Textbook of
Soil Mechanics and
Foundation Engineering
Geotechnical Engineering Series

Geotechnical Engineering Series

- **Textbook of Soil Mechanics and Foundation Engineering**

- **Advanced Foundation Engineering**

Textbook of
Soil Mechanics and
Foundation Engineering
Geotechnical Engineering Series

VNS Murthy
Consulting Geotechnical Engineer
Bangalore, India

CBSPD

CBS Publishers & Distributors Pvt Ltd

New Delhi • Bengaluru • Chennai • Kochi • Kolkata • Lucknow • Mumbai
Hyderabad • Jharkhand • Nagpur • Patna • Pune • Uttarakhand

Textbook of
Soil Mechanics and
Foundation Engineering
Geotechnical Engineering Series

ISBN-13: 978-81-239-1362-9
ISBN-10: 81-239-1362-1

First Edition: 2007
Reprint: 2008, 2009, 2010, 2011, 2012, 2013, 2014, 2015, 2016, 2017, 2018, 2022,
    2025

Published by **Satish Kumar Jain** and produced by **Varun Jain** for

CBS Publishers & Distributors Pvt Ltd

4819/XI Prahlad Street, 24 Ansari Road, Daryaganj, New Delhi 110 002, India.
Ph: 011-23266838, 23289259   Website: www.cbspd.com
    e-mail: delhi@cbspd.com

Corporate Office: 204 FIE, Industrial Area, Patparganj, Delhi 110 092
Ph: 011-4934 4934   Fax: 011-4934 4935
    e-mail: publishing@cbspd.com; publicity@cbspd.com

Branches

- **Bengaluru:** Seema House 2975, 17th Cross, KR Road, Banasankari 2nd Stage, Bengaluru 560 070; Karnataka, India
  Ph: +91-80-26771678/79   Fax: +91-80-26771680   e-mail: bangalore@cbspd.com
- **Chennai:** 18/8B, Subbarayan Street, Shenoy Nagar, Chennai 600 030, Tamil Nadu, India
  Ph: +91-44-42032115, 26681266   e-mail: chennai@cbspd.com
- **Kochi:** 42/1325, 1326, Power House Road, Opp KSEB, Power House, Ernakulum Kochi 682 018, Kerala, India
  Ph: +91-484-4059061-65,67   Fax: +91-484-4059065   e-mail: kochi@cbspd.com
- **Kolkata:** 147, Hind Ceramics Compound, 1st Floor, Nilgunj Road, Belghoria, Kolkata-700056, West Bengal, India
  Ph: +033-25633055, 033-25633056   e-mail: kolkata@cbspd.com
- **Lucknow:** Basement, Khushnuma Complex, 7 Meerabai Marg (Behind Jawahar Bhawan), Lucknow-226001, UP, India
  Ph: +0522-4000032   e-mail: tiwari.lucknow@cbspd.com
- **Mumbai:** PWD Shed, Gala no 25/26, Ramchandra Bhatt Marg, Next to JJ Hospital Gate no. 2, Opp. Union Bank of India, Noorbaug, Mumbai-400009, Maharashtra, India
  Ph: 022-66661880/89   e-mail: mumbai@cbspd.com

Representatives

- Hyderabad · 0-9885175004   · Jharkhand   0-9811541605   · Nagpur   0-8692091830
- Patna   0-9334159340   · Pune   0-9664372571   · Uttarakhand 0-9716462459

Printed at Glorious Printers, Jhilmil Industrial Area, Delhi, India

Dedicated

to

the Cause of Students

Department of Civil Engineering

Indian Institute of Science

Bangalore - 560 012, India

Date : 16-11-2005

FOREWORD

IT gives me great pleasure to write the Foreword to this book *Textbook of Soil Mechanics and Foundation Engineering* by Dr. V.N.S. Murthy. Eversince he wrote the first book in 1972, he has been revising and enlarging the book to improve and include the current topics and interests in the subject. During the last century, the subject of geotechnical engineering grew at an unprecedented rate, thanks to the great contributions of researchers and professional engineers. To keep pace with the advancements and to appreciate the merits of such advancements, a sound knowledge of the basics is essential; and good books, as many in number, are the need of the hour. Dr. Murthy's book is one such.

The book offers a comprehensive coverage of topics of study in undergraduate curriculum and also to some extent in the post graduate curriculum. Each topic is developed in a logical sequence and helps the students to grasp the fundamentals very clearly. In this, Dr. Murthy has brought to bear his rich experience of teaching for three decades. Major strength of the book is in the style of presentation, which is lucid.

The book contains 26 chapters. The first five chapters cover the basics of soil formation, phase relationships, index properties and classification. The next four chapters deal with permeability, seepage, pore water and effective pressure. Stress distribution, compressibility, shear strength and soil improvement are covered in the next four chapters. This is followed by chapters on lateral earth pressures, sheet piles, and slope stability. Shallow foundations and the aspects of bearing capacity and settlement are presented in a long chapter after adequately introducing a chapter on soil exploration. A fairly elaborate treatment of deep foundations — vertical and laterally loaded piles, caissons and drilled piers, is given in the next three chapters. The final chapters of the book deal with the aspects of foundations in expansive and collapsible soils, concrete gravity retaining walls, mechanically stabilized earth retaining walls, braced cuts and cellular cofferdams, to give a wider coverage of interest to the readers.

In each chapter, a number of carefully chosen example problems have been worked out to illustrate the concepts and to help in proper understanding of the subject. At the end, a number of problems have been set for solution by the students, which when completed will certainly enable them to develop a grasp on the subject.

This excellent *Textbook* should be of great service to the students and teachers of geotechnical engineering, and also to the professional engineers.

K.S. Subba Rao

PREFACE

ALL structures have to be built on soils. Our main objective in the study of **Soil Mechanics and Foundation Engineering** is to lay down certain basic principles, theories and procedures for the design and construction of safe and sound structures. The following requirements are considered essential in any **textbook** on the subject.

1. A book which deals with the fundamentals of the subject and explains the basic principles in a logical way.

2. A book which is comprehensive and covers the requirements of students at undergraduate level.

3. A book which serves as a foundation course for the postgraduate students for further development of advanced knowledge in the subject.

4. A book which is practical and pragmatic and serves as a reference book for practising engineers.

The *textbook* that is being published by CBS Publishers and Distributors, New Delhi, satisfies all the above requirements. The subject matter is updated up to the year 2000. There is no need for me to elucidate further on the merits of my book. Prof. K.S. Subba Rao in his *Foreword* has clearly amplified the quality of my book and as such I do not want to add anything further.

V.N.S. Murthy

Acknowledgements

I thank sincerely Prof. K.S. Subba Rao for writing the Foreword to my book. He is an eminent well-known foundation engineer. Now he is Emeritus Professor in the Department of Civil Engineering, Indian Institute of Science, Bangalore. He graduated from Mysore University in 1961, obtained the Master of Engineering and Doctor of Philosophy degrees from the Indian Institute of Science in 1963 and 1969 respectively. He served on the faculty of IISc till retirement in 2004.

Prof. Subba Rao has 35 years of experience in teaching, research and consulting in geotechnical engineering. His research interests span from foundation engineering and structure soil interaction to expansive soils. He has guided 19 doctoral and several masters degree theses and has authored more than 120 publications in refereed journals and conferences.

Prof. Subba Rao has delivered the prestigious annual lecture of IGS, and is a recipient of IGS Kucckelman award for his contributions to geotechnical engineering. He is elected as Life Fellow of the Indian National Academy of Engineering, the Indian Geotechnical Society, and the Institution of Engineers (India).

The typesetting for my book has been done by Chaitanya Graphics, who are experts in total CAD/CAM/CAE solutions. They have done an excellent job.

Last but not the least, I wish to express my gratitude to my wife Sharadamani for extending all cooperation during the preparation of the book.

V.N.S. Murthy

REVIEWS

"*Textbook of Soil Mechanics and Foundation Engineering* is a long title befitting a major work. I am pleased to introduce this superb volume destined for a readership of students, professors, and consultants. What makes this text different from other books on these subjects that appear each year and why am I recommending it to you? I have been working and teaching in the area of geotechnical engineering for 25 years. I have read and used scores of textbooks in my classes and practice. Dr. Murthy's text is by far the most comprehensive text I have found. You will find that his organization of the subject matter follows a logical progression. His example problems are numerous and, like the text, start from fundamental principles and progressively develop into more challenging material. They are the best set of example problems I have seen in a textbook. Dr. Murthy has included ample homework problems with a range of difficulty meant to help the student new to the subject to develop his/her confidence and to assist the experienced engineer in his/her review of the subject and in professional development.

I have been impressed by the coverage, the clarity of the presentation, and the insights into the hows and whys of soil and foundation behavior. Often I have been astonished at Dr. Murthy's near-conversational approach to sharing helpful insights. You get the impression he's right there with you guiding you along, anticipating your questions, and providing instruction and necessary information as the next steps in the learning process. I believe you will enjoy this book and that it will receive a warm welcome wherever it is used. I thank Dr. Murthy for his commitment to write this textbook and for sharing his professional experience with us."

— **Mark T. Bowers**, PhD, PE
Associate Professor of Civil Engineering, University of Cincinnati, USA

"This comprehensive, pertinent and up-to-date volume is well suited for use as a textbook for graduate students as well as a reference book for consulting geotechnical engineers and contractors. The book is well written with numerous examples on applications of basic principles to solve practical problems.

This textbook by Prof. V. N. S. Murthy is highly recommended for students specializing in geotechnical engineering and for practicing civil engineers in the United States and Europe. The book includes recent developments such as soil improvement and stabilization methods and applications of geotextiles to control settlements and lateral earth pressure. Numerous graphs and examples illustrate the most important concepts in geotechnical engineering. This textbook should serve as a valuable reference book for many years to come."

Bengt B. Broms, Ph D
Nanyang Technical University, Singapore (retired)

CONTENTS

CHAPTER 11   COMPRESSIBILITY AND CONSOLIDATION   283

CHAPTER 12 SHEAR STRENGTH   329

CHAPTER 18   SHALLOW FOUNDATION   619

PART A—BEARING CAPACITY OF FOUNDATIONS

PART B—SAFE BEARING PRESSURE AND SETTLEMENT OF FOUNDATIONS

PART C—COMBINED FOOTINGS AND MAT FOUNDATIONS

CHAPTER 19   PILE FOUNDATION   723

PART A–TYPES OF PILES AND INSTALLATION

PART B–VERTICAL LOAD BEARING CAPACITY OF SINGLE VERTICAL PILE

PART C–PILE GROUPS SUBJECTED TO VERTICAL LOADS

PART D–BEHAVIOUR OF LATERALLY LOADED VERTICAL AND BATTER PILES

CHAPTER 20   CAISSON FOUNDATIONS   819

CHAPTER 21   DRILLED PIER FOUNDATIONS   837

CHAPTER 22   BRACED CUTS AND DRAINAGE   879

PART A—BRACED CUTS

PART B—DRAINAGE

CHAPTER 23   CELLULAR COFFERDAMS   899

CHAPTER 24   FOUNDATIONS ON COLLAPSIBLE AND EXPANSIVE SOILS   915

PART A—COLLAPSIBLE SOILS

PART B—EXPANSIVE SOILS

CHAPTER 25   CONCRETE RETAINING WALLS   957

CHAPTER 26   MECHANICALLY STABILISED EARTH
RETAINING WALLS   975

APPENDIX A   SI UNITS IN GEOTECHNICAL ENGINEERING   1007

APPENDIX B   SLOPE STABILITY CHARTS AND TABLES   1013

REFERENCES 1027

INDEX   1039

CHAPTER 1

INTRODUCTION

1.1 GENERAL REMARKS

Karl Terzaghi writing in 1951, on 'The Influence of Modern Soil Studies on the Design and Construction of Foundations' commented on foundations as *follows:*

> *Foundations can appropriately be described as a necessary evil. If a building is to be constructed on an outcrop of sound rock, no foundation is required. Hence, in contrast to the building itself which satisfies specific needs, appeals to the aesthetic sense, and fills its matters with pride, the foundations merely serve as a remedy for the deficiencies of whatever whimsical nature has provided for the support of the structure at the site which has been selected. On account of the fact that there is no glory attached to the foundations, and that the sources of success or failures are hidden deep in the ground, building foundations have always been treated as step children; and their acts of revenge for the lack of attention can be very embarrassing.*

The comments made by Terzaghi are very significant and should be taken note of by all the practising architects and engineers. Architects or engineers who do not wish to make use of the growing knowledge of foundation design are not rendering true service to their profession. Since substructures are as important as superstructures, persons who are well qualified in the design of substructures should always be consulted and the old proverb that a 'stitch in time saves nine' should always be kept in mind.

The design of foundations is a branch of civil engineering. Experience has shown that most of these branches have passed in succession through two stages, the empirical and the scientific, before they reached the present one which may be called the stage of maturity.

The stage of scientific reasoning in the design of foundations started with the publication of the book *Erdbaumechanik* (means *Soil Mechanics*) by Karl Terzaghi in 1925. This book represents the first attempt to treat soil mechanics on the basis of the physical properties of soils. Terzaghi's contribution for the development of soil mechanics and foundation engineering is so vast that he may truly be called as the *father of soil mechanics*. His activity extended over a period of about 50 years starting from the year 1913. He was born on October 2, 1883 in Prague and died on October

25, 1963 in Winchester, Massachusetts, USA. His amazing career is well documented in the book *From Theory to Practice in Soil Mechanics* (Wiley, 1960).

Many investigators in the field of soil mechanics were inspired by Terzaghi. Some of the notable personalities who followed his footsteps are Ralph B. Peck, Arthur Casagrande, A. W. Skempton, etc. Because of the unceasing efforts of these and other innumerable investigators, soil mechanics and foundation engineering has come to stay as a very important part of civil engineering profession.

The transition of foundation engineering from the empirical stage to that of scientific stage started almost at the commencement of the 20th century. The design of foundations during the empirical stage was based mostly on intuition and experience. There used to be many occasional failures since the procedure of design was only by trial and error method.

However, in the present scientific age, the design of foundations based on scientific analysis has received a lot of impetus. Theories have been developed based on fundamental properties of soils. Still one can witness unsatisfactory performance of structures constructed even on scientific principles. The reasons for such performance are many. The soil mass on which a structure is to be built is heterogeneous in character and no theory can simulate field conditions. The fundamental properties of soil which we determine in the laboratories may not reflect truly the properties of the soil *in-situ*. A judicial combination of theory and experience is very essential for successful performance of any structure built on earth. Another method that is gaining popularity is the *observational approach*. This procedure consists in making appropriate observations soon enough during construction to detect signs of departure of the real conditions from those assumed by the designer and in modifying either the design or the method of construction in accordance with the findings.

1.2   A BRIEF HISTORICAL DEVELOPMENT

Many structures that were built centuries ago are still the monuments of curiosity even today. Egyptian temples built three or four thousand years ago are still existing though the design of the foundations were not based on any presently known principles. Romans built notable engineering structures such as harbours, breakwaters, aqueducts, bridges, large public buildings and a vast network of durable and excellent roads. The *leaning tower of Pisa* in Italy built during the 14th century is still a centre of tourists attraction. The tower is 179 ft. in height and its top was out of plumb by about 16.5 ft in the year 1970. Many bridges were also built during 15th to 17th centuries. Timber piles were used for many of the foundations.

Another marvel of engineering achievement is the construction of the famed mausoleum Taj Mahal outside the city of Agra. This was constructed in 17th century by the Mogul Emperor of Delhi, Shahjahan, to commemorate his favourite wife Mumtaz Mahal. The mausoleum is built on the bank of the river Jamuna. The proximity of the river required special attention in the building of the foundations. It is reported that masonry cylindrical wells have been used for the foundations. It goes to the credit of the engineers who designed and constructed this grand structure which is still quite sound even after a lapse of about three centuries.

The first rational approach for the computation of earth pressures on retaining walls was formulated by Charles Augustin Coulomb (1736-1806), a famous French scientist. He proposed a theory in 1770 and this theory is called as "Classical Earth Pressure Theory". Poncelet (1788-1887) extended Coulomb's theory by giving an elegant graphical method for finding the magnitude of earth pressure on walls. Later, in 1866 Karl Culmann (1821-1881) gave the Coulomb-Poncelet Theory a geometrical formulation, thus supplying the method with a broad scientific basis. William J. Macqom Rankine (1820-1872) who was a Professor of Civil Engineering at the University of Glasgow, proposed a new earth pressure theory in 1857, which is also called as a *Classical Earth Pressure Theory.*

Darcy, on the basis of his experiments on filter sands, proposed a law in 1856 for the flow of water in permeable materials and in the same year G. G. Stokes gave an equation for determining the terminal velocity of solid particles falling in liquids. The rupture theory of Otto Mohr (1835-1918) was made known in 1871. Mohr's Stress Circles are extensively used in the study of shear strengths of soils. One of the most important contributions to the engineering science was made by J. V. Boussinesq (1842-1929) who proposed a theory for determining stress distribution under loaded areas in a semi-infinite, elastic, homogeneous, and isotropic medium.

In 1911, A. Atterberg, a Swedish scientist, proposed simple tests for determining the consistency limits of cohesive soils. In 1913, W. Fellenius headed the Swedish Geotechnical Commission for determining the causes of failure of many railway and canal embankments. The so-called *swedish circle method* or otherwise termed as the *slip circle method* was the outcome of his investigation which was published in 1922.

The development of the science of soil mechanics and foundation engineering from the year 1925 onwards was phenomenal. Terzaghi laid down definite procedures in his book published in 1925 for determining properties and the strength characteristics of soils. The modem soil mechanics was born in 1925. The present stage of knowledge in Soil Mechanics and the design procedures of foundations are mostly due to the works of Terzaghi and his band of devoted collaborators.

1.3   SOIL MECHANICS AND FOUNDATION ENGINEERING

Terzaghi defined soil mechanics as follows:

Soil mechanics is the application of the laws of mechanics and hydraulics to engineering problems dealing with sediments and other unconsolidated accumulations of solid particles produced by the mechanical and chemical disintegration of rocks regardless of whether or not they contain an admixture of organic constituents.

The term *soil mechanics* is now accepted quite generally to designate that discipline of engineering science which deals with the properties and behaviour of soil as a structural material.

All structures have to be built on soils. Our main objective in the study of soil mechanics is to lay down certain principles, theories and procedures for the design of a safe and sound structure. The subject of *foundation engineering* deals with the design of various types of substructures under different soil and environmental conditions.

During the design, the designer has to make use of the properties of soils, the theories pertaining to the design and his own practical experience to adjust the design to suit the field conditions. He has to deal with natural soil deposits which perform the engineering function of supporting the foundation and the superstructure above it. The soil deposits in nature are available in an extremely erratic manner producing thereby an infinite variety of possible combinations which would affect the choice and design of foundations. The foundation engineer must have the ability to interpret the principles of soil mechanics to suit the field conditions. The success or failure of his design depends upon as to how much he is in tune with Nature.

CHAPTER 2

FORMATION AND STRUCTURE OF SOILS

2.1 INTRODUCTION

The word 'soil' has different meaning for different professions. To the agriculturist, soil is the top thin layer of earth within which organic forces are predominant and which is responsible for the support of plant life. To the geologist also, soil is the material in the top thin zone within which roots occur. But from the point of view of an engineer, soil includes all earth materials, organic and inorganic, occurring in the zone overlying the rock crust.

The behaviour of a structure depends upon the properties of the soil materials on which the structure rests. The properties of the soil materials depend upon the properties of the rocks from which they are derived. A brief discussion of the parent rocks is, therefore, quite essential in order to understand the properties of soil materials.

2.2 ROCK CLASSIFICATION

Rock can be defined as a compact, semi-hard to hard mass of natural material composed of one or more minerals. The rocks that are encountered at the surface of the earth or beneath, are commonly classified into three groups according to their modes of origin. They are igneous, sedimentary and metamorphic rocks.

2.3 IGNEOUS ROCKS

Igneous rocks are considered to be the primary rocks formed by the cooling of molten magmas, or by the recrystallisation of older rocks under heat and pressure great enough to render them fluid. They have been formed on or at various depths below the earth surface. There are two main classes of igneous rocks. They are:

1. Extrusive (poured out at surface), and

2. Intrusive (large rock masses which have not been formed in contact with the atmosphere).

Initially both classes of rocks were in a molten state. Their present state results directly from the way in which they solidified. Due to violent volcanic eruptions in the past, some of the molten materials emitted into the atmosphere with gaseous extrusions, these cooled quickly and eventually fell on the earth's surface as volcanic ash and dust. Extrusive rocks are distinguished, in general, by their glass-like structure.

Intrusive rocks, cooling and so solidifying at great depths and under pressure containing entrapped gases, are wholly crystalline in texture. Such rocks occur in masses of great extent, often going to unknown depths. Some of the important rocks that belong to the igneous group are *granite and basalt*. Granite is primarily composed of feldspar, quartz and mica and possesses a massive structure. Basalt is a dark-coloured fine-grained rock. It is characterised by the predominance of plagioclase, the presence of considerable amounts of pyroxene and some olivine and the absence of quartz. The colour varies from dark-grey to black. Both granite and basalt are used as building stones.

2.4 SEDIMENTARY ROCKS

When the products of the disintegration and decomposition of any rock type are transported, redeposited, and partly or fully consolidated or cemented into a new rock type, the resulting material is classified as a sedimentary rock. The sedimentary rocks generally are formed in quite definitely arranged beds, or strata, which can be seen to have been horizontal at one time although sometimes displaced through angles up to 90 degrees. Sedimentary rocks are generally classified on the basis of grain size, texture and structure. From our engineering point of view, the most important rocks that belong to the group are *sandstones, limestones,* and *shales.*

2.5 METAMORPHIC ROCKS

Rocks formed by the complete or incomplete recrystallisation of igneous or sedimentary rocks by high temperatures, high pressures, and/or high shearing stresses are metamorphic rocks. The rocks so produced may display features varying from complete and distinct foliation of a crystalline structure to a fine fragmentary partially crystalline state caused by direct compressive stress, including also the cementation of sediment particles by siliceous matter. Metamorphic rocks formed without intense shear action have a massive structure; Some of the important rocks that belong to this group are *gneiss, schist, slate* and *marble*. The characteristic feature of gneiss is its structure, the mineral grains are elongated, or platy, and banding prevails. Generally gneiss is a good engineering material. Schist is a finely foliated rock containing a high percentage of mica. Depending upon the amount of pressure applied by the metamorphic forces, schist may be a very good building material. Slate is a dark coloured, platy rock with extremely fine texture and easy cleavage. Because of this easy cleavage, slate is split into very thin sheets and used as a roofing material. Marble is the end product of the metamorphism of limestone and other sedimentary rocks composed of calcium or magnesium carbonate. It is very dense and exhibits a wide variety of colours. In construction, marble is used for facing concrete or masonry exterior and interior walls and floors.

2.6 ROCK MINERALS

It is very essential to examine the properties of the rock forming minerals since all soils are derived through the disintegration or decomposition of some parent rock. A *'mineral'* is a natural inorganic substance of a definite structure and chemical composition. Some of the very important physical properties of minerals are crystal form, colour, hardness, cleavage lustre, fracture, and specific gravity. Out of these only two, specific gravity and hardness, are of foundation engineering interest. The specific gravity of the minerals affects the specific gravity of soils derived from them. The specific gravity of the most rock and soil taking minerals varies from 2.50 (some feldspars) and 2.65 (quartz) to 3.5 (augite or olivine). Gypsum has a smaller value of 2.3 and salt (NaCl) has 2.1. Some iron minerals may have higher values, for instance, magnetite has 5.2.

The hardness of a mineral (symbol H) is expressed by its number in Mohs' scale of hardness (Table 2.1). Each mineral listed in that scale can scratch all minerals of smaller numbers but in turn can be scratched only by minerals of a higher numbers than its own.

It is reported that about 95 percent of the known part of the lithosphere consists of igneous rocks and only 5 percent of sedimentary rocks. The soil formation is mostly due to the disintegration of igneous rock which may be termed as a parent rock.

The average mineral composition of igneous rocks is given in Table 2.2. Feldspars are the most common rock minerals, which account for the abundance of clays derived from the feldspars on the earth's surface. Quartz comes next in order of frequency. Most sands are composed of quartz.

TABLE 2.1

Mohs' scale of hardness

Standard mineral	Hardness, H	Approximate identification method
Talc	1	Will mark cloth.
Gypsum	2	Can be scratched by a finger nail.
Calcite	3	Can be scratched by a copper coin.
Fluorite	4	Can be scratched by a pocket-knife.
Apatite	5	Can be scratched by a pocket-knife.
Orthoclase	6	Will scratch window glass.
Quartz	7	Cannot be scratched by a steel file.
Topaz	8	
Corundum	9	Will scratch most metals, but not diamonds.
Diamond	10	Will scratch any substance except other diamonds.

TABLE 2.2

Mineral composition of igneous rocks

Mineral	Percent
Quartz	12-20
Feldspar	60-50
Ca, Fe and Mg, Silicates	17-14
Micas	4-8
Others	7-8

2.7 FORMATION OF SOILS

Soil is defined as a natural aggregate of mineral grains, with or without organic constituents, that can be separated by gentle mechanical means such as agitation in water. Whereas rock is considered to be a natural aggregate of mineral grains connected by strong and permanent cohesive forces. The process of weathering of the rock decreases the cohesive forces binding the mineral grains and leads to the disintegration of bigger masses to smaller and smaller particles. Soils are formed by the process of weathering of the parent rock. The weathering of the rocks might be mechanical (disintegration, and/or chemicals (decomposition).

Mechanical Weathering

Mechanical weathering of rocks to smaller particles are due to the action of such agents as the expansive forces of freezing water in fissures, due to sudden changes of temperature or due to the abrasion of rock by moving water or glaciers. Temperature changes of sufficient amplitude and frequency bring about changes in the volume of the rocks in the superficial layers of the earth's crust in terms of expansion and contraction. Such a volume change-sets up tensile and shear stresses in the rock and ultimately leading to the fracture of even large rocks. This type of rock weathering takes place in a very significant manner in arid climates where free, extreme atmospheric radiation brings about considerable variation in temperature at sunrise and sunset.

Chemical Weathering

Chemical weathering (decomposition) can transform hard rock minerals into soft, easily erodable matter. The principal types of decomposition are *hydration, oxidation, carbonation, desilication and leaching*. Oxygen and carbondioxide which are always present in the air readily combine with the elements of rock in the presence of water.

Oxidation

Oxidation occurs frequently in rocks containing iron, which decomposes in a manner similar to the rusting of steel when in contact with moist-air.

Carbonation

The minerals containing iron, calcium, magnesium, sodium, or potassium can be decomposed by carbonic acid which is formed by carbondioxide with water. Thus practically all igneous rocks may be decomposed in this manner. Particularly limestones can easily be dissolved by water containing carbonic acid. Dolomites are much more resistant to carbonation being composed of less soluble magnesium carbonate. It is interesting to note that silica (SiO_2) is not decomposed by carbonation and as such quartz, which is composed of pure silica, is one of the most stable minerals. Therefore acid rocks (rocks containing much silica), such as granite are much more resistant to weathering than are basic igneous rocks such as basalt and gabbro. However, orthoclase feldspar can easily be decomposed by carbonic acid with the resulting formation of a new soft clay mineral, kaolinite, and other bye-products.

Hydration

Hydration is a common process of rock decay by which water is combined with some other soil substances thus producing certain new minerals. The process of hydration are more intensive in humid than in the arid climates.

Desilication

Desilication consists in leaching out of dissolved or colloidal silica freed in the course of other chemical processes.

Leaching

Leaching is the process whereby water-soluble parts (calcium carbonate, for example) are dissolved and washed out from the soil by rainfall, percolating water, subsurface flow or other water.

2.8 GENERAL TYPES OF SOILS

It has been discussed earlier that soil is formed by the process of physical and chemical weathering. The individual size of the constituent parts of even the weathered rock might range from the smallest state (colloidal) to the largest possible (boulders). This implies that all the weathered constituents of a parent rock cannot be called as soil. According to their grain size, soil particles are classified as cobbles, gravel, sand, silt and clay. Grains having diameters in the range of 4.75 to 80 mm are called as gravel. If the grains are visible to the naked eye, but are less than about 4.75 mm in size the soil is described as sand. The lower limit of visibility of grains for the naked eyes is about 0.075 mm. Soil grains ranging from 0.075 to 0.002 mm in termed as silt and those that are finer than 0.002 mm as clay. This classification is purely based on size which does not indicate the properties of fine grained materials.

Residual and Transported Soils

On the basis of origin of their constituents, soils can be divided into two large groups:

1. Residual soils, and
2. Transported soils.

Residual soils are those that remain at the place of their formation as a result of the weathering of parent rocks. The depth of residual soils depends primarily on climatic conditions and the time of exposure: In some areas, this depth might be considerable. In temperate zones residual soils are commonly stiff and stable. An important characteristic of residual soil is that the sizes of grains are indefinite. For example, when a residual sample is sieved, the amount passing any given sieve size depends greatly on the time and energy expended in shaking, because of the partially disintegrated condition.

Transported soils are soils that are found at locations far removed from their place of formation. The transporting agencies of such as soils are glaciers, wind and water. The soils are named according to the mode of transportation. *Alluvial* soils are those that have been transported by running water. The soils that have been deposited in quiet lakes, are *lacustrine* soils. *Marine soils* are those deposited in sea water. The soils transported and deposited by wind are *aeolin* soils. Those deposited primarily through the action of gravitational force, as in land slides, are *colluvial* soils. *Glacial* soils are those deposited by glaciers. Many of these transported soils are loose and soft to a depth of several hundred feet. Therefore, difficulties with foundations and other types of construction are generally associated with transported soils.

Organic and Inorganic Soils

Soils in general are further classified as *Organic* or *Inorganic*. Soils of organic origin are chiefly formed either by growth and subsequent decay of plants such as peat mosses, or by the accumulation of fragments of the inorganic skeletons or shells of organisms. Hence a soil of organic origin can be either organic or inorganic. The term organic soil ordinarily refers to a transported soil consisting of the products of rock weathering with a more or less conspicuous admixture of decayed vegetable matter.

2.9 NAMES OF SOME SOILS THAT ARE GENERALLY USED IN PRACTICE

Sand and Gravel

They are cohesionless aggregates of rounded subangular or angular fragments of more or less unaltered rocks or minerals. Particles of size from 0.075 to 4.75 mm are referred to as . sand, and those with a size from 4.75 to 80 mm as gravel. Fragments with diameters more than 80 mm and less than 300 mm are known as Cobbles.

Inorganic Silt

Fine-grained soil with little or no plasticity. The least plastic varieties generally consist of more or less uniform size grains of quartz and are sometimes called *rock flour.* On the other hand, the plastic types contain a considerable percentage of flake shaped particles and are referred to as plastic silt.

Organic Silt

A fine-grained more or less plastic soil with an admixture of finely divided particles of organic matter.

Clay

It is an aggregate of mineral particles of microscopic and submicroscopic range. The soil may be inorganic or organic. Inorganic clays are generally more plastic than the organic, whereas the organic clays are more compressible because of the presence of finely divided organic matter.

Bentonite

It is a clay formed by the decomposition of volcanic ash with a high content of montmorillonite. It exhibits the properties of clay to an extreme degree.

Varved Clays

These clays are made of thin alternate layers of silt and fat clays of glacial origin. They possess the undesirable properties of both silt and clay. The constituents of varved clays were transported into fresh water lakes by the melted water at close of the ice age.

Kaolin, China Clay

A very pure form of white clay used in the ceramic industry.

Boulder Clay

It is a mixture of an unstratified sedimented deposit of glacial clay, containing unsorted rock fragments of all sizes ranging from boulders, cobbles, and gravel to finely pulverized clay material.

Calcareous Soil

It is a soil containing calcium carbonate. Such soil effervescences when tested with weak hydrochloric acid.

Marl

It is loosely attributed to deposits which consist of mixture of calcareous sands, or clays, or loam.

Hardpan

A relatively hard, densely cemented soil layer, like rock which does not soften when wet. Boulder clays or glacial till is also sometimes named as hardpan.

Caliche

An admixture of clay, sand, and gravel cemented by calcium carbonate deposited from ground water.

Peat

A fibrous aggregate of finer fragments of decayed vegetable matter. Peat is very compressible and one should be cautious while using for supporting foundations of structures.

Loam

A mixture of combined sand, silt and clay.

Loess

This is a fine-grained, air-borne deposit characterised by a very uniform grain size, and high void ratio. The size of particles range between about 0.01 to 0.05 mm. The soil can stand deep vertical cuts because of slight cementation between particles. It is formed in dry continental regions and its colour is yellowish light brown.

Shale

This is a material in the state of transition from clay to slate. Shale itself is sometimes considered a rock but, when it is exposed to the air or has a chance to take in water it may rapidly decompose.

Black Cotton Soil

Black cotton soil is a material characterised by its high expansive and shrinkage properties. Its colour varies from dark grey to black. This soil is very common in many parts of India. Great care is required when structures are to be built on black cotton soil.

2.10 STRUCTURE OF SOILS

The formation of soil materials from the weathering of the parent rock has been considered in the previous section. The size of the soil materials in a mass of soil may range from the finest to the coarsest. The behaviour of a soil mass under stress is a function of many factors such as the size and shape of grains, gradation of grains, mineralogical composition of the particles, arrangement of the grains in relation to each other, the interparticle forces, the adsorption complexes and possibly many other factors. A study of these various factors will lead to a better understanding of the behaviour of soils under different conditions of loading. The word *structure* used herein can be applied to the structure of individual grains as well as the structure of the soil media.

Soil structure may be defined as the physical constitution of a soil material as expressed by the size, shape, and arrangement of the solid particles to form compound particles and the compound particles themselves.

This concept of soil structure by Brewer (1964) encompasses all the divergent definitions of different workers and therefore can be accepted. In soil materials it can be applied satisfactorily to compound particles, to primary particles (single grains) and to even more specific levels such as the structure of individual minerals (size, shape and arrangement of atoms) and even the structure of atoms (size, shape and arrangement of electrons, protons and neutrons).

The soil structure as defined above may be discussed in detail under two headings. They are:

1. Soil particle structure, and

2. Soil mass structure.

The soil particle structure deals with the structure of individual atoms and minerals whereas the soil mass structure discusses the pattern of arrangement of soil particles in a mass of soil.

2.11 SOIL PARTICLE SIZE AND SHAPE

The size of particles as explained earlier, may range from gravel to the finest size possible. Their characteristics vary with the size. The soil particles coarser than 0.075 mm are visible to the naked eye or may be examined by means of a hand lens. They constitute the coarser fractions of the soils. Grains finer than 0.075 mm can constitute the finer fractions of soils. It is possible to distinguish the grains lying between 0.075 mm and 2 μ (1 μ = 1 micron = 0.001 mm) under a microscope. Grains having size between 2 μ and 0.1 μ can be observed under a microscope but their shapes cannot be made out. The shape of grains smaller than 1 μ can be determined by means of an electron microscope. The molecular structure of particles can be investigated by means of X-ray analysis.

The coarser fractions of soils consist of gravel and sand. The individual particles of gravel, which is nothing but fragments of rock, is composed of one or more minerals, whereas sand grains contain mostly one mineral which is quartz. The individual grains of gravel and sand may be angular, subangular, subrounded or well-rounded as shown in Fig. 2.1. Gravel may contain grains which may be flat also. Some sands contain a fairly high percentage of mica flakes that give them the property of elasticity.

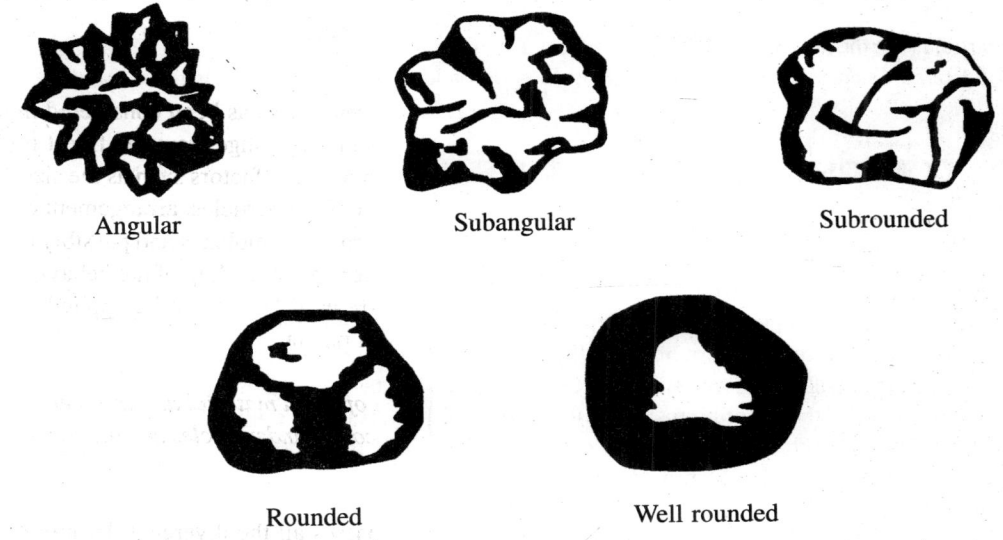

Angular   Subangular   Subrounded

Rounded   Well rounded

Fig. 2.1   Shapes of coarser fractions of soils

Silt and clay constitute the finer fractions of the soil. Any one grain of this fraction generally consists of only one mineral. The particles may be angular, flake-shaped or sometimes needle-like.

2.12 SPECIFIC SURFACE

The soil is essentially a disperse system, that is, a system in which the particles are in a fine state of subdivision or dispersion. In soils, the dispersed or the solid phase predominates and the dispersion medium, soil water, only helps to fill the pores between the solid particles. The significance of the concept of dispersion becomes more apparent when the relationship of surface to particle size is considered. In the case of silt, sand and larger size particles the ratio of the area of surface of the particles to the volume of the sample is relatively small. Whereas this ratio becomes increasingly large as size decreases from 2 μ which is the upper limit for clay-sized particles. An useful index of relative importance of surface effects is the *specific surface* of grain. The specific surface is defined as the total area of the surface of grains expressed in square centimeters per gram or per cubic centimetre of dispersed phase.

The shape of the clay particles is an important property from the physical point of view. The amount of surface per unit mass or volume *v* varies with the shape of the particles. Moreover, the amount of contact area per unit surface also changes with shape. It is a fact that a sphere has the smallest surface area per unit volume whereas a plate exhibits the maximum. Ostwald (Von Buzagh, 1937) has emphasized the importance of shape in determining specific surface of colloidal systems. Table 2.3 gives the effect of shape upon the specific surface of colloidal particles as per the calculations of Ostwald. Since disc-shaped particles can be brought more in intimate contact with each other, this shape has a pronounced effect upon the mechanical properties of the system. The interparticle forces between the surfaces of particles have a significant effect on the property of the soil mass if the particles in the media belong to the clay fraction. The surface activity depends not only on the specific surface but also on the chemical and mineralogical composition of the solid particles. Since clay particles are the active portions of a soil because of their high specific surface and their chemical constitution, a discussion on the chemical composition and structure of minerals is essential.

2.13 COMPOSITION OF CLAY MINERALS

The word 'clay' is generally understood to refer to a material composed of a mass of a small mineral particles which, in association with certain quantities of water, exhibits property of plasticity. According to the clay mineral concept, clay materials are essentially composed of extremely small crystalline particles of one or more members of a small group of minerals that are commonly known as clay minerals. These minerals are essentially hydrous aluminium silicates, with magnesium or iron replacing wholly or in part for the aluminium, in some minerals. Many clay materials may contain organic material and water-soluble salts. Organic materials occur either as discrete particles of wood, leaf matter, spores, etc., or it may be present as organic molecules adsorbed on the surface of the clay mineral particles.

TABLE 2.3

The effect of shape on the specific surface of colloidal particles (after Ostwald)

Shape	Dimensions	Specific surface per cc
Sphere	Radius = 0.6204	4.836 sq.cm
Cube	Length of side = 1.0	6.0 sq.cm
Disc	Radius = $\sqrt{v/\pi h}$	Specific surface = $(2/h) + 2\sqrt{\pi h}$
$h = 0.1$ cm (1mm)	1.728	21.121 sq.cm
$h = 0.0001$ cm *(I μ)*	56.42	2×10^4 sq.cm
$h = 0.00005$ cm *(0.5 μ)*	79.8	4×10^4 sq.cm
$h = 0.00001$ cm *(0.1 μ)*	178.4	20×10^4 sq.cm

TABLE 2.4

Clay minerals

	Name of mineral	Structural formula
I.	Kaolin group	
	1. Kaolinite	$Al_4Si_4O_{10}(OH)_8$
	2. Halloysite	$Al_4Si_6O_6(OH)_{16}$
II.	Montmorillonite group	
	Montmorillonite	$Al_4Si_8O_{20}(OH)_4 nH_2O$
III.	Illite group	
	Illite	$K_y(Al_4Fe_2.Mg_4.Mg_6)Si_{8-y}$ $Al_y(OH)_4O_{20}$

The water-soluble salts that are present in clay materials must have been entrained in the clay at the time of accumulation or may have developed subsequently as a consequence of ground water movement and weathering or alteration processes.

Clays can be divided into three general groups on the basis of their crystalline arrangement and it is observed that roughly similar engineering properties are connected with all the clay minerals belonging to the same group. An initial study of the crystal structure of clay minerals leads to a better understanding of the behaviour of clays under different conditions of loading. Table 2.4 gives the groups of minerals and some of the important minerals under each group.

2.14   ATOMS AND ATOMIC BONDS

Atoms

Molecules of minerals are composed of atoms of chemical elements and these are the basic building blocks of all matter. The atoms of an element are formed of basic particles, viz., protons, neutrons and electrons. An atom consists of a nucleus with one or more protons, each carrying a positive electromagnetic charge, and mayor may not contain one or more neutrons carrying no charge. The number of satellite electrons revolving about the nucleus is the same as the protons but carry the opposite charge, that is, the negative charge. The individual elements are determined by the atomic number which is equal to the number of protons in the nucleus or the number of electrons attached to the nucleus of their atom. Each atom of every element is electrically balanced since the number of protons and electrons are equal in number and carry opposite charges.

There are six known major energy levels in which an electron or a group of electrons revolve around the nucleus of an atom. Each energy level is termed as a shell, only a limited number of electrons can exist in anyone energy level or shell. Table 2.5 gives the number of electrons that can exist at each shell. The closest and the farthest shells to the nucleus are termed as K and P respectively.

TABLE 2.5

Electrons that can exist at or below different shell or energy levels

Shell	No. of electrons at each shell level	Total number of electrons at and below each shell level
K	2	2
L	8	10
M	8	18
N	18	36
0	18	54
P	32	86

TABLE 2.6

Valence of some elements

Elements	No. of electrons in atom	Valence	Elements	No. of electrons in atom	Valence
H	1	± 1	Na	11	1
O	8	− 2	Ca	20	2
Cl	17	− 1	K	19	1
Si	14	± 4	Mg	12	2
Al	13	+ 3	P	15	− 3
Fe	26	Varies	S	16	± 2
			C	6	− 4

Valence

Valence of an element is the number of electrons that is in excess or deficient at a shell level. If the electrons are in excess it is termed as positive valence, and negative if deficient. For example, magnesium atom contains 12 electrons and protons with 10 of the electrons below shell level M, and, at shell level M. It is an excess of two above the shell level L or lacks 6 to complete the shell level M. The least of the two is 2 which is taken as the valence of Mg and is positive. Table 2.6, gives the valence of some of the common elements that compose the rocks of the earth crust.

Atomic Bonds

Atoms unite together to form molecules and molecules to aggregates. There are certain forces that bind them together. The nature of these bonding forces is not completely understood. However, some of the important bonding forces that are recognised as useful in the field of soil mechanics are:

1. Primary valence bond

2. Secondary valence bond

Primary Valence Bond

Primary valence bond is a chemical combination of two or more elements because of the lack of a complete complement of electrons in their outermost shells. One atom joins with another atom by adding electrons to its outer shell or by losing them to arrive at a stable compound. The number of electrons an atom gains or loses depends upon the valence of the element given in Table 2.6. Atoms which lose or gain electrons in this manner are called *ions* and the forces binding them together are called *Ionic Bonds*. For example in the formation of a molecule, two ions of Al join with three ions of 0 to form Al_2O_3. This is possible because each Al ion has an excess of 3 electrons in its outer ring and 0 lacks 2 electrons in its outer ring. A stable compound can be formed since the two Al ions are in excess of 6 electrons and the three 0 ions lack 6 electrons. The ionic bonds are the strongest of the bonds that hold atoms together.

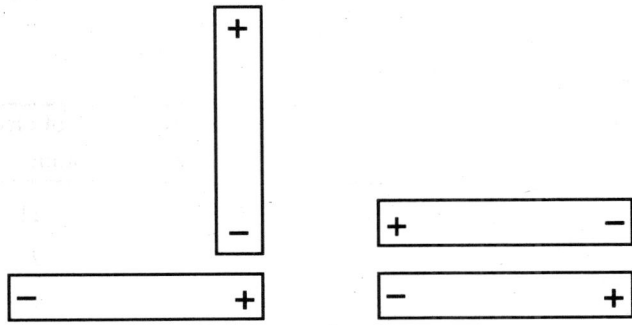

Fig. 2.2   Water molecules as oriented dipoles

Secondary Valence Bonds

Atoms in one molecule bonding to atoms in another molecule (intermolecular bonds) are called *Secondary Bonds*. These bonds are of two types. They are:

1.   Van der Waals forces

2.   Hydrogen bonds

Van der Waals forces are the attractive forces between the surface of two parallel clay mineral particles separated by water. These forces depend entirely on the crystal structure of the minerals and on the distance of separation. This force increases with the decrease in the distance between the surfaces of the minerals. Van der Waals postulated in 1873 the existence of a common attractive force acting between all atoms and molecules of matter. Subsequently it was proposed that this force was caused by two oriented dipoles. A common example of such a bond is the attractive force between molecules of water. The water molecule is a permanent dipole similar to a bar magnet. When these dipoles are oriented as shown in Fig. 2.2 the molecules are held by the interaction of the magnetic fields.

The hydrogen atom contains only one electron in its shell and as such it may be considered as having one excess or lacking one electron. The hydrogen atom can therefore enter into combination with two atoms of negative charge. This bond between the hydrogen proton and two anions is called the *hydrogen bond* which is a weak bond as compared to the ionic bond.

2.15   STRUCTURE OF CLAY MINERALS

Clay Minerals are essentially crystalline in nature though some clay minerals do contain material which is non-crystalline (for example allophane). Two fundamental building blocks are involved in the formation of clay mineral structures. They are:

1.   Tetrahedral unit.

2.   Octahedral unit.

The tetrahedral unit consists of four oxygen atoms (or hydroxyls, if needed to balance the structure) placed at the apices of a tetrahedron enclosing a silicon atom which combine together to form a shell-like structure with all the tips pointing in the same direction. The oxygen at the bases of all the units lie in a common plane.

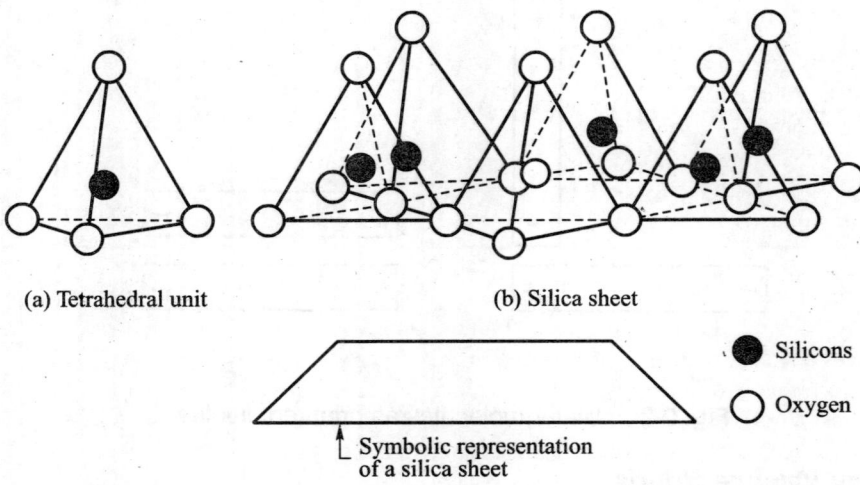

(a) Tetrahedral unit   (b) Silica sheet

● Silicons

○ Oxygen

Symbolic representation
of a silica sheet

Fig. 2.3   Basic structural units in the silicon sheet

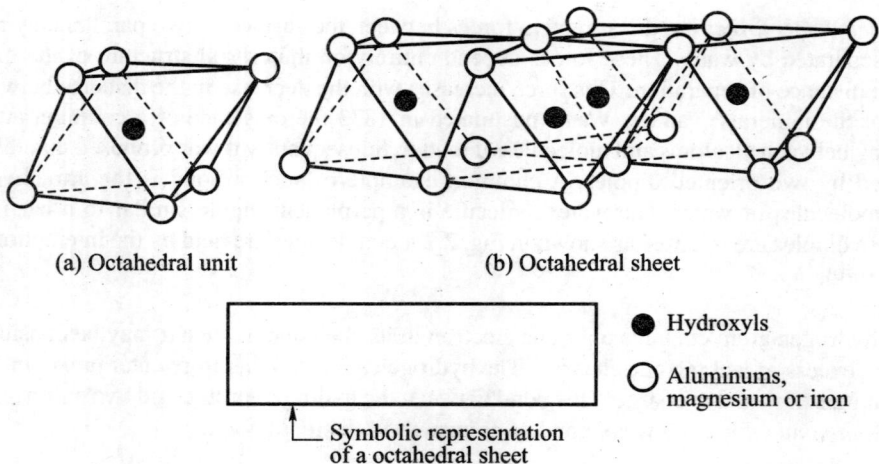

(a) Octahedral unit   (b) Octahedral sheet

● Hydroxyls

○ Aluminums,
   magnesium or iron

Symbolic representation
of a octahedral sheet

Fig. 2.4   Basic structural units in octahedral sheet

Each of the oxygen ions at the base is common to two units. The arrangement is shown in Fig. 2.3. The oxygen atoms are negatively charged with two negative charges each and the silicon with four positive charges. Each of the three oxygen ions at the base shares its charges with the adjacent tetrahedral unit. The sharing of charges leaves three negative charges at the base per tetrahedral unit and this along with two negative charges at the apex makes a total of 5 negative charges to balance the 4 positive charges of the silicon ion. The process of sharing the oxygen ions at the base with neighboring units leaves a net charge of –1 per unit.

The second building block is an octahedral unit with six hydroxyl ions at apices of an octahedral enclosing an aluminum ion at the centre. Iron or magnesium ions may replace aluminium ions in some units. These octahedral units are bound together in a sheet structure with each hydroxyl ion common to three octahedral units. This sheet is sometimes termed as *gibbsite* sheet. The Al ion has 3 positive charges and each hydroxyl ion divides its −1 charge with two other neighboring units. This sharing of negative charge with other units leaves a total of 2 negative charges per unit $[(1/3) \times 6]$. The net charge of a unit with an aluminium ion at the centre is +1. Figure 2.4 gives the structural arrangements of the units.

Formation of Minerals

The combination of two sheets of silica and gibbsite in different arrangements and conditions lead to the formation of different clay minerals as given in Table 2.4.

Kaolinite Mineral

This is the most common mineral of the Kaolin group. The building blocks of gibbsite and silica sheets are arranged as shown in Fig. 2.5 to give the structure of the kaolinite layer. The structure is composed of a single tetrahedral sheet and a single alumina octahedral sheet combined in units so that the tips of the silica tetrahedrons and one of the layers of the octahedral sheet form a common layer. All the tips of the silica tetrahedrons point in the same direction and towards the centre of the unit made of the silica and octahedral sheets. This gives rise to strong ionic bonds between the silica and gibbsite sheets. The thickness of the layer is about 7 Å (one angstrom = 10^{-8} cm) thick. The kaolinite mineral is formed by stacking the layers of 7 Å thick one above the other with the base of the silica sheet bonding to hydroxyls of the gibbsite sheet by hydrogen bond. Since hydrogen bonds are comparatively strong, the kaolinite crystals consist of many sheet stackings that are difficult to dislodge. The mineral is therefore, stable, and water cannot enter between the sheets to expand the unit cells. The lateral dimensions of kaolinite particles range from 1000 to 20,000 Å and the thickness varies from 100 to 1000 Å.

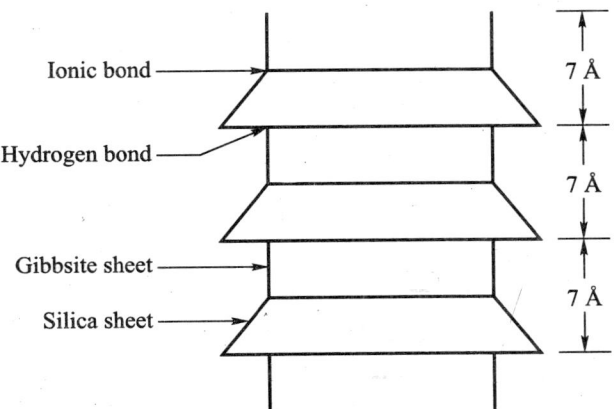

Fig. 2.5   Structure of kaolinite layer

Halloysite Mineral

Halloysite minerals are made up of successive layers with the same structural composition as those composing kaolinite. In this case, however, the successive units are randomly packed and may be separated by a single molecular layer of water. The dehydration of the interlayers by the removal of the water molecules leads to changes in the properties of the mineral. An important structural feature of halloysite is that the particles appear to take tubular forms as opposed to the platy shape of kaolinite.

Montmorillonite Mineral

It is the most common mineral of the montmorillonite group. The structural arrangement of this mineral is composed of units made of two silica tetrahedral sheets with a central alumina octahedral sheet. All the tips of the tetrahedron point in the same direction and toward the centre of the unit. The silica and gibbsite sheets are combined in such a way that the tips of the tetrahedrons of each silica sheet and one of the hydroxyl layer, of the octahedral sheet form a common layer. The atoms common to both the silica and gibbsite layer become oxygen instead of hydroxyls. The thickness of the silica-gibbsite-silica unit is about 10 Å (Fig. 2.6). In stacking these combined units one above the other, oxygen layers of each unit are adjacent to oxygen of the neighbouring units with a consequence that there is a very weak bond and an excellent cleavage between them. Water can enter between the sheets, causing them to expand significantly and thus the structure can break into 10 Å thick structural units. Thus the soils containing a considerable amount of montmorillonite minerals will exhibit high swelling and shrinkage characteristics. The lateral dimensions of montmorillonite particles range from 1000 to 5000 Å with thickness varying from 10 to 50 Å. Bentonite clay and the black cotton soil belong to the montmorillonite group.

Illite

The basic structural unit of illite is similar to that of montmorillonite except that some of the silicons are always replaced by aluminium atoms and the resultant charge deficiency is balanced by potassium ions. The potassium ions occur between unit layers. The bonds with the nonexchangeable K^+ ions are weaker than the hydrogen bonds, but stronger than the water bond of montmorillonite. Illite,

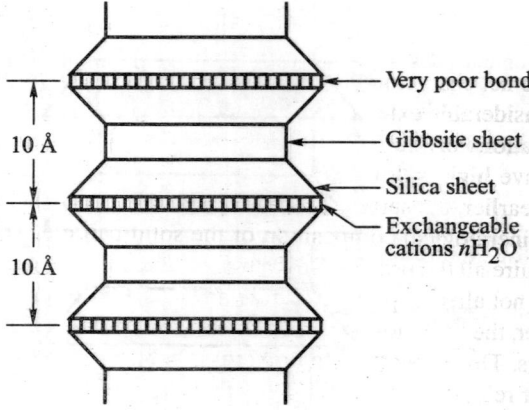

10 Å

10 Å

Very poor bond

Gibbsite sheet

Silica sheet

Exchangeable cations $n\mathrm{H_2O}$

Fig. 2.6 Structure of montmorillonite layer

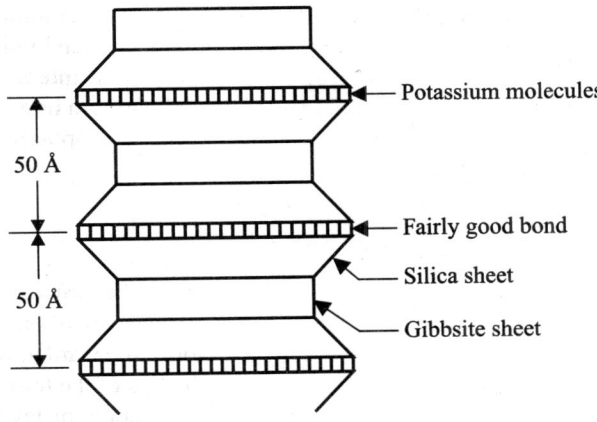

Fig. 2.7   Structure of illite layer

therefore, does not swell as much in the presence of water as does montmorillonite. The lateral dimensions of illite clay particles are about the same as of montmorillonite, 1000 to 5000 Å, but the thickness of illite particles is greater than that of montmorillonite particles, 50 to 500 Å. The arrangement of silica and gibbsite sheets are as shown in Fig. 2.7.

2.16   CLAY PARTICLE-WATER RELATIONS

Introduction

The property of a soil mass depends upon the behaviour of the discrete particles composing the mass and the pattern of particle arrangement. In all these cases water plays an important part. The behaviour of the soil mass is profoundly influenced by the inter-particle-water relationships, the ability of the soil particles to adsorb exchangeable cations and the amount of water present.

Adsorbed Water

We have seen earlier the net charge on the surface of every particle is negative. The intensity of the charge depends to a considerable extent on the mineralogical character of the particle. The physical and chemical manifestations of the surface charge constitute the surface activity of the mineral. Minerals are said to have high or low surface activity, depending on the intensity of the surface charge. As pointed out earlier, the surface activity depends not only on the specific surface but also on the chemical and mineralogical composition of the solid particle. The surface activity of sand, therefore, shall not acquire all the properties of a true clay, even if it is ground to a fine powder. The presence of water does not alter its properties of coarser fractions considerably excepting changing its unit weight. However, the behaviour of a saturated soil mass consisting of fine sand might change under dynamic loadings. This aspect of the problem is not considered here. This article deals only with clay particle-water relations.

In nature every soil particle is surrounded by water. Since the centres of positive and negative charges of water molecules do not coincide, the molecules behave like dipoles. The negative charge

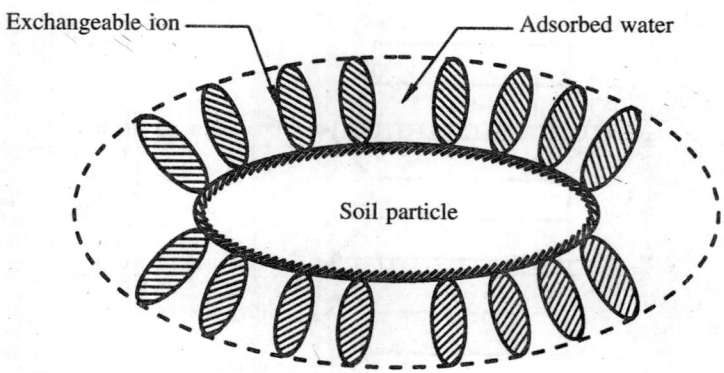

Exchangeable ion ———   ——— Adsorbed water

Soil particle

Fig. 2.8   Adsorbed water layer surrounding a soil particle

on the surface of the soil particle, therefore, attracts the positive (hydrogen) end of the water molecules. The water molecules are arranged in a definite pattern in the immediate vicinity of the boundary between solid and water. More than one layer of water molecules sticks on the surface with considerable force and this attractive force decreases with the increase in the distance of the water molecule from the surface. The electrically attracted water that surrounds the clay particle is known as the *diffused double-layer of water*. The water located within the zone of influence is known as the *Adsorbed Layer* as shown in Fig. 2.8. Within the zone of influence the physical properties of the water are very different from those of free or normal water at the same temperature. Near the surface of the particle the water has the property of a solid. At the middle of the layer it resembles a very viscous liquid and beyond the zone of influence, the properties of the water become normal. The adsorbed water affects the behaviour of clay particles when subjected to external stresses, since it comes between the particle surfaces. To drive off the adsorbed water, the clay particle must be heated to more than 200 °C, which would indicate that the bond between the water molecules and the surface is considerably greater than that between normal water molecules.

The adsorbed film of water on coarse particles is thin in comparison with the diameter of the particles. In fine grained soils, however, this layer of adsorbed water is relatively much thicker and might even exceed the size of the grain. The forces associated with the adsorbed layers therefore play an important part in determining the physical properties of the very fine-grained soils, but have little effect on the coarser soils.

Soils in which the adsorbed film is thick compared to the grain size have properties quite different from other soils having the same grain sizes but smaller adsorbed films. The most pronounced characteristic of the former is their ability to deform plastically without cracking when mixed with varying amounts of water. This is due to the grains moving across one another supported by the viscous interlayers of the films. Such soils are called *cohesive soils*, for they do not disintegrate with pressure but can be rolled into threads with ease. Here the cohesion is not due to direct molecular interaction between soil particles at the points of contact but to the shearing strength of the adsorbed layers that separate the grains at these points.

Base Exchange

Electrolytes dissociate when dissolved in water into positively charged cations and negatively charged anions. Acids break up into cations of hydrogen and anions such as Cl or SO_4. Salts and bases split into metallic cations such as Na, K or Mg, and nonmetallic anions. Even water itself is an electrolyte, because of very small fraction of its molecules always dissociate into hydrogen ions H^+ and hydroxyl ions OH^-. These positively charged H^+ ions migrate to the surface of the negatively charged particles and form what is known as adsorbed layer. These H^+ ions can be replaced by the other cations like Na, K or Mg. These cations enter the adsorbed layers and constitute what is termed as an *Adsorption complex*. The process of replacing cations of one kind by those of another in an adsorption complex is known as *base exchange*. By base exchange is meant the capacity of colloidal particles to change the cations adsorbed on their surface. Thus a hydrogen clay (colloid with adsorbed H cations) can be changed to sodium clay (colloid with adsorbed Na cations) by a constant percolation of water containing dissolved Na salts. Such changes can be used to decrease the permeability of a soil. Not all adsorbed cations are exchangeable. The quantity of exchangeable cations in a soil is termed *Exchange Capacity*.

The base exchange capacity is generally defined in terms of the mass of a cation which may be held on the surface of 100 gm dry mass of mineral. It is generally more convenient to employ a definition of base exchange capacity in milli-equivalents (meq) per 100 gm dry soil. One meq is one milligram of hydrogen or the portion of any ion which will combine with or displace 1 milligram of hydrogen. This can be explained with examples.

Example 2.1

It is required to determine the exchange capacity of clay if 100 gm of dry clay adsorbs 200 mgm of calcium.

Solution

Unit mass of Ca = 40; Unit mass of H = 1

Ca^{++} ion has two positive charges whereas the H^+ ion has one positive charge.

Therefore one meq of Ca = 40/2 = 20 mgm of Ca.

Since 100 gm of dry clay adsorbs 200 mgm of Ca, the clay's cation exchange capacity in meq = 200/20 = 10 meq/100 gm of clay.

Example 2.2

If one cubic metre of clay has a dry unit mass of 1600 kg, find the quantity of H and Ca that can be adsorbed by this soil.

Solution

Since 100 gm of soil can adsorb 10 mgm of hydrogen (as per Example 1),

1600 kg of soil can adsorb = (10/100 x 100) x 1600 = 0.16 kg of H.

The quantity of Ca that can be adsorbed is 20 times of H, since one meq of Ca = 20) µg of Ca. Therefore the quantity of Ca that can be adsorbed is 3.2 kg.

The relative exchange capacity of some of the clay minerals is given in Table 2.7.

TABLE 2.7

Exchange capacity of some clay minerals

Mineral group	Exchange capacity (meq per 100 g)
Kaolinites	3.8
Illites	40
Montmorillonites	80

TABLE 2.8

Cations arranged in the order of decreasing shear strength of clay

$$NH_4^+ > H^+ > K^+ > Fe^{+++} > Al^{+++} > Mg^+ > Ba^{++} > Ca^{++} > Na^+ > Li^+$$

If one element, such as H, Ca, or N a prevails over the other in the adsorption complex of a clay, the clay is sometimes given the name of this element, for example H-clay or Ca-clay. The thickness and the physical properties of the adsorbed film surrounding a given particle, depend to a large extent on the character of the adsorption complex. These films are relatively thick in the case of strongly water-adsorbent cations like Li^+ and Na^+ cations but very thin for H^+. The films of other cations have intermediate values. Soils with adsorbed Li^+ and Na^+ cations are relatively more plastic at low water contents and possess smaller shear strength because the particles are separated by thicker viscous film. Some of the cations in Table 2.8 are arranged in the order of decreasing shear strength of clay.

Sodium clays in nature are a product either of the deposition of clays in sea water or of their saturation by saltwater flooding or capillary action. Calcium clays are formed essentially by fresh water sediments. Hydrogen clays are a result of prolonged leaching of a clay by pure or acidic water, with the resulting removal of all other exchangeable bases.

2.17   SOIL MASS STRUCTURE

The orientation of particles in a mass depends on the size and shape of the grains as well as upon the minerals of which the grains are formed. The structure of soils that is formed by natural deposition could be altered due to external forces. Figures 2.9 to 2.11 give the various types of structures of soil. Figure 2.9(a) is a *single grained structure* which is formed by the settlement of coarse grained soils in suspension in water. Figure 2.9(b) is a *flocculent structure* formed by the deposition of fine soil fraction in water. Figure 2.9(c) is a *honeycomb structure* which is formed by the disintegration of flocculent structure under superimposed load. The particles oriented in flocculent structure will have edge to face contact as shown in Fig. 2.10(a) whereas in honeycomb structure, the particles will have face to face contact as shown in Fig. 2.10(b). Natural clay sediments will have more or less flocculated particle orientations. Marine clays generally have more open structure than fresh water clays. Figure 2.11 shows the schematic view of salt water and fresh water deposits.

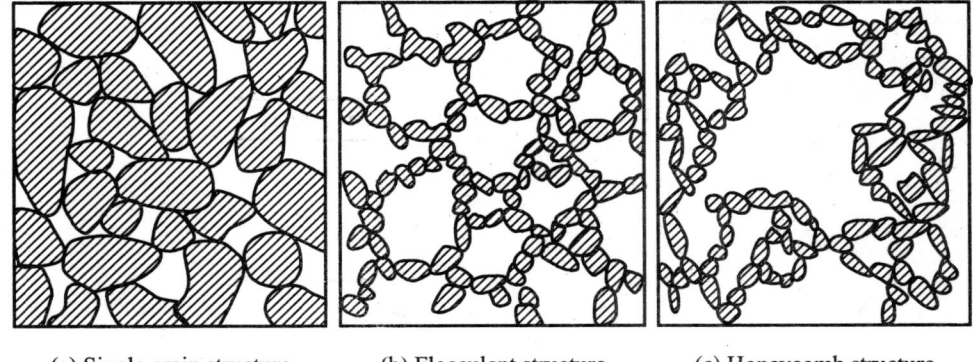

(a) Single grain structure   (b) Flocculent structure   (c) Honeycomb structure

Fig. 2.9

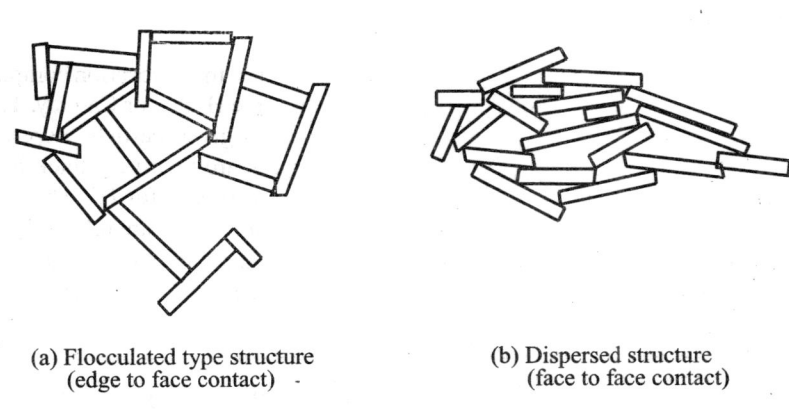

(a) Flocculated type structure   (b) Dispersed structure
(edge to face contact) -   (face to face contact)

Fig. 2.10

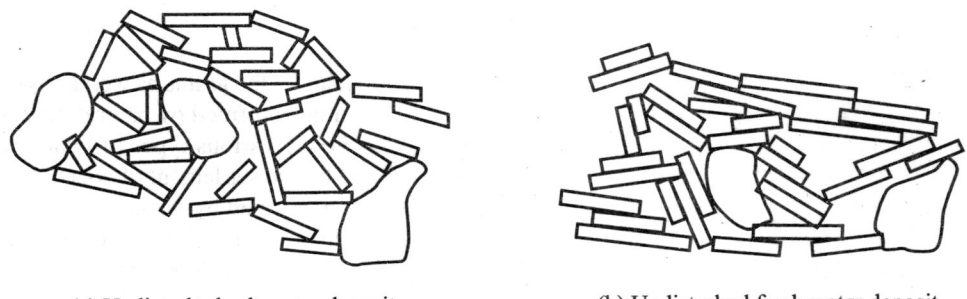

(a) Undisturbed salt water deposit   (b) Undisturbed fresh water deposit

Fig. 2.11   Schematic diagrams of salt water and fresh water deposits

CHAPTER 3

SOIL PHASE RELATIONSHIP

3.1 INTRODUCTION

Soil mass is generally a three phase system. It consists of solid particles, liquid and gas. For all practical purposes, the liquid may be considered to be water (although in some cases, the water may contain some dissolved salts) and the gas as air. The phase system may be expressed in SI units either in terms of mass-volume or weight-volume relationships. The inter relationships of the different phases are important since they help to define the condition or the physical make-up of the soil.

Mass-Volume Relationship

In SI units, the mass M, is normally expressed in kg and the density ρ in kg/m^3. Sometimes, the mass and densities are also expressed in gm and gm/cm^3 or Mg and Mg/m^3 respectively. The density of water ρ_o at 4 °C is exactly 1.00 g/cm^3 (= 1000 kg/m^3 = 1 Mg/m^3). Since the variation in density is relatively small over the range of temperatures encountered in ordinary engineering practice, the density of water ρ_w at other temperatures may be taken the same as that at 4 °C. The volume is expressed either in cm^3 or m^3.

Weight-Volume Relationship

Unit weight or weight per unit volume is still the common measurement in geotechnical engineering practice. The density ρ, may be converted to unit weight, γ by using the relationship.

$$\gamma = \rho g \tag{3.1a}$$

The 'standard' value of g is 9.807 m/s^2 (= 9.81 m/s^2 for all practical purposes).

Conversion of Density of Water ρ_w to Unit Weight γ_w

From Eq. (3.1a)

$$\gamma_w = \rho_w g \tag{3.1b}$$

Substituting ρ_w = 1000 kg/m^3 and g = 9.81 m/s^2, we have

$$\gamma_w = 1000 \frac{kg}{m^3}\left(9.82 \frac{m}{s^2}\right) = 9810 \frac{kg.m}{m^3 s^2} \tag{3.1c}$$

Since $1N$ (newton) $= \dfrac{kg.m}{s^2}$, we have,

$$\gamma_w = 9810 \frac{N}{m^3} = 9.81 \frac{kN}{m^3} \tag{3.1d}$$

or $\gamma_w = \left(1 \times \dfrac{g}{cm^3}\right) \times 9.81 = 9.81 \, kN/m^3 \tag{3.1e}$

In general, the unit weight of a soil mass may be obtained from the equation

$$\gamma = 9.81\rho \ kN/m^3 \tag{3.1f}$$

where in Eq. (3.1f), ρ is in g/cm^3. For example, if a soil mass has a dry density, $\rho_d = 1.7$ g/cm^3, the dry unit weight of the soil is

$$\gamma_d = 9.81 \times 1.7 = 16.68 \ kN/m^3 \tag{3.1g}$$

3.2   MASS-VOLUME RELATIONSHIPS

The phase-relationships in terms of mass-volume and weight-volume for a soil mass are shown by a block diagram in Fig. 3.1. A block of unit sectional area is considered. The volumes of the different constituents are shown on the right side and the corresponding mass/weights on the right and left sides of the block. The mass/weight of air may be assumed as zero.

Volumetric Ratios

There are three volumetric ratios, that are very useful in geotechnical engineering and these can be determined directly from the phase diagram, Fig. 3.1.

1.   The *void ratio*, e, is defined as

$$e = \frac{V_v}{V_s} \tag{3.2}$$

where, V_v = volume of voids, and V_s = volume of the solids.

The void ratio e is normally expressed as a *decimal*.

2.   The *porosity n* is defined as

$$n = \frac{V_v}{V} \times 100\% \tag{3.3}$$

where, V = total volume of the soil sample.

The *porosity n* is always expressed as a *percentage*.

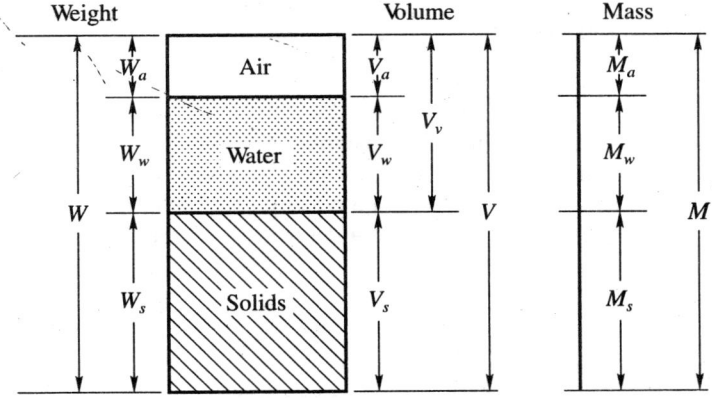

Fig. 3.1 Block diagram

3. The *degree of saturation S* is defined as

$$S = \frac{V_w}{V_v} \times 100\%$$ (3.4)

where, V_w = volume of water

It is always expressed as a *percentage* of the total volume of the voids V_v. When $S = 0\%$, the soil is completely dry, and when $S = 100\%$, the soil is fully saturated.

4. The air void ratio, A_v in a sample is expressed as

$$A_v = \frac{V_a}{V}\% = n(1-S)100$$ (3.4a)

Air content, A_c is expressed as

$$A_c = \frac{V_a}{V_v}\% = (1-S)\times 100$$ (3.4b)

where S is expressed as a fraction.

Mass-Volume Relationships

The other aspects of the phase diagram connected with mass or weight can be explained with reference to Fig. 3.1.

Water Content, *w*

The water content, w, of a soil mass is defined as the ratio of the mass of water, M_w, in the voids to the mass of solids, M_s, as

$$w = \frac{M_w}{M_s} \times 100$$ (3.5)

The water content, which is usually expressed as a percentage, can range from zero (dry soil) to several hundred percent. The natural water content for most soils is well under 100%, but for the soils of volcanic origin (for example bentonite) it can range up to 500% or more.

Density

Another very useful concept in geotechnical engineering is *density* which is expressed as mass per unit volume. There are several commonly used densities. These may be defined as total (or bulk), or moist density, ρ_t; the dry density, ρ_d; the saturated density, ρ_{sat}; the density of the particles, solid density, ρ_s; and density of water ρ_w. Each of these densities is defined as follows with respect to Fig. 3.1.

Total Density, $$\rho_t = \frac{M}{V} \tag{3.6}$$

Dry density, $$\rho_d = \frac{M_s}{V} \tag{3.7}$$

Saturated density, $$\rho_{sat} = \frac{M}{V} \qquad \text{for } S = 100\% \tag{3.8}$$

Density of solids, $$\rho_s = \frac{M_s}{V_s} \tag{3.9}$$

Density of water, $$\rho_w = \frac{M_w}{V_w} \tag{3.10}$$

Specific Gravity

The specific gravity of a substance is defined as the ratio of its mass in air to the mass of an equal volume of water at reference temperature, 4 °C. The specific gravity of a mass of soil (including air, water and solids) is termed as mass specific gravity G_m. It is expressed as

$$G_m = \frac{\rho_t}{\rho_w} = \frac{M}{V\rho_w} \tag{3.11}$$

The specific gravity of solids, G, (excluding air and water) is expressed by

$$G = \frac{\rho_s}{\rho_w} = \frac{M}{V_s\rho_w} \tag{3.12}$$

Interrelationships of different parameters

We can establish relationships between the different parameters defined by equations from (3.2) through (3.12). In order to develop the relationships, the block diagram (Fig. 3.2) is made use of. Since the sectional area perpendicular to the plane of the paper is assumed as unity, the heights of

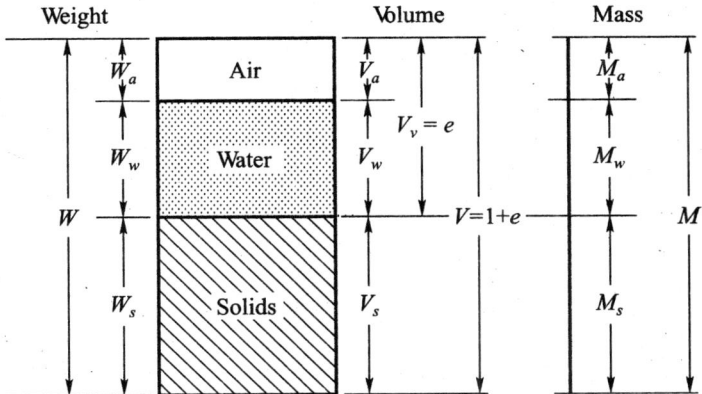

Fig. 3.2 Block diagram

the blocks will represent the volumes. The volume of solids may therefore be represented as $(V_s = 1)$. When the soil is fully saturated, the voids are completely filled with water.

Relationship between e and n (Fig. 3.2)

$$e = \frac{V_v}{V_s} = \frac{V_v}{1} = V_v, \ n = \frac{V_v}{V} = \frac{e}{1+e}$$

or $e = \dfrac{n}{1-n}$ (3.13)

Relationship between e, G and S

Case 1: When partially saturated $(S < 100\%)$

$$S = \frac{V_w}{V_v} = \frac{V_w}{e}; \text{But } V_w = \frac{M_w}{\rho_w} = \frac{wM_s}{\rho_w} = \frac{wGV_s\rho_w}{\rho_w} = wG$$

Therefore, $S = \dfrac{wG}{e}$ or $e = \dfrac{wG}{S}$ (3.14a)

Case 2: When saturated $(S = 100\%)$

From Eq. (3.14a), we have (for $S = 1$)

$e = wG$ (3.14b)

Relationships between density ρ and other parameters

The density of soil can be expressed in terms of other parameters for cases of soil (1) partially saturated $(S < 100\%)$; (2) fully saturated $(S = 100\%)$; (3) Fully dry $(S = 0)$; and (4) submerged.

Case 1: For $S < 100\%$

$$\rho_t = \frac{M}{V} = \frac{M_s(1+w)}{1+e} = \frac{G\rho_w(1+w)}{1+e} \qquad (3.15)$$

From Eq. (3.13) $w = eS/G$; substituting for w in Eq. (3.15), we have

$$\rho_t = \frac{\rho_w(G+eS)}{1+e} \qquad (3.16)$$

Case 2: For $S = 100\%$

From Eq. (3.16)

$$\rho_t = \rho_{sat} = \frac{\rho_w(G+e)}{1+e} \qquad (3.17)$$

Case 3: For $S = 0\%$

From Eq. (3.16)

$$\rho_t = \rho_d = \frac{\rho_w G}{1+e} \qquad (3.18)$$

Case 4: When the soil is submerged

If the soil is submerged, the density of the submerged soil ρ_b, is equal to the density of the saturated soil reduced by density of water.

That is

$$\rho_b = (\rho_{sat} - \rho_w) = \frac{\rho_w(G+e)}{1+e} - \rho_w = \frac{\rho_w(G-1)}{1+e}$$

$$(3.19)$$

Relative Density (or Density Index I_D)

The looseness or denseness of sandy soils can be expressed numerically by index density I_D, defined by the equation

$$I_D = \frac{e_{max} - e}{e_{max} - e_{min}} \times 100 \qquad (3.20)$$

in which

e_{max} = void ratio of sand in its loosest state having a dry density of ρ_{dm}

e_{min} = void ratio in its densest state having a dry density of ρ_{dM}

e = void ratio under in-situ condition having a dry density of ρ_d

From Eq. (3.18), a general equation for e may be written as

$$e = \frac{\rho_w G - \rho_d}{\rho_d}$$

Now substituting the corresponding dry densities for e_{max}, e_{min} and e in Eq. (3.20) and simplifying, we have

$$I_D = \frac{\rho_{dM}}{\rho_d} \times \frac{\rho_d - \rho_{dm}}{\rho_{dM} - \rho_{dm}} \times 100 \tag{3.21}$$

The loosest state for a granular material can usually be created by allowing the dry material to fall into a container from a funnel held in such a way that the free fall is about one centimeter. The densest state can be established by a combination of static pressure and vibration of soil packed in a container.

3.3   WEIGHT-VOLUME RELATIONSHIPS

The weight-volume relationships can be established from the earlier equations by substituting γ for ρ and W for M. The various equations are tabulated below.

1. Water content, $$w = \frac{W_w}{W_s} \times 100 \tag{3.5a}$$

2. Total unit weight, $$\gamma_t = \frac{W}{V} \tag{3.6a}$$

3. Dry unit weight, $$\gamma_d = \frac{W_s}{V} \tag{3.7a}$$

4. Saturated unit weight, $$\gamma_{sat} = \frac{W}{V} \tag{3.8a}$$

5. Unit weight of solids, $$\gamma_s = \frac{W_s}{V_s} \tag{3.9a}$$

6. Unit weight of water, $$\gamma_w = \frac{W_w}{V_w} \tag{3.10a}$$

7. Mass specific gravity, $$G_m = \frac{W}{V\gamma_w} \tag{3.11a}$$

8. Specific gravity of solids, $$G = \frac{W_s}{V_s\gamma_w} \tag{3.12a}$$

9. Total unit weight for $S < 100$, $$\gamma_t = \frac{G\gamma_w(1+w)}{1+e} \tag{3.15a}$$

or $$\gamma_t = \frac{\gamma_w(G+eS)}{1+e} \tag{3.16a}$$

10. Saturated unit weight, $\gamma_{sat} = \dfrac{\gamma_w(G+e)}{1+e}$ (3.17a)

11. Dry unit weight, $\gamma_d = \dfrac{\gamma_w G}{1+e}$ (3.18a)

12. Submerged unit weight, $\gamma_b = \dfrac{\gamma_w(G-1)}{1+e}$ (3.19a)

13. Density index, $I_D = \dfrac{\gamma_{dM}}{\gamma_d} \times \dfrac{\gamma_d - \gamma_{dm}}{\gamma_{dM} - \gamma_{dm}}$ (3.21a)

3.4   GENERAL REMARKS

The void ratios of natural sand deposits depend upon the shape of the grains, the uniformity of grain size, and the conditions of sedimentation. The void ratios of clay soils range from anything less than unity to 5 or more. The soils with higher void ratios have a loose structure and generally belong to the montmorillonite group. The specific gravity of solid particles of most of the soils varies from 2.5 to 2.9. For most of the calculations, G can be assumed as 2.65 for cohesionless soils and 2.70 for clay soils. The dry unit weights (γ_d) of granular soils range from 14 to 18 kN/m^3, whereas, the saturated unit weights of fine grained soils can range from 12.5 to 22.7 kN/m^3. Table 3.1 gives typical values of porosity, void ratio, water content (when saturated) and unit weights of various types of soils.

TABLE 3.1

Porosity, void ratio, water content and unit weights of typical soils in Natural State

S. no.	Description	Porosity n %	Void ratio e	Water content w%	Unit weight kN/m^3 γ_d	γ_{sat}
1	2	3	4	5	6	7
1	Uniform sand, loose	46	0.85	32	14.0	18.5
2	Uniform sand, dense	34	0.51	19	17.0	20.5
3	Mixed-grained sand, loose	40	0.67	25	15.6	19.5
4	Mixed-grained sand, dense	30	0.43	16	18.2	21.2
5	Glacial till, mixed grained	20	0.25	9	20.8	22.7
6	Soft glacial clay	55	1.20	45	11.9	17.3
7	Soft glacial clay	37	0.60	22	16.7	20.3
8	Soft slightly organic clay	66	1.90	70	9.1	15.5
9	Soft highly organic clay	75	3.00	110	6.8	14.0
10	Soft bentonite	84	5.20	194	4.2	12.4

3.5 DETERMINATION OF SPECIFIC GRAVITY

Specific gravity as such does not indicate the behaviour of a soil mass under external loads, but it is an important factor which is used in computing other soil properties, for example, the void ratio of a soil, its unit weight, soil particle size determination by means of the hydrometer method. It is also used in consolidation studies of clays, in calculating the degree of saturation of a soil, and in other calculation. Therefore, greater attention should be given for a precise determination of the specific gravity of a soil.

Laboratory Method of Specific Gravity Determination

A pycnometer or constant volume method has been found to be the most reliable for the determination of specific gravity. Commonly, about 200 g of dry mass of sample, a 500 cc constant volume bottle, and distilled water are used. Figure 3.3(A) gives a specific gravity bottle.

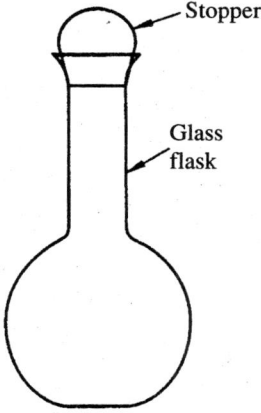

Fig. 3.3(A)   Specific gravity bottle

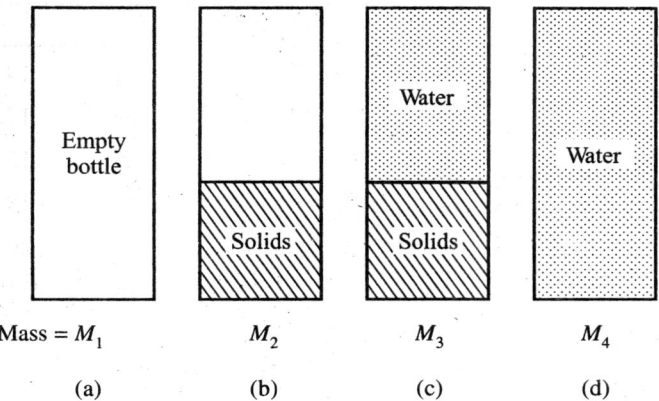

Fig. 3.3(B)   Block diagram for specific gravity determination

    In the computation of the specific gravity of a soil from laboratory data, the mass of the specific gravity bottle filled with distilled water at the test temperature will be needed. This value is usually taken from a plot of temperature versus weight of bottle plus water. The plot or calibration curve, can be determined either by experimental or by theoretical means (the students should refer to any book on Soil Testing for details). Next it is required to determine the mass of bottle with a known mass of soil and water filled just to the constant volume mark. Greatest-care should be taken to expel all air from the mixture. The air may be expelled by a gentle boiling or by suction by applying a partial vacuum to the suspension or by both. The mass is taken after the bottle with the suspension is cooled to the room temperature. The measurements that are used in the computation of specific gravity are:

(a) Mass of empty dry bottle $= M_1$,

(b) Mass of bottle + dry soil sample $= M_2$,

(c) Mass of bottle + soil + water $= M_3$, and

(d) Mass of bottle + water $= M_4$,

Now the following equations can be written,

From (a) and (b)

Mass of dry soil mass $= M_s = M_2 - M_1$ (e)

From (b) and (c)

Mass of water in (c) $= \overline{M}_w = M_3 - M_2$ (f)

From (d) and (a)

Mass of water in (d) $= M_w = (M_4 - M_1)$ (g)

From (g) and (f)

Mass of water equivalent to the volume of soil solids $= M'_w = M_w - \overline{M}_w$ (h)

or $M'_w = (M_4 - M_1) - (M_3 - M_2)$

The specific gravity of solids is defined as

$$G = \frac{\text{mass of dry soil}}{\text{mass of an equi. vol. of water}}$$

or $G = \dfrac{M_s}{M'_w} = \dfrac{M_2 - M_1}{(M_4 - M_1) - (M_3 - M_2)}$

or $G = \dfrac{M_s}{M_s - (M_3 - M_4)}$ (3.22)

3.6   DETERMINATION OF WATER CONTENT

Two methods may be used for determining the water content of a wet sample of soil. They are

1. Oven dry method

2. Pycnometer method

Oven Dry Method

The following steps describe briefly the method.

1.   Take a clean container with lid. Let the mass of container with lid $= M_1$.

2.   Take about 300 g of the wet sample (whose water content is required to be determined) in the container.

3.   Keep the container with the wet sample (lid removed) in a thermostatically controlled oven and dry the same with the temperature of the oven kept at about 110 °C. The drying period depends upon the type of soil. For example, clay sample may take 10 to 15 hours, whereas sandy sample take about 4 hours. After drying, cool the container with the sample in a desiccator. Replace the lid on the container. Now let the mass of container with lid + dry sample $= M_3$.

The water content of the sample may be expressed as follows

$$\text{Water content,} \quad w = \frac{M_2 - M_3}{M_3 - M_1} \times 100 \text{ percent} \qquad (3.23)$$

Pycnometer Method

A pycnometer is a glass bottle of about 1 litre capacity, fitted with a brass conical cap by means of a screw type cover as shown in Fig. 3.4(A). The cap has a small hole at the top.
The various steps to be followed for determining water content, w, are as follows,

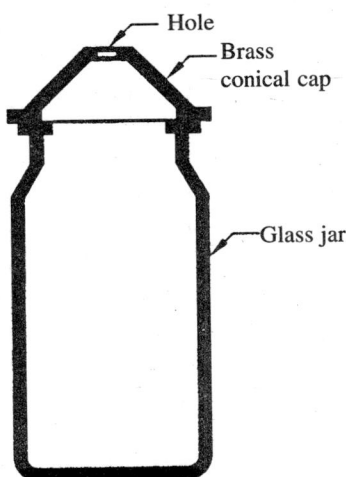

Fig. 3.4(A)   Pycnometer bottle for water content determination

(a) Find the mass of pycnometer (M_1).

(b) Fill the pycnometer with about 300 g of wet sample.

Mass of pycnometer + wet soil = M_2

(c) Fill water into the pycnometer upto about half the height. Stir water with the sample by using a glass rod. Continue stirring as more and more water is added. Finally, bring the water level flush with the hole in the conical cap. Now

Mass of pycnometer + soil + water = M_3

(d) Empty the pycnometer of all its contents and clean it. Next fill it with water upto the top level of the hole in the cap.

Mass of pycnometer + water = M_3

Let,

ρ_w = density of water, ρ_s = density of solids,

V_s = volume of dry solids of the sample, G = specific gravity of the solids.

The equation for the water content of the sample may be obtained as follows [refer Fig. 3.4(B)].

Let M_s = mass of solid particles in Fig. 3.4B(c).

Volume of $M_s = V_s = \dfrac{M_s}{G}$

The equation for the water content of the sample may also be obtained with reference to Fig. 3.4(B).

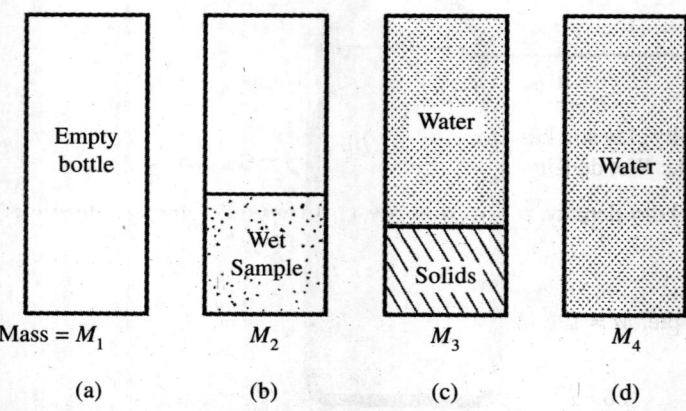

Fig. 3.4(B)   Block diagram for water content determination

In Fig. 3.4(B), M_s is the mass of soil particles and its volume is equal to M_s/G. Thus, if the soil particles from (c) are replaced with water of mass M_s/G, we get the mass M_4 in (d) of Fig. 3.4(B). We may now write

$$M_4 = M_3 - M_s + \frac{M_s}{G}$$

$$M_s\left(\frac{G-1}{G}\right) = M_3 - M_4$$

From which

$$M_s = (M_3 - M_4)\left(\frac{G}{G-1}\right)$$

Now the mass of water in the wet sample is

$$M_w = (M_2 - M_1) - M_s$$

Since water content is

$$w = \frac{M_w}{M_s} \times 100$$

we may write

$$w = \frac{(M_2 - M_1) - M_s}{M_s} \times 100 = \left(\frac{(M_2 - M_1)}{M_s} - 1\right) \times 100$$

or $\qquad w = \left(\frac{M_2 - M_1}{M_3 - M_4} \times \frac{G-1}{G} - 1\right) \times 100$ $\hspace{2cm}$ (3.24)

Example 3.1

A sample of wet silty clay soil has a mass of 126 kg. The following data were obtained from laboratory tests on the sample: Wet density, $\rho_t = 2.1$ g/cm^3, $G = 2.7$, water content, $w = 15\%$.

Determine (i) dry density, ρ_d, (ii) porosity, (iii) void ratio, and (iv) degree of saturation.

Solution

Mass of sample, $M = 126$ kg.

Volume $\quad V = \dfrac{126}{2.1 \times 10^3} = 0.06 \text{ m}^3$

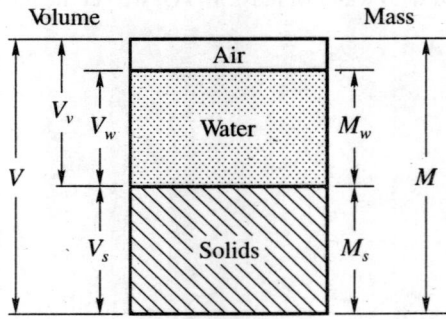

Fig. Ex. 3.1

Now, $M_s + M_w = M$, or $M_s + wM_s = M_s(1+w) = M$

Therefore, $M_s = \dfrac{M}{1+w} = \dfrac{126}{1.15} = 109.57 \text{ kg}; M_w = M - M_s = 16.43 \text{ kg}$

Now, $V_w = \dfrac{W_w}{\rho_w} = \dfrac{16.43}{1000} = 0.01643 \text{ m}^3$

$V_s = \dfrac{M_s}{G\rho_w} = \dfrac{109.57}{2.7 \times 1000} = 0.04058 \text{ m}^3$

$V_v = V - V_s = 0.06000 - 0.04058 = 0.01942 \text{ m}^3.$

(i) Dry density, $\rho_d = \dfrac{M_s}{V} = \dfrac{109.57}{0.06} = 1826.2 \text{ kg/m}^3$

(ii) Porosity, $n = \dfrac{V_v}{V} \times 100 = \dfrac{0.01942 \times 100}{0.06} = 32\%$

(iii) Void ratio, $e = \dfrac{V_v}{V_s} = \dfrac{0.01942}{0.04058} = 0.47$

(iv) Degree of saturation, $S = \dfrac{V_w}{V_v} \times 100 = \dfrac{0.01643}{0.01942} \times 100 = 84.5\%$

Example 3.2

Earth is required to be excavated from borrow pits for building an embankment. The wet unit weight of undisturbed soil in wet condition is 18 kN/m³ and its water content is 8%. In order to build a 4 m high embankment with top width 2 m and side slopes 1 : 1, estimate the quantity of earth required to be excavated per meter length of embankment. The dry density required in the embankment is 15 kN/m³ with a moisture content of 10%. Assume the specific gravity of solids as 2.67. Also determine the void ratios and the degree of saturation of the soil in both the undisturbed and remoulded states.

Solution

The dry unit weight of soil in the borrow pit is

$$\gamma_d = \frac{\gamma_t}{1+w} = \frac{18}{1.08} = 16.7 \text{ kN/m}^3$$

Volume of embankment per meter length V_e

$$V_e = \frac{2+10}{2} \times 4 = 24 \text{ m}^3$$

The dry density of soil in the embankment is 15 kN/m³

Volume of earth required to be excavated V_{ex} per metre

$$V_{ex} = 24 \times \frac{15}{16.7} = 21.55^3 / \text{m}$$

Undisturbed state

$$V_s = \frac{\gamma_d}{G\gamma_w} = \frac{16.7}{2.67} \times \frac{1}{9.81} = 0.64 \text{ m}^3; \ V_v = 1 - 0.64 = 0.36 \text{ m}^3$$

$$e = \frac{0.36}{0.64} = 0.56, \ W_w = 18.0 - 16.7 = 1.3 \text{ kN}$$

Degree of saturation, $S = \dfrac{W_w}{V_v} \times 100$, where

$$V_w = \frac{W_w}{\gamma_w} = \frac{1.3}{9.81} = 0.133 \text{ m}^3$$

Now, $\quad S = \dfrac{0.133}{0.36} \times 100 = 36.9\%$

Remoulded state

$$V_s = \frac{\gamma_d}{G\gamma_w} = \frac{15}{2.67 \times 9.81} = 0.57 \text{ m}^3$$

$$V_v = 1 - 0.57 = 0.43 \text{ m}^3$$

$$e = \frac{0.43}{0.57} = 0.75; \quad \gamma_t = \gamma_d(1+w) = 15 \times 1.1 = 16.5 \text{ kN/m}^3$$

Therefore, $W_w = 16.5 - 15.0 = 1.5 \text{ kN}$

$$V_w = \frac{1.5}{9.81} = 0.153 \text{ m}^3$$

$$S = \frac{0.153}{0.43} \times 100 = 35.6\%$$

Example 3.3

The moisture content of an undisturbed sample of clay belonging to a volcanic region is 265% under 100% saturation. The specific gravity of the solids is 2.5. The dry unit weight is 3.3 kN/m³. Determine (i) the saturated unit weight, (ii) the submerged unit weight, and (iii) void ratio.

Solution

(i) Saturated unit weight, $\gamma_{sat} = \gamma_t$

$$W = \gamma_t = W_w + W_s = wW_s + W_s = W_s(1 + w)$$

$$= 3.3(1 + 2.65) = 3.3 \times 3.65 = 12 \text{ kN/m}^3$$

(ii) Submerged unit weight, γ_b

$$\gamma_b = \gamma_{sat} - \gamma_w = 12.0 - 9.81 = 2.19 \text{ kN/m}^3$$

(iii) Void ratio, e

$$V_s = \frac{\gamma_d}{G\gamma_w} = \frac{3.3}{2.5 \times 9.81} = 0.132 \text{ m}^3$$

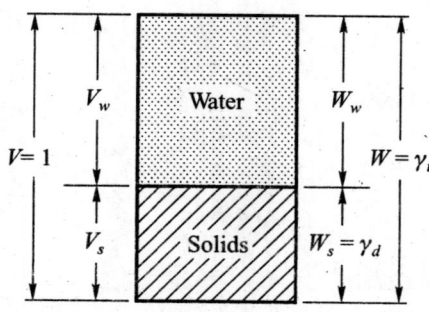

Fig. Ex.3.3

$$V_v = V_w = \frac{w \times W_s}{\gamma_w} = 2.65 \times \frac{3.3}{9.81} = 0.89 \, m^3$$

$$e = \frac{V_v}{V_s} = \frac{0.89}{0.132} = 6.74$$

Example 3.4

A sample of saturated clay from a consolidometer test has a total mass of 1.526 kg and a dry mass of 1.053 kg: the specific gravity of the solid particles is 2.7. For this sample, determine the water content, void ratio, porosity and total density.

Solution

$$w = \frac{M_w}{M_s} \times 100\% = \frac{1526 - 1053}{1053} = 44.9\%$$

$$e = \frac{wG}{S} = \frac{0.45 \times 2.7}{1} = 1.212$$

$$n = \frac{e}{1+e} = \frac{1.212}{1+1.212} = 0.548$$

$$\rho_t = \frac{(2.7 + 1 \times 1.212)(1)}{1 + 1.212} = 1.768 \, g/cm^3 = 17.68 \, kN/m^3$$

Example 3.5

A sample of silty clay has a volume of 14.88 cm³, a total mass of 28.81 g, a dry mass of 24.83 g, and a specific gravity of solids 2.7. Determine the void ratio and the degree of saturation.

Solution

Void ratio

$$V_s = \frac{M_s}{G\rho_w} = \frac{24.83}{2.7(1)} = 9.2 \, cm^3$$

$$V_v = V - V_s = 14.88 - 9.2 = 5.68 \, cm^3$$

$$e = \frac{V_w}{V_s} = \frac{5.68}{9.2} = 0.618$$

Degree of saturation

$$w = \frac{M_w}{M_s} = \frac{28.81 - 24.83}{24.83} = 0.16$$

$$S = \frac{Gw}{e} = \frac{2.7(0.16)}{0.618} = 0.70 \text{ or } 70\%$$

Example 3.6

A soil sample in its natural state has a mass of 2.29 kg and a volume of 1.15×10^{-3} m^3. Under an oven dried state, the dry mass of the sample is 2.035 kg. The specific gravity of the solids is 2.68. Determine the total density, water content, void ratio, porosity, and degree of saturation and the air void ratio.

Solution

$$\rho_t = \frac{W}{V} = \frac{2.29}{1.15 \times 10^{-3}} = 1990 \text{ kg/m}^3$$

$$w = \frac{W_w}{M_s} = \frac{2.29 - 2.035}{2.035} = 0.125 \text{ or } 12.5\%$$

$$e = \frac{V_v}{V_s}, \quad V_s = \frac{M_s}{G\rho_w} = \frac{2035}{2.68(1)} = 759.33 \text{ cm}^3$$

$$V_v = V - V_s = 1150 - 759.33 = 390.67 \text{ cm}^3$$

$$e = \frac{390.67}{759.33} = 0.51$$

$$n = \frac{e}{1+e} = \frac{0.51}{1+0.51} = 0.34 \text{ or } 34\%$$

$$S = \frac{wG}{e} = \frac{0.125 \times 2.68}{0.51} = 0.656 = 65.6\%$$

$$A_v = n(1-S) = 0.34(1-0.656) = 0.12 \text{ or } 12\%$$

Example 3.7

A soil sample has a total unit weight of 16.97 kN/m^3 and a void ratio of 0.84. The specific gravity of the soil particles is 2.70. Determine the moisture content, dry unit weight and degree of saturation of the sample.

Solution

Degree of saturation [from Eq. (3.16a)]

$$\gamma_t = \frac{\gamma_w(G+eS)}{1+e} \quad \text{or} \quad 16.97 = \frac{(9.81)(2.7+0.84S)}{1+0.84} \quad \text{or} \quad S = 58\%$$

Dry unit weight (Eq. 3.18a)

$$\gamma_d = \frac{\gamma_w G}{1+e} = \frac{(9.81)2.7}{1+0.84} = 14.4 \text{ kN/m}^3$$

Water content (Eq. 3.14a)

$$w = \frac{Se}{G} = \frac{0.58 \times 0.84}{2.7} = 0.18 \text{ or } 18\%$$

Example 3.8

A soil sample in its natural state has, when fully saturated, a water content of 32.5%. Determine the void ratio, dry and total unit weights. Calculate the total weight of water required to saturate a soil mass of volume 10 m³. Assume $G = 2.69$.

Solution

Void ratio (Eq. 3.14a)

$$e = \frac{wG}{S} = \frac{32.5 \times 2.69}{(1) \times 100} = 0.874$$

Total unit weight (Eq. 3.15a)

$$\gamma_t = \frac{G\gamma_w(1+w)}{1+e} = \frac{2.69(9.81)(1+0.325)}{1+0.874} = 18.7 \text{ kN/m}^3$$

Dry unit weight (Eq. 3.18a)

$$\gamma_d = \frac{\gamma_w G}{1+e} = \frac{2.69 \times 9.81}{1+0.874} = 14.08 \text{ kN/m}^3$$

From Eq. (3.6a), $W = \gamma_t V = 18.66 \times 10 = 186.6$ kN

From Eq. (3.7a), $W_s = \gamma_d V = 14.08 \times 10 = 140.8$ kN

Weight of water $= W - W_s = 186.6 - 140.8 = 45.8$ kN

Example 3.9

A soil mass in its natural state is partially saturated having a water content of 17.5% and a void ratio of 0.87. Determine the degree of saturation, total unit weight, and dry unit weight. What is the weight of water required to make a mass of 10 m³ volume to get saturated? Assume $G = 2.69$.

Solution

Degree of saturation, $S = \dfrac{wG}{e} = \dfrac{0.175 \times 2.69 \times 100}{0.87} = 54.11\%$

Total unit weight, $\gamma_t = \dfrac{G\gamma_w(1+w)}{1+e} = \dfrac{2.69(1+0.175) \times 9.81}{1+0.87} = 16.58 \, kN/m^3$

Dry unit weight, $\gamma_d = \dfrac{\gamma_t}{1+w} = \dfrac{16.58}{1+0.175} = 14.11 \, kN/m^3$

Extra weight of water required for full saturation

$$\gamma_t = \frac{\gamma_w(G+e)}{1+e} = \frac{(2.69 + 0.87)9.81}{1.87} = 18.68 \, kN/m^3$$

For saturation of 10 m³ of mass,

extra water required = $(18.68 - 16.58) \times 10 = 21 \, kN$

Example 3.10

The void ratio of a clay sample is 0.5 and the degree of saturation is 70%. Compute the water content, dry and wet unit weights of the soil. Assume $G = 2.7$.

Solution

Water content, $w = \dfrac{eS}{G} = \dfrac{0.5 \times 0.7}{2.7} = 0.13 = 13\%$

Bulk unit weight, $\gamma_t = \dfrac{G\gamma_w(1+w)}{1+e} = \dfrac{2.7 \times 9.81(1+0.13)}{1.5} = 19.95 \, kN/m^3$

Dry unit weight, $\gamma_d = \dfrac{\gamma_t}{1+w} = \dfrac{19.95}{1.13} = 17.65 \, kN/m^3$

Example 3.11

A sample of soil compacted according to standard proctor test has a unit weight of 20.58 kN/m³ at 100% compaction and at optimum water content of 14%. What is the dry unit weight? What is the dry unit weight at zero air voids? If the voids become filled with water what would be the saturated unit weight? Assume $G = 2.67$.

Solution

As per Eq. (3.5a), $W_w = wW_s = 0.14\,W_s$

Assume in Fig. Ex. 3.1, $V = 1\ m^3$.

Hence, $W_w + W_s = \gamma_t = 20.58\ kN$

or $014\,W_s + W_s = 20.58\ kN$

Simplifying, $W_s = 18.052\ kN$, $W_w = 20.58 - 18.052 = 2.528\ kN$

As per Eq. (3.7a) $\gamma_d = \dfrac{W_s}{V} = \dfrac{18.052}{1} = 18.052\ kN/m^3$

The volume of the solids, $V_s = \dfrac{W_s}{G\gamma_w} = \dfrac{18.052}{9.81} \times \dfrac{1}{2.67} = 0.69\ m^3$

The volume of the voids $= V - V_s = 1.00 - 0.69 = 0.31\ m^3$

The volume of water $= \dfrac{2.528}{9.81} = 0.25\ m^3$

The dry unit weight at zero-air void $= \dfrac{W_s}{V_s + V_w} = \dfrac{18.052}{0.69 + 0.25} = 19.2\ kN/m^3$

If the air voids are filled with water, $W_w = 0.31 \times 9.81 = 3.04\ kN$

The satured unit weight $\gamma_{sat} = \dfrac{W}{V} = \dfrac{W_s + W_w}{V} = \dfrac{18.052 + 3.04}{1} = 21.1\ kN/m^3$

Example 3.12

A sample of sand above water table was found to have a natural moisture content of 15% and a unit weight of 18.84 kN/m³. Laboratory tests on a dried sample indicated values $e_{min} = 0.50$ and $e_{max} = 0.85$ for the densest and loosest states respectively. Compute the degree of saturation and the density index. Assume $G = 2.65$.

Solution

Refer block diagram of Fig. 3.1

Assume $V = 1\ m^3$, hence $\gamma_t = 18.84\ kN/m^3$

This is also the weight of solids plus water. Since the water content is 15%, we may write

$$w = \dfrac{W_w}{W_s} = 0.15 \text{ or } W_w = 0.15\,W_s$$

Since, $W_w + W_s = 18.84$ kN,   we have, $0.15\,W_s + W_s = 18.84$

Therefore, $W_s = \dfrac{18.84}{1.15} = 16.38$ kN, $W_w = 18.84 - 16.38 = 2.46$ kN

$V_s = \dfrac{W_s}{G\gamma_w} = \dfrac{16.38}{2.65 \times 9.81} = 0.63\,\text{m}^3$, $V_w = \dfrac{2.46}{9.81} = 0.25\,\text{m}^3$

The volume of air, $V_a = V - (V_s + V_w) = 1 - (0.63 + 0.25) = 0.12\,\text{m}^3$.

The volume of void, $V_v = V_w + V_a = 0.25 + 0.12 = 0.37\,\text{m}^3$.

The degree of saturation, $S = \dfrac{V_w}{V_v} \times 100 = \dfrac{0.25}{0.37} \times 100 = 68\%$

The void ratio, $e = \dfrac{V_v}{V_s} = \dfrac{0.37}{0.63} = 0.59$

The density index, $I_D = \dfrac{e_{max} - e}{e_{max} - e_{min}} \times 100 = \dfrac{0.85 = 0.59}{0.85 - 0.50} \times 100 = 74\%$

Example 3.13

How many cubic meters of fill can be constructed at a voids ratio of 0.7 from 1,19,000 m³ of borrow material that has a void ratio of 1.2?

Solution

(Refer Fig. 3.2)

Consider the block diagram Fig. 3.2. Here $V_s = 1$, $V = 1 + e$. The sectional area of the block 1×1 units. When the soil is in the borrow pit, $V_s = 1$ m³, $V = 1 + 1.2 = 2.2$ m³. When the soil is put in the fill, $V_s = 1$ m³, $V = 1 + 0.7 = 1.7$ m³. This means that for every 2.2 m³ of material excavated we can construct a fill of 1.7 m³. Therefore, the total quantity of fill that can be constructed is

$$= \dfrac{1.7}{2.2} \times 1,91,000 = 1,47,591\,\text{m}^3$$

Example 3.14

The natural water content of a sample taken from a soil deposit was found to be 11.5%. It has been calculated that the maximum density for the soil will be obtained when the water content reaches 21.5%. Compute how much of water must be added to each 100 kN of soil (in its natural state) in order to increase the water to 21.5%. Assume that the degree of saturation in its natural state was 40% and $G = 2.7$.

Solution

(Refer Fig. 3.2)

Consider a block of a sample of sectional area 1×1 m.

Let $V_s = 1$ m³.

For the soil in its natural state the void ratio is (Eq. 3.14a)

$$e = \frac{wG}{S} = \frac{0.115 \times 2.7}{0.4} = 0.776$$

For the same void ration of 0.776, the water content has to be increased to 21.5% to attain the maximum density. The degree of saturation required is (Eq 3.14a)

$$S = \frac{wG}{e} = \frac{0.215 \times 2.7}{0.776} = 0.748 \quad \text{or} \quad 74.8\%$$

Now for Eq. (3.4), volume of water

at 40% saturation, $V_{w1} = SV_v = S_1 e = 0.4 \times 0.776 = 0.31$ m³

at 78.4% saturation, $V_{w2} = S_{2e} = 0.784 \times 0.776 = 0.58$ m³

Excess water required $= (V_{w2} - V_{w1})\,\gamma_w = 0.27$ m³ $\times 9.81 = 2.65$ kN

Volume of mass of section 1 m $\times$ 1 m and height $(1 + e) = 1.766$ m³

Weight of mass at 40% saturation

weight of solids $= V_s G \gamma_w = 1 \times 2.7 \times 9.81 = 26.49$ kN

Weight of water $= V_{w1}\,\gamma_w = 0.31 \times 9.81 \quad = 3.04$ kN

_____

Total 29.53 kN

For every 29.53 kN of mass, the excess weight of water = 2.65 kN

For 100 kN of mass, the excess water required $= \dfrac{2.65}{29.53} \times 100 = 8.98$ kN

Example 3.15

In an oil well drilling project, drilling mud was used to retain the sides of the borewell. In one litre of suspension in water, the drilling mud fluid consists of the following material:

Material	Weight (gm)	Sp. gr
Clay	410	2.81
Sand	75	2.69
Iron fillings	320	7.13

Find the unit weight of the drilling fluid of uniform suspension.

Solution

Volume of suspension = 1000 cm^3

$$\text{Volume of clay} = \frac{410}{2.81} = 145.91 \, \text{cm}^3$$

$$\text{Volume of sand} = \frac{75}{2.69} = 27.88 \, \text{cm}^3$$

$$\text{Volume of iron fillings} = \frac{320}{7.13} = 44.88 \, \text{cm}^3$$

Total volume of solids = 218.67 cm^3

Volume of water = 1000 - 218.67 = 781.33 cm^3

Total weight of solids = 805.00 gm

Weight of water = 781.33 × 1 = 781.33 gm

Total weight of drilling fluid = 1586.33 gm

$$\text{Unit weight of suspension} = \frac{1586.33}{1000} = 1.586 \, \text{gm/cm}^3$$

Example 3.16

In a field exploration, a soil sample was collected in a sampling tube of internal diameter 5.0 cm below ground water table. The length of the extracted sample was 10.2 cm and its weight was 387 gm. If $G = 2.7$, and the weight of the dried sample is 313 gm, find the porosity void ratio, degree of saturation, and the dry density of the sample.

Solution

Internal diameter of sampling tube = 5.0 cm

Length of sample = 10.2 cm

Moist-weight of sample = 387.0 gm

Dry weight of the sample = 313.0 gm

$G = 2.7$

$$\text{Water content} = \frac{W_w}{W_s} = \frac{387-313}{313} \times 100 = 23.64\%$$

$$\text{Area of sample} = \frac{\pi}{4} \times 5^2 = 19.63 \, \text{cm}^2$$

Volume of sample = 1963 × 10.2 = 200.28 cm^3

$$\text{Bulk unit weight, } \gamma_t = \frac{387}{200.25} = 1.93 \, \text{gm/cm}^3$$

$$\text{Dry unit weight, } \gamma_d = \frac{313}{200.25} = 1.56 \, \text{gm/cm}^3$$

$$\text{or} \quad \gamma_d = \frac{\gamma_t}{1+w} = \frac{1.93}{1+0.2364} = 1.56 \, \text{gm/cm}^3$$

$$\text{since} \quad \gamma_d = \frac{G\gamma_w}{1+e} = \frac{G}{1+e}$$

$$\text{or} \quad e = \frac{G}{\gamma_d} - 1 = \frac{2.7}{1.56} - 1 = 0.73 \, \text{gm/cm}^3$$

$$n = \frac{e}{1+e} = \frac{0.73}{1.73} = 0.42 = 42\%$$

$$Se = wG, \quad \text{or} \quad S = \frac{wG}{e} = \frac{23.64 \times 2.7}{0.73} = 87.45\%$$

Example 3.17

A saturated sample of undisturbed clay has a volume of 19.2 cm^3 and weights 32.5 gm. After oven drying, the weight reduces to 20.2 gm. Determine the following

(a) water content, (b) specific gravity, (c) void ratio, (d) saturated unit weight of the clay sample.

$$\text{(a)} \quad w = \frac{32.5-20.2}{20.2} \times 100 = 60.89\%$$

$$\text{(b)} \quad \gamma_d = \frac{W_s}{V} = \frac{20.2}{19.2} = 1.05 \, \text{gm/cm}^3$$

$$e = \frac{wG}{S} = wG = 0.6089G \qquad\qquad (i)$$

From equation, $\quad \gamma_d = \dfrac{G\gamma_w}{1+e} = \dfrac{G}{1+e}, \quad$ we have

$$e = \frac{G}{\gamma_d} - 1 \qquad\qquad (ii)$$

Equating (i) and (ii), we have

$$0.6089G = \frac{G}{1.05} - 1$$

or $\qquad G = \dfrac{1}{0.361} = 2.77$

(c) $\ e = wG = 0.6089 \times 2.77 = 1.688$

(d) $\ \gamma_{sat} = \dfrac{\gamma_w(G+e)}{1+e} = \dfrac{1(2.77+1.688)}{1+1.688} = 1.6585 \ \text{gm/cm}^3$

Example 3.18

The natural bulk density of sandy stratum is 18.5 kN/m³ and has a water content of 8%. For determining of density index, dried sand from the stratum was filled loosely into a 300 cm³ mould and vibrated to give a maximum density. The loose weight of the sample in the mould was 480 gm, and the dense weight was 570 gm. If $G = 2.66$, find the density index of the sand in its natural state.

Solution

Use Eq. (3.21a)

$$I_D = \frac{\gamma_{dM}}{\gamma_d} \cdot \frac{\gamma_d - \gamma_{dm}}{\gamma_{dM} - \gamma_{dm}}$$

Given: $V = 300 \ \text{cm}^3$, $G = 2.66$, weight (loose) $= 480$ gm, weight (dense) $= 570$ gm; $\gamma_t = 18.5 \ \text{kN/m}^3$, $w = 8\%$, we have

$$\gamma_d = \frac{\gamma_t}{1+w} = \frac{18.5}{1.08} = 17.12 \ \text{kN/m}^3$$

$$\gamma_{dM} = \frac{570}{300} = 1.9 \ \text{gm/cm}^3 = 1.9 \times 9.81 = 18.64 \ \text{kN/m}^3$$

$$\gamma_{dm} = \frac{480}{300} = 1.6 \text{ gm/cm}^3 = 1.6 \times 9.81 = 15.70 \text{ kN/m}^3$$

$$I_D = \frac{18.64}{17.12} \left[\frac{17.12 - 15.70}{18.64 - 15.70} \right] \times 100 = 52.6\%$$

Example 3.19

An earth embankment is to be compacted to a density of 19 kN/m³ at a moisture content of 14 percent. The *in-situ* bulk density and water content of the borrow pit are 18 kN/m³ and 8% respectively. How much excavation should be carried out from the borrow pit for each m³ of the embankment?

Solution

Borrow pit	Embankment
$\gamma_t = 18 \text{ kN/m}^3$	$\gamma_t = 19.0 \text{ kN/m}^3$
$w = 8\%$	$w = 14\%$
$\gamma_d = \dfrac{18}{1+0.08} = 16.67 \text{ kN/m}^3$	$\gamma_d = \dfrac{19}{1+0.14} = 16.67 \text{ kN/m}^3$

The excavation from borrow pit per m³ of embankment is

$$\frac{\gamma_d \text{ (embankment)}}{\gamma_d \text{ (borrow pit)}} = \frac{16.67}{16.67} = 1$$

or $\gamma_d(\text{emb}) = \gamma_d \text{ (borrow)}$

or   1 m³ (emb) = 1 m³ (borrow)

Example 3.20

An undisturbed sample of soil has a volume of 29 cm³ and weighs 48 gm. The dry weight of the sample is 32 gm. The value of $G = 2.66$. Determine (a) natural water content, (b) the *in-situ* void ratio, (c) degree of saturation, and (d) the saturated unit weight of soil.

Solution

Given $V = 29$ cm³, W (moist) = 48 gm, W_s (dry) = 32 gm. $G = 2.66$

(a) $w_n = \dfrac{W - W_s}{W_s} = \dfrac{48 - 32}{32} \times 100 = 50\%$

(b) $\gamma_t = \dfrac{W}{V} = \dfrac{48}{29} = 1.655 \text{ g/cm}^3$

$\gamma_d = \dfrac{\gamma_t}{1+w} = \dfrac{1.655}{1+0.5} \times 9.81 = 10.82 \text{ kN/m}^3$

$e = \dfrac{G\gamma_w}{\gamma_d} - 1 = \dfrac{2.66 \times 9.81}{10.81} - 1 = 1.41$

(c) From Eq. (3.14a)

$S = \dfrac{wG}{e} = \dfrac{0.5 \times 2.66}{1.41} = 0.937 = 93.7\%$

(d) From Eq. (3.17a)

$\gamma_{\text{sat}} = \dfrac{\gamma_w(G+e)}{1+e} = \dfrac{9.81(2.66+1.41)}{1+1.41} = 16.56 \text{ kN/m}^3$

Example 3.21

A mass of soil coated with a thin layer of paraffin weighs 4.37×10^{-3} kN. When immersed in water it displaces 3.2×10^{-4} m^3 of water. The paraffin is peeled of and found to weigh 1.77×10^{-4} kN. The specific gravity of the soil particles is 2.7 and that of paraffin is 0.9. Determine the void ratio of the soil if its water content is 10%.

Solution

Wt. of soil + paraffin $= 4.76 \times 10^{-4}$ kN

Wt. of paraffin $= 1.77 \times 10^{-4}$ kN

Wt. of soil mass $= 4.58 \times 10^{-3}$ kN

Vol. of soil + paraffin $= 3.2 \times 10^{-4}$ m^3

Vol. of paraffin $= \dfrac{1.77 \times 10^{-4}}{0.9 \times 9.81} = 2.0 \times 10^{-5}$ m^3 $= 0.2 \times 10^{-4}$ m^3

Vol. of soil $= 3.2 \times 10^{-4} - 0.2 \times 10^{-4} = 3.0 \times 10^{-4}$ m^3

Bulk unit weight, γ_t $= \dfrac{4.58 \times 10^{-3}}{3 \times 10^{-4}} = 15.27 \text{ kN/m}^3$

or $\quad \gamma_t = \dfrac{(1+w)G\gamma_w}{1+e} = \dfrac{(1+0.1)\times 2.7 \times 9.81}{1+e} = 15.27$

Simplifying, we have,

$$e = \frac{1.1 \times 2.7 \times 9.81}{15.27} - 1 = 0.908$$

Example 3.22

225 g of soil was oven dried and placed in a specific gravity bottle and then filled with water upto a constant volume mark made on the bottle. The mass of bottle with water and soil is 1650 g. The specific gravity bottle was filled with water alone upto the constant volume mark and weighed. It's mass was found to be 1510 g. Determine the sp. gravity of the soil.

Solution

Given $M_s = 225$ g, $M_3 = 1650$ g and $M_4 = 1510$ g.

From Eq. (3.22)

$$G = \frac{M_s}{M_s - (M_3 - M_4)} = \frac{225}{225 - (1650 - 1510)} = 2.65$$

Example 3.23

It is required to determine the water content of a wet silty sandy sample of soil weighing 400 g. This mass of soil was placed in a pyenometer and water filled upto the top of the conical cap and weighed (M_3). Its mass was found to be 2350 g. The pycnometer was next filled with clean water and weighed and its mass was found to be 2200 g (M_4). Assuming $G = 2.67$, determine the water content of the soil sample.

Solution

From Eq. (3.24), the equation for water content is

$$w = \left(\frac{M_2 - M_1}{M_3 - M_4} \times \frac{G-1}{G} - 1 \right) \times 100$$

where, mass of wet soil $= M_s = (M_2 - M_1) = 400$ g, $M_3 = 2350$ g and $M_4 = 2200$ g.

By substituting, we have

$$w = \left(\frac{400}{2350 - 2200} \times \frac{2.67 - 1}{2.67} - 1 \right) \times 100 = 66.8\%$$

3.7   QUESTIONS AND PROBLEMS

3.1   A soil sample has a volume of 160 cc and a mass of 304 g when partially saturated and
      269.28 g when dry. The specific gravity of the solid particles is 2.64. Determine the porosity,
      the void ratio, the water content and the degree of saturation.

3.2   A clay sample is found to have a mass of 423.53 g in its natural state. It is then dried in an
      electric oven at 105 °C. The dried mass is found to be 337.65 g. The specific gravity of the
      solid is 2.70 and the density of the soil mass in its natural stale is 1700 kg/m³. Determine
      the water content, degree of saturation and the dry density of the mass in its natural state.

3.3   A sample of sand in its natural state has a relative density of 65 percent. The dry unit
      weight of the sample at its densest and loosest states are respectively 18.0 and 14 kN/m³.
      Assuming the specific gravity of the solids as 2.64, determine (i) its dry density (ii) wet
      density when full saturated and (iii) submerged density.

3.4   A fully saturated sample of clay soil has a void ratio of 2.3 and a moisture content of 85
      percent in its natural state. Assuming the specific gravity of the solids as 2.7, determine its
      dry, bulk and the submerged unit weights of the clay.

3.5   A clay fill has bulk unit weight of 19 kN/m³. If the moisture content is 25 percent and the
      specific gravity of the particles is 2.7, determine the degree of saturation.

3.6   A soil sample has a bulk unit weight of 21 kN/m³ and the degree of saturation is 80
      percent. Determine its void ratio and water content if the specific gravity of the solids is
      2.70.

3.7   The maximum and minimum void ratios that could be obtained from a sample of sand are
      0.85 and 0.53 respectively. Determine the relative density of the sample in its natural state
      if its dry unit weight is 15.3 kN/m³. Assume a suitable value for the specific gravity of the
      solids.

3.8   The saturated unit weigt of a clay sample is 20.7 kN/m³. Its water content is 25 percent.
      Determine the void ratio, the dry and the submerged unit weights of the sample. Assume
      a suitable value for the specific gravity of the solids.

3.9   The mass of wet sample of soil in a drying dish is 462 g. The sample and the dish have a
      mass of 364 g after drying in an oven at 110 °C overnight. The mass of dish alone is 39 g.
      Determine the water content of the soil.

3.10 The total density, ρ_t of a mass of soil is 1.76 Mg/m³ and its water content is 10%. Calculate
      ρ_d, e, n, S, and ρ_{sat}. Draw a phase diagram. Assume density of solids $\rho_s = 2.70$ Mg/m³.

3.11 A soil sample, extracted from the field has $e = 0.62$, $w_n = 15$ percent and $\rho_s = 2.65$ Mg/m³.
      Determine: ρ_d, ρ_t, w for $S = 100\%$, and ρ_{sat} for S 100%.

3.12 The following information were known about a sample of soil brought to the laboratory
      for testing: $\rho_s = 2700$ kg/m³, $S = 100\%$, and $w = 46\%$. It is required to know e, ρ_{sat}, ρ_b.

3.13 A sample of soil which is fully saturated has a mass of 1350 g in its natural state and 975
      g after drying. Determine the natural water content of the soil.

3.14 Determine the water content of a fully saturated soil with dry density, equal to 1.70 Mg/m³.
      Assume $\rho_s = 2.71$ Mg/m³.

3.15 A sample of soil has a dry density of 1.65 Mg/m^3 and the value of ρ_s is 2.68 Mg/m^3. Determine w, e, P_{sat}.

3.16 The void ratio of a clay sample is 0.5 and $S = 70\%$. Assuming $\rho_s = 2750$ kg/m^3, calculate, w, ρ_d and ρ_t.

3.17 The volume of water in a sample of moist soil is 0.056 m^3. The volume of solids $V_s = 0.28$ m^3. Determine w. Assume $\rho_s = 2590$ kg/m^3.

3.18 A sample of dry quartz sand has a density of 1.68 Mg/m^3. Determnine ρ_t for $S = 75\%$. Assume $\rho_s = 2.65$ Mg/m^3.

3.19 A sample of saturated clay has a natural water content of 43%. The specific gravity of the solid matter is 2.70. What are the void ratio, e the porosity, n; and the saturated unit weight of the soil, γ_{sat}?

3.20 A sample of sand above the water table was found to have a natural moisture content of 15% and a unit weight of 18.85 kN/m^3. Laboratory tests on a dried sample indicated values $e_{min} = 0.50$, and $e_{max} = 0.85$ for the densest and loosest states respectively. Compute the degree of saturation, S, and the relative density D_r: Assume $G = 2.65$.

3.21 A 50 cm^3 sample of moist clay was obtained by pushing a sharpened hollow cylinder into the wall of a test pit. The extruded sample had a mass 85 g, and after oven drying a mass of 60 g. Compute w, e, S, and ρ_d if $\rho_s = 2.7$ Mg/m^3.

3.22 A sample of saturated clay has a water content of 56%. Compute e, γ_{sat} and n if $G = 2.72$.

3.23 A soil sample has a mass of 5 kg, and a volume of 2605.5 cm^3. If $w = 13.4\%$, and $\rho_s = 2.65$ Mg/m^3, compute ρ_t, ρ_d, e, n and S.

3.24 A pit sample of soft saturated clay has a volume of 100 cm^3 and a mass of 175 g. If the oven-dry mass is 120 g, compute w, e, and ρ_s.

3.25 A pit sample of moist quartz sand was obtained from a pit by core-cutter method. The volume of the sample obtained was 150 cm^3 and its total mass was found to be 250 g. In the laboratory the dry mass of the sand alone was found to be 240 g. Tests on dry sand indicated $e_{max} = 0.80$ and $e_{min} = 0.48$. Estimate ρ_s, w, e, S, ρ_d and D_r of the sand in the field.

3.26 The water content of a saturated clay of volcanic origin is 400%. If $\rho_s = 2.4$ Mg/m^3, compute ρ_{sat}, e, and n.

3.27 The dry density of a compacted sand sample is 20 kN/m^3. Estimate G and determine w at $S = 100\%$. What would be the unit weight if $S = 20\%$?

3.28 Laboratory tests on a sample of uniform sand give the following data: $n = 43\%$ and $w_n = 12\%$; $e_{max} = 0.85$, $e_{min} = 0.55$. Assume that $G = 2.65$ and compute the void ratio, relative density, degree of saturation, and dry unit weight.

3.29 A laboratory test on a saturated clay soil gave $e = 2.5$. Calculate its water content and dry density if $G = 2.70$.

3.30 800 m^3 of earth fill is to be constructed at a void ratio of 0.75. How many cubic metre of earth is to be excavated from a borrow pit in which the void ratio is 1.1?

3.31   An earthen embankment under construction has a bulk unit weight of 15.7 kN/m³ and a moisture content of 10 percent. Compute the quantity of water in litres required to be added per cubic metre of earth to raise its moisture content to 14 percent at the same void ratio.

3.32   The wet unit weight of a glacial outwash soil is 19.2 kN/m³, the specific gravity of the solid particles of the soil is $G = 2.67$, and the moisture content of the soil is $w = 12\%$ by dry weight. Calculate (a) dry unit weight, (b) porosity, (c) void ratio, (d) degree of saturation, and (e) percent of air voids. Draw the phase diagram of this soil showing the absolute and relative volumetric and gravimetric relationships.

3.33   The moisture content of a saturated clay sample is 345%. The unit weight of the solids is 23.4 kN/m³. Determine the saturated unit weight of clay. Draw a phase diagram.

3.34   The unit of the solids of a given sand is 25.6 kN/m³. Its void ratio is 0.572. Calculate (a) the unit weight of dry sand (b) the unit weight of sand when saturated, and (c) the submerged unit weight of the sand.

3.35   Derive the equation, $e = wG/100$ which expresses the relationship between the void ratio e the specific gravity, G and the moisture content, w for full saturation of voids.

3.36   A clay sample, originally 2.50 cm thick and at initial void ratio of 1.120 was subjected to a compressive load. After the clay sample was completely consolidated, the thickness of it was measured to be 2.426 cm. Compute the final void ratio.

CHAPTER 4

INDEX PROPERTIES OF SOILS

4.1 INTRODUCTION

The various properties of soils which would be considered as index properties are:

1. The size and shape of particles.

2. The relative density (or density index) or consistency of soil.

The index properties of soils can be studied in general under two classes. They are:

1. Soil grain properties.

2. Soil aggregate properties.

The principle soil grain properties are the specific gravity, size and shape of grains and the mineralogical character of the finer fractions (applied to clay soils). The most significant aggregate property of cohesionless soils is the relative density, whereas that of cohesive soils is the consistency. Water content can also be studied as an aggregate property as applied to cohesive soils. The strength and compressibility characteristics of cohesive soils are functions of water content. As such water content is an important factor in understanding the aggregate behaviour of cohesive soils. Whereas cohesionless soils are concerned, water content does not alter its properties significantly except when the mass is submerged, in which case only its unit weight gets reduced.

4.2 THE SHAPE AND SIZE OF PARTICLES

Shape of Particles

The shapes of particles as conceived by visual inspection gives only a qualitative idea with regard to the behaviour of a soil mass composed of such particles. Since particles finer than 0.075 mm diameter cannot be seen by naked eye, one can visualize the nature of the coarse grained particles only. Coarser fractions composing of, angular grains are capable of supporting heavier static loads and can be compacted to a dense mass by vibration. The influence of shape of particles on the compressibility characteristics of soils are:

1.   Reduction in the volume of mass upon the application of pressure depends upon the shape of particles.

2.   A small mixture of mica to sand will result in a large increase in its compressibility.

Size of Grains

The classification according to size divides the soil broadly into two distinctive groups, namely, cohesionless and cohesive. Since the properties of cohesionless soils are to a considerable extent, based on grain size distribution, classification of cohesionless soil according to size would therefore be helpful. Fine grained soils are so much affected by structures, shape of grains, geological origin, and other factors that their grains size distribution alone tells little about their physical properties. However, one can assess the nature of a mixed soil on the basis of the percentage of fine grained soil present in it. It is, therefore, essential to classify the soil according to grain size.

4.3   THE DETERMINATION OF GRAIN SIZE DISTRIBUTION IN SOILS

The classification of soils as gravel, sand, silt and clay as per IS: 1498-1970 scale is given in Fig. 4.1. Soil particles which are coarser than 0.075 mm are generally termed as *cohesionless* and the finer ones as silt, clay and peat (organic soil) are considered *fine grained*. From the engineering point of view, these two types of soils have distinctive characteristics. In the case of cohesionless soils, gravitational forces determine the engineering characteristics. Whereas interparticle forces are predominant in the case of fine grained soils. The dependence of the behaviour of a soil mass on the size of particles had led the investigators to classify soils according to their size. Many classification systems are prevalent. The Indian soil textural classification as per IS: 1498-1959 and IS 1498-1970 are given Fig. 4.1

The physical separation of a sample of soil by any method into two or more fractions, each containing only particles of certain sizes, is termed *fractionation*. The determination of the mass of material in fractions containing only particles of certain sizes is termed *mechanical analysis*. Mechanical analysis is one of the oldest and most common forms of soil analysis, It provides the basic information for revealing the uniformity or gradation of the materials within established size ranges and for textural classifications. A good mechanical analysis is not equally valuable in different branches of engineering. The size of the soil grains is of importance in such cases as construction of

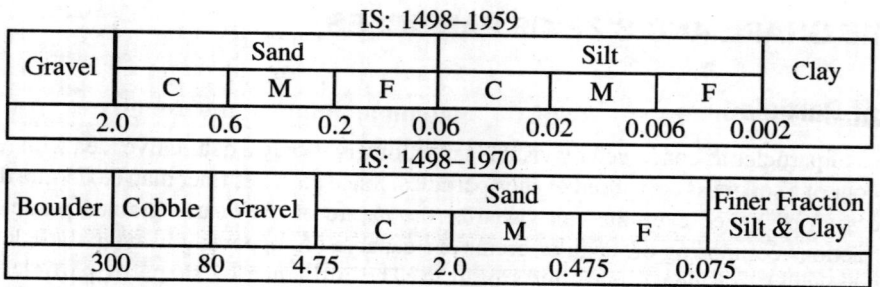

Fig. 4.1   Soil classifications (size of particles in nm)

the earth dams or rail road and highway embankments where earth is used as a material that should satisfy definite specifications. In foundations of structures, data from mechanical analysis are generally illustrative; other properties such as compressibility and shearing resistance, are of more importance. Mechanical analysis can be divided into:

1. Sieve Analysis.

2. Wet Mechanical Analysis.

The grain size distributions of coarse materials can easily be made by passing a sample through a set of sieves and weighing the amount retained on each sieve. The use of good standard set of sieves, well made and in good condition, will give sufficiently accurate results. The finest screen, which is practicable, is a sieve with mesh openings of 0.075 mm. If the sample being analysed contains a large proportion of grains below this diameter, the finer material will have to be analysed by wet mechanical analysis.

4.4 SIEVE ANALYSIS

Sieve analysis is carried out by using a set of standard sieves. Sieves are made by weaving two sets of wires at right angles to one another, The square holes thus formed between the wires provide the limit which determines the size of the particles retained on a particular sieve. The sieve sizes are given in terms of the number of openings per inch. The number of openings per inch vary according to different standards. put according to the Indian Standard, the sieve number is the mesh width expressed in mm or microns, μ (1 μ = 0.001 mm). Thus, an ASTM 60 sieve has 60 openings per inch width with each opening of 0.251 mm. The corresponding IS number for the same opening is 230 μ. Table 4.1 gives a comparison of British Standard, ASTM and the Indian Standard Sieves.

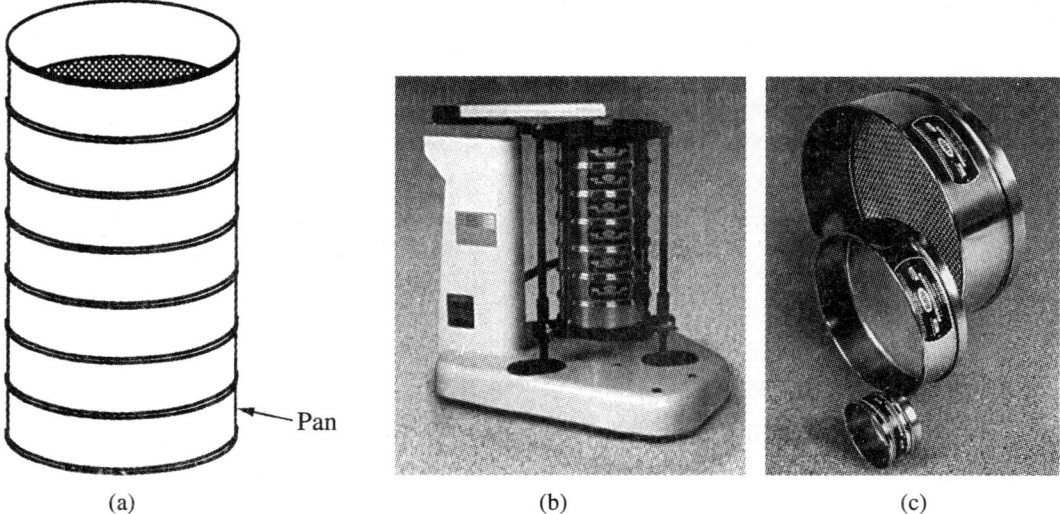

(a)   (b)   (c)

Fig. 4.2 (a) Sieves arrangement, (b) sieve shaker and (c) a set of sieves for a test in the laboratory (Courtesy: Soiltest, USA)

The usual procedure is to use a set of sieves which will yield equal grain, size intervals on a logarithmic scale. A good spacing of soil particle diameters on the grain size distribution curve will be obtained if a nest of sieves is used in which each sieve has an opening approximately one-half of the coarser sieve above it in the nest. If the soil contains gravel, the coarsest sieve that can be used to separate out gravel from sand is the 4.75 mm of IS sieve or No.4 of ASTM sieve. To separate out the silt-clay fractions from the sand fractions, 75 μ. of IS Sieve or No. 200 of BS or ASTM sieve may be used. The intermediate sieves between the coarsest and the finest may be selected on the basis of the principle explained earlier. The IS sieve of size 4.75 mm, 2.0 mm, 1.00 mm, 425 μ, 212 μ, 125 μ, and 75 μ may be used.

TABLE 4.1

Comparison of different standard sieves

| B S Sieves BS: 410–1962 | | US standard sieves sieves | | IS sieves IS: 460–1962 | |
| Designation | Aperture mm | Designation | Aperture mm | Designation | Aperture mm |
(1)	(2)	(3)	(4)	(5)	(6)
2 in	50.80	2 in	50.80	50 mm	50.00
1½ in	38.10	1½ in	38.10	40 mm	40.00
3/4 in	19.05	¾ in	19.00	20 mm	20.00
3/8 in	9.52	3/8 in	9.51	10 mm	10.00
3/16 in	4.76	4	4.75	4.75 mm	4.75
6	2.80	8	2.36	2.80 mm	2.80
8	2.00	10	2.00	2.00 mm	2.00
12	1.40	14	1.40	1.40 mm	1.40
14	1.20	16	1.18	1.18 mm	1.18
16	1.00	18	1.00	1.00 mm	1.00
25	0.600	30	0.60	600 μ	0.600
30	0.500	35	0.50	500 μ	0.50
36	0.420	40	0.425	425 μ	0.425
44	0.355	50	0.300	355 μ	0.355
60	0.250	60	0.250	250 μ	0.250
72	0.210	70	0.210	212 μ	0.212
85	0.180	80	0.180	180 μ	0.180
100	0.150	100	0.150	150 μ	0.150
120	0.125	120	0.125	125 μ	0.125
170	0.090	170	0.088	90 μ	0.090
200	0.075	200	0.075	75 μ	0.075
350	0.045	270	0.053	45 μ	0.045

The sieve analysis is carried out by sieving a known dry mass of sample through the nest of sieves placed one below the other so that the openings decrease in size from the top sieve downwards, with a pan at the bottom of the stack as shown in Fig. 4.2. The whole nest of sieves is given a horizontal shaking for about 10 minutes (if required, more) till the mass of soil remaining on each sieve reaches a constant value (the shaking can be done by hand Or using a mechanical shaker, if available). The amount of shaking required depends on the shape and number of particles. If a sizable portion of soil is retained on the No. 200 ASTM sieve (or 75 μ IS Sieve), it should be washed. This is done, by placing the sieve with a pan at the bottom and pouring clean water on the screen. A spoon or glass rod is to be used to stir the slurry. The soil which is washed through is recovered, dried and weighed. The mass of soil recovered is subtracted from the mass retained on No. 200 ASTM sieve before washing and added to the soil that has passed through No. 200 by dry sieving. The mass of soil required for sieve analysis is about 500 g of oven-dried soil with all the particles separated out by some means. By determining the mass of soil sample left on each sieve, the following calculations can be made.

1. Percentage retained on any sieve $\quad = \dfrac{\text{mass of soil retained}}{\text{Total soil mass}} \times 100$

2. Cumulative percentage retained on any sieve $\quad =$ Sum of percentages retained on all coarser sieves

3. Percentage finer than any sieve size, N $\quad =$ 100 per cent minus cumulative percentage retained.

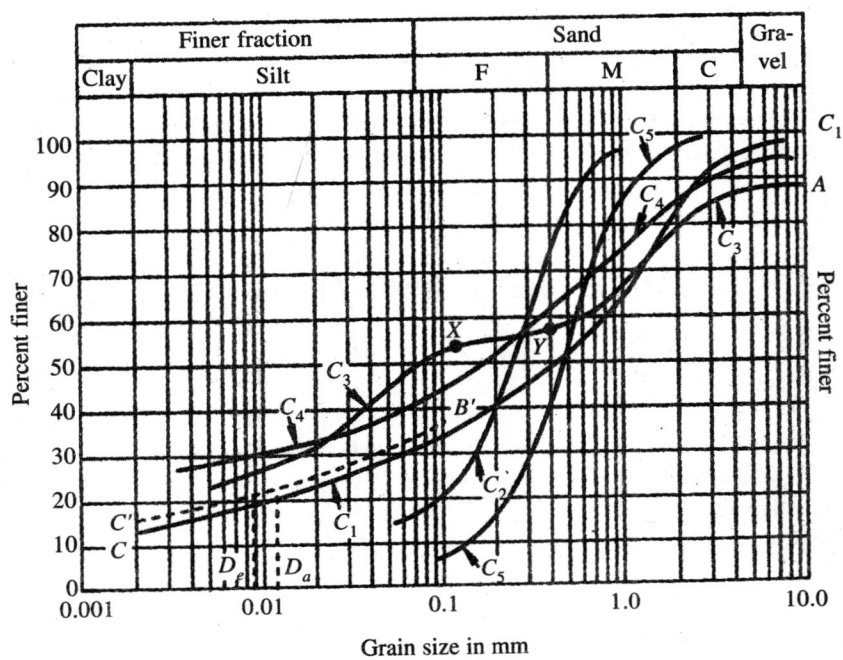

Fig. 4.3 Grain size distribution curves

The results may be plotted in the form of a graph on a semi-log paper with the percentage finer on the arithmetic scale and the particle diameter on the log scale as shown in Fig. 4.3. A smooth curve can be drawn through them depending on the distribution of sizes in the soil.

4.5 WET MECHANICAL ANALYSIS

The smallest opening that can be used in a sieve depends upon the smallest diameter of wire that can be used for the construction of mesh. Sieves with the smallest opening of 0.037 mm, are manufactured but the smallest sieve commonly used in soil mechanics work is the No. 200 sieve which has an opening of 0.075 mm. Therefore, grain size analysis of materials finer than the No. 200 sieve is made by other means known as wet mechanical analysis. The wet mechanical analysis is based on sedimentation principle of Stokes' Law. A number of methods have been proposed and numerous procedures used for determining the grain distribution curve of fine grained materials on the sedimentation principle. Some of the methods are:

1.   Successive sedimentation

2.   Sedimentation into clear water

3.   Observations of amount sediment per unit-volume at a given point in the sedimentation tube

4.   Observation of the total amount of soil in suspension above a given elevation

5.   Observation of total sedimented soil

6.   Elutriation

Of all the methods mentioned above, method 3 is the one that is generally used in the field of soil mechanics. In this method the amount of sediment per unit of volume at a given point can be determined either by a hydrometer or a pipette. Since hydrometer is quite popular amongst soil engineers, and is accurate enough for all practical purposes, only this method is described in this book.

4.6 THE HYDROMETER METHOD OF ANALYSIS

The hydrometer method was originally proposed in 1926 by Prof. Bouyoucos of Michigan Agricultural College, and later modified by Arthur Casagrande of Harvard. This method depends upon variations in the density of a soil suspension contained in a litre graduate jar 1000 cc). The density of the suspension is measured with a hydrometer at determined time Intervals; then the coarsest diameter of particles in suspension at a given time and the percentage of particles finer than that coarsest (suspended) diameter are computed. These computations are based on the Stokes' formula which is described below.

Stokes' Law

G. G. Stokes (1856), an English physicist, proposed an equation for determining the terminal velocity of a failing sphere in a liquid. According to him if a single sphere is allowed to fall through a liquid of indefinite extent, the terminal velocity, v can be expressed as,

$$v = \frac{\gamma_s - \gamma_w}{18\mu} D^2 \, \text{m/sec} = CD^2 \, \text{m/sec} \tag{4.1}$$

in which,

   v = Terminal velocity of fall of sphere through liquid in m/sec

   γ_s = Unit weight of solid sphere in kN/m³

   γ_w = Unit weight of liquid sphere in kN/m³

   μ = Absolute viscosity of liquid in kN-s/m³

   D = Diameter of sphere in m.

   Normally it is more convenient to express D in mm. A convenient form of expression for D may be developed as follows.

   The terminal velocity of a particle of diameter D falling a height H_e m in t secs may be written as

   From Eq. (4.1),

$$v = \frac{H_e}{t} = CD^2 \text{ or } D = \sqrt{\frac{1}{C}} \sqrt{\frac{H_e}{t}} \, m \tag{4.1b}$$

Now C may be express as

$$C = \frac{\gamma_s - \gamma_w}{18\mu} = \frac{\gamma_w(G-1)}{18\mu} = \frac{10.00(G-1)}{18\mu} = \frac{(G-1)}{1.8\mu} \tag{4.1c}$$

since $\gamma_s = \gamma_w G$, and $\gamma_w = 10$ kN/m³ for all practical purposes.

   Now the Eq. (4.1b) gets reduced to by expressing t in min

$$D = \sqrt{\frac{1.8\mu}{(G-1)}} \sqrt{\frac{H_e}{60t}}$$

If H_e is expressed in centimetres, and t in minutes and D in mm the equation for D is

$$\frac{D \, (\text{mm})}{1000} = \sqrt{\frac{1.8\mu}{G-1}} \sqrt{\frac{H_e}{100 \times 60t}} \tag{4.2}$$

or $$D \, (\text{mm}) = \sqrt{\frac{300\mu}{G-1}} \sqrt{\frac{H_e}{t}} = 10^{-3} F \sqrt{\frac{H_e}{t}} \tag{4.3}$$

here, $$F = 10^3 \times \sqrt{\frac{300\mu}{G-1}} \tag{4.3a}$$

TABLE 4.2

Values of factor _F_

Temp °C	μ (poise)	G				
		2.50	2.60	2.65	2.70	2.80
20	0.0101	14.31	13.86	13.65	13.44	13.07
22	0.0096	13.97	13.53	13.32	13.12	12.76
24	0.00916	13.65	13.21	13.01	12.82	12.46
26	0.00875	13.34	12.91	12.72	12.53	12.18
28	0.00836	13.04	12.64	12.44	12.25	11.91
30	0.00800	12.76	12.36	12.17	11.99	11.69
32	0.00767	12.50	12.10	11.91	11.75	11.41
34	0.00736	12.26	11.85	11.69	11.52	11.18
36	0.00706	11.98	11.61	11.45	11.27	10.96
38	0.00679	11.75	11.40	11.72	11.06	10.75
40	0.00654	11.57	11.19	11.02	11.085	10.56

The factor F is a function of G and μ. Table 4.2 gives the values of F for various values of G and μ.

The time t in minutes for a particle of diameter D (mm) to fall a height H_e (cm) may be expressed as

$$t = \frac{300\mu}{(G-1)} \frac{H_e}{D^2} \text{ min} \qquad (4.4)$$

where μ is in (kN-sec)/m^2

Assumptions of Stokes Law and Its Validity

The Stokes Law assumes spherical particles falling in a liquid of infinite extent, and all the Particles have the same unit weight g_s. The particles reach constant terminal velocity within a few seconds after it is allowed to fall.

However, the above assumptions are not strictly valid since the particles used in the analysis are not truly spherical, and sedimentation is done in a jar of limited cross-section. There will be influence of one particle over the other. The specific gravity of the solids. When particles of diameter larger than 0.2 mm are used there will be turbulence in the water immediately surrounding the falling particles which vitiates the results. If particles are smaller than 0.0002 mm, Brownian movement occurs, and hence the velocity of settlement will be too small for dependable measurement.

Since Particles are not spherical, the concept of an equivalent diameter has been introduced. A particle is said to have an equivalent diameter, D_e, if a sphere of diameter D having the same unit weight as the particle, has the same velocity of fall as the particle, For bulky grams, $D_e \approx D$, whereas for flaky particles $D/D_e = 4$ or more.

TABLE 4.3
Viscosity of water ($\times 10^{-7}$ kN-sec/m^2)

°C	0	1	2	3	4	5	6	7	8	9
0	17.94	17.32	16.74	16.19	15.68	15.19	14.73	14.29	13.87	13.48
10	13.10	12.74	12.39	12.06	11.75	11.45	11.16	10.88	10.60	10.34
20	10.09	9.84	9.61	9.38	9.16	8.95	8.75	8.55	8.36	8.18
30	8.00	7.83	7.67	7.51	7.36	7.21	7.06	6.92	6.79	6.66
40	6.54	6.42	6.30	6.18	6.08	5.97	5.87	5.77	5.68	5.58
50	5.49	5.40	5.24	5.24	5.15	5.07	4.99	4.92	4.84	4.77
60	4.70	4.63	4.50	4.50	4.43	4.37	4.31	4.24	4.19	4.13
70	4.07	4.02	3.91	3.91	3.86	3.81	3.76	3.71	3.66	3.62
80	3.57	3.53	3.44	3.44	3.40	3.36	3.32	3.28	3.24	3.20
90	3.13	3.13	3.06	3.06	3.03	2.99	2.96	2.93	2.90	2.87
100	2.84	2.82	2.76	2.76	2.73	2.70	2.67	2.64	2.62	2.59

The effect of influence of one particle over the other is minimised by limiting the mass of soil for sedimentation analysis to 50 g in the sedimentation jar of 10^3 cm^3 capacity.

Effect of Temperature on Velocity of Fall

The unit weight of water γ_w at 4°C is taken as equal to 9.81 kN/m^3. There will be variation in the density of water if the room temperature of the test varies considerably. Normally, the variation in density is quite negligible and can be neglected. However, the variation in viscosity, μ, with temperature is quite considerable, and as such the viscosity of water should be taken as applicable to the temperature of the test. The viscosity of water expressed in units of kN-sec/m^2 is given in Table 4.3. [It may be noted here 1 poise = 10^{-4} kN-sec/m^2 and 1 milli poise = 10^{-7} kN-sec/m^2]

Description of a Hydrometer

Figure 4.4 shows a streamlined type of hydrometer that is generally used in hydrometer analysis. The hydrometer possesses a long stem and a bulb. The total length of the hydrometer varies from about 30 to 40 cm. The stem is graduated to give the specific gravity of soil suspensions. The calibration may be in three specific gravity ranges, namely, 0.995 to 1.030, 0.995 to 1.040, and 1.000 to 1.060. The first of these range is preferable because the larger space between 0.001 division makes it easier to obtain the desired accuracy. Also, the maximum hydrometer reading in a soil suspension is generally less than 1.030, for 50 gm of soil.

Principle of the Method

Hydrometer is used for the determination of unit weight of suspensions at different depths and particular intervals of time. A unit volume of soil suspension at a depth H_e and at any time t contain particles finer than a particular diameter D. The value of this diameter is determined by applying Stokes' Law whereas the percentage finer than this diameter is determined by the use of hydrometer. The principle of the method is that the reading of the hydrometer gives the unit weight of suspension at the center of volume of hydrometer. The first step in the presentation of this method is to calibrate the hydrometer.

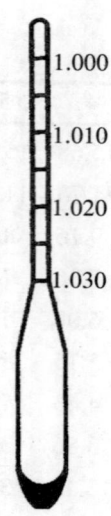

Fig. 4.4 Hydrometer

Calibration of Hydrometer and Corrections

Calibration

The readings r_h on the stem of a hydrometer gives the density of suspension at the centre of immersion which coincides with the centre of bulb for all practical purposes. It is however convenient to express the readings r_h as R_h where

$$R_h = (r_h - 1)10^3 \tag{4.5}$$

The distance of any reading R_h (or r_h) on the stem to the centre of volume may be designated as H_e as shown in Fig. 4.5. The distance H_e varies linearly with the reading R_h. An expression for H_e may be written as follows for any reading R_h.

$$H_e = H_{e_1} - \frac{(H_{e_1} - H_{e_2})R_h}{30} \tag{4.6}$$

where H_{e_1} = maximum depth to the centre of volume from $r_h = 1.00$ or $R_h = 0$.

H_{e_2} = minimum depth to the centre of volume from $r_h = 1.030$ or $R_h = 30$.

A straight line C_1 may be drawn giving the relationship between H_e and R_h as shown in Fig. 4.5. This curve C_1 can be used for determining H_e for readings within the first 4 minutes of hydrometer test starting from time $t = 0$. But if hydrometer is inserted into the suspension at any time t after 4 min, there will be an immersion correction to be applied to H_e.

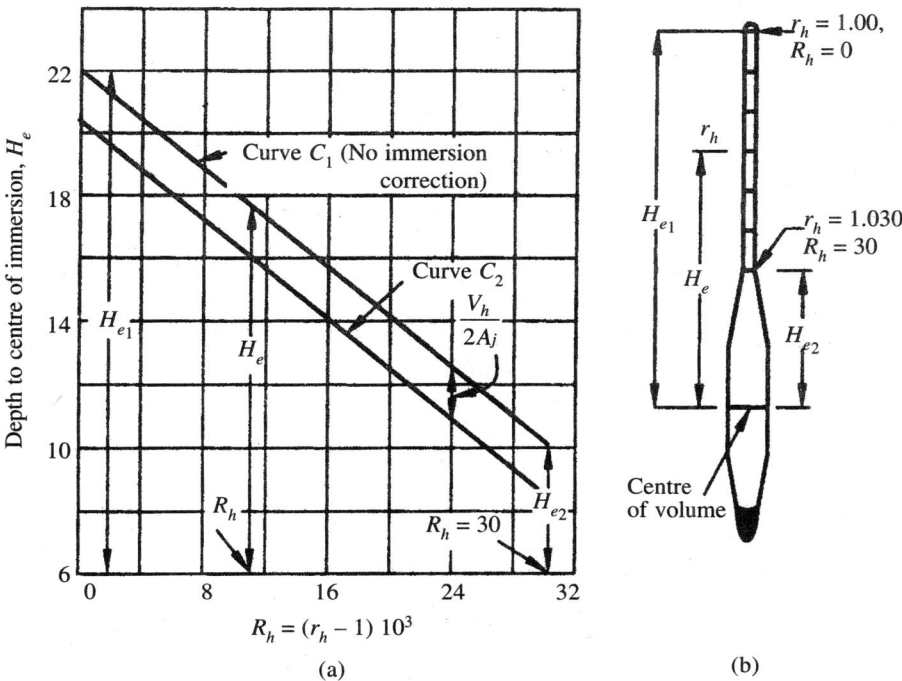

Fig. 4.5 Calibration chart for hydrometer

When the hydrometer is inserted into the suspension, the surface of suspension rises as shown in Fig. 4.6. The distance H_e in Fig. 4.6 is the actual distance through which particle of diameter D has fallen. The point at level A_1 at depth H_e occupies the position A_2 (which coincides with the centre of volume of hydrometer) in the figure after the immersion of the hydrometer and correspondingly the surface of suspension rises from B_1 to B_2. The depth H'_e is therefore greater than H_e through which the particle of diameter D has fallen the value of H_e can be obtained from the equation

$$H_e = H'_e - \frac{V_h}{2A_j} \tag{4.7}$$

where, V_h = volume of hydrometer; A_j = cross-sectional area of hydrometer jar.

The volume of the hydrometer can be determined by immersing it in a graduated jar of water and noting the increase of volume as read on the measuring jar. The cross-sectional area of the jar, A_j, is obtained by dividing the volume between two calibration marks by the distance between the same two marks.

Corrected calibration curve C_2 for the distance H_e may be obtained by subtracting the value $V_h/2A_j$ from the curve C_1 given in the Fig. 4.5. The curve C_2 should be used for observations after 4 min when the hydrometer has to be inserted and taken out after each reading. The equation for curve C_2 may be written as

$$H_e = H_{e_1} - \frac{(H_{e_1} - H_{e_2})R_h}{30} - \frac{V_h}{2A_j} \tag{4.8}$$

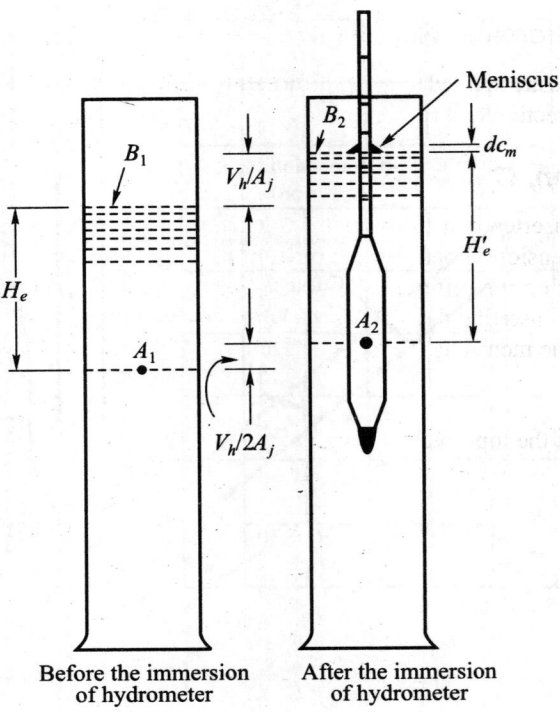

Before the immersion   After the immersion
of hydrometer   of hydrometer

Fig. 4.6   Immersion correction

Test Procedure

A very brief outline of the test procedure is given here. The students should read some books on laboratory testing of oils for details.

A soil of about 50 g dry mass is mixed with distilled water and made into a thin paste. The paste is mixed well with a suitable quantity of deflocculating agent. A mechanical stirrer may be used for mixing the paste well. The paste is then transferred to a graduated cylinder and enough distilled water is added to bring the level to the 1000 cc mark. The suspension is mixed well by placing the palm of the hand over the open end turning the jar upside down and back. The jar is next placed on a table, the hydrometer is inserted carefully into suspension and the timer started. Readings are taken at the total elapsed times of ½, 1, 2 and 4 minutes without removing the hydrometer.

The hydrometer is removed after the 4-minute reading. The suspension is remixed as before, the hydrometer is inserted only after the lapse of 4 minutes and the reading is taken. The hydrometer is removed after this reading and reinserted just before the subsequent reading. The subsequent readings are taken at total elapsed time intervals of 8, 15, 30, 60, 120, 240, 480, and 1440 minutes.

After each reading in the suspension, the hydrometer should taken out and inserted into another jar of distilled water kept close by, which is also filled up to 1000 cc mark. The distilled water also contains the same quantity of deflocculating agent as in the suspension. Temperature observations and hydrometer reading is distilled water are taken at intervals.

In all the cases the readings are taken to the top of the meniscus.

Corrections to Hydrometer Readings

The following corrections are applied to the hydrometer readings. They are: (a) meniscus corrections, (b) dispersing agent correction and (c) temperature change correction.

Meniscus correction, C_m

When a hydrometer is inserted in a suspension, a meniscus is formed around the stem as shown in Fig. 4.6. Since the suspension is opaque, the reading is taken to the top of the meniscus and this should corrected by adding the difference in height between the top and bottom of the meniscus. This can be found out by inserting the hydrometer into another jar containing clear water and noting the difference between the meniscus. This is a constant for each hydrometer. The correction C_m may be found out as follows.

Let R'_h = reading at the top-level of meniscus; R_h = reading at the bottom level of meniscus.

We may write

$$R_h - R'_h = C_m$$

or $\quad R_h = R'_h + C_m$ (4.9)

C_m is always positive.

Equation (4.9) is used for determining the depth H_e in the sedimentation analysis.

Dispersing agent correction, C_d

To prevent formation of flocs in the suspension, a dispersing agent is always added to the suspension which increases the density of suspension to a certain extent. Due to increase in density, the hydrometer floats higher in the suspension giving thereby a higher reading. The correction should therefore be negative. This correction can be determined by noting the hydrometer reading first in clear water and then in the same water with the dispersing agent added to it (the same quantity as is used in the suspension). This correction is designated as C_d.

Temperature change correction, C_t

The hydrometer is calibrated to read unity at the calibration temperature of 27°C. If the temperature of test T is more than 27°C, the hydrometer reading will be less than what it should be and hence C_t is *positive*. If the temperature of test $T°$ is less than 27°C the reading of the hydrometer is more than what it should be and hence C_t is *negative*.

The final corrected hydrometer reading, R

The corrections are required to be applied to the observed hydrometer readings $R'_h [= (r'_h - 1) \times 10^3]$. The final corrected reading R may now be expressed as

$$R = R'_h + C_m \pm C_t - C_d$$ (4.10a)

or $\quad R = R'_h \pm C$ (4.10b)

where as $\quad C = C_m \pm C_t - C_d$ (4.10c)

is called the *composite correction*. The composite correction C can be found out directly by taking readings in a comparison cylinder which contains the same quantity of distilled water and dispersing agent and has the same temperature. As the hydrometer has been calibrated at 27°C to indicate a specific gravity 1:00, the difference between the reading taken at the top of the meniscus and 1.00 is in magnitude equal to the composite correction. For example if the hydrometer reading $r_c = 1.002$, the composite correction $C = -(1.002 - 1.00) \times 10^3 = -2.0$. If $r_c = 0.997$, $C = -(0.997 - 1.00) \times 10^3$ $= +3.0$.

The composite correction has to be found out before the start of sedimentation test and afterward at every 30 minute intervals.

Equation (4.10) is used for determining the percent finer than diameter D in a given soil mass in the sedimentation analysis.

Percent Finer

Let, V = volume of suspension = 10^3 cc [= 1 litre]; M_s = mass of dry soil taken for sedimentation analysis (passing through 75 μ sieve)

The mass of soil M_s is mixed thoroughly in distilled water and suspension of volume $V = 1000$ cc is made. When the solids in the suspension are uniformly dispersed, the mass of solids m_s in a unit volume of suspension at any depth may be written as (at time $t = 0$).

$$m_s = \frac{M_s}{V}$$

The volume of solids v_s is

$$v_s = \frac{m_s}{G\rho_w} = \frac{M_s}{VG\rho_w}$$

Therefore, the volume of water in unit volume of suspension is

$$v_w = (1 - v_s) = (1 - \frac{M_s}{VG\rho_w})$$

Initial density ρ_i of suspension at time $t = 0$ is

$$\rho_i = m_s + m_w = \frac{M_s}{V} + \left(1 - \frac{M_s}{VG\rho_w}\right)\rho_w$$

Since $\rho_w = 1$ g/cm^3, we have

$$\rho_i = \frac{M_s}{V} + \left(1 - \frac{M_s}{VG}\right) \tag{4.11}$$

After a lapse of time t, the unit volume of suspension at any depth H_e contains only particles finer than a particular diameter D, since particles coarser than this diameter have fallen a distance greater than H_e as per Stokes' Law. The coarsest diameter of the particle in unit volume of suspension at depth H_e and at time t is given Eq. (4.3). Since all particles coarser than diameter D in Eq. (4.3) have fallen a distance greater than H_e, within the unit volume at depth H_e, all particles coarser than D have disappeared and the content of particles finer than D is unchanged.

Let, M_D = the mass of all particles finer than D in the sample taken for analysis. The density of suspension ρ_f after time t at depth H_e can be derived from Eq. (4.11) as

$$\rho_f = \frac{M_D}{V} + \left(1 - \frac{M_D}{GV}\right) \tag{4.12a}$$

where M_D/V = mass particles of diameter smaller than D in the unit volume of suspension at depth He after time t.

As per the calibration of the hydrometer, the corrected reading of the hydrometer, r is equal to the density of suspension at the centre of volume of the hydrometer. As such, we may write,

$$\rho_f = r = \frac{M_D}{V} + (1 - \frac{M_D}{GV}) \tag{4.12b}$$

From Eq. (4.12b), we have

$$M_D = \frac{G}{G-1} V(r-1) \tag{4.13}$$

Since $V = 10^3$, we can write

$(r - 1)\, 10^3 = R$ = corrected hydrometer reading.

Now Eq.(4.13) reduces to

$$M_D = \frac{G}{G-1} R = \frac{G}{G-1}(R'_h \pm C) \tag{4.14}$$

where R in Eq. (4.14) is same as that given in Eq. (4.10b) and C is the composite correction given in Eq. (4.10c).

Since the mass of dry soil taken for sedimentation analysis is M_s, the percentage N' of particles finer than diameter D may be written as

$$\text{Percent finer, } N' = \frac{M_D}{M_s} \times 100 = \frac{G}{G-1}\frac{(R'_h \pm C)}{M_s} \tag{4.15}$$

Let, M_p = the total mass of dry soil passing through 75 μ sieve.

M = total mass of soil taken for combined sieve and hydrometer analysis

The percentage finer N for the combined analysis may be written as

$$\text{Percent finer, } N = N'\% \times \frac{M_p}{M} \tag{4.16}$$

Now Eq. (4.16) with Eq. (4.3) given points for drawing grain size distribution curve.

Example 4.1

Determine the times t required in minutes for particles of diameters 0.2, 0.02, 0.01 and 0.005 mm to fall a height of 10 cm from surface in water.

Given: $\mu = 8.15 \times 10^{-3}$ poises, $G = 2.65$.

Solution

$\mu = 8.15 \times 10^{-3} \times 10^{-4}$ (kN-sec)/m² $= 8.15 \times 10^{-7}$ (kN-sec)/m²

Use Eq. (4.4)

$$t = \frac{300\mu}{(G-1)} \frac{H_e}{D^2}$$

substituting for H_e, G and m and simplifying, we have

$$t = \frac{1.49 \times 10^{-3}}{D^2} \text{ min}$$

The times required for the various of D are as given below.

D (mm)	t
0.2	2.24 sec
0.02	3 m 43.5 sec
0.01	14 m 54 sec
0.05	59 m 36 sec

Example 4.2

A sand sample comprising of particles ranging from 0.20 to 0.05 mm is put on the surface of a still water tank of 3 m depth. Find the time required for the coarsest and the finest of the particles to reach the bottom of the tank.

Given: $G = 2.7$, $\mu = 0.51$ poise.

Solution

Use Eq.(4.4)

$\mu = 0.51$ poise $= 0.15 \times 10^{-4}$ (kN-sec)/m²

$H_e = 300$ cm, $G = 2.7$

Substituting and simplifying Eq.(4.4), we have,

$$t = \frac{2.7}{D^2} \text{ min}$$

For $D = 0.20$ mm, $t = 67$ min 30 sec.

$D = 0.05$ mm, $t = 1080$ min.

Example 4.3

A soil particle falls a depth of 100 cm from the surface of water in a tank. The time taken for the fall is 10 min. Determine the diameter of the particle, if the temperature of water is 30 °C.

Given $G = 2.65$.

Solution

From Table 4.2, for $T = 30°C$, $G = 2.65$ the value of $F = 12.17$

Use Eq. (4.3)

$$D = 10^{-3} \times F \sqrt{\frac{H_e}{t}} \text{ mm}$$

By substituting for $H_e = 100$ cm, we have

$$D = 12.17 \times 10^{-3} \sqrt{\frac{100}{10}} = 0.0385 \text{ mm}$$

Example 4.4

In a hydrometer analysis, particles of different sizes were mixed to from a uniform suspension of volume 1000 cm³. It is required the time required to determine the time required for the fall of all the particles of size ranging from 0.05 mm to 0.001 mm and also the time required for the coarsest of the particles to fall. The depth of fall given is 20 cm. The other data available are $G = 2.70$, and $\mu = 8.1 \times 10^{-3}$ poises.

Solution

The time required for the fall may be found out from Eq.(4.4)

$$t(\text{min}) = \frac{300\mu}{(G-1)} \cdot \frac{H_e}{D^2}$$

where $\mu = 8.1 \times 10^{-3}$ poises $= 8.1 \times 10^{-7}$ (kN- sec)/m². The depth of fall $H_e = 20$ cm. The time required for the coarsest particle 0.05 mm is

$$t = \frac{300 \times 8.1 \times 10^{-7}}{(2.7-1)} \times \frac{20}{(0.05)^2} = 1.1414 \text{ min or 1 min 8.6 sec.}$$

The time required for all the particles to settle is the time required for the finest 0.001 mm particle. Therefore

$$t = \frac{300 \times 8.1 \times 10^{-7}}{1.7} \times \frac{20}{(0.001)^2} = 2858.8 \text{ min}$$

or $t = 47$ h 38 min 48 sec.

Example 4.5

A sedimentation analysis by hydrometer method was done with 50 g of oven dried sample. The volume of soil suspension is $V = 10^3$. The hydrometer reading $R'_h = 19.50$ after a lapse of time of 60 minutes after the commencement of the best.

Given : 0.52, H_e (effective) = 14.0 cm, $C = 2.50$, and $\mu = 0.01$ poise.

Calculate the smallest particle size, which would have settled a depth of 14.0 cm and the percentage finer than this size.

Solution

From Eq. (4.3)

$$D \,(\text{mm}) = \sqrt{\frac{300\mu}{G-1}} \sqrt{\frac{H_e}{D^2}}$$

where $\mu = 0.01 \times 10^{-4}$ (kN-sec)/m^2.

Substitutimg

$$D = \sqrt{\frac{300 \times 0.01 \times 10^{-4}}{2.7-1}} \sqrt{\frac{14}{60}} = 0.0064 \,\text{mm}.$$

Percent finer, N' use Eq. (4.15),

$$N'\% = \frac{G}{G-1} \frac{(R'_h \pm C)}{M_s} \times 100$$

or $$N'\% = \frac{2.7}{2.7-1} \frac{(19.5-2.5)}{50} \times 100 = 54\%$$

Example 4.6

A 500 g of dry soil was used for a combined sieve and hydrometer analysis. The soil mass passing through 75 μ sieve = 120 g. Hydrometer analysis was carried out on a mass of 40 g that passed through 75 μ sieve. The average temperature recorded during the test was 30°C.

Given : $G = 2.65$, $C_m = 0.50$, $C_d = 0.6$, $C_t = 0.915$, $\mu = 8.15 \times 10^{-3}$ poises. $H_{e_1} = 22.0$ cm for $R_h = 0$, $H_{e_2} = 10.0$ cm for $R_h = 30$, $A_j = 30$ cm^2, $V_h = 40$ cm.

The hydrometer reading $R'_h = 15.00$ after a lapse of time of time of 120 min after the start of the test. Determine the particles size D and percent timer $N'\%$ and $N\%$.

Solution

From Eq.(4.6) we have

$$H_e = H_{e_1} = \frac{(H_{e_1} - H_{e_2})R_h}{30}$$

Where, $R_h = R'_h + C_m$

$$H_e = 22.0 = \frac{(22.0-10)}{30} \times (15.0+0.5) - \frac{40}{2\times30} = 15.16 \, \text{cm}$$

From Eq.(4.3)

$$D = \sqrt{\frac{300\times8.15\times10^{-7}}{2.65-1}} \sqrt{\frac{15.16}{120}} = 0.0043 \, \text{mm}$$

From Eq.(4.15)

Percent finer, $\quad N'\% = \dfrac{G}{(G-1)M_s}(R'_h + C_m + C_t - C_d)$

or $\quad\quad\quad\quad N'\% = \dfrac{G}{(G-1)M_s}(R'_h + C)$

Now, $C = C_m + C_t - C_d = 0.5 + 0.975 - 0.6 = +0.875$

Substituting, we have

$$N\% = \frac{2.65}{2.65-1} \times \frac{(15+0.875)}{40} \times 100 = 63.7\%$$

From Eq.(4.16)

$$N\% = N'\% \times \frac{M_p}{M} = 63.7 \times \frac{120}{500} = 15.3\%$$

Example 4.7

400 g of dry soil was taken for combined sieve and hydrometer analysis. The mass of soil passing through 75 μ sieve was found to be 165 g. Hydrometer analysis was carried out on 40 g dry soil taken from the mass passing through 75 μ sieve. The hydrometer readings r'_h and R'_h taken in a suspension of 1000 cc are given below.

Elapsed time	r'_h	$R'_h[=(r'_h-1)\times10^3]$	Elapsed time	r'_h	$R'_h[=(r'_h-1)\times10^3]$
30 sec	1.0215	21.5	30 min	1.0167	16.7
1 min	1.0210	21.0	62 min	1.0160	16.0
2 min	1.0203	20.3	120 min	1.0150	15.0
4 min	1.0193	19.3	244 min	1.0140	14.0
8 min	1.0189	18.9	1459 min	1.0120	12.0
15 min	1.0179	17.9			

The average temperature recorded was 25°C. The specific gravity of the particles $G = 2.65$. The other data are : $C_m = 0.5$, $C_d = 0.6$, and $C_i = 0.975$. $H_{e_1} = 22.00$ cm for $R'_h = 0$ and $H_{e_2} = 10.00$ cm for $R_h = 30$. Sectional area of jar $= 30$ cm²; $V_h = 40$ cm³.

Determine : (a) $N'\%$ and $N\%$ (b) plot D vs $N\%$.

Solution

From Eq.(4.3), the expression for D is

$$D = 10^{-3} F \sqrt{\frac{H_e}{t}}$$

where, $F = 10^3 \sqrt{\frac{300\mu}{G-1}}$

From Table 4.2, $F = 12.86$ for $G = 2.65$ at $T = 25°C$. Now D may be expressed as

$$D = 12.86 \times 10^{-3} \sqrt{\frac{H_e}{t}} = \frac{1}{78} \sqrt{\frac{H_e}{t}} \tag{a}$$

where H_e is in cm and t in min

Percent finer $N'\%$ from Eq.(4.15) is

$$N'\% = \frac{G}{(G-1)} \frac{R}{M_s} \times 100$$

where, $R = R'_h \pm C$,   where C = composite correction.

$$C = C_m - C_t - C_d = 0.5 - 0.975 - 0.6 = -1.075$$

Therefore, $R = R'_h - 1.075$

Substituting for R, $G = 2.65$ and simplifying , we have

$$N\% = 4.015 (R'_h - 1.075) \tag{b}$$

Equation for H_e is

$$H_e = H_{e_1} - \frac{\left(H_{e_1} - H_{e_2}\right) R_h}{30} \tag{c}$$

Substituting for $H_{e_1} = 22.0$ cm and $H_{e_2} = 10$ cm, and simplifying, we have,

$$H_e = 22.0 - 0.4 R_h$$

for reading within 4 minutes. For reading greater than 4 min

$$H_e = (22.0 - 0.4R_h) - \frac{V_h}{2A_j} \qquad\qquad\text{(d)}$$

where, $V_h = 40 \text{ cm}^3$ and $A_j = 30 \text{ cm}^2$. Substituting and simplifying, we have

$$H_e = (22.0 - 0.4\,R_h) - 0.67 = 21.33 - 0.4\,R_h$$

Since $R_h = R'_h + C_m$, the values of R_h can be found for various values of R'_h observed at various time intervals. Hence, the values of H_e may be computed by making use of Eq. (c) upto 4 min amd Eq.(d) for beyond 4 min . Since H_e and T are known D may be found out from Eq.(a) and also $N'\%$ from Eq.(b). We may now calculate $N\%$ from equation.

$$N\% = N'\% \times \frac{M_p}{M}$$

where, M_p = total mass of soil passing through 75μ sieve = 165 g ; M = total mass of soil taken for combined analysis = 400 g.

Therefore, $N\% = N'\% \times \dfrac{165}{400} = 0.412N'\%$

The results of calculations are given in a tabular form in Table 4.4. The percentage finer $N\%$ versus D is plotted as shown in Fig. Ex. 4.7.

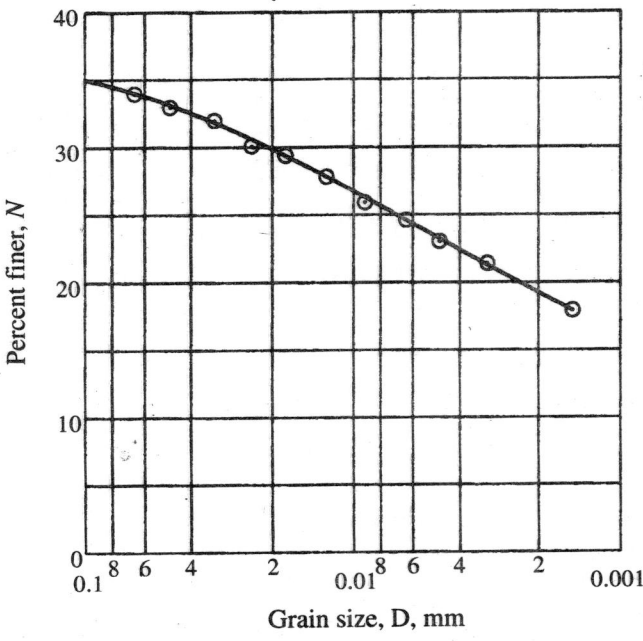

Fig. Ex. 4.7 Hydrometer analysis

TABLE 4.4

Hydrometer analysis

Elapsed time t (min)	Hydro reading R'_h	R_h $= R'_h + C_m$ $= R'_h + 0.5$	H_e cm	$D = (1/78)$ $\sqrt{H_e/t}$ mm	$R =$ $R'_h - 1.075$	$N'\%$ $4.015\ R$	$N\%$ $0.4125\ N'\%$
1	2	3	4	5	6	7	8
1/2	21.5	22.0	13.2	0.066	20.43	82.0	33.83
1	21.0	21.5	13.4	0.047	19.93	80.0	33.0
2	20.3	20.8	13.68	0.034	19.23	77.21	31.85
4	19.3	19.8	13.41	0.024	18.22	73.15	30.17
8	18.9	19.4	13.57	0.017	17.83	72.21	29.79
15	17.9	18.4	14.07	0.012	16.83	67.57	27.87
30	16.7	17.2	14.45	0.009	15.63	62.75	25.88
60	16.0	16.5	14.73	0.0062	14.93	59.94	24.73
120	15.0	15.5	15.15	0.0046	13.93	55.92	23.06
244	14.0	14.5	15.53	0.0032	12.93	57.91	21.41
1459	12.0	12.5	16.33	0.0014	1093	43.88	18.10

4.7 GRAIN SIZE DISTRIBUTION CURVES

A typical set of grain distribution curves are given in Fig. 4.3 with the grain size D as abscissa on the logarithmic scale and the per cent finer N as ordinate on the arithmetic scale. On the curve C_1 the section AB represents the portion obtained by sieve analysis and the section $B'C'$ by hydrometer analysis. Since hydrometer analysis gives equivalent diameters which are generally less than the actual sizes, the section $B'C'$ will not be a continuation of AB and would occupy a position shown by the dotted curve. If we assume that the curve BC is the actual curve obtained by sketching it parallel to the $B'C'$, then at any percentage finer, say 20 per cent, the diameters D_a and D_e, represent the actual and equivalent diameters respectively. The ratio of D_a to D_e can be quite high for flaky grains.

The shapes of the curves indicate the nature of the soil tested. On the basis of the shapes we can classify soils as:

1. Uniformly graded or poorly graded.

2. Well graded.

3. Gap graded.

Uniformly graded soils are represented by nearly vertical lines as shown by curve C_2 in Fig. 4.3. Such soils possess particles of almost the same diameter. A well graded soil, represented by curve C_1, possesses a wide range of particle sizes ranging from gravel to clay size particles. A gap graded soil, as shown by curve C_3 has some of the sizes of particles missing. On this curve the soil particles falling within the range of XY are missing.

The grain distribution curves as shown in Fig. 4.3 can be used to understand certain grain size characteristics of soils. Allen Hazen has shown that the permeability of clean filter sands in loose state can be correlated with numerical values designated D_{10} the effective grain size. The effective

TABLE 4.5

Soil grading according to uniformity coefficient C_u

C_u	Type of soil
< 5	Uniform size particles
5–15	Medium graded soil
> 15	Well graded soil

grain size is corresponding to 10 per cent finer particles. Hazen found that the sizes smaller than the effective size affected the functioning of filters more than did the remaining 90 percent of the sizes. To determine whether a material is uniformly graded or well graded, Hazen proposed the following equation:

$$C_u = \frac{D_{60}}{D_{10}} \tag{4.17}$$

where D_{60} is the diameter of the particle at 60 percent finer on the grain size distribution curve. The uniformity coefficient, C_u, is about one, if the grain size distribution curve is almost vertical, and the value increases with gradation. For all practical purposes we can consider the following values for granular soils given in Table 4.5.

There is another procedure, which is sometimes used to determine the gradation of particles. This is based on the term used called the *coefficient of curvature* which is expressed as

$$C_c = \frac{D_{30}^2}{D_{10} \times D_{60}} \tag{4.18}$$

wherein D_{30} is the size of particle at 30 percent finer on the gradation curve. The soil is said to be well graded if C_c, lies between 1 and 3.

Two samples of soils are said to be similarly graded, if their grain size distribution curves are almost parallel to each other on a semilogarithmic plot. For example curve C_1 and C_4, and C_2 and C_5 are similarly graded. When the curves are almost parallel to each other the ratios of their diameters at any percentage finer approximately remains constant. Such curves are useful in the design of filter materials around drainage pipes.

4.8 DENSITY INDEX OF COHESIONLESS SOILS

The density of granular soils varies with the shape and size of grains, the gradation and the manner in which the mass is compacted. If all the grains are assumed to be spheres of uniform size and packed as shown in Fig. 4.7(a) the void ratio of such a mass amounts to about 0.90. However, if the grains are packed as shown in Fig. 4.7(b) the void ratio of the mass is about 0.35. The soil corresponding to the higher void ratios is called loose and that corresponding to the lower void ratio is called dense. If the soil grains are not uniform, then smaller grains, fill in the space between the bigger ones and the void ratios of such soils get reduced to as low as 0.25 in the densest state. Further if the grains are angular, they tend to form looser structures than rounded grains because their sharp edges and points hold the grains farther apart. But if the mass with angular grains is compacted by vibration, it forms a dense structure. Static load alone will not alter the density of grains significantly but if it is accompanied by vibration, there will be considerable change in the density. The water present in voids may act as a lubricant to a certain extent for an increase in the density under vibration. Since the change in void ratio would change the density and this in turn changes the strength characteristics of granular soils. Void ratio or the unit weight of soil can be used to compare the strength characteristics of samples of granular soils of the same origin. The term used to indicate the strength characteristics in a qualitative manner is termed as *density index* which is already expressed by Eq. (3.20). On the basis of relative density, we can classify sandy soils as loose, medium or dense as in Table 4.6.

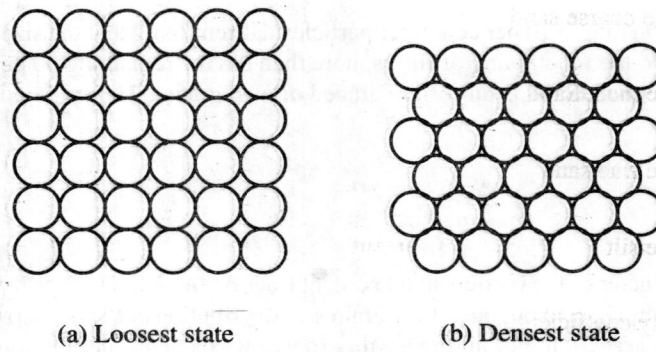

(a) Loosest state   (b) Densest state

Fig. 4.7   Packing of grains of uniform size

TABLE 4.6

Classification of sandy soils

Density index, I_D, %	Type of soil
0–15	Very loose
15–50	Loose
50–70	Medium dense
70–85	Dense
85–100	Very dense

Example 4.8

500 g of dry soil was used for sieve analysis. The masses of soil retained on each sieve is given below:

IS sieve	Mass in g	IS sieve	Mass in g
2.00 mm	10	500 μ	135
1.40 mm	18	250 μ	145
1.00 mm	60	125 μ	56
		75 μ	45

Plot a grain distribution curve and computed the following:

(a) Percentages of gravel, coarse sand, medium sand, fine and silt, as per IS: 1498–1959, (b) Uniformity coefficient and (c) Coefficient of curvature.

Comment on the type of soil

Solution

(a) Percentage gravel $\quad= 100 - 98 = 2$ percent

Percentage coarse sand $\quad= 98 - 61.5 = 36.5$ percent

Percentage med. Sand $\quad= 61.5 - 22 = 39.5$ percent

Percentage fine sand $\quad= 22.0 - 3.0 = 19.0$ percent

Percentage silt $\quad= 3$ percent.

(b) Uniformity coefficient $U = \dfrac{0.61}{0.098} = 6$

(c) Coefficient of curvature $= \dfrac{(D_{30})^2}{D_{10} \times D_{60}} = \dfrac{(0.28)^2}{0.098 \times 0.61} = 1.3.$

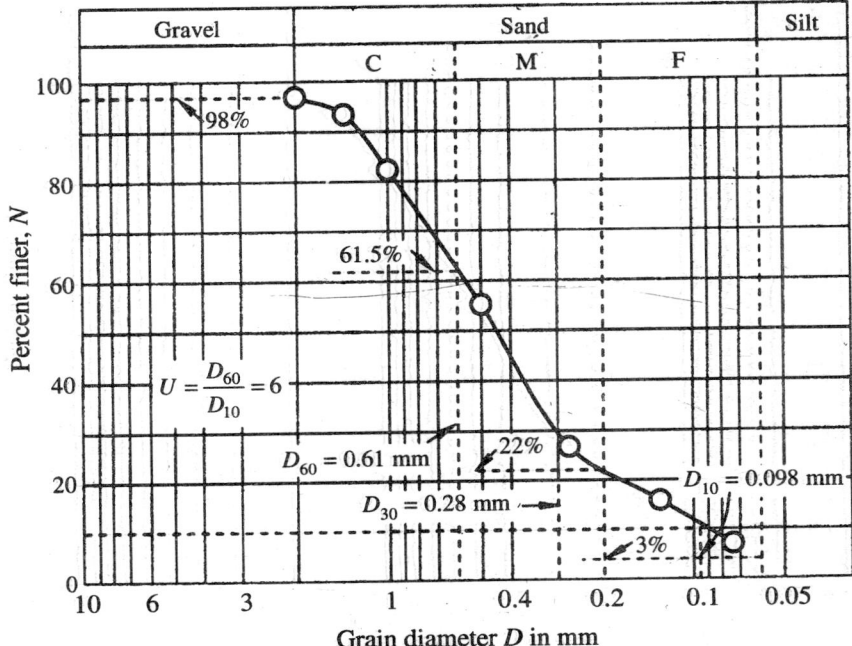

Fig. Ex. 4.8

IS Sieve	Diameter, D of grains in mm	Mass retained in g	% retained	Cumulative % retained	% finer N
2.00 mm	2.00	10	2.0	2.0	98.0
1.40 mm	1.40	18	3.6	5.6	94.4
1.00 mm	1.00	60	12.0	17.6	82.4
500 μ	0.500	135	27.0	44.6	55.4
250 μ	0.25	145	29.0	73.6	26.4
125 μ	0.125	56	11.2	84.8	15.2
75 μ	0.075	45	9.0	93.8	6.2

Comments: As per the value of uniformity coefficient, the soil is medium graded, but as per the value of C_c, the soil is said to be well graded since the value lies between 1 and 3.

4.9 CONSISTENCY OF CLAY SOIL

Consistency is a term used to indicate the degree of firmness of cohesive soils. The consistency of natural cohesive soil deposits is expressed qualitatively by such terms as very soft, stiff, very stiff and hard. The physical properties of clays greatly differ at different water contents. A soil which is very soft at a higher percentage of water content becomes very hard with the decrease of water content. However, it has been found that at the same water content, two samples of clay of different origins may possess different consistency. One clay may be relatively soft while the other may be hard. Further, certain decrease in water content may have little effect on one sample of clay but may transform the other sample from almost a liquid to a very firm condition. Water content alone, therefore, is not an adequate index of consistency for engineering and many other purposes. Consistency of a soil can be expressed in terms of:

1. Atterberg limits of soils.

2. Unconfined compressive strengths of soils.

Consistency of Soils as Per Atterberg Limits

Atterberg, a Swedish scientist, considered the consistency of soils in 1911, and proposed a series of tests for defining the properties of cohesive soils. These tests indicate the range of plastic state (plasticity is defined as the property of cohesive soils which possess the ability to undergo changes of shape without rupture) and other states. He showed that if the water content of a thick suspension of clay is gradually reduced, the clay water mixture undergoes changes from liquid state through a plastic state and finally into a solid state. The different states through which the soil sample passes through with the decrease in the moisture content is depicted in Fig. 4.8. The water content corresponding to the transition from one state to another is termed as *Atterberg Limit* and the tests required to determine the limits are the *Atterberg Limit Tests*. The testing procedures of Atterberg were subsequently improved by A. Casagrande.

States	Limit	Consistency	Volume change
Liquid		Very soft	↑
.... w_l	*Liquid limit*	*Soft*	
Plastic		Stiff	Decrease in volume
.... w_p	Plastic limit	Very stiff	
Semi solid			
.... w_s	Shrinkage limit	Extremely	↓
Solid		Hard	Constant volume

Fig. 4.8 Different states and consistency of soils with Atterberg limits

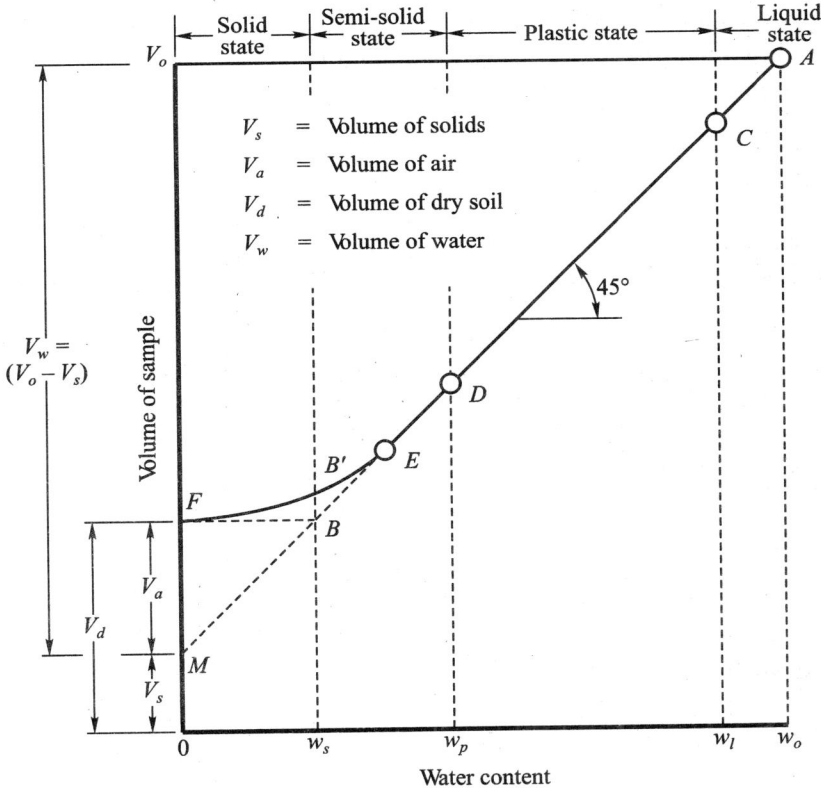

Fig. 4.9 Curve showing transition stages from the liquid to solid state

The transition state from the liquid stated to a plastic stated is called the *liquid limit* w_l. At this stage all soil posses a certain small shear strength . This arbitrarily chosen shear strength is probably the smallest value that is feasible to measure in a standardized procedure. The transition from the plastic stated to the semisolid state is termed as *plastic limit*, w_p. At this state the soil rolled into threads of about 3 mm diameter just crumbles. Further decrease of the water contents of the same will lead finally to the point where the sample can decrease in volume no further. At this point the sample begins to dry at the surface, saturation is no longer complete, and further decrease m water m the voids occurs without change in the void volume. The colour of the soil begins to change from dark to light. This water content is called the *shrinkage limit*, w_s. The limits expressed above are all expressed by their percentage of water contents. The range of water content between the liquid and plastic limits, which is an important measure of plastic behaviour, is called the *plasticity index*, I_p, i.e.

$$I_p = w_l - w_p$$

There is decrease in the volume of the soil sample as its moisture content decreases. This decrease in volume takes place up to shrinkage limit w_s, (theoretically) and there will be no further decrease in volume beyond this limit with the decrease in moisture content. The soil remains saturated up to the shrinkage limit and when once this limit is crossed, the soil becomes partially saturated. The air takes the place of the moisture that is lost due to evaporation. At about 105° to 110°C, there will not be any normal water left in the pores and soil at this temperature is said to be *Oven-dry*. The decrease in volume of the soil mass with the decrease in moisture content is depicted in the form of a diagram shown in Fig. 4.9. A soil sample of volume V_o and water content w_o is represented by point A in the figure.

As the soil loses moisture content there is a corresponding change in the volume of soils. The volume change of soil is equal to volume of moisture lost. The straight line, *AE,* therefore, gives the volume of soil at different water contents. The points C and *D* represent the transition stages of soil sample at liquid and plastic limits respectively. As the moisture content is reduced further beyond the point *D,* the decrease in volume of the soil sample will not be linear with the decrease in moisture beyond a point E due to many causes. One possible cause is that air might start entering into the voids of the soil. This can happen only when the normal water between the particles are removed. If the normal water between some particles are removed, the soil particles surrounded by adsorbed water will come in contact with each other. Greater pressure is required if these particles are to be brought still closer. As such the change in volume is less than the change in moisture content. Therefore, the curve *DEB'F* depicts the transition from plastic limit to the dry condition of soil represented by point F. However, for all practical purposes, the abscissa of the point of intersection B of the tangents *FB* and *EB* may be taken as the shrinkage limit, w_s. The straight line AB when extended meets the ordinate at point *M*. The ordinate of M gives the volume of the solid particles V_s. Since the ordinate of F is the dry volume, V_d of the sample, the volume of air V_a, is given by $(V_d - V_s)$.

Determination of Atterberg Limits

Liquid Limit

The apparatus shown in Fig. 4.10 is the Casagrande's Liquid Limit Device used for determining the liquid limits of soils. A photograph is shown in Fig. 4.11. The device contains a brass cup which could be raised and allowed to fall on a hard rubber base by turning the handle. The cup is raised by one cm. The limits are determined on that portion of soil finer than a No. 40 sieve (ASTM). About 100 g of soil is mixed thoroughly with distilled water into a uniform paste. A portion of the paste is placed in the cup and levelled to a maximum depth of 12 mm. A channel of the dimensions of 11 mm wide and 8 mm depth is cut, as shown in Fig. 4.10(a) through the sample along the symmetrical axis of the cup. The grooving tool should always be held normal to the cup at the point of contact.

The handle next turned at a rate of about two revolutions per second and the number of blows necessary to close the-groove along the bottom for a distance of 12 mm is counted. The groove should be closed by a flow of the soil and not by slippage between the soil and the cup. The water content of the soil in the cup is altered and the tests repeated. At least four tests should be carried out by adjusting the water contents in such a way that the number of blows required to close the groove may fall within the range of 5 to 40 blows. A plot of water content against the log of blows is made as shown in Fig. 4.12. Within the range of 5 to 40 blows, the plotted points lie almost on a straight line. The curve so obtained is known as a *flow curve*. The water content corresponding to 25 blows is termed as *liquid limit*. The equation of the flow curve can be written as

$$w = -I_f \log N + C \qquad\qquad (4.19)$$

where, w = water content

I_f = Slope of the flow curve, termed as flow index

N = Number of blows

C = a constant.

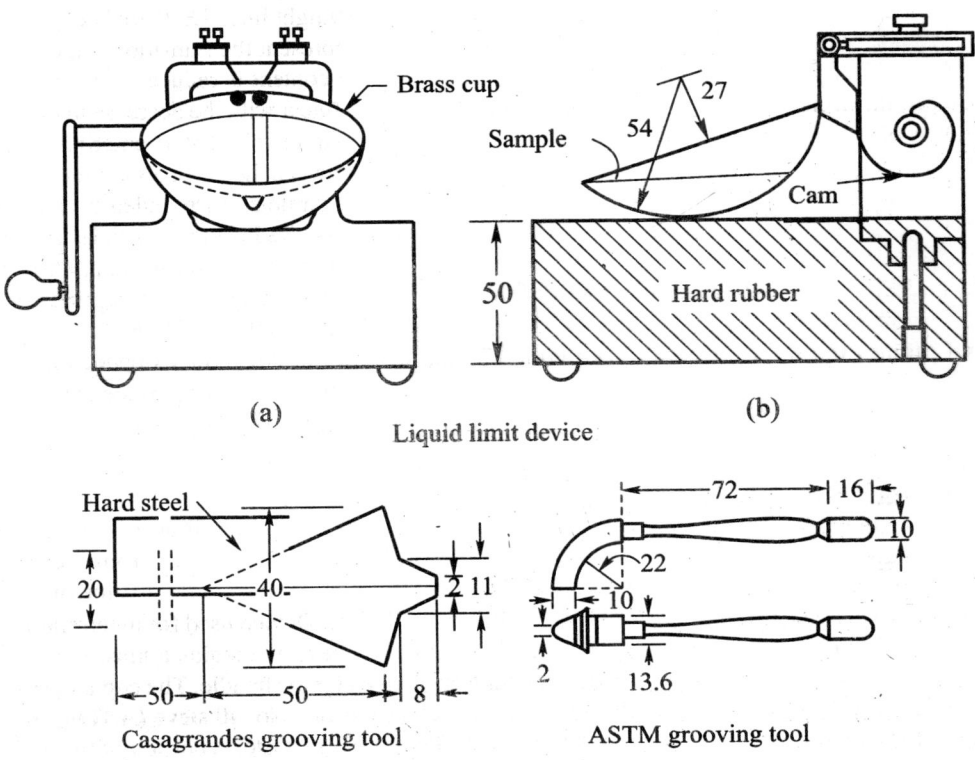

Fig. 4.10   Casagrande's liquid limit apparatus

Two types of grooving tools generally in use, one is the Casagrande type and the other ASTM type. The Casagrande type is usually recommended for cohesive soil and the other for soil with more of granular materials in it. For a derailed study, students are recommended to read a book on Laboratory Testing of Soils.

Fig. 4.11 Hand-operated liquid limit device

In order to draw the mean 'flow line' it is better to have two groups of points as far apart as is convenient. The ranges for each group are chosen from 5 to 20 and from about 30 to 40 blows respectively. The force that resists the deformation of the sides of the groove, is the shearing resistance of the soil. The number of blows required to close the groove for a specified length is indirectly a relative measure of the shearing resistance of the soil at the corresponding water content. Twenty-five blows has been fixed on the basis, that if the groove closes the specified distance at 25 blows, all plastic soils at liquid limits possess a constant value of shearing resistance. It has been found by means of direct shear tests on different types of clays that the liquid limit corresponds to a shearing strength of about 2.7 kN/m^2. This applies to all plastic soils where the impact forces produce only the deformation of the soil. However, most non-plastic soils are much more pervious than clay and the impact forces tend to cause the pore-water to flow towards the groove, thereby, actually softening the soil and reducing the shear strength near the groove. For such soils, the liquid limit no longer represents the water content at which the soil has a certain definite but very small shearing strength. Liquid and other limits have therefore no meaning for cohesionless soils which are non-plastic.

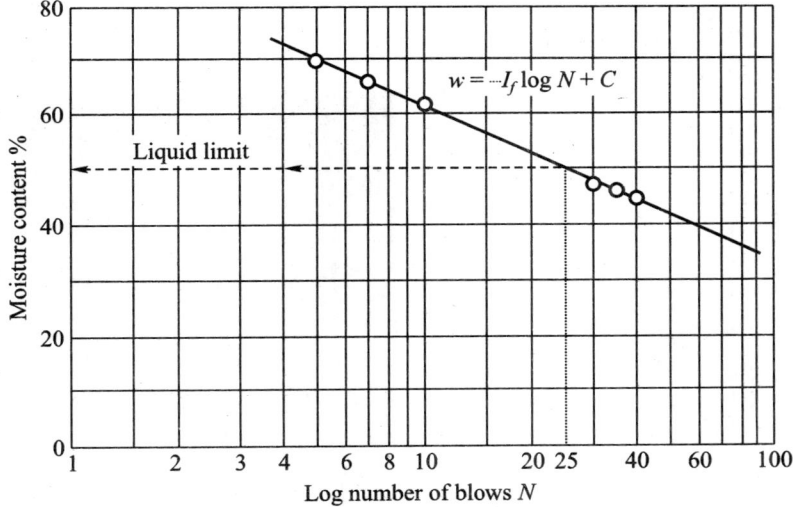

Fig. 4.12 Determination of liquid limit

Since the number of blows represent a relative measure of the shear strength of soil, the same semi-log plot can also show a relationship between water content and shearing strength of a soil. If this plot is extended from a semi-liquid stage, through the entire plastic range to the semisolid stage, the plot is not a straight line. This plot gradually curves and asymptotically approaches the zero water content coordinate. However, the plot is almost a straight line within the test range of 5 to 40 blows.

Liquid Limit by One-point Method

The determination of liquid limit as explained earlier requires considerable amount of time and labour. We can however use what is termed as 'one-point method' if an approximate value of the limit is required. The formula used for this purpose is

$$w_l = w\left(\frac{N}{25}\right)^n \qquad (4.20)$$

where w is the water content corresponding to the number of blows N, and n, an index whose value has been found to vary from 0.068 to 0.121. An average value of 0.10 may be useful for all practical purposes. It is, however, a good practice to check this method with the conventional method as and when possible.

Liquid Limit by the Use of Static Cone Penetrometer

IS: 2720 (Part V)-1965 describes a method of determining liquid limit by making use of a static cone penetrometer. This method is based on a similar method developed in Russia. The procedure is as follows:

Figure 4.13(a) shows the arrangement of the apparatus. The soil whose liquid limit is to be found out is mixed well into a soft consistency and filled into the cylindrical mould of 5 cm diameter and 5 cm high. The cone which has a central angle of 31° and a total mass of 148 g will be kept free

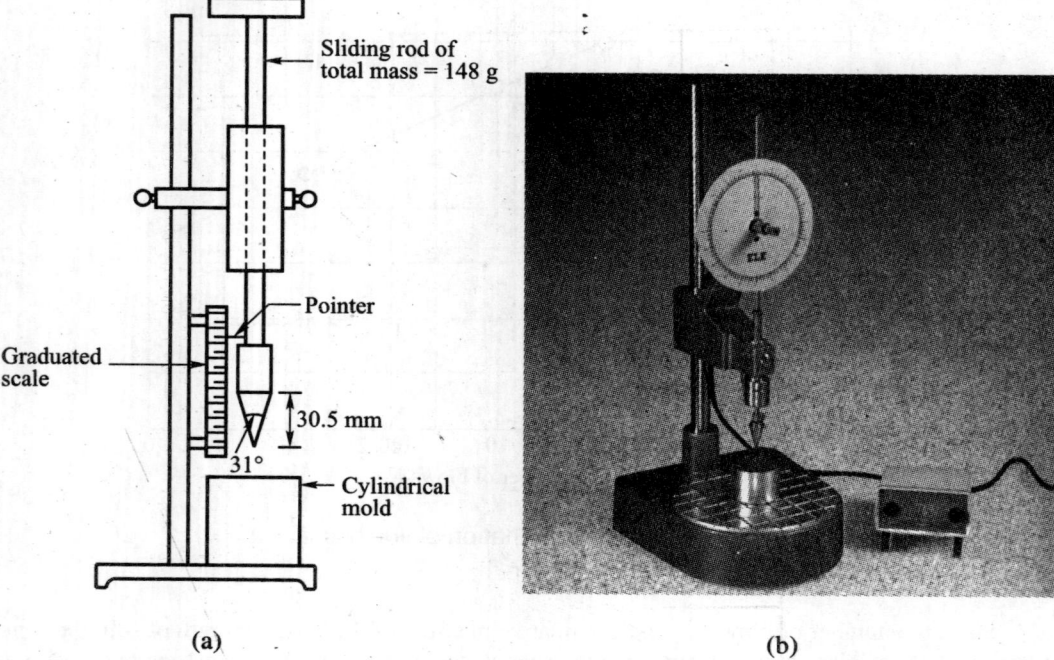

Fig. 4.13 Liquid limit by the use of the static cone penetrometer

on the surface of the soil. The depth of penetration y of the cone is measured in mm on the graduated scale after 30 sec of penetration. The liquid limit w_l, may be computed by using the formula,

$$w_l = w_y + 0.01(25 - y)(w_y + 15)$$

where, w_y is the water content corresponding to the penetration y.

The procedure is based on the assumption that the penetration lies between 20 to 30 mm. Even this method has to be used with caution.

Figure 4.13(b) shows BS-1377 static cone penetrometer.

Example 4.9

Liquid limit tests on a given sample of clay were carried. The data obtained are as given below.

Test No.	1	2	3	4
Water content %	70	64	47	44
Number of blows, N	5	8	30	45

Draw the flow curve on a semi-log paper and determine the liquid limit and flow index of the soil.

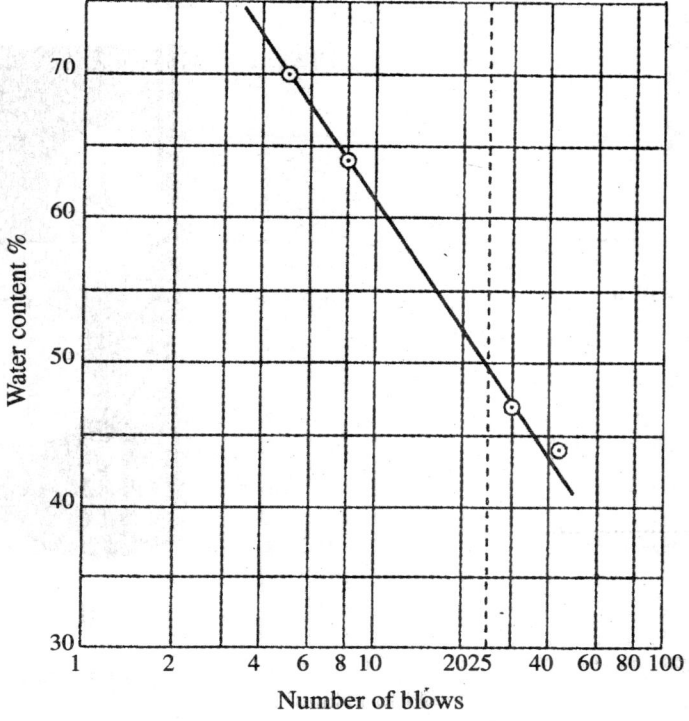

Fig. Ex. 4.9

Solution

Figure Ex. 4.9 gives the flow curve for the given sample of clay soil. As per the curve, we have liquid limit, $w_l = 50\%$

Flow index, $I_f = 29$

Plastic Limit

About 15 g of soil passing through No. 40 sieve (ASTM), is mixed thoroughly. The soil is rolled on a glass plate with the hand, until it is about 3 mm in diameter. This procedure of mixing and rolling is repeated till the soil shows signs of crumbling when the diameter is 3 mm. The water content of the crumbled portion of the thread is determined. This is called as plastic limit.

Shrinkage Limit

Shrinkage limit of a soil can be determined by anyone of the following methods:

1. Determination of w_s, when the specific gravity of the soils G is unknown.

2. Determination of w_s, when the specific gravity of the solids, G is known.

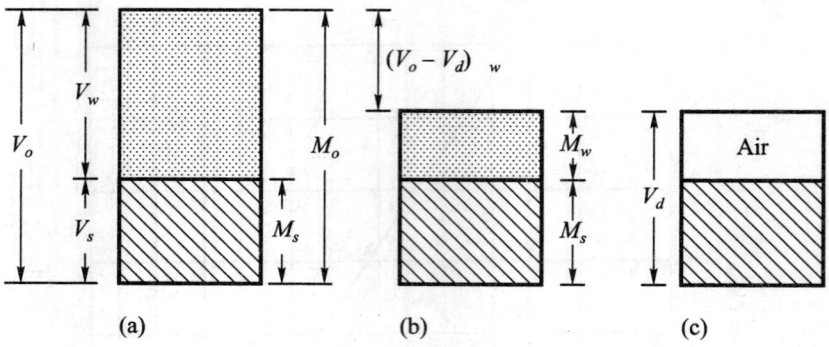

Fig. 4.14   Determination of shrinkage limit

Method I, when G is unknown

Three block diagrams of a sample of soil having the same mass of solids M_s, are given in Fig. 4.14. Block diagram (a) represents a specimen in the plastic state, which just fills a container of known volume, V_o. The mass of the specimen is M_o. The specimen is then dried gradually, and as it reaches the shrinkage, limit, the specimen is represented by the block diagram (b). The specimen remains saturated up to this limit but reaches a constant volume V_d. When the specimen is completely dried, its mass will be M_s whereas, its volume remains as V_d.

These different states are also represented on Fig. 4.9. The shrinkage limit can be written as

$$w_s = \frac{M_w}{M_s} \times 100$$

where, $M_w = M_o - M_s - (V_o - V_d)\rho_w$

Therefore, $w_s = \dfrac{(M_o - M_s) - (V_o - V_d)p_w}{M_s} \times 100\%$ (4.22)

The volume of the dry specimen can be determined by displacement of mercury as explained below.

Determination of Dry Volume V_d of Sample by Displacement in Mercury

Place a small dish filled with mercury up to the top in a big dish.

Cover the dish with a glass plate containing three metal prongs in such a way that the plate is entrapped. Remove the mercury spilt over into the big dish and take out the cover plate from the small dish. Place the soil sample on the mercury. Submerge the sample with the pronged glass plate and make the glass plate flush with the top of the dish. Weigh the mercury that is spilt over due to displacement. The volume of the sample is obtained by dividing the weight of the mercury by its specific gravity which may be taken as 13.6. Fig. 4.15 shows the apparatus used for the determination of volume.

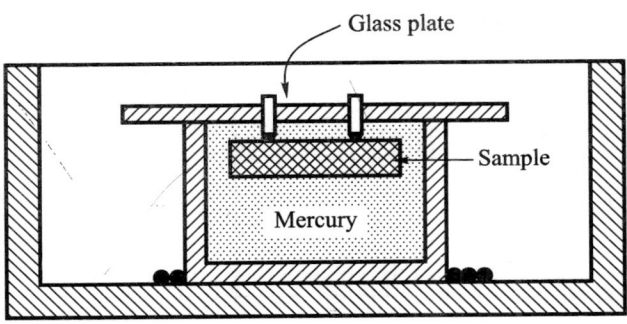

Fig. 4.15   Determination of dry volume by mercury displacement method

Method II, when G is known

$$w_s = \frac{M_w}{M_s} \times 100 \text{ But, } M_w = (V_d - V_s)\rho_w = \left(V_d - \frac{M_s}{G\gamma\rho_o}\right)\rho_w$$

Therefore, $$w_s = \frac{\left(V_d - \dfrac{M_s}{G\rho_o}\right)\rho_w}{M_s} \times 100 = \left[\frac{V_d\rho_w}{M_s} - \frac{\rho_w}{G\rho_o}\right] \times 100$$

or $$w_s = \left[\frac{V_d\rho_w}{M_s} - \frac{G_w}{G}\right] \times 100 \qquad\qquad (4.23)$$

If $\rho_w \approx \rho_o = 1$ gm/cm^3, we have $G_w = 1$ Eq. (4.23) reduces to

$$w_s = \left[\frac{V_d}{M_s} - \frac{1}{G}\right] \times 100 \qquad\qquad (4.24)$$

where, G_w = specific gravity of water at room temperature. The specific gravity of solids, G, may be determined as explained in Chapter 3.

Discussion on Limits and Indices

Plasticity index and liquid limit are the important factors that help an engineer to understand the consistency or plasticity of a clay. Shearing strength, though constant at liquid limits, varies at plastic limits for all clays. A high plastic clay (some times called fat clay) has higher shearing strength at plastic limit and the threads at this limit are rather hard to roll whereas a lean clay can be rolled easily at plastic limit and thereby possesses low shearing strength.

There are some fine grained soils that appear similar to clays but they cannot be rolled into threads so easily. Such materials are not really plastic. They may be just at the border line between plastic and non-plastic soils. In such soils, one finds the liquid limit practically identical with the plastic limit. There are cases where the liquid limit is even lower than the plastic limit giving thereby a negative plasticity index.

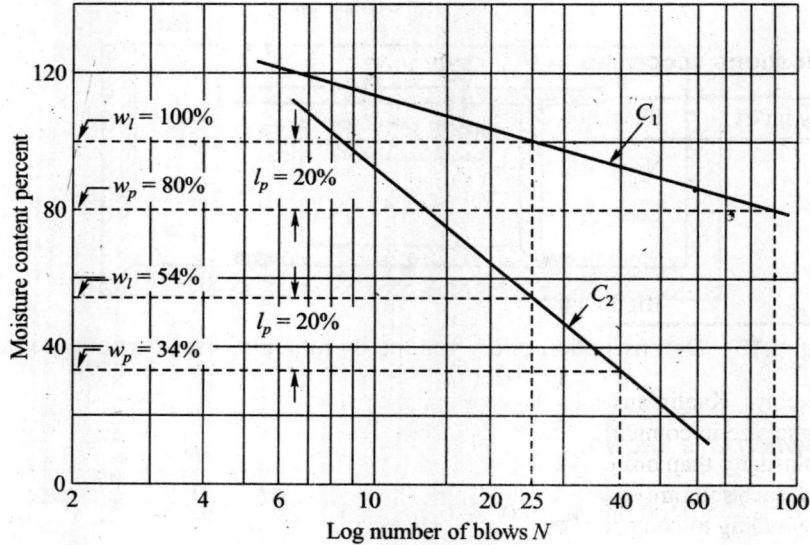

Fig. 4.16 Two samples of soils with different flow indices

Two soils may differ in their physical properties even though they have the same liquid and plastic limits or the same plasticity index. Such soils possess different flow indices. For example in Fig. 4.16 is shown two flow curves C_1 and C_2 of two samples of soils, C_1 is flatter than C_2. It may be assumed for the sake of explanation that both the curves are straight lines even in the plastic range and the same liquid limit device is used to determine their plastic limits also. The plasticity index I_p is taken to be the same for both the soils. I_p is the decrease in water content from liquid limit to plastic limit. It can be seen from the figure that the sample of flow curve C_1 has liquid and plastic limits of 100 and 80 percents respectively, giving thereby a plasticity index, I_p of 20 percent. The sample of flow curve C_2 has liquid and plastic limits of 54 and 34 percents giving thereby the same plasticity index value of 20 percent. Though the plasticity indices of the two samples remain the same the resistance offered by the two samples for slippage at their plastic limits are different. The sample one takes 90 blows for slippage whereas the second one takes only 40 blows. This indicate that at plastic limits, the cohesive strength of sample 1 with a lower flow index is larger than the sample 2 with a higher flow index.

Plasticity Index I_p

Plasticity index I_p indicates the degree of plasticity of a soil. Greater the difference between liquid and plastic limits greater is the plasticity of the soil. A cohesionless soil has zero plasticity index. Such soils are termed as non-plastic. Fat clays are highly plastic and possess a high plasticity index. Soil possessing large values of w_l and I_p are said to be highly plastic or fat. Those with low values are described as slightly plastic or lean. Atterberg classifies the soil according to their plasticity indices as in Table 4.7.

A liquid limit greater than 100 is uncommon for inorganic clays of non-volcanic origin. However, for clays containing considerable quantities of organic matter and clays of volcanic origin, the liquid limit may considerably exceed 100. Bentonite, a material consisting of chemically disintegrated volcanic ash, has a liquid limit ranging from 400 to 600. It contains approximately 70 percent of scale like particles of colloidal size as compared with about 30 percent for ordinary

TABLE 4.7

Soil classifications according to Plasticity Index

Plasticity index	Plasticity
0	Non-plastic
<7	Low plastic
7–17	Medium plastic
>17	Highly plastic

highly plastic clays. Kaolin and mica powder consist partially or entirely of scale like particles of relatively coarse size in comparison with highly colloidal particles in plastic clays. They therefore possess less plasticity than ordinary clays. Organic clays possess liquid limits greater than 50. The plastic limits of such soils are equally higher. Therefore, soils with organic content have low plasticity indices corresponding to comparatively high liquid limits.

Toughness Index, I_t

The shearing strength of a clay at plastic limit is a measure of its toughness. Two clays having the same plasticity index possess toughness which is inversely proportional to the flow indices. An approximate numerical value for the toughness can be derived as follows.

Let, $s_l =$ Shearing strength corresponding to liquid limit, w_l, which is assumed to be constant for all plastic clays.

$s_p =$ Shearing strength at plastic limit, which can be used as a measure of toughness of a clay.

Now $w_l = -I_f \log N_l + C$, $w_p = -I_f \log N_p + C$

where N_l and N_p are the number of blows at liquid and plastic limits respectively. The flow curve is assumed to be a straight line extending into the plastic range as shown in Fig. 4.16.

Let, $N_l = ms_l$, $N_p = ms_p$, where m is a constant.

We can write

$$w_l = I_f \log ms_l + C, \quad w_p = -I_f \log ms_p + C$$

Therefore, $I_f = w_l - w_p = I_f (\log ms_p - \log ms_l) = I_f \log \dfrac{s_p}{s_l}$

or $I_t = \dfrac{I_p}{I_f} = \log \dfrac{s_p}{s_l}$ (4.25)

Since we are interested only in a relative measure of toughness, I_t can be obtained from Eq. (4.25) as the ratio of plasticity index and flow index. The value of I_t generally falls between 0 and 3 for most clay soils. When I_t is less than one, the soil is friable at the plastic limit. I_t is quite an useful index to distinguish soils of different physical properties.

TABLE 4.8

Values of I_l and I_c according to consistency of soil

Consistency	I_l	I_c
Semisolid or solid state	Negative	>1
Very stiff state ($w_n = w_p$)	0	1
Very soft state ($w_n = w_l$)	1	0
Liquid state (when disturbed)	>1	Negative

Liquidity Index I_l

The Atterberg limits are found for remoulded soil samples. These limits as such do not indicate the consistency of undisturbed soils. The index that is used to indicate the consistency of undisturbed soils is called as the *liquidity index* or water plasticity ratio. The liquidity index IS expressed as

$$I_l = \frac{w_n - w_p}{I_p} \tag{4.26}$$

where, w_n is the natural moisture content of soil in the undisturbed state. The liquidity index of undisturbed soil can vary from less than zero to greater than 1. The value of I_l varies according to the consistency of soil as in Table 4.8.

The liquidity index indicates the state of the soil in the field. If the natural moisture content of the soil is closer to the liquid limit the soil can be considered as soft, and the soil is stiff if the natural moisture content is closer to the plastic limit. There are some soils whose natural moisture contents are higher than the liquid limits. Such soils generally belong to the montmorillonite group and possess brittle structure. The soil of this type when disturbed by vibration flow like a liquid. The liquidity index values of such soils are greater than unity. One has to be cautious in using such soils for foundations of structures.

Consistency Index, I_c

The consistency index may be defined as

$$I_c = \frac{w_l - w_n}{I_p} \tag{4.27}$$

The index I_c also reflect the state of the clay soil condition in the field in an undisturbed state just in the same way as I_l described earlier. The values of I_c for different states of consistency are given in Table 4.8 along with the values I_p. It may be seen that values of I_c and I_c are just opposite to each other for the same consistency of soil.

From Eqs (4.26) and (4.27) we have

$$\therefore I_l + I_c = \frac{w_l - w_p}{I_p} = 1 \tag{4.28}$$

Effect of Drying on Plasticity

Drying produces an invariable change in the colloidal characteristics of the organic matter in a soil. The distinction between organic and inorganic soils can be made by performing two liquid limit tests on the same material. One test is made on an air-dried sample and the other on an oven dried one. If the liquid limit of the oven dried sample is less than about 0.7 times that for the air-dried sample, the soils may be classed as organic. Oven-drying also lowers plastic limits of organic soils, but the drop in plastic limit is less than in liquid limit.

Effect of Admixtures on Plasticity

The effect of admixtures on the limits has been investigated in soil stabilisation studies. It has been found that chemicals having a sodium base affect the limits most, usually increasing both the liquid and plastic limits considerably. However, ordinary table salt, NaCl, produces a decrease, in the liquid limit. The effect of base exchange on plasticity is explained in Sect. 3.7 under base exchange.

The Shrinkage and Swelling of Clay Soils

As explained earlier if a moist cohesive soils is subjected to drying it loses moisture and shrinks. The degree of shrinkage, S_r is expressed as

$$S_r = \frac{V_o - V_d}{V_o} \times 100 \tag{4.29}$$

where, V_o, is the initial volume and V_d the final volume at the shrinkage limit after drying.

On the basis of the degree of shrinkage, Scheidig (1948) classified soils as in Table 4.9.

The soils that belong to the montmorillonite group shrink more than the soils of the other groups. When these soils shrink, because of the heterogeneous nature of the soils, cracks develop on the surface. The cracks are sometimes as wide as 5 cm at the surface and extend to a great depth. These are desiccation cracks. In tropical countries the depths of cracks may go up to about 3 to 4 metres. These dried soils swell considerably when they come in contact with water. The natural soils that are dried in summer swell considerably during the rainy season. The soils that shrink and swell considerably are called as expansive soils. The black-cotton soil of India belongs to this category. Light structures that are built directly on expansive soils suffer considerable damage because of the shrinking and swelling characteristics of the soil.

Skempton considers that the significant change in the volume of a clay soil during shrinking or swelling is a function of plasticity index and the quantity of colloidal clay particles present in soil. According to him, the clay soil can be classified *inactive*, *normal* or active as per the following equation. The *activity* of clay is expressed by an *activity* number A_c.

TABLE 4.9

Soil classification according to degree of shrinkage S_r

$S_r \%$	Plasticity
<5	Good
5–10	Medium good
10–15	Poor
>15	Very poor

Activity number $A_c = \dfrac{\text{Plasticity index, } I_p}{\text{Percent finer than 2 micron}}$ (4.30)

Table 4.10 gives the type of soil according to the value of A_c.

The clay soil which has an activity value greater than 1-4 can be considered as belonging to the swelling type.

The relationship between plasticity index and clay fraction is shown in Fig. 4.17.

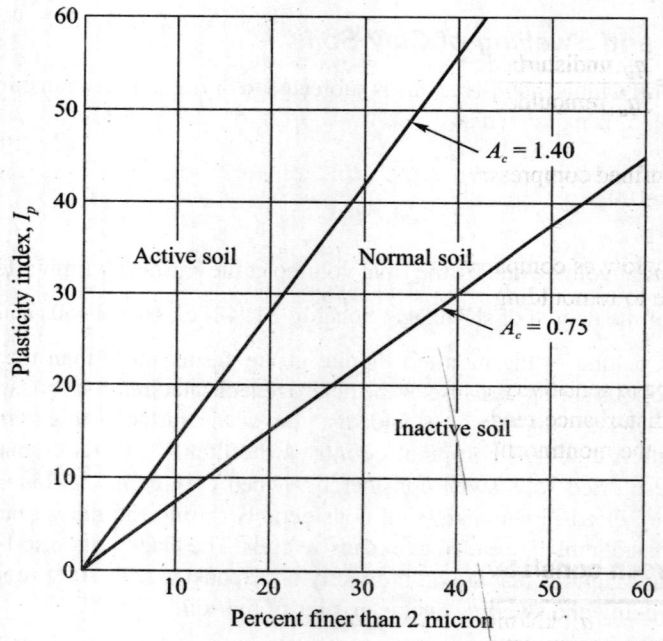

Fig. 4.17 Classification of soil according to activity number

TABLE 4.10

Soil classification according to activity number

A_c	Soil type
<0.75	Inactive
0.75–1.40	Normal
>1.40	Active

Consistency of Soils as per the Unconfined Compressive Strength

The consistency of a natural soil is different from that of a remoulded soil at the same water content. Remoulding destroys the structure of the soil and the particle orientation. The liquidity index value which is an indirect measure of consistency is only qualitative. Measurement of consistency of undisturbed soils on certain specific values vary quantitatively on the basis of its unconfined compressive strength. The unconfined compressive strength, q_u, is defined as the ultimate load per unit cross sectional area that a cylindrical specimen of soil (with height to diameter ratio of 2) can take under compression without any lateral pressure. Water content of the soil is assumed to remain constant during the duration of the test which generally takes only a few minutes. Table 4.11 indicates the relationship between consistency and q_u.

As explained earlier, remoulding of undisturbed sample of clay at the same water content alters its consistency, because of the destruction of its original structure. The degree of disturbance of undisturbed clay sample due to remoulding can be expressed as

$$\text{Sensitivity, } S_t = \frac{q_u \text{, undisturbed}}{q'_u \text{, remoulded}} \tag{4.31}$$

where q'_u is the unconfined compressive strength of remoulded clay at the same water content as the undisturbed clay.

When q'_u is very low as compared to q_u the clay is highly sensitive. When $q_u = q'_u$ the clay is said to be insensitive to remoulding. On the basis of the values of S_t clays can be classified as in Table 4.12.

The clays that have sensitivity greater than 8 should be treated with care during construction operations because disturbance tends to transform them, at least temporarily, into viscous fluids. Such clays belong to the montmorillonite group and possess flocculent structure.

TABLE 4.11

Relationship between consistency of clays and q_u

Consistency	q_u, kN/m^2	Consistency	q_u, kN/m^2
Very soft	<25	Stiff	100–200
Soft	25–50	Very stiff	200–400
Medium	50–100	Hard	>400

TABLE 4.12

Soil classification on the basis of sensitivity (after Skempton and Northey)

Sensitivity, S_t	Nature of clay	Sensitivity, S_t	Nature of clay
1	Insensitive clays	4–8	Sensitive clays
1–2	Low-sensitive clays	8–16	Extra sensitive clays
2–4	Medium sensitive clays	>16	Quick clays

Thixotropy

If a remoulded clay sample h, with sensitivity greater than one is allowed to stand without further disturbance and change in water content, it may regain at least part of its original strength and stiffness. This increases in strength is due to the gradual reorientation of the adsorbed molecules of water, and is known as *thixotropy* (from the Greek *thix*, meaning *touch* and *tropein*, meaning *to change*). The regaining of a part of the strength after remoulding has important application in connection with pile-driving operations, and other types of construct in which disturbance of natural clay formations is inevitable.

Example 4.10

The laboratory tests on sample of soil gave the following results:

$w_n = 24\%$, $w_l = 62\%$, $w_p = 28\%$, percentage of particle less than 2 m = 23%

Determine: (a) The liquidity index, (b) Activity number, (c) Consistency and nature of soil.

Solution

(a) Plasticity index, $I_p = w_l - w_p = 62 - 28 = 34\%$

Liquidity index, $I_l = \dfrac{w_n - w_p}{I_p} = \dfrac{24 - 28}{34} = -0.12$

(b) Activity number, $A_c = \dfrac{I_p}{\% \text{ of particles} < 2\mu} = \dfrac{34}{23} = 1.48.$

Comments

(i)   *Since the I_t is negative, the consistency of the soil is very stiff to extremely stiff (semisolid state).*

(ii) *Since the I_p is greater than 17% it is highly plastic.*

(iii) *Since A_c value greater than 1.40, the soil is active and is subjected to significant volume change (shrinkage and swelling) just like black cotton soil of India.*

Example 4.11

Two soil samples tested in soil mechanics laboratory gave the following results:

	Sample no. 1	Samples no. 2
Liquid limit	50%	40%
Plastic limit	30%	20%
Flow indices, I_f	27	17

Determine: (a) the toughness indices I_t, (b) Comment on the types of soils.

Solution

(a) $I_t = \dfrac{w_l - w_p}{I_f}$

Sample 1, $I_t = \dfrac{50-30}{27} = \dfrac{20}{27} = 0.74$; Sample 2, $I_l = \dfrac{40-20}{17} = \dfrac{20}{27} = 1.18$

(b)

(i)   Both the soils are clay soils as their toughness indices lie between 0 and 3.

(ii) Soil one is friable at the plastic limit since its I_t value is less than one.

(iii) Soil two is stiffer than soil one at the plastic limit since I_t value of the latter is higher.

Example 4.12

The natural moisture content of an excavated soil is 32%. Its liquid limit is 60% and plastic, limit is 27%. Determine the plasticity index of the soil and comment about the nature of the soil. .

Solution

Plasticity index, $I_p = w_t - w_p = 60 - 27 = 33\%$

The nature of the soil can be judged by determining its liquidity index, I_t from Eq. (4.26)

$$I_l = \frac{w_n - w_p}{I_p} = \frac{32-37}{33} = 0.15$$

since the value of I_l is very close to 0, the nature of soil as per Table 4.8 is very stiff.

Example 4.13

A soil with a liquidity index of –0.20 has a liquid limit of 56% and a plasticity index of 20%. What is its natural water content? What is the nature of this soil?

Solution

As per Eq. (4.26)

Liquidity index, $I_l = \dfrac{w_n - w_p}{I_p}$

$w_p = w_l - I_p = 56 - 20 = 36$, $w_n = I_t I_p + w_p = -0.20 \times 20 + 36 = 32$.

Since I_l is negative, the soil is in a semisolid or solid state as per Table 4.8.

Example 4.14

Four different types of soils were encountered in a big project. The liquid limits, w_l, plastic limits w_p, and the natural moisture contents w_n of the soils are given below:

Soil type	$w_l\%$	$w_p\%$	$w_n\%$
1	120	40	150
2	80	35	70
3	60	30	30
4	65	32	25

Required (a) the liquidity indices I_l of the soils, (b) the consistency of the natural soils and (c) the possible behaviour of the soils under vibrating loads.

Solution

(a) $I_l = \dfrac{w_n - w_p}{I_p}$

By substituting the appropriate values in this equation, we have

Type	I_l	Type	I_l
1	1.375	3	0
2	0.750	4	-0.21

(b) From Table 4.8, type I is in a liquid state, type 2 in a very soft state, type 3 in very soft state and type 4 in a semisolid state.

(c) The soil types 3 and 4 are not much affected by vibrating loads. Type 1 is very sensitive even for small disturbance and as such not suitable for any foundation. Whereas type 2 is also very soft, greater settlement of foundation or failure of foundation due to development of pore water pressure under saturated condition takes place due to any sudden application of loads.

Example 4.15

A shrinkage limit test on a clay soil gave the following data. Compute the shrinkage limit. Assume that the total volume of dry soil cake is equal to its total volume at the shrinkage limit what is the degree of shrinkage? Comment on the nature of soil

Mass of shrinkage dish and saturated soil $M_1 = 38.78$ g

Mass of shrinkage dish and oven dry soil $M_2 = 30.46$ g

Mass of shrinkage dish $\qquad M_3 = 10.65$ g

Volume of shrinkage dish $\qquad V_o = 16.29$ cm³

Total volume of oven-dry soil cake $\qquad V_d = 10.00$ cm³

Solution

Refer Fig. 4.14

The equation for shrinkage limit $w_s = \dfrac{M_w}{M_s}$

where M_w = mass of water in the voids at shrinkage limit.

M_o = mass of sample at plastic state = $M_1 - M_3 = 38.78 - 10.65 = 28.13$ g

Volume of water lost from plastic state to shrinkage limit $\Delta V = (V_o - V_d)$

or $\Delta V = 16.29 - 10.00 = 6.29$ cm³

Mass of dry soil = $M_s = M_2 - M_3 = 30.46 - 10.65 = 19.81$ g

Now; $M_w = M_o - M_s - (V_o - V_d)\, \rho_w = 28.13 - 19.81 - (6.29)(1) = 2.03$ g

From Eq. (4.22), $w_s = \dfrac{(M_o - M_s) - (V_o - V_d)\rho_w}{M_s} = \dfrac{M_w}{M_s} = \dfrac{2.03}{19.81} = 0.102 = 10.2\%$

As per Eq. (4.29), the degree of shrinkage, S_r is

$$S_r = \frac{V_o - V_d}{V_o} \times 100 = \frac{(16.29 - 10.1) \times 100}{16.29} = 38\%$$

From Table 4.9 the soil is of very poor quality.

Example 4.16

A sample of clay soil has liquid limit of 62%, and its plasticity index is 32%.

(a) What is the state of consistency of the soil if the soil in its natural state has a water content of 34%.

(b) Calculate the shrinkage limit if the void ratio of the sample at the shrinkage limit is 0.70. Assume $G = 2.70$.

Solution

(a) The consistency of a natural soil is judged by its liquidity index I_l.

$$w_p = w_l - I_p = 62 - 32 = 30, \quad I_l = \frac{34 - 30}{32} = 0.125.$$

As per Table 4.8, the consistency of soil is very stiff as $I_l \to 0$.

(b) Refer to Fig. 4.14

Assume $\qquad V_s = 1\ m^3,\ M_s = GV_s\,\rho_w = 2.7 \times 1 \times 1 = 2.7\ g$

$\qquad\qquad\qquad M_w = (V_d - V_s)\,\rho_w,\ \text{where}\ V_d = V_s + e = 1 + e$

Therefore, $\qquad M_w = (1 + e - 1)\,\rho_w = e\,\rho_w = 0.7 \times (1) = 0.7\ g$

shrinkage limit $w_s = \dfrac{M_w}{M_s} = \dfrac{0.7}{2.7} = 0.259\ \text{or}\ 25.9\%.$

Example 4.17

A sample of clay soil has a water content of 40 percent at full saturation. Its shrinkage limit is 15%. Assuming $G = 2.70$, determine the degree of shrinkage S_r. Comment on the nature, the soil.

Solution

Refer Fig. 4.14

$\qquad$ Shrinkage limit $w_s = \dfrac{M_w}{M_s}$

$\qquad\qquad M_s = GV_s\,\rho_w,\ \text{where}\ V_s = 1cm^3,\ \rho_w = 1\ g/cm^3$

Therefore, $\qquad M_s = 2.7 \times 1 \times (1) = 2.7\ g$

Now, $\qquad\qquad M_w = w_s M_s = 0.15 \times 2.7 = 0.40\ g$

and $\qquad\qquad V_w = \dfrac{0.40}{(1)} = 0.40\ cm^3$

Therefore, $\qquad V_d = V_s + V_w = 1 + 0.40\ cm^3$

From Fig. 4.14

$\qquad\qquad M_w = w_n M_s = 0.4 \times 2.7 = 1.08\ g$

or $\quad V_w = \dfrac{1.08}{(1)} = 1.08\ cm^3$

$\qquad\qquad V_o = V_s + V_w = 1 + 1.08 = 2.08\ cm^3$

Degree of shrinkage $S_r = \dfrac{V_o - V_d}{V_o} = \dfrac{2.08 - 1.40}{2.08} \times 100 = 32.7\%$

From Table 4.9, the soil is of very poor quality.

Example 4.18

A mass of an over dried soil pat is 0.78 N. When immersed in mercury the dry soil displaces 4.75 N of mercury. If the specific gravity of soil $G = 2.72$, what is the shrinkage limit of the soil. Assume the specific gravity G_m of mercury = 13.6.

Solution

Use Eq. (4.22)

$$w_s = \left(\frac{V_d \gamma_w}{M_s} - \frac{G_w}{G} \right) = \left(\frac{V_d \gamma_w}{M_s} - \frac{1}{G} \right)$$

by assuming $G_w \approx 1$

Since $\dfrac{M_s}{V_d} = \gamma_d$, we may write $w_s = \left(\dfrac{\gamma_w}{\gamma_d} - \dfrac{1}{G} \right)$

Volume of dry soil $V_d = \dfrac{4.75 \times 10^{-3}}{13.6 \times 9.81} = 0.356 \times 10^{-4} \, \text{m}^3$

Therefore, $\gamma_d = \dfrac{0.78 \times 10^{-3}}{0.356 \times 10^{-4}} = 22.28 \, \text{kN} / \text{m}^3$

By substituting,

$$w_s = \frac{9.81}{22.28} - \frac{1}{2.72} = 0.081 \text{ or } 8.1\%.$$

Example 4.19

The soil types given in Ex. 4.14 contain soil particles finer than 2 microns as given below:

Soil type	1	2	3	4
Percent finer than 2 micron	50	55	45	50

Classify the soils according to their activity values.

Solution

As per Eq. (4.30), activity A_c is expressed as

$$A_c = \frac{\text{Plasticity Index}, I_p}{\%\text{finer than } 2\mu}$$

substituting the appropriate values, the activity numbers and the types of soils from Table 4.10 as follows.

Soil type	1	2	3	4
A_c	1.6	0.82	0.67	0.67
Type of soil	Active	Normal	Inactive	Inactive

The soil type 1 is a swelling type as its A_c value is higher than 1.4.

4.10 QUESTIONS AND PROBLEMS

4.1   What are the index properties of soils? Why are they required? Discuss.

4.2   Explain and discuss a laboratory method of determining specific gravity of solids. Give the probable values of specific gravity we can use for sand and clay materials.

4.3   Does the size and shape of particles reflect the physical properties of cohesionless soils? Discuss.

4.4   Explain a method of determining the grain size distribution of cohesionless soils.

4.5   What is Stoke's Law? Under what conditions is this law valid?

4.6   Explain the principle of hydrometer method for determining the grain size distribution of fine grained soils.

4.7   A soil sample is a mixture of cohesionless and cohesive soils. Explain and discuss a method of determining the grain size distribution of the soil.

4.8   Explain and discuss the various corrections required to be applied to hydrometer readings in the sedimentation analysis.

4.9   Explain and discuss how some of the corrections to the hydrometer readings can be eliminated by following a particular procedure in the sedimentation analysis.

4.10   Sketch grain size distribution curves for uniformly graded, well graded and gap graded soils and discuss their characteristics.

4.11   Explain and discuss the significance of the values of uniformity coefficient and the coefficient of curvature.

4.12   What do you understand by the term 'consistency' of a soil? How do you express the consistency of a soil? Discuss.

4.13   Explain and discus the use of plasticity index to understand the nature of a soil.

4.14   Explain and discuss the use of liquidity index, activity number and sensitivity of clay.

4.15 In a specific gravity test the following data were obtained at the room temperature of 40 °C.

Mass of pycnometer + Water	=	720.36 g
Mass of pycnometer + Solid + Water	=	750.46 g
Mass of solids	=	47.50 g

Determine the specific gravity of the solids. Assume specific gravity of water at 40 °C, $G_T = 0.985$. [Ans: $G = 2.69$]

4.16 In a sieve analysis of a given sample of sand the following informations were obtained. Effective grain size = 0.25 mm, uniformity coefficient 6.0, coefficient of curvature = 1.0.

Sketch the curve on a semilog paper.

4.17 Sieve analysis of a given sample of sand were carried out by making use of IS sieves. The total weight of sand used for the analysis was 522 g. The following informations were obtained.

Sieve Size in mm	4.750	2.000	1.000	0.500	0.355	0.180	0.125	0.075
Weight retained in gm	25.75	61.75	67.00	126.0	57.75	78.75	36.75	36.75
Pan								31.5

Plot the grain distribution curve on a semi-log paper and compute the following (use IS: 1498-1970).

(i) Percent gravel

(ii) Percent of coarse, medium and fine sand

(iii) Percent of finer fraction

(iv) Uniformity coefficient

(v) Coefficient of curvature

4.18 Combined mechanical analysis of a given sample of soil was carried out. The total weight of soil used in the analysis was 350 g. The sample was divided into coarser and finer fractions of 125 g by washing it through IS-75 microns sieve. The coarser fraction was used for the sieve analysis and 50 g of the finer fraction was use for the hydrometer analysis. The test results were as given below:

Sieve analysis:

IS sieve size	Mass retained g	IS sieve size	Mass retained g
4.75 mm	9.0	355μ	24.5
2.00 mm	15.5	180μ	49.0
1.40 mm	10.5	125μ	28.0
1.00 mm	10.5	75μ	43.0
500μ	35.0		

The hydrometer was inserted into the suspension at the start of the test and readings were taken up to 2 min. It was next removed and introduced just before each of the subsequent readings.

Temperature of suspension   = 30 °C

Volume of hydrometer bulb   = 40 cc

Temperature of hydrometer calibration   = 27 °C

The values of H_e

when $r_h = 1.00$ H_{e1} = 22.50 cm

when $r_h = 1.030$, H_{e2} = 9.50 cm

Area of sedimentation jar, A_j = 30 cm^2

Volume of suspension   = 10^3 cc

Meniscus correction C_m = +0.4

Correction for the change in density

due to the addition of dispersing agent, C_d = –0.2

Temperature correction C_t = 0.75

(i)   Show step by step all the computations required for combined analysis.

(ii)   Plot the grain size distribution curve on a semi-log paper.

(iii) Determine the percentages of gravel, sand, and fine fractions present in the sample (use IS: 1498-1970)

(iv)   Compute the uniformity coefficient and the coefficient of curvature.

(v)   Comment on the basis of test results whether the soil is well graded or not.

Hydrometer analysis: Reading In suspension

Time, min	Reading, R'_h	Time, min	Reading, R'_h
1/4	28.00	30	5.10
1/2	24.00	60	4.25
1	20.50	120	3.10
2	17.20	240	2.30
4	12.00	480	1.30
8	8.50	1440	0.70
15	6.21		

4.19   Liquid limit tests were carried out on two given samples of clay. The test data are as given
       below.

Test Nos	1	2	3	4
Sample no. 1 Water content %	120	114	98	86
Number of blows, N	7	10	30	40
Sample no. 2 Water content %	96	74	45	30
Number of blows, N	9	15	32	46

The plastic limit of Sample No.1 is 40 percent and that of Sample No.2 is 32 percent.

Required:

(i)   The flow indices of the two samples,

(ii)   The toughness indices of the samples,

(iii) Comment on the type of soils on the basis the toughness index values.

4.20   Four different types of soils were encountered in a big project. Their liquid limits (w_l),
       plastic limits (w_p) and their natural moisture contents w_n were as given below:

Soil type	w_l %	w_p %	w_n %
1	120	40	150
2	80	35	70
3	60	30	30
4	65	32	25

Required:

(i)   The liquidity indices of the soils. (ii) The consistency of the natural soils (i.e. whether
       soft, stiff, etc.)

(ii)   The possible behaviour of the soils under vibrating loads.

4.21   The soil types as given in problem 4.20 contained soil particles finer than 2 microns as
       given below:

Soil type	1	2	3	4
Percent finer than 2 micron	50	55	45	50

Classify the soils according to their activity values.

4.22   A sample of clay has a water content of 40 percent at full saturation. Its shrinkage limit is 15 percent. Assuming $G = 2.70$, determine Its shrinkage ratio. Comment on the quality of the soil.

4.23   A sample of clay soil has a liquid limit of 62%, and its plasticity index is 32 percent.

   (i)   What is the state of consistency of the soil if the soil in its natural state has a water content of 34 percent.

   (ii)   Calculate the shrinkage limit if the void ratio of the sample at the shrinkage limit is 0.7.0.

   Assume $G = 2.70$

4.24   A soil with a liquidity index of –0.20 has a liquid limit 56 percent and plasticity index of 20 percent. What is its natural water content?

4.25   In a sedimentation test 20 g of soil of specific gravity 2.69 and passing 0.2 mm sieve were dispersed in 1000 ml of water having a viscosity of 0.001 Ns/m². One hour after the commencement of sedimentation, 20 ml of the suspension were taken by means of a pipette from a depth of 100 mm. The amount of solid particles (in the sample of 20 ml taken by pipette) obtained on drying was 0.07 g. Compute:

   (a)   The largest size of particle remaining in suspension at a depth of 100 mm, 60 mins after the commencement of sedimentation.

   (b)   The percentage of particles finer than this size in the original sample.

   (c)   the time interval from the commencement after which the largest particle remaining in suspension at 100 mm depth is one-quarter of this size.

4.26   A sample of soil weighing 50 g is dispersed in 1000 ml of water. How long after the commencement of sedimentation should the hydrometer reading be taken in order to estimate the percentage of particles less than 0.002 mm effective diameter, if the centre of the hydrometer is 150 mm below the surface of the water.

   Assume: $G_s = 2.7$; $\mu = .0..0.01$ Ns/m².

4.27   The results of sieving analysis of a soil were as follows:

Sieve size (mm)	Mass retained (g)	Sieve size (mm)	Mass retained (g)
20	0	2	3.5
12	1.7	1.4	1.1
1.0	2.3	0.5	30.5
6.3	8.4	0.355	45.3
5.6	5.7	0.180	25.4
2.8	12.9	0.063	7.4

   The total mass of the sample was 147.2 g.

   (a)   Plot the particle-size distribution curve and describe the soil. Comment on the flat part of the curve;

    (b)   state the effective grain size;

    (c)   find Allen Hazen's uniformity coefficient; (d) design a filter suitable for protecting this soil.

4.28   A liquid limit test carried out on a sample of inorganic soil taken from below the water table gave the following results;

Cone penetration y (mm)	15.5	18.2	21.4	23.6
Moisture content wy %	34.6	40.8	48.2	53.4

A plastic limit test gave a value of 33%. Determine the average liquid limit and plasticity index of this soil and give its classification.

4.29   For a particle of 0.005 mm diameter, how many hours are required to settle a depth of 3 m in a lake from its surface. Assume G = 2.7, $\mu = 0.001$ Ns/m^2.

4.30   The oven dry mass of a sample of clay was 11.26 g. Volume of the dry sample was determined by immersing in mercury and the mass of the displaced liquid was 80.29 g. Find the shrinkage limit, w_s, of clay by assuming G = 2.70.

4.31   Particles of five different sizes are mixed in the proportion shown below and enough water is added to make 1000 cm^3 of the suspension

Particle size (mm)	Mass (g)	
0.050	6	
0.020	20	
0.010	15	
0.005	5	
0.001	4	Total 50 g

It is ensured that the suspension is thoroughly mixed so as to have a uniform distribution of particles. All particles have specific gravity of 2.7.

    (a)   What is the largest particle size present at a depth of 6 cm after 5 mins of start of sedimentation,

    (b)   What is the specific gravity of the suspension at a depth of 6 cm after 5 mins of start of sedimentation?

    (c)   How long should the sedimentation be allowed so that all the particles have settled below 6 cm. Assume $\mu = 0.9 \times 10^{-6}$ kN-s/m^2.

4.32   A sample of clayey silt is mixed at about its liquid limit of 40%. It is placed carefully in a small porcelain dish with a volume of 19.3 cm^3 and weighs 34.67 g. After oven drying, the soil pat displaced 216.8 g of mercury.

    (a)   Determine the shrinkage limit, w_s, of the soil sample,

    (b)   Estimate the r_s of the soil.

4.33   During the determination of the shrinkage limit of a sandy clay, the following laboratory data was obtained:

Wet wt. of soil + dish	=	87.85 g
Dry wt. of soil + dish	=	76.91 g
Wt. of dish	=	52.70 g

The volumetric determination of soil pat

Wt. of dish + mercury	=	430.8 g
Wt. of dish	=	244.62 g
Calculate the shrinkage limit, assuming ρ_s =		2.65 Mg/m^3

4.34   A sedimentation analysis by a hydrometer was done with 50 g of oven dried soil sample. The hydrometer reading in 1000 cm^3 soil suspension 60 mins after the commencement of sedimentation is 19.5. The meniscus correction is 0.5. For R_h = 20.0, the value of H_e = 12 cm from the calibration chart of the hydrometer. The combined correction for density changes and variation in temperature is –2.00. Assuming G = 2.70 and $\mu = 1 \times 10^{-6}$ kN-s m^2 for water, calculate the smallest particle size which would have settled during the time of 60 mins and percentage of particles finer than this size.

CHAPTER 5

CLASSIFICATION OF SOILS

5.1 INTRODUCTION

The behaviour of a soil mass under load depends upon many factors such as the properties of the various constituents present in the mass, the degree of density, the degree of saturation, the environmental conditions, etc. If soils are grouped on the basis of certain definite principles and rated according to its performance, the properties of a given soil can be understood to a certain extent, on the basis of some simple tests. The objectives of this chapter is to discuss the following:

1. Field identification of soils.

2. Classification of soils.

5.2 FIELD IDENTIFICATION OF SOILS

The methods of field identification of soils can conveniently be discussed under the headings of coarse-grained and fine-grained -soil materials.

Coarse-Grained Soil Materials

The coarse-grained soil materials are mineral fragments, that may be identified primarily on the basis of grain size. The different constituents of coarse-grained materials are sand and gravel. As described in the earlier chapters the size of sand varies from 0.075 mm to 4.75 mm and that of gravel from 4.75 mm to 80 mm. The sand can further be classified as coarse, medium and fine (Fig. 4.2). The engineer should have an idea of the relative sizes of the grains in order to identify the various fractions. The description of sand and gravel should include an estimate of the quantity of material in the different size ranges as well as a statement of the shape and mineralogical composition of the grains. The mineral grains can be rounded, subrounded, angular or subangular. The presence of mica or a weak material such as shale affects the durability or compressibility of the deposit. A small magnifying glass can be used to identify the small fragments of shale or mica. The property of a coarse grained material mass depends also on the uniformity of the sizes of the grains. A well-graded

113

sand is more stable for a foundation base as compared to a uniform or poorly graded material.

Fine-Grained Soil Materials

Inorganic Soils: The constituent parts of fine-grained materials are silt and clay fractions. The properties of these materials have been discussed in detail in the earlier chapter. Since both these materials are microscopic in size, physical properties other than grain size must be used as criteria for field identification. The classification tests used in the field for preliminary identification are:

1. Dry strength test
2. Shaking test
3. Plasticity test
4. Dispersion test

Dry Strength

The strength of a soil in a dry state is an indication of its cohesion and hence of its nature. It can be estimated by crushing a 3 mm size of a dried fragment between thumb and forefinger. A clay fragment can be broken only with a great effort, whereas a silt fragment crushes easily.

Shaking Test

Shaking test is also called as dilatancy test. It helps to distinguish silt from clay since silt is more permeable than clay. In this test a part of soil mixed with water to a very soft consistency is placed in the palm of the hand. The surface of the soil is smoothed out with a knife and the soil pat is shaken by tapping the back of the hand. If the soil is silt, the water will rise quickly to the surface and give it a shiny glistening appearance. If the pat is then deformed either by squeezing or by stretching, the water will flow back into the soil and leave the surface with a dull appearance. Since clay soils contain much smaller voids than silts and are much less permeable, the appearance of the surface of pat does not change during the shaking test. An estimate of the relative proportions of silt and clay in an unknown soil mixture can be made by noting whether the reaction is rapid, slow or nonexistent.

Plasticity Test

If a sample of moist soil can be manipulated between the palms of the hands and fingers and rolled into a long thread of about 3 mm diameter, the soil then contains a significant amount of clay. Whereas silt cannot be rolled into a thread of 3 mm diameter without severe cracking.

Dispersion Test

This test is useful for making a rough estimate of sand, silt and clay present in a material. The procedure consists in dispersing a small quantity of the soil in water taken in a glass cylinder and allowing the particles to settle. The coarser particles settle first followed by finer ones. Ordinarily sand particles settle within about 30 seconds if the depth of water is about 10 cm. Silt particles settle in about 1/2 to 240 minutes, whereas particles of clay size remain in suspension for at least several hours and sometimes several days. The time required for some of the particle sizes to settle through 10 cm is given in Table 5.1.

TABLE 5.1

Dispersion test-time to settle 10 cm depth

Particle size (mm)	Vel. of fall cm/sec	Time to settle 10 cm depth
2.0	360	0.03 sec
0.6	32	0.31 sec
0.2	36	2.78 sec
0.06	324×10^{-3}	30.80 sec
0.02	36×10^{-3}	4^m 38 sec
0.002	36×10^{-5}	7^h 43 m
0.0002	36×10^{-7}	32^d 3^h 36 m

The table indicates that colloidal particles finer than 0.0002 mm will not generally settle because of *Brownian movement*.

Organic Soils

Surface soils and many underlying formations may contain significant amounts of solid matter derived from organisms. While shell fragments and similar solid matter are found at some locations, organic material in soil is usually derived from plant or root growth and consists almost completely disintegrated matter, such as muck or more fibrous material, such as peat. The soils with organic matter are weaker and more compressible than soils having the same mineral composition but lacking in organic matter. The presence of an appreciable quantity of organic material can usually be recognised by the dark-grey to black colour and the odour of decaying vegetation which it lends to the soil.

Organic Slit

It is fine grained more or less plastic soil containing mineral particles of silt size and finely divided particles of organic matter. Shells and visible fragments of partly decayed vegetative matter may also be present.

Organic Clay

It is a clay soil which owes some of its significant physical properties to the presence of finely divided organic matter. Highly organic soil deposits such as peat or muck may be distinguished by a dark-brown to black colour, by the presence of fibrous particles of vegetable matter in varying states of decay. The organic odour is a distinguishing characteristic of the soil. The organic odour can sometimes be distinguished by a slight amount of heat.

5.3 CLASSIFICATION OF SOILS

The soils in nature rarely exist separately as gravel, sand, silt, clay or organic matter, but are usually found as mixtures with varying proportions of these components. Grouping of soils on the basis of certain definite principles would help the engineer to rate the performance of a given soil either as a sub-base material for roads and airfield pavements, foundations of structures, etc. The classification or grouping of soils is mainly based on one or two index properties of soil which are described in detail in Chapter 4. The methods that are used for classifying soils are based on one or the other of the following two broad systems:

1. Textural system which is based only on grain size distribution.

2. The systems that are based on grain size distribution and limits of soil.

There are many classification systems under the Textural System alone. However, the important ones are the IS: 1498-1970 and the U.S. Department of Agriculture. Only these two systems are described in this book.

Many systems are in use that are based on grain size distribution and limits of soil. The systems that are quite popular amongst engineers, are the U.S. Bureau of Public Roads Administration System (PRA System) and the Unified Classification System.

5.4 TEXTURAL SYSTEM

IS: 1498-1970 System: Figure 4.2 gives IS grain size classification system. In the use of the IS system, the grain size distribution of the soil is determined. Then the mass of the soil grains within each size group is computed. Finally the percentage of the soil represented by each size group is obtained. Thus, a mixed-grained soil might be described as 10 percent gravel, 52 percent sand, 38 percent silt and clay, according to the IS classification.

U.S. Department of Agriculture System

The boundaries between the various soil fractions as per this systems are given in Table 5.2.

By making use of the grain size limits mentioned above, for sand, silt and clay, a triangular classification chart, has been developed as shown in Fig. 5.1 for classifying mixed soils. The first step in the classification of soil is to determine the percentages of sand, silt and clay-size materials in a given sample by mechanical analysis. With the given relative percentages of the sand, silt and clay, a point is located on the triangular chart as shown in Fig. 5.1. The designation given on the chart for the area in which the point falls is then used as the classification of the sample. This method of classification does not reveal any properties of the soil other than grain-size distribution. Because of its simplicity, it is widely used by workers in the fields of agriculture and highway engineering. One significant disadvantage of this. method is that the textural name as derived from the chart does not always correctly express the physical characteristics of the soil. For example, since some clay size particles are much less active than others. a soil described as clay on the basis of this system may have physical properties more typical of silt.

TABLE 5.2

Soil fractions as per U.S. department of agriculture

Soil fraction	Diameter in mm
Gravel	>2.00
Sand	2–0.05
Silt	0.05–0.002
Clay	<0.002

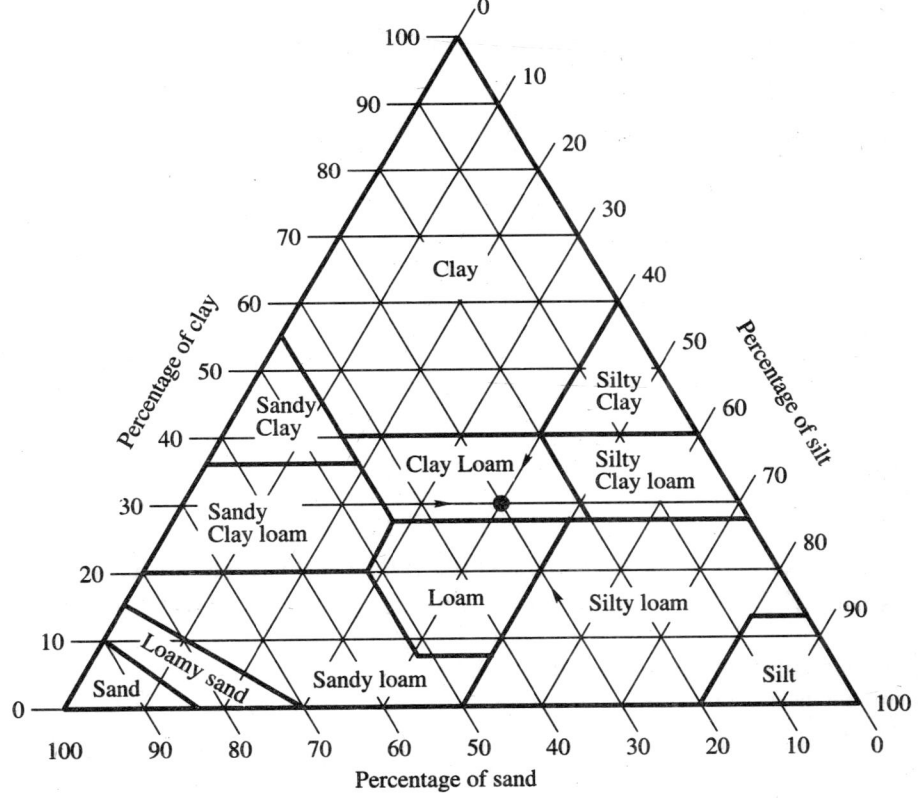

Fig. 5.1 U.S. Department of agriculture textural classification

The triangular chart of the Department of Agriculture uses the word *loam* to describe the various types of mixed soils, such as clay loam, silt loam, sandy clay loam, etc. A loam is a mixture of sand, silt and clay particles in varying proportions. The term *loam* originated in agricultural soil work and was taken over by highway engineers who have to deal with surface soil layers. In order to eliminate this ambiguous term *loam* from engineering use, the Mississippi River Commission. U.S.A. has proposed a new triangular chart, Fig. 5.2, which does not use the word loam and replaces it by such terms as silty clay. sandy silt. etc. This appears logical, since the whole classification is based only on the sand, silt and clay content.

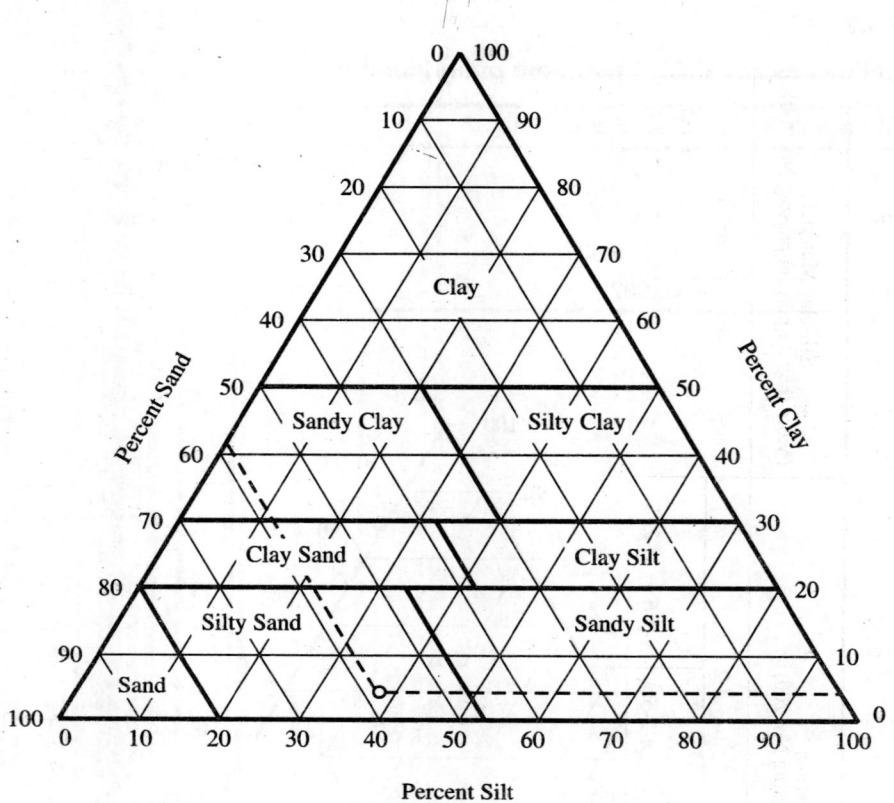

Fig. 5.2 Triangular Classification Chart of the Mississippi River Commission, U.S.A.

5.5   PUBLIC ROADS ADMINISTRATION CLASSIFICATION SYSTEM (PRA SYSTEM) ALSO KNOWN AS AASHTO (1978) SYSTEM

This system was originally proposed in the year 1928 by the U.S. Bureau of Public Roads for the use of highway engineers. A committee of highway engineers for the Highway Research Board, U.S.A. met in 1945 and made, an extensive revision of (the PRA system. This system in USA is known as AASHTO (American Association of State Highway and Transport Officials) System. The revised system comprises seven groups of inorganic soils, A-1 to A-7 with 12 subgroups in all. The system is based on the following three soil properties:

1.   Particle-size distribution

2.   Liquid limit

3.   Plasticity index.

Besides, a Group Index is introduced to identify completely soils containing appreciable fine-grained materials. The characteristics of various groups are defined in Table 5.3. The Group Index may be determined from the equation.

Group Index $(GI) = 0.2a + 0.005ac + 0.01bd$ (5.1)

TABLE 5.3

Public roads administration classification (1945) highway research board modification

General classification	Granular Materials (35 percent or less passing No. 200)							Silt-clay Materials (More than 35 percent passing No. 200)			
Group classification	A-1-a	A-1-b	A-3	A-2-4	A-2-5	A-2-6	A-2-7	A-4	A-5	A-6	A-7 (A-7-5, A-7-6)
Sieve analysis percent passing											
No. 10	50 max										
No. 40	30 max	50 max	51 min								
No. 200	15 max	25 max	10 max	35 max	35 max	35 max	35 max	36 min	36 min	36 min	36 min
Characteristics of fraction passing No. 40											
Liquid limit				40 max	41 min	40 max	41 min	40 max	41 min	40 max	41 min
Plasticity Index	6 max		N.P.	10 max	10 max	11 min	11 max	10 max	10 max	11 min	11 min
Group Index	0		0	0	0	0	0	8 max	12 max	16 max	20 max
Usual types of significant constituent materials	Stone fragments—gravel and sand		Fine sand	Silty or clayey gravel and sand				Silty soils		Clayey soils	
General rating as subgrade	Excellent to good							Fair to poor			

in which,

a = That portion of percentage of soil particles passing No. 200 (ASTM) sieve greater than 35 and not exceeding 75, expressed as a positive whole number (0 to 40).

b = That portion of percentage of soil particles passing No. 200 sieve, greater than 15 and not exceeding 55 expressed as a positive whole number (0 to 40).

c = That portion of the liquid limit greater than 40 and not exceeding 60, expressed as a positive whole number (0 to 20).

d = That portion of the plasticity index greater than 10 and not exceeding 30, expressed as a positive whole number (0 to 20).

The group index is a means of rating the value of a soil as a subgrade material within its own group. It is not used in order to place a soil in a particular group, that is done directly from the results of sieve analysis, the liquid limit and plasticity index. The higher the value of the group index, the poorer is the quality of the material. The group index is a function of the amount of material passing the No. 200 sieve, the liquid limit and the plasticity index.

If the pertinent index value for a soil falls below the minimum limit associated with a, b, c or d, the value of the corresponding term is zero, and the term drops out of the group index equation. When the values of a, b, c, d, are more than the prescribed maximum values, then the highest value of 40 or 20 should be assigned. The group index value should be shown in parenthesis after group symbol as A-6(12) where 12 is the group index.

Classification Procedure

The classification procedure can be explained with the help of a simple example.

Classification procedure: With required data in mind, proceed from left to right in the chart; correct group will be found by process of elimination. The first group from the left consistent with the test data is the correct classification. The A-7 group is subdivided in to A-7-5 or A-7-6 depending on the plastic limit. For $w_p < 30$, the classification is A-7-6; for $w_p \geq 30$, it is A-7-5.

Example 5.1

A sample of inorganic soil has the following grain size characteristics

Size (mm)	Percent passing
2.0	95
0.074 (No. 200)	75

The liquid limit is 56 percent, and the plasticity index 25 percent. Classify the soil according to the PRA system.

Solution

Percent of fine grained soil = 75

Computation of Group Index:

$a = 75 - 35 = 40$

$b = 40$ (since $75 - 15 = 60$ which is greater than the limiting value 40).

$c = 56 - 40 = 16, d = 25 - 10 = 15$

Group Index, $GI = 0.2 \times 40 + 0.005 \times 40 \times 16 + 0.01 \times 40 \times 15 = 17$

On the basis of percent of fine-grained soils, liquid limit and plasticity index values, the soil is either A-7-5 or A-7-6. Since $w_p = w_i - I_p = 56 - 25 = 31$, *which is greater than 30, classification is A-7-5(17).*

5.6   UNIFIED SOIL CLASSIFICATION SYSTEM (USCS)

The Unified Classification System is based on the recognition of the type and predominance of the constituents considering grains-size, gradation, plasticity and compressibility. It divides soil into three major divisions: Coarse-grained soils, fine grained soils, and highly organic (peaty) soils. In the field, identification is accomplished by visual examination for the coarse grained and a few simple hand tests for the fine-grained soils. In the laboratory, the grain-size curve and the Atterberg Limits can be used. The peaty soils are readily identified by colour, odour, spongy feel and fibrous texture.

The Unified Soil Classification System is a modified version of A. Casagrande's Airfield Classification (AC) System developed by him in 1942 for the Corps of Engineers, U.S.A. Since 1942 the original classification has been expanded and revised in cooperation with the Bureau of Reclamation, USA, so that it applies not only to airfields but also to embankments, foundations, and other engineering features. This system was adopted in 1952. In 1969 the American Society for Testing and Materials (ASTM) adopted the Unified System as a standard method for classification for engineering purposes (ASTM D-2487). The various symbols used in this system are:

1.   *Coarse-grained Soils*

G = Gravel and gravelly soils, S = Sand and sandy soils. The gravel and sand are further divided into subgroups as:

    W = Well graded, fairly clean material

    C = Well graded with excellent clay binder

    P = Poorly graded, fairly clean material

    F = Coarse materials containing fines not included in preceding groups

2.   *Fine-grained Soils*

    M = Inorganic silty and very fine sandy soils

    C = Inorganic clays

    O = Organic silts and clays

    P = peat

The above groups of the fine-grained soils (excluding peat) are further subdivided according to their liquid limits into:

L = Fine-grained soils with liquid limits below 50, indicating low to medium compressibility or plasticity characteristics

H = Fine-grained soils with liquid limits above 50 indicating high compressibility or plasticity characteristics.

The types of soil is indicated by a suitable combination of symbols such as:

GW = Well graded gravels or gravel-sand mixtures with little or no fines

GC = Clayey gravels or sand-clay mixtures

SC = Clayey sands or sand-clay mixtures

ML = Inorganic silt with low to medium compressibility

CH = Inorganic clays of high plasticity.

Table 5.4a gives the groupings for coarse grained soils and Table 5.4b for fine-grained soils. The classification of fine grained soils is easier to understand on Casagrande's plasticity chart which is described in the next paragraph.

Plasticity Chart as per Unified Classification System

A. Casagrande devised a plasticity chart which is useful for identifying and classifying fine-grained soils. In this chart the ordinate represents the values of the plasticity index and the abscissa the values of the liquid limit. The equation of the inclined line (which is termed as A-line) is I_p= 0.73 $(w_i - 20)$. All soils represented by points below the line may be inorganic silts, organic silts or organic clays. Inorganic clays may be of low, medium or high plasticity depending on whether liquid limit is below 30, between 30 and 50 or above 50. Similarly, the inorganic silts are said to be of low, medium or high compressibility according to the region they fall below the A-line. Organic silts are represented by points within the region bounded by liquid limits of 30 and 50, and organic clays by points in the region with liquid limits greater than 50.

The procedure for distinguishing between organic and inorganic soils is explained in Chapter 4. The soils are considered as fine grained when more than 50 percent of the particles pass through No. 200 sieve (ASTM). The chart proposed by Casagrande has been modified and used under Unified Classification System. As per the Unified Classification System, the whole region is divided into four parts by drawing a vertical line through liquid limit 50 percent. The points lying to the left of the line come under low to medium compressibility or plasticity and those lying on the right come under high compressibility or plasticity. The grouping of fine-grained soils are:

ML = Inorganic silts and very fine sands with slight plasticity, having liquid limit less than 50.

CL = Inorganic clays of low to medium plasticity, gravelly clays, sandy clays, silty clays, lean clays with $w_l < 50$.

OL = Organic silts and organic silt-clays of low plasticity with $w_l < 50$.

TABLE 5.4a

Unified Classification System—Coarse-Grained Soils (More than half of material is larger than No. 200 sieve size[1])

Soil (1)	Major divisions (2)	Groups symbols[2] (3)	Typical names (4)	Field identification procedures[3] (5)	Laboratory classification criteria (6)
Gravels	Clean gravels	GW	Well graded gravels, gravel-sand mixtures, little or no fines	Wide range in grain sizes and substantial amounts of all intermediate particles sizes	$C_u = \dfrac{D_{60}}{D_{10}}$ greater than 4 where C_u = uniformity coefficient $C_c = \dfrac{(D_{30})^2}{D_{10} \times D_{60}}$ between one and three
		GP	Poorly graded gravels or gravel-sand mixtures, little or no fines	Predominantly one size or a range of sizes with some intermediate sizes missing	GP does not meet all gradation requirement for GW
	Gravels[4] with fines[6]	GM	Silty gravels, gravel-sand-silt mixture	Nonplastic fines or fines with low plasticity	Atterberg limits below A line or PI less than 4
		GC	Clayey gravels, gravel-sand-clay mixture	Plastic fines	Atterberg limits above A line with PI greater than 7

For the GM/GC rows, the far-right column notes: Above A-line with PI between 4 and 7 are borderline cases requiring use of dual symbols

contd.

TABLE 5.4a contd.

Unified classification system—coarse-grained soils (More than half of material is larger than No. 200 sieve size[1])

Soil (1)	Major divisions (2)	Groups symbols[2] (3)	Typical names (4)	Field identification procedures[3] (5)	Laboratory classification criteria (6)	
Sands[7]	Clean sands[5]	SW	Well graded sands, gravelly sand, little or no fines	Wide range in grain sizes and substantial amounts of all intermediate particles sizes	$C_u = \dfrac{D_{60}}{D_{10}}$ greater than 6	
		SP	Poorly graded sands or gravelly sands, little or no fines	Predominantly one size or a range of sizes with some intermediate sizes missing	SP does not meet all gradation requirements	
	Sands with fines[6]	SM	Silty sands, sand-silt mixture	Nonplastic fines or fines with low plasticity	Atterberg limits below A line or PI less than 4	Limits plotting in hatched zone with PI between 4 and 7 are borderline cases requiring use of dual symbols
		SC	Clayey sand, sand-clay mixture	Plastic fines	Atterberg limits with PI greater than 7	

Footnotes to the Table 5.4a

1. All sieve sizes in this table are U.S. Standard.

2. Soils possessing characteristics of two groups are designated by a combination of group symbols; for example, *GW-GC* is well-graded gravel-sand mixture with clay binder.

3. Excluding particles larger than 3" and basing fractions on estimated weights.

4. More than half of coarse fraction is larger than No.4 sieve size.

5. Little or no fines.

6. Appreciable amount of fines.

7. More than half of coarse fraction is smaller than No.4 sieve size.

Information Required for Describing Soils

Give typical name; indicate approximate percentages of sand and gravel, maximum size; angularity, surface condition, and hardness of the coarse grains; local or geologic name and other pertinent descriptive information; and symbol in parentheses.

For undisturbed soils add information on stratification, degree of compactness, cementation, moisture conditions, and drainage characteristics.

Example: Silty sand, gravelly; about 20 percent, hard, angular gravel particles 1/2" maximum size, rounded and subangular sand grains, coarse to fine; about IS percent non-plastic fines with low dry strength, well compacted and moist in place; alluvial sand (SM).

Determine percentages of gravel and sand from grain-size curve. Depending on percentage of fines (fraction smaller than No. 200 sieve size) coarse grained soils are classified as follows:

Less than 5 percent *GW, GP, SW, SP*

More than 12 percent *GM, GC, SM, SC*

5 to 12 percent borderline cases requiring use of dual symbols

From: U.S. Army Corps of Engineers Water Ways Experiment Station, The Unified Classification System Tech. Memo 3-357, March 1953.

MH = Inorganic silts micaceous or diatomaceous fine sandy or silty soils, and elastic silts with $w_t > 50$.

CH = Inorganic clays of high plasticity (fat clays) with $w_t > 50$.

OH = Organic clays of medium to high plasticity with $w_t > 50$.

Pt = Peat and highly organic soils.

The soil groups are shown in the Plasticity Chart, Fig. 5.3.

The limits as obtained for the black cotton soil near Indore (India) is also plotted on the Plasticity Chart. As per the unified classification system, this soil is classified as *CH*.

TABLE 5.4b

Unified classification system–fine-grained soils
(More than half of material is smaller than No. 200 sieve size)

Soil	Major divisions	Group symbols	Identification procedures on fraction smaller than No, 40 sieve size		
			Dry strength	Dilatancy	Toughness
Silt and clays	Liquid limit less than 50	ML	None to slight	Quick to slow	None
		CL	Medium to high	None to very slow	Medium
		OL	Slight to medium	Slow	Slight
	Liquid limit less than 50	ML	Slight to medium	Slow to none	Slight to medium
		CH	High to very high	None	High
		OH	Medium to high	None to very slow	Slight to medium
Highly organic soil		Pt	Readily identified by colour, odour, spongy feel and frequently by fibrous texture		

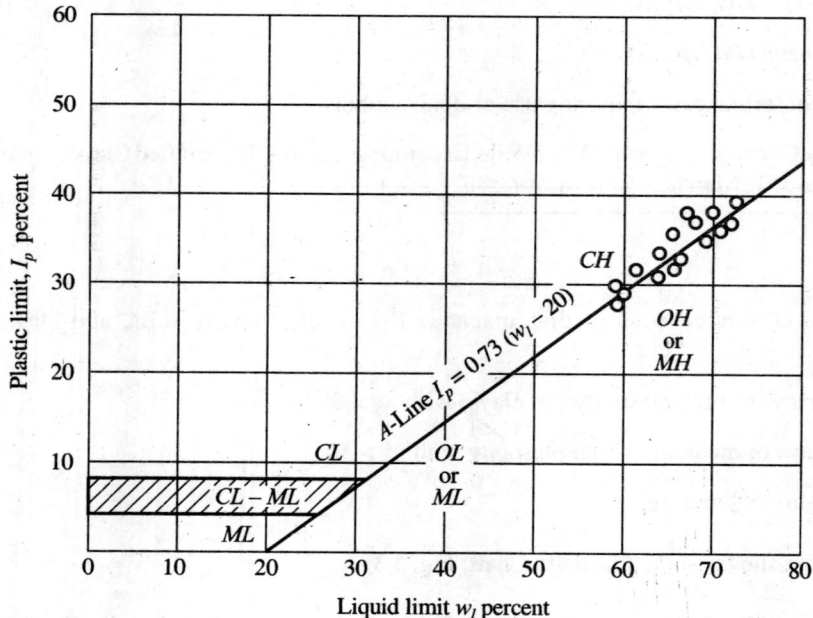

Fig. 5.3   Plasticity chart for fine-grained soils

5.7 THE INDIAN SOIL CLASSIFICATION SYSTEM (IS: 1498-1970)

The Indian Bureau of Standards have adopted the Unified Classification System (USCS) as their code of practice (IS: 1498–1970) with some modification of the plasticity chart for the classification and identification of soils. The modified form of the plasticity chart is explained below.

Plasticity Chart as per Indian Classification System (IS: 1498-1970)

In the case of Indian Soil Classification chart the plasticity chart given in Fig. 5.4 is divided into six regions by drawing another vertical line at liquid limit w_l equal to 35. The plasticity (clays) or compressibility (sills) are designated as low (L), medium (I) and high (H) according to the regions. The designation of the types of soil falling in the various regions areas down in Fig. 5.4 and the various soil groups are given below.

All soil lying above the A-line may be clays of low ($w_l < 35$), medium ($35 < w_l < 50$) and high ($50 > w_l$) plasticity. Those lying below the A-line in the corresponding liquid limit zones are normally inorganic silts or organic clays of low, medium and high compressibility. In accordance with this general classification, the following group symbols are used.

CL, CI and *CH* = Inorganic clays of low, medium and high plasticity.

ML = Inorganic silt of low plasticity.

OL = Organic silts or organic silt-clays of low plasticity

OI = Organic clays of medium plasticity

MI = Inorganic silts or silty sands of medium plasticity.

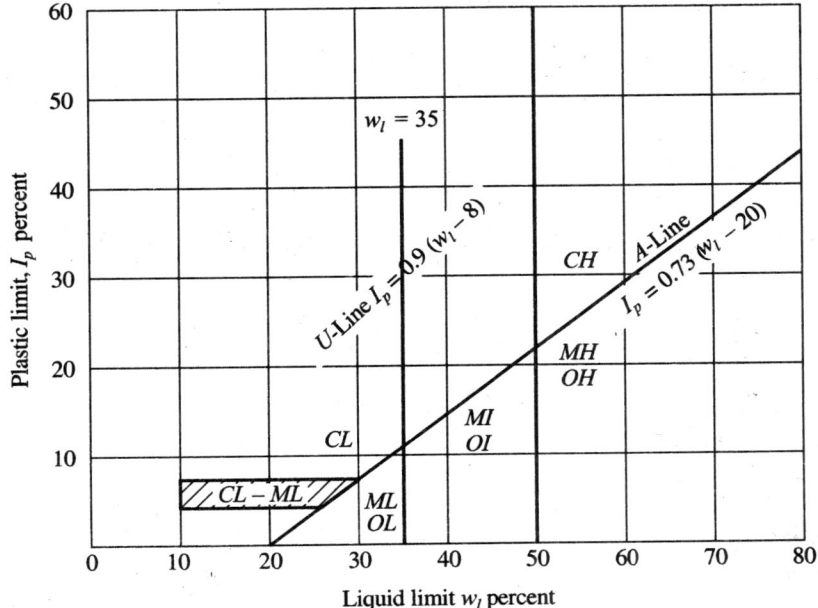

Fig. 5.4 Plasticity chart as per IS: 1498-1970

MH = Inorganic silts of high compressibility.

OH = Organic clays of high compressibility or plasticity.

The soils of a particular group may be admixtures of one or more components of the other soils including fine sand.

5.8   GENERAL COMMENTS ON THE SYSTEMS OF SOIL CLASSIFICATION

The various classification systems described earlier are based on

1. The properties of soil grains.
2. The properties applicable to remoulded soils.

The systems do not take into account the properties of intact materials as found in nature. Since the foundation materials of most of the engineering structure are undisturbed, the properties of intact materials only determine the soil behaviour during and after construction. The classification of a soil according to any of the accepted systems does not in itself enable detailed studies of soils to be dispensed with altogether. Solving flow, compression and stability problems merely on the basis or soil classification can lead to disastrous results. However, soil classification has been found to be a valuable tool to the engineer. It helps the engineer by giving him general guidance through making available in an empirical manner the results of field experience.

Example 5.2

Mechanical analysis on four different samples designated as A, B, C and D were carried out in a soil Laboratory. The results of tests are given below. Hydrometer analysis was carried out on sample D. Soil is non-plastic.

Sample D: Liquid Limit = 42, Plastic Limit = 24, Plasticity Index = 18

Classify the Soil as per the Unified Classification System.

Samples ASTM Sieve Designation	A	B	C	D
	Percentage finer than			
63.0 mm	100		93	
20.0 mm	64		76	
6.3	39	100	65	
2.0 mm	24	98	59	
600 μ	12	90	54	
212 μ	5	9	47	100
63 μ	1	2	34	95
20 μ			23	69
6 μ			4	46
2 μ			7	31

Solution

Grain size distribution curves of samples A, B, C and D are given in Fig. Ex. 5.2. The values of C_u and C_c are obtained from the curves as given below.

Sample A: Well graded sandy gravel classified as G_w. Gravel size particles more than 50%, fine grained soil less than 5%. C_u, greater than 4, and C_c lies between 1 and 3.

Sample B: Poorly-graded sand, classified as *SP* with 96% of particles being of sand. Finer fraction less than 5%. $C_u = 1.8$, C_c is not between 1 and 3.

Sample C: Gravel-sand-silt mixture, classified as *GM*. Coarse grained fraction greater than 66% and fine grained fraction less than 34%. The soil is non-plastic. C_u is very high but C_c is 0.25 only.

Sample D: Silty-clay of low plasticity, classified as *CL*. Finer fraction 95% with clay size particles 31 %. The point plots just above the A-line in the *CL* zone on the plasticity chart.

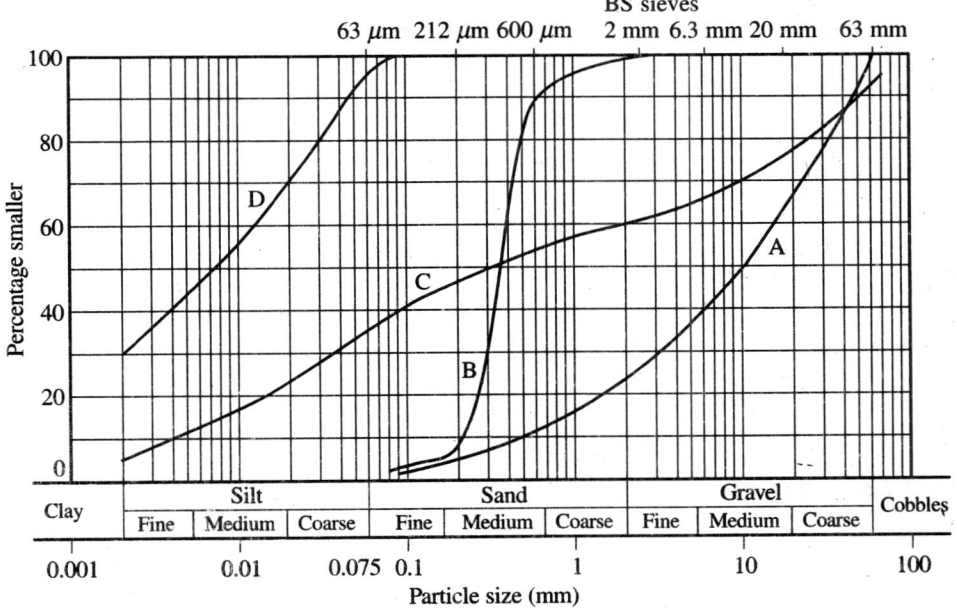

Fig. Ex. 5.2

Sample	D_{10}	D_{30}	D_{60}	C_u	C_c
A	0.47	3.5	16.00	34.0	1.60
B	0.23	0.30	0.41	1.8	0.95
C	0.003	0.042	2.40	800.0	0.25

Example 5.3

The following data refers to a light silty clay that was assumed; to be saturated in the undisturbed condition. On the basis of these data determine the liquidity index, sensitivity, and void ratio of the saturated soil. Classify the soil according to the Unified and AASHTO (PRA) systems. Assume $G = 2.7$.

Index property	Undisturbed	Remoulded
Unconfined compressive strength, q_u kN/m²	244 kN/m²	144 kN/m²
Water content, %	22	22
Liquid limit, %		45
Plastic limit, %		20
Shrinkage limit, %		12
% passing no. 200 sieve		90

Solution

Liquidity Index (Eq. 4.26),

$$I_l = \frac{W_n - W_p}{W_l - W_p} = \frac{22 - 20}{45 - 20} = 0.08$$

Sensitivity (Eq. 4.31),

$$S = \frac{q_u \text{ undisturbed}}{q_u \text{ disturbed}} = \frac{244}{144} = 1.7$$

Void ratio,

$$e = \frac{V}{V_s} \quad \text{but}$$

As per Eq. (3.14b) for $S = 1$, $e = wG = 0.22 \times 2.7 = 0.594$.

Unified Classification

Use plasticity chart Fig. 5.3. $w_l = 45$, $I_p = 25$. The point falls above the A-line in the CL-zone, that is the soil is inorganic clay of low to medium plasticity.

AASHTO System

Group Index $GI = 0.2a + 0.005ac + 0.01bd$

$a = 90 - 35 = 55$ but limited to 40

$b = 90 - 15 = 75$ but limited to 40

$c = 45 - 40 = 5$

$d = 25 - 10 = 15$ (here $I_p = w_l - w_n = 45 - 20 = 25$)

Group index $GI = 0.2 \times 40 + 0.005 \times 40 \times 5 + 0.01 \times 40 \times 15$

$$= 8 + 1 + 6 = 15$$

We can enter Table 5.3 with the following data

% passing 200 sieve = 90%

Liquid limit   = 45%

Plasticity index   = 25%

With this, the soil is either A-7-5 or A-7-6. Since plastic limit $w_p < 30$, the soil is classified as A-7-6. According to this system the soil is clay.

5.9 QUESTIONS AND PROBLEMS

5.1 What are the tests you carry out to identify soils in the field? Explain.

5.2 What are the different systems of classification of soil? Discuss the merits and demerits of each of the systems.

5.3 Explain and discuss the U.S. Department of Agriculture System of Classification of soil.

5.4 Explain and discuss the Mississippi River Commission System of Classification of soils.

5.5 Explain and discuss the PRA Classification System.

5.6 Explain and discuss the Unified Soil Classification System.

5.7 What is a plasticity chart? How is this chart useful for classifying fine-grained soils? Discuss.

5.8 The sieve analysis of a given sample of soil gave information that 57 percent of the particles passed through IS 75 micron sieve. The liquid and plastic limits of the soil were 62 and 28 percents respectively. Classify the soil as per the PRA and the Unified Classification Systems.

5.9 The mechanical analysis of a sample of a soil gave the following information:

$$\begin{array}{lll} \text{Sand} & = & 35 \text{ percent} \\ \text{Silt} & = & 40 \text{ percent} \\ \text{Clay} & = & 25 \text{ percent} \end{array}$$

Classify the soil as per the Triangular Classification System.

CHAPTER 6

PERMEABILITY OF SOIL

6.1   INTRODUCTION

A material is permeable if it contains continuous voids. All materials such as rocks, concrete, soils, etc. are permeable. The flow of water through all of them obeys approximately the same laws. Hence, the difference between the flow of water through rock or concrete is one of degree. The permeability of soils has a decisive effect on the stability of foundations, seepage loss through embankments of reservoirs, drainage of subgrades, excavation of open cuts in water bearing sand, rate of flow of water into wells and many others.

Flow of Water through a Soil Mass

Hydraulic Gradient

When water flows through a saturated soil mass there is certain resistance for the flow because of the presence of solid matters. However, the laws of *fluid mechanics,* which are applicable for the flow of fluids through pipes are also applicable to flow of water through soils. As per *Bernoulli's equation,* the total head, at any point in water under steady flow condition may be expressed as:

Total head = pressure head + velocity head + elevation head

This principle can be understood with regards to the flow of water through a sample of soil of length L and cross-sectional area A as shown in Fig. 6.1(a). As per this figure the heads of water at points A and B as the water flows from A to B are given as follows (with respect to a datum)

Total head at A, $\quad H_A = Z_A + \dfrac{P_A}{\gamma_w} + \dfrac{V_A^2}{2g}$

Total head at B, $\quad H_B = Z_B + \dfrac{P_B}{\gamma_w} + \dfrac{V_B^2}{2g}$

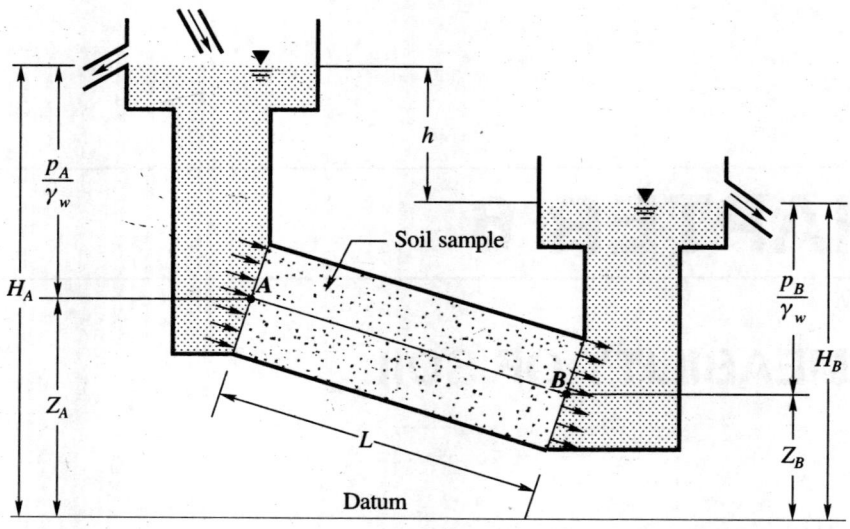

Fig. 6.1(a)   Flow of water through a sample of soil

As the water flows from A to B, there is an energy loss which is represented by the difference in the total heads H_A and H_B

$$\text{or} \quad H_A - H_B = \left(Z_A + \frac{p_A}{\gamma_w} + \frac{V_A^2}{2g} \right) - \left(Z_B + \frac{p_B}{\gamma_w} + \frac{V_B^2}{2g} \right) = h$$

where, p_A and p_B = pressure heads, V_A and V_B = velocity, g = acceleration due to gravity, γ_w = unit weight of water, h = loss of head.

For all practical purposes the velocity head is a small quantity and may be neglected. The loss of head of h units is effected as the water flows from A to B. The loss of head per unit length of flow may be expressed as

$$i = \frac{h}{L} \tag{6.1}$$

where i is called the *hydraulic gradient*.

Laminar and Turbulent Flow

Problems relating to the flow of fluids in general may be divided into two main classes:

1. Those in which the flow is laminar.

2. Those in which the flow is turbulent.

There is a certain velocity, v_c, below which for a given diameter of a straight tube and for a given fluid at a particular temperature, the flow will always remain laminar. Likewise there is a higher velocity, v_T, above which the flow will always be turbulent. The lower bound there is a

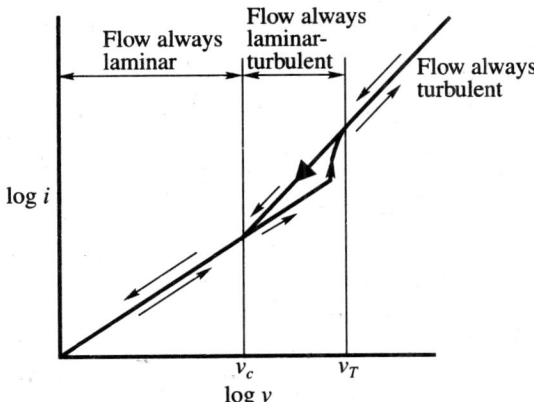

Fig. 6.1(b) Relationship between velocity of flow and hydraulic gradient for flow of liquids in a pipe

higher velocity, v_T, and above which the flow will always be turbulent. The lower bound velocity, v_T, of turbulent flow is about 6.5 times the upper bound velocity v_c of laminar flow as shown in Fig. 6.1(b). The upper bound velocity of laminar flow is called the *lower critical velocity*. The fundamental laws that determine the state existing for any given case were determined by Osborne Reynolds. He found the lower critical velocity is inversely proportional to the diameter of the pipe and gave the following general expression applicable for any fluid and for any system of unit.

$$N_R = \frac{v_c D \gamma_0}{\mu g} = 2000$$

where, N_R = reynolds number taken as 2000 as the maximum value for the flow to remain always laminar, D = diameter of pipe, v_c = critical velocity below which the flow always remains laminar, γ_0 = unit weight of fluid at 4 °C, μ = viscosity of fluid, g = acceleration due to gravity.

The principal difference between laminar flow and turbulent flow is that in the former case the velocity is proportional to the first power of the hydraulic gradient, i, whereas in the latter case it is 4/7 the power of i. According to Hagen-Poiseuille's' law the flow through a capillary tube may be expressed as

$$q = \frac{\gamma_w R^2 a i}{8\mu} \tag{6.2a}$$

or $$v = \frac{q}{a} = \frac{\gamma_w R^2 i}{8\mu} \tag{6.2b}$$

where, R = radius of capillary tube of sectional area a, q = discharge through the tube, v = average velocity through the tube, μ = coefficient of viscosity.

6.2   DARCY'S LAW

Darcy in 1856 derived an empirical formula to understand the behaviour of flow through saturated soils after a great deal of tests on soils. He found that the quantity of water q per sec flowing through a cross-sectional area of soil under hydraulic gradient i can be expressed by the formula

$$q = kiA \qquad\qquad (6.3)$$

or the velocity of flow can be written as

$$v = \frac{q}{a} = ki \qquad\qquad (6.4)$$

where k is termed the *hydraulic conductivity* (or coefficient of permeability) has the units of velocity. A in Eq. (6.4) is the cross-sectional area of soil normal to the direction of flow which includes the area of the solids and the voids, whereas the area a in Eq. (6.2) is the area of a capillary tube. The essential point in Eq. (6.3) is that the flow through the soils is also proportional to the first power of the hydraulic gradient i as propounded by Poiseuille's law. From this, we are justified in concluding that the flow of water through the pores of a soil is laminar. It is found that, on the basis of extensive investigations made since Darcy introduced his law in 1856, this law is valid strictly for fine grained types of soils.

6.3   DISCHARGE AND SEEPAGE VELOCITIES

Figure 6.2 shows a soil sample of length L and cross-sectional area A. The sample is placed in a cylindrical horizontal tube between screens. The tube is connected to two reservoirs R_1 and R_2 in which the water levels are maintained constant. The difference in head between R_1 and R_2 is h. This difference in head is responsible for the flow of water. Since Darcy's law assumes no change in the volume of voids and the soil is saturated, the quantity of flow past sections AA, BB and CC should remain the same for steady flow conditions. We may express the equation of continuity as follows

$$q_{aa} = q_{bb} = q_{cc}$$

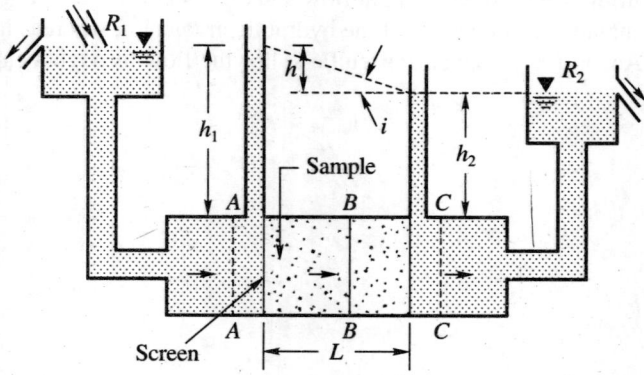

Fig. 6.2   Flow of water through a sample of soil

If the soil be represented as divided into solid matter and void space, then the area available for the passage of water is only A_v. If v_s is the velocity of flow in the voids, and v, the average velocity across the section then, we have

$$A_v V_s = Av \quad \text{or} \quad v_s = \frac{A}{A_v} v$$

Since, $\quad \dfrac{A}{A_v} = \dfrac{1}{n} = \dfrac{1+e}{e}, \quad v_s = \dfrac{v}{n} = \left(\dfrac{1+e}{e}\right)v$ $\hspace{3cm}$ (6.5)

Since $(1 + e)/e$ is always greater than unity, v_s is always greater than v. Here, v_s is called the *seepage velocity* and v the *discharge velocity*.

6.4 METHODS OF DETERMINATION OF COEFFICIENTS OF PERMEABILITY OF SOILS

The flow of water through soils depends upon its permeability. Greater the value of the coefficient of permeability, greater is the flow. The water retaining capacity and the stability of earth dams, the capacity of pumping installations for the lowering of the ground water level during excavations, the rate of settlement of buildings and many other depend upon the value of coefficient of permeability of soils. Methods that are in common use for determining the coefficient of permeability k can be classified under laboratory and field methods.

Laboratory methods

1. Constant Head Permeability method.
2. Falling Head Permeability method.

Indirect method

Computation from the grain size distribution.

Field methods

1. Pumping Tests
2. Bore hole Tests

The various types of apparatus which are used in soil laboratories for determining the coefficients of permeability of soils are called *Permeameters*. The apparatus used for the constant head permeability test is called a *Constant Head Permeameter* and the one used for the falling head test is a *Falling Head Permeameter*. All the methods except the direct testing of soils in place are laboratory methods. The soil samples used in laboratory methods are either undisturbed or disturbed. Since it is not possible to obtain undisturbed samples of cohesionless soils, the laboratory tests on cohesionless materials are always conducted on samples which are reconstructed to the same density as they exist in nature. The results of tests on such reconstructed soils are often misleading since it is impracticable to obtain representative samples and place them in the test apparatus to give exactly the same density and structural arrangement of particles. Direct testing of soils in place is generally preferred in cases where it is not possible to procure undisturbed samples. Since this method is quite costly, it is generally carried out in connection with major projects such as foundation investigation for dams and large bridges or building foundation jobs where lowering of water table is involved. In place of pumping tests, bore hole tests as proposed by the U.S. Bureau of Reclamation are quite inexpensive as these tests eliminate the use of observation wells.

Constant Head Permeability Test

Figure 6.3(a) shows a constant head permeameter which consists of a vertical tube of lucite (or any other material) containing soil sample which is reconstructed or undisturbed as the case may be. The diameter and height of the tube can be of any convenient dimensions. The head and tail water levels are kept constant by overflows. The sample of length L and cross-sectional area A is subjected to a head h which is constant during the progress of a test. A test is performed by allowing water to flow through the sample and measuring the quantity of discharge Q in time t.

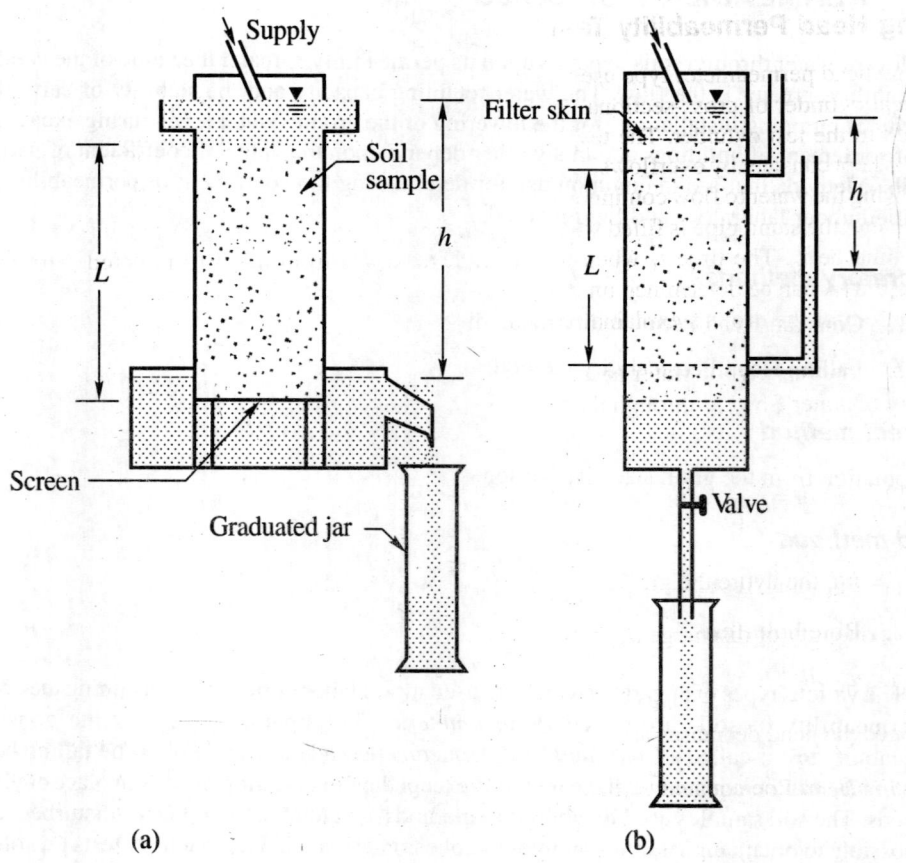

Fig. 6.3   Constant head permeability test

The coefficient of permeability k can be computed directly from Darcy's law expressed as follows

$$Q = k\frac{h}{L}At \qquad (6.6)$$

or $\quad k = \dfrac{QL}{hAt} \qquad (6.7)$

The constant head permeameter test is more suited for coarse grained soils such as gravelly sand and coarse and medium sand. Permeability tests in the laboratory are generally subjected to various types of experimental errors. One of the most important of these arises from the formation of a filter skin of fine materials on the surface of the sample. The constant head permeameter of the type shown in Fig. 6.3(b) can eliminate the effect of the surface skin. In this apparatus the loss of head is measured through a distance in the interior of the sample, and the drop in head across the filter skin has no effect on the results.

Falling Head Permeability Test

A falling head permeameter type used for the test is shown in Fig. 6.4(a). The soil sample is kept in a vertical cylinder of cross-sectional area A. A transparent stand pipe of cross sectional area, a, is attached to the test cylinder. The test cylinder is kept in a container filled with water, the level of which is kept constant by overflows. Before the commencement of the test the soil sample is saturated by allowing the water to flow continuously through the sample from the stand pipe. After the saturation is complete, the stand pipe is filled with water up to a height of h_0 and a stop watch is started. Let the initial time be t_0. The time t_1 when the water level drops from h_0 to h_1 is noted. The hydraulic conductivity k can be determined on the basis of the drop in head $(h_0 - h_1)$ and the elapsed time $(t_1 - t_0)$ required for the drop as explained below.

Let h be the head of water at any time t. Let in time dt the head drop by an amount dh. The quantity of water flowing through the sample in time dt from Darcy's law is

$$dQ = kiA\,dt = k\frac{h}{L}A\,dt \qquad (6.8)$$

where, $i = h/L$ the hydraulic gradient.

The quantity of discharge dQ can also be expressed as

$$dQ = -a\,dh \qquad (6.9)$$

Since the head decreases as the time increases dh is a negative quantity in Eq. (6.9).

Eq. (6.8) can be equated to Eq. (6.9)

$$-a\,dh = k\frac{h}{L}A\,dt \qquad (6.10)$$

The discharge Q in time $(t_1 - t_0)$ can be obtained by integrating the Eq. (6.8) or (6.9). Therefore, Eq. (6.10) can be rearranged and integrated as follows:

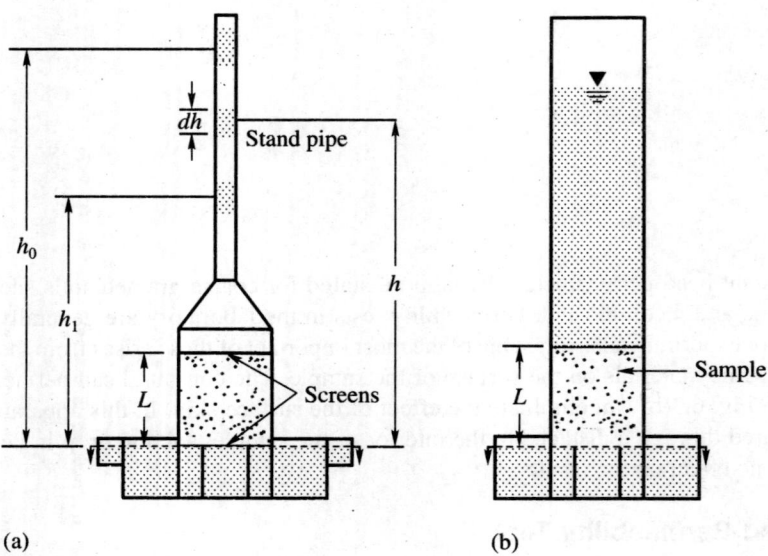

Fig. 6.4   Falling head permeability test

$$-a\int_{h_0}^{h_1}\frac{dh}{h}=\frac{kA}{L}\int dt \text{ or } a\log_e\frac{h_0}{h_1}=\frac{kA}{L}\cdot(t_1-t_0)$$

The general expression for k is

$$k=\frac{aL}{A(t_1-t_0)}\log_e\frac{h_0}{h_1} \text{ or } k=\frac{2.3aL}{A(t_1-t_0)}\log_{10}\frac{h_0}{h_1} \qquad (6.11)$$

The setup shown in Fig. 6.4(a) is generally used for comparatively fine materials such as fine sand and silt where the time required for the drop in head from h_0 to h_1 is neither unduly too long nor too short for accurate recordings. If the time is too long evaporation of water from the surface of water might take place and also temperature variations might affect the volume of the sample. These would introduce serious errors in the results. The setup is suitable for soils having permeabilities ranging from 10^{-3} to 10^{-6} cm per sec. Sometimes, falling head permeameters are used for coarse grained soils also. For such soils, the cross-sectional area of the stand pipe is made the same as the test cylinder so that the drop in head can conveniently be measured. Fig. 6.4(b) shows the test setup for coarse grained soils. When $a = A$, Eq. (6.11) is reduced to

$$k=\frac{2.3L}{(t_1-t_0)}\log_{10}\frac{h_0}{h_1} \qquad (6.12)$$

Example 6.1

A constant head permeability test was carried out on a cylindrical sample of sand 10 cm diameter and 15 cm height. 160 cm³ of water was collected in 1.75 min, under a head of 30 cm. Compute the coefficient of permeability in m/year and the velocity of flow in m/sec.

Solution

The formula for determining k is

$$k = \frac{Q}{Ait}$$

$$Q = 160\,\text{cm}^3, \quad A = 3.14 \times \frac{10^2}{4} = 78.5\,\text{sq. cm}$$

$$i = \frac{h}{L} = \frac{30}{15} = 2, \quad t = 105\,\text{sec}$$

Therefore, $k = \dfrac{160}{78.5 \times 2 \times 105} = 9.7 \times 10^{-3}$ cm/sec $= 9.7 \times 10^{-5}$ m/sec $= 3060$ m/year

Velocity of flow $= ki = 9.7 \times 10^{-5} \times 2 = 1.94 \times 10^{-4}$ m/sec

Example 6.2

A sand sample of 35 cm² cross-sectional area and 20 cm long was tested in a constant head permeameter. Under a head of 60 cm, the discharge was 120 ml in 6 min. The dry weight of sand used for the test was 1120 g, and $G = 2.68$. Determine (a) the coefficient of permeability in cm/sec, (b) the discharge velocity, and (c) the seepage velocity.

Solution

Use Eq. (6.7), $k = \dfrac{QL}{hAt}$

where $Q = 120$ ml, $t = 6$ min, $A = 35$ cm², $L = 20$ cm, and $h = 60$ cm. Substituting, we have

$$k = \frac{120 \times 120}{60 \times 35 \times 6 \times 60} = 3.124 \times 10^{-3}\ \text{cm/sec}$$

Discharge velocity, $v = ki = 3.1214 \times 10^{-3} \times \dfrac{60}{20} = 9.52 \times 10^{-3}$ cm/sec

Seepage velocity v_s

$$\gamma_d = \frac{W_s}{V} = \frac{1120}{35 \times 20} = 1.6\ \text{g/cm}^3$$

From Eq. (3.18), $\gamma_d = \dfrac{\gamma_w G}{1+e}$ or $e = \dfrac{G}{\gamma_d} - 1$ since $\gamma_w = 1\,\text{g/cm}^3$

Substituting, $\quad e = \dfrac{2.68}{1.6} - 1 = 0.675$

$$n = \frac{e}{1+e} = \frac{0.675}{1+0.675} = 0.403$$

Now, $\quad v_s = \dfrac{v}{n} = \dfrac{9.52 \times 10^{-3}}{0.403} = 2.36 \times 10^{-2}\ \text{cm/sec}$

Example 6.3

Calculate the value of k of a sample of 6 cm height and 50 cm^2 cross-sectional area, if a quantity of water of 430 cm^3 flows down in 10 min under an effective constant head of 40 cm. On oven drying, the test specimen weighed 498 g. Assuming $G = 2.65$, calculate the seepage velocity of water during the test.

Solution

From Eq. (6.7), $\quad k = \dfrac{QL}{hAt} = \dfrac{430 \times 6}{40 \times 50 \times 10 \times 60} = 2.15 \times 10^{-3}\ \text{cm/sec}$

Discharge velocity, $\quad v = ki = k = \dfrac{h}{L} = 2.15 \times 10^{-3} \times \dfrac{40}{6} = 14.33 \times 10^{-3}\ \text{cm/sec}$

$$\gamma_d = \frac{W_s}{V} = \frac{498}{50 \times 6} = 1.66\ \text{g/cm}^3$$

$$e = \frac{G}{\gamma_d} - 1 = \frac{2.65}{1.66} - 1 = 0.596$$

$$n = \frac{e}{1+e} = \frac{0.596}{1+0.596} = 0.373$$

Seepage velocity, $\quad v_s = \dfrac{v}{n} = \dfrac{14.33 \times 10^{-3}}{0.373} = 38.37 \times 10^{-3}\ \text{cm/sec}$

Example 6.4

The coefficient of permeability of a soil sample was determined in a soil mechanics laboratory by making use of a falling head permeameter. The data used and the test results obtained were as follows: diameter of sample = 6 cm, height of sample = 15 cm, diameter of stand pipe = 2 cm, initial head $h_0 = 45$ cm. Find head h_1 after a time of 2 min = 30 cm. Determine the coefficient of permeability in m/day.

Solution

The formula for determining k is

$$k = \frac{2.3aL}{At} \log_{10} \frac{h_0}{h_1} \quad \text{where } t \text{ is the time elapsed.}$$

Area of stand pipe, $\quad a = \dfrac{3.14 \times 2 \times 2}{4 \times 100 \times 100} = 3.14 \times 10^{-4}$ sq. m

Area of sample, $\quad A = \dfrac{3.14 \times 6 \times 6}{4 \times 100 \times 100} = 28.26 \times 10^{-4}$ sq. m

Height of sample, $\quad L = \dfrac{15}{100} = 15 \times 10^{-2}$ m

Head, $\quad h_0 = \dfrac{45}{100} = 45 \times 10^{-2}$ m, $\quad h_1 = \dfrac{30}{100} = 30 \times 10^{-2}$ m

Elapsed time, $\quad t = 105 \sec = \dfrac{105}{60 \times 60 \times 24} = 12.15 \times 10^{-2}$ days

$$k = \frac{2.3 \times 3.14 \times 10^{-4} \times 15 \times 10^{-2}}{28.26 \times 10^{-4} \times 12.15 \times 10^{-2}} = \log_{10} \frac{45}{30} = 5.5 \text{ m/day}$$

Computation of *k* from the Grain Size Distriction

If the cross-section of a tube is circular, the flow in the tube as per Poiseuille's Law, Eq. (6.2a) is

$$q = \frac{\gamma_w R^2}{8\mu} ai \qquad (6.13)$$

The average velocity in the tube from Eq. (6.2b) is

$$v_a = \frac{q}{a} = \frac{\gamma_w R^2 i}{8\mu} = \frac{\gamma_w d^2}{32\mu} i \qquad (6.14)$$

where d = diameter of the tube = $2R$.

Equation (6.14) expresses that for a given hydraulic gradient, the velocity of water in a circular pipe is proportional to the square of the diameter of the tube. Since the average diameter of the voids in soil at a given porosity increases practically in proportion to the grain size, D, it is possible to express k on the basis of Poiseulle's Law as

$$k = CD^2 \qquad (6.15)$$

where C = some constant.

From his experiments with clean filter sands of high uniformity ($U < 5$), Allen Hazen in 1911 obtained the empirical equation

$$k = CD_{10}^2 \qquad\qquad\qquad (6.16)$$

in which k is in cm per sec. D_{10} is effective size of grains in cm, C a factor which varies from 100 to 150. Sometimes a value of 100 for C is assumed for all practical purposes.

It is necessary to bear in mind that Eq. (6.16) is very empirical in nature and should not be relied on too much except where a rough idea of coefficient of permeability is required.

Example 6.5

A loose uniform sand with rounded grains has an effective grain size D_{10} equal to 0.3 mm. Estimate the coefficient of permeability.

Solution

Use Eq. (6.16)

$k = CD_{10}^2$ where C may be taken equal to 100 and D_{10} in centimetre. Substituting,

$k = 100 \times (0.003)^2 = 9 \times 10^{-2}$ cm/sec

6.5   FIELD METHODS OF DETERMINATION OF *k*

Direct Determination of *k* of Soils in Place by Pumping Test

The most reliable information concerning the permeability of a deposit of coarse grained material below the water table can usually be obtained by conducting pumping tests in the field. Although such tests have their most extensive application in connection with dam foundations, they may also prove advisable on large bridge or building foundation jobs where the water table must be lowered.

The arrangement consists of a test well and a series of observation wells. The test well is sunk through the permeable stratum up to the impermeable layer. A well sunk into a water bearing stratum, termed an *aquifer*, and tapping free flowing ground water having a free ground water table under atmospheric pressure, is termed a *gravity* or *unconfined well*. A well sunk into an aquifer where the ground water flow is confined between two impermeable soil layers, and is under pressure greater than atmospheric, is termed as *artesian* or *confined well*. Observation wells are drilled at various distances from the test or pumping well along two straight lines, one oriented approximately in the direction of ground water flow and the other at right angles to it. A minimum of two observation wells and their distances from the test well are needed. These wells are to be provided on one side of the test well in the direction of the ground water flow.

The test consists of pumping out water continuously at a uniform rate from the test well until the water levels in the test and observation wells remain stationary. When this condition is achieved the water pumped out of the well is equal to the inflow into the well from the surrounding strata. The water levels in the observation wells and the rate of water pumped out of the well would provide the necessary additional data for the determination of k.

As the water from the test well is pumped out, a steady state will be attained when the water pumped out will be equal to the inflow into the well. At this stage the depth of water in the well will remain constant. The drawdown resulting due to pumping is called the *cone of depression*. The maximum drawdown D_0 is in the test well. It decreases with the increase in the distance from the test well. The depression dies out gradually and forms theoretically, a circle around the test well called the *circle of influence*. The radius of this circle, R_i, is called the *radius of influence* of the depression cone.

Assumptions

The development of equations for the determination of k which is based on Dupuit-Thiem's theory of well hydraulics involve the following assumptions:

1. The soil surrounding the well is assumed as homogeneous.

2. The flow towards the test well is assumed as steady, laminar, radial and horizontal.

3. The horizontal velocity is independent of depth.

4. The ground water table is assumed as horizontal in all directions.

5. The hydraulic gradient at any point on the drawdown is equal to the slope of the tangent at the point.

Equation for *k* for an Unconfined Aquifer

Figure 6.5 gives the arrangement of test and observation wells for an unconfined aquifer. Only two observation wells at radial distances of r_1 and r_2 from the test well are shown. When the inflow of water into the test well is steady, the depths of water in these observation wells are h_1 and h_2 respectively.

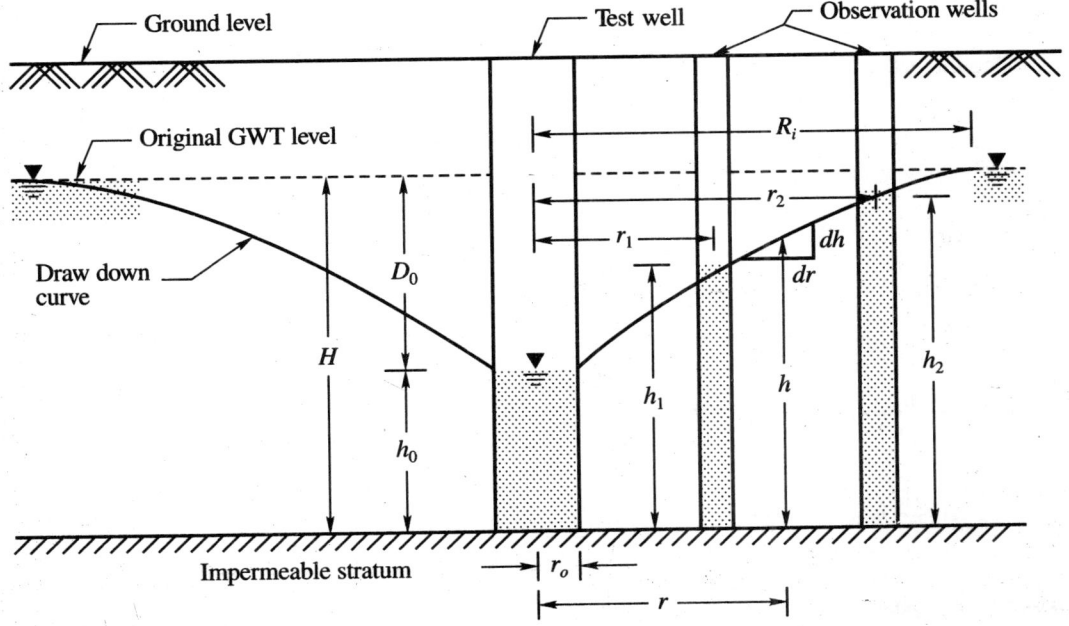

Fig. 6.5   Pumping test in an unconfined aquifer

Let h be the depth of water at radial distance r. The area of the vertical cylindrical surface of radius r and depth h through which water flows is

$$A = 2\pi rh \tag{6.13}$$

The hydraulic gradient is $\quad i = \dfrac{dh}{dr}$

As per Darcy's law the rate of inflow into the well when the water levels in the wells remain stationary is

$$q = kiA$$

Substituting for A and i the rate of inflow across the cylindrical surface is

$$q = k\frac{dh}{dr} 2\pi rh$$

Rearranging the terms, we have

$$\frac{dr}{r} = \frac{2\pi khdh}{q} \tag{6.14}$$

The integral of Eq. (6.14) within the boundary limits is

$$\int_{r_1}^{r_2} \frac{dr}{r} = \frac{2\pi k}{q} \int_{h_1}^{h_2} h\,dh \tag{6.15}$$

The integral of Eq. (6.15) is

$$\log_e \frac{r_1}{r_2} = \frac{\pi k}{q}(h_2^2 - h_1^2) \tag{6.16}$$

The equation for k is

$$k = \frac{q}{\pi(h_2^2 - h_1^2)} \log_e \frac{r_1}{r_2} \tag{6.17}$$

When reduced to common logarithms,

$$k = \frac{2.3q}{\pi(h_2^2 - h_1^2)} \log_{10} \frac{r_1}{r_2} \tag{6.18}$$

Alternate Equations

If r_0 is the radius of the test well, h_0 is the depth of water in the well, Eq. (6.15) may also be written as

$$\int_{r_0}^{r_1} \frac{dr}{r} = \frac{2\pi k}{q} \int_{h_0}^{h_1} h\, dh$$

on integration, we get another equation for k as

$$k = \frac{2.3q}{\pi(h_1^2 - h_0^2)} \log_{10} \frac{r_1}{r_0} \tag{6.19}$$

Again Eq. (6.15) can be integrated between limits $r = r_0$, $r = R_i$, and between $h = r_0$ and $h = H$ where, R_i is the radius of influence and H is depth of water before pumping. We have

$$\int_{r_0}^{R_1} \frac{dr}{r} = \frac{2\pi k}{q} \int_{h_0}^{H} h\, dh$$

on integration and rearranging, we have

$$k = \frac{2.3q}{\pi(H^2 - h_0^2)} \log_{10} \frac{R_i}{r_0} \tag{6.20}$$

If we write $h_0 = (H - D_0)$ in Eq. (6.20), where D_0 is the depth of maximum drawdown in the test well, we have

$$k = \frac{2.3q}{\pi D_0 (H - D_0)} \log_{10} \frac{R_i}{r_0} \tag{6.21}$$

Now from Eq. (6.21), the maximum yield from the well may be written as,

$$q = \frac{\pi D_0 k (2H - D_0)}{2.3} \frac{1}{\log_{10}(R_i / r_0)} \tag{6.22}$$

Radius of Influence R_i: Based on experience, Sichardt (1930) gave an equation for estimating the radius of influence for the stabilised flow condition as

where, $R_i = 3000 D_0 \sqrt{k}$ meters $\tag{6.23}$

D_0 = maximum drawdown in meters

k = hydraulic conductivity in m/sec

Equation for k in a Confined Aquifer

Figure 6.6 shows a confined aquifer with the test and observation wells. The water in the observation wells rises above the top of the aquifer due to artesian pressure. When pumping from such an artesian well two cases might arise. They are:

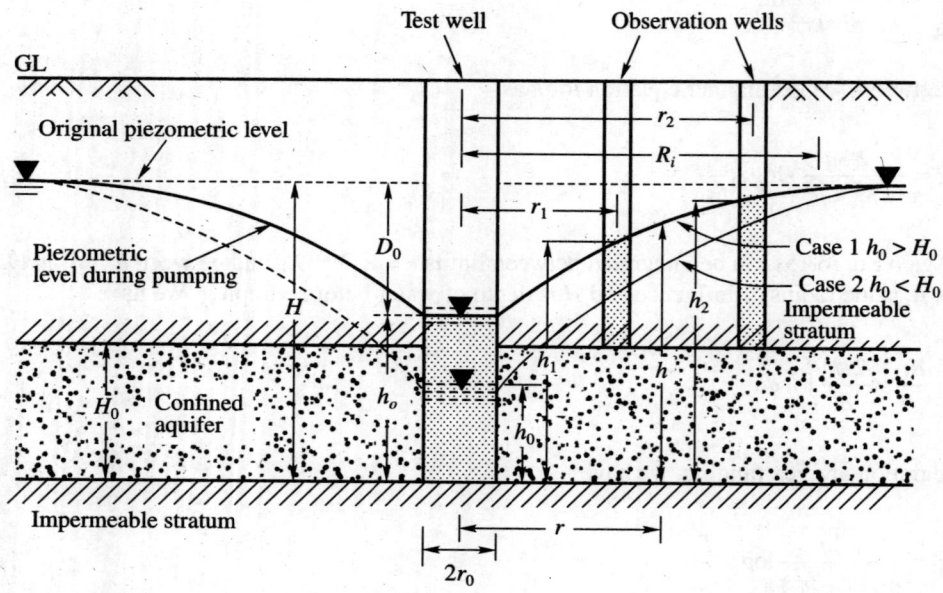

Fig. 6.6 Pumping test in confined aquifer

Case 1. The water level in the test well might remain above the roof level of the aquifer at steady flow condition.

Case 2. The water level in the test well might fall below the roof level of the aquifer at steady flow condition.

If H_0 is the thickness of the confined aquifer and h_0 is the depth of water in the test well at the steady flow condition Case 1 and Case 2 may be stated as:

Case 1. When $h_0 > H_0$. **Case 2.** When $h_0 < H_0$.

Case 1. When $h_o > H_o$

In this case, the area of a vertical cylindrical surface of any radius r does not change as in Eq. (6.13), since the depth of the water bearing strata is limited to the thickness H_0. Therefore, the discharge surface area is

$$A = 2\pi r H_0 \tag{6.24}$$

Again writing $i = \dfrac{dh}{dr}$, the flow equation as per Darcy's law is

$$q = kiA = k\frac{dh}{dr}2\pi r H_0 \tag{6.25}$$

arranging the terms in Eq. (6.25), we have

$$dh = \frac{q}{2\pi k H_0} \frac{dr}{r} \tag{6.26}$$

Integration of Eq. (6.26) yields,

$$\int_{h_1}^{h_2} dh = \frac{q}{2\pi k H_0} \int_{r_1}^{r_2} \frac{dr}{r} \tag{6.27}$$

or $\quad (h_2 - h_1) = \frac{q}{2\pi H_0} \log_e \frac{r_1}{r_2}$ $\hspace{2cm}$ (6.27a)

The equation for k is

$$k = \frac{q}{2\pi H_0 (h_2 - h_0)} \log_e \frac{r_2}{r_1} \tag{6.28}$$

reducing to common logarithms, the equation for k is

$$k = \frac{2.3q}{2\pi H_0 (h_2 - h_0)} \log_{10} \frac{r_2}{r_1} \tag{6.29}$$

Alternate Equations

As before we can write the following equation for determining k

$$k = \frac{2.3q}{2\pi H_0 (h_1 - h_0)} \log_{10} \frac{r_1}{r_2} \tag{6.30}$$

$$k = \frac{2.3q}{2\pi H_0 (H - h_0)} \log_{10} \frac{R_i}{r_0} \tag{6.31}$$

or $\quad k = \frac{2.3q}{2\pi H_0 D_0} \log_{10} \frac{R_i}{r_0}$ $\hspace{2cm}$ (6.32)

Case 2. When $h_o < H_o$

Under the condition when h_0 is less than H_0, the flow pattern close to the well is similar to that of an unconfined aquifer whereas at distances farther from the well the flow is artesian. Musket has developed an equation which could be used to determine the coefficient of perm·ability. The equation is

$$k = \frac{2.3q}{\pi (2HH_0 - h_0^2 - h_0^2)} \log_{10} \frac{R_i}{r_0} \tag{6.33}$$

Example 6.6

A pumping test was carried out for determining the coefficient of permeability of soil in place. A well of diameter 40 cm was drilled up to the impermeable stratum. The depth of the water bearing stratum was 8 m. The yield from the well was 4 m³/min at a steady drawdown of 4.5 m. Determine the coefficient of permeability of the soil in m/day if the observed radius of influence was 150 m.

Solution

The formula for determining k is

$$k = \frac{2.3q}{\pi D_0 (H - D_0)} \log_{10} \frac{R_i}{r_0}$$

$q = 4 \text{ m}^3/\text{min} = 4 \times 60 \times 24 \text{ m}^3/\text{day}$

$D_0 = 4.5\text{m}, \quad H = 8 \text{ m}, \quad R_i = 150 \text{ m}, \quad r_0 = 0.2 \text{ m}$

$$k = \frac{2.3 \times 4 \times 60 \times 24}{3.14 \times 4.5(2 \times 8 - 4.5)} \log_{10} \frac{150}{0.2} = 234 \text{ m/day}$$

Example 6.7

Calculate the yield per hour from a well driven into a confined aquifer. The following data are available:

height of original piezometric level from the bed of aquifer, $H = 9.0$ m,

thickness of aquifer, $H_a = 5.0$ m,

the depth of water in the well at steady state, $h_0 = 5.5$ m,

coefficient of permeability of soil $= 0.024$ m/min,

radius of well, $r_0 = 10$ cm, radius of influence, $R_i = 175$ m.

Solution

Since h_0 is greater than H_a the equation for q is

$$q = \frac{2\pi k H_0 (H - h_0)}{2.3 \log_{10} (R_i / r_0)}$$

where, $k = 0.024$ m/min $= 1.44$ m/hr. Now

$$q = \frac{2 \times 3.14 \times 1.44 \times 5(9.0 - 5.5)}{2.3 \times 3.24} = 2.12 \text{ m}^3/\text{hour}$$

Example 6.8

A pumping test was made in pervious gravels and sands extending to a depth of 15.24 m, where a bed of clay was encountered. The normal ground water level was at the ground surface. Observation wells were located at distances of 3.05 and 7.62 m from the pumping well. At a discharge of 216 litres per minute from the pumping well, a steady state was attained in about 24 hr. The draw-down at a distance of 3.05 m was 1.68 m and at 7.62 m was 0.37 m. Compute the coefficient of permeability.

Solution

Use Eq. (6.18) where

$$k = \frac{2.3q}{\pi(h_2^2 - h_1^2)} \log_{10} \frac{r_1}{r_2}$$

where $q = \dfrac{216}{60} = 3.6 \text{ litres/sec} = 3.6 \times 10^3 \text{ cm}^3/\text{sec}$

$r_1 = 3.05 \text{ m}, \quad r_2 = 7.62 \text{ m}, \quad h_2 = 15.24 - 0.37 = 14.87 \text{ m}, \quad h_1 = 15.24 - 1.68 = 13.56 \text{ m}$

$$k = \frac{2.3 \times 3.6 \times 10^3}{3.14(14.87^2 - 13.56^2) \times 10^4} \log_{10} \frac{7.62}{3.05} = 2.8 \times 10^{-3} \text{ cm/sec}$$

Example 6.9

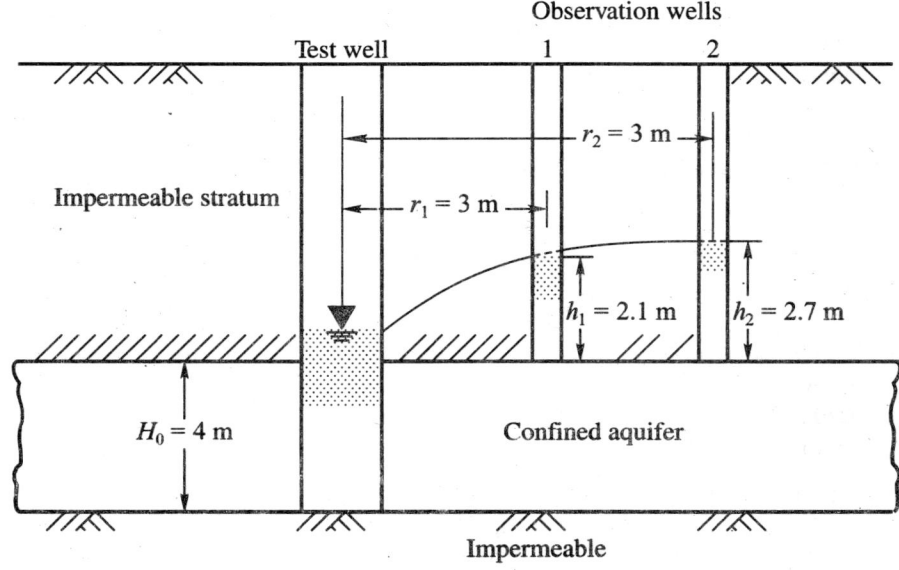

Fig. Ex. 6.9

A field pumping test was conducted from an aquifer of sandy soil of 4 m thick confined between two impervious strata. When equilibrium was established, 90 litres of water was pumped out per hour. The water elevation in an observation well 3.0 m away from the test well was 2.1 m and another 6.0 m way was 2.7 m from the rool level of the impervious stratum of the aquifer. Find the value of k of the soil in m/sec (Fig. Ex. 6.9).

Solution

Use Eq. (6.29)

$$k = \frac{2.3q}{2\pi H_0(h_2 - h_0)} \log_{10} \frac{r_2}{r_1}$$

We have, $q = 90$ l/hr $= 90 \times 10^3$ cm³/hr $= 25 \times 10^{-6}$ m³/sec

$$k = \frac{2.3 \times 25 \times 10^{-6}}{2 \times 3.14 \times 4(2.7 - 2.1)} \log_{10} \frac{6}{3} = 1.148 \times 10^{-6} \text{ m/sec}$$

Borehole Permeability Tests

The borehole tests may sometimes be used in place of pumping tests which are relatively costly. These tests are usually carried out in boreholes made for site investigation. The tests are classified as:

1. Constant water level method
2. Falling water level method
3. Rising water level method

Constant water level tests are performed when the soil strata to be tested lied above or below the water table, when the layers are below the water level, the tests may be any of the above three type.

1. Constant Water Level Method (USBR Method)

Figure 6.7 shows the arrangement for a constant water level test through the open end of a pipe casing which has been drilled to the desired depth. The hole stops above the water table in (a), and goes below the water table in (c) of Fig. 6.7. When the hole goes below the water table, it is necessary to keep it filled with water to prevent heaving of soil at the bottom. The test is performed by pumping water into the hole and adjusting remains constant. When the water level in the hole remains constant, the rate of inflow of water q is equal to the rate of outflow from the hole through its bottom. The differential head Hg above the bottom of the hole in (a) and above water table in (c) of Fig. 6.7 is responsible for the outflow.

In case the soil to be tested is of low permeability the gravity head Hg provided in (a) and (c) of [Fig. 6.7] will not be sufficient to induce significant outflow. In such circumstannces additional pressure head Hp can be added to the gravity head Hg to maintain a constant rate of flow as shown in (b) and (d) of [Fig. 6.7].

The coefficient of permeability is calculated by making use of the formula

$$k = \frac{0.18q}{r_0 H} \qquad (6.34)$$

In Eq. (6.34), q is the constant rate of flow into the hole, r_0 is the inside radius of the casing, and H is the differential head of water used in maintaining the steady rate. Any consistent units may be used in computing the permeability. The borehole tests may lead to errors in the determination of k. The possible causes of errors are:

1. Leakage along casing,

2. Clogging due to presence of fines or sediment in the test water,

3. Air locking due to gas bubbles in soil or water, and

4. Flow of water into pockets that are opened by excessive head in the test holes.

Considerable care and judgement is very essential if borehole test are to be successful.

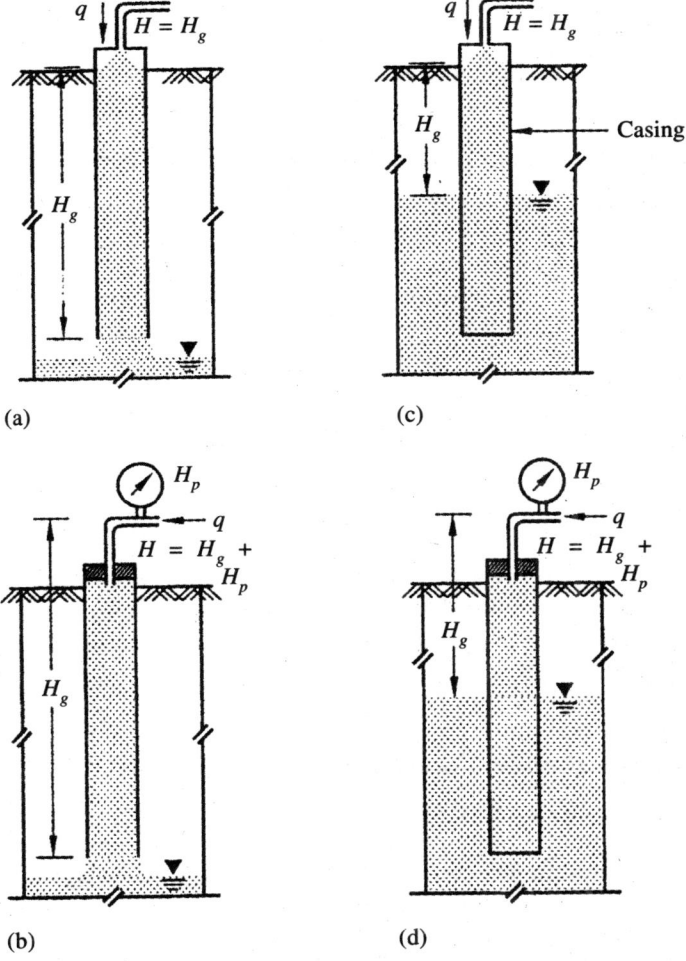

Fig. 6.7   Constant water level method of permeability test

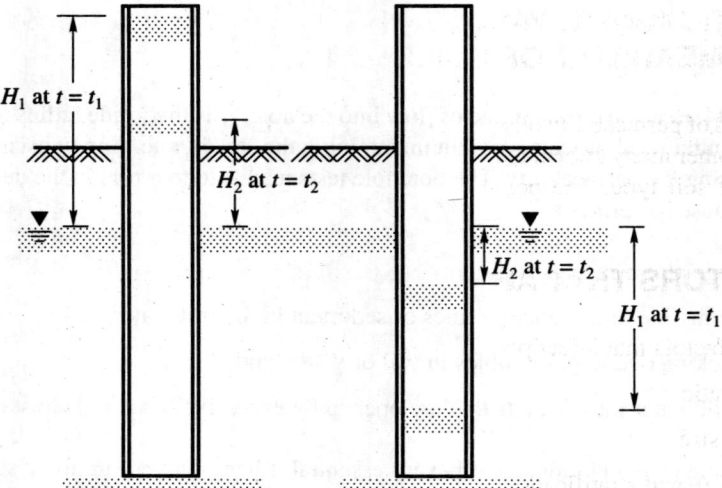

(a) Falling water head method   (b) Rising water head method

Fig. 6.8   Falling and rising water method of determining k

2. *Falling Water level, Method: (Cased hole and soil flush with bottom)*

In this test, the casing is filled with water which is then allowed to seep into the soil. The rate of drop of the water level in the casing is observed by measuring the depth of the water surface below the top of the casing at 1, 2 and 5 minutes after the start of the test and at 5 minutes intervals thereafter. These observations are made until the rate of drop becomes negligible or until sufficient readings have been obtained. The coefficient of permeability is computed as [Fig. 6.8(a)].

$$k = \frac{\pi r_0}{5.5(t_2 - t_1)} \log_e \frac{H_1}{H_2} \tag{6.35}$$

where, H_1 = Piezometric head at $t = t_1$, H_2 = Piezometric head at $t = t_2$.

If coefficient of permeability k is to be found out in fine grained soils, the time required to conduct this test will be very considerable since the rate of drop of water level in the casing is very small. Whereas, the time required to conduct a test in coarse grained soils is not much as the rate of drop of water will be quite high. It is therefore, essential to decide the type of test to be carried out in advance to avoid possible errors in the tests.

3. *Rising Water Level Method (Cased Hole and Soil Flush with Bottom)*

This method, most commonly referred to as the time-lag method, consists of bailing the water out of the casing and observing the rate of rise of the water level in the casing at intervals until the rise in water level becomes negligible. The rate is observed by measuring the elapsed time and the depth of the water surface below the top of the casing. The intervals at which the readings are required will vary somewhat with the permeability of the soil. Eq. (6.35) is applicable for this case, [Fig. 6.8(b)]. A rising water level test should always be followed by sounding the bottom of the holes to determine whether the test created a quick condition.

6.6   APPROXIMATE VALUES OF THE COEFFICIENTS OF PERMEABILITY OF SOILS

The coefficients of permeability of soils vary according to their type, textural composition, structure, void ratio and other many known factors. Therefore, no single value can be assigned to a soil purely on the basis of soil type. The possible coefficients of permeability of some soils are given in Table 6.1.

6.7   FACTORS THAT AFFECT PERMEABILITY

The important factors that affect permeability of soils are:

1.   Void ratio

2.   Grain size

3.   Structure and stratification

4.   Temperature

Some of the factors that affect permeability are interrelated such as grain size and void ratio. Smaller the grain size, smaller the voids which leads to reduced size of flow channels and lower permeability.

Effect of Void Ratio

To estimate k at void ratios other than the test void ratios, Taylor (1948) proposed:

$$k_1 : k_2 = \frac{c_1 e_1^3}{1+e_1} : \frac{c_2 e_2^3}{1+e_2} \tag{6.36}$$

where the coefficients c_1 and c_2, depend on the soil structure. Very approximately for sands $c_1 \approx c_2$. Another relationship which has been found to be useful for sand is

$$k_1 : k_2 = c_2' e_1^2 : c_1' e_2^2 \tag{6.37}$$

where again $c_1' \approx c_2'$ for sands approximately.

For fine grained soils, the expression for k recommended is

$$k = k'(e - 0.1)^2 \tag{6.38}$$

TABLE 6.1
Coefficient of permeability of some soils (after Casagrande and Fadum)

k (cm/sec)	Soils type	Drainage conditions
10^1 to 10^2	Clean gravels	Good
10^1	Clean sand	Good
10^{-1} to 10^{-4}	Clean sand and gravels mixtures	Good
10^{-5}	Very fine sand	Poor
10^{-6}	Silt	Poor
10^{-7} to 10^{-9}	Clay soils	Practically impervious

Effect of Structure and Stratification on Permeability

The coefficient of permeability of a disturbed sample may be different from that of the undisturbed sample even though the void ratio is the same. This may be due to a change in the structure or due to the stratification of the undisturbed soil or a combination of both of these factors. In nature we may find fine grained soils having structures either flocculated or dispersed. Two fine-grained soils at the same void ratios, one dispersed and the other flocculated, with exhibit different coefficients of permeability.

Soils may be stratified by the deposition of different materials in layers which possess different permeability characteristics. In such stratified soils engineers desire to have the average permeability either in the horizontal or vertical directions. The average permeability can be computed if the permeabilities of each layers are found out in the laboratory. The procedure is as follows:

$k_1, k_2, ..., k_n$ = Coefficients of permeability of individual strata of soil either in the vertical or horizontal directions.

$z_1, k_2, ..., k_n$ = Thickness of the corresponding strate.

$Z = z_1 + z_2 + \cdots + z_n$

k_h = Average coefficient of permeability parallel to the bedding planes (usually horizontal).

k_v = Average coefficient of permeability perpendicular to the bedding planes (usually vertical).

Flow in the Horizontal Direction (Fig. 6.9)

When the flow is in the horizontal direction the hydraulic gradient i remains the same for all the layers. Let $v_1, v_2, ..., v_n$ be the discharge velocities in the corresponding strata. Then

$$Q = kiZ = (v_1 z_1 + v_1 z_1 + ... + v_n z_n)$$

$$= (k_1 i z_1 + k_2 i z_2 + ... + k_n i z_n) \tag{6.39}$$

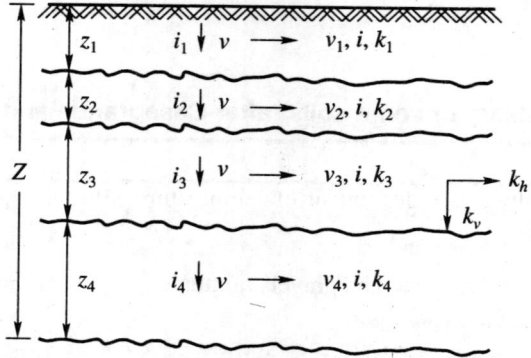

Fig. 6.9   Flow through stratified layers of soil

Therefore,

$$k_h = \frac{1}{Z}(k_1 z_1 + k_2 z_2 + \cdots + k_n z_n) \tag{6.40}$$

Flow in the Vertical Direction

When flow is in the vertical direction, the hydraulic gradients for each of the layers are different and let these be denoted by i_1, i_2, ..., i_n. Let h be the total loss of head as the water flows from the top layer to the bottom through a distance of Z. The average hydraulic gradient is h/Z. The principle of continuity of flow requires that the downward velocity be the same in each layer. Therefore,

$$v = k_v \frac{h}{Z} = k_1 i_1 = k_2 i_2 = \cdots = k_n i_n \tag{6.41}$$

If h_1, h_2, ..., h_n, are the head losses in each of the layers, we have

$$h = h_1 + h_2 + \cdots + h_n \tag{6.42}$$

or $\quad h = z_1 i_1 + z_2 i_2 + z_n i_n \tag{6.43}$

From Eqs (6.41) and (6.42), we obtain

$$k_v = \frac{Z}{\dfrac{z_1}{k_1} + \dfrac{z_2}{k_2} + \cdots + \dfrac{z_n}{k_n}} \tag{6.44}$$

It should be noted that in all stratified layers of soils the horizontal permeability is generally greater than the vertical permeability.

Effect of Temperature on Permeability

As per Eq. (6.36) the coefficient of permeability k is a product of a quantity k' which is dependent on temperature and a function of the void ratio e. The quantity k' is expressed as

$$k' = \frac{C\gamma_w}{\mu} \tag{6.45}$$

where, C is a constant which is independent of temperature. The expression for coefficient of permeability may now be written as

$$k = CF(e)\frac{\gamma_w}{\mu} \tag{6.46}$$

Equation (6.46) clearly indicates that for a soil of given structure and void ratio, k varies as γ_w/μ. Since the variation of μ with temperature is more significant than the variation of density of water, we can consider for all practical purposes that k varies as the inverse of μ. It is a common practice to reduce coefficient of permeability obtained at different temperatures to a standard temperature which is taken as 20 °C. The coefficient of permeability at temperature T, k_T, can be reduced to that at 20 °C, k_{20}, by using

$$k_{20} = \frac{\mu_T}{\mu_{20}} k_T \qquad\qquad (6.47)$$

in which μ_T and μ_{20} are the viscosities of water at temperatures T °C and 20 °C respectively.

Example 6.10

The following details refer to a test to determine the value of k of a soil sample: sample thickness = 2.5 cm, diameter of soil sample = 7.5 cm, diameter of stand pipe = 10 mm, initial head of water in the stand pipe = 100 cm, water level in the stand pipe after 3 h 20 min = 80 cm. Determine the value of k if $e = 0.75$. What is the value of k of the same soil at a void ratio $e = 0.90$?

Solution

Use Eq. (6.11) where, $\quad k = \dfrac{2.3aL}{A(t_1 - t_2)} \log_{10} \dfrac{h_0}{h_1}$

We have, $\quad a = \dfrac{3.14}{4}(1)^2 = 0.785 \text{ cm}^2$

$A = \dfrac{3.14}{4}(7.5)^2 = 44.17 \text{ cm}^2$

$t = 12000 \text{ sec}$

By substituting the value of k for $e_1 = 0.75$ is

$$k = k_1 = \frac{2.3 \times 0.785 \times 2.5}{44.17 \times 12000} \times \log \frac{100}{80} = 0.826 \times 10^{-6} \text{ cm/sec}$$

For determining k at any other void ratio, use Eq. (6.36)

$$\frac{k_1}{k_2} = \frac{e_1^3}{1+e_1} \Big/ \frac{e_2^3}{1+e_2} = \frac{1+e_2}{1+e_1} \times \left(\frac{e_1}{e_2}\right)^3$$

Now, $\quad k_2 = \dfrac{1+e_2}{1+e_1} \times \dfrac{e_1}{e_2}^3 \times k_1$

By substituting we have k_2 for $e_2 = 0.90$ is

$$k_2 = \frac{1.75}{1.90} \times \left(\frac{0.9}{0.75}\right)^3 \times 0.826 \times 10^{-6} = 1.3146 \times 10^{-6} \text{ cm/sec}$$

Example 6.11

A sand deposit contains three distinct horizontal layers of equal thickness. (Refer Fig. 6.9). The coefficient of permeability of the upper and lower layers is 10^{-3} cm/sec and that of the middle is 10^{-2} cm/sec. What are the values of the horizontal and vertical coefficients of permeability of the three layers, and what is their ratio.

Solution

Horizontal flow

$$k_h = \frac{1}{2}(k_1 z_1 + k_2 z_2 + k_3 z_3) = \frac{1}{3}(k_1 + k_2 + k_3) \text{ since } z_1 = z_2 = \frac{z}{3}$$

$$k_h = \frac{1}{3}(10^{-3} + 10^{-2} + 10^{-3}) = \frac{1}{3}(2 \times 10^{-3} + 10^{-2}) = 4 \times 10^{-3} \text{ cm/sec}$$

Vertical flow

$$k_v = \frac{Z}{\dfrac{z_1}{k_1} + \dfrac{z_2}{k_2} + \dfrac{z_3}{k_3}} = \frac{1}{\dfrac{1}{k_1} + \dfrac{1}{k_2} + \dfrac{1}{k_3}} = \frac{3}{\dfrac{2}{k_1} + \dfrac{1}{k_2}}$$

$$= \frac{3k_1 k_2}{2k_2 + k_1} = \frac{3 \times 10^{-3} \times 10^{-2}}{1.4 \times 10^{-2} \times 10^{-3}} = 1.4 \times 10^{-3} \text{ cm/sec}$$

$$\frac{k_h}{k_v} = \frac{4 \times 10^{-3}}{1.4 \times 10^{-3}} = 2.86$$

Example 6.12

In a falling head permeameter, the sample used is 20 cm long having a cross-sectional area of 24 cm^2. Calculate the time required for a drop of head from 25 to 12 cm if the cross-sectional area of the stand pipe is 2 cm^2. The sample of soil is made of three layers. The thickness of the first layer from the top is 8 cm and has a value of $k_1 = 2 \times 10^{-4}$ cm/sec, the second layer of thickness 8 cm has $k_2 = 5 \times 10^{-4}$ cm/sec and the bottom layer of thickness 4 cm has $k_3 = 7 \times 10^{-4}$ cm/sec. Assume that the flow is taking place perpendicular to the layers (Fig. Ex. 6.12).

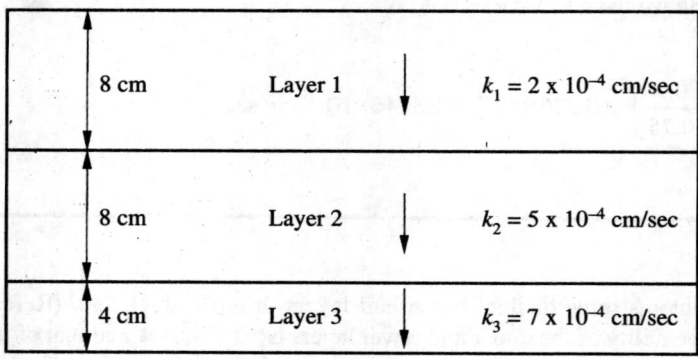

Fig. Ex. 6.12

Solution

Use Eq. (4.28)

$$k_v = \frac{Z}{\dfrac{z_1}{k_1} + \dfrac{z_2}{k_2} + \dfrac{z_3}{k_3}} = \frac{20}{\dfrac{8}{2\times10^{-4}} + \dfrac{8}{5\times10^{-4}} + \dfrac{4}{7\times10^{-4}}}$$

$$= 32.4 \times 10^{-4} \text{ cm/sec}$$

Now for Eq. (6.11)

$$k = \frac{2.3aL}{A(t_1 - t_0)} \log_{10} \frac{h_0}{h_1}$$

or $$(t_1 - t_0) = t = \frac{2.3aL}{Ak} \log_{10} \frac{h_0}{h_1} = \frac{2.3 \times 2 \times 20}{24 \times 3.24 \times 10^{-4}} \log_{10} \frac{25}{12}$$

$$= 3775.37 \text{ sec} = 1^{\text{h}} 2^{\text{min}} 55.37 \text{ sec}$$

Example 6.13

The data given below relate to two falling head permeameter tests performed on two different soil samples:

(a) stand pipe area = 4 cm², (b) sample area = 28 cm², (c) sample height = 5 cm, (d) initial head in the stand pipe = 100 cm, (e) final head = 20 cm, (f) time required for the fall of water level in test 1, t = 500 sec, (g) for test 2, t = 15 sec.

Determine the values of k for each of the samples. If these two types of soils form adjacent layers in a natural state with flow (a) in the horizontal direction, and (b) flow in the vertical direction.

Determine the average permeability for both the cases by assuming that the thickness of each layer as equal to 150 cm.

Solution

Use Eq. (6.11)

$$k = \frac{2.3aL}{At}\log_{10}\frac{h_1}{h_2}$$

For test 1

$$k_1 = \frac{2.3\times4\times5}{28\times500}\log_{10}\frac{100}{20} = 2.3\times10^{-3} \text{ cm/sec}$$

For test 2

$$k_2 = \frac{2.3\times4\times5}{28\times15}\log_{10}\frac{100}{20} = 2.3\times10^{-3} \text{ cm/sec}$$

Flow in the horizontal direction

Use Eq. (6.40)

$$k_h = \frac{1}{Z}(k_1z_1 + k_2z_2) = \frac{1}{300}(2.3\times500 + 76.7\times150)\times10^{-3}$$

$$= 39.5 \times 10^{-3} \text{ cm/sec}$$

Flow in the vertical direction

Use Eq. (6.44)

$$k_v = \frac{Z}{\dfrac{z_1}{k_1} + \dfrac{z_2}{k_2}} = \frac{300}{\dfrac{150}{2.3\times10^{-3}} + \dfrac{150}{76.7\times10^{-3}}} = 4.46\times10^{-3} \text{ cm/sec}$$

6.8 QUESTIONS AND PROBLEMS

6.1 Prove as per Poiseuille's law that velocity in laminar flow is proportional to the first power of the hydraulic gradient.

6.2 Explain and discuss the validity of Darcy's law.

6.3 Explain with a neat diagram a method for determining coefficient of permeability of medium sand in the laboratory.

6.4 Explain with a neat diagram a method for determining coefficient of permeability of sandy silt in the laboratory.

6.5 Explain with diagram a method for determining k for clay soils in the laboratory.

6.6 Explain the difference between unconfined and confined aquifer.

6.7 Explain and discuss the validity of laboratory methods for determining k.

6.8 What are the various factors that affect coefficient of permeability? Discuss.

6.9 Explain and discuss the borehole method of determining k.

6.10 A falling head permeability test was performed on a sample of silty sand. The time required for the head to fall in the stand pipe from 60 cm to 30 cm mark was 70 min. The sectional area of the stand pipe was 1.25 sq. cm. If the height and diameter of the sample were respectively 10 and 9 cm, determine the value of k.

6.11 A constant head permeability test was carried out on a sample of sand. The diameter and length of the sample were 10 cm and 20 cm respectively. The head of water was maintained at 35 cm. If 110 cm^3 of water is collected in $1^m 20^s$, compute the coefficient of permeability of sand.

6.12 Determine the maximum yield from a well driven into an unconfined aquifer. The following data are given:

Maximum drawdown, $D_0 = 4.5$ m, radius of the well, $r_0 = 20$ cm, the depth of water table = 8 m, coefficient of permeability, $k = 8 \times 10^{-2}$ cm/sec.

6.13 A pumping test was carried out to determine the coefficient of permeability of soil at a site which was selected for the construction of an earth dam. Observation wells were established at distances of 3 and 6 m from the test well. The following data were obtained:

Depth of water table = 16 m, discharge under steady condition = 2.3 m^3/min, draw down at outer well = 0.5 m, drawdown at inner well = 1.5 m.

Determine the coefficient of permeability of the soil.

6.14 A pumping test was carried out in a confined aquifer to determine the coefficient of permeability of the aquifer. The thickness of the aquifer was 5 m. The draw down in the test well from the original piezometric level at a steady discharge of 1.5 m^3/min was found to be 4.5 m. The radius of influence may be assumed as 100 m. The radius of the test well was 10 cm. If the original piezometric level is at height of 10 m above the bed of the aquifer, compute the coefficient of permeability.

CHAPTER 7

SEEPAGE THROUGH SOIL

7.1   INTRODUCTION

The interaction between soils and percolating water has an important influence on:

1.   The design of foundations and earth slopes.

2.   The quantity of water that will be lost by percolation through a dam or its subsoil.

Foundation failures due to *piping* are quite common. Piping is a phenomenon by which the soil on the downstream sides of some hydraulic structures get lifted up due to excess pressure of water. The pressure that is exerted on the soil due to the seepage of water is called the *seepage force or pressure*. In the stability of slopes, the seepage force is a very important factor. Shear strengths of soils are reduced due to the development of neutral stress or pore pressures. A detailed understanding of the hydraulic conditions is therefore essential for a satisfactory design of structures.

The computation of seepage loss under or through a dam, the uplift pressures caused by the water on the base of a concrete dam and the effect of seepage on the stability of earth slopes can be studied by constructing flow nets.

Flow Net

Flow net is a network of flow lines and equipotential lines intersecting at right angles to each other.

The path which a particle of water follows in its course of seepage through a saturated soil mass is called a *flow line*.

Equipotential lines are lines that intersect the flow lines at right angles. At all points along an equipotential line, the water would rise in piezometric tubes to the same elevation known as the *piezometric head*. Figure 7.1 gives a typical example of a flow net for the flow below a sheet pile wall. The head of water on the upstream side of the sheet pile is h_t and on the downstream side h_d. The head lost as the water flows from the upstream to the downstream side is h.

7.2   LAPLACE EQUATION

Figure 7.1(a) illustrates the flow of water along curved lines which are parallel to the section shown. The figure represents a section through an impermeable diaphragm extending to a depth D below the horizontal surface of a homogeneous stratum of soil of depth H.

It is assumed that the difference h between the water levels on the two sides of the diaphragm is constant. The water enters the soil on the upstream side of the diaphragm, flows in a downward direction and rises on the downstream side towards the surface.

Consider a prismatic element P shown shaded in Fig. 7.1(a) which is shown on a larger scale in (b). The element is a parallelopiped with sides dx, dy and dz. The x and z directions are as shown in the figure and the y direction is normal to the section. The velocity v of water which is tangential to the stream line can be resolved into components v_x and v_z in the x and z directions respectively.

Let,

$i_x = \dfrac{\partial h}{\partial x}$, the hydraulic gradient in the horizontal direction.

$i_z = \dfrac{\partial h}{\partial z}$, the hydraulic gradient in the vertical direction.

k_x = hydraulic conductivity in the horizontal direction.

k_z = hydraulic conductivity in the vertical direction.

If we assume that the water and soil are perfectly incompressible, and the flow is steady, then the quantity of water that enters the element must be equal to the quantity that leaves it.

The quantity of water that enters the side ab

$= v_x dz dy$

The quantity of water that leaves the side cd

$= \left(v_x + \dfrac{\partial v_x}{\partial x} dx \right) dz dy$

The quantity of water that enters the side bc

$= v_z dx dy$

The quantity of water that leaves the side ad

$= \left(v_z + \dfrac{\partial v_z}{\partial z} dz \right) dx dy$

Therefore, we have the equation,

$$v_x dz dy + v_z dx dy = \left(v_x + \frac{\partial v_x}{\partial x} dx \right) dz dy + \left(v_z + \frac{\partial v_z}{\partial z} dz \right) dx dy$$

Upon simplifying, we get

$$\frac{\partial v_x}{\partial x} + \frac{\partial v_z}{\partial z} = 0 \qquad\qquad (7.1)$$

Equation (7.1) expresses the necessary condition for continuity of flow. According to Darcy's law we may write,

$$v_x = -k_x \frac{\partial h}{\partial x} \qquad\qquad (7.2)$$

$$v_z = -k_z \frac{\partial h}{\partial z} \qquad\qquad (7.3)$$

Substituting for v_x and v_z in Eq. (7.1), we get

$$\frac{\partial}{\partial x}\left(-k_x \frac{\partial h}{\partial x} \right) + \frac{\partial}{\partial z}\left(-k_z \frac{\partial h}{\partial z} \right) = 0$$

or $\quad k_x \dfrac{\partial^2 h}{\partial x^2} + \dfrac{\partial^2 h}{\partial z^2} = 0 \qquad\qquad (7.4)$

When $k_z = k_x$, i.e. when the permeability is the same in all directions, Eq. (7.4) reduces to

$$\frac{\partial^2 h}{\partial x^2} + \frac{\partial^2 h}{\partial z^2} = 0 \qquad\qquad (7.5)$$

Equation (7.5) is the *Laplace's Equation* for homogeneous soil. It says that the change of gradient in the x-direction plus the change of gradient in the z-direction is zero. The solution of this equation gives a family of curves meeting at right angles to each other. One family of these curves represents flow lines and the other equipotential lines. For the simple case shown in Fig. 7.1, the flow lines represent a family of semi-ellipses and the equipotential lines semi-hyperbolas.

Anisotropic Soil

Soils in nature do possess permeabilities which are different in the horizontal and vertical directions. The permeability in the horizontal direction is greater than in the vertical direction in sedimentary deposits and in most earth embankments. In loess deposits the vertical permeability is greater than the horizontal permeability. The study of flow nets would be of little value if this variation in the permeability is not taken into account. Eq. (7.4) applies for a soil mass where anisotropy exists. This equation may be written in the form.

$$\frac{\partial^2 h}{\left(\frac{k_z}{k_x}\right)\partial x^2} + \frac{\partial^2 h}{\partial z^2} = 0$$

(7.6)

If we consider a new coordinate variable x_c measured in the same direction as x multiplied by a constant, expressed by

$$x_c + x_x \sqrt{\frac{k_z}{k_x}}$$

(7.7)

Equation (7.6) may be written as

$$\frac{\partial^2 h}{\partial x_c^2} + \frac{\partial^2 h}{\partial z^2} = 0$$

(7.8)

Now Eq. (7.8) is a *Laplace equation* in the coordinates x_c and z. This equation indicates that a cross-section through an anisotropic soil can be transformed to an imaginary section which possesses the same permeability in all directions. The transformation of the section can be effected as per Eq. (7.7) by multiplying the x-coordinates by $\sqrt{k_z/k_x}$ and keeping the z-coordinates at the natural scale. The flow net can be sketched on this transformed section. The permeability to be used with the transformed section is

$$k_e = \sqrt{k_x k_z}$$

(7.9)

An alternative method of transforming a stratified section is to keep the natural dimensions in the x direction and change the z coordinates by multiplying them by $\sqrt{k_x/k_z}$. If k_x is greater than k_z this method would extend the section in the z direction.

7.3   FLOW NET CONSTRUCTION

Properties of a Flow Net

The properties of a flow net can be expressed as given below:

1. Flow and equipotential lines are smooth curves.

2. Flow lines and equipotential lines meet at right angles to each other.

3. No two flow lines cross each other.

4. No two flow or equipotential lines start from the same point.

Boundary Conditions

Flow of water through earth masses is in general three dimensional. Since the analysis of three-dimensional flow is too complicated, the flow problems are solved on the assumption that the flow is two-dimensional. All flow lines in such a case are parallel to the plane of the figure, and the condition is therefore known as two-dimensional flow. All flow studies dealt with herein are for the steady state case. The expression for boundary conditions consists of statements of head or flow conditions at all boundary points. The boundary conditions are generally four in number though there are only three in some cases. The boundary conditions for the case shown in Fig. 7.1 are as follows:

1. Line AB is a boundary equipotential line along which the head is h_t.

2. The line BCB' along the sheet pile wall is a flow boundary.

3. The line $B'N$ is a boundary equipotential line along which the head is equal to h_d.

4. The line FG is a flow line.

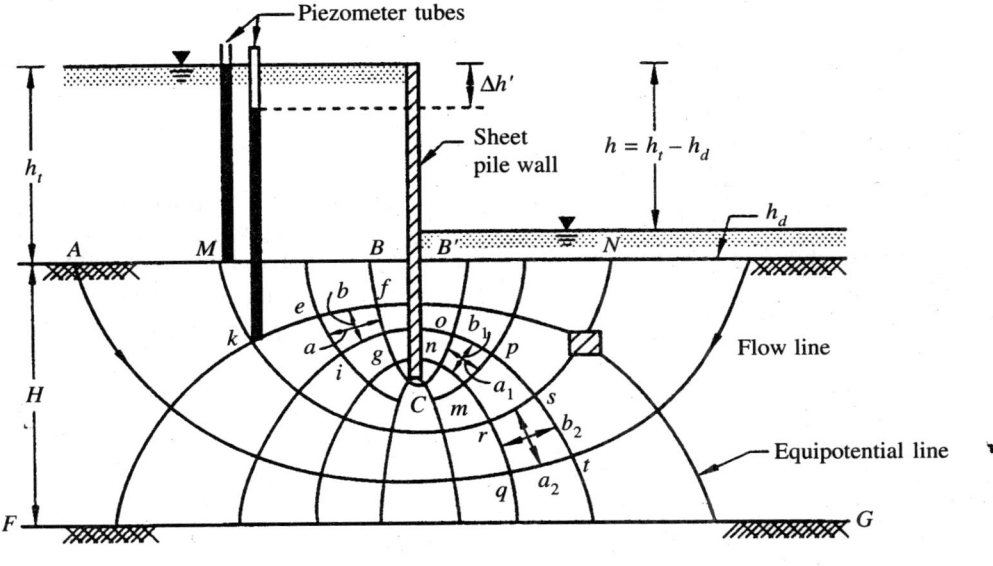

(a) Flow net

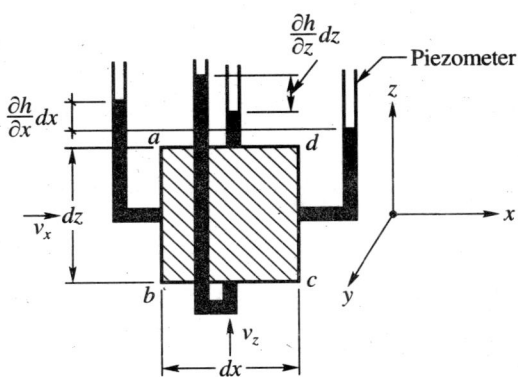

(b) Flow through a prismatic element

Fig. 7.1   Flow through a homogeneous stratum of soil

If we consider any flow line, say, MkN in Fig. 7.1, the potential head at M is h_t and at N is h_d. The total head lost as the water flows along the line is h which is the difference between the upstream and downstream heads of water. The head lost as the water flows from M to k is Δh which is the difference between the heads shown by the piezometers. This loss of head Δh is a fraction of the total head lost.

Flow Net Construction

Flow nets are constructed in such a way as to keep the ratio of the sides of each block bounded by two flow lines and two equipotential lines a constant. If all the sides of one such block are equal, then the flow net must consist of squares. The square block referred to here does not constitute a square according to the strict meaning of the word, it only means that the average width of the square blocks are equal. For example, in Fig. 7.1, the width a of figure $efgi$ is equal to its length b.

The area bounded by any two neighboring flow lines is called a *flow channel*. If the flow net is constructed in such a way that the ratio a/b remains the same for all figures, then it can be shown that there is the same quantity of seepage in each flow channel. In order to show this consider two figures $efgi$ and $nopm$ in one flow channel and another figure $rstq$ in another flow channel as shown in Fig. 7.1. The figure $rstq$ is chosen in such a way that it lies within the same equipotential lines that bound the figure $nopm$. Darcy's law for the discharge through any figure such as $efgi$ per unit length of the section may be written as

$$\Delta q = kia = \frac{\Delta h}{b}a = k\Delta h\frac{a}{b}$$

where, h represents the head loss in crossing the block. The expressions in this form for each of the three figures under consideration are

$$\Delta q = k\Delta h\frac{a}{b}, \ \Delta q_1 = k\Delta h_1\frac{a_1}{b_1}, \ \Delta q_2 = k\Delta h_2\frac{a_2}{b_2} \qquad (7.10)$$

In the above equation the value of k remains the same for all the figures. If the blocks are all squares then

$$\frac{a}{b} = \frac{a_1}{b_1} = \frac{a_2}{b_2} = 1$$

Since figures $efgi$ and $nopm$ are in the same flow channel, we have $\Delta q = \Delta q_1$. Since the figures $efgi$ and $rstq$ are within the same equipotential lines we have $\Delta h_1 = \Delta h_2$. If these equations are inserted we obtain the following relationship:

$$\Delta q = \Delta q_2 \text{ and } \Delta h = \Delta h_2$$

This proves that the same quantity flows through each figure and there is the same head drop in crossing each figure if all the blocks are squares or possess the same ratio a/b. Flow nets are constructed by keeping the ratio a/b the same in all figures. Square flow nets are generally used in practice as this is easier to construct.

There are many methods that are in use for the construction of flow nets. Some of the important methods are

1. Analytical method,

2. Electrical analog method,

3. Scaled model method,

4. Graphical method.

The analytical method, based on the Laplace equation although rigorously precise, is not universally applicable in all cases because of the complexity of the problem involved. The mathematics involved even in some elementary cases is beyond the comprehension of many design engineers. Although this approach is sometimes useful in the checking of other methods, it is largely of academic interest.

The electrical analogy method has been extensively made use of in many important design problems. However, in most of the cases in the field of soil mechanics where the estimation of seepage flows and pressures are generally required, a more simple method such as the graphical method is preferred.

Scaled models are very useful to solve seepage flow problems. Soil models can be constructed to depict flow of water below concrete dams or through earth dams. These models are very useful to demonstrate the fundamentals of fluid flow, but their use in other respects is limited because of the large amount of time and effort required to construct such models.

Graphical method that has been developed by Forchheimer has been found to be very useful in solving complicated flow problems. A. Casagrande has improved this method by incorporating many suggestions. The main drawback of this method is that a good deal of practice and aptitude are essential to produce a satisfactory flow net. In spite of these drawbacks, the graphical method is quite popular amongst engineers as fairly accurate flownets can be produced with some practice.

Graphical Method

The usual procedure for obtaining flow nets is a graphical, trial sketching method, sometimes called the Forchheimer solution. This method of obtaining flow nets is the quickest and the most practical of all the available methods. A. Casagrande has offered many suggestions to the beginner who is interested in flow net construction. Some of his suggestions are summarized below:

1. Study carefully the flow net pattern of well-constructed flow nets.

2. Try to reproduce the same flow nets without seeing them.

3. As a first trial, use not more than four to five flow channels. Too many flow channels would confuse the issue.

4. Follow the principle of 'whole to part', i.e. one has to watch the appearance of the entire flow net and when once the whole net is found approximately correct, finishing touches can be given to the details.

5. All flow and equipotential lines should be smooth and there should not be any sharp transitions between straight and curved lines.

The above suggestions, though quite useful for drawing flow nets, are not sufficient for a beginner. In order to overcome this problem, taylor has proposed a procedure known as the procedure by explicit trials. Some of the salient features of this procedure are given below:

1. As a first step in the explicit trial method, one trial flow line or one trial equipotential line is sketched adjacent to a boundary flow line or boundary equipotential.

2. After choosing the first trial line (say it is a flow line), the flow path between the line and the boundary flow line is divided into a number of squares by drawing equipotential lines. These equipotential lines are extended to meet the bottom flow line at right angles keeping in view that the lines drawn should be smooth without any abrupt transitions.

3. The remaining flow lines are next drawn, adhering rigorously to square figures.

4. If the first trial is chosen property, the net drawn satisfies all the necessary conditions. Otherwise, the last drawn flow line will cross the bottom boundary flow line, indicating that the trial line chosen is incorrect and needs modification.

5. In such a case, a second trial line should be chosen and the procedure repeated.

A typical example of a flownet under a concrete dam is given in Fig. 7.4(a). It should be understood that the number of flow channels will be an integer only by chance. That means, the bottom flow line sketched might not produce full squares with the bottom boundary flow line. In such a case the bottom flow channel will be a fraction of a full flow channel. It should also be noted that the figure formed by the first sketched flow line with the last equipotential line in the region is of irregular form. This figure is called a *singular square*. The basic requirement for such squares, as for all the other squares, is that continuous sub-division of the figures give an approach to true squares. Such singular squares are formed at the tips of sheet pile walls also. Squares must be thought of as valid only where the Laplace equation applies. The Laplace equation applies to soils which are homogeneous and isotropic. When the soil is anisotropic, the flow net should be sketched as before on the transformed section. The transformed section can be obtained from the natural section explained earlier.

7.4   DETERMINATION OF QUANTITY OF SEEPAGE

Flow nets are useful for determining the quantity of seepage through a section. The quantity of seepage q is calculated per unit length of the section. As explained earlier, the flow through any square can be written as

$$\Delta q = k\Delta h \qquad (7.11)$$

Let the number of flow channel and equipotential drops in a section be N_f and N_d, respectively. Since all drops are equal, we can write

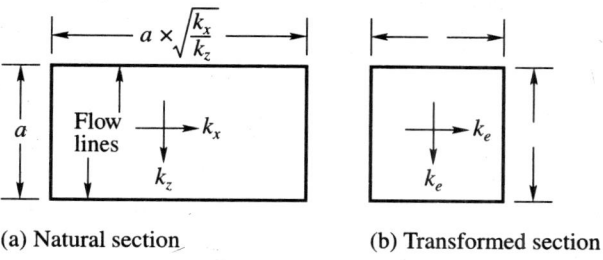

(a) Natural section   (b) Transformed section

Fig. 7.2 Flow through anisotropic soil

$$\Delta h = \frac{h}{N_d} \tag{7.12}$$

Since the discharge in each flow channel is the same we can write,

$$q = N_f \Delta q \tag{7.13}$$

Substituting for Δq and Δh, the Eq. (7.13) can be written as

$$q = kh \frac{N_f}{N_d} \tag{7.14}$$

Equation (7.14) can also be used to compute the seepage through anisotropic sections by writing k_e in place of k. As per Eq. (7.9), k_e is equal to $\sqrt{k_x k_z}$, where k_x and k_z are the coefficients of permeability in the x and z directions respectively. The validity of this relationship can be proved as follows. Consider a figure bounded by flow and equipotential lines in which the flow is parallel to the x direction. In Fig. 7.2 the figure in question is drawn to a transformed scale in (b) and the same to the natural scale in (a).

In Fig. 7.2(b) the permeability has the effective value k_e in both the x and z directions and the flow through the square according to Eq. (7.11), is

$$\Delta q = k_e \frac{\Delta h a}{a} = k_e \Delta h \tag{7.15}$$

In Fig. 7.2(a) the coefficients of permeability k_x in the horizontal section must apply because the flow is horizontal and the sketch is to the natural scale. The flow equation is, therefore,

$$\Delta q = k_x iA = k_x \frac{\Delta h a}{a \times \sqrt{\dfrac{k_x}{k_z}}} \tag{7.16}$$

Equating Eq. (7.15) and (7.16), we get

$$k_e = \sqrt{k_x k_z} \tag{7.17}$$

Flownets in Anisotropic Soils

To obtain a flow net for anisotropic soil conditions, the natural cross-section has to be redrawn to satisfy the condition of Laplace Eq. (7.5).

The transformed section may be obtained by multiplying either the natural horizontal distances by $\sqrt{k_z/k_x}$ or the vertical distances by $\sqrt{k_x/k_z}$ keeping the other dimension unaltered. Normally the vertical dimensions are kept as they are but the horizontal dimensions are multiplied by $\sqrt{k_z/k_x}$. The natural section gets shortened or lengthened in the x-direction in accordance with the condition that k_x is greater or less than k_z.

Figure 7.3(a) is a natural section with flow taking place around a sheet pile wall. The horizontal permeability is assumed to be 4 times that of the vertical permeability. Fig. 7.3(b) is transformed section with the horizontal dimensions multiplied by a factor equal to $\sqrt{k_z/k_x} = \sqrt{1/4} = 1/2$. This section is now assumed to possess the same permeability of $k_e = \sqrt{4k_z^2} = 2k$ in all directions. The

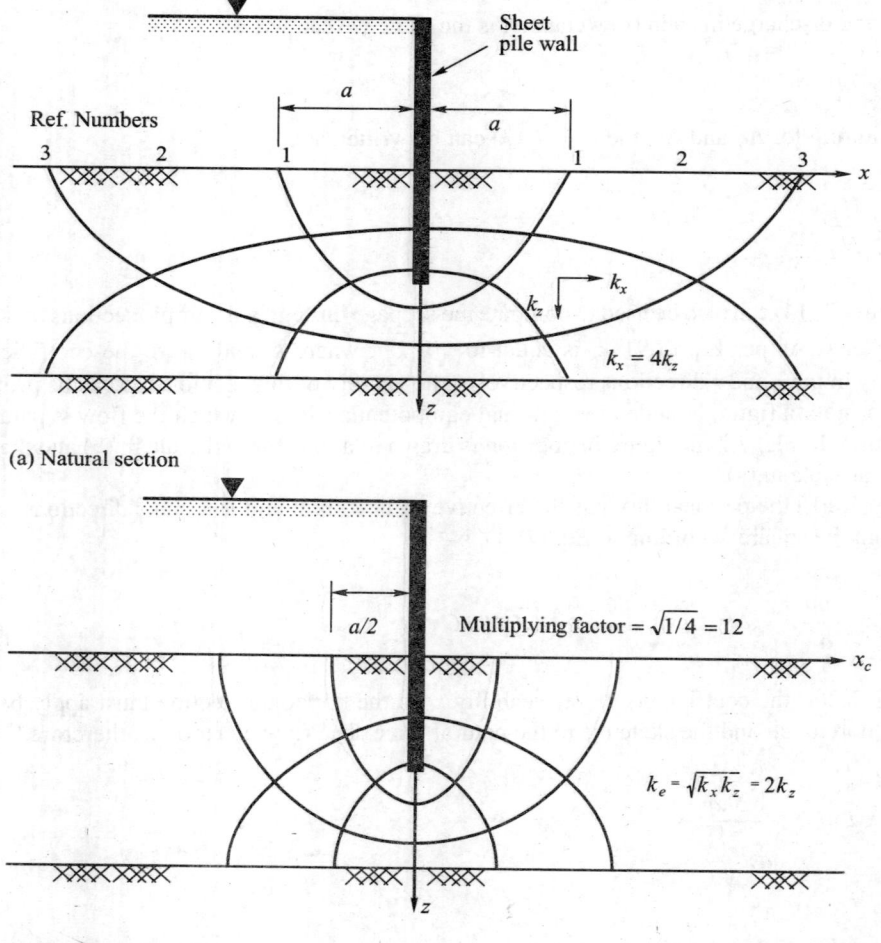

(a) Natural section

(b) Transformed section

Fig. 7.3 Flownet in anisotropic soil

flownets are constructed on this section in the usual way. The same flow net is transferred to the natural section in (a) of Fig. 7.3, by multiplying the x-coordinates of points on the flow and equipotential lines by the factor 2.

On the natural cross-section the flow net will not be composed of squares but of rectangles elongated in the direction of greater permeability.

7.5   DETERMINATION OF SEEPAGE PRESSURE

Flow nets are useful in the determination of the seepage pressure at any point along the flow path. Consider the cubical element $efgi$ in Fig. 7.1 with all the sides equal to a. Let h_1 be the piezometric head acting on the face ef and h_2 on face gi.

The total force on face $ef = P_1 = a^2 \gamma_w h_1$

The total force on face $gi = P_2 = a^2 \gamma_w h_2$

The differential force acting on the element is

$P_1 - P_2 = P_3 = a^2 \gamma_w (h_1 - h_2)$

Since $(h_1 - h_2)$ is the head drop h, we can write

$$P_3 = a^2 \gamma_w \Delta h = a^3 \frac{\Delta h}{a} \gamma_w = a^3 i \gamma_w$$

where a^3 is the volume of the element. The force per unit volume of the element is, therefore,

$$P_s = i \gamma_w \tag{7.18}$$

This force which exerts a drag on the element known as the *seepage pressure*. It has the dimension of unit weight, and at any point its line of action is tangent to the flow line. The seepage pressure is a very important factor in the stability analysis of earth slopes. If the line of action of the seepage force acts in the vertical direction upward as on an element adjacent to point B' in Fig. 7.1, the force that is acting downward to keep the element stable is the buoyant unit weight of the element. When these two forces balance, the soil will just be at the point of being lifted up, and there will be effectively no grain-to-grain pressures. The gradient at which this occurs can be computed from the balance of forces given by Eqs (7.19) and (7.18). Therefore, we can write

$$i_c \gamma_w = \frac{(G-1)\gamma_w}{1+e} \quad \text{or} \ i_c = \frac{G-1}{1+e} \tag{7.19}$$

The soil will be in *quick condition* at this gradient, which is therefore called i_c, the *critical gradient*.

Example 7.1

In order to compute the seepage loss through the foundation of a cofferdam, flownets were constructed. The result of the flownet study gave $N_f = 6$, $N_d = 16$. The head of water lost during seepage was 6 m. If the coefficient of permeability of the soil is $k = 4 \times 10^{-5}$ m/min, compute the seepage loss per metre length of dam per day.

Solution

The equation for seepage loss is

$$q = kh\frac{N_f}{N_d}$$

Substituting the given values, we have

$$q = 4\times10^{-5}\times6\times\frac{6}{16} = 9\times10^{-5}\,\text{m}^3/\text{min} = 13\times10^{-2}\,\text{m}^3/\text{day}$$

Example 7.2

Two lines of sheet piles were driven in a river bed as shown in Fig. Ex. 7.2. The depth of water over the river bed is 2.5 m. The trench level within the sheet piles is 2.0 m below the river bed. The water level within the sheet piles is kept at trench level by resorting to pumping. If a quantity of water flowing into the trench from outside is 0.30 m³/hour per foot length of sheet pile, what is the hydraulic coefficient of permeability of sand? What is the hydraulic gradient immediately below the trench bed?

Solution

Figure Ex. 7.2 gives the flow net and other details. The differential head between the bottom of trench and the water level in the river is 4.5 m.

Number of channels = 6

Number of equipotential drops = 10

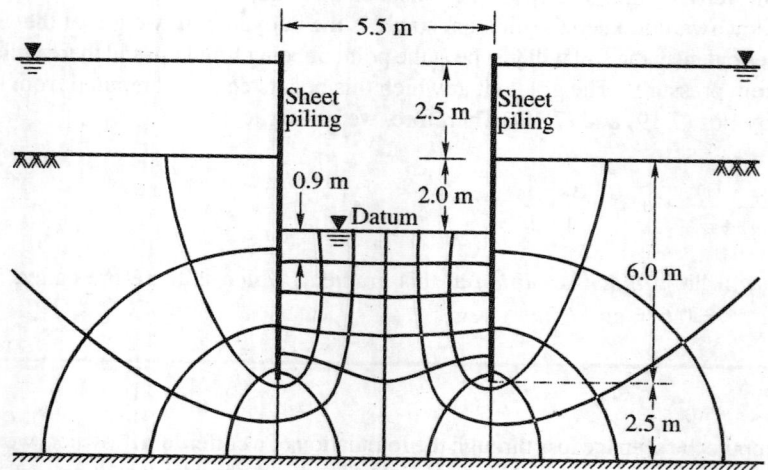

Fig. Ex. 7.2

$$q = kh \frac{N_f}{N_d} \text{ or } 0.30 = 4.5 \times \frac{6}{10} \times k$$

$$\text{or } k = \frac{0.30 \times 10}{4.5 \times 6} \times \frac{1}{60 \times 60} = 3.08 \times 10^{-5} \text{ m/s}$$

The distance between the last two equipotentials given is 0.9 m. The calculated hydraulic gradient is

$$i = \frac{\Delta h}{\Delta s} = \frac{4.5}{10 \times 0.9} = 0.50$$

Example 7.3

A concrete dam is constructed across a river over a permeable stratum of soil of limited thickness. The water heads are upstream side 16 m and 2 m on the downstream side. The flow net constructed under the dam gives $N_f = 7$ and $N_d = 12$. Calculate the seepage loss through the subsoil if the average value of the hydraulic conductivity is 6×10^{-3} cm/sec horizontally and 3×10^{-4} cm/sec vertically. Calculate the exit gradient if the average length of the last field is 0.9 m.

Solution

Upstream side $h_1 = 16$ m

Downstream side $h_2 = 2$ m

$N_f = 7, N_d = 21$

$k_h = 6 \times 10^{-3}$ cm/sec, $k_v = 3 \times 10^{-4}$ cm/sec

$$k_e = \sqrt{k_h k_v} = \sqrt{6 \times 10^{-3} \times 3 \times 10^{-4}}$$

$$= 1.34 \times 10^{-3} \text{ cm / sec}$$

$$q = k_e h \frac{N_f}{N_d} = 14 \times 100 \times 1.34 \times 10^{-3} \times \frac{7}{21}$$

$$= 0.626 \text{ cm}^3 / \text{sec}$$

The head loss per potential drop $= \dfrac{h}{N_d} = \dfrac{14}{21} = 0.67$

The exit gradient $i = \dfrac{\Delta h}{l} = \dfrac{0.67}{0.9} = 0.74$

7.6   DETERMINATION OF UPLIFT PRESSURES

Water that seeps below masonry dams or weirs founded on permeable soils exerts pressures on the bases of structures. These pressures are called *uplift pressures*. Uplift pressures reduce the effective weight of the structure and thereby cause instability. It is therefore very essential to determine the uplift pressures on the base of dams or weirs accurately. Accurate flow nets should be constructed in cases where uplift pressures are required to be determined. The method of determining the uplift pressures can be explained with an example.

Consider a concrete dam founded on a permeable foundation as shown in Fig. 7.4(a). A sheet pile wall is provided on the upstream side of the base of the dam in order to reduce uplift pressure on the base of the dam.

Assume the tail water level as the datum. The pressure head is the same as the uplift pressure or the piezometric head at any point on the base of dam. For any point p the driving residual head h_p is the sum of the pressure head H_p, and the elevation head z_p, we may therefore write

$$h_p = H_p + z_p \qquad\qquad (7.20a)$$

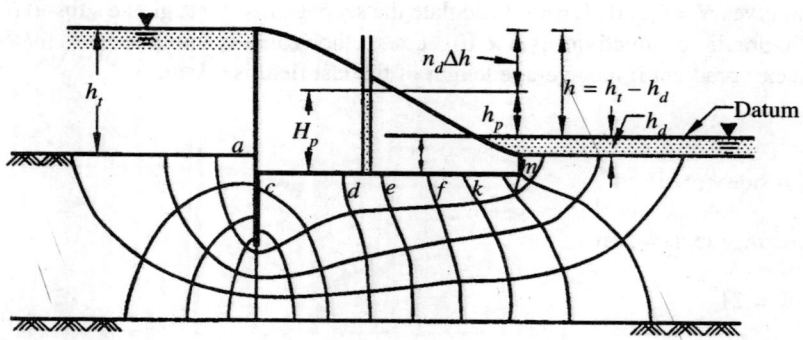

(a) A typical flownet under a weir

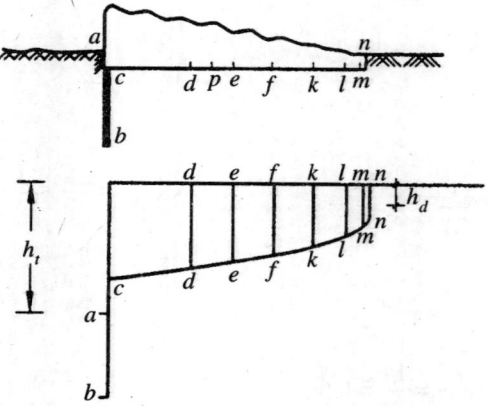

(b) Uplift pressure on the base of concrete dam

Fig. 7.4   Uplift pressure on the base of concrete dam

Assuming the tail water level as datum and the elevation head is negative for points below the datum, the pressure head at point p is

$$H_p = h_p - (-z_p) = h_p + z_p \qquad\qquad\qquad (7.20b)$$

where $h_p = (h - n_d \Delta h)$, n_d is the number of potential drops of value Δh up to the point p, and z_p is the elevation head of the point p. The uplift pressure head distribution on the base of the dam as per Eq. (7.18) is shown in Fig. 7.4(b). The uplift pressure at any points a, b, c, d, etc. marked in the figure are as shown in (b). The uplift pressure at any point is the product of H_p and γ_w, the unit weight of water.

7.7   SEEPAGE THROUGH HOMOGENEOUS EARTH DAMS

In almost all problems concerning seepage beneath a sheet pile wall or through the foundation of a concrete dam all boundary conditions are known. However, in the case of seepage through an earth dam the upper boundary or the uppermost flow line is not known. This upper boundary is a free water surface and will be referred to as the *line of seepage or phreatic line*. The seepage line may therefore be defined as the line above which there is no hydrostatic pressure and below which there is hydrostatic pressure. In the design of all earth dams, the following factors are very important.

1.   The seepage line should not cut the downstream slope.

2.   The seepage loss through the dam should be the minimum possible.

The two important problems that are required to be studied in the design of earth dams are:

1.   The prediction of the position of the line of seepage in the cross-section.

2.   The computation of the seepage loss.

If the line of seepage is allowed to intersect the downstream face much above the toe, more or less serious sloughing may take place and ultimate failure may result. This mishap can be prevented by providing suitable drainage arrangements on the downstream side of the dam.

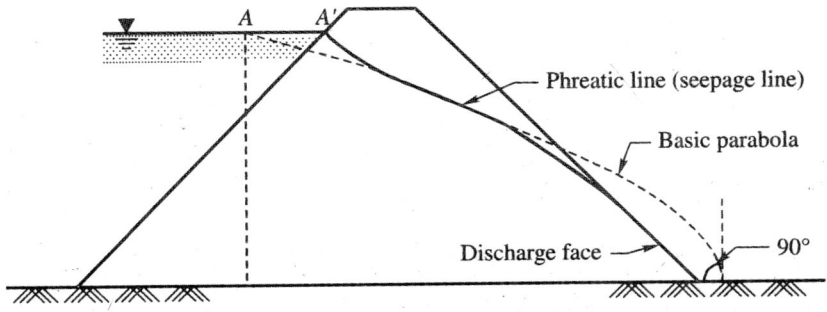

Fig. 7.5   Basic parabola and the phreatic line for a homogeneous earth dams

The section of an earth dam may be homogeneous or non-homogeneous. A homogeneous dam contains the same material over the whole section and only one coefficient of permeability may be assumed to hold for the entire section. In the non homogeneous or the composite section, two or more permeability coefficients may have to be used according to the materials used in the section. When a number of soils of different permeabilities occur in a cross-section, the prediction of the position of the line of seepage and the computation of the seepage loss become quite complicated.

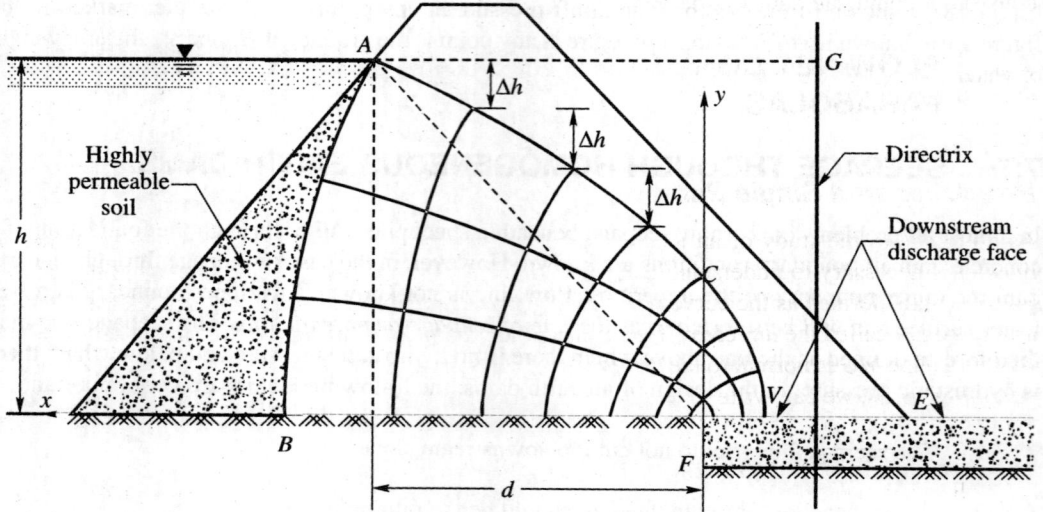

(a) Conjugate confocal parabolas

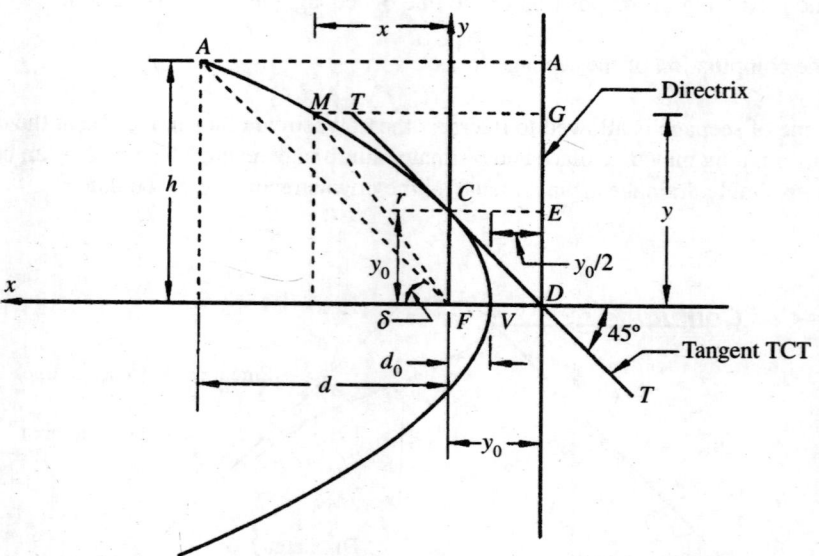

(b) Properties of a parabolas

Fig. 7.6   Ideal flownet comprising of conjugate confocal parabolas

It has been noticed from experiments on homogeneous earth dam models that the line of seepage assumes more or less the shape of a parabola as illustrated in Fig. 7.5. In some sections a little divergence from a regular parabola is required at the surfaces of entry and discharge of the line of seepage. In some ideal sections where conditions are favourable the entire seepage line may be considered as a parabola. When the entire seepage line is a parabola, all the other flow lines will be confocal parabolas. The equipotential lines for this ideal case will be conjugate confocal parabolas as shown in Fig. 7.6(a). As a first step it is necessary to study the ideal case where the entire flow net consists of conjugate confocal parabolas.

7.8   FLOW NET CONSISTING OF CONJUGATE CONFOCAL PARABOLAS

Properties of a Single Parabola

As a prelude to the study of an ideal flow net comprising of parabolas as flow and equipotential lines, it is necessary to understand the properties of a single parabola. The parabola, illustrated in Fig. 7.6(b), is defined as the curve whose every point is equidistant from a point F called the *focus* and a line DA called the *directrix*. If we consider any point M on the curve, we can write $FM = MG$, where the line MG is normal to the directrix. If F is the origin of coordinates, and the coordinates of point M is (x, y), we can write

$$FM = \sqrt{x^2 + y^2} = MG = x + y_0$$

$$\text{or } x = \frac{y^2 - y_0^2}{2y_0} \tag{7.21}$$

Equation (7.21) is the equation of the basic parabola. If the parabola intersects the y-axis at C, we can write as before for the point C

$$FC = CE = y_0$$

Similarly for the vertex point V, the focal distance a_0 is

$$FV = VD = a_0 = y_0/2 \tag{7.22}$$

Properties of Conjugate Confocal Parabolas

Figure 7.7(a) illustrates the ideal flow net consisting of conjugate confocal parabolas. All the parabolas have a common focus F.

The boundary lines of such an ideal flow net are:

1.   The upstream face AB, an equipotential line, is a parabola.

2.   The downstream discharge face FV, an equipotential line, is horizontal.

3. ACV, the flow line, is a parabola.

4. BF, the bottom flow line, is horizontal.

The known boundary conditions are only three in number. They are, the two equipotential lines AB and FV, and the bottom flow line BF. The top flow line ACV is the one that is unknown. The theoretical investigation of Prof. Kozeny i n 1931 revealed that the flow net for such an ideal condition mentioned above with a horizontal discharge face FV consists of two families of confocal parabolas with a common focus F. Since the conjugate confocal parabolas should intersect at right angles to each other, all the parabolas crossing the vertical line FC should have their intersection points lie on this line.

Since the seepage line is a line of atmospheric pressure only, the only type of head that can exist along it is the elevation head. Therefore, there must be constant drops in elevation between the points at which successive equipotentials meet the top flow line, as shown in Fig. 7.6(a).

In all seepage problems connected with flow through earth dams, the focus F of the basic parabola is assumed to lie at the intersection of the downstream discharge face FV and the bottom flow line BF as shown in Fig. 7.6(a). The point F is therefore known. The point A, which is the intersection point of the top flow line of the basic parabola and the upstream water level, is also supposed to be known. When the point A is known, its coordinates (d, h) with respect to the origin F can be determined. With these two known points, the basic parabola can be constructed as explained below. We may write from Fig. 7.7(b).

$$FA = AA = \sqrt{d^2 + h^2}$$

$$FD = 2a_0 = \sqrt{d^2 + h^2} - d$$

$$a_0 = \frac{1}{2}\left(\sqrt{d^2 + h^2} - d\right) \tag{7.23}$$

Seepage Loss through the Dam

The seepage flow q across any section can be expressed according to Darcy's law as

$$q = kiA \tag{7.24}$$

considering the section FC in Fig. 7.6(a), where the sectional area A is equal to y_0, the hydraulic

gradient i can be determined analytically as follows:

From Eq. (7.21), the equation of the parabola can be expressed as

$$y = \sqrt{2xy0 + y_0^2} \tag{7.25}$$

The hydraulic gradient i at any point on the seepage line in Fig. 7.6(a) can be expressed as

$$\frac{dy}{dx} = \frac{y_0}{\sqrt{2xy0 + y_0^2}} \tag{7.26}$$

For the point C which has coordinates $(0, y_0)$, the hydraulic gradient from Eq. (7.26) is

$$\frac{dy}{dx} = \frac{y_0}{\sqrt{y_0^2}} = 1$$

Therefore, the seepage quantity across section FC is

$$q = k\frac{dy}{dx} y_0 = ky_0 \qquad\qquad (7.27)$$

7.9 SEEPAGE THROUGH HOMOGENEOUS AND ISOTROPIC EARTH DAMS

Types of Entry and Exit of Seepage Lines

The flow net comprising of conjugate confocal parabolas as flow equipotential lines is an ideal case which is not generally met in practice. Though the top flow line resembles a parabola for most of its length, the departure from the basic parabola takes place at the faces of entry and discharge of the flow line. The departure from the basic parabola depends upon the conditions prevailing at the points of entrance and discharge of the flow line as illustrated in Fig. 7.7 from (a) to (e).

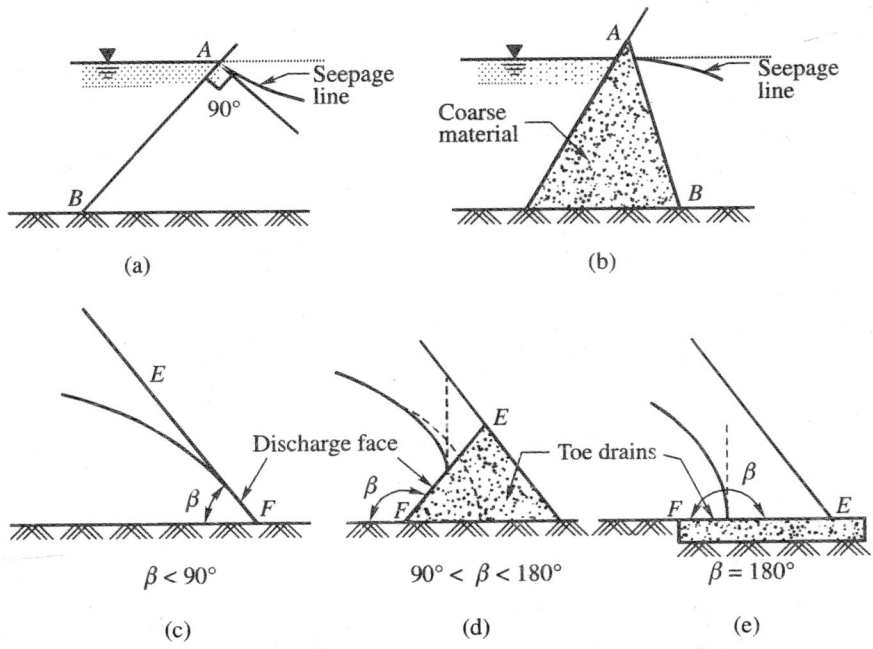

Fig. 7.7 Types of entry and exit of seepage lines

The seepage line should be normal to the equipotential line at the point of entry as shown in Fig. 7.7(a). However, this condition is violated in Fig. 7.7(b), where the angle made by the upstream face AB with the horizontal is less than 90°. It can be assumed in this case the coarse material used to support the face AB is highly permeable and does not offer any resistance for flow. In such cases AB taken as the upstream equipotential line. The top flow line cannot therefore be normal to the equipotential line. However, this line possesses zero gradient and velocity at the point of entry. This zero condition relieves the apparent inconsistency of deviation from a normal intersection.

The conditions prevailing at the downstream toe of the dam affect the type of exit of the flow line at the discharge face. In Fig. 7.7(c) the material at the toe is the same as in the other parts of the dam whereas in (d) and (e) rock toe drains are provided. This variation in the soil condition at the toe affects the exit pattern of the flow line. The flow line will meet the discharge face FE tangentially in 7.8(c). This has to be so because the particles of water as they emerge from the pores at the discharge face have to conform as nearly as possible to the direction of gravity. But in cases where rock toe drains are provided, the top flow line becomes tangential to the vertical line drawn at the point of exit on the discharge face as shown in (d) and (e) of Fig. 7.7.

7.10 METHOD OF LOCATING SEEPAGE LINE

The general method of locating the seepage line in any homogeneous dam resting on an impervious foundation may be explained with reference to Fig. 7.8(a). As explained earlier, the focus F of the basic parabola is taken as the intersection point of the bottom flow line BF and the discharge face EF. In this case the focus coincides with the toe of the dam. One more point is required to construct the basic parabola. Analysis of the location of seepage lines by A. Casagrande has revealed that the basic parabola with focus F intersects the upstream water surface at A such that AA' = 0.3 m, where m is the projected length of the upstream equipotential line A'B on the water surface. Point A is called the corrected entrance point. The parabola APSV may now be constructed as per Eq. (7.21). The divergence of the seepage line from the basic parabola is shown as A'P and SD in Fig. 7.8(a). For dams with flat slopes, the divergences may be sketched by eye keeping in view the boundary requirements. The error involved in sketching by eye, the divergence on the downstream side, might be considerable if the slopes are steeper. Procedures have therefore been developed to sketch the downstream divergence as explained below. As shown in Fig. 7.6(a), E is the point at which the basic parabola intersects the discharge face. Let the distance ED be designated as Δa and the distance DF as a. The values of Δa and $a + \Delta a$ vary with the angle, β, made by the discharge face with the horizontal measured clockwise. The angle may vary from 30° to 180°. The discharge face is horizontal as shown in Fig. 7.7(e). Casagrande (1937) determined the ratios of $\Delta a / (a + \Delta a)$ for a number of discharge slopes varying from 30° to 180° and the relationship is shown in a graphical form Fig. 7.8(b).

The distance $(a + \Delta a)$ can be determined by constructing the basic parabola with F as the focus. With the known $(a + \Delta a)$ and the discharge face angle β, Δa can be determined from Fig. 7.8(b). The point D may therefore be marked out at a distance of .a from E. With the point D known, the divergence DS may be sketched by eye.

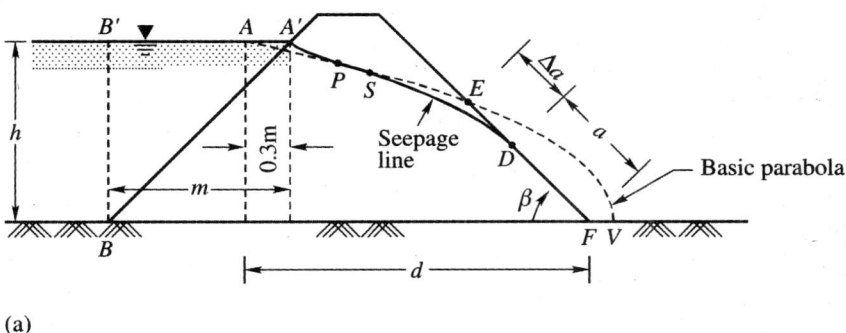

(a)

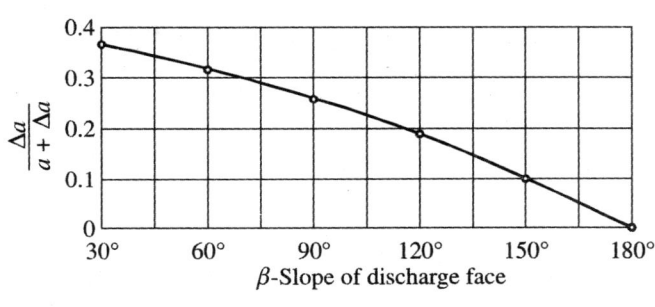

(b)

Fig. 7.8   Construction of seepage line

It should be noted that the discharge length a, which is boundary of the flownet, is neither an equipotential nor a flow line, since it is at atmospheric pressure. It is a boundary along which the head at any point is equal to the elevation.

Analytical Solutions for Determining a and q

A. Casagrande has proposed the following equation for determining a for $\beta < 30°$

$$a = \frac{d}{\cos \beta} - \sqrt{\frac{d^2}{\cos^2 \beta} - \frac{h^2}{\sin^2 \beta}} \qquad (7.28)$$

L. Casagrande (1932) gave the following equation for a when β lies between 30° and 90°.

$$a = \sqrt{h^2 + d^2} - \sqrt{d^2 - h^2 \cot^2 \beta} \qquad (7.29)$$

The discharge q per unit length through any cross-section of the dam may be expressed as follows:

For $\beta < 30°$, $q = ka \sin \beta \tan \beta$ (7.30)

For $30° < \beta < 90°$, $q = ka \sin^2 \beta$ (7.31)

Example 7.4

Figure Ex. 7.4 gives the section of homogeneous dam having coefficient of permeability $k = 2 \times 10^{-6}$ cm/sec. Draw the seepage line and compute the seepage loss per metre length of the dam.

Solution

The depth h if water on up side = 10 m.

The projected length of slope $A'B$ on the water surface = 10 m.

The point A on the water level is a point on the basic parabola. Therefore, $AA' = 0.3 \times 32.81 = 3$ m.

F is the focus of the parabola. The distance of the directrix from the focus F is

$$y_0 = \sqrt{d^2 + h^2} - d$$

where, $d = 21$ m, $h = 10$ m.

Therefore, $y_0 = \sqrt{21^2 + 10^2} - 21 = 2.26$ m

The distance of the vertex of the parabola from F is

$$FV = a_0 = \frac{y_0}{2} = \frac{2.26}{2} = 1.13 \text{ m}$$

The (x, y) coordinates of the basic parabola may be obtained from Eq. (7.21).

$$x = \frac{y^2 - y_0^2}{2y_0} = \frac{y^2 - (2.26)^2}{2 \times 2.26} = \frac{y^2 - 5.11}{4.52}$$

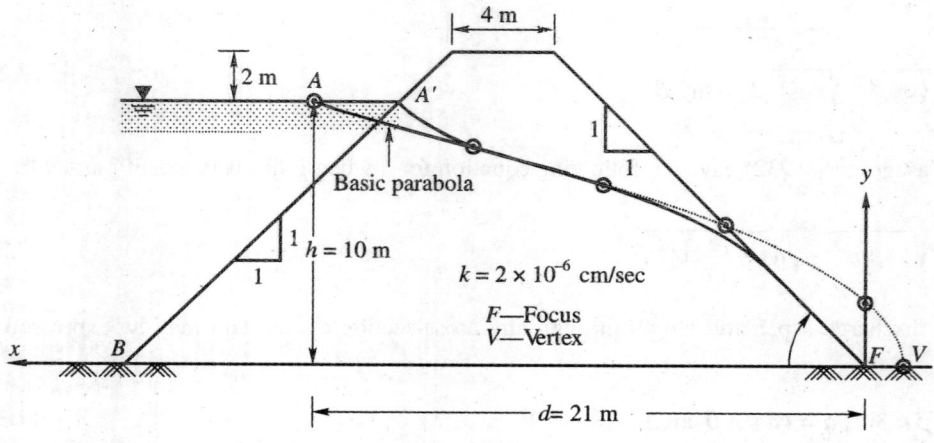

Fig. Ex. 7.4

or $y = \sqrt{4.52x + 5.11}$

Given below are values of y for various values of x

x	0	5	10	15	21
y	2.26	5.26	7.1	8.54	10.0

The parabola has been constructed with the above coordinates as shown in Fig. Ex. 7.4.

From Fig. Ex. 7.4 $\Delta a + a = 7.5$ m

From Fig. 7.8, for a slope angle $\beta = 45°$

$$\frac{\Delta a}{a + \Delta a} = 0.35$$

or $\Delta a = 0.35 (a + \Delta a) = 0.35 \times 7.5 = 2.63$ m

From Eq. (7.27) $q = ky_0$

where $k = 2 \times 10^{-6}$ cm/sec or 2×10^{-8} m/sec and $y_0 = 2.26$ m

$q = 2 \times 10^{-8} \times 2.26 = 4.52 \times 10^{-8}$ m³/sec per metre length of dam.

Example 7.5

An earth dam which is anisotropic is given in Fig. Ex. 7.5(a). The coefficients of permeability k_x and k_z in the horizontal and vertical directions are respectively 4.5×10^{-8} m/s and 1.6×10^{-8} m/s. Construct the flow net and determine the quantity of seepage through the dam. What is the porewater pressure at point P?

Solution

The transformed section is obtained by multiplying the horizontal distances by $\sqrt{k_z/k_x}$ and by keeping the vertical dimensions unaltered. Figure Ex. 4.17(a) is a natural section of the dam. The scale factor for transformation in the horizontal direction is

$$\text{Scale factor} = \sqrt{\frac{k_z}{k_x}} = \sqrt{\frac{1.6 \times 10^{-8}}{4.5 \times 10^{-8}}} = 0.6$$

The transformed section of the dam is given in Fig. Ex. 7.5(b). The isotropic equivalent coefficient of permeability is

$$k_e = \sqrt{k_x k_z} = \left(\sqrt{4.5 \times 1.6}\right) \times 10^{-8} = 2.7 \times 10^{-8} \text{ m/s}$$

Confocal parabolas can be constructed with the focus of the parabola at A. The basic parabola passes through point G such that

$GC = 0.3\, HC = 0.3 \times 27 = 8.10$ m

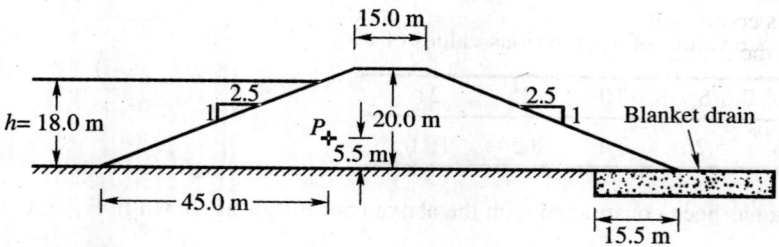

(a) Natural section

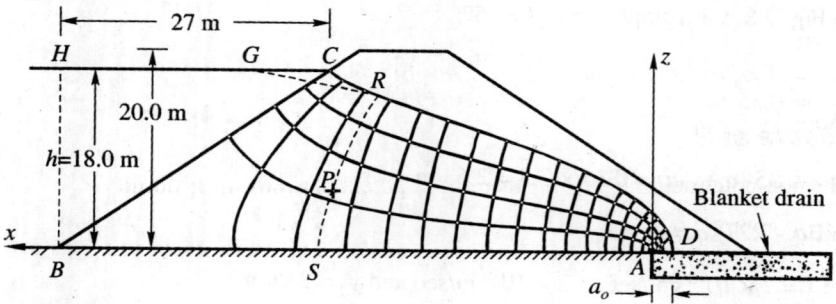

(b)Transformed section

Fig. Ex. 7.5

The coordinates of G are:

$x = +40.80$ m, $z = +18.0$ m

As per Eq. (7.21) $x = \dfrac{z^2 - 4a_0^2}{4a_0}$ (a)

Substituting for x and z, we get $40.80 = \dfrac{18\,a^2 - 4\,a_0^2}{4\,a_0}$

Simplifying we have, $4a_0^2 + 163.2a_0 - 324 = 0$

Solving, $a_0 = 1.9$ m

Substituting for a_0 in Eq. (a) above, we can write

$x = \dfrac{z^2 - 14.4}{7.6}$ (b)

By using Eq. (b), the coordinates of a number of points on the basic parabola may be calculated.

x(m)	1.9	0.0	5.0	10.0	20.0	30.0
z(m)	0.0	3.8	7.24	9.51	12.9	15.57

The basic parabola is shown in Fig. Ex. 7.5(b).

The flownet is completed by making the entry corrections by ensuring that the potential drops are equal between the successive equipotential lines at the top seepage line level.

As per Fig. Ex. 7.5(b), there are 3.8 flow channels and 18 equipotential drops. The seepage per unit length of dam is

$$q = k_e h \frac{N_f}{N_d} = 2.7 \times 10^{-8} \times 18 \times \frac{3.8}{18} = 1 \times 10^{-7} \text{ m}^3/\text{s}$$

The quantity of seepage across section Az can also be calculated without the flownet by using Eq. (7.27)

$$q = k_e y_0 = 2k_e a_0 = 2 \times 2.7 \times 10^{-8} \times 1.9 \approx 1 \times 10^{-7} \text{ m}^3/\text{sec}$$

Pore pressure at P

Let RS be the equipotential line passing through P, potential drops up to point P equals $2.4\Delta h = 2.4 \times 1 = 2.4$ m

$$\text{since} \quad \Delta h = \frac{N_d}{h} = \frac{18}{18} = 1$$

where h = headloss = 18 m.

Residual head at $P = 18 - 2.4 = 15.6$ m

Assuming the base of the dam as datum, the elevation head = 5.50 m.

Therefore the head at point $P = 15.6 - 5.5 = 10.1$ m.

The pore pressure at P is, therefore, $u_w = 10.1 \times 9.81 = 99 \text{ kN/m}^2$

7.11   PIPING FAILURE

Piping failures caused by heave can be expected to occur on the downstream side of a hydraulic structure when the uplift forces of seepage exceed the downward forces due to the submerged weight of the soil.

The mechanics of failure due to seepage was first presented by Terzaghi. The principle of this method may be explained with respect to seepage flow below a sheet pile wall. Fig. 7.10(a) is a sheet pile wall with the flow net drawn. The uplift pressures acting on a horizontal plane ox can be determined as explained in Sec. 7.6. The ordinates of curve C in Fig. 7.9(b) represent the uplift pressure at any point on the line ox. It is seen that the uplift pressure is greatest close to the wall and gradually becomes less with an increase in the distance from the wall. When the upward forces of seepage on a portion of ox near the wall become equal to the downward forces exerted by the submerged soil, the surface of the soil rises as shown in Fig. 7.9(a). This heave occurs simultaneously with an expansion of the volume of the soil, which causes its permeability to increase. Additional seepage causes the sand to boil, which accelerates the flow of water and leads to complete failure. Terzaghi determined from model tests that heave occurs within a distance of about $D/2$ (where D is the depth of penetration of the pile) from the sheet pile and the critical section ox passes through the lower edge of the sheet pile.

Factor of Safety Against Heave

The prism *aocd* in Fig. 7.9(b) subjected to the possible uplift has a depth of D and width $D/2$.

The average uplift pressure on the base of prism is equal to $\tilde{a}wha$. The total uplift force per unit length of wall is

$$U = \frac{1}{2}\gamma_w h_a D \tag{7.33}$$

The submerged weight of the prism *aocd* is

$$W_b = \frac{1}{2}\gamma_b D^2$$

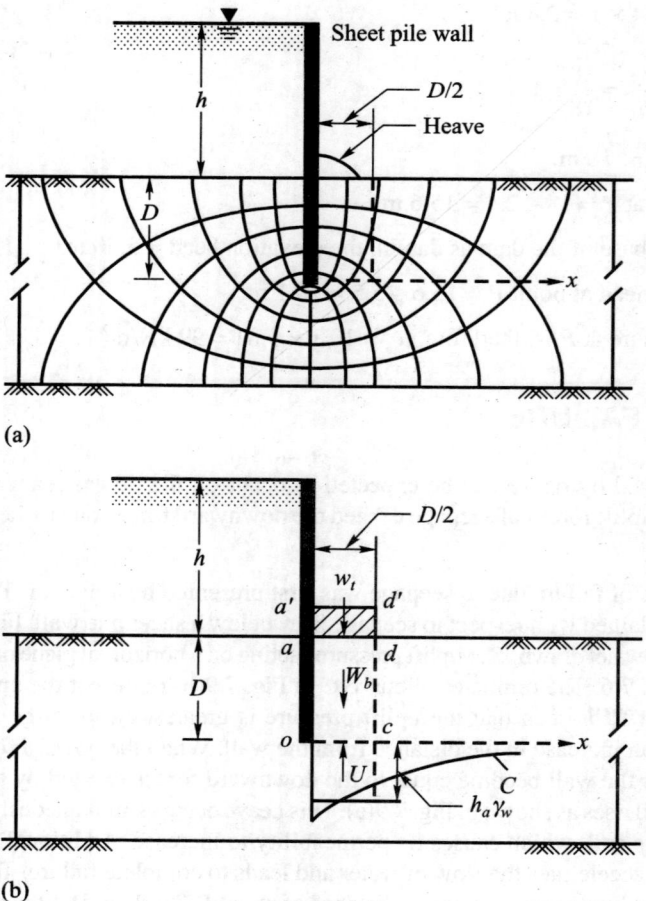

(a)

(b)

Fig. 7.9 Piping failure

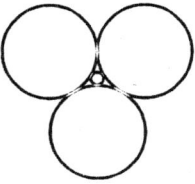

(a) Size of smallest spherical particle which just fits the space of bigger size

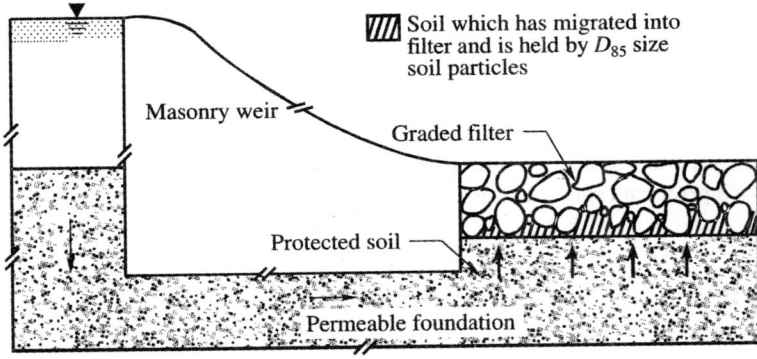

(b) Condition of the boundary between protected soil and the filter material

Fig. 7.10   Requirements of a filter

where γ_b is the submerged unit weight of the material. The factor of safety with respect to piping can therefore be expressed as

$$F_s = \frac{W_b}{U} = \frac{D\gamma_b}{h_a\gamma_w} \tag{7.34}$$

It is not economical to drive the sheet piles deeply enough to prevent heave, the factor of safety can be increased by placing a weighted filter over the prism *aocd* as shown by the prism *aa'd'd*. If the weight of such a filter is W_t, the new factor of safety can be written as

$$F'_s = \frac{W_b + W_t}{U} \tag{7.35}$$

Filter Requirements to Control Piping

Filter drains are required on the downstream sides of hydraulic structures and around drainage pipes. A properly graded filter prevents the erosion of soil in contact with it due to seepage forces. To prevent the movement of erodible soils into or through filters, the pore spaces between the filter particles should be small enough to hold some of the protected materials in place. Taylor (1948) shows that if three perfect spheres have diameters greater than 6.5 times the diameter of a small sphere, the small spheres can move through the larger as shown in Fig. 7.10(a). Soils and aggregates are always composed of ranges of particle sizes, and if pore spaces in filters are small enough to hold the 85 percent size (D_{85}) of the protected soil in place, the finer particles will also be held in place as exhibited schematically in Fig. 7.10(b).

The requirements of a filter to keep the protected soil particles from invading the filter significantly are based on particle size. These requirements were developed from tests by Terzaghi which were later extended by the U.S. Army Corps of Engineers (1953). The resulting filter specifications relate the grading of the protective filter to that of the soil being protected by the following:

$$\frac{D_{15\,filter}}{D_{85\,soil}} \leq 5, \ 4 < \frac{D_{15\,filter}}{D_{15\,soil}} \leq 20, \ \frac{D_{50\,filter}}{D_{50\,soil}} \leq 25 \tag{7.36}$$

The criteria may be explained as follows:

1. The 15 percent size (D_{15}) of filter material must be less than 4 times the 85 percent size (D_{85}) of a protected soil. The ratio of D_{15} of a filter to D_{85} of a soil is called the *piping ratio*.

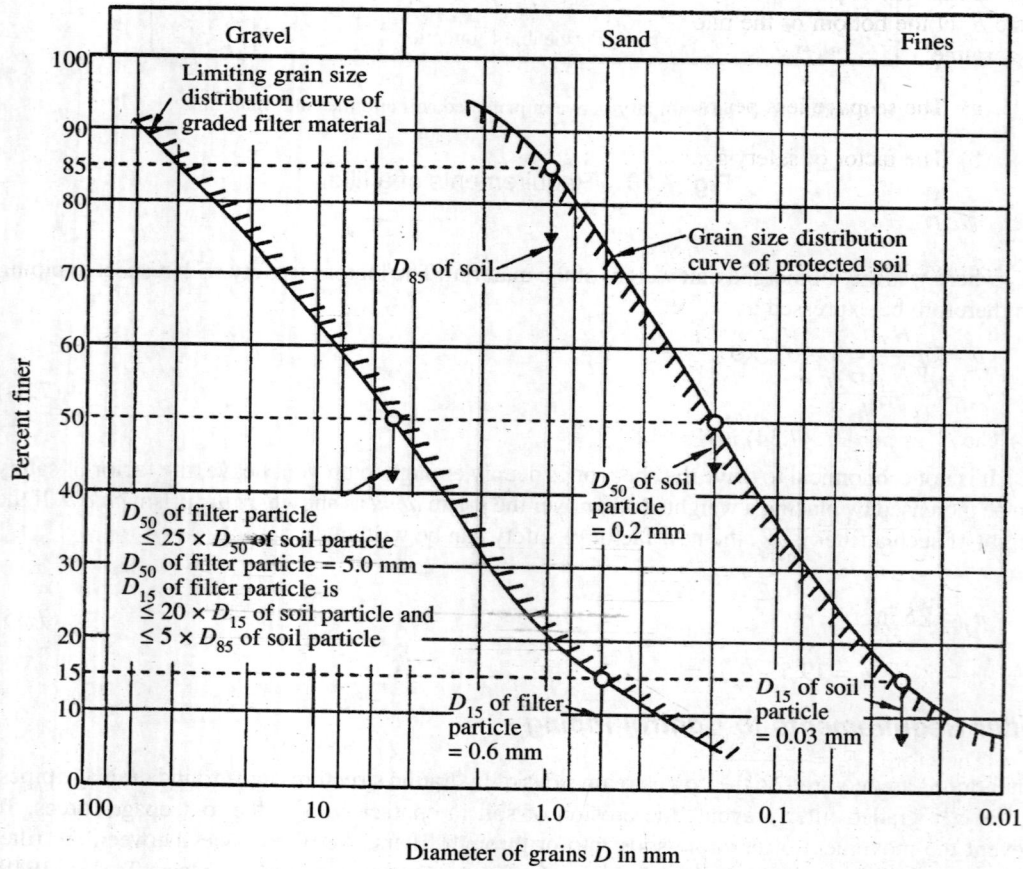

Fig. 7.11 Grain size distribution curves of graded filter and protected materials

2. The 15 percent size (D_{15}) of a filter material should be at least 4 times the 15 percent size (D_{15}) of a protected soil but not more than 20 times of the latter.

3. The 50 percent size (D_{50}) of filter material should be less than 25 times the 50 percent size (D_{50}) of protected soil.

Experience indicates that if the basic filter criteria mentioned above are satisfied in every part of a filter, piping cannot occur under even extremely severe conditions.

A typical grain size distribution curve of a protected soil and the limiting sizes of filter materials for constructing a graded filter is given in Fig. 7.11. The size of filter materials must fall within the two curves and to satisfy the requirements given.

Example 7.6

A sheet pile wall was driven across a river. It retains a head of water of 12.0 m depth and driven 6.0 m depth below river bed. The soil below the river bed is silty sand and extends up to a depth of 12.0 m where it meets an impermeable stratum of clay. Flow net analysis gave $N_f = 6$ and $N_d = 12$. The coefficient of permeability of the sub-soil is $k = 8 \times 10^{-5}$ m /min. The average uplift pressure head h_a at the bottom of the pile is 3.5 m. The saturated unit weight of the soil $\gamma_{sat} = 19.5$ kN/m^3. Determine:

(a) The seepage less per meter length of pile per day.

(b) The factor of safety against heave on the downstream side of the pile.

Solution

(a) Seepage loss,

$$q = kh \frac{N_f}{N_d} = 8 \times 10^{-5} \times 9 \times \frac{6}{12} = 36 \times 10^{-5} \text{ m}^3/\text{min} = 51.84 \times 10^{-2} \text{ m}^3/\text{day}$$

(b) The F_s as per Eq. (7.34) is (Ref. Fig 7.9)

$$F_s = \frac{W_b + W_t}{U} = \frac{D\gamma_b}{h_a \gamma_w}$$

$h_a = 3.5$ m

$\gamma_b = \gamma_{sat} - \gamma_w = 19.5 - 9.81 = 9.69$ kN / m^3

Therefore, $F_s = \dfrac{6 \times 9.69}{3.5 \times 9.81} = 1.69$

7.12   QUESTIONS AND PROBLEMS

7.1   Explain the practical applications of a flow net.

7.2   Prove mathematically that there is the same quantity of seepage flow between any two adjacent flow channels.

7.3   Explain and discuss the merits and demerits of different methods of sketching flow nets.

7.4   What is the function of a drainage gallery in an earth dam? Discuss.

7.5   Explain a graphical method of determining uplift pressures on the base of a weir founded on permeable foundation. Assume that the soil is anisotropic.

7.6   Explain and discuss the properties of conjugate confucal parabolas.

7.7   Explain with neat diagrams a method of locating the phreatic line in a homogeneous earth dam. Assume that there is no toe drain.

7.8   Explain the boundary conditions for drawing flow nets for a homogeneous earth dam.

7.9   Explain the method of designing a graded flow nets for a homogeneous earth dam.

7.10   A homogeneous earth dam is shown in Fig. Prob. 7.10. Sketch the phreatic line and estimate the quantity of seepage. What would be the effect of the seepage if the length of the horizontal drain is 20 m?

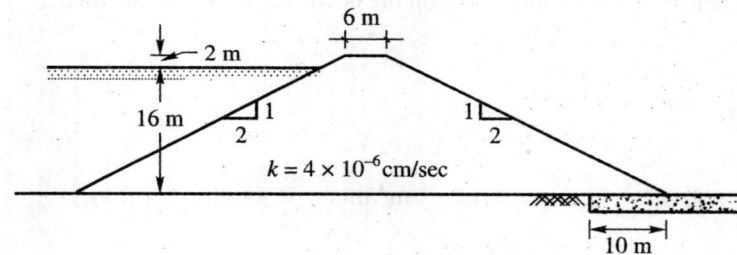

Fig. Prob. 7.10

7.11   Sketch the phreatic line in the Fig. Prob. 7.10 by assuming that there is no drainage gallery. Estimate the seepage loss.

7.12   A graded filter is required to be provided on the downstream side of a hydraulic structure. The protected soil has the following grain size distribution:

ASTM Sieve No.	8	10	18	40	70	120	200
Percent finer	100	85	60	45	35	25	10

Draw the grain size distribution curves of the protected soil and the graded filter on a semilog paper. Discuss the significance of the relationship between the two.

7.13 (a) Show graphically the pressure distribution on the base of the weir given in Fig. Prob. 7.13. If the specific gravity of the masonry is 2.3, is the weir safe against uplift?

(b) Estimate the seepage loss:

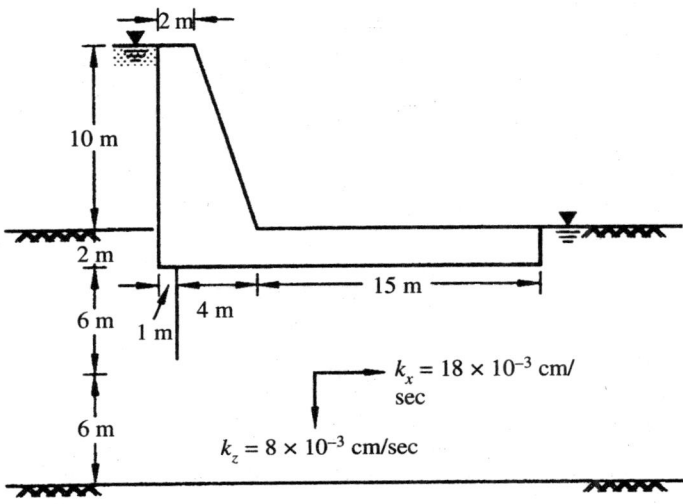

Fig. Prob. 7.13

7.14 The section through a long cofferdam is shown in Fig. Prob. 7.14. The coefficient of permeability of the soil being 4×10^{-7} m/s. Draw the flow net and determine the quantity of seepage entering the Cofferdam. Calculate the seepage pressures at C and D.

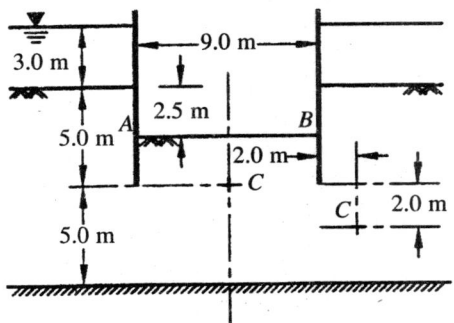

Fig. Prob. 7.14

7.15 The foundation soil at the toe of a masonry dam has a porosity of 41% and a ρ_s of 2.68 Mg/m³. To assume safety against piping, the specifications state that the upward gradient must not exceed 25% of the gradient at which quick condition occurs. What is the maximum permissible upward gradient.

7.16 Figure Prob. 7.16 gives a section of dam. Construct flow net and compute the flow under the dam per metre of dam. The value of $k = 3.5 \times 10^{-4}$ cm/sec.

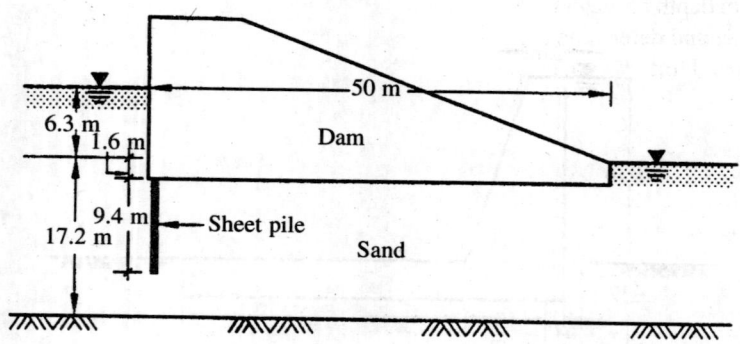

6.3 m
1.6 m
50 m
Dam
9.4 m — Sheet pile
17.2 m
Sand

Fig. Prob. 7.16

7.17 Figure Prob. 7.17 gives sheet pile driven into a permeable stratum. Construct flow net and determine the quantity of seepage per metre length of piling. Assume $k = 8 \times 10^{-4}$ cm/sec.

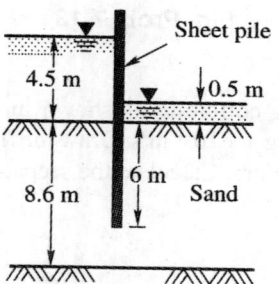

Sheet pile
4.5 m
0.5 m
6 m
8.6 m
Sand

Fig. Prob. 7.17

7.18 Figure Prob. 7.18 gives a cross-section of a dam. The subsoil is anisotropic and has permeabilities $k_h = 2 \times 10^{-6}$ cm/sec and $k_v = 5 \times 10^{-7}$ cm/sec. Find the rate of flow beneath the dam per metre length of the dam. Assume $N_f = 4$, and $N_d = 8$.

7.19 Figure Prob. 7.19 gives the section of a dam founded on permeable strata of 12.5 m thick having a value of $k = 1.3 \times 10^{-4}$ cm/sec. The dam is 90 m long. Construct a flow net and determine the quantity of water that will be lost per day be seepage. In this case, there will be no impervious apron or sheet pile wall. Hint: There will be 5 flow channels and 8 potential drops.

7.20 In order to decrease the seepage loss sheet pile of depth 5.8 m below the ground level on the d/s of the dam was driven at the toe of the dam with an impermeable apron of 6 m wide was constructed on the u/s side as shown by dotted lines in the Fig. Prob. 7.19. Construct flow nets for this case and compute the seepage loss. Hint: Flow channels = 5; potential drops = 14.

7.21   A sheet pile wall is driven to a depth of 6 m into permeable stratum which extends to a depth of 13.5 m below the ground level. Below this there is an impermeable stratum. There is 4.5 m depth of water on one side and no water on the other side. Make a neat sketch of the flow net and determine the seepage loss in litres per day, assuming the value of $k = 6 \times 10^{-4}$ cm/sec. Hint: $N_f = 5$, and $N_d = 10$.

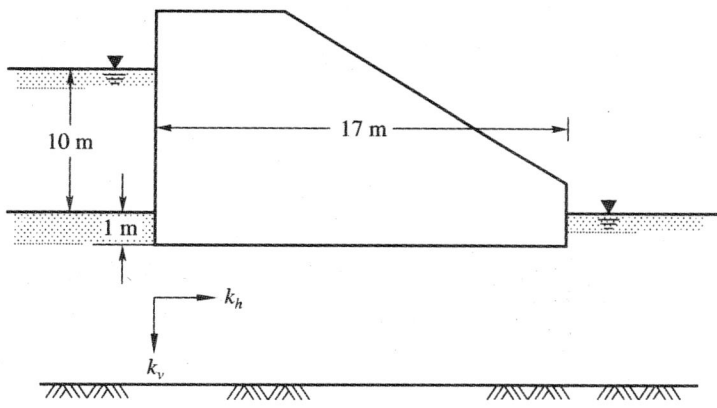

Fig. Prob. 7.18

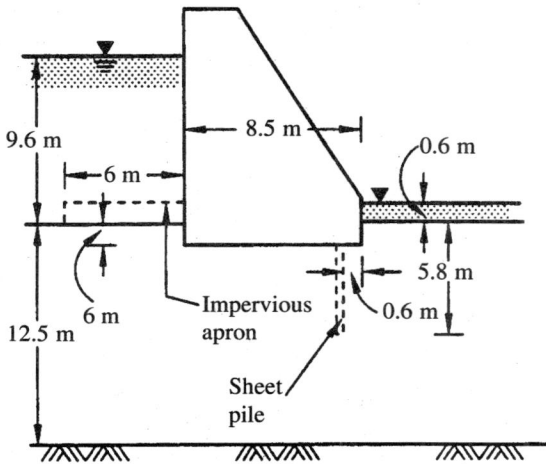

Fig. Prob. 7.19

CHAPTER 8

EFFECTIVE AND POREWATER PRESSURES

8.1   DEVELOPMENT OF EQUATIONS

The pressure transmitted through grain to grain at the contact points through a soil mass is termed as *intergranular* or *effective pressure*. It is known as effective pressure since this pressure is responsible for the decrease in the void ratio or increase in the frictional resistance of a soil mass.

If the pores of a soil mass are filled with water and if a pressure induced into the pore water, tries to separate the grains, this pressure is termed as *pore water pressure* or *neutral stress*. The effect of this pressure is to increase the volume or decrease the frictional resistance of the soil mass.

The effects of the intergranular and pore water pressures on a soil mass can be illustrated by means of simple practical examples.

Consider a rigid cylindrical mold, Fig. 8.1(a), in which dry sand is placed. Assume that there is no side friction. Load Q is applied at the surface of the soil through a piston. The load applied at the surface is transferred to the soil grains in the mold through their points of contact. If the load is quite considerable, it would result in the compression of the soil mass in the mould. The compression might be partly due to the elastic compression of the grains at their points of contact and partly due to relative sliding between particles. If the sectional area of the cylinder is A, the average stress at any level XY may be written as

$$\sigma_a = \frac{Q}{A} \tag{8.1}$$

The stress σ_a is the average stress and not the actual stress prevailing at the grain to graincontacts which is generally very high. Any plane such as XY will not pass through all the points of contact and many of the grains are cut by the plane as shown in Fig. 8.1(b). The actual points of contact exhibit a wavy form. However, for all practical purposes the average stress is considered. Since this stress is responsible for the deformation of the soil mass, it is termed the *intergranular* or *effective stress*. We may therefore write,

$$\sigma_a = \sigma' \tag{8.2}$$

where σ' is the *effective stress*.

Consider now another experiment. Let the soil in the mould Fig. 8.1(a) be fully saturated and made completely watertight. If the same load Q is placed on the piston, this load will not be transmitted to the soil grains as in the earlier case. If we assume that water is incompressible, the external load Q will be transmitted to the water in the pores. This pressure that is developed in the water is called the *pore water* or *neutral stress* u_w as shown schematically in Fig. 8.1(c). This pore water pressure u_w prevents the compression of the soil mass. The value of this pressure is

$$u_w = \frac{Q}{A} \tag{8.3}$$

If the valve V provided in the piston is opened, immediately there will be expulsion of water through the hole in the piston. The flow of water continues for some time and then stops.

The expulsion of water from the pores decreases the pore water pressure and correspondingly increases the intergranular pressure. At any stage the total pressure Q/A is divided between water

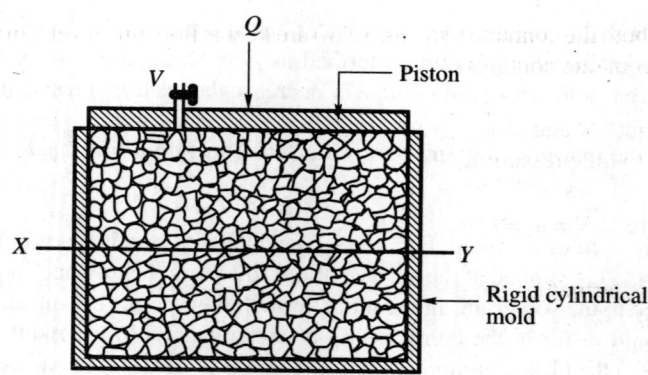

(a) Soil under load in a rigid container

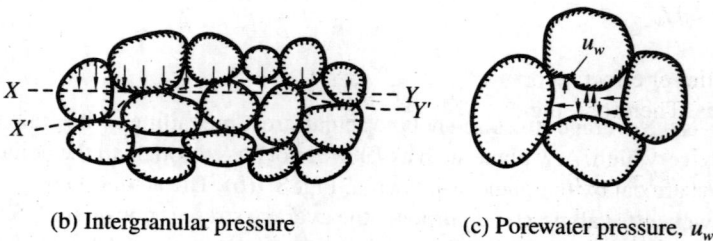

(b) Intergranular pressure   (c) Porewater pressure, u_w

Fig. 8.1   Effective and porewater pressures

and the points of contact of grains. A new equation may therefore be written as

Total pressure $\sigma_t = \dfrac{Q}{A} =$ Intergranular pressure + Porewater pressure

or $\sigma_t = \sigma' + u_w$ (8.4)

Final equilibrium will be reached when there is no expulsion of water. At this stage the pore water pressure $u_w = 0$. All the pressure will be carried by the soil grains. Therefore, we can write,

$\sigma_t = \sigma'$ (8.5)

The porewater pressure u_w can be induced in the pores of a soil mass by a head of water over it. When there is no flow of water through the pores of the mass, the intergranular pressure remains constant at any level. But if there is flow, the intergranular pressure increases or decreases according to the direction of flow. This aspect of the problem may be discussed with some simple illustrations as given below.

In Fig. 8.2 the container A is filled with sand upto a depth z_1 and water up to a depth z_2 above the sand surface. A flexible tube C connects the bottom of the container A to another container B. The water levels are kept constant in these two containers.

Case 1.   *When no flow takes place through the soil*

The water surfaces in both the containers in Fig. 8.2(a) are kept at the same level. Under this condition, no flow takes place from one container to another.

Consider two points M and N as shown in the figure on a horizontal plane. The water pressure at M should be equal to the pressure at N according to the laws of hydraulics. Therefore,

the water pressure at $N = u_z = (z + z_2)\, \gamma_w$ (8.6)

The pressure u_z is termed as the pore water pressure acting on the grains at depth z from the surface of the sample.

However, the total pressure at point N is due to the water head plus the weight of the submerged soil above N. If γ_b is the submerged unit weight of the soil, the total pressure at N is

$\sigma_z = z\gamma_b + (z + z_2)\gamma_w$ (8.7)

The intergranular or effective pressure at the point N is the difference between the total and the pore water pressures. Therefore, the effective pressure σ_z' is

$\sigma_z' = \sigma_z - u_z = z\gamma_b + (z + z_2)\gamma_w - (z + z_2)\gamma_w = z\gamma_b$ (8.8a)

Equation (8.8a) clearly demonstrates that the effective pressure σ_z' is independent of the depth of water z_2 above the submerged soil surface. The total pore water and effective pressures at the bottom of the soil sample are as follows:

Total pressure $\sigma_t = \sigma_B = (z_1 + z_2)\gamma_w + z_1\gamma_b$ (8.8b)

Porewater pressure $u_B = (z_1 + z_2)\gamma_w$ (8.8c)

Effective pressure $\sigma_B' = (\sigma_B - u_B) = z_1\gamma_b$ (8.8d)

The stress diagrams are shown in Fig. 8.2(a).

Case 2.   *When flow takes place through the soil from top to bottom*

In Fig. 8.2(b) the water surface in container B is kept at h units below the surface in A. This difference in head permits water to flow from container A to B.

Since container B with the flexible tube can be considered as a piezometer tube freely communicating with the bottom of container A, the piezometric head or the porewater pressure head at the bottom of container A is $(z_1 + z_2 - h)$. Therefore, the porewater pressure u_B at the bottom level is

$$u_B = (z_1 + z_2 - h)\gamma_w \qquad (8.9)$$

As per Fig. 8.2(a), the pore water pressure at the bottom of container A when no flow takes place through the soil sample is

$$u_B = (z_1 + z_2)\gamma_w \qquad (8.10)$$

It is clear from Eqs (8.9) and (8.10) that there is a decrease in pore water pressure to the extent of $h\gamma_w$ when water flows through the soil sample from top to bottom. It may be understood that this decrease in pore water pressure is not due to velocity of the flowing water. The value of the velocity head $V^2/2g$ is a negligible quantity even if we take the highest velocity of flow that is encountered in natural soil deposits. For example, a velocity of 0.3 in per sec, which is an enormous value as compared to typical seepage velocities, gives a velocity head of only 0.45 cm which may be disregarded in all seepage problems. As in Fig. 8.2(a), the total pressure σ_B at the bottom of the container in this case also remains the same. Therefore,

$$\sigma_B = z\gamma_b + (z_1 + z_2)\gamma_w + z_1\gamma_b \qquad (8.11)$$

The effective pressure σ_B' at the bottom of the container is

$$\sigma_B' = (\sigma_B - u_B) = (z_1 + z_2)\gamma_w + z_1\gamma_b - (z_1 + z_2 - h)\gamma_w = z_1\gamma_b + h\gamma_w \qquad (8.12)$$

Equation (8.12) indicates that in case 2 there is an increase in the effective pressure by $h\gamma_w$ at the bottom of the container A as compared to Case 1. The effective pressure at the top surface of the sample is zero as before. Therefore, the effective pressure σ_z' at any depth z can be written as

$$\sigma_z' = \sigma_B' \frac{z}{z_1} = (z_1\gamma_b + h\gamma_w)\frac{z}{z_1} = z\gamma_b + \frac{hz\gamma_w}{z_1} \qquad (8.13)$$

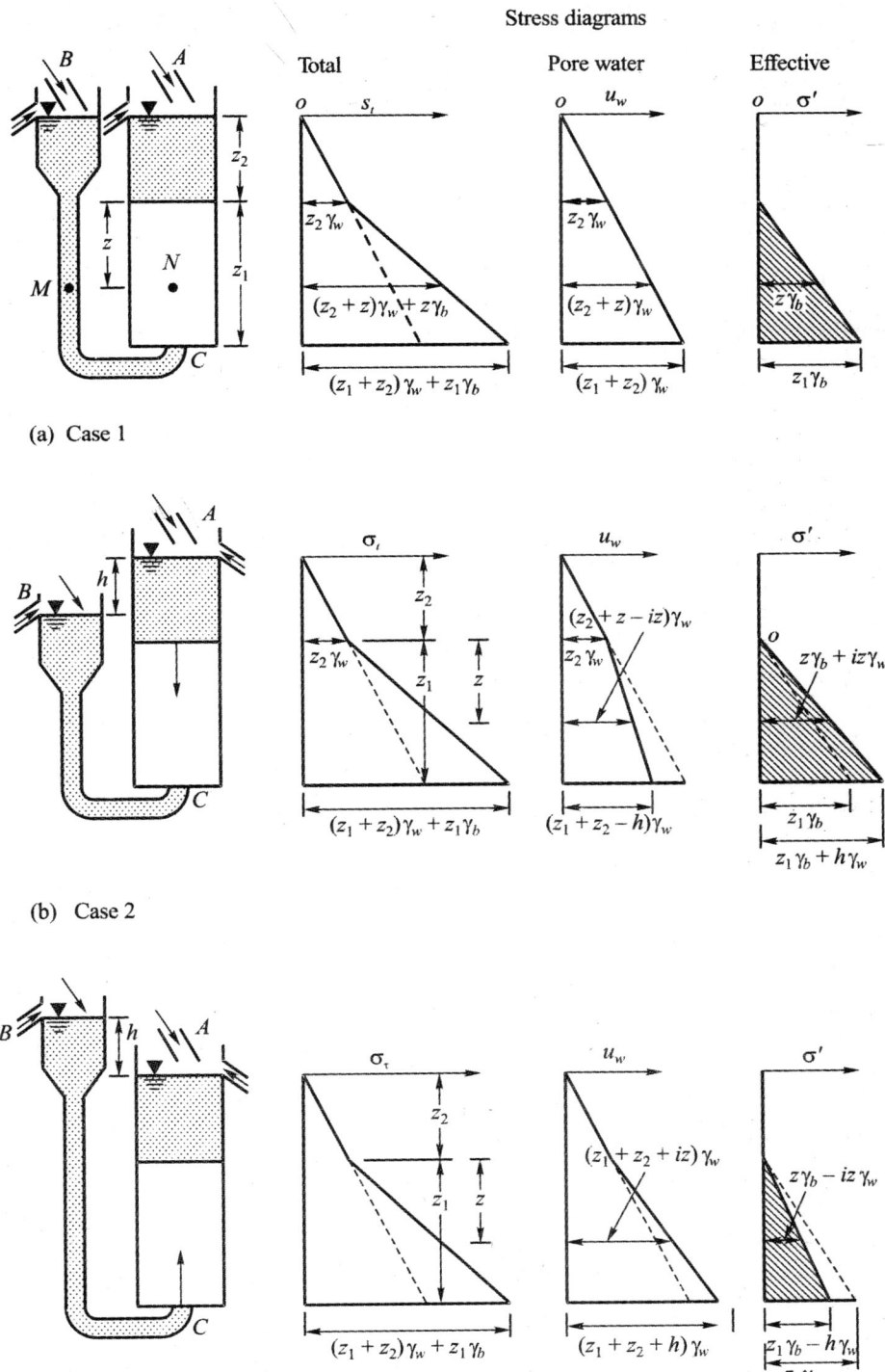

Stress diagrams

(a) Case 1

(b) Case 2

(c) Case 3

Fig. 8.2 Concept of total pressure, pore pressure and effective pressure under different conditions of flow

Equation (8.13) indicates that $hz\gamma_w/z_1$ is the increase in the effective pressure as the water flows from the surface to a depth z. This increase in effective pressure due to the flow of water through the pores of the soil is known as *seepage pressure*. It may be noted that h is the total loss of head as the water flows from the top surface of the sample to a depth z_1.

The corresponding loss of head at depth z is $(z/z_1)h$. Since $(h/z_1) = i$, the *hydraulic gradient*, the loss of head at depth z can be expressed as iz. Therefore, the seepage pressure at any depth may be expressed as $iz\gamma_w$. The effective pressure at depth z can be written as

$$\sigma_z' = z\gamma_b + iz\gamma_b \tag{8.14}$$

The distribution of porewater and effective pressures are shown in Fig. 8.2(b). In normal soil deposits when flow takes place in the direction of gravity there will be an increase in the effective pressure.

Case 3. *When flow takes place through the soil from bottom to top*

In Fig. 8.2(c), the water surface in container B is kept above that of A by h units. This arrangement permits water to flow upwards through the sample in container A. The total piezometric or the pore water head at the bottom of the sample is given by

$$(z_1 + z_2 + h)$$

Therefore, the porewater pressure u_B at the bottom of the sample is

$$u_B = (z_1 + z_2 + h)\gamma_w \tag{8.15}$$

As before the total pressure head σ_B at the bottom of the sample is

$$\sigma_B = (z_1 + z_2)\gamma_w + z_1\gamma_b \tag{8.16}$$

The effective pressure σ_B' at the bottom of sample is, therefore,

$$\sigma_B' = (\sigma_B - u_B) = (z_1 + z_2)\gamma_w + z_1\gamma_b - (z_1 + z_2 + h)\gamma_w = z_1\gamma_b - h\gamma_w \tag{8.17}$$

As in Eq. (8.14) the effective pressure at any depth z can be written as

$$\sigma_z' = z\gamma_b - iz\gamma_w \tag{8.18}$$

Equation (8.18) indicates that there is a decrease in the effective pressure due to upward flow of water. At any depth z, $z\gamma_b$, is the pressure of the submerged soil acting downward and $iz\gamma_b$ is the seepage pressure acting upward. The effective pressure σ_z', reduces to zero when these two pressures balance. This happens when

$$\sigma_z' = z\gamma_b + iz\gamma_w = 0 \text{ or } i = i_c = \frac{\gamma_b}{\gamma_w} \tag{8.19}$$

Equation (8.19) indicates that the effective pressure reduces to zero when the hydraulic gradient attains a maximum value which is equal to the ratio of the submerged unit weight of soil and the unit weight of water. This gradient is known as the *critical hydraulic gradient ic*. In such cases, cohesionless soils lose all of their shear strength and bearing capacity and a visible agitation of soil grains is observed. This phenomenon is known as *boiling* or a *quick sand condition*. By substituting in Eq. (8.19) for γ_b

$$\gamma_b = \frac{\gamma_w(G-1)}{1+e}, \quad \text{we have}$$

$$i_c = \frac{(G-1)}{1+e} \tag{8.20}$$

The critical gradient of natural granular soil deposits can be calculated if the void ratios of the deposits are known. For all practical purposes the specific gravity of granular materials can be assumed as equal to 2.65. Table 8.1 gives the critical gradients of granular soils at different void ratios ranging from 0.5 to 1.0.

TABLE 8.1

Critical gradients of granular

Sl. No.	Void ratio	i_c
1	0.5	1.10
2	0.6	1.03
3	0.7	0.94
4	0.8	0.92
5	1.0	0.83

It can be seen from Table 8.1 that the critical gradient decreases from 1.10 by about 25 percent only as the void ratio increases by 100 percent from an initial value of 0.5 to 1.0. The void ratio of granular deposits generally lies within the range of 0.6 to 0.7 and as such a critical gradient of unity can justifiably be assumed for all practical purposes. It should be remembered that a quick condition does not occur in clay deposits since the cohesive forces between the grains prevent the soil from boiling.

Quick conditions are common in excavations below the ground water table. This can be prevented by lowering the ground water elevation by pumping before excavation. Quick conditions occur most often in fine sands or silts and cannot occur in coarse soils. The larger the particle size, the greater is the porosity. To maintain a critical gradient of unity, the velocity at which water must be supplied at the point of inflow varies as the permeability. Therefore a quick condition cannot occur in a coarse soil unless a large quantity of water can be supplied.

8.2 EXAMPLES

Example 8.1

The depth of water in a well is 3 m. Below the bottom of the well lies a layer of sand 5 meters thick overlying a clay deposit. The specific gravity of the solids of sand and clay are respectively 2.64

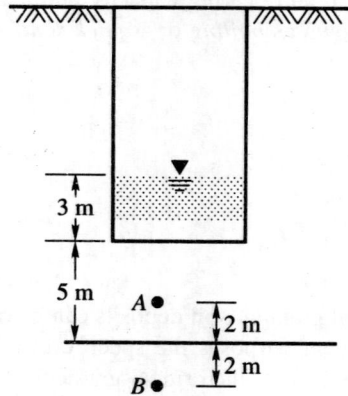

Fig. Ex. 8.1

and 2.70. Their water contents are respectively 25 and 20 percent. Compute the total, intergranular and pore water pressures at points A and B shown in Fig. Ex. 8.1.

Solution

The formula for the submerged unit weight is

$$\gamma_b = \frac{\gamma_w(G-1)}{1+e}$$

Since the soil is saturated,

$$e = wG, \gamma_b = \frac{\gamma_w(G-1)}{1+wG}$$

For sand, $\gamma_b = \dfrac{9.81(2.64-1)}{1+0.25\times2.64} = 9.7\ \text{kN/m}^3$

For clay, $\gamma_b = \dfrac{9.81(2.70-1)}{1+0.20\times2.70} = 10.8\ \text{kN/m}^3$

Pressure at point A

(i)   Total pressure $= 3 \times 9.7$ (sand) $+ 6 \times 9.81 = 29.1 + 58.9 = 88\ \text{kN/m}^2$

(ii)   Effective pressure $= 3 \times 9.7 = 29.1\ \text{kN/m}^2$

(iii) Porewater pressure $= 6 \times 9.81 = 58.9\ \text{kN/m}^2$

Pressure at point B

(i)   Total pressure $= 5 \times 9.7 + 2 \times 10.83 + 10 \times 9.81 = 168.3\ \text{kN/m}^2$

(ii) Intergranular pressure = $5 \times 9.7 + 2 \times 10.83 = 70.2$ kN/m²

(iii) Porewater pressure = $10 \times 9.81 = 98.1$ kN/m²

Example 8.2

If water in the well in example 8.1 is pumped out up to the bottom of the well, estimate the change in the pressures at points A and B given in Fig. Ex. 8.1.

Solution

Change in pressure at points A and B

(i) Change in total pressure = decrease in water pressure due to pumping

$$= 3 \times 9.81 = 29.43 \text{ kN/m}^2$$

(ii) Change in effective pressure = 0

(iii) Change in porewater pressure = decrease in water pressure due to pumping

$$= 3 \times 9.81 = 29.43 \text{ kN/m}^2$$

Example 8.3

A trench is excavated in fine sand for a building foundation, up to a depth of 4 m. The excavation was carried out by providing the necessary side supports for pumping water. The water levels at the sides and the bottom of the trench are as given Fig. Ex. 8.3. Examine whether the bottom of the trench is subjected to a quick condition if $G = 2.64$ and $e = 0.7$. If so, what is the remedy?

Solution

As per Fig. Ex. 8.3 the depth of the water table above the bottom of the trench = 3 m. The sheeting is taken 2 m below the bottom of the trench to increase the seepage path.

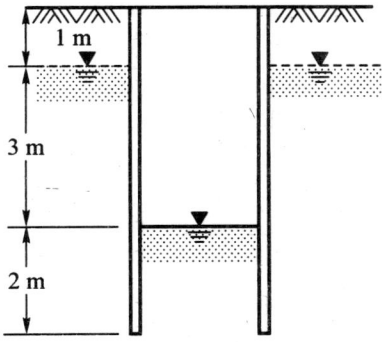

Fig. Ex. 8.3

The equation for the critical gradient is $i_c = \dfrac{(G-1)}{1+e}$

If the trench is to be stable, the hydraulic gradient, i, prevailing at the bottom should be less than i_c. The hydraulic gradient i is

$$i = \frac{h}{L},$$

There will be no quick condition if,

$$\frac{h}{L} < \frac{G-1}{1+e}$$

From the given data

$$i_c = \frac{2.64-1}{1+0.7} = \frac{1.64}{1.7} = 0.96$$

$$\frac{h}{L} = \frac{3}{2} = 1.50$$

It is obvious that $h/L > i_c$. There will be quick condition.

Remedy:

(i) Increase L to at least a 4 m depth below the bottom of trench so that $h/L = 0.75$ which gives a margin of factor of safety.

(ii) Keep the water table outside the trench at a low level by pumping out water. This reduces the head h.

(iii) Do not pump water up to the bottom level of the trench. Arrange the work in such a way that the work may be carried out with some water in the trench.

Any suggestion given above should be considered by keeping in view the site conditions and other practical considerations.

Example 8.4

A clay layer 3.66 m thick rests beneath a deposit of submerged sand 7.92 m thick. The top of the sand is located 3.05 m below the surface of a lake. The saturated unit weight of the sand is 19.62 kN/m³ and of the clay is 18.36 kN/m³.

Compute (a) the total vertical pressure, (b) the pore water pressure, and (c) the effective vertical pressure at mid height of the clay layer (Refer to Fig. Ex. 8.4).

Solution

(a) Total pressure

The total pressure σ_t over the midpoint of the clay is due to the saturated weights of clay and sand layers plus the weight of water over the bed of sand, that is

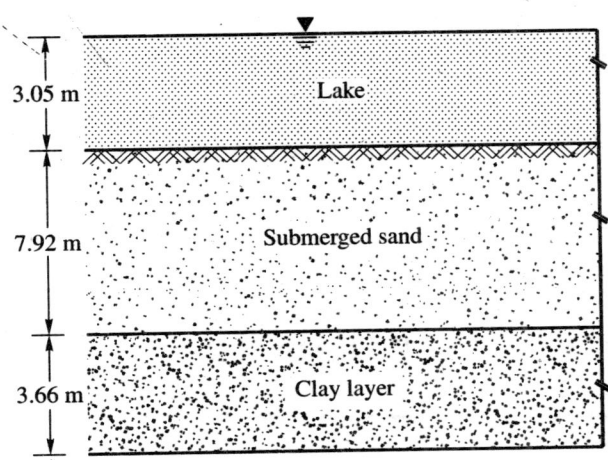

Fig. Ex. 8.4

$$\sigma_t = \frac{3.66}{2} \times 18.36 + 7.92 \times 19.62 + 3.05 \times 9.81 = 33.6 + 155.4 + 29.9 = 218.9 \text{ kN/m}^2$$

(b) Porewater pressure is due to the total water column above the midpoint.
That is

$$u_w = \frac{3.66}{2} \times 9.81 + 7.92 \times 9.81 + 3.05 \times 9.81 = 125.6 \text{ kN/m}^2$$

(c) Effective vertical pressure

$$\sigma_t - u_w = \sigma' = 218.9 - 125.6 = 93.3 \text{ kN/m}^2$$

Example 8.5

The surface of a saturated clay deposit is located permanently below a body of water. Laboratory tests have indicated that the average natural water content of the clay is 47% and that the specific gravity of the solid matter is 2.74. What is the vertical effective pressure at a depth of 11.28 m below the top of the clay.

Solution

To find the effective pressure, we have to find first the submerged unit weight of soil expressed as

$$\gamma_b = \frac{(G-1)\gamma_w}{1+e}$$

Now from Eq. (3.13), $e = \dfrac{wG}{S} = wG$

or $e = 0.47 \times 2.74 = 1.29$

Therefore,

$$\gamma_b = \frac{(2.74-1.00)\times 9.81}{1+1.29} = 7.45 \text{ kN/m}^3$$

Effective pressure, $\sigma' = 11.28 \times 7.45 = 84.1 \text{ kN/m}^2$

Example 8.6

If the water level in Ex. 8.5 remains unchanged and an excavation is made by dredging, what depth of clay must be removed to reduce the effective pressure at point A at a depth of 11.28 m by 48 kN/m²?

Solution

As in Ex. 8.5, $\gamma_b = 7.45 \text{ kN/m}^3$, let the depth of excavation be D. The effective depth over the point is $(11.28 - D)$ m. The depth of D must be such which gives an effective pressure of $84.1 - 48.0 = 36.1 \text{ kN/m}^2$

or $(11.28 - D) \times 7.45 = 36.1$

or $D = \dfrac{11.28 \times 7.45 - 36.1}{7.45} = 6.43$ m

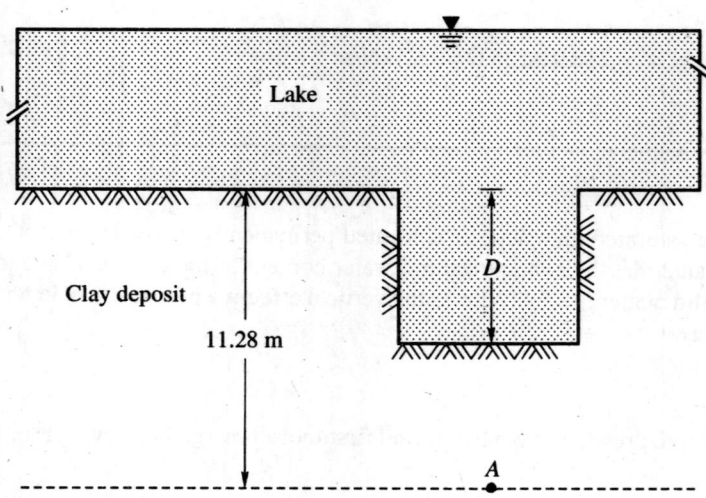

Fig. Ex. 8.6

Example 8.7

The water table is lowered from a depth of 3.05 m to a depth of 6.10 m in a deposit of silt. All the silt is saturated even after the water table is lowered. Its water content is 26%. Estimate the increase in the effective pressure at a depth of 10.4 m on account of lowering the water table. Assume $G = 2.7$.

Solution

Effective pressure before lowering the water table

The water table is at a depth of 3.05 m and the soil above this depth remains saturated but not submerged. The soil from 3.05 m to 10.4 m remains submerged. Therefore, the effective pressure at 10.4 m depth is

$$\sigma_1' = 3.05\gamma_{sat} + (10.40 - 3.05)\gamma_b$$

Now, $\gamma_{sat} = \dfrac{\gamma_w(G+e)}{1+e}$ $\gamma_b = \dfrac{\gamma_w(G-1)}{1-e}$ $\gamma_w = 9.81$ kN/m², $e = wG$ for $S = 1$

Therefore, $e = 0.26 \times 2.7 = 0.70$

$$\gamma_{sat} = \frac{9.81(2.7+0.7)}{1+0.7} = 19.62 \text{ kN/m}^3$$

$$\gamma_b = \frac{9.81(2.7-1)}{1+0.7} = 9.81 \text{ kN/m}^3$$

$$\sigma_1' = 3.05 \times 19.62 + 7.35 \times 9.81 = 131.90 \text{ kN/m}^2$$

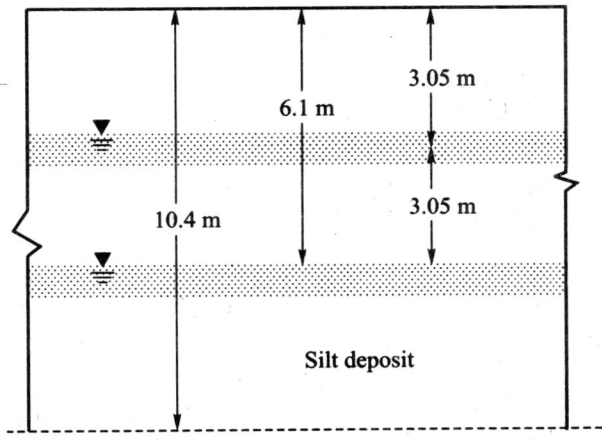

Silt deposit

Fig. Ex. 8.7

Effective pressure after lowering of water table

After lowering the water table to a depth of 6.1 m, the soil above this level remains saturated but effective and below this submerged. Therefore, the altered effective pressure is

$$\sigma_2' = 6.1\gamma_{sat} + (10.4 - 6.1)\gamma_b = 6.1 \times 19.62 + 473 \times 9.81 = 161.9 \, kN/m^2$$

The increase in the effective pressure is

$$\sigma_2' - \sigma_1' = \Delta\sigma' = 161.9 - 131.90 = 30 \, kN/m^2$$

Example 8.8

Compute the critical hydraulic gradients for the following materials: (a) Coarse gravel, $k = 10$ cm/sec, $G = 2.67$, $e = 0.65$ (b) Sandy silt, $k = 10^{-6}$ cm/sec, $G = 2.67$, $e = 0.80$.

Solution

As per Eq. (8.20), the critical gradient i_c may be expressed as

$$i_c = \frac{(G-1)}{1+e}$$

(a) Coarse gravel

$$i = \frac{2.67 - 1}{1 + 0.65} = 10.1$$

(b) Sandy silt

$$i = \frac{2.67 - 1}{1 + 0.80} = 0.93$$

Example 8.9

A large excavation is made in a stiff clay whose saturated unit weight is 17.27 kN/m³. When the depth of excavation reaches 7.5 m, cracks appear and water begins to flow upward to bring sand to the surface. Subsequent borings indicate that the clay is underlain by sand at a depth of 11 m below the original ground surface.

What is the depth of the water table outside the excavation below the original ground level?

Solution

Making an excavation in the clay creates a hydraulic gradient between the top of the sand layer and the bottom of the excavation. As a consequence, water starts seeping in an upward direction from the sand layer towards the excavated floor. Because the clay has a very low permeability, flow equilibrium can only be reached after a long period of time. The solution must be considered over a short time interval.

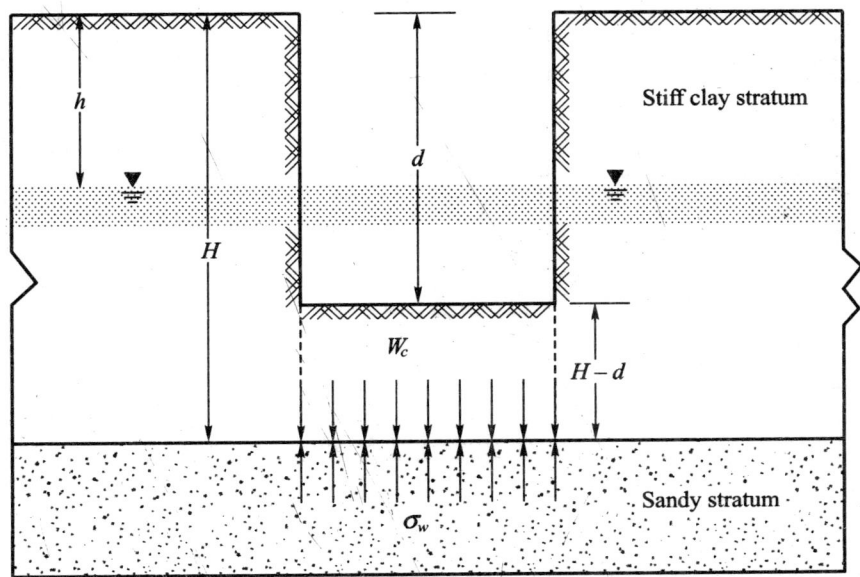

Fig. Ex. 8.9

The floor of the excavation at depth d is stable only if the water pressure σ_w at the top of the sand layer at a depth of 11 m is counter balanced by the saturated weight W per unit area of the clay above it disregarding the shear strength of the clay.

Let H = total thickness of clay layer = 11 m, d = depth of excavation in clay = 7.5 m, h = depth of water table from ground surface, γ_{sat} = saturated unit weight of the clay.

Let $(H - d) = 11 - 7.5 = 3.5$ m, the thickness of clay strata below the bottom of the trench.

$$W_c = \gamma_{sat} = (h - d) = 17.27 \times 3.5 = 60.24 \text{ kN/m}^2$$

$$\sigma_w = \gamma_w (H - h) = 9.81(11 - h) \text{ kN/m}^2$$

cracks may develop when $W_c = \sigma_w$

or $60.24 = 9.81(11 - h)$ or $h = 11 - \dfrac{60.24}{9.81} = 4.84$ m

---

Example 8.10

The water table is located at a depth of 3.0 m below the ground surface in a deposit of sand 11.0 m thick (Fig. Ex. 8.10). The sand is saturated above the water table. The total unit weight of the sand is 20 kN/m³. Calculate the (a) the total pressure, (b) the pore water pressure and (c) the effective pressure at depths 0, 3.0, 7.0, and 11.0 m from the ground surface, and draw the pressure distribution diagram.

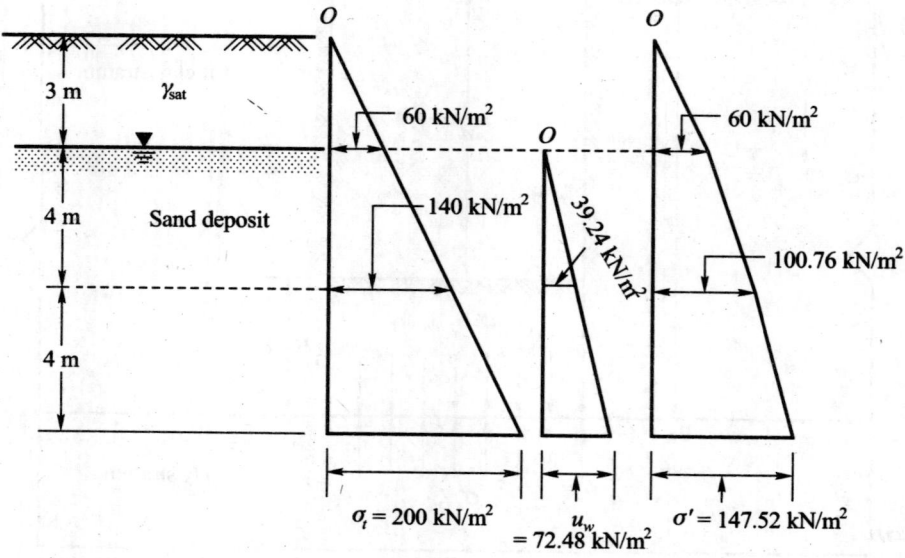

Fig. Ex. 8.10

Solution

$$\gamma_{sat} = 20 \text{ kN/m}^3, \ \gamma_b = 20 - 9.81 = 10.19 \text{ kN/m}^3$$

Depth (m)	Total pressure σ_t (kN/m²)	Pore water pressure u_w (kN/m²)	Effective pressure σ' (kN/m²)
0	0	0	0
3	3 × 20 = 60	0	60
7	7 × 20 = 140.00	4 × 9.81 = 39.24	100.76
11	11 × 20 = 220.00	8 × 9.81 = 78.48	141.52

The pressure distribution σ_t, u_w and σ' are given in Fig. Ex. 8.10.

Example 8.11

A clay stratum 8.0 m thick is located at a depth of 6 m from the ground surface. The natural moisture content of the clay is 56% and $G = 2.75$. The soil stratum between the ground surface and the clay consists of fine sand. The water table is located at a depth of 2 m below the ground surface. The submerged unit weight of fine sand is 10.5 kN/m³, and its moist unit weight above the water table is 18.68 kN/m³. Calculate the effective stress at the center of the clay layer.

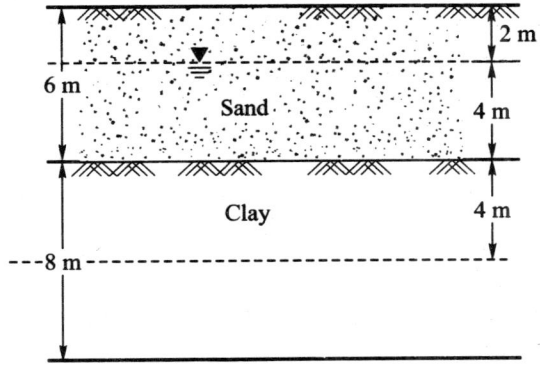

Fig. Ex. 8.11

Solution

Fine sand

Above water table: $\gamma_t = 18.68$ kN/m³

Below water table: $\gamma_b = 10.5$ kN/m³

$$\gamma_{sat} = 10.5 + 9.81 = 20.31 \text{ kN/m}^3$$

Clay stratum

For $S = 1.0$,

$$e = wG = 0.56 \times 2.75 = 1.54$$

$$\gamma_{sat} = \frac{\gamma_w(G+e)}{1+e} = \frac{9.81(2.75+1.54)}{1+1.54} = 16.57 \text{ kN/m}^3$$

$\gamma_b = 16.57 - 9.81 = 6.76$ kN/m³

At a depth 10.0 m from GL

$\sigma_t = 2 \times 18.68 + 4 \times 20.31 + 4 \times 16.57$

$ = 37.36 + 81.24 + 66.28 = 184.88$ kN/m²

$u_w = 4 \times 9.81 + 4 \times 9.81 = 39.24 + 39.24 = 78.48$ kN/m²

Effective stress, $\sigma' = \sigma_t - u_w = 184.88 - 78.44 = 106.40$ kN/m²

Example 8.12

A 12 m thick layer of relatively impervious saturated clay lies over a gravel aquifer. Piezometer tubes introduced to the gravel layer show an artesian pressure condition with the water level standing in the tubes 3 m above the top surface of the clay stratum. The properties of the clay are $e = 1.2$ and $G = 2.7$.

Determine (a) the effective stress at the top of the gravel strata, (b) the depth of excavation that can be made in the clay stratum without bottom heave.

Solution

(a) At the top of the gravel stratum

$$\sigma_t = 12 \times 17.39 = 208.68 \text{ kN/m}^2$$

The pore water pressure at the top of gravel strata

$$u_w = 9.81 \times 15 = 147.15 \text{ kN/m}^2$$

The effective stress at the top of gravel strata

$$\sigma' = \sigma_t - u_w = 208.68 - 147.15 = 61.53 \text{ kN/m}^2$$

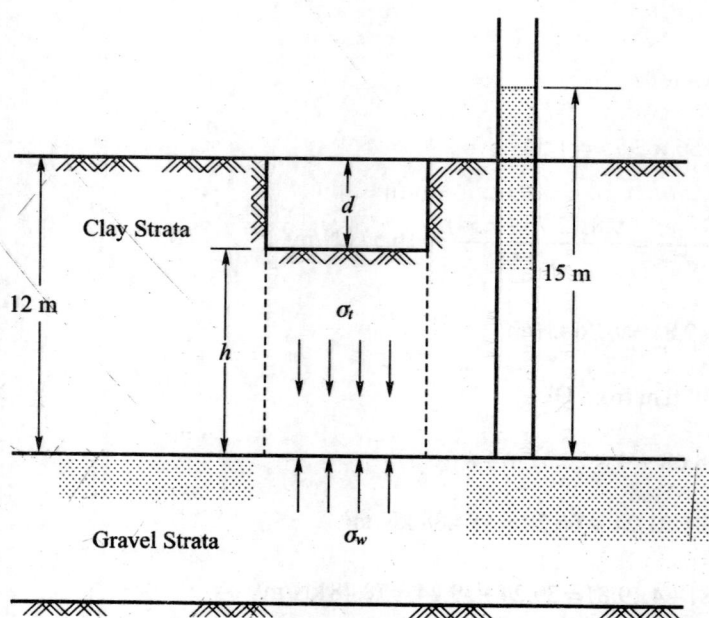

Fig. Ex. 8.12

(b) If an excavation is made into the clay stratum as shown in Fig. Ex. 8.12, the depth must be such that

$$\sigma_t = u_w$$

Let the bottom of the excavation be h m above the top of gravel layer. Now the downward pressure acting at the top of the gravel layer is

$$\sigma_t = \gamma h = 17.39h \text{ kN/m}^2$$

$$u_w = 147.15 \text{ kN/m}^2$$

Now, $17.39h = 147.15$

or $\qquad h = \dfrac{147.15}{17.39} = 8.46 \text{ m} = 8.5 \text{ m}$

Depth of excavation, $d \neq 12 - 9.30 = 2.7 \text{ m}$

This is just the depth of excavation with a factor of safety $F = 1.0$. If we assume a minimum $F_s = 1.10$

$$h = \dfrac{147.15 \times 1.1}{17.39} = 9.3 \text{ m}$$

Depth of excavation $= 12 - 9.30 = 2.7 \text{ m}$

8.3 QUESTIONS AND PROBLEMS

8.1 The depth of water in a lake is 3 m. The soil properties as obtained from soil exploration below the bed of the lake are as given below.

Depth from bed of lake (m)	Type of soil	Void ratio e	Sp. gr. G
0–4	Clay	0.9	2.70
4–9	Sand	0.75	2.64
9–15	Clay	0.60	2.70

Calculate the following pressures at a depth of 12 m below the bed level of the lake.

(i) The total pressure, (ii) The porewater pressure and (iii) The intergranular pressure.

8.2 The water table in a certain deposit of soil is at a depth of 2 m below the ground surface. The soil consists of clay up to a depth of 4 m from the ground and below which lies sand. The clay stratum is saturated above the water table.

Given: Clay stratum: $w = 30$ percent, $G = 2.72$; Sandy stratum: $w = 26$ percent, $Gs = 2.64$.

Required

(i)   The total pressure, pore pressure and effective pressure at a depth of 8 m below the ground surface.

(ii)   The change in the effective pressure if the water table is brought down to a level of 4 m below the ground surface by pumping.

8.3   Water flows from container B to A as shown in Fig. 8.2(c). The piezometric head at the bottom of container A is 2.5 m and the depth of water above the sand deposit is 0.25 m. Assuming the depth of the sand deposit is 1.40 m, compute the effective pressure at the middle of the sand deposit. Assume $e = 0.65$ and $G = 2.64$ for the sand. Sketch the pressure distribution with respect to depth in container A.

8.4   In order to excavate a trench for the foundation of a structure, the water table level was lowered from a depth of 1.20 metre to a depth of 4.5 m in a silty sand deposit. Assuming that the soil above the water table remained saturated at a moisture content of 28 percent, estimate the following:

(i)   The increase in effective stress at a depth of 5 metre.

(ii)   The effective stress at a depth of 4 metre assuming the specific gravity of the solids as 2.68.

8.5   Soil is placed in the containers shown in Fig. Prob. 8.5. The saturated unit weight of soil is 20 kN/m³. Calculate the pore pressure, and the effective stress at elevation A, when (a) the water table is at elevation A, and (b) when the water table rises to E1.B.

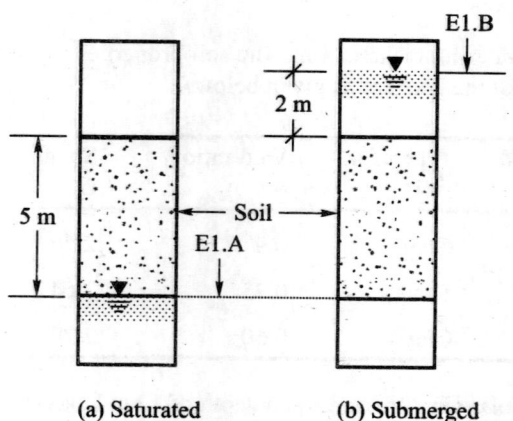

(a) Saturated   (b) Submerged

Fig. Prob. 8.5

8.6 Figure Prob. 8.6 gives a soil profile. Calculate the total and effective stresses at point A.

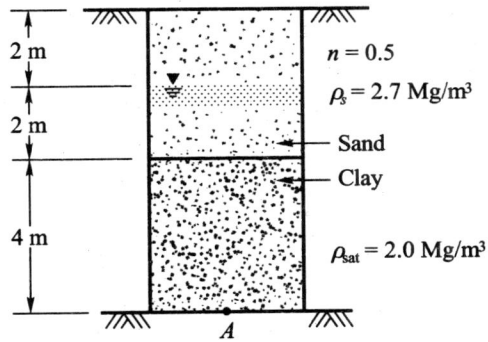

Fig. Prob. 8.6

8.7 For the soil profile given in Fig. Prob. 8.6, plot the total, neutral, and effective stresses with depth for the entire soil profile.

8.8 For the soil profile given Fig. Prob. 8.6, determine the effective stress at point A for the following conditions: (a) Water table at ground level, (b) water table at E1.A. (assume the soil above this level remains saturated and (c) water table 2 m above ground level.

8.9 A soil profile is shown in Fig. Prob. 8.9. Complete the effective stresses at the middle of clay layer for positions of water table at (a) A initially and, then (b) changed to position B. In the second position of water table, assume that soil above this level remain saturated.

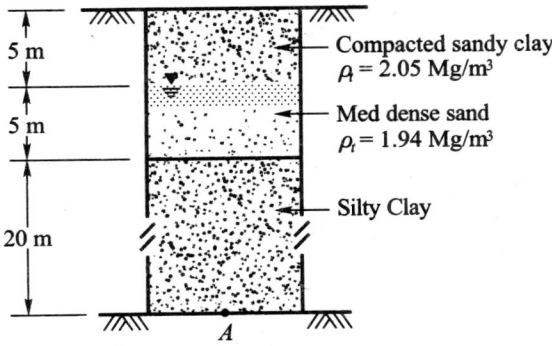

Fig. Prob. 8.9

CHAPTER 9

CAPILLARY WATER RISE IN SOIL

9.1 INTRODUCTION

The term water level, water table and phreatic surface designate the locus of the levels to which water rises in observation wells in free communication with the voids of the soil in site. The water table can also be defined as the surface at which the neutral stress u_w, in the soil is equal to zero.

If the water contained in the soil were subjected to no force other than gravity, the soil above the water table would be perfectly dry. In reality, every soil in the field is completely saturated above this level up to a certain height. The water that occupies the voids of the soil located above the water table constitutes *soil moisture.*

If the lower part of mass of dry soil comes into contact with water, the water rises in the voids to a certain height above the free water surface. The upward flow into the voids of the soil is attributed to the *surface tension* of the water. The height to which water rises above the water table against the force of gravity is called as *capillary rise.* The height of capillary rise is greatest for very fine grained soil materials. The water that rises above the water table attains the maximum height h_c only in the smaller voids. A few large voids may effectively stop capillary rise in certain parts of the soil mass. As a consequence, only a portion of the capillary zone above the free water surface remains fully saturated and the remaining partially saturated.

The seat of the surface tension is located at the boundary between air and water. Within the boundary zone the water is in a state of tension comparable to that in a stretched rubber membrane attached to the walls of the voids of the soil. However, in contrast to the tension in a stretched membrane, the surface tension in the boundary film of water is entirely unaffected by either the contraction or stretching of the film. The water held in the pores of soil above the free water surface is retained in a state of reduced pressure. This reduced pressure is called as *capillary pressure* or *soil moisture suction pressure.*

The existence of surface tension can be demonstrated as follows:

A greased sewing needle, Fig. 9.1, can be made to float on water because water has no affinity to grease, and, therefore, the water surface curves down under the needle until the upward component of the surface tension is large enough to support the weight of the needle. In Fig. 9.1, T_s is the surface tension per unit length of the needle and W_n the weight of the needle. The upward vertical

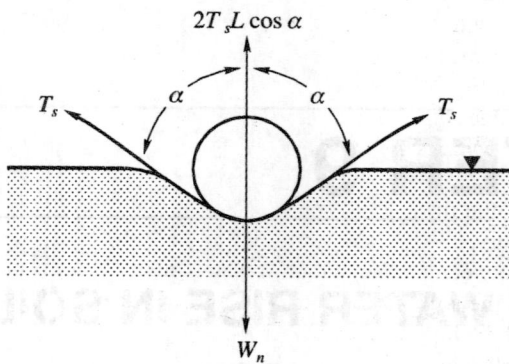

Fig. 9.1 Needle smeared with grease floating on water

force due to the surface tension is $2T_sL \cos \alpha$ where L the length of the needle. The needle floats when this vertical weight of the needle W_n acting downwards.

9.2   RISE OF WATER IN CAPILLARY TUBES

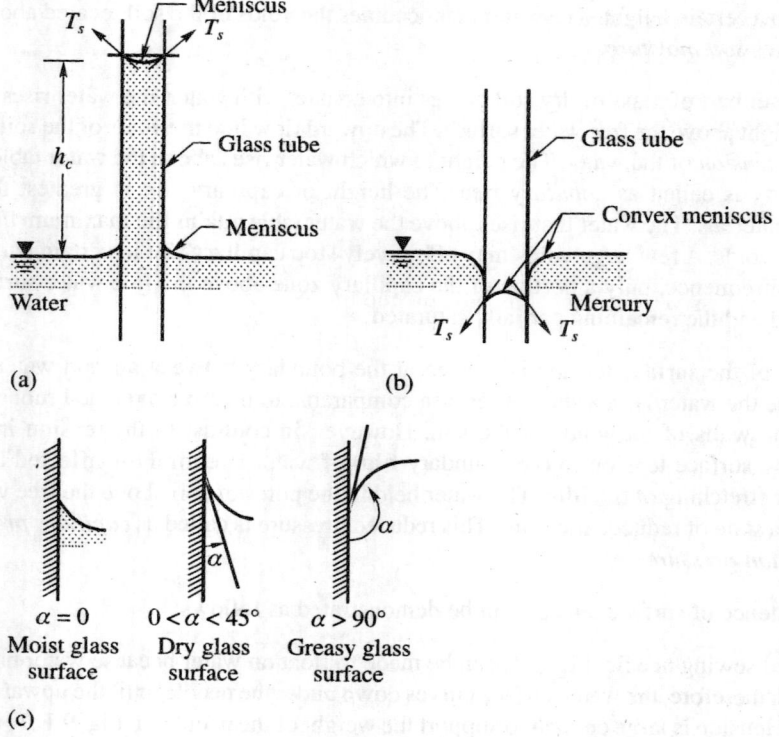

Fig. 9.2 Capillary rise and meniscus

The phenomenon of capillary rise can be demonstrated by immersing the lower end of a very small diameter glass tube into water. Such a tube is known as capillary tube. As soon as the lower end of the tube comes into contact with the water, the attraction between the glass and the water molecules combined with the surface tension of the water pulls the water up into the tube to a height he above the water level as shown in Fig. 9.2(a). The height h_c is known as the height of capillary rise. The upper surface of water assumes the shape of a cup, called the *meniscus* that joins the walls of the tube at an angle a known as the *contact angle*.

On the other hand, if the tube is dipped into mercury a depression of the surface in the tube below the surface of mercury, and the formation of a convex meniscus as shown in Fig. 9.2(b). The reason for the difference between the behaviour of water and mercury resides in the different affinity between the molecules of the solid and water or mercury. If there is a strong affinity between the molecules of the solid and the liquid, the surface of the liquid will climb up on the wall of the solid until a definite contact angle a is established. The contact angle between a clean moist glass surface and water is zero; that is, the water surface and water, α is not a constant. It may be as high as 45° at first gradually reducing there from to much smaller values. Probably the inevitable contamination of surfaces cleaned by ordinary methods, and the humidity of air are responsible for such variations. Figure 9.2(c) shows the contact angles between water and the surfaces under different conditions.

9.3 SURFACE TENSION

Surface tension is a force that exists at the surface of the meniscus. Along the line of contact between the meniscus in a tube and the walls of the tube itself, the surface tension, T_s is expressed as the force per unit length acting in the direction of the tangent as shown in Fig. 9.3(a). The components of this force along the wall and perpendicular to the wall are

Along the wall = $T_s \cos \alpha$ per unit length of wall.

Normal to the wall = $T_s \sin \alpha$ per unit length of wall.

The force normal to the wall tries to pull the wall of the tube together and the one along the wall produces a compressive force in the tube below the line of contact.

The meniscus can be visualised as a suspension bridge in three dimensions which is supported on the walls of the tube. The column of water of height he below the meniscus is suspended from this bridge by means of the molecular attraction of the water molecules. If the meniscus has stopped moving upward in the tube, then there must be equilibrium between the weight of the column of water suspended from the meniscus and the force with which the meniscus is clinging to the wall of the tube. We can write the following equation of equilibrium

$$\pi d T_s \cos\alpha = \frac{\pi d^2 h_c \gamma_w}{4} \text{ or } h_c = \frac{4 T_s \cos\alpha}{d\gamma_w} \qquad (9.1)$$

The surface tension T_s for water at 20 °C can be taken as equal to 75×10^{-8} kN per cm. The surface tensions of some of the common liquids are given in Table 9.1.

Equation (9.1) can be simplified by, assuming $\alpha = 0$ for moist glass and by substituting for T_s. Therefore, for the case of water, the capillary height he can be written as

$$h_c = \frac{4T_s}{d\gamma_w} = \frac{4 \times 75 \times 10^{-8} \times 10^{-6}}{d \times 9.81} = \frac{0.3}{d} \tag{9.2}$$

In Eq. (9.2) h_c and d are expressed in cm, wherein, $\gamma_w = 9.81$ kN/m³.

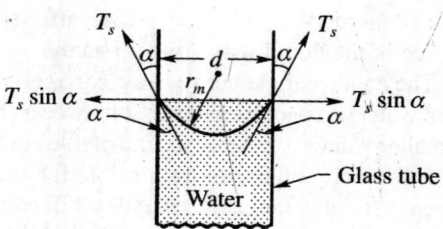

(a) Forces due to surface tension

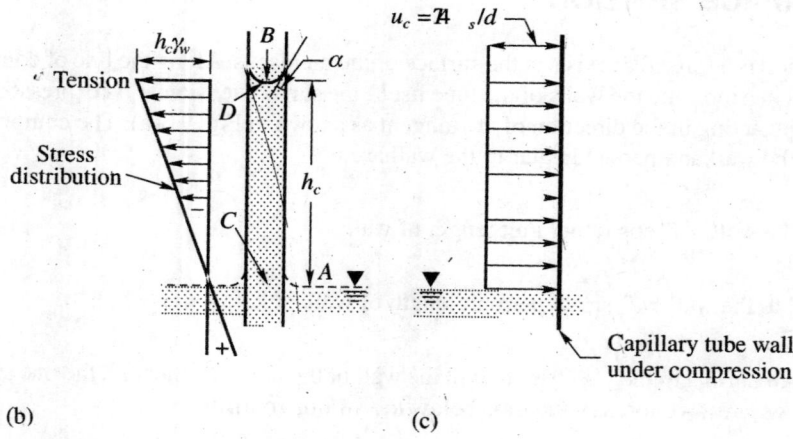

(b)

(c)

Fig. 9.3 Capillary pressure

TABLE 9.1

Surface tension of some Liquids at 20 °C

Liquids	T_s kN/cm x 10^{-8}
Ethyl Alcohol	22.03
Benzene	28.90
Carbon Tetra Chloride	26.90
Mercury	573.00
Petroleum	26.00
Water	75.00

9.4   STRESS DISTRIBUTION IN WATER BELOW THE MENISCUS

Figure 9.3(b) shows a capillary tube with its bottom end immersed in water. The pressure is atmospheric at points A and B. Since the point C is at the same level as A, according to the laws of hydraulics, the pressure at C is also atmospheric. Since the point D which is just below the meniscus is higher than point C by the head he, the pressure at D must be less than atmospheric by the amount $h\gamma_w$. Therefore, the pressure at any point in water between C and D is less than atmospheric. That means, the water above the point C is in tension if we refer to atmospheric pressure as zero pressure. The tension in water at any height h above C is given by $h_c\gamma_w$. Whereas the pressure in water below the free surface A is above atmospheric and therefore it is in compression. The stress distribution in water is given in Fig. 9.3(b).

Thus, the tension u_w in water immediately below the meniscus is given by

$$u_w = -h_c\gamma_w = -\frac{4T_s\cos\alpha}{d} \tag{9.3}$$

If r_m is the radius of the meniscus, Fig. 9.3(a), we can write,

$$r_m = \frac{d}{2\cos\alpha} \text{ or } d = 2r_m\cos\alpha$$

Substituting for d in Eq. (9.3), we have

$$u_m = -\frac{4T_s\cos\alpha}{2r_m\cos\alpha} = -\frac{2T_s}{r_m} \tag{9.4}$$

It may be noted here that at the level of the meniscus the magnitude of capillary pressure ue that compresses the wall of the tube is also equal to the capillary tension in the water just below the meniscus. The magnitude of capillary pressure ue remains constant with depth as shown in Fig. 9.3(c) whereas the capillary tension, u_w, in water varies from a maximum of $h_c\gamma_w$ at the meniscus level to zero at the free water surface level as shown in Fig. 9.3(b).

9.5   SHAPES AND POSITION OF CAPILLARY TUBES ON CAPI LLARY TENSION

Capillary Tubes in a Horizontal Position

Capillary tension in a liquid can be produced not only by allowing the liquid to rise in a capillary tube but in various other ways as well. For example, if a capillary tube of diameter d is filled with water, and then left on a table in a horizontal position as shown in Fig. 9.4(a), water will evaporate from both ends and gradually form concave surface. These menisci will cling to the edges of the tube until the minimum contact angle is reached, which for moist glass and water is zero. Thus, an increase in curvature of the meniscus at each end will proceed until semi-spherical menisci are formed. The different position of the menisci are shown as 1, 2, 3 and 4 in the Fig. 9.4. During the process as the menisci reaches from position 1 to 2, 3 and 4 (when at 4, has reached the minimum contact angle), the tension in the water must increase to the maximum for this size of the tube. This maximum tension is identical with the weight of a column of water the height of which is equal to the maximum height of capillary rise for this diameter. If evaporation continues, the menisci will withdraw into the interior part of the tube and occupy positions such as 5, during which process the radius of curvature, and the corresponding tension in the water will remain unchanged. In contrast to the variable stress in the water in the vertical tubes, the tension in horizontal tubes remains the same throughout the tube.

Capillary Tubes of Different Lengths in Vertical Position

One of the factors affecting the capillary tension in a vertical column of water is the length of the column with respect to the maximum height of capillary rise, h_c. If h represents the length of the water column, the stress conditions for various relationships of h and h_c can be explained as below.

If a short capillary tube is filled with water and held free in vertical position, complete meniscus will be developed as each end of the tube is wet and occupies the position shown in Fig. 9.4(b). The column of water will not be stable and flows down due to its weight W_w, since the surface tension T_s at the two ends of the column is equal and there is no resultant force to oppose gravity.

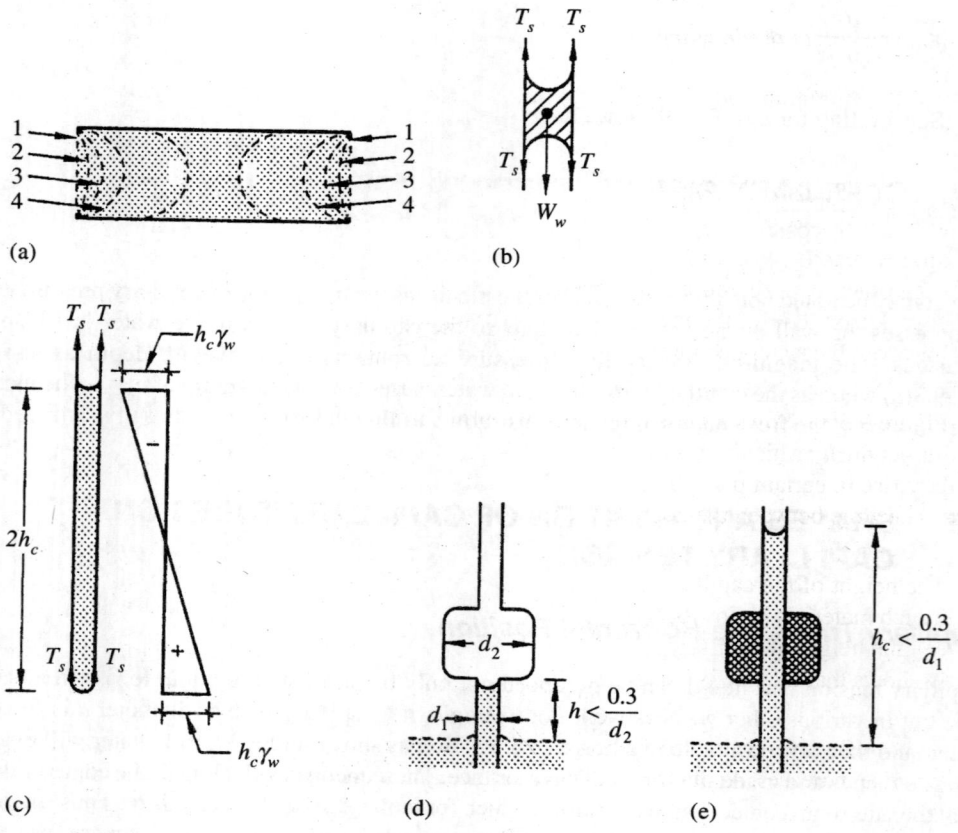

Fig. 9.4 The capillary height in tubes of different sizes and position

Figure 9.4(c) shows a long capillary tube filled with water and held free in a vertical position. If the top inside surface of the tube is clean and wet, and the bottom of the walls of the tube is oily, fully developed menisci will be formed as shown in the figure. The bottom meniscus holds the water in the bottom half of the tube in compression and the top meniscus holds the top half in tension. The total length of the column of water held in the tube is, therefore, $2h_c$. The pressure distribution is as shown in the figure.

Capillary Tubes of Different Sizes and Shapes

Consider two tubes of variable cross-sections as shown in Fig. 9.4(d) and (e). Both the tubes possess the same sectional properties. When the bottom of one tube is immersed in water keeping the base of the bulb at height h above the free water surface, the water will rise in the tube of diameter d_1 only up to a limiting height h, if h is less than $0.3/d_2$ [refer to Eq. (9.2)]. On the other hand if the tube (e) is filled with water from the top, water can stand above the bulb provided the bulb is within the height $0.3/d_1$ from the free water surface.

In the case of the second bulb it is only the column of water of diameter d_1 that is supported by the meniscus and the water shown hatched is supported by the bottom of the bulb. The same principle explained above holds good for tubes of other sizes and shapes. In all these cases, the rise of water beyond a given point in a capillary opening depends on two factors. They are

(a) The height of the point above free water surface.

(b) The minimum radii of curvature which the meniscus can assume at the point.

9.6   CAPILLARY RISE OF WATER IN SOILS

In contrast to capillary tubes the continuous voids in soils have a variable width. They communicate with each other in all directions and constitute an intricate network of voids. When water rises into the network from below, the lower part of the network becomes completely saturated. In the upper part, however, the water occupies only the narrowest voids and the wider areas remain filled with air.

Figure 9.5(a) shows a glass tube filled with fine sand. Sand would remain fully saturated only up to a height h_c which is considerably smaller than h_c. A few large voids may effectively stop capillary rise in certain parts. The water would rise, therefore, to a height of hc only in the smaller voids. The zone between the depths $(h_c - h_c)$ will remain partially saturated.

The height of the capillary rise is greatest for very fine grained soils materials, but the rate of rise in such materials is slow because of their low permeability. Figure 9.5(b) shows the relationship between the height of capillary rise in 24 hours and the grain size of a uniform quartz powder. This clearly shows that the rise is a maximum for materials falling in the category of silts and fine sands.

As the effective-grain size 'decreases', the size of the voids also decreases, and the height of capillary rise increases. A rough estimation of the height of capillary rise can be determined from the equation.

$$h_c = \frac{C}{eD_{10}} \tag{9.5}$$

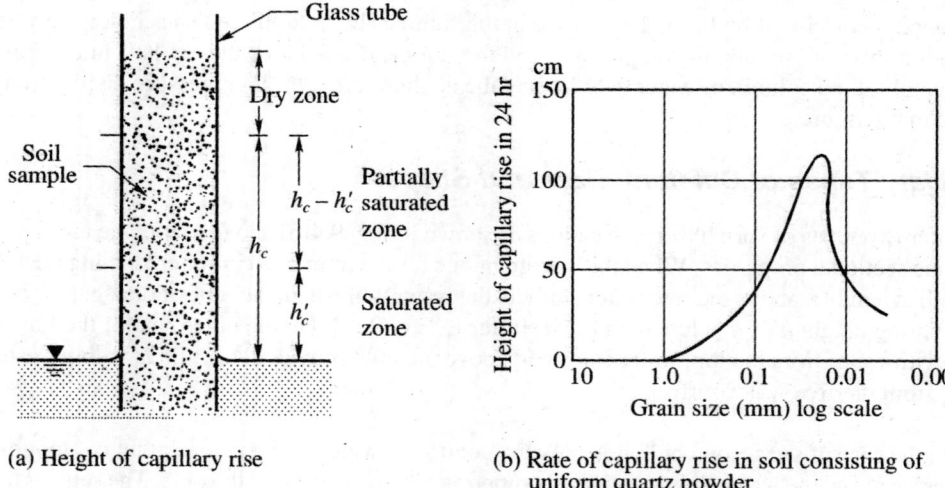

(a) Height of capillary rise

(b) Rate of capillary rise in soil consisting of
uniform quartz powder

Fig. 9.5   Capillary rise in soils

in which e is the void ratio, D_{10} is Hazens' effective diameter in centimetres and C is an empirical constant which can have a value between 0.1 and 0.5 sq. cm.

Capillary Siphoning

Capillary forces are able to raise water against the force of gravity not only into capillary tubes or the voids in columns of dry soil, but also into narrow open channels or V-shaped grooves. If the highest point of the groove is located below the level, to which the surface tension can lift the water, the capillary forces will pull the water into the descending part of the groove and will slowly empty the vessel. This process is known as capillary siphoning. The same process may also occur in the voids of soil. For example, water may flow over the crest of an impermeable core in a dam in spite of the fact that the elevation of the free water surface is below the crest of the core as shown in Fig. 9.6.

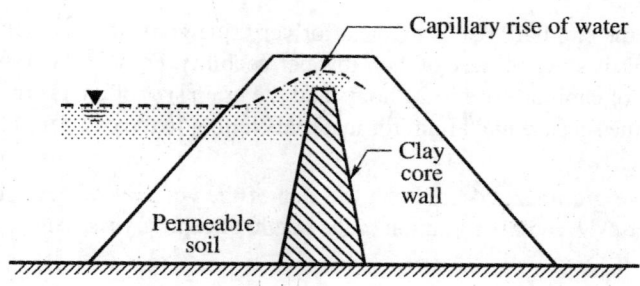

Fig. 9.6   Capillary siphoning

9.7   CAPILLARY PRESSURE IN SOILS

The tension U_w in water just below the meniscus as given by the Eq. (9.3) is

$$u_m = -\frac{4T_s \cos\alpha}{d}$$

Since this pressure is below atmospheric pressure, it draws the grains of soils closer to each other at all points where the menisci touch the soil grains. Intergranular pressure of this type is called capillary pressure. The effective or intergranular pressure at any point in a soil mass can be expressed by

$$\sigma' = \sigma_t - u_m \qquad (9.6)$$

where σ_t is the total pressure, σ' is the effective or the integranular pressure and u_w is the porewater pressure. When the water is in compression u_w is positive, and when it is in tension u_w is negative. Since u_w is negative, in the capillary zone, the intergranular pressure is increased by u_w. The equation therefore, can be written as

$$\sigma' = \sigma_t - (-u_m) = \sigma_t + u_m \qquad (9.7)$$

The increase in the intergranular pressure due to capillary pressure acting on the grains leads to greater strength of the soil mass. The effect of the capillary pressure on the volume of a soil mass can be explained as below.

Consider a sample of fine grained soil which is saturated, and exposed to drying. First the shining surface of the sample will change to a dull appearance as curved mensici are formed at each opening. As water is removed by evaporation, the menisci which are suspended from every opening decrease their radius of curvature, thereby increasing the capillary pressure [refer to Eq. (9.3)] in the skeleton of mineral grains. As a result the skeleton gets compressed. Evaporation will continue

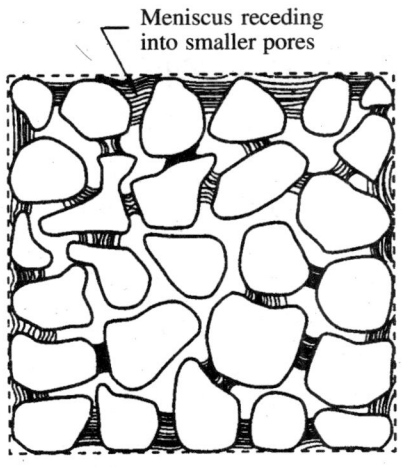

Fig. 9.7   The effect of capillary pressure on a soil mass

TABLE 9.2

Capillary rise and pressure in soil

Soil	Size particles (mm)	Capillary rise h_c (cm)	Capillary pressure u_w (kPa)
Sand, coarse	2.00–0.6	1.5–5	0.15–0.50
Sand, medium	0.06–0.2	5.0–15	0.5–1.50
Sand, fine	0.20–0.06	15–50	1.5–5.00
Silt	0.06–0.002	50–1500	5.00–150.00
Clay, coarse	0.002–0.0002	1500–15000	150.00–1500
Clay, colloid	< 0.0002	> 15000	> 1500

to decrease the radius of curvature of all menisci, Fig. 9.7, and increase the pressure in the skeleton, until a point is reached when the capillary pressure can no longer compress the skeleton further, and then the menisci will withdraw into the interior. When this occurs there is a change in appearance of the surface from dark to light colour. At this stage in the drying out process, the sample attains a volume which remains practically constant with further evaporation. This limit is called as the shrinkage limit. Any further evaporation must result in a gradual retreat of the menisci into the interior of the sample. 1f a daied lump of clay soil is immersed in water, water will be drawn into the voids due to suction with a great force. Thus, results in crumbling of the soil sample.

The *capillary pressure* of some soils rises from zero at saturation to a very high value in oven dry soil. Table 9.2 may serve to give an idea of the maximum height of capillary rise and the maximum capillary pressure that may develop in a soil mass on the assumption that the size of the opening is equal to that of the particles.

9.8   STRESS CONDITION IN SOIL DUE TO SURFACE TENSION FORCES

It is to be assumed here that the soil above the ground water table remains dry prior to the rise of capillary water. The stress condition in the dry soil mass get changed due to the rise of capillary water.

Now consider the soil profile given in Fig. 9.8(a). When a dry soil mass above the GWT comes in contact with water, water rises by capillary action. Let the height of rise is h_c and assume that the soil within this zone gets saturated due to capillary water. Assume that the menisci formed at height h_c coincide with the ground surface. The plane of the menisci is called as the *capillary fringe*.

The vertical stress distribution of the dry soil mass is shown in Fig. 9.8(b). The vertical stress distribution of the saturated mass of soil is grven in Fig. 9.8(d). The tension in water is maximum at the menisci level, say equal to u_w and zero at GWT level as shown in Fig. 9.8(e).

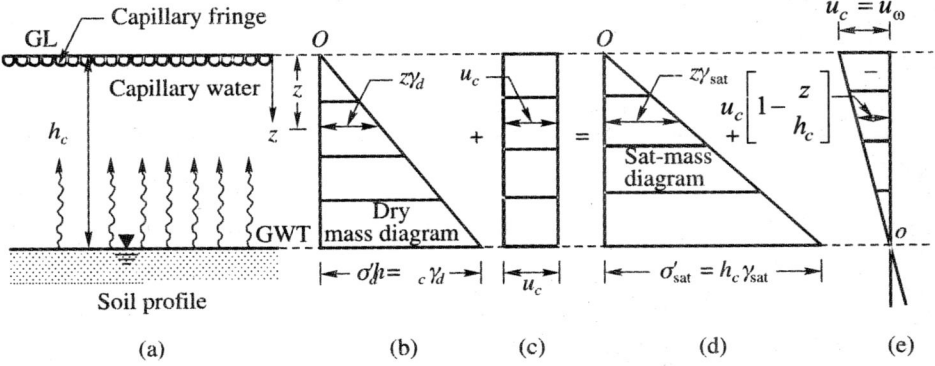

Fig. 9.8 Effect of capillary pressure Uc on soil vertical stress diagram

Prior to capillary rise the maximum pressure of the dry mass, σ'_d, at GWT level is

$$\sigma'_d = \gamma_a h_c$$

where, $\gamma_d = $ *dry* unit weight of soil.

After the capillary rise the maximum pressure of the saturated weight of soil at GWT level is

$$\sigma'_{sat} = \gamma_{sat} h_c$$

Since the pore water pressure at GWT level is zero, it is obvious that the difference between the two pressures σ'_{sat} and σ'_d represent the increase in pressure due to capillary rise which is actually the capillary pressure, which may be expressed as

$$u_c = h_c (\gamma_{sat} - \gamma_d) \qquad\qquad\qquad\qquad\qquad (a)$$

By substituting for

$$\gamma_{sat} = \frac{(G+e)\gamma_w}{1+e}, \text{ and } \gamma_d = \frac{Ge}{1+e}$$

in Eq. (a), we have, after simplifying

$$u_c = \frac{e}{1+e} h_c \gamma_w = n h_c \gamma_w \qquad\qquad\qquad\qquad (9.8)$$

where, $e = $ void ratio,

   $n = $ porosity

It is clear from Eq. (9.8) that the capillary pressure for soil is directly proportional to the porosity of the soil and this pressure is very much less than $h_c \gamma_w$ which is used only for a fine bore and uniform diameter capillary tube.

The distribution of capillary pressure u_c (constant with depth) is given the Fig. 9.8(c). The following equation for the pressure at any depth z may be written as per Fig. 9.8.

$$z\gamma_d + u_c = z\gamma_{sat} + u_c \left[1 - \frac{z}{h_c}\right] \tag{9.9}$$

Example 9.1

The diameter of a clean capillary tube is 0.08 mm. Determine the expected rise of water in the tube.

Solution

As per Eq. (9.2) the expected rise, h_c, in the capillary is

$$h_c = \frac{0.3}{d} = \frac{0.3}{0.008} = 196 \text{ cm}$$

where, d is centimetre

Example 9.2

Water table is at a depth of 10 m in a silty soil mass. The sieve analysis indicate the effective diameter D_{10} of the soil mass is 0.05 mm. Determine the capillary rise of water above the water table and the maximum capillary pressure (a) by using Eq. (9.3) and (b) by using Eq. (9.8) Assume void ratio $e = 0.51$.

Solution

Use Eq. (9.5) and assuming $C = 0.5$ the capillary rise of water is now

$$h_c = \frac{C}{eD_{10}} = \frac{0.5}{0.51 \times 0.005} = 196 \text{ cm}$$

(a) As per Eq. (9.3)

the capillary pressure, is $u_w = -h_c\gamma_w = -1.96 \times 9.81 = -19.2 \text{ kN/m}^2$

(b) As per Eq. (9.8)

Porosity, $n = \dfrac{e}{1+e} = \dfrac{0.51}{1.15} = 0.34$

$u_w = u_c = -nh_c\gamma_w = -0.34 \times 19.2 = 6.53 \text{ kN/m}^2$

Example 9.3

A layer of silty soil of thickness 5 m lies below the ground surface at a particular site and below the silt layer lies a clay starta. Ground water table is at a depth of 4 m below ground level. The following data are available for both the silt and clay layers of soil.

Silt layer: Particle size $D_{10} = 0.018$ mm, $e = 0.7$, and $G = 2.7$

Clay layer: $e = 0.8$ and $G = 2.75$

Required: (a) Height of capillary water rise, (b) capillary pressure, (c) the efiedive pressure at the ground surface level, at GWT level, at the bottom of the silt layer and at a depth of $H = 6$ m below ground level, and (d) At a depth 2 m below ground level.

Solution

For silty soil:

$$\gamma_d = \frac{G\gamma_w}{1+e} = \frac{2.7 \times 9.81}{1.7} = 15.6 \text{ kN/m}^3$$

$$\gamma_{sat} = \frac{(G+e)\gamma_w}{1+e} = \frac{(2.7+0.7)9.81}{1.7} = 19.62 \text{ kN/m}^3$$

$$\gamma_b = \gamma_{sat} - \gamma_w = 19.62 - 9.81 = 9.81 \text{ kN/m}^3$$

Clay strata:

$$\gamma_{sat} = \frac{(2.75+0.8)9.81}{1.8} = 19.35 \text{ kN/m}^3$$

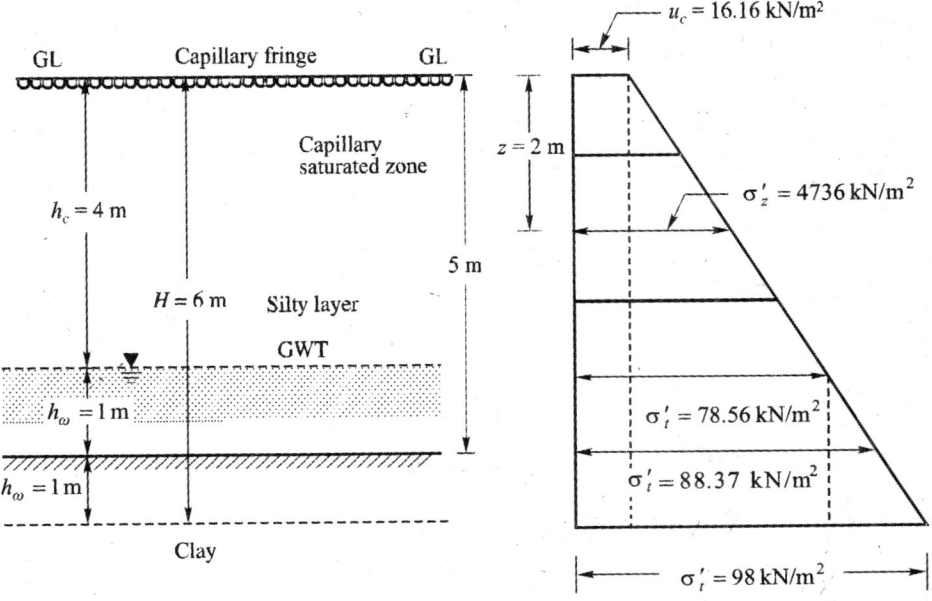

Effective pressure distribution diagram

Fig. Ex. 9.3

$$\gamma_b = 19.35 - 9.81 = 9.54 \, \text{kN/m}^3$$

(a) *Height of capillary water rise*

$$h_c = \frac{C}{eD_{10}} \quad \text{as per Eq. (9.5)}$$

Assume $C = 0.5$ sq. cm.

We have, $h_c = \dfrac{0.5}{0.7 \times 0.018 \times 10} = 397$ cm or say 4.0 m

It is clear from h_c, that the plane of menisci formed by the capillary water with the ground surface as the water table is also at a depth of 4 m form GL.

(b) *Capillary pressure* u_c

As per Eq. (9.8), $u_c = nh_c\gamma_w = \dfrac{e}{1+e} h_c\gamma_w$

or $u_c = \dfrac{0.7}{1.7} \times 4 \times 9.81 = 16.16 \, \text{kN/m}^2$

(c) *The effective pressure at GL*

Since the plane of meniscus coincides with the ground surface, the effective at GL is equal to the capillary pressure u_c

Total effective pressure at GWT level, σ'_{sat}

As per Fig. Ex. 9.3

$$\sigma'_{sat} = \sigma'_d + u_c = \gamma_d h_c + u_c$$

$$\sigma'_{sat} = 15.6 \times 4 \times 16.16 = 78.56 \, \text{kN/m}^2$$

Total effective pressure at the bottom level of silt layer

Bottom of silt layer is at a depth of 1 m below GWT level. The effective press this depth is

$$\sigma' = \gamma_b h_w = 9.81 \times 1 = 9.81 \, \text{kN/m}^2$$

Total effective pressure, $\sigma'_t = \sigma'_{sat} + \sigma' = 78.56 + 9.81 = 88.37 \, \text{kN/m}^2$

Total effective pressure at a depth of 6 m below GL

This lies in clay strata at a depth of 1 m below the bottom of silty layer.

The increase in effective pressure within this depth is

$$\sigma' = \gamma_b h_w = 9.54 \, \text{kN/m}^2$$

The total effective pressure, $\sigma'_t = 88.37 + 9.54 = 97.91 \, \text{kN/m}^2 \approx 98 \, \text{kN/m}^2$

(d) σ'_z at 2 m below GL

$$\sigma'_z = u_c + z\gamma_d = 16.16 + 2 \times 15.6 = 47.36 \, \text{kN/m}^2$$

The pressure distribution diagram is given in Fig. Ex. 9.3.

Example 9.4

At a particular site, lies a layer of fine sand of 8 m thick below ground surface and having a void ratio of 0.7 the GWT is at a depth of 4 m below ground surface. The average saturation of sand above the capillary fringe is 50%. The soil gets saturated due to capillary action to a height of 2.0 m above GWT level. Assuming $G = 2.65$, calculate the total effective pressure on a horizontal plane at a depth of 6 m below ground level and at depth 3 m below ground surface.

Solution

$$\gamma_d \frac{G\gamma_w}{1+e} = \frac{2.65 \times 9.81}{1.7} = 15.29 \, \text{kN/m}^3$$

$$\gamma_{sat} = \frac{(e+G)\gamma_w}{1+e} = \frac{(0.7+2.65) \times 9.81}{1.7} = 19.33 \, \text{kN/m}^3$$

$$\gamma_b = \gamma_{sat} - \gamma_w = 19.33 - 9.81 = 9.52 \, \text{kN/m}^3$$

moist weight of soil above capillary fringe

$$\gamma_m = \frac{(G+es)\gamma_w}{1+e} = \frac{(2.65+0.7 \times 0.5) \times 9.81}{1.7} = 17.31 \, \text{kN/m}^3$$

Capillary pressure,

$$u_c = nh_c\gamma_w = \frac{e}{1+e} h_c\gamma_w = \frac{0.7}{1.7} \times 2 \times 9.81 = 8.08 \, \text{kN/m}^2$$

Effective stresses at different levels

(a) At Ground level $\sigma' = 0$

(b) Over burden pressure at fringe level $= \sigma'_o = \overline{h_c}\gamma_m = 2 \times 17.31 = 34.62 \, \text{kN/m}^2$

(c) Effective pressure at fringe level $= \sigma'_c = \sigma'_o + u_c = 34.62 + 8.08 = 42.70 \, \text{kN/m}^2$

(d) Effective pressure at GWT level $= \sigma'_{sat} = \sigma'_c + \sigma'_d = 42.70 + 2 \times 15.29$

$$= 42.70 + 30.58 = 73.28 \, \text{kN/m}^2$$

(e) Effective pressure at 6 m below GL

$$\sigma'_t = \sigma'_{sat} + h_w\gamma_b = 73.28 + 2 \times 9.52 = 73.28 + 10.04 = 92.32 \, \text{kN/m}^2$$

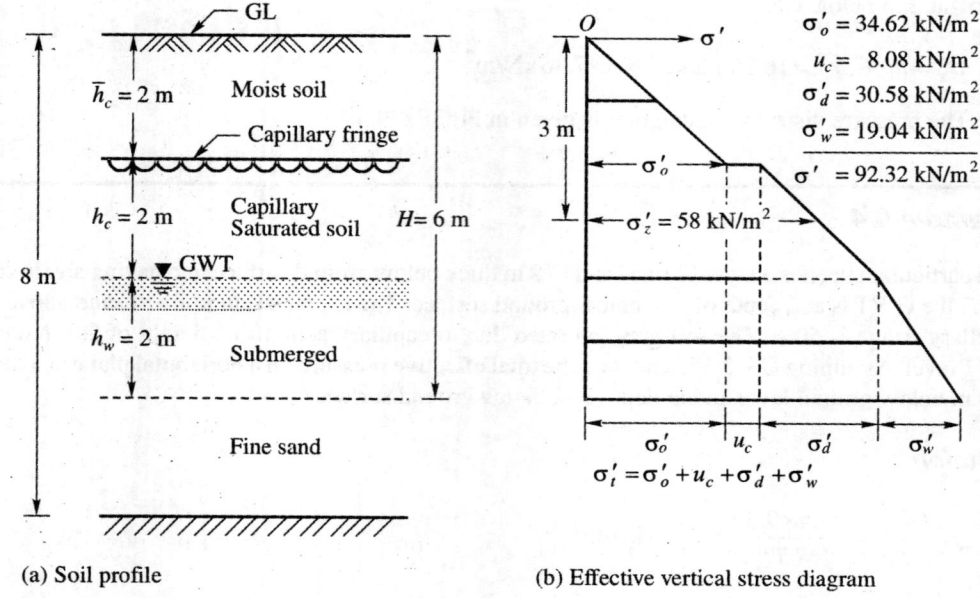

(a) Soil profile   (b) Effective vertical stress diagram

Fig. Ex. 9.4

Effective stress depth 3 m below GL

Refer Fig. Ex. 9.4.

$$\sigma'_z = \sigma'_o + u_c + (z - h_c)\gamma_d = 34.62 + 8.08 + (3-2) \times 15.29 \approx 58 \text{ kN/m}^2$$

9.9   QUESTIONS AND PROBLEMS

9.1   Why does water rise and mercury gets depressed in a capillary tube? Discuss.

9.2   Develop an equation for determining the rise of water in a capillary tube.

9.3 · Explain and discuss the effect of size, shape and disposition of capillary tubes on capillary rise and pressure.

9.4   Explain and discuss the relation between grain size and rate of rise of capillary water in soil.

9.5   Explain the meaning of capillary siphoning and its practical application.

9.6   What are the practical advantages of capillary pressure in soil? Discuss.

9.7   A capillary tube of variable diameters d1 and d2 is filled with water and kept on a table with its axis horizontal as shown in Fig. Prob. 9.7. Explain the process of formation of menisci in the small and big ends of the tube and the relationship between their contact angles. Which of the contact angles, α_1, or α_2 reach zero value first and why? Discuss. Develop an equation for computing the capillary pressure when one of the menisci reaches zero angle first.

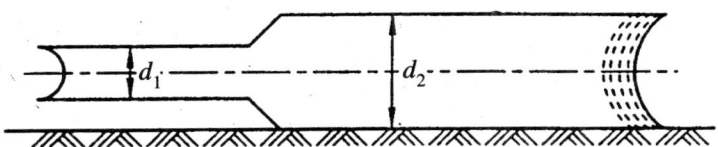

Fig. Prob. 9.7

9.8 A glass tube, open both ends, has an internal diameter of 0.0002 mm. The tube is held vertically and water is added from the top end. What is the maximum height h, of the column of water that will be supported.

9.9 Calculate (a) the theoretical height of capillary rise h_c and (b) the capillary pressure, u_w, in a silty soil with D_{10} 0.004 mm. Assume void ratio as equal to 0.50. Use Eq. (9.8) and $C = 0.5$.

9.10 Calculate the height to which water will rise in a soil deposit consisting of fine silt of uniform is size. The depth of water below the ground surface is 20 m. Assume the surface tension is 75 × 10⁻⁸ kN/cm and contact angle is zero. The average size of the pores is 0.004 mm.

CHAPTER 10

STRESS DISTRIBUTION IN SOILS

10.1   INTRODUCTION

Estimation of vertical stresses at any point in a soil-mass due to external vertical loadings are of great significance in the prediction of settlements of buildings, bridges, embankments and many other structures. Equations have been developed to compute stresses at any point in a soil mass on the basis of the theory of elasticity. According to elastic theory, constant ratios exist between stresses and strains. For the theory to be applicable, the real requirement is not that the material necessarily be elastic, but there must be constant ratios between stresses and the corresponding strains. Therefore, in non-elastic soil masses, the elastic theory may be assumed to hold so long as the stresses induced in the soil mass are relatively small. Since the stresses in the subsoil of a structure having adequate factor of safety against shear failure are relatively small in comparison with the ultimate strength of the material, the soil may be assumed to behave elastically under such stresses.

When a load is applied to the soil surface, it increases the vertical stresses within the soil mass. The increased stresses are greatest directly under the loaded area, but extend indefinitely in all directions. Many formulae based on the theory of elasticity have been used to compute stresses in soils. They are all similar and differ only in the assumptions made to represent the elastic conditions of the soil mass. The formulae that are most widely used are the Boussinesq and Westergaard formula. These formulae were first developed for point loads acting at the surface. These formulae have been integrated to give stresses below uniform strip loads and rectangular loads.

The extent of elastic layer below the surface loadings may be any one of the following:

1.   Infinite in the vertical and horizontal directions.

2.   Limited thickness in the vertical direction underlain with a rough rigid base such as a rocky bed.

The loads at the surface may act on flexible or rigid footings. The stress conditions in the elastic layer below vary according to the rigidity of the footings and the thickness of elastic layer. All the external loads considered in this book are vertical loads only as the vertical loads are of practical importance for computing settlements of foundations.

10.2 BOUSSINESQ'S FORMULA FOR POINT LOADS

Figure 10.1 shows a load Q acting at a point O on the surface of a semi-infinite solid. A semi-infinite solid is the one bounded on one side by a horizontal surface, here the surface of the earth, and infinite in all the other directions. The problems limited to the determination at any point P at a depth z a vertical stress σ_z, horizontal stress σ_x, and shear stress τ_{xz}, on planes parallel to the limiting plane. The problem was solved by Prof. J. Boussinesq in 1885 on the following assumptions.

1. The soil mass is elastic, isotropic, homogeneous and semi-infinite.

2. The soil is weightless.

3. The load is a point load acting on the surface.

The soil is said to be isotropic if there are identical elastic properties throughout the mass and in every direction through any point of it. The soil is said to be homogeneous if there are identical elastic properties at every point of the mass in identical directions.

The expression obtained by Boussinesq for computing vertical stress σ_z, at point P (Fig. 10.1) due to a point load Q is

$$\sigma_z = \frac{3Q}{2\pi z^2} \frac{1}{\left[1+(r/z)^2\right]^{5/2}} = \frac{Q}{z^2} I_B \qquad (10.1)$$

where, r = the horizontal distance between an arbitrary point P below the surface and the vertical axis through the point load Q.

z = the vertical depth of the point P from the surface.

$$I_B = \text{Boussinesq stress coefficient} = \frac{3}{2\pi} \frac{1}{\left[1+(r/z)^2\right]^{5/2}} \qquad (10.1a)$$

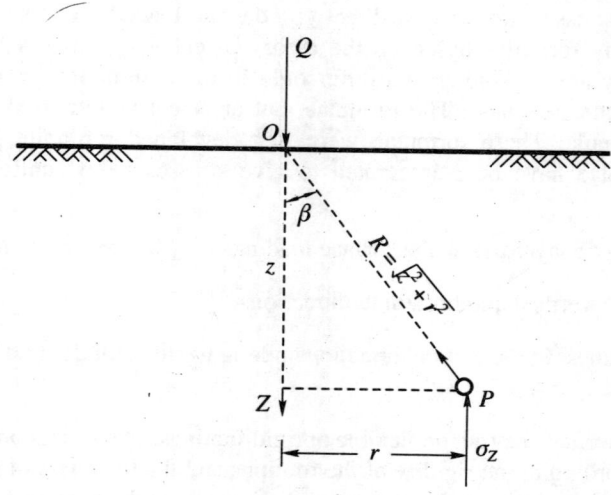

Fig. 10.1 Vertical pressure within an earth mass

The values of the Boussinesq coefficient I_B can be determined for a number of values of r/z. The variation of I_B with r/z in a graphical form is given in Fig. 10.2. It can be seen from this figure that I_B has a maximum value of 0.48 at $r/z = 0$, i.e. the value of I_B is a maximum value of 0.48 at $r/z = 0$, i.e. the value of I_B is a maximum on the Z-axis, indicating thereby that the stress is a maximum below the point load. Experience indicates that the computed values might be in error as much as ± 25 percent or more.

10.3   WESTERGAARD'S FORMULA FOR POINT LOADS

Boussinesq assumed that the soil is elastic, isotropic and homogeneous for the development of a point load formula. However, the soil is neither isotropic nor homogeneous. The most common type of soils that are met in nature are the water deposited sedimentary soils.

When the soil particles get deposited in water, typical clay strata usually have their lenses of coarser materials within them. The soils of this type can be assumed as laterally reinforced by numerous, closely spaced, horizontal sheets of negligible thickness but of infinite rigidity, which prevent the mass as a whole from undergoing lateral movement of soil grains. Westergaard, a British Scientist, proposed in 1938 a formula for the computation of vertical stress σ_z by a point load, Q, at the surface as

$$\sigma_z = \frac{3Q}{2\pi z^2} \frac{\sqrt{(1-2\mu)/(2-2\mu)}}{\left[(1-2\mu)/(2-\mu)+(r/z)^2\right]^{3/2}} = \frac{Q}{z^2} I_W \qquad (10.2)$$

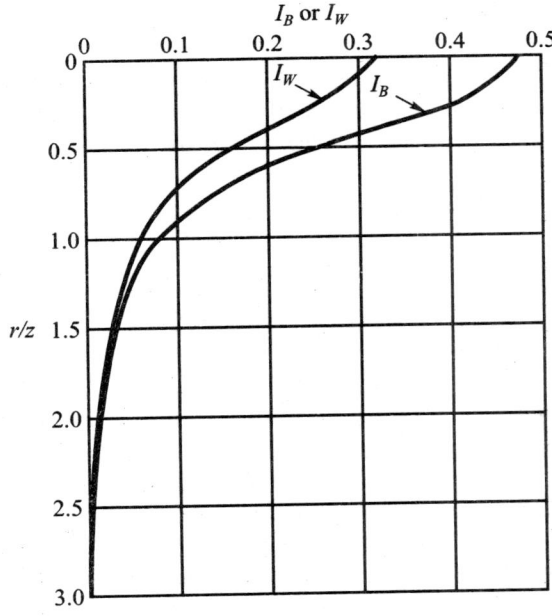

Fig. 10.2   Values of I_B or I_W for use in Boussinesq or Westergaard formula

in which μ is Poisson's ratio. If μ is taken as zero for all practical purposes, Eq. (10.2) simplifies to

$$\sigma_z = \frac{Q}{\pi z^2} \cdot \frac{1}{[1+2(r/z)^2]^{3/2}} = \frac{Q}{z^2} I_w \tag{10.3}$$

where $I_w = \dfrac{(1/\pi)}{[1+2(r/z)^2]^{3/2}}$ is the Westergaard stress coefficient. The variation of I_w with the

ratios of (r/z) is shown graphically in Fig. 10.2 along with the Boussinesq's coefficient I_B. The value of I_w at r/z = 0 is 0.32 which is less than that of I_B by 33 percent.

10.4   COMPARISON OF BOUSSINESQ AND WESTERGAARD EQUATIONS

We may plot the vertical stress distribution on horizontal planes at different depths z from the surface due to the load Q acting at the surface, shown in Fig. 10.3(a), by making use of Boussinesq's or Westergaard's equation. It is clear from the figure that the stress is greatest directly under the load but extend indefinitely in all directions. The greater the depth, smaller the concentration of stress directly beneath the load, but at any depth, if the incremental stresses were integrated over the area, the total force would equal the applied load Q. Fig. 10.3(a) illustrates the difference in the distribution of stress as per the Boussinesq's and Westergaard's equations.

The vertical stress distribution with depth along the vertical plane below the load is shown in Fig. 10.3b. The stress at the surface directly beneath the load though tends to infinity theoritically but in reality reaches a finite value as the load occupies a finite area.

It is clear from the curves in Fig. 10.2 that the Westergaard equation gives values consistently less than the Boussinesq for the same point load up to a ratio r/z equal to 1.5. When the ratio exceeds 1.5, Westergaard formula gives a greater stress. For all ratios of r/z less than about 0.8, the vertical stresses as per Westergaard formula is approximately equal to two-thirds of the values given by the Boussinesq's formula.

Experimental measurements of pressures developed in soils due to boundary loads are not sufficient to take a final decision as to which of the two formulae are applicable to field condition. However, the stratified condition upon which the Westergaard's equation is based is definitely nearer to the condition existing in sedimentary soils than the isotropic condition assumed by Boussinesq. It has been noticed in some cases that the settlements estimated by making use of the stresses computed from Boussinesq's equation are larger than the observed ones. Engineers still prefer to use Boussinesq's equation for the computation of vertical stresses, since the settlement computed from these stresses give conservative values. As such, all further discussions are restricted to Boussinesq theory only.

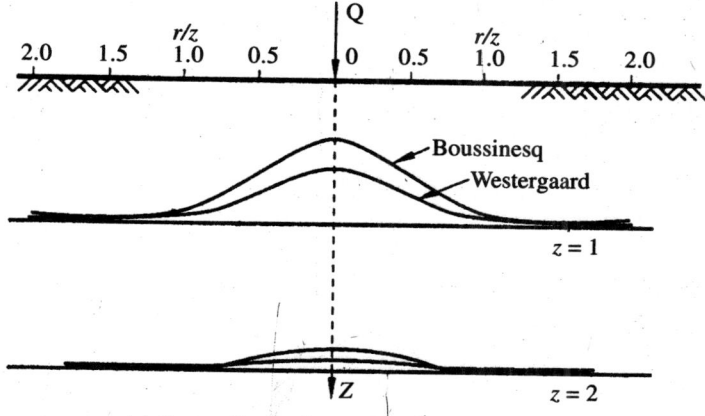

(a) Stress distribution on horizontal planes

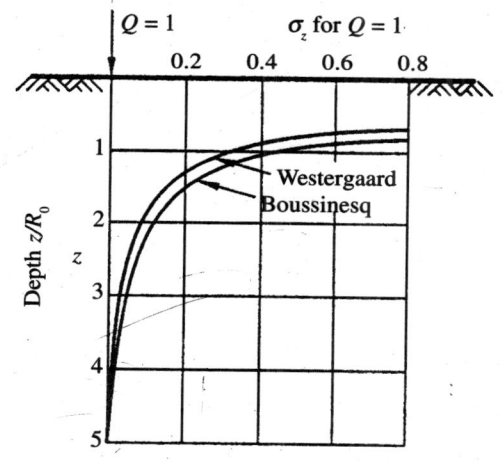

(b) Variation of stress with depth under a unit load

Fig. 10.3 Stress distribution with respect to depth

Example 10.1

A concentrated of load of 1000 kN is applied at the ground surface. Compute the vertical pressure (i) at a depth of 4 m below the load, (ii) at a distance of 3 m at the same depth. Use Boussinesq's equation.

Solution

The equation is

$$\sigma_z = \frac{Q}{z^2} I_B, \text{ where } I_B = \frac{3/2\pi}{\left[1+(r/z)^2\right]^{5/2}}$$

(i) When $r/w = 0$, $I_B = 3/2\pi = 0.48$, $\sigma_z = 0.48\dfrac{Q}{z^2} = 0.48\times\dfrac{1000}{4\times4} = 30$ kN/m^2

(ii) When $r/z = 3/4 = 0.75$

$$I_B = \frac{3/2\pi}{\left[1+(0.75)^2\right]^{5/2}} = 0.156, \sigma_z = \frac{0.156\times1000}{4\times4} = 9.8 \text{ kN/m}^2$$

Example 10.2

A concentrated load of 200 kN act at foundation level at a depth of 2 m below ground surface. Find the vertical stress along the axis of the load at a depth of 10 m and at a radial distance of 16.4 ft at the same depth by (a) Boussinesq, and (b) Westergaard formulae for $\mu = 0$. Neglect the depth of the foundation.

Solution

(a) Boussinesq Eq. (10.1)

$$\sigma_z = \frac{Q}{z^2}I_B, \ I_B = \frac{3}{2\pi}\left[\frac{1}{1+(r/z)^2}\right]^{5/2}$$

Substituting the known values, and simplifying

$I_B = 0.2733$ for $r/z = 0.5$

$$\sigma_z = \frac{200}{10^2}\times0.2733 = 0.55 \text{ kN/m}^2$$

(b) Westergaard (Eq. 10.3)

$$\sigma_z = \frac{Q}{z^2}I_w, I_w = \frac{1}{\pi}\left[\frac{1}{1+2(r/z)^2}\right]^{3/2}$$

Substituting the known values and simplifying, we have,

$I_w = 0.1733$ for $r/z = 0.5$

therefore,

$$\sigma_z = \frac{200}{10^2}\times0.1733 = 0.346 \text{ kN/m}^2$$

Example 10.3

A rectangular raft of size 30×12 m founded at a depth of 2.5 m below the ground surface is subjected to a uniform pressure of 150 kPa. Assume the centre of the area is the origin of coordinates $(0, 0)$, and the corners have coordinates $(6, 15)$. Calculate stresses at a depth of 20 m below the foundation level by the methods of (a) Boussinesq, and (b) Westergaard at coordinates of $(0, 0)$, $(0, 15)$, $(6, 0)$ $(6, 15)$ and $(10, 25)$. Also determine the ratios of the stresses as obtained by the two methods. Neglect the effect of foundation depth on the stresses.

Solution

Equations (a) Boussinesq

$$\sigma_z = \frac{Q}{z^2} I_B, \quad I_B = \frac{0.48}{1 + (r/z)^{2}} {}^{5/2}$$

(b) Westergaard: $\quad \sigma_z = \frac{Q}{z^2} I_w, I_w = \frac{0.48}{1 + (r/z)^{2}} {}^{3/2}$

The ratios of r/z at the given locations for $z = 20$ m are as follows

Location	r/z	Location	r/z
$(0, 0)$	0	$(6, 15)$	$\left(\sqrt{6^2 + 15^2}\right)/20 = 0.81$
$(6, 0)$	$6/2 = 0.3$	$(10, 25)$	$\left(\sqrt{10^2 + 25^2}\right)/20 = 1.35$
$(0, 15)$	$15/20 = 0.75$		

The stresses at the various locations at $z = 20$ m may be calculated by using the equations given above. The results are tabulated below for the given total load $Q = qBL = 150 \times 12 \times 30 = 54000$ kN acting at $(0, 0)$ coordinate. $Q/z^2 = 135$.

Location	r/z	Boussinesq		Westergaard		σ_B/σ_W
		I_B	σ_B(kPa)	I_W	σ_W(kPa)	
$(0, 0)$	0	0.48	65	0.32	43	1.51
$(6, 0)$	0.3	0.39	53	0.25	34	1.56
$(0, 15)$	0.75	0.16	22	0.10	14	1.57
$(6,15)$	0.81	0.14	19	0.09	12	1.58
$(10, 25)$	1.35	0.036	5	0.03	4	1.25

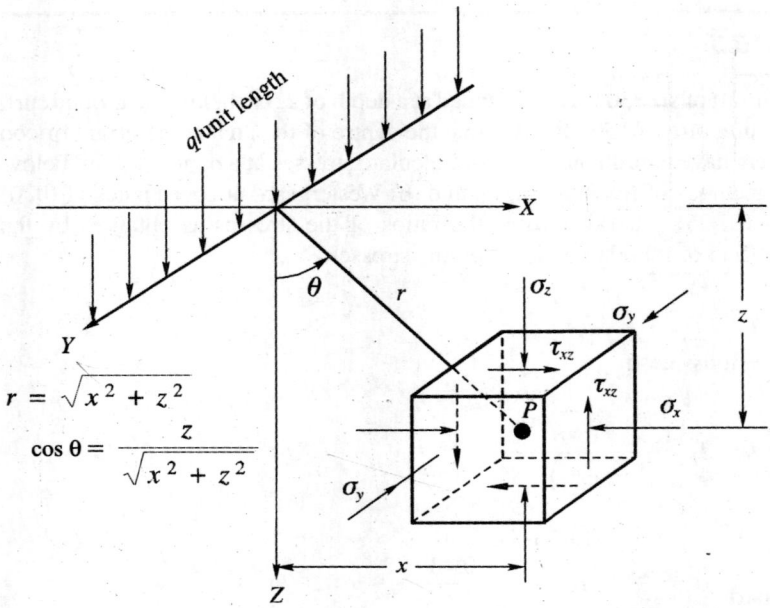

Fig. 10.4 Stresses due to vertical line load in rectangular coordinates

10.5 LINE LOADS

The basic equation used for computing σ_z, at any point P in an elastic semi-infinite mass is Eq. (10.1) of Boussinesq. By applying the principle of his theory, the stresses at any point in the mass due to a line load of infinite extent acting at the surface may be obtained. The state of stress encountered in this case is that of a plane strain condition. The strain at any point P in the Y-direction parallel to the line load is assumed equal to zero. The stress σ_y normal to the xz-plane (Fig. 10.4) is the same at all sections and the shear stresses on these sections are zero. By applying the theory of elasticity, stresses at any point P (Fig. 10.4) may be obtained either in polar coordinates or in rectangular coordinates. The vertical stress σ_z at point P may be written in rectangular coordinates as

$$\sigma_z = \left(\frac{q}{z}\right)\frac{2/\pi}{\left[1+(x/z)^2\right]^2} = \frac{q}{z}I_z \tag{10.4}$$

where, I_z is the influence factor equal to 0.637 at $x/z = 0$.

Example 10.4

Three parallel strip footings 3 m wide each and 5 m apart centre to centre transmit contact pressures of 200, 150 and 100 kN/m² respectively. Calculate the vertical stress due to the combined loads beneath the centers of each footing at a depth of 3 m below the base. Assume the footings are placed at a depth of 2 m below the ground surface. Use Boussinesq's method for line loads.

Solution

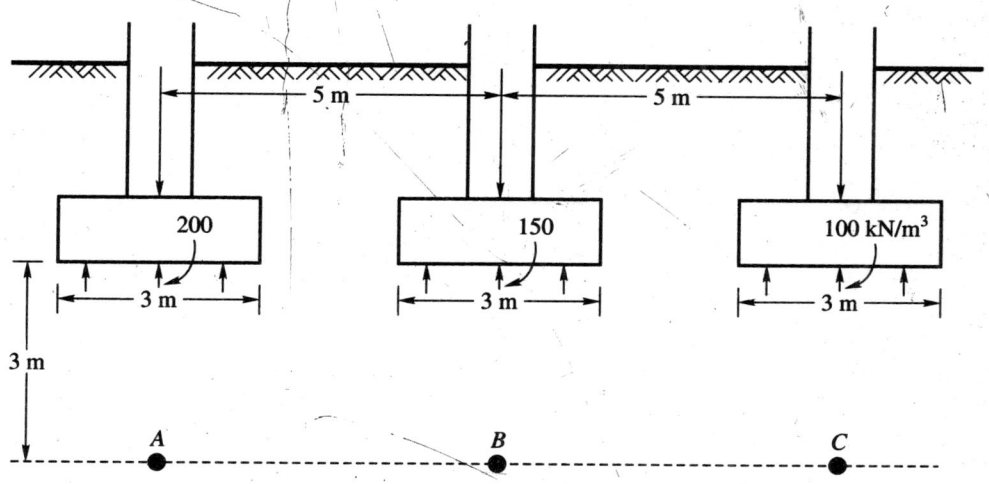

Fig. Ex. 10.4 Three parallel footings

From Eq. (10.4), we have

$$\sigma_z = \left(\frac{q}{z}\right)\frac{2/\pi}{\left[1+(x/2)^2\right]^2} = \frac{q}{z}I_z$$

The stress at A (Fig. Ex. 10.4), we have

$$(\sigma_z)_A = \frac{2\times 200}{3.14\times 3}\left[\frac{1}{1+(0/3)^2}\right]^2 + \frac{2\times 150}{3.14\times 3}\left[\frac{1}{1+(5/3)^2}\right]^2$$

$$+ \frac{2\times 100}{3.14\times 3}\left[\frac{1}{1+(10/3)^2}\right]^2 = 45\,\text{kN/m}^2$$

The stress a B

$$(\sigma_z)_B = \frac{2\times 200}{3\pi}\left[\frac{1}{1+(5/3)^2}\right]^2 + \frac{2\times 150}{3\pi}\left[\frac{1}{1+(0/3)^2}\right]^2$$

$$+ \frac{2\times 100}{3\pi}\left[\frac{1}{1+(5/3)^2}\right]^2 = 36.3\,\text{kN/m}^2$$

The stress a C

$$(\sigma_z)_C = \frac{2\times 200}{3\pi}\left[\frac{1}{1+(10/3)^2}\right]^2 + \frac{2\times 150}{3\pi}\left[\frac{1}{1+(5/3)^2}\right]^2 + \frac{2\times 100}{3\pi} = 23.74\,\text{kN/m}^2.$$

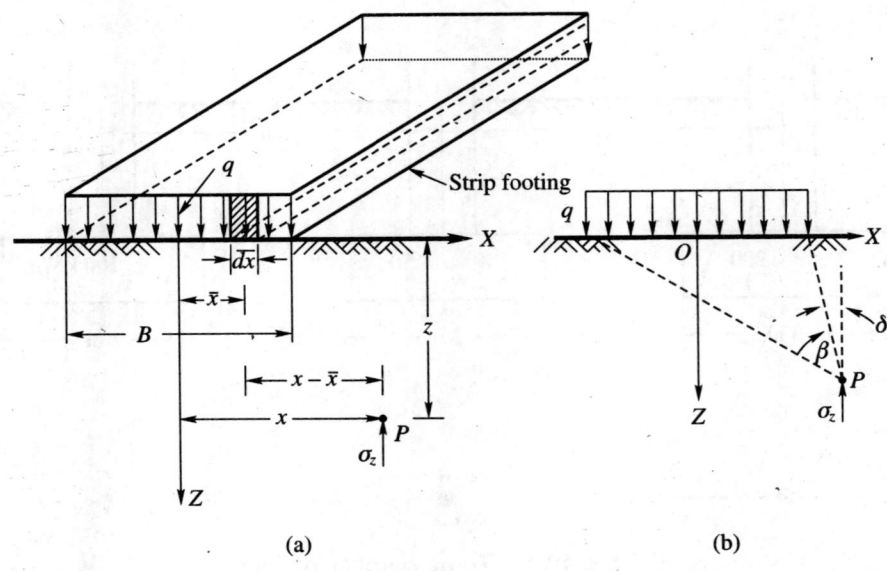

(a)   (b)

Fig. 10.5   Strip load

10.6   STRIP LOADS

The state of stress encountered in this case also is that of a plane strain condition. Such conditions are found for structures extended very much in one direction, such as strip and wall foundations, foundations of retaining walls, embankments, dams and the like. For such structures the distribution of stresses in any section (except for the end portions of 2 to 3 times the widths of the structures from its end) will be the same as in the neighboring sections, provided that the load does not change in directions perpendicular to the plane considered.

Figure 10.5(a) shows a load q per unit area acting on a strip of infinite length and of constant width B. The vertical stress at any arbitrary point P due to a line load of $q\,d\overline{x}$ acting at $x = \overline{x}$ can be written from the Eq. (10.4) as

$$d\sigma_z = \frac{2q}{\pi} \frac{z^3}{[(x-\overline{x})^2 + z^2]^2} \tag{10.5}$$

Applying the principle of superposition, the total stress σ_z at point P due to a strip load distributed over a width $B(= 2b)$ may be written as

$$\sigma_z = \frac{2q}{\pi} \int_{-b}^{+b} \frac{z^3}{[(x-\overline{x})^2 + z^2]^2} dx$$

or $\sigma_z = \dfrac{q}{\pi} \left[\tan^{-1} \dfrac{z}{x-b} - \tan^{-1} \dfrac{z}{x+b} - \dfrac{2bz(x^2 - b^2 - z^2)}{(x^2 - b^2 + z^2)^2 + 4b^2 z^2} \right]$ $\tag{10.6}$

The non-dimensional values of σ_z/q are given graphically in Fig. 10.6. Eq. (10.6) can be expressed in a more convenient form as

$$\sigma_z = \frac{q}{\pi}\left[\beta + \sin\beta\cos(\beta + 2\delta)\right] \tag{10.7}$$

where β and δ are the angles as shown in Fig. 10.5(b). Equation (10.7) is very convenient for computing σ_z, since the angles β and δ can be obtained graphically for any point P. The principal stresses σ_1 and σ_3 at any point P may be obtained from the equations.

$$\sigma_1 = \frac{q}{\pi}(\beta + \sin\beta) \tag{10.8}$$

$$\sigma_1 = \frac{q}{\pi}(\beta - \sin\beta) \tag{10.9}$$

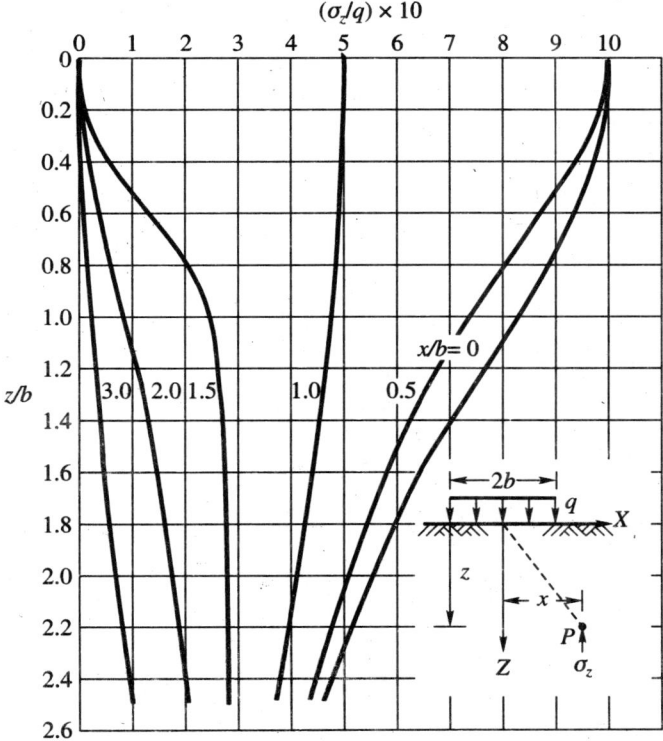

Fig. 10.6   Non-dimensional values of σ_z/q for strip load

Example 10.5

The wall of a building foundation of width 3 m, length 30 m carries a uniformly distributed load 300 kN/m². Compute the vertical stress along the centre of the foundation at a depth of 3.0 m below the bottom of the foundation. Ignore the surcharge effect of the soil over the base level of the foundation.

Solution

The state of stress encountered in this case is that of a plane strain condition. The equation applicable to strip loads is to be used here. Use chart given in Fig. 10.6, for computing vertical stress σ_z.

Half width b of foundation $3/2 = 1/5$ m, $z/b = 3.0/1.5 = 2.0$, $x/b = 0$

From Fig. 10.6 $(\sigma_z/q) \times 10 = 5.6$ or $(\sigma_z/q) = 0.56$

$$(\sigma_z/q) = 0.56q = 0.5 \times 300 = 169 \text{ kPa.}$$

10.7 STRESSES BENEATH THE CORNER OF A RECTANGULAR FOUNDATION

Consider an infinitely small unit of area of size $db \times dl$, shown in Fig. 10.7. The pressure acting on the small area may be replaced by a concentrated load dQ applied to the center of the area.

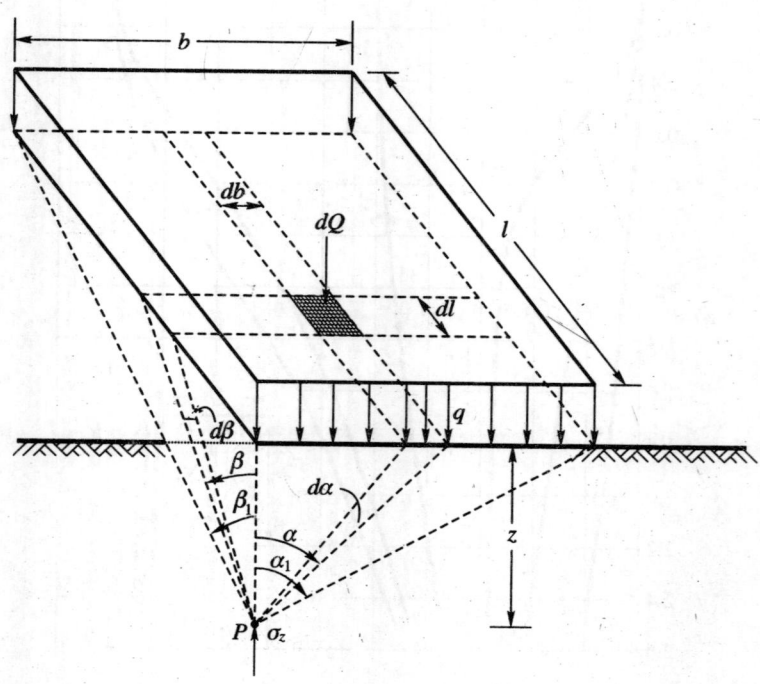

Fig. 10.7 Vertical stress under the corner of rectangular foundation

Hence

$$dQ = q\, db\, dl \qquad\qquad (10.10)$$

The increase of the vertical stress σ_z due to the load dQ can be expressed per Eq. (10.11) as

$$d\sigma_z = \frac{dQ}{2\pi} \frac{3z^3}{(z^2 + r^2)^{5/2}} \qquad\qquad (10.11)$$

The stress produced by the pressure q over the entire rectangle $b \times l$ can then be obtained by expressing dl, db and r in terms of the angles α and β, and integrating

$$\sigma_z = \int_{\alpha=0}^{\alpha=\alpha_1} \int_{\beta=0}^{\beta=\beta_1} d\sigma_z \qquad\qquad (10.12)$$

There are several forms of solution for Eq. (10.12). The one that is normally used is of the following form

$$\sigma_z = q\, \frac{1}{4\pi} \left[\frac{2mn(m^2 + n^2 + 1)^{1/2}}{m^2 + n^2 + m^2 n^2 + 1} \frac{m^2 + n^2 + 2}{m^2 + n^2 + 1} + \tan^{-1} \frac{2mn(m^2 + n^2 + 1)^{1/2}}{m^2 + n^2 - m^2 n^2 + 1} \right] \qquad (10.13)$$

or $\quad \sigma_z = ql \qquad\qquad (10.14)$

wherein, $m = b/z$, $n = l/z$, are pure numbers. I is a dimensionless factor and represents the influence of a surcharge covering a rectangular area on the vertical stress at a point located at a depth z below one of its corners.

Equation (10.14) is presented in graphical form in Fig. 10.8. This chart helps to compute pressures beneath loaded rectangular areas. The chart also shows that the vertical pressure is not materially altered if the length of the rectangle is greater than ten times its width. Fig. 10.9 may also be used for computing the influence value I based on the values of m and n and may also be used to determine stresses below points that lie either inside or outside the loaded areas as follows.

When the Point is Inside

Let O be an interior point of a rectangular loaded area $ABCD$ shown in Fig. 10.10(a). It is required to compute the vertical stress σ_z below this point O at a depth z from the surface. For this purpose, divide the rectangle $ABCD$ into four rectangles marked 1 to 4 in the Fig. 10.10(a) by drawing lines through O. For each of these rectangles, compute the ratios z/b. The influence value I may be obtained from Fig. 10.8 or 10.9 for each of these ratios and the total stress at P is therefore

$$\sigma_z = q(I_1 + I_2 + I_3 + I_4) \qquad\qquad (10.15)$$

When the Point is Outside

Let O be an exterior point of loaded rectangular area $ABCD$ shown in Fig. 6.9(b). It is required to compute the vertical stress σ_z below point O at a depth z from the surface.

Construct rectangles as shown in the figure. The point O is the corner point of the rectangle OB_1CD_1. From the figure it can be seen that

$$\text{Area } ABCD = OB_1CD_1 - OB_1BD_2 - OD_1DA_1 + OA_1AD_2 \qquad (10.16)$$

The vertical stress at point P located at a depth z below point O due to a surcharge q per unit area of $ABCD$ is equal to the algebraic sum of the vertical stresses produced by loading each one of the areas listed on the right hand side of the Eq. (10.16) with q per unit of area. If I_1 to I_4 are the influence factors of each of these areas, the total vertical stress is

$$\sigma_z = q(I_1 - I_2 - I_3 + I_4) \qquad (10.17)$$

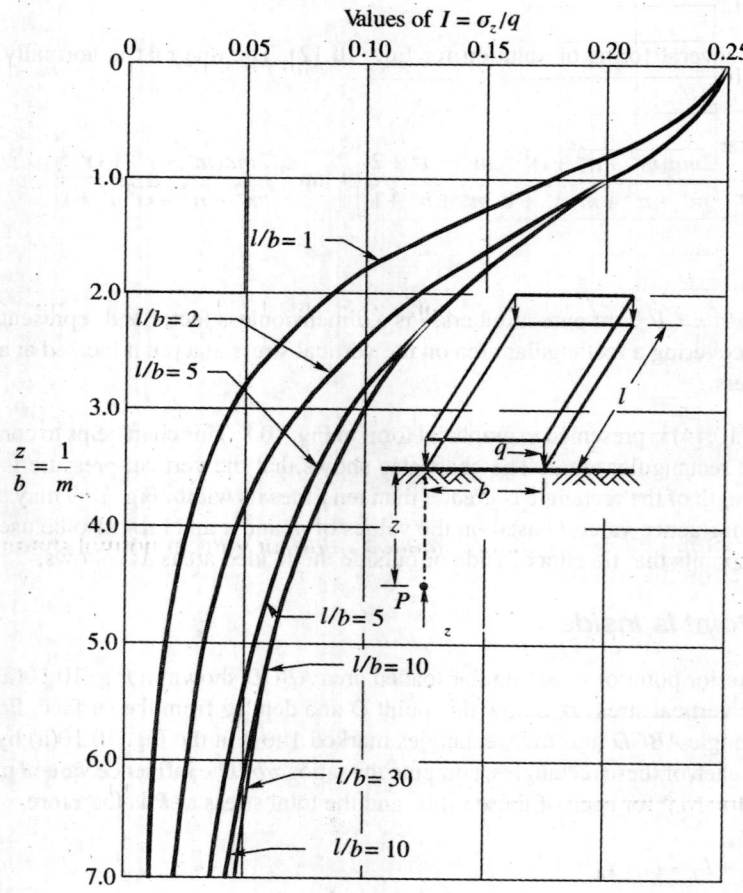

Fig. 10.8 Chart for computing σ_z below the corner of a rectangular foundation
(after Steinbrenner)

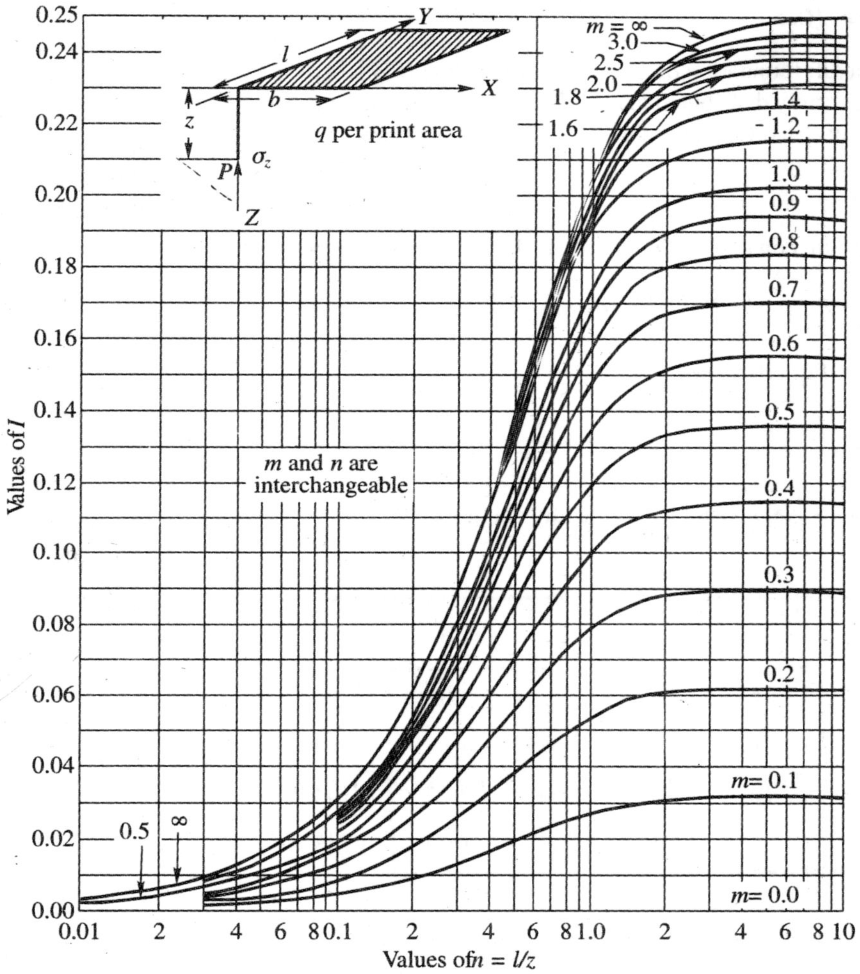

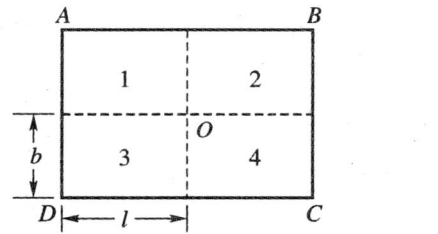

Fig. 10.9 Graph for determining influence value for vertical normal stress σ_z at point P located beneath one corner of a uniformly loaded rectangular area. (After Fadum)

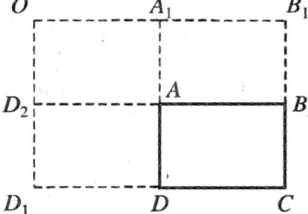

(a)When the point 'O' is within the rectangle   (b)When the point 'O' is outside the rectangle

Fig. 10.10 Computation of vertical stress below a point

Example 10.6

ABCD is a raft foundation of a multi-story building [Fig. 10.10(b)] wherein $AB = 20$ m, and $BC = 12$ m. The uniformly distributed load q over the raft is 350 kN/m². Determine σ_z at a depth of 6 m below point O [Fig. 10.10(b)] wherein $AA_1 = 4$ m and $A_1O = 6$ m. Use Fig. 10.9.

Solution

Rectangles are constructed as shown in [Fig. 10.10(b)]. As per this figure.

$$\text{Area } ABCD = OB_1CD_1 - OB_1BD_2 - OD_1DA_1 + OA_1AD_2$$

As per Eq. (10.17)

$$\sigma_z = q(I_1 - I_2 - I_3 + I_4) = 350(0.245 - 0.168 - 0.194 + 0.145) = 9.8 \text{ kNm}^2$$

The same value can be obtained by using Fig. 10.8.

Rectangle	l (m)	b (m)	m	n	I
OB_1CD_1	26	16	2.67	4.33	0.245
OB_1BD_2	26	4	2 0.67	4.33	0.168
OD_1DA_1	16	6	1.00	2.67	0.194
OA_1AD_2	6	4	0.67	1.00	0.145

Example 10.7

A rectangular raft of size 30×12 m founded on the ground surface is subjected to a uniform pressure of 150 kN/m². Assume the center of the area as the origin of coordinates (0, 0), and corners with coordinates (6, 15). Calculate the induced stress at a depth of 20 m by the exact method at location (0, 0).

Solution

Divide the rectangle 12×30 m into four equal parts of size 6×15 m.

The stress below the corner of each footing may be calculated by using charts given in Fig. 10.8 or Fig. 10.9. Here Fig. 10.8 is used.

For a rectangle 6×15 m, $z/b = 20/6 = 3.34$, $l/b = 15/6 = 2.5$.

For $z/b = 3.34$, $l/b = 2.5$, $\overline{\sigma}_z/q = 0.07$

Therefore, $\sigma_z = 4\,\overline{\sigma}_z = 4 \times 0.07q = 4 \times 0.07 \times 150 = 42 \text{ kN/m}^2$.

10.8 STRESSES UNDER UNIFORMLY LOADED CIRCULAR FOOTING

Stresses Along the Vertical Axis of Symmetry

Figure 10.11 shows a plan and section of the loaded circular footing. The stress required to be determined at any point P along the axis is the vertical stress σ_z and the radial stress σ_r.

Let dA be an elementary area considered as shown in Fig. 10.11. dQ may be considered as the point load acting on this area which is equal to $q\,dA$. We may write

$$dQ = q\,dA = qr\,d\theta\,dr \tag{10.18}$$

The vertical stress $d\sigma$ at point P due to point load dQ may be expressed [Eq. (10.1a)] as

$$d\sigma_z = \frac{3q}{2\pi}\frac{z^3 r\,d\theta\,dr}{(r^2+z^2)^{5/2}} \tag{10.19}$$

The integral form of the equation for the entire circular area may be written as

$$\sigma_z = \int_{\theta=0}^{\theta=2\pi}\int_{r=0}^{r=R_0} d\sigma_z = \frac{3qz^3}{2\pi}\int_{\theta=0}^{\theta=2\pi}\int_{r=0}^{r=R_0}\frac{r\,d\theta\,dr}{(r^2+z^2)^{5/2}}$$

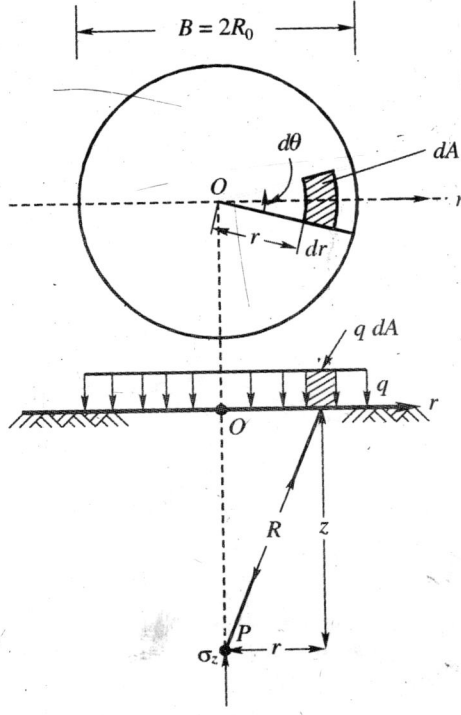

Fig. 10.11 Vertical stress under uniformly loaded circular footing

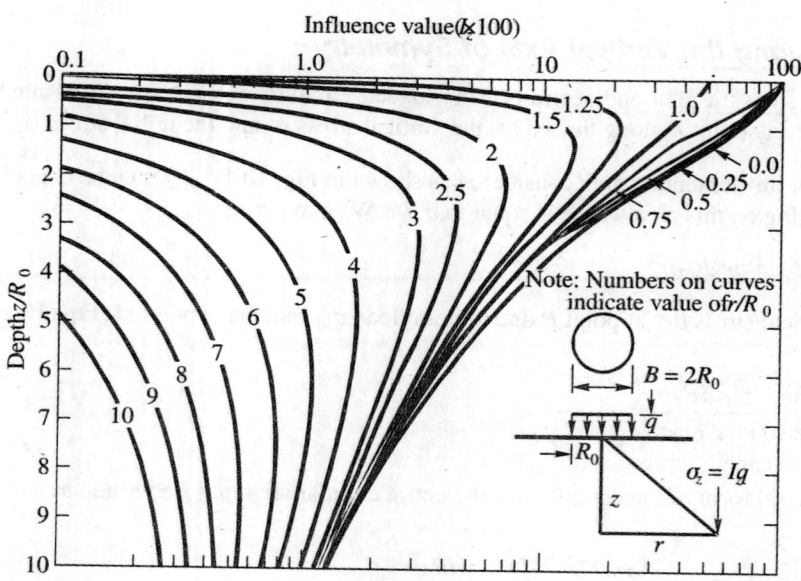

Fig. 10.12 Influence diagram for vertical normal stress at various points within an elastic half-space under a uniformly loaded circular area. (After Foster and Ahlvin, 1954)

On integration we have, $\sigma_z = q\left[1 - \dfrac{z^3}{(R_0^2 + z^2)^{3/2}}\right]$ (10.20)

or $\sigma_z / q = 1 - \left[\dfrac{1}{1 + (R_0 / z)^2}\right]^{3/2} = I_z$ (10.21)

where, I_z is the influence coefficient. The stress at any point P on the axis of symmetry of a circular loaded area may be calculated by the use of Eq. (10.21), vertical stresses σ_z may be calculated by using the influence coefficient diagram given in Fig. 10.12.

Example 10.8

A water tank is required to be constructed at a place with a circular foundation having a diameter of 16 m founded at a depth of 2 m below the ground surface. The estimated distributed load on the foundation is 325 kN/m². Assuming that the subsoil extends to a great depth and is isotropic and homogeneous, determine the stresses σ_z at points (i) $z = 8$ m, $r = 0$, (ii) $z = 8$ m, $r = 8$ m, (iii) $z = 16$ m, $r = 0$ and (iv) $z = 16$ m, $r = 8$ m, where r is the radial distance from the central axis. Neglect the effect of the depth of the foundation on the stresses (Use Fig. 10.12).

Solution

$q = 325$ kN/m², $R_0 = 8$ m. The results are given in a tabular form as follows:

Point		z/R_0	r/R_0	I	σ_z kN/m²
(i)	(8, 0)	1	0	0.7	227.5
(ii)	(8, 8)	1	1.0	0.33	107.25
(iii)	(16, 0)	2	0	0.3	97.5
(iv)	(16, 8)	2	1.0	0.2	65

Example 10.9

For a raft of size 30 × 12 m, compute the stress at 20 m depth below the centre of the raft by assuming that the rectangle can be represented by an equivalent circle.

Solution

The radius of a fictitious circular footing of area equal to the rectangular footing of size 30 × 12 m is

$$\pi R_0^2 = 30 \times 12 = 360 \text{ sq.m or} \qquad R_0 = \sqrt{\frac{360}{\pi}} = 10.70 \text{ m}$$

Use Eq. (10.12) for computing σ_z at 20 m depth

Now, $z/R_0 = \dfrac{20}{10.7} = 1.9$, and $r/R_0 = 0$. From Fig. 10.12, $I_z = 0.3$

Therefore, $\sigma_z = 0.3\, q = 0.3 \times 150 = 45 \text{ kN/m}^2$

The value for σ_z obtained by this method agrees very well with the point load method given in Ex. 10.14.

Example 10.10

A ring footing of external diameter 8 m and internal diameter 4 m rests at a depth 2 m below the ground surface. It carries a load intensity of 150 kN/m². Find the vertical stress at depths of 2, 4 and 8 m along the axis of the footing below the footing base. Neglect the effect of the excavation on the stress.

Solution

From Eq. (10.21) we have,

$$\frac{\sigma_z}{q} = 1 - \left[\frac{1}{1 + (R_0/z)^2} \right]^{3/2} = I_z$$

where, q = contact pressure 150 kN/m², I_z = Influence coefficient.

The stress σ_z at any depth z on the axis of the ring is expressed as

$$\sigma_z = \sigma_{z_1} - \sigma_{z_2} = q(I_{z_1} - I_{z_2})$$

where σ_{z_1} = stress due to the circular footing of diameter 8 m, and $I_z = I_{z_1}$ and $R_0 / z = (R_1 / z)$

σ_{z_2} = stress due to the footing of diameter 4 m, $I_z = I_{z_2}$ and $R_0 / z = (R_2 / z)$

The values of I_z may be obtained from Table 10.1A for various values of R_0/z. The stress σ_z at depths 2, 4 and 8 m are given below:

Depth	R_1/z	I_{z_1}	R_2/z	I_{z_2}	$(I_{z_1} - I_{z_2}) q = \sigma_z$ kN/m²
2	2	0.911	1.0	0.692	39.7
4	1.0	0.647	0.5	0.287	54.0
8	0.5	0.285	0.25	0.087	30.0

10.9   VERTICAL STRESS BENEATH LOADED AREAS OF IRREGULAR SHAPE

Newmark's Influence Chart

When the foundation consists of a large number of footings or when the loaded mats or rafts are not regular in shape, a chart developed by Newmark (1942) is more practical than the methods explained before. It is based on the following procedure. The vertical stress σ_z below the centre of a circular area of radius R which carries uniformly distributed load q is determined per Eq. (10.21).

$$\frac{\sigma_z}{q} = 1 - \left[\frac{1}{1 + (R/z)^2} \right]^{3/2} \qquad (10.22)$$

It may be seen from Eq. (6.21) that when $R/z = \infty$, $\sigma_z/q = 1$, that is $\sigma_z = q$. This indicates that if the loaded area extends to infinity, the vertical stress in the semi-infinite solid at any depth z is the same as unit load q at the surface. If the loaded area is limited to any given radius R_1 it is possible to determine from Eq. (10.21) the ratios R/z for which the ratio of σ_z/q may have any specified value, say 0.8 or 0.6. Table 10.1A gives the ratios of R/z for different values of σ_z/q.

Table 10.1A may be used for the computation of vertical stress σ_z at any depth z below the centre of a circular loaded area of radius R. For example, at any depth z, the vertical stress $\sigma_z = 0.8\ q$ if the radius of the loaded area at the surface is $R = 1.387\ z$. At the same depth, the vertical stress is $\sigma_z = 0.7q$ if $R = 1.110\ z$. If instead of loading the whole area, if only the annular space between the circles of radii 1.387 z and 1.110 z are loaded, the vertical stress at z at the centre of the circle is $\Delta\sigma_z = 0.8\ q - 0.7\ q = 0.1q$. Similarly, if the annular space between circles of radii 1.110 z and 0.917 z are loaded, the vertical stress at the same depth z is $\Delta\sigma_z = 0.7\ q - 0.6\ q = 0.1\ q$. We may therefore draw a series of

TABLE 10.1A

Values of R/z for different values of σ_z/q

$\sigma_z/qR/z$	σ_z/q	R/z
0.00	0.000	0.80	1.387
0.10	0.270	0.90	1.908
0.20	0.401	0.92	2.094
0.30	0.518	0.94	2.351
0.40	0.637	0.96	2.748
0.50	0.766	0.98	3.546
0.60	0.917	1.00	∞
0.70	1.110		

concentric circles on the surface of the ground in such a way that when the annular space between any two consecutive circles is loaded with a load q per unit area, the vertical stress $\Delta\sigma_z$ produced at any depth z below the center remains a constant fraction of q. We may write, therefore,

$$\Delta\sigma_z = Cq \tag{10.23a}$$

where C is constant. If an annular space between any two consecutive concentric circles is divided into n equal blocks and if any one such block is loaded with a distributed load q, the vertical stress produced at the center is, therefore,

$$\frac{\Delta\sigma_z}{n} = \frac{C}{n}q = C_i q \tag{10.23b}$$

$$\frac{\Delta\sigma_z}{n} = C_i \text{ when } q = 1.$$

That is, a load $q = 1$ covering one of the blocks will produce a vertical stress C_i. In other words, the *influence value* of each loaded block is C_i. If the number of loaded blocks is N, and if the intensity of load is q per unit area, the total vertical stress at depth z below the center of the circle is

$$\sigma_z = C_i Nq \tag{10.24}$$

The graphical procedure for computing the vertical stress σ_z due to any surface loading is as follows.

Select some definite scale to represent depth z. For instance a suitable length AB in cm as shown in Fig. 10.13Aa to represent depth z in meters. In such a case, the scale is 1 cm = z/AB metres. The length of the radius $R_{0.8}$ which corresponds to $\sigma_z/q = 0.8$ is then equal to $1.387 \times AB$ cm, sand a circle of that radius may be drawn. This procedure may be repeated for other ratios of σ_z/q, for instance, for $\sigma_z/q = 0.7, 0.5$, etc. shown in Fig. 10.13A.

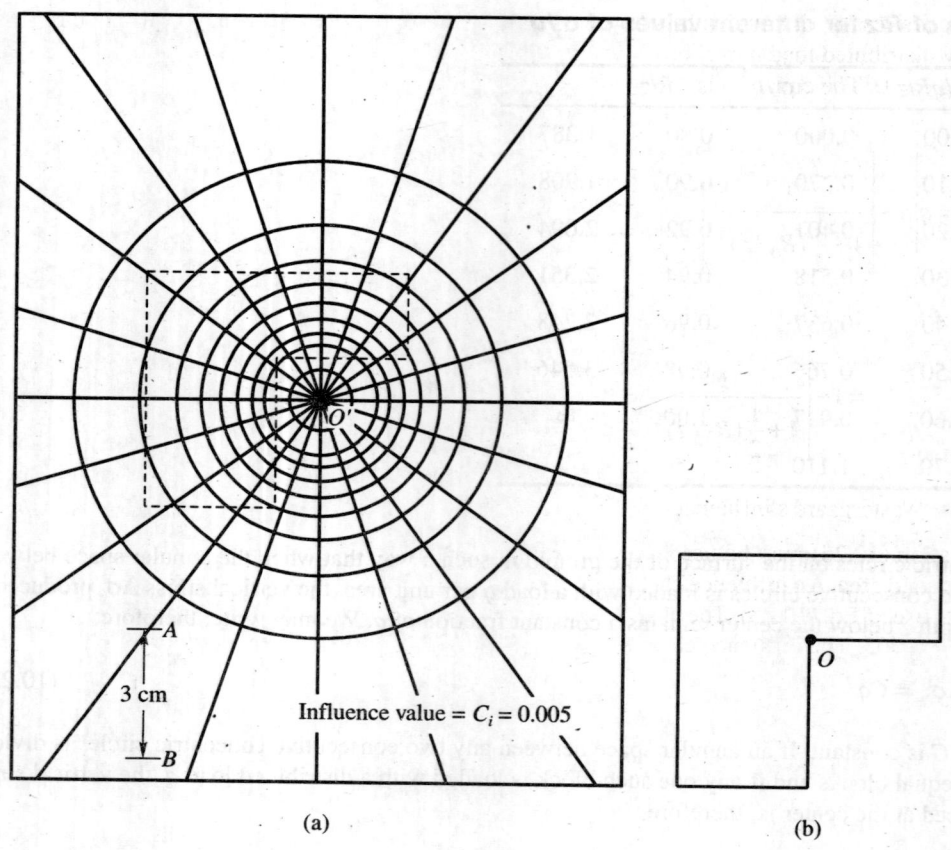

Influence value = $C_i = 0.005$

3 cm

(a)

(b)

Fig. 10.13A   Newmark's influence chart

The annular space between the circles may be divided into n equal blocks, and in this case $n = 20$. The influence value C_i is therefore equal to $0.1/20 = 0.005$. A plan of the foundation is drawn on a tracing paper to a scale such that the distance AB on the chart corresponds to the depth z at which the stress σ_z is to be computed. For example, if the vertical stress at a depth of 9 m is required, and if the length AB chosen is 3 cm, the foundation plan is drawn to a scale of 1 cm = 9/3 = 3 m. In case the vertical stress at a depth 12 m is required, a new foundation plan on a separate tracing paper is required. The scale for this plan is 1 cm = 12/AB = 12/3 = 4 m.

This means that a different tracing has to be made for each different depth whereas the chart remains the same for all. Fig. 10.13A(b) gives a foundation plan, which is loaded with a uniformly distributed load q per unit area. It is now required to determine the vertical stress σ_z at depth vertically below point O shown in the figure. In order to determine σ_z, the foundation plan is laid over the chart in such a way that the surface point O coincides with the center O' of the chart as shown in Fig. 10.13A(a). The number of small blocks covered by the foundation plan is then counted. Let this number be N. Then the value of σ_z at depth z below O is

$$\sigma_z = C_i Nq, \text{ which is the same as Eq. (10.24)}.$$

Westergaard Influence Chart

The vertical stress σ_z at any depth below the centre of a circular area of radius R_0 which carries a uniformly distributed load q may be obtained from Westergaard's equation for point load (Eq. 10.3) by putting $\mu = 0$. The equation is

$$\sigma_z = q \left[1 - \left[\frac{1}{1 + \frac{1}{2}(R_0/z)^2} \right]^{1/2} \right] \qquad (10.25a)$$

or

$$\text{or} \quad \frac{\sigma_z}{q} = I_w = 1 - \left[\frac{1}{1 + \frac{1}{2}(R_0/z)^2} \right]^{1/2} \qquad (10.25b)$$

where I_w = Westergaard's influence value.

Equation (10.25) may be used to compute stress at any point P on the axis of symmetry of a circular loaded area. An influence chart may be constructed in the same way as the Newmark's chart by making use of Eq. (10.25). The chart is shown in Fig. 10.13B for $\mu = 0$. The influence value C_i for the chart = 0.005. The influence chart is constructed from the values given in Table 10.1B.

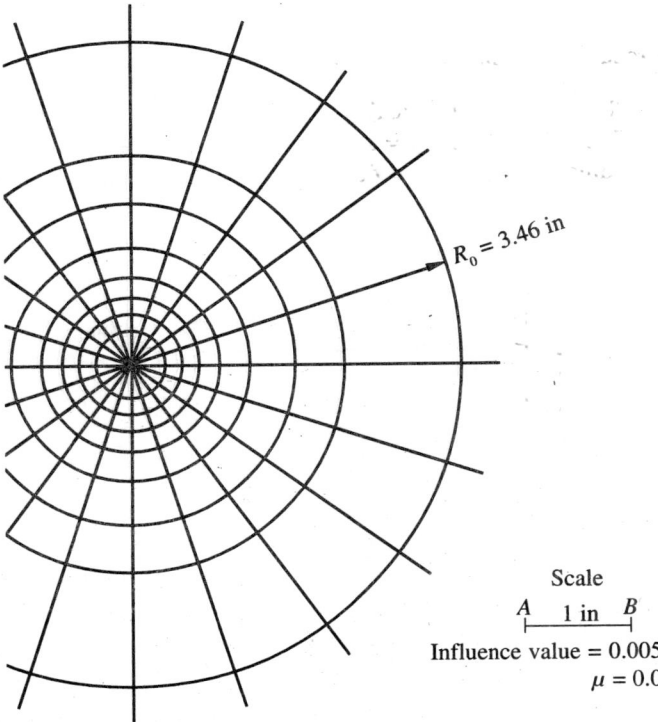

Fig. 10.13B Influence chart for vertical pressure at any depth $z = AB$ in the soil, based on the Westergaard theory

TABLE 10.1B

Values of R/z for various values σ_z/q for Westergaard's Influence chart for $\mu = 0$

σ_z/q	σ_z/q	σ_z/q	r/z
0.00	0.000	0.76	2.860
0.10	0.343	0.80	3.464
0.20	0.530	0.82	3.864
0.30	0.721	0.84	4.363
0.40	0.943	0.88	5.850
0.60	1.620	0.90	7.036
0.70	2.249	0.94	11.764
0.72	2.424	0.98	35.348

Example 10.11

A raft foundation of the size given in Fig. Ex. 10.11 carries a uniformly distributed load of 300 kN/ m². Estimate the vertical pressure at a depth 9 m below the point O marked in the figure.

Solution

The depth at which σ_z required is 9 m.

As per Fig. 10.13A, the scale of the foundation plan is $AB = 9$ m or 1 cm = 3 m. The foundation plan is required to be made to a scale of 1 cm = 3 m on tracing paper. This plan is superimposed on Fig. 10.13A with O coinciding with the centre of the chart. The plan is shown in dotted lines in Fig. 10.13A.

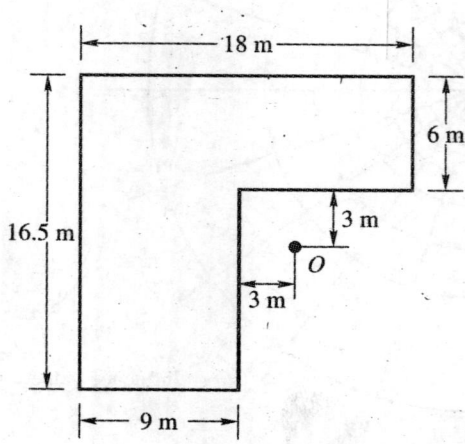

Fig. Ex. 10.11

Number of loaded blocks occupied by the plan, $N = 62$

Influence value, $C_i = 0.005$, $q = 300$ kN/m²

The vertical stress, $\sigma_z = C_i \, Nq = 0.005 \times 62 \times 300 = 93$ kN/m².

10.10 EMBANKMENT LOADINGS

Long earth embankments with sloping sides represent trapezoidal loads. When the top width of the embankment reduces to zero, the load becomes a triangular strip load. The basic problem is to determine stresses due to a linearly increasing vertical loading on the surface.

Linearly Increasing Vertical Loading

Figure 10.14(a) shows a linearly increasing vertical loading starting from zero at A to a finite value q per unit length at B. Consider an elementary strip of width db at a distance b from A. The load per unit length may be written as

$$dq = (q/a) \, b \, db$$

If dq is considered as a line load on the surface, the vertical stress d 10.14A at P [Fig. 10.14(a)] due to dq may be written from Eq. (10.4) as

$$d\sigma_z = \left(\frac{1}{a}\right)\left(\frac{2q}{\pi}\right)\frac{z^3 b \, db}{\left[(x-b)^2 + z^2\right]^2}$$

Therefore,

$$\sigma_z = \int d\sigma_z = \left(\frac{1}{a}\right)\left(\frac{2q}{\pi}\right)\int_{b=0}^{b=a}\frac{z^3 b \, db}{\left[(x-b)^2 + z^2\right]^2}$$

on integration, $\sigma_z = \dfrac{q}{2\pi}\left(\dfrac{2x}{a}\alpha - \sin 2\beta\right) = qI_z$ (10.26)

where I_z is non-dimensional coefficient whose values for various values of x/a and z/a are given in Table 10.2.

If the point P lies in the plane BC [Fig. 6.13(a)], then $\beta = 0$ at $x = a$. Eq. (6.25) reduces to

$$\sigma_z = \frac{q}{\pi}(\alpha)$$ (10.27)

The equations for σ_z and τ_{xz} can be developed in the same way as for σ_z. The final equations are

$$\sigma_x = \frac{q}{2\pi}\left(\frac{x}{a}\alpha - 2.3\frac{z}{a}\log\frac{R_1^{\,2}}{R_2^{\,2}} + \sin 2\beta\right)$$ (10.28)

$$\tau_{xz} = \frac{q}{2\pi}\left(1 + \cos 2\beta - \frac{z}{\bar{a}}\alpha\right)$$ (10.29)

where $\bar{a} = a/2$

Figures 10.14(b) and (c) show the distribution of stress σ_z on vertical and horizontal sections under the action of a triangular loading as a function of q. The maximum vertical stress occurs below the centre of gravity of the triangular load as shown in Fig. 10.14(c).

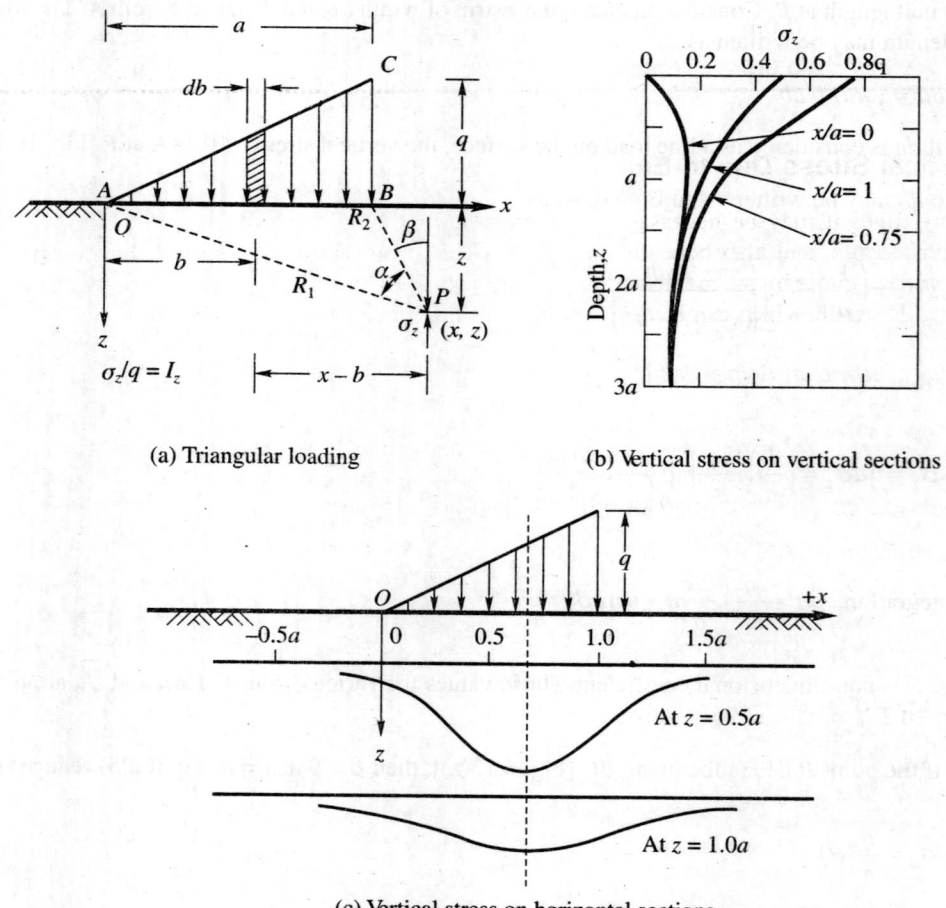

(a) Triangular loading

(b) Vertical stress on vertical sections

(c) Vertical stress on horizontal sections

Fig. 10.14   Stresses in a semi-infinite mass due to triangular loading on the surface

TABLE 10.2

I_z for Triangular load Eq. (10.26)

x/a				z/a			
	0.00	0.5	1.0	1.5	2	4	6
−1.500	0.00	0.002	0.014	0.020	0.033	0.051	0.041
−1.00	0.00	0.003	0.025	0.048	0.061	0.060	0.041
0.00	0.00	0.127	0.159	0.145	0.127	0.075	0.051
0.50	0.50	0.410	0.275	0.200	0.155	0.085	0.053
0.75	0.75	0.477	0.279	0.202	0.163	0.082	0.053
1.00	0.50	0.353	0.241	0.185	0.153	0.075	0.053
1.50	0.00	0.056	0.129	0.124	0.108	0.073	0.050
2.00	0.00	0.017	0.045	0.062	0.069	0.060	0.050
2.50	0.00	0.003	0.013	0.041	0.050	0.049	0.045

Vertical Stress Due to Embankment Loading

Many times it may be necessary to determine the vertical stress σ_z beneath road and railway embankments, and also beneath earth dams. The vertical stress beneath embankments may be determined either by the method of superposition by making use of Eq. (10.27) or by making use of a single formula which can be developed from first principle

σ_z by method of Superposition

Consider an embankment given in Fig. 10.15. σ_z at P may be calculated as follows:

The trapezoidal section of the embankment $ABCD$, may be divided into triangular sections by drawing a vertical line through the point P as shown in Fig. 10.15. We may write

$$ABCD = AGE + FGB - EDJ - FJC$$

If σ_{z_1}, σ_{z_2}, σ_{z_3}, and σ_{z_4} are the vertical stresses at point P due to the loadings of figures AGE, FGB, EDJ and FJC respectively, the vertical stress σ_z due to loading of figure $ABCD$ may be written as

$$\sigma_z = \sigma_{z_1} + \sigma_{z_2} - \sigma_{z_3} - \sigma_{z_4}$$

By applying the principle of superposition for each of the triangle by making use of Eq. (10.27), we get

$$\sigma_z = \frac{q}{\pi}\left[(\alpha_1 + \alpha_2 + \alpha_3 + \alpha_4) + (b_1/a_1)\alpha_1 + (b_2/a_2)\alpha_4\right]$$

$$\sigma_z = qI_z = \frac{q}{\pi}f(a/z, b/z) \tag{10.30}$$

where I_z is influence factor for trapezoidal load which is a function of a/z and b/z.

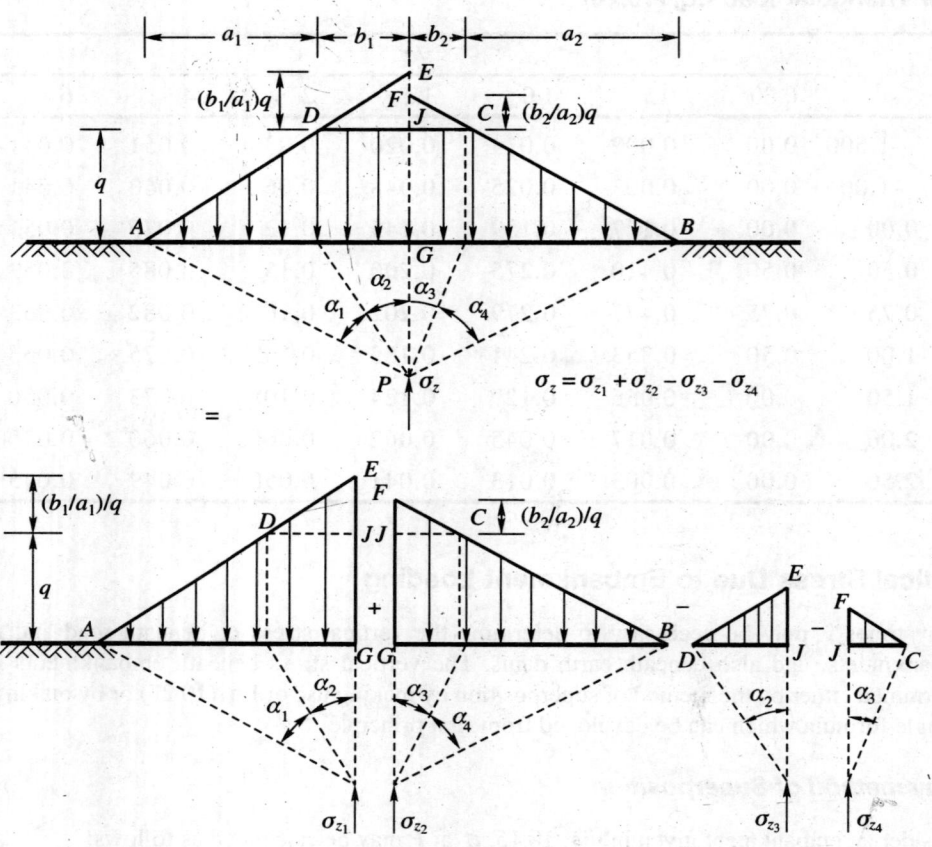

Fig. 10.15 Vertical stress due to embankent

The values of I_z for various values of a/z and b/z are given in Fig. 10.16. (After J. D. Osterberg)

σ_z from a single formula for asymmetrical trapezoidal loading

A single formula can be developed for trapezoidal loading for computing σ_z at a point P (Fig. 10.17) by applying the basic Eq. (10.30). The origin of coordinates is as shown in the figure. The final equation may be expressed as

$$\sigma_z = \frac{q}{\pi}\left[(\alpha_1 + \alpha_2 + \alpha_3) + \frac{b_1}{a_1}(\alpha_1 + R\alpha_3) + \frac{x}{a_1}(\alpha_1 - R\alpha_3)\right] \tag{10.31}$$

where α_1, α_2, and α_3 are the angles subtended at the point P in the supporting medium by the loading and $R = a_1/a_2$. When $R = 1$, the stresses are due to that of a symmetrical trapezoidal loading. When the top width is zero, i.e. when $b = 0$, $\alpha_2 = 0$, the vertical stress σ_z will be due to triangular loading. the expression for triangular loading is

$$\sigma_z = \frac{q}{\pi}\left[(\alpha_1 + \alpha_3) + \frac{x}{a_1}(\alpha_1 - R\alpha_3)\right]$$

(10.32)

Equation (10.31) and Eq. (10.32) can be used to compute σ_z at any point in the supporting medium. The angles α_1, α_2, and α_3 may conveniently be obtained by graphical procedure where these angles are expressed as radians in the equations.

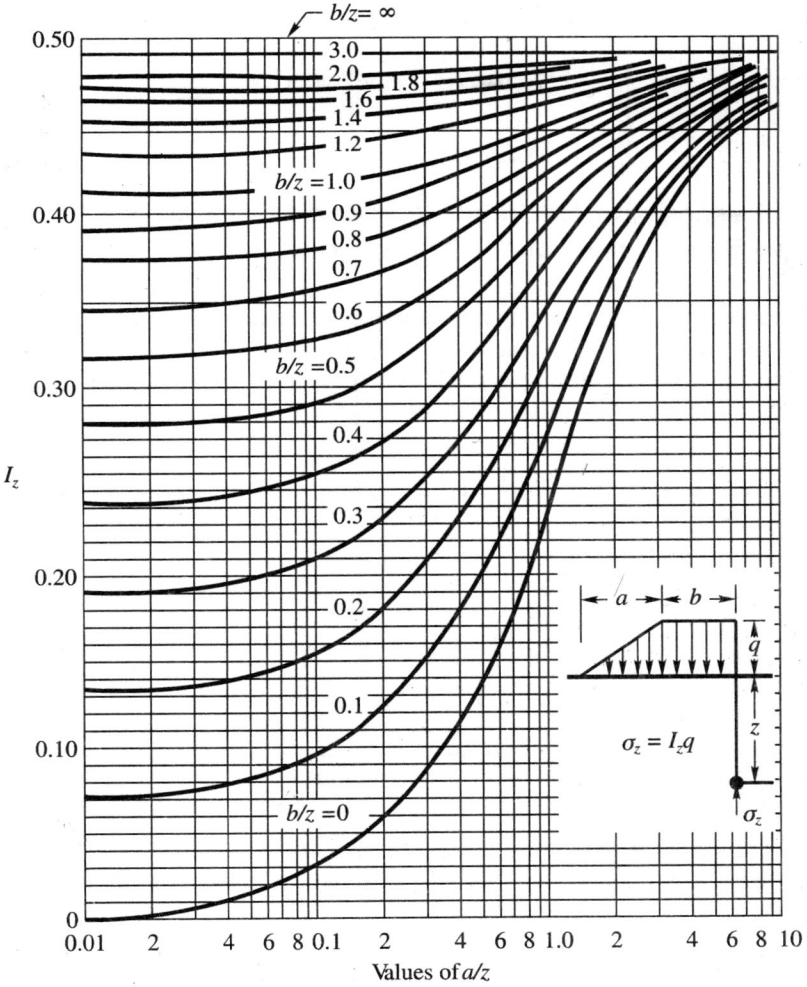

Fig. 10.16 A graph to determine compressive stresses from a load varying by straight line law (After Osterberg)

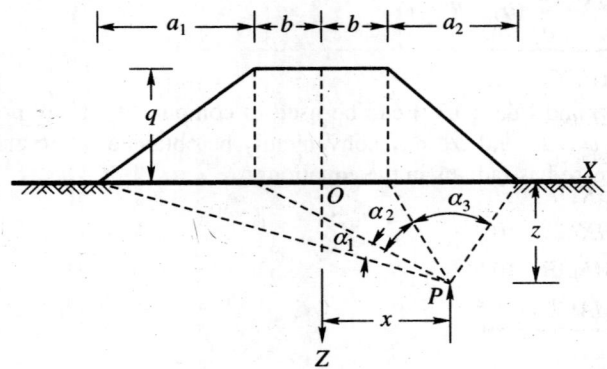

Fig. 10.17 Trapezoidal loads

Principal Stresses and Maximum Shearing Stress at any Point P in Two-dimensional Stress Space due to Trapezoidal Loading

σ_z can be calculated at a point P for any type of trapezoidal loading. By using the same principle of superposition and using Eq. (10.28) amd Eq. (10.29) and putting $\beta = 0$, the stresses σ_x and τ_{xz} can be calculated. The following relationships can be used for computing σ_1, σ_3, and τ_{max}.

$$\sigma_1, \sigma_3 = \frac{\sigma_z + \sigma_x}{2} \pm \frac{1}{2}\sqrt{(\sigma_z - \sigma_x)^2 + 4\tau^2_{zx}} \qquad (10.33)$$

$$\tau_{max} = \frac{1}{2}\sqrt{(\sigma_z - \sigma_x)^2 + 4\tau^2_{zx}} \qquad (10.34)$$

Example 10.12

A 3 m high embankment is to be constructed as shown in Fig. Ex. 10.18. If the unit weight of soil used in the embankment is 19.0 kN/m³, calculate the vertical stress due to the embankment loading at points P_1, P_2, and P_3.

Solution

$q = \gamma H = 19 \times 3 = 57$ kN/m², $z = 3$ m

The embankment is divided into blocks as shown in Fig. 10.18 for making use of the graph given in Fig. 10.16. The calculations are arranged as follows:

Point	Block	b (m)	a (m)	b/z	a/z	I
P_1	ACEF	1.5	3	0.5	1	0.396
	EDBF	4.5	3	1.5	1	0.477
P_2	AGH	0	1.5	0	0.5	0.15
	GKDB	7.5	3	2.5	1.0	0.493
	HKC	0	1.5	0	0.5	0.15
P_3	MLDB	10.5	3.0	3.5	1.0	0.498
	MACL	1.5	3.0	0.5	1.0	0.39

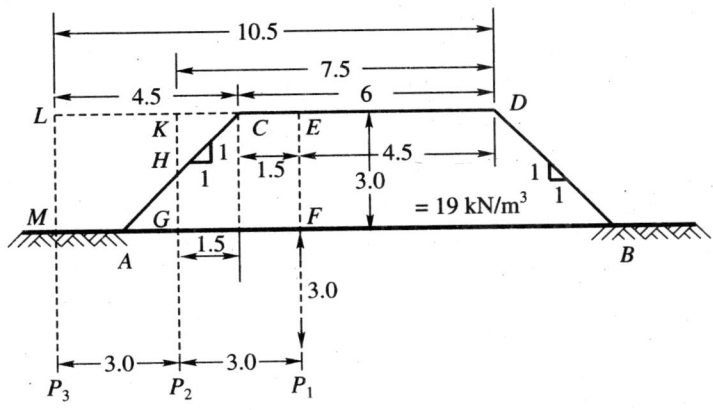

Fig. 10.18 Vertical stresses at P_1, P_2 and P_3

Vertical stress σ_z

At point P_1, $\sigma_z = (0.396+0.477)\times 57 = 55.47\,\text{kN/m}^2$

At point P_2, $\sigma_z = 0.15\times(57/2)+0.493\times 57-0.15\times(57/2)=28.1\,\text{kN/m}^2$

At point P_3, $\sigma_z = (0.498-0.385)57=6.4\,\text{kN/m}^2$

10.11 APPROXIMATE METHODS FOR COMPUTING σ_z

Two approximate methods are generally used for computing stresses in a soil mass below loaded areas. They are:

1. Use of the point load formulas such as Boussinesq's equation.

2. 2 : 1 method which gives an average vertical stress σ_z at any depth z. This method assumes that the stresses distribute from the loaded edge points at an angle of 2 (vertical) to 1 (horizontal).

The first method if properly applied gives the point stress at any depth which compares fairly well with exact methods, whereas the second does not give any point stress but only gives an average stress σ_z at any depth. The average stress computed by the second method has been found to be in error depending upon the depth at which the stress is required.

Point Load Method

Equation (10.1) may be used for the computation of stresses in a soil mass due to point loads acting at the surface. Since loads occupy finite areas, the point load formula may still be used if the footings are divided into smaller rectangles or squares and a series of concentrated loads of value $q\,dA$ are assumed to act at the center of each square or rectangle. Here dA is the area of the smaller blocks and q the pressure per unit area. The only principle to be followed in dividing a bigger area into smaller blocks is that the width of the smaller block should be less than one-third the depth z of the point at which the stress is required to be computed. The loads acting at the centers of each smaller area may be considered as point loads and Boussinesq's formula may then be applied. The difference between the point load method and the exact method explained earlier is clear from Fig. 10.19. In this figure the abscissa of the curve C_1 represents the vertical stress σ_z at different depths z below the center of a square area $B \times B$ which carries a surcharge q per unit area or a total surcharge load of $B^2 q$. This curve is obtained by the exact method explained under Sect. 6.6. The abscissa of the curve C_2 represents the corresponding stresses due to a concentrated load $Q = B^2 q$ acting at the center of the square area. The figure shows that the difference between the two curves becomes very small for values of z/B in excess of three. Hence in a computation of the vertical stress σ_z at a depth z below an area, the area should be divided into convenient squares or rectangles such that the least width of any block is not greater than $z/3$.

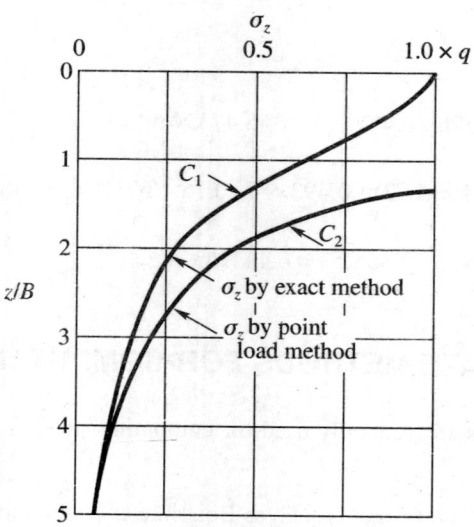

Fig. 10.19 σ_z by point load method

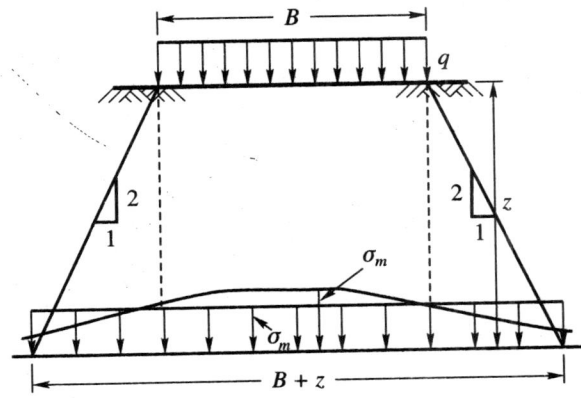

Fig. 10.20 σ_m 2 : 1 method

2 : 1 Method

In this method, the stress is assumed to be distributed uniformly over areas lying below the foundation. The size of the area at any depth is obtained by assuming that the stresses spread out at an angle of 2 (vertical) to 1 (horizontal) from the edges of the loaded areas shown in Fig. 10.20. The average stress at any depth z is

$$\sigma_a = \frac{Q}{(B+z)(L+z)} \tag{10.35}$$

The maximum stress σ_m by an exact method below the loaded area is different from the average stress σ_a at the same depth. The value of σ_m/σ_a reaches a maximum of about 1.6 at $z/b = 0$-5, where $b = $ half width.

Example 10.13

A footing of size 4 m × 4 m carries a uniformly distributed load of 150 kN/m². Compute vertical pressure by point load method at a depth of 6 m below the centre of the footing.

Solution

Point load formula,

$$\sigma_z = \frac{Q}{z^2}I_B, \text{ the depth factor, } \frac{z}{B} = 1.5 < 3$$

Divide the footing into four equal footings of size 2 m × 2 m.

The depth factor $\dfrac{z}{B} = \dfrac{6}{2} = 3$. This satisfies the conditon.

Concentrated load at the centres of each footing $= 2 \times 2 \times 150 = 600$ kN

The radial distance from the centre of footing 2 m × 2 m to the centre of footing 4 m × 4 m is

$$r = \sqrt{1^2 + 1^2} = 1.414, \quad r/z = \frac{1.414}{6} = 0.236$$

$$I_B = \frac{3/2\pi}{\left[1 + 0.236^2\right]^{5/2}} = 0.405$$

Vertical pressure vertically below the centre of footing at depth 6 m is

$$\sigma_z = \frac{4 \times 600}{6^2} \times 0.405 = 27.0 \text{ kN/m}^2$$

Example 10.14

Calculate the stress by point load method of Boussinesq at a depth of 20 m below th centre of a raft of a size 30 × 12 m founded at a depth of 2.5 m subjected to a uniform pressure of 150 kN/m². Neglect the effect of foundation depth. Assume the coordinate of the centre of the raft is (0, 0).

Solution

The stress as calculated in Ex. 10.3 for the same size of raft do not give accurate results as the ratio of depth to least width $z/B = 20/12 = 1.67$ is less than 3 which is necessary for accurate results. In order to achieve this objective, the rectangle 12 × 30 m may be divided into four equal rectangles of size 6 × 15 in which case, the depth to width ratio of each works out to 3.34 which is more than 3. The loads acting at the centres of each of the rectangles may be considered as point loads having coordinates (7.5, 3.0). The sum of the stresses acting at coordinate (0, 0) due to point loads at the centres of each rectangle gives the necessary value which is more accurate than the method given in Ex. 10.3.

The radial distance from the coordinate (7.5, 3.0) to (0, 0) is

$$r = \sqrt{(7.5)^2 + 3^2} = 0.08 \text{ m, and } \frac{r}{z} = \frac{8.08}{20} = 0.404$$

$$I_B = \frac{0.48}{\left[1 + (0.404)^2\right]^{5/2}} = 0.329$$

$$Q = \frac{150 \times 30 \times 12}{4} = \frac{54000}{4} = 13,500 \text{ kN}$$

$$\frac{Q}{z^2} = \frac{13.500}{20^2} = 33.75$$

Therefore, total stress at (0, 0) is,

$$\sigma_B = 4 \times 0.329 \times 33.75 = 44.4 \text{ kN/m}^2$$

The value of σ_z by the above method is very much less than that given in Ex. 10.3 for the same location.

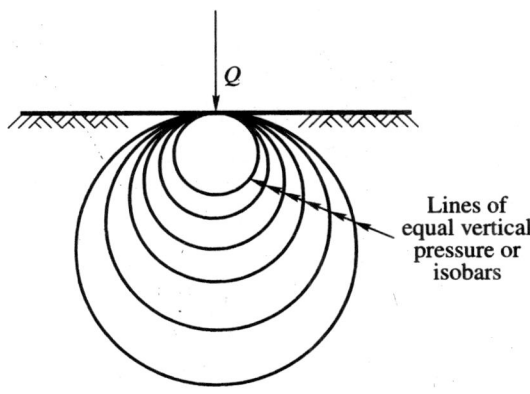

Fig. 10.21   Bulb of pressure

10.12 PRESSURE ISOBARS

Definition

An *Isobar* is a line which connects all points of equal stress below the ground surface. In other words, an isobar is a stress contour. We may draw any number of isobars as shown in Fig. 10.21 for any given load system. Each isobar represents a fraction of the load applied at the surface. Since these isobars form closed figures and resemble the form of a bulb, they are also termed *bulb of pressure* or simply the *pressure bulb*. Normally isobars are drawn for vertical, horizontal and shear stresses. The one that is most important in the calculation of settlements of footings is the vertical pressure isobar.

Significant Depth

In his opening discussion on settlement of structures at the First International Conference on Soil Mechanics and Foundation Engineering (held in 1936 at Harvard University in Cambridge, Mass, USA), Terzaghi stressed the importance of the bulb of pressure and its relationship with the seat of settlement. As stated earlier we may draw any number of isobars for any given load system, but the one that is of practical significance is the one which encloses a soil mass which is responsible for the settlement of the structure. The depth of this stressed zone may be termed as the *significant depth D_s* which is responsible for the settlement of the structure. Terzaghi recommended that for all practical purposes one can take a *stress contour* which represents 20 percent of the foundation contact pressure q, i.e. equal to $0.2q$. The depth of such an isobar can be taken as the *significant depth D_s* which represents the seat of settlement forhe foundation. Terzaghi's recommendation was based on his observation that direct stresses are considered of negligible magnitude when they are smaller than 20 percent of the intensity of the applied stress from structural loading, and that most of the settlement, approximately 80 percent of the total, takes place at a depth less than D_s. The depth D_s is approximately equal to 1.5 times the width of square or circular footings [Fig. 10.22(a)].

If several loaded footings are spaced closely enough, the individual isobars of each footing in question would combine and merge into one large isobar of the intensity as shown in [Fig. 10.22(b)]. The combined significant depth D_s is equal to about $1.5B$.

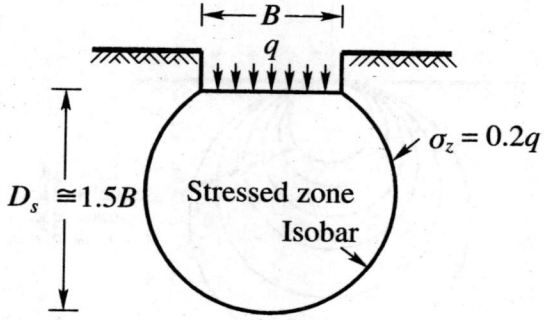

(a) Significant depth of stressed zone

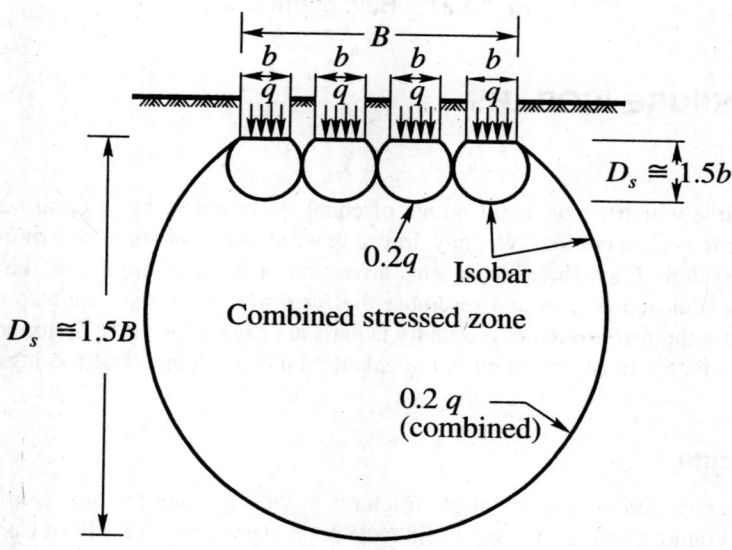

(b) Effect of closely placed footings

Fig. 10.22 Significant depth of stressed zone

Pressure Isobars for Footings

Pressure isobars of square, rectangular and circular footings may conveniently be used for determining vertical pressure, σ_z, at any depth, z, below the base of the footings. The depths z from the ground surface, and the distance r (or x) from the center of the footing are expressed as a function of the width of the footing B. In the case of circular footing B represents the diameter.

The following pressure isobars are given based on either Boussinesq or Westergaard's equations.

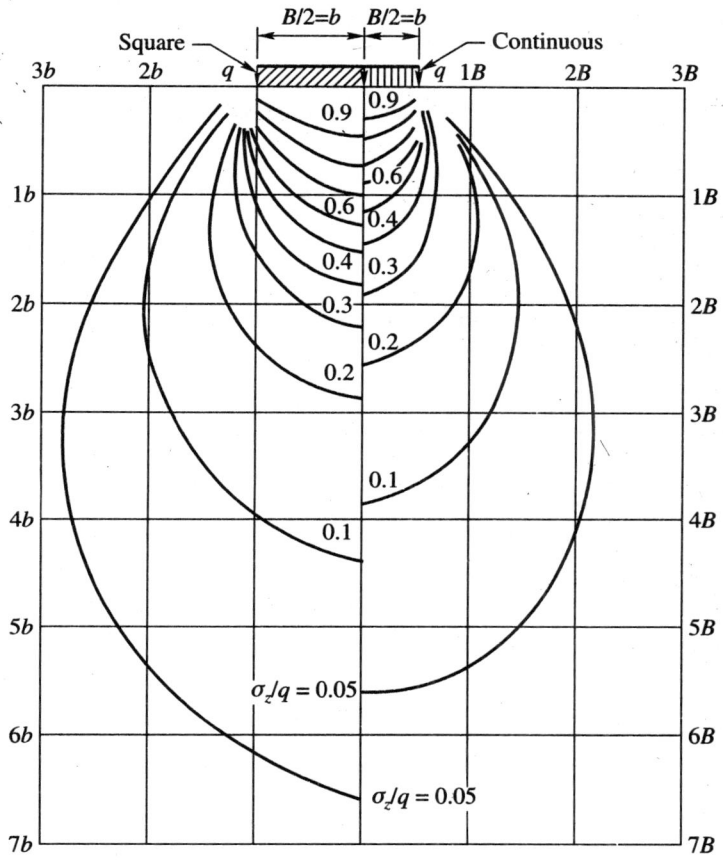

Fig. 10.23   Pressure isobars based on Boussinesq equation for square and continuous footings

1.   Boussinesq isobars for square and continuous footings, Fig. 10.23.

2.   Boussinesq isobar for circular footings, Fig. 10.24.

3.   Westergaard isobars for square and continuous footings, Fig. 10.25.

Example 10.15

A single concentrated load of 1000 kN acts at the ground surface. Construct an isobar for $\sigma_z = 40$ kN/m² by making use of the Boussinesq equation.

Solution

From Eq. (10.1) we have

$$\sigma_z = \frac{3Q}{2\pi z^2}\left[\frac{1}{1+(r/z)^2}\right]^{\frac{5}{2}}$$

We may now write by rearranging an equation for the radial distance r as

$$r = \sqrt{z}\sqrt{\left(\frac{3Q}{2\pi z^2 \sigma_z}\right)^{\frac{2}{5}} - 1}$$

Now for $Q = 1000$ kN, $\sigma_z = 40$ kN/m^2, we obtain the values of r_1, r_2, r_3, etc. for different depths z_1, z_2, z_3, etc. The values so obtained are

z (m)	r (m)
1.0	1.30
2.0	1.04
3.0	0.60
3.45	0.00

The isobar for $\sigma_z = 40$ kN/m^2 may be obtained by plotting z against r as shown in Fig. Ex. 10.15.

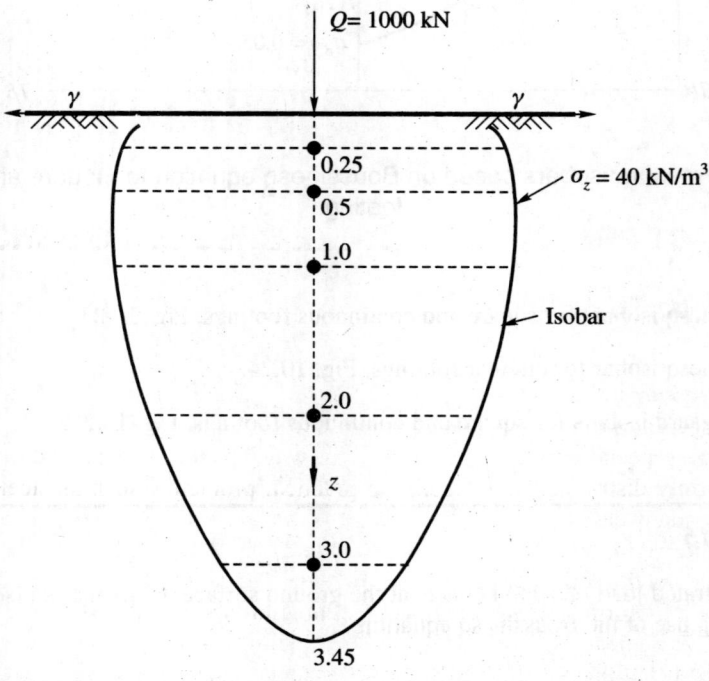

Fig. Ex. 10.15

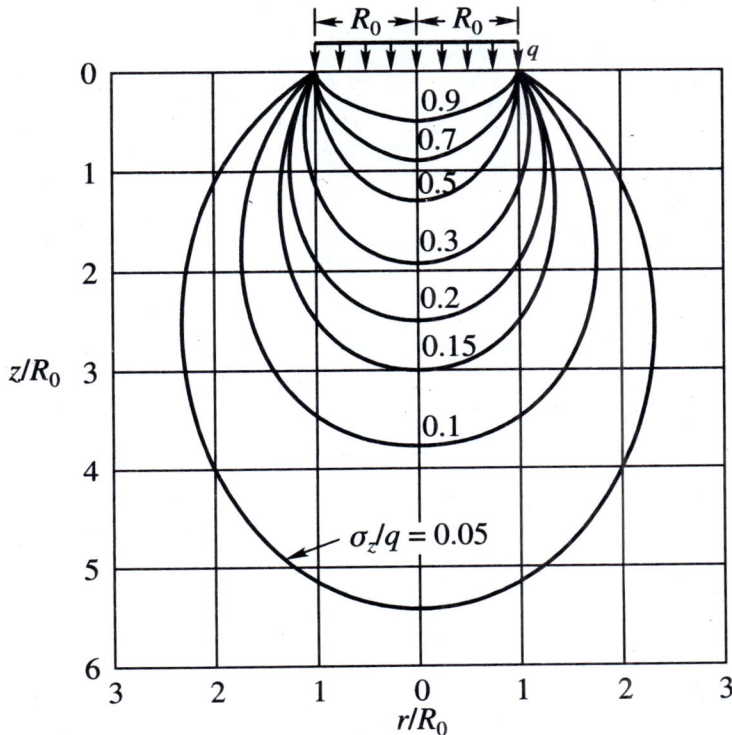

Fig. 10.24 Pressure isobars based on Boussinesq equation for uniformly loaded circular footings

10.13 CONTACT PRESSURE OVER THE BASE OF FOOTINGS

Theoretical distribution of contact pressure over the base of circular and strip footings

The term *contact pressure* indicates the normal stress at the surface of contact between a footing and the supporting earth. The problem of pressure distribution over the base of structures is of high practical importance, especially for flexible foundations which are to be calculated for bending. The condition of uniformly distributed load occurs in practical problems, such as steel tanks for the storage of fluids. In many cases, however, the structural member (such as a footing) in contact with the soil will be quite rigid, and the settlement is more or less uniform over the area of contact between the footing and the soil. Since a uniform stress causes a dish-shaped pattern of settlement, in order to produce a uniform settlement, the contact pressure must increase on the outside of the loaded area and decrease near the central line provided that the material is perfectly elastic. For an elastic footing the distribution of contact pressure depends on the elastic properties of the supporting medium, on the depth of the elastic layer, on the flexural rigidity of the footing and on, the distribution of loads on the footing.

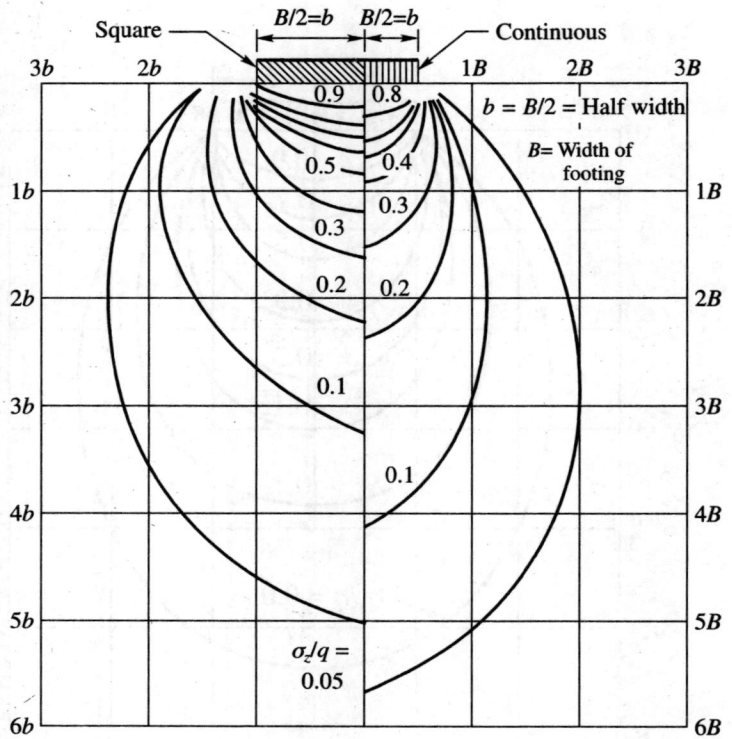

Fig. 10.25 Pressure isobars based on Westergaard equation for square and long continuous footing.

Borowicka (1936, 38) analysed the problem of distribution of pressure on the base of uniformly loaded circular and long strip footings resting on the surface of semi-infinite mass. He found the pressure varying on the base according to the rigidity of the footings. the stiffer the footing, the less uniform is the distribution of contact pressure. The flexibility of the footing is presented in the form of an equation as

$$K_r = \frac{1-\mu_s^{\ 2}}{6(1-\mu_f)^2} \frac{E_f}{E_s} \left(\frac{h}{b}\right)^2 \tag{10.36}$$

where, μ_s, μ_f = Poisson's ratios of subgrade soil and the footing material respectively,

E_s, E_f = Young's modulus of soil and footing material respectively

h = Thickness of footing

b = radius of circular footing or half width of strip footing.

The factor K_r in Eq. (10.36) can be considered a measure for the relative stiffness of the footings with K_r ranging from 0 to ∞. When $K_r = 0$, the footing is said to be perfectly flexible, when $K_r = ∞$, it is said to be perfectly rigid.

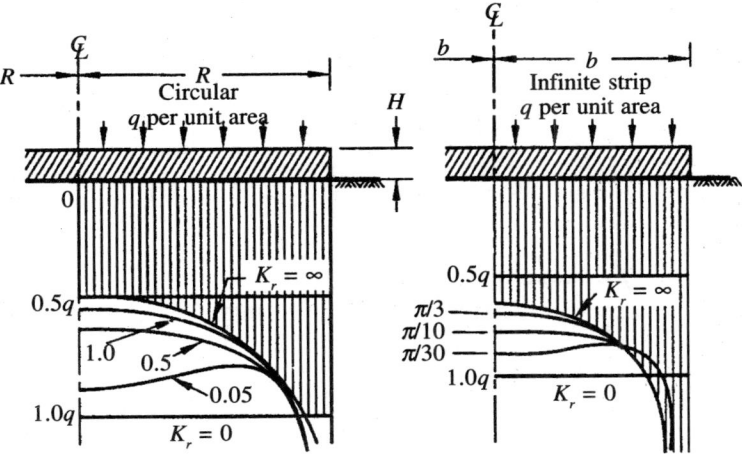

Fig.10.26 Stress distribution under circular and strip footings of varying rigidity
(After Borowicka)

If q_c is the contact pressure, the ratio $q_c/q = 0.5$ for a perfectly rigid circular foundation and $q_c/q = 0.67$ for a perfectly rigid strip foundation as shown in Fig. 10.26. At $K_r = \infty$, the edge pressure reaches infinity theoretically for both the circular and strip foundations. It should be noted here that this pressure distribution is valid only when the imposed pressure q is sufficiently less than the ultimate bearing capacity q_d of the soil and also the soil is perfectly elastic. As the load, q on the footing is increased gradually, the stress condition in the subsoil changes from one state of elastic to plastic equilibrium. This transition influences not only the intensity and the distribution of the stresses in the loaded material, but it also changes the distribution of the contact pressure on the base of the footing. As a consequence, the contact pressures at the edges of footings get reduced from infinity to a finite value as the load increases from elastic to plastic state. Since the soil met in nature is either plastic (cohesive) or non-plastic (cohesionless), the contact pressure at the base of the footing varies according to the type of soil met with and also depth of embedment of the footing as described below.

Contact pressure distribution in cohesionless soils

Figure 10.27(a) is a flexible footing on cohesionless soil at the ground surface subjected to a uniformly distributed load. Since the footing is completely flexible, uniform distribution of pressure also acts on the surface of the soil. The soil outside the edge of the footing is not under pressure and has no strength. Therefore, when the given intensity of load is applied, the outer edge of the footing undergoes a relatively larger settlement. Below the centre of the footing the soil develops strength and rigidity, and because of this, the settlement is relatively small.

The settlement of a rigid footing on cohesion less soil is uniform [Fig. 10.27(b)]. Under, uniform settlement, the soil pressure at the edges is zero and maximum at the centre. The distribution of pressure is approximately parabolic. If the footing is very large such as a mat or a raft, the pressure distribution is ellipsoidal in shape [Fig. 10.27(c)].

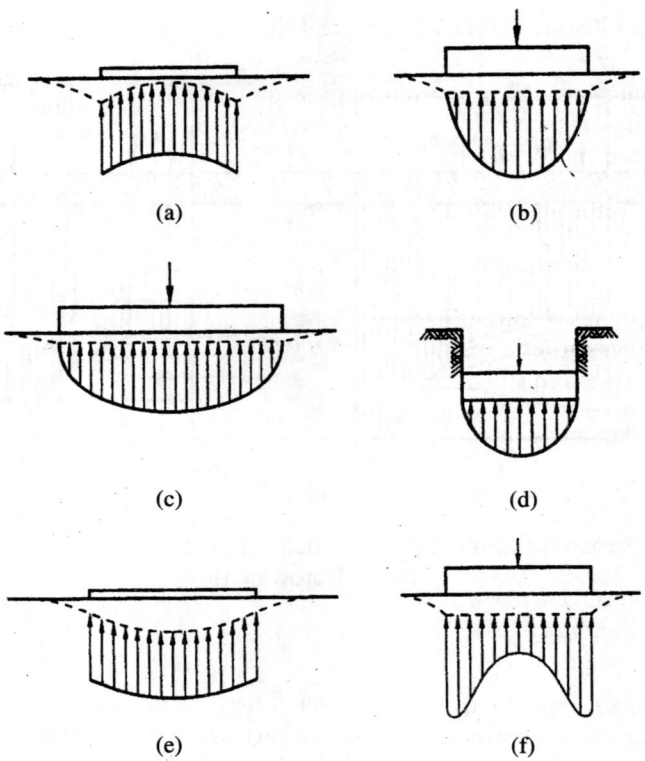

(a) Flexible footing on cohesionless soil at ground level, (b) Rigid footing on cohesionless soil at ground level, (c) Mat or raft on cohesion less soil, (d) Rigid footing on cohesion less soil below ground level, (e) Flexible footing on cohesive soil, (f) Rigid footing on cohesive soil.

Fig. 10.27   Contact pressure distribution

The pressure distribution below rigid footings founded below ground surface is shown in Fig. 10.27(d). The pressure at the edges is not zero because of the effect of overburden.

Contact pressure distribution in highly cohesive soils

The pressure is uniform below flexible footings founded on cohesive soil [Fig. 10.27(e)]. The footing settlement is more at the centre and less at the edges under uniform distribution of pressure. Whereas, when the footing is rigid, the settlement is uniform but the pressure is more at the edges and less at the centre [Fig. 10.27(f)].

Effect of non-uniform distribution of contact pressure on the design of footings

When a footing rests on cohesionless soil, the actual pressure distribution is parabolic with zero pressures at the edges. However, uniform pressure distribution is assumed in the design of footings. The assumption of uniform pressure distribution is conservative. Whereas, when a footing rests on clay soil, high pressures at outer edges result in both shear and bending moments higher than those computed on the basis of uniform pressure. Ordinarily no correction is made for the error, because the safety of the design is sufficiently large to make up for it.

10.14 QUESTIONS AND PROBLEMS

10.1   A column of a building transfers a concentrated load of 1000 kN to the soil in contact with the footing. Estimate the vertical pressure at the following points by making use of the Boussinesq and Westergaard equations.

     (i)   Vertically below the column load at depths of 1, 2, and 4 metres.

     (ii)   At radial distances of 2, 4 and 6 metres and at a depth of 4 m.

10.2   Three footings are placed at locations forming an equilateral triangle of 4 m sides. Each of the footings carries a vertical load of 500 kN. Estimate the vertical pressures by means of the Boussinesq equation at a depth of 3 metres at the following locations:

     (i)   Vertically below the centres of the footings.

     (ii)   Below the centre of the triangle.

     (iii)   Below the centre of the triangle.

10.3   A reinforced concrete water tank of size 8 m × 8 m and resting on the ground surface, carries a uniformly distributed load of 250 kN/m². Estimate the maximum vertical pressures at depths of 12 and 24 metres by point load approximation.

10.4   What would be the vertical pressures for the Prob. 10.3 by an exact method?

10.5   Two footings of sizes 4 m × 4 m and 3 m × 3 m are placed 6 m centre to centre apart at the same level and carry concentrated loads of 1500 and 1250 kN respectively. Compute the vertical pressures at the following points:

     (i)   Midway between footings at depth of 4 m.

     (ii)   Vertical below the centres of footings at the same depths given in (i) by point load formula.

10.6   A water tower is founded on a circular ring type foundation. The width of the ring is 4 m and its internal radius is 8 m. Assuming the distributed load per unit area as 300 kN/m², determine the vertical pressure at a depth of 6 m below the centre of the foundation.

10.7   A rectangular footing of size 12 m × 8 m carries a uniformly distributed load of 250 kN/m². Determine the vertical pressure 6 m below a point which is located at a distance of 10 m from the centre of the footing on its longitudinal axis by making use of curves in Fig. 10.8.

10.8   An embankment for road traffic is required to be constructed with the following dimensions:

Top width = 8 m, Height = 4 m, Side slopes= 1 : 1½

The unit weight of soil under the worst condition is 21 kN/m³. The surcharge load on the road surface may be taken as 50 kN/m². Compute the vertical pressure at a depth of 6 m below the ground surface at the following locations:

     (i)   On the central longitudinal plane of the embankment.

     (ii)   Below the toes of the embankment.

10.9   If the top width of the road given in Prob. 10.8 is reduced to zero, what would be the change in the vertical pressure at the same points?

10.10   A square footing founded of size 4×4 m founded on the surface carried a distributed load of 100 kN/m². Determine the increase in pressure at a depth of 4 m by 2: 1 method

10.11   A long masonry wall footing carries a uniformly distributed load of 200 kN/m². If the width of the footing is 4 m, determine the vertical pressures at a depth of 3 m below the (i) centre and (ii) edge of the footing.

10.12   A and B are two footings of size 1.5×1.5 m each placed in position as shown in Fig. Prob. 10.12. Each of the footings carry a column load of 400 kN. Determine by Boussinesq method, the excess load, the footing B carries due to the effect of the load on A. Assume that the excess vertical stress $(s_z$ at the centre of the base of footing B represent the average for the whole area.

10.13   For the Prob. 10.12, determine the total vertical stress s_z, at a depth of 1.0 m below the central line of footing B by Boussinesq method.

10.14   For the Prob. 10.12, determine the maximum vertical pressure at the base of footing B.

10.15   Three concentrated loads $Q_1 = 1000$ kN, $Q_2 = 2000$ kN and $Q_3 = 3000$ kN act in one vertical plane and they are placed in the order Q_1–Q_2–Q_3. Their spacing is 4 m⁻³ m. Plot the vertical stress distribution diagram on a horizontal plane at a depth $Z = 1.5$ m.

10.16   A long foundation of 0.6 m wide carries a uniform linear load of 100 kN/m. Assuming that the foundation load system represent a linear load, calculate the vertical stress, s_z, at a point N_1 the coordinates of which are $x = 2.75$ m, and $z = 1.5$ m, where x-coordinate is normal to the line load-from the central line of the footing.

10.17   A circular ring foundation for an overhead tank transmits a contact pressure of 300 kN/m². Its internal diameter is 6 m and external diameter 10 m. Compute the vertical stress on the central line of the footing at a depth of 6.5 m below the ground level. The footing is founded at a depth of 2.5 m.

10.18   In Prob. 10.17, if the foundation for the tank is a raft of diameter 10 m, determine the vertical stress at 6.5 m depth on the central line of the footing. All the other data remain the same.

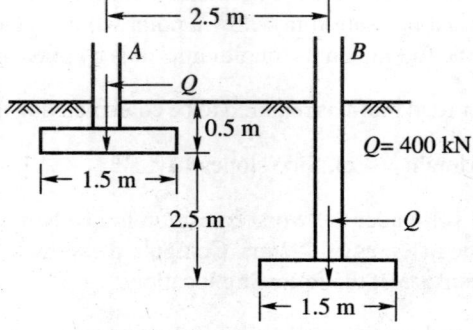

Fig. Prob.10.12

10.19   A square footing of 4 × 4 m is founded at a depth of 1.5 m below the ground level. The imposed pressure at the base is 375 kN/m². Determine the vertical pressure at a depth of 7.5 m below the ground surface on the central line of the footing by Boussinesq point load method.

10.20   The centre of a rectangular area at the ground surface has cartesian coordinate (0,0) and the corners have coordinates (6,15). All dimensions are in metres. The area carries a uniform pressure of 150 kPa. Estimate the stresses at a depth of 20 m below ground surface at each of the following locations: (0;0), (0,15), (6,0). Obtain the values by Boussinesq point load method.

10.21   Calculate the vertical stress at a depth of 15 m below a point 3 m from the comer (along the longest side) of a rectangular loaded area 10 × 30 m carrying a uniform load of 120 kPa by Boussinesq point load method.

10.22   How far apart must two 20 m diameter tanks be placed such that their stress overlap is not greater than 10% of contact stress at a depth of 10 m.

10.23   A strip footing 3 m wide is loaded on the ground surface with a pressure equal to 200 kPa. Calculate the vertical stresses at depths of 3, 6, and 12 m under the centre of the footing.

10.24   A rectangular footing 6 × 3 m carries a uniform pressure of 300 kN/m² on the surface of a soil mass. Determine the vertical stress at a depth of 4.5 m below the surface on the centre line 1.5 m outside the long edge of the foundation by using Boussinesq point load method.

10.25   A load of 1500 kN is imposed on a foundation 2 m square at a shallow depth in a soil mass. Determine the vertical stress at a point 5 m below the centre of the foundation (a) assuming the load is uniformly distributed over the foundation, (b) assuming the load acts as a point load at the centre of foundation.

CHAPTER 11

COMPRESSIBILITY AND CONSOLIDATION

11.1 CONSOLIDATION

When a saturated clay-water system is subjected to an external pressure, the pressure applied is initially taken by the water in the pores resulting thereby in an excess pore water pressure. If drainage is permitted, the resulting hydraulic gradients initiate a flow of water out of the clay mass and the mass begins to compress. A portion of the applied stress is transferred to the soil skeleton, which in turn causes a reduction in the excess pore pressure. This process, involving a gradual compression occurring simultaneously with a flow of water out of the mass and with a gradual transfer of the applied pressure from the pore water to the mineral skeleton is called *consolidation*. The process opposite to consolidation is called *swelling*, which involves an increase in the water content due to an increase in the volume of the voids.

Consolidation may be due to one or more of the following factors:

1. Due to external static loads from structures.
2. Due to self-weight of the soil such as recently placed fills.
3. Due to lowering of the ground water table.
4. Due to desiccation.

The total compression of a saturated clay strata under excess effective pressure may be considered as the sum of:

1. Immediate compression,°
2.   Primary consolidation, and
3. Secondary compression.

The portion of the settlement of a structure which occurs more or less simultaneously with the applied loads is referred to as the *initial* or *immediate settlement*. This settlement is due to the immediate compression of the soil layer under undrained condition and is calculated by assuming the soil mass to behave as an elastic soil.

If the rate of compression of the soil layer is controlled solely by the resistance of the flow of water under the induced hydraulic gradients, the process is referred to as *primary consolidation*.

The portion of the settlement that is due to the primary consolidation is called *primary consolidation settlement* or *compression*. At the present time the only theory of practical value for estimating time-dependent settlement due to volume changes, that is under primary consolidation is *the one-dimensional theory*.

The third part of the settlement is due to secondary consolidation or compression of the clay layer. This compression is supposed to start after the primary consolidation ceases, that is after the excess pore water pressure approaches zero. It is often assumed that secondary compression proceeds linearly with the logarithm of time. However, a satisfactory treatment of this phenomenon has not been formulated for computing settlement under this category.

Principle of Consolidation

The process of consolidation of a clay-soil-water system may be explained with the help of a mechanical model as described by Terzaghi and Frohlich.

The model consists of a cylinder with a frictionless piston as shown in Fig. 11.1. The piston is supported on one or more helical metallic springs. The space underneath the piston is completely filled with water. The springs represent the mineral skeleton in the actual soil mass and the water below the piston is the pore water under saturated conditions in the soil mass. When a load per unit area is placed on the piston, this stress is fully transferred to the water (as water is assumed to be incompressible) and the water pressure increases. The pressure in the water is

$$u = p$$

This is analogous to pore water pressure, u, that would be developed in a clay-water system under external pressures. If the whole model is leakproof without any holes in the piston, there is no chance for the water to escape. Such a condition represents a highly impermeable clay-water system in which there is a very high resistance for the flow of water. It has been found out by Reltov (1947) and Rosa (1950) that in the case of compact plastic clays that the minimum initial gradient required to cause flow may be as high as 20 to 30.

If a few holes are made in the piston, the water will immediately escape through the holes. With the escape of water through the holes a part of the load carried by the water is transferred to the

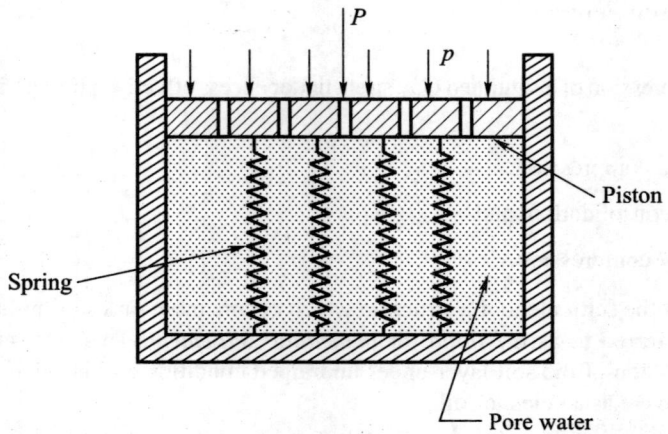

Fig. 11.1 Mechanical model to explain the principle of consolidation

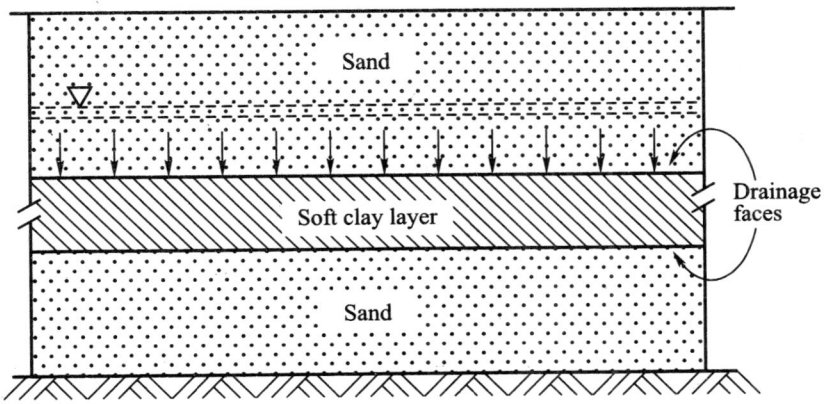

Fig. 11.2 Clay layer sandwiched between sand layers

springs. This process of transference of load from water to spring goes on until the flow stops when all the load will be carried by the spring and none by the water. The time required to attain this condition depends upon the number and size of the holes made in the piston. A few small holes represents a clay soil with poor drainage characteristics.

After the spring-water system attains equilibrium condition under the imposed load, the settlement of the piston is analogous to the compression of the clay-water system under external pressures.

One-Dimensional Consolidation

In many instances the settlement of a structure is due to the presence of one or more layers of soft clay located between layers of sand or stiffer clay as shown in Fig. 11.2. The adhesion between the soft and stiff layers almost completely prevents the lateral movement of the soft layers. The theory that was developed by Terzaghi on the basis of this assumption is called the *one-dimensional onsolidation theory*. In the laboratory this condition is simulated most closely by the *confined compression* or *consolidation test*.

11.2   CONSOLIDOMETER

The compressibility of a saturated, clay-water system is determined by means of the apparatus shown diagrammatically in Fig. 11.3 and was devised by Terzaghi. This apparatus is also known as an *Oedometer*.

The consolidation test is usually performed at room temperature, in floating or fixed rings of diameter from 5 to 11 cm and from 2 to 4 cm in height. Fig. 11.3 is a fixed ring type. In a floating ring type, the ring is free to move in the vertical direction.

The soil sample is contained in the brass ring between two porous stones about 1.25 cm thick. By means of the porous stones water has free access to and from both surfaces of the specimen. The compressive load is applied to the specimen through a piston, either by means of a hanger and dead weights or by a system of levers. The compression is measured on a dial gauge.

At the bottom of the soil sample the water expelled from the soil flows through the filter stone into the water container. At the top, a well-jacket filled with water is placed around the stone in order to prevent excessive evaporation from the sample during the test. Water from the sample also flows into the jacket through the upper filter stone. The soil sample is kept submerged in a saturated condition during the test.

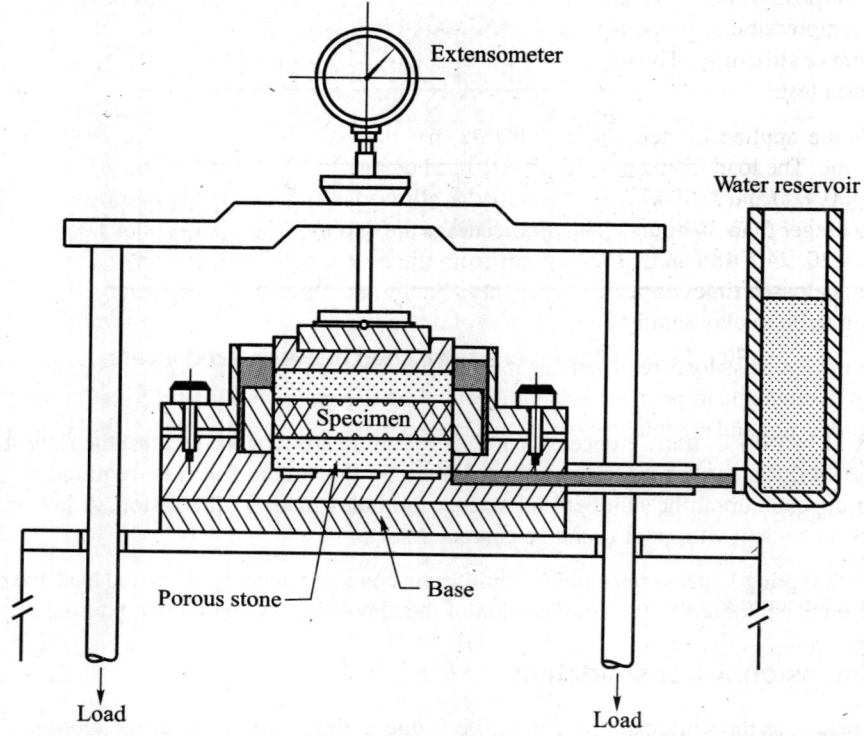

Fig. 11.3(a)   A Schematic diagram of a consolidometer

Fig. 11.3(b)   Tabletop consolidation apparatus (Courtesy : Soiltest, USA)

11.3   THE STANDARD ONE-DIMENSIONAL CONSOLIDATION TEST

The main purpose of the consolidation test on soil samples is to obtain the necessary information about the compressibility properties of a saturated soil for use in determining the magnitude and rate of settlement of structures. The following test procedure is applied to any type of soil in the standard consolidation test.

Loads are applied in steps in such a way that the successive load intensity, p, is twice the preceding one. The load intensities commonly used being 1/4, 1/2, 1, 2, 4, 8, and 16 kg/cm^2 (25, 50, 100, 200, 400, 800 and 1600 kN/m^2). Each load is allowed to stand until compression has practically ceased (no longer than 24 hours). The dial readings are taken at elapsed times of 1/4, 1/2, 1, 2, 4, 8, 15, 30, 60, 120, 240, 480 and 1440 minutes from the time the new increment of load is put on the sample (or at elpased times as per requirements). Sandy samples are compressed in a relatively short time as compared to clay samples and the use of one day duration is common for the latter.

After the greatest load required for the test has been applied to the soil sample, the load is removed in decrements to provide data for plotting the expansion curve of the soil in order to learn its elastic properties and magnitudes of plastic or permanent deformations. The following data should also be obtained:

1.   Moisture content and weight of the soil sample before the commencement of the test.

2.   Moisture content and weight of the sample after completion of the test.

3.   The specific gravity of the solids.

4.   The temperature of the room where the test is conducted.

11.4   PRESSURE-VOID RATIO CURVES

The pressure-void ratio curve can be obtained if the void ratio of the sample at the end of each increment of load is determined. Accurate determinations of void ratio are essential and may be computed from the following data:

1.   The cross-sectional area of the sample A, which is the same as that of the brass ring.

2.   The specific gravity, G_s, of the solids.

3.   The dry weight, W_s, of the soil sample.

4.   The sample thickness, h, at any stage of the test.

Let V_s = volume of the solids in the sample

where

$$V_s = \frac{W}{G\gamma_0}, \quad \gamma_0 = \text{unit of water at 4 °C.}$$

We can also write

$$V_s = h_s A \text{ or } h_s = \frac{V_s}{A}$$

where, h_s = thickness of solid matter.

If e is the void ratio of the sample, then

$$e = \frac{Ah - Ah_s}{Ah_s} = \frac{h - h_s}{h_s} \tag{11.1}$$

In Eq. (11.1) h_s, is a constant and only h is a variable which decreases with increment load. If the thickness h of the sample is known at any stage of the test, the void ratio at all the stages of the test may be determined.

The equilibrium void ratio at the end of any load increment may be determined by the change of void ratio method as follows:

Change of Void-Ratio Method

In one-dimensional compression the change in height Δh per unit of original height h equals the change in volume ΔV per unit of original volume V.

$$\frac{\Delta h}{h} = \frac{\Delta V}{V} \tag{11.2}$$

V may now be expressed in terms of void ratio e.

We may write (Fig. 11.4),

$$V_v = eV_s, V = V_s(1+e), V_v' = e'V_s$$

$$V' = V_s(1+e')$$

$$\frac{\Delta V}{V} = \frac{V - V'}{V} = \frac{V_s(1+e) - V_s(1+e')}{V} = \frac{e - e'}{1+e} = \frac{\Delta e}{1+e}$$

Therefore,

$$\frac{\Delta h}{h} = \frac{\Delta e}{1+e}$$

or

$$\Delta e = \frac{1+e}{h}\Delta h \tag{11.3}$$

wherein, Δe = change in void ratio under a load, h = initial height of sample, e = initial void ratio of sample, e' = void ratio after compression under a load, Δh = compression of sample under the load which may be obtained from dial gauge readings.

Normally the void ratios are worked out backwards starting from the final void ratio after completion of the test. The final sample thickness h_f and the final void e_f should be known for this purpose.

A typical pressure-void ratio curves for an undisturbed clay sample are shown in Fig. 11.5, plotted both on arithmetic and on semilog scales. The curve on the log scale indicates clearly two branches, a fairly horizontal initial portion and a nearly straight inclined portion. The coordinates of point A in the figure represent the void ratio e_0 and effective overburden pressure p_0 corresponding to a state of the clay in the field as shown in the inset of the figure. When a sample is extracted by

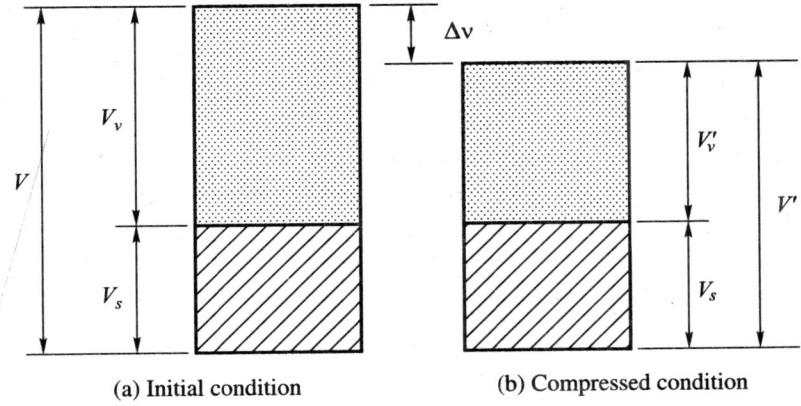

(a) Initial condition   (b) Compressed condition

Fig. 11.4   Change of void ratio

means of the best of techniques, the water content of the clay does not change significantly. Hence, the void ratio e_0 at the start of the test is practically identical with that of the clay in the ground. When the pressure on the sample in the consolidometer reaches p_0, the e-log p curve should pass through the point A unless the test conditions differ in some manner from those in the field. In reality the curve always passes below point A, because even the best sample is at least slightly disturbed.

The curve that passes through point A is generally termed as a *field curve* or *virgin curve*. In settlement calculations, the field curve is to be used.

Pressure-Void Ratio Curves for Sand

Normally, no consolidation tests are conducted on samples of sand as the compression of sand under external load is almost instantaneous as can be seen in Fig. 11.6(a), which gives a typical curve showing the time versus the compression caused by an increment of load.

In this sample more than 90 percent of the compression has taken place within a period of less than 2 minutes. The time lag is largely of a frictional nature. The compression is about the same whether the sand is dry or saturated. The shape of typical e-p curves for loose and dense sands are

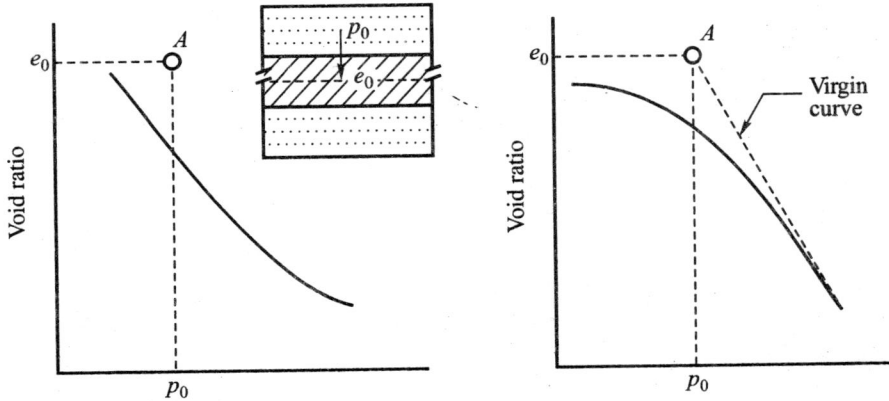

Fig. 11.5   Pressure-void ratio curves

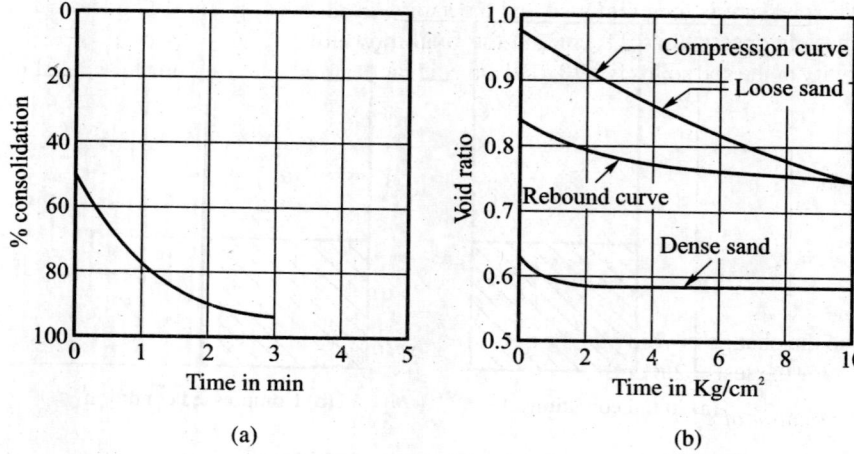

Fig. 11.6 Pressure-void ratio curves for sand

shown in Fig. 11.6(b). The amount of compression even under a high load intensity is not significant as can be seen from the curves.

Pressure-Void Ratio Curves for Clays

The compressibility characteristics of clays depend on many factors. The most important factors are:

1. Whether the clay is normally consolidated or overconsolidated.
2. Whether the clay is sensitive or insensitive.

Normally Consolidated and Overconsolidated Clays

A clay is said to be normally consolidated if the present effective overburden pressure p_0 is the maximum pressure to which the layer has ever been subjected at any time in its history, whereas a clay layer is said to be overconsolidated if the layer was subjected at one time in its history to a greater effective overburden pressure, p_c, than the present pressure, p_0. The ratio p_c / p_0 is called the *over consolidation ratio* (O_{CR}).

The overconsolidation of a clay stratum may have been caused due to some of the following factors:

1. Due to weight of an overburden of soil which has eroded.
2. Due to weight of a continental ice sheet that melted.
3. Due to desiccation of layers close to the surface.

Experience indicates that the natural moisture content, w_n, is commonly close to the liquid limit, w_e, for normally consolidated clay soil whereas for the overconsolidated clay, w_n is close to plastic limit w_p.

Example 11.1

During a consolidation test, a sample of fully saturated clay 3 cm thick is consolidated under a pressure increment of 200 kN/m². When equilibrium is reached, the sample thickness is reduced to

2.60 cm. The pressure is then removed and the sample is allowed to expand and absorb water. The final thickness is observed as 2.8 cm and the final moisture content is determined as 24%. If the specific gravity of the soil solids is 2.70, find the void ratio of the sample before and after consolidation.

Solution

Use equation

$$\Delta e = \frac{(1+e_f)}{h_f} \Delta h$$

to determine the change in void ratios where in, e_f = final void ratio after the test; h_f = final thickness of sample after the test = 2.8 cm.

1. *Determination of* e_f

   Weight of solids = $M_s = V_s G \rho_0 = 1 \times 2.70 \times 1 = 2.70$ g.

   $\dfrac{M_w}{M_s} = 0.24$ or $M_w = 0.24 \times 2.70 = 0.672$ gm, $e_f = V_w = 0.672$.

2. *Changes in thickness from final stage to equilibrium stage with load on*

   $\Delta h = 2.80 - 2.60 = 0.20$ cm, $\Delta e = \dfrac{(1+0.672)0.20}{2.80} = 0.119.$

   Void ratio after consolidation = $e_f - \Delta e = 0.672 - 0.119 = 0.553.$

3. *Change in void ratio from the commencement to the end of consolidation is*

   $\Delta e = \dfrac{1+0.553}{2.6}(3.00-2.60) = \dfrac{1.553}{2.6} \times 0.40 = 0.238.$

   Void ratio at the start of consolidation = 0.553 + 0.238 = 0.791.

Example 11.2

A recently completed fill was 10 m thick and its initial average void ratio was 1.0. The fill was loaded on the surface by constructing an embankment covering a large area of the fill. Some months after the embankment was constructed, measurements of the fill indicated an average void ratio of 0.8. Estimate the compression of the fill.

Solution

As per Eq. (11.3), the compression of the fill may be calculated as by the expression

$$\Delta H = \frac{\Delta e}{1+e_0} H_0$$

where ΔH = the compression, Δe = change in void ratio, e_0 = initial void ratio, H_0 = thickness of fill.

Substituting, $\Delta H = \dfrac{1.0 - 0.8}{1 + 1.0} \times 10\,\text{m} = 1.0\,\text{m}.$

11.5 DETERMINATION OF PRECONSOLIDATION PRESSURE

Several methods have been proposed for determining the value of the maximum consolidation pressure. They fall under the following categories. They are

1. Field method,

2. Graphical procedure based on consolidation test results.

Field Method

The field method is based on geological evidence. The geology and physiography of the site may help to locate the original ground level. The overburden pressure in the clay structure with respect to the original ground level may be taken as the preconsolidation pressure p_c. Usually the geological estimate of the maximum consolidation pressure is very uncertain. In such instances, the only remaining procedure for obtaining an approximate value of p_c is to make an estimate based on the results of laboratory tests or on some relationships established between p_c and other soil parameters.

Graphical Procedure

There are a few graphical methods for determining the preconsolidation pressure based on laboratory test data. No suitable criteria exists for-appraising the relative merits of the various methods. The methods that are suggested are

1. Casagrande method, 2. Burmister method and 3. Schmertmann method.

The method that is commonly used is the Casagrande's Method which is described here.

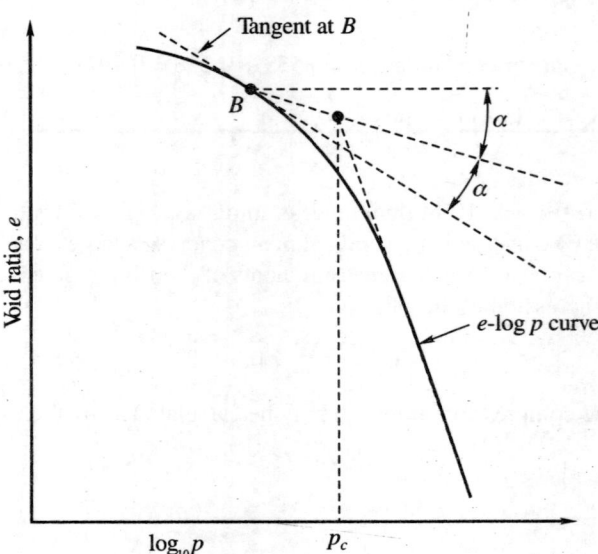

Fig. 11.7 Method of determining p_c by Casagrande method

Casagrande Method

The earliest and the most widely used method was the one proposed by A. Casagrande (1936). The method involves locating the point of maximum curvature, B, on the laboratory e-log p curve of an undisturbed sample as shown in Fig. 11.7. From B, a tangent is drawn to the curve and a horizontal line is also constructed. The angle between these two lines is then bisected. The abscissa of the point of intersection of this bisector with the upward extension of the inclined straight part corresponds to the preconsolidation pressure p_c.

11.6 e-*log* p FIELD CURVES FROM NORMALLY AND PRE-CONSOLIDATED UNDISTRIBUTED SAMPLES OF CLAY OF LOW SENSITIVITY

It has been explained earlier with reference to Fig. 11.5, that the laboratory e-log p curve of an undisturbed sample does not pass through point A and always passes below the point. It has been found from investigation that the inclined straight portion of e-log p curves of undisturbed or remoulded samples of clay soil intersect at one point at a low void ratio and corresponds to $0.4e_0$ shown as point C in Fig. 11.8 (Schmertmann, 1955). It is therefore, logical to assume the field curve labelled as K_f should also pass through this point. The field curve can be drawn from point A, having coordinates (e_0, p_0), which corresponds to the *in-situ* condition of the soil. The straight line AC in Fig. 11.8a gives the field curve K_f for normally consolidated clay soil of low sensitivity.

The field curve for overconsolidated clay soil consists of two straight lines, represented by AB and BC in Fig. 11.8b. Schmertmann (1955) has shown that the initial section AB of the field curve is parallel to the mean slope MN of the rebound laboratory curve. Point B is the intersection point of the vertical line passing through the preconsolidation pressure pc on the abscissa and the sloping line

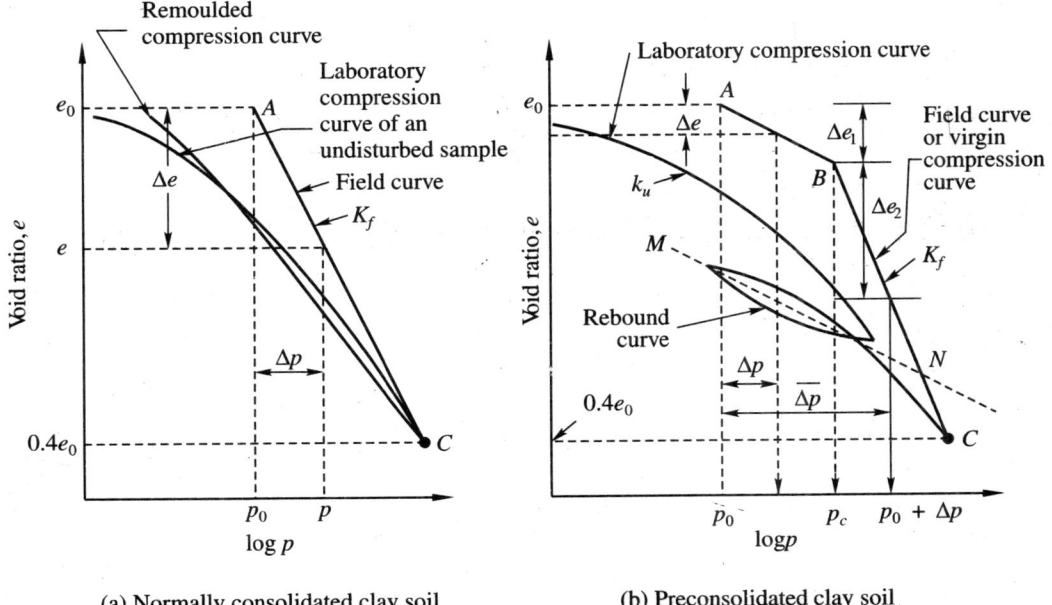

(a) Normally consolidated clay soil (b) Preconsolidated clay soil

Fig. 11.8 Field e-log p curves

AB. Since point C is the intersection of the laboratory compression curve and the horizontal line at void ratio $0.4e_0$, line BC can be drawn. The slope of line MN which is the slope of the rebound curve is called the *swell index* C_s.

11.7 COMPUTATION OF ULTIMATE SETTLEMENT

Settlement Equations for Normally Consolidated Clays

For computing the ultimate settlement of a structure founded on clay the following data are required

1. The thickness of the clay stratum, H
2. The initial void ratio, e_0
3. The consolidation pressure p_0 or p_c
4. The field consolidation curve K_f

The slope of the field curve K_f on a semilogarithmic diagram is designated as the *compression index* C_c (Fig. 11.8).

For normally loadedd clays, the equation for C_c may be written as

$$C_c = \frac{e_0 - e}{\log_{10} p - \log_{10} p_0} \frac{e_0 - e}{\log_{10} p/p_0} \frac{\Delta e}{\log_{10} p/p_0} \tag{11.4}$$

In one-dimensional compression, as per Eq. (11.2), the change in height ΔH per unit of original H may be written as equal to the change in volume ΔV per unit of original volume V (Fig. 11.9).

$$\frac{\Delta H}{H} = \frac{\Delta V}{V} \tag{11.5}$$

Considering a unit sectional area of the clay stratum, we may write

$$V = H, \quad V_1 = H_1$$

$$\Delta V = (H - H_1) = H_s(1 + e_0) - H_s(1 + e_1) = H_s(e_0 - e_1)$$

Therefore,

$$\frac{\Delta V}{V} = \frac{H_s(e_0 - e_1)}{H_s(1 + e_0)} = \frac{(e_0 - e_1)}{(1 + e_0)} = \frac{\Delta e}{1 + e_0} \tag{11.6}$$

Substituting for $\Delta V/V$ in Eq. (11.5)

$$\Delta H = H \frac{\Delta e}{1 + e_0} \tag{11.7}$$

If we designate the compression ΔH of the clay layer as the total settlement S_t of the structure built on it, we have

$$\Delta H = S_t = H \frac{\Delta e}{1 + e_0} \tag{11.8}$$

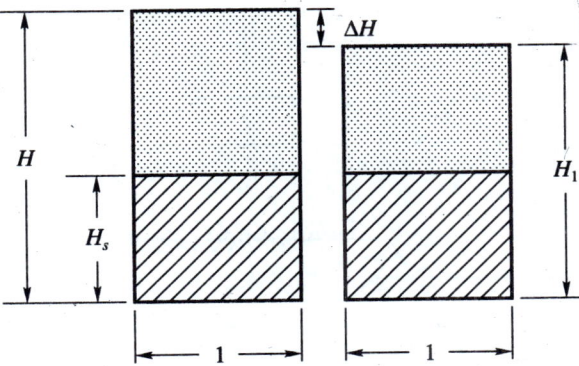

Fig. 11.9 Change of height due to one-dimensional compression

Settlement Calculation from e-log p Curves

Substituting for Δe in Eq. (11.8) we have

$$S_t = \frac{C_c}{1+e_0} H \log_{10} \frac{p_0 + \Delta p}{p_0} \tag{11.9}$$

or $\quad S_t = \frac{C_c}{1+e_0} H \log_{10} \frac{p_0 + \Delta p}{p_0} \tag{11.10}$

Net change in pressure Δp produced by the structure at the middle of a clay stratum is calculated from the Boussinesq or Westergaard theories as explained in Chapter 10.

If the thickness of the clay stratum is too large, the stratum may be divided into layers of smaller thickness not exceeding 3 m. The net change in pressure Δp at the middle of each layer will have to be calculated. Consolidation tests will have to be completed on samples taken from the middle of each of the strata and the corresponding compression indices will have to be determined. The equation for the total consolidation settlement may be written as

$$S_t = \sum H_i \frac{C_c}{1+e_0} \log_{10} \frac{p_0 + \Delta p}{p_0} \tag{11.11}$$

where the subscript i refers to each layer in the subdivision. If there is a series of clay strata of thickness H_1, H_2, etc. separated by granular materials, the same Eq. (11.10) may be used for calculating the total settlement.

Settlement Calculation from e-p Curves

We can plot the field e-p curves from the laboratory test data and the field e-log p curves. The weight of a structure or of a fill increases the pressure on the clay stratum from the overburden pressure p_0 to the value $p_0 + \Delta p$ (Fig. 11.10). The corresponding void ratio decreases from e_0 to e. Hence, for the range in pressure from p_0 to $(p_0 + \Delta p)$, we may write

$$e_0 - e = \Delta e = a_v \Delta p$$

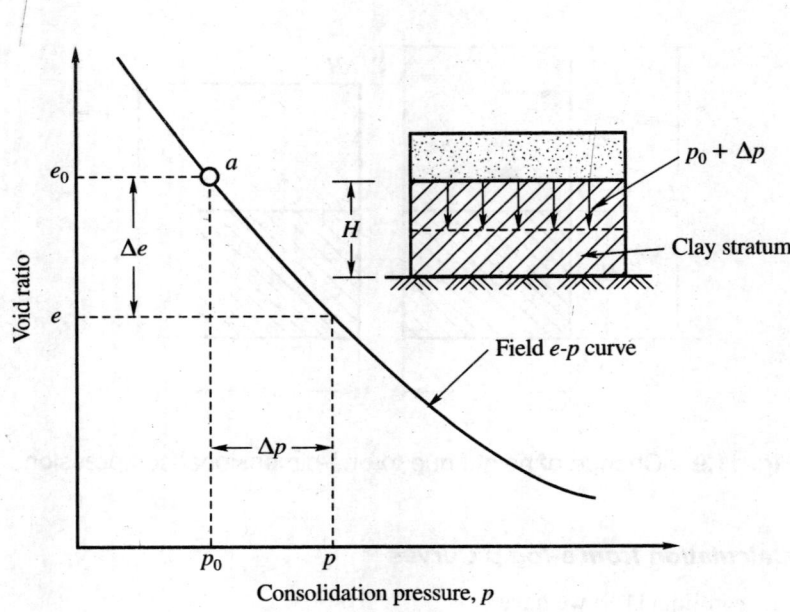

Fig. 11.10 Settlement calculation from *e-p* curve

or $a_v(\text{cm}^2/\text{gm}) = \dfrac{\Delta e}{p(\text{cm}^2/\text{gm})}$ (11.12)

where, a_v is called the *coefficient of compressibility*.

For a given difference in pressure, the value of the coefficient of compressibility decreases as the pressure increases. Now substituting for Δe in Eq. (11.8) from Eq. (11.12), we have the equation for settlement

$$S_t = \frac{a_v H}{1+e_0}\Delta p = m_v H \Delta p$$ (11.13)

where, $m_v = a_v/(1 + e_0)$ is known as the *coefficient of volume compressibility*.

It represents the compression of the clay per unit of original thickness due to a unit increase of the pressure.

Settlement Calculation from *e*-log *p* Curve for Overconsolidated Clay Soil

Figure 11.8b gives the field curve K_f for preconsolidated clay soil. The settlement calculation depends upon the excess foundation pressure Δp over and above the existing overburden pressure p_0.

Settlement Computation, if $p_0 + \Delta p \le p_c$ (Fig. 11.8b)

In such a case, use the sloping line *AB*. If C_s = slope of this line (also called the swell index), we have

$$C_s = \frac{\Delta e}{\log(p_0 + \Delta p)p_0}$$ (11.14a)

or $\quad \Delta e = C_s \log \dfrac{p_0 + \Delta p}{p_0}$ (11.14b)

By substituting for Δe in Eq. (11.8), we have

$$S_t = \frac{C_s H}{1 + e_0} \log \frac{p_0 + \Delta p}{p_0}$$ (11.15a)

Settlement Computation, if $p_0 < p_0 < p_0 + \overline{\Delta p}$

We may write from Fig. 11.8b

$$\Delta e = \Delta e_1 + \Delta e_2 = C_s \log_{10} \frac{p_c}{p_0} + C_c \log_{10} \frac{p_0 + \overline{\Delta p}}{p_c}$$ (11.15b)

In this case the slope of both the lines AB and BC in Fig. 11.8b are required to be considered. Now the equation for S_t may be written as [from Eq. (11.8) and Eq. (11.15b)]

$$S_t = \frac{C_s H}{1 + e_0} \log_{10} \frac{p_c}{p_0} + \frac{C_c H}{1 + e_0} \log_{10} \frac{p_0 + \overline{\Delta p}}{p_c}$$ (11.15c)

The swell index $C_s \approx 4\ 1/5$ to $1/10\ C_c$.

Nagaraj and Murthy (1985) have proposed the following equation for C_s as

$$C_s = 0.0463 \left(\frac{w_l}{100} \right) G$$

where w_l = liquid limit, G_s = specific gravity of solids.

Compression Index C_c—Empirical Relationships

Research workers in different parts of the world have established empirical relationships between the *compression index* C_c and other soil parameters. A few of the important relationships are given below.

Skempton's Formula

Skempton (1944) established a relationship between C_c and liquid limits for remolded clays as

$$C_c = 0.007(w_l - 10)$$ (11.16)

where w_l is in percent.

Terzaghi and Peck Formula

Based on the work of Skempton and others, Terzaghi and Peck (1948) modified Eq. (11.16) applicable to normally consolidated clays of low to moderate sensitivity as

$$C_c = 0.009(w_l - 10)$$ (11.17)

Azzouz et al. Formula

Azzouz; et al. (1976) proposed a number of correlations based on the statistical analysis of a number of soils. The one of the many which is reported to have 86 percent reliability is

$$C_c = 0.37(e_0 + 0.003w_l + 0.0004w_n - 0.34) \tag{11.18}$$

where e_0 = in situ void ratio, w_l and w_n are in percent. For organic soil they proposed

$$C_c = 0.115 \, w_n \tag{11.19}$$

Hough's Formula

Hough (1957), on the basis of experiments on precompressed soils, has given the following equation

$$C_c = 0.3 \, (e_0 - 0.27) \tag{11.20}$$

Nagaraj and Srinivasa Murthy Formula

Nagaraj and Srinivasa Murthy (1985) have developed equations based on their investigation as follows

$$C_c = 0.2343 \, e_l \tag{11.21}$$

$$C_c = 0.39 \, e_0 \tag{11.22}$$

where e_l is the void ratio at the liquid limit, and e_0 is the in-situ void ratio.

In the absence of consolidation test data, one of the formulae given above may be used for computing C_c according to the judgment of the engineer.

Example 11.3

A stratum of normally consolidated clay 7 m thick is located at a depth 12 m below ground level. The natural moisture content of the clay is 43 percent and its liquid limit is 48 percent. The specific gravity of the solid particles is 2.76. The water table is located at a depth 5 m below ground surface. The soil is sand above the clay stratum. The submerged unit weight of the sand is 11 kN/m³ and the same weighs 18 kN/m³ above the water table. The average increase in pressure at the centre of the clay stratum is 120 kN/m² due to the weight of a building that will be constructed on the sand above the clay stratum. Estimate the expected settlement of the structure.

Solution

1. Determination of e and γ_b for the clay [Fig. Ex. 11.3]

$$\frac{M_w}{M_s} = w_n, M_s = V_s G \rho_0 = 1 \times 2.76 \times 1 = 2.76 \text{ gms}$$

$$w_n = 0.43, M_w = 0.43 \times 2.76 = 1.19$$

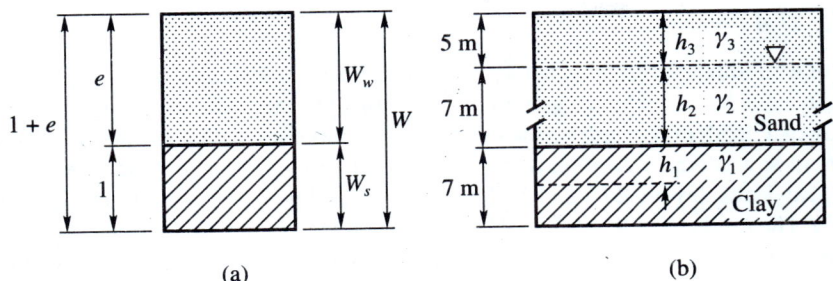

(a)   (b)

Fig. Ex. 11.3

$$e_0 = \frac{V_w}{V_s} = \frac{1.19}{1} 1.19 = 1.19$$

$$M = M_w + M_s = 1.19 \times 2.76 = 3.95 \text{ gms}$$

$$\rho_t = \frac{M}{1+e_0} = \frac{3.95}{2.19} = 1.8 \text{ gm/cm}^3$$

$$\rho_b = (1.8 - 1) = 0.8 \text{ gm/cm}^3$$

2.   Determination of overburden pressure p_0

$$p_0 = \gamma_1 h_1 + \gamma_2 h_2 + \gamma_3 h_3$$

or

$$p_0 = 0.8 \times 9.81 \times 3.5 + 11 \times 7 \times 1.8 \times 9.81 \times 5 = 193 \text{ kN/m}^2$$

3.   Compression index [Eq. 11.17]

$$C_c = 0.009(w_t - 10) = 0.009 \times (48 - 10) = 0.34$$

4.   Excess pressure

$$\Delta p = 120 \text{ kN/m}^2$$

5.   Total settlement

$$S_t = \frac{C_c}{1+e_0} H \log_{10} \frac{p_0 + \Delta p}{p_0}$$

$$= \frac{0.34}{2.19} \times 700 \log_{10} \frac{193 + 120}{193} = 22.6 \text{ cm}$$

Estimated settlement = 22.6 cm.

Example 11.4

A column of a building carries a load of 4000 kN. The load is transferred to sub soil through a square footing of size 5 × 5 m founded at a depth of 2 m below ground level. The soil below the footing is fine sand up to à depth of 5 m and below this is a soft compressible clay of thickness 5 m. The water table is found at a depth of 2 m below the base of the footing. The specific gravities of the solid particles of sand and clay are 2.64 and 2.72 and their natural moisture contents are 25 and 40 percent respectively. The sand above the water table may be assumed to remain saturated. If the plastic limit and the plasticity index of the clay are 30 and 40 percent respectively, estimate the probable settlement of the footing (see Fig. Ex. 11.4).

Solution

1. Required Δp at the middle of the clay layer using the Boussinesq equation

$$\frac{D}{B} = \frac{7.5}{5} = 1.5 < 3.0$$

Divide the footing into 4 equal parts so that $D/B > 3$

The concentrated load at the center of each part = 1000 kN

Radial distance, $r = 1.77$ m

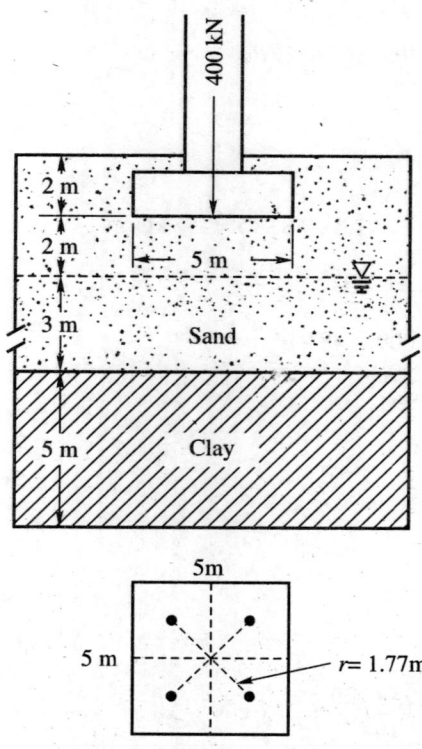

Fig. Ex. 11.4

By the Boussinesq equation the excess pressure Δp at depth 7.5 m is ($I_B = 0.41$)

$$\Delta p = 4 \times \frac{Q}{Z_2} I_B = \frac{4 \times 1000}{7.5^2} \times 0.41 = 29 \text{ kN/m}^2$$

2. Void ratio and unit weights

As per the procedure explained in Ex. 11.3

For sand $\gamma_t = 19.8 \text{ kN/m}^3$ $\gamma_b = 9.8 \text{ kN/m}^3$

For clay $\gamma_b = 8.2 \text{ kN/m}^3$ $e_0 = 1.09$

3. Overburden pressure p_0

$p_0 = 2.5 \times 8.2 + 3 \times 9.8 + 2 \times 19.8 = 89.5 \text{ kN/m}^2$

4. Compression index

$w_l = I_p + w_p = 40 + 30 = 70\%,$ $\qquad C_c = 0.009(70 - 10) = 0.54$

Settlement $\quad S_t = \dfrac{0.54}{1+1.09} \times 500 \times \log_{10} \dfrac{89.5 + 29}{89.5} = 15.7 \text{ cm.}$

Example 11.5

Soil investigation at a site gave the following information. Fine sand exists to a depth of 10.6 m and below this lies a soft clay layer 7.60 m thick. The water table is at 4.60 m below the ground surface. The submerged unit weight of sand γ_b is 10.4 kN/m³, and the wet unit weight above the water table is 17.6 kN/m³. The water content of the normally consolidated clay $w_n = 40\%$, its liquid limit $w_l = 45\%$, and the specific gravity of the solid particles is 2.78. The proposed construction will transmit a net stress of 120 kN/m² at the centre of the clay layer. Find the average settlement of the clay layer.

Solution

For calculating settlement (Eq. 11.10)

$$S_t = \frac{C_c}{1 + e_0} H \log \frac{p_0 + \Delta p}{p_0} \quad \text{where } \Delta p = 120 \text{ kN/m}^2$$

From Eq. (11.17), $C_c = 0.009 (w_l - 10) = 0.009 (45 - 10) = 0.32$

From Eq. (3.13), $e_0 = \dfrac{wG}{S} = wG = 0.40 \times 2.78 = 1.11 \qquad$ since $S = 1$

γ_b, the submerged unit weight of clay is

From Eq. (3.17), $\gamma_{sat} = \dfrac{\gamma_w (G + e_0)}{1 + e_0} = \dfrac{9.81 (2.78 + 1.11)}{1 + 1.11} = 18.1 \text{ kN/m}^3$

$$\gamma_b = \gamma_{sat} - \gamma_w = 18.1 - 9.81 = 8.29 \, \text{kN/m}^3$$

The effective vertical stress p_0 at the mid height of the clay layer is

$$p_0 = 4.60 \times 17.6 + 6 \times 10.4 + \frac{7.60}{2} \times 8.29 = 174.9 \, \text{kN/m}^3$$

Now,
$$S_t = \frac{0.32 \times 7.60}{1 + 1.11} \log \frac{174.9 + 120}{174.9} = 0.26 \, \text{m} = 26 \, \text{cm}$$

Average settlement = 26 cm.

Example 11.6

A soil sample has a compression index of 0.3. If the void ratio e at a stress of $1.4 \, \text{kg/m}^2$ is 0.5, compute (i) the void ratio if the stress is increased to $2 \, \text{kg/m}^2$, and (ii) the settlement of a soil stratum 4 m thick.

Solution

Given : $C_c = 0.3, e_1 = 0.50, p_1 = 1.4 \, \text{kg/cm}^2, p_2 = 2 \, \text{kg/cm}^2$

(i) Now from Eq. (11.4),

$$C_c = \frac{e_1 - e_2}{\log p_2 - \log_{10} p_1} = \frac{e_1 - e_2}{\log_{10} p_2 / p_1}$$

or $e_2 = e_1 - C_c \log_{10} p_2 / p_1$

substituting the known values, we have

$$e_2 = 0.5 - 0.3 \log_{10} \frac{2}{1.4} = 0.454$$

(ii) The settlement per Eq. (11.10) is

$$S = \frac{C_c}{1 + e_1} H \log \frac{p_2}{p_1} = \frac{0.3 \times 4 \times 100}{1.5} \log \frac{2}{1.4} = 12.3 \, \text{cm}.$$

Example 11.7

Two points on a curve for a normally consolidated clay have the following coordinates.

Point 1: $e_1 = 0.7,$ $p_1 = 1.0 \, \text{kg/cm}^2$

Point 2: $e_2 = 0.6,$ $p_2 = 3.0 \, \text{kg/cm}^2$

If the average overburden pressure on a 6 m thick clay layer is $1.5 \, \text{kg/cm}^2$, how much settlement will the clay layer experience due to additional load intensity of $1.6 \, \text{kg/cm}^2$.

Solution

From Eq. (11.4) we have

$$C_c = \frac{e_1 - e_2}{\log p_2 / p_1} = \frac{0.7 - 0.6}{\log 3.0 / 1.0} = 0.21$$

We need the initial void ratio e_0 at an overburden pressure of 1.5 kg/cm².

Now, we have write,

$$C_c = \frac{e_0 - e_2}{\log p_2 / p_0} = 0.21$$

or

$$(e_0 - 0.6) = 0.21 \log 3 / 1.5 = 0.063$$

or

$$e_0 = 0.6 + 0.063 = 0.663.$$

$$\text{Settlement, } S = \frac{C_c}{1 + e_0} H \log_{10} \frac{p_0 + \Delta p}{p_0}$$

Substituting the known values, we have, (for $\Delta p = 1.6$ kg/cm²)

$$S = \frac{0.21 \times 6 \times 100}{1.633} \log_{10} \frac{1.5 + 1.6}{1.5} = 23.9 \text{ cm.}$$

11.8   ONE-DIMENSIONAL CONSOLIDATION THEORY OF TERZAGHI

One dimensional consolidation theory as proposed by Terzaghi is generally applicable in all cases that arise in practice where

1.   The secondary compression is not very significant,

2.   The clay stratum is drained on one or both the surfaces,

3.   The clay stratum is deeply buried, and

4.   The clay stratum is thin compared with the size of the loaded areas.

The following assumptions are made in the development of the theory:

1.   The voids of the soil are completely filled with water,

2.   Both water and solid constituents are incompressible,

3.   Darcy's law is strictly valid,

4.   The coefficient of permeability is a constant,

5.   The time lag of consolidation is due entirely to the low permeability of the soil, and

6.   The clay is laterally confined.

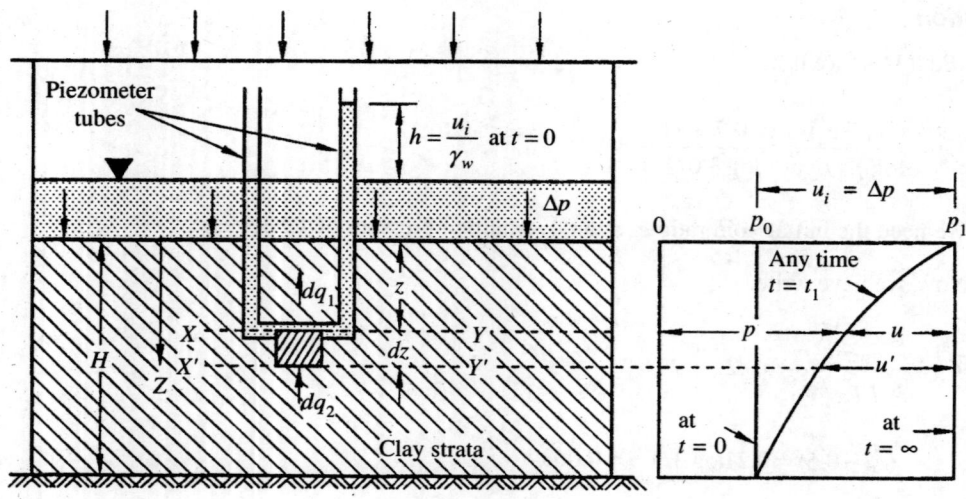

Fig. 11.11   One-dimensional consolidation

Differential Equation for One-Dimensional Flow

Consider a stratum of soil infinite in extent in the horizontal direction (Fig. 11.11) but of such thickness H, that the pressures created by the weight of the soil itself may be neglected in comparison to the applied pressure.

Assume that drainage takes place only at the top and further assume that the stratum has been subjected to a uniform pressure of p_0 for such a long time that it is completely consolidated under that pressure and that there is a hydraulic equilibrium prevailing, i.e., the water level in the piezometric tube at any section XY in the clay stratum stands at the level of the water table (piezometer tube in Fig. 11.14).

Let an increment of pressure Δp be applied. The total pressure to which the stratum is subjected is

$$p_1 = p_0 + \Delta p \tag{11.23}$$

Immediately after the increment of load is applied the water in the pore space throughout the entire height, H, will carry the additional load and there will be set up an excess hydrostatic pressure u_i throughout the pore water equal to Δp as indicated in Fig. 11.11.

After an elapsed time $t = t_1$, some of the pore water will have escaped at the top surface and as a consequence, the excess hydrostatic pressure will have been decreased and a part of the load transferred to the soil structure. The distribution of the pressure between the soil and the pore water, p and u respectively at any time t, may be represented by the curve as shown in the figure. It is evident that

$$p_1 = p + u \tag{11.24}$$

at any elapsed time t and at any depth z, and u is equal to zero at the top. The pore pressure u, at any depth, is therefore a function of z and t and may be written as

$$u = f(z, t) \tag{11.25}$$

Consider an element of volume of the stratum at a depth z, and thickness dz (Fig. 11.11). Let the bottom and top surfaces of this element have unit area.

The consolidation phenomenon is essentially a problem of non-steady flow of water through a porous mass. The difference between the quantity of water that enters the lower surface at level $X'Y'$ and the quantity of water which escapes the upper surface at level XY in time element dt must equal the volume change of the material which has taken place in this element of time. The quantity of water is dependent on the hydraulic gradient which is proportional to the slope of the curve t.

The hydraulic gradients at levels XY and $X'Y'$ of the element are

$$i = \frac{1}{\gamma_w} \frac{\partial u}{\partial z} \tag{11.26}$$

$$i' = \frac{1}{\gamma_w} \frac{\partial}{\partial z}\left(u + \frac{\partial u}{\partial z} dz\right) = \frac{1}{\gamma_w} \frac{\partial u}{\partial z} + \frac{1}{\gamma_w} \frac{\partial^2 u}{\partial u \partial z^2} dz \tag{11.27}$$

If k is the hydraulic conductivity (coefficient of permeability) the outflow from the element at level XY in time dt

$$dq_1 = ikdt = \frac{k}{\gamma_w} \frac{\partial u}{\partial z} dt \tag{11.28}$$

The inflow at level $X'Y'$ is

$$dq_2 = ikdt = \frac{k}{\gamma_w} \frac{\partial u}{\partial z} dt + \frac{\partial^2 u}{\partial z^2} dz \, dt \tag{11.29}$$

The difference in flow is therefore

$$dq = dq_1 - dq_2 = -\frac{k}{\gamma_w} \frac{\partial^2 u}{\partial z^2} dz \, dt \tag{11.30}$$

From the consolidation test performed in the laboratory, it is possible to obtain the relationship between the void ratios corresponding to various pressures to which a soil is subjected. This relationship is expressed in the form of a pressure-void ratio curve which gives the relationship as expressed in Eq. (11.12)

$$de = a_v dp \tag{11.31}$$

The change in volume Δdv of the element given in Fig. 11.11 may be written as per Eq. (11.7)

$$\Delta dv = \Delta dz = \frac{de}{1+e} dz \tag{11.32}$$

Substituting for de, we have

$$\Delta dv = \frac{a_v}{1+e} dp \, dz \tag{11.33}$$

Here dp is the change in effective pressure at depth z during the time element dt. The increase in effective pressure dp is equal to the decrease in the pore pressure, du.

Therefore,
$$dp = -du = \frac{\partial u}{\partial t} dt \qquad (11.34)$$

Hence,
$$\Delta dv = -\frac{a_v}{1+e}\frac{\partial u}{\partial t} dt\, dz = -m_v \frac{\partial u}{\partial t} dt\, dz \qquad (11.35)$$

Since the soil is completely saturated, the volume change Δdv of the element of thickness dz in time dt is equal to the change in volume of water dq in the same element in time dt.

Therefore, $dq = \Delta dv$ \qquad (11.36)

or
$$\frac{k}{\gamma_w}\frac{\partial^2 u}{\partial z^2} dz\, dt = -\frac{a_v}{1+e}\frac{\partial u}{\partial t} dt\, dz$$

or
$$\frac{k(1+e)}{\gamma_w a_v}\frac{\partial^2 u}{\partial z^2} = \frac{\partial u}{\partial t} = c_v \frac{\partial^2 u}{\partial z^2} \qquad (11.37)$$

where
$$c_v = \frac{k(1+e)}{\gamma_w a_v} = \frac{k}{\gamma_w m_v} \qquad (11.38)$$

is defined as the *coefficient of consolidation*.

Equation (11.37) is the differential equation for one-dimensional flow. The differential equation for three-dimensional flow may be developed in the same way. The equation may be written as

$$\frac{\partial u}{\partial t} = \frac{1+e}{\gamma_w a_v}\left(k_x \frac{\partial^2 u}{\partial x^2} + k_y \frac{\partial^2 u}{\partial y^2} + k_z \frac{\partial^2 u}{\partial z^2} \right) \qquad (11.39)$$

where k_x, k_y and k_z are the coefficients of permeability (hydraulic conductivity) in the coordinate directions of x, y and z respectively.

As consolidation proceeds, the values of k, e and a_v all decrease with time but the ratio expressed by Eq. (11.38) may remain approximately constant.

Mathematical Solution for the One-Dimensional Consolidation Equation

To solve the consolidation Eq. (11.37) it is necessary to set up the proper boundary conditions. For this purpose, consider a layer of soil having a total thickness $2H$ and having drainage facilities at both the top and bottom faces as shown in Fig. 11.12. Under this condition no flow will take place across the center line at depth H. The center line can therefore be considered as an impervious barrier. The boundary conditions for solving Eq. (11.37) may be written as

    1. $u = 0$ when $z = 0$ 2. $u = 0$ when $z = 2H$

    3. $u = \Delta p$ for all depths at time $t = 0$

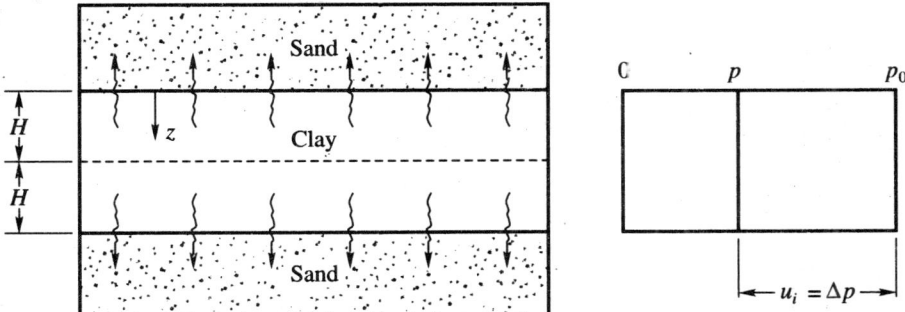

Fig. 11.12 Boundary conditions

On the basis of the above conditions, the solution of the differential Eq. (11.37) can be accomplished by means of Fourier Series.

The solution is

$$u = \sum_{N=0}^{N=\infty} \frac{4\Delta p}{(2N+1)\pi} \sin\left[\frac{(2N+1)\pi z}{2H}\right] \varepsilon^{-\{(2N+1)^2 \pi^2 T\}/4}$$

or $\quad u = \sum_{N=0}^{N=\infty} \frac{2\Delta p}{m} \sin\left[\frac{mz}{H}\right] \varepsilon^{-m^2 T}$ (11.40)

where $\quad m = \dfrac{(2N+1)\pi}{2}, \quad T = \dfrac{C_v t}{H^2}$, a non-dimensional factor.

Eq. (11.40) can be expressed in a general form as

$$\frac{u}{\Delta p} = f\left(\frac{z}{H}, T\right)$$ (11.41)

If Equation (11.41) can be solved by assuming T constant for various values of z/H, curves corresponding to different values of the *time factor T* may be obtained as given in Fig. 11.13. It is of interest to determine how far the consolidation process under the increment of load Δp has progressed at a time t corresponding to the *time factor T* at a given depth z. The term U_z is used to express this relationship. It is defined as the ratio of the amount of consolidation which has already taken place to the total amount which is to take place under the load increment.

The curves in Fig. 11.13 shows the distribution of the pressure Δp between solid and liquid phases at various depths. At a particular depth, say $z/H = 0.5$, the stress in the soil skeleton is represented by AC and the stress in water by CB. AB represents the original excess hydrostatic pressure $u_i = \Delta p$. The degree of consolidation U_z percent at this particular depth is then

$$U_z\% = 100 \times \frac{AC}{AB} = \frac{\Delta p - u}{\Delta p} = 100\left(1 - \frac{u}{\Delta p}\right)$$ (11.42)

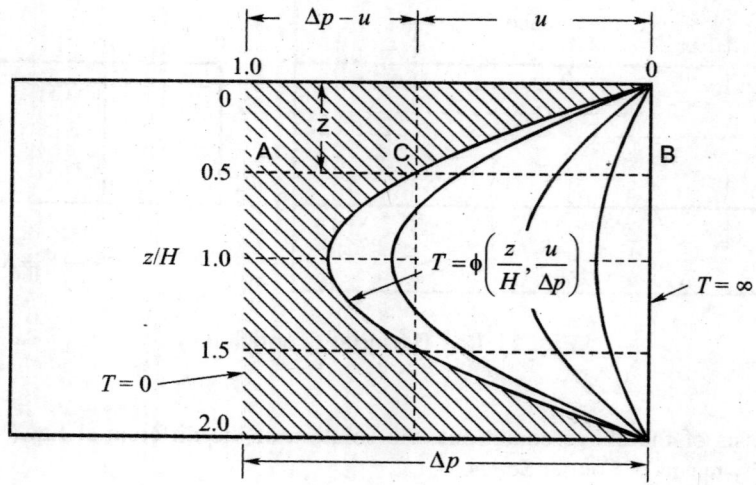

Fig. 11.13 Consolidation of clay layer as a function of T

Following a similar reasoning, the average degree of consolidation $U\%$ for the entire layer at a time factor T is equal to the ratio of the shaded portion (Fig. 11.13) of the diagram to the entire area which is equal to $2H\,\Delta p$. Therefore

$$U\% = \frac{\int_0^{2H}(\Delta p - u)dz}{2H\Delta p}\times 100$$

or $$U\% = \frac{100}{2H}\left[2H - \frac{1}{\Delta p}\int_0^{2H}u\,dz\right] \tag{11.43}$$

Hence, Eq. (11.43) after integration reduces to

$$U\% = 100\left[1 - \sum_{N=0}^{N=\infty}\frac{2}{m^2}\varepsilon^{-m^2 T}\right] \tag{11.44}$$

It can be seen from Eq. (11.44) that the degree of consolidation is a function of the time factor T only which is a dimensionless ratio. The relationship between T and $U\%$ may therefore be established once and for all by solving Eq. (11.44) for various values of T. Values thus obtained are given in Table 11.1 and also plotted on a semilog plot as shown in Fig. 11.14.

For values of $U\%$ between 0 and 60%, the curve in Fig. 11.14 can be represented almost exactly by the equation

$$T = \frac{\pi}{4}\left(\frac{U\%}{100}\right)^2 \tag{11.45}$$

TABLE 11.1

Relationship between *U* and *T*

U%	T	U%	T	U%	T
0	0	40	0.126	75	0.477
10	0.008	45	0.159	80	0.565
15	0.018	50	0.197	85	0.684
20	0.031	55	0.238	90	0.848
25	0.049	60	0.287	95	1.127
30	0.071	65	0.342	100	∞
35	0.096	70	0.405		

which is the equation of a parabola. Substituting for T, Eq. (11.45) may be written as

$$\frac{U\%}{100} = \sqrt{\frac{4c_v}{\pi H^2}} \sqrt{t}$$

(11.46)

In Eq. (11.46), the values of c_v and H are constants. One can determine the time required to attain a given degree of consolidation by using this equation. It should be noted that H represents half the thickness of the clay stratum when the layer is drained on both sides, and it is the full thickness when drained on one side only.

For values of $U\%$ greater than 53%, the curve in Fig.11.14 may be represented by an equation

$$T = 1.781 - 0.933 \log_{10}(100 - U\%)$$

(11.47)

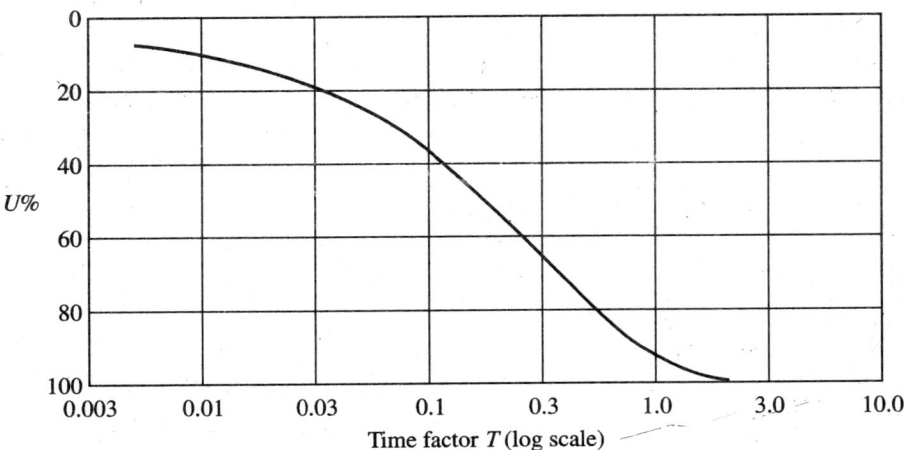

Fig. 11.14 *U* verses *T*

Effect of Boundary Conditions on Consolidation

A layer of clay which permits drainage through both surfaces is called an *open layer*. The thickness of such a layer is always represented by the symbol 2H, in contrast to the symbol H used for the thickness of half-closed layers which can discharge their excess water only through one surface.

The relationship expressed between T and U given in Table 11.1 applies to all the following cases:

1. Where the clay stratum is drained on both sides and the initial consolidation pressure distribution is uniform or linearly increasing or decreasing with depth.

2. Where the clay stratum is drained on one side but the consolidation pressure is uniform with depth.

TABLE 11.2

Relation between U% and T (Special Cases)

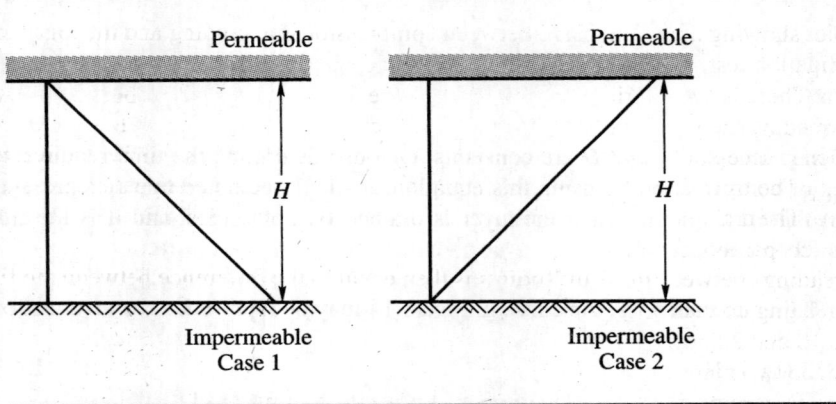

U%	Consolidation pressure increase with depth Time Factors, T	Consolidation pressure decrease with depth
00	0	0
10	0.047	0.003
20	0.100	0.009
30	0.158	0.024
40	0.221	0.048
50	0.294	0.092
60	0.383	0.160
70	0.500	0.271
80	0.665	0.44
90	0.94	0.72
95	1	0.8
100	∞	∞

Separate relationship between T and U are required for half closed layers with thickness H where the consolidation pressures increase or decrease with depth. Such cases are only exceptional and as such not dealt with in detail here. However, the relation between $U\%$ and T for these two cases are given in Table 11.2.

11.9 DETERMINATION OF THE COEFFICIENT OF CONSOLIDATION

The coefficient of consolidation c_v can be evaluated by means of laboratory tests by fitting the experimental curve with the theoretical.

There are two laboratory methods that are in common use for the determination of c_v. They are:

1. Casagrande Logarithm of Time Fitting Method.

2. Taylor Square Root of Time Fitting Method.

Logarithm of Time Fitting Method

Figure 11.15 is a plot showing the relationship between compression dial reading and the logarithm of time of a consolidation test. The theoretical consolidation curve using the log scale for the time factor is also shown. There is a similarity of shape between the two curves. On the laboratory curve, the intersection formed by the final straight line produced backward and the tangent to the curve at the point of inflection is accepted as the 100 percent primary consolidation point and the dial reading is designated as R_{100}. The time-compression relationship in the early stages is also parabolic just as the theoretical curve. The dial reading at zero primary consolidation R_0 can be obtained by selecting any two points on the parabolic portion of the curve where times are in the ratio of 1 : 4. The difference in dial readings between these two points is then equal to the difference between the first point and the dial reading corresponding to zero primary consolidation. For example, two points A and B whose times 10 and 2.5 minutes respectively, are marked on the curve. Let z_1 be the ordinate difference between the two points. A point C is marked vertically over B such that $BC = z_1$. Then the point C corresponds to zero primary consolidation. The procedure is repeated with several points. An average horizontal line is drawn through these points to represent the theoretical zero percent consolidation line.

The interval between 0 and 100% consolidation is divided into equal intervals of percent consolidation. Since it has been found that the laboratory and the theoretical curves have better correspondence at the central portion, the value of c_v is computed by taking the time t and time factor T at 50 percent consolidation. The equation to be used is

$$T_{50} = \frac{c_v t_{50}}{H_{50}^2} \text{ or } c_v = \frac{T_{50}}{t_{50}} H_{50}^2 \tag{11.48}$$

From Table 11.1, we have at $U = 50\%$, $T = 0.197$. From the initial height H_i of specimen and compression dial reading at 50% consolidation, H_{50} for double drainage is

$$H_{50} = \frac{H_i - \Delta H}{2} \tag{11.49}$$

where ΔH = Compression of sample up to 50% consolidation.

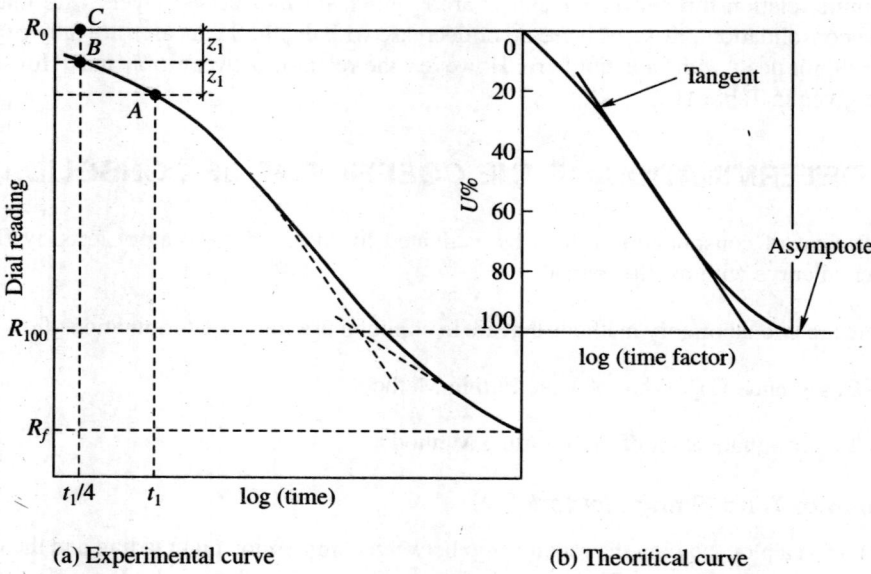

(a) Experimental curve   (b) Theoritical curve

Fig. 11.15   Log of time fitting method

Now the equation for c_v may be written as

$$c_v = 0.197 \frac{H_{50}^2}{t_{50}}$$ (11.50)

Square Root of Time Fitting Method

This method was devised by Taylor. In this method, the dial readings are plotted against the square root of time as given in Fig. 11.16(a). The theoretical curve U versus $\sqrt{T}$ is also plotted and shown in Fig. 11.16(b). On the theoretical curve a straight line exists up to 60 percent consolidation while at 90 percent consolidation the abscissa of the curve is 1.15 times the abscissa of the straight line produced.

The fitting method consists of first drawing the straight line which best fits the early portion of the laboratory curve. Next a straight line is drawn which at all points has abscissa 1.15 times as great as those of the first line. The intersection of this line and the laboratory curve is taken as the 90 percent consolidation point. Its value may be read and is designated as t_{90}.

Usually the straight line through the early portion of the laboratory curve intersects the zero time line at a point differing somewhat from the initial point. This intersection point is called the *corrected zero point*. If one-ninth of the vertical distance between the corrected zero point and the 90 per cent point is set off below the 90 percent point, the point obtained is called the *100 percent primary compression point* (R_{100}). The compression between zero and 100 percent point is called *primary compression*.

At the point of 90 percent consolidation, the value of $T = 0.848$. The equation of c_v may now be written as

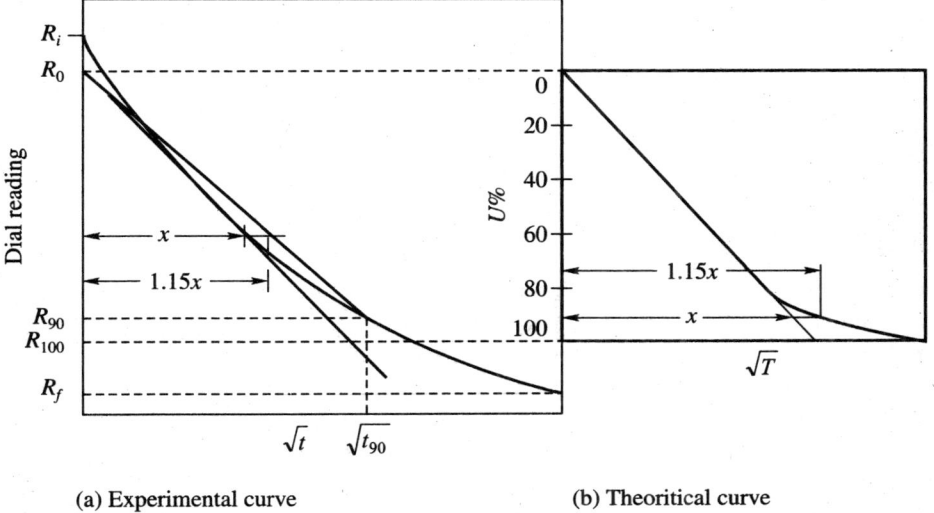

(a) Experimental curve (b) Theoritical curve

Fig. 11.16 Square root of time fitting method

$$c_v = 0.848\frac{H_{90}^2}{t_{90}}$$ (11.51)

where, $H_{90} = \dfrac{H_i - \Delta H}{2}$ for double drainage. ΔH = Compression of sample up to 90% consolidation.

Compression Ratios

In Figs (11.15) or (11.16).

R_i = initial reading of the dial gauge, R_0 = Reading at zero per cent consolidation, R_{90} = Reading at 90% consolidation, R_{100} = Reading at 100% consolidation, R_f = One day dial reading.

We may write the following relationships:

$$R_{100} = R_{90} - \frac{1}{9}(R_0 - R_{90})$$

Total compression	=	$R_i - R_f$
Primary compression	=	$R_0 - R_{100}$
Initial compression	=	$R_i - R_0$
Secondary compression	=	$R_{100} - R_f$

$$\text{Primary compression ratio} \quad = \quad \frac{10}{9}\frac{R_0 - R_{100}}{R_i - R_f}$$ (11.52)

Initial compression ratio $\qquad = \dfrac{R_i - R_0}{R_i - R_f}$ $\hspace{3cm}$ (11.53)

Secondary compression ratio $\qquad = \dfrac{R_{100} - R_f}{R_i - R_f}$ $\hspace{2.5cm}$ (11.54)

Secondary Compression

In certain types of clays the secondary time effects are very pronouned to the extesnt that in some cases the entire time compression curve has the shape of an almost straight sloping line when plotted to a semi-logarithmic scale, instead of the typical inverted S-shape of the curve of clays with pronouned primary consolidation effects. These so called secondary time effects are a phenomenon somewhat analogous to the creep of other overstressed material in a plastic state. A delayed progressive slippage of grain upon grain as the particles adjust themselves to dense condition, appears to be responsible for the secondary effects. When the rate of plastic deformations of the individual soil particles or of their slippage on each other, is slower than the rate of decreasing volume of voids between the particles, then secondary effects predominate and this is reflected by the shape of the time compression of soils are not yet fully understood, and no satisfactory method has yet been developed for a rigorous and reliable analysis and forecasts of the magnitude of these effects. Highly organic soils are normally subjected to considerable secondary consolidation.

However, the rate of secondary consolidation may be expressed by the *coefficient of secondary consolidation.*

$$C_\alpha = \frac{\Delta e}{1 + e_0} \frac{1}{\log_{10}(t_2 / t_1)} = \frac{C_t}{1 + e_0} \hspace{3cm} (11.55)$$

where C_t, the slope of the sraight-line portion of the *e*-log *t* curve, is known as the *secondary compression index.*

Numberically, C_t is equal to the value of Δe of a single cycle of time on the curve (Fig. 11.17)

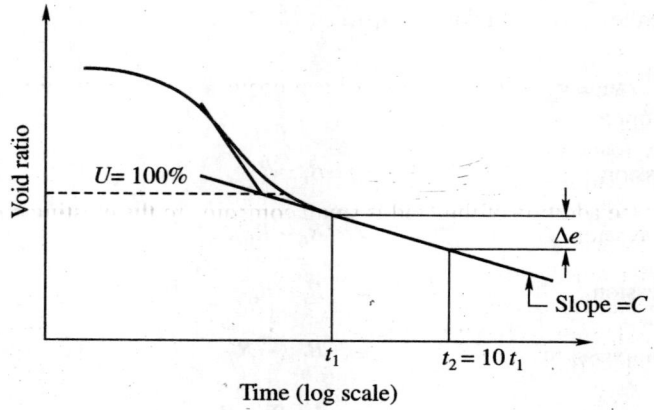

Fig. 11.17 *e*-log time curve representing secondary compression

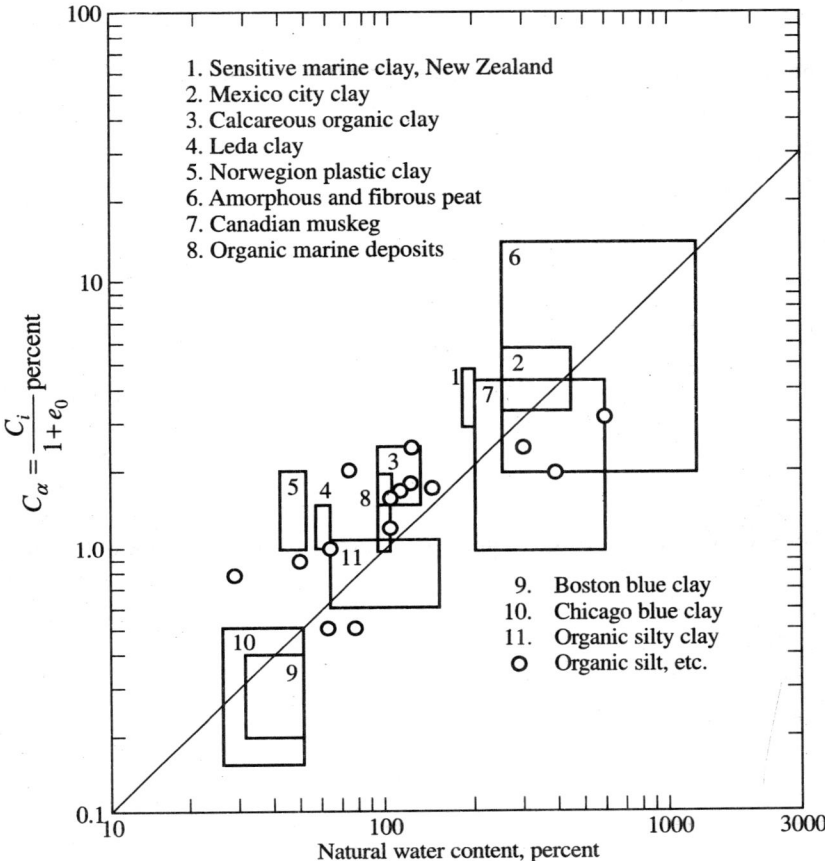

Fig. 11.18 Relationship between coefficient of secondary consolidation and natural water content of normally loaded deposits of clays and various compression organic soils (after Mesri 1972)

The value of C_α for normally loaded compressible soils increases in a general way with the compressibility and hence, with the natural water content, in the manner shown in Fig. 11.18 (Mesri 1972). Although the range in values for a given water content is extremely large, the relation gives a conception of the upper limit of the rate of secondary settlement that may be anticipated if the deposit is normally loaded or if the stress added by the proposed construction will appreciably exceed the preconsolidation load. The rate is likely to be much less if the clay is strongly preloaded or if the stress after the addition of the load is small compared to the existing overburden pressure. The rate is also influenced by the length of time the preload may have acted, by the existence of shearing stresses and by the degree of disturbance of the samples. The effects of these various factors have not yet been evaluated. Secondary compression is high in plastic clays and organic soils. Table 11.3 provides a classification of soil based on secondary compressibility. If an 'young, normally loaded clay', having an effective overburden of p_0 is left undisturbed for thousands of years, there will be creep or secondary consolidation. This will reduce the void ratio and consequently increases the preconsolidation pressure which will be much greater than the existing effective overburden pressure p_0. Such a clay may be called an aged, normally consolidated clay.

TABLE 11.3

Classification of soil based on secondary compressibility

C_a	Secondary compressibility
< 0.002	Very low
0.004	Low
0.008	Medium
0.016	High
0.032	Very high
0.064	Extremely high

Example 11.8

A 2.5 cm thick sample of clay was taken from the field for predicting the time of settlement for a proposed building which exerts a uniform pressure of 100 kN/m² over the clay stratum. The sample was loaded to 100 kN/m² and proper drainage was allowed from top and bottom. It was seen that 50 percent of the total settlement occurred in 3 minutes. Find the time required for 50 percent of the total settlement of the building, if it is to be constructed on a 6 m thick layer of clay which extends from the ground surface and is underlain by sand.

Solution

The data applied to sample,

Given $U\% = 50\%$. Therefore, T for 50% consolidation $= 0.197$.

The sample is drained on both sides.

$$c_v = \frac{TH^2}{t} = 0.197 \times \frac{(2.5)^2}{4} \times \frac{1}{3} = 10.25 \times 10^{-2} \text{ cm}^2/\text{min}.$$

The time t for 50% consolidation is,

$$t = \frac{0.197 \times 300 \times 300 \times 100}{10.25 \times 60 \times 24} = 120 \text{ days}.$$

Example 11.9

The void ratio of a clay sample A decreased from 0.572 to 0.505 under a change in pressure from 122 to 180 kN/m². The void ratio of another sample B decreased from 0.61 to 0.557 under the same increment of pressure. The thickness of sample A was 1.5 times that of B. Nevertheless the time taken for 50% consolidation was 3 times larger for sample B than for A. What is the ratio of coefficient of permeability of sample A to that of B?

Solution

Let H_a = thickness of sample A, H_b = thickness of sample B, m_{va} = coefficient of volume compressibility of sample A, m_{vb} = coefficient of volume compressibility of sample B, c_{va} = coefficient of consolidation for sample A, c_{vb} = coefficient of consolidation for sample B, Δp_a = increment of load for sample A, Δp_b = increment of load for sample B, k_a = coefficient of permeability for sample A, and k_b = coefficient of permeability of sample B.

We may write the following relationship

$$m_{va} = \frac{\Delta e_a}{1+e_a}\frac{1}{\Delta p_a}, \quad m_{vb} = \frac{\Delta e_b}{1+e_b}\frac{1}{\Delta p_b}$$

where e_a is the void ratio of sample A at the commencement of the test and Δe_a is the change in void ratio. Similarly e_b and Δe_b apply to sample B.

$$\frac{m_{va}}{m_{vb}} = \frac{\Delta p_b}{\Delta p_a}\frac{\Delta e_a}{\Delta e_b}\frac{1+e_b}{1+e_a}, \quad \text{and} \quad T_a = \frac{c_{va}t_a}{H_a^2}, \quad T_b = \frac{c_{vb}t_b}{H_b^2}$$

wherein T_a, t_a, T_b and t_b correspond to samples A and B respectively. We may write

$$\frac{c_{va}}{c_{vb}} = \frac{T_a}{T_b}\frac{H_a^2}{H_b^2}\frac{t_b}{t_a}, \quad k_a = c_{va}m_{va}\gamma_w, \quad k_b = c_{vb}m_{vb}\gamma_w$$

Therefore, $\dfrac{k_a}{k_b} = \dfrac{c_{va}}{c_{vb}}\dfrac{m_{va}}{m_{vb}}$

Given in this problem $e_a = 0.572$, and $e_b = 0.61$

$$\Delta e_a = 5.72 - 0.505 = 0.067, \Delta e_b = 0.610 - 0.557 = 0.053$$

$$\Delta p_a = \Delta p_b = 180 - 122 = 58 \text{ kN/m}^2, \quad H_a = 1.5\, H_b$$

But $\qquad t_b = 3t_a$

We have, $\qquad \dfrac{m_{va}}{m_{vb}} = \dfrac{0.067}{0.053} \times \dfrac{1+0.61}{1+0.572} = 1.29$

$$\frac{c_{va}}{c_{vb}} = 1.5^2 \times 3 = 6.75$$

Therefore, $\dfrac{k_a}{k_b} = 6.75 \times 1.29 = 8.7$

The ratio is 8.7 : 1.

Example 11.10

A strata of normally consolidated clay of thickness 3 m is drained on one side only. It has a hydraulic conductivity of $k = 5 \times 10^{-8}$ cm/sec and a coefficient of volume compressibility $m_v = 125 \times 10^{-2}$ cm²/sec. Determine the ultimate value of the compression of the stratum by assuming a uniformly distributed load of 250 kN/m² and also determine the time required for 20 percent and 80 percent consolidation.

Solution

Total compression,

$$S_t = m_v H \Delta p = 125 \times 10^{-4} \times 300 \times 250 \times 10^{-4} = 9.40 \, \text{cm}.$$

For determining the relationship between $U\%$ and T for 20% consolidation use the equation

$$T = \frac{\pi}{4} \left(\frac{U\%}{100} \right)^2 \quad \text{or} \quad T = \frac{3.14}{4} \times \left(\frac{20}{100} \right)^2 = 0.0314$$

For 80% consolidation use the equation

$$T = 1.781 - 0.933 \log_{10} (100 - U\%)$$

Therefore $T = 1.781 - 0.933 \log_{10} (100 - 80) = 0.570$.

The coefficient of consolidation is

$$c_v = \frac{k}{\gamma_w m_v} = \frac{5 \times 10^{-8}}{9.81 \times 10^{-6} \times 125 \times 10^{-2}} = 4 \times 10^{-3} \, \text{cm}^2/\text{sec}$$

The time required for 20% and 80% consolidation is

$$t_{20} = \frac{H^2 T}{c_v} = \frac{300 \times 300 \times 0.0314 \times 10^3}{4 \times 60 \times 60 \times 24} = 8.18 \, \text{days}$$

$$t_{80} = \frac{300 \times 300 \times 0.57 \times 10^3}{4 \times 60 \times 60 \times 24} = 156 \, \text{days}$$

Example 11.11

The loading period for a new building extended from May 1995 to May 1997. In May 1960, the average measured settlement was found to be 11.43 cm. It is known that the ultimate settlement will be about 35.56 cm. Estimate the settlement in May 1965. Assume double drainage to occur.

Solution

For the majority of practical cases in which loading is applied over a period, acceptable accuracy is obtained when calculating time-settlement relationships by assuming the time datum to be midway through the loading or construction period.

$S_t = 11.43$ cm when $t = 4$ years and $S = 35.56$ cm.

The settlement is required for $t = 9$ years, that is, 1965. Assuming as a starting point that at $t = 9$ years, the degree of consolidation will be $= 0.53$. Under these conditions per Eq. (11.45),

$$U = 1.13 \sqrt{T}.$$

If S_{t_1} = settlement at time t_1, S_{t_2} = settlement at time t_2

$$\frac{S_{t_1}}{S_{t_2}} = \frac{U_1}{U_2} = \sqrt{\frac{T_1}{T_2}} = \sqrt{\frac{t_1}{t_2}} \quad \text{since } T = \frac{c_v t}{H^2}$$

where $\dfrac{c_v t}{H^2}$ is a constant. Therefore, $\dfrac{11.43}{S_{t_2}} = \dfrac{\sqrt{4}}{\sqrt{9}}$ or $S_{t_2} = 17.15$ cm

Therefore, at $t = 9$ years, $U = \dfrac{17.5}{35.56} = 0.48$

Since the value of U is less than 0.60 the assumption is valid. Therefore, the estimated settlement is 17.15 cm. In the event of the degree of consolidation exceeding 0.60, Eq. (11.47) has to be used to obtain the relationship between T and U.

Example 11.12

An oedometer test is performed on a 2 cm thick clay sample. After 5 minutes, 50% consolidation is reached. After how long a time would the same degree of consolidation be achieved in the field where the clay layer is 3.70 m thick? Assume the sample and the clay layer have the same drainage boundary conditions (double drainage).

Solution

The time factor T is defined as $T = \dfrac{c_v t}{H^2}$

where H = half the thickness of the clay for double drainage.

Here, the time factor T and coefficient of consolidation are the same for both the sample and the field clay layer. The parameter that changes is the time t. Let t_1 and t_2 be the times required to reach 50% consolidation both in the oedometer and field respectively. $t_1 = 5$ min

Therefore, $\dfrac{c_v t_1}{H_1^2} = \dfrac{c_v t_2}{H_2^2}$;

Now, $t_2 = \left(\dfrac{H_2}{H_1}\right)^2 t_1 = \left(\dfrac{370}{2}\right)^2 \times 5 \times \dfrac{1}{60} \times \dfrac{1}{24}$ days ≈ 119 days.

Example 11.13

A laboratory sample of clay 2 cm thick took 15 min to attain 60 percent consolidation under a double drainage condition. What time will be required to attain the same degree of consolidation for a clay layer 3 m thick under the foundation of a building for a similar loading and drainage condition, what is the value of c_v.

Solution

Use Eq. (11.47) for $U > 53\%$ for determining T

$$T = 1.781 - 0.933 \log_{10} (100 - U\%)$$

$$= 1.781 - 0.933 \log_{10} (100 - 60) = 0.286.$$

From Eq. (11.48) the coefficient of consolidation, c_v is

$$C_v = \frac{TH^2}{t} = \frac{0.286 \times (1)^2}{15} = 1.91 \times 10^{-2} \text{ cm}^2/\text{min}.$$

The value of c_v remains constant for both the laboratory and field conditions. As such, we may write,

$$\left(\frac{TH^2}{t} \right)_{lab} = \left(\frac{TH^2}{t} \right)_{field}$$

where H = half the thickness = 1 cm for the lab and 150 cm for field strata, and t = 15 min. for lab.

Therefore, we have

$$\left(\frac{T(1)^2}{0.25} \right)_{lab} = \left(\frac{T(150)^2}{t} \right)_{field}$$

or $t_f = (150)2 \times 0.25 = 5625$ hr

for the field strata to attain the same degree of consolidation.

11.10 RATE OF SETTLEMENT DUE TO CONSOLIDATION

It has been explained that the ultimate settlement S_t of a clay layer due to consolidation may be computed by using either Eq. (11.10) or Eq. (11.13). If S is the settlement at any time t after the imposition of load on the clay layer, the degree of consolidation of the layer in time t may be expressed as

$$U\% = \frac{S}{S_t} \times 100 \text{ percent} \tag{11.56}$$

Since U is a function of the time factor T, we may write

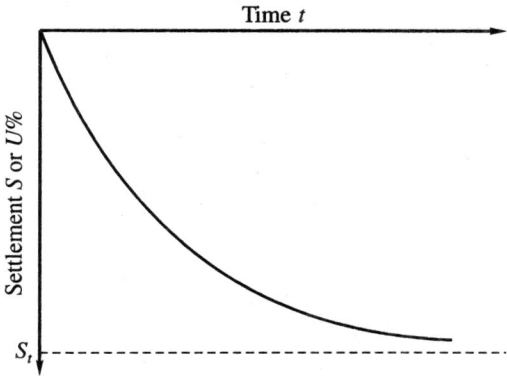

Fig. 11.19 Time-settlement curve

$$U\% = 100 f(T) = \frac{S}{S_t} \times 100 \qquad\qquad (11.57)$$

The rate of settlement curve of a structure built on a clay layer may be obtained by the following procedure:

1. From consolidation test data, compute m_v and c_v.

2. Compute the total settlement S_t that the clay stratum would experience with the increment of load Δp.

3. From the theoretical curve giving the relation between U and T, find T for different degrees of consolidation, say 5, 10, 20, 30 percent, etc.

4. Compute from equation $t = \dfrac{TH^2}{c_v}$ the values of t for different values of T. It may be noted here that for drainage on both sides H is equal to half the thickness of the clay layer.

5. Now a curve can be plotted giving the relation between t and $U\%$ or t and S as shown in Fig. 11.19.

11.11 TWO AND THREE-DIMENSIONAL CONSOLIDATION, PROBLEMS

When the thickness of a clay stratum is great compared with the width of the loaded area, the consolidation of the stratum is three-dimensional. In a three-dimensional process of consolidation the flow occurs either in radial planes or else the water particles travel along flow lines which do not lie in planes. The problem of this type is complicated though a general theory of three-dimensional consolidation exists (Biot 1941). But a simple example of three-dimensional consolidation is the consolidation of a stratum of soft clay or silt by providing sand drains and surcharge for accelerating consolidation.

The most important example of two dimensional consolidation in engineering practice is the consolidation of the case of a hydraulic fill dam. In two-dimensional flow, the excess water drains out of the clay in parallel planes. Gilboy (1934) has analysed the two dimensional consolidation of a hydraulic fill dam.

11.12 QUESTIONS AND PROBLEMS

11.1 A bed of sand 10 m thick is underlain by a compressible of clay 3 m thick under which lies sand. The water table is at a depth of 4 m below the ground surface. The total unit weights of sand below and above the water table are 20.5 and 17.7 kN/m³ respectively. The clay has a natural water content of 42%, liquid limit 46% and specific gravity 2.76. Assuming the clay to be normally consolidated, estimate the probable final settlement under an average excess pressure of 100 kN/m².

11.2 The effective overburden pressure at the middle of a saturated clay layer 4 m thick is 10 T/m² and is drained on both sides. The overburden pressure at the middle of the clay stratum is expected to be increased by 150 kN/m² due to the load from a structure at the ground surface. An undisturbed sample of clay 20 mm thick is tested in a consolidometer. The total change in thickness of the specimen is 0.80 mm when the applied pressure is 100 kN/m². The final water content of the sample is 24 percent and the specific gravity of the solids is 2.72. Estimate the probable final settlement of the proposed structure.

11.3 The following observations refer to a standard laboratory consolidation test on an undisturbed sample of clay.

Pressure kN/m²	Final dial gauge Reading × 10⁻² mm	Pressure kN/m²	Final dial gauge Reading × 10⁻² mm
0	0	400	520
50	180	100	470
100	250	0	355
200	360		

The sample was 75 mm in diameter and had an initial thickness of 18 mm. The moisture content at the end of the test was 45.5%; the specific gravity of solids was 2.53.

Compute the void ratio at the end of each loading increment and also determine whether the soil was overconsolidated or not. If it was overconsolidated, what was the overconsolidation ratio if the effective overburden pressure at the time of sampling was 60 kN/m²?

11.4 The results of a consolidation test on a soil sample for a load increased from 200 to 400 kN/m² are given below:

The thickness of the sample corresponding to the dial reading 1255 is 1.561 cm. Find the value of coefficient of consolidation by square root of time fitting method in sq. m/year. One division of dial gauge corresponds to 2.5 × 10⁻⁴ cm. The sample is drained on both faces.

11.5 A 2.5 cm thick sample was tested in a consolidometer under saturated conditions with drainage both sides. 30 percent of consolidation was reached under a load in 15 minutes.

For the same conditions of stressing but with only one way drainage, estimate the amount of time in days it would take for a 6 m thick layer of the same soil to consolidate in the field to attain the same degree of consolidation.

11.6   A layer of normally consolidated clay is 7 m thick and lies under a recently constructed building. The pressure of sand overlying the clay layer is 300 kN/m^2, and the new construction increases the overburden pressure at the middle of the clay layer by 100 kN/m^2. If the compression index is 0.5, compute the final settlement assuming $w_n = 45\%$, $G = 2.70$, and the clay is submerged with the water table at the top of the clay stratum.

11.7   A consolidation test was made on a sample of saturated marine clay. The diameter and thickness of the sample were 5.5 cm and 3.75 cm respectively. The sample weighed 650 g at the start of the test and 480 g in the dry state after the test. The specific gravity of solids was 2.72. The dial readings corresponding to the final equilibrium condition under each load are given below.

(a)   Compute the void ratios and plot the e-log p curve.

(b)   Estimate the maximum preconsolidation pressure by the Casagrande method.

(c)   Draw the field curve and determine the compression index.

(d)   Compute the values of a_v and m_v for every load increment and plot the same against load.

Pressure, kN/m^2	DR cm × 10^{-4}	Pressure, kN/m^2	DR cm × 10^{-4}
0	0	106	1880
6.7	175	213	3340
11.3	275	426	5000
26.6	540	852	6600
53.3	965		

11.8   The dial readings recorded during a consolidation test at a certain load increment are given below.

Time min	Dial Reading cm × 10^{-4}	Time min	Dial Reading cm × 10^{-4}
0	240	15	622
0.10	318	30	738
0.25	340	60	842
0.50	360	120	930
1.00	385	240	975
2.00	415	1200	1070
4.00	464	1620	1090
8.00	530	-	-

Determine c_v by both the square root of time and log of time fitting methods.

11.9   A structure built on a layer of clay settled 5 cm in 60 days after it was built. If this settlement corresponds to 20 percent average consolidation of the clay layer, plot the time settlement curve of the structure for a period of 3 years from the time it was built.

11.10   The following table gives dial readings against time as obtained during a consolidation test
         on a soft clay ($w_l = 43\%$, $w_p = 21\%$, $w_n = 39\%$), when the pressure was increased from 166
         to 333 kN/m². The void ratio after 100% consolidation under 166 kN/m² was 0.945 and
         that under 333 kN/m² was 0.812. The dial was set to zero at the beginning of the test and the
         initial height of the sample was 1.87 cm. Drainage was permitted at both faces of the sample.
         Plot the curves by

   (i)   The square root of time fitting method

   (ii)   The logarithm of time fitting method

   Determine by both the methods.

   (i)   Coefficient of consolidation

   (ii)   Initial compression ratio

   (iii)   Primary compression ratio

   (iv)   Secondary compression ratio

   (v)   Coefficient of permeability in m/year.

Time min	Dial reading cm × 10⁻⁴	Time min	Dial reading cm × 10⁻⁴
0	122	8.00	765
0.10	220	15.00	906
0.25	252	30.00	1025
0.50	300	70.00	1130
1.00	356	140.00	1178
2.00	452	260.00	1216
4.00	590	455.00	1250
		1440.00	1320

11.11   In a laboratory consolidation test a sample of clay with a thickness of 2.50 cm reached 50%
         consolidation in 8 minutes. The sample was drained top and bottom. The clay layer from
         which the sample was taken is 8 m thick. It is covered by a layer of sand through which
         water can escape and is underlain by a practically impervious bed of intact shale. How long
         will the clay layer require to reach 50 per cent consolidation?

11.12   The following points are coordinates on a pressure-void ratio curve for an undisturbed
         clay.

p	40	80	150	320	kN/m²
e	1.202	1.098	0.991	0.893	

         Find (i) C_c, (ii) m_v, at load intervals of 100 kN/m² starting from $p = 100$ kN/m² and (iii) the
         magnitude of compression in a 3 m thick layer of this clay for a load increment from 100 to
         200 kN/m².

11.13   The following data were obtained from a consolidation test performed on an undisturbed clay sample 3 cm in thickness:

(i) $p_1 = 165$ kN/m², $e = 0.895$

(ii) $p_2 = 310$ kN/m², $e_2 = 0.782$

Assume that the average value of its coefficient of percentage is 3.5×10^{-9} cm/sec. By utilizing the known theoritical relationship between per cent consolidation and time factor, compute and plot the decrease in thickness with time for:

(a)   9 m thick layer of this clay, which is drained on the upper surface only.

(b)   Identical with (a) except that at depth at 3 m, it is separated by a thin horizontal sand layer providing free drainage.

11.14   Assuming that the clay strata in Prob. 11.10 is 6 m thick drained on both sides, plot a time-settlement curve for the increment of load given in the same problem.

11.15   8 m depth of sand overlies a 6 m layer of clay, below which is an impermeable stratum (Fig. Prob. 11.15); the water table is 2 m below the surface of the sand. Over a period of 1 year a 3 m depth of fill (unit weight 20 kN/m³) is to be dumped on the surface over an extensive area. The saturated unit weight of the sand is 19 kN/m³ and that of the clay 20 kN/m³, and above the water table the unit weight of the sand is 17 kN/m³. For the clay, the relationship between void ratio and effective stress (unit kN/m²) can be represented by the equation

$$e = 0.88 - 0.32 \log \sigma' /1100$$

and the coefficient of consolidation is 1.26 m²/year.

(a) Calculate the final settlement of the area due to consolidation of the clay and the settlement after a period of 3 years from the start of dumping. (b) If a very thin layer of sand, freely draining, existed 1.5 m above the bottom of the clay layer, what would be the values of the final and 3 year settlements.

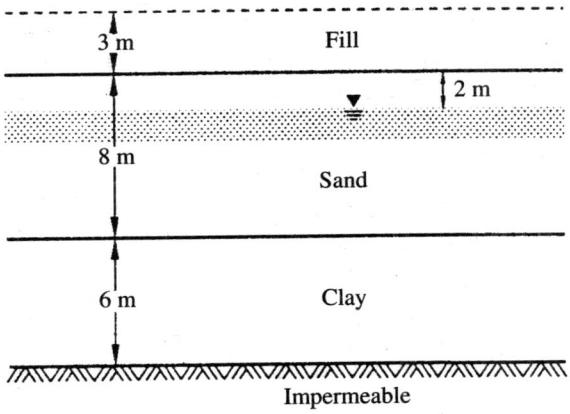

Fig. Prob. 11.15

11.16   In a consolidation test a specimen of saturated clay 19 mm thick reaches 50% consolidation
         in 5 mins. How long would it take a layer of this clay 5 m thick to reach the same degree of
         consolidation under the same stress and drainage conditions? How long would it take the
         layer to reach 30% consolidation? Assume drainage on both sides.

11.17   The thickness of a compressible layer, prior to placing of fill over it covering a large area,
         is 10 m. Its original void ratio was 1.0. Sometime after the fill was constructed measurements
         indicated that the average void ratio was 0.8. Determine the compression of the soil layer.

11.18   A 10 m thick clay layer with single drainage settles 9 cm. In 3.5 year, the coefficient of
         consolidation for this clay was found to be 0.544×10^{-2} cm^2/sec. Compute the ultimate
         consolidation settlement and find out how long it will take to settle to 90% of this amount.

11.19   The time factor T for a clay layer undergoing consolidation is 0.2. What is the average
         degree of consolidation (consolidation ratio) for the layer?

11.20   If the final consolidation settlement for the clay layer in Prob. 11.19 is expected to be 1.0
         m, how much settlement has occurred when the time factor is (a) 0.2 and (b) 0.7?

11.21   A certain compressible layer has a thickness of 4 m. After 1 yr. when the clay is 50%
         consolidated, 8 cm of settlement has occurred. For similar clay and loading conditions,
         how much settlement would occur at the end of 1 yr. and 4 yr. if the thickness of this new
         layer were 8 m.

11.22   A layer of normally consolidated clay 3.5 m thick has an average void ratio of 1.3. Its
         compression index is 0.6 and its coefficient of consolidation is 1 m^2/yr. When the existing
         vertical pressure on the clay layer is doubled, what change in thickness of the clay layer
         will result?

11.23   The settlement analysis for a proposed structure indicates that 6 cm of settlement will occur
         in 4 yr. and that the ultimate total settlement will be 25 cm. The analysis is based on the
         assumptions that the compressible clay layer is drained on both sides. However, it is suspected
         that there may not be drainage at the bottom surface. For the case of single drainage, estimate
         (a) the ultimate total settlement and (b) the time required for 6 cm of settlement.

11.24   The water content of a soft clay is 54.2% and the liquid limit is 57.3%. Estimate the
         compression index.

11.25   The time to reach 60% consolidation is 32.5 sec for a sample 1.27 cm thick tested in a
         laboratory under conditions of double drainage. How long will the corresponding layer in
         nature requires to reach the same degree of consolidation if it is 4.57 m thick and drained
         on one side only.

11.26   A certain clay layer 9 m which is expected to have an ultimate settlement of 40.6 cm. If the
         settlement was 10.2 cm in four years how much longer will it take to obtain a total of
         150 mm.

11.27   If the coefficient of consolidation of a 3 m duck layer of clay is $c_v = 0.0003$ cm^2/sec, what
         is the average consolidation of that layer of clay, (a) in one year with two-way drainage,
         and (b) the same as above for one-way drainage.

11.28   The coefficient of consolidation of a normally loaded clay has been found to be 5.57×10^{-4} sq. m/min. The average natural moisture content of the deposit is 40%; the unit weight of the solid matter is 2.8 Mg/m³, and compression index C_c is 0.36. If the clay deposit is 6.1m thick drained both sides, calculate the final consolidation settlement S_c.

11.29   A rigid foundation block, circular in plan and 6 m diameter rests on a bed of compact sand 5 m deep. Below the sand is 1.6 m thick of clay overlying impervious bed rock. Ground water level is 1.5 m below the surface of the sand. The density of sand above WT is 1920 kg/m³, the saturated density of sand is 2080 kg/m³, and the saturated density of the clay is 1990 kg/m³.

A laboratory consolidation test on an undisturbed sample of the clay, 20 mm thick and drained top and bottom, gave the following results

Pressure (kN/m²)	50	100	200	300	400
Void ratio	0.73	0.68	0.625	0.58	0.54

If the contact pressure at the underside of the foundation is 200 kN/m²,

(a)   Calculate the final average settlement of the foundation assuming 2:1 method for the spread of load.

(b)   If the consolidation test sample reached 90% consolidation in 106 mins, how long will it take the foundation to reach 90% of its final settlement.

CHAPTER 12

SHEAR STRENGTH

12.1 INTRODUCTION

One of the most important and the most controversial engineering properties of soil is its shear strength or ability to resist sliding along internal surfaces within a mass. The stability of a cut, the slope of an earth dam, the foundations of structures, the natural slopes of hillsides and other structures built on soil depend upon the shearing resistance offered by the soil along the probable surfaces of slippage. There is hardly a problem in the field of engineering which does not involve the shear properties of the soil in some manner or the other.

Basic Concept of Shearing Resistance and Shearing Strength

The basic concept of shearing resistance and shearing strength can be made clear by studying first the basic principles of friction between solid bodies. Consider a prismatic block B resting on a plane surface MN as shown in Fig. 12.1. Block B is subjected to the force P_n which acts at right angles to the surface MN, and the force F_a that acts tangentially to the plane. The normal force P_n remains constant whereas F_a gradually increases from zero to a value which will produce sliding. If the tangential force F_a is relatively small, block B will remain at rest, and the applied horizontal force will be balanced by an equal and opposite force F_r on the plane of contact. This resisting force is developed as a result of roughness characteristics of the bottom of block B and plane surface MN. The angle δ formed by the resultant R of the two forces F_r and P_n with the normal to the plane MN is known as the *angle of obliquity*.

If the applied horizontal force F_a is gradually increased, the resisting force F_r will likewise increase, always being equal in magnitude and opposite in direction to the applied force. Block B will start sliding along the plane when the force F_a reaches a value which will increase the angle of obliquity to a certain maximum value δ_m. If block B and plane surface MN are made of the same material, the angle δ_m is equal to ϕ which is termed the *angle of friction*, and the value $\tan \phi$ is termed the *coefficient of friction*. If block B and plane surface MN are made of dissimilar materials, the angle ä is termed the *angle of wall friction*. The applied horizontal force F_a on block B is a shearing force and the developed force is friction or *shearing resistance*. The maximum shearing resistance which the materials are capable of developing is called the *shearing strength*.

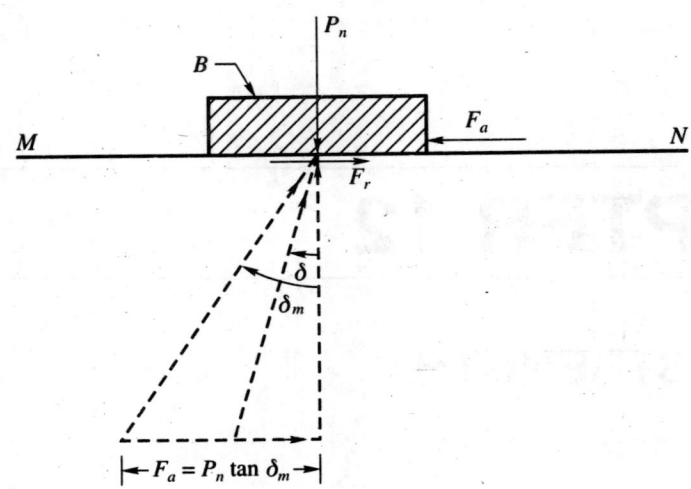

Fig. 12.1 Basic concept of shearing resistance and strength

If another experiment is conducted on the same block with a higher normal load P_n the shearing force F_a will correspondingly be greater. A series of such experiments would show that the shearing force F_a is proportional to the normal load P_n, that is

$$F_a = P_n \tan\phi \tag{12.1}$$

If A is the overall contact area of block B on plane surface MN, the relationship may be written as

$$\text{shear strength, } s = \frac{F_a}{A} = \frac{P_n}{A}\tan\phi$$

$$\text{or } s = \sigma \tan\phi \tag{12.2}$$

12.2 THE COULOMB EQUATION

The basic concept of friction as explained in Sect. 12.1 applies to soils which are purely granular in character. Soils which are not purely granular exhibit an additional strength which is due to the cohesion between the particles. It is, therefore, still customary to separate the shearing strength s of such soils into two components, one due to the cohesion between the soil particles and the other due to the friction between them. The fundamental shear strength equation proposed by the French engineer Coulomb is

$$s = c + \sigma \tan\phi \tag{12.3}$$

This equation expresses the assumption that the cohesion c is independent of the normal pressure σ acting on the plane of failure. At zero normal pressure, the shear strength of the soil is expressed as

$$s = c \tag{12.4}$$

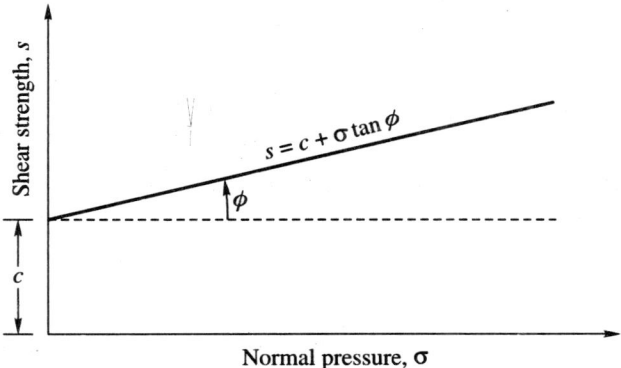

Fig. 12.2 Coulomb's law

According to Eq. (12.4), the cohesion of a soil is defined as the shearing strength at zero normal pressure on the plane of rupture.

In Coulomb's equation c and σ are empirical parameters, the values of which for any soil depend upon several factors; the most important of these are:

1. The past history of the soil.

2. The initial state of the soil, i.e. whether it is saturated or unsaturated.

3. The permeability characteristics of the soil.

4. The conditions of drainage allowed to take place during the test.

Since c and σ in Coulomb's Eq. (12.3) depend upon many factors, c is termed as *apparent cohesion* and ϕ the angle of shearing resistance. For cohesionless soil $c = 0$, then Coulomb's equation becomes

$$s = \sigma \tan\phi \qquad (12.5)$$

The relationship between the various parameters of Coulomb's equation is shown diagrammatically in Fig. 12.2.

12.3   METHODS OF DETERMINING SHEAR STRENGTH PARAMETERS

Methods

The shear strength parameters c and ϕ of soils either in the undisturbed or remolded states may be determined by any of the following methods:

1. *Laboratory methods*

   (a) Direct or box shear test

   (b) Triaxial compression test

   (c) Laboratory Vane Shear test

2.   *Field method:* Vane shear test or by any other indirect methods.

Shear Parameters of Soils *in-situ*

The laboratory or the field method that has to be chosen in a particular case depends upon the type of soil and the accuracy required. Wherever the strength characteristics of the soil in-situ are required, laboratory tests may be used provided undisturbed samples can be extracted from the stratum. However, soils are subject to disturbance either during sampling or extraction from the sampling tubes in the laboratory even though soil particles possess cohesion. It is practically impossible to obtain undisturbed samples of cohesionless soils and highly pre-consolidated clay soils. Soft sensitive clays are nearly always remolded during sampling. Laboratory methods may, therefore, be used only in such cases where fairly good undisturbed samples can be obtained. Where it is not possible to extract undisturbed samples from the natural soil stratum, any one of the following methods may have to be used according to convenience and judgment:

1.   Laboratory tests on remolded samples which could at best simulate field conditions of the soil.

2.   Any suitable field test.

The present trend is to rely more on field tests as these tests have been found to be more reliable than even the more sophisticated laboratory methods.

Shear Strength Parameters of Compacted Fills

The strength characteristics of fills which are to be constructed, such as earth embankments, are generally found in a laboratory. Remoulded samples simulating the proposed density and water content of the fill materials are made in the laboratory and tested. However, the strength characteristics of existing fills may have to be determined either by laboratory or field methods keeping in view the limitations of each method.

Direct or Box Shear Test

The original form of apparatus for the direct application of shear force is the shear box. The box shear test, though simple in principle, has certain shortcomings which will be discussed later on. The apparatus consists of a square brass box split horizontally at the level of the center of the soil sample, which is held between metal grilles and porous stones as shown in Fig. 12.3a. Vertical load is applied to the sample as shown in the figure and is held constant during a test. A gradually increasing horizontal load is applied to the lower part of the box until the sample fails in shear. The shear load at failure is divided by the cross-sectional area of the sample to give the ultimate shearing strength. The vertical load divided by the area of the sample gives the applied vertical stress σ. The test may be repeated with a few more samples having the same initial conditions as the first sample. Each sample is tested with a different vertical load.

The horizontal load is applied at a constant rate of strain. The lower half of the box is mounted on rollers and is pushed forward at a uniform rate by a motorized gearing arrangement. The upper half of the box bears against a steel proving ring, the deformation of which is shown on the dial gauge indicating the shearing force. To measure the volume change during consolidation and during the shearing process another dial gauge is mounted to show the vertical movement of the top platen. The horizontal displacement of the bottom of the box may also be measured by another dial gauge which is not shown in the figure.

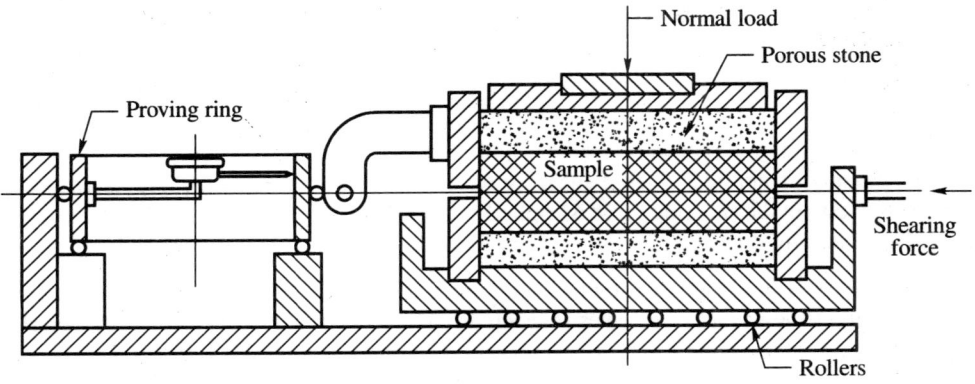

Fig. 12.3a Constant rate of strain shear box

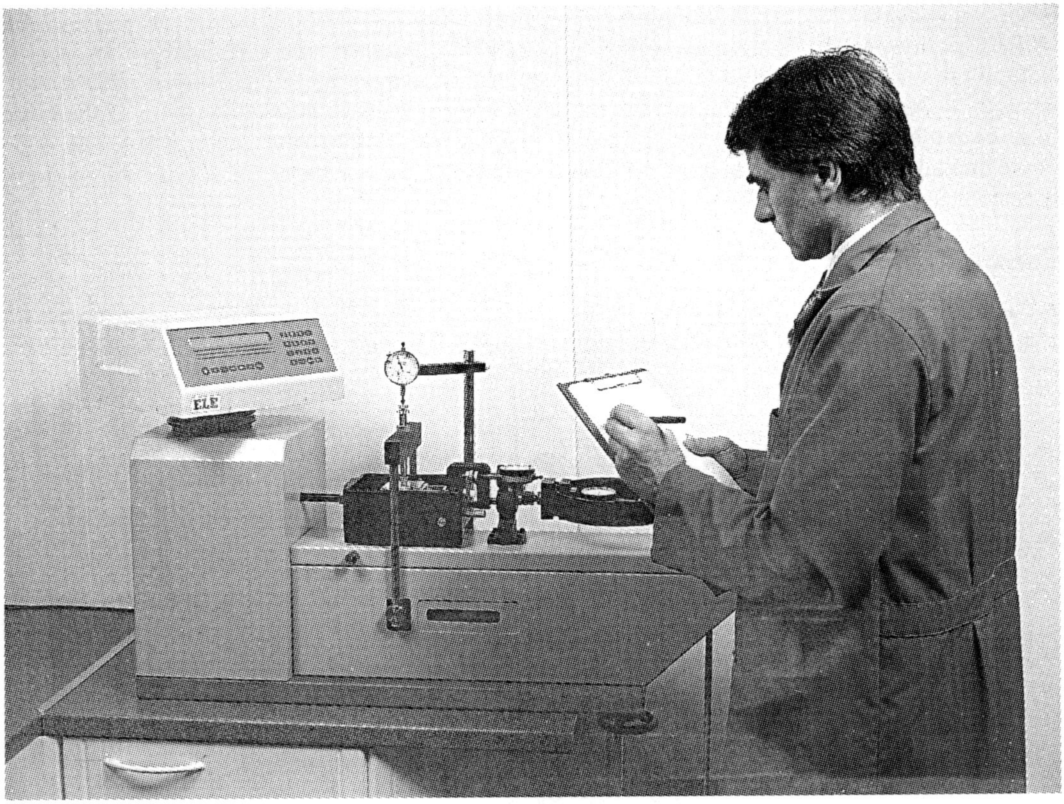

Fig. 12.3b Strain controlled shear apparatus (Courtesy : Soiltest)

Procedure for Determining Shearing Strength of Soil

In the direct shear test, a sample of soil is placed into the shear box. The size of the box normally used for clays and sands is 6 × 6 cm and the sample is 2 cm thick. A large box of size 30 × 30 cm with sample thickness of 15 cm is sometimes used for gravelly soils.

The soils used for the test are either undisturbed samples or remoulded. If undisturbed, the specimen has to be carefully trimmed and fitted into the box. If remolded samples are required, the soil is placed into the box in layers at the required initial water content and tamped to the required dry density.

After the specimen is placed in the box, and all the other necessary adjustments are made, a known normal load is applied. Then a shearing force is applied. The normal load is held constant throughout the test but the shearing force is applied at a constant rate of strain (which will be explained later on). The shearing displacement is recorded by a dial gauge.

Dividing the normal load and the maximum applied shearing force by the cross-sectional area of the specimen at the shear plane gives respectively the unit normal pressure σ and the shearing strength s at failure of the sample. These results may be plotted on a shearing diagram where σ is the abscissa and s the ordinate. The result of a single test establishes one point on the graph representing the Coulomb formula for shearing strength. In order to obtain sufficient points to draw the Coulomb graph, additional tests must be performed on other specimens which are exact duplicates of the first. The procedure in these additional tests is the same as in the first, except that a different normal stress is applied each time. Normally, the plotted points of normal and shearing stresses at failure of the various specimens will approximate a straight line. But in the case of saturated, highly cohesive clay soils in the undrained test, the graph of the relationship between the normal stress and shearing strength is usually a curved line, especially at low values of normal stress. However, it is the usual practice to draw the best straight line through the test points to establish the Coulomb law. The slope of the line gives the angle of shearing resistance and the intercept on the ordinate gives the apparent cohesion (See. Fig. 12.2).

Triaxial Compression Test

A diagrammatic layout of a triaxial test apparatus is shown in Fig. 12.4a. In the triaxial compression test, three or more identical samples of soil are subjected to uniformly distributed fluid pressure

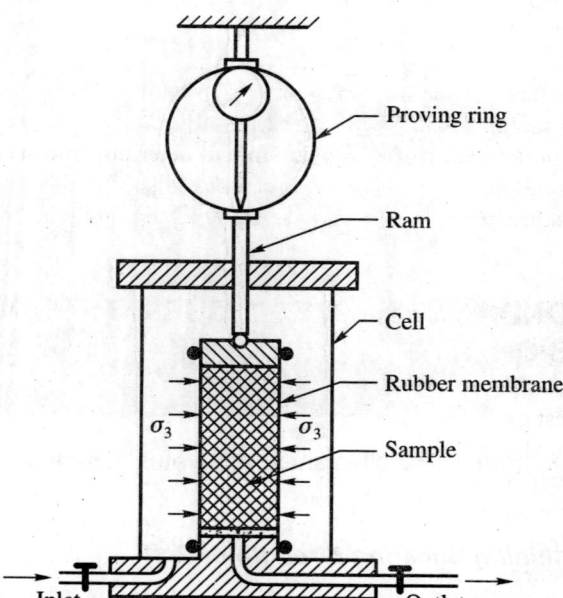

Fig. 12.4a   Diagrammatic layout of triaxial test apparatus

Fig. 12.4b Shows a photograph of a triaxial test apparatus

around the cylindrical surface. The sample is sealed in a watertight rubber membrane. Then axial load is applied to the soil sample until it fails. Although only compressive load is applied to the soil sample, it fails by shear on internal faces. It is possible to determine the shear strength of the soil from the applied loads at failure. In order to interpret the results of a triaxial compression test, it is necessary to analyse the stress relationships which exist at a point in the interior of a stressed body.

12.4 STRESS CONDITIONS IN A SOIL DURING TRIAXIAL COMPRESSION TEST

In triaxial compression test a cylindrical specimen is subjected to a constant all-round fluid pressure which is the minor principal stress σ_3 since the shear stress on the surface is zero. The two ends are subjected to axial stress which is the major principal stress σ_1. The stress condition in the specimen goes on changing with the increase of the major principal stress σ_1. It is of interest to analyze the state of stress along inclined sections passing through the sample at any stress level σ_1 since failure occurs along inclined surfaces.

Consider the cylindrical specimen of soil in Fig. 12.5(a) which is subjected to principal stresses σ_1 and σ_3 ($\sigma_2 = \sigma_3$).

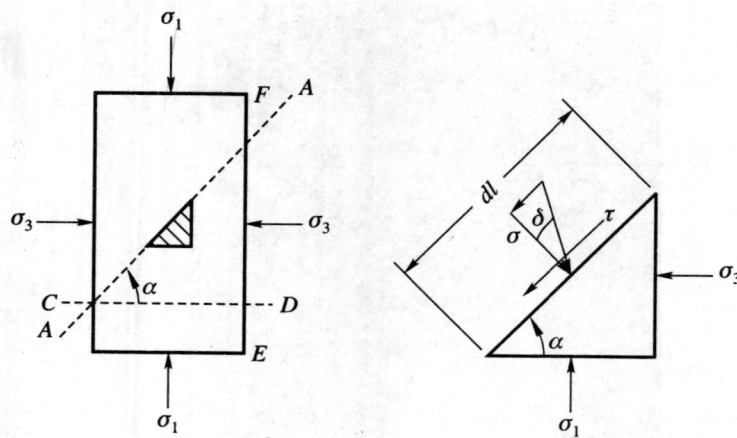

Fig. 12.5 Stress condition in a triaxial compression test specimen

Now CD, a horizontal plane, is called a principal plane since it is normal to the principal stress σ_1 and the shear stress is zero on this plane. EF is the other principal plane on which the principal stress σ_3 acts. AA is the inclined section on which the state of stress is required to be analyzed.

Consider as before a small prism of soil shown shaded in Fig. 12.5(a) and the same to an enlarged scale in Fig. 12.5(b). All the stresses acting on the prism are shown. The equilibrium of the prism requires

$$\Sigma \text{ Horizontal forces} = \sigma_3 \sin \alpha \, dl - \sigma \sin \alpha \, dl + \tau \cos \alpha \, dl = 0$$

$$\Sigma \text{ Vertical forces} = \sigma_1 \cos \alpha \, dl - \sigma \cos \alpha \, dl - \tau \sin \alpha \, dl = 0$$

Solving the Equations, we have

$$\sigma = \frac{\sigma_1 + \sigma_3}{2} + \frac{\sigma_1 - \sigma_3}{2} \cos 2\alpha \tag{12.6}$$

$$\tau = \frac{1}{2}(\sigma_1 - \sigma_3) \sin 2\alpha \tag{12.7}$$

Let the resultant of σ and τ make an angle δ with the normal to the inclined plane. One should remember that when α is less than $90°$, the shear stress τ is positive, and the angle δ is also positive.

Relationship between the Principal Stresses and Cohesion c

If the shearing resistance s of a soil depends on both friction and cohesion, sliding failure occurs in accordance with the Coulomb Eq. (12.3), that is, when

$$\tau = s = c + \tan\phi$$

Substituting for the values of σ and τ from Eqs (12.6) and (12.7), we obtain

$$\sigma_3 \sin \alpha \cos \alpha - \sigma_3 \sin \alpha \cos \alpha = c + \sigma_3 \tan\phi + \cos^2\alpha \, \sigma_1 \tan\phi - \cos^2\alpha \, \sigma_3 \tan\phi$$

Solving for σ_1

$$\sigma_1(\sin \alpha \cos \alpha - \cos^2 \alpha \tan\phi) = c + \sigma_3 \tan\phi + \sigma_3(\sin \alpha \cos \alpha - \cos^2 \alpha \tan\phi)$$

or $\quad \sigma_1 = \sigma_3 + \dfrac{c + \sigma_3 \tan\phi}{\sin \alpha \cos \alpha - \cos^2 \alpha \tan\phi}$ $\qquad\qquad$ (12.8)

The plane with the least resistance to shearing along it will correspond to the minimum value of σ_1 which can produce failure in accordance with Eq. (12.8). σ_1 will be at a minimum when the denominator in the second member of the equation is at a maximum, that is, when

$$\frac{d}{d\alpha}(\sin \alpha \cos \alpha - \cos^2 \alpha \tan\phi) = 0$$

Differentiating, and writing $\alpha = \alpha_c$, we get

$$\cos^2 \alpha_c - \sin^2 \alpha_c + 2\tan\phi \cos \alpha_c \sin \alpha_c = 0$$

or $\cos 2\alpha_c = -\sin 2\alpha_c \tan\phi$

$\cot 2\alpha_c = -\tan\phi = \cot(90 + \phi)$

$\alpha_c = 45° + \phi/2$ $\qquad\qquad$ (12.9)

Substituting for α in Eq. (12.8) and remembering that

$$\tan\phi = \cot(90 - \phi) = -\cot(90 + \phi) = -\cot 2\alpha_c = \frac{1}{2}(\tan\alpha_c - \cot\alpha_c)$$

we get,

$$\sin \alpha_c \cos \alpha_c - \cos^2 \alpha_c \tan\phi = \sin \alpha_c \cos \alpha_c - \frac{\cos^2 \alpha_c}{2}\left(\frac{\sin \alpha_c}{\cos \alpha_c} - \frac{\cos \alpha_c}{\sin \alpha_c}\right)$$

$$= \frac{\cos \alpha_c}{2\sin \alpha_c}(\sin^2 \alpha_c + \cos^2 \alpha_c) = \frac{1}{2\tan\alpha_c}$$

Substituting in Eq. (12.8) and simplifying, we get

$$\sigma_1 = \sigma_3 \tan^2(45° + \phi/2) + 2c \tan(45° + \phi/2)$$

or $\quad \sigma_1 = \sigma_3 N_\phi + 2c\sqrt{N_\phi}$ $\qquad\qquad$ (12.10)

where $N_\phi = \tan^2(45° + \phi/2)$ is called the *flow value*.

If the cohesion $c = 0$, we have

$\sigma_1 = \sigma_3 N_\phi$ $\qquad\qquad$ (12.11)

If $\phi = 0$, we have

$$\sigma_1 = \sigma_3 + 2c \tag{12.12}$$

If the sides of the cylindrical specimen are not acted on by the horizontal pressure σ_3, the load required to cause failure is called the unconfined compressive strength q_u. It is obvious that an unconfined compression test can be performed only on a cohesive soil. According to Eq. (12.10), the unconfined compressive strength q_u is equal to

$$\sigma_1 = q_u = 2c\sqrt{N_\phi} \tag{12.13a}$$

If $\phi = 0$, then

$$q_u = 2c \tag{12.13b}$$

or the shear strength

$$s = c = \frac{q_u}{2} \tag{12.13c}$$

Equation (12.13c) shows one of the simplest ways of determining the shear strength of cohesive soils.

Mohr Circle of Stress

Squaring Eqs (12.6) and (12.7) and adding, we have

$$\tau^2 + \left[\sigma - \frac{\sigma_1 + \sigma_3}{2}\right]^2 = \left(\frac{\sigma_1 - \sigma_3}{2}\right)^2 \tag{12.14}$$

Now, Eq. (12.14) is the equation of a circle

The centre of the circle has coordinates $\tau = 0$, and $\sigma = (\sigma_1 + \sigma_3)/2$, and its radius is $(\sigma_1 - \sigma_3)/2$. The coordinates of points on the circle represent the normal and shearing stresses on inclined planes at a given point. The circle is called the *Mohr circle of stress*, after O. Mohr, who first recognized this useful relationship around 1870. Again from diagram, the normal and shearing stresses on any plane passing through a point in a stressed body (Fig. 12.5) may be determined if the principal stresses σ_1 and σ_3 are known. Since σ_1 and σ_3 are always known in a cylindrical compression test, Mohr diagram is a very useful tool to analyse stresses on failure planes.

Consider a cylindrical specimen of soil in Fig. 12.6 subjected to normal stresses σ_1 and σ_3 which are major and minor principal stresses respectively. A Mohr cirle of stress may be drawn for this specimen with a radius of $(\sigma_1 - \sigma_3)/2$ having its centre at C at a distance of $(\sigma_1 + \sigma_3)/2$ on the abscissa from the origin of coordinates O. The origin of planes or the pole P_0 may be obtained from points E and F (Fig. 12.6) parallel to planes on which the minor and major principal stresses act. In this case, the pole P_0 lies on the abscissa and coincides with the point E. The normal stress σ and shear τ on any arbitary plane AA making an angle α with the major principal plane may be determined as follows.

From the pole P_0 draw a line P_0P_1 parallel to the plane AA (Fig. 12.6). The coordinates of the point P_1 gives the stresses σ and τ. From the stress circle we may write $\angle P_1CF = 2\alpha$,

$$\sigma = \frac{\sigma_1 + \sigma_3}{2} + \frac{\sigma_1 - \sigma_3}{2}\cos 2\alpha, \quad \tau = \frac{\sigma_1 - \sigma_3}{2}\sin 2\alpha$$

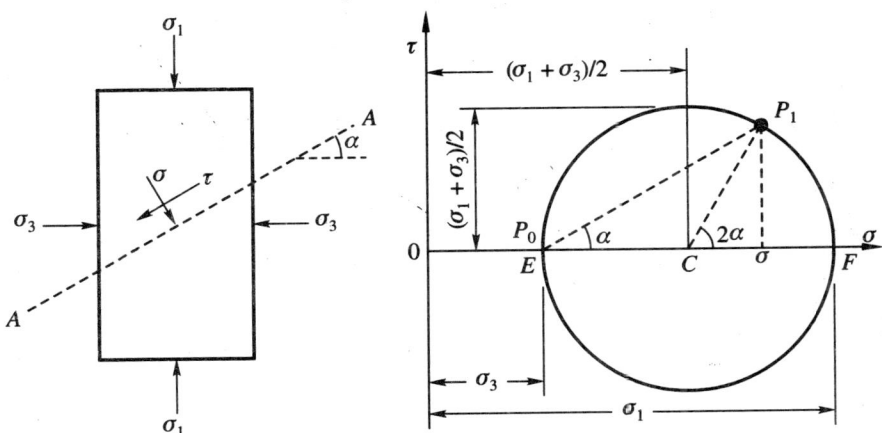

Fig. 12.6 Mohr stress circle for a cylindrical specimen

The equations for σ and τ, obtained from Mohr circle of stress, are the same as Eqs (12.6) and (12.7) obtained analytically.

12.5 MOHR-COULOMB FAILURE THEORY

Various theories relating to the stress condition in engineering materials at the time of failure are available in the engineering literature. Each of these theories may explain satisfactorily the actions of certain kinds of materials at the time they fail, but no one of them is applicable to all materials. The failure of a soil mass is more nearly in accordance with the tenets of the Mohr theory of failure than with those of any other theory and the interpretation of the triaxial compression test depends to a large extent on this fact. The Mohr theory is based on the postulate that a material will fail when the shearing stress on the plane along which the failure is presumed to occur is a unique function of the normal stress acting on that plane. The material fails along the plane only when the angle between the resultant of the shearing stress and the normal stress is a maximum, that is, where the combination of normal and shearing stresses produces the maximum obliquity angle δ.

According to Coulomb's law, the condition of failure is that the shear stress

$$\tau \leq c + \sigma \tan\alpha$$

Mohr Diagram for Triaxial Compression Test at Failure

Consider a cylindrical specimen of soil possessing both cohesion and friction is subjected to a conventional triaxial compression test. In the conventional test the lateral pressure σ_3 is held constant and the vertical pressure σ_1 is increased at a constant rate of stress or strain until the sample fails. If σ_1 is the peak value of the vertical pressure at which the sample fails, the two principal stresses that are to be used for plotting the Mohr circle of rupture are σ_3 and σ_1. In Fig. 12.7 the values of σ_1 and σ_3 are plotted on the σ-axis and a circle is drawn with ($\sigma_1 - \sigma_3$) as diameter. The centre of the circle lies at a distance of $(\sigma_1 + \sigma_3)/2$ from the origin. As per Eq. (12.9), the soil fails along a plane which makes an angle $\alpha = 45° + \phi/2$ with the major principal plane. In Fig. 12.7 the two lines P_0P_1 and P_0P_2 (where P_0 is the origin of planes) are the conjugate rupture planes. The two lines M_0N and M_0N_1 drawn tangential to the rupture circle at points P_1 and P_2 are called Mohr envelopes. If the Mohr envelope can be drawn by some other means, the orientation of the failure planes may be determined.

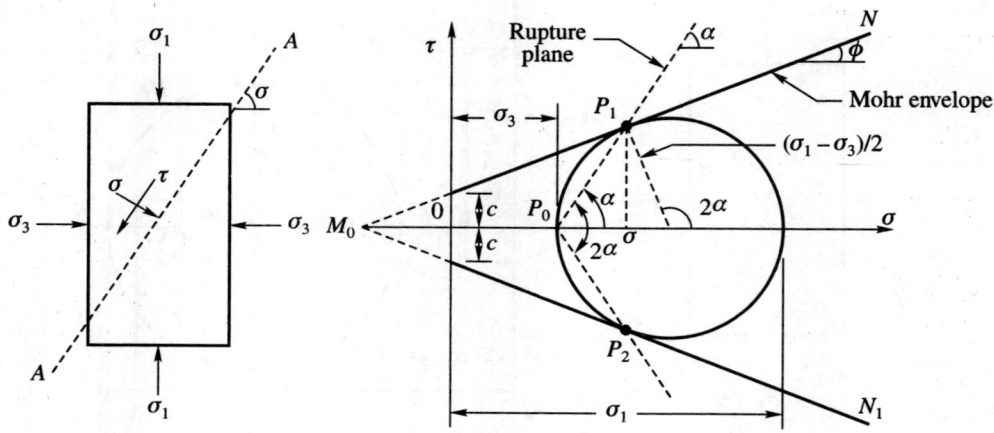

Fig. 12.7   Mohr diagram for triaxial test at failure for c-ϕ soil

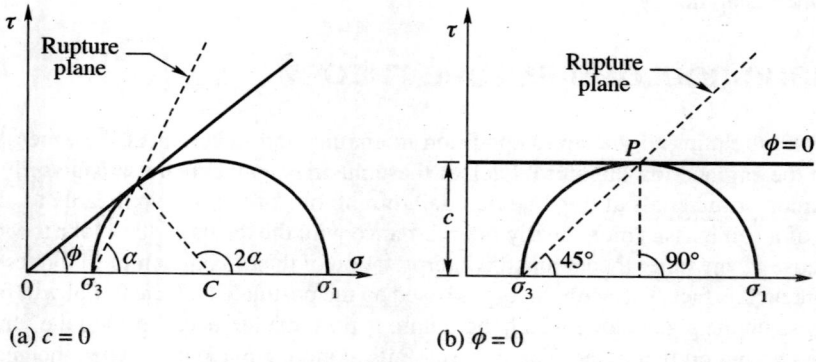

(a) $c = 0$ (b) $\phi = 0$

Fig. 12.8   Mohr diagram for soils with $c = 0$ and $\phi = 0$

The results of analysis of triaxial compression tests as explained in Sect. 12.4 are now presented in a graphical form in Fig. 12.7. The various information that can be obtained from the figure are

1.   The angle of shearing resistance ϕ = the slope of the Mohr envelope.

2.   The apparent cohesion c = the intercept of the Mohr envelope on the τ-axis.

3.   The inclination of the rupture plane = α.

4.   The angle between the conjugate planes = 2α.

If the soil is cohesionless with $c = 0$ the Mohr envelopes pass through the origin, and if the soil is purely cohesive with $\phi = 0$ the Mohr envelope is parallel to the abscissa. The Mohr envelopes for these two types of soils are shown in Fig. 12.8.

Mohr Diagram for a Direct Shear Test at Failure

In a direct shear test the sample is sheared along a horizontal plane. This indicates that the failure

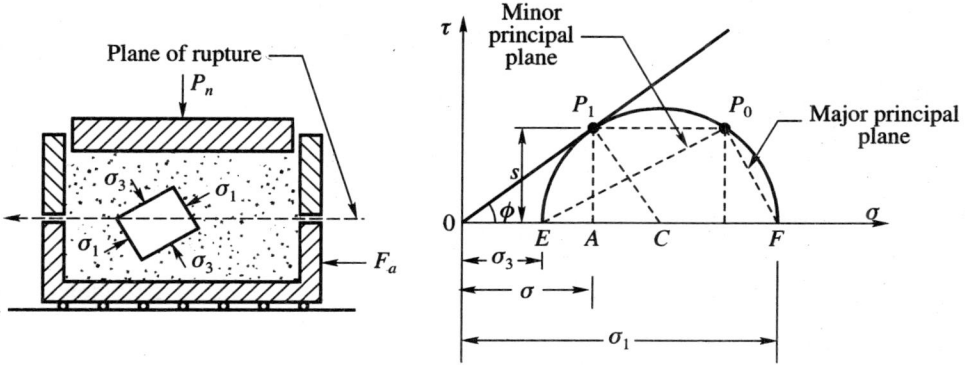

Fig. 12.9 Mohr diagram for a direct shear test at failure

plane is horizontal. The normal stress σ on this plane is the external vertical load divided by the area of the sample. The shear stress at failure is the external lateral load divided by the area of the sample.

The point P_1 on the stress diagram in Fig. 12.9 represents the stress condition on the failure plane. The coordinates of the point are

normal stress $= \sigma$, shear stress $\tau = s$.

If it is assumed that the Mohr envelope is a straight line passing through the origin (for cohesionless soil or normally consolidated clays), it follows that the maximum obliquity δ_m occurs on the failure plane and $\delta_m = \phi$. Therefore, the line OP_1 must be tangent to the Mohr circle, and the circle may be constructed as follows:

Draw P_1C normal to OP_1. Point C which is the intersection point of the normal with the abscissa is the centre of the circle. CP_1 is the radius of the circle. The Mohr circle may now be constructed which gives the major and minor principal stresses σ_1 and σ_3 respectively.

Since the failure is on the horizontal plane, the origin of planes P_0 may be obtained by drawing a horizontal line through P_1 giving P_0. P_0F and P_0E give the directions of the major and minor principal planes respectively.

12.6 EFFECTIVE STRESSES

So far, the discussion has been based on consideration of total stresses. It is to be noted that the strength and deformation characteristics of a soil can be understood better by visualizing it as a compressible skeleton of solid particles enclosing voids. The voids may completely be filled with water or partly with water and air. Shear stresses are to be carried only by the skeleton of solid particles. However, the total normal stresses on any plane are, in general, the sum of two components.

Total normal stress   = component of stress carried by solid particles

+ pressure in the fluid in the void space.

This visualization of the distribution of stresses between solid and fluid has two important consequences:

1. When a specimen of soil is subjected to external pressure, the volume change of the specimen is not due to the total normal stress but due to the difference between the total normal stress and the pressure of the fluid in the void space. The pressure in the fluid is the pore pressure u. The difference which is called the effective stress σ' may now be expressed as

$$\sigma' = \sigma - u \tag{12.15}$$

2. The shear strength of soils, as of all granular materials, is largely determined by the frictional forces arising during slip at the contacts between the soil particles. These are clearly a function of the component of normal stress carried by the solid skeleton rather than of the total normal stress. For practical purposes the shear strength equation of Coulomb is given by the expression

$$s = c' + (\sigma - u) \tan \phi' = c' + \sigma' \tan \phi' \tag{12.16}$$

where c' = Apparent cohesion in terms of effective stresses

ϕ' = Angle of shearing resistance in terms of effective stresses

σ = Total normal pressure to the plane considered

u = Pore-pressure.

The effective stress parameters c' and ϕ' of a given sample of soil may be determined provided the pore pressure u developed during the shear test is measured. The pore pressure u is developed when the testing of the soil is done under undrained conditions. However, if free drainage takes place during testing, there will not be any development of pore pressure. In such cases, the total stresses themselves are effective stresses.

The principal stresses may also be expressed either as total stresses or as effective stresses if the values of pore pressures are known.

If u is the pore pressure developed during a triaxial test, we may write as before

$$\sigma_1' = \sigma_1 - u, \quad \sigma_3' = \sigma_3 - u \tag{12.17}$$

where σ_1' and σ_3' are the effective principal stresses.

Example 12.1

What is the shearing strength of soil along a horizontal plane at a depth of 4 m in a deposit of sand having the following properties:

Angle of internal friction, $\phi = 35°$

Dry unit weight, $\gamma_d = 17$ kN/m³

Specific gravity, $G = 2.7$.

Assume the ground water table is at a depth of 2.5 m from the ground surface. Also find the change in shear strength when the water table rises to the ground surface.

Solution

The effective vertical stress at the plane of interest is

$$\sigma' = 250 \times \gamma_d + 150 \times \gamma_b$$

Given $\gamma_d = 17$ kN/m³ and $G = 2.7$

We have, $\gamma_d = 17 = \dfrac{G}{1+e} \gamma_w = \dfrac{2.7}{1+e} \times 9.81$

or $17e = 26.5 - 17 = 9.49$ or, $e = \dfrac{9.49}{17} = 0.56$

Therefore, $\gamma_b = \dfrac{G-1}{1+e} \gamma_w = \dfrac{2.7-1.0}{1+0.56} \times 9.81 = 10.7$ kN/m³

Hence, $\sigma' = 2.5 \times 17 + 1.5 \times 10.7 = 58.55$ kN/m²

Hence, the shearing strength of the sand is

$$s' = \sigma' \tan \phi = 58.55 \times \tan 35° = 41 \text{ kN/m}^2$$

If the water table rises to the ground surface, i.e. by a height of 2.5 m, the change in the effective stress will be,

$$\Delta \sigma' = \gamma_d \times 2.5 - \gamma_b \times 2.5 = 17 \times 2.5 - 10.7 \times 2.5 = 15.75 \text{ kN/m}^2 \text{(negative)}$$

Hence, the decrease in shear strength will be,

$$= \Delta \sigma' \tan 35° = 15.75 \times 0.70 = 11 \text{ kN/m}^2$$

Example 12.2

An unconfined cylindrical specimen of clay fails under an axial stress of 240 kN/m². The failure plane was inclined at an angle of 55° to the horizontal. Determine the shear strength parameters of the soil.

Solution

From Eq. (12.10), we have

$$\sigma_1 = \sigma_3 N_\phi + 2c\sqrt{N_\phi}, \text{ where, } N_\phi = \tan^2\left(45° + \frac{\phi}{2}\right)$$

since $\sigma_3 = 0$, we have

$$\sigma_1 = 2c\sqrt{N_\phi} = 2c \tan\left(45° + \frac{\phi}{2}\right), \text{ where } \sigma_1 = 240 \text{ kN/m}^2 \qquad \text{(a)}$$

From Eq. (12.9), the failure angle α is

$$\alpha = 45° + \frac{\phi}{2}, \text{ since } \alpha = 55°, \text{ we have}$$

$$\phi = (55-45)\times 2 = 20°$$

From Eq. (a),

$$c = \frac{\sigma_1}{2\tan\left(45° + \dfrac{\phi}{2}\right)} = \frac{240}{2\tan 55°} = 84\text{kN/m}^2$$

Example 12.3

A direct box test when conducted on a remoulded sample of sand gave the following observations at the time of failure: Normal load, $\sigma = 288$ N; shear load = 173 N. The cross-sectional area of the sample = 36 cm².

Determine: (i) the angle of internal friction, (ii) the magnitude and direction of the principal stresses in the zone of failure.

Solution

Such problems can be solved in two ways, namely graphically and analytically. The analytical solution has been left as an exercise for the students.

Graphical Solution

(i) Shear stress $\tau = \dfrac{173}{36} = 4.8\,\text{kN/cm}^2 = 48\,\text{kN/m}^2$

Normal stress, $\sigma = \dfrac{288}{36} = 8.0\,\text{kN/m}^2$

we now know one point on the Mohr envelope. Plot point A (Fig. Ex. 12.3) with coordinates $\tau = 48$ kN/m², and $\sigma = 80$ kN/m². Since cohesion $c = 0$ for sand, the Mohr envelope OM passes through the origin. The slope of OM gives the angle of internal friction $\phi = 31°$.

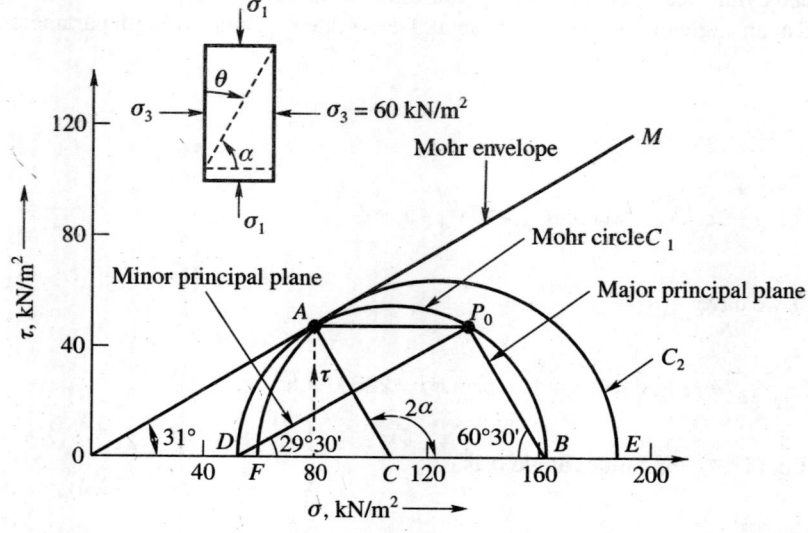

Fig. Ex. 12.3

(ii) In Fig. Ex. 12.3, draw line AC normal to the envelope OM cutting the abscissa at point C. With C as centre, and AC as radius, draw Mohr circle C_1 which cuts the abscissa at points B and D, which gives

major principal stress $= OB = \sigma_1 = 163.5$ kN/m^2

minor principal stress $= OD = \sigma_3 = 53.5$ kN/m^2

Now, $\angle ACB = 2\alpha =$ twice the angle between the failure plane and the major principal plane. Measurement gives

$2\alpha = 121°$ or $\alpha = 60°30'$

Since in a direct shear test the failure plane is horizontal, the angle made by the major principal plane with the horizontal will be 60.30". The minor principal plane should be drawn at a right angle to the major principal plane.

The directions of the principal planes may also be found out by determining the pole P_o. P_o is obtained by drawing a horizontal line from point A which is parallel to the failure plane in the direct shear test. Now $P_o B$ and $P_o D$ give the directions of the major and minor principal planes respectively.

Example 12.4

Determine the magnitude of the deviator stress if a sample of the same sand with the same void ratio as given in Ex. 12.3 was tested in a triaxial apparatus with a confining pressure of 60 kN/m^2.

Solution

In the case of a triaxial test on an identical sample of sand as given in Ex. 12.3, use the same Mohr envelope OM (Fig. Ex. 12.3). Now the point F on the abscissa gives the confining pressure $\sigma_3 = 60$ kN/m^2. A Mohr circle C_2 may now be drawn passing through point F and tangential to the Mohr envelope OM. The point E gives the major principal stress σ_1 for the triaxial test.

Now, $\sigma_1 = OE = 188$ kN/m^2, $\sigma_1 = 60$ kN/m^2

Therefore, $\sigma_1 - \sigma_3 = 188 - 60 = 128$ kN/m^2 = Deviator stress

Example 12.5

Boreholes reveal that a thin layer of alluvial silt exists at a depth of 15.25 m below the surface of the ground. The soil above this level has an average dry unit weight of 15.23 kN/m^3 and an average water content of 30%. The water table is approximately at the surface. Tests on undisturbed samples give the following data: $c_u = 48$ kN/m^2, $\phi = 13°$, $c_d = 41$ kN/m^2, $\phi_d = 23°$. Estimate the shearing resistance of the silt on a horizontal plane (a) when the shear stress builds up rapidly, and (b) when the shear stress builds up very slowly.

Solution

Bulk density $\gamma_t = \gamma_d(1+w) = 15.23 \times 1.3 = 19.8$ kN/m^2

Submerged density $\gamma_b = 19.80 - 9.81 = 9.99$ kN/m^2

Total normal pressure at 15.25 m depth $= 15.25 \times 19.80 = 302 \text{kN/m}^2$

Effective pressure at 15.25 m depth $= 15.25 \times 9.99 = 152.3 \text{kN/m}^2$

(a) For rapid build-up, use the properties of the undrained state and total pressure.

At a total pressure of 302 kN/m^2

shear strength, $s = c + \sigma \tan\phi = 48 + 302 \tan 13° = 117.72 \text{kN/m}^2$

(b) For slow build-up, use effective stress properties:

At an effective stress of 152.3kN/m^2,

shear strength $= 41 + 152.3 \tan 23° = 105.5 \text{kN/m}^2$

12.7 TYPES OF LABORATORY TESTS

The laboratory tests on soils may be on:

1. Undisturbed samples, or
2. Remoulded samples.

Further, the tests may be conducted on soils that are:

1. Fully saturated, or
2. Partially saturated.

The type of test to be adopted depends upon how best we can simulate the field conditions. Generally speaking, the various shear tests for soils may be classified as follows:

1. *Unconsolidated-Undrained Tests(UU):* The samples are subjected to an applied pressure under conditions in which drainage is prevented, and then sheared under conditions of no drainage.

2. *Consolidated-Undrained or Quick Tests(CU):* The samples are allowed to consolidate under an applied pressure and then sheared under conditions of no drainage.

3. *Consolidated-Drained or Slow Tests(CD):* The samples are consolidated as in the previous test, but the shearing is carried out slowly under conditions of no excess pressure in the pore space.

The drainage condition of a sample is generally the deciding factor in choosing a particular type of test in the laboratory. The purpose of carrying out a particular test is to simulate field conditions as far as possible. Because of the high permeability of sand, consolidation occurs relatively rapidly and is usually completed during the application of the load. Tests on sand are therefore generally carried out under drained conditions (drained or slow test).

For soils other than sands the choice of test conditions depends upon the purpose for which the shear strength is required. The guiding principle is that drainage conditions of the test should conform as closely as possible to the conditions under which the soils will be stressed in the field.

Undrained or quick tests are generally used for foundations on clay soils, since during the period of construction only a small amount of consolidation will have taken place and consequently the moisture content will have undergone little change. For clay slopes or cuts undrained tests are used both for design and for the investigation of failures.

Consolidated-undrained tests are used where changes in moisture content are expected to take place due to consolidation before the soil is fully loaded. An important example is the condition known as "sudden drawdown" such as that occurs in an earth dam behind which the water level is lowered at a faster rate than at which the material of the dam can consolidate. In the consolidated-undrained tests used in this type of problem, the consolidation pressures are chosen to represent the initial conditions of the soil, and the shearing loads correspond to the stresses called into play by the action of sudden drawdown.

As already stated, drained tests are always used in problems relating to sandy soils. In clay soils drained tests are sometimes used in investigating the stability of an earth dam, an embankment or a retaining wall after a considerable interval of time has passed.

Very fine sand, silts and silty sands also have poor drainage qualities. Saturated soils of these categories are likely to fail in the field under conditions similar to those under which consolidated quick tests are made.

12.8 SHEARING TEST APPARATUS FOR THE VARIOUS TYPES OF TESTS

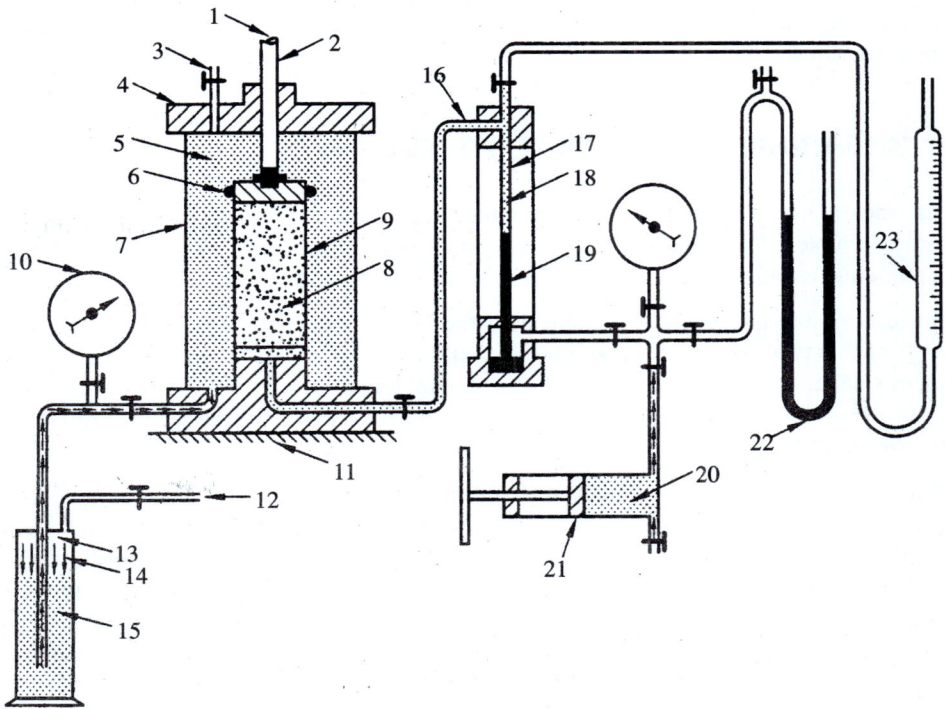

1.	Axial load	7.	Pressure cylinder	13.	Air reservoir	19.	Mercury
2.	Ram	8.	Sample	14.	Air pressure	20.	Water
3.	Air release valve	9.	Rubber membrane	15.	Water	21.	Screw control
4.	Top cap	10.	Pressure gauge	16.	Flexible copper tube	22.	Mercury manometer
5.	Water	11.	Base	17.	Glass capillary tube	23.	Burette
6.	Rubbering	12.	Air from compressor	18.	Water		

Fig. 12.10 A typical arrangement of triaxial cell for compression test

The various types of shear tests mentioned earlier may be carried out either by the box shear test or the triaxial compression test apparatus. Tests that may be made by the two types of apparatus are:

Box Shear Test Apparatus

1. Quick and consolidated quick tests on clay samples only.

2. Slow tests on any soil.

Triaxial Compression Test Apparatus

All types of tests can conveniently be carried out in this apparatus.

Box shear test apparatus is not suited for undrained or consolidated-undrained tests on samples other than clay samples, because the other soils are so permeable that even a rapid increase of the stresses in the sample may cause at least a noticeable change of the water content.

Figure 12.10 gives a typical arrangement of a triaxial cell for compression test. The pore-pressure in the sample is measured either by the 'Null Method' or by the use of pressure gauges. Samples of diameter 3.75, 5 or 10 cm may be used for the tests. The standard triaxial tests use 3.75 cm diameter samples. The height to diameter ratio used varies from 1.5 to 2.5, but normally 2 us used. The reader may refer to some book or laboratory tests for understanding the details of the equipment and for tests.

12.9   SHEARING STRENGTH TESTS ON SAND

In all the shearing tests on sand, only the moulded samples are used as it is not pracitable to get undisturbed samples. The soil samples are to be made approximately to the same dry density as it exists *in-situ* and tested either by direct shear or triaxial compression tests.

The tests on soils are generally carried out by the strain-controlled type apparatus. The principal advantage of this type of test on dense sand is that its peak-shear resistance, as well as the shear resistances smaller than the peak, can be observed and plotted.

Direct Shear Test

Only the drained or the slow shear tests on sand may be carried out by using the box shear test apparatus. The box is filled with sand to the required density. The sample is sheared at a constant vertical pressure σ. The shear stresses are calculated at various displacements of the shear box. The test is repeated with different pressures σ.

If the sample consists of loose sand, the shearing stress increases with increasing displacement until failure occurs. If the sand is dense, the shear failure of the sample is preceded by a decrease of the shearing stress from a peak value to an ultimate value (also known as residual value) lower than the peak value.

Typical stress-strain curves for loose and dense sands are shown in Fig. 12.11(a). By plotting the shear strengths corresponding to the state of failure in the different shear tests against the normal pressure a straight line is obtained for loose sand and a slightly curved line for dense sand [Fig. 12.11(c)]. However, for all practical purposes, the curvature for the dense sand can be disregarded and an average line may be drawn. The slopes of the lines give the corresponding angles of friction ϕ of the sand. The general equation for the lines may be written as

$$s = \sigma \tan \phi$$

For a given sand, the angle ϕ increases with increasing relative density. For loose sand it is roughly equal to the *angle of repose*, defined as the angle between the horizontal and the slope of a heap produced by pouring clean dry sand from a small height. The angle of friction varies with the shape of the grains. Sand samples containing well graded angular grains give higher values of ϕ as compared to uniformly graded sand with rounded grains. The angle of friction ϕ for dense sand at peak shear stress is higher than that at ultimate shear stress. Table 12.1 gives some typical values of ϕ (at peak) and $.\phi_u$ (at ultimate).

Triaxial Compression Test

Reconstructed sand samples at the required density are used for the tests. The procedure of making samples should be studied separately (refer to any book on Soil Testing). Tests on sand may be conducted either in a saturated state or in a dry state. Slow or consolidated undrained tests may be carried out as required.

Slow Tests

At least three identical samples having the same initial conditions are to be used. For slow tests under saturated conditions the drainage valve should always be kept open. Each sample should be

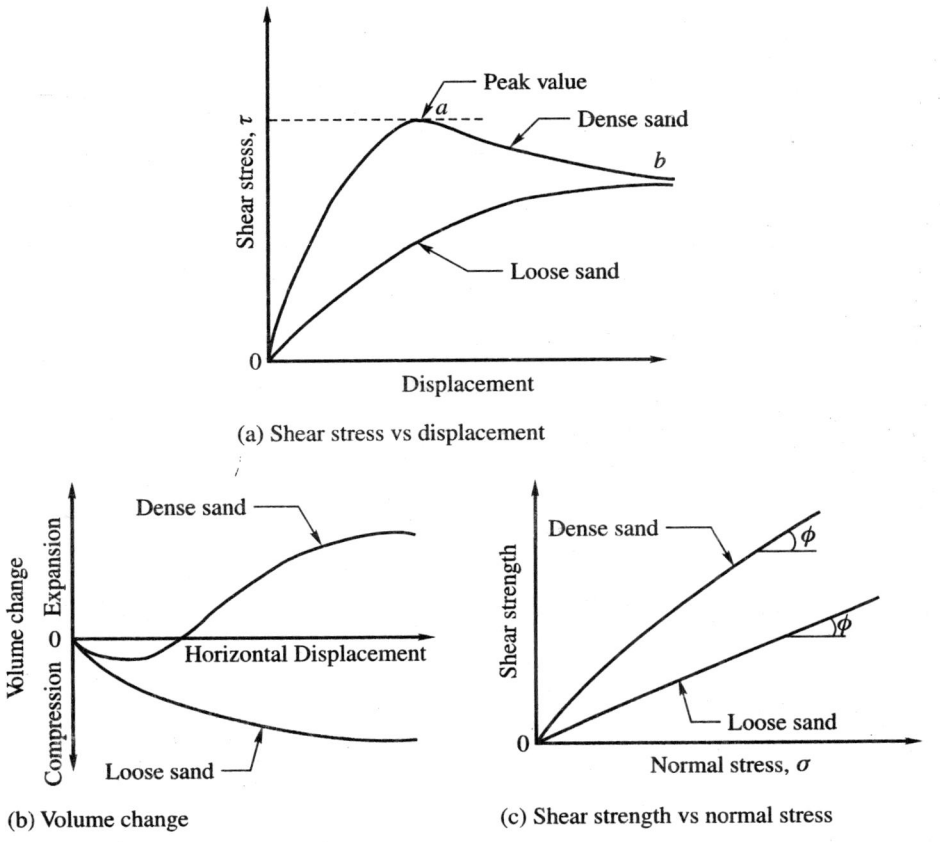

(a) Shear stress vs displacement

(b) Volume change

(c) Shear strength vs normal stress

Fig. 12.11   Direct shear test on sand

tested under different constant all-round pressures for example, 1, 2 and 3 kg/cm². Each sample is sheared to failure by increasing the vertical load at a sufficiently slow rate to prevent any build up of excess pore pressures.

TABLE 12.1
Typical values of ϕ and ϕ_u for granular soils

Types of soil	ϕ deg	ϕ_u deg
Sand: rounded grains		
Loose	28 to 30	
Medium	30 to 35	26 to 30
Dense	35 to 38	
Sand: angular grains		
Loose	30 to 35	
Medium	35 to 40	30 to 35
Dense	40 to 45	
Sandy gravel	34 to 48	33 to 36

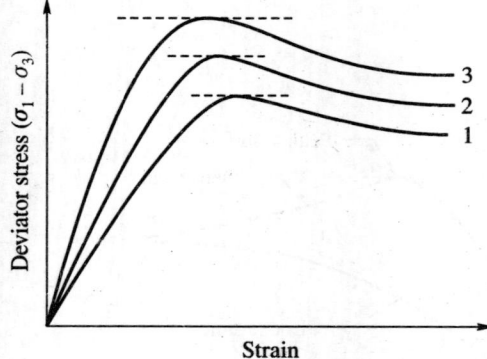

(a) Stress-strain curves for three samples at dense state

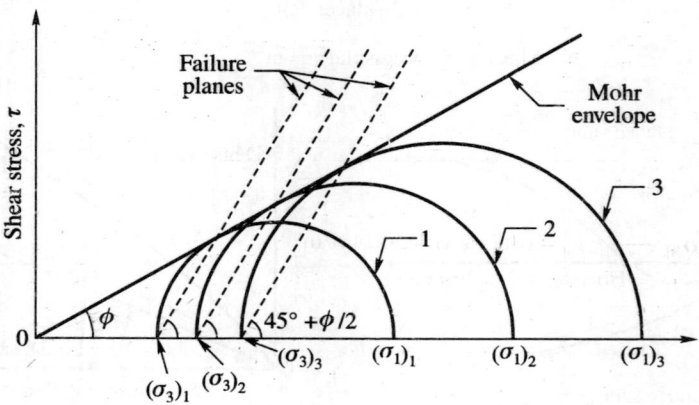

(b) Mohr envelope

Fig. 12.12   Mohr envelope for dense sand

At any stage of loading the major principal stress is the all-round pressure σ_3 plus the intensity of deviator stress ($\sigma_1 - \sigma_3$). The actually applied stresses are the effective stresses in a slow test, that is $\sigma_1 = \sigma'_1$ and $\sigma_3 = \sigma'_3$. Dense samples fail along a clearly defined rupture plane whereas loose sand samples fail along many planes which result in a symmetrical bulging of the sample. The compressive strength of a sample is defined as the difference between the major and minor principal stresses at failure ($\sigma_1 - \sigma_3)_f$.

The typical stress-strain curves for three samples in a dense state and the Mohr circles for these samples at peak strength are shown in Fig. 12.12(a).

If the experiment is properly carried out there will be one common tangent to all these three circles and this will pass through the origin. This indicates that the Mohr envelope is a straight line for sand and the sand has no cohesion. The angle made by the envelope with the σ-axis is called the angle of internal friction. The failure planes for each of these samples are shown in Fig. 12.12(b). Each of them make an angle α with the horizontal which is approximately equal to

$$\alpha = 45° + \phi/2$$

From Fig. 12.12(b) an expression for the angle of internal friction may be written as

$$\sin\phi = \frac{\sigma_1 - \sigma_3}{\sigma_1 + \sigma_3} = \frac{\sigma_1/\sigma_3 - 1}{\sigma_1/\sigma_3 + 1} \tag{12.18}$$

Example 12.6

A consolidated drained triaxial test was conducted on a granular soil. At failure $\sigma'_1/\sigma'_3 = 4.0$. The effective minor principal stress at failure was 100 kN/m^2. Compute ϕ' and the principal stress difference at failure.

Solution

We have

$$\sin\phi = \frac{\sigma'_1/\sigma'_3 - 1}{\sigma'_1/\sigma'_3 + 1} = \frac{4-1}{4+1} = 0.6 \text{ or } \phi' = 37°$$

The principal stresses difference at failure is

$$(\sigma'_1 - \sigma'_3) = \sigma'_3\left(\frac{\sigma'_{1f}}{\sigma'_{3f}} - 1\right) = 100(4-1) = 300 \text{ kN/m}^2$$

Example 12.7

A drained triaxial test on sand with $\sigma'_3 = 150$ kN/m^2 gave $(\sigma'_1/\sigma'_3)_f = 3.7$. Compute (a) σ'_{1f}, (b) ($\sigma_1 - \sigma_3)_f$ and (c) ϕ'.

Solution

(a) We have, $\dfrac{\sigma_1'}{\sigma_3'} = 3.7$

Therefore, $\sigma_1' = 3.7\sigma_3' = 3.7 \times 150 = 555 \text{ kN/m}^2$

(b) $(\sigma_1 - \sigma_3)_f = (\sigma_1'/\sigma_3')_f = 555 - 150 = 405 \text{ kN/m}^2$

(c) $\sin\phi = \dfrac{\sigma_1'/\sigma_3' - 1}{\sigma_1'/\sigma_3' + 1} = \dfrac{3.7 - 1}{3.7 + 1} = 0.574$ or $\phi' = 35°$

Example 12.8

Assume the test specimen in Ex. 12.7 was sheared undrained at the same total cell pressure of 150 kN/m². The induced excess pore water pressure at failure u_f was equal to 70 kN/m². Compute:

(a) σ'_{1f}

(b) $(\sigma_1 - \sigma_3)_f$

(c) ϕ in terms of total stress

(d) the angle of the failure plane α_f.

Solution

(a) and (b): Since the void ratio after consolidation would be the same for this test as for Ex. 12.7, assume ϕ' is the same.

As before, $(\sigma_1 - \sigma_3)_f = \sigma'_{3f}\left(\dfrac{\sigma_1'}{\sigma_3'} - 1\right)$

$\sigma'_{3f} = \sigma_{3f} - u_f = 150 - 70 = 80 \text{ kN/m}^2$

So, $(\sigma_1 - \sigma_3)_f = 80(3.7 - 1) = 216 \text{ kN/m}^2$

$\sigma'_{1f} = (\sigma_1 - \sigma_3)_f + \sigma'_{3f} = 216 + 80 = 296 \text{ kN/m}^2$

(c) $\sin\phi_{total} = \dfrac{\sigma_1 - \sigma_3}{\sigma_1 + \sigma_3} = \dfrac{216}{296 + 70} = 0.42$ or $\phi_{total} = 24°.8$

(d) From Eq.(12.9),

$$\alpha_f = 45° + \dfrac{\phi'}{2} = 45° + \dfrac{35}{2} = 62°.5$$

where ϕ' is taken from Ex. 12.7.

Example 12.9

A saturated specimen of cohesionless sand was tested under drained conditions in a triaxial compression test apparatus and the sample failed at a deviator stress of 482 kN/m^2 and the plane of failure made an angle of 60° with the horizontal. Find the magnitudes of the principal stresses. What would be the magnitudes of the deviator stress and the major principal stress at failure for another identical specimen of sand if it is tested under a cell pressure of 200 kN/m^2?

Solution

As per Eq. (12.9), the failure α is expressed as equal to

$$\alpha = 45° + \frac{\phi}{2}$$

Since $\alpha = 60°$, we have $\phi = 30°$

From Eq.(12.18), we have, $\sin\phi = \dfrac{\sigma_1 - \sigma_3}{\sigma_1 + \sigma_3}$

we have, $\phi = 30°$, $\sigma_1 - \sigma_3 = 482\,\text{kN/m}^2$. Substituting we have

$$\sigma_1 + \sigma_3 = \frac{\sigma_1 - \sigma_3}{\sin\phi} = \frac{482}{\sin 30°} = 964\,\text{kN/m}^2 \tag{a}$$

$$\sigma_1 - \sigma_3 = 482\,\text{kN/m}^2 \tag{b}$$

solving (a) and (b), we have

$$\sigma_1 = 723\,\text{kN/m}^2, \text{ and } \sigma_3 = 241\,\text{kN/m}^2$$

For the identical sample

$$\phi = 30°, \quad \sigma_3 = 200\,\text{kN/m}^2$$

From Eq. (12.8), we have,

$$\sin 30° = \frac{\sigma_1 - 200}{\sigma_1 + 200}$$

on solving, we have $\sigma_1 = 600\,\text{kN/m}^2$ and $(\sigma_1 - \sigma_3) = 400\,\text{kN/m}^2$

Consolidated-quick Tests on Sands and Critical Void Ratio

Consolidated-quick shear test

A consolidated-quick shear test on samples of sand will have to be carried only in a triaxial shear test apparatus as the box-shear test apparatus is not suitable for this purpose. A sample of sand completely saturated and having a known initial void ratio is subjected to an all-round pressure σ_3 and consolidated under this pressure with the drainage valve kept open. The samples are next

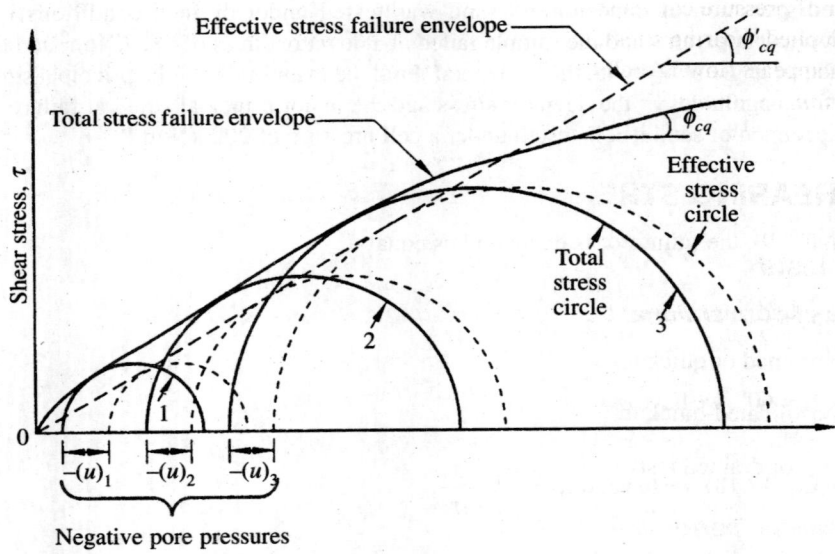

Fig. 12.13   Consolidated quick tests on saturated sand

sheared by keeping the drainage valve closed and increasing the axial load at a constant rate of strain. The tests are repeated on identical specimens and under different all-round pressure σ_3.

Mohr circles which correspond to peak strengths of three samples tested with different all-round pressures and the corresponding Mohr envelope are shown in Fig. 12.13. The Mohr envelope is curved at low pressure σ_3 and approaches a straight line with a slope-angle ϕ_{cq}. The value ϕ_{cq} is referred to as the *consolidated-quick value* of the angle of *shearing resistance*. Coulomb law is not strictly valid at low pressures as is evident in Fig. 12.13. However, if pore-pressures are measured and effective Mohr stress circles are drawn as shown by dashed lines, the Mohr envelope of the effective stress circles will remain a straight line with a slope angle ϕ'_{cq}.

Critical void ratio

Negative pore water pressure increases the shear strength and positive pore water pressure decreases the strength of the sample. However, at a particular all-round pressure σ_3, there is practically no change in the pore water pressure. This indicates that the strength of the sample at this all-round pressure under consolidated-quick test is the same as that of the slow test and there will be no change in the volume of the sample during shearing in both the cases. The initial void ratio which corresponds to this state of the stress is called as the *critical void ratio*.

Liquefaction of Sand

If a mass of sand in a saturated condition has a void ratio greater than the critical void ratio and if it is subjected to a suddenly applied shearing stress, as from an earthquake, heavy blasting, pipe driving or any other dynamic force, the sand tends to decrease in volume. As a result, the pore water is subjected to a suddenly applied hydrostatic excess pressure, and a portion of the weight of overlying material is transferred from intergranular pressure to pore water pressure. The effective stress in the soil is thus reduced. Since the shearing strength of a granular soil depends upon the effective stress, this transfer of pressure causes sudden decrease in the shearing strength. If it is reduced to a value below the applied shearing stress, the mass will fail in shear. Such a failure occurs suddenly, and the whole mass appears flow laterally as if it were a liquid. This type of failure is sometimes referred to as *liquefaction.*

12.10 SHEARING STRENGTH TESTS ON CLAY SOIL

Types of tests

The types of conventional tests on clay soils may be

1. Undrained or quick tests,

2. Consolidated-quick tests, and

3. Slow or drained tests.

The test may be performed either in

1. Box-shear test apparatus, or

2. Triaxial compression test apparatus.

Undrained Strength of Saturated Soils

The tests may be carried out either on undisturbed or on remoulded soil samples. The procedure of the test is the same in both the cases. A series of samples (at least a minimum of three) having the same initial conditions are tested under undrained condition. With σ_3, the all-round pressure, acting on a sample under conditions of no drainage, the axial pressure is increased until failure occurs at a deviator stress $(\sigma_1 - \sigma_3)$. From the deviator stress, the major principal stress σ_1 is determined. If the other samples are tested in the same way but with different values of σ_3, it is found that for all types of saturated clay, the deviator stress at failure (compressive strength) is entirely independent of the magnitude of σ_3 as shown in Fig. 12.14. The diameter of all the Mohr circles are equal and the Mohr envelope is parallel to σ-axis indicating that the angle of shearing resistance $\phi_u = 0$. The symbol ϕ_u represents the angle of shearing resistance under undrained condition. Thus, saturated clays behave as purely cohesive materials with the following properties:

$$\phi_u = 0, \; c_u = \frac{1}{2}(\sigma_1 - \sigma_3) \tag{12.19}$$

where c_u is the symbol used for cohesion under undrained condition.

Equation (12.19) holds good for the particular case of an unconfined compression test in which $\sigma_3 = 0$. Since this test requires very simple apparatus, it is often used, especially for field work, as a ready means of measuring the shearing strength of saturated clay, in this case

$$c_u = \frac{q_u}{2}, \; \text{where } q_u = (\sigma_1 - \sigma_3)_f = (\sigma_1)_f \tag{12.20}$$

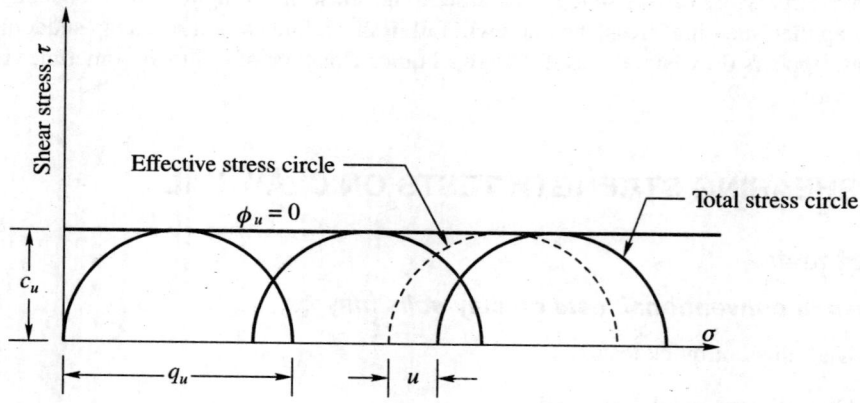

Fig. 12.14   Mohr circle for undrained shear test on saturated clay soil

Effective Stresses

If during the test, pore-pressures are measured, the effective principal stresses may be written as

$$\sigma_1' = \sigma_1 - u, \ \sigma_3' = \sigma_3 - u$$

where, u is the pore water pressure measured during the test. The effective deviator stress at failure may be written as

$$(\sigma_1' - \sigma_3')_f = (\sigma_1 - u)_f - (\sigma_3 - u)_f = (\sigma_1 - \sigma_3)_f \tag{12.21}$$

Equation (12.21) shows that the deviator stress is not affected by the pore water pressure. As such the effective stress circle is only shifted from the position of the total stress circle as shown in Fig. 12.14

Undrained Tests on Partially Saturated Clay

The tests may be carried out either on undisturbed or on remoulded soil samples. All the samples shall have the same initial condition before the test i.e. they should possess the same water content and dry density. The tests are conducted in the same way as for saturated samples. Each sample is tested under undrained condition with different all-round pressures σ_3. Mohr circles for three soil samples and the Mohr envelope are shown in Fig. 12.15. Though all the samples had the same initial conditions, the deviator stress increases with the increase in the all-round pressure σ_3 as shown in the figure. This indicates that the strength of the soil increases with the increasing values of σ_3. The degree of saturation also increases with the increase of σ_3. The Mohr envelope which is curved at lower values of σ_3 becomes almost parallel to the σ-axis as full saturation is reached. Thus, it is not strictly possible to quote single values for the parameters c_u' and ϕ_u for partially saturated clays, but over any range of normal pressure σ_n encountered in a practical example, the envelope can be approximated to a straight line and the approximate values of c_u and ϕ_u can be used in the analysis.

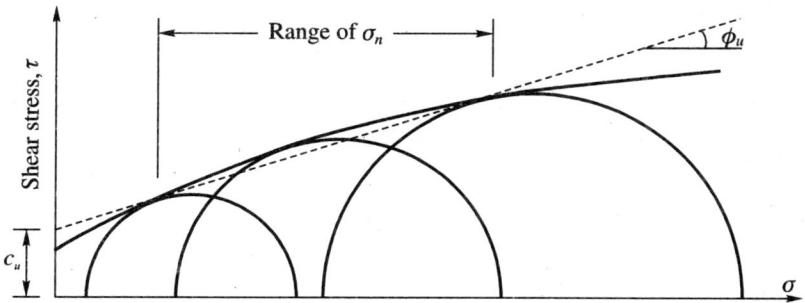

Fig.12.15   Mohr circle for undrained shear tests on partially saturated clay soil

TABLE 12.2

Undrained shear strength parameters for partially saturated soils

Types of soil	c_u	ϕ_u	c'	ϕ'
		Over range $\sigma_3 = 3.5$ kg/cm^2		
Sand with clay binder	0.80	23°	0.70	40°
Lean silty clay	0.87	13°	0.45	31°
Clay, moderate plasticity	0.93	9°	0.60	28°
Clay, very plastic	0.87	8°	0.67	22°

Effective Stresses

If the pore-pressures are measured during the test, the effective circles can be plotted as shown in Fig. 12.16 and the parameters c' and ϕ' obtained. The envelope to the circles, when plotted in terms of effective stresses, is linear.

Typical undrained shear strength parameters for partially saturated compacted samples are shown in Table 12.2.

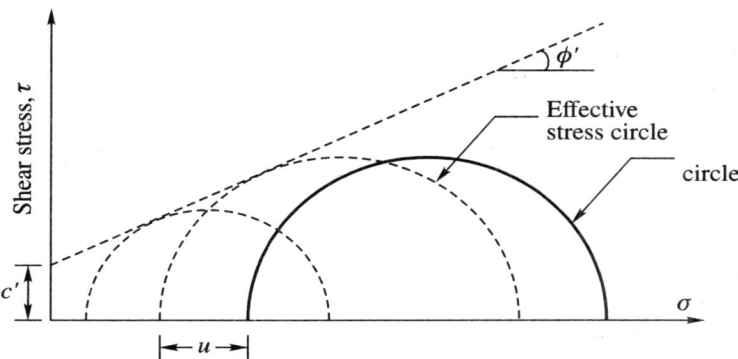

Fig.12.16   Effective stress circles for drained shear tests on partially saturated
clay soils

Normally Consolidated Saturated Clay under Undrained test

If two clay samples 1 and 2 are consolidated under ambient pressures of p_1 and p_2 and are then subjected to undrained triaxial tests without further change in cell pressure, the results may be expressed by the two Mohr circles C_1 and C_2 respectively as shown in Fig. 12.17(b). The failure envelope tangential to these circles passes through the origin and its slope is defined by ϕ_{cu}, the angle of shearing resistance in consolidated undrained tests. If the pore pressures are measured the effective stress Mohr circles C'_1 and C'_2 can also be plotted and the slope of this envelope is ϕ_{cu}. The effective principal stresses are:

$$\sigma'_{11} = \sigma_{11} - u_1; \ \sigma'_{12} = \sigma_{12} - u_2$$

$$\sigma'_{31} = \sigma_{31} - u_1; \ \sigma'_{32} = \sigma_{32} - u_2$$

where u_1 and u_2 are the pore water pressures for the samples 1 and 2 respectively.

It is an experimental fact that the envelopes to the total and effective stress circles are linear. Figure 12.17(b) shows the nature of the variation on the deviator stress ($\sigma_1 - \sigma_3$) and the porewater pressure u in the specimen during the test with the axial strain. The pore water pressure builds up during shearing with a corresponding decrease in the volume of the sample.

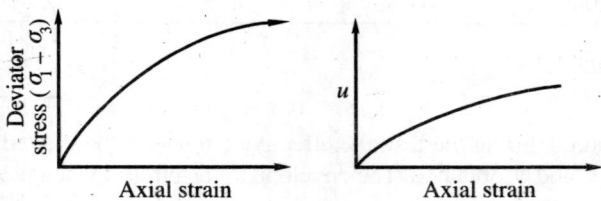

(a) Variation of ($\sigma_1 - \sigma_3$) and u with axial strain

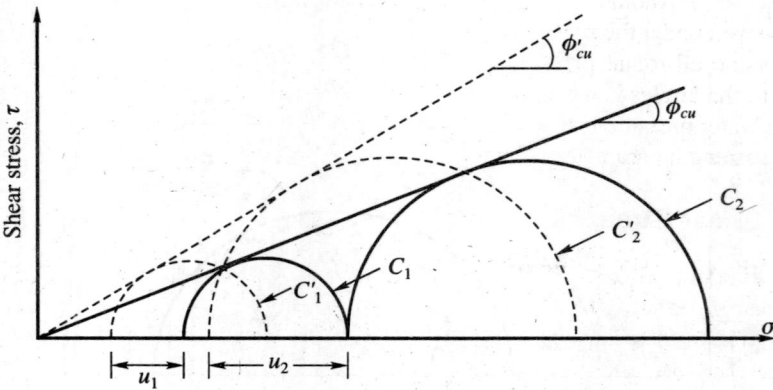

(b) Mohr envelope

Fig. 12.17 Normally consolidated clay under undrained triaxial test

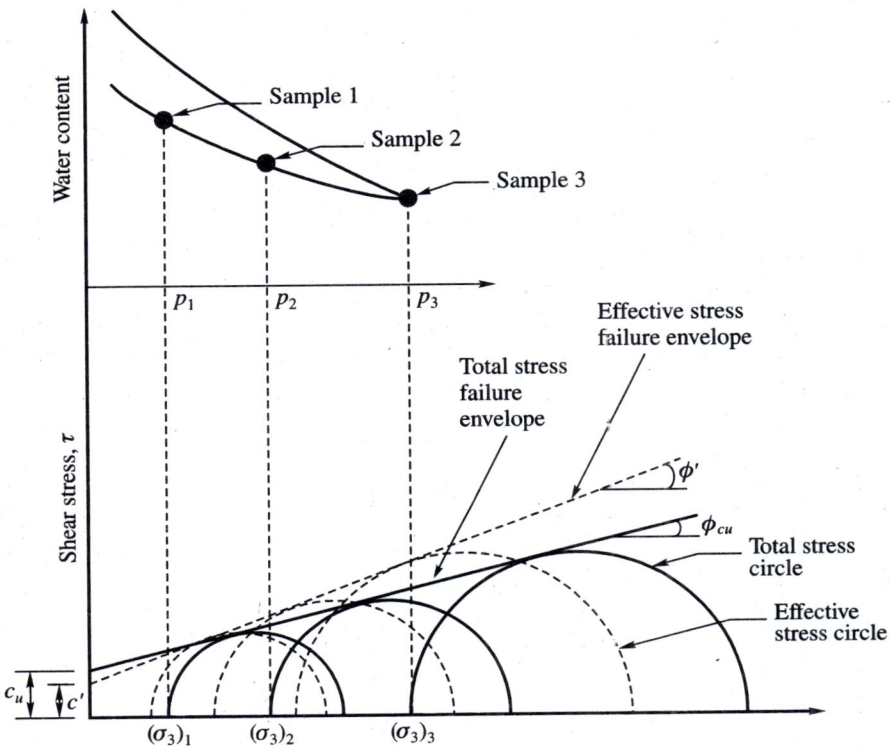

Fig. 12.18 Consolidated quick tests on saturated overconsolidated clay soil

Over-consolidated Saturated Clay under Undrained Test

Let a saturated sample 1 be consolidated under an ambient pressure p_a and then allowed to swell under the pressure p_1. An undrained triaxial test is carried out on this sample under the all-round pressure $p_1(= \sigma_{31})$. Another sample 2 is also consolidated under the same ambient pressure p_a and allowed to swell under the pressure $p_2(= \sigma_{32})$. An undrained triaxial test is carried out on this sample under the same all-round pressure p_2. The two Mohr circles are plotted and the Mohr envelope tangential to the circles is drawn as shown in Fig. 12.18. The shear strength parameters are c_u and ϕ_{cu}. If pore water pressure is measured, effective stress Mohr circles may be plotted as shown in the figure. The strength parameters for effective stresses are represented by c' and ϕ'.

Drained Shear Strength

In drained triaxial tests the soil is first consolidated under an ambient pressure p_a and then subjected to an increasing deviator stress until failure occurs, the rate of strain being controlled in such a way that at no time is there any appreciable pore-pressure in the soil. Thus, at all times the applied stresses are effective, and when the stresses at failure are plotted in the usual manner, the failure envelope is directly expressed in terms of effective stresses. For normally consolidated clays and for sands the envelope is linear for normal working stresses and passes through the origin as shown in Fig. 12.19. The failure criterion for such soils is therefore the angle of shearing resistance in the drained condition ϕ_d.

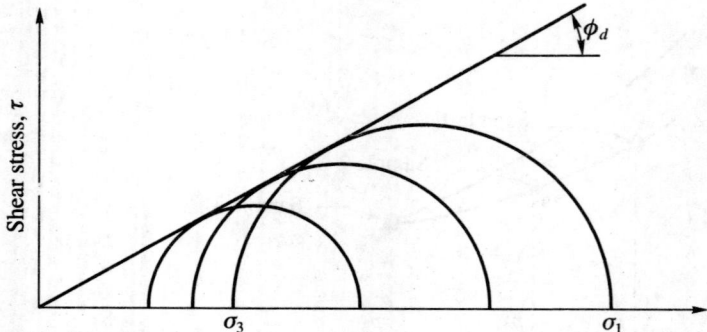

Fig. 12.19   Drained tests on normally consolidated clay samples

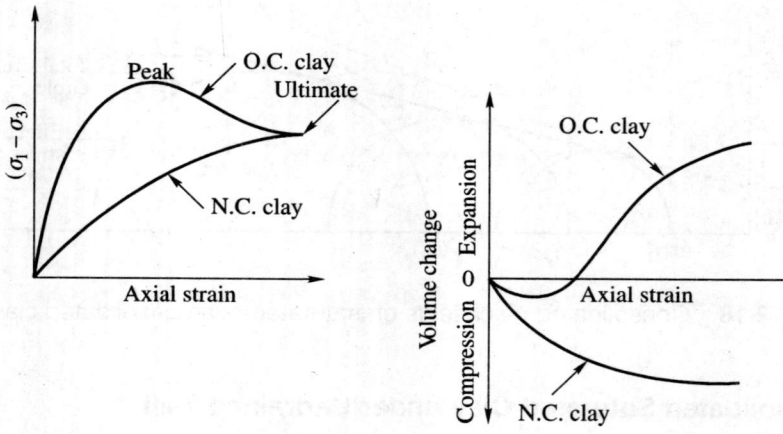

(a) Variation of $(\sigma_1 - \sigma_3)$ with axial strain

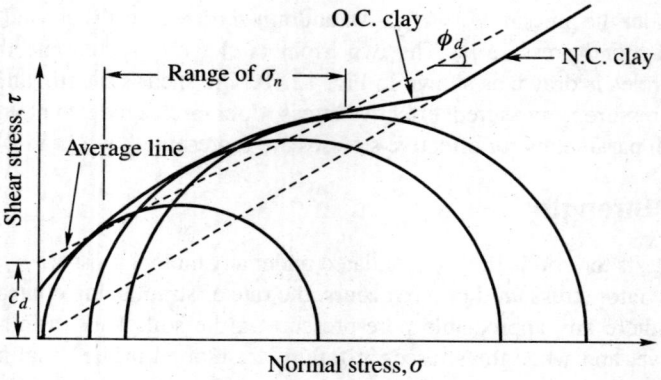

(b) Mohr envelope

Fig. 12.20   Drained tests on over-consolidated clays

For overconsolidated clays, the envelope intersects the axis of zero pressure at a value c_d, the apparent cohesion in the drained test. The Mohr envelope for over-consolidated clays is not linear as may be seen in Fig. 12.20(b). An average line is to be drawn within the range of normal pressure σ_n. The shear strength parameters c_d and ϕ_d are referred to this line.

Since the stresses in a drained test are effective, it might be expected that a given ϕ_d would be equal to ϕ' as obtained from undrained tests with pore-pressure measurement. In normally consolidated clays and in loose sands the two angles of shearing resistance are in fact closely equal since the rate of volume change in such materials at failure in the drained test is approximately zero and there is no volume change throughout an undrained test on saturated soils. But in dense sands and heavily overconsolidated clays there is typically a considerable rate of positive volume change at failure in drained tests, and work has to be done not only in overcoming the shearing resistance of the soils, but also in increasing the volume of the specimen against the ambient pressure. Yet in undrained tests on the same soils, the volume change is zero and consequently ϕ_d for dense sands and heavily over-consolidated clays is greater than ϕ'. Fig. 12.20(a) shows the nature of variation of the deviator stress with axial strain. During the application of the deviator stress, the volume of the specimen gradually reduces for normally consolidated clays. However, overconsolidated clays go through some reduction of volume initially but then expand.

12.11 BASIC SHEAR STRENGTH PARAMETERS OF CLAY SOILS (HVORSLEV PARAMETERS)

Failure Criterion

The values of cohesion and friction in terms of effective stresses are found to vary with the stress history of the sample, and though useful for practical purpose cannot be considered as basic physical properties of the soil. The true cohesion is a function of the water content of the soil (i.e. void ratio) whereas the true angle of friction is a constant for the soil. In any conventional series of tests, an increase in effective stress is accompanied by a decrease in water content and the resultng failure envelope represents the combined influence of the two parameters.

It should be remembered that the values c and ϕ in the Coulomb equation represent merely two empirical coefficients. The term cohesion is retained only for historical reasons. It is used as an abbreviation of the term *apparent cohesion*. In contrast to the apparent cohesion, the *true cohesion* represents that part of the shearing resistance of a soil which is a function only of the water content. There is no relation between apparent and true cohesion other than the name.

Determination of True Cohesion and True Friction

Consolidation test results show that if a sample of saturated soil is consolidated and then unloaded, the pressure void ratio relationship on the unloading cycle is different from that obtained on first loading. Two samples can, therefore, exist, one normally consolidated and the other over-consolidated, each having the same void ratio, but in equilibrium under different effective stresses [points A and B in Fig. 12.21(a)].

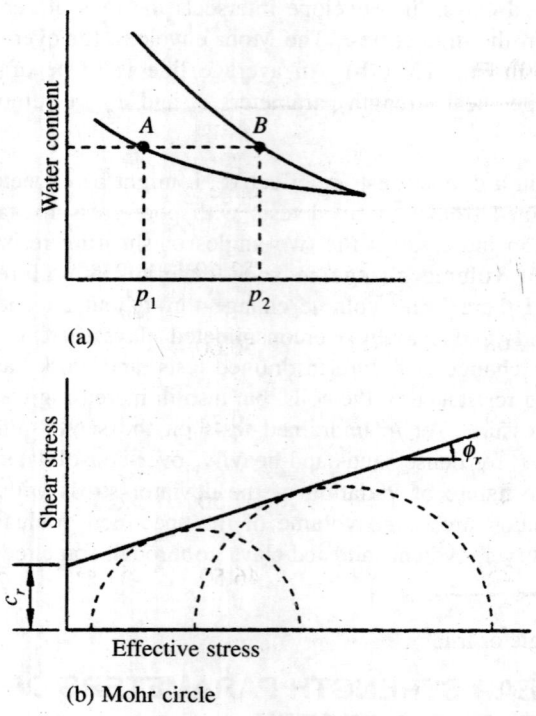

Fig. 12.21   True shear parameters

If undrained tests are carried out on the two samples, and the pore-pressures are measured, the Mohr circle for the two tests can be plotted in terms of effective stresses as shown if Fig. 12.21(b). But since the void ratio is identical in both the tests, the true cohesion is also identical, and the difference in strength between the samples must be an expression of the true internal friction. Thus, the failure envelope to the two Mohr circles can be expressed by the equation

$$s = c_r + (\sigma - u)\tan\phi_r \tag{12.22}$$

where, c_r = true cohesion, ϕ_r = angle of true internal friction

The value of ϕ_r for a particular clay at a given water content can be found by the tests indicated earlier, and also by measuring the angle α in the compression specimen. Numerous tests have shown that ϕ_r varies only slightly with water content and an average value can therefore be given for a clay over normal ranges of water content as shown in Table 12.3.

It will, be noted that for clays of moderate or high plasticity the angle of friction ϕ_r has appreciable values. This shows that even in clays that the particles are effectively in contact and that solid body friction persists at the intergranular 'contacts'. It may be seen from the table that if the actual angle made by slip planes α is measured during experiment, the true angle may ϕ_r also be determined.

TABLE 12.3

Basic strength parameters

Soil	w_1	I_p	Activity	Water content range %	Drained ϕ_d	ϕ_r	$2\alpha - 90°$ $= \phi_r$
Undisturbed clay	123	87	1.42	52.60	23°	18°	21.0
Remoulded clay	98	68	1.11	50.60	22°	13.5°	14.0
Remoulded London clay	74	49	0.98	29.34	20.50°	12.5°	12.5
Remoulded Illite clay	73	45	0.90	37.50	22.0°	16.3°	16.0
Remoulded Kaolinite clay	63	25	0.32	46.50	21.5°	21°	22.5

In contrast with the angle of true internal friction the true cohesion varies greatly with water content.

12.12 UNCONFINED COMPRESSION TESTS

The unconfined compression test is a special case of a triaxial compression test in which the allround pressure $\sigma_3 = 0$. The tests are carried out only on saturated samples which can stand without any lateral support. The test, is, therefore, applicable to cohesive soils only. The test is an undrained test and is based on the assumption that there is no moisture loss during the test. The unconfined compression test is one of the simplest and quickest tests used for the determination of the shear strength of cohesive soils. These tests can also be performed in the field by making use of simple loading equipments.

Any compression testing apparatus with arrangement for strain control may be used for testing the samples . The axial load σ_1 may be applied mechanically or pneumatically.

Specimens of height to diameter ratio of 2 are normally used for the tests. The sample fails either by shearing on an inclined plane (if the soil is of brittle type) or by bulging. The vertical stress at any stage of loading is obtained by dividing the total vertical load by the cross-sectional area. The cross-sectional area of the sample increases with the increase in compression. The cross-sectional area A at any stage of loading of the sample may be computed on the basic assumption that the total volume of the sample remains the same. That is

$$A_0 h_0 = A_h$$

where A_0, h_0 = initial cross-sectional area and height of sample respectively.

A, h = cross-sectional area and height respectively at any stage of loading.

If Δh is the compression of the sample, the strain is

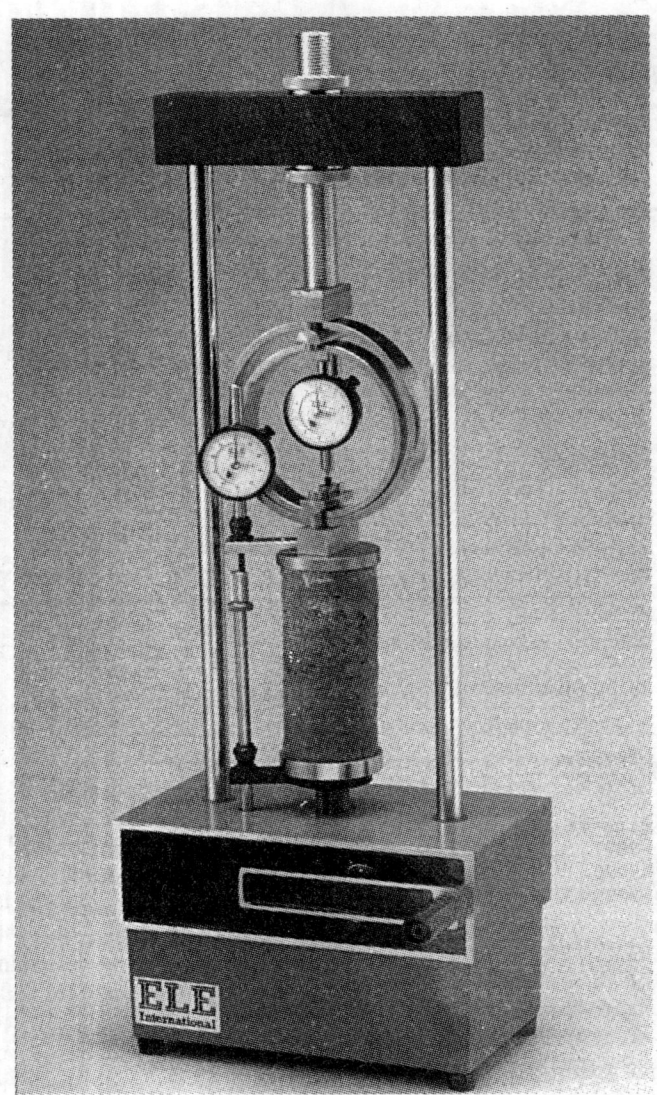

Fig. 12.22 Unconfined compression test equipment (Courtesy : Soiltest)

$$\varepsilon = \frac{\Delta h}{h_0}$$

Since, $\Delta h = h_0 - h$, we may write $A_0 h_0 = A(h_0 - \Delta h)$

Therefore, $A = \dfrac{A_0\, h_0}{h_0 - \Delta h} = \dfrac{A_0}{1 - \Delta h / h_0} = \dfrac{A_0}{1 - \varepsilon}$ \qquad\qquad\qquad (12.23)

The average vertical stress at any stage of loading may be written as

$$\sigma_1 = \frac{P}{A} = \frac{P(1-\varepsilon)}{A_0} \tag{12.24}$$

where P is the vertical load at the strain ε.

Using the relationship given by Eq. (12.24) stress-strain curves may be plotted. The peak value is taken as the unconfined compressive strength q_u, that is

$$(\sigma_1)_f = q_u \tag{12.25}$$

Example 12.10

When an undrained triaxial compression test was conducted on specimens of clayey silt, the following results were obtained:

Specimen No.	1	2	3
σ_3 (kN/m²)	17	44	56
σ_1 (kN/m²)	157	204	225
u (kN/m²)	12	20	22

Determine the values of shear parameters considering (a) total stresses and (b) effective stresses.

Solution

(a) Total stresses

For a solution with total stresses, draw Mohr circles C_1, C_2 and C_3 for each of the specimens using the corresponding principal stresses σ_1 and σ_3.

Draw a Mohr envelope tangent to these circles as shown in Fig. Ex. 12.10. Now from the figure, we have

$c = 48$ kN/m², $\phi = 15°$

(b) With effective stresses

The effective principal stresses may be found by subtracting the pore pressures u from the total principal stresses as given below.

Specimen No.	1	2	3
$\sigma'_3 = (\sigma_3 - u)$ kN/m²	5	24	34
$\sigma'_1 = (\sigma_1 - u)$ kN/m²	145	184	203

As before draw Mohr circles C'_1, C'_2 and C'_3 for each of the specimens as shown in Fig. Ex. 12.10. Now from the figure

$c' = 46$ kN/m², $\phi' = 20°$

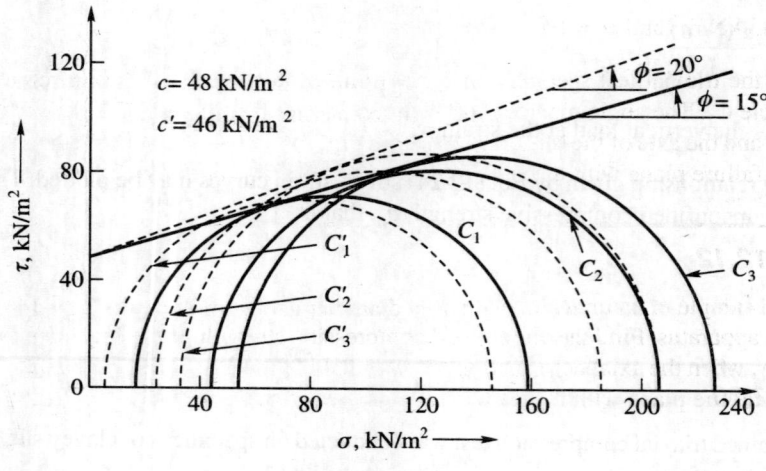

Fig. Ex. 12.10

Example 12.11

A soil has an unconfined compressive strength of 120 kN/m². In a triaxial compression test a specimen of the same soil when subjected to a chamber pressure of 40 kN/m² failed at an additional stress of 160 kN/m². Determine:

(i) The shear strength parameters of the soil, (ii) the angle made by the failure plane with the axial stress in the triaxial test.

Solution

There is one unconfined compression test result and one triaxial compression test result. Hence, two Mohr circles, C_1, and C_2 may be drawn as shown in Fig. Ex. 12.11. For Mohr circle C_1, $\sigma_3 = 0$ and $\sigma_1 = 120$ kN/m², and for Mohr circle C_2, $\sigma_3 = 40$ kN/m² and $\sigma_1 = (40 + 160) = 200$ kN/m². A common tangent to these two circles is the Mohr envelope which gives

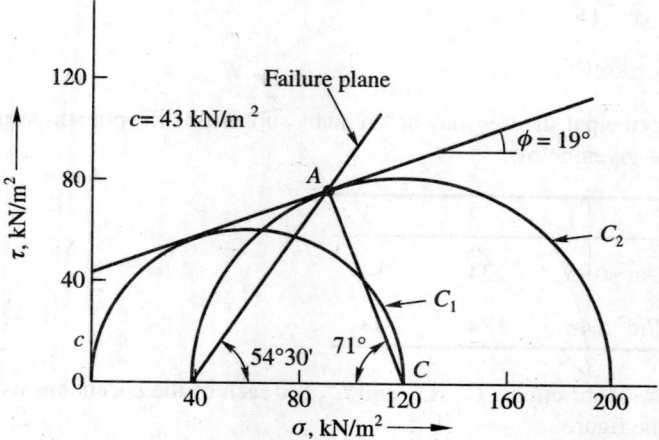

Fig. Ex. 12.11

(i) $c = 43$ kN/m² and $\phi = 19°$.

(ii) For the triaxial test specimen, A is the point of tangency for Mohr circle C_2 and C is the centre of circle C_2. The angle made by AC with the abscissa is equal to twice the angle between the failure plane and the axis of the sample $= 2\theta$. From Fig. Ex. 12.11, $2\theta = 71°$ and $\theta = 35.5°$. The angle made by the failure plane with the σ-axis is $\alpha = 90° - 35.5° = 54.5°$.

Example 12.12

A cylindrical sample of saturated clay 4 cm in diameter and 8 cm high was tested in an unconfined compression apparatus. Find the unconfined compression strength, if the specimen failed at an axial load of 360 N, when the axial deformation was 8 mm. Find the shear strength parameters if the angle made by the failure plane with the horizontal plane was recorded as 50°.

Solution

As per Eq. (12.24), the unconfined compression strength of the soil is given by

$$\sigma_1 = \frac{P(1-\varepsilon)}{A_0}, \text{ where } P = 360\ N$$

$$A_0 = \frac{3.14}{4} \times (4)^2 = 12.56 \text{cm}^2, \ \varepsilon = \frac{0.8}{8} = 0.1$$

Therefore, $\sigma_1 = \dfrac{360(1-0.1)}{12.56} = 25.8 \text{N/cm}^2 \text{ or } 258 \text{ kN/m}^2$

Now, $\phi = 2\alpha\text{-}90°$ (Refer Fig. 12.7) where, $\alpha = 50°$. Therefore, $\phi = 2\times50\text{-}90° = 10°$.

Hence, draw the Mohr circle as shown in Fig. Ex. 12.12 ($\sigma_3 = 0$ and $\sigma_1 = 258$ kN/m²) and from the centre C of the circle, draw CA at $2\alpha = 100°$. At point A, draw a tangent to the circle. The tangent is the Mohr envelope which gives

$c = 106$ kN/m², and $\phi = 10°$

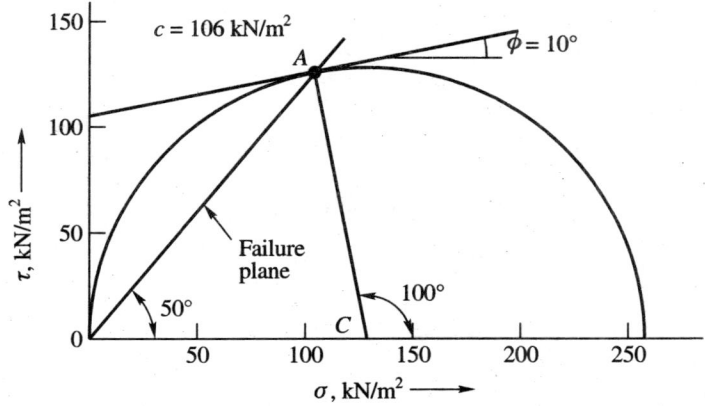

Fig. Ex. 12.12

Example 12.13

A cylindrical sample of soil having a cohesion of 80 kN/m² and an angle of internal friction of 20° is subjected to a cell pressure of 100 kN/m².

Determine: (i) the maximum deviator stress $(\sigma_1 - \sigma_3)$ at which the sample will fail, and (ii) the angle made by the failure plane with the axis of the sample.

Graphical solution

$\sigma_3 = 100$ kN/m², $\phi = 20°$, and $c = 80$ kN/m².

A Mohr circle and the Mohr envelope can be drawn as shown in Fig. Ex. 12.13(a). The circle cuts the σ-axis at B (= σ_3), and at E (= σ_1). Now $\sigma_1 = 433$ kN/m², and $\sigma_3 = 100$ kN/m².

Analytical solution

As per Eq. (12.10)

$$\sigma_1 = \sigma_3 \tan^2\left(45° + \frac{\phi}{2}\right) + 2c \tan\left(45° + \frac{\phi}{2}\right)$$

Substituting the known values, we have

$$\tan(45° + \phi/2) = \tan(45° + 10) = \tan 55° = 1.428$$

$$\tan^2(45° + \phi/2) = 2.04.$$

Therefore,

$$\sigma_1 = 100 \times 2.04 + 80 \times 1.428 = 433 \, \text{kN/m}^2.$$

$$(\sigma_1 - \sigma_3) = (433 - 100) = 333 \, \text{kN/m}^2.$$

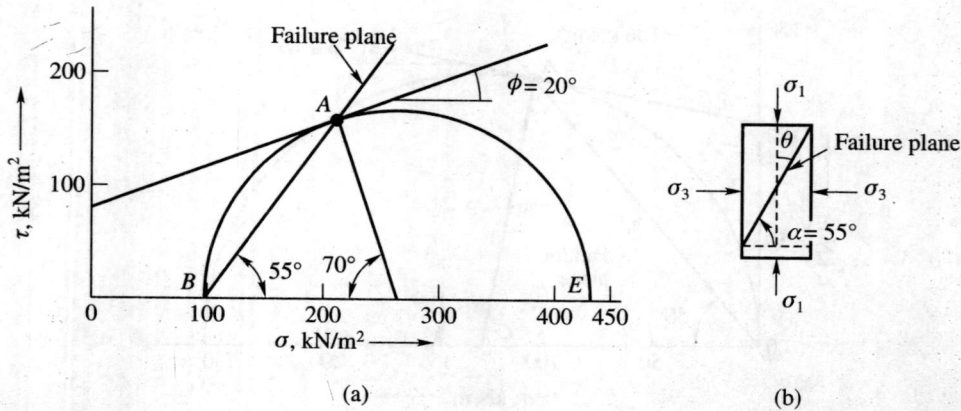

(a)

(b)

Fig. Ex. 12.13

And, if θ = angle made by failure planes with the axis of the sample, (Fig. Ex. 12.13(b))

$$2\theta = 90 - \phi = 90 - 20 = 70° \quad \text{or} \quad \theta = 35°.$$

Therefore, the angle made by the failure plane with σ-axis.

$$\alpha = 90 - 35 = 55°$$

Example 12.14

A normally consolidated clay was consolidated under a stress of 150 kN/m², then sheared undrained in axial compression. The principal stress difference at failure was 100 kN/m², and the induced pore pressure at failure was 88 kN/m². Determine (a) the Mohr-Coulomb strength parameters, in terms of both total and effective stresses analytically, (b) compute $(\sigma_1 / \sigma_3)_f$ and $(\sigma'_1 / \sigma'_3)_f$, and (c) etermine the theoretical angle of the failure plane in the specimen.

Solution

The parameters required are effective parameters c' and ϕ', and total parameters c and ϕ.

(a) Given σ_{3f} = 150 kN/m², and $(\sigma_1 / \sigma_3)_f$ = 100 kN/m². The total principal stress at failure σ_{1f} is obtained from

$$\sigma_{1f} = (\sigma_1 - \sigma_3)_f + \sigma_{3f} = 100 + 150 = 250 \text{ kN/m}^2$$

Effective $\sigma'_{1f} = \sigma_{1f} - u_f = 250 - 88 = 162 \text{ kN/m}^2$

$\sigma'_{3f} = \sigma_{3f} - u_f = 150 - 88 = 62 \text{ kN/m}^2$

Now $\sigma_1 = \sigma_3 \tan^2(45° + \phi/2) + 2c \tan(45° + \phi/2)$

Since the soil is normally consolidated $c = 0$ on the Mohr-Coulomb envelope. As such

$$\frac{\sigma_1}{\sigma_3} = \tan^2(45° + \phi/2) = \frac{1 + \sin\phi}{1 - \sin\phi}, \quad \text{or} \quad \sin\phi = \frac{\sigma_1 - \sigma_3}{\sigma_1 + \sigma_3}$$

$$\text{Total } \phi = \sin^{-1}\left[\frac{100}{250 + 150}\right] = \sin^{-1}\frac{100}{400} = 14°.5$$

$$\text{Effective } \phi' = \sin^{-1}\left[\frac{100}{162 + 62}\right] = \sin^{-1}\frac{100}{224} = 26°.5$$

(b) The stress ratios to failure are

$$\frac{\sigma_1}{\sigma_3} = \frac{250}{150} = 1.67, \quad \frac{\sigma'_1}{\sigma'_3} = \frac{162}{62} = 2.16$$

(c) From Eq. (12.9)

$$\alpha_f = 45° + \frac{\phi'}{2} = 45° + \frac{26.5}{2} = 58°.25$$

The above problem can be solved graphically also by constructing Mohr-Coulomb envelope.

Example 12.15

The following results were obtained at failure in a series of consolidated-undrained tests, with pore pressure measurement, on specimens of saturated clay. Determine the values of the effective stress parameters c' and ϕ' by drawing Mohr circles.

σ_3 kN/m²	$\sigma_1 - \sigma_3$ kN/m²	u_w kN/m²
150	192	80
300	341	154
450	504	222

Solution

The values of the effective principal stresses σ_1 and σ_3 at failure are tabulated below:

σ_3	σ_1	σ'_3	σ'_1
150	342	70	262
300	641	146	487
450	954	228	732

The Mohr circles in terms of effective stresses and the failure envelope are drawn in Fig. Ex. 12.15. The shear strength parameters as measured are:

$$c' = 16\,\text{kN/m}^2; \quad \phi' = 29°$$

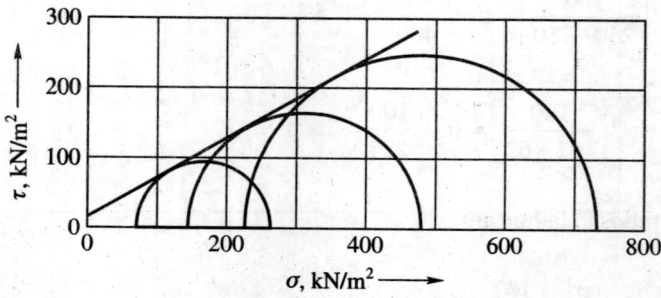

Fig. Ex. 12.15

Example 12.16

The following results were obtained at failure in a series of triaxial tests on specimens of a saturated clay initially 38 mm in diameter and 76 mm long. Determine the values of the shear strength parameters with respect to (a) total stress, and (b) effective stress.

Type of test	σ_3 kN/m²	Axial load (N)	Axial compression (mm)	Volume change (ml)
(a) Undrained	200	222	9.83	–
	400	215	10.06	–
	600	226	10.28	–
(b) Drained	200	467	10.81	6.6
	400	848	12.26	8.2
	600	1265	14.17	9.5

Solution

The principal stress difference at failure in each test is obtained by dividing the axial load by the cross-sectional area of the specimen at failure. The corrected cross-sectional area is calculated from Eq. (12.23). There is, of course, no volume change during an undrained test on a saturated clay. The initial values of length, area and volume for each specimen are $h_0 = 76$ mm, $A_0 = 11.35$ cm²; $V_0 = 86.0$ cm³ respectively.

The Mohr circles at failure and the corresponding failure envelopes for both series of tests are shown in Fig. Ex. 12.16. In both cases the failure envelope is the line nearest to the common tangent to the Mohr circles. The total stress parameters representing the undrained strength of the clay are:

$$c_u = 85 \, \text{kN/m}^2; \phi_u = 0$$

The effective stress parameters, representing the drained strength of the clay, are:

$$c' = 20 \, \text{kN/m}^2; \phi' = 26°$$

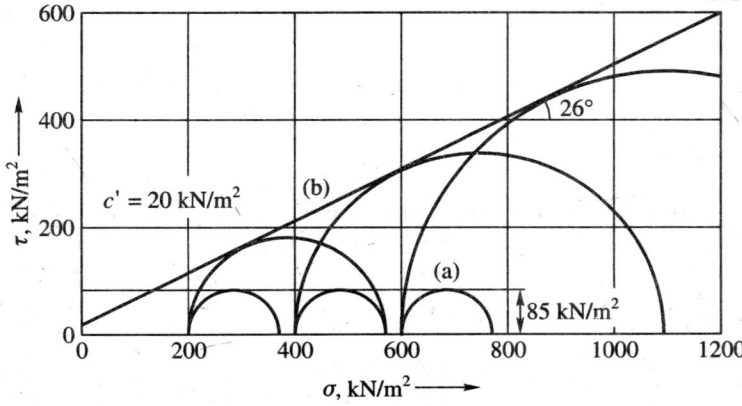

Fig. Ex. 12.16

σ_3 kN/m^2	$\Delta h/h_0$	$\Delta V/V_0$ cm^2	Area kN/m^2	$\sigma_1 - \sigma_3$ kN/m^2	σ_1
a	200	0.129		13.04	170	370
	400	0.132		13.09	160	564
	600	0.135	-	13.12	172	772
b	200	0.142	0.077	12.22	382	582
	400	0.161	0.095	12.25	691	1091
	600	0.186	0.110	12.40	1020	1620

12.13 PORE PRESSURE PARAMETERS UNDER UNDRAINED LOADING

Soils in nature are at equilibrium under their overburden pressure. If the same soil is subjected to an instantaneous additional loading, there will be development of pore pressure if drainage is delayed under the loading. The magnitude of the pore pressure depends upon the permeability of the soil, the manner of application of load, the stress history of the soil, and possibly many other factors. If a load is applied slowly and drainage takes place with the application of load, there will practically be no increase of pore pressure. However, if the hydraulic conductivity of the soil is quite low, and if the loading is relatively rapid, there will not be sufficient time for drainage to take place. In such cases, there will be an increase in the pore pressure in excess of the existing hydrostatic pressure. It is therefore necessary many times to determine or estimate the excess pore pressure for the various types of loading conditions. Pore pressure parameters are used to express the response of pore pressure to changes in total stress *under undrained conditions*. Values of the parameters may be determined in the laboratory and can be used to predict pore pressures in the field under similar stress conditions.

Pore Pressure Parameters Under Triaxial Test Conditions

A typical stress application on a cylindrical element of soil under triaxial test conditions is shown in Fig. 12.23(a) ($\Delta\sigma_1 > \Delta\sigma_3$). Δu is the increase in the pore pressure without drainage. From Fig. 12.23(a), we may write

$$\Delta u_3 = B\Delta\sigma_3, \Delta u_1 = AB(\Delta\sigma_1 - \Delta\sigma_3), \text{ therefore,}$$

$$\Delta u = \Delta u_1 + \Delta u_3 = B[\Delta\sigma_3 + A(\Delta\sigma_1 - \Delta\sigma_3)] \tag{12.26}$$

$$\text{or } \Delta u = B\Delta\sigma_3 + \overline{A}(\Delta\sigma_1 - \Delta\sigma_3) \tag{12.27}$$

where, $\overline{A} = AB$

for saturated soils $B = 1$, so

$$\Delta u = \Delta\sigma_3 + A(\Delta\sigma_1 - \Delta\sigma_3) \tag{12.28}$$

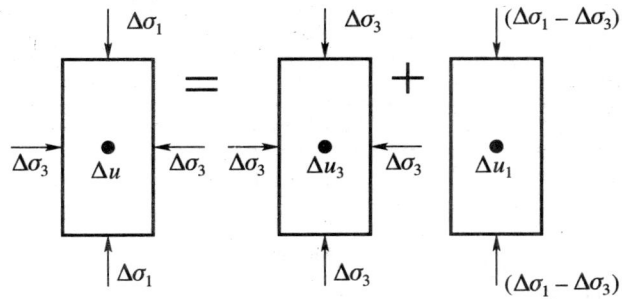

Fig. 12.23(a) Excess water pressure under triaxial test conditions

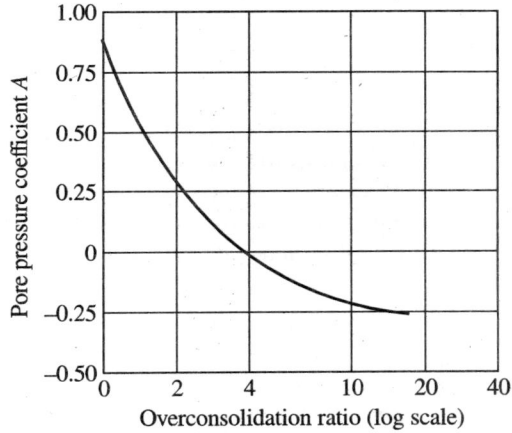

Fig. 12.23(b) Relationship between overconsolidation ratio and pore pressure coefficient A

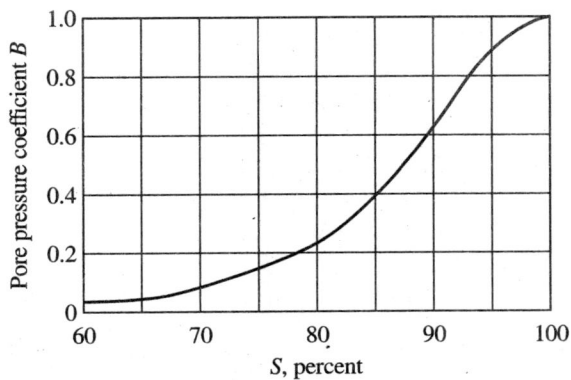

Fig. 12.23(c) Typical relationship between B and degree of saturation S

where A and B are called pore pressure parameters. The variation of A under a failure condition (A_f) with the overconsolidation ratio, O_{CR}, is given in Fig. 12.23(b). Some typical values of A_f are given in Table 12.4. The value of B varies with the degree of saturation as shown in Fig. 12.23(c).

TABLE 12.4

Typical values of A_f

Type of soil	Volume change	A_f
Highly sensitive clay	large contraction	+ 0.75 to + 1.5
Normally consolidated clay	contraction	+ 0.5 to + 1.0
Compacted sandy clay	slight contraction	+ 0.25 to + 0.75
Lightly overconsolidated clay	none	+ 0.00 to + 0.5
Compacted clay gravel	expansion	− 0.25 to + 0.25
Heavily overconsolidated clay	expansion	− 0.5 to 0

Example 12.17

An embankment is being constructed of soil whose properties are $c' = 51$ kN/m^2, $\phi' = 21°$ (all effective stresses), and $\gamma = 15.7$ kN/m^3. The pore pressure parameters as determined from triaxial tests are $A = 0.5$, and $B = 0.9$. Find the shear strength of the soil at the base of the embankment just after the height of fill has been raised from 3 m to 6 m. Assume that the dissipation of pore pressure during this stage of construction is negligible, and that the lateral pressure at any point is one-half of the vertical pressure.

Solution

The equation for pore pressure is

$$\Delta u = B[\Delta\sigma_3 + A(\Delta\sigma_1 - \Delta\sigma_3)]$$

$$\Delta\sigma_1 = \text{Vertical pressure due to 3 m of fill} = 3 \times 15.7 = 47.1\,\text{kN/m}^2$$

$$\Delta\sigma_3 = \frac{47.1}{2} = 23.55\,\text{kN/m}^2$$

Therefore, $\Delta u = 0.9[23.55 + 0.5 \times 23.55] = 31.8\,\text{kN/m}^2$

Original pressure, $\sigma_1 = 3 \times 15.7 = 47.1\,\text{kN/m}^2$

Therefore, $\sigma' = \sigma_1 + \Delta\sigma_1 - \Delta u$

$$= 47.1 + 47.1 - 31.8 = 62.4\,\text{kN/m}^2$$

Strear strength, $s = c' + \sigma' \tan\phi' = 51 + 62.4\tan 21° = 75$ kN/m^2

12.14 VANE SHEAR TESTS

From experience it has been found that the vane test can be used as a reliable *in-situ* test for determining the shear strength of soft-sensitive clays. It is in deep beds of such material that the vane test is most valuable, for the simple reason that there is at present no other method known by which the shear strength of these clays can be measured. Repeated attempts, particularly in Sweden, have failed to obtain undisturbed samples from depths of more than about 10 meters in normally consolidated clays of high sensitivity even using the most modern form of thin-walled piston samplers. In these soils the vane is indispensable. The vane should be regarded as a method to be used under the following conditions:

1. The clay is normally consolidated and sensitive.

2. Only the undrained shear strength is required.

It has been determined that the vane gives results similar to those obtained from unconfined compression tests on undisturbed samples.

The soil mass should be in a saturated condition if the vane test is to be applied. The vane test cannot be applied to partially saturated soils to which the angle of shearing resistance is not zero.

Description of the Vane

The vane consists of a steel rod having at one end four small projecting blades or vanes parallel to its axis, and situated at 90° intervals around the rod. A post hole borer is first employed to bore a hole up to a point just above the required depth. The rod is pushed or driven carefully until the vanes are embedded at the required depth. At the other end of the rod above the surface of the ground a torsion head is used to apply a horizontal torque and this is applied at a uniform speed of about 0.1° per sec until the soil fails, thus generating a cylinder of soil. The area consists of the peripheral surface of the cylinder and the two round ends. The first moment of these areas divided by the applied moment gives the unit shear value of the soil. Fig. 12.24(a) gives a diagrammatic sketch of a field vane.

Determination of Cohesion or Shear Strength of Soil

Consider the cylinder of soil generated by the blades of the vane when they are inserted into the undisturbed soil *in-situ* and gradually turned or rotated about the axis of the shaft or vane axis. The turning moment applied at the torsion head above the ground is equal to the force multiplied by the eccentricity.

Let the force applied = P eccentricity (lever arm) = x units

Turning moment = Px

The surface resisting the turning is the cylindrical surface of the soil and the two end faces of the cylinder.

Therefore,

resisting moment $= (2\pi r \times L \times c_u \times r + 2\pi r^2 \times c_u \times 0.67\, r) = 2\pi r^2\, c_u\, (L + 0.67\, r)$

where r = radius of the cylinder and c_u the undrained shear strength.

At failure the resisting moment of the cylinder of soil is equal to the turning moment applied at the torsion head.

Therefore, $Px = 2\pi r^2 c_u (L + 0.67r)$

$$c_u = \frac{Px}{2\pi r^2 (L + 0.67r)} \qquad (12.30)$$

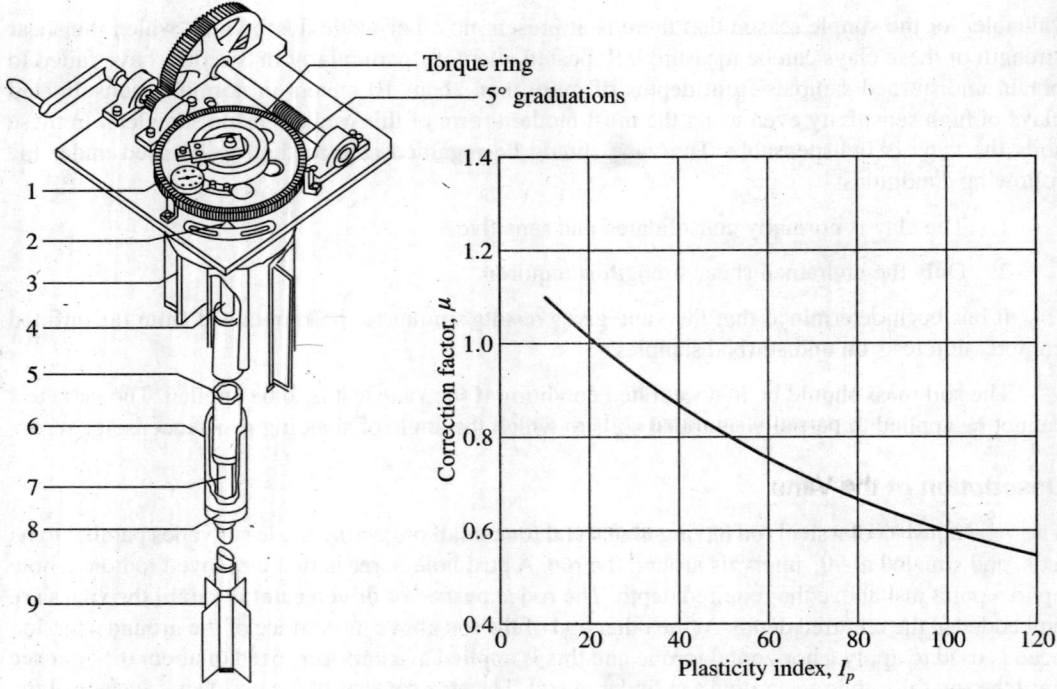

Torque ring
5° graduations

1. Straingauge for
   reading torque

2. Rotation indicator

3. 8-in casing with side fins for
   anchoring torque assembly

4. Torque rod

5. A-rod for applying torque to
   vane. Made up in 5-ft lengths

   (a)

6. BX casing for housing
   torque rod and A rod

7. Vane rod

8. BX-casing-point containing
   bearing and water seals for vane rod

9. Vane varying sizes
   2 in dia by 4 in
   3 in dia by 6 in
   4 in dia by 8 in

   (b)

Fig. 12.24 Vane shear test (a) diagrammatid sketch of a field vane,
(b) correction factor μ (Bjerrum, 1973)

The standard dimensions of field vanes as recommended by ASTM (1994) are given in
Table 12.5.

Some investigators believe that vane shear tests in cohesive soil gives a values of the shear
strength about 15 percent greater than in unconfined compression tests. There are others who believe
that vane tests give lower values.

TABLE 12.5

Recommended dimensions of field vanes (ASTM, 1994)

Casing size	Height, mm (L)	Diameter, mm (d)	Blade thickness mm	Diameter of rod mm
AX	76.2	38.1	1.6	12.7
BX	101.6	50.8	1.6	12.7
NX	127.0	63.5	3.2	12.7

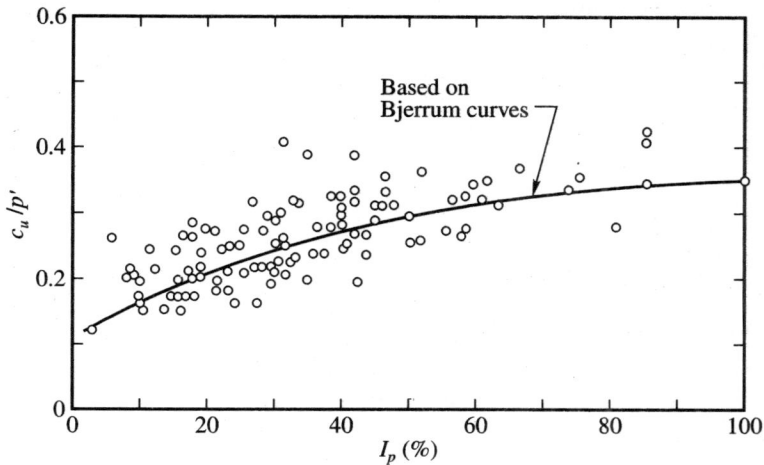

Fig. 12.24(c) Undrained shear strengths from field vane tests on inorganic soft clays and silts (after Tavenas and Leroueil, 1987)

Bjerrum (1973) back computed a number of embankment failures on soft clay and concluded that the vane shear strength tended to be too high. Figure 12.24(b) gives correction factors for the field vane test as a function of plasticity index, I_p (Ladd et al., 1977). We may write

$$c_u \text{ (field)} = \mu c_u \text{ (vane)} \tag{12.31}$$

where μ is the correction factor (Fig. 12.24(b)).

Figure 12.24(c) give relationships between plasticity index I_p and c_u/p' where c_u is the undrained shear strength obtained by field vane and p' the effective overburden pressure. This plot is based on comprehensive test data compiled of Tavenas and Leroueil (1987). Necessary correction factors have been applied to the data as per Fig. 12.24 (b) before plotting.

Example 12.18

At a depth of 6 m below the ground surface at a site, a vane shear test gave a torque value of 6040 N-cm. The vane was 10 cm high and 7 cm across the blades. Estimate the shear strength of the soil.

Solution

Per Eq. (12.30)

$$c_u = \frac{\text{Torque } (T)}{2\pi r^2 \left[L + (2/3) r\right]}$$

where, $T = 6040$ N-cm, $L = 10$ cm, $r = 3\text{-}5$ cm.

substituting, we have

$$c_u = \frac{6040}{2 \times 3.14 \times 3.5^2 (10 + 0.67 \times 3.5)} = 6.4 \text{ N/cm}^2 = 64 \text{ kN/m}^2$$

Example 12.19

A vane 11.25 cm long, and 7.5 cm in diameter was pressed into soft clay at the bottom of a borehole. Torque was applied to cause failure of soil. The shear strength of clay was found to be 37 kN/m². Determine the torque that was applied.

Solution

From Eq. (12.30),

$$\text{Torque,} \quad T = c_u \left[2\pi r^2 (L + (2/3)r) \right], \text{ where, } c_u = 37 \text{ kN/m}^2 = 3.7 \text{ N/cm}^2$$

$$= 3.7 \left[2 \times 3.14 \times (3.75)^2 (11.25 + 0.67 \times 3.75) \right] = 4500 \text{ N cm}$$

12.15 OTHER METHODS FOR DETERMINING UNDRAINED SHEAR STRENGTH OF COHESIVE SOILS

We have discussed in earlier sections three methods for determining the undrained shear strength of cohesive soils. They are

1. Unconfined compression test

2. UU triaxial test

3. Vane shear test

In this section two more methods are discussed. The instruments used for this purpose are

1. Torvane (TV)

2. Pocket penetrometer (PP)

Torvane

Torvane, a modification of the vane, is convenient for investigating the strength of clays in the walls of test pits in the field or for rapid scanning of the strength of tube or split spoon samples. Fig 12.25(a) gives a diagrammatic sketch of the instrument. Figure 12.25(b) gives a photograph of the same. The vanes are pressed to their full depth into the clay below a flat surface, whereupon a torque is applied through a calibrated spring until the clay fails along the cylindrical surface circumscribing the vanes and simultaneously along the circular surface constituting the base of the cylinder. The value of the shear strength is read directly from the indicator on the calibrated spring.

Specification for three sizes of vanes are given below (Holtz et al., 1981)

Diameter (mm)	Height of vane (mm)	Maximum shear strength (kPa)
19	3	250
25	5	100 (standard)
48	5	20

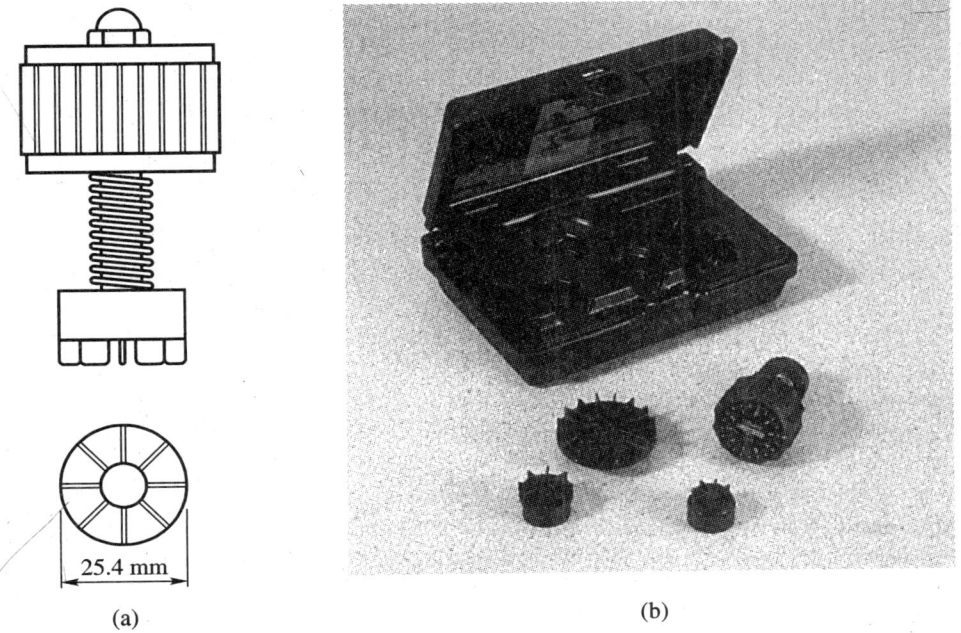

25.4 mm

(a)   (b)

Fig. 12.25   Torvane shear device (a) a diagrammatic sketch, and (b) a photograph
(Courtesy: Soiltest)

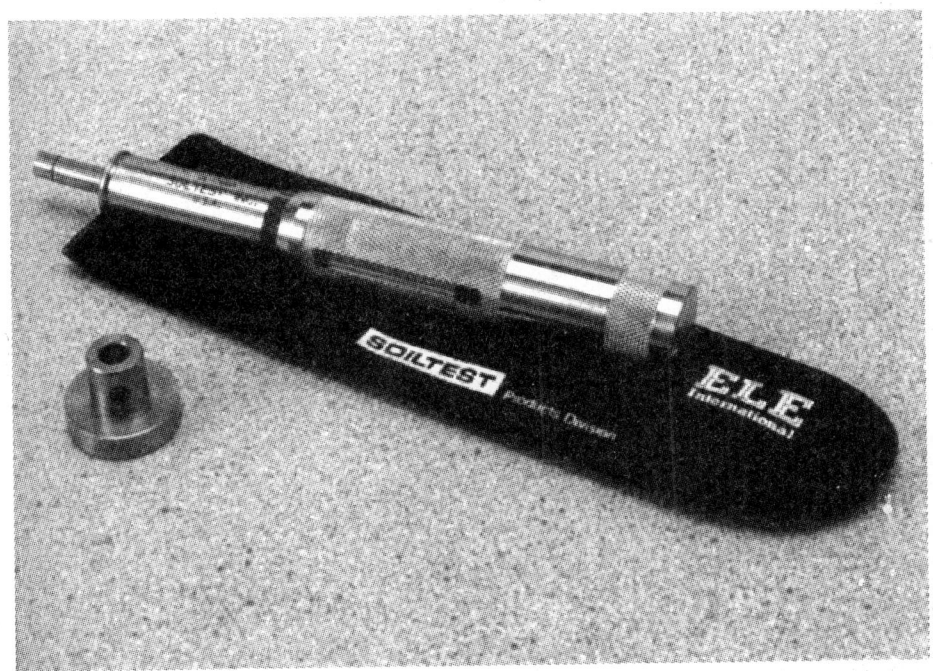

Fig. 12.26   Pocket penetrometer (PP), a hand-held device which indicates unconfined
compressive strength (Courtesy: Soiltest, USA)

Pocket Penetrometer

Figure 12.26 shows a pocket penetrometer (Holtz et al., 1981) which can be used to determine undrained shear strength of clay soils both in the laboratory and in the field. The procedure consists in pushing the penetrometer directly into the soil and noting the strength marked on the calibrated spring.

12.16 THE RELATIONSHIP BETWEEN UNDRAINED SHEAR STRENGTH AND EFFECTIVE OVERBURDEN PRESSURE

It has been discussed in previous sections that the shear strength is a function of effective consolidation pressure. If a relationship between undrained shear strength, c_u, and effective consolidation pressure p' can be established, we can determine c_u if p' is known and vice versa. If a soil stratum in nature is normally consolidated the existing effective overburden pressure p'_0 can be determined from the known relationship. But in overconsolidated natural clay deposits, the preconsolidation pressure p'_c is unknown which has to be estimated by any one of the available methods. If there is a relationship between p'_c and c_u, c_u can be determined from the known value of p'_c. Alternatively, if c_u is known, p'_c can be determined. Some of the relationships between c_u and p' are presented below. A typical variation of undrained shear strength with depth is shown in Fig. 12.27 for both normally consolidated and heavily overconsolidated clays. The higher shear strength as shown in Fig. 12.27(a) for normally consolidated clays close to the ground surface is due to desiccation of the top layer of soil.

Skempton (1957) established a relationship which may be expressed as

$$\frac{c_u}{p'} = 0.10 + 0.004 I_p \qquad\qquad (12.32)$$

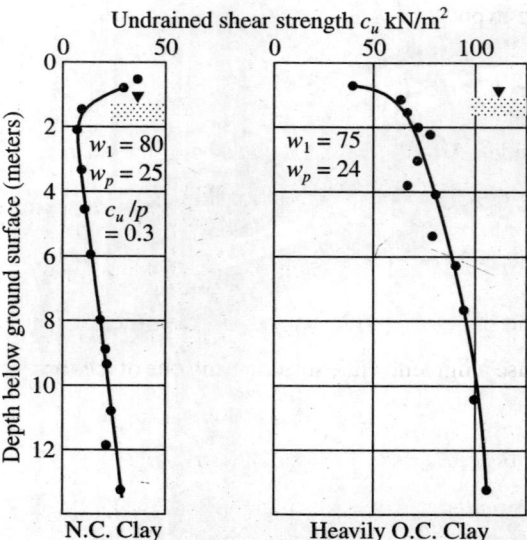

Fig. 12.27 Typical variations of undrained shear strength with depth (After Bishop and Henkel, 1962)

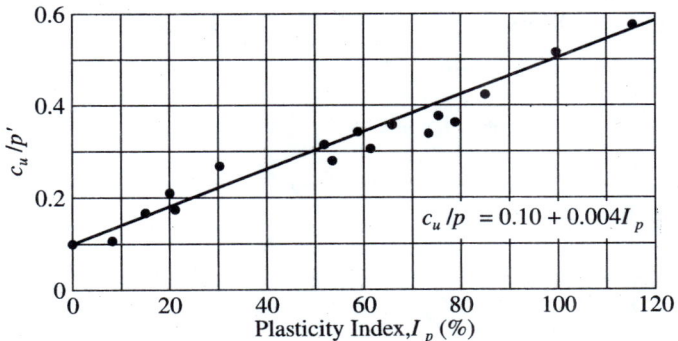

Fig. 12.28 Relation between c_u/p' and plasticity index

He found a close correlation between c_u/p' and I_p as illustrated in Fig. 12.28. Though the Eq. (12.32) was originally meant for normally consolidated clays, it has been used for overconsolidated clays also, p' may be replaced by p'_0 as the existing effective overburden pressure for normally consolidated clays, and by p'_c for overconsolidated clays. Peck et al., (1974) has extensively used this relationship for determining preconsolidation pressure p'_c. Equation (12.32) may also be used for determining p'_c indirectly. If p'_c can be determined independently, the value of the undrained shear strength c_u for overconsolidated clays can be obtained from Eq. (12.32). The values of c_u so obtained may be checked with the values determined in the laboratory on undisturbed samples of clay.

Bjerrum and Simons (1960) proposed a relationship between c_u/p' and plasticity index I_p as

$$\frac{c_u}{p'} = 0.45(I_p)^{\frac{1}{2}} \text{ for } I_p > 5\%$$ (12.33)

The scatter is expected to be of the order of ± 25 percent of the computed value.

Another relationship expressed by them is

$$\frac{c_u}{p'} = 0.18(I_l)^{-\frac{1}{2}} \text{ for } I_l > 0.5$$ (12.34)

where I_l is the liquidity index. The scatter is found to be of the order of ± 30 percent.

Karlsson and Viberg (1967) proposed a relationship as

$$\frac{c_u}{p'} = 0.005w_l \text{ for } w_l > 20 \text{ percent}$$ (12.35)

where w_l is the liquid limit in percent. The scatter is of the order of ± 30 percent.

The engineer has to use judgment while selecting any one of the forms of relationships mentioned above.

c_u/p' Ratio Related to Overconsolidation Ratio p'_c/p'_0

Ladd and Foott (1974) presented a non-dimensional plot (Fig. 12.29) giving a relationship between a nondimensional factor N_f and overconsolidation ratio OCR. Figure 12.29 is based on direct simple shear tests carried out on five clays from different origins. The plot gives out a trend but requires further investigation.

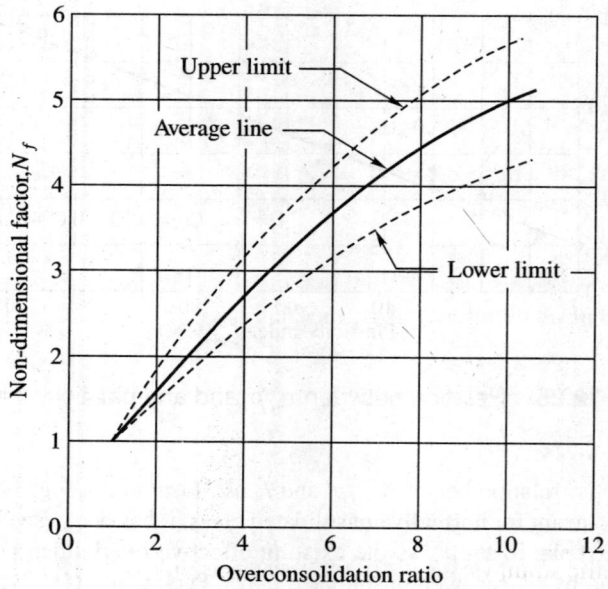

Fig. 12.29 Relationship between N_f and overconsolidation ratio OCR (Ladd and Foott, 1974)

The non-dimensional factor N_f is defined as

$$N_f = \frac{(c_u / p_0')_{NC}}{(c_u / p_0')_{OC}} \qquad (12.36)$$

where p_0' = existing overburden pressure

   OC = overconsolidated

   NC = normally consolidated

From the plot in Fig. 12.29 the shear strength c_u of overconsolidated clay can be determined if p_0' and $(c_u/p_0')_{NC}$ are known.

12.17 QUESTIONS AND PROBLEMS

12.1   What is Coulomb's equation for shear strength of soil? Discuss the factors that affect the shear strength parameters of soil.

12.2   Derive the equations for normal and shear failure stresses in terms of applied stresses or, σ_1 and σ_3 on a cylindrical specimen in a triaxial compression test.

12.3   Explain the method of drawing a Mohr circle for a cylindrical sample in a triaxial test. Establish the geometrical relationships between the stresses on the failure plane and externally applied principal stresses. Compare these relations with those obtained analytically in question 12.2.

12.4   Derive the equation giving major principal stress in terms of minor principal stress and shear strength parameters of soil.

12.5   Classify the shear tests based on drainage conditions. Explain how the pore pressure variation and volume change take place during these tests. Enumerate the field conditions which necessitate each of these tests.

12.6   Enlist features of a triaxial compression test apparatus and describe them briefly.

12.7   What are the advantages and disadvantages of a triaxial compression test in comparison with a direct shear test?

12.8   For which types of soils, will the unconfined compression test give reliable results? Draw a Mohr circle for this test. How do you consider the change in the area of the specimen which takes place during the test in the final results?

12.9   What types of field tests are necessary for determining the shear strength parameters of sensitive clays? Derive the relationships that are useful for analyzing the observations of this test.

12.10   For loose and dense sands, draw the following typical diagrams:

(i) Deviator stress vs. linear strain, and

(ii) Volumetric strain vs. linear strain. Discuss them.

12.11   What is 'critical void ratio'? How would you determine it in the laboratory? Also explain the conditions causing liquefaction of sand.

12.12   Briefly discuss the effects of drainage conditions on the shear strength parameters of clay soil.

12.13   A direct shear test on specimens of fine sand gave the following results:

Normal stress (kN/m^2)   100   175

Shearing stress (kN/m^2)   46   81

Determine :

(i) the angle of internal friction of the soil, and

(ii) shear strength of the soil at a depth of 5 metres from the ground surface.

Assume specific gravity of solids is 2.65, void ratio 0.7 and the ground water table is at a depth of 2 m from the ground surface.

12.14   A specimen of clean sand when subjected to a direct shear test failed at a stress of 120 kN/m^2 when the normal stress was 160 kN/m^2.

Determine:

(i) the angle of internal friction, and

(ii) the deviator stress at which the failure will take place, if a cylindrical sample of the same sand is subjected to a triaxial test with a cell pressure of 100 kN/m^2. Find the angle made by the failure plane with the horizontal.

12.15   A specimen of fine sand, when subjected to a drained triaxial compression test, failed at a deviator stress of 400 kN/m^2. It was observed that the failure plane made an angle of 30° with the axis of the sample. Estimate the value of the cell pressure to which this specimen would have been subjected to.

12.16   A specimen of sandy silt, when subjected to a drained triaxial test failed at major and minor principal stresses of 120 kN/m^2 and 50 kN/m^2 respectively. At what value of deviator stress would another sample of the same soil fail, if it were subjected to a confining pressure of 75 kN/m^2?

12.17   When an unconfined compression test was conducted on a specimen of silty-clay, it showed a strength of 150 kN/m^2. Determine the shear parameters of the soil, if the angle made by the failure plane with the axis of specimen was 35°.

12.18   When an unconfined compression test was conducted on a soil specimen, it failed at an axial pressure of 350 kN/m^2. If the soil has cohesion of 50 kN/m^2 and angle of internal friction of 24°, what was the cell pressure of the test? Also find the angle made by the failure plane with the direction of σ_3. Use both the graphical and analytical methods.

12.19   Given the following triaxial test data, plot the results in a Mohr diagram and determine ϕ.

σ_3 kN/m^2	Peak σ_1 kN/m^2	σ_3 kN/m^2	Peak σ_1 kN/m^2
69	190	276	759
138	376	345	959
207	580	414	1143

12.20   In a triaxial test a soil specimen was consolidated under an allround pressure of 800 kN/m^2 and a back pressure of 400 kN/m^2. Thereafter, under undrained conditions, the all-round pressure was raised to 900 kN/m^2, resulting in a pore water pressure of 495 kN/m^2, then (with all-round pressure remaining at 900 kN/m^2) axial load was applied to give a principal stress difference of 585 kN/m^2 and a pore water pressure of 660 kN/m^2. Calculate the values of pore-pressure coefficients A and B.

12.21   A normally consolidated clay was consolidated under a stress of 150 kPa, then sheared undrained in axial compression. The principal stress difference at failure was 100 kPa, and the induced pore pressure at failure was 88 kPa. Determine analytically (a) the slopes of the total and effective Mohr stress envelops and (b) the theoritical angle of failure plane.

12.22   A sand is hydrostatically consolidated in a triaxial test apparatus to 420 kPa and then sheared with the drainage valves open. At failure, $(\sigma_1 - \sigma_3)$ is 1046 kPa. Determine the major and minor principal stresses at failure and the angle of shearing resistance.

12.23   The same sand as in Prob. 12.22 is tested in a direct shear apparatus under a normal pressure of 420 kPa. The sample fails when a shear stress of 280 kPa is reached. Determine the major and minor principal stresses at failure and the angle of shearing resistance. Plot the Mohr diagram.

12.24   A sample of dense sand tested in a triaxial CD test failed along a well defined failure plane at an angle of 66° with the horizontal. Find the effective confining pressure of the test if the principal stress difference at failure was 100 kPa.

12.25   A drained triaxial test is performed on a sand with $\sigma'_{3f} = 500$ kPa. At failure, $\tau_{max} = 660$ kPa. Find σ'_{1f}, $(\sigma_1 - \sigma_3)_f$ and ϕ'.

12.26   If the test of Prob. 12.25 had been conducted undrained, determine $(\sigma_1 - \sigma_3)_f$, ϕ', ϕ_{total} and the angle of the failure plane in the specimen. The pore water pressure $u = 100$ kPa.

12.27   If the test of Prob. 12.26 is conducted at an initial confining pressure of 1000 kPa, estimate the principal stress difference and the induced pore pressure at failure.

12.28   A sample of silty sand was tested consolidated drained in a triaxial cell where $\sigma_1 = \sigma_3$ = 475 kPa. If the total axial stress at failure was 1600 kPa while $\sigma_3 = 475$ kPa, compute the angle of shearing resistance and the theoretical orientation of the failure plane with respect to the horizontal.

12.29   A *CD* axial compression triaxial test on a normally consolidated clay failed along a clearly defined failure plane of 57°. The cell pressure during the test was 200 kPa. Estimate ϕ' the maximum σ'/σ'_3, and the principal stress difference at failure.

12.30   Two identical samples of soft saturated normally consolidated clay were consolidated to 150 kPa in a triaxial apparatus. One specimen was sheared under drained condition, and the principal stress difference at failure was 300 kPa. The other specimen was sheared undrained, and the principal stress difference at failure was 200 kPa. Determine ϕ_d and ϕ_{cu}.

12.31   A normally consolidated clay sample was consolidated in a triaxial apparatus at an all-round pressure of 1000 kPa and then sheared under undrained condition. The $(\sigma_1 - \sigma_3)$ at failure was equal to 1 Mpa. Determine, ϕ_d and α.

12.32   Two sets of triaxial tests were carried out on two samples of glacial silt. The results are (a) $\sigma_{11} = 400$ kN/m², $\sigma_{31} = 100$ kN/m², (b) $\sigma_{12} = 680$ kN/m², $\sigma_{32} = 200$ kN/m².

The angle of rupture in both the tests is measured to be the same and is equal to 59°. Determine the magnitude of ϕ and c.

12.33   A triaxial compression test on a cylindrical cohesive sample gave the following effective stresses

(a) Major principal stress, $\sigma'_1 = 6895$ kN/m²

(b) Minor principal stress, $\sigma'_3 = 1379$ kN/m²

(c) The angle of inclination of rupture plane = 60° with the horizontal.

Determine analyitically the (i) normal stress, (ii) the shear stress, (iii) the resultant stress on the rupture plane through a point, and (iv) the angle of obliquity of the resultant stress with the shear plane.

12.34   Given the results of two sets of triaxial stress tests:

$\sigma_{11} = 1800$ kN/m²; $\sigma_{31} = 1000$ kN/m²

$\sigma_{12} = 2800$ kN/m²; $\sigma_{32} = 2000$ kN/m²

Compute ϕ and c, and draw the stress diagram

12.35   What is the shear strength in terms of effective stress on a plane within a saturated soil mass at a point where the normal stress is 295 kN/m² and the porewater pressure 120 kN/m². The effective stress parameters for the soils are $c' = 12$ kN/m², $\phi' = 30°$.

12.36   The effective stress parameters for a fully saturated clay are known to be $c' = 15$ kN/m^2, $\phi' = 29°$. In an unconsolidated-undrained triaxial test on a sample of the same clay the all-round pressure was 250 kN/m^2 and the principal stress difference at failure 134 kN/m^2. What was the value of pore water pressure in the same at failure?

12.37   In an *in-situ* vane test on a saturated clay a torque of 35 Nm is required to shear the soil. The vane is 50 mm by 100 mm long. What is the undrained strength of the clay?

12.38   A drained triaxial test is to be performed on a uniform dense sand with rounded grains. The all-round pressure is to be 200 kN/m^2. At about what vertical pressure should the sample fail?

12.39   Compute the shearing resistance against sliding along a horizontal plane at a depth of 6.1 m in a deposit of sand. The water table is at a depth of 2.13 m. The unit weight of moist sand above the water table is 18.54 kN/m^3 and the saturated weight of submerged sand is 20.11 kN/m^3. Assume that the sand is drained freely and ϕ_d for the wet sand is 32°.

12.40   It is believed that the shear strength of a soil under certain conditions in the field will be governed   by Coulomb's law, wherein $c = 19.15$ kN/m^2, and $\phi = 22°$. What minimum lateral pressure would be required to prevent failure of the soil at a given point if the vertical pressure were 431 kN/m^2?

12.41   A sample of dry sand was tested in a direct shear device under a vertical pressure of 137.9 kN/m^2. Compute the angle of internal friction of the sand.

12.42   The sand in a deep natural deposit has an angle of internal friction of 40° in the dry state and dry unit weight of 17.28 kN/m^3. If the water table is at a depth of 6.1 m, what is the shearing resistance of the material along a horizontal plane at a depth of 3.05 m?

12.43   Compute the shearing resistance under the conditions specified in Prob. 12.42, if the water table is at the ground surface.

12.44   A drained triaxial test was conducted on dense sand with a confining pressure of 143.6 kN/m^2. The sample failed at an added vertical pressure of 520 kN/m^2. Compute the angle of internal friction ϕ and the angle of inclination α of the failure planes on the assumption that Coulomb's law is valid.

12.45   A saturated sample of dense sand was consolidated in a triaxial test apparatus at a confining pressure of 143.6 kN/m^2. Further drainage was prevented. During the addition of vertical load, the pore pressure in the sample was measured. At the instant of failure, it amounted to 115 kN/m^2. The added vertical pressure at this time was 138.85 kN/m^2. What was the value of ϕ for the sand?

12.46   An undrained triaxial test was carried out on a sample of saturated clay with a confining pressure of 95.76 kN/m^2. The unconfined compressive strength obtained was 346.65 kN/m^2. Determine the excess vertical pressure in addition to the all-round pressure required to make the sample fail.

12.47   The following data refer to three triaxial tests performed on representative undisturbed samples:

Test	Cell pressure kN/m²	Axial dial reading (division) at failure
1	50	66
2	150	106
3	250	147

The load dial calibration factor is 1.4 N per division. Each sample is 75 mm long and 37.5 mm diameter. Find by graphical means, the value of apparent cohesion and the angle of internal friction for this soil.

12.48   A series of undrained triaxial tests on samples of saturated soil gave the following results

σ_3, kN/m²	100	200	300
u, kN/m²	20	70	136
$(\sigma_1 - \sigma_3)$, kN/m²	290	400	534

Find the values of the parameters c and ϕ.

(a) with respect to total stress, and (b) with respect to effective stress.

12.49   (a) In a vane test a torque of 46 Nm is required to cause failure of the vane in a clay soil. The vane is 150 mm long and has a diameter of 60 mm. Calculate the apparent shear strength of the soil from this test. (b) When a vane of 200 mm long and 90 mm in diameter is used in the same soil and the torque at failure was 138 Nm. Calculate the ratio of the shear strength of the clay in a vertical direction to that in the horizontal direction.

Hint: Use torque $T = Px = c_v (\pi dL) d/2 + (c_h \pi d^2/4) (1/3d) \times 2$ where c_v = cohesion in the vertical plane, and c_v = cohesion in the horizontal plane.

CHAPTER 13

SOIL IMPROVEMENT

13.1 INTRODUCTION

General practice is to use shallow foundations for the foundations of buildings and other such structures, if the soil close to the ground surface possesses sufficient bearing capacity. However, where the top soil is either loose or soft, the load from the superstructure has to be transferred to deeper firm strata. In such cases, pile or pier foundations are the obvious choice.

There is also a third method which may in some cases prove more economical than deep foundations or where the alternate method may become inevitable due to certain site and other environmental conditions. This third method comes under the heading *foundation soil improvement*. In the case of earth dams, there is no other alternative than compacting the remoulded soil in layers to the required density and moisture content. The soil for the dam will be excavated at the adjoining areas and transported to the site. There are many methods by which the soil at the site can be improved. Soil improvement is frequently termed *soil stabilization*, which in its broadest sense is alteration of any property of a soil to improve its engineering performance. Soil improvement

1. Increases shear strength

2. Reduces permeability, and

3. Reduces compressibility

The methods of soil improvement considered in this chapter are

1. Mechanical compaction

2. Dynamic compaction

3. Vibroflotation

4. Preloading

5. Sand and stone columns

389

6. Use of admixtures
7. Injection of suitable grouts
8. Use of geotextiles

13.2 MECHANICAL COMPACTION

Mechanical compaction is the least expensive of the methods and is applicable in both cohesionless and cohesive soils. The procedure is to remove first the weak soil up to the depth required, and refill or replace the same in layers with compaction. If the soil excavated is cohesionless or a sand-silt clay mixture, the same can be replaced suitably in layers and compacted. If the soil excavated is a fine sand, silt or soft clay, it is not advisable to refill the same as these materials, even under compaction, may not give sufficient bearing capacity for the foundations. Sometimes it might be necessary to transport good soil to the site from a long distance. The cost of such a project has to be studied carefully before undertaking the same.

The compaction equipment to be used on a project depends upon the size of the project and the availability of the compacting equipment. In projects where excavation and replacement are confined to a narrow site, only tampers or surface vibrators may be used. On the other hand, if the whole area of the project is to be excavated and replaced in layers with compaction, suitable roller types of heavy equipment can be used. Cohesionless soils can be compacted by using vibratory rollers and cohesive soils by sheepsfoot rollers.

The control of field compaction is very important in order to obtain the desired soil properties. Compaction of a soil is measured in terms of the dry unit weight of the soil. The dry unit weight, γ_d, may be expressed as

$$\gamma_d = \frac{\gamma_t}{1+w}$$ (13.1)

where,

γ_t = total unit weight

w = moisture content

Factors Affecting Compaction

The factors affecting compaction are

1. The moisture content
2. The compactive effort

The compactive effort is defined as the amount of energy imparted to the soil. With a soil of given moisture content, increasing the amount of compaction results in closer packing of soil particles and increased dry unit weight. For a particular compactive effort, there is only one moisture content which gives the maximum dry unit weight. The moisture content that gives the maximum dry unit weight is called the *optimum moisture content*. If the compactive effort is increased, the maximum dry unit weight also increases, but the optimum moisture content decreases. If all the desired qualities of the material are to be achieved in the field, suitable procedures should be adopted to compact the earthfill. The compactive effort to the soil is imparted by mechanical rollers or any other compacting device. Whether the soil in the field has attained the required maximum dry unit weight can be determined by carrying out appropriate laboratory tests on the soil. The following tests are normally carried out in a laboratory.

1. Standard proctor test (ASTM Designation D-698), and
2. Modified proctor test (ASTM Designation D-1557)

13.3   LABORATORY TESTS ON COMPACTION

Standard Proctor Compaction Test

Proctor (1933) developed this test in connection with the construction of earth fill dams in California. The standard size of the apparatus used for the test is given in Fig 13.1. Table 13.1 gives the standard specifications for conducting the test (ASTM designation D-698). Three alternative procedures are provided based the soil material used for the test.

Test Procedure

A soil at a selected water content is placed in layers into a mould of given dimensions (Table 13.1 and Fig. 13.1), with each layer compacted by 25 or 56 blows of a 5.5 lb (2.5 kg) hammer dropped from a height of 12 in (305 mm), subjecting the soil to a total compactive effort of about 12,375 fl-lb/ft^3 (600 kNm/m^3). The resulting dry unit weight is determined. The procedure is repeated for a sufficient number of water contents to establish a relationship between the dry unit weight and the water content of the soil. This data, when, plotted, represents a curvilinear relationship known as the compaction curve or moisture-density curve. The values of water content and standard maximum dry unit weight are determined from the compaction curve as shown in Fig. 13.2.

TABLE 13.1

Specification for standard proctor compaction test

	Procedure		
Item	A	B	C
1. Diameter of mould	4 in. (101.6 mm)	4 in. (101.6 mm)	6 in. (152.4 mm)
2. Height of mould	4.584 in. (116.43 mm)	4.584 in. (116.43 mm)	4.584 in. (116.43 mm)
3. Volume of mould	0.0333 ft^3 (944 cm^3)	0.0333 ft^3 (944 cm^3)	0.075 ft^3 (2124 cm^3)
4. Weight of hammer	5.5 lb (2.5 kg)	5.5 lb (2.5 kg)	5.5 lb (2.5 kg)
5. Height of drop	12.0 in. (304.8 mm)	12.0 in. (304.8 mm)	12.0 in. (304.8 mm)
6. No. of layers	3	3	3
7. Blows per layer	25	25	56
8. Energy of compaction	12,375 ft-lb/ft^3 (600 kN-m/m^3)	12,375 ft-lb/ft^3 (600 kN-m/m3)	12,375 ft-lb/ft^3 (600 kN-m/m^3)
9. Soil material	Passing No. 4 sieve (4.75 mm). May be used if 20% or less retained on No. 4 siere	Passing No 4 sieve (4.75 mm). Shall be used if 20% or more retained on No. 4 sieve and 20% or less retained on 3/8 in (9.5 mm) sieve	Passing No. 4 sieve (4.75 mm). Shall be used if 20% or more retained on 3/8 in. (9.5 mm) sieve and less than 30% retained on 3/4 in. (19 mm) sieve

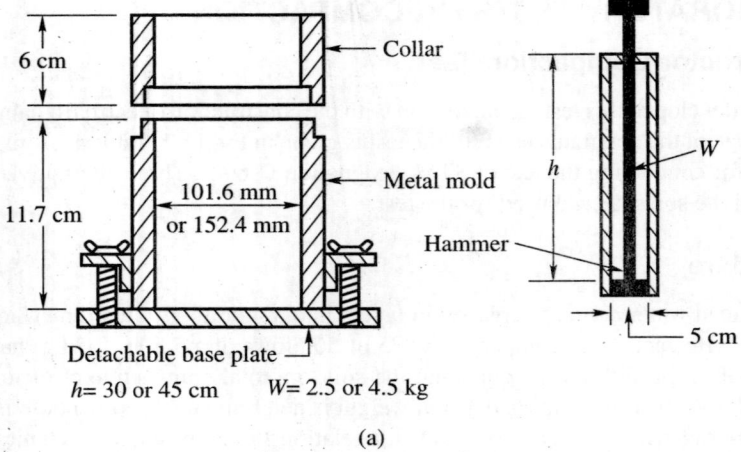

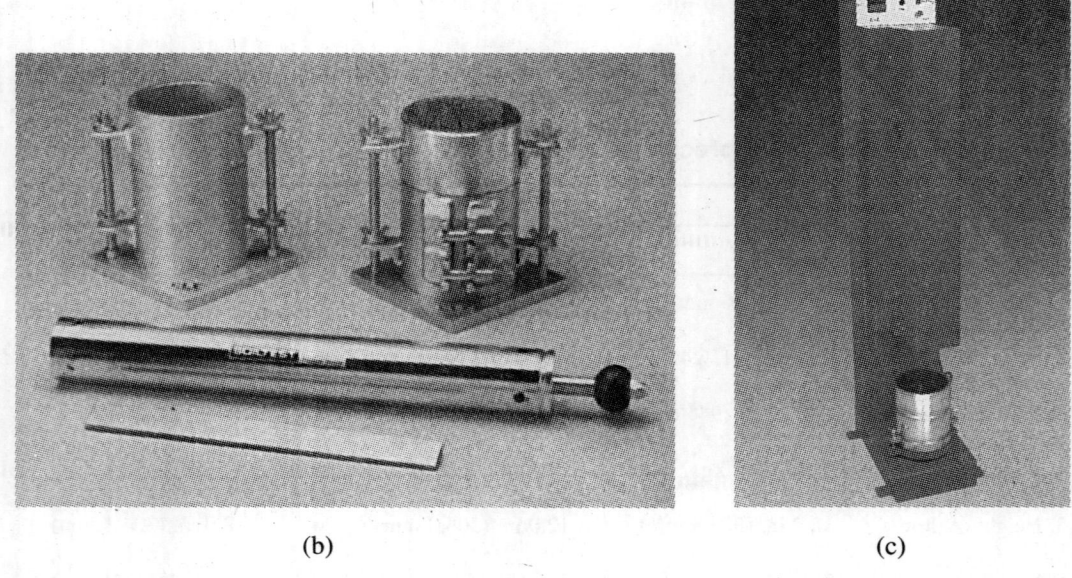

Fig. 13.1   Proctor compaction apparatus: (a) diagrammatic sketch,
(b) photograph of mould, and (c) automatic soil compactor (Courtesy: Soilest)

Modified Proctor Compaction Test (ASTM Designation: D1557)

This test method covers laboratory compaction procedures used to determine the relationship between water content and dry unit weight of soils (compaction curve) compacted in a 4 in. or 6 in. diameter mould with a 10 lb (5 kg) hammer dropped from a height of 18 in. (457 mm) producing a compactive effort of 56,250 ft-lb/ft^3 (2,700 kN-m/m^3). As in the case of the standard test, the code provides three alternative procedures based on the soil material tested. The details of the procedures are given in Table 13.2.

TABLE 13.2 Specification for modified proctor compaction test

Item		A	B	C
			Procedure	
1.	Mould diameter	4 in. (101.6mm)	4 in. (101.6 mm)	6 in. (101.6 mm)
2.	Volume of mould	0.0333 ft³ (944 cm3)	0.0333 ft³ (944 cm³)	0.075 ft³ (2124 cm³)
3.	Weight of hammer	10 lb (4.54 kg)	10 lb (4.54 kg)	10 lb (4.54 kg)
4.	Height of drop	18 in. (457.2mm)	18 in. (457.2mm)	18 in. (457.2mm)
5.	No. of layers	5	5	5
6.	Blows / layer	25	25	56
7.	Energy of compaction	56,250 ft lb/ft³ (2700 kN-m/m³)	56,250 ft lb/ft³ (2700 kN-m/m³)	56,250 ft lb/ft³ (2700 kN-m/m³)
8.	Soil material	May be used if 20% or less retained on No. 4 sieve.	Shall be used if 20% or more retained on No. 4 sieve and 20% or less retained on the 1/8 in. sieve	Shall be used if more-than 20% retained on 3/8 in. sieve and less-than 30% retained on the 3/4 in. sieve (19 mm)

Test Procedure

A soil at a selected water content is placed in five layers into a mould of given dimensions, with each layer compacted by 25 or 56 blows of a 10 lb (4.54 kg) hammer dropped from a height of 18 in. (457 mm) subjecting the soil to a total compactive effort of about 56,250 ft-lb/ft³ (2700 kN-m/m³). The resulting dry unit weight is determined. The procedure is repeated for a sufficient number of water contents to establish a relationship between the dry unit weight and the water content for the soil. This data, when plotted, represents a curvilinear relationship known as the compaction curve or moisture-dry unit weight curve. The value of the optimum water content and maximum dry unit weight are determined from the compaction curve as shown in Fig. 13.2.

Determination of Zero Air Voids Line

Referring to Fig. 13.3, we have

Degree of saturation $\qquad S = \dfrac{V_w}{V_v}$

Water content, $\qquad w = \dfrac{W_w}{W_s}$

Dry weight of solids, $\qquad W_s = V_s G \gamma_w = G \gamma_w \quad$ since $V_s = 1$

$$V_w = \frac{W_w}{\gamma_w} = \frac{w G \gamma_w}{\gamma_w} = wG$$

Therefore $\qquad S = \dfrac{wG}{V_v}$ (13.2)

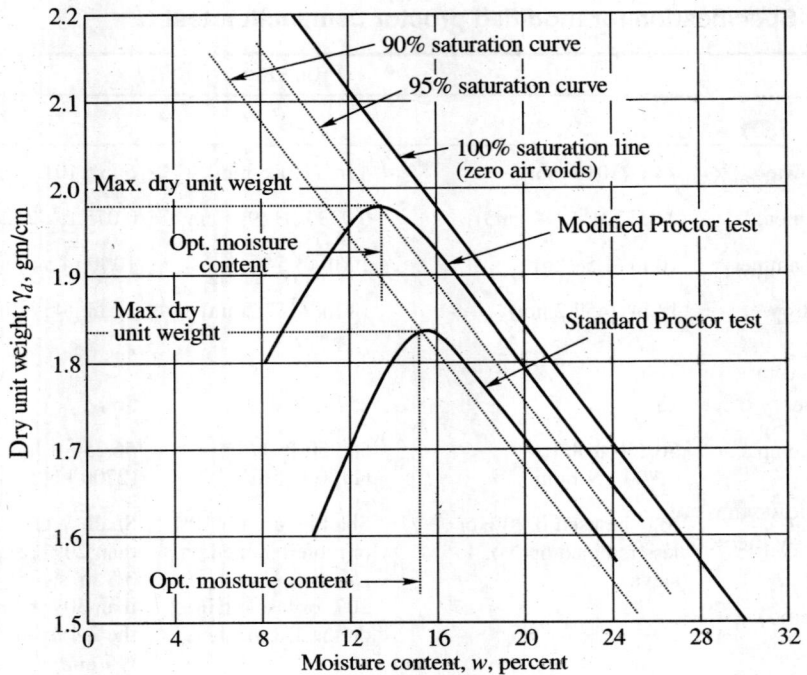

Fig. 13.2   Moisture-dry unit weight relationship

or
$$V_v = \frac{wG}{S}$$

Dry unit weight
$$\gamma_d = \frac{W_s}{1+V_v} = \frac{G\gamma_w}{1+\dfrac{wG}{S}} \qquad (13.3)$$

In Eq. (13.3), since G and γ_w, remain constant for a particular soil, the dry unit weight is a function of water content for any assumed degree of saturation. If $S = 1$, the soil is fully saturated (zero air voids). A curve giving the relationship between γ_d and w may be drawn by making use of Eq. (13.3) for $S = 1$. Curves may be drawn for different degrees of saturation such as 95, 90, 80, etc.

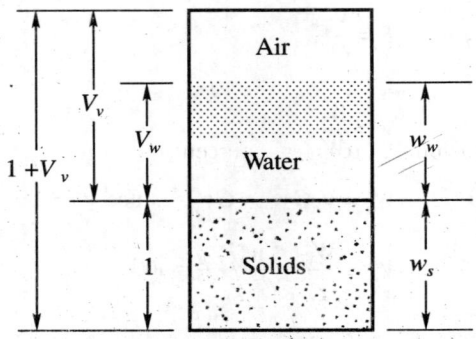

Fig.13.3   Block diagram for determining zero air voids line

percents. Fig. 13.2 gives typical curves for different degrees of saturation along with moisture-dry unit weight curves obtained by different compactive efforts.

Example 13.1

A proctor compaction test was conducted on a soil sample, and the following observations were made:

Water content, percent	7.7	11.5	14.6	17.5	19.7	21.2
Mass of wet soil, g	1739	1919	2081	2033	1986	1948

If the volume of the mould used was 950 cm³ and the specific gravity of soils grains was 2.65, make necessary calculations and draw, (i) compaction curve and (ii) 80% and 100% saturation lines.

Solution

From the known mass of the wet soil sample and volume of the mould, wet density or wet unit weight is obtained by the equations,

$$\rho_t (g/cm^3) = \frac{M}{V} = \frac{\text{Mass of wet sample in gm}}{950 \, cm^3} \text{ or } \gamma_t \approx 9.81 \times \rho_t (g/cm^3)$$

Then from the wet density and corresponding moisture content, the dry density or dry unit weight is obtained from,

$$\rho_d = \frac{\rho_t}{1+w} \text{ or } \gamma_d = \frac{\gamma_t}{1+w}$$

Thus for each observation, the wet density and then the dry density are calculated and tabulated as follows:

Water content, percent	7.7	11.5	14.6	17.5	19.7	21.2
Mass of wet sample, g	1739	1919	2081	2033	1986	1948
Wet density, g/cm³	1.83	2.02	2.19	2.14	2.09	2.05
Dry density, g/cm³	1.70	1.81	1.91	1.82	1.75	1.69
Dry unit weight kN/m³	16.7	17.8	18.7	17.9	17.2	16.6

Hence, the compaction curve, which is a plot between the dry unit weight and moisture content can be plotted as shown in the Fig. Ex. 13.1. The curve gives,

Maximum dry unit weight, $MDD = 18.7 \, kN/m^3$

Optimum moisture content, $OMC = 14.7$ percent

For drawing saturation lines, make use of Eq. (13.3), viz.,

$$\gamma_d = \frac{G\gamma_w}{1 + \frac{wG}{S}}$$

where, $G = 2.65$, given, S = degree of saturation 80% and 100%, w = water content, may be assumed as 8%, 12%, 16%, 20% and 24%.

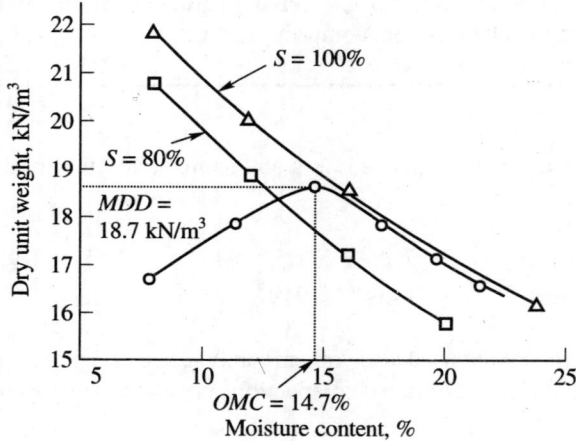

Fig. Ex. 13.1

Hence for each value of saturation and water content, find γ_d, and tabulate:

Water content, percentage	8	12	16	20	24
γ_d kN/m³ for $S = 100\%$	21.45	19.73	18.26	17.0	15.69
γ_d kN/m³ for $S = 80\%$	20.55	18.61	17.00	15.64	14.49

With these calculations, saturation lines for 100% and 80% are plotted, as shown in the Fig. Ex. 13.1.

Also the saturation, corresponding to $MDD = 18.7$ kN/m³ and $OMC = 14.7\%$ can be calculated as,

$$18.7 = \gamma_d = \frac{G\gamma_w}{1 + \dfrac{wG}{S}} = \frac{2.65 \times 9.81}{1 + \dfrac{0.147 \times 2.65}{S}}$$

which gives $S = 99.7\%$

Example 13.2

A small cylinder having volume of 600 cm³ is pressed into a recently compacted fill of embankment filling the cylinder. The mass of the soil in the cylinder is 1100 g. The dry mass of the soil is 910 g. Determine the void ratio and the saturation of the soil. Take the specific gravity of the soil grains as 2.7.

Solution

Wet density of soil

$$\rho_t = \frac{1100}{600} = 1.83 \, g/cm^3 \text{ or } \gamma_t = 17.99 \, kN/m^3$$

Water content, $$w = \frac{1100 - 910}{910} = \frac{190}{910} = 0.209 = 20.9\%$$

Dry unit weight, $\quad \gamma_d = \dfrac{\gamma_t}{1+w} = \dfrac{17.99}{1+0.209} = 14.88 \text{ kN/m}^3$

From, $\quad\quad\quad\quad \gamma_d = \dfrac{G\gamma_w}{1+e}$ we have $e = \dfrac{G\gamma_w}{\gamma_d} - 1$

Substituting and simplifying

$$e = \dfrac{2.7 \times 9.81}{14.88} - 1 = 0.78$$

From, $\quad\quad\quad\quad Se = wG, \quad \text{or} \quad S = \dfrac{wG}{e} = \dfrac{0.209 \times 2.7 \times 100}{0.78} = 72.35\%$

13.4 EFFECT OF COMPACTION ON ENGINEERING BEHAVIOUR

Effect of Moisture Content on Dry Density

The moisture content affects the behaviour of the soil. When the moisture content is low, the soil is stiff and difficult to compress. Thus, low unit weight and high air contents are obtained (Fig. 13.2). As the moisture content increases, the water acts as a lubricant, causing the soil to soften and become more workable. This results in a denser mass, higher unit weights and lower air contents under compaction. The water and air combination tend to keep the particles apart with further compaction, and prevent any appreciable decrease in the air content of the total voids, however, continue to increase with moisture content and hence the dry unit weight of the soil falls.

To the right of the peak of the dry unit weight-moisture content curve (Fig. 13.2), lies the saturation line. The theoretical curve relating dry density with moisture content with no air voids is approached but never reached since it is not possible to expel by compaction all the air entrapped in the voids of the soil.

Effect of Compactive Effort on Dry Unit Weight

For all types of soil with all methods of compaction, increasing the amounts of compaction, that is, the energy applied per unit weight of soil, results in an increase in the maximum dry unit weight and a corresponding decrease in the optimum moisture content as can be seen in Fig. 13.4.

Shear Strength of Compacted Soil

The shear strength of a soil increases with the amount of compaction applied. The more the soil is compacted, the greater is the value of cohesion and the angle of shearing resistance. Comparing the shearing strength with the moisture content for a given degree of compaction, it is found that the greatest shear strength is attained at a moisture content lower than the optimum moisture content for maximum dry unit weight. Fig. 13.5 shows the relationship between shear strength and moisture-dry unit weight curves for a sandy clay soil. It might be inferred from this that it would be an advantage to carry out compaction at the lower value of the moisture content. Experiments, however, have indicated that soils compacted in this way tend to take up moisture and become saturated with a consequent loss of strength.

Effect of Compaction on Structure

Figure 13.6 illustrates the effects of compaction on clay structure (Lambe, 1958a). Structure (or fabric) is the term used to describe the arrangement of soil particles and the electric forces between adjacent particles.

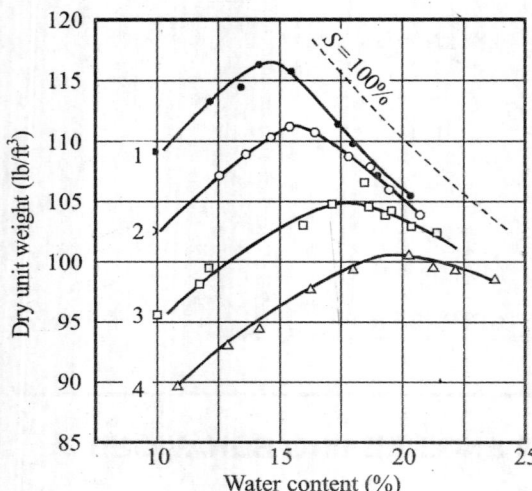

No.	Layers	Blows per layer	Hammer weight (lb)	Hammer drop (in.)
1	5	55	10	18 (mod. AASHTO)
2	5	26	10	18
3	5	12	10	18 (std. AASHTO)
4	3	25	5½	12

Note: 6 in. diameter mould used for all tests

Fig. 13.4 Dynamic compaction curves for a silty clay (from Turnbull, 1950)

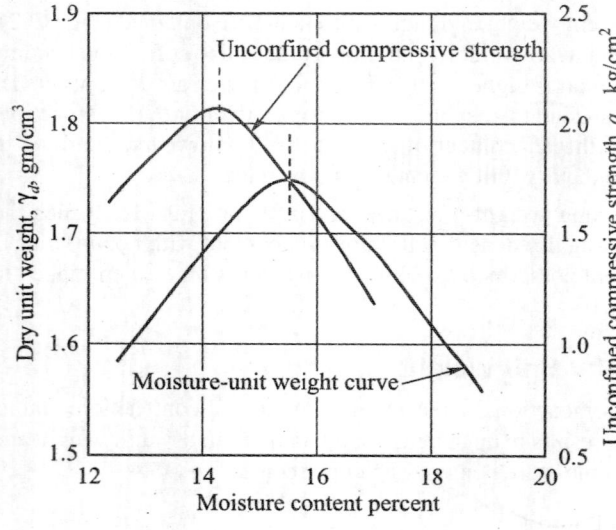

Fig. 13.5 Relationshipbetween compaction and shear strength curves

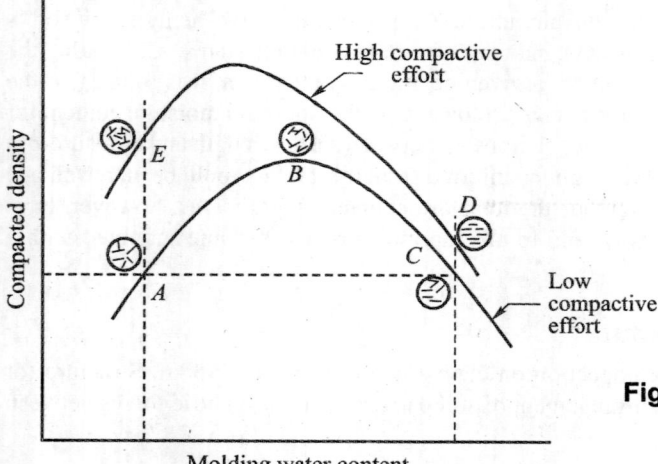

Fig. 13.6 Effects of compaction on structure(from Lambe, 1958a)

\ The effects of compaction conditions on soil structure, and thus on the engineering behaviour of the soil, vary considerably with soil type and the actual conditions under which the behaviour is determined.

At low water content, w_A in Fig. 13.6, the repulsive forces between particles are smaller than the attractive forces, and as such the particles flocculate in a disorderly array. As the water content increases beyond w_A, the repulsion between particles increases, permitting the particles to disperse, making particles arrange themselves in an orderly way. Beyond w_B the degree of particle parallelism increases, but the density decreases. Increasing the compactive effort at any given water content increases the orientation of particles and therefore gives a higher density as indicated in Fig. 13.6.

Effect of Compaction on Permeability

Figure 13.7 depicts the effect of compaction on the permeability of a soil. The figure shows the typical marked decrease in permeability that accompanies an increase in moulding water content on the dry side of the optimum water content. A minimum permeability occurs at water contents slightly above optimum moisture content (Lambe, 1958a), after which a slight increase in permeability occurs. Increasing the compactive effort decreases the permeability of the soil.

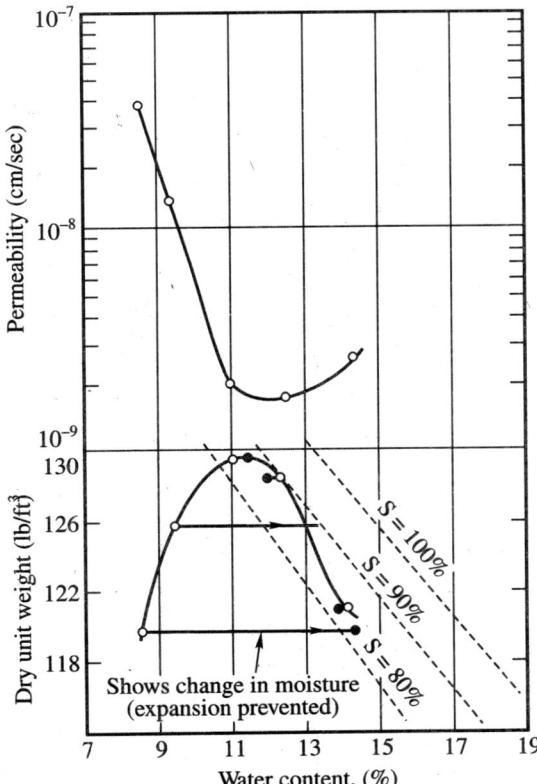

Fig. 13.7   Compaction-permeability tests on Siburua clay (from Lambe, 1962)

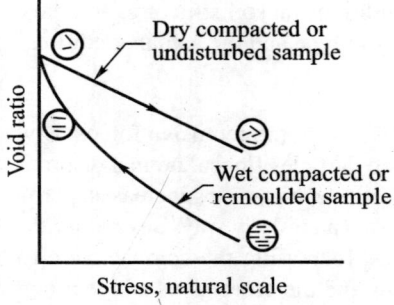

(a) Low-stress consolidation

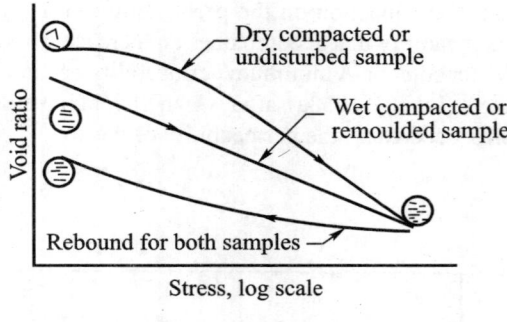

(b) High-stress consolidation

Fig. 13.8   Effect of one-dimensional compression on structure (Lambe, 1958b)

Effect of Compaction on Compressibility

Figure 13.8 illustrates the difference in compaction characteristics between two saturated clay samples at the same density, one compacted on the dry side of optimum and one compacted on the wet side (Lambe, 1958b). At low stresses the sample compacted on the wet side is more compressible than the one compacted on the dry side. However, at high applied stresses the sample compacted on the dryside is more compressible than the sample compacted on the wet side.

13.5   FIELD COMPACTION AND CONTROL

The necessary compaction of subgrades of roads, earth fills, and embankments may be obtained by mechanical means. The equipment that are normally used for compaction consists of

1.   Smooth wheel rollers

2.   Rubber tired rollers

3.   Sheepsfoot rollers

4.   Vibratory rollers

Laboratory tests on the soil to be used for construction in the field indicate the maximum dry density that can be reached and the corresponding optimum moisture content under specified methods of compaction. The field compaction method should be so adjusted as to translate laboratory condition

Fig. 13.9 Smooth wheel roller (Courtesy: Caterpillar, USA)

into practice as far as possible. The two important factors that are necessary to achieve the objectives in the field are

1. The adjustment of the natural moisture content in the soil to the value at which the field compaction is most effective.

2. The provision of compacting equipment suitable for the work at the site.

The equipment used for compaction are briefly described below:

Smooth Wheel Roller

There are two types of smooth wheel rollers. One type has two large wheels, one in the rear and a similar single drum in the front. This type is generally used for compacting base courses. The equipment weighs from 50 to 125 kN (Fig.13.9). The other type is the tandem roller normally used for compacting paving mixtures. This roller has large single drums in the front and rear and the weights of the rollers range from 10 to 200 kN.

Fig. 13.10   Sheepsfoot roller (Courtesy: Vibromax America Inc.)

Rubber Tired Roller

The maximum weight of this roller may reach 2000 kN. The smaller rollers usually have 9 to 11 tires on two axles with the tires spaced so that a complete coverage is obtained with each pass. The tire loads of the smaller roller are in the range of 7.5 kN and the tire pressures in the order of 200 kN/m². The larger rollers have tire loads ranging from 100 to 500 kN per tire, and tire pressures range from 400 to 1000 kN/m².

Sheepsfoot Roller

Sheepsfoot rollers are available in drum widths ranging from 120 to 180 cm and in drum diameters ranging from 90 to 180 cm. Projections like a sheepsfoot are fixed on the drums. The lengths of these projections range from 17.5 cm to 23 cm. The contact area of the tamping foot ranges from 35 to 56 sq. cm. The loaded weight per drum ranges from about 30 kN for the smaller sizes to 130 kN for the larger sizes (Fig. 13.10).

Vibratory Roller

The weights of vibratory rollers range from 120 to 300 kN. In some units vibration is produced by weights placed eccentrically on a rotating shaft in such a manner that the forces produced by the rotating weights are essentially in a vertical direction. Vibratory rollers are effective for compacting granular soils (Fig. 13.11).

Fig. 13.11   Vibratory drum on smooth wheel roller (Courtesy: Caterpillar, USA)

election of Equipment for Compaction in the Field

ne choice of a roller for a given job depends on the type of soil to be compacted and percentage of ompaction to be obtained. The types of rollers that are recommended for the soils normally met are:

Type of soil	Type of roller recommended
Cohesive soil	Sheepsfoot roller, or Rubber tired roller
Cohesionless soils	Rubber-tired roller or Vibratory roller.

1ethod of Compaction

he first approach to the problem of compaction is to select suitable equipment. If the compaction is :quired for an earth dam, the number of passes of the roller required to compact the given soil to the :quired density at the optimum moisture content has to be determined by conducting a field trial :st as follows:

The soil is well mixed with water which would give the optimum water content as determined in ie laboratory. It is then spread out in a layer. The thickness of the layer normally varies from 15 to 2.5 cm. The number of passes required to obtain the specified density has to be found by determining ie density of the compacted material after every definite number of passes. The density may be hecked for different thickness in the layer. The suitable thickness of the layer and the number of asses required to obtain the required density will have to be determined.

In cohesive soils, densities of the order of 95 percent of standard Proctor can be obtained wit
practically any of the rollers and tampers; however, vibrators are not effective in cohesive soil:
Where high densities are required in cohesive soils in the order of 95 percent of modified Procto
rubber tired rollers with tire loads in the order of 100 kN and tire pressure in the order of 600 kN/m
are effective.

In cohesionless sands and gravels, vibrating type equipment is effective in producing densitie
up to 100 percent of modified Proctor. Where densities are needed in excess of 100 percent c
modified Proctor such as for base courses for heavy duty air fields and highways, rubber tired roller
with tire loads of 130 kN and above and tire pressure of 1000 kN/m² can be used to produce densitie
up to 103 to 104 percent of modified Proctor.

Field Control of Compaction

Methods of Control of Density

The compaction of soil in the field must be such as to obtain the desired unit weight at the optimur
moisture content. The field engineer has therefore to make periodic checks to see whether th
compaction is giving desired results. The procedure of checking involves:

1. Measurement of the dry unit weight, and
2. Measurement of the moisture content.

There are many methods for determining the dry unit weight and/or moisture content of the soi
in-situ. The important methods are:

1. Core-Cutter method,
2. Sand cone method,
3. Rubber balloon method,
4. Nuclear method, and
5. Proctor needle method.

Core-Cutter Method

The apparatus consists of a cutter, dolly and a rammer as shown in Fig. 13.12. The cutter is driver
into the soil.

The dolly is placed over the cutter while driving to prevent damage to the edges of the cutter
When the cutter is just at the level of the ground, the cutter with the soil is then dug out of the groune
and any soil extruding from its ends is trimmed off so that the cutter contains volume of soil jus
equal to its internal volume. The mass of the contained soil is found out and its moisture conten
determined. The dry density of the soils is found out as follows:

Let

M_1 = Mass of cutter with soil

M_2 = Mass of cutter

V = Internal volume of cutter

w = Water content of the soil

The dry density ρ_d is

$$\rho_d = \frac{M}{V} \times \frac{1}{1+w} = \frac{\rho_t}{1+w}$$

where, $\dfrac{M}{V} = \rho_t$, the bulk density of the soil

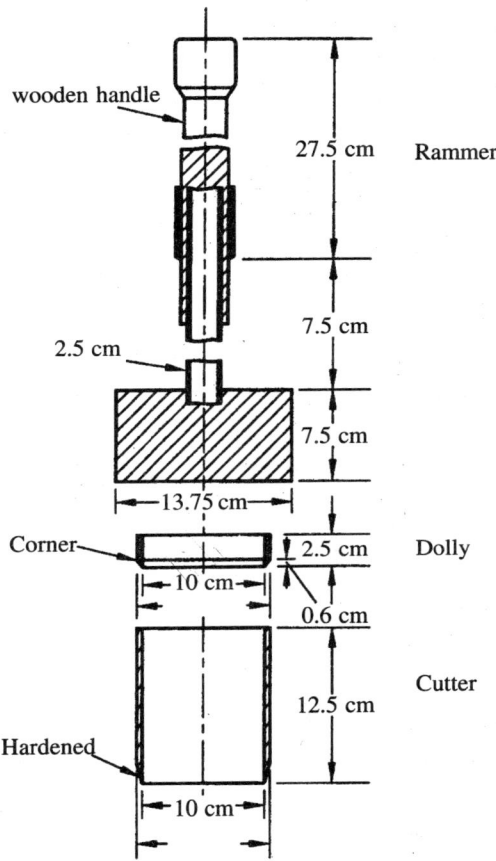

Fig. 13.12 Core-cutter apparatus

Sand Cone Method (ASTM Designation D-1556)

The sand for the sand cone method consists of a sand pouring jar shown in Fig. 13.13a. The jar contains uniformly graded clean and dry sand. A hole about 10 cm in diameter is made in the soil to be tested up to the depth required. The weight of soil removed from the hole is determined and its water content is also determined. Sand is run into the hole from the jar by opening the valve above the cone until the hole and the cone below the valve is completely filled. The valve is closed. The jar is calibrated to give the weight of the sand that just fills the hole, that is, the difference in weight of the jar before and after filling the hole after allowing for the weight of sand contained in the cone is the weight of sand poured into the hole.

Let

W_s = weight of dry sand poured into the hole
G = specific gravity of sand particles
W = weight of soil taken out of the hole
w = water content of the soil

Volume of sand in the hole = volume of soil taken out of the hole

that is, $V = \dfrac{W_s}{G\gamma_w}$ (13.4a)

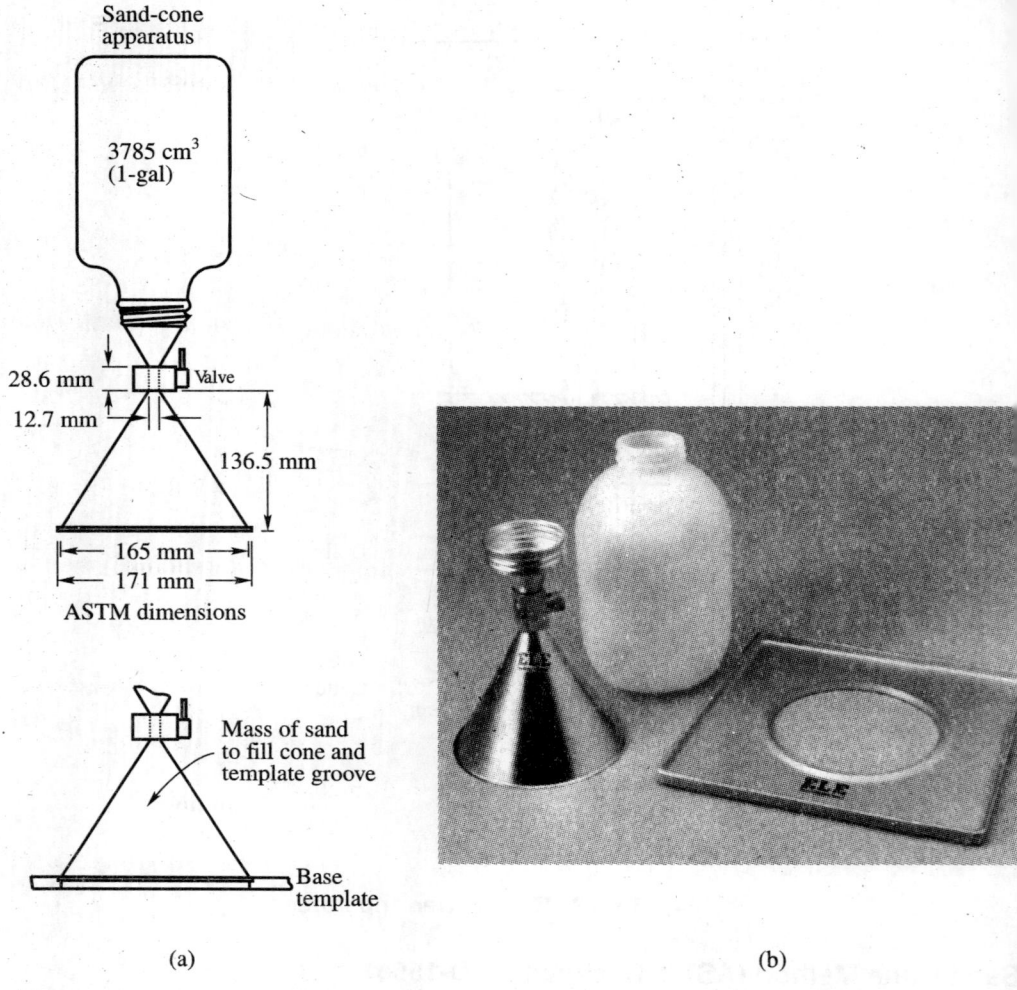

Sand-cone
apparatus

3785 cm³
(1-gal)

28.6 mm — Valve

12.7 mm

136.5 mm

165 mm

171 mm

ASTM dimensions

Mass of sand
to fill cone and
template groove

Base
template

(a)

(b)

Fig. 13.13   Sand-cone apparatus: (a) Schematic diagram, and (b) Photograph

The bulk unit weight of soil, $\gamma_t = \dfrac{W}{V} = \dfrac{WG\gamma_w}{W_s}$ (13.4)

The dry unit weight of soil, $\gamma_d = \dfrac{\gamma_t}{1+w}$

Rubber Balloon Method (ASTM Designation: D 2167)

The volume of an excavated hole in a given soil is determined using a liquid-filled calibrated cylinder
for filling a thin rubber membrane. This membrane is displaced to fill the hole. The inplace un
weight is determined by dividing the wet mass of the soil removed by the volume of the hole. Th
water (moisture) content and the in-place unit weight are used to calculate the in-place dry un
weight. The volume is read directly on the graduated cylinder. Fig. 13.13(c) shows the equipment

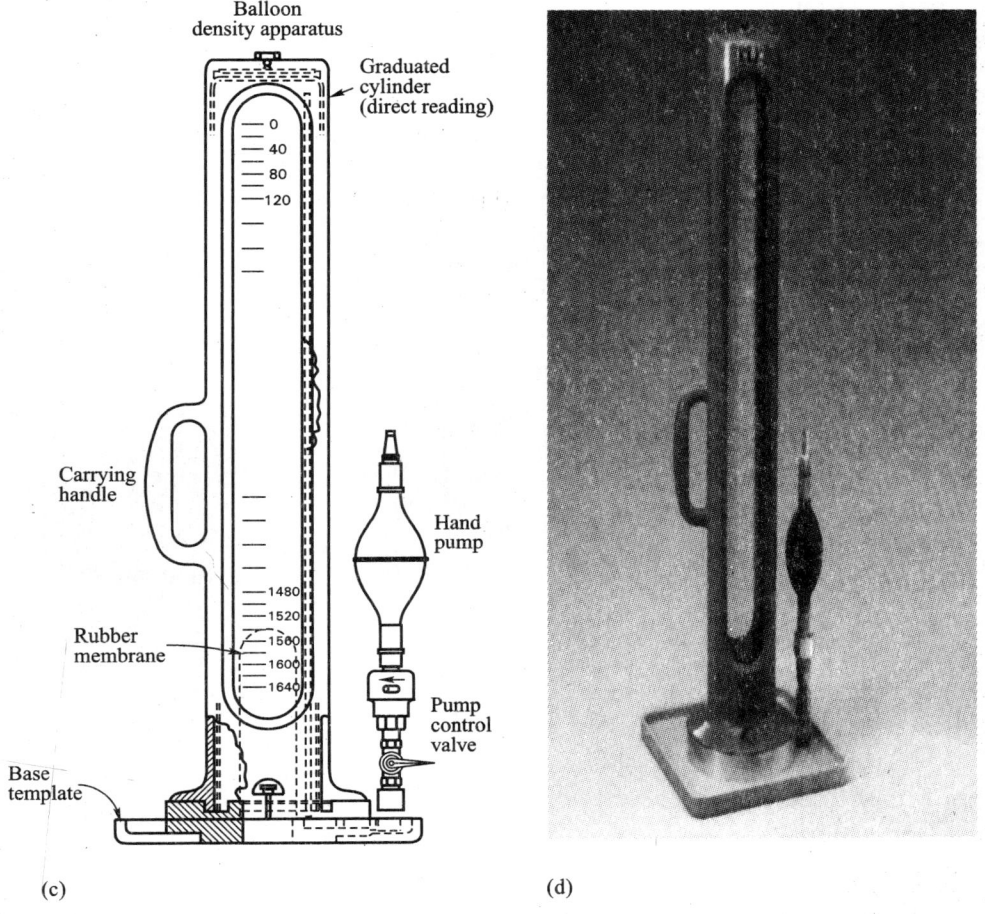

Fig. 13.13   Rubber balloon density apparatus, (c) diagrammatic sketch, and
(d) a photograph

Nuclear Method

The modern instrument for rapid and precise field measurement of moisture content and unit weight is the Nuclear density/Moisture meter. The measurements made by the meter are non-destructive and require no physical or chemical processing of the material being tested. The instrument may be used either in drilled holes or on the surface of the ground. The main advantage of this equipment is that a single operator can obtain an immediate and accurate determination of the *in-situ* dry density and moisture content.

Proctor Needle Method

The Proctor needle method is one of the methods developed for rapid determination of moisture contents of soils *in-situ*. It consists of a needle attached to a spring loaded plunger, the stem of which is calibrated to read the penetration resistance of the needle in lbs/in² or kg/cm². The needle is supplied with a series of bearing points so that a wide range of penetration resistances can be measured. The bearing areas that are normally provided are 0.05, 0.1, 0.25, 0.50 and 1.0 sq. in. The apparatus is shown in Fig. 13.14. A Proctor penetrometer set is shown in Fig. 13.15 (ASTMD–1558).

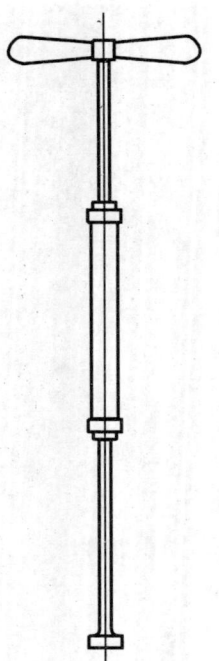

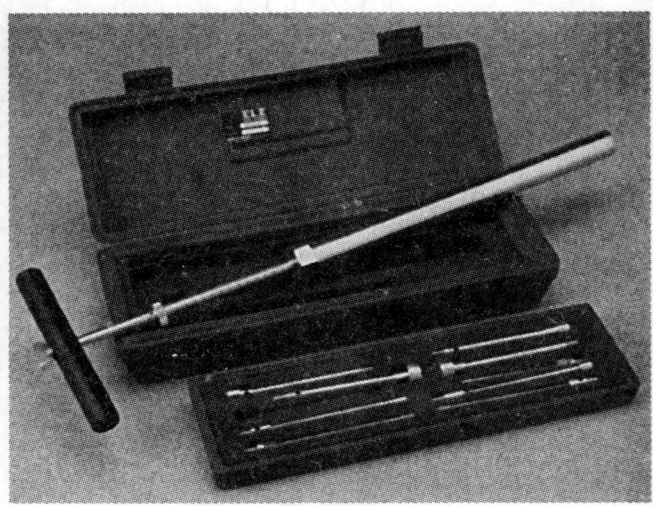

Fig. 13.14   Proctor needle   **Fig. 13.15**   Proctor penetrometer set (Courtesy: Soiltest)

Laboratory Penetration Resistance Curve

A suitable needle point is selected for a soil to be compacted. If the soil is cohesive, a needle with a larger bearing area is selected. For cohesionless soils, a needle with a smaller bearing area will be sufficient. The soil sample is compacted in the mold.

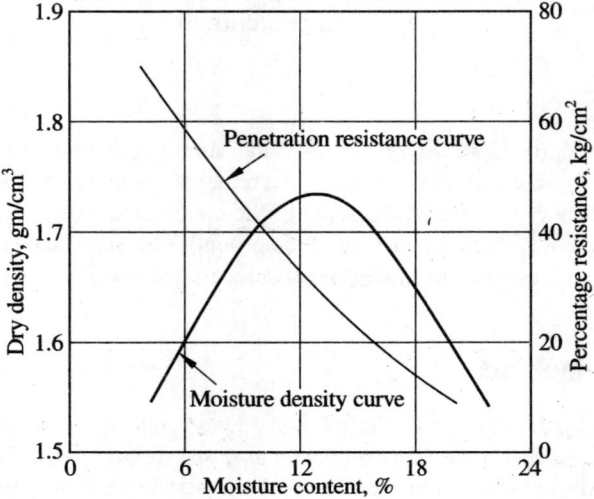

Fig. 13.16   Field method of determining water content by Proctor needle method

The penetrometer with a known bearing area of the tip is forced with a gradual uniform push at a rate of about 1.25 cm per sec to a depth of 7.5 cm into the soil. The penetration resistance in kg/cm^2 is read off the calibrated shaft of the penetrometer. The water content of the soil and the corresponding dry density are also determined. The procedure is repeated for the same soil compacted at different moisture contents. Curves giving the moisture-density and penetration resistance-moisture content relationship are plotted as shown in Fig. 13.16.

To determine the moisture content in the field, a sample of the wet soil is compacted into the mould under the same conditions as used in the laboratory for obtaining the penetration resistance curve. The Proctor needle is forced into the soil and its resistance is determined. The moisture content is read from the laboratory calibration curve.

This method is quite rapid, and is sufficiently accurate for fine-grained cohesive soils. However, the presence of gravel or small stones in the soil makes the reading on the Proctor needle less reliable. It is not very accurate in cohesionless sands.

Example 13.3

The following observations were recorded when a sand cone test was conducted for finding the unit weight of a natural soil:

Total density of sand used in the test = 1.4 g/cm^3

Mass of the soil excavated from hole = 950 g.

Mass of the sand filling the hole = 700 g.

Water content of the natural soil = 15 percent.

Specific gravity of the soil grains = 2.7

Calculate: (i) the wet unit weight, (ii) the dry unit weight, (iii) the void ratio, and (iv) the degree of saturation.

Solution

Volume of the hole
$$V_p = \frac{700}{1.4} = 500 \text{ cm}^3$$

Wet density of natural soil,
$$\rho_t = \frac{950}{500} = 1.9 \text{ g/cm}^3 \text{ or } \gamma_t = 18.64 \text{ kN/m}^3$$

Dry density
$$\rho_d = \frac{\rho_t}{1+w} = \frac{1.9}{1+0.15} = 1.65 \text{ g/cm}^3$$

$$\rho_d = \frac{G}{1+e}\rho_w = \frac{2.7}{1+e} \times 1 \text{ or } 1.65 + 1.65e = 2.7$$

Therefore
$$e = \frac{2.7-1.65}{1.65} = 0.64$$

And
$$S = \frac{wG}{e} = \frac{0.15 \times 2.7 \times 100}{0.64} = 63\%$$

Example 13.4

Old records of a soil compacted in the past gave compaction water content of 15% and saturation 85%. What might be the dry density of the soil?

Solution

The specific gravity of the soil grains is not known, but as it varies in a small range of 2.6 to 2.7, it can suitably be assumed. An average value of 2.65 is considered here.

Hence

$$e = \frac{wG}{S} = \frac{0.15 \times 2.65}{0.85} = 0.47$$

and dry density $\rho_d = \frac{G}{1+e} \rho_w = \frac{2.65}{1+0.47} \times 1 = 1.8 \text{ g/cm}^3$ or dry unit weight $= 17.66 \text{ kN/m}^3$

Example 13.5

The following data are available in connection with the construction of an embankment:

(a) **Soil from borrow pit:** Natural density = 1.75 Mg/m³, Natural water content = 12%

(b) **Soil after compaction:** density = 2 Mg/m³, water content = 18%.

For every 100 m³ of compacted soil of the embankment, estimate:

(i) The quantity of soil to be excavated from the borrow pit, and

(ii) The amount of water to be added

Note: 1 g/cm³ = 1000 kg/m³ = $10^3 \times 10^3$ g/m³ = 1 Mg/m³ where Mg stands for Megagram = 10^6 g.

Solution

The soil is compacted in the embankment with density of 2 Mg/m³ and with 18% water content.

Hence, for 100 m³ of soil

Mass of compacted wet soil = 100 × 2.0 = 200 Mg = 200 × 10³ kg.

Mass of contacted dry soil $= \frac{200}{1+w} = \frac{200}{1+0.18} = 169.5 \text{ Mg} = 169.5 \times 10^3 \text{ kg}$

Mass of wet soil to be excavated = 169.5 (1+ w) = 169.2 (1+ 0.12) = 189.84 Mg

Volume of the wet soil to be excavated $= \frac{189.84}{1.75} = 108.48 \text{ m}^3$

Now, in the natural state, the moisture present in 169.5 × 10³ kg of dry soil would be

169.5 × 10³ × 0.12 = 20.34 × 10³ kg

and the moisture which the soil will possess during compaction is

169.5 × 10³ × 0.18 = 30.51 × 10³ kg

Hence, mass of water to be added for every 100 m³ of compacted soil is

(30.51 − 20.34) 10³ = 10.17 × 10³ kg.

Example 13.6

A sample of soil compacted according to the standard Proctor test has a density of 2.06 g/cm³ at 100% compaction and at an optimum water content of 14%. What is the dry unit weight? What is the dry unit weight at zero air-voids? If the voids become filled with water what would be the saturated unit weight? Assume $G = 2.67$.

Solution

Refer to Fig. Ex. 13.6. Assume $V =$ total volume = 1 cm³. Since water content is 14% we may write,

$$\frac{M_w}{M_s} = 0.14$$

and since, $M_w + M_s = 2.06$ g

$$0.14 M_w + M_s = 0.14 M_w = 2.06$$

or $\quad M_s = \dfrac{2.06}{1.14} = 1.807$ g

$$M_w = 0.14 \times 1.807 = 0.253 \text{ g.}$$

By definition, $\rho_d = \dfrac{M_s}{V} = \dfrac{1.87}{1} = 1.807$ g/cm³ or $\gamma_d = 1.807 \times 9.81 = 17.73$ kN/m³

The volume of solids (Fig. 13.6) is

$$V_s = \frac{1.807}{2.67} = 0.68 \text{ g/cm}^3$$

The volume of voids = $1 - 0.68 = 0.32$ cm³

The volume of water = 0.253 cm³

The volume of air = $0.320 - 0.253 = 0.067$ cm³

If all the air is squeezed out of the samples the dry density at zero air voids would be, by definition,

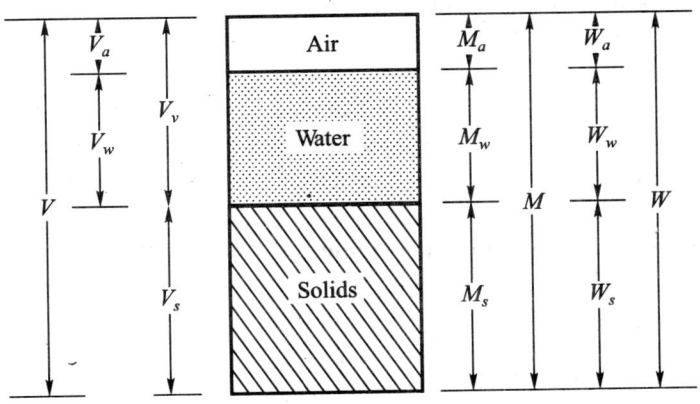

Fig. Ex. 13.6

$$\rho_d = \frac{1.807}{0.68 + 0.253} = 1.94 \text{ g/cm}^3 \quad \text{or} \quad \gamma_d = 1.94 \times 9.81 = 19.03 \text{ kN/m}^3$$

on the other hand, if the air voids also were filled with water,

The mass of water would be = $0.32 \times 1 = 0.32$ g

The saturated density is

$$\rho_{sat} = \frac{1.807 + 0.32}{1} = 2.13 \text{ g/cm}^3$$

13.6 COMPACTION FOR DEEPER LAYERS OF SOIL

Three types of dynamic compaction for deeper layers of soil are discussed here. They are:

1. Vibroflotation.
2. Dropping of a heavy weight.
3. Blasting.

Vibroflotation

The vibroflotation technique is used for compacting granular soil only. The vibroflot is a cylindrical tube containing water jets at top and bottom and equipped with a rotating eccentric weight, which develops a horizontal vibratory motion as shown in Fig. 13.17. The vibroflot is sunk into the soil using the lower jets and is then raised in successive small increments, during which the surrounding material is compacted by the vibration process. The enlarged hole around the vibroflot is backfilled with suitable granular material. This method is very effective for increasing the density of a sand deposit for depths up to 30 m. Probe spacings of compaction holes should be on a grid pattern of about 2 m to produce relative densities greater than 70 percent over the entire area. If the sand is coarse, the spacings may be somewhat larger.

In soft cohesive soil and organic soils the vibroflotation technique has been used with gravel as the backfill material. The resulting densified stone column effectively reinforces softer soils and acts as a bearing pile for foundations.

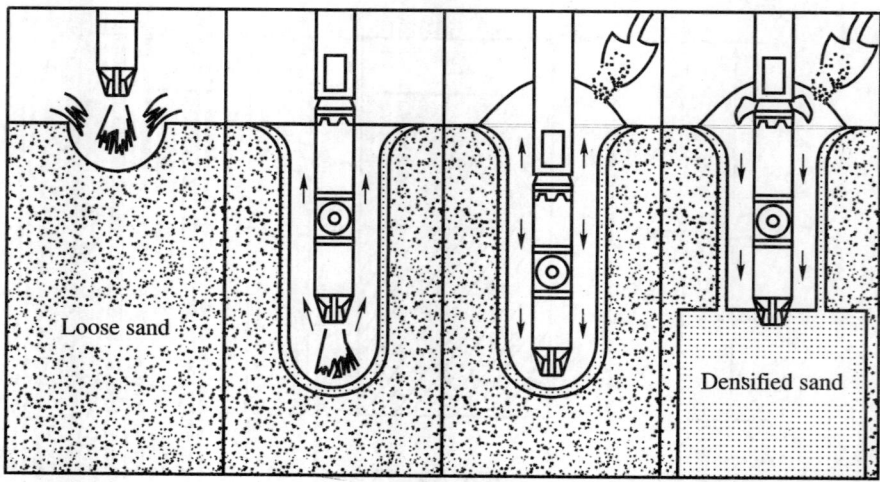

Fig. 13.17   Compaction by using vibroflot (Brown, 1977)

Dropping of a Heavy Weight

The repeated dropping of a heavy weight on to the ground surface is one of the simplest of the methods of compacting loose soil.

The method, known as deep dynamic compaction or deep dynamic consolidation may be used to compact cohesionless or cohesive soils. The method uses a crane to lift a concrete or steel block, weighing up to 500 kN and up to heights of 40 to 50 m, from which height it is allowed to fall freely on to the ground surface. The weight leaves a deep pit at the surface. The process is then repeated either at the same location or sequentially over other parts of the area to be compacted. When the required number of repetitions is completed over the entire area, the compaction at depth is completed. The soils near the surface, however, are in a greatly disturbed condition. The top soil may then be levelled and compacted, using normal compactiing equipment. The principal claims of this method are:

1. Depth of recompaction can reach up to 10 to 12 m.

2. All soils can be compacted.

3. The method produces equal settlements more quickly than do static (surcharge type) loads.

The depth of recompaction, D, in meters is approximately given by Leonards, et al., (1980) as

$$D \approx \frac{1}{2}(Wh)^{1/2} \tag{13.5}$$

where W = weight of falling mass in metric tons,

h = height of drop in meters.

Blasting

Blasting, through the use of buried, time-delayed explosive charges, has been used to densify loose, granular soils. The sands and gravels must be essentially cohesionless with a maximum of 15 percent of their particles passing the No. 200 sieve size and 3 percent passing 0.005 mm size. The moisture condition of the soil is also important for surface tension forces in the partially saturated state limit the effectiveness of the technique. Thus the soil, as well as being granular, must be dry or saturated, which requires sometimes prewetting the site via construction of a dike and reservoir system.

The technique requires careful planning and is used at a remote site. Theoretically, an individual charge densifies the surrounding adjacent soil and soil beneath the blast. It should not lift the soil situated above the blast, however, since the upper soil should provide a surcharge load. The charge should not create a crater in the soil. Charge delays should be timed to explode from the bottom of the layer being densified upward in a uniform manner. The uppermost part of the stratum is always loosened, but this can be surface-compacted by vibratory rollers. Experience indicates that repeated blasts of small charges are more effective than a single large charge for achieving the desired results.

13.7 PRELOADING

Preloading is a technique that can successfully be used to densify soft to very soft cohesive soils. Large-scale construction sites composed of weak silts and clays or organic materials (particularly marine deposits), sanitary land fills, and other compressible soils may often be stabilized effectively and economically by preloading. Preloading compresses the soil. Compression takes place when the water in the pores of the soil is removed which amounts to artificial consolidation of soil in the field. In order to remove the water squeezed out of the pores and hasten the period of consolidation, horizontal and vertical drains are required to be provided in the mass. The preload is generally in

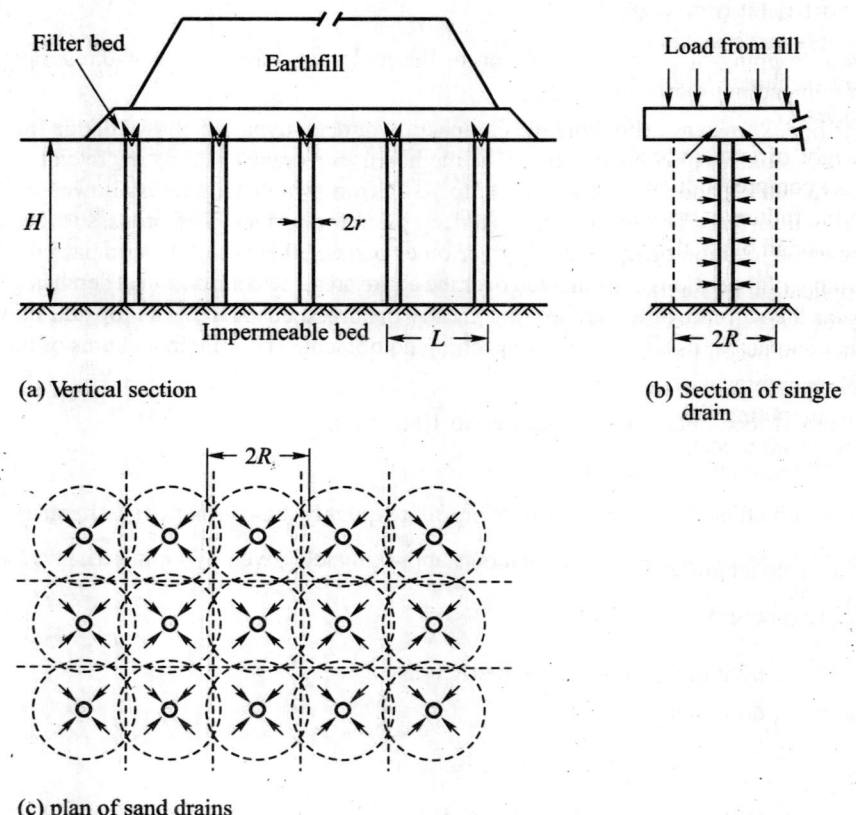

Filter bed

Earthfill

Load from fill

H

$2r$

Impermeable bed $\leftarrow L \rightarrow$

$\leftarrow 2R \rightarrow$

(a) Vertical section

(b) Section of single drain

$\leftarrow 2R \rightarrow$

(c) plan of sand drains

Fig. 13.18   Consolidation of soil using sand drains

the form of an imposed earth fill which must be left in place long enough to induce consolidation. The process of consolidation can be checked by providing suitable settlement plates and piezometers. The greater the surcharge load, shorter the time for consolidation. This is a case of three-dimension consolidation.

Two types of vertical drains considered are

1.   Cylindrical sand drains
2. · Wick (prefabricated vertical) drains

Sand Drains

Vertical and horizontal and drains are normally used for consolidating very soft clay, silt and other compressible materials. The arrangement of sand drains shown in Fig. 13.18 is explained below:

1.   It consists of a series of vertical sand drains or piles. Normally medium to coarse sand is used.
2.   The diameter of the drains are generally not less than 30 cm and the drains are placed in a square grid pattern at distances of 2 to 3 meters apart. Economy requires a careful study of the effect of spacing the sand drains on the rate of consolidation.
3.   Depth of the vertical drains should extend up to the thickness of the compressible stratum.

4. A horizontal blanket of free draining sand should be placed on the top of the stratum and the thickness of this may be up to a meter, and

5. Soil surcharge in the form of an embankment is constructed on top of the sand blanket in stages.

The height of surcharge should be so controlled as to keep the development of pore water pressure in the compressible strata at a low level. Rapid loading may induce high pore water pressures resulting in the failure of the stratum by rupture. The lateral displacement of the soil may shear off the sand drains and block the drainage path.

The application of surcharge squeezes out water in radial directions to the nearest sand drain and also in the vertical direction to the sand blanket. The dashed lines shown in Fig. 13.18(b) are drawn midway between the drains. The planes passing through these lines may be considered as impermeable membranes and all the water within a block has to flow to the drain at the centre. The problem of computing the rate of radial drainage can be simplified without appreciable error by assuming that each block can be replaced by a cylinder of radius R such that

$$\pi R^2 = L^2$$

where L is the side length of the prismatic block.

The relation between the time t and degree of consolidation $U_z\%$ is determined by the equation

$$U_z\% = 100\, f(T)$$

wherein,

$$T = \frac{c_v t}{H^2} \tag{13.6}$$

If the bottom of the compressible layer is impermeable, then H is the full thickness of the layer.

For radial drainage, Rendulic (1935) has shown that the relation between the time t and the degree of consolidation $U_r\%$ can be expressed as

$$U_r\% = 100\, f(T) \tag{13.7}$$

wherein,

$$T_r = \frac{c_v r}{4R^2} t \tag{13.8}$$

is the time factor. The relation between the degree of consolidation $U_r\%$ and the time factor T_r depends on the value of the ratio R/r. The relation between T_r and $U_r\%$ for ratios of R/r equal to 1, 10 and 100 in Fig. 13.19 are expressed by curves C_1, C_{10} and C_{100} respectively.

Installation of Vertical Sand Drains

The sand drains are installed as follows

1. A casing pipe of the required diameter with the bottom closed with a loose-fit-cone is driven up to the required depth,

2. The cone is slightly separated from the casing by driving a mandrel into the casing, and

3. The sand of the required gradation is poured into the pipe for a short depth and at the same time the pipe is pulled up in steps. As the pipe is pulled up, the sand is forced out of the pipe by applying pressure on to the surface of the sand. The procedure is repeated till the holes is completely filled with sand.

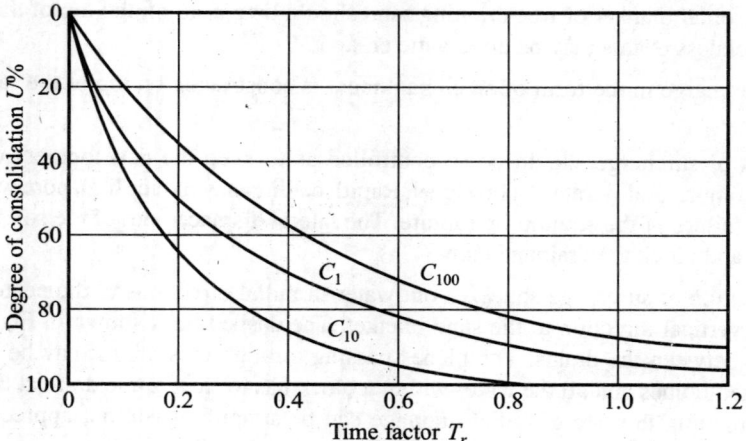

Fig. 13.19 T_r versus U_r

The sand drains may also be installed by jetting a hole in the soil or by driving an open casing into the soil, washing the soil out of the casing, and filling the hole with sand afterwards.

Sand drains have been used extensively in many parts of the world for stabilizing soils for port development works and for foundations of structures in reclaimed areas on the sea coasts. It is possible that sand drains may not function satisfactorily if the soil surrounding the well gets remoulded. This condition is referred to as *smear*. Though theories have been developed by considering different thickness of smear and different permeability, it is doubtful whether such theories are of any practical use since it would be very difficult to evaluate the quality of the smear in the field.

Wick (Prefabricated Vertical) Drains

Geocomposites used as drainage media have completely taken over certain geotechnical application areas. Wick drains, usually consisting of plastic fluted or nubbed cores that are surrounded by a geotextile filter, have considerable tensile strength. Wick drains do not require any sand to transmit flow. Most synthetic drains are of a strip shape. The strip drains are generally 100 mm wide and 2 to 6 mm thick. Fig. 13.20 shows typical core shapes of strip drains (Hausmann, 1990).

Wick drains are installed by using a hollow lance. The wick drain is threaded into a hollow lance, which is pushed (or driven) through the soil layer, which collapses around it. At the ground surface the ends of the wick drains (typically at 1 to 2 m spacing) are interconnected by a granular soil drainage layer or geocomposite sheet drain layer. There are a number of commercially available wick drain manufacturers and installation contractors who provide information on the current products, styles, properties, and estimated costs (Koerner, 1999).

With regards to determining wick drain spacings, the initial focal point is on the time for the consolidation of the subsoil to occur. Generally the time for 90% consolidation ($t90$) is desired. In order to estimate the time t, it is first necessary to estimate an equivalent sand drain diameter for the wick drain used. The equations suggested by Koerner (1999) are

$$d_{sd} = \sqrt{\frac{d_v^2}{n_s}}$$

(13.9a)

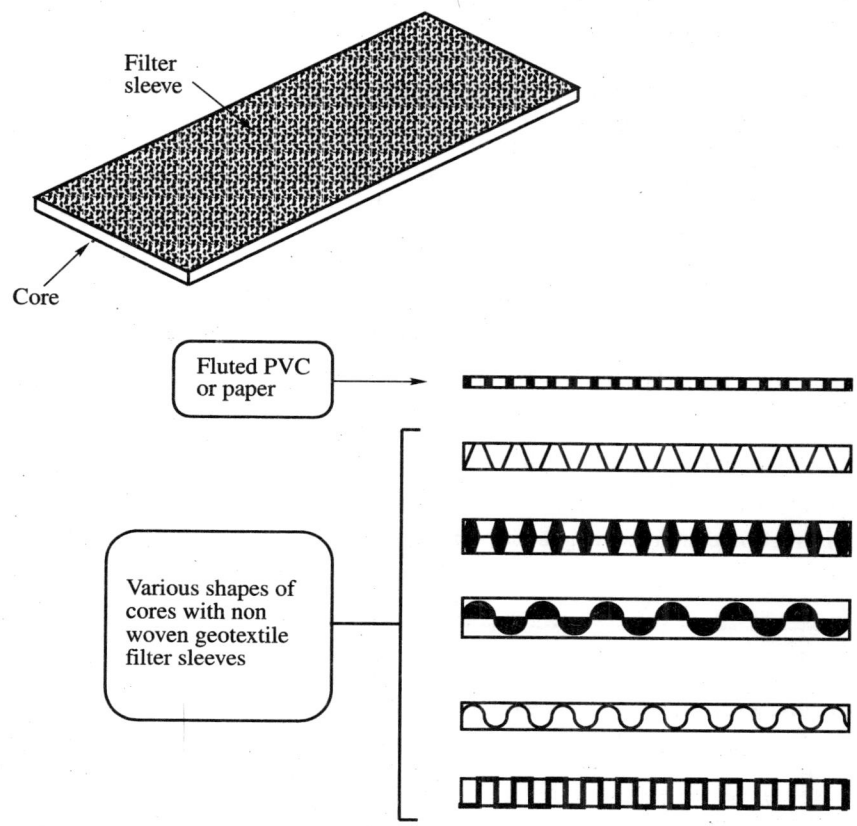

Fig. 13.20 Typical core shapes of strip drains (Hausmann, 1990)

$$d_v = \sqrt{\frac{4btn_d}{\pi}}$$

(13.9b)

where d_{sd} = equivalent sand drain diameter

d_v = equivalent void circle diameter

b, t = width and thickness of the wick drain

n_s = porosity of sand drain

$$n_d = \frac{\text{Void area of wick drain}}{\text{total cross sectional area of strip}} = \frac{\text{Void area of wick drain}}{b \times t}$$

It may be noted here that equivalent sand drain diameters for various commercially available wick drains vary from 30 to 50 mm (Korner, 1999).

The equation for estimating the time t for consolidation is (Koerner, 1999)

$$t = \frac{D^2}{8c_h}\left[\ln\frac{D}{d} - 0.75\right]\ln\frac{1}{1-U}$$

(13.10)

where t = time for consolidation

c_h = coefficient of consolidation of soil for horizontal flow

d = equivalent diameter of strip drain

$$= \frac{\text{circumference}}{\pi}$$

D = sphere of influence of the strip drain;
      (a) for a triangular pattern, $D = 1.05 \times$ spacing D_t
      (b) for a square pattern, $D = 1.13 \times$ spacing D_s

D_t = distance between drains in triangular spacing and
D_s = distance for square pattern
U = average degree of consolidation

Advantages of Using Wick Drains (Koerner, 1999)

1. The analytic procedure is available and straightforward in its use.
2. Tensile strength is definitely afforded to the soft soil by the installation of the wick drains.
3. There is only nominal resistance to the flow of water if it enters the wick drain.
4. Construction equipment is generally small.
5. Installation is simple, straightforward and economic.

Example 13.7

What is the equivalent sand drain diameter of a wick drain measuring 96 mm wide and 2.9 mm thick that is 92% void in its cross section? Use an estimated sand porosity of 0.3 for typical sand in a sand drain.

Solution

Total area of wick drain = $b \times t = 96 \times 2.9 = 279$ mm^2
    Void area of wick drain = $n_d \times b \times t = 0.92 \times 279 = 257$ mm^2
    The equivalent circle diameter (Eq. 13.9b) is

$$d_v = \sqrt{\frac{4btn_d}{\pi}} = \sqrt{\frac{4 \times 257}{3.14}} = 18.1 \text{ mm}$$

the equivalent sand drain diameter (Eq. 13.9a)

$$d_{sd} = \sqrt{\frac{d_v^2}{n_s}} = \sqrt{\frac{18.1^2}{0.3}} = 33 \text{ mm}$$

Example 13.8

Calculate the times required for 50, 70 and 90% consolidation of a saturated clayey silt soil using wick drains at various triangular spacings. The wick drains measure 100×4 mm and the soil has a c_h = 6.5 10^{-6} m^2/min.

Solution

In the simplified formula the equivalent diameter d of a strip drain is

$$d = \frac{\text{circumference}}{\pi} = \frac{100 + 100 + 4 + 4}{3.14} = 66.2 \text{ mm}$$

Using Eq. (13.10)

$$t = \frac{D^2}{8c_h}\left[\ln\frac{D}{d} - 0.75\right]\ln\left(\frac{1}{1-U}\right)$$

substituting the known values

$$t = \frac{D^2}{8\,(6.5\times10^{-6})}\left[\ln\frac{D}{0.0062} - 0.75\right]\ln\left(\frac{1}{1-U}\right)$$

The times required for the various degrees of consolidation are tabulated below for assumed theoretical spacings of wick drains.

Wick drain spacings D(m)	Time in days for various degrees of consolidation (U)		
	50%	70%	90%
2.1	110	192	367
1.8	77	133	254
1.5	49	86	164
1.2	29	50	95
0.9	14	24	46
0.6	4.8	8.4	16
0.3	0.6	1.1	2.1

For the triangular pattern, the spacing D_t is

$$D_t = \frac{D}{1.05}.$$

13.8 SAND COMPACTION PILES AND STONE COLUMNS

Sand Compaction Piles

Sand compaction piles consists of driving a hollow steel pipe with the bottom closed with a collapsible plate down to the required depth; filling it with sand, and withdrawing the pipe while air pressure is directed against the sand inside it. The bottom plate opens during withdrawal and the sand backfills the voids created earlier during the driving of the pipe. The in-situ soil is densified while the pipe is being withdrawn, and the sand backfill prevents the soil surrounding the compaction pipe from collapsing as the pipe is withdrawn. The maximum limits on the amount of fines that can be present are 15 percent passing the No. 200 sieve (0.075 mm) and 3 percent passing 0.005 mm. The distance between the piles may have to be planned according to the site conditions.

Stone Columns

The method described for installing sand compaction piles or the vibroflot described earlier can be used to construct stone columns. The size of the stones used for this purpose range from about 6 to 40 mm. Stone columns have particular application in soft inorganic, cohesive soils and are generally inserted on a volume displacement basis.

The diameter of the pipe used either for the construction of sand drains or sand compaction piles can be increased according to the requirements. Stones are placed in the pipe instead of sand, and the technique of constructing stone columns remains the same as that for sand piles.

Stone columns are placed 1 to 3 m apart over the whole area. There is no theoretical procedure for predicting the combined improvement obtained, so it is usual to assume the foundation loads are carried only by the several stone columns with no contribution from the intermediate ground (Bowles, 1996).

Bowles (1996) gives an approximate formula for the allowable bearing capacity of stone columns as

$$q_a = \frac{K_P}{F_s}(4c + \sigma'_r) \qquad \qquad (13.11)$$

where $K_P = \tan^2(45° + \phi'/2)$,

$\phi' =$ drained angle of friction of stone,

$c =$ either drained cohesion (suggested for large areas) or the undrained shear strength c_u,

$\sigma'_r =$ effective radial stress as measured by a pressuremeter (but may use $2c$ if pressuremeter data are not available),

$F_s =$ factor of safety, 1.5 to 2.0.

The total allowable load on a stone column of average cross-section area A_c is

$$Q_a = q_a A_c \qquad \qquad (13.12)$$

Stone columns should extend through soft clay to firm strata to control settlements. There is no end bearing in Eq. (13.11) because the principal load carrying mechanism is local perimeter shear.

Settlement is usually the principal concern with stone columns since bearing capacity is usually quite adequate (Bowles, 1996). There is no method currently available to compute settlement on a theoretical basis.

Stone columns are not applicable to thick deposits of peat or highly organic silts or clays (Bowles, 1996). Stone columns can be used in loose sand deposits to increase the density.

13.9   SOIL STABILIZATION BY THE USE OF ADMIXTURES

The physical properties of soils can often economically be improved by the use of admixtures. Some of the more widely used admixtures include lime, portland cement and asphalt. The process of soil stabilization first involves mixing with the soil a suitable additive which changes its property and then compacting the admixture suitably. This method is applicable only for soils in shallow foundations or the base courses of roads, airfield pavements, etc.

Soil-lime Stabilization

Lime stabilization improves the strength, stiffness and durability of fine grained materials. In addition, lime is sometimes used to improve the properties of the fine grained fraction of granular soils. Lime has been used as a stabilizer for soils in the base courses of pavement systems, under concrete foundations, on embankment slopes and canal linings.

Adding lime to soils produces a maximum density under a higher optimum moisture content than in the untreated soil. Moreover, lime produces a decrease in plasticity index.

Lime stabilization has been extensively used to decrease swelling potential and swelling pressures in clays. Ordinarily the strength of wet clay is improved when a proper amount of lime is added. The improvement in strength is partly due to the decrease in plastic properties of the clay and partly to the pozzolanic reaction of lime with soil, which produces a cemented material that increases in strength with time. Lime-treated soils, in general, have greater strength and a higher modulus of elasticity than untreated soils.

Recommended percentages of lime for soil stabilization vary from 2 to 10 percent. For coarse soils such as clayey gravels, sandy soils with less than 50 per cent silt-clay fraction, the per cent of lime varies from 2 to 5, whereas for soils with more than 50 percent silt-clay fraction, the percent of lime lies between 5 and 10. Lime is also used with fly ash. The fly ash may vary from 10 to 20 per cent, and the percent of lime may lie between 3 and 7.

Soil-Cement Stabilization

Soil-cement is the reaction product of an intimate mixture of pulverized soil and measured amounts of portland cement and water, compacted to high density. As the cement hydrates, the mixture becomes a hard, durable structural material. Hardened soil-cement has the capacity to bridge over local weak points in a subgrade. When properly made, it does not soften when exposed to wetting and drying, or freezing and thawing cycles.

Portland cement and soil mixed at the proper moisture content has been used increasingly in recent years to stabilize soils in special situations. Probably the main use has been to build stabilized bases under concrete pavements for highways and airfields. Soil cement mixtures are also used to provide wave protection on earth dams. There are three categories of soil-cement (Mitchell and Freitag, 1959). They are:

1. Normal soil-cement usually contains 5 to 14 percent cement by weight and is used generally for stabilizing low plasticity soils and sandy soils.

2. Plastic soil-cement has enough water to produce a wet consistency similar to mortar. This material is suitable for use as water proof canal linings and for erosion protection on steep slopes where road building equipment may not be used.

3. Cement-modified soil is a mix that generally contains less than 5 percent cement by volume. This forms a less rigid system than either of the other types, but improves the engineering properties of the soil and reduces the ability of the soil to expand by drawing in water.

The cement requirement depends on the gradation of the soil. A well graded soil containing gravel, coarse sand and fine sand with or without small amounts of silt or clay will require 5 percent or less cement by weight. Poorly graded sands with minimal amount of silt will require about 9 percent by weight. The remaining sandy soils will generally require 7 percent. Non-plastic or moderately plastic silty soils generally require about 10 percent, and plastic clay soils require 13 percent or more.

Bituminous Soil Stabilization

Bituminous materials such as asphalts, tars, and pitches are used in various consistencies to improve the engineering properties of soils. Mixed with cohesive soils, bituminous materials improve the bearing capacity and soil strength at low moisture content. The purpose of incorporating bitumen into such soils is to water proof them as a means to maintain a low moisture content. Bituminous materials added to sand act as a cementing agent and produces a stronger, more coherent mass. The amount of bitumen added varies from 4 to 7 percent for cohesive materials and 4 to 10 percent for sandy materials. The primary use of bituminous materials is in road construction where it may be the primary ingredient for the surface course or be used in the subsurface and base courses for stabilizing soils.

13.10 SOIL STABILIZATION BY INJECTION OF SUITABLE GROUTS

Grouting is a process whereby fluid like materials, either in suspension, or solution form, are injected into the subsurface soil or rock.

The purpose of injecting a grout may be any one or more of the following:

1. To decrease permeability.
2. To increase shear strength.
3. To decrease compressibility.

Suspension-type grouts include soil, cement, lime, asphalt emulsion, etc., while the solution type grouts include a wide variety of chemicals. Grouting proves especially effective in the following cases:

1. When the foundation has to be constructed below the ground water table. The deeper the foundation, the longer the time needed for construction, and therefore, the more benefit gained from grouting as compared with dewatering.

2. When there is difficult access to the foundation level. This is very often the case in city work, in tunnel shafts, sewers, and subway construction.

3. When the geometric dimensions of the foundation are complicated and involves many boundaries and contact zones.

4. When the adjacent structures require that the soil of the foundation strata should not be excavated (extension of existing foundations into deeper layers).

Grouting has been extensively used primarily to control ground water flow under earth and masonry dams, where rock grouting is used. Since the process fills soil voids with some type of stabilizing material grouting is also used to increase soil strength and prevent excessive settlement.

Many different materials have been injected into soils to produce changes in the engineering properties of the soil. In one method a casing is driven and injection is made under pressure to the soil at the bottom of the hole as the casing is withdrawn. In another method, a grouting hole is drilled and at each level in which injection is desired, the drill is withdrawn and a collar is placed at the top of the area to be grouted and grout is forced into the soil under pressure. Another method is to perforate the casing in the area to be grouted and leave the casing permanently in the soil.

Penetration grouting may involve portland cement or fine grained soils such as bentonite or other materials of a particulate nature. These materials penetrate only a short distance through most soils and are primarily useful in very coarse sands or gravels. Viscous fluids, such as a solution of sodium silicate, may be used to penetrate fine grained soils. Some of these solutions form gels that restrict permeability and improve compressibility and strength properties.

Displacement grouting usually consists of using a grout like portland cement and sand mixture which when forced into the soil displaces and compacts the surrounding material about a central core of grout. Injection of lime is sometimes used to produce lenses in the soil that will block the flow of water and reduce compressibility and expansion properties of the soil. The lenses are produced by hydraulic fracturing of the soil.

The injection and grouting methods are generally expensive compared with other stabilization techniques and are primarily used under special situations as mentioned earlier. For a detailed study on injections, readers may refer to Caron et al., (1975).

13.11 SOIL STABILIZATION BY ELECTRICAL AND THERMAL METHODS.

Electrical Method

The electrical method is used to densify the in situ cohesive soils. The method is called as electro-osmosis. It consists of placing in the soil to be stabilized a number of electrodes and then passing a direct current between them. The electric current induces a flow of water from the anode to the cathode (This is because of the attraction of cations, and of the unbalanced, negatively charged clay particles themselves, to the anode). The cathode is generally a perforated metal pipe which is used as a well point for removing the water. The anode can be any type of metal rod. Typical electro-osmotic stabilization configuration are shown in Fig.13.21. In general, both the cathodes and anodes should be placed about 2 m apart beneath the lowest elevation to be stablized. Typical spacings of the cathodes (the well points) are 6 to 9 m apart with the anodes being placed midway between them.

The flow rate to a cathode well point can be estimated using a modification of Darav's law as follows:

$$q = k_e i_e A \tag{13.13}$$

where, q = flow rate, m^3/s,

k_e = electro-asmotic coefficient of permeability based on voltage, 1×10^{-9} to 7×10^{-9} m/s per V/m,

i_e = electrical potential gradient, V/m,

A = cross-sectional area, m^2.

The consumption of power varies from 1 to 10 kW per m^3 of stabilized soil.

Thermal Methods

Heat is very rarely used to stabilize soils. However it is technically possible to stabilize saturated clays by heat. Russians have stabilised deep deposits of partly saturated loses by burning a mixture of liquid fuel and air injected into the ground through a network of pipes. A temperature of 100 °C

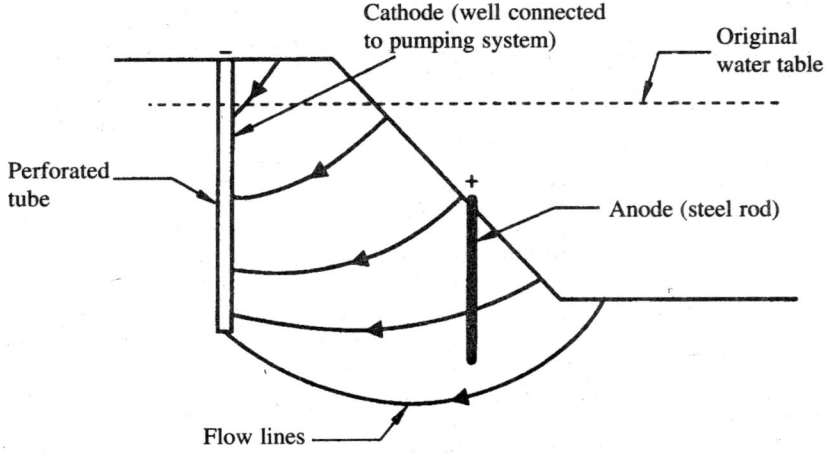

Fig. 13.21   Electro-osmotice stabilization of soils

causes drying and increase in the strength of clays. A permanent change in the structure of clays is possible at a temperature of 500 °C, and at 1000 °C there will be fusion of clay particles transforming clay into a solid substance much like brick. However, the economics of using heat precludes its use in construction projects.

Ground freezing appears to be gaining popularity in some cases. It is accomplished by bringing a refrigerant into the proximity of soil porewater that is stationary. The porewater around the refrigerant pipe begins to freeze, and with continued exposure the ice layers expand in all directions. A series refrigerant pipes layed close to each other will help to form a continuous wall of ice. The freezed soil possesses high strength and low permeability. It can stabilize a wide range of soil types. Freezing technique has been successfully used in sinking tunnel shafts, advancing tunnels in running ground, providing lateral restriant for excavations, etc.

13.12 QUESTIONS AND PROBLEMS

13.1   Differentiate: (i) Compaction and consolidation, and (ii) Standard Proctor and modified Proctor tests.

13.2   Draw an ideal 'compaction curve' and discuss the effect of moisture on the dry unit weight of soil.

13.3   Explain: (i) the unit, in which the compaction is measured, (ii) 95 percent of Proctor density, (iii) zero air-voids line, and (iv) effect of compaction on the shear strength of soil.

13.4   What are the types of rollers used for compacting different types of soils in the field? How do you decide the compactive effort required for compacting the soil to a desired density in the field?

13.5   What are the methods adopted for measuring the density of the compacted soil? Briefly describe the one which will suit all types of soils.

13.6   A soil having a specific gravity of solids $G = 2.75$, is subjected to proctor compaction test in a mould of volume $V = 945$ cm^3. The observations recorded are as follows:

Observation number	1	2	3	4	5
Mass of wet sample, g	1389	1767	1824	1784	1701
Water content, percentage	7.5	12.1	17.5	21.0	25.1

What are the values of maximum dry unit weight and the optimum moisture content? Draw 100% saturation line.

13.7   A field density test was conducted by sand cone method. The observation data are given below:

(a) Mass of jar with cone and sand (before use) = 4950 g, (b) mass of jar with cone and sand (after use) = 2280 g, (c) mass of soil from the hole = 2925 g, (d) dry density of sand = 1.48 g/cm^3, (e) water content of the wet soil = 12%. Determine the dry unit weight of compacted soil.

13.8   If a clayey sample is saturated at a water content of 30%, what is its density? Assume a value for specific gravity of solids.

13.9   A soil in a borrow pit is at a dry density of 1.7 Mg/m^3 with a water content of 12%. If a soil mass of 2000 cubic meter volume is excavated from the pit and compacted in an embankment with a porosity of 0.32, calculate the volume of the embankment which can be constructed out of this material. Assume $G = 2.70$.

13.10 In a Proctor compaction test, for one observation, the mass of the wet sample is missing. The oven dry mass of this sample was 1800 g. The volume of the mould used was 950 cm³. If the saturation of this sample was 80 percent, determine (i) the moisture content, and (ii) the total unit weight of the sample. Assume $G = 2.70$.

13.11. A field-compacted sample of a sandy loam was found to have a wet density of 2.176 Mg/m³ at a water content of 10%. The maximum dry density of the soil obtained in a standard Proctor test was 2.0 Mg/m³. Assume $G = 2.65$. Compute ρ_d, S, n and the percent of compaction of the field sample.

13.12 A proposed earth embankment is required to be compacted to 95% of standard Proctor dry density. Tests on the material to be used for the embankment give $\rho_{max} = 1.984$ Mg/m³ at an optimum water content of 12%. The borrow pit material in its natural condition has a void ratio of 0.60. If $G = 2.65$, what is the minimum volume of the borrow required to make 1 cu.m of acceptable compacted fill?

13.13 The following data were obtained from a field density test on a compacted fill of sandy clay. Laboratory moisture density tests on the fill material indicated a maximum dry density of 1.92 Mg/m³ at an optimum water content of 11%. What was the percent compaction of the fill? Was the fill water content above or below optimum.

Mass of the moist soil removed from the test hole	= 1038 g
Mass of the soil after oven drying	= 914 g
Volume of the test hole	= 478.55 cm³

13.14 A field density test performed by sand-cone method gave the following data.

Mass of the soil removed + pan	= 1590 g
Mass of the pan	= 125 g
Volume of the test hole	= 750 cm3
Water content information	
Mass of the wet soil + pan	= 404.9 g
Mass of the dry soil + pan	= 365.9 g
Mass of the pan	= 122.0 g

Compute: ρ_d, γ_d, and the water content of the soil. Assume $G = 2.67$

CHAPTER 14

LATERAL EARTH PRESSURE

14.1 INTRODUCTION

Structures that are built to retain vertical or nearly vertical earth banks or any other material are called *retaining walls*. Retaining walls may be constructed of masonry or sheet piles. Some of the purposes for which retaining walls are used are shown in Fig. 14.1.

The retaining walls may retain water also. The earth retained may be natural soil or fill. The principal types of retaining walls are given in Figs 14.1 and 14.2.

Whatever may be the type of wall, all the walls listed above have to withstand lateral pressures either from earth or any other material on their faces. The pressures acting on the walls try to move the walls from their position. The walls should be so designed as to keep them stable in their position. Gravity walls resist movement because of their heavy sections. They are built of mass concrete or stone or brick masonry. No reinforcement is required in these walls. Semi-gravity walls are not as heavy as gravity walls. A small amount of reinforcement is used for reducing the mass of concrete. The stems of cantilever walls are thinner in section. The base slab is the cantilever portion. These walls are made of reinforced concrete. Counterfort walls are similar to cantilever walls except that the stem of the walls span horizontally between vertical brackets known as counterforts. The counterforts are provided on the backfill side. Buttressed walls are similar to counterfort walls except the brackets or buttress walls are provided on the opposite side of the backfill.

In all these cases, the backfill tries to move the wall from its position. The movement of the wall is partly resisted by the wall itself and partly by soil in front of the wall.

The sheet pile walls are more flexible than the other types. The earth pressure on these walls is dealt with in Chapter 15. This chapter deals with earth pressures on rigid walls only.

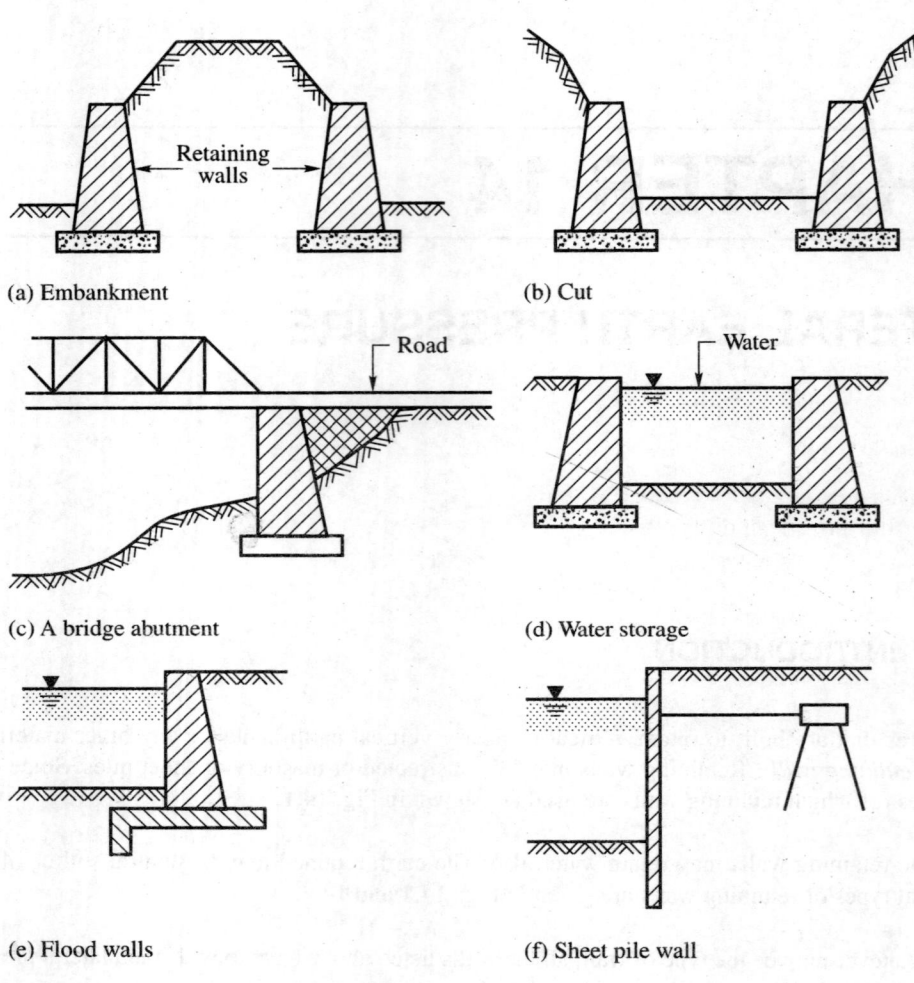

(a) Embankment

(b) Cut

(c) A bridge abutment

(d) Water storage

(e) Flood walls

(f) Sheet pile wall

Fig. 14.1   Use of retaining walls

14.2   LATERAL EARTH PRESSURE THEORY

There are two classical earth pressure theories. They are

1.   Coulomb's earth pressure theory.

2.   Rankine's earth pressure theory.

The first rigorous analysis of the problem of lateral earth pressure was published by Coulomb in (1776). Rankine (1857) proposed a different approach to the problem in 1857. These theories propose to estimate the magnitudes of two pressures called *Active earth pressure* and *Passive earth pressure*.

These pressures are defined as follows:

Consider a rigid retaining wall with a plane vertical face, as shown in Fig. 14.3(a), is backfilled
with cohesionless soil. If the wall does not move even after back filling, the pressure exerted on the wall
termed as pressure for the *at rest condition* of the wall. If suppose the wall gradually rotates about
point A and moves away from the backfill, the unit pressure on the wall is gradually reduced and after a
particular displacement of the wall at the top, the pressure reaches a constant value. The pressure is the
minimum possible. This pressure is termed the *active pressure* since the weight of the backfill is
responsible for the movement of the wall. If the wall is smooth, the resultant pressure acts normal to the
face of the wall. If the wall is rough, it makes an angle δ with the normal on the wall. The angle δ is
called the *angle of wall friction*. As the wall moves away from the backfill, the soil tends to move
forward. When the wall movement is sufficient, a soil mass of weight W ruptures along surface AC'C
shown in Fig. 14.3(a). This surface is slightly curved. If the surface is assumed to be a plane surface AC,
analysis would indicate that this surface would make an angle of $45° + \phi/2$ with the horizontal.

If the wall is now rotated about A *towards the backfill, the actual failure plane AC'C* is also a
curved surface [Fig. 14.3(b)]. However, if the failure surface is approximated as a plane AC, this
makes an angle $45° – \phi/2$ with the horizontal and the pressure on the wall increases from the value of
the at rest condition to the maximum value possible. The maximum pressure P_p that is developed is
termed the *passive earth pressure*. The pressure is called passive because the weight of the backfill
opposes the movement of the wall. It makes an angle δ with the normal if the wall is rough.

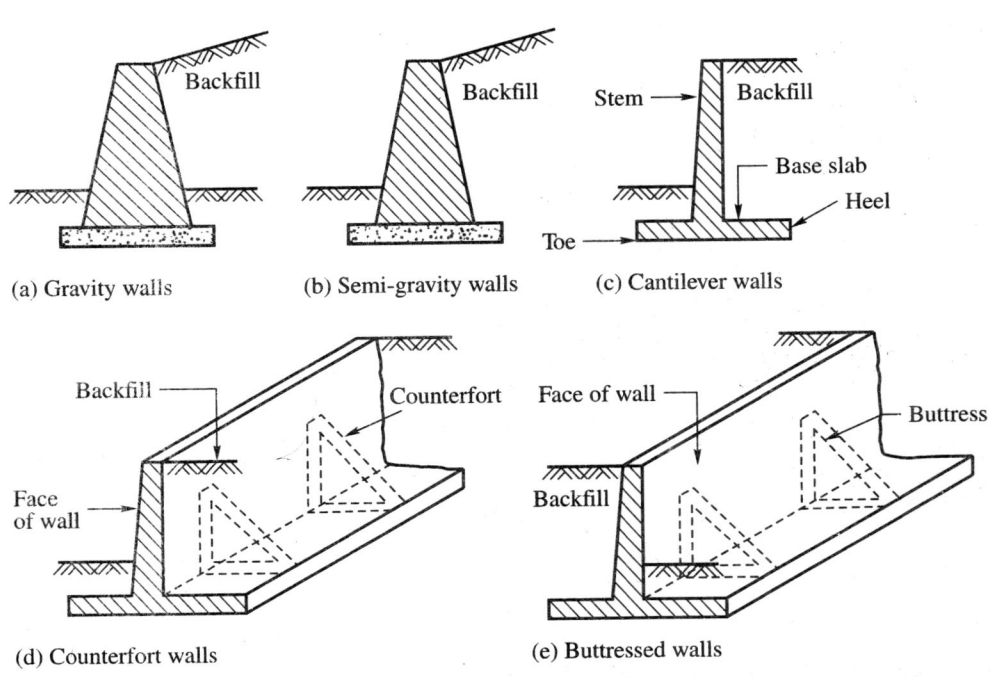

(a) Gravity walls   (b) Semi-gravity walls   (c) Cantilever walls

(d) Counterfort walls   (e) Buttressed walls

Fig. 14.2   Principal types of rigid retaining walls

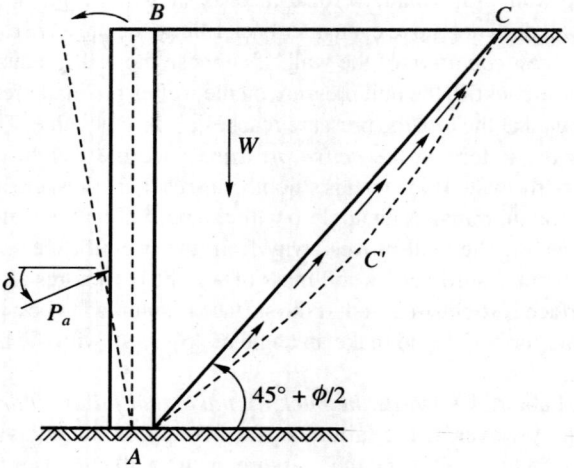

(a) Active earth pressure

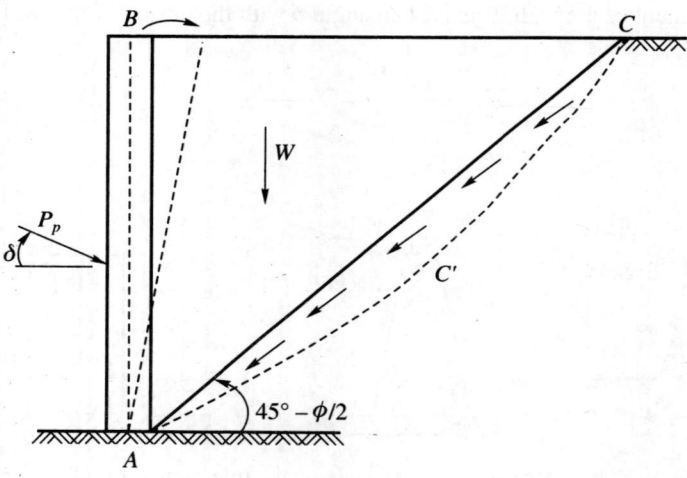

(b) Passive earth pressure

Fig. 14.3   Wall movement for the development of active and passive earth pressures

The gradual decrease or increase of pressure on the wall with the movement of the wall from the at rest condition may be depicted as shown in Fig. 14.4.

The movement Δ_p required to develop the passive state is considerably larger than Δ_a required for the active state.

14.3   LATERAL EARTH PRESSURE FOR AT REST CONDITION

Figure 14.1 gives different types of retaining walls for retaining earth. The earth retained may be natural earth or filled up soil. These backfill materials exert certain lateral pressure on the wall. If the wall is rigid and does not move with the pressure exerted on the wall, the soil behind the wall will be in a state of *elastic equilibrium*. Consider a prismatic element E in the backfill at depth z shown in Fig. 14.5.

The element E is subjected to the following pressures.

Vertical pressure = $\sigma_v = \gamma z$; lateral pressure = σ_h

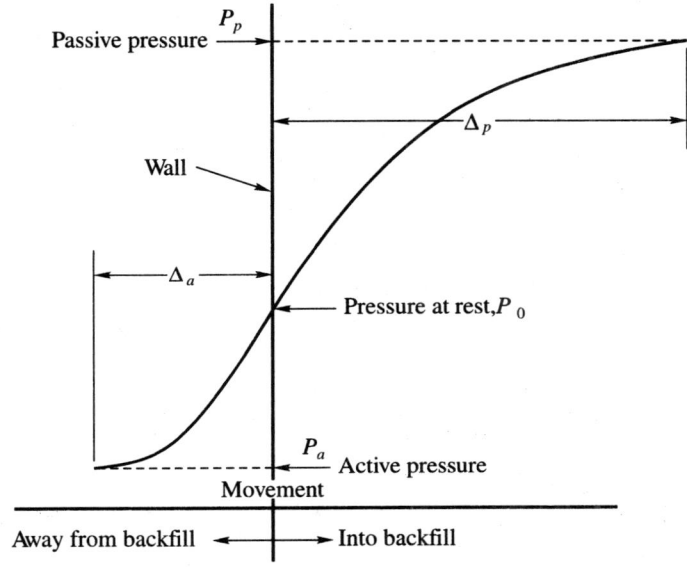

Fig. 14.4 Development of active and passive earth pressures

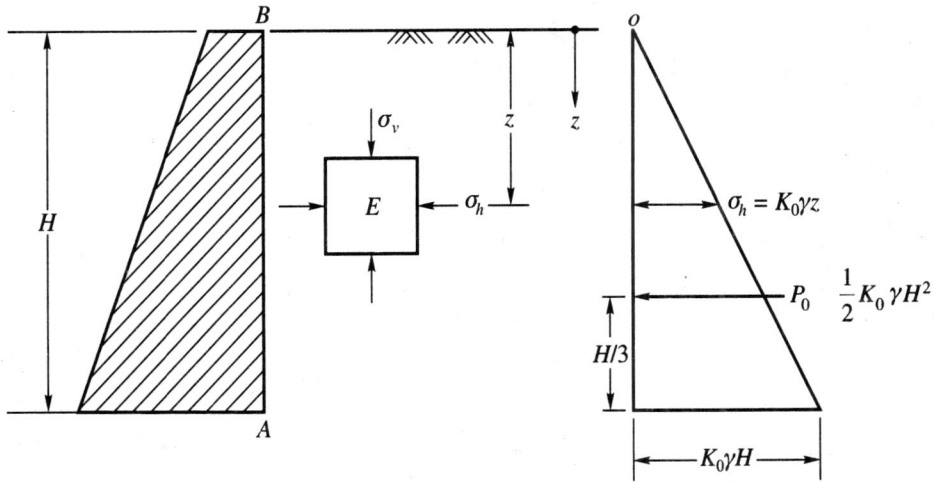

Fig. 14.5 Lateral earth pressure for at rest condition

where γ is the effective unit weight of the soil. If we consider the backfill is homogeneous then both σ_v and σ_h increase linearly with depth z. In such a case, the ratio of σ_h to σ_v remains constant with respect to depth, that is

$$\frac{\sigma_h}{\sigma_v} = \frac{\sigma_h}{\gamma z} = \text{constant} = K_0$$

where, K_0 is called the *coefficient of earth pressure for the at rest condition.*

The lateral earth pressure σ_h acting on the wall at any depth z may be expressed as

$$\sigma_h = K_0 \gamma z \tag{14.1a}$$

The expression for σ_h at depth H, the height of the wall, is

$$\sigma_h = K_0 \gamma H \tag{14.1b}$$

The distribution of σ_h on the wall is given in Fig. 14.5(b).

The total pressure P_0 for the soil for the at rest condition is

$$P_0 = \frac{1}{2} K_0 \gamma H^2 \tag{14.1c}$$

The value of K_0 depends upon the relative density of the sand and the process by which the deposit was formed. If this process does not involve artificial tamping the value of K_0 ranges from about 0.40 for loose sand to 0.6 for dense sand. Tamping the layers may increase it to 0.8.

The value of K_0 may also be obtained on the basis of elastic theory. If a cylindrical sample of soil is acted upon by vertical stress σ_v and horizontal stress σ_h, the lateral strain ε_1 may be expressed as

$$\varepsilon_1 = \frac{1}{E} [\sigma_h - \mu(\sigma_h + \sigma_v)] \tag{14.2}$$

where E = Young's modulus, μ = Poisson's ratio.

The lateral strain $\varepsilon_1 = 0$ when the earth is in the at rest condition. For this condition, we may write

$$\frac{1}{E} [\sigma_h - \mu(\sigma_h + \sigma_v)] = 0 \text{ or } \frac{\sigma_h}{\sigma_v} = \frac{\mu}{1-\mu} \tag{14.3}$$

$$\text{or } \sigma_h = \left(\frac{\mu}{1-\mu}\right) \sigma_v = K_0 \sigma_v = K_0 \gamma z$$

$$\text{where } \frac{\mu}{1-\mu} = K_0, \ \sigma_v = \gamma z \tag{14.4}$$

TABLE 14.1

Coefficients of earth pressure for at rest K_0 condition

Type of soil	I_p	K_0
Loose sand, saturated	–	0.46
Dense sand, saturated	–	0.36
Dense sand, dry ($e = 0.6$)	–	0.49
Loose sand, dry ($e = 0.8$)	–	0.64
Compacted clay	9	0.42
Compacted clay	31	0.60
Organic silty clay, undisturbed ($w_1 = 74\%$)	45	0.57

According to Jaky (1944), a good approximation for K_0 is given by Eq. (14.5).

$$K_0 = 1 - \sin\phi \qquad (14.5)$$

which fits most of the experimental data.

Numerical values of K_0 for some soils are given in Table 14.1.

Example 14.1

If a retaining wall 5 m high is restrained from yielding, what will be the at-rest earth pressure per meter length of the wall? Given: the backfill is cohesionless soil having $\phi = 30°$ and $\gamma = 18$ kN/m^3. Also determine the resultant force for the at-rest condition.

Solution

From Eq. (14.5)

$$K_0 = 1 - \sin\phi = 1 - \sin 30° = 0.5$$

From Eq. (14.1c)

$$P_0 = \frac{1}{2} K_0 \gamma H^2 = \frac{1}{2} \times 0.5 \times 18 \times 5^2 = 112.5 \text{ kN/m length of wall}$$

14.4 RANKINE'S STATES OF PLASTIC EQUILIBRIUM FOR COHESIONLESS SOILS

Let XY in Fig. 14.6(a) represent the horizontal surface of a semi-infinite mass of cohesionless soil with a unit weight γ. The soil is in an initial state of elastic equilibrium. Consider a prismatic block $ABCD$. The depth of the block is z and the cross-sectional area of the block is unity. Since the element is symmetrical with respect to a vertical plane, the normal stress on the base AD is

$$\sigma_v = \gamma z \tag{14.6}$$

σ_v is a principal stress. The normal stress σ_v on the vertical planes AB or DC at depth z may be expressed as a function of vertical stress.

$$\sigma_h = f(\sigma_v) = K_0 \gamma z \tag{14.7}$$

where K_0 is the coefficient of earth pressure for the at rest condition which is assumed as a constant for a particular soil. The horizontal stress σ_h varies from zero at the ground surface to $K_0 \gamma z$ at depth z.

If we imagine that the entire mass is subjected to horizontal deformation, such deformation is a plane deformation. Every vertical section through the mass represents a plane of symmetry for the entire mass. Therefore, the shear stresses on vertical and horizontal sides of the prism are equal to zero.

Due to the stretching, the pressure on vertical sides AB and CD of the prism decreases until the conditions of *plastic equilibrium* are satisfied, while the pressure on the base AD remains unchanged. Any further stretching merely causes a plastic flow without changing the state of stress. The transition from the state of *plastic equilibrium* to the state of *plastic flow* represents the failure of the mass. Since the weight of the mass assists in producing an expansion in a horizontal direction, the subsequent failure is called *active failure*.

If, on the other hand, the mass of soil is compressed, as shown in Fig. 14.6(b), in a horizontal direction, the pressure on vertical sides AB and CD of the prism increases while the pressure on its base remains unchanged at γz. Since the lateral compression of the soil is resisted by the weight of the soil, the subsequent failure by plastic flow is called a *passive failure*.

The problem now consists of determining the stresses associated with the states of plastic equilibrium in the semi-infinite mass and the orientation of the surface of sliding. The problem was solved by Rankine (1857).

Therefore, the plastic states which are produced by stretching or by compressing a semi-infinite mass of soil parallel to its surface are called *active* and *passive Rankine states* respectively. The orientation of the planes may be found by Mohr's diagram.

Horizontal stretching or compressing of a semi-infinite mass to develop a state of plastic equilibrium is only a concept. However, local states of plastic equilibrium in a soil mass can be created by rotating a retaining wall about its base either away from the backfill for an active state or into the backfill for a passive state in the way shown in Fig. 14.3. In both cases, the soil within wedge ABC will be in a state of plastic equilibrium and line AC represents the rupture plane.

Mohr Circle for Active and Passive States of Equilibrium in Granular Soils

Point P_1 on the σ-axis in Fig. 14.6(c) represents the state of stress on base AD of prismatic element $ABCD$ in Fig. 14.6(a). Since the shear stress on AD is zero, the vertical stress on the base

$$\sigma_v = \gamma z \tag{14.8}$$

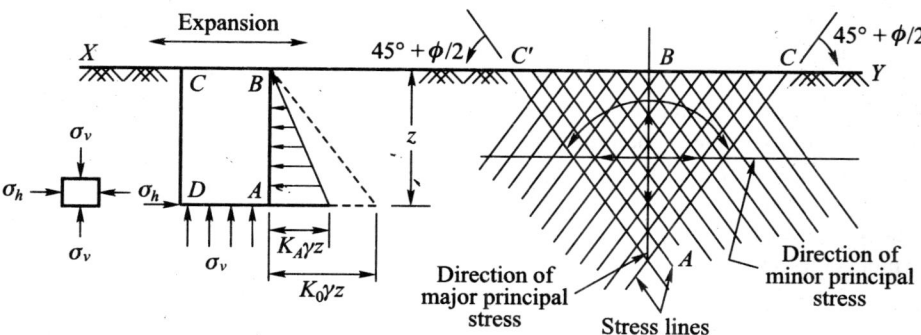

(a) Active state

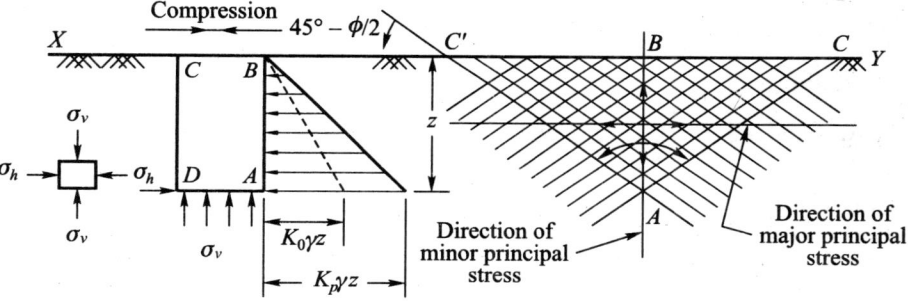

(b) Passive state

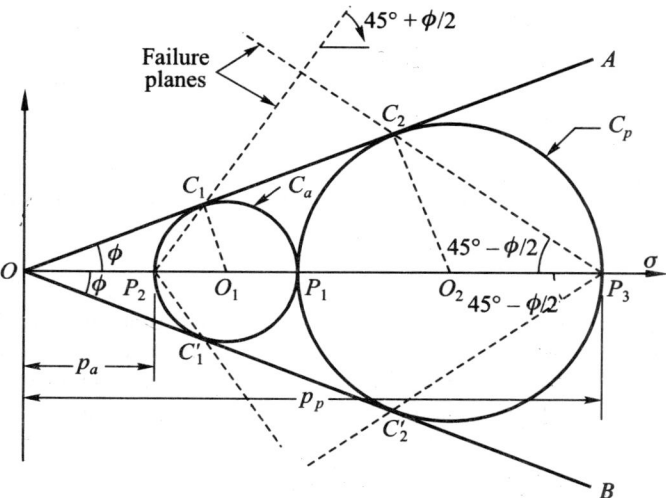

(c) Mohr stress diagram

Fig. 14.6 Rankine's condition for active and passive failures in a semi-infinite mass of cohesionless soil

is a principal stress. OA and OB are the two Mohr envelopes which satisfy the Coulomb equation of shear strength

$$s = \sigma \tan \phi \tag{14.9}$$

Two circles C_a and C_p can be drawn passing through P_1 and at the same time tangential to the Mohr envelopes OA and OB. When the semi-infinite mass is stretched horizontally, the horizontal stress on vertical faces AB and CD Fig. 14.6(a) at depth z is reduced to the minimum possible and this stress is less than vertical stress σ_v. Mohr circle C_a gives the state of stress on the prismatic element at depth z when the mass is in active failure. The intercepts OP_1 and OP_2 are the major and minor principal stresses respectively.

When the semi-infinite mass is compressed Fig. 14.6(b), the horizontal stress on the vertical face of the prismatic element reaches the maximum value OP_3 and circle C_p is the Mohr circle which gives that state of stress.

Active State of Stress

From Mohr circle C_a

Major principal stress $= OP_1 = \sigma_1 = \gamma z$

Minor principal stress $= OP_2 = \sigma_3$

$$OO_1 = \frac{\sigma_1 + \sigma_3}{2}, \quad O_1 C_1 = \frac{\sigma_1 - \sigma_3}{2}$$

From triangle $OO_1 C_1$, $\dfrac{\sigma_1 - \sigma_3}{2} = \dfrac{\sigma_1 + \sigma_3}{2} \sin \phi$

$$\text{or } \sigma_1 = \sigma_3 \left(\frac{1 + \sin \phi}{1 - \sin \phi} \right) = \sigma_3 \tan^2 (45° + \phi/2) = \sigma_3 N_\phi \tag{14.10}$$

$$\text{Therefore, } p_a = \sigma_3 = \frac{\sigma_1}{N_\phi} = \gamma z K_A \tag{14.11}$$

where $\sigma_1 = \gamma z$, K_A = coefficient of earth pressure for active state = $\tan^2 (45° - \phi/2)$.

From point P_1, draw a line parallel to the base AD on which σ_1 acts. Since this line coincides with the σ-axis, point P_2 is the origin of planes. Lines $P_2 C_1$ and $P_2 C'_1$ give the orientations of the failure planes. They make an angle of $45° + \phi/2$ with the σ-axis. The lines drawn parallel to the lines $P_2 C_1$ and $P_2 C'_1$ in Fig. 14.6(a) give the shear lines along which the soil slips in the plastic state. The angle between a pair of conjugate shear lines is $(90° - \phi)$.

Passive State of Stress

C_p is the Mohr circle in Fig. 14.6(c) for the passive state and P_3 is the origin of planes.

Major principal stress $= \sigma_1 = p_p = OP_3$

Minor principal stress $= \sigma_3 = OP_1 = \gamma z$

From the triangle OO_2C_2, as before $\sigma_1 = \gamma z N_\phi$

Since $\sigma_1 = p_p$ and $\sigma_3 = \gamma z$, we have

$$p_p = \gamma z N_\phi = \gamma z K_p \qquad (14.12)$$

where K_p = coefficient of earth pressure for passive state $= \tan^2(45° + \phi/2)$.

The shear failure lines are P_3C_2 and $P_3C'_2$ and they make an angle of $45° - \phi/2$ with the horizontal. The shear failure lines are drawn parallel to P_3C_2 and $P_3C'_2$ in Fig. 14.6(b). The angle between any pair of conjugate shear lines is $(90° + \phi)$.

14.5 RANKINE'S EARTH PRESSURE AGAINST A VERTICAL SECTION WITH THE SURFACE HORIZONTAL IN COHESIONLESS BACKFILL

Active Earth Pressure

The section AB in Fig. 14.6(a) in a semi-infinite mass is replaced by a smooth wall AB in Fig. 14.7(a).

The lateral pressure acting against smooth wall AB is due to the mass of soil ABC above failure line AC which makes an angle of $45° + \phi/2$ with the horizontal. The lateral pressure distribution on wall AB of height H increases in simple proportion to depth. The pressure acts normal to the wall AB [Fig. 14.7(b)].

The lateral active pressure at A is

$$p_a = \gamma H K_A \qquad (14.13)$$

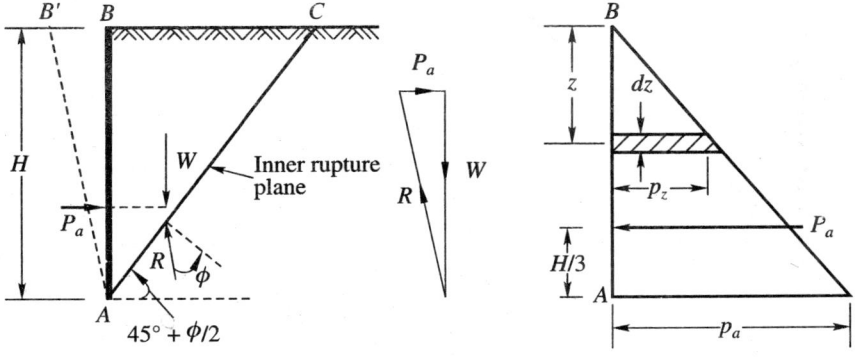

Fig. 14.7 Rankine's active earth pressure in cohesionless soil

The total pressure on AB is therefore

$$P_a = \int_0^H p_z dz = K_A \int_0^H \gamma z dz = \frac{1}{2} K_A \gamma H^2 \qquad (14.14)$$

where, $K_A = \tan^2(45° - \phi 2) = \dfrac{1-\sin\phi}{1+\sin\phi}$ (14.14a)

P_a act at a height $H/3$ above the base of the wall.

Passive Earth Pressure

If wall AB is pushed into the mass to such an extent as to impart uniform compression throughout, the mass, soil wedge ABC in Fig. 14.8(a) will be in Rankine's passive state of plastic equilibrium. The inner rupture plane AC makes an angle $45° + \phi/2$ with the vertical AB. The pressure distribution on wall AB is linear as shown in Fig. 14.8(b).

The passive pressure p_p at A is

$$p_p = \gamma H K_p$$

the total pressure against the wall is

$$P_p = \int_0^H p_z dz = K_p \int_0^H \gamma z dz = \frac{1}{2} K_p \gamma H^2 \qquad (14.15)$$

where, $K_p = \tan^2(45° + \phi/2) = \dfrac{1+\sin\phi}{1-\sin\phi}$ (14.15a)

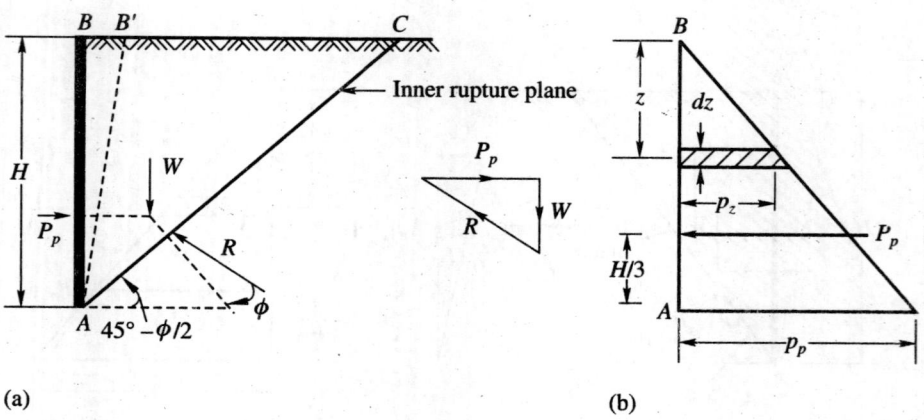

(a)   (b)

Fig. 14.8 Rankine's passive earth pressure

Relationship between K_P and K_A

The ratio of K_P and K_A may be written as

$$\frac{K_p}{K_A} = \frac{\tan^2(45° + \phi/2)}{\tan^2(45° - \phi/2)} = \tan^4(45° + \phi/2) \tag{14.16}$$

For example, if $\phi = 30°$, we have,

$$\frac{K_p}{K_A} = \tan^4 60° = 9, \text{ or } K_p = 9K_A$$

This simple demonstration indicates that the value of K_P is quite large compared to K_A.

Backfill Soil Submerged with the Surface Horizontal

When the backfill is fully submerged, two types of pressures act on wall AB. (Fig. 14.9) They are

1.   The active earth pressure due to the submerged weight of soil

2.   The lateral pressure due to water

At any depth z, the total unit pressure on the wall is

$$\overline{p_a} = p_a + p_w = \gamma_b z K_A + \gamma_w z$$

At depth $z = H$, we have

$$\overline{p_a} = \gamma_b H K_A + \gamma_w H$$

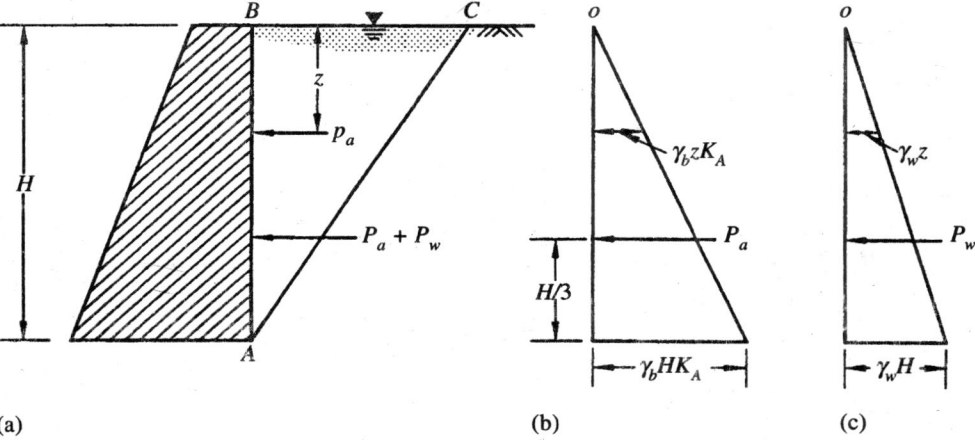

(a)   (b)   (c)

Fig. 14.9 Rankine's active pressure under submerged condition in cohesionless soil

where γ_b is the submerged unit weight of soil and γ_w the unit weight of water. The total pressure acting on the wall at a height $H/3$ above th e base is

$$\overline{P_a} = P_a + P_w = \frac{1}{2}\gamma_b H^2 K_A + \frac{1}{2}\gamma_w H^2 \tag{14.17}$$

When the Backfill is Partly Submerged with a Uniform Surcharge Load

The ground water table is at a depth of H_1 below the surface and the soil above this level has an effective moist unit weight of γ. The soil below the water table is submerged with a submerged unit weight γ_b. In this case, the total unit pressure may be expressed as given below.

At depth H_1 at the level of the water table

$$\overline{p_a} = qK_A + \gamma H_1 K_A$$

At depth H we have

$$\overline{p_a} = qK_A + \gamma H_1 K_A + \gamma_b H_2 K_A + \gamma_w H_2$$

$$\text{or} \quad \overline{p_a} = qK_A + (\gamma H_1 + \gamma_b H_2)K_A + \gamma_w H_2 \tag{14.18}$$

The pressure distribution is given in Fig. 14.10(b). It is assumed that the value of ϕ remains the same throughout the depth H.

From Fig. 14.10(b), we may say that the total pressure $\overline{P_a}$ acting per unit length of the wall may be written as equal to

$$\overline{P_a} = \text{Area of the diagram} (1+2+3) \tag{14.19}$$

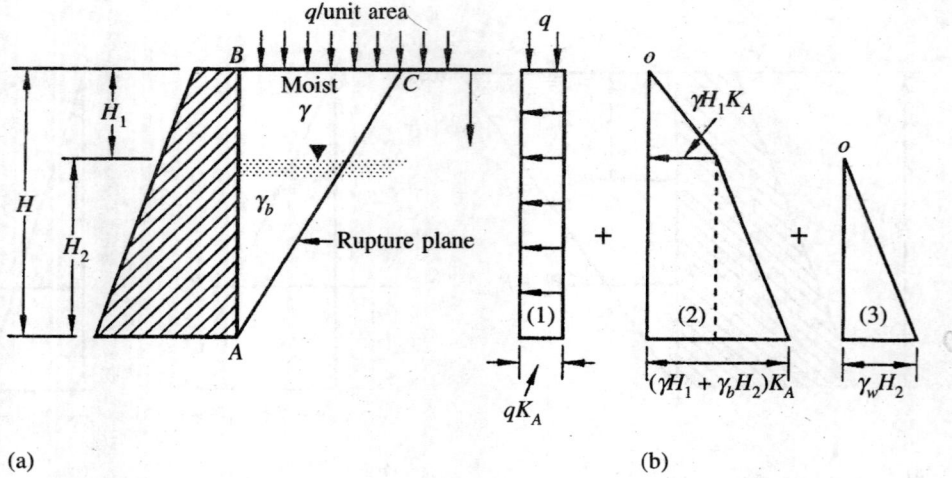

(a)   (b)

Fig. 14.10 Rankine's active pressure in cohesionless backfill under party submerged condition with surcharge load

The point of application of $\overline{P}_a$ above the base of the wall can be found by taking moments of all the forces acting on the wall about A.

14.6 RANKINE'S ACTIVE EARTH PRESSURE WITH A SLOPING COHESIONLESS BACKFILL SURFACE

Figure 14.11(a) shows a smooth vertical wall with a sloping backfill of cohesionless soil. As in the case of a horizontal backfill, the active state of plastic equilibrium can be developed in the backfill by rotating the wall about A away from the backfill. Let AC be the rupture line and the soil within the wedge ABC be in an active state of plastic equilibrium.

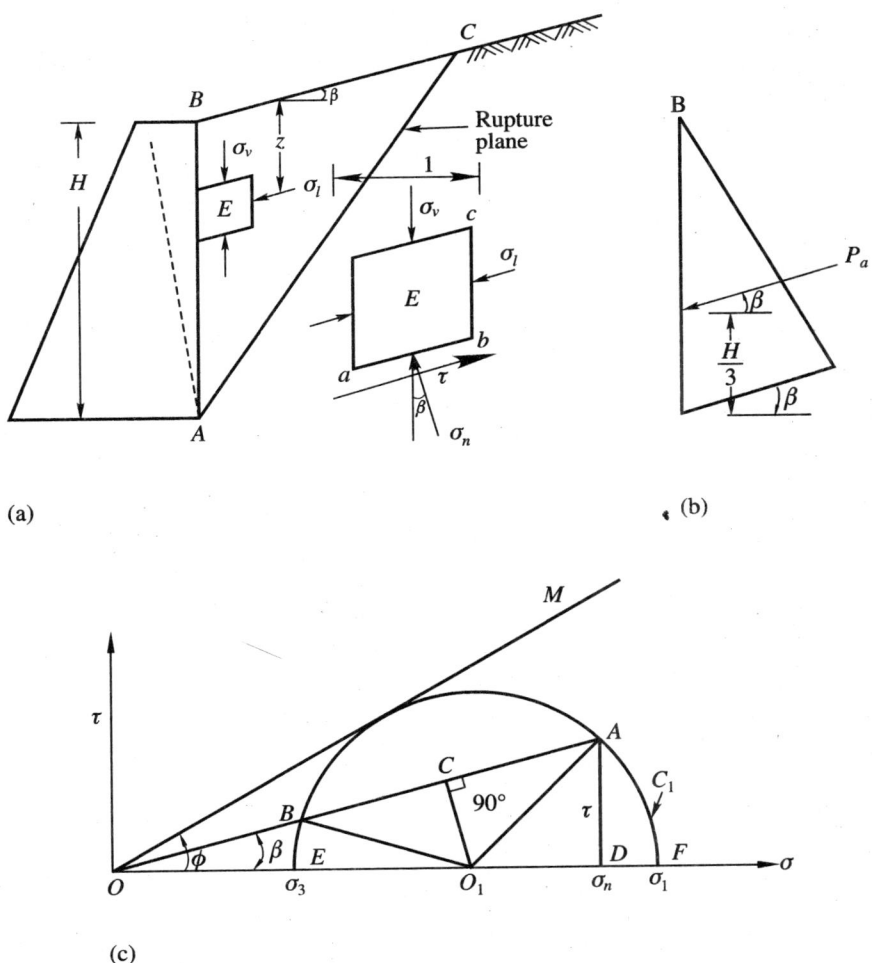

(a)   (b)

(c)

Fig. 14.11   Rankine's active pressure for a sloping cohesionless backfill

Consider a rhombic element E within the plastic zone ABC which is shown to a larger scale outside. The base of the element is parallel to the backfill surface which is inclined at an angle β to the horizontal. The horizontal width of the element is taken as unity.

Let σ_v = the vertical stress acting on an elemental length $ab = \gamma z \cos \beta$

σ_l = the lateral pressure acting on vertical surface bc of the element

The vertical stress σ_v can be resolved into components σ_n the normal stress and τ the shear stress on surface ab of element E. We may now write

$$\sigma_n = \sigma_v \cos \beta = \gamma z \cos \beta \cos \beta = \gamma z \cos^2 \beta \tag{14.20}$$

$$\tau = \sigma_v \sin\beta = \gamma z \cos \beta \sin\beta \tag{14.21}$$

A Mohr diagram can be drawn as shown in Fig. 14.11(c). Here, length $OA = \gamma z \cos \beta$ making an angle β with the σ-axis. $OD = \sigma_n = \gamma z \cos^2 \beta$ and $AD = \tau = \gamma z \cos \beta \sin \beta$. OM is the Mohr envelope making an angle ϕ with the σ-axis. Now Mohr circle C_1 can be drawn passing through point A and at the same time tangential to envelope OM. This circle cuts line OA at point B and the σ-axis at E and F.

Now OB = the lateral pressure $\sigma_l = p_a$ in the active state.

The principal stresses are

$$OF = \sigma_1 \text{ and } OE = \sigma_3$$

The following relationships can be expressed with reference to the Mohr diagram.

$$BC = CA = \frac{\sigma_1 + \sigma_3}{2} \sqrt{\sin^2 \phi - \sin^2 \beta}$$

$$\sigma_v = OA = OC + CA = \frac{\sigma_1 + \sigma_3}{2} \cos \beta + \frac{\sigma_1 + \sigma_3}{2} \sqrt{\sin^2 \phi - \sin^2 \beta}$$

$$\sigma_l = p_a = OC - BC = \frac{\sigma_1 + \sigma_3}{2} \cos \beta - \frac{\sigma_1 + \sigma_3}{2} \sqrt{\sin^2 \phi - \sin^2 \beta} \tag{14.22}$$

Now we have (after simplification)

$$\frac{\sigma_l}{\sigma_v} = \frac{p_a}{\gamma z \cos \beta} = \frac{\cos \beta - \sqrt{\cos^2 \beta - \cos^2 \phi}}{\cos \beta + \sqrt{\cos^2 \beta - \cos^2 \phi}}$$

or $\quad p_a = \gamma z \cos \beta \times \dfrac{\cos \beta - \sqrt{\cos^2 \beta - \cos^2 \phi}}{\cos \beta + \sqrt{\cos^2 \beta - \cos^2 \phi}} = \gamma z K_A \tag{14.23}$

where, $K_A = \cos \beta \times \dfrac{\cos \beta - \sqrt{\cos^2 \beta - \cos^2 \phi}}{\cos \beta + \sqrt{\cos^2 \beta - \cos^2 \phi}}$ $\qquad\qquad$ (14.24)

is called as *the coefficient of earth pressure* for the active state.

The pressure distribution on the wall is shown in Fig. 14.11(b). The active pressure at depth H is

$$p_a = \gamma H K_A$$

which acts parallel to the surface. The total pressure P_a per unit length of the wall is

$$P_a = \frac{1}{2}\gamma H^2 K_A \qquad\qquad (14.25)$$

which acts at a height $H/3$ from the base of the wall and parallel to the sloping surface of the backfill.

14.7 APPLICATION OF RANKINE'S ACTIVE PRESSURE TO CANTILEVER WALLS WITH COHESIONLESS BACKFILLS

Figure 14.12 shows a cantilever wall. In this case the vertical section AB of height H is taken as the wall on which the total active pressure P_a acts at a height $H/3$ from the base point A. AC is the rupture plane. W is the weight of soil between the wall and the section AB. The resultant of P_a and W is the total active earth pressure P_R per unit length of the wall as shown in Fig. 14.12.

Similar method can be applied for a sloping pressure surface of a gravity retaining wall.

14.8 RANKINE'S PASSIVE EARTH PRESSURE FOR A SLOPING COHESIONLESS BACKFILL SURFACE

An equation for P_p for a sloping backfill surface can be developed in the same way as for an active case. The equation for P_p may be expressed as

$$P_p = \frac{1}{2}\gamma H^2 K_p \qquad\qquad (14.26)$$

where, $K_p = \cos\beta \times \dfrac{\cos\beta + \sqrt{\cos^2 \beta - \cos^2 \phi}}{\cos\beta - \sqrt{\cos^2 \beta - \cos^2 \phi}}$ $\qquad\qquad$ (14.27)

P_p acts at a height $H/3$ above point A and parallel to the sloping surface.

The method of application of the principal of Rankine to cantilever walls is shown in Fig. 14.13. The resultant of P_p and W can be found out in the same was as for an active case explained in Fig. 14.12.

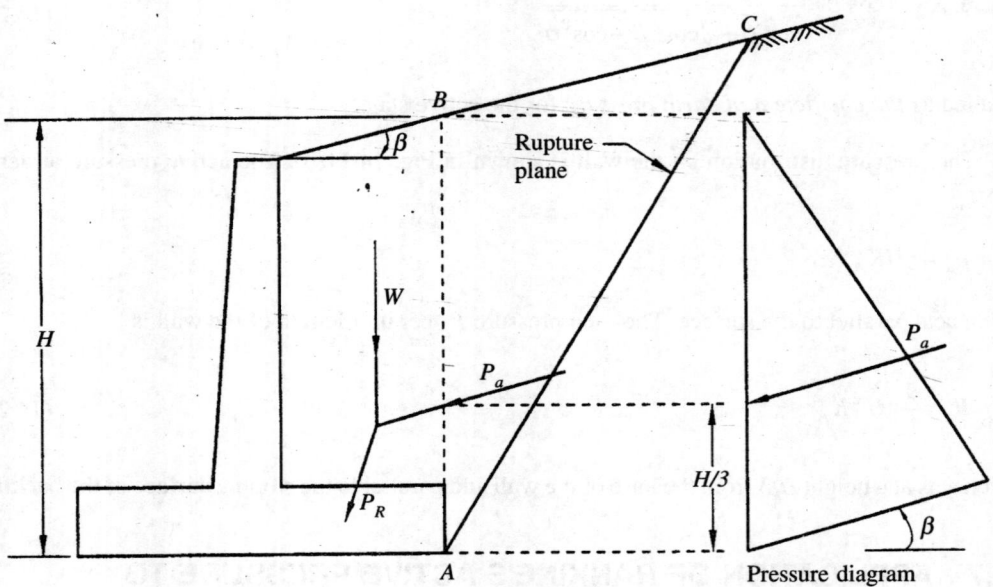

Fig. 14.12   Rankine's active condition applied to cantilever wall with cohesionless backfill material

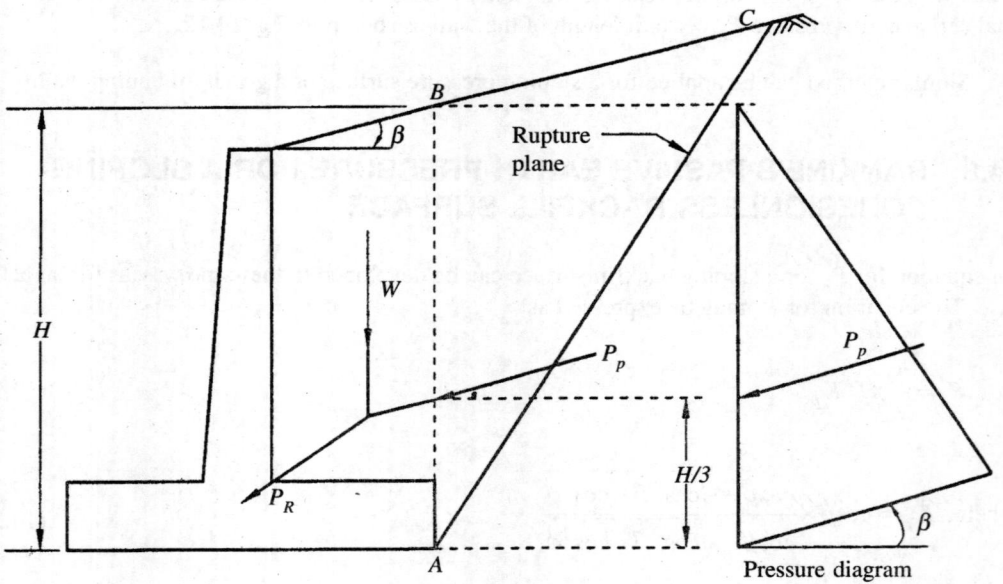

Fig. 14.13   Rankine's passtive state applied to cantilever wall with cohesionless backfill material

Example 14.2

A cantilever retaining wall of 7 meter height retains sand. The properties of the sand are $e = 0.5$, $\phi = 30°$ and $G = 2.7$. Using Rankine's theory determine the active earth pressure at the base when the backfill is (i) dry, (ii) saturated and (iii) submerged, and also the resultant active force in each case. In addition determine the total water pressure under the submerged condition.

Solution

$$e = 0.5 \text{ and } G = 2.7, \gamma_d = \frac{G\gamma_w}{1+e} = \frac{2.7}{1+0.5} \times 9.81 = 17.66 \text{ kN/m}^3$$

Saturated unit weight

$$\gamma_{sat} = \frac{(G+e)\gamma_w}{1+e} = \frac{2.7+0.5}{1+0.5} \times 9.81 = 20.92 \text{ kN/m}^3$$

Submerged density

$$\gamma_b = \gamma_{sat} - \gamma_w = 20.92 - 9.81 = 11.1 \text{ kN/m}^3$$

For $\phi = 30$, $\quad K_A = \dfrac{1-\sin\phi}{1+\sin\phi} = \dfrac{1-\sin30°}{1+\sin30°} = \dfrac{1}{3}$

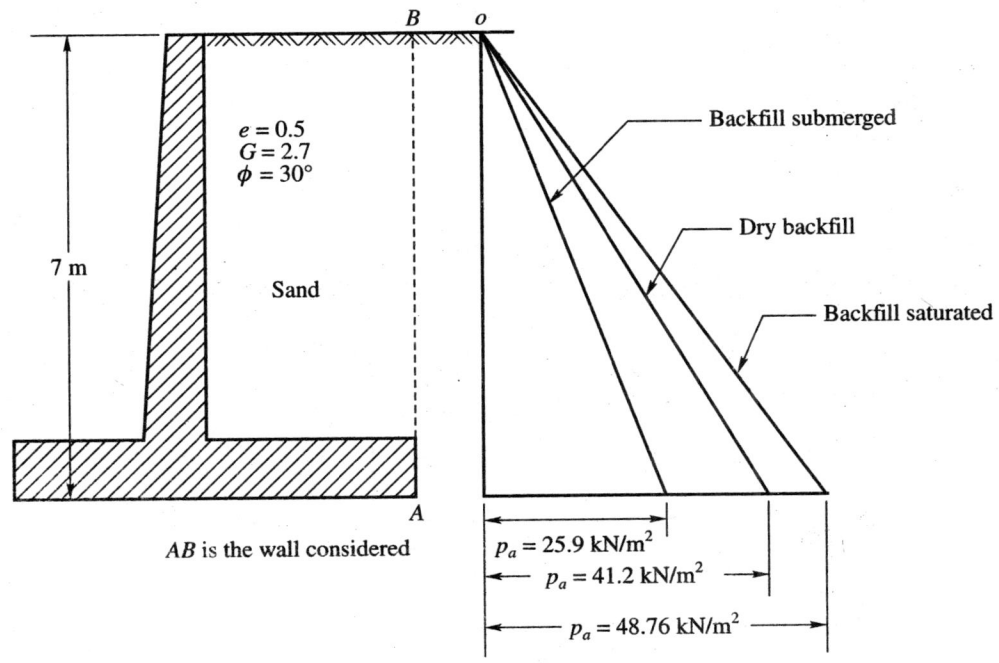

$p_a = 25.9 \text{ kN/m}^2$

$p_a = 41.2 \text{ kN/m}^2$

$p_a = 48.76 \text{ kN/m}^2$

$e = 0.5$
$G = 2.7$
$\phi = 30°$

7 m

Sand

AB is the wall considered

Backfill submerged

Dry backfill

Backfill saturated

Fig. Ex. 14.2

Active earth pressure at the base is

(i) for dry backfill,

$$P_a = K_A \gamma_d H = \frac{1}{3} \times 17.66 \times 7 = 41.2 \text{ kN/m}^2$$

(ii) for saturated backfill,

$$P_a = K_A \gamma_{sat} H = \frac{1}{3} \times 20.9 \times 7 = 48.76 \text{ kN/m}^2$$

(iii) for submerged backfill,

$$P_a = K_A \gamma_b H = \frac{1}{3} \times 11.1 \times 7 = 25.9 \text{ kN/m}^2$$

Example 14.3

For the earth retaining structure shown in Fig. Ex. 14.3, construct the earth pressure diagram for the active state and determine the total thrust per unit length of the wall.

Solution

For $\phi = 30°$, $K_A = \dfrac{1 - \sin 30°}{1 + \sin 30°} = \dfrac{1}{3}$

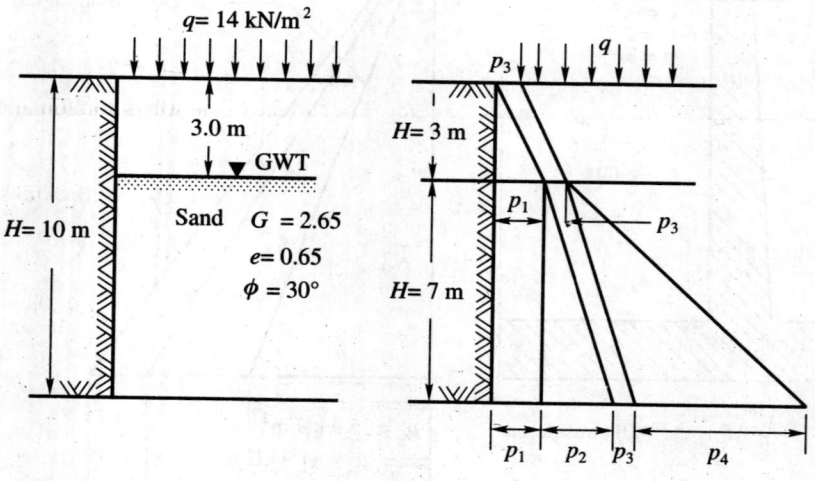

(a) Given system   (b) Pressure diagram

Fig. Ex. 14.3

Dry unit weight $\gamma_d = \dfrac{G\gamma_w}{1+e} = \dfrac{2.65}{1+0.65} \times 9.81 = 15.7 \text{ kN/m}^3$

$$\gamma_d = \dfrac{(G-1)\gamma_w}{1+e} = \dfrac{2.65-1}{1+0.65} \times 9.81 = 9.81 \text{ kN/m}^3$$

Assuming the soil above water table be dry, [Refer Fig. Ex. 14.3(b)].

$p_1 = K_A \gamma_d H_1 = \dfrac{1}{3} \times 15.7 \times 3 = 15.7 \text{ kN/m}^2$

$p_2 = K_A \gamma_b H_2 = \dfrac{1}{3} \times 9.81 \times 7 = 22.9 \text{ kN/m}^2$

$p_3 = K_A \times q = \dfrac{1}{3} \times 14 = 4.66 \text{ kN/m}^2$

$p_4 = (K_A)_w \gamma_w H_2 = 1 \times 9.81 \times 7 = 68.7 \text{ kN/m}^2$

Total thrust = summation of area of different parts of pressure diagram

$$= \dfrac{1}{2} p_1 H_1 + p_1 H_2 + \dfrac{1}{2} p_2 H_2 + p_3 (H_1 + H_2) + \dfrac{1}{2} p_4 H_4$$

$$= \dfrac{1}{2} \times 15.7 \times 3 + 15.7 \times 7 + \dfrac{1}{2} \times 22.9 \times 7 + 4.66(7+3) + \dfrac{1}{2} \times 68.7 \times 7$$

$$\approx 501 \text{ kN/m}$$

Example 14.4

A retaining wall with a vertical back of height 7.32 m supports a cohesionless soil of unit weight 17.3 kN/m³ and an angle of shearing resistance $\phi = 30°$. The surface of the soil is horizontal. Determine the magnitude and direction of the active thrust per meter of wall using Rankine's theory.

Solution

For the condition given here, Rankine's theory disregards the friction between the soil and the back of the wall.

The coefficient of active earth pressure K_A is

$$K_A = \dfrac{1-\sin\phi}{1+\sin\phi} = \dfrac{1-\sin 30°}{1+\sin 30°} = \dfrac{1}{3}$$

The lateral active earth pressure P_a is

$$P_a = \dfrac{1}{2} K_A \gamma H^2 = \dfrac{1}{2} \times \dfrac{1}{3} \times 17.3 \times (7.32)^2 = 154.5 \text{ kN}$$

Example 14.5

A wall of 8 m height retains sand having a density of 1.936 Mg/m^3 and an angle of internal friction of 34°. If the surface of the backfill slopes upwards at 15° to the horizontal, find the active thrust per unit length of the wall. Use Rankine's conditions.

Solution

There can be two solutions: analytical and graphical. The analytical solution can be obtained from Eqs (14.25) and (14.24), viz.

$$P_a = \frac{1}{2} K_A \gamma H^2$$

where $K_A = \cos \beta \times \dfrac{\cos \beta - \sqrt{\cos^2 \beta - \cos^2 \phi}}{\cos \beta + \sqrt{\cos^2 \beta - \cos^2 \phi}}$

where, $\beta = 15°$, $\cos \beta = 0.9659$ and $\cos^2 \beta = 0.933$

and $\phi = 34°$ gives $\cos^2 \phi = 0.688$

Hence $K_A = 0.966 \times \dfrac{0.966 - \sqrt{0.933 - 0.688}}{0.966 + \sqrt{0.933 - 0.688}} = 0.311$

$$\gamma = 1.936 \times 9.81 = 19.0 \, \text{kN/m}^3$$

Hence $P_a = \dfrac{1}{2} \times 0.311 \times 19(8)^2 = 189 \, \text{kN/m wall}$

Graphical Solution

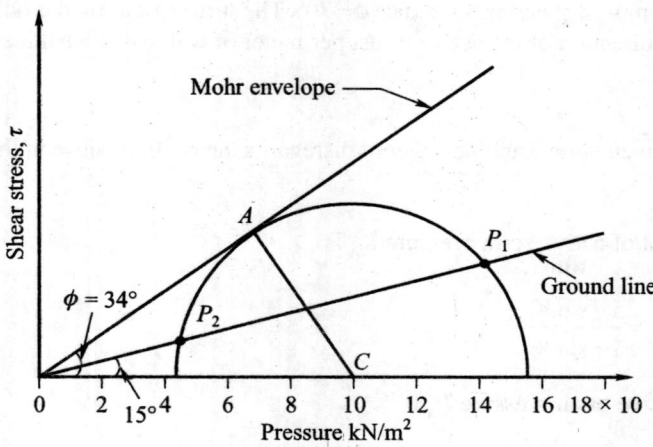

Fig. Ex. 14.5

Vertical stress at a depth $z = 8$ m is

$$\gamma H \cos\beta = 19 \times 8 \times \cos 15° = 147 \text{ kN/m}^2$$

Now draw the Mohr envelope at an angle of 34° and the ground line at an angle of 15° with the horizontal axis as shown in Fig. Ex. 14.5.

Using a suitable scale plot $OP_1 = 147$ kN/m².

(i) Centre of circle C lies on the horizontal axis,

(ii) Circle passes through point P_1, and

(iii) Circle is tangent to the Mohr envelope.

The point P_2, at which the circle cuts the ground line represents earth pressure. The length OP_2 measures 47.5 kN/m².

Hence, thrust per unit length, $P_a = \dfrac{1}{2} \times 47.5 \times 8 = 190 \text{ kN/m}^2$

Example 14.6

A rigid retaining wall 5 m high supports a backfill of cohesionless soil with $\phi = 30°$. The water table is below the base of the wall. The backfill is dry and has a unit weight of 18 kN/m³. Determine Rankine's passive earth pressure per meter length of the wall.

Solution

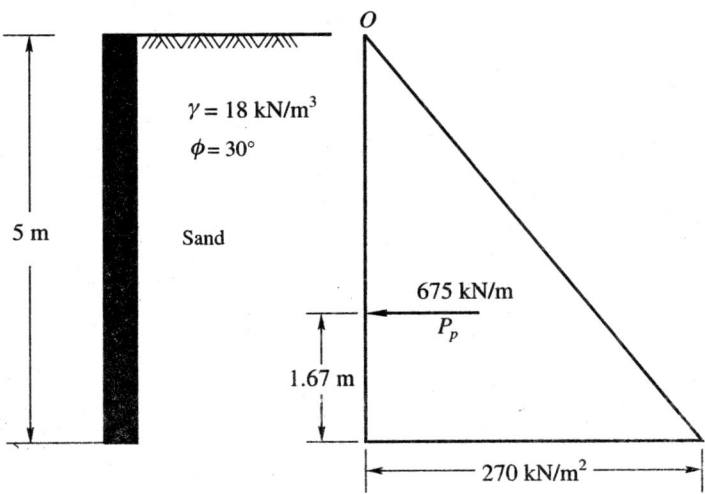

Fig. Ex. 14.6

From Eq. (14.15a)

$$K_p = \frac{1+\sin\phi}{1-\sin\phi} = \frac{1+\sin30°}{1-\sin30°} = \frac{1+0.5}{1-0.5} = 3$$

p_p at $A = K_p \gamma H = 3 \times 18 \times 5 = 270$ kN/m^2

From Eq. (14.15)

$$P_P = \frac{1}{2}K_P \gamma H^2 = \frac{1}{2} \times 3 \times 18 \times 5^2 = 675 \text{ kN/m length of wall}$$

The pressure distribution is given in Fig. Ex. 14.6.

Example 14.7

A counterfort wall of 10 m height retains a non-cohesive backfill. The void ratio and angle of internal friction of the backfill respectively are 0.70 and 30° in the loose state and they are 0.40 and 40° in the dense state. Calculate and compare active and passive earth pressures for both the cases. Take the specific gravity of solids as 2.7.

Solution

(i) In the loose state, $e = 0.70$ which gives

$$\gamma_d = \frac{G\gamma_w}{1+e} = \frac{2.7}{1+0.7} \times 9.81 = 15.6 \text{ kN/m}^3$$

For $\phi = 30°$, $K_A = \dfrac{1-\sin\phi}{1+\sin\phi} = \dfrac{1-\sin30°}{1+\sin30°} = \dfrac{1}{2}$ and $K_p = 3$

$$\text{Max.} p_a = K_A \gamma_d H = \frac{1}{3} \times 15.6 \times 10 = 52 \text{ kN/m}^2$$

$$\text{Max.} p_p = K_P \gamma_d H = 3 \times 15.6 \times 10 = 468 \text{ kN/m}^2$$

(ii) In the dense state, $e = 0.40$, which gives,

$$\gamma_d = \frac{2.7}{1+0.4} \times 9.81 = 18.93 \text{ kN/m}^3$$

For $\phi = 40°$, $K_A = \dfrac{1-\sin40°}{1+\sin40°} = 0.217$, $K_P = \dfrac{1}{K_A} = 4.6$

$$\text{Max.} p_a = K_A \gamma_d H = 0.217 \times 18.93 \times 10 = 41.1 \text{ kN/m}^2$$

and Max. $p_p = 4.6 \times 18.93 \times 10 = 870.8$ kN/m^2

Comment: The comparison of the results indicates that densification of soil decreases the active earth pressure and increases the passive earth pressure. This is advantageous in the sense hat active earth pressure is a disturbing force and passive earth pressure is a resisting force.

14.9 RANKINE'S ACTIVE EARTH PRESSURE WITH COHESIVE SOILS WITH HORIZONTAL BACKFILL ON SMOOTH VERTICAL WALLS

In Fig. 14.14(a) is shown a prismatic element in a semi-infinite mass with a horizontal surface. The vertical pressure on the base AD of the element at depth z is

$$\sigma_v = \gamma z$$

The horizontal pressure on the element when the mass is in a state of plastic equilibrium may be determined by making use of Mohr's stress diagram [Fig. 14.14(b)].

The Mohr envelopes $O'A$ and $O'B$ for cohesive soils are expressed by Coulomb's equation

$$s = c + \tan\phi \qquad (14.28)$$

The point P_1 on the σ-axis represents the state of stress on the base of the prismatic element. When the mass is in the active state σ_v is the major principal stress σ_1. The horizontal stress σ_h is the minor principal stress σ_3. The Mohr circle of stress C_a passing through P_1 and tangential to the Mohr envelopes $O'A$ and $O'B$ represents the stress conditions in the active state. The relation between the two principal stresses may be expressed by the expression

$$\sigma_1 = \sigma_3 N_\phi + 2c\sqrt{N_\phi} \qquad (14.29)$$

Substituting $\sigma_1 = \gamma z$, $\sigma_3 = p_a$ and transposing we have

$$p_a = \frac{\gamma z}{N_\phi} - \frac{2c}{\sqrt{N_\phi}} = \gamma z K_A - 2c\sqrt{K_A} \qquad (14.30)$$

The active pressure $p_a = 0$ when

$$\frac{\gamma z}{N_\phi} - \frac{2c}{\sqrt{N_\phi}} = 0 \qquad (14.31)$$

That is, p_a is zero at depth z, such that

$$z = z_0 = \frac{2c}{\gamma}\sqrt{N_\phi} \qquad (14.32)$$

At depth $z = 0$, the pressure p_a is

$$p_a = -\frac{2c}{\sqrt{N_\phi}} \qquad (14.33)$$

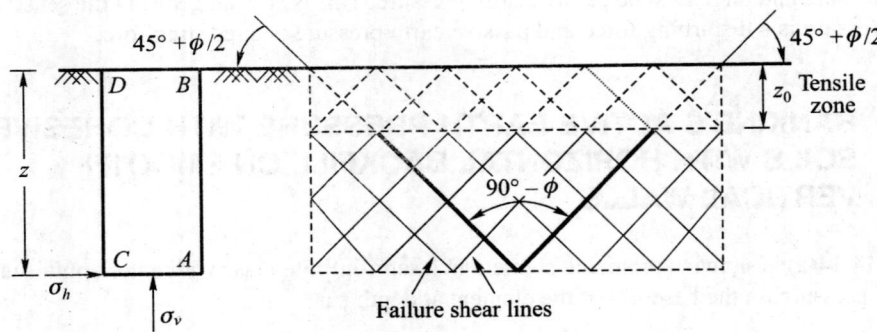

(a) Semi-infinite mass

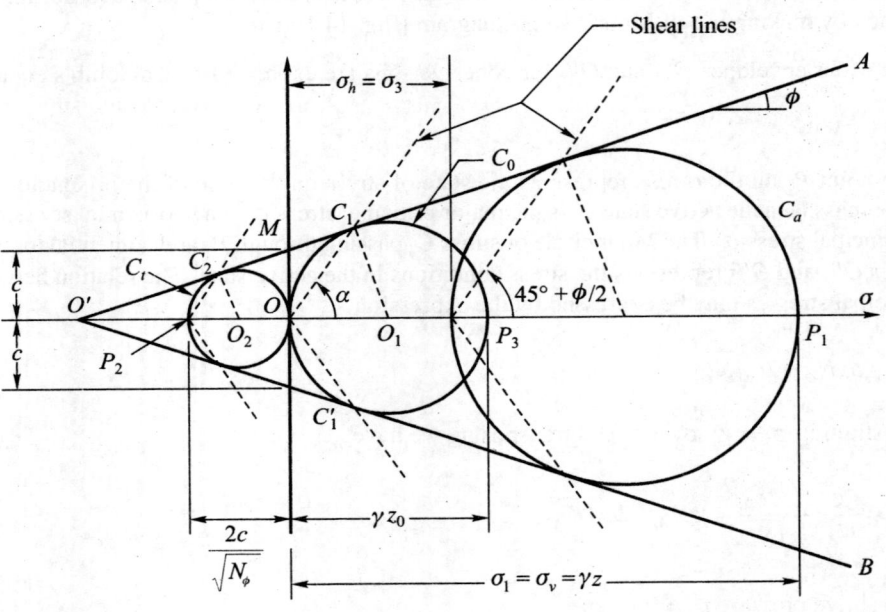

(b) Mohr diagram

Fig. 14.14 Active earth pressure of cohesive soils with horizontal backfills on vertical walls

Equations (14.33) and (14.32) indicate that the active pressure p_a is tensile between depth 0 and z_0. The Eqs (14.32) and (14.33) can also be obtained from Mohr circles C_0 and C_t respectively.

Shear Lines Pattern

The shear lines are shown in Fig. 14.14(a). Up to depth z_0 they are shown dotted to indicate that this zone is in tension.

Total Active Earth Pressure on a Vertical Section

If AB is the vertical section [14.15(a)], the active pressure distribution against this section of height H is shown in Fig. 14.15(b) as per Eq. (14.30). The total pressure against the section is

$$P_a = \int_0^H pz\, dz = \int_0^H \frac{\gamma z}{N_\phi}\, dz - \int_0^H \frac{2c}{\sqrt{N_\phi}}\, dz$$

$$= \frac{1}{2}\gamma H^2 \frac{1}{N_\phi} - 2c\frac{H}{\sqrt{N_\phi}} \tag{14.34}$$

The shaded area in Fig. 14.15(b) gives the total pressure P_a. If the wall has a height

$$H = H_c = \frac{4c}{\gamma}\sqrt{N_\phi} = 2z_0 \tag{14.35}$$

the total earth pressure is equal to zero. This indicates that a vertical bank of height smaller than H_c can stand without lateral support. H_c is called the *critical depth*. However, the pressure against the wall increases from $-2c/\sqrt{N_\phi}$ at the top to $+2c/\sqrt{N_\phi}$ at depth H_c, whereas on the vertical face of an unsupported bank the normal stress is zero at every point. Because of this difference, the greatest depth of which a cut can be excavated without lateral support for its vertical sides is slightly smaller than H_c.

For soft clay, $\phi = 0$, and $N_\phi = 1$

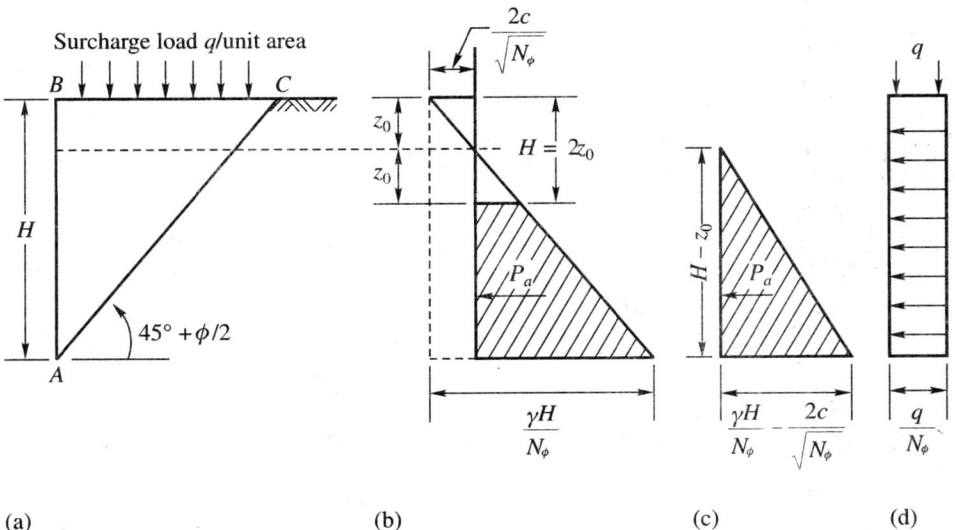

(a)   (b)   (c)   (d)

Fig. 14.15 Active earth pressure on vertical sections in cohesive soils

Therefore, $P_a = \frac{1}{2}\gamma H^2 - 2cH$ (14.36)

and $H_c = \frac{4c}{\gamma}$ (14.37)

Soil does not resist any tension and as such it is quite unlikely that the soil would adhere to the wall within the tension zone of depth z_0 producing cracks in the soil. It is therefore, commonly assumed that the active earth pressure is represented by the shaded area in Fig. 14.15(c).

The total pressure on wall AB is equal to the area of the triangle in Fig. 14.15(c) which is equal to

$$P_a = \frac{1}{2}\left(\frac{\gamma H}{N_\phi} - \frac{2c}{\sqrt{N_\phi}}\right)(H - z_0)$$ (14.38a)

or $P_a = \frac{1}{2}\left(\frac{\gamma H}{N_\phi} - \frac{2c}{\sqrt{N_\phi}}\right)\left(H - \frac{2c}{\gamma}\sqrt{N_\phi}\right)$ (14.38b)

Simplifing, we have

$$P_a = \frac{1}{2}\gamma H^2 \frac{1}{N_\phi} - 2cH \frac{1}{\sqrt{N_\phi}} + \frac{2c^2}{\gamma}$$ (14.38c)

For soft clay, $\phi = 0$

$$P_a = \frac{1}{2}\gamma H^2 - 2cH + \frac{2c^2}{\gamma}$$ (14.39)

It may be noted that here $K_A = 1/N_\phi$

Effect of Surcharge and Water Table

Effect of Surcharge

When a surcharge load q per unit area acts on the surface, the lateral pressure on the wall due to surcharge remains constant with depth as shown in Fig. 14.15(d) for the active condition. The lateral pressure due to a surcharge under the active state may be written as

$$p_{aq} = \frac{q}{N_\phi}$$

The total active pressure due to surcharge load is, therefore

$$P_{aq} = \frac{qH}{N_\phi}$$ (14.40)

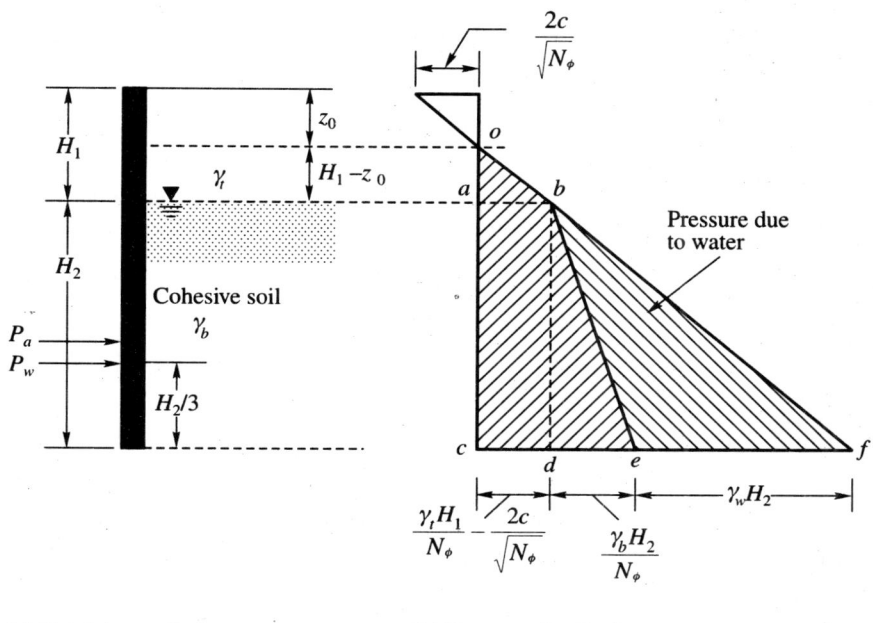

(a) Retaining wall   (b) Pressure distribution

Fig. 14.16   Effect of water table on lateral earth pressure

Effect of Water Table

If the soil is partly submerged, the submerged unit weight below the water table will have to be taken into account in both the active and passive states.

Figure 14.16(a) shows the case of a wall in the active state with cohesive material as backfill. The water table is at a depth of H_1 below the top of the wall. The depth of water is H_2.

The lateral pressure on the wall due to partial submergence is due to soil and water as shown in Fig. 14.16(b). The pressure due to soil = area of the figure *oceb*.

The total pressure due to soil

$$P_a = oab + acdb + bde$$

or $$P_a = \frac{1}{2}(H_1 - z_0)\left(\frac{\gamma_t H_1}{N_\phi} - \frac{2c}{\sqrt{N_\phi}}\right) + \left(\frac{\gamma_t H_1}{N_\phi} - \frac{2c}{\sqrt{N_\phi}}\right)H_2 + \frac{1}{2}\frac{\gamma_b H_2^{\,2}}{N_\phi}$$
(14.41)

After substituting for $z_0 = \dfrac{2c}{\gamma_t}\sqrt{N_\phi}$

and simplifing we have

$$P_a = \frac{1}{2N_\phi}\left(\gamma_t H_1^2 + \gamma_b H_2^2\right) - \frac{2c}{\sqrt{N_\phi}}(H_1 + H_2) + \frac{\gamma_t H_1 H_2}{N_\phi} + \frac{2c^2}{\gamma_t} \qquad (14.42)$$

It is to be noted here that the water table is assumed to lie below the depth H_1 from the ground level.

The total pressure on the wall due to water is

$$P_w = \frac{1}{2}\gamma_w H_2^2 \qquad (14.43)$$

The point of application of P_a can be determined without any difficulty. The point of application P_w is at a height of $H_2/3$ from the base of the wall.

If the backfill material is cohesionless, the terms containing cohesion c in Eq. (14.42) reduces to zero.

14.10 RANKINE'S PASSIVE EARTH PRESSURE OF COHESIVE SOILS WITH HORIZONTAL BACKFILL ON SMOOTH VERTICAL WALLS

If the wall AB in Fig. 14.17(a) is pushed towards the backfill, the horizontal pressure p_h on the wall increases and becomes greater than the vertical pressure σ_v. When the wall is pushed sufficiently inside, the backfill attains Rankine's state of plastic equilibrium. The pressure distribution on the wall may be expressed by the equation

$$\sigma_1 = \sigma_3 N_\phi + 2c\sqrt{N_\phi}$$

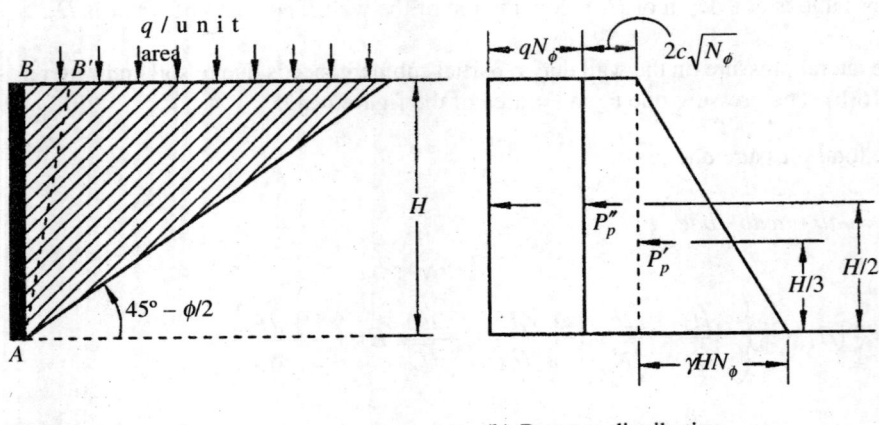

(a) Wall

(b) Pressure distribution

Fig. 14.17 Passive earth pressure on vertical sections in cohesive soils

In the passive state, the horizontal stress σ_h is the major principal stress σ_1 and the vertical stress σ_v is the minor principal stress σ_3. Since $\sigma_3 = \gamma z$, the passive pressure at any depth z may be written as

$$\sigma_1 = \sigma_h = P_p = \gamma z N_\phi + 2c\sqrt{N_\phi} = \gamma z K_P + 2c\sqrt{K_P} \qquad (14.44a)$$

At depth $z = 0$, $\quad p_p = 2c\sqrt{N_\phi} = 2c\sqrt{K_P}$

At depth $z = H$, $\quad p_p = \gamma H N_\phi + 2c\sqrt{N_\phi} = \gamma H K_p + 2c\sqrt{K_P} \qquad (14.44b)$

The distribution of pressure with respect to depth is shown in Fig. 14.17(b). The pressure increases hydrostatically. The total pressure on the wall may be written as a sum of two pressures P'_p and P''_p and we have

$$P'_p = \int_0^H \gamma z N_\phi \, dz = \frac{1}{2}\gamma H^2 N_\phi = \frac{1}{2}\gamma H^2 K_P \qquad (14.45a)$$

This acts at a height $H/3$ from the base.

$$P''_p = \int_0^H 2c\sqrt{N_\phi} \, dz = 2cH\sqrt{N_\phi} = 2cH\sqrt{K_P} \qquad (14.45b)$$

This acts at a height of $H/2$ from the base.

The passive pressure due to surcharge load of q per unit area is

$$P_{pq} = qN_\phi = qK_P$$

The total passive pressure due to surcharged load is

$$P_{pq} = qHN_\phi = qHK_P \qquad (14.46)$$

which acts at mid-height of the wall.

It may be noted here that $N_\phi = K_P$

Example 14.8

A retaining wall has a vertical back and is 7.32 m high. The soil is sandy loam of unit weight 17.3 kN/m³. It has a cohesion of 12 kN/m² and $\phi = 20°$. Neglecting wall friction, determine the active thrust on the wall. The upper surface of the fill is horizontal.

Solution

(Refer to Fig. 14.15)

When the material exhibits cohesion, the pressure on the wall at depth z is given by (Eq. 14.30)

$$p_a = \gamma z K_A + 2c\sqrt{K_A}$$

where $K_A = \dfrac{1-\sin\phi}{1+\sin\phi} = \dfrac{1-\sin20°}{1+\sin20°} = 0.49,\ \sqrt{K_A} = 0.7$

when the depth is small the expression for z is negative because of the effect of cohesion as per theory up to a depth z_0. The soil is in tension and the soil draws away from the wall

$$z_0 = \frac{2c}{\gamma}\sqrt{N_\phi} = \frac{2c}{\gamma}\sqrt{K_P}$$

where $K_P = \dfrac{1+\sin\phi}{1-\sin\phi} = 2.04,$ and $\sqrt{K_P} = 1.43$

Therefore $z_0 = \dfrac{2\times12}{17.3}\times1.43 = 1.98$ m

Pressure at surface ($z = 7.32$ m) is

$$p_a = -2c\sqrt{K_A} = -2\times12\times0.7 = -16.8\ \text{kN/m}^2$$

The negative sign indicates tension.

Pressure at the base of the wall ($z = 7.32$ m) is

$$p_a = 17.3\times7.32\times0.49 - 16.8 = 45.25\ \text{kN/m}^2$$

As per theory, the area of the upper triangle in Fig. 14.15(b) to the left of the pressure axis represents a tensile force which should be subtracted from the compressive force on the lower part of the wall below the depth z_0. Since tension cannot be applied physically between the soil and the wall, this tensile force is neglected. It is therefore commonly assumed that the active earth pressure is represented by the shaded area in Fig. 14.15(c). The total pressure on the wall is equal to the area of the triangle in Fig. 14.15(c).

$$P_a = \frac{1}{2}\left(\gamma H K_A - 2c\sqrt{K_A}\right)(H - z_0)$$

$$= \frac{1}{2}(17.3\times7.32\times0.49 - 2\times12\times0.7)(7.32 - 1.98) = 120.8\ \text{kN/m}$$

Example 14.9

Find the resultant thrust on the wall in Ex. 14.8 if the drains are blocked and water builds up behind the wall until the water table reaches a height of 2.75 m above the bottom of the wall.

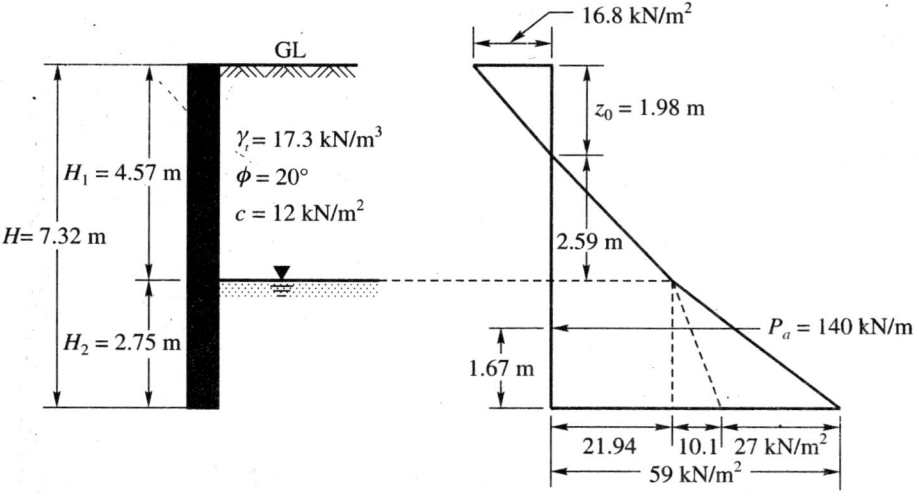

Fig. Ex. 14.9

Solution

For details refer to Fig. 14.16.

As per this figure,

$H_1 = 7.32 - 2.75 = 4.57$ m, $H_2 = 2.75$ m, $H_1 - z_0 = 4.57 - 1.98 = 2.59$ m

The base pressure as per Fig. 14.16(b)

1. $\gamma_t H_1 K_A - 2c\sqrt{K_A} = 17.3 \times 4.57 \times 0.49 - 2 \times 12 \times 0.7 = 21.94$ kN/m²

2. $\gamma_b H_2 K_A = (17.3 - 9.81) \times 2.75 \times 0.49 = 10.1$ kN/m²

3. $\gamma_w H_2 = 9.81 \times 2.75 = 27$ kN/m²

The total pressure $= P_a =$ Pressure due to soil + water

From Eqs (14.41), (14.43), and Fig. 14.16(b)

$P_a = oab + acdb + bde + bef$

$$= \frac{1}{2} \times 2.59 \times 21.94 + 2.75 \times 21.94 + \frac{1}{2} \times 2.75 \times 10.1 + \frac{1}{2} \times 2.75 \times 27$$

$$= 28.84 + 60.3 + 13.89 + 37.13 = 140 \text{ kN/m}$$

Point of application of P_a may be found by taking moments of each area and P_a about the base. Let h be the height of P_a above the base. Now

$$140 \times h = 28.84 \left(\frac{1}{3} \times 2.59 + 2.75 \right) + 60.34 \times \frac{2.75}{2} + 13.89 \times \frac{2.75}{3} + \frac{37.13 \times 2.75}{3}$$

$$= 104.20 + 83.0 + 12.7 + 34.0 = 233.9$$

or $\quad h = \dfrac{233.9}{140} = 1.67 \text{ m}$

Example 14.10

A rigid retaining wall 6 m high has a saturated backfill of soft clay soil. The properties of the clay soil are $\gamma_{sat} = 17.56$ kN/m³, and unit cohesion $c_u = 18$ kN/m². Determine (a) the expected depth of the tensile crack in the soil (b) the active earth pressure before the occurrence of the tensile crack, and (c) the active pressure after the occurrence of the tensile crack.

Solution

At $z = 0$, $p_a = -2c = -2 \times 18 = -36$ kN/m² since $\phi = 0$

At $z = H$, $p_a = \gamma H - 2c = 17.56 \times 6 - 2 \times 18 = 69.36$ kN/m²

(a) From Eq. (14.32), the depth of tensile crack z_0 is (for $\phi = 0$)

$$z_0 = \frac{2c}{\gamma} = \frac{2 \times 18}{17.56} = 2.05 \text{ m}$$

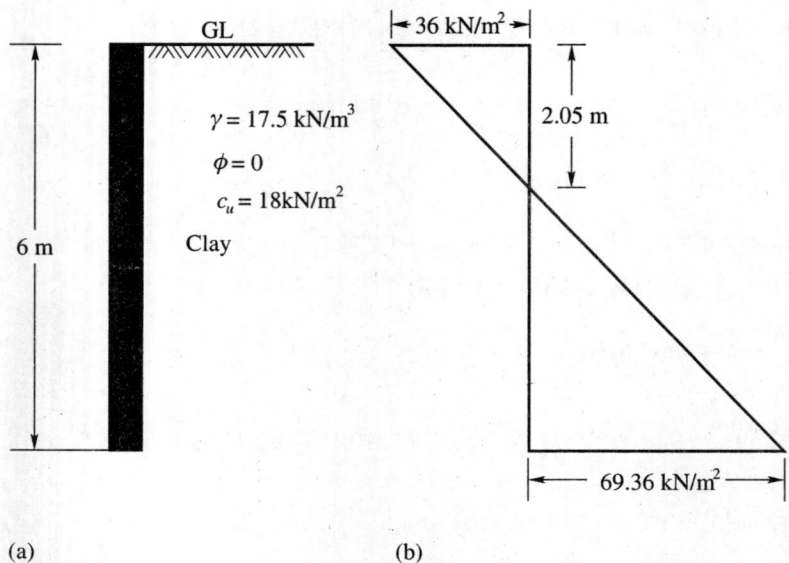

(a)   (b)

Fig. Ex. 14.10

(b) The active earth pressure before crack occurs.

Use Eq. (14.36) for computing P_a

$$P_a = \frac{1}{2}\gamma H^2 - 2cH$$

since $K_A = 1$ for $\phi = 0$. Substituting, we have

$$P_a = \frac{1}{2} \times 7.56 \times 6^2 - 2 \times 18 \times 6 = 316 - 216 = 100 \text{ kN/m}$$

(c) P_a after the occurrence of tensile crack.

Use Eq. (14.38a),

$$P_a = \frac{1}{2}(\gamma H - 2c)(H - z_0)$$

Substituting

$$P_a = \frac{1}{2}(17.56 \times 6 - 2 \times 18)(6 - 2.05) = 137 \text{ kN/m}$$

Example 14.11

A rigid retaining wall of 6 m height (Fig. Ex. 14.11) has two layers of backfill. The top layer to a depth of 1.5 m is sandy clay having $\phi = 20°$, $c = 12.15$ kN/m^2 and $\gamma = 16.4$ kN/m^3. The bottom layer is sand having $\phi = 30°$, $c = 0$, and $\gamma = 17.25$ kN/m^3.

Determine the total active earth pressure acting on the wall and draw the pressure distribution diagram.

Solution

For the top layer,

$$K_A = \tan^2\left(45° - \frac{20}{2}\right) = 0.40, \quad K_P = \frac{1}{0.40} = 2.5$$

The depth of tensile zone, z_0 is

$$z_0 = \frac{2c}{\gamma}\sqrt{K_P} = \frac{2 \times 12.15\sqrt{2.5}}{16.4} = 2.34 \text{ m}$$

Since the depth of the sandy clay layer is 1.5 m, which is less than z_0, the tensile crack develops up to $z = 1.5$ m only.

Chapter 1

K_A for the sandy layer is

$$K_A = \tan^2\left(45° - \frac{\phi}{2}\right) = \tan^2\left(45° - \frac{30}{2}\right) = \frac{1}{3}$$

At depth $z = 1.5$, the vertical pressure σ_v is

$$\sigma_v = \gamma z = 16.4 \times 1.5 = 24.6 \text{ kN/m}^2$$

The active pressure is

$$p_a = K_A \gamma z = \frac{1}{3} \times 24.6 = 8.2 \text{ kN/m}^2$$

At depth 6 m, the effective vertical pressure is

$$\sigma_v = 1.5 \times 16.4 + 4.5 \times 17.25 = 24.6 + 77.62 = 102.23 \text{ kN/m}^2$$

The active pressure p_a is

$$p_a = K_A \sigma_v = \frac{1}{3} \times 102.23 = 34.1 \text{ kN/m}^2$$

The pressure distribution diagram is given in Fig. Ex. 14.11.

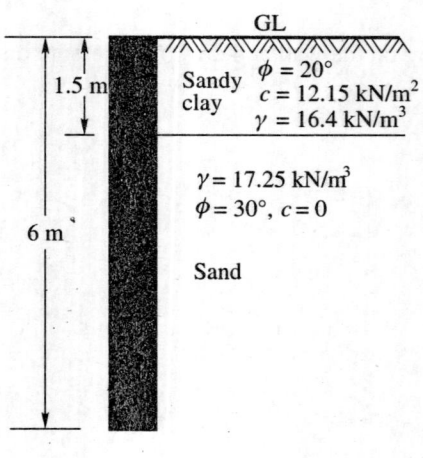
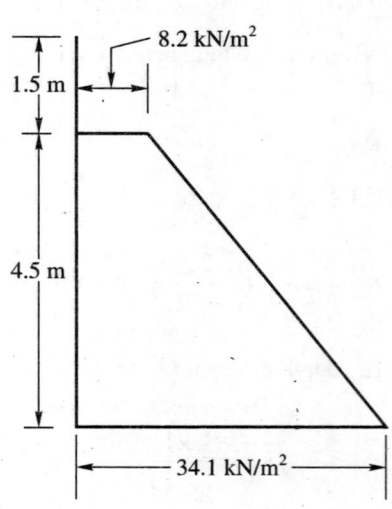

Fig. Ex. 14.11

Example 14.12

A smooth rigid retaining wall 6 m high carries a uniform surcharge load of 12 kN/m². The backfill is clayey sand with the following properties: $\gamma = 16.0$ kN/m³, $\phi = 25°$, and $c = 6.5$ kN/m².

Determine the passive earth pressure and draw the pressure diagram.

Solution

For $\phi = 25°$, the value of K_p is

$$K_P = \frac{1+\sin\phi}{1-\sin\phi} = \frac{1+0.423}{1-0.423} = \frac{1.423}{0.577} = 2.47$$

From Eq. (14.44a), p_p at any depth z is

$$p_p = \gamma z K_P + 2c\sqrt{K_P} = \sigma_v K_P 2c\sqrt{K_P}$$

At depth $z = 0$, $\sigma_v = 12$ kN/m²

$$p_p = 12\times2.47 + 2\times6.5\sqrt{2.47} = 50\,\text{kN/m}^2$$

At $z = 6$ m, $\sigma_v = 12 + 6 \times 16 = 50$ kN/m²

$$p_p = 108\times2.47 + 2\times6.5\sqrt{2.47} = 287.19\,\text{kN/m}^2$$

The pressure distribution is shown in Fig. Ex. 14.12.

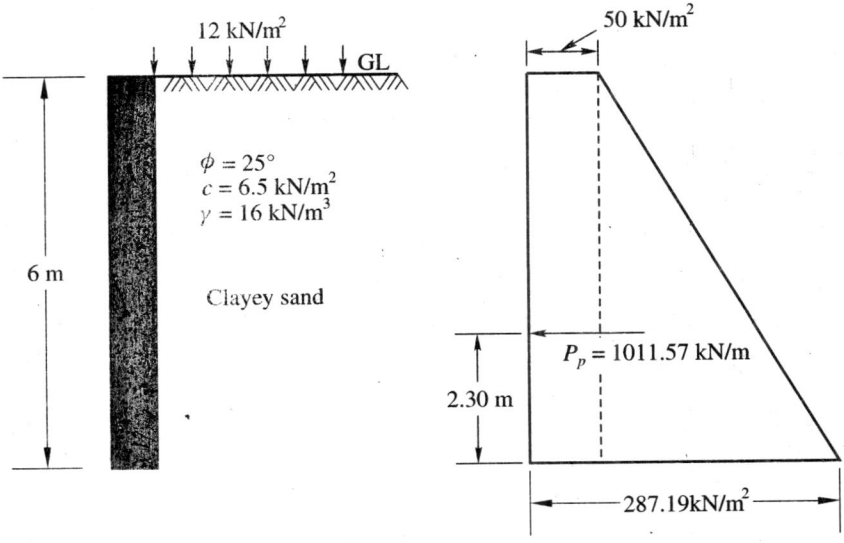

Fig. Ex. 14.12

The total passive pressure P_p acting on the wall is

$$P_p = 50 \times 6 + \frac{1}{2} \times 6(287.19 - 50) = 300 + 711.57 = 1011.57 \text{ kN/m length of wall}$$

Location of resultant

Taking moments about the base

$$P_p \times h = \frac{1}{2} \times 6 \times 50 \times 6 + \frac{1}{2} \times 6 \times 237.195 \times \frac{6}{3}$$

$$= 900 + 1423.14 = 2323.14$$

$$\text{or} \quad h = \frac{2323.14}{P_p} = \frac{2323.14}{1011.57} = 2.30 \text{ m}$$

14.11 COULOMB'S EARTH PRESSURE THEORY FOR SAND FOR ACTIVE STATE

Coulomb made the following assumptions in the development of his theory:

1. The soil is isotropic and homogeneous
2. The rupture surface is a plane surface
3. The failure wedge is a rigid body
4. The pressure surface is a plane surface
5. There is wall friction on the pressure surface
6. Failure is two-dimensional and
7. The soil is cohesionless

Consider Fig. 14.18.

1. AB is the pressure face
2. The backfill surface BE is a plane inclined at an angle β with the horizontal
3. α is the angle made by the pressure face AB with the horizontal
4. H is the height of the wall
5. AC is the assumed rupture plane surface, and
6. θ is the angle made by the surface AC with the horizontal

If AC in Fig. 18(a) is the probable rupture plane, the weight of the wedge W per unit length of the wall may be written as

$W = \gamma A$, where $A = $ area of wedge ABC

Area of wedge $ABC = A = 1/2 \ AC \times BD$

where BD is drawn perpendicular to AC.

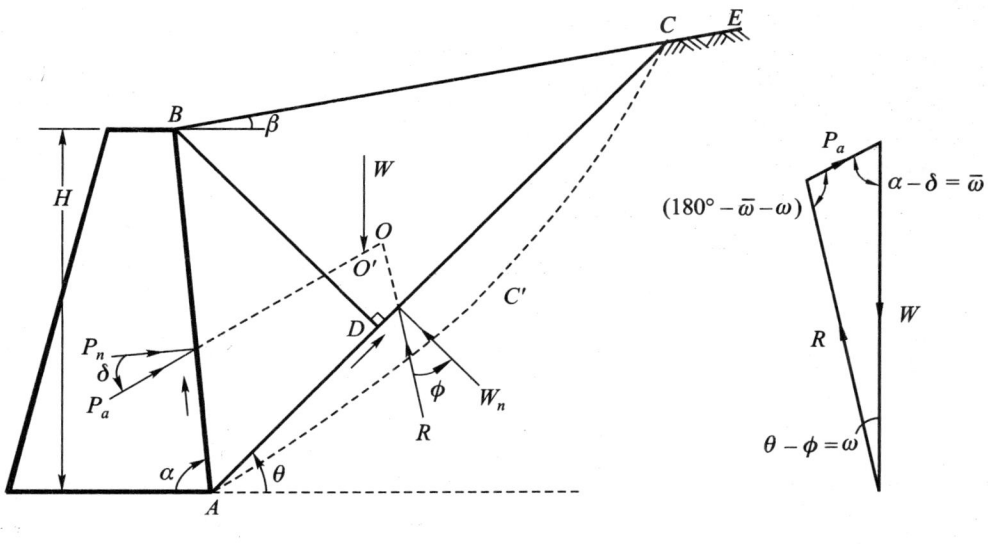

(a) Retaining wall

(b) Polygon of forces

Fig. 14.18 Conditions for failure under active conditions

From the law of sines, we have

$$AC = AB \frac{\sin(\alpha + \beta)}{\sin(\theta - \beta)}, \quad BD = AB \sin(\alpha + \theta), \quad AB = \frac{H}{\sin \alpha}$$

Make the substitution and simplifying we have,

$$W = \gamma A = \frac{\gamma H^2}{2\sin^2 \alpha} \sin(\alpha + \theta) \frac{\sin(\alpha + \beta)}{\sin(\theta - \beta)} \tag{14.47}$$

The various forces that are acting on the wedge are shown in Fig. 14.18(a). As the pressure face AB moves away from the backfill, there will be sliding of the soil mass along the wall from B towards A. The sliding of the soil mass is resisted by the friction of the surface. The direction of the shear stress is in the direction from A towards B. If P_n is the total normal reaction of the soil pressure acting on face AB, the resultant of P_n and the shearing stress is the active pressure P_a making an angle δ with the normal. Since the shearing stress acts upwards, the resulting P_a dips below the normal. The angle δ for this condition is considered *positive*.

As the wedge ABC ruptures along plane AC, it slides along this plane. This is resisted by the frictional force acting between the soil at rest below AC, and the sliding wedge. The resisting shearing stress is acting in the direction from A towards C. If W_n is the normal component of the weight of wedge W on plane AC, the resultant of the normal W_n and the shearing stress is the reaction R. This makes an angle ϕ with the normal since the rupture takes place within the soil itself. Statical equilibrium requires that the three forces P_a, W, and R meet at a point. Since AC is not the actual rupture plane, the three forces do not meet at a point. But if the actual surface of failure $AC'C$ is considered, all the

three forces meet at a point. However, the error due to the nonconcurrence of the forces is ve:
insignificant and as such may be neglected.

The polygon of forces is shown in Fig. 14.18(b). From the polygon of forces, we may write

$$\frac{P_a}{\sin(\theta-\phi)} = \frac{W}{\sin\left(180° - \alpha - \theta + \phi + \delta\right)}$$

$$\text{or } P_a = \frac{W\sin(\theta-\phi)}{\sin(180° - \alpha - \theta + \phi + \delta)} \tag{14.4}$$

In Eq. (14.48), the only variable is θ and all the other terms for a given case are constan
Substituting for W, we have

$$P_a = \frac{\gamma H^2}{2\sin^2\alpha} = \frac{\sin(\theta-\phi)}{\sin\left(180° - \alpha - \theta + \phi + \delta\right)}\left(\sin(\alpha+\theta)\frac{\sin(\alpha+\beta)}{\sin(\theta-\beta)}\right) \tag{14.49}$$

The maximum value for P_a is obtained by differentiating Eq. (14.49) with respect to θ an
equating the derivative to zero, i.e.

$$\frac{dP_a}{d\theta} = 0$$

The maximum value of P_a so obtained may be written as.

$$P_a = \frac{1}{2}\gamma H^2 K_A \tag{14.50}$$

where K_A is the active earth pressure coefficient.

$$K_A = \frac{\sin^2(\alpha+\phi)}{\sin^2\alpha\sin(\alpha-\delta)\left[1+\sqrt{\frac{\sin(\phi+\delta)\sin(\phi-\beta)}{\sin(\alpha-\delta)\sin(\alpha+\beta)}}\right]^2} \tag{14.51}$$

TABLE 14.2

Active earth pressure coefficients K_A for $\beta = 0$ and $\alpha = 90°$

$\phi =$	15	20	25	30	35	40
$\delta = 0$	0.59	0.49	0.41	0.33	0.27	0.22
$\delta = +\phi/2$	0.55	0.45	0.38	0.32	0.26	0.22
$\delta = +2/3\phi$	0.54	0.44	0.37	0.31	0.26	0.22
$d = +\phi$	0.53	0.44	0.37	0.31	0.26	0.22

The total normal component P_n of the earth pressure on the back of the wall is

$$P_n = P_a \cos\delta = \frac{1}{2}\gamma H^2 K_A \cos\delta \qquad (14.52)$$

If the wall is vertical and smooth, and if the backfill is horizontal, we have

$$\beta = \delta = 0 \text{ and } \alpha = 90°$$

Substituting these values in Eq. (14.51), we have

$$K_A = \frac{1-\sin\phi}{1+\sin\phi} = \tan^2\left(45° - \frac{\phi}{2}\right) = \frac{1}{N_\phi} \qquad (14.53)$$

where $N_\phi = \tan^2\left(45° - \frac{\phi}{2}\right)$

The coefficient K_A in Eq. (14.53) is the same as Rankine's. The effect of wall friction is frequently neglected where active pressures are concerned. Table 14.2 makes this clear. It is clear from this table that K_A decreases with an increase of δ and the maximum decrease is not more than 10 percent.

14.12 COULOMB'S EARTH PRESSURE THEORY FOR SAND FOR PASSIVE STATE

In Fig. 14.19, the notations used are the same as in Fig. 14.18. As the wall moves into the backfill, the soil tries to move up on the pressure surface AB which is resisted by friction of the surface. Shearing stress on this surface therefore acts downward. The passive earth pressure P_p is the resultant of the normal pressure P_{pn} and the shearing stress. The shearing force is rotated upward with an angle δ which is again the angle of wall friction. In this case δ is *positive*.

As the rupture takes place along assumed plane surface AC, the soil tries to move up the plane which is resisted by the frictional force acting on that line. The shearing stress therefore, acts downward. The reaction R makes an angle ϕ with the normal and is rotated upwards as shown in the figure.

The polygon of forces is shown in (b) of the Fig. 14.19. Proceeding in the same way as for active earth pressure, we may write the following equations:

$$W = \frac{\gamma H^2}{2\sin^2\alpha}\sin(\alpha+\theta)\frac{\sin(\alpha+\beta)}{\sin(\theta-\beta)} \qquad (14.54)$$

$$P_p = \frac{W\sin(\theta+\phi)}{\sin\left(180° - \theta - \phi - \delta - \alpha\right)} \qquad (14.55)$$

Differentiating Eq. (14.55) with respect to δ and setting the derivative to zero, gives the minimum value of P_p as

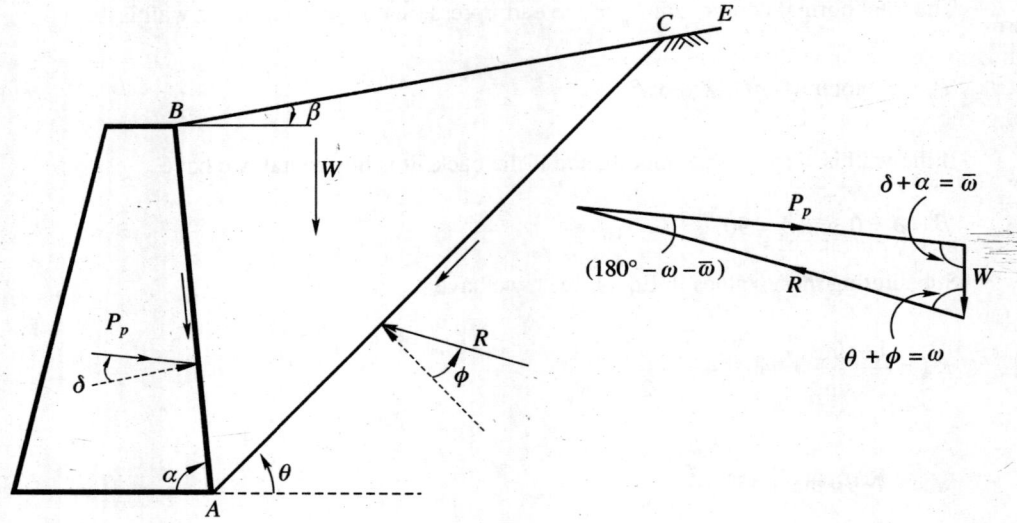

(a) Forces on the sliding wedge   (b) Polygon of forces

Fig. 14.19 Conditions for failure under passive state

$$P_p = \frac{1}{2}\gamma H^2 K_p \tag{14.56}$$

where K_P, which is called as the *passive earth pressure coefficient* is expressed as

$$K_P = \frac{\sin^2(\alpha - \phi)}{\sin^2\alpha \sin(\alpha + \delta)\left[1 - \sqrt{\dfrac{\sin(\phi + \delta)\sin(\phi + \beta)}{\sin(\alpha + \delta)\sin(\alpha + \beta)}}\right]^2} \tag{14.57}$$

Equation (14.57) is valid for both the positive and negative values β and δ.

The total normal component of the passive earth pressure P_p on the back of the wall is

$$P_{pn} = \frac{1}{2}\gamma H^2 K_P \cos\delta \tag{14.58}$$

For a smooth vertical wall with a horizontal backfill, we have

$$K_P = \frac{1 + \sin\phi}{1 - \sin\phi} = \tan^2\left(45° + \frac{\phi}{2}\right) = N_\phi \tag{14.59}$$

Equation (14.59) is also the Rankine's passive earth pressure coefficient. We can see from Eqs (14.53) and (14.59) that

$$K_P = \frac{1}{K_A}$$

Coulomb sliding wedge theory of plane surfaces of failure is valid with respect to passive pressure, i.e., to the resistance of non-cohesive soils only. If wall friction is zero for a vertical wall and horizontal backfill, the value of K_p may be calculated using Eq. (14.59). If wall friction is considered in conjunction with plane surfaces of failure, much too high, and therefore unsafe values of earth resistance will be obtained, especially in the case of high friction angles ϕ. For example for $\phi = \delta = 40°$, and for plane surfaces of failure, $K_p = 92.3$, whereas for curved surfaces of failure $K_p = 17.5$. However, if δ is smaller than $\phi/3$, the difference between the real surface of sliding and Coulomb's plane surface is very small and we can compute the corresponding passive earth pressure coefficient by means of Eq. (14.57). If δ is greater than $\phi/3$, the values of K_p should be obtained by analyzing curved surfaces of failure.

14.13 CONDITIONS UNDER WHICH RANKINE'S AND COULOMB'S FORMULAE ARE APPLICABLE UNDER ACTIVE STATE

Conjugate Rupture Planes Under Active State

When a backfill of cohesionless soil is under an active state of plastic equilibrium due to the stretching of the soil mass at every point in the mass two rupture planes called as conjugate rupture planes are formed called as inner rupture plane and outer rupture plane as shown in Fig. 14.20 at point A. These rupture planes make angles of α_i and α_0 with the vertical. The equations for these angles may be written as (for a sloping backfill)

$$\alpha_i = \frac{90-\phi}{2} + \frac{\varepsilon-\beta}{2}$$

$$\alpha_0 = \frac{90-\phi}{2} - \frac{\varepsilon-\beta}{2}$$

where, $\quad \sin\varepsilon = \frac{\sin\beta}{\sin\phi}$

when $\beta = 0$, $\quad \alpha_i = \frac{90-\phi}{2} = 45° - \frac{\phi}{2}$, $\quad \alpha_0 = \frac{90-\phi}{2} = 45° - \frac{\phi}{2}$

The angle between the two rupture planes $= 90 - \phi$ which is the same as given the Fig. 14.6(a) for an active state for a horizontal backfill surface.

Conditions for Rankine's Formula

1. Wall should be vertical and smooth

2. When walls are inclined, it should not come in the way of the formation of outer rupture plane and the soil should not slide on the wall.

If a retaining wall with a plane back is located with its back AB' (Fig. 14.20) to the left of the vertical AB so that it does not interfere with the formation of the outer plane of rupture, and if the soil

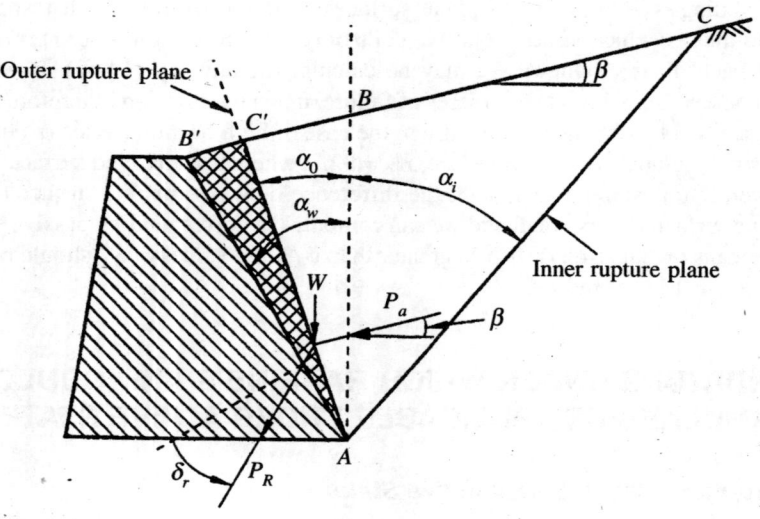

Fig. 14.20   Application of Rankine's active condition to gravity walls

between this plane of rupture and the back of the wall does not slide along the back of the wall but moves with the wall, the wedge is bounded on two sides by planes of rupture AC and AC'.

As shown in Fig. 14.20 plastic state exists within the wedge ACC'. The retaining wall does not interfere with the formation of this wedge since the angle α_w made by the pressure face AB' with the vertical is greater than α_0. In such cases the method of calculating the lateral pressure on AB' is as follows.

1. Apply Rankine's formula (Eq. 14.25) for the vertical section AB.

2. Combine P_a with W, the weight of soil within the wedge ABB' to give the resultant P_R.

Let the resultant P_R in this case make an angle δ_r with the normal to the face of the wall. Let the maximum angle of wall friction be δ_m. If $\delta_r > \delta_m$, the soil slides along the face AB' of the wall. In such an eventuality, Rankine's formula is not recommended but Coulomb's formula may be used.

Conditions for Coulomb's Formula

1. The back of the wall must be plane or nearly place.

2. Coulomb's formula may be applied under all other conditions where the surface of the wall is not smooth and where the soil slides along the surface.

In general the following recommendations may be made for the application of Rankine's or Coulomb's formula without the introduction of significant errors:

1. Use Rankine formula for cantilever and counterfort walls (Fig. 14.21).

2. Use Coulomb's formula for solid and semisolid gravity walls.

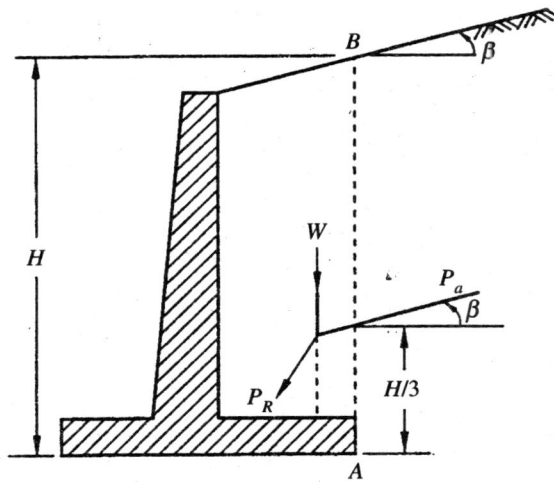

Fig. 14.21   Lateral earth pressure on cantilever walls under active condition

In the case of cantilever walls (Fig. 14.21), P_a is the active pressure acting on the vertical ection AB passing through the heel of the wall. The pressure is parallel to the backfill surface and cts at a height $H/3$ from the base of the wall where H is the height of the section AB. The resultant ressure P_R is obtained by combining the lateral pressure P_a with the weight of the soil W between he section AB and the wall.

14.14 GRAPHICAL SOLUTIONS TO EARTH PRESSURE THEORIES

Analytical solutions to lateral earth pressure problems have been presented so far. Coulomb's equations or both active and passive cases may be applied directly to gravity walls when they satisfy the ollowing basic requirements:

1.   When the surface of wall is plane

2.   When the backfill surface is plane

3.   When the soil is cohesionless

If any of the above requirements is not satisfied, one has to resort to graphical methods for solving the problems.

Similarly, Rankine's expressions may be applied directly for plane backfill surfaces provided the other conditions enunciated in Section 14.13 are also satisfied.

Graphical solutions are very handy and any type of complicated problems can be solved. Two of the graphical methods are described in this chapter.

The methods that are briefly described here are

1.   Culmann's method

2.   Poncelet's or Rebhann's construction

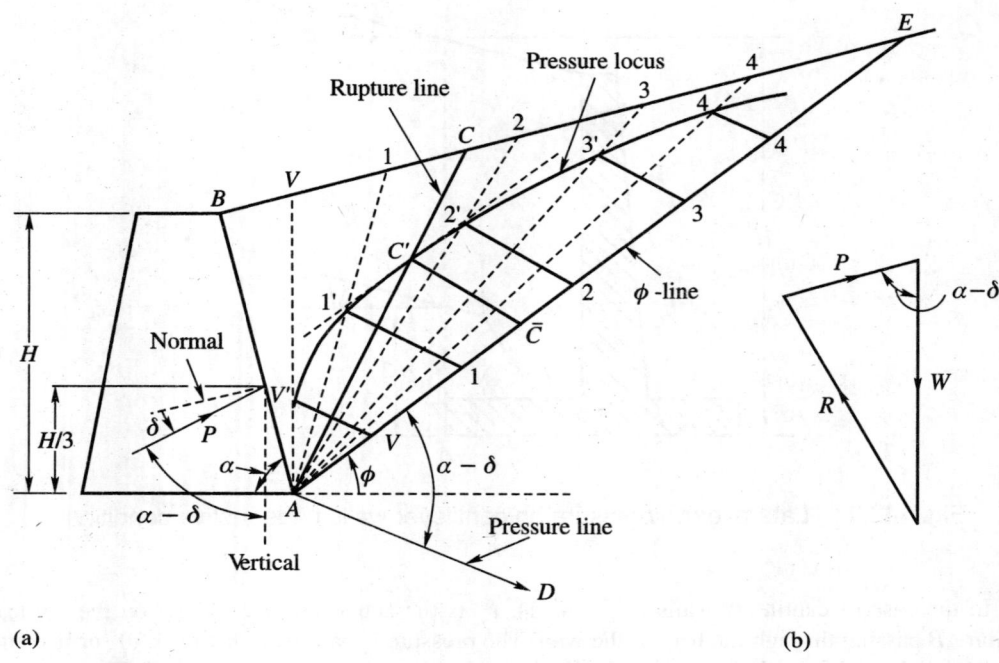

Fig. 14.22   Active pressure by Culmann's method for cohesionless soils

14.15 ACTIVE PRESSURE BY CULMANN'S METHOD FOR COHESIONLESS SOILS WITHOUT SURCHARGE LINE LOAD

The Culmann's method is the same as the trial wedge method. In Culmann's method, the force polygons are constructed directly on the ϕ-line AE taking AE as the load line. The procedure is as follows:

In Fig. 14.22(a) AB is the retaining wall drawn to a suitable scale. The various steps in the construction of the pressure locus are:

1. Draw ϕ-line AE at an angle ϕ to the horizontal.

2. Lay off on AE distances, $AV, A1, A2, A3$, etc. to a suitable scale to represent the weights of wedges $ABV, AB1, AB2, AB3$, etc. respectively.

3. Draw lines parallel to AD from points $V, 1, 2, 3$ to intersect assumed rupture lines $AV, A1, A2, A3$ at points $V', 1', 2', 3'$, etc. respectively.

4. Join points $V', 1', 2', 3'$, etc. by a smooth curve which is the pressure locus.

5. Select point C' on the pressure locus such that the tangent to the curve at this point is parallel to the ϕ-line AE.

6. Draw $C'\overline{C}$ parallel to the pressure line AD. The magnitude of $C'\overline{C}$ in its natural units gives the active pressure P_a.

7.   Join AC' and produce to meet the surface of the backfill at C. AC is the rupture line.

For the plane backfill surface, the point of application of P_a is at a height of $H/3$ from the base of the wall.

Passive Earth Pressure by Culamann's Method for Cohesionless Soils

Figure 14.23 gives Culamann's construction procedure. The various steps in the construction are:

1.   Draw ϕ-line AE at an angle ϕ to the horizontal.

2.   Draw pressure line AD making an angle $(\alpha + \delta)$ with the line AE.

3.   Mark on AE distances $A1$, $A2$, $A3$, etc. to represent weights W_1, W_2, W_3, etc. of the trial wedges $AB1$, $AB2$, $AB3$, etc. respectively.

4.   Draw from points 1, 2, 3, etc. on the ϕ-line, lines parallel to AD to interest the assumed rupture lines $A1$, $A2$, $A3$, etc. at points $1'$, $2'$, $3'$, etc. respectively.

5.   Draw a smooth curve through points $1'$, $2'$, $3'$, etc. to give the pressure locus.

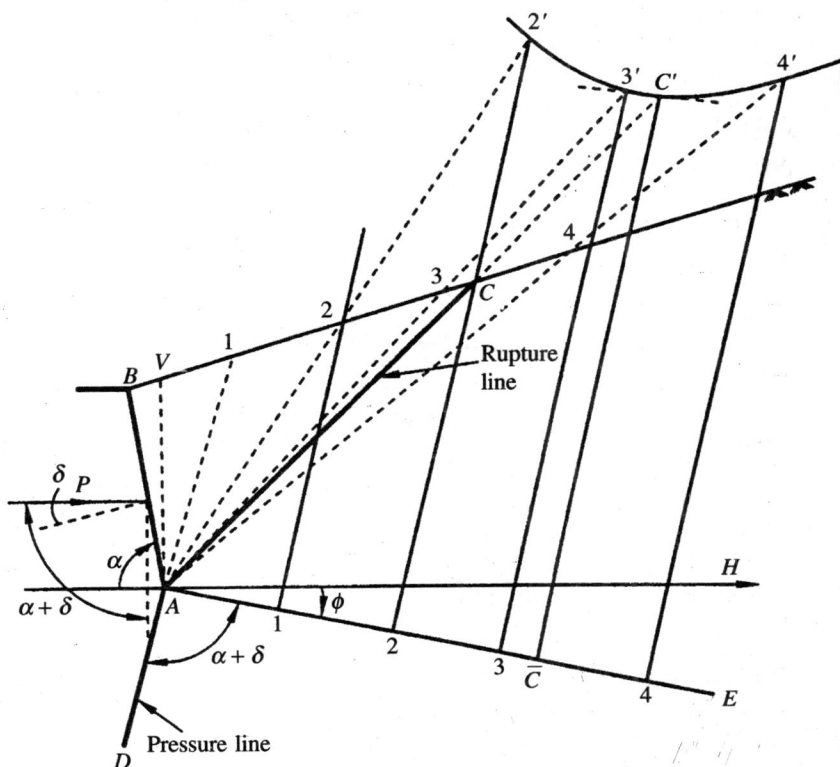

Fig. 14.23   Passive earth pressure by Culamann's method for cohesionless soils

6. Select point C' on the pressure locus such that the tangent to the curve at this point is parallel to the ϕ-line.

7. Draw from C' a line $C'\,\overline{C}$ parallel to AD. The magnitude of $C'\,\overline{C}$ in its natural units gives the active pressure P_p.

8. Join AC' which meets the ground surface at C. AC is the rupture line.

Example 14.13

For a retaining wall system, the following data were available: (i) Height of wall = 7 m, (ii) Properties of backfill: $\gamma_d = 16$ kN/m³, $\phi = 35°$, (iii) angle of wall friction, $\delta = 20°$, (iv) back of wall is inclined at 20° to the vertical (positive batter), and (v) backfill surface is sloping at 1 : 10.

Determine the magnitude of the active earth pressure by Culmann's method.

Solution

(a) Figure Ex. 14.13 shows the wall drawn to a scale of 1 cm = 2 m. ϕ line and pressure lines are also drawn.

(b) The trial rupture lines Bc_1, Bc_2, Bc_3, etc. are drawn by making $Ac_1 = c_1c_2 = c_2c_3$, etc. = 1 cm.

(c) The length of perpendicular from B to the backfill surface = 3.6 cm.

(d) The areas of wedges BAc_1, BAc_2, BAc_3, etc. are respectively equal to 1/2(base lengths Ac_1, Ac_2, Ac_3, etc.) × perpendicular length.

(e) The weights of the wedges in (d) above per meter length of wall may be determined by multiplying the areas by the unit weight of the soil. The results are tabulated below:

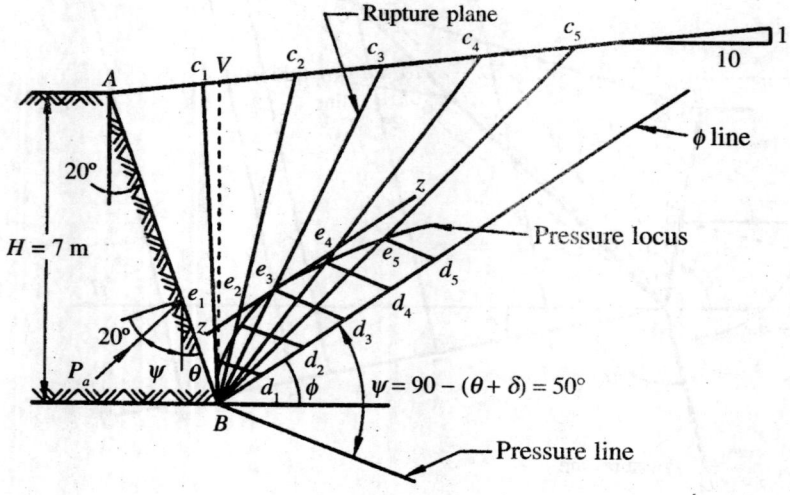

Fig. Ex. 14.13

Wedge	Weight, kN	Wedge	Weight, kN
BAc_1	115	BAc_4	460
BAc_2	230	BAc_5	575
BAc_3	345		

(f) The weights of the wedges BAc_1, BAc_2, etc. are respectively plotted are Bd_1, Bd_2, etc. on the ϕ-line using, a scale 1 cm = 200 kN.

(g) Lines are drawn parallel to the pressure line from points d_1, d_2, d_3 etc. to meet respectively the trial rupture lines Bc_1, Bc_2, Bc_3 etc. at points e_1, e_2, e_3, etc.

(h) The pressure locus is drawn passing through points e_1, e_2, e_3, etc.

(i) Line zz is drawn tangential to the pressure locus at a point at which zz is parallel to the ϕ line. This point coincides with the point e_3.

(j) $e_3 d_3$ gives the active earth pressure when converted to force units.

$P_a = 0.9$ cm $\times$ 200 = 180 kN per meter length of wall.

(k) Bc_3 is the critical rupture plane.

14.16 ACTIVE EARTH PRESSURE FOR COHESIONLESS SOILS BY PONCELET CONSTRUCTION

The Poncelet construction is also called sometimes as Rebhann's construction since the construction is based on the principles enunciated by Rebhann. The procedure of construction is explained below with reference to Fig. 14.24.

1. Draw the retaining wall to a suitable scale and the backfill surface at the required angle.

2. Draw ϕ-line AE to intersect the surface at E.

3. Draw a semicircle on AE as diameter.

4. Draw pressure line AD at an angle $(\alpha - \delta)$ to the ϕ-line.

5. Draw BF parallel to AD.

6. Erect a perpendicular to AE at F to cut the semicircle in G.

7. With centre A and radius AG, draw an arc to cut AE in J.

8. Draw JC parallel to BF.

9. AC is now the rupture plane.

10. With J as centre and JC as the radius draw an arc to cut AE at M.

Now the area of the triangle MJC in its natural units multiplied by the unit weight of the soil gives the active earth pressure P_a, that is,

$$P_a = \frac{1}{2}\gamma l l_n \tag{14.60}$$

where, l = length of side $JC = JM$, l_n = height of perpendicular CN.

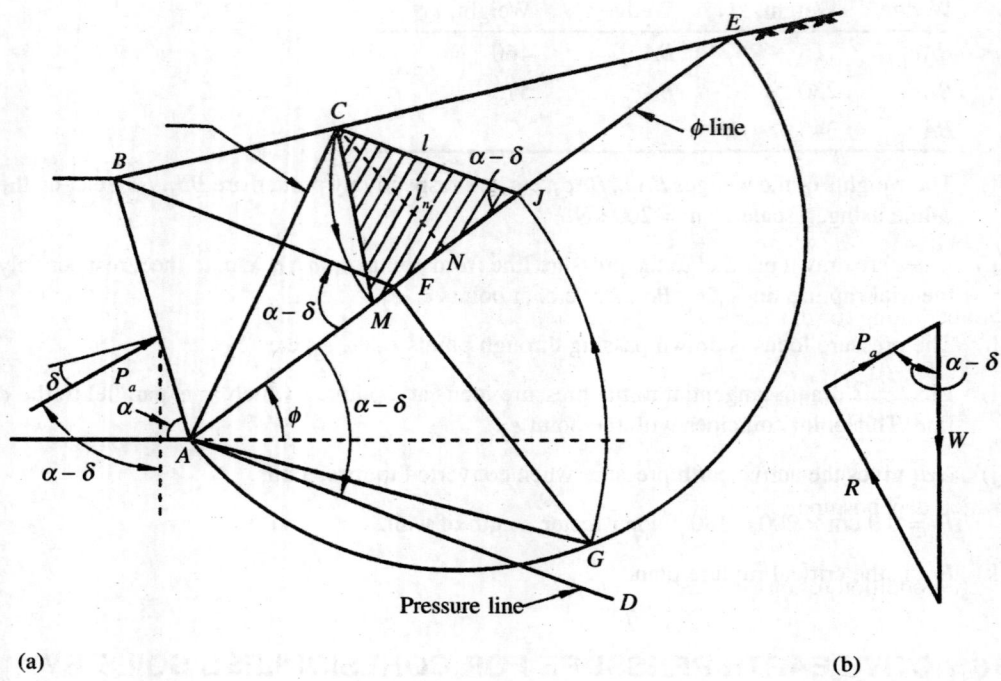

Fig. 14.24 Active earth pressure by Poncelet construction for cohesionless soils

Proof: Let AC in Fig. 14.24(a) be the assumed rupture plane. The polygon of forces acting on the trial wedge ABC is shown in Fig. 14.24(b). The angle between the direction of W and the earth pressure P_a is $(\alpha - \delta)$ in the polygon of forces. The angle between AJ and JC in Fig. 14.24(a) is also $(\alpha - \delta)$. We can therefore consider that the polygon of forces is similar to the triangle AJC. We may therefore write

$$\frac{P_a}{W} = \frac{CJ}{AJ}$$

or $P_a = W\dfrac{CJ}{AJ} = W\dfrac{1}{d}$ (14.61)

where, W is the weight of the assumed failure wedge ABC and $AJ = d$.

The weight W may be written as

W = weight of wedge ABE − weight of wedge ACE

or $W = \dfrac{1}{2}\gamma AE(BF - CJ)\sin(\alpha - \delta) = \dfrac{1}{2}\gamma AE \times BF \times \sin(\alpha - \delta) - \dfrac{1}{2}\gamma AE \times CJ \sin(\alpha - \delta)$

$$= \frac{1}{2}\gamma a(b-l)\sin(\alpha - \delta)$$ (14.62)

where $AE = a$, $BF = b$, $CJ = l$

Substituting for W in Eq. (14.61), we have

$$P_a = \frac{1}{2}\gamma al \frac{(b-l)}{d} \sin(\alpha - \delta)$$

(14.63)

Let $d = (a - kl)$ where $k = \dfrac{EJ}{CJ} = \text{constant}$

Substituting for d, we have

$$P_a = \frac{\gamma al(b-l)}{2(a-kl)} \sin(\alpha - \delta)$$

(14.64)

In Eq. (14.64), the terms a, b, α, and δ are constants for a given problem and the only variables is l which depends upon the position of the rupture plane.

The condition for maximum P_a is $\dfrac{dP_a}{dl} = 0$, that is,

$$\frac{dP_a}{dl} = (a - kl)(b - 2l) + kl(b - l) = 0$$

Simplifying and substituting for k, we have

$$(b - l)a = ld$$

(14.65)

Multiplying both sides of the Eq. (14.65) by $(1/2)\sin(\alpha - \delta)$ we have

$$\frac{1}{2}ba \sin(\alpha - \delta) - \frac{1}{2}al \sin(\alpha - \delta) = \frac{1}{2}ld \sin(\alpha - \delta)$$

or   triangle ABE – triangle ACE = triangle ABC = triangle AJC

(14.66)

Equation (14.66) indicates that the criterion for maximum pressure P_a is that the triangles ABC and AJC should be equal in area and also

$$d = \frac{a}{l}(b - l)$$

(14.67)

Substituting the value of d in the expression for P_a in Eq. (14.63), we have

$$P_a = \frac{1}{2}\gamma al^2 \sin(\alpha - \delta)$$

(14.68)

Since from Fig. 14.24(a), $l_n = l \sin(\alpha - \delta)$, the equation for P_a may be written as

$$P_a = \frac{1}{2}\gamma al_n$$

(14.69)

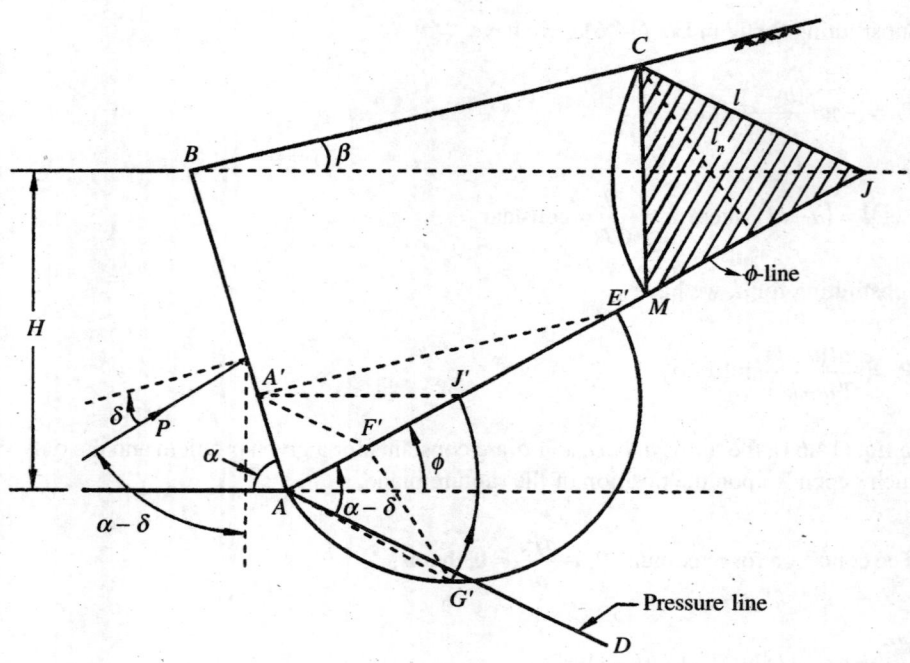

Fig. 14.25 Special case of Poncelet construction when the angle β is approximately equal to ϕ

Special Cases of Poncelet Construction

Two special cases may arise in Poncelet construction. They are

Case 1. When the slope of the ground surface β is approximately equal to the angle of friction ϕ of the backfill material.

Case 2. When the slope of the ground surface β is equal to the angle of friction ϕ of the backfill material.

Case 1. When β and ϕ are nearly equal

In this the ϕ-line may not intersect the ground within the limits of the paper. The construction should therefore be modified so as to get the pressure triangle within the limits of the paper. The construction shown in Fig. 14.25 is explained below:

1. Choose any arbitrary point E' on ϕ-line and construct a semicircle with AE' as diameter.

2. Draw $E'A'$ parallel to ground surface.

3. Draw $A'F'$ parallel to the pressure line.

4. Draw $F'G'$ perpendicular to the ϕ-line.

5. Make $AJ' = AG'$.

6. Join $A'J'$.

7. Draw BJ parallel to $A'J'$.

8. Draw JC parallel to the pressure line.

9. Make $JC = JM$.

P_a = area of triangle $JCM \times$ unit weight of material $= \dfrac{1}{2}\mathcal{H}l_n$

Case 2. When the ground line and the ϕ line are parallel to each other

When the two lines meet at infinity, the pressure triangle may be constructed at any point on the ϕ-line. The construction procedure is (Fig. 14.26).

1. From any point J on the ϕ-line draw line JC parallel to the pressure line AD.

2. Make JM = JC. The pressure triangle is JMC.

$$P_a = \frac{1}{2}\mathcal{H}l_n$$

In this case, the ϕ-line itself is the rupture line.

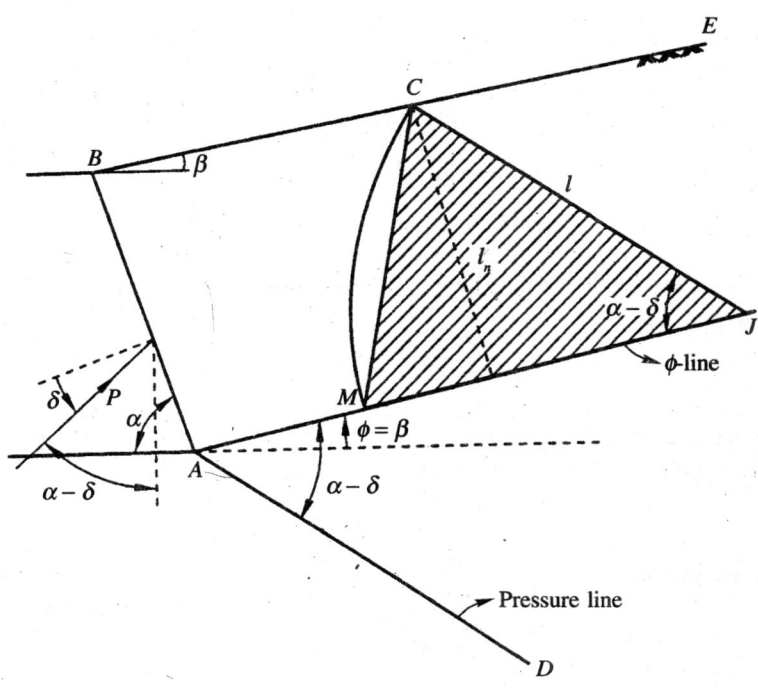

Fig. 14.26 Special cases of Poncelet construction when $\beta = \phi$ for active case

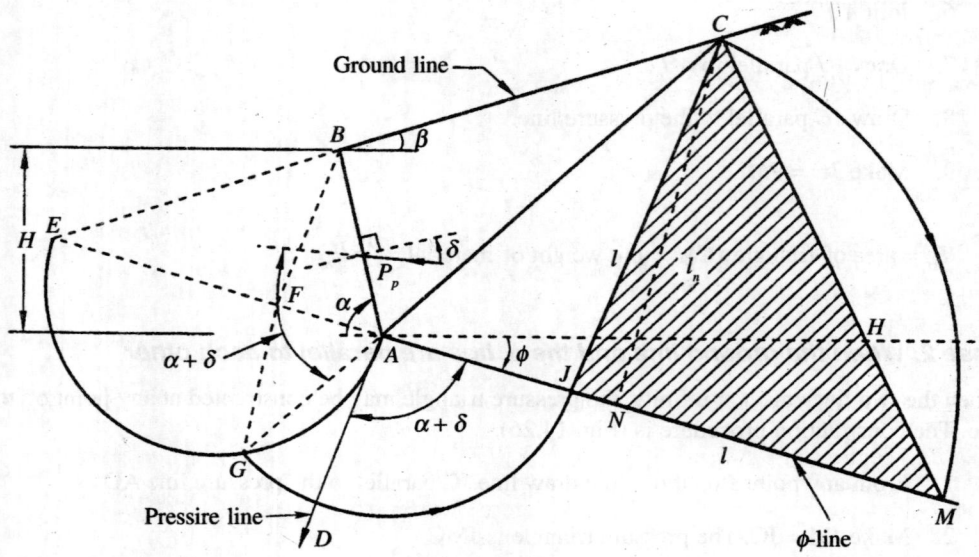

Fig. 14.27 Passive earth pressure by Poncelet construction for cohesionless soils

Poncelet Construction for Passive Pressure

The construct is shown in Fig. 14.27 and the various steps are explained below:

1.   Draw ϕ-line below the horizontal line AH and produce the line backward to meet the ground line at E.

2.   Draw the pressure line AD making angle $\alpha + \delta$ with the ϕ-line.

3.   Construct a semicircle on AE with AE as diameter.

4.   Draw BF parallel to line AD.

5.   Erect perpendicular FG to AE at F.

6.   Mark $AJ = AG$.

7.   Draw JC parallel to line AD.

8.   Join AC. AC is the rupture line.

9.   Mark $JM = JC$. The triangle JCM is the pressure triangle.

The passive pressure $P_p = \dfrac{1}{2}\gamma l l_n$

where, $l = JC$, $l_n = CN$, perpendicular to AM.

The point of application of P_p depends upon the pressure distribution on the surface AB. For plane ground surface, the point of application is at a height $H/3$ from the base of wall.

Example 14.14

Solve Example 14.13 by Poncelet construct. (Same as Rebhann's construction)

Solution

Draw the wall and the backfill as shown in Fig. Ex. 14.14.

Draw ϕ line and pressure line.

With BC as diameter, draw a semicircle BEC.

From A draw a line parallel to pressure line, cutting the ϕ-line at point D.

From D draw DE perpendicular to ϕ-line.

With B as centre and BE as radius, draw an arc EF.

From F draw a line parallel to the pressure line and cutting the ground line at G.

The line BG is the critical failure plane.

Measure GF, which is equal to this 5.42 m in natural units.

Hence, earth pressure is given by,

$$P_a = \frac{1}{2}\gamma(GF)^2 \sin\psi = \frac{1}{2}\times16\times(5.42)^2 \times \sin50° = 180 \text{ kN/m of wall}$$

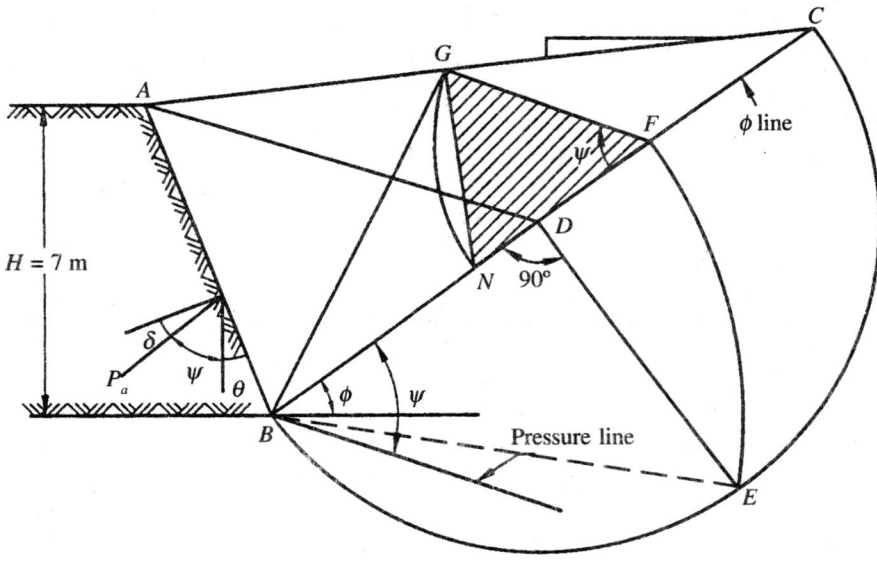

Fig. Ex. 14.14

14.17 LATERAL PRESSURES BY THEORY OF ELASTICITY FOR SURCHARGE LOADS ON THE SURFACE OF BACKFILL

The surcharges on the surface of a backfill parallel to a retaining wall may be any one of the followin

1. A concentrated load
2. A line load
3. A strip load

Lateral Pressure at a Point in a Semi-Infinite Mass due to a Concentrate Load on the Surface

Tests by Spangler (1938), and others indicate that lateral pressures on the surface of rigid walls ca be computed for various types of surcharges by using modified forms of the theory of elasticit equations. Lateral pressure on an element in a semi-infinite mass at depth z from the surface may b calculated by Boussinesq theory for a concentrated load Q acting at a point on the surface. Th equation may be expressed

$$p_h = \frac{Q}{2\pi z^2}\left[3\sin^2\beta\cos^2\beta - \frac{(1-2\mu)\cos^2\beta}{1+\cos\beta}\right] \qquad (14.70)$$

If we write $r = x$ in Fig. 10.1 and redefine the terms as

$x = mH$ and, $z = nH$

where H = height of the rigid wall and take Poisson's ratio $\mu = 0.5$, we may write Eq. (14.70) a:

$$p_h = \frac{3Q}{2\pi H^2}\frac{m^2 n}{\left(m^2 + n^2\right)^{5/2}} \qquad (14.71)$$

Equation (14.71) is strictly applicable for computing lateral pressures at a point in a semiinfinite mass. However, this equation has to be modified if a rigid wall intervenes and breaks the continuity of the soil mass. The modified forms are given below for various types of surcharge loads.

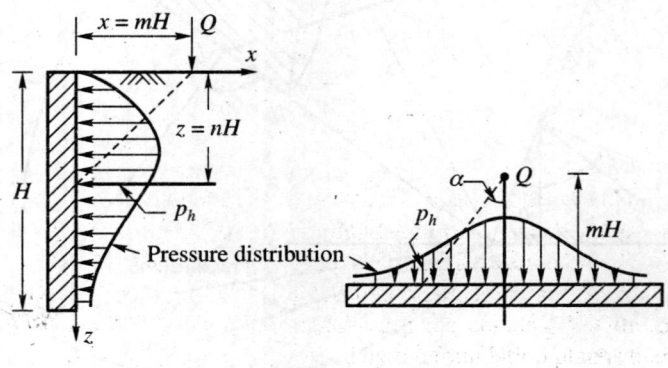

(a) Vertical section   (b) Horizontal section

Fig. 14.28 Lateral pressure against a rigid wall due to a point load

Lateral Pressure on a Rigid Wall Due to a Concentrated Load on the Surface

Let Q be a point load acting on the surface as shown in Fig. 14.28. The various equations are

(a) For $m > 0.4$

$$p_h = \frac{1.77Q}{H^2} \frac{n^2}{\left(m^2+n^2\right)^3}$$

(14.72)

(b) For $m \le 0.4$

$$p_h = \frac{0.28Q}{H^2} \frac{n^2}{\left(0.16+n^2\right)^3}$$

(14.73)

(c) Lateral pressure at points along the wall on each side of a perpendicular from the concentrated load Q to the wall (Fig. 14.28b)

$$p'_h = P_h \cos^2(1.1\alpha)$$

(14.74)

Lateral Pressure on a Rigid Wall due to Line Load

A concrete block wall, conduit laid on the surface, or wide strip loads may be considered as a series of parallel line loads as shown in Fig. 14.29. The modified equations for computing p_h are as follows:

(a) For $m > 0.4$

$$p_h = \frac{4}{\pi} \frac{q}{H} \left[\frac{m^2 n}{\left(m^2+n^2\right)^2} \right]$$

(14.75)

(b) For $m \le 0.4$

$$p_h = \frac{q}{H} \left[\frac{0.203n}{\left(0.16+n^2\right)^2} \right]$$

(14.76)

Lateral Pressure on a Rigid Wall due to Strip Load

A strip load is a load intensity with a finite width, such as a highway, railway line or earth embankment which is parallel to the retaining structure. The application of load is as given in Fig. 14.30.

The equation for computing p_h is

$$p_h = \frac{2q}{\pi} (\beta - \sin\beta \cos2\alpha)$$

(14.77)

The total lateral pressure per unit length of wall due to strip loading may be expressed as (Jarquio, 1981).

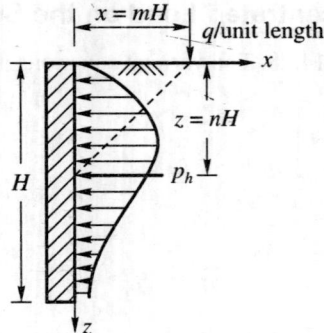

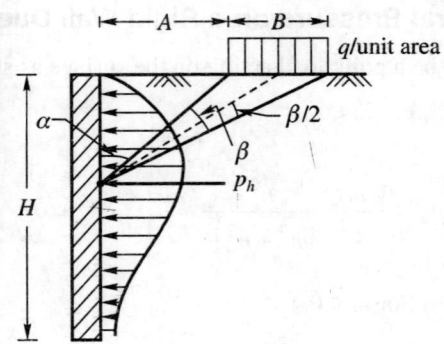

Fig. 14.29 Lateral pressure against a
rigid wall due to line load

Fig. 14.30 Lateral pressure against a
rigid wall due to a strip load

$$p_h = \frac{q}{90}[H(\alpha_2 - \alpha_1)] \tag{14.78}$$

where $\alpha_1 = \tan^{-1}\left(\dfrac{A}{H}\right)$ and $\alpha_2 = \tan^{-1}\left(\dfrac{A+B}{H}\right)$

Example 14.15

A railway line is laid parallel to a rigid retaining wall as shown in Fig. Ex. 14.15. The width of the railway track and its distance from the wall is shown in the figure. The height of the wall is 10 m. Determine

(a) The unit pressure at a depth of 4m from the top of the wall due to the surcharge load

(b) The total pressure acting on the wall due to the surcharge load

Solution

(a) From Eq. (14.77)

The lateral earth pressure p_h at depth 4 m is

$$p_h = \frac{2q}{\pi}(\beta - \sin\beta \cos2\alpha)$$

$$= \frac{2\times60}{3.14}\left[\frac{18.44}{180}\times3.14 - \sin18.44° \cos2\times36.9\right] = 8.92 \text{ kN/m}^2$$

(b) From Eq. (14.78)

$$p_h = \frac{q}{90}[H(\alpha_2 - \alpha_1)]$$

where, $q = 60 \text{ kN/m}^2, H = 10 \text{ m}$

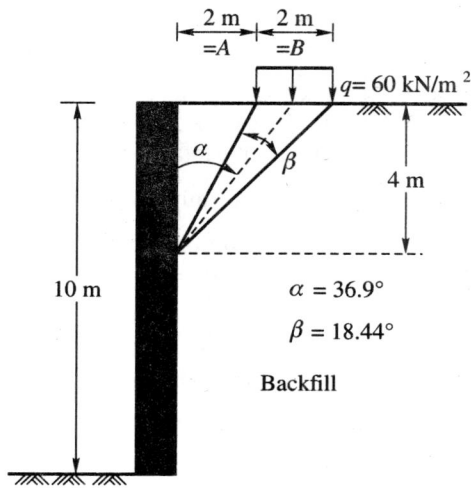

Fig. Ex. 14.15

$$\alpha_1 = \tan^{-1}\frac{A}{H} = \tan^{-1}\frac{2}{10} = 11.31°$$

$$\alpha_2 = \tan^{-1}\left(\frac{A+B}{H}\right) = \tan^{-1}\left(\frac{2+2}{10}\right) = 21.80°$$

$$P_h = \frac{60}{90}[10(21.80-11.31)] \approx 70 \text{ kN/m}$$

14.18 QUESTIONS AND PROBLEMS

14.1   What do you understand by the "state of general plastic equillibrium"? Explain the concept of active and passive earth pressures with the help of Mohr circle and shear strength envelope.

14.2   What is Coulomb's wedge theory of earth pressure? Explain the conditions for obtaining the maximum active earth pressure.

14.3   What are the 'conjugate' stresses? How does Rankine use them for the problem of inclined surface of the backfill?

14.4   What is 'earth pressure at rest'? Derive an equation for determining the magnitude of earth pressure for at rest backfill?

14.5   Differentiate critically the classical earth pressure theories of Rankine and Coulomb.

14.6   State the procedure of Poncelet's graphical construction. Give the proof underlying the construction.

14.7   What earth pressure theories would you recommend in determining the earth pressures in the case of (i) cantilever retaining wall, and (d) gravity retaining wall? Why?

14.8   With suitable illustration, describe the Culmann's trial wedge method of graphical construction for non-cohesive backfill under active earth pressure condition.

14.9   A wall of 6 m height retains non-cohesive backfill of dry unit weight 18 kN/m³ and angle of internal friction 30°. Use Rankine's theory and find total active earth pressure per metre length of the wall. Estimate the change in the total pressure in the following circumstances:

   (i)   The top of backfill carrying a uniformly distributed load of 6 kN/m².

   (ii)   The backfill under submerged condition with the water table at an elevation of 2 m below the top of the wall. Assume $G = 2.65$, and soil above water table remains saturated.

14.10   For an earth retaining wall shown in Fig. Prob. 14.10, find the total active earth pressure per metre length of wall. Use Culmann's graphical construction. Take the angle of wall friction as 20°.

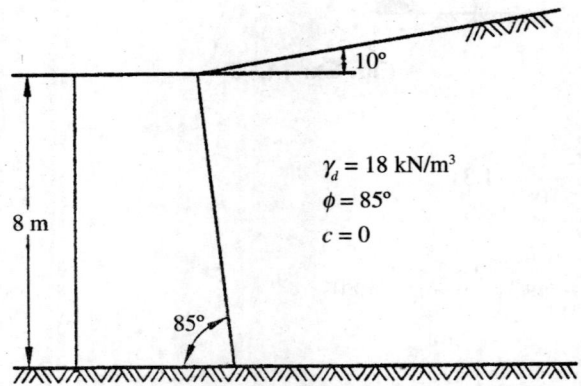

$\gamma_d = 18$ kN/m³
$\phi = 85°$
$c = 0$

8 m

10°

85°

Fig. Prob. 14.10

14.11   Solve Prob. 14.10 by Poncelet's graphical construction. What will be the point of application and the direction of the total pressure?

14.12   A retaining wall of 6 m height having a smooth back, retains a backfill made up of two stratas

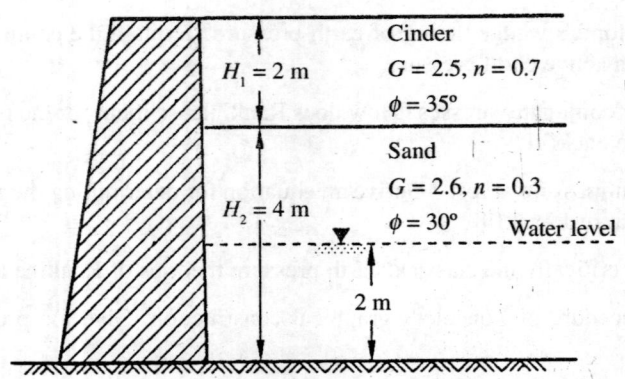

Cinder
$H_1 = 2$ m $G = 2.5, n = 0.7$
                     $\phi = 35°$

Sand
                     $G = 2.6, n = 0.3$
$H_2 = 4$ m $\phi = 30°$ Water level

2 m

Fig. Prob. 14.12

shown in Fig. Prob. 14.12. Construct the active earth pressure diagram and find the magnitude and point of action of the resultant thrust. Comment on the suitability of cinder and sand as backfill.

4.13 A retaining wall of 5.5 m height retains silty-clay backfill, having cohesion $c = 30$ kN/m² and angle of internal friction $\phi = 25°$. Draw the diagram of earth pressure along the height of the wall and determine the total pressure per metre length of wall. Consider the back of wall as smooth and top of backfill as horizontal.

4.14 (a) Calculate the total active pressure on a vertical wall 5 m high retaining sand of unit weight 17 kN/m³ for which $\phi = 35°$. The surface is horizontal and the water table is below the bottom of the wall. (b) Determine the thrust on the wall if the water table rises to a level 2 m below the surface of the sand. The saturated unit weight of the sand is 20 kN/m³.

4.15 The soil conditions adjacent to a rigid retaining wall are shown in Fig. Prob. 14.15, a surcharge pressure of 50 kN/m² is carried on the surface behind the wall. For soil (A) above the water table, $c' = 0$, $\phi' = 38°$, $\gamma' = 18$ kN/m³. For soil (B) below the WT, $c' = 10$ kN/m², $\phi' = 28°$, and $\gamma_{sat} = 20$ kN/m³. Calculate the maximum unit active pressure behind the wall, and the resultant thrust per unit length of the wall.

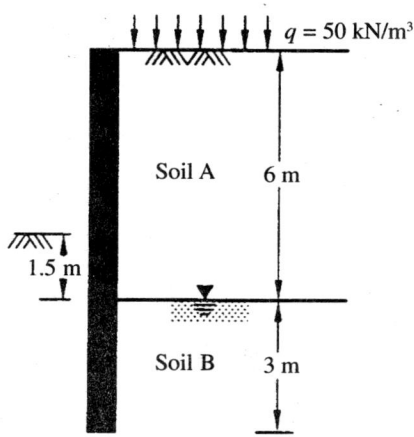

$q = 50$ kN/m³

Soil A 6 m

1.5 m

Soil B 3 m

Fig. Prob. 14.15

4.16 A retaining wall of 6 m high with the surface horizontal retains a back fill of sand of $\phi = 37°$; having a unit weight of 17 kN/m³. Determine the total active thrust on the wall according to Rankine's Theory. If the wall is prevented from yielding, what is the approximate value of the thrust of the wall?

14.17 The depths of soil behind and in front of a rigid retaining wall are 8 m and 3 m respectively, both the soil surfaces being horizontal. The appropriate shear strength parameters for the soil are $c_u = 30$ kN/m², and $\phi_u = 22°$, and the unit weight is 18 kN/m³. Using Rankine theory, determine the total active thrust behind the wall and the total passive resistance in front of the wall.

14.18 Estimate the total at-rest lateral earth pressure at a depth of 5 m in a dense sand deposit (a) with water table much deeper than 5 m, (b) and also for the case of the water table at the ground surface. Assume the saturated unit weight of the sand to the 20.5 kN/m³, and weight of the sand above the water table to be 18.4 kN/m³.

14.19   A smooth vertical wall 3.5 m high retains a mass of dry loose sand. The dry unit weight of the sand is 15.6 kN/m^3 and the angle of friction ϕ is 32°. Estimate the total force per meter acting against the wall (a) if the wall is prevented from yielding and (b) if the wall is allowed to yield.

14.20   The backfill material of a retaining wall of 6 m high is sand having $\phi = 35°$. The water table is 2.5 m below the top of the wall and supports a uniform surcharge of 30 kN/m^2. The unit weight of sand above water table is 18.7 kN/m^3 and the saturated unit weight below the water table is 21.2 kN/m^3. by using Rankine's Theory determine the magnitude of the resultant active force.

14.21   A smooth retaining wall is 4 m high and supports a cohesive backfill with a unit weight of 17 kN/m^3. The shear strength parameters of the soil are cohesion = 10 kPa and $\phi = 10°$. Calculate the total active thrust acting against the wall and the depth to the point of zero lateral pressure.

CHAPTER 15

SHEET PILE WALL

15.1   INTRODUCTION

Sheet pile walls are retaining walls constructed to retain earth, water or any other fill material. These walls are thinner in section as compared to masonry walls described in Chapter 14. Sheet pile walls are generally used in,

1.   Water front structures, for example, in building wharfs, quays, and piers

2.   Building diversion dams, such as cofferdams

3.   River bank protection

4.   Retaining the sides of cuts made in earth

Sheet piles may be of timber, reinforced concrete or steel. Timber piling is used for short spans and to resist light lateral loads. They are mostly used for temporary structures such as braced sheeting in cuts. If used in permanent structures above the water level, they require preservative treatment and even then, their span of life is relatively short. Timber sheet piles are joined to each other by tongue-and-groove joints as indicated in Fig. 15.1. Timber piles are not suitable for driving in soils consisting of stones as the stones would dislodge the joints.

Reinforced concrete sheet piles are precast concrete members, usually with a tongue-and-groove joint. Typical section of piles are shown in Fig. 15.2. These piles are relatively heavy and bulky. They displace large volumes of solid during driving. This large volume displacement of soil tends

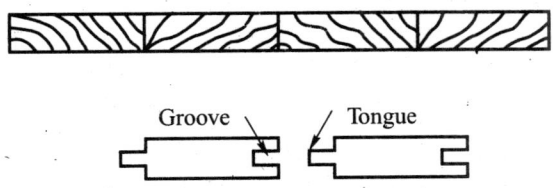

Groove   Tongue

Fig. 15.1   Timber pile wall section

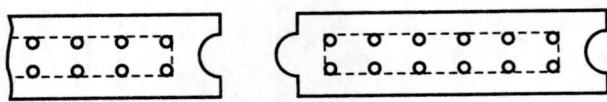

Fig. 15.2 R.C. Sheet pile wall section

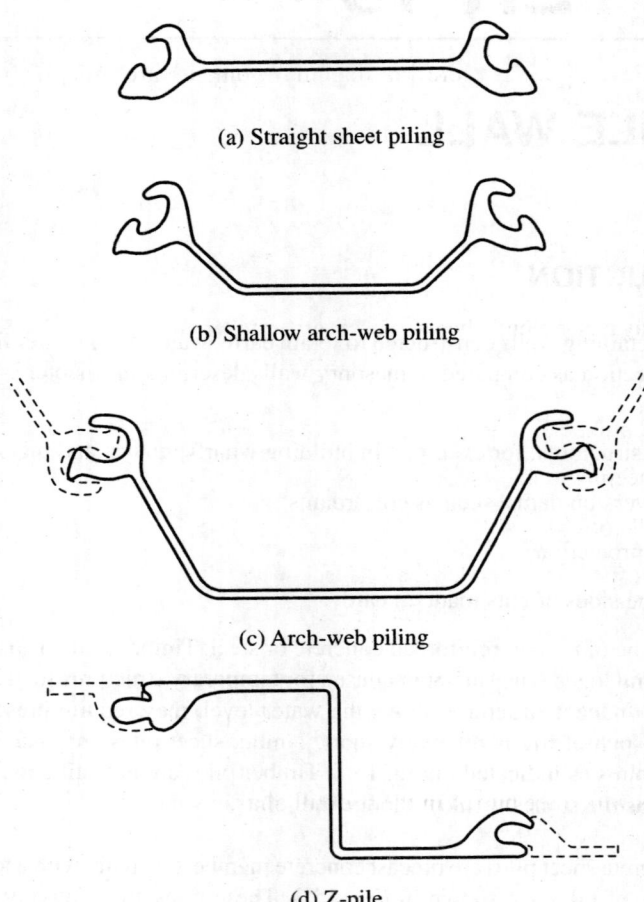

(a) Straight sheet piling

(b) Shallow arch-web piling

(c) Arch-web piling

(d) Z-pile

Fig. 15.3 Sheel pile sections

to increase the driving resistance. The design of piles has to take into account the large driving stresses and suitable reinforcement has to be provided for this purpose.

The most common types of piles used are steel sheet piles. Steel piles possess several advantages over the other types. Some of the important advantages are:

1. They are resistant to high driving stresses as developed in hard or rocky material

2. They are lighter in section

3. They may be used several times.

4. They can be used either below or above water and possess longer life.

5. Suitable joints which do not deform during driving can be provided to have a continuous wall.

6. The pile length can be increased either by welding or bolting.

Steel sheet piles are available in the market in several shapes. Some of the typical pile sections are shown in Fig. 15.3. The archweb and Z-piles are used to resist large bending moments, as in anchored or cantilever walls. Where the bending moments are less, shallow-arch piles with corresponding smaller section moduli can be used. Straight-web sheet piles are used where the web will be subjected to tension, as in cellular cofferdams. The ball-and-socket type of joints, Fig. 15.3(d), offer less driving resistance than the thumb-and-finger joints, Fig. 15.3(c).

15.2   SHEET PILE STRUCTURES

Steel sheet piles may conveniently be used in several civil engineering works. They may be used as:

1. Cantilever sheet piles

2. Anchored bulkheads

3. Braced sheeting in cuts

4. Single cell cofferdams

5. Cellular cofferdams, circular type

6. Cellular cofferdams (diaphragm)

Anchored bulkheads Fig. 15.4(b) serve the same purpose as retaining walls. However, in contrast to retaining walls whose weight always represent an appreciable fraction of the weight of the sliding wedge, bulkheads consist of a single row of relatively light sheet piles of which the lower ends are driven into the earth and the upper ends are anchored by tie or anchor rods. The anchor rods are held in place by anchors which are buried in the backfill at a considerable distance from the bulkhead.

Anchored bulkheads are widely used for dock and harbor structures. This construction provides a vertical wall so that ships may tie up alongside, or to serve as a pier structure, which may jet out into the water. In these cases sheeting may be required to laterally support a fill on which railway lines, roads or warehouses may be constructed so that ship cargoes may be transferred to other areas. The use of an anchor rod tends to reduce the lateral deflection, the bending moment, and the depth of the penetration of the pile.

The cantilever sheet piles depend for their stability on an adequate embedment into the soil below the dredge line. Since the piles are fixed only at the bottom and are free at the top, they are called *cantilever sheet piles*. These piles are economical only for moderate wall heights, since the required section modulus increases rapidly with an increase in wall height, as the bending moment increases with the cube of the cantilevered height of the wall. The lateral deflection of this type of wall, because of the cantilever action, will be relatively large. Erosion and scour in front of the wall, i.e., lowering the dredge line, should be controlled since stability of the wall depends primarily on the developed passive pressure in front of the wall.

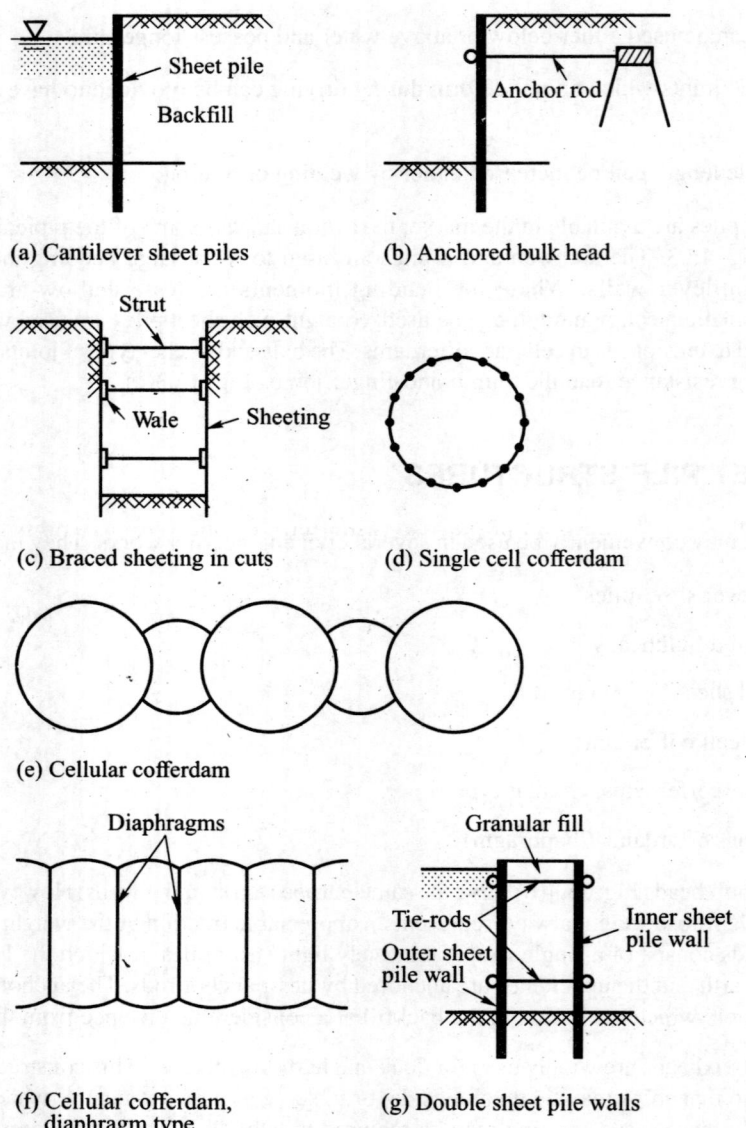

(a) Cantilever sheet piles   (b) Anchored bulk head

(c) Braced sheeting in cuts   (d) Single cell cofferdam

(e) Cellular cofferdam

(f) Cellular cofferdam,   (g) Double sheet pile walls
    diaphragm type

Fig. 15.4 Use of sheet piles

15.3 CANTILEVER SHEET PILE WALLS

When the height of earth to be retained by sheet piling is small, the piling acts as a cantilever. The forces acting on sheet pile walls include:

1. The active earth pressure on the back of the wall which tries to push the wall away from the backfill.

2.   The passive pressure in front of the wall below the dredge line. The passive pressure resists the movements of the wall

The active and passive pressure distributions on the wall are assumed hydrostatic. In the design of the wall, although the Coulomb approach considering wall friction tends to be more realistic, the Rankine approach (with the angle of wall friction $\delta = 0$) is normally used. The pressure due to water may be neglected if the water levels on both sides of the wall are the same. If the difference in level is considerable, the effect of the difference on the pressure will have to be considered. Effective unit weights of soil should be considered in computing the active and passive pressures.

General Principle of Design of Cantilever Sheet Piling

The action of the earth pressure against cantilever sheet piling can be best illustrated by a simple case shown in Fig. 15.5(a). In this case, the sheet piling is assumed to be perfectly rigid. When a horizontal force P is applied at the top of the piling, the upper portion of the piling tilts in the direction of P and the lower portion moves in the opposite direction as shown by a dashed line in the figure. Thus the piling rotates about a stationary point O'. The portion above O' is subjected to a passive earth pressure from the soil on the left side of the piles and an active pressure on the right side of the piling, whereas the lower portion $o'g$ is subjected to a passive earth pressure on the right side and an active pressure on the left side of the piling. At point O' the piling does not move and therefore is subjected to equal and opposite earth pressures (at-rest pressure from both sides) with a net pressure equal to zero. The net earth pressure (the difference between the passive and the active) is represented by $abO'c$ in Fig. 15.5(b). For the purpose of design, the curve $bO'c$ is replaced by a straight line dc. Point d is located at such a location on the line af that the sheet piling is in static equilibrium under the action of force P and the earth pressures represented by the areas ade and ecg. The position of point d can be determined by a trial and error method.

This discussion leads to the conclusion that cantilever sheet piling derives its stability from passive earth pressure on both sides of the piling. However, the distribution of earth pressure is

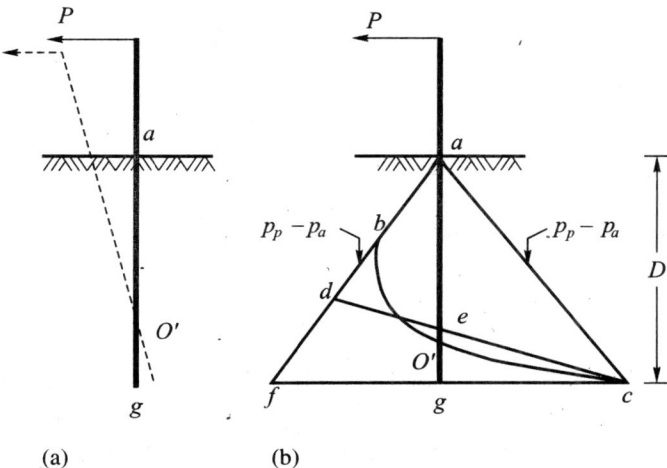

(a)   (b)

Fig. 15.5   Example illustrating earth pressure on cantilever sheet piling

different between sheet piling in granular soils and sheet piling in cohesive soils. The pressure distribution is likely to change with time for sheet pilings in clay.

15.4 DEPTH OF EMBEDMENT OF CANTILEVER WALLS IN SANDY SOILS

The active pressure acting on the back of the wall tries to move the wall away from the backfill. If the depth of embedment is adequate the wall rotates about a point O' situated above the bottom of the wall as shown in Fig. 15.6(a). The types of pressure that act on the wall when rotation is likely to take place about O' are:

1. Active earth pressure at the back of wall from the surface of the backfill down to the point of rotation, O'. The pressure is designated as P_{a_1}.

2. Passive earth pressure in front of the wall from the point of rotation O' to the dredge line. This pressure is designated as P_{p_1}.

3. Active earth pressure in front of the wall from the point of rotation to the bottom of the wall. This pressure is designated as P_{a_2}.

4. Passive earth pressure at the back of wall from the point of rotation O' to the bottom of the wall. This pressure is designated as P_{p_2}.

The pressures acting on the wall are shown in Fig. 15.6(a).

   If the passive and active pressures are algebraically combined, the resultant pressure distribution below the dredge line will be as given in Fig. 15.6(b). The various notations used are:

D	=	Minimum depth of embedment with a factor of safety equal to 1
K_A	=	Rankine active earth pressure coefficient
K_P	=	Rankine passive earth pressure coefficient
K	=	$K_P - K_A$
$\bar{p}_a$	=	Effective active earth pressure acting against the sheet pile at the dredge line level $= \gamma H K_A$
$\bar{p}_p$	=	Effective passive earth pressure at the base of the pile wall and acting towards the backfill $= \gamma D K - \gamma H K_A$
$\bar{p}'_p$	=	Effective passive earth pressure at the base of the sheet pile wall acting against the backfill side of the wall $= \gamma D K + \gamma H K_A$
$\bar{p}''_p$	=	Effective passive earth pressure at level of $O = \gamma y_0 K + \gamma H K_P$
γ	=	Effective unit weight of the soil assumed the same below and above dredge level
y_0	=	Depth of point O below dredge line where the active and passive pressures are equal
$\bar{y}$	=	Height of point of application of the total active pressure P_a above point O
h	=	Height of point G above the base of the wall
D_0	=	Height of point O above the base of the wall

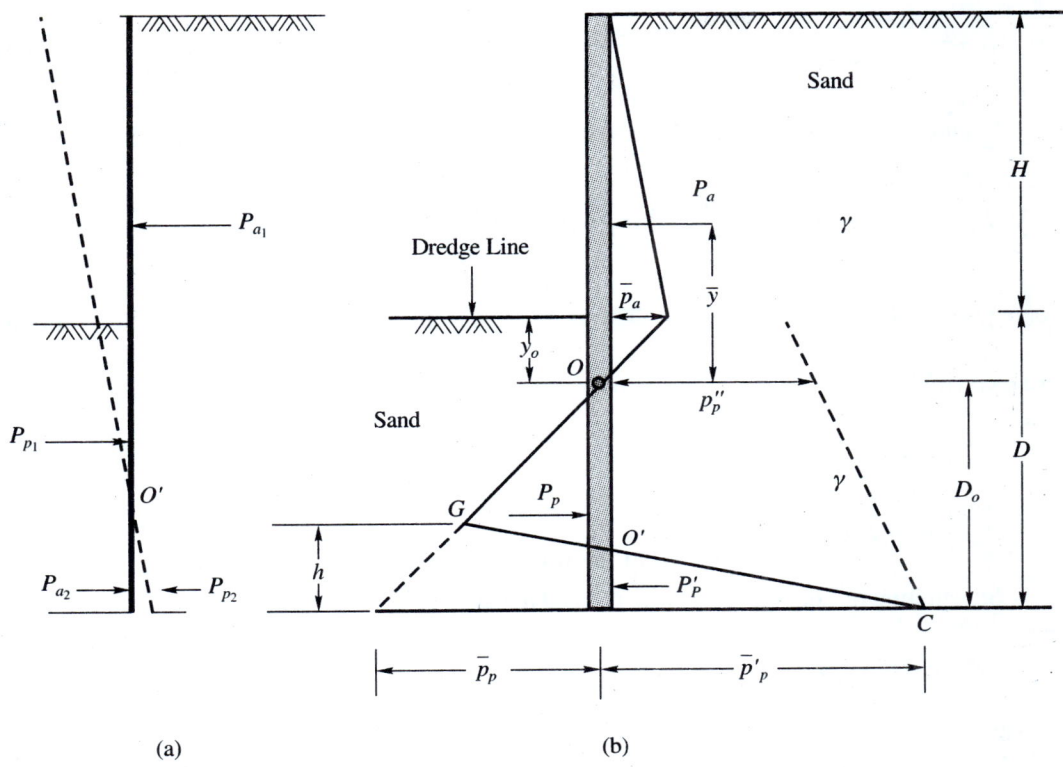

Fig. 15.6 Pressure Distribution on a cantilever wall

Expression for y_0

At point O, the passive pressure acting towards the right should equal the active pressure acting towards the left, that is

$$\gamma y_0 K_p = \gamma (H + y_0) K_A$$

Therefore, $\gamma y_0 (K_p - K_A) = \gamma H K_A$

$$y_0 = \frac{\gamma H K_A}{\gamma (K_p - K_A)} = \frac{\bar{P}_a}{\gamma K} \qquad (15.1)$$

Expression for h

For statical equilibrium, the sum of all the forces in the horizontal direction shall be equal to zero. That is

$$P_a - \frac{1}{2} \bar{P}_p (D - y_0) + \frac{1}{2} \left(\bar{P}_p + \bar{P}'_p \right) h = 0$$

Solving for h, we obtain

$$h = \frac{\bar{P}_p(D - y_0) - 2P_a}{\bar{P}_p + \bar{P}'_p} \tag{15.2}$$

Taking moments of all the forces about the bottom of pile, and equating to zero, we have,

$$P_a(D_0 + \bar{p}) - \frac{1}{2}\bar{P}_p \times D_0 \times \frac{D_0}{3} + \frac{1}{2}(\bar{P}_p + \bar{P}'_p) \times h \times \frac{h}{3} = 0$$

$$\text{or } 6P_a(D_0 + \bar{y}) - \bar{P}_p D_0^2 + (\bar{P}_p + \bar{P}'_p)h^2 = 0 \tag{15.3}$$

we can write

$$\bar{P}_p = \gamma K D_0$$

$$\bar{P}'_p = \bar{P}''_p + \gamma K D_0$$

Substituting in Eq. (15.3) for $\bar{p}_p$, $\bar{p}'_p$ and h and simplifying, we have,

$$D_0^4 + C_1 D_0^3 + C_2 D_0^2 + C_3 D_0 + C_4 = 0 \tag{15.4}$$

where

$$C_1 = \frac{\bar{P}''_p}{\gamma K}$$

$$C_2 = -\frac{8P_a}{\gamma K}$$

$$C_3 = -\left[\frac{6P_a}{(\gamma K)^2}(2\bar{y}\gamma K + p''_p)\right]$$

$$C_4 = -\left[\frac{6P_a\bar{y}p''_p + 4P_a^2}{(\gamma K)^2}\right] \tag{15.4a}$$

The solution of Eq. (15.4) gives the depth D_0. The method of trial and error is generally adopted to solve this equation. The minimum depth of embedment D with a factor of safety equal to 1 is, therefore

$$D = D_0 + y_0 \tag{15.4b}$$

A minimum factor of safety of 1.5 to 2 may be obtained by increasing the minimum depth D by 20 to 40 percent.

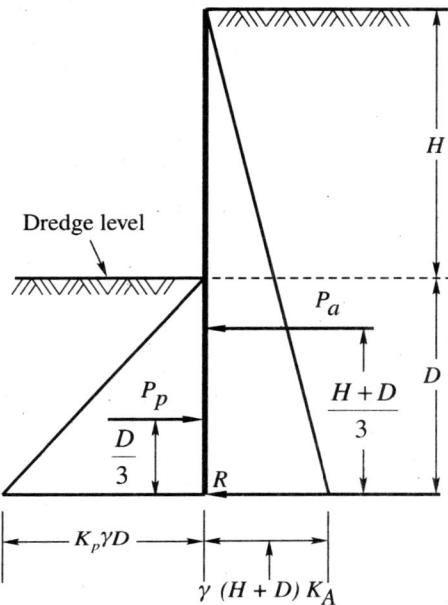

Fig. 15.7 Simplified method of determining D for cantilever sheet pile

(a) Simplified Method

The solution of the fourth degree equation is quite laborious and the problem can be simplified by assuming the passive pressure $\overline{p}'_p$ (Fig. 15.6) as a concentrated force R acting at the foot of the pile. The simplified arrangement is shown in Fig. 15.7.

For equilibrium, the moments of the active pressure on the right and passive resistance on the left about the point of reaction R must balance

$$\frac{1}{3}P_p D - \frac{P_a}{3}(H+D) = 0$$

Now $P_p = \frac{1}{2}K_p \gamma D^2$ and $P_a = \frac{1}{2}K_A \gamma (H+D)^2$

Therefore, $K_p D^3 - K_A (H+D)^3 = 0$

or $KD^3 - 3HD(H+D)K_A = 0$ (15.5)

The solution of Eq. (15.5) gives a value for D which is at least a guide to the required depth. The depth calculated should be increased by at least 20 percent to provide a factor of safety and to allow extra length to develop the passive pressure R. The approximate figures, applicable to cohesionless soils with no hydrostatic pressure, given by Henry, are given in Table 15.1.

TABLE 15.1

Approximate depths of embedment as given by Henry

ϕ	D
20°	2.0 H
25°	1.5 H
30°	1.2 H
35°	0.9 H
40°	0.7 H

Example 15.1

Find the depth of embedment for the sheet-piling shown in Fig.Ex. 15.1(a). The soil has effective unit weight of 17 kN/m³ and an angle of internal friction of 30°.

Solution

(a) Rigorous Analysis

$$\text{For } \phi = 30°, \ K_A = \tan^2(45° - \phi/2) = \tan^2 30 = \frac{1}{3}$$

$$K_p = \frac{1}{K_A} = 3, \ K = K_p - K_A = 3 - \frac{1}{3} = 2.67.$$

The pressure distribution along the sheet pile is assumed as shown in the Fig.Ex.15.1(b)

$$\bar{p}_a = \gamma H K_A = 17 \times 6 \times \frac{1}{3} = 34 \text{ kN/m}^2$$

$$y_0 = \frac{\bar{p}_a}{\gamma(K_p - K_A)} = \frac{34}{17 \times 2.67} = 0.75 \text{ m}$$

$$P_a = \frac{1}{2}\bar{p}_a H + \frac{1}{2}\bar{p}_a y_0 = \frac{1}{2} \times 34 \times 6 + \frac{1}{2} \times 34 \times 0.75$$

$$= 102 + 13 = 115 \text{ kN/metre length of wall.}$$

$$\bar{p}_p = \gamma D(K_p - K_A) - \bar{p}_a = 17 \times D \times 2.67 - 34 = 45.3D - 34$$

$$\bar{p}'_p = \gamma H K_p + \gamma D(K_p - K_A) = 17 \times 6 \times 3 + 17 \times D \times 2.67 = 306 + 45.3D$$

$$\bar{p}''_p = \gamma H K_p + \gamma y_0(K_p - K_A) = 17 \times 6 \times 3 + 17 \times 0.75 \times 2.67 = 340 \text{ kN/m}^2$$

To find $\bar{y}$, $P_a\bar{y} = \frac{1}{2}\bar{P}_aH\left(\frac{H}{3}+y_0\right)+\frac{1}{2}\bar{P}_ay_0\left(\frac{2}{3}y_0\right)$

$= \frac{1}{2}\times34\times6\times(2+0.75)+\frac{1}{2}\times34\times0.75\times\frac{2}{3}\times0.75 = 286.9$

Therefore, $\bar{y} = \dfrac{286.9}{115} = 2.50$ m.

Now D_0 can be found from Eq.(15.4), namely

$$D_0^4 + C_1D_0^3 + C_2D_0^2 + C_3D_0 + C_4 = 0$$

$C_1 = \dfrac{\bar{P}_p''}{\gamma K} = \dfrac{34}{17\times2.67} = 7.4$

$C_2 = -\dfrac{8P_a}{\gamma K} = \dfrac{8\times115}{17\times2.67} = -20.3$

$C_3 = -\dfrac{6P_a}{(\gamma K)^2}\left(2\bar{y}\gamma K + \bar{P}_p''\right)$

$= -\dfrac{6\times115}{(17\times2.67)^2}(2\times2.50\times17\times2.67+340) = -192$

$C_4 = -\dfrac{6P_a\bar{y}\,p_p'' + 4P_a^2}{(\gamma K)^2}$

$= -\dfrac{6\times115\times2.50\times340+4\times(115)^2}{(17\times2.67)^2} = -314$

Substituting for C_1, C_2, C_3 and C_4, we have

$$D_0^4 + 7.4D_0^3 - 20.3D_0^2 - 192D_0 - 314 = 0$$

This equation when solved by the method of trial and error gives

$D_0 \approx 5.4$ m

Depth of Embedment

$D = D_0 + y_0 = 5.4 + 0.75 = 6.15$ m

Increasing D by 40%, we have

$D = 6.15 + 2.46 = 8.61$ m

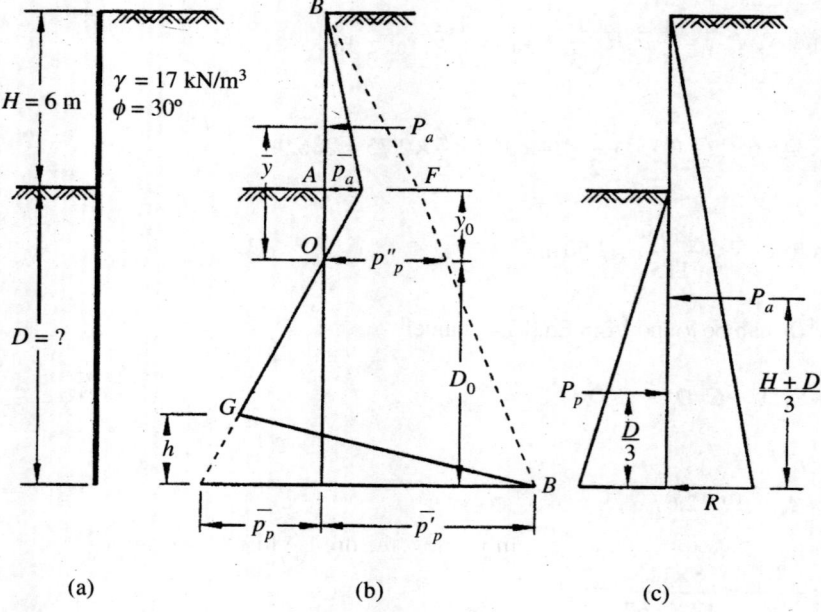

Fig.Ex. 15.1

(b) Simplified Analysis

The pressure distribution along the wall is assumed as shown in Fig.Ex.15.1(c)

$$P_a = \frac{1}{2}K_a\gamma(H+D)^2 = \frac{1}{2} \times \frac{1}{3} \times 17(6+D)^2$$

$$= 2.8D^2 + 34D + 102$$

$$P_p = \frac{1}{2}K_a\gamma D^2 = \frac{1}{2} \times 3 \times 17 \times D^2 = 25.5D^2$$

Taking moment of these forces about the base of wall, we have,

$$P_a \times \frac{(H+D)}{3} - P_p \times \frac{D}{3} = 0$$

$$\text{or } (2.8D^2 + 34D + 102)\left(2 + \frac{D}{3}\right) - 25.5D^2 \times \frac{D}{3} = 0$$

Simplifying, we have

$$0.76D^3 - 3.66D^2 - 10.2D - 20.4 = 0$$

On solving the equation by trial and error method we have

$$D = 7.1\,\text{m}$$

This depth has a factor of safety of 1. To increase this factor of safety from 1.5 to 2, the depth should be increased by 20 to 40 percent.

Depth for a factor of safety of 2 is = 7.10 + 2.84 = 9.94 m.

15.5 DEPTH OF EMBEDMENT OF CANTILEVER WALLS IN COHESIVE SOILS

The pressure distribution on a sheet pile wall is shown in Fig. 15.8.

The active pressure p_a at any depth z may be expressed as

$$p_a = \sigma_v K_A - 2c\sqrt{K_A}$$

where σ_v = vertical pressure, γz

z = depth from the surface of the backfill.

The passive pressure p_p at any depth y below the dredge line may be expressed as

$$p_p = \sigma_v K_p + 2c\sqrt{K_p}$$

The active pressure distribution on the wall from the backfill surface to the dredge line is shown in Fig. 15.8. The soil is supposed to be in tension up to a depth of z_0 and the pressure on the

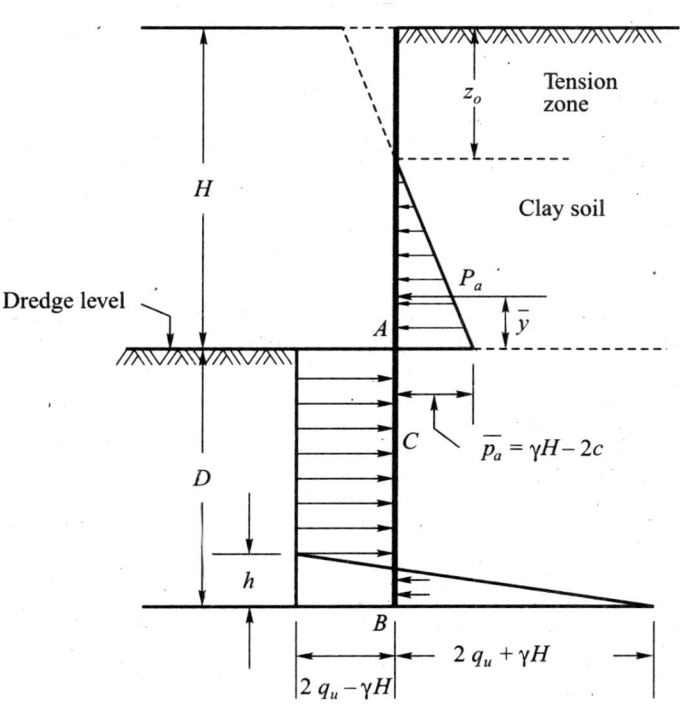

Fig. 15.8 Depth of embedment of cantilever walls in cohesive soils

wall is zero in this zone. The net pressure distribution on the wall is shown by the shaded triangle.

At the dredge line (at point A)

(a) The active pressure $\bar{p}_a$ acting towards the left is

$$\bar{p}_a = \gamma H K_A - 2c\sqrt{K_A}$$

when $\quad \phi = 0$

$$\bar{p}_a = \gamma H - 2c = \gamma H - q_u$$

where q_u = unconfined compressive strength of the clay soil = $2c$.

(b) The passive pressure acting towards the right at the dredge level,

$$\bar{p}_p = 2c, \quad \text{since}, \phi = 0$$

or $\quad \bar{p}_p = q_u$

The resultant of passive and active pressures at the dredge level is

$$\bar{p}_p - \bar{p}_a = q_u - (\gamma H - q_u)$$

or $\qquad = 2q_u - \gamma H \qquad\qquad\qquad\qquad\qquad\qquad (15.6)$

The resultant of the passive and active pressures at any depth y below the dredge line is

Passive pressure, $p_p = \gamma y + q_u$

Active pressure $p_a = \gamma(H + y) - q_u$

The resultant pressure is

$$p_p - p_a = (\gamma y + q_u) - [\gamma(H + y) - q_u]$$

$$= 2q_u - \gamma H \qquad\qquad\qquad\qquad\qquad\qquad (15.7)$$

Equations (15.6) and (15.7) indicate that the resultant pressure remains constant at $(2q_u - \gamma H)$ at all depths.

If passive pressure is developed on the backfill side at the bottom of the pipe, at point B, then

$$p_p = \gamma(H + D) + q_u \quad \text{acting towards the left}$$

$$p_a = \gamma D - q_u \quad \text{acting towards the right}$$

The resultant is

$$p_p - p_a = \gamma(H + D) + q_u - \gamma D + q_u$$

$$= \gamma H + 2q_u \qquad\qquad\qquad\qquad\qquad\qquad (15.8)$$

For statical equilibrium, the sum of all the horizontal forces should be equal to zero.

$$P_a - (2q_u - \gamma H)D + \frac{1}{2}(2q_u + 2q_u)h = 0$$

Simplifying,

$$P_a + 2q_u h - 2q_u D + \gamma HD = 0$$

$$h = \frac{D(2q_u - \gamma H) - P_a}{2q_u} \tag{15.9}$$

Also, for equilibrium, the sum of the moments at any point should be zero. Taking moments about the base, we have

$$P_a(\bar{y} + D) + \frac{h^2}{3}(2q_u) - \frac{(2q_u - \gamma H)D^2}{2} = 0 \tag{15.10}$$

Substituting for h in (Eq.15.10) and simplifying, we have

$$C_1 D^2 + C^2 D + C_3 = 0 \tag{15.11}$$

where $C_1 = (2q_u - \gamma H)$

$C_2 = 2P_a$

$$C_3 = -\frac{P_a(6q_u\bar{y} + P_a)}{(q_u + \gamma H)} \tag{15.11a}$$

The depth computed from Eq. (15.11) should be increased by 20 to 40 percent so that a factor of safety of 1.5 to 2.0 may be obtained. Alternatively the unconfined compressive strength q_u may be divided by a factor of safety.

Limiting Height of Wall

Equation (15.7) indicates that when $(2q_u - \gamma H) = 0$ the resultant pressure is zero. The wall will not be stable. In order that the wall may be stable, the condition that must be satisfied is

$$\frac{2q_u}{F} \geq \gamma H \tag{15.12}$$

where F = Factor of safety.

Example 15.2

Solve Example 15.1, if the soil is clay having an unconfined compressive strength of 70 kN/m² and a unit weight of 17 kN/m³.

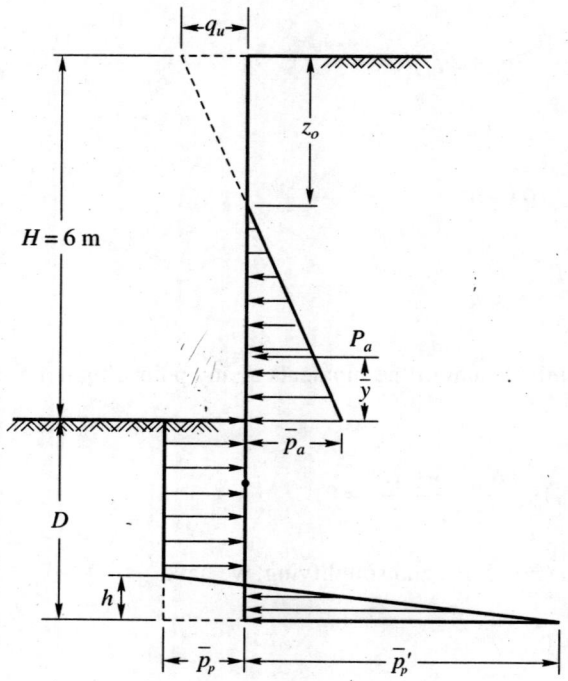

Fig. Ex. 15.2

Solution

The pressure distribution is assumed as shown in Fig. Ex. 15.2.

For $\phi_u = 0$, $\bar{p}_a = \gamma H - q_u = 17 \times 6 - 7 = 32 \text{ kN/m}^2$

$$z_0 = \frac{q_u}{\gamma} = \frac{70}{17} \text{ m} = 4.12 \text{ m}$$

$$P_a = \frac{1}{2}\bar{p}_a(H - z_0) = \frac{1}{2} \times 32 \times (6.0 - 4.12) = 30 \text{ kN/m of wall}$$

$$\bar{p}_p = 2q_u - \gamma H = 2 \times 70 - 17 \times 6 = 38 \text{ kN/m}^2$$

$$\bar{p}'_p = 2q_u + \gamma H = 2 \times 10 + 17 \times 6 = 242 \text{ kN/m}^2$$

$$\bar{y} = \frac{1}{2}(H - z_0) = \frac{1}{3}(6.0 - 4.12) = 0.63 \text{ m}$$

For the determination of h, equate the summation of all horizontal forces to zero, we have

$$P_a - \bar{P}_p \times D + \frac{1}{2}(\bar{P}_p + \bar{P}'_p)h = 0$$

or $\quad 30 - 38 \times D + \frac{1}{2}(38 + 242)h = 0$

Therefore $h = \dfrac{3.8D - 3}{14}$

For the determination of D, taking moments of all the forces about the base of the wall, we have

$$P_a \times (D + \bar{y}) - \bar{P}_p \times \frac{D^2}{2} + (\bar{P}_p + \bar{P}'_p) \times \frac{h}{2} \times \frac{h}{3} = 0$$

or $\quad 30(D + 0.63) - 38 \times \dfrac{D^2}{2} + (38 + 242) \times \dfrac{h^2}{6} = 0$

Substituting for h we have,

$$3D + 1.89 - 1.9D^2 + 4.7\left[\frac{3.8D - 3}{14}\right]^2 = 0$$

Simplifying, we have

$$D^2 - 0.57D - 1.35 = 0$$

Solving D, $D = 2.2$ m; Increasing D by 40%, we have $D = 2.2 + 0.90 = 3.1$ m.

Example 15.3

Solve Example 15.1 if the soil below the dredge line is clay having a cohesion of 35 kN/m² and the backfill is sand having an angle of internal friction of 30°. The unit weight of both the soils may be assumed as 17 kN/m³.

Solution

Refer Fig. Ex. 15.3

$$\bar{P}_a = \gamma H K_A = 17 \times 6 \times \frac{1}{3} = 34 \text{ kN/m}^2$$

$$\bar{P}_p = 2q_u - \gamma H = 2 \times 2 \times 35 - 17 \times 6 = 38 \text{ kN/m}^2$$

Taking moments about the base we have

$$\frac{1}{2}P_a H\left(\frac{H}{3} + D\right) - \bar{P}_p \times D \times \frac{D}{2} = 0$$

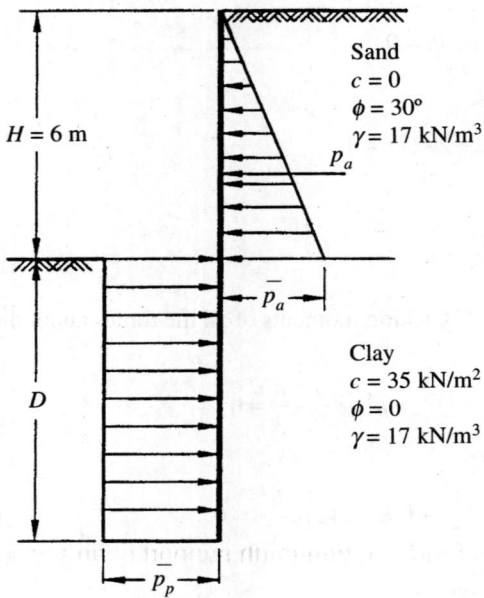

Fig. Ex. 15.3

or $\quad \dfrac{1}{2} \times 34 \times 6(2+D) - 38 \times \dfrac{D^2}{2} = 0$

Simplifying, we have,

$D^2 - 5.4D - 10.7 = 0$

Solving, we have $D = 7$ m

Increase D by 40% we have,

$D = 7 + 2.8 = 9.8$ m.

15.6 FREE-EARTH SUPPORT METHOD: DEPTH OF EMBEDMENT OF ANCHORED SHEET PILES IN GRANULAR SOILS

If the sheet piles have been driven to a shallow depth, the deflection of a bulkhead is somewhat similar to that of a vertical elastic beam whose lower end B is simply supported and the other end is fixed as shown in Fig. 15.9. Bulkheads which satisfy this condition are called bulkheads with *free earth support*. There are two methods of applying the factor of safety in the design of bulkheads.

1. Compute the minimum depth of embedment and increase the value by 20 to 40 percent to give a factor of safety of 1.5 to 2.

2. The alternative method is to apply the factor of safety to K_p and determine the depth of embedment.

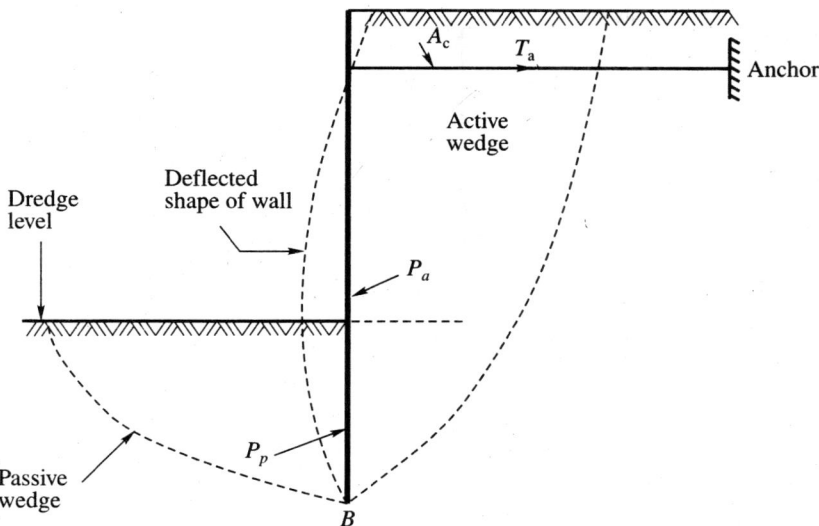

Fig. 15.9 Conditions for free-earth support of an anchored bulk head

Method 1: Minimum Depth of Embedment

The water table is assumed to be at a depth h_1 from the surface of the backfill. The anchor rod is fixed at a height h_2 above the dredge line. The sheet pile is held in position by the anchor rod and the tension in the rod is T_a. The forces that are acting on the sheet pile are

1. Active pressure due to the soil behind the pile,

2. Passive pressure due to the soil in front of the pile, and

3. The tension in the anchor rod.

The problem is to determine the minimum depth of embedment D. The forces that are acting on the pile wall are shown in Fig. 15.10.

The resultant of the passive and active pressures acting below the dredge line is shown in Fig. 15.10. The distance y_0 to the point of zero pressure is

$$y_0 = \frac{\overline{p}_a}{\gamma_b K} \tag{15.13}$$

The system is in equilibrium when the sum of the moments of all the forces about any point is zero. For convenience if the moments are taken about the anchor rod, we have

$$P_p h_4 = P_a \overline{y}_a \tag{15.14}$$

But $P_p = \dfrac{1}{2} \gamma_b K D_0^2$

$$h_4 = h_2 + y_0 + \frac{2}{3} D_0$$

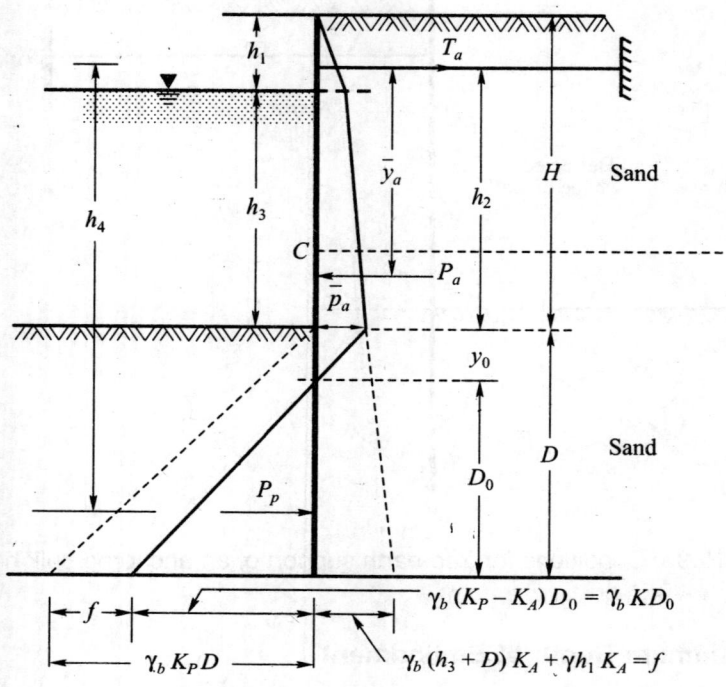

Fig. 15.10   Depth of embedment of an anchored bulkhead by free-earth support method
(method 1)

Therefore,

$$P_a \bar{y}_a = \frac{1}{2} \gamma_b K D_0^2 \left(h_2 + y_0 + \frac{2}{3} D_0 \right)$$

Simplifying the equation, we have

$$C_1 D_0^3 + C_2 D_0^2 + C_3 = 0 \qquad (15.15)$$

where $C_1 = \dfrac{\gamma_b K}{3}$

$$C_2 = \left(\frac{\gamma_b K}{3} \right)(h_2 + y_0) \qquad (15.15a)$$

$$C_3 = -P_a \bar{y}_a$$

γ_b = Submerged unit weight of soil

$K = K_p - K_A$

The force in the anchored rod, Ta, is found by summing the horizontal forces as

$$T_a = P_a - P_p \tag{15.16}$$

The minimum depth of embedment is

$$D = D_0 + y_0 \tag{15.17}$$

Increase the depth D by 20 to 40% to give a factor of safety of 1.5 to 2.0 respectively.

Example 15.4

Determine the depth of embedment and the force in the tie rod of the anchored bulkhead shown in Fig. Ex. 15.4(a). The backfill and the soil below the dredge line is sand, having the following properties.

$$G = 2.6, \ e = 1.0 \text{ and } \phi = 30°$$

Solve the problem by the free-earth support method.

Solution

Assume the soil above the water level is dry

For $\phi = 30°$, $K_A = \dfrac{1}{3}$, $K_p = 3.0$

and $K = K_p - K_A = 3 - \dfrac{1}{3} = 2.67$

For $G = 2.6$ and $e = 1.0$.

$\gamma_{sat} = 18 \text{ kN/m}^3$, and $\gamma_d = 13 \text{ kN/m}^3$.

$\gamma_d = \gamma_{sat} - \gamma_w = 18 - 9.81 = 8.19 \text{ kN/m}^3$.

where $\gamma_w = 9.81 \text{ kN/m}^3$.

The pressure distribution along the bulkhead is as shown in Fig. Ex. 15.4(b).

$$\bar{p}_1 = \gamma_d h_1 K_A = 13 \times 2 \times \frac{1}{3} = 8.67 \text{ kN/m}^2.$$

$$\bar{p}_a = \bar{p}_1 + \gamma_b h_3 K_A = 8.67 + 8.19 \times 3 \times \frac{1}{3} = 16.86 \text{ kN/m}^2$$

$$y_0 = \frac{\bar{p}_a}{\gamma_d \times K} = \frac{16.86}{8.19 \times 2.67} = 0.77 \text{ m}$$

$$P_a = \frac{1}{2} \times \bar{p}_1 \times h_1 + \bar{p}_1 \times h_3 + \frac{1}{2}(\bar{p}_a - \bar{p}_1)h_3 + \frac{1}{2}\bar{p}_a y_0$$

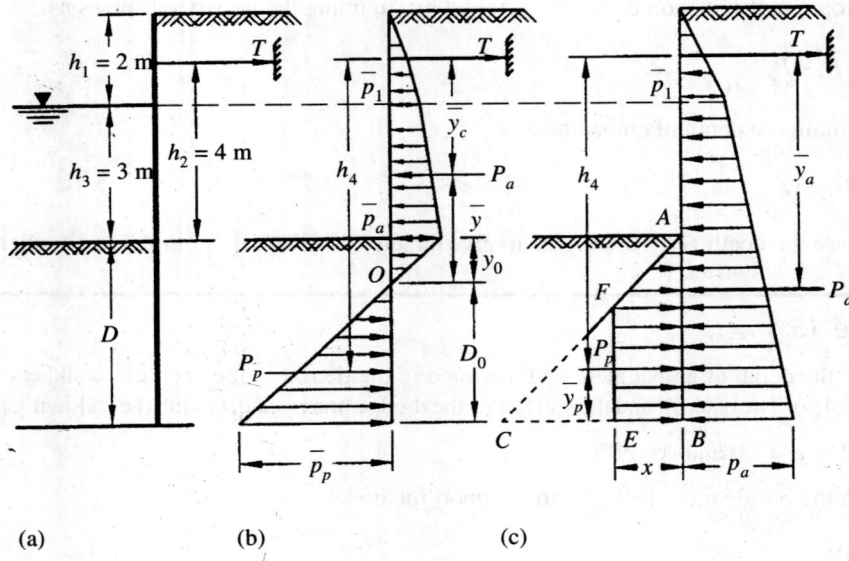

(a)　　　　　　(b)　　　　　　(c)

Fig. Ex. 15.4

$$= \frac{1}{2} \times 8.67 \times 2 + 8.67 \times 3 + \frac{1}{2}(16.7 - 8.67)3$$

$$+ \frac{1}{2} \times 1.67 \times 0.77 = 53.1 \, \text{kN/m of wall}$$

To find $\bar{y}$, taking moments of areas about 0, we have

$$53.1 \times \bar{y} = \frac{1}{2} \times 8.67 \times 2\left(\frac{1}{2} + 3 + 0.77\right) + 8.67 \times 3(3/2 + 0.77)$$

$$+ \frac{1}{2}(16.7 - 8.67) \times 3(3/3 + 0.77) + \frac{1}{2} \times 16.7 \times \frac{2}{3} \times 0.77^2 = 126.7$$

We have $\quad \bar{y} = \dfrac{126.7}{53.1} = 2.4 \, \text{m}, \ \bar{y}_c = 4 + 0.77 - 2.4 = 2.37 \, \text{m}$

Now $\quad P_p = \dfrac{1}{2} \times \gamma_b \times K \times D_0^2 = \dfrac{1}{2} \times 8.19 \times 2.67 D_0^2 = 10.93 D_0^2$

and its distance from the anchor rod is

$$h_4 = h_2 + y_0 + 2/3D_0 = 4 + 0.77 + 2/3D_0 = 4.77 + 0.67D_0$$

Now, taking the moments of forces about the tie rod, we have

$$P_a \times \bar{y} = P_p \times h_4$$

$$53.1 \times 2.37 = 10.93 D_0^2 \times (4.77 + 0.67 D_0)$$

Simplifying, we have

$$D_0 \approx 1.5 \text{ m}, \ D = y_0 + D_0 = 0.77 + 1.5 = 2.27 \text{ m}$$

For finding tension in the anchor rod, we have

$$P_a - P_p - T_a = 0$$

Therefore, $T_a = P_a - P_p = 53.1 - 10.93(1.5)^2 = 28.51 \text{ kN/m of wall.}$

Note: The depth of embedment may be increased from 2.28 m to 2.75 m or 3.20 m, in order to provide a factor of safety of 1.5 or 2.0 respectively.

Method 2: Depth of Embedment by Applying a Factor of Safety to K_p

(a) Granular Soil Both in the Backfill and Below the Dredge Line

The forces that are acting on the sheet pile wall are as shown in Fig. 15.11. The maximum passive pressure that can be mobilized is equal to the area of triangle *ABC* shown in the figure. The passive pressure that has to be used in the computation is the area of figure *ABEF* (shaded). The triangle *ABC* is divided by a vertical line *EF* such that

$$\text{Area } ABEF = \frac{\text{Area } ABC}{\text{Factor of safety}} = P'_p$$

The width of figure *ABEF* and the point of application of P'_p can be calculated without any difficulty.

The equilibrium of the system requires that the sum of all the horizontal forces and moments about any point, for instance, about the anchor rod, should be equal to zero.

Hence $P'_p + T_a - P_a = 0$ (15.18)

$P_a \bar{y}_a - P'_p h_4 = 0$ (15.19)

where $P'_p = \dfrac{1}{2} \gamma_b K_p D^2 \times \dfrac{1}{F_s}$

and F_s = assumed factor of safety.

The anchor pull may be found out from Eq. (15.18) and the Eq. (15.19) be solved for D.

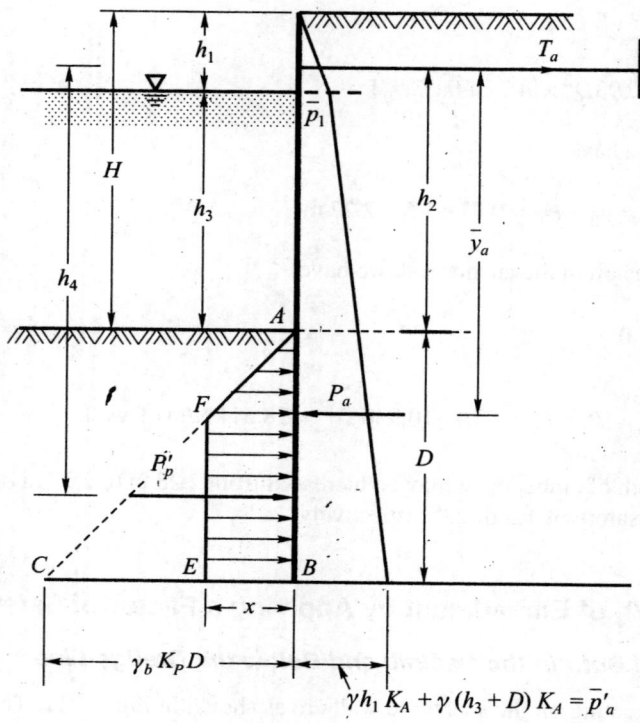

Fig. 15.11   Depth of embedment by free-earth support method (method 2)

Example 15.5

Solve Ex. 15.4 by applying directly a factor of safety of 2 to the passive pressure.

Solution

[Refer Fig. Ex. 15.4(c)]

$$P_p = \text{area } ABEF = \frac{\text{Area } ABC}{F_s}$$

$$P_p = \frac{1}{2}\gamma_b K_p D^2 \times \frac{1}{F_s} = \frac{1}{2} \times 8.0 \times 3 \times D^2 \times \frac{1}{2} = 6D^2$$

To find x we may write, Area $ABEF = 6.0D^2 = \dfrac{(AB + EF)x}{2}$

Since $AB = D$, we have, $EF = D - \dfrac{x}{\gamma_b K_p} = D - \dfrac{x}{24}$

Therefore, $6D^2 = \dfrac{(D + D - x/2.4)x}{2}$

Simplifying, we have

$x^2 - 4.8Dx + 2.9D^2 = 0$, or $x = 0.7D$.

Taking moments of area $ABEF$ about the base, we have

$\bar{y}_p = 0.44D$, $h_4 = 4 + (D - \bar{y}_p) = 4 + 0.56D$.

Now, from the active earth pressure diagram, we have

$$\bar{P}_1 = \gamma_b h_1 K_A = 13 \times 2 \times \frac{1}{3} = 8.67 \text{ kN/m}^2$$

$$P_a = \bar{P}_1 + \gamma_b (h_3 + D) K_A = 8.67 + \frac{8 \times (3 + D)}{3} = 16.67 + 2.67D$$

$$P_a = \frac{1}{2} P_1 h_1 + \frac{P_1 + P_a}{2}(h_3 + D)$$

$$= \frac{1}{2} \times 8.67 \times 2 + \frac{8.67 + 16.67 + 2.67D}{2}(3 + D) = 1.3D^2 + 16.8D + 46.7.$$

Taking moments of forces about the anchor rod, we have

$$P_a \bar{y}_a = \frac{1}{2} \times 8.67 \times 2 \left(\frac{2}{3} \times 2 - 1 \right) + 0.867 \frac{(3 + D)}{2} \left(1 + \frac{3 + D}{2} \right)$$

$$+ \frac{1}{2}\left(16.67 + 2.67D^2 - 8.67\right)(3 + D)\left[\frac{2}{3}(3 + D) + 1 \right]$$

$$= 0.89D^3 + 13.7D^2 + 66.7D + 104$$

$P_p h_4 = 6D^2(4 + 0.56D) = 24D^2 + 3.36D^3$

Since summation of moments about the anchor rod should be zero, we have

$0.89D^3 + 13.7D^2 + 66.7D + 104 = 24D^2 + 3.36D^3$

Simplifying we have, $D^3 + 4.17D^2 - 27D - 105 = 0$

On solving by trial and error method, $D = 3.8$ m (with a factor of safety of 2)

Now, force in the anchor rod is

$T_a = P_a - P_p$

where $\qquad P_a = 1.3 \times 3.8^2 + 16.8 \times 3.8 + 46.7$

$\qquad\qquad = 129.31 \,\text{kN/m length of wall}$

and $\qquad P_p = 6 D^2 = 6(3.8)^2 = 86.64 \,\text{kN/m length of wall}.$

Therefore, $\quad T_a = 129.31 - 86.64 = 42.67 \,\text{kN/m length of wall}.$

(b) Depth of Embedment when the Soil Below Dredge Line is Cohesive and the Backfill Granular

Figure 15.12 shows the pressure distribution.

The surcharge at the dredge line due to the backfill may be written as

$$q = \gamma h_1 + \gamma_b h_3 = \gamma_e H$$

where h_3 = depth of water above the dredge line, γ_e effective equivalent unit weight of the soil, and $H = h_1 + h_3$.

The active earth pressure acting towards the left at the dredge line is (when $\phi = 0$)

$$\overline{P}_a = q - q_u$$

The passive pressure acting towards the right is

$$\overline{P}_p = q_u$$

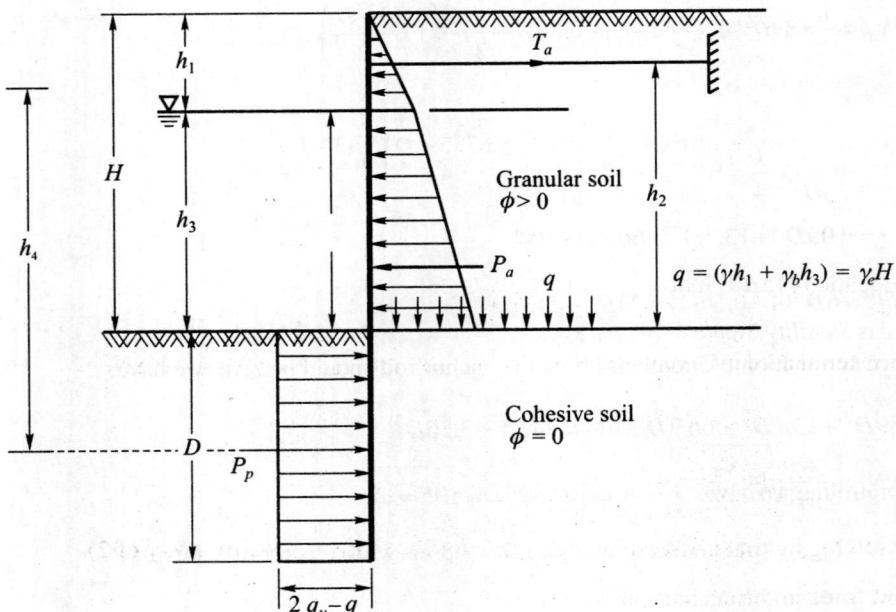

Fig. 15.12 Depth of embedment when the soil below the dredge line is cohesive

The resultant of the passive and active earth pressures is

$$\overline{P}_p - \overline{P}_a = 2q_u - q \tag{15.20}$$

The pressure remains constant with depth. Taking moments of all the forces about the anchor rod, we have

$$P_a \overline{y} - D(2q_u - q)(h_2 + D/2) = 0 \tag{15.21}$$

where $\overline{y}$ = the distance of the anchor rod from P_a.

Simplifying Eq. (15.21), we have

$$D^2 + C_1 D + C_2 = 0 \tag{15.22}$$

where $C_1 = 2h_2$

$$C_2 = -\frac{2\overline{y}P_a}{2q_u - q} \tag{15.22a}$$

The force in the anchor rod is given by Eq. (15.16).

It can be seen from Eq. (15.20) that the wall will be unstable if

$$2q_u - q = 0$$

or $\quad 4c - q = 0$

If we write for all practical purposes $q = \gamma_e H = \gamma H$, then Eq. (15.20) may be written as

$$4c - \gamma H = 0$$

or $\quad N_s = \dfrac{c}{\gamma H} = \dfrac{1}{4} = 0.25 \tag{15.23}$

Equation (15.23) indicates that the wall is unstable if the ratio $\dfrac{c}{\gamma H}$ is equal to 0.25. The N_s is termed is *Stability Number*. The stability is a function of the wall height H, but is relatively independent of the material used in developing q. If the wall adhesion c_a is taken into account the stability number N_S becomes

$$N_s = \frac{c}{\gamma H}\sqrt{1 + \frac{c_a}{c}} \tag{15.24}$$

At passive failure $\sqrt{1 + \dfrac{c_a}{c}}$ is approximately equal to 1.25.

The stability number for sheet pile walls embedded in cohesive soils may be written as

$$N_s = \frac{1.25c}{\gamma H}$$ (15.25)

When factor of safety $F_s = 1$ and $\dfrac{c}{\gamma H} = 0.25$, we have $N_s = 0.30$.

The stability number N_s required in determining the depth of sheet pile walls is therefore

$$N_s = 0.30 \times F_s$$ (15.26)

Example 15.6

Solve Ex. 15.4, if the backfill is sand with $\phi = 30°$ and the soil below the dredge line is clay having $c = 20$ kN/m². For both the soils, assume $G = 2.6$.

Solution

The pressure distribution along the bulkhead is as shown in Fig. Ex. 15.6.

$\bar{p}_1 = 8.67$ kN/m² as in Ex.15.4, $\bar{p}_a = 16.7$ kN/m²

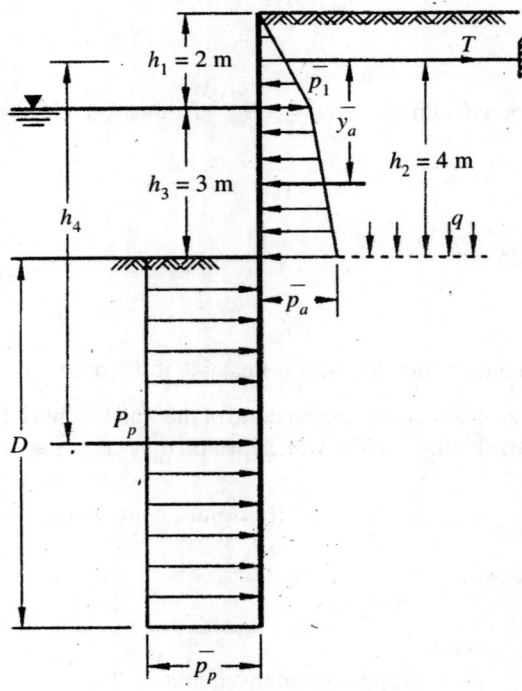

Fig. Ex. 15.6

$$P_a = \frac{1}{2}\bar{p}_1 h_1 + \bar{p}_1 h_3 + \frac{1}{2}(\bar{p}_a - \bar{p}_1)h_3$$

$$= \frac{1}{2}\times 8.67 \times 2 + 8.67 \times 3 + \frac{1}{2}(16.7 - 8.67)\times 3 = 46.7 \text{ kN/m}$$

To determine $\bar{y}_a$, take moments about the tie rod.

$$P_a \times \bar{y}_a = \frac{1}{2}\times 8.67 \times 2\left(\frac{2}{3}\times 2 - 1\right) + 8.67 \times 3 \times 2.5$$

$$+ \frac{1}{2}(16.7 - 8.67)\times 3 \times 3 = 104$$

Therefore, $\bar{y}_a = \dfrac{104}{P_a} = \dfrac{104}{46.7} = 2.33 \text{ m}$

Now, $q = \gamma_d h_1 + \gamma_b h_3 = 13\times 2 + 8\times 3 = 50 \text{ kN/m}^2$

$$\bar{p}_p = 2\times 20^2 - 50 = 30 \text{ kN/m}^2$$

Therefore, $P_a = \bar{p}_p \times D = 30 \text{ kN/m}$

and $\qquad h_4 = \left(4 + \dfrac{D}{2}\right)$

Taking moments of forces about the anchor rod, we have

$$P_a \times \bar{y}_a - P_p \times h_4 = 0$$

or $\quad 46.7 \times 2.23 - 30D\left(4 + \dfrac{D}{2}\right) = 0$

or $\quad 1.5D^2 + 12D - 10.36 = 0$ or $D \approx 0.8$ m.

$D = 0.8$ m is obtained with a factor of safety equal to one. It should be increased by 20 to 40 percent to increase the factor of safety from 1.5 to 2.0. For a factor of safety of 2, the depth of embedment should be at least 1.12 m.

$$P_P = 30 \times 1.12 = 33.6 \text{ kN/m length of wall}$$

The tension in the tie rod is

$$T_a = P_a - P_p = 46.7 - 33.6 = 13.1 \text{ kN/m length of wall.}$$

15.7   FIXED EARTH SUPPORT METHOD (GRANULAR SOIL)

If the sheet piles are driven to a considerable depth, as shown in Fig. l5.11(a) the lower end of the bulkhead is practically fixed in position, because the resistance of sand adjoining the end does not permit more than an insignificant deviation of the wall from its initial vertical position. Therefore anchored sheet pile walls of this type will be called as *bulkheads with fixed earth support*. An adequately anchored bulkhead with free earth support can fail by bending or on account of the failure of the sand adjoining the wall at the passive side by shear along a curved surface of sliding (Fig. 15.9). A securely anchored bulkhead with fixed earth support can fail only by bending.

The deflected shape of the pile under fixed support condition is shown in Fig. 15.13(a). The actual active pressure distribution on such a pile is not linear as can be seen by the dashed line in the figure. However, for all practical purposes, a linear distribution is assumed. The point *b* may be considered as the point of fixity of the sheet pile and the point I on the pile is the point of inflection (zero bending moment). The point *O* is the point of zero shear. Fig. 15.13(b) and (c) show the pressure and moment distribution on the wall respectively.

The problem of bulkheads with fixed earth support may be solved by anyone of the following methods.

1. Elastic line methods

2. Equivalent beam method.

The elastic line method is quite complicated and time consuming. Whereas the equivalent beam method represents a simplification of the elastic line method and the method involves a considerable saving in time and labour at a small sacrifice of accuracy. Only the equivalent beam method is therefore described in this book.

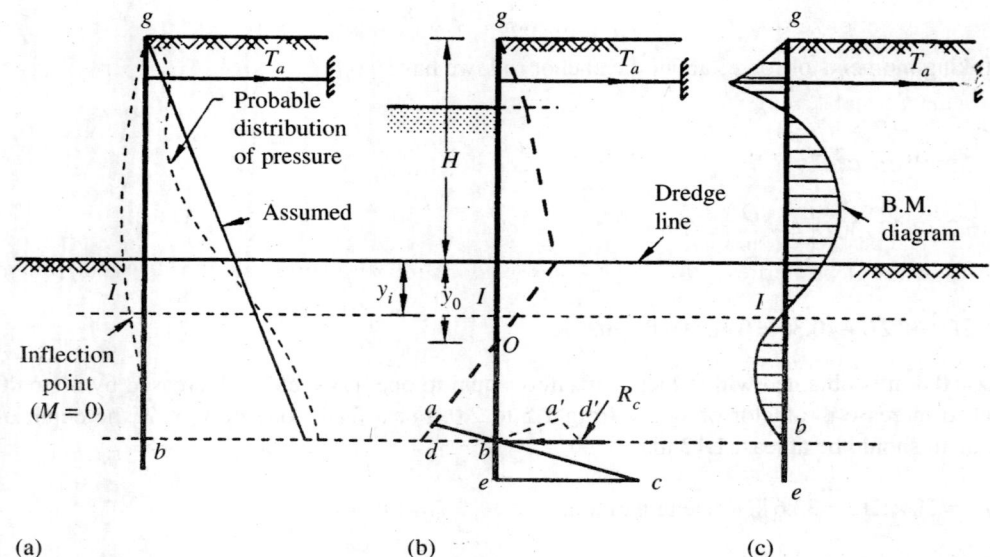

Fig. 15.13   Fixed-earth support condition

Equivalent Beam Method

The equivalent beam method is based on the following principle:

1.   The pressure diagram triangle Oab is transformed into triangle obd Fig. 15.13(b), thus increasing the area by the amount adb. This increase is however counterbalanced by adding an equivalent area $ba'd'$ on the back of the wall. The pressure due to the area bce and $ba'd'$ may be replaced by a concentrated load R_c acting at point b to the left Fig. 15.13(b).

2.   The sheet pile wall gob is assumed as a beam. This beam is supported freely at one end at anchbr rod level by the reaction T_a (pull in the anchor rod) and fixed at the other end at point b with reaction R_c Fig. 15.13(b).

3.   Since the moment at the inflection point I is zero, the beam gob can be splitted into two parts (i) An equivalent, freely supported beam gI of the original beam gb with reactions R_I and T_a acting towards the right. (ii) Another beam Ib, supported at I by the reaction R_I acting towards the left and the bottom b with reaction R_c Fig. 15.14.

The solution to the problem by equivalent beam method involves the following.

1.   The determination of the reaction R_I.

2.   Determination of the depth of embedment.

The solution to the problem depends upon the depth y_i of the inflection point I below the dredge line. It has been found that the depth y_i is a function of ϕ. The values of y_i for different values of ϕ as obtained by elastic line method of analysis are (Terzaghi).

$\phi = 20°$	$30°$	$40°$
$y_i = 0.25\ H$	$0.08\ H$	$-0.007\ H$

The angle of internal friction ϕ of sandy backfills is approximately $30°$ corresponding to a value of y_i of about $0.1\ H$. Hence if both the backfill and the earth below the dredge line are sandy, $y_i = 0.1$ H may be used without significant error. Another approach that is normally recommended is to consider the point of zero shear the same as the point of zero moment (Anderson).

Thus, the value of y_i may be taken as equal to y_0 without significant error. Fig. 15.14(b) gives the equivalent beam section with the assumption $y_i = y_0$.

Determination of Reaction R_I and T_a

The pressure distribution diagram on the sheet pile in Fig. 15.14(a) is shown in Fig. 15.14(b). The total active and passive earth pressures P_a and P_p acting on the wall can be calculated by making use of the known soil properties in the same way as a free-earth support method. The unknown quantities are

1.   The reaction R_I

2.   The tension T_a in the anchor rod

3.   The depth D_b

1. The reaction R_I can be found out by taking moments of all the forces acting on the beam go about the anchor point P.

2. The algebraic sum of all the horizontal forces acting on the beam $g0$ equated to zero gives the value of T_a.

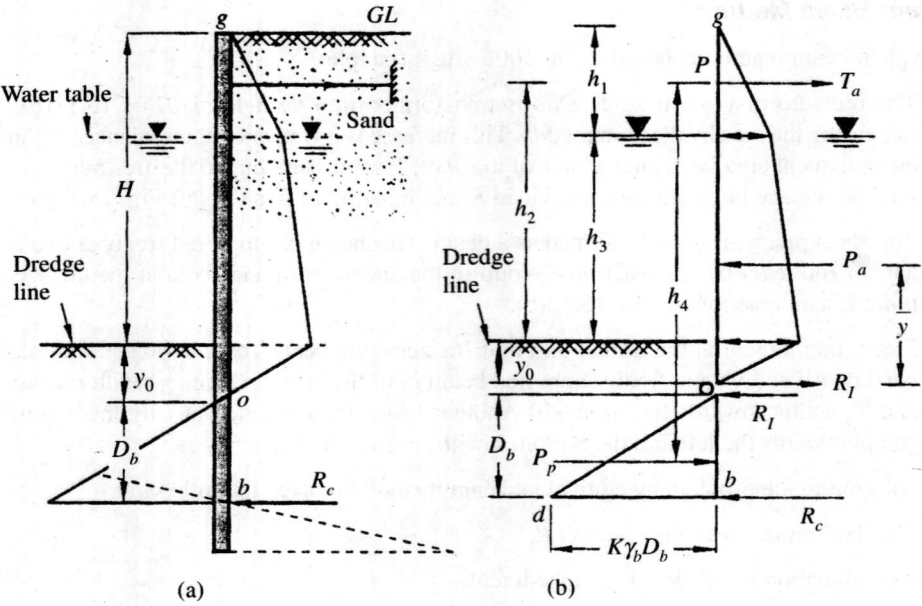

Fig. 15.14   The principles of equivalent beam method

Determination of Depth of Embedment

Let D_b = Depth of point b from point O on the beam.

The depth y_0 can be found out by making use of Eq. (15.1)

Now consider the lower part of the pile Ob, Fig. 15.14(b). Summing moments about b, we have

$$\frac{1}{2}\gamma_b K D_b \times D_b \times \frac{D_b}{3} = R_I D_b$$

or $$\frac{1}{6}\gamma_b K D_b^2 = R_I$$

or $$D_b = \sqrt{\frac{6R_I}{\gamma_b K}} \qquad (15.27)$$

Let $\bar{D}_b$ = depth of the point of fixity b from dredge line

Therefore $\bar{D}_b = D_b + y_0$ \qquad (15.28)

The depth of embedment D of sheet pile wall below the dredge line for complete fixity is taken as

$$D = 1.2\bar{D}_b \qquad (15.29)$$

Example 15.7

Solve Ex. 15.4, if the bulkhead has a fixed support. Use equivalent beam method with the following concepts:

(i) The point of zero moment being assumed as lying at a depth of 0.1 H from the level of dredge line, and

(ii) The point of zero moment being assumed as coinciding with the point of zero shear.

Solution

(i) With the first concept,

$$y_i = 0.1\,H = 0.5 \text{ m}$$

The pressure distribution and the separation of bulkhead into two beams are as shown in Fig. Ex. 15.7a

Consider the upper part gI,

$$\left.\begin{array}{l}\bar{p}_1 = 8.7 \text{ kN/m}^2 \\ \bar{p}_2 = 16.7 \text{ kN/m}^2\end{array}\right\} \text{ as in Ex.15.4.}$$

$$\bar{p}_3 = \bar{p}_2 - \gamma_b K y_i = 16.7 - 8.0 \times 2.67 \times 0.5 = 6 \text{ kN/m}^2$$

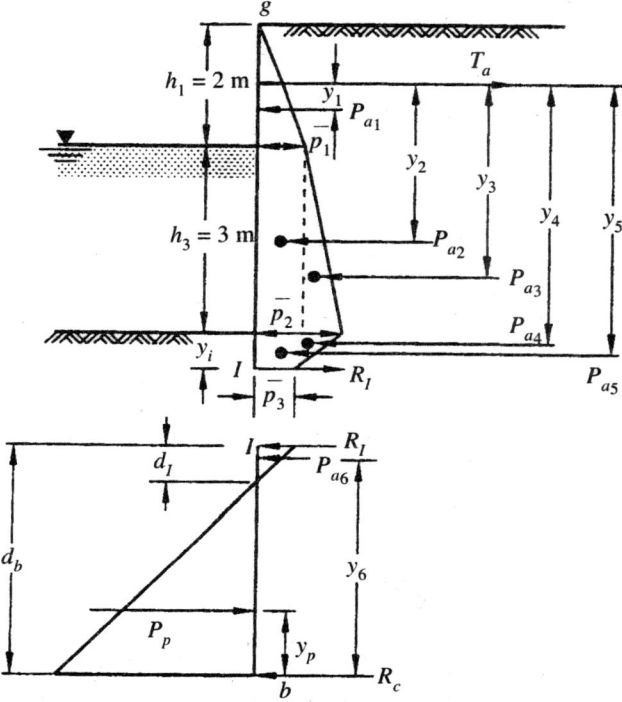

Fig. Ex. 15.7a

Therefore, $P_{a_1} = \frac{1}{2} \times 8.67 \times 2 = 8.67$ kN/m, $y_1 = \frac{2'}{3} \times 2 - 1 = 0.33$ m

$$P_{a_2} = 8.67 \times 3 = 26 \text{ kN/m}, \quad y_2 = 1.5 + 1 = 2.5 \text{ m}$$

$$P_{a_3} = \frac{1}{2} \gamma_b K_A h_3^2 = \frac{1}{2} \times 8.0 \times \frac{1}{3} \times (3)^2 = 12 \text{ kN/m}$$

$$y_3 = 2.0 + 1 = 3.0 \text{ m}, \quad P_{a_5} = 6.0 \times 0.5 = 3.0 \text{ kN/m}$$

$$y_5 = 4 + 0.25 = 4.25 \text{ m}$$

$$P_{a_4} = \frac{1}{2} (\bar{p}_2 - \bar{p}_3) y_i = \frac{1}{2} (16.7 - 6.0) \times 0.5 = 2.7 \text{ kN/m}$$

$$y_4 = 4 + \frac{1}{3} \times \frac{1}{2} = 4.16 \text{ m}$$

Taking moments about the anchor rod, we hve

$$R_I \times 4.5 = P_{a_1} y_1 + P_{a_2} y_2 + P_{a_3} y_3 + P_{a_4} y_4 + P_{a_5} y_5$$

$$R_I \times 4.5 = 8.7 \times 0.33 + 26 \times 2.5 + 12 \times 3 + 2.7 \times 4.16 + 3 \times 4.25 = 128$$

Therefore, $R_I = \dfrac{128}{4.5} = 28.5$ kN/m

and Force $T_a = P_{a_1} + P_{a_2} + P_{a_3} + P_{a_4} + P_{a_5} - R_I$

$$= 8.7 + 26 + 12 + 2.7 + 3 - 2.85 = 23.9 \text{ kN/m}$$

For beam Ib, $d_1 = \dfrac{\bar{p}_3}{\gamma_b K} = \dfrac{6.0}{8.0 \times 2.67} = 0.28$ m,

$$P_{a_6} = \frac{1}{2} \times 6.0 \times 0.28 = 0.84 \text{ kN/m}$$

$$y_6 = d_b - \frac{1}{3} \times 0.28 = (d_b - 0.093) \text{ m}$$

$$P_p = \frac{1}{2} \gamma_b K (d_b - d_1)^2$$

$$= \frac{1}{2} \times 8 \times 2.67 \times (d_b - 0.28)^2 = 10.67 (d_b - 0.28)^2$$

$$y_p = \frac{1}{3} (d_b - 0.28) \text{ m}$$

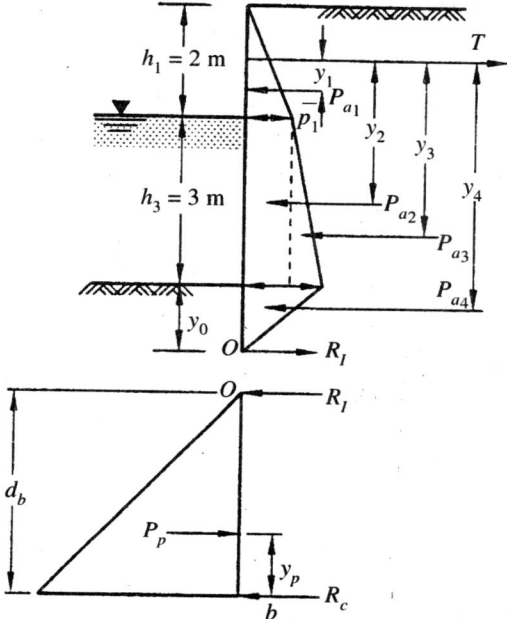

Fig. Ex. 15.7b

Taking moments of the forces about point b,

$$R_I \times d_b + P_{a6} \times y_6 - P_p \times y_p = 0$$

$$28.5d_b - 0.84(d_b - 0.093) - \frac{10.67}{3}(d_b - 0.28)^3 = 0$$

On simplification, we have, $0.36d_b^2 - 0.3d_b - 2.85 = 0$

Solving for d_b, we have $d_b = 3.3$ m

Depth of embedment

$D = 0.5 + 3.3 = 3.8$ m

Increasing D by 20% for fixity,

$D_{final} = 3.8 + 0.8 = 4.6$ m

(ii) With the second concept, the pressure distribution diagram is as shown in Fig. Ex. 15.7b.

Values of $\bar{p}_1, \bar{p}_2, P_{a_1}, y_1, P_{a_2}, y_2, P_{a_3}$ and y_3 remain the same as given under the first concept.

$y_0 = y_i + d_1 = 0.5 + 0.28 = 0.78$ m

Therefore, $P_{a_4} = \frac{1}{2} p_2 y_0 = \frac{1}{2} \times 16.7 \times 0.78 = 6.5$ kN/m

and $\qquad y_4 = 4 + \frac{1}{2} y_0 = 4 + 0.26 = 4.26$ m

Taking moments of all the forces about the anchor rod,

$$R_1(4 + y_0) = P_{a_1} y_1 + P_{a_2} y_2 + P_{a_3} y_3 + P_{a_4} y_4$$

By substituting and simplifying, we have

$$R_1 = \frac{132}{4.78} = 27.5 \text{ kN/m}$$

The tension in the tie rod is $T_a = P_{a_1} + P_{a_2} + P_{a_3} + P_{a_4} - R_1$

$$= 8.7 + 26 + 12 + 6.5 - 27.5 = 25.8 \text{ kN/m}$$

For beam ob $P_p = \frac{1}{2} \gamma K d_b^2 = \frac{1}{2} \times 8 \times 2.67 \times d_b^2 = 10.7 d_b^2$

Taking moments of P_p and R_1 about b

$$P_p y_p - R_1 d_b = 0$$

Therefore, $10.7 d_b^2 - 27.5 d_b = 0$, or $12.8 d_b^2 - 99 = 0$

which gives $d_b = 2.8$ m

Depth of embedment $D = y_0 + d_b = 0.78 + 2.8 = 3.58$ m

Increasing D by 20% for fixity,

$D_{\text{final}} = 3.58 + 0.7 = 4.28$ m

15.8 QUESTIONS AND PROBLEMS

15.1 Classify the sheet piles according to their materials. Discuss their relative advantages and disadvantages.

15.2 With suitable illustration, describe the analysis used for finding the depth of embedment.

15.3 Design the depth of embedment for a sheet pile wall shown in Fig. Prob. 15.3.

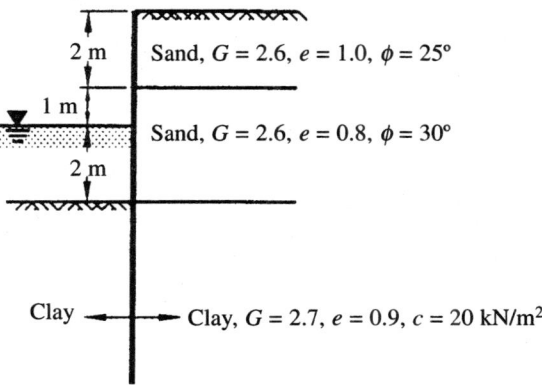

Fig. Prob. 15.3

15.4 Describe the 'equivalent beam' and 'the fixed-earth-support' methods of designing anchored bulkheads.

15.5 Design the anchored bulkhead shown in Fig. Prob. 15.5 by (a) Free-earth-support and (b) fixed-earth-support methods.

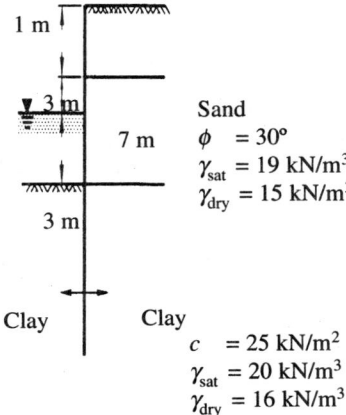

Fig. Prob. 15.5

CHAPTER 16

STABILITY OF SLOPES

16.1 INTRODUCTION

Slopes of earth are of two types:

1. Natural slopes

2. Man made slopes

Natural slopes are those that exist in nature and are formed by natural causes. Such slopes exist in hilly areas. The sides of cuttings, the slopes of embankments constructed for roads, railway lines, canals etc. and the slopes of earth dams constructed for storing water are examples of man made slopes. The slopes whether natural or artificial may be

1. Infinite slopes

2. Finite slopes

The term infinite slope is used to designate a constant slope of infinite extent. The long slope of the face of a mountain is an example of this type, whereas finite slopes are limited in extent. The slopes of embankments and earth dams are examples of finite slopes. The slope length depends on the height of the dam or embankment.

Slope Stability: Slope stability is an extremely important consideration in the design and construction of earth dams. The stability of a natural slope is also important. The results of a slope failure can often be catastrophic, involving the loss of considerable property and many lives.

Causes of Failure of Slopes: The important factors that cause instability in a slope and lead to failure are

1. Gravitational force

2. Force due to seepage water

3. Erosion of the surface of slopes due to flowing water

4. The sudden lowering of water adjacent to a slope

5. Forces due to earthquakes

The effect of all the forces listed above is to cause movement of soil from high points to low points. The most important of such forces is the component of gravity that acts in the direction of probable motion. The various effects of flowing or seeping water are generally recognized as very important in stability problems, but often these effects have not been properly identified. It is a fact that the seepage occurring within a soil mass causes seepage forces, which have much greater effect than is commonly realized.

Erosion on the surface of a slope may be the cause of the removal of a certain weight of soil, and may thus lead to an increased stability as far as mass movement is concerned. On the other hand, erosion in the form of undercutting at the toe may increase the height of the slope, or decrease the length of the incipient failure surface, thus decreasing the stability.

When there is a lowering of the ground water or of a freewater surface adjacent to the slope, for example in a sudden drawdown of the water surface in a reservoir there is a decrease in the buoyancy of the soil which is in effect an increase in the weight. This increase in weight causes increase in the shearing stresses that may or may not be in part counteracted by the increase in shearing strength,

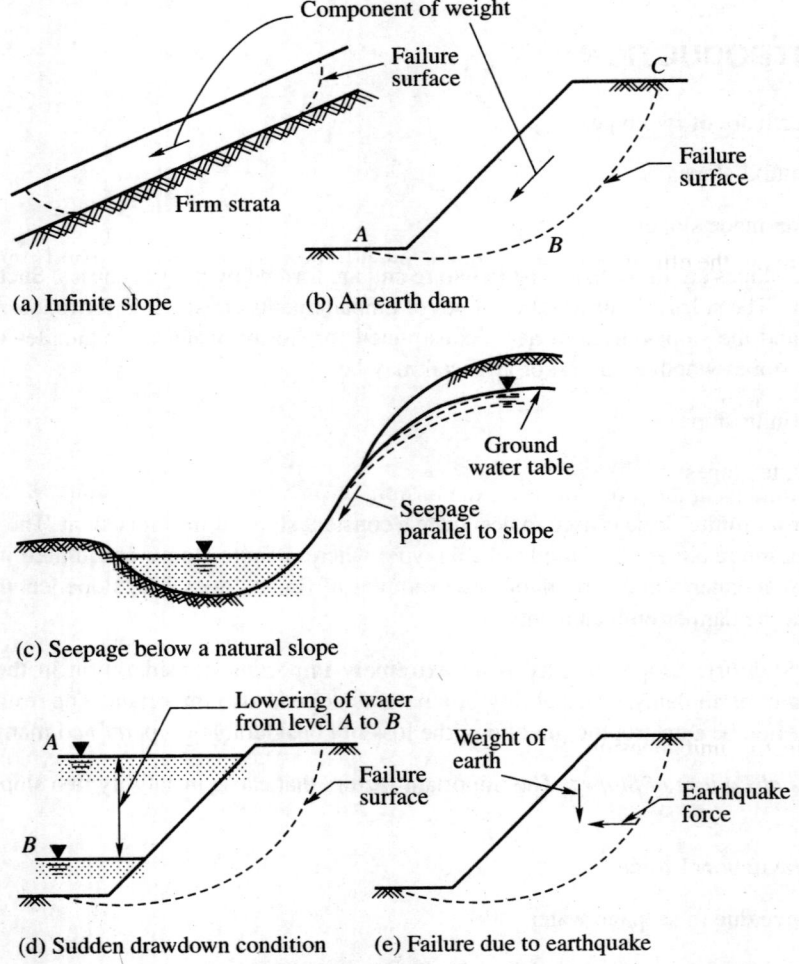

(a) Infinite slope   (b) An earth dam

(c) Seepage below a natural slope

(d) Sudden drawdown condition   (e) Failure due to earthquake

Fig. 16.1   Forces that act on earth slopes

depending upon whether or not the soil is able to undergo compression which the load increase tends to cause. If a large mass of soil is saturated and is of low permeability, practically no volume changes will be able to occur except at a slow rate, and in spite of the increase of load the strength increase may be inappreciable.

Shear at constant volume may be accompanied by a decrease in the intergranular pressure and an increase in the neutral pressure. A failure may be caused by such a condition in which the entire soil mass passes into a state of liquefaction and flows like a liquid. A condition of this type may be developed if the mass of soil is subject to vibration, for example, due to earthquake forces.

The various forces that act on slopes are illustrated in Fig. 16.1.

16.2   GENERAL CONSIDERATIONS AND ASSUMPTIONS IN THE ANALYSIS

There are three distinct parts to an analysis of the stability of a slope. They are:

1. Testing of samples to determine the cohesion and angle of internal friction

If the analysis is for a natural slope, it is essential that the sample be undisturbed. In such important respects as rate of shear application and state of initial consolidation, the condition of testing must represent as closely as possible the most unfavorable conditions ever likely to occur in the actual slope.

2. The study of items which are known to enter but which cannot be accounted for in the computations

The most important of such items is progressive cracking which will start at the top of the slope where the soil is in tension, and aided by water pressure, may progress to considerable depth. In addition, there are the effects of the non-homogeneous nature of the typical soi and other variations from the ideal conditions which must be assumed.

3. Computation

If a slope is to fail along a surface, all the shearing strength must be overcome along that surface which then becomes a surface of rupture. Any one such as ABC in Fig. 16.1(b) represents one of an infinite number of possible traces on which failure might occur.

It is assumed that the problem is two dimensional, which theoretically requires a long length of slope normal to the section. However, if the cross section investigated holds for a running length of roughly two or more times the trace of the rupture, it is probable that the two dimensional case holds within the required accuracy.

The shear strength of soil is assumed to follow Coulomb's law

$$s = c' + \sigma' \tan\phi' \qquad (16.1)$$

where,

c' = effective unit cohesion

σ' = effective normal stress on the surface of rupture = $(\sigma - u)$

σ = total normal strees on the surface of rupture

u = pore water pressure on the surface of rupture

ϕ' = effective angle of internal friction.

The item of great importance is the loss of shearing strength which many clays show when subjected to a large shearing strain. The stress-strain curves for such clays show the stress rising with increasing strain to a maximum value, after which it decreases and approaches an ultimate value which may be much less than the maximum. Since a rupture surface tends to develop progressively rather than with all the points at the same state of strain, it is generally the ultimate value that should be used for the shearing strength rather than the maximum value.

16.3   FACTOR OF SAFETY

In stability analysis, two types of factors of safety are normally used. They are

1. Factor of safety with respect to shearing strength.

2. Factor of safety with respect to cohesion. This is termed the factor of safety with respect to height.

Let,

F_s = factor of safety with respect to strength

F_c = factor of safety with respect to cohesion

F_H = factor of safety with respect to height

F_ϕ = factor of safety with respect to friction

c'_m = mobilized cohesion

ϕ'_m = mobilized angle of friction

τ = average value of mobilized shearing strength

s = maximum shearing strength.

The factor of safety with respect to shearing strength, F_s, may be written as

$$F_s = \frac{s}{\tau} = \frac{c' + \sigma' \tan\phi'}{\tau}$$

The shearing strength mobilized at each point on a failure surface may be written as

$$\tau = \frac{c'}{F_s} + \sigma' \frac{\tan\phi'}{F_s}$$

or $\tau = c'_m + \sigma' \tan\phi'_m$ (16.2)

where $c'_m = \dfrac{c'}{F_s}$, $\tan\phi'_m = \dfrac{\tan\phi'}{F_s}$

Actually the shearing resistance (mobilized value of shearing strength) does not develop to a like degree at all points on an incipient failure surface. The shearing strains vary considerably and the shearing stress may be far from constant. However the above expression is correct on the basis of average conditions.

If the factors of safety with respect to cohesion and friction are different, we may write the equation of the mobilized shearing resistance as

$$\tau = \frac{c'}{F_c} + \sigma' \frac{\tan\phi'}{F_\phi}$$ (16.3)

It will be shown later on that F_c depends on the height of the slope. From this it may be concluded that the factor of safety with respect to cohesion may be designated as the *factor of safety with respect to height*. This factor is denoted by F_H and it is the ratio between the critical height and the actual height, the critical height being the maximum height at which it is possible for a slope to be stable. We may write from Eq. (16.3)

$$\tau = \frac{c'}{F_H} + \sigma' \tan\phi' \qquad\qquad (16.4)$$

where F_ϕ is arbitrarily taken equal to unity.

Example 16.1

The shearing strength parameters of a soil are

$c' = 26.7 \text{ kN/m}^2$

$\phi' = 15°$

$c'_m = 17.8 \text{ kN/m}^2$

$\phi'_m = 12°$

Calculate the factor of safety (a) with respect to strength, (b) with respect to cohesion and (c) with respect to friction. The average intergranular pressure σ' on the failure surface is 102.5 kN/m^2.

Solution

On the basis of the given data, the average shearing strength on the failure surface is

$$s = 26.7 + 102.5 \tan 15°$$

$$= 26.7 + 102.5 \times 0.268 = 54.2 \text{ kN/m}^2$$

and the average value of mobilized shearing resistance is

$$\tau = 17.8 + 102.5 \tan 12°$$

$$= 17.8 + 102.5 \times 0.212 = 39.6 \text{ kN/m}^2$$

$$F_s = \frac{54.2}{39.6} = 1.37; \quad F_c = \frac{26.70}{17.80} = 1.50; \quad F_\phi = \frac{\tan\phi'}{\tan\phi_m} = \frac{0.268}{0.212} = 1.27$$

The above example shows the factor of safety with respect to shear strength, F_s is 1.27, whereas the factors of safety with respect to cohesion and friction are different. Consider two extreme cases:

1. When the factor of safety with respect to cohesion is unity.
2. When the factor of safety with respect to friction is unity.

Case 1

$$\tau = 39.60 = 26.70 + \frac{102.50}{F_\phi} \tan 15° = 26.7 + \frac{102.5 \times 0.268}{F_\phi} = 26.70 + \frac{27.50}{F_\phi}$$

$$F_\phi = \frac{27.50}{12.90} = 2.13$$

Case 2

$$\tau = 39.60 = \frac{26.70}{F_c} + 102.50 \tan 15°$$

$$= \frac{26.70}{F_c} + 27.50$$

$$F_c = \frac{26.70}{12.10} = 2.20$$

We can have any combination of F_c and F_ϕ between these two extremes cited above to give the same mobilized shearing resistance of 39.6 kN/m². Some of the combinations of F_c and F_ϕ are given below.

Combination of F_c and F_ϕ

F_c	1.00	1.26	1.37	1.50	2.20
F_ϕ	2.12	1.50	1.37	1.27	1.00

Under Case 2, the value of $F_c = 2.20$ when $F_\phi = 1.0$. The factor of safety $F_c = 2.20$ is defined as the *factor of safety with respect to cohesion.*

Example 16.2

What will be the factors of safety with respect to average shearing strength, cohesion and internal friction of a soil, for which the shear strength parameters obtained from the laboratory tests are $c' = 32$ kN/m² and $\phi' = 18°$; the expected parameters of mobilized shearing resistance are $c'_m = 21$ kN/m² and $\phi'_m = 13°$ and the average effective pressure on the failure plane is 110 kN/m². For the same value of mobilized shearing resistance determine the following:

1. Factor of safety with respect to height;
2. Factor of safety with respect to friction when that with respect to cohesion is unity; and
3. Factor of safety with respect to strength.

Solution

The available shear strength of the soil is

$$s = 32 + 110 \tan 18° = 32 + 35.8 = 67.8 \text{ kN/m}^2$$

The mobilized shearing resistance of the soil is

$$\tau = 21 + 110 \tan 13° = 21 + 25.4 = 46.4 \text{ kN/m}^2$$

Factor of safety with respect to average strength, $\quad F_s = \dfrac{67.8}{46.4} = 1.46$

Factor of safety with respect to cohesion, $\quad F_c = \dfrac{32}{21} = 1.52$

Factor of safety with respect to friction, $\quad F_\phi = \dfrac{\tan \phi'}{\tan \phi'_m} = \dfrac{\tan 18°}{\tan 13°} = \dfrac{0.3249}{0.2309} = 1.41$

Factor of safety with respect to height, $\quad F_H (= F_c)$ will be at $F_\phi = 1.0$

$$\tau = 46.4 = \frac{32}{F_c} + \frac{110 \tan 18°}{1.0}, \quad \text{therefore, } F_c = \frac{32}{46.4 - 35.8} = 3.0$$

Factor of safety with respect to friction at $F_c = 1.0$ is

$$\tau = 46.4 = \frac{32}{1.0} + \frac{110 \tan 18°}{F_\phi}, \quad \text{therefore, } F_\phi = \frac{35.8}{46.4 - 32} = 2.49$$

Factor of safety with respect to strength F_s is obtained when $F_c = F_\phi$. We may write

$$\tau = 46.4 = \frac{32}{F_s} + \frac{110 \tan 18°}{F_s} \quad \text{or } F_s = 1.46$$

16.4 STABILITY ANALYSIS OF INFINITE SLOPES IN SAND

As an introduction to slope analysis, the problem of a slope of infinite extent is of interest. Imagine an infinite slope, as shown in Fig. 16.2, making an angle β with the horizontal. The soil is cohesionless and completely homogeneous throughout. Then the stresses acting on any vertical plane in the soil are the same as those on any other vertical plane. The stress at any point on a plane EF parallel to the surface at depth z will be the same as at every point on this plane.

Now consider a vertical slice of material $ABCD$ having a unit dimension normal to the page. The forces acting on this slice are its weight W, a vertical reaction R on the base of the slice, and two lateral forces P_1 acting on the sides. Since the slice is in equilibrium, the weight and reaction are equal in magnitude and opposite in direction. They have a common line of action which passes through the centre of the base AB. The lateral forces must be equal and opposite and their line of action must be parallel to the sloped surface.

The normal and shear stresses on plane AB are

$$\sigma'_n = \gamma z \cos^2 \beta$$

$$\tau = \gamma z \cos \beta \sin \beta$$

where σ'_n = effective normal stress,
γ = effective unit weight of the sand.

If full resistance is mobilized on plane AB, the shear strength, s, of the soil per Coulomb's law is

$$s = \sigma'_n \tan\phi'$$

when $\tau = s$, substituting for s and σ'_n, we have

$$\gamma z \cos \beta \sin \beta = \gamma z \cos^2 \beta \tan\phi'$$

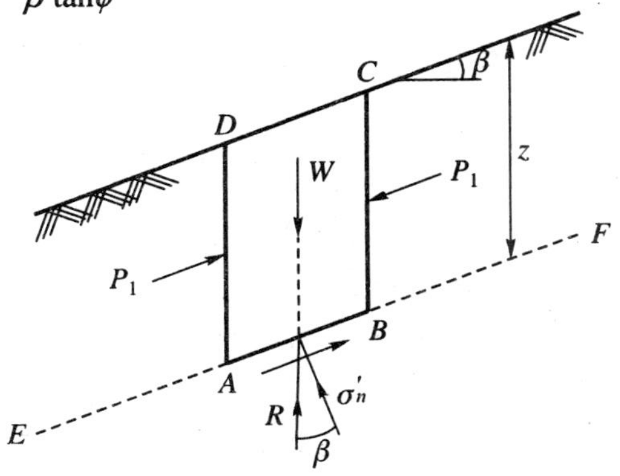

Fig. 16.2 Stability analysis of infinite slope in sand

or $\tan\beta = \tan\phi'$ (16.5a)

Equation (16.5a) indicates that the maximum value of β is limited to ϕ'. if the slope is to be stable. This condition holds true for cohesionless soils whether the slope is completely dry or completely submerged under water.

The factor of safety of infinite slopes in sand may be written as

$$F_s = \frac{\tan\phi'}{\tan\beta}$$ (16.5b)

16.5 STABILITY ANALYSIS OF INFINITE SLOPES IN CLAY

The vertical stress σ_v acting on plane AB (Fig. 16.3) where

$$\sigma_v = \gamma z \cos\beta$$

is represented by OC in Fig. 16.3 in the stress diagram. The normal stress on this plane is OE and the shearing stress is EC. The line OC makes an angle β with the σ-axis.

The Mohr strength envelope is represented by line FA whose equation is

$$s = c' + \sigma' \tan\phi'$$

According to the envelope, the shearing strength is ED where the normal stress is OE.

When β is greater than ϕ'. the lines OC and FD meet. In this case the two lines meet at A. As long as the shearing stress on a plane is less than the shearing strength on the plane, there is no danger of failure. Figure 16.3 indicates that at all depths at which the direct stress is less than OB, there is no possibility of failure. However at a particular depth at which the direct stress is OB, the shearing

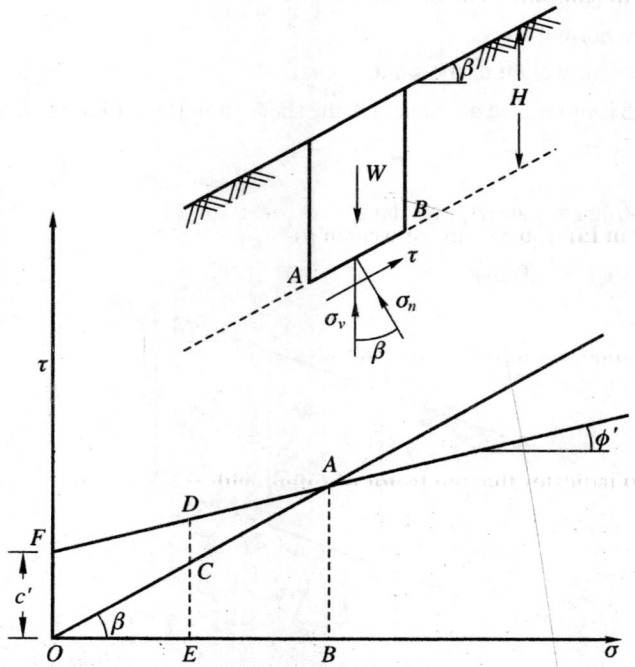

Fig. 16.3   Stability analysis of infinite slopes in clay soils

strength and shearing stress values are equal as represented by AB, failure is imminent. This depth at which the shearing stress and shearing strength are equal is called the *critical depth*. At depths greater than this critical value, Fig. 16.3 indicates that the shearing stress is greater than the shearing strength but this is not possible. Therefore it may be concluded that the slope may be steeper than ϕ' as long as the depth of the slope is less than the critical depth.

Expression for the Stability of an Infinite Slope of Clay of Depth H

Equation (16.2) gives the developed shearing stress as

$$\tau = c'_m + \sigma' \tan\phi'_m \tag{16.6}$$

Under conditions of no seepage and no pore pressure, the stress components on a plane at depth H and parallel to the surface of the slope are

$$\tau = \gamma H \sin \beta \cos \beta$$

$$\sigma' = \gamma H \cos^2 \beta$$

Substituting these stress expressions in the equation above and simplifying, we have

$$c'_m = \gamma H \cos^2 \beta (\tan\beta - \tan \phi'_m)$$

$$\text{or } N_s = \frac{c'_m}{\gamma H} = \cos^2 \beta (\tan\beta - \tan \phi'_m) \tag{16.7}$$

where H is the allowable height and the term $c'_m /\gamma H$ is a dimensionless expression called the *stability number* and is designated as N_s. This dimensionless number is proportional to the required cohesion and is inversely proportional to the allowable height. The solution is for the case when no seepage is occurring. If in Eq. (16.7) the factor of safety with respect to friction is unity, the stability number with respect to cohesion may be written as

$$N_s = \frac{c'}{F_c \gamma H} = \cos^2 \beta (\tan\beta - \tan \phi') \tag{16.8}$$

where $c'_m = \dfrac{c'}{F_c}$

The stability number in Eq. (16.8) may be written as

$$N_s = \frac{c'}{F_c \gamma H} = \frac{c'}{\gamma H_c} \tag{16.9}$$

where H_c = critical height. From Eq. (16.9), we have

$$F_c = \frac{H_c}{H} = F_H \tag{16.10}$$

Equation (16.10) indicates that the factor of safety with respect to cohesion, F_c, is the same as the factor of safety with respect to height F_H.

If there is seepage parallel to the ground surface throughout the entire mass, with the free water surface coinciding with the ground surface, the components of effective stresses on planes parallel to the surface of slopes at depth H are given as [Fig. 16.4(a)].

Normal stress

$$\sigma'_n = (\gamma_{sat} - \gamma_w)H \cos^2 \beta = \gamma_b H \cos^2 \beta \tag{16.11a}$$

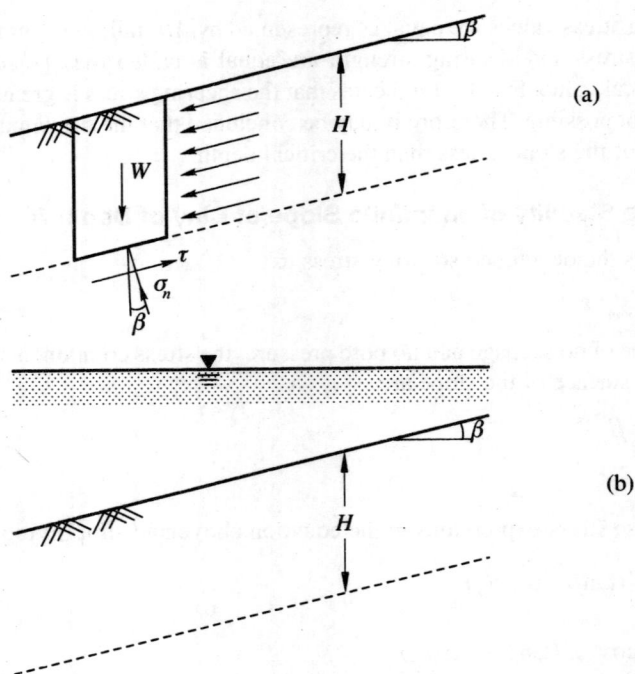

Fig. 16.4 Analysis of infinite slope (a) with seepage flow through the entire mass, and (b) with completely submerged slope.

the shearing stress

$$\tau = \gamma_{sat} H \sin \beta \cos \beta \qquad (16.11b)$$

Now substituting Eqs (16.11a) and (16.11b) into equation

$$\tau = c'_m + \sigma'_n \tan \phi'_m$$

and simplifying, the stability expression obtained is

$$\frac{c'_m}{\gamma_{sat} H} = \cos^2 \beta \tan \beta - \frac{\gamma_b}{\gamma_{sat}} \tan \phi'_m \qquad (16.12)$$

As before, if the factor of safety with respect to friction is unity, the stability number which represents the cohesion may be written as

$$N_s = \frac{c'}{F_c \gamma_{sat} H} = \frac{c'}{\gamma_{sat} H_c} = \cos^2 \beta \tan \beta - \frac{\gamma_b}{\gamma_{sat}} \tan \phi' \qquad (16.13)$$

If the slope is completely submerged, and if there is no seepage as in Fig. 16.4(b), then Eq. (16.13) becomes

$$N_s = \frac{c'}{F_c \gamma_b H} = \frac{c'}{\gamma_b H_c} = \cos^2 \beta (\tan \beta - \tan \phi') \qquad (16.14)$$

where γ_b = submerged unit weight of the soil.

Example 16.3

Find the factor of a slope of infinite extent having a slope angle $= 25°$. The slope is made of cohesionless soil with $\phi = 30°$.

Solution

Factor of safety

$$F_s = \frac{\tan\phi'}{\tan\beta} = \frac{\tan 30°}{\tan 25°} = \frac{0.5774}{0.4663} = 1.238$$

Example 16.4

Analyze the slope of Example 16.3 if it is made of clay having $c' = 30$ kN/m², $\phi' = 20°$, $e = 0.65$ and $G = 2.7$ and under the following conditions: (i) when the soil is dry, (ii) when water seeps parallel to the surface of the slope, and (iii) when the slope is submerged.

Solution

For $e = 0.65$ and $G = 2.7$

$$\gamma_d = \frac{2.7 \times 9.81}{1+0.65} = 16.05 \text{ kN/m}^3, \quad \gamma_{sat} = \frac{(2.7+0.65) \times 9.81}{1+0.65} = 19.9 \text{ kN/m}^3,$$

$$\gamma_b = 10.09 \text{ kN/m}^3$$

(i) For dry soil the stability number N_s is

$$N_s = \frac{c'}{\gamma_b H_c} = \cos^2 \beta(\tan\beta - \tan\phi') \quad \text{when } F_\phi = 1$$

$$= (\cos 25°)(\tan 25° - \tan 20°) = 0.084.$$

Therefore, the critical height $H_c = \dfrac{c'}{\gamma_d \times N_s} = \dfrac{3}{16.05 \times 0.084} = 22.25$ m

(ii) For seepage parallel to the surface of the slope [Eq. (16.13)]

$$N_s = \frac{c'}{\gamma_t H_c} = \cos^2 25°(\tan 25° - \frac{10.09}{19.9}\tan 20° = 0.2315$$

$$H_c = \frac{c'}{\gamma_t N_s} = \frac{30}{19.9 \times 0.2315} = 6.51 \text{ m}$$

(iii) For the submerged slope [Eq. (16.14)]

$$N_s = \cos^2 25°(\tan 25° - \tan 20°) = 0.084$$

$$H_c = \frac{c'}{\gamma_b N_s} = \frac{30}{10.09 \times 0.084} = 35.4 \text{ m}$$

16.6 METHODS OF STABILITY ANALYSIS OF SLOPES OF FINITE HEIGHT

The stability of slopes of infinite extent has been discussed in previous sections. A more common problem is the one in which the failure occurs on curved surfaces. The most widely used method of analysis of homogeneous, isotropic, finite slopes is the *Swedish method* based on circular failure surfaces. Petterson (1955) first applied the circle method to the analysis of a soil failure in connection with the failure of a quarry wall in Goeteberg, Sweden. A Swedish National Commission, after studying a large number of failures, published a report in 1922 showing that the lines of failure of most such slides roughly approached the circumference of a circle. The failure circle might pass above the toe, through the toe or below it. By investigating the strength along the arc of a large number of such circles, it was possible to locate the circle which gave the lowest resistance to shear. This general method has been quite widely accepted as offering an approximately correct solution for the determination of the factor of safety of a slope of an embankment and of its foundation. Developments in the method of analysis have been made by Fellenius (1947), Terzaghi (1943), Gilboy (1934), Taylor (1937), Bishop (1955), and others, with the result that a satisfactory analysis of the stability of slopes, embankments and foundations by means of the circle method is no longer an unduly tedious procedure.

There are other methods of historic interest such as the *Culmann method* (1875) and the *logarithmic spiral method*. The Culmann method assumes that rupture will occur along a plane. It is of interest only as a classical solution, since actual failure surfaces are invariably curved. This method is approximately correct for steep slopes. The logarithmic spiral method was recommended by Rendulic (1935) with the rupture surface assuming the shape of logarithmic spiral. Though this method makes the problem statically determinate and gives more accurate results, the greater length of time required for computation overbalances this accuracy.

There are several methods of stability analysis based on the circular arc surface of failure. A few of the methods are described below

Methods of Analysis

The majority of the methods of analysis may be categorized as limit equilibrium methods. The basic assumption of the limit equilibrium approach is that Coulomb's failure criterion is satisfied along the assumed failure surface. A free body is taken from the slope and starting from known or assumed values of the forces acting upon the free body, the shear resistance of the soil necessary for equilibrium is calculated. This calculated shear resistance is then compared to the estimated or available shear strength of the soil to give an indication of the factor of safety.

Methods that consider only the whole free body are the (a) slope failure under undrained conditions, (b) friction-circle method (Taylor, 1937, 1948) and (c) Taylor's stability number (1948).

Methods that divide the free body into many vertical slices and consider the equilibrium of each slice are the Swedish circle method (Fellenius, 1927), Bishop method (1955), Bishop and Morgenstern method (1960) and Spencer method (1967). The majority of these methods are in chart form and cover a wide variety of conditions.

16.7 PLANE SURFACE OF FAILURE

Culmann (1875) assumed a plane surface of failure for the analysis of slopes which is mainly of interest because it serves as a test of the validity of the assumption of plane failure. In some cases this assumption is reasonable and in others it is questionable.

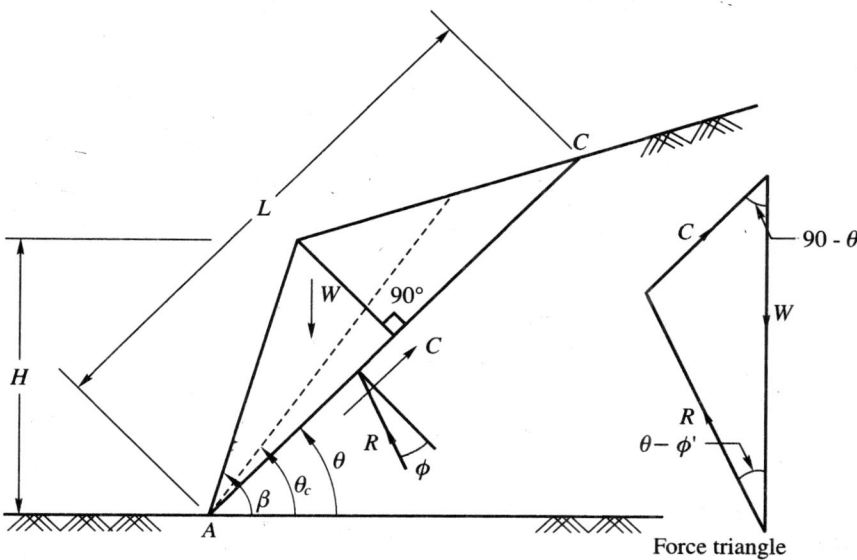

Fig. 16.5 Stability of slopes by Culmann method

The method as indicated above assumes that the critical surface of failure is a plane surface passing through the toe of the dam as shown in Fig. 16.5.

The forces that act on the mass above trial failure plane AC inclined at angle θ with the horizontal are shown in the figure. The expression for the weight, W, and the total cohesion C are respectively,

$$W = \frac{1}{2}\gamma LH \cos ec\beta \sin(\beta - \theta)$$

and $C = c'L$

The use of the law of sines in the force triangle, shown in the figure, gives

$$\frac{C}{W} = \frac{\sin(\theta - \phi')}{\cos \phi'}$$

Substituting herein for C and W, and rearranging we have

$$\frac{c'}{\gamma H_\theta} = \frac{1}{2}\cos ec\beta \sin(\beta - \theta)\sin(\theta - \phi')\sec \phi'$$

in which the subscript θ indicates that the stability number is for the trial plane at inclination θ.

The most dangerous plane is obtained by setting the first derivative of the above equation with respect to θ equal to zero. This operation gives

$$\theta'_c = \frac{1}{2}(\beta + \phi')$$

where θ'_c is the critical angle for limiting equilibrium and the stability number for limiting equilibrium may be written as

$$\frac{c'}{\gamma H_c} = \frac{1 - \cos(\beta - \phi')}{4\sin \beta \cos \phi'} \tag{16.15}$$

where H_c is the critical height of the slope.

If we write

$$F_c = \frac{c'}{c'_m}, \quad F_\phi = \frac{\tan\phi'}{\tan\phi'_m}$$

where F_c and F_ϕ are safety factors with respect to cohesion and friction respectively, Eq. (16.15) may be modified for chosen values of c'_m and ϕ'_m as

$$\frac{c'_m}{\gamma H} = \frac{1-\cos(\beta-\phi'_m)}{4\sin\beta\cos\phi'_m} \tag{16.16}$$

The critical angle for any assumed values of c'_m and ϕ'_m is

$$\theta_c = \frac{1}{2}(\beta+\phi'_m) \tag{16.17}$$

From Eq. (16.16), the allowable height of a slope is

$$H = \frac{4c'_m \sin\beta\cos\phi'_m}{\gamma[1-\cos(\beta-\phi'_m)]} \tag{16.18}$$

Example 16.5

Determine by Culmann's method the critical height of an embankment having a slope angle of 40° and the constructed soil having $c' = 630$ psf, $\phi' = 20°$ and effective unit weight $= 114$ lb/ft³. Find the allowable height of the embankment if $F_c = F_\phi = 1.25$.

Solution

$$H_c = \frac{4c'\sin\beta\cos\phi'}{\gamma[1-\cos(\beta-\phi')]} = \frac{4\times630\times\sin40°\cos20°}{114(1-\cos20°)} = 221\,\text{ft}$$

For $F_c = F_\phi = 125$, $\quad c'_m = \dfrac{c'}{F_c} = \dfrac{630}{1.25} = 504\,\text{lb/ft}^2$

and $\tan\phi'_m = \dfrac{\tan\phi'}{F_\phi} = \dfrac{\tan20°}{1.25} = 0.291$, $\phi'_m = 16.23°$

Allowable height, $H = \dfrac{4\times504\sin40°\cos16.23°}{114[1-\cos(40-16.23°)]} = 128.7\,\text{ft}$

16.8   CIRCULAR SURFACES OF FAILURE

The investigations carried out in Sweden at the beginning of this century have clearly confirmed that the surfaces of failure of earth slopes resemble the shape of a circular arc. When soil slips along a circular surface, such a slide may be termed as a rotational slide. It involves downward and outward movement of a slice of earth as shown in Fig. 16.6(a) and sliding occurs along the entire surface of contact between the slice and its base. The types of failure that normally occur may be classified as

1.   Slope failure

2. Toe failure

3. Base failure

In slope failure, the arc of the rupture surface meets the slope above the toe. This can happen when the slope angle β is quite high and the soil close to the toe possesses high strength. Toe failure occurs when the soil mass of the dam above the base and below the base is homogeneous. The base failure occurs particularly when the base angle β is low and the soil below the base is softer and more plastic than the soil above the base. The various modes of failure are shown in Fig. 16.6.

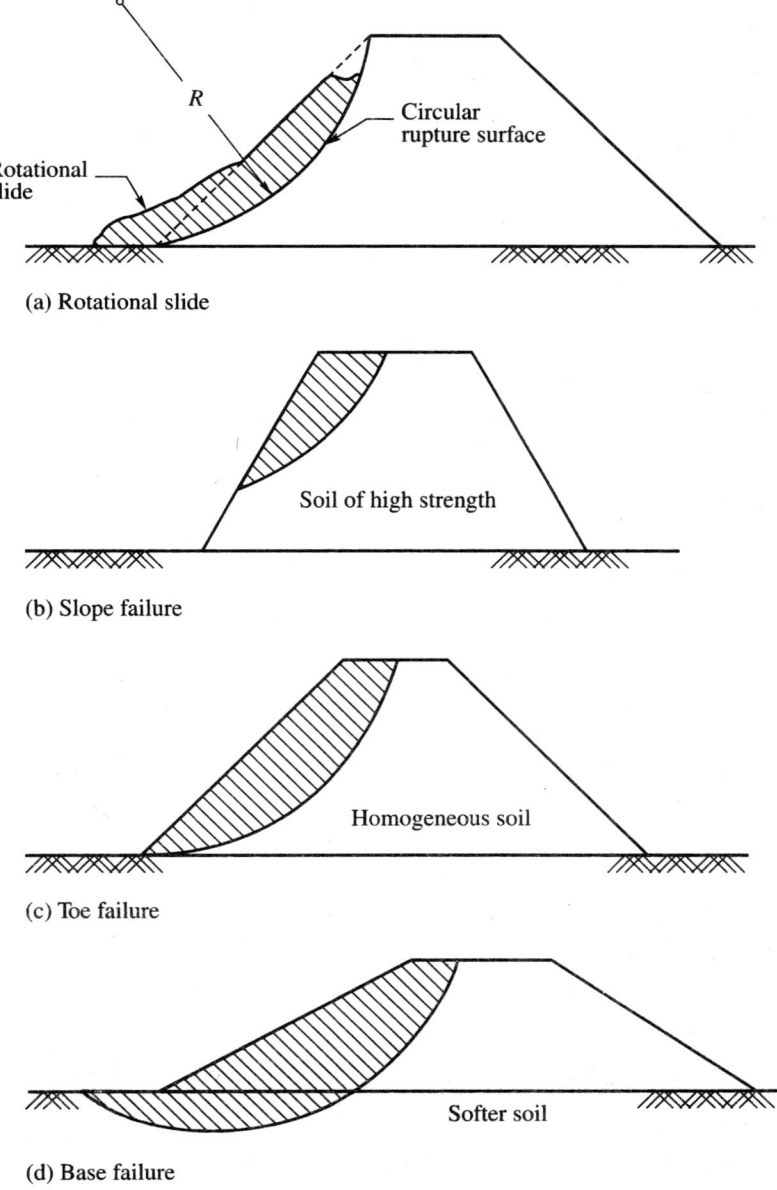

(a) Rotational slide

(b) Slope failure

(c) Toe failure

(d) Base failure

Fig. 16.6   Types of failure of earth dams

16.9   FAILURE UNDER UNDRAINED CONDITIONS ($\phi_u = 0$)

A fully saturated clay slope may fail under undrained conditions ($\phi_u = 0$) immediately after construction. The stability analysis is based on the assumption that the soil is homogeneous and the potential failure surface is a circular arc. Two types of failures considered are:

1.   Slope failure
2.   Base failure

The undrained shear strength c_u of soil is assumed to be constant with depth. A trial failure circular surface AB with centre at O and radius R is shown in Fig. 16.7(a) for a toe failure. The slope AC and the chord AB make angles β and α with the horizontal respectively. W is the weight per unit

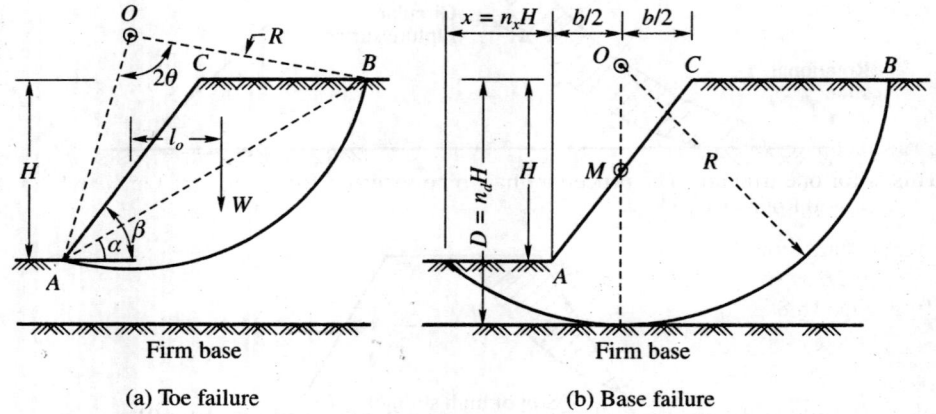

(a) Toe failure   (b) Base failure

Fig. 16.7   Critical circle positions for (a) slope failure (after (Fellenius, 1927), (b) base failure

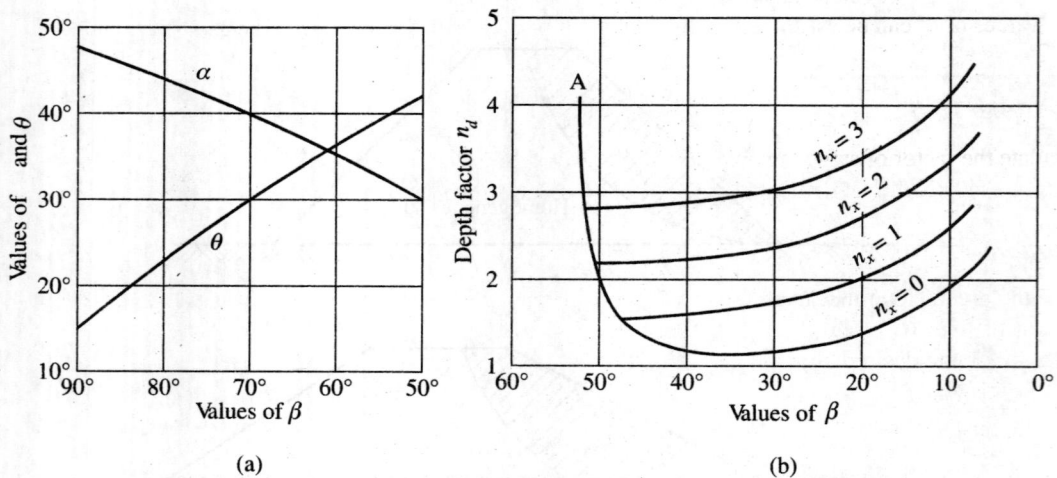

(a)   (b)

Fig. 16.8   (a) Relation between slope angle β and parameters α and θ for location of critical toe circle when β is greater than 53°, (b) relation between slope angle β and depth factor n_d for various values of parameter n_x (after Fellenius, 1927)

length of the soil lying above the trial surface acting through the centre of gravity of the mass. l_o is the lever arm, L_a is the length of the arc, L_c the length of the chord AB and c_m the mobilized cohesion for any assumed surface of failure.

We may express the factor of safety F_s as

$$F_s = \frac{c_u}{c_m} \tag{16.19}$$

For equilibrium of the soil mass lying above the assumed failure surface, we may write resisting moment M_r = actuating moment M_a

The resisting moment $M_r = L_a c_m R$

Actuating moment, $M_a = W l_o$

Equation for the mobilized c_m is

$$c_m = \frac{W l_o}{L_a R} \tag{16.20}$$

Now the factor of safety F for the assumed trial arc of failure may be determined from Eq. (16.19). This is for one trial arc. The procedure has to be repeated for several trial arcs and the one that gives the least value is the critical circle.

If failure occurs along a toe circle, the centre of the critical circle can be located by laying off the angles α and 2θ as shown in Fig. 16.7(a). Values of α and θ for different slope angles β can be obtained from Fig. 16.8(a).

If there is a base failure as shown in Fig. 16.7(b), the trial circle will be tangential to the firm base and as such the centre of the critical circle lies on the vertical line passing through midpoint M on slope AC. The following equations may be written with reference to Fig. 16.7(b).

$$\text{Depth factor, } n_d = \frac{D}{H}, \quad \text{Depth factor, } n_x = \frac{x}{H} \tag{16.21}$$

Values of n_x can be estimated for different values of n_d and β by means of the chart Fig. 16.8(b).

Example 16.6

Calculate the factor of safety against shear failure along the slip circle shown in Fig. Ex. 16.6 Assume cohesion = 40 kN/m², angle of internal friction = zero and the total unit weight of the soil = 20.0 kN/m³.

Solution

Draw the given slope $ABCD$ as shown in Fig. Ex. 16.6. To locate the centre of rotation, extend the bisector of line BC to cut the vertical line drawn from C at point O. With O as centre and OC as radius, draw the desired slip circle.

Radius $OC = R = 36.5$ m, Area $BECFB = \frac{2}{3} \times EF \times BC$

$$= \frac{2}{3} \times 4 \times 32.5 = 86.7 \text{ m}^2$$

Therefore $W = 86.7 \times 1 \times 20 = 1734$ kN

W acts through point G which may be taken as the middle of FE.

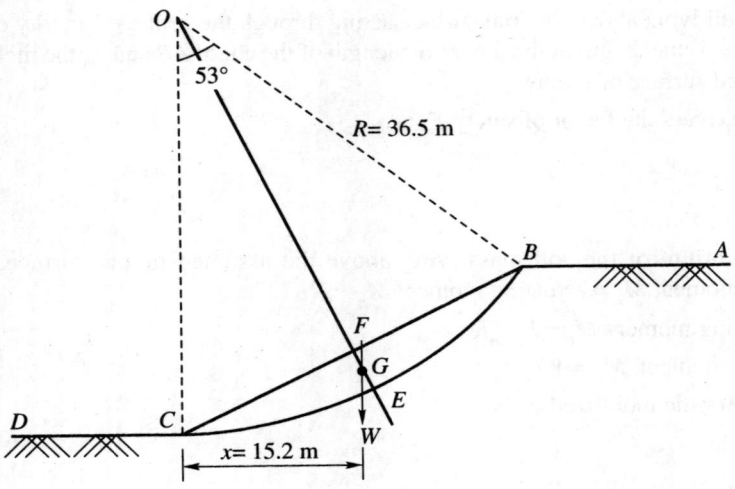

Fig. Ex. 16.6

From the figure we have, $x = 15.2$ m, and $\theta = 53°$

$$\text{Length of arc } BEC = R\theta = 36.5 \times 53° \times \frac{3.14}{180} = 33.8 \text{ m}$$

$$F_s = \frac{\text{length of arc} \times \text{cohesion} \times \text{radius}}{Wx} = \frac{33.8 \times 40 \times 36.5}{1734 \times 15.2} = 1.87$$

16.10 FRICTION-CIRCLE METHOD

Physical Concept of the Method

The principle of the method is explained with reference to the section through a dam shown in Fig. 16.9. A trial circle with centre of rotation O is shown in the figure. With centre O and radius $R \sin \phi'$

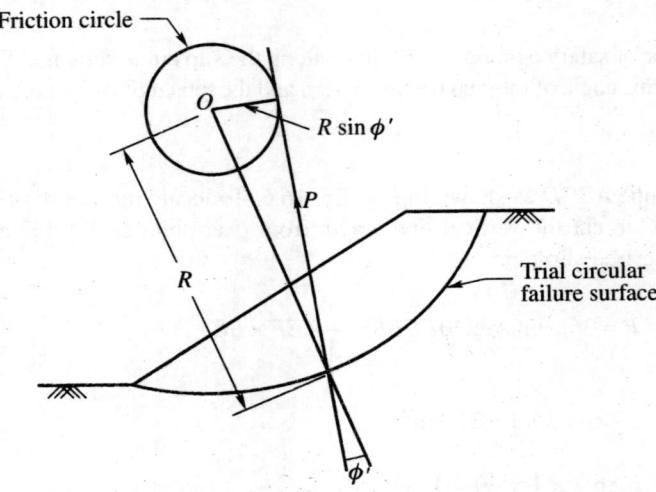

Fig. 16.9   Principle of friction circle method

where R is the radius of the trial circle, a circle is drawn. Any line tangent to the inner circle must intersect the trial circle at an angle ϕ' with R. Therefore, any vector representing an intergranular pressure at obliquity ϕ' to an element of the rupture arc must be tangent to the inner circle. This inner circle is called the *friction circle* or ϕ-*circle*. The friction circle method of slope analysis is a convenient approach for both graphical and mathematical solutions. It is given this name because the characteristic assumption of the method refers to the ϕ-circle.

The forces considered in the analysis are

1. The total weight W of the mass above the trial circle acting through the centre of mass. The centre of mass may be determined by any one of the known methods.

2. The resultant boundary neutral force U. The vector U may be determined by a graphical method from flownet construction.

3. The resultant intergranular force, P, acting on the boundary.

4. The resultant cohesive force C.

Actuating Forces

The actuating forces may be considered to be the total weight W and the resultant boundary force U as shown in Fig. 16.10.

The boundary neutral force always passes through the centre of rotation O. The resultant of W and U, designated as Q, is shown in the figure.

Resultant Cohesive Force

Let the length of arc AB be designated as L_a, the length of chord AB by L_c. Let the arc length L_a be divided into a number of small elements and let the mobilized cohesive force on these elements be designated as C_1, C_2, C_3, etc. as shown in Fig. 16.11. The resultant of all these forces is shown by the force polygon in the figure. The resultant is $A'B'$ which is parallel and equal to the chord length AB. The resultant of all the mobilized cohesional forces along the arc is therefore

$$C = c'_m L_c$$

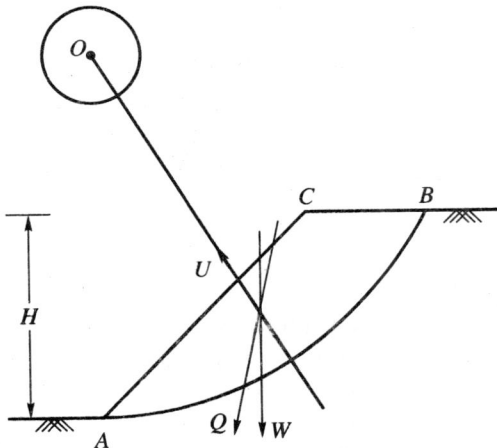

Fig. 16.10 Actuating forces

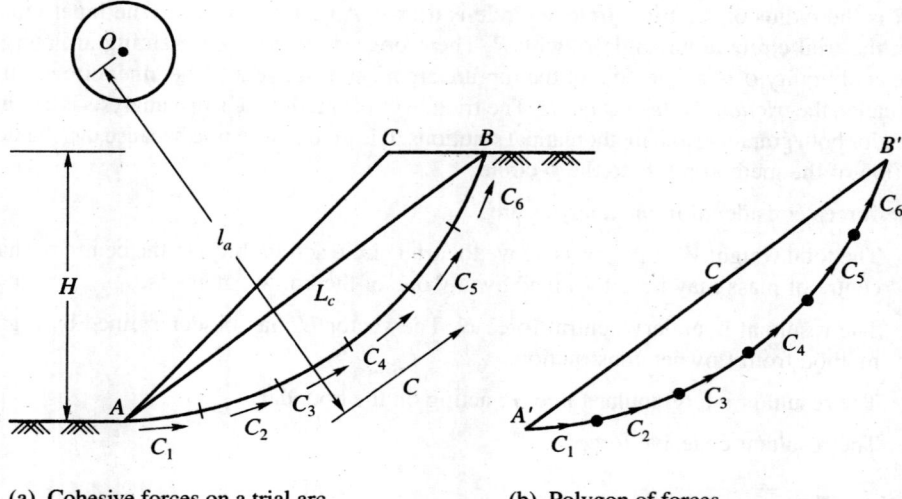

(a) Cohesive forces on a trial arc   (b) Polygon of forces

Fig. 16.11   Resistant cohesive forces

We may write $c'_m = \dfrac{c'}{F_c}$

wherein c' = unit cohesion, F_c = factor of safety with respect to cohesion.

The line of action of C may be determined by moment consideration. The moment of the total cohesion is expressed as

$$c'_m L_a R = c'_m L_c l_a$$

where l_a = moment arm. Therefore,

$$l_a = R\frac{L_a}{L_c} \tag{16.22}$$

It is seen that the line of action of vector C is independent of the magnitude of c'_m.

Resultant of Boundary Intergranular Forces

The trial arc of the circle is divided into a number of small elements. Let P_1, P_2, P_3, etc. be the intergranular forces acting on these elements as shown in Fig. 16.12. The friction circle is drawn with a radius of $R \sin \phi'_m$

where $\tan\phi'_m = \dfrac{\tan\phi'}{F_\phi}$

The lines of action of the intergranular forces P_1, P_2, P_3, etc. are tangential to the friction circle and make an angle of ϕ'_m at the boundary. However, the vector sum of any two small forces has a line of action through point D, missing tangency to the ϕ'_m-circle by a small amount. The resultant of all granular forces must therefore miss tangency to the ϕ'_m-circle by an amount which is not considerable. Let the distance of the resultant of the granular force P from the centre of the circle be designated as $KR \sin \phi'_m$ (as shown in Fig. 16.12). The magnitude of K depends upon the type of intergranular

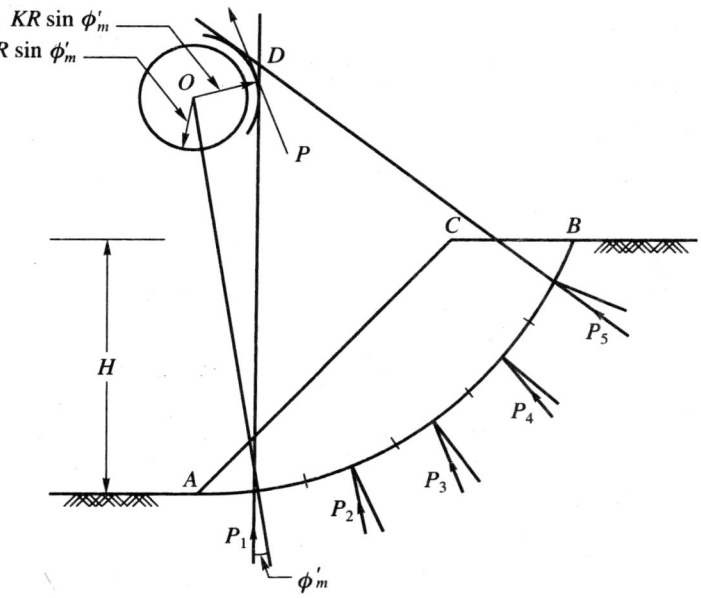

Fig. 16.12 Resultant of intergranular forces

pressure distribution along the arc. The most probable form of distribution is the sinusoidal distribution.

The variation of K with respect to the central angle α' is shown in Fig. 16.13. The figure also gives relationships between α' and K for a uniform stress distribution of effective normal stress along the arc of failure.

The graphical solution based on the concepts explained above is simple in principle. For the three forces Q, C and P of Fig. 16.14 to be in equilibrium, P must pass through the intersection of the

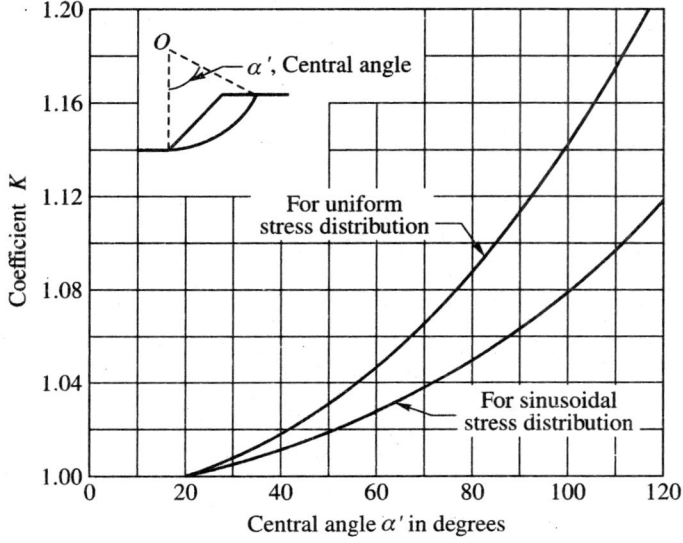

Fig. 16.13 Relationship between K and central angle α'

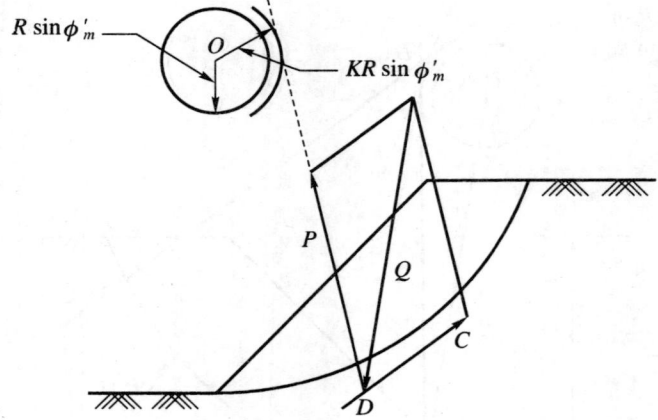

Fig. 16.14   Force triangle for the friction-cirle method

known lines of action of vectors Q and C. The line of action of vector P must also be tangent to the circle of radius $KR \sin \phi'_m$. The value of K may be estimated by the use of curves given in Fig. 16.13, and the line of action of force P may be drawn as shown in Fig. 16.14. Since the lines of action of all three forces and the magnitude of force Q are known, the magnitude of P and C may be obtained by the force parallelogram construction that is indicated in the figure. The circle of radius of $KR \sin \phi'_m$ is called the *modified friction circle*.

Determination of Factor of Safety With Respect to Strength

Figure 16.15(a) is a section of a dam. AB is the trial failure arc. The force Q, the resultant of W and U is drawn as explained earlier. The line of action of C is also drawn. Let the forces Q and C meet at

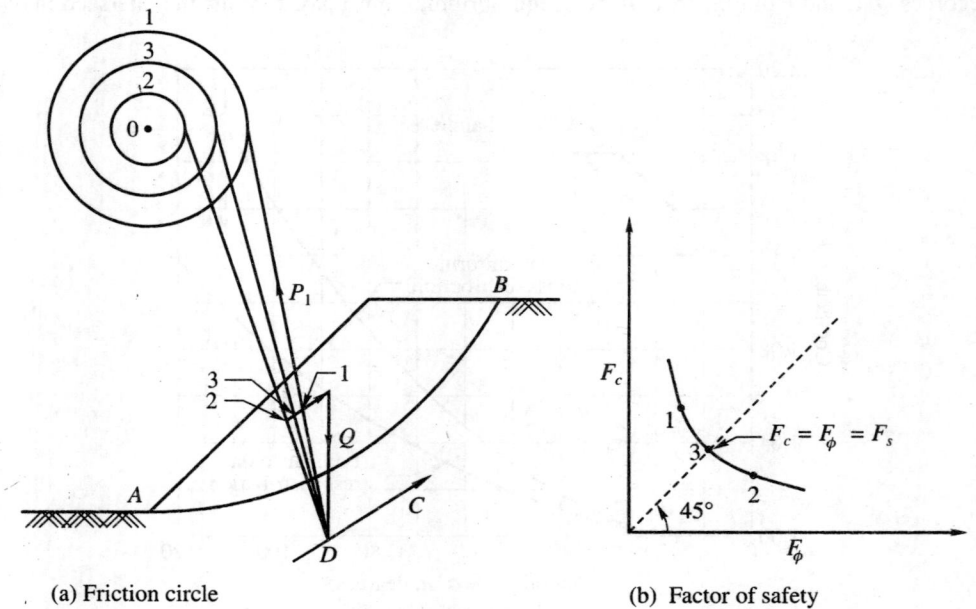

(a) Friction circle   (b) Factor of safety

Fig. 16.15   Graphical method of determining factor of safety with respect to strength

point D. An arbitrary first trial using any reasonable ϕ'_m value, which will be designated by ϕ'_{m1} is given by the use of circle 1 or radius $KR \sin \phi'_{m1}$. Subscript 1 is used for all other quantities of the first trial. The force P_1 is then drawn through D tangent to circle 1. C_1 is parallel to chord and point 1 is the intersection of forces C_1 and P_1. The mobilized cohesion is equal $c'_{m1} L_c$. From this the mobilized cohesion c'_{m1} is evaluated. The factors of safety with respect to cohesion and friction are determined from the expressions

$$F'_c = \frac{c'}{c'_{m1}}, \text{ and } F_{\phi 1} = \frac{\tan \phi'}{\tan \phi'_{m1}}$$

These factors are the values used to plot point 1 in the graph in Fig. 16.15(b). Similarly other friction circles with radii $KR \sin \phi'_{m2}$, $KR \sin \phi'_{m3}$. etc. may be drawn and the procedure repeated. Points 2, 3, etc. are obtained as shown in Fig. 16.15(b). The 45° line, representing $F_c = F_\phi$, intersects the curve to give the factor of safety F_s for this trial circle.

Several trial circles must be investigated in order to locate the critical circle, which is the one having the minimum value of F_s.

Example 16.7

An embankment has a slope of 2 (horizontal) to 1 (vertical) with a height of 10 m. It is made of a soil having a cohesion of 30 kN/m², an angle of internal friction of 5° and a unit weight of 20 kN/m³. Consider any slip circle passing through the toe. Use the friction circle method to find the factor of safety with respect to cohesion.

Solution

Refer to Fig. Ex. 16.7. Let EFB be the slope and AKB be the slip circle drawn with centre O and radius R = 20 m.

Length of chord $AB = L_c = 27$ m

Take J as the midpoint of AB, then

Area $AKBFEA$ = area $AKBJA$ + area $ABEA$

$$= \frac{2}{3} AB \times JK + \frac{1}{2} AB \times EL$$

$$= \frac{2}{3} \times 27 \times 5.3 + \frac{1}{2} 27 \times 2.0 = 122.4 \text{ m}^2$$

Therefore, the weight of the soil mass $= 122.4 \times 1 \times 20 = 2448$ kN

It will act through point G, the centroid of the mass which can be taken as the mid point of FK.

Now, $\theta = 85°$,

Length of arc $AKB = L = R\theta = 20 \times 85 \times \frac{3.14}{180} = 29.7$ m

Moment arm of cohesion, $l_a = R \frac{L}{L_c} = 20 \times \frac{29.7}{27} = 22$ m

From centre O, at a distance l_a, draw the cohesive force vector C, which is parallel to the chord AB. Now from the point of intersection of C and W, draw a line tangent to the friction circle drawn at

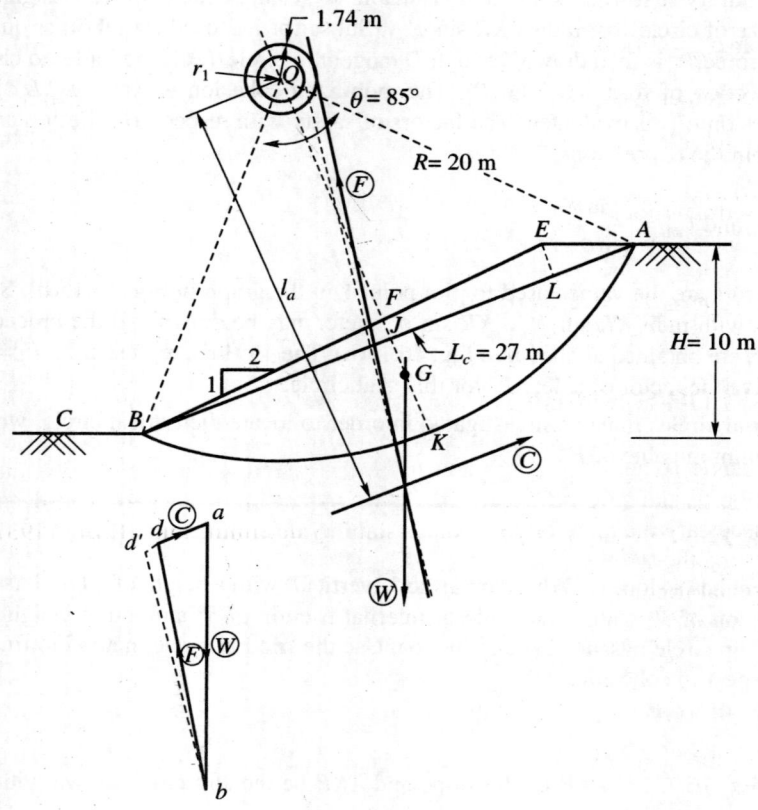

Fig. Ex. 16.7

O with a radius of $R \sin \phi' = 20 \sin 5° = 1.74$ m. This line is the line of action of the third force F.

Draw a triangle of forces in which the magnitude and the direction for W is known and only the directions of the other two forces C and F are known.

Length ad gives the cohesive force $C = 520$ kN

Mobilized cohesion,

$$c'_m = \frac{C}{L} = \frac{520}{29.7} = 17.51 \, \text{kN/m}^2$$

Therefore, the factor of safety with respect to cohesion, F_c, is

$$F_c = \frac{c'}{c'_m} = \frac{30}{17.51} = 1.713$$

F_c will be 1.713 if the factor of safety with respect to friction, $F_\phi = 1.0$

If, $F_s = 1.5$, then $\phi'_m = \dfrac{\tan 5°}{F_s} = 0.058$ rad; or $\phi'_m = 3.34°$

The new radius of the friction circle is

$$r_1 = R\sin \phi'_m = 20 \times \sin 3.3° = 1.16 \text{ m}.$$

The direction of F changes and the modified triangle of force abd' gives,

cohesive force $= C = $ length $ad' = 600$ kN

Mobilised cohesion, $c'_m = \dfrac{C}{L} = \dfrac{600}{29.7} = 20.2 \text{ kN/m}^2$

Therefore, $F_c = \dfrac{c'}{c'_m} = \dfrac{30}{20.2} \approx 1.5$

16.11 TAYLOR'S STABILITY NUMBER

If the slope angle β, height of embankment H, the effective unit weight of material γ, angle of internal friction ϕ', and unit cohesion c' are known, the factor of safety may be determined. In order to make unnecessary the more or less tedious stability determinations, Taylor (1937) conceived the idea of analyzing the stability of a large number of slopes through a wide range of slope angles and angles of internal friction, and then representing the results by an abstract number which he called the "*stability number*". This number is designated as N_s. The expression used is

$$N_s = \frac{c'}{F_c \gamma H} \tag{16.23}$$

From this the factor of safety with respect to cohesion may be expressed as

$$F_c = \frac{c'}{N_s \gamma H} \tag{16.24}$$

Taylor published his results in the form of curves which give the relationship between N_s and the slope angles β for various values of ϕ'. as shown in Fig. 16.16. These curves are for circles passing through the toe, although for values of β less than 53°, it has been found that the most dangerous circle passes below the toe. However, these curves may be used without serious error for slopes down to $\beta = 14°$. The stability numbers are obtained for factors of safety with respect to cohesion by keeping the factor of safety with respect to friction (F_ϕ) equal to unity.

In slopes encountered in practical problems, the depth to which the rupture circle may extend is usually limited by ledge or other underlying strong material as shown in Fig. 16.17. The stability number N_s for the case when $\phi' = 0$ is greatly dependent on the position of the ledge. The depth at which the ledge or strong material occurs may be expressed in terms of a depth factor n_d which is defined as

$$n_d = \frac{D}{H} \tag{16.25}$$

where $D = $ depth of ledge below the top of the embankment, $H = $ height of slope above the toe.

For various values of n_d and for the $\phi = 0$ case the chart in Fig. 16.17 gives the stability number N_s for various values of slope angle β. In this case the rupture circle may pass through the toe or below the toe. The distance x of the rupture circle from the toe at the toe level may be expressed by a distance factor n_x which is defined as

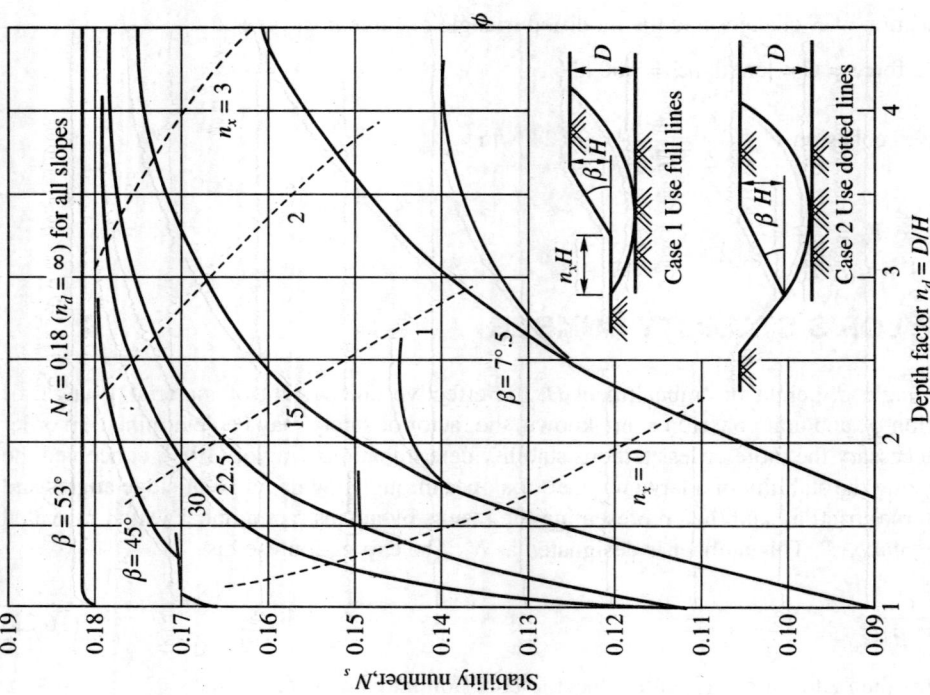

Fig. 16.17　Taylor's stability numbers $\phi' = 0$ case
(after Taylor, 1937)

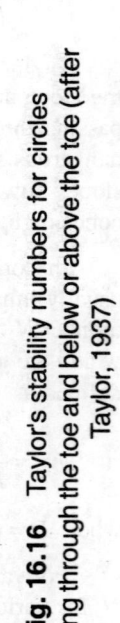

Fig. 16.16　Taylor's stability numbers for circles
passing through the toe and below or above the toe (after
Taylor, 1937)

$$n_x = \frac{x}{H} \tag{16.26}$$

The chart in Fig. 16.17 shows the relationship between n_d and n_x. If there is a ledge or other stronger material at the elevation of the toe, the depth factor n_d for this case is unity.

Factor of Safety with Respect to Strength

The development of the stability number is based on the assumption that the factor of safety with respect to friction F_ϕ, is unity. The curves give directly the factor of safety F_c with respect to cohesion only. If a true factor of safety F_s with respect to strength is required, this factor should apply equally to both cohesion and friction. The mobilized shear strength may therefore be expressed as

$$s_m = \frac{s}{F_s} = \frac{c'}{F_s} + \frac{\sigma' \tan \phi'}{F_s}$$

In the above expression, we may write

$$\frac{c'}{F_s} = c'_m, \quad \tan \phi'_m = \frac{\tan \phi'}{F_s}, \quad \text{or} \quad \phi'_m = \frac{\phi'}{F_s} \text{(approx.)} \tag{16.27}$$

c'_m and ϕ'_m may be described as average values of mobilized cohesion and friction respectively.

Example 16.8

The following particulars are given for an earth dam of height 39 ft. The slope is submerged and the slope angle $\beta = 45°$.

$$\gamma_b = 69 \text{ lb/ft}^3$$

$$c' = 550 \text{ lb/ft}^2$$

$$\phi' = 20°$$

Determine the factor of safety F_s.

Solution

Assume as a first trial $F_s = 2.0$

$$\phi'_m = \frac{20}{2} = 10° \text{ (approx.)}$$

For $\phi'_m = 10°$ and $\beta = 45°$ the value of N_s from Fig. 16.16 is 0.11, we may write

from Eq. (16.23) $N_s = \dfrac{c'}{F_c \gamma H}$, substituting

$$0.11 = \frac{550}{2 \times 69 \times H}$$

$$\text{or } H = \frac{550}{2 \times 69 \times 0.11} = 36.23 \text{ ft}$$

If $F_s = 1.9, \phi'_m = \dfrac{20}{1.9} = 10.53°$ and $N_s = 0.105$

$$H = \frac{550}{1.9 \times 69 \times 0.105} = 40\,\text{ft}$$

The computed height 40 ft is almost equal to the given height 39 ft. The computed factor of safety is therefore 1.9.

Example 16.9

An excavation is to be made in a soil deposit with a slope of 25° to the horizontal and to a depth of 25 meters. The soil has the following properties:

$$c' = 35\,\text{kN/m}^2,\ \phi' = 15°\ \text{and}\ \gamma = 20\,\text{kN/m}^3$$

1. Determine the factor of safety of the slope assuming full friction is mobilized.

2. If the factor of safety with respect to cohesion is 1.5, what would be the factor of safety with respect to friction?

Solution

1. For $\phi' = 15°$ and $\beta = 25°$, Taylor's stability number chart gives stability number $N_s = 0.03$.

$$F_c = \frac{c'}{N_s \gamma H} = \frac{35}{0.03 \times 20 \times 25} = 2.33$$

2. For $F_c = 1.5$, $N_s = \dfrac{c'}{F_c \times \gamma \times H} = \dfrac{35}{1.5 \times 20 \times 25} = 0.047$

For $N_s = 0.047$ and $\beta = 25°$, we have from Fig. 16.16, $\phi'_m = 13°$

Therefore, $F_\phi = \dfrac{\tan \phi'}{\tan \phi'_m} = \dfrac{\tan 15°}{\tan 13°} = \dfrac{0.268}{0.231} = 1.16$

Example 16.10

An embankment is to be made from a soil having $c' = 420\,\text{lb/ft}^2$, $\phi' = 18°$ and $\gamma = 121\,\text{lb/ft}^3$. The desired factor of safety with respect to cohesion as well as that with respect to friction is 1.5. Determine

1. The safe height if the desired slope is 2 horizontal to 1 vertical.

2. The safe slope angle if the desired height is 50 ft.

Solution

$$\tan \phi' = \tan 18° = 0.325,\ \phi'_m = \tan^{-1} \frac{0.325}{1.5} = 12.23°$$

1. For $\phi' = 12.23°$ and $\beta = 26.6°$ (i.e. 2 horizontal and 1 vertical) the chart gives $N_s = 0.055$

Therefore, $0.055 = \dfrac{c'}{F_c \gamma H} = \dfrac{420}{1.5 \times 121 \times H}$

$$\text{Therefore, } H_{safe} = \frac{420}{1.5 \times 121 \times 0.055} = 42 \text{ ft}$$

2.   Now, $N_s = \dfrac{c'}{F_c \gamma H} = \dfrac{420}{1.5 \times 121 \times 50} = 0.046$

For $N_s = 0.046$ and $\phi'_m = 12.23°$, slope angle $\beta = 23.5°$

16.12 TENSION CRACKS

If a dam is built of cohesive soil, tension cracks are usually present at the crest. The depth of such cracks may be computed from the equation

$$z_0 = \frac{2c'}{\gamma} \qquad\qquad\qquad (16.28)$$

where z_0 = depth of crack, c' = unit cohesion, γ = unit weight of soil.

The effective length of any trial arc of failure is the difference between the total length of arc minus the depth of crack as shown in Fig. 16.18.

16.13 STABILITY ANALYSIS BY METHOD OF SLICES FOR STEADY SEEPAGE

The stability analysis with steady seepage involves the development of the pore pressure head diagram along the chosen trial circle of failure. The simplest of the methods for knowing the pore pressure head at any point on the trial circle is by the use of flownets which is described below.

Determination of Pore Pressure with Seepage

Figure 16.19 shows the section of a homogeneous dam with an arbitrarily chosen trial arc. There is steady seepage flow through the dam as represented by flow and equipotential lines. From the equipotential lines the pore pressure may be obtained at any point on the section. For example at point a in Fig. 16.19 the pressure head is h. Point c is determined by setting the radial distance ac

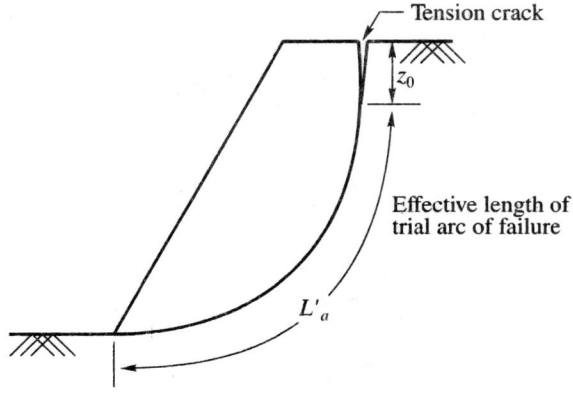

Fig. 16.18   Tension crack in dams built of cohesive soils

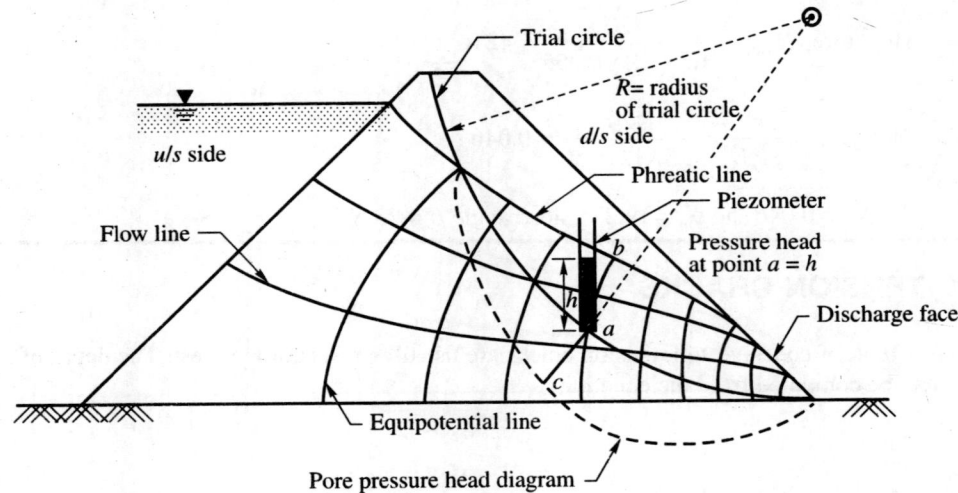

Fig. 16.19 Determination of pore pressure with steady seepage

equal to h. A number of points obtained in the same manner as c give the curved line through c which is a pore pressure head diagram.

Method of Analysis (graphical method)

Figure 16.20(a) shows the section of a dam with an arbitrarily chosen trial arc. The centre of rotation of the arc is 0. The pore pressure acting on the base of the arc as obtained from flow nets is shown in Fig. 16.20(b).

When the soil forming the slope has to be analyzed under a condition where full or partial drainage takes place the analysis must take into account both cohesive and frictional soil properties based on *effective stresses*. Since the effective stress acting across each elemental length of the assumed circular arc failure surface must be computed in this case, the method of slices is one of the convenient methods for this purpose. The method of analysis is as follows.

The soil mass above the assumed slip circle is divided into a number of vertical slices of equal width. The number of slices may be limited to a maximum of eight to ten to facilitate computation. The forces used in the analysis acting on the slices are shown in Figs. 16.20(a) and (c). The forces are:

1. The weight W of the slice.

2. The normal and tangential components of the weight W acting on the base of the slice. They are designated respectively as N and T.

3. The pore water pressure U acting on the base of the slice.

4. The effective frictional and cohesive resistances acting on the base of the slice which is designated as S.

The forces acting on the sides of the slices are statically indeterminate as they depend on the stress deformation properties of the material, and we can make only gross assumptions about their relative magnitudes.

In the conventional slice method of analysis the lateral forces are assumed equal on both sides of the slice. This assumption is not strictly correct. The error due to this assumption on the mass as a whole is about 15 percent (Bishop, 1955).

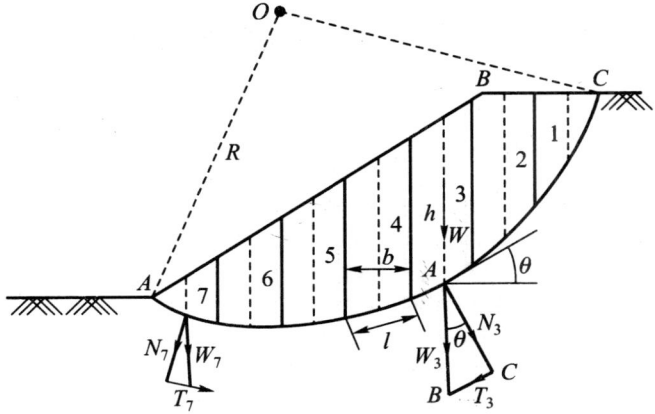

(a) Total normal and tangential components

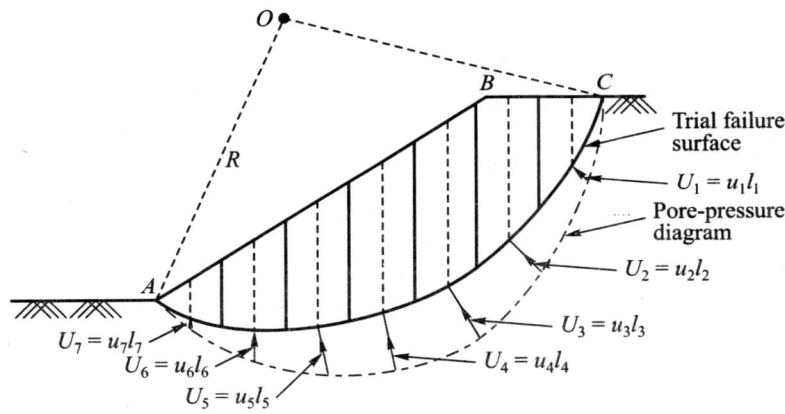

(b) Pore-pressure diagram

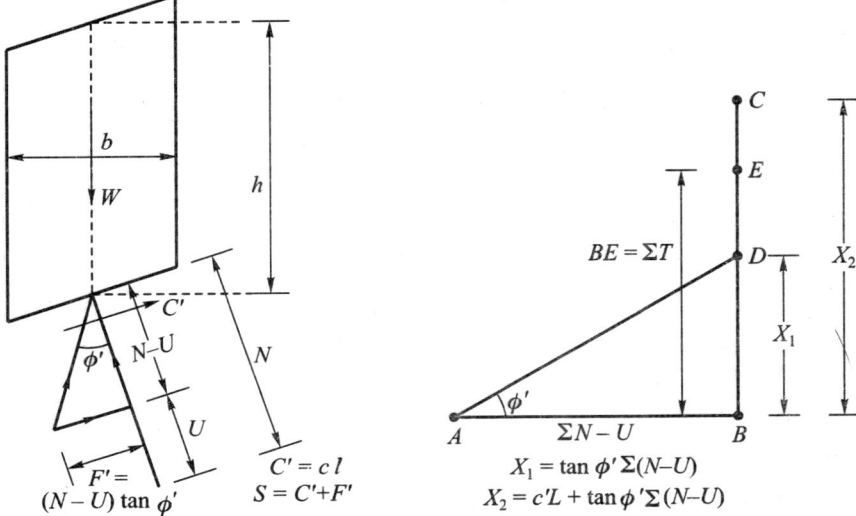

(c) Resisting forces on the base of slice

(d) Graphical representation of all the forces

Fig. 16.20 Stability analysis of slope by the method of slices

The forces that are actually considered in the analysis are shown in Fig. 16.20(c). The various components may be determined as follows:

1. The weight, W, of a slice per unit length of dam may be computed from

$$W = \gamma h b$$

where, γ = total unit weight of soil, h = average height of slice, b = width of slice.

If the widths of all slices are equal, and if the whole mass is homogeneous, the weight W can be plotted as a vector AB passing through the centre of a slice as in Fig. 16.20(a). AB may be made equal to the height of the slice.

2. By constructing triangle ABC, the weight can be resolved into a normal component N and a tangential component T. Similar triangles can be constructed for all slices. The tangential components of the weights cause the mass to slide downward. The sum of all the weights cause the mass to slide downward. The sum of all the tangential components may be expressed as $\overline{T} = \Sigma T$. If the trial surface is curved upward near its lower end, the tangential component of the weight of the slice will act in the opposite direction along the curve. The algebraic sum of T should be considered.

3. The average pore pressure u acting on the base of any slice of length l may be found from the pore pressure diagram shown in Fig. 16.20(b). The total pore pressure, U, on the base of any slice is

$$U = ul$$

4. The effective normal pressure N' acting on the base of any slice is

$$N' = N - U[\text{Fig.16.20(c)}]$$

5. The frictional force F' acting on the base of any slice resisting the tendency of the slice to move downward is

$$F = (N - U)\tan\phi'$$

where ϕ' is the effective angle of friction. Similarly the cohesive force C' opposing the movement of the slice and acting at the base of the slice is

$$C' = c'l$$

where c' is the effective unit cohesion. The total resisting force S acting on the base of the slice is

$$S = C' + F' = c'l + (N - U)\tan\phi'$$

Figure 16.20(c) shows the resisting forces acting on the base of a slice.

The sum of all the resisting forces acting on the base of each slice may be expressed as

$$S_s = c'\Sigma l + \tan\phi'\Sigma(N - U) = c'L + \tan\phi'\Sigma(N - U)$$

where $\Sigma l = L$ = length of the curved surface.

The moments of the actuating and resisting forces about the point of rotation may be written as follows:

Actuating moment = $R\,\Sigma T$

Resistant moment = $R[c'L + \tan\phi'\Sigma(N - U)]$

The factor of safety F_s may now be written as

$$F_s = \frac{[c'L + \tan\phi'\Sigma(N - U)]}{\Sigma T} \qquad (16.29)$$

The various components shown in Eq. (16.29) can easily be represented graphically as shown in Fig. 16.20(d). The line AB represents to a suitable scale $\Sigma(N - U)$. BC is drawn normal to AB at B and equal to $c'L + \tan\phi'\,\Sigma(N - U)$. The line AD drawn at an angle ϕ' to AB gives the intercept BD on BC equal to $\tan\phi'\,\Sigma(N - U)$. The length BE on BC is equal to ΣT. Now

$$F_s = \frac{BC}{BE} \tag{16.30}$$

Centres for Trial Circles Through Toe

The factor of safety F_s as computed and represented by Eq. (16.29) applies to one trial circle. This procedure is followed for a number of trial circles until one finds the one for which the factor of safety is the lowest. This circle that gives the least F_s is the one most likely to fail. The procedure is quite laborious. The number of trial circles may be minimized if one follows the following method.

For any given slope angle β (Fig. 16.21), the centre of the first trial circle centre O may be determined as proposed by Fellenius (1927). The direction angles α_A and α_B may be taken from Table 16.1. For the centres of additional trial circles, the procedure is as follows:

Mark point C whose position is as shown in Fig. 16.21. Join CO. The centres of additional circles lie on the line CO extended. This method is applicable for a homogeneous $(c - \phi)$ soil. When the soil is purely cohesive and homogeneous the direction angles given in Table 16.1 directly give the centre for the critical circle.

Centres for Trial Circles Below Toe

Theoretically if the materials of the dam and foundation are entirely homogeneous, any practicable earth dam slope may have its critical failure surface below the toe of the slope. Fellenius found that the angle intersected at 0 in Fig. 16.22 for this case is about 133.5°. To find the centre for the critical circle below the toe, the following procedure is suggested.

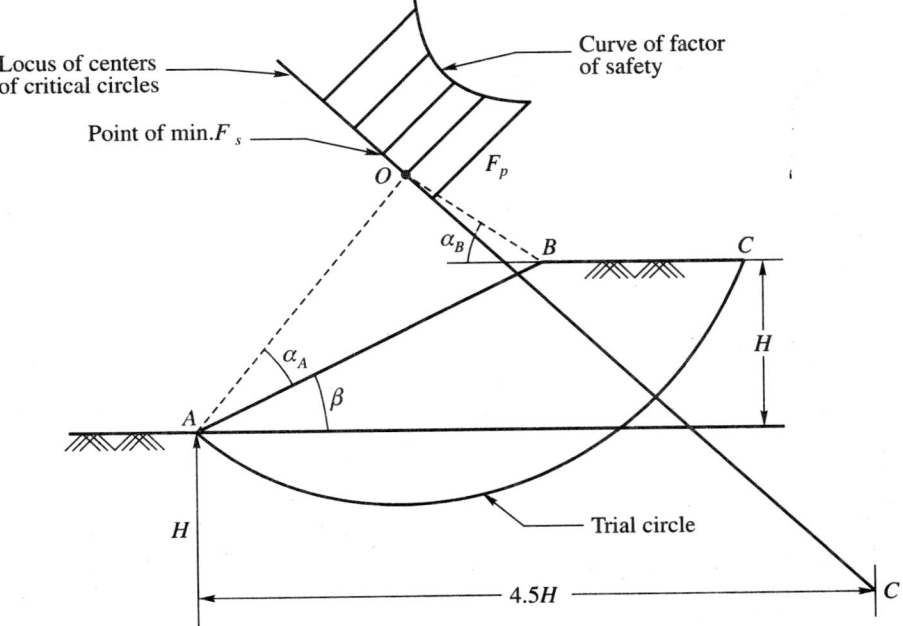

Fig. 16.21 Location of centres critical circles passing through toe of dam

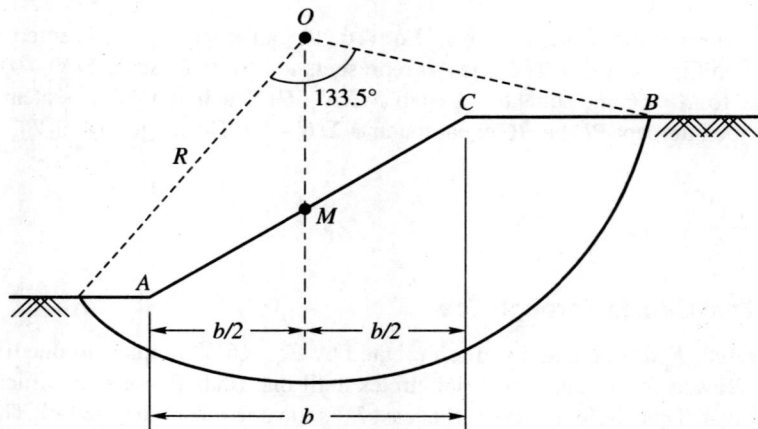

Fig. 16.22   Centres of trial circles for base failure

TABLE 16.1
Direction angles α°_A and α°_B for centres of critical circles

Slope	Slope angle	Direction angles	
	β	α°_A	α°_B
0.6 : 1	60	29	40
1 : 1	45	28	37
1.5 : 1	33.8	26	35
2 : 1	26.6	25	35
3 : 1	18.3	25	35
5 : 1	11.3	25	37

Erect a vertical at the midpoint M of the slope. On this vertical will be the centre O of the first trial circle. In locating the trial circle use an angle (133.5°) between the two radii at which the circle intersects the surface of the embankment and the foundation. After the first trial circle has been analyzed the centre is some what moved to the left, the radius shortened and a new trial circle drawn and analyzed. Additional centres for the circles are spotted and analyzed.

Example 16.11

An embankment is to be made of a sandy clay having a cohesion of 30 kN/m², angle of internal friction of 20° and a unit weight of 18 kN/m³. The slope and height of the embankment are 1.6 : 1 and 10 m respectively. Determine the factor of safety by using the trial circle given in Fig. Ex. 16.11 by the method of slices.

Solution

Consider the embankment as shown in Fig. Ex.16.11. The centre of the trial circle O is selected by taking $\alpha_A = 26°$ and $\alpha_B = 35°$ from Table 16.1. The soil mass above the slip circle is divided into 13 slices of 2 m width each. The weight of each slice per unit length of embankment is given by $W = h_a b \gamma_t$, where h_a = average height of the slice, b = width of the slice, γ_t = unit weight of the soil.

The weight of each slice may be represented by a vector of height h_a if b and γ_t remain the same for the whole embankment. The vectors values were obtained graphically. The height vectors may be

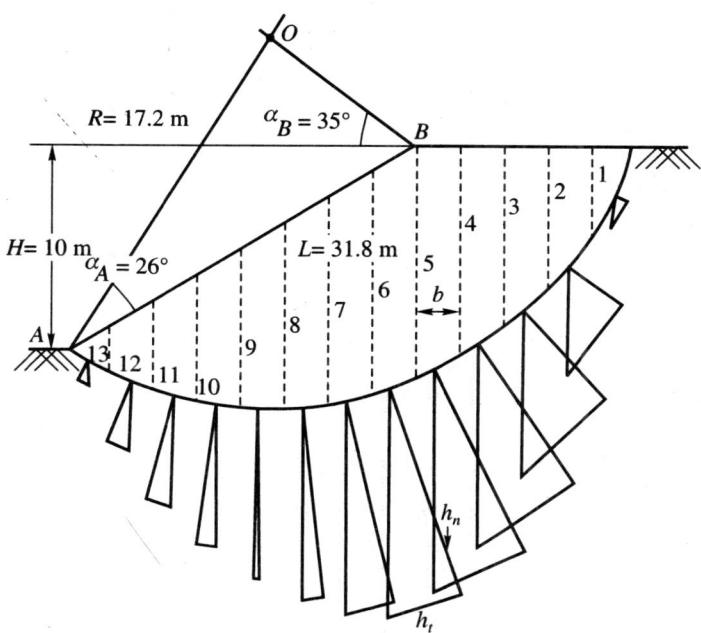

Fig. Ex. 16.11

resolved into normal components h_n and tangential components h_t. The values of h_a, h_n and h_t for the various slices are given below in a tabular form.

Values of h_a, h_n and h_t

Slice No.	$h_a(m)$	$h_n(m)$	$h_t(m)$	Slice No.	$h_a(m)$	$h_n(m)$	$h_t(m)$
1	1.8	0.80	1.72	8	9.3	9.25	1.00
2	5.5	3.21	4.50	9	8.2	8.20	−0.20
3	7.8	5.75	5.30	10	6.8	6.82	−0.80
4	9.5	7.82	5.50	11	5.2	5.26	−1.30
5	10.6	9.62	4.82	12	3.3	3.21	−1.20
6	11.0	10.43	3.72	13	1.1	1.0	−0.50
7	10.2	10.20	2.31				

The sum of these components h_n and h_t may be converted into forces ΣN and ΣT respectively by multiplying them as given below

$$\Sigma h_n = 81.57 \text{ m}, \quad \Sigma h_t = 24.87 \text{ m}$$

Therefore, $\quad \Sigma N = 81.57 \times 2 \times 18 = 2937 \text{ kN}$

$$\Sigma T = 24.87 \times 2 \times 18 \text{ kN}$$

Length of arc $= L = 31.8 \text{ m}$

$$\text{Factory of safety} = \frac{c'L + \tan \phi' \Sigma N}{\Sigma T} = \frac{30 \times 31.8 + 0.364 \times 2937}{895} = 2.26$$

16.14 BISHOP'S SIMPLIFIED METHOD OF SLICES

Bishop's method of slices (1955) is useful if a slope consists of several types of soil with different values of c and ϕ and if the pore pressures u in the slope are known or can be estimated. The method of analysis is as follows:

Figure 16.23 gives a section of an earth dam having a sloping surface AB. ADC is an assumed trial circular failure surface with its centre at O. The soil mass above the failure surface is divided into a number of slices. The forces acting on each slice are evaluated from limit equilibrium of the slices. The equilibrium of the entire mass is determined by summation of the forces on each of the slices.

Consider for analysis a single slice $abcd$ (Fig. 16.23a), which is drawn to a larger scale in Fig. 16.23(b). The forces acting on this slice are:

W = weight of the slice

N = total normal force on the failure surface cd

U = pore water pressure = ul on the failure surface cd

F_R = shear resistance acting on the base of the slice

E_1, E_2 = normal forces on the vertical faces bc and ad

T_1, T_2 = shear forces on the vertical faces bc and ad

θ = the inclination of the failure surface cd to the horizontal

The system is statically indeterminate. An approximate solution may be obtained by assuming that the resultant of E_1 and T_1 is equal to that of E_2 and T_2, and their lines of action coincide. For equilibrium of the system, the following equations hold true.

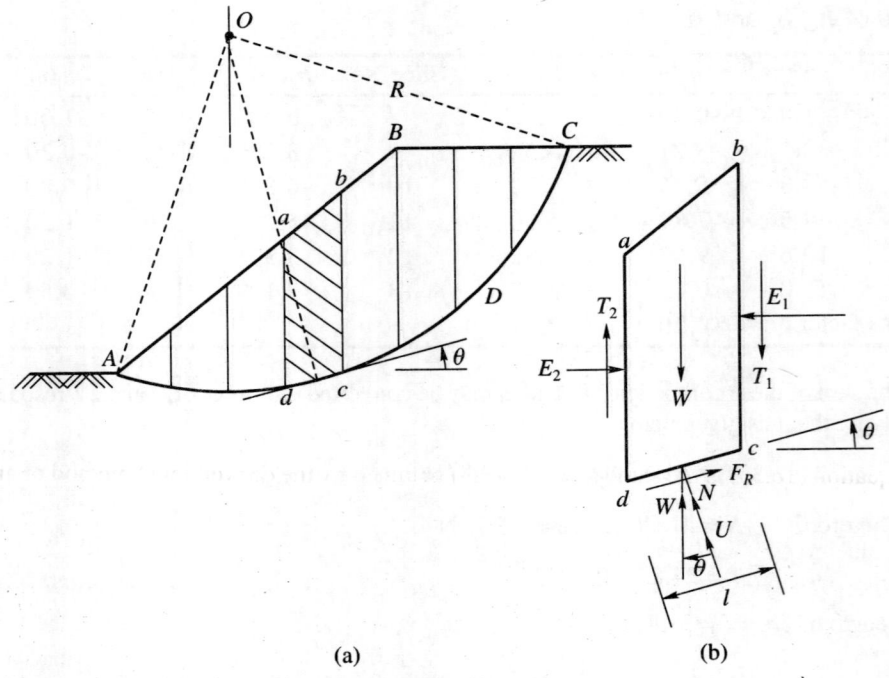

(a) (b)

Fig. 16.23 Bishop's simplified method of analysis

$$N = W \cos\theta$$

$$F_t = W \sin\theta \tag{16.31}$$

where F_t = tangential component of W

The unit stresses on the failure surface of length, l, may be expressed as

normal stress, $\sigma_n = \dfrac{W \cos\theta}{l}$

shear stress, $\tau_n = \dfrac{W \sin\theta}{l} \tag{16.32}$

The equation for shear strength, s, is

$$s = c' + \sigma' \tan\phi' = c' + (\sigma - u) \tan\phi'$$

where σ' = effective normal stress

c' = effective cohesion

ϕ' = effective angle of friction

u = unit pore pressure

The shearing resistance to sliding on the base of the slice is

$$sl = c'l + (W \cos\theta - ul) \tan\phi'$$

where $ul = U$, the total pore pressure on the base of the slice (Fig 16.23b)

$$sl = F_R$$

The total resisting force and the actuating force on the failure surface ADC may be expressed as

Total resisting force F_R is

$$\Sigma F_R = \Sigma\left[c'l + (W \cos\theta - ul) \tan\phi'\right] \tag{16.33}$$

Total actuating force F_t is

$$\Sigma F_t = \Sigma W \sin\theta \tag{16.34}$$

The factor of safety F_s is then given as

$$F_s = \frac{\Sigma F_R}{\Sigma F_t} = \frac{\Sigma\left[c'l + (W \cos\theta - ul) \tan\phi'\right]}{\Sigma W \sin\theta} \tag{16.35}$$

Equation (16.35) is the same as Eq. (16.29) obtained by the conventional method of analysis.

Bishop (1955) suggests that the accuracy of the analysis can be improved by taking into account the forces E and T on the vertical faces of each slice. For the element in Fig. 16.23(b), we may write an expression for all the forces acting in the vertical direction for the equilibrium condition as

$$N' \cos\theta = W + (T_1 - T_2) - ul \cos\theta - F_R \sin\theta \tag{16.36}$$

If the slope is not on the verge of failure ($F_s > 1$), the tangential force F_t is equal to the shearing resistance F_R on cd divided by F_s.

$$F_R = \frac{c'l}{F_s} + N'\frac{\tan\phi'}{F_s} \tag{16.37}$$

where, $N' = N - U$, and $U = ul$

Substituting Eq. (16.37) into Eq. (16.36) and solving for N', we obtain

$$N' = \frac{\left(W + \Delta T - U\cos\theta - \dfrac{c'l}{F_s}\sin\theta\right)}{\cos\theta + \dfrac{\tan\phi'\sin\theta}{F_s}} \tag{16.38}$$

where, $\Delta T = T_1 - T_2$.

For equilibrium of the mass above the failure surface, we have by taking moments about O

$$\Sigma W\sin\theta R = \Sigma F_R R \tag{16.39}$$

By substituting Eqs (16.37) and (16.38) into Eq. (16.39) and solving we obtain an expression for F_s as

$$F_s = \frac{\Sigma\{c'l\cos\theta + [(W - U\cos\theta) + \Delta T]\tan\phi'\}\frac{1}{m_\theta}}{\Sigma W\sin\theta} \tag{16.40}$$

where, $\quad m_\theta = \cos\theta + \dfrac{\tan\phi'\sin\theta}{F_s} \tag{16.41}$

The factor of safety F_s is present in Eq. (16.40) on both sides. The quantity $\Delta T = T_1 - T_2$ has to be evaluated by means of successive approximation. Trial values of E_1 and T_1 that satisfy the equilibrium of each slice, and the conditions

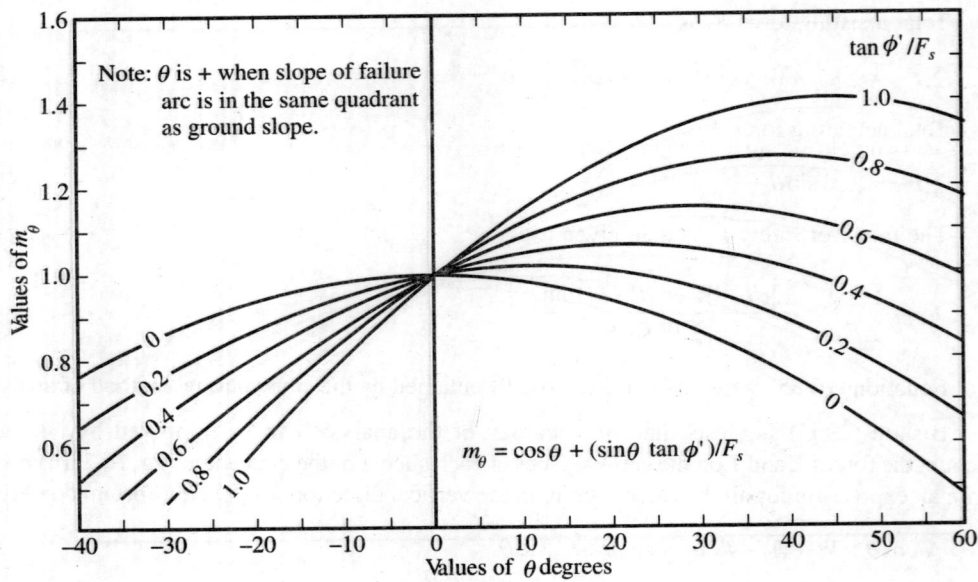

Fig. 16.24 Values of m_θ (after Janbu et al., 1956)

$(E_1 - E_2) = 0$ and $(T_1 - T_2) = 0$

are used. The value of F_s may then be computed by first assuming an arbitrary value for F_s. The value of F_s may then be calculated by making use of Eq. (16.40). If the calculated value of F_s differs appreciably from the assumed value, a second trial is made and the computation is repeated. Figure 16.24 developed by Janbu et al. (1956) helps to simplify the computation procedure.

It is reported that an error of about 1 percent will occur if we assume $\Sigma(T_1 - T_2) \tan \phi' = 0$. But if we use the conventional method of analysis using Eq. (16.35) the error introduced is about 15 percent (Bishop, 1955).

16.15 BISHOP AND MORGENSTERN METHOD FOR SLOPE ANALYSIS

Equation (16.40) developed based on Bishop's analysis of slopes, contains the term pore pressure u. The Bishop and Morgenstern method (1960) proposes the following equation for the evaluation of u

$$r_u = \frac{u}{\gamma h} \tag{16.42}$$

where, u = pore water pressure at any point on the assumed failure surface

γ = unit weight of the soil

h = the depth of the point in the soil mass below the ground surface

The pore pressure ratio r_u is assumed to be constant throughout the cross-section, which is called a *homogeneous pore pressure distribution*. Figure 16.25 shows the various parameters used in the analysis.

The factor of safety F_s is defined as

$$F_s = m - n r_u \tag{16.43}$$

where, m, n = stability coefficients.

The m and n values may be obtained either from charts in Figs. B.1 to B.6 or Tables B1 to B6 in Appendix B. The depth factor given in the charts or tables is as per Eq. (16.25), that is $n_d = D/H$, where H = height of slope, and D = depth of firm stratum from the top of the slope. Bishop and Morgenstern (1960) limited their charts (or tables) to values of $c'/\gamma H$ equal to 0.000, 0.025, and 0.050.

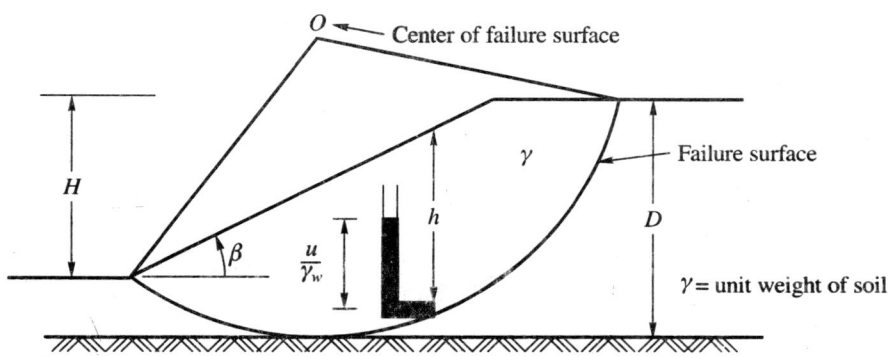

Fig. 16.25 Specifications of parameters for Bishop-Morgenstern method of analysis

Extension of the Bishop and Morgenstern Slope Stability Charts

As stated earlier, Bishop and Morgenstern (1960) charts or tables cover values of $c'/\gamma H$ equal to 0.000, 0.025, and 0.050 only. These charts do not cover the values that are normally encountered in natural slopes. O' Connor and Mitchell (1977) extended the work of Bishop and Morgenstern to cover values of $c'/\gamma H$ equal to 0.075 and 0.100 for various values of depth factors n_d. The method employed is essentially the same as that adopted by the earlier authors. The extended values are given in the form of charts and tables from Figs. B.7 to B.14 and Tables B7 to B14 respectively in Appendix B.

Method of Determining F_s

1. Obtain the values of r_u and $c/\gamma H$

2. From the tables in Appendix B, obtain the values of m and n for the known values of $c/\gamma H$, ϕ and β, and for $n_d = 0$, 1, 1.25 and 1.5.

3. Using Eq. (16.43), determine F_s for each value of n_d.

4. The required value of F_s is the lowest of the values obtained in step 3.

Example 16.12

Figure Ex. 16.12 gives a typical section of a homogeneous earth dam. The soil parameters are: $\phi' = 30°$, $c' = 590$ lb/ft², and $\gamma = 120$ lb/ft³. The dam has a slope 4:1 and a pore pressure ratio $r_u = 0.5$. Estimate the factor of safety F_s by Bishop and Morgenstern method for a height of dam $H = 140$ ft.

Solution

Height of dam $H = 140$ ft

$$\frac{c'}{\gamma h} = \frac{590}{120 \times 140} = 0.035$$

Given: $\phi' = 30°$, slope 4:1 and $r_u = 0.5$.

Since $c'/\gamma H = 0.035$, and $n_d = 1.43$ for $H = 140$ ft, the F_s for the dam lies between $c'/\gamma H = 0.025$ and 0.05 and n_d between 1.0 and 1.5. The equation for F_s is

$$F_s = m - nr_u$$

Using the Tables in Appendix B, the following table can be prepared for the given values of $c/ \gamma H$, ϕ, and β.

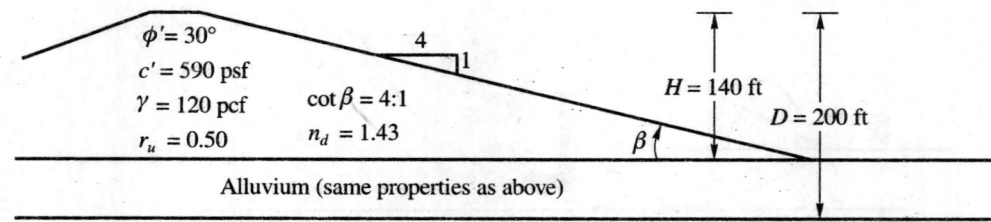

$\phi' = 30°$
$c' = 590$ psf
$\gamma = 120$ pcf $\cot \beta = 4{:}1$
$r_u = 0.50$ $n_d = 1.43$

$H = 140$ ft

$D = 200$ ft

β

Alluvium (same properties as above)

Fig. Ex. 16.12

From Tables B2 and B3 for $c'/\gamma H = 0.025$

n_d	m	n	F_s	
1.0	2.873	2.622	1.562	
1.25	2.953	2.806	1.55	Lowest

From Tables B4, B5 and B6 for $c'/\gamma H = 0.05$

n_d	m	n	F_s	
1.0	3.261	2.693	1.915	
1.25	3.221	2.819	1.812	Lowest
1.50	3.443	3.120	1.883	

Hence, $n_d = 1.25$ is the more critical depth factor. The value of F_s for $c'/\gamma H = 0.035$ lies between 1.55 (for $c'/\gamma H = 0.025$) and 1.812 (for $c'/\gamma H = 0.05$). By proportion $F_s = 1.655$.

16.16 MORGENSTERN METHOD OF ANALYSIS FOR RAPID DRAWDOWN CONDITION

Rapid drawdown of reservoir water level is one of the critical states in the design of earth dams. Morgenstern (1963) developed the method of analysis for rapid drawdown conditions based on the Bishop and Morgenstern method of slices. The purpose of this method is to compute the factor of safety during rapid drawdown, which is reduced under no dissipation of pore water pressure. The assumptions made in the analysis are

1. Simple slope of homogeneous material
2. The dam rests on an impermeable base
3. The slope is completely submerged initially
4. The pore pressure does not dissipate during drawdown

Morgenstern used the pore pressure parameter $\bar{B}$ as developed by Skempton (1954) which states

$$\bar{B} = \frac{u}{\sigma_1} \tag{16.44}$$

where $\sigma_1 = \gamma h$

γ = total unit weight of soil or equal to twice the unit weight of water

h = height of soil above the lower level of water after drawdown

The charts developed take into account the drawdown ratio which is defined as

$$R_d = \frac{\bar{H}}{H} \tag{16.45}$$

where R_d = drawdown ratio

$\bar{H}$ = height of drawdown

H = height of dam (Fig. 16.26)

All the potential sliding circles must be tangent to the base of the section.

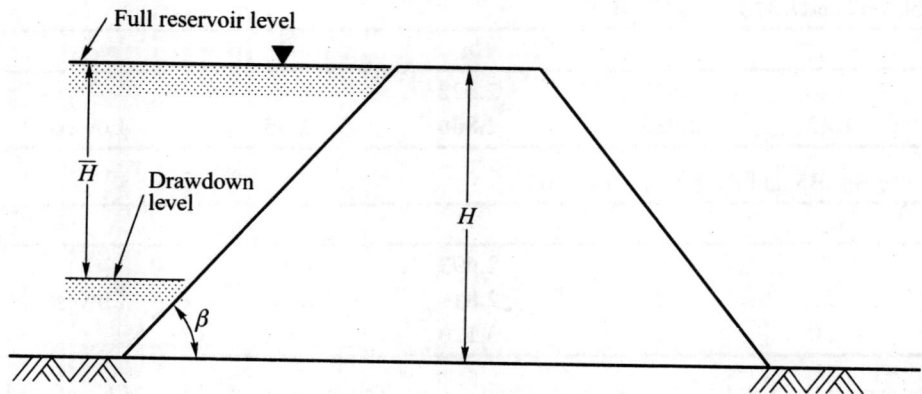

Fig. 16.26 Dam section for drawdown conditions

The stability charts are given in Figs 16.27 to 16.29 covering a range of stability numbers $c'/\gamma H$ from 0.0125 to 0.050. The curves developed are for the values of ϕ' of 20°, 30°, and 40° for different values of β.

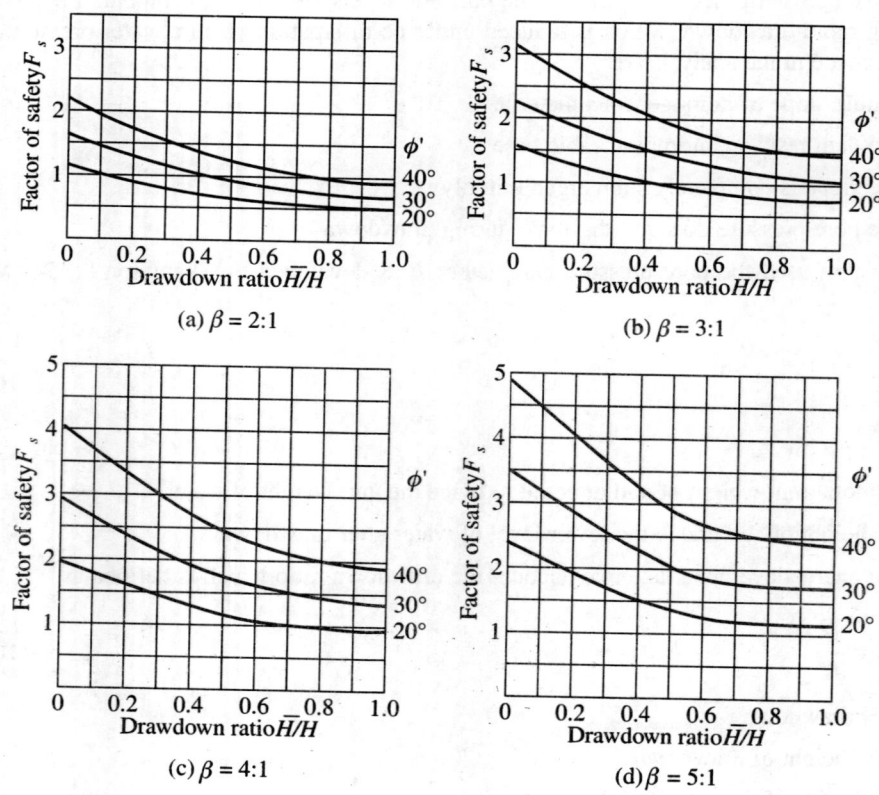

Fig. 16.27 Drawndown stability chart for $c'/\gamma H = 0.0125$ (after Morgenstern, 1963)

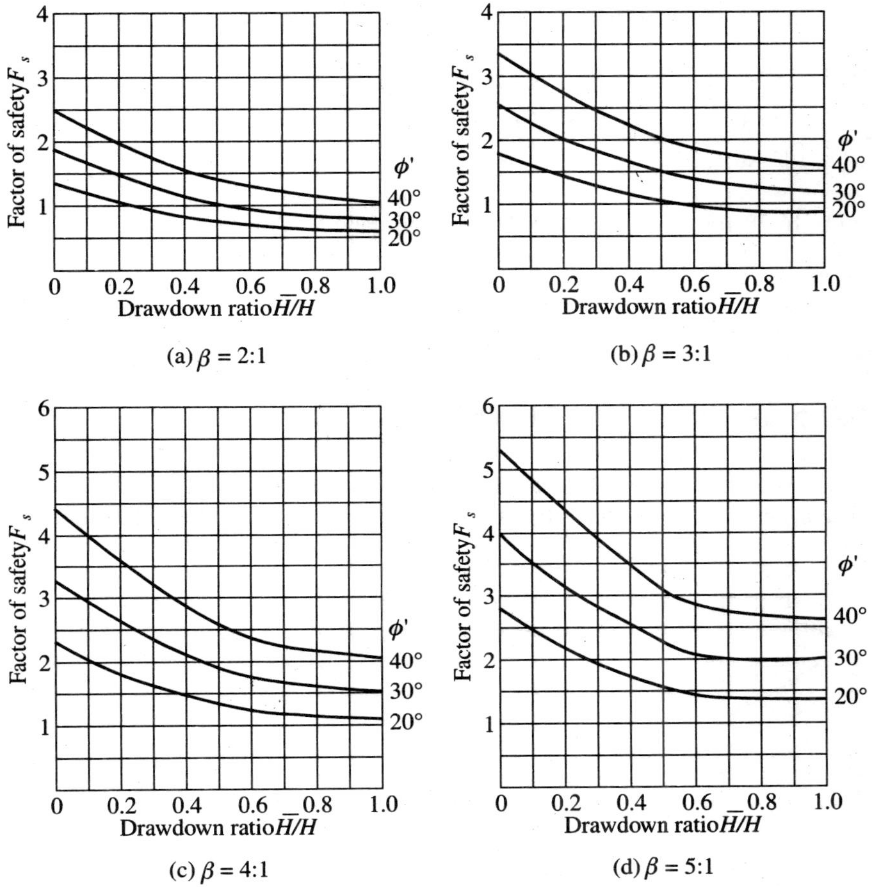

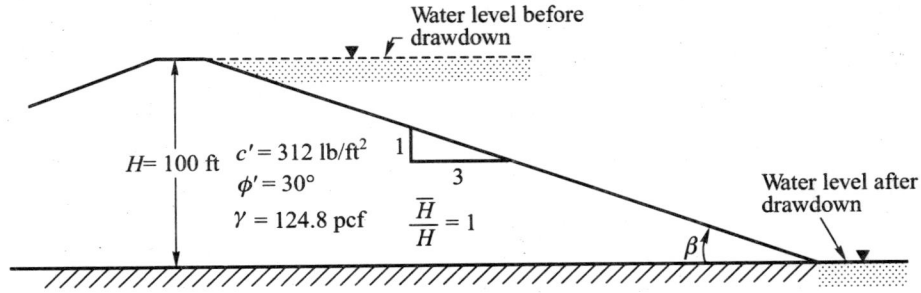

Fig. 16.28 Drawdown stability chart for $c'/\gamma H = 0.025$ (after Morgenstern, 1963)

Example 16.13

It is required to estimate the minimum factor of safety for the complete drawdown of the section shown in Fig. Ex. 16.13 (Morgenstern, 1963)

Water level before drawdown

$H = 100$ ft $c' = 312$ lb/ft^2

$\phi' = 30°$

$\gamma = 124.8$ pcf

$\dfrac{\overline{H}}{H} = 1$

Water level after drawdown

β

Fig. Ex. 16.13

Solution

From the data given in the Fig. Ex. 16.13

$$N_s = \frac{c'}{\gamma h} = \frac{312}{124.8 \times 100} = 0.025$$

From Fig.16.28, for $N_s = 0.025$, $\beta = 3:1$, $\phi' = 30°$, and $\overline{H}/H = 1$,

$$F_s = 1.20$$

It is evident that the critical circle is tangent to the base of the dam and no other level need be investigated since this would only raise the effective value of N_s resulting in a higher factor of safety.

16.17 SPENCER METHOD OF ANALYSIS

Spencer (1967) developed his analysis based on the method of slices of Fellenius (1927) and Bishop (1955). The analysis is in terms of effective stress and satisfies two equations of equilibrium, the first

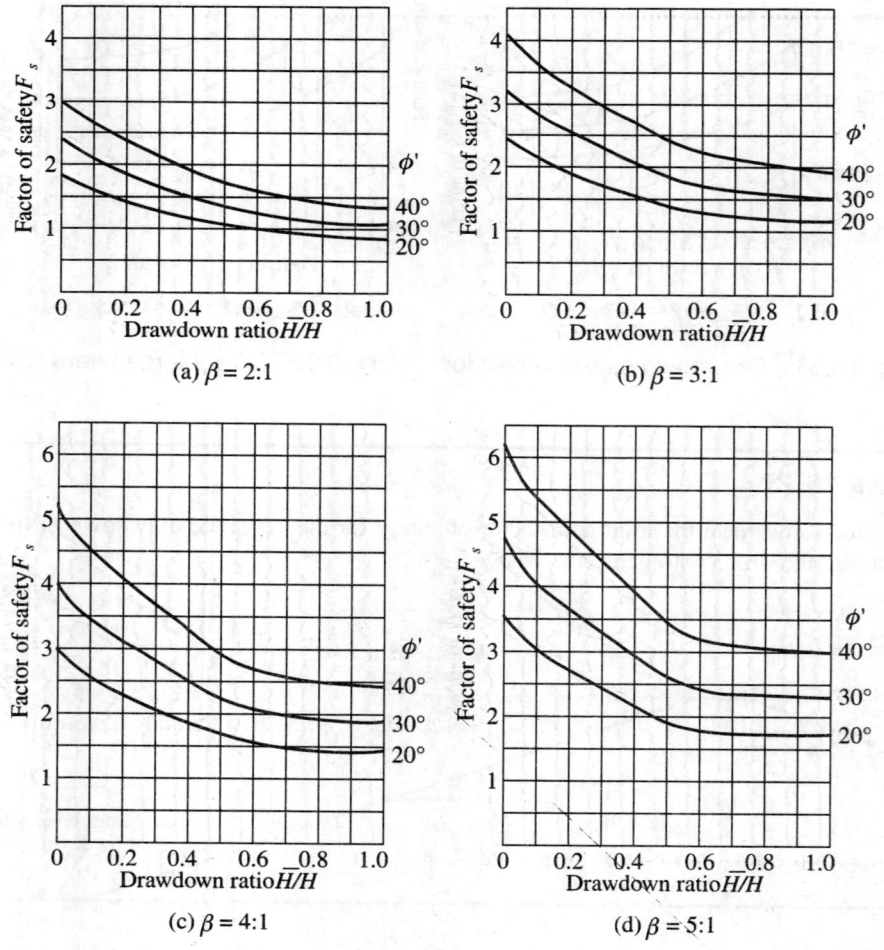

Fig. 16.29 Drawdown stability char for $c'/\gamma H = 0.05$ (after Morgenstern, 1963)

with respect to forces and the second with respect to moments. The interslice forces are assumed to be parallel as in Fig. 16.23. The factor of safety F_s is expressed as

$$F_s = \frac{\text{Shear strength available}}{\text{Shear strength mobilized}} \qquad (16.46)$$

The mobilized angle of shear resistance and other factors are expressed as

$$\tan\phi'_m = \frac{\tan\phi'}{F_s} \qquad (16.47)$$

pore pressure ratio, $r_u = \dfrac{u}{\gamma h}$ \qquad (16.48)

Stability factor, $N_s = \dfrac{c'}{F_s \gamma H}$ \qquad (16.49)

The charts developed by Spencer for different values of N_s, ϕ'_m and r_u are given in Fig. 16.30. The use of these charts will be explained with worked out examples.

Example 16.14

Find the slope corresponding to a factor of safety of 1.5 for an embankment 100 ft high in a soil whose properties are as follows:

$$c' = 870\,\text{Ib/sq ft},\ \gamma = 120\,\text{Ib/ft}^3,\ \phi' = 26°,\ r_u = 0.5$$

Solution (by Spencer's Method)

$$N_s = \frac{c'}{F_s \gamma H} = \frac{870}{1.5\times120\times100} = 0.048$$

$$\tan\phi'_m = \frac{\tan\phi'}{F_s} = \frac{0.488}{1.5} = 0.325$$

$$\phi'_m = 18°$$

Referring to Fig. 16.30c, for which $r_u = 0.5$, the slope corresponding to a stability number of 0.048 is 3:1.

Example 16.15

What would be the change in strength on sudden drawdown for a soil element at point P which is shown in Fig. Ex. 16.15? The equipotential line passing through this element represents loss of water head of 1.2 m. The saturated unit weight of the fill is 21 kN/m³.

Solution

The data given are shown in Fig. Ex. 16.15. Before drawdown,
The stresses at point P are:

$$\sigma_0 = \gamma_w h_w + \gamma_{sat} h_c = 9.81\times3 + 21\times4 = 113\,\text{kN/m}^2$$

$$u_0 = \gamma_w (h_w + h_c - h') = 9.81(3+4-1.2) = 57\,\text{kN/m}^2$$

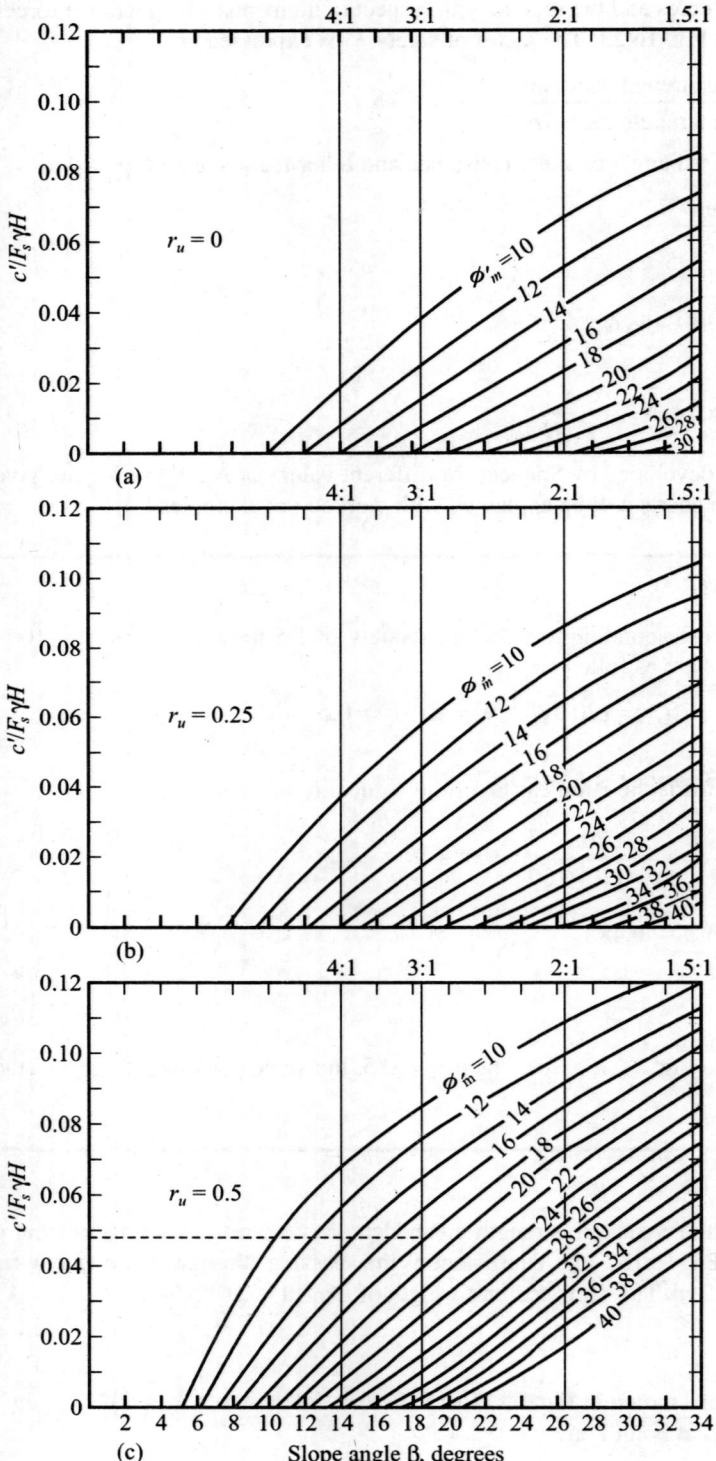

Fig. 16.30 Stability charts (after Spencer, 1967)

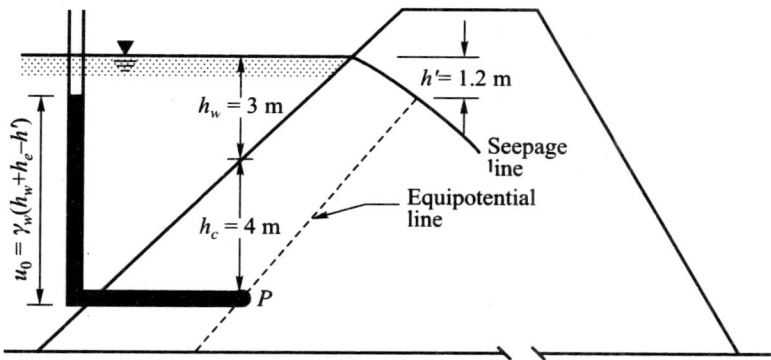

Fig. Ex. 16.15

Therefore $\sigma_0' = \sigma_0 - u_0 = 113 - 57 = 56\,\text{kN/m}^2$

After drawdown,

$\sigma = \gamma_{sat}h_c = 21 \times 4 = 84\,\text{kN/m}^2$

$u = \gamma_w(h_c - h') = 9.81(4 - 1.2) = 27.5\,\text{kN/m}^2$

$\sigma' = \sigma - u = 84 - 27.5 = 56.5\,\text{kN/m}^2$

The change in strength is zero since the effective vertical stress does not change.

Note: There is no change in strength due to sudden drawdown but the direction of forces of the seepage water changes from an inward direction before drawdown to an outward direction after drawdown and this is the main cause for the reduction in stability.

16.18 QUESTIONS AND PROBLEMS

16.1   Find the critical height of an infinite slope having a slope angle of 30°. The slope is made of stiff clay having a cohesion 20 kN/m², angle of internal friction 20°, void ratio 0.7 and specific gravity 2.7. Consider the following cases for the analysis.

(a) the soil is dry.

(b) the water seeps parallel to the surface of the slope.

(c) the slope is submerged.

16.2   An infinite slope has an inclination of 26° with the horizontal. It is underlain by a firm cohesive soil having $G = 2.72$ and $e = 0.52$. There is a thin weak layer 20 ft below and parallel to the slope ($c' = 525$ lb/ft², $\phi' = 16°$). Compute the factors of safety when (a) the slope is dry, and (b) ground water flows parallel to the slope at the slope level.

16.3   An infinite slope is underlain with an overconsolidated clay having $c' = 210$ lb/ft², $\phi' = 8°$ and $\gamma_{sat} = 120$ lb/ft³. The slope is inclined at an angle of 10° to the horizontal. Seepage is parallel to the surface and the ground water coincides with the surface. If the slope fails parallel to the surface along a plane at a depth of 12 ft below the slope, determine the factor of safety.

16.4   A deep cut of 10 m depth is made in sandy clay for a road. The sides of the cut make an angle of 60° with the horizontal. The shear strength parameters of the soil are $c' = 20$ kN/m², $\phi' = 25°$, and $\gamma = 18.5$ kN/m³. If AC is the failure plane (Fig Prob. 16.4), estimate the factor of safety of the slope.

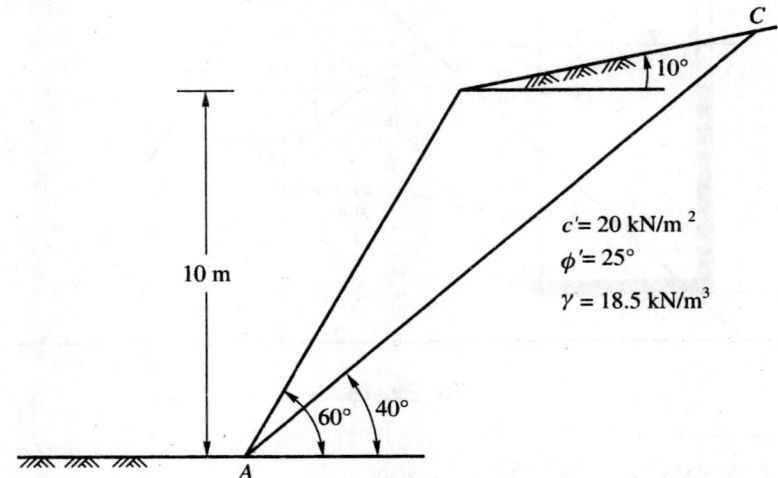

Fig. Prob. 16.4

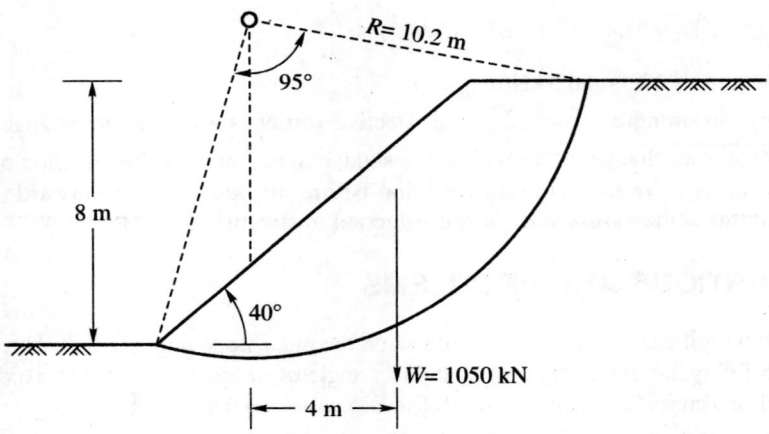

Fig. Prob. 16.5

16.5   A 40° slope is excavated to a depth of 8 m in a deep layer of saturated clay having strength parameters $c = 60$ kN/m², $\phi = 0$, and $\gamma = 19$ kN/m³. Determine the factor of safety for the trial failure surface shown in Fig. Prob. 16.5.

16.6   An excavation to a depth of 8 m with a slope of 1:1 was made in a deep layer of saturated clay having $c_u = 65$ kN/m² and $\phi_u = 0$. Determine the factor of safety for a trial slip circle passing through the toe of the cut and having a centre as shown in Fig. Prob. 16.6. The unit weight of the saturated clay is 19 kN/m³. No tension crack correction is required.

16.7   A 45° cut was made in a clayey silt with $c = 15$ kN/m², $\phi = 0$ and $\gamma = 19.5$ kN/m³. Site exploration revealed the presence of a soft clay stratum of 2 m thick having $c = 25$ kN/m² and $\phi = 0$ as shown in Fig. Prob. 16.7. Estimate the factor of safety of the slope for the assumed failure surface.

16.8   A cut was made in a homogeneous clay soil to a depth of 8 m as shown in Fig. Prob. 16.8. The total unit weight of the soil is 18 kN/m³, and its cohesive strength is 25 kN/m². Assuming

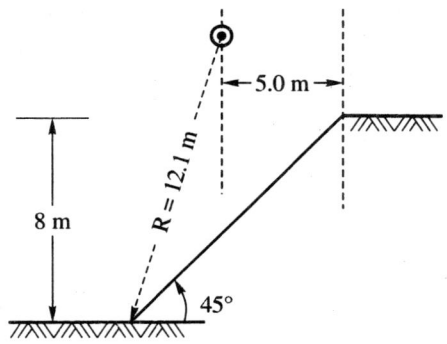

Fig. Prob. 16.6

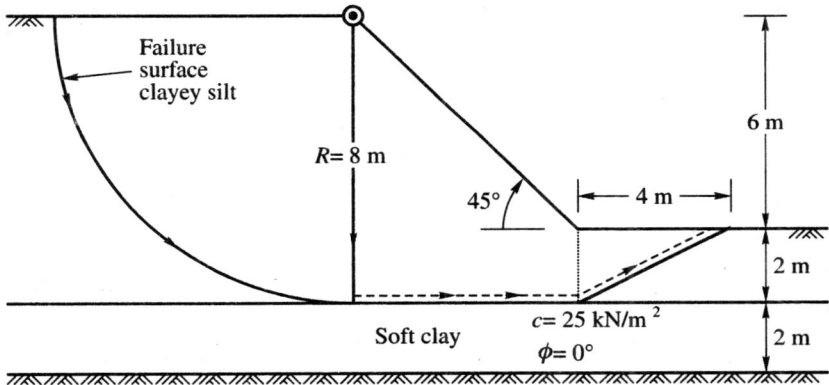

Fig. Prob. 16.7

a $\phi = 0$ condition, determine the factor of safety with respect to a slip circle passing through the toe. Consider a tension crack at the end of the slip circle on the top of the cut.

16.9 A deep cut of 10 m depth is made in natural soil for the construction of a road. The soil parameters are: $c' = 35$ kN/m², $\phi' = 15°$ and $\gamma = 20$ kN/m³.

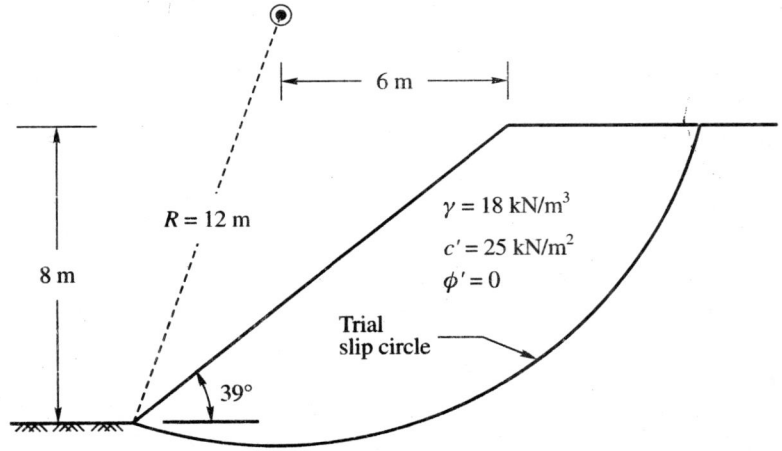

Fig. Prob. 16.8

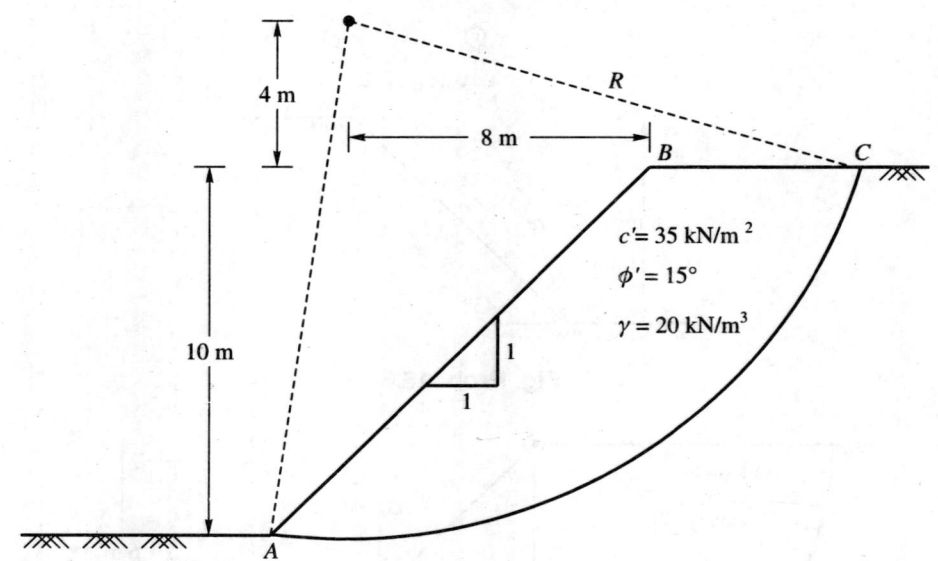

Fig. Prob. 16.9

The sides of the cut make angles of 45° with the horizontal. Compute the factor of safety using friction circle method for the failure surface AC shown in Fig. Prob. 16.9.

16.10   An embankment is to be built to a height of 50 ft at an angle of 20° with the horizontal. The soil parameters are: $c' = 630$ lb/ft², $\phi' = 18°$ and $\gamma = 115$ lb/ft³.

Estimate the following;

1. Factor of safety of the slope assuming full friction is mobilized.

2. Factor of safety with respect to friction if the factor of safety with respect to cohesion is 1.5.

Use Taylor's stability chart.

16.11   A cut was made in natural soil for the construction of a railway line. The soil parameters are: $c' = 700$ lb/ft², $\phi' = 20°$ and $\gamma = 110$ lb/ft³.

Determine the critical height of the cut for a slope of 30° with the horizontal by making use of Taylor's stability chart.

16.12   An embankment is to be constructed by making use of sandy clay having the following properties: $c' = 35$ kN/m², $\phi' = 25°$ and $\gamma = 19.5$ kN/m³.

The height of the embankment is 20 m with a slope of 30° with the horizontal as shown in Fig. Prob. 16.12. Estimate the factor of safety by the method of slices for the trial circle shown in the figure.

16.13   If an embankment of 10 m height is to be made from a soil having $c' = 25$ kN/m2, $\phi' = 15°$, and $\gamma = 18$ kN/m³, what will be the safe angle of slope for a factor of safety of 1.5?

16.14   An embankment is constructed for an earth dam of 80 ft high at a slope of 3:1. The properties of the soil used for the construction are: $c' = 770$ lb/ft², $\phi' = 30°$, and $\gamma = 110$ lb/ft³. The estimated pore pressuer ratio $r_u = 0.5$. Determine the factor of safety by Bishop and Morgenstern method.

16.15   For the Prob. 16.14, estimate the factor of safety for $\phi' = 20°$. All the other data remain the same.

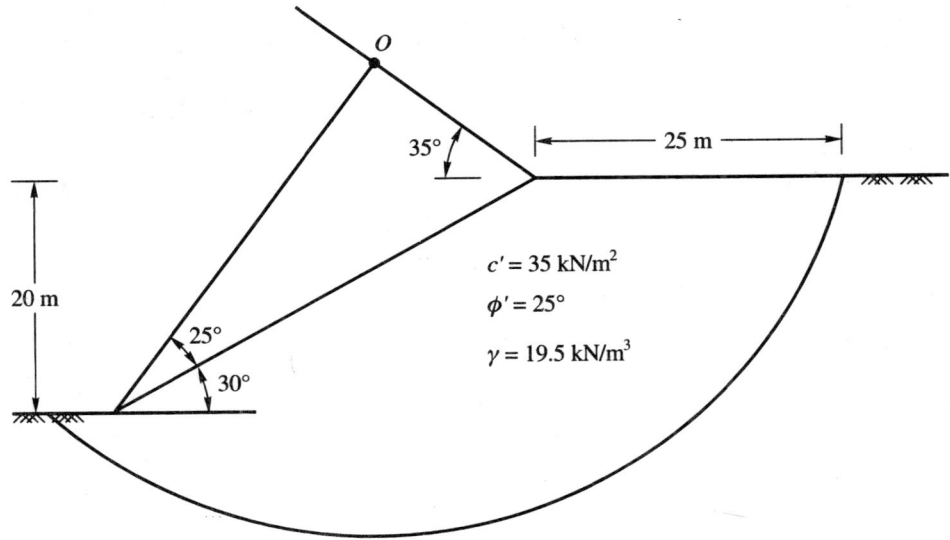

Fig. Prob. 16.12

16.16 For the Prob. 16.14, estimate the factor of safety for a slope of 2:1 with all the oother data remain the same.

16.17 A cut of 25 m dopth is made in a compacted fill having shear strength parameters of $c' = 25$ kN/m², and $\phi' = 20°$. The total unit weight of the material is 19 kN/m³. The pore pressuer ratio has an average value of 0.3. The slope of the sides is 3:1. Estimate the factor of safety using the Bishop and Morgenstern method.

16.18 For the Prob. 16.17, estimate the factor of safety for $\phi' = 30°$, with all the other data remain the same.

16.19 For the Prob. 16.17, esatimate the factor of safety for a slope of 2:1 with all the other data remaining the same.

16.20 Estimate the minimum factor of safety for a complete drawdown condition for the section of dam in Fig. Prob. 16.20. The full reservoir level of 15 m depth is reduced to zero after drawdown.

16.21 What is the safety factor if the reservoir level is brought down from 15 m to 5 m depth in the Prob. 16.20?

16.22 An earth dam to be constructed at a site has the following soil parameters: $c' = 600$ lb/ft², $\gamma = 110$ lb/ft³, and $\phi' = 20°$. The height of of dam $H = 50$ ft.

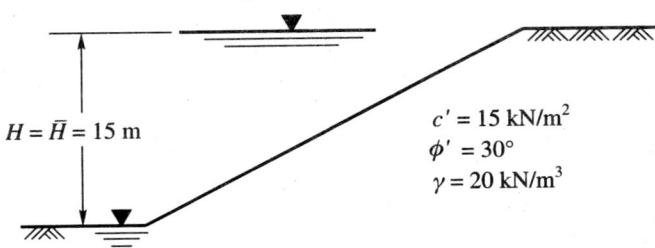

Fig. Prob. 16.20

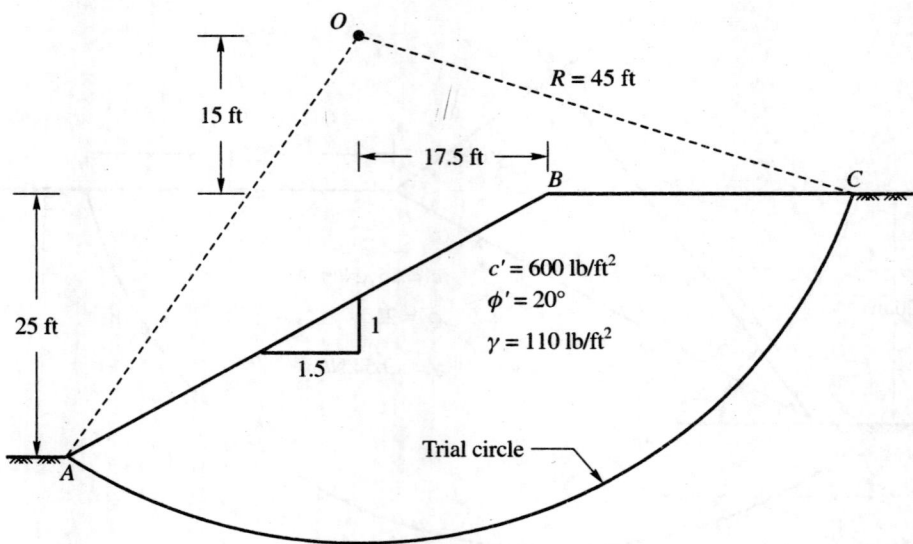

Fig. Prob. 16.24

The pore pressure ratio $r_u = 0.5$. Determine the slope of the dam for a factor of safety of 1.5 using Spencer's method (1967).

16.23   If the given pore pressure ratio is 0.25 in Prob. 16.22, what will be the slope of the dam?

16.24   An embankment has a slope of 1.5 horizontal to 1 vertical with a height of 25 feet. The soil parameters are:

$c' = 600$ lb/ft², $\phi' = 20°$, and $\gamma = 110$ lb/ft³.

Determine the factor of safety using friction circle method for the failure surface AC shown in Fig. Prob. 16.24.

16.25   It is required to construct an embankment for a reservoir to a height of 20 m at a slope of 2 horizontal to 1 vertical. The soil parameters are:

$c' = 40$ kN/m², $\phi' = 18°$, and $\gamma = 17.5$ kN/m³.

Estimate the following:

1. Factor of safety of the slope assuming full friction is mobilized.

2. Factor of safety with respect to friction if the factor of safety with respect to cohesion is 1.5.

Use Taylor's stability chart.

16.26   A cutting of 40 ft depth is to be made for a road as shown in Fig. Prob. 16.26. The soil properties are:

$c' = 500$ lb/ft², $\phi' = 15°$, and $\gamma = 115$ lb/ft³.

Estimate the factor of safety by the method of slices for the trial circle shown in the figure.

16.27   An earth dam is to be constructed for a reservior. The height of the dam is 60 ft. The properties of the soil used in the construction are:

$c' = 400$ lb/ft², $\phi° = 20°$, and $\gamma = 115$ lb/ft³, and $\beta = 2:1$.

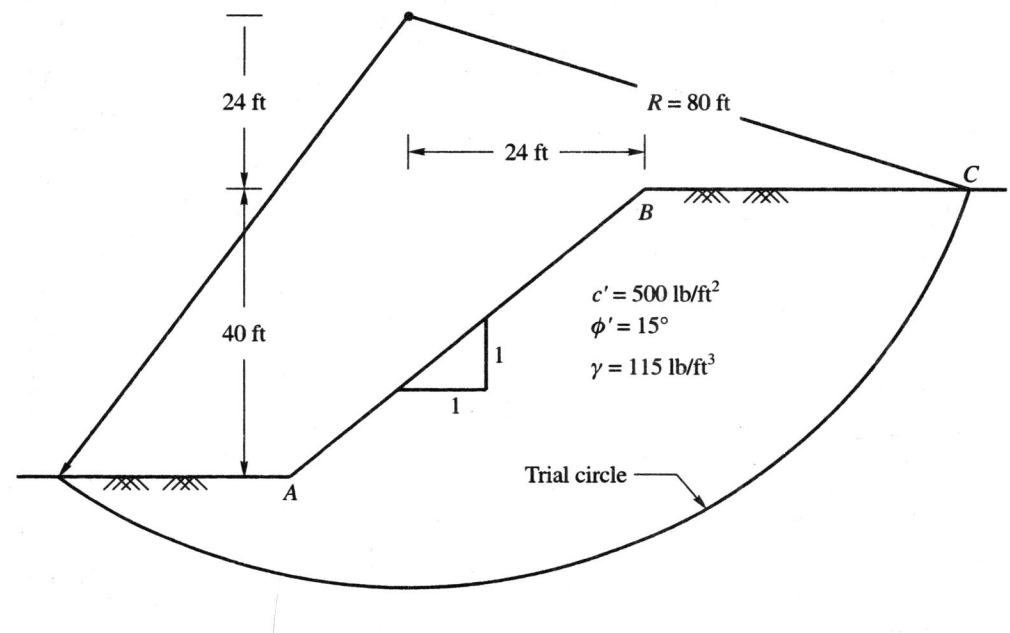

Fig. Prob. 16.26

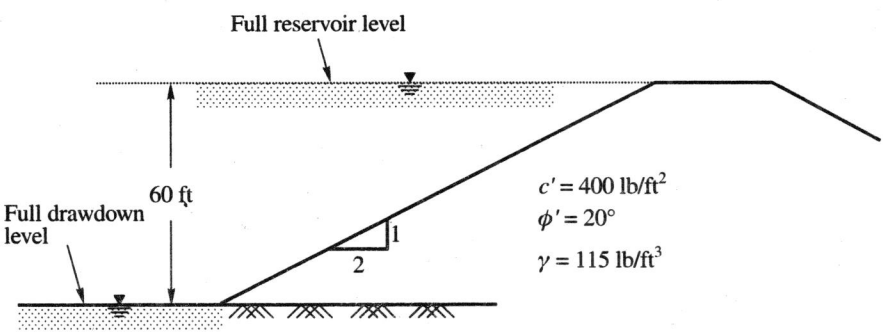

Fig. Prob. 16.27

Estimate the minimum factor of safety for the complete drawn from the full reservior level as shown in Fig. Prob. 16.27 by Morgenstern method.

16.28   What is the factor of safety if the water level is brought down from 60 ft to 20 ft above the bed level of reservoir in Prob. 16.27?

16.29   For the dam given in Prob. 16.27, determine the factor of safety for ru = 0.5 by Spencer's method.

CHAPTER 17

SOIL EXPLORATION

17.1 INTRODUCTION

The elements of soil exploration depend mostly on the importance and magnitude of the project, but generally should provide the following:

1. Information to determine the type of foundation required such as a shallow or deep foundation.

2. Necessary information with regards to the strength and compressibility characteristics of the subsoil to allow the Design Consultant to make recommendations on the safe bearing pressure or pile load capacity.

Soil exploration involves broadly the following:

1. Planning of a program for soil exploration.

2. Collection of disturbed and undisturbed soil or rock samples from the holes drilled in the field. The number and depths of holes depend upon the project.

3. Conducting all the necessary *in-situ* tests for obtaining the strength and compressibility characteristics of the soil or rock directly or indirectly.

4. Study of ground-water conditions and collection of water samples for chemical analysis.

5. Geophysical exploration, if required.

6. Conducting all the necessary tests on the samples of soil /rock and water collected.

7. Preparation of drawings, charts, etc.

8. Analysis of the data collected.

9. Preparation of report.

The *in-situ* field tests for a project may consist of any one or more of the following tests:

1. Standard penetration tests in boreholes (SPT),

2. Static cone penetration tests (CPT),

3. Vane shear test (VST),

4. Plate load test (PLT),

5. Permeability test (PT).

Each of the elements of soil exploration is discussed briefly in the following sections.

17.2 BORING OF HOLES

There a many methods of boring holes. They are:

1. Hand operated or power driven augers, 2. A combination of shell and auger, 3. Wash boring, 4. Rotary drilling, 5. Percussion drilling.

The first three of the above methods are normally used in soils and the last two in rock. The types of *hand operated augers* are shown in Fig. 17.1. These augers are used up to a maximum depth of 10 m and generally suitable for all types of soils above water table but suitable only below water table in clay soils. The diameters of the holes vary from 10 to 20 cm. In the case of *power driven augers,* a continuous flight of augers are used for borings (Fig. 17.2). These augers are suitable in all types of soils. The hollow stem can be used for sampling or conducting standard penetration tests and plugged when not in use.

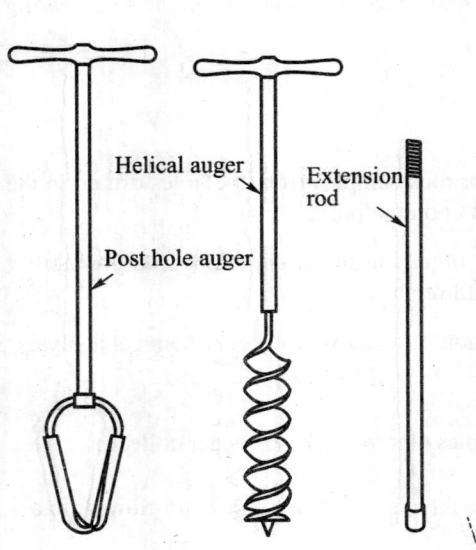

Fig. 17.1   Hand augers

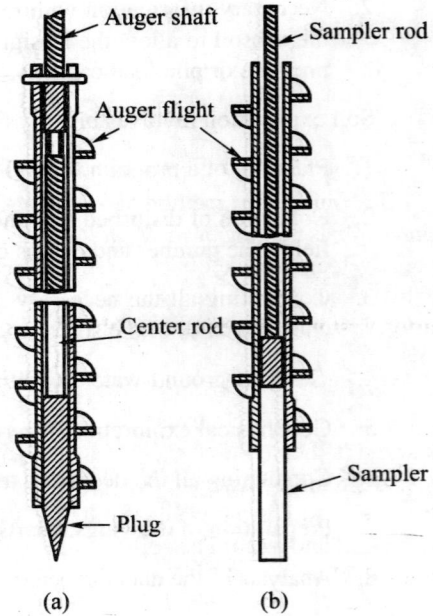

Fig. 17.2   Hollow-stem auger.
(a) Plugged while advancing the auger
(b) Plug removed and sampler inserted to sample soil below auger

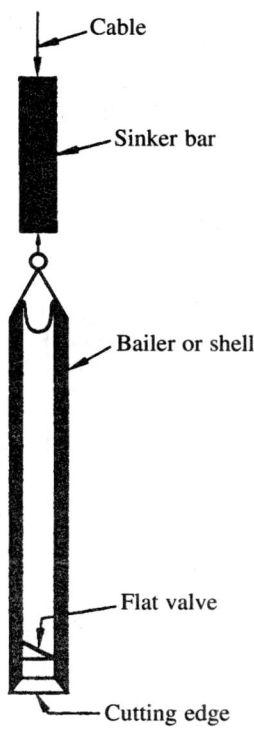

Fig. 17.3 Shell with sinker bar

Shell and auger method is quite popular in India. A shell, (Fig. 17.3), which is also called as a *sand bailer,* is nothing but a heavy duty pipe with a cutting edge. The length and weight vary according to requirements. Sinker bars are sometimes added to add weight to the bailer. Raising and dropping of shell in a hole cuts the soil which is pushed into the tube. It is emptied when full. Boring is always started first with angering, and shell is used when not in use.

Wash boring (Fig. 17.4) is commonly used for boring in difficult soil. To start with, the hole is advanced a short depth by auger and then a casing pipe is pushed to prevent the sides from caving in. The hole is then continued by the use of a chopping bit fixed at the end of a string of hollow drill rods. A stream of water under pressure is forced through the rod and the bit into the hole, which loosens the soil as the water flows up around the pipe. The loosened soil in suspension in water is discharged into a tub. The soil in suspension settles down in the tub and the clean water flows into a sump which is reused for circulation. The motive power for wash boring is either mechanical or man power.

Rotary drilling is also another method of wash boring normally used in rocky strata. In this method a cutter bit or a core barrel with a coring bit (Fig. 17.6) attached to the end of a string of drill rods is rotated by a power rig. (Fig. 17.5). The core barrel is used primarily in rocky strata to get rock samples. The types of coring bits are shown in Fig. 17.6. Diamond coring bits are the most versatile of all the coring bits.

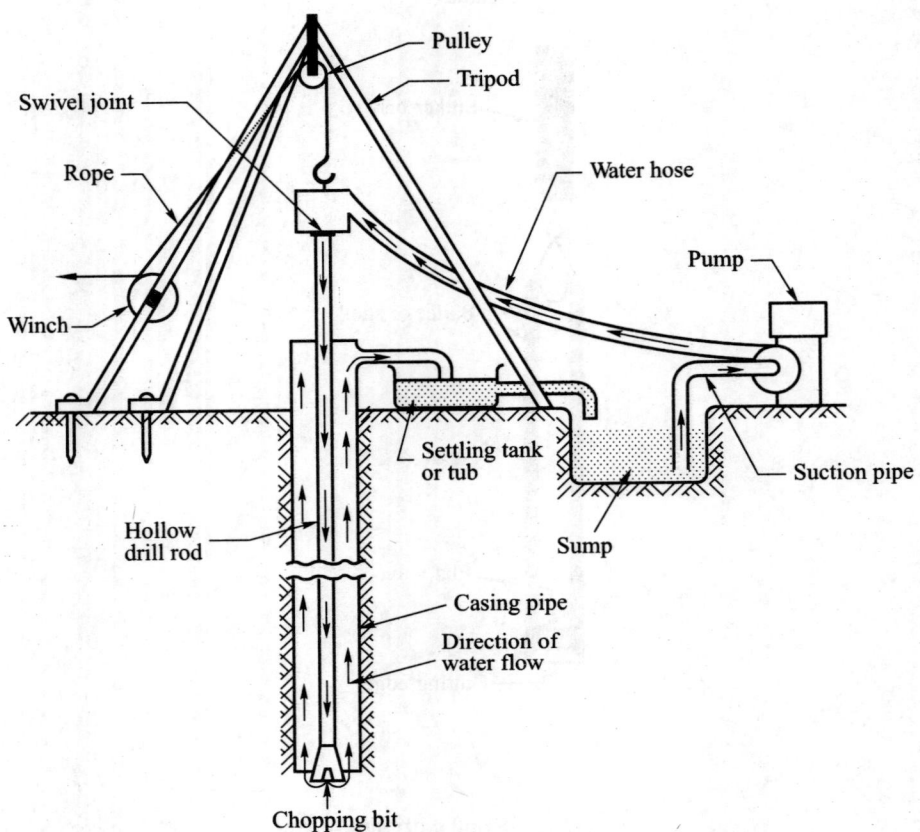

Fig. 17.4   Wash boring

There is another type of rotary drilling which is called as *Calyx* or *shot core drilling*. In this case, cutting action is provided by a slotted bit of mild steel and by very hard steel shots which are fed into the drill hole with the wash water and reaches the bit via the annular space between the core and the wall of the barrel. Slots cut into the bit at its lower end facilitate movement of the shot to the bottom and outside of the bit. This method of drilling is generally used in tube well boring.

Percussion drilling is another method of drilling hole in which a heavy drilling bit is alternately raised and dropped in such a manner that it powders the underlying material and forms into a slurry in water. This slurry is removed out of the hole by means of bailers or sand pumps.

In all types of drilling the sides of the holes may be stabilised, if required, by the use of drilling mud or casing pipes A drilling mud is nothing but bentonite clay mixed in water.

The machinery used to advance holes and take sample is called as a *drill rig* (Fig. 17.5) The rig may be power-driven or hand driven. The power driven is normally provided with a water pump or air compressor which provides water or air under pressure for the removal of cuttings from the drill hole and cooling of the rotary bits. A *winch* is also provided to raise and lower the drilling tools and casing pipes. In hand operated rig a tripod is normally used.

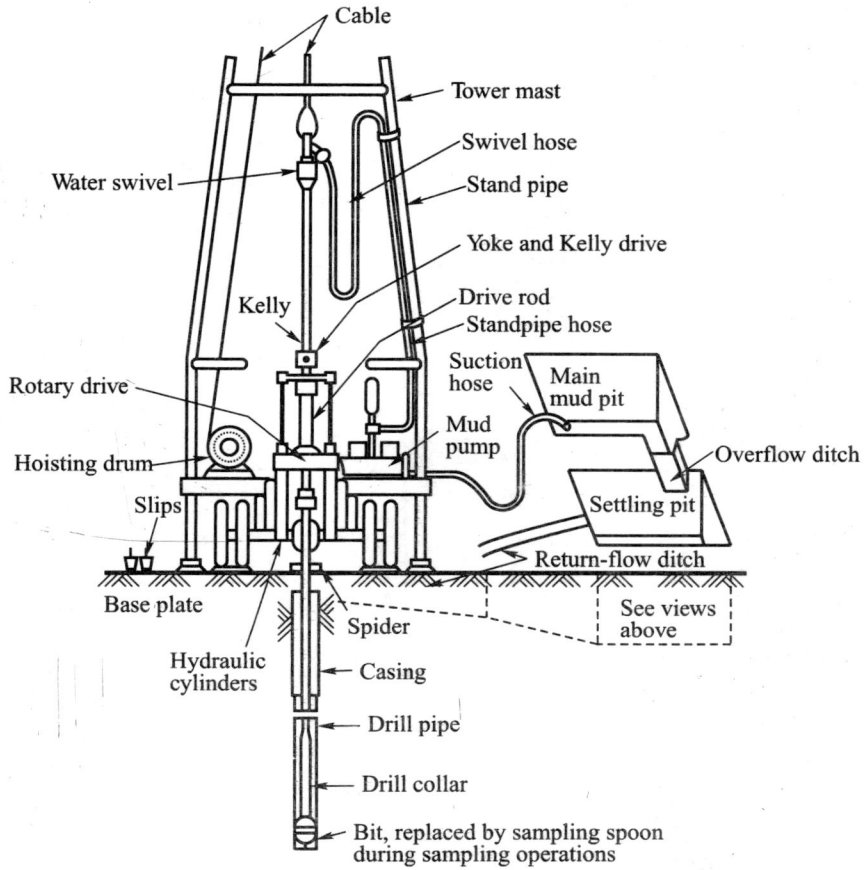

Fig. 17.5 Rotary drilling rig (After Hvorslev, 1948)

17.3   SAMPLING IN SOIL

The two types of samples that are extracted from bore holes are (i) disturbed samples, and (ii) undisturbed samples

Auger samples are disturbed samples, and are used to identify and classify soils. The standard penetration test sample obtained in a *split-spoon sampler* (Fig. 17.7) is also a disturbed sample.

Open Drive Sampler

An open drive sampler (Fig. 17.8) is normally a thin-walled seamless steel tube sampler (called as Shelby type) with a hard cutting edge, and connected to a sampler head. The sampler head contains a ball check valve and ports which permit the easy escape of water or air from the sample tube as the sample enters it. The thickness of tube wall is normally 1/16 to 1/8 inch which is governed by the *area ratio, A_r*, which is defined as

$$A_r = \frac{d_0^2 - d_i^2}{d_i^2}$$

(17.1)

Fig. 17.6 Coring bits: (a) diamond with conventional waterways, (b) diamond with bottom discharge waterways, (c) carbide insert, blade type, (d) carbide insert, pyramid type, (e) saw tooth (Courtesy of Sprague & Henwood, Inc. and Acker Drill Co., Inc.)

where, d_i = inside diameter, and d_0, = outside diameter. A_r is a measure of the volume of the collected sample. The ratio of A_r varies from 10 to 20 percent depending upon the type of soil. A higher ratio is required in very stiff to hard clay strata, and smaller ratio in softy clay. The diameters of sampling tubes vary from 50 to 100 mm and length from 450 to 600 mm. The procedure of sampling involves attaching a string of drill rods to the sampler tube adopter and lowering the sampler on to the bottom of the bore hole. The sampler is either pushed or driven to the required depth, and then sheared off by giving a twist to the drill rod at the top. The sampling tube is then taken out of the hole, and the tube is taken out of the sampler head. The top and bottom of the sample is suitably sealed with molten wax. The degree of disturbance of a soil during extraction is expressed as

$$R_r = \frac{L_a}{L_t} \tag{17.2}$$

where, R_r = Recovery ratio, L_a = actual length of sample in the tube, and L_t = total length of the sampling tube driven below the bottom of the bore hole. R_r = 1 indicates no disturbance to the sample, whereas $R_r < 1$ indicates the sample is compressed and $R_r > 1$ indicates that there is expansion of sample within the tube.

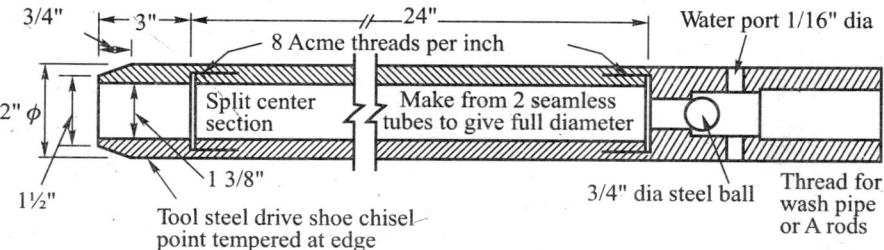

Fig. 17.7   Split spoon sampler for standard penetration test

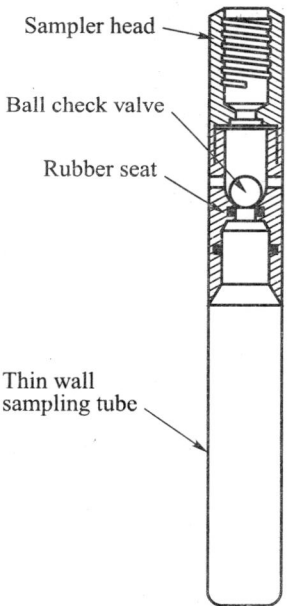

Fig. 17.8   Thin wall "shelly tube" sampler

Piston Sampler (After Osterberg)

To improve the quality of samples and to increase the recovery of soft or slightly cohesive soils, a *piston sampler* is normally used. Such a sampler consists of a thin walled tube fitted with a piston that closes the end of the sampling tube until the apparatus is lowered to the bottom of the bore hole (Fig. 17.9a). The sampling tube is pushed into the soil hydraulically by keeping the piston stationary (Fig. 17.9b). The presence of the piston prevents the soft soils from squeezing rapidly into the tube and thus eliminates most of the distortion of the sample. The piston also helps to increase the length of sample that can be recovered by creating a slight vacuum that tends to retain the sample if the top of the column of soil begins to separate from the piston. During the withdrawal"of the sampler, the piston also prevents water pressure from acting on the top of the sample and thus increases the chances of recovery. The design of piston samplers has been refined to the extent that it is sometimes

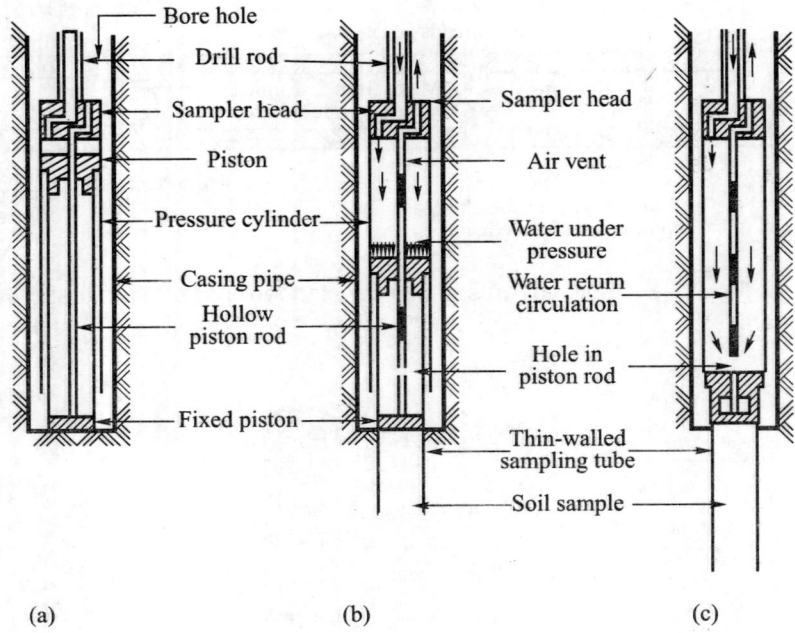

Fig. 17.9 Osterberg Piston Sampler, (a) Sampler is set in drilled hole, (b) Sample tube is pushed hydraulically into the soil, (c) Pressure is released through hole in piston rod.

possible to take undisturbed samples of sand from below the water table. However, piston sampling is relatively a costly procedure and may be adopted only where its use is justified.

Samples from Test Pits

Excavated test pits are one of the oldest methods of soil exploration since they permit a detailed visual examination of the subsurface material in-situ condition. Disturbed and practically undisturbed samples can be obtained from the pits. The depths of pits depend upon the soil condition and the position of the water table. Normally the depths are limited to a maximum of 3 m or up to the water table level whichever is earlier. Since the undisturbed samples obtained above the water table are not fully saturated, the test results obtained from such samples have to be analysed with caution.

Example 17.1

Determine the area ratio of a shelby tube type sampler of 51 mm external diameter.

Solution

Let the internal diameter be 48 mm. Area ratio from Eq. (17.1) is

$$A_r = \frac{d_0^2 - d_i^2}{d_i^2} = \frac{51^2 - 48^2}{48^2} \times 100 = 12.9 \text{ percent.}$$

Example 17.2

What is the probable wall thickness of a sampling tube of 75 mm external diameter which is required for sampling in stiff to very stiff clay soil.

Solution

Experience indicates that the minimum area ratio of a sampling tube required for sampling in stiff to very stiff clay strata is 20 percent. Now the thickness of wall for the minimum area ratio may be calculated as follows,

$$\frac{75^2 - d_i^2}{d_i^2} = 0.20$$

Simplifying $d_i = 68.465$ mm; The wall thickness $= \dfrac{75 - 68.465}{2} = 3.267$ mm.

17.4 ROCK CORE SAMPLING

Rock coring is the process in which a sampler consisting of a tube (core barrel) with a cutting bit at its lower end cuts an annular hole in a rock mass, thereby creating a cylinder or core of rock which is recovered in the core barrel. Rock cores are normally obtained by rotary drilling.

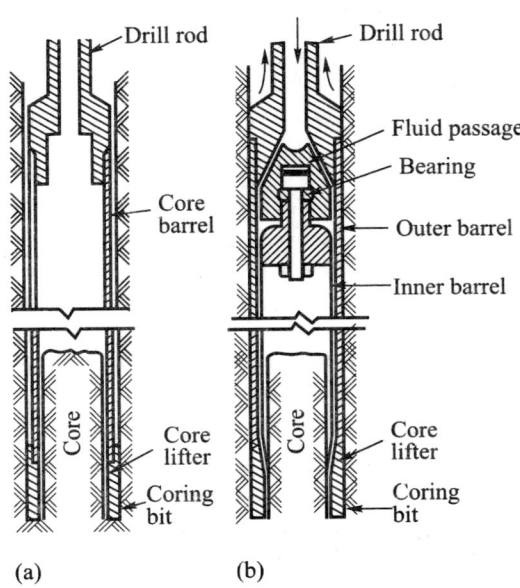

Fig. 17.10 Schematic diagram of core barrels (a) Single tube, (b) Double tube.

TABLE 17.1

Standard sizes of core barrels, drill rods, and compatible casing (Peck *et al* 1974)

Symbol	Core barrel		Symbol	Drill rod	Symbol	Casing	
	Hole dia (in)	Core dia (in)		Outside dia (in)		Outside dia (in	Inside dia (in)
EWX, EWM	$1\frac{1}{2}$	$1\frac{3}{16}$	E	$1\frac{15}{16}$	–	–	–
AWX, AWM	$1\frac{15}{16}$	$1\frac{3}{16}$	A	$1\frac{5}{8}$	EX	$1\frac{13}{16}$	$1\frac{1}{2}$
BWX, BWM	$2\frac{3}{8}$	$1\frac{5}{8}$	B	$1\frac{7}{8}$	AX	$2\frac{1}{4}$	$1\frac{29}{32}$
NWX, NWM	3	$2\frac{1}{8}$	N	$2\frac{3}{8}$	BX	$2\frac{7}{8}$	$2\frac{3}{8}$
$2\frac{3}{4}\times3\frac{7}{8}$	$3\frac{7}{8}$	$2\frac{11}{16}$	–		NX	$3\frac{1}{2}$	3

Note: Symbol *X* indicates single barrel, *M* indicates double barrel.

The primary purpose of core drilling is to obtain intact samples. The behaviour of a rock mass is affected more by the presence of fractures in the rock. The size and spacing of fractures, the degree of weathering of fractures, and the presence of soil within the fractures are critical items. Fig. 17.10 gives a schematic diagram of core barrels with coring bits at the bottom. As discussed earlier, the cutting element may consist of diamonds, tungsten carbide inrts or chilled shot. The core barrel may consist of a single tube or a double tube. Samples taken in a single tube barrel are likely to experience considerable disturbance due to torsion to swelling and contamination with the drilling fluid, but these disadvantages are not there if the coring apparatus to be in hard, intact, rocky strata. However, if double tube barrel is used, the core is protected by the circulating fluid. Most core barrels are capable of retaining cores up to a length of 2 m. Single barrel is used in Calyx drilling. Standard rock cores range from about 1¼ inches to nearly 6 inches in diameter. The more common sizes are given in Table 17.1

The *recovery ratio R_r* defined as the percentage ratio between the length of the core recovered and the length of the core drilled on a given run, is related to the quality of rock encountered in boring, but it is also influenced by the drilling technique and the type and size of core barrel used. Generally the use of a double tube barrel results in higher recovery ratios than can be obtained with single tube barrels. A better estimate of in-situ rock quality is obtained by a modified core recovery ratio known as the *Rock Quality Designation (RQD)* which is expressed as

$$RQD = \frac{\overline{L}_a}{L_t} \qquad (17.3)$$

where, $\overline{L}_a$ = total length of intact hard and sound pieces of core of length greater than 100 mm arranged in its proper position,

L_t = total length of drilling.

TABLE 17.2

Relation of RQD and *in-situ* Rock Quality and allowable contact pressure q_a (Peck *et al* 1974)

RQD %	Rock quality	q_a (tons/ft^2)
90–100	Excellent	200–300
75–90	Good	120–200
50–75	Fair	65–120
25–50	Poor	30–65
0–25	Very Poor	10–25

Breaks obviously caused by drilling are ignored. The diameter of the core should preferably not less than $2\,\frac{1}{8}$ inches. Table 17.2 gives the rock quality description as related to *RQD*.

17.5 STANDARD PENETRATION TEST

The SPT is the most commonly used *in situ* test in a bore hole for soil in USA. The test is made by making use of a split spoon sampler shown in Fig. 17.7. The method has been standardized as ASTM D-1586 (1997) with periodic revision since 1958. The method of carrying out this test is as follows:

1. The split spoon sampler is connected to a string of drill rods and is lowered into the bottom of the bore hole which was drilled and cleaned in advance.

2. The sampler is driven into the soil strata to a maximum depth of 18 in by making use of a 140 lb (63.5 kg) weight falling freely from a height of 30 in (760 mm) on to an anvil fixed on the top of drill rod. The weight is guided to fall along a guide rod. The weight is raised and allowed to fall by means of a manila rope, one end tied to the weight and the other end passing over a pulley on to a hand operated winch or a motor driven cathead.

3. The number of blows required to penetrate each of the successive 6 in (150 mm) depths is counted to produce a total penetration of 18 in (450 mm).

4. To avoid seating errors, the blows required for the first 6 in of penetration are not taken into account; those required to increase the penetration from 6 in (150 mm) to 18 in (450 mm) constitute the *N*-value.

As per some codes of practice if the *N*-value exceeds 100, it is termed as refusal, and the test is stopped even if the total penetration falls short of the last 300 mm depth of penetration. Standardization of refusal at 100 blows allows all the drilling organizations to standardize costs so that higher blows if required may be eliminated to prevent the excessive wear and tear of the equipment. The SPT is conducted normally at 1 to 2 m intervals. The intervals may be increased at greater depths if necessary.

Standardization of SPT

The validity of the SPT has been the subject of study and research by many authors for the last many years. The basic conclusion is that the best results are difficult to reproduce. Some of the important factors that affect reproducibility are

1. Variation in the height of fall of the drop weight (hammer) during the test

2. The number of turns of rope around the cathead, and the condition of the manila rope

3. Length and diameter of drill rod

4. Diameter of bore hole

5. Overburden pressure

There are many more factors that hamper reproducibility of results. Normally corrections used to be applied for a quick condition in the hole bottom due to rapid withdrawal of the auger. ASTM 1586 has stipulated standards to avoid such a quick condition. Discrepancies in the input driving energy and its dissipation around the sampler into the surrounding soil are the principal factors for the wide range in N values. The theoretical input energy may be expressed as

$$E_{in} = Wh \tag{17.4}$$

where $W =$ weight or mass of the hammer

$h =$ height of fall

Investigation has revealed (Kovacs and Salomone, 1982) that the actual energy transferred to the driving head and then to the sampler ranged from about 30 to 80 percent. It has been suggested that the SPT be standardized to some energy ratio R_e keeping in mind the data collected so far from the existing SPT. Bowles (1996) suggests that the observed SPT value N be reduced to a standard blow count corresponding to 70 percent of standard energy. Terzaghi, et al., (1996) suggest 60 percent. The standard energy ratio may be expressed as

$$R_{es} = \frac{\text{Actual hammer energy to sampler, } E_a}{\text{Input energy, } E_{in}}$$

Corrections to the Observed SPT Value

Three types of corrections are normally applied to the observed N values. They are:

1. Hammer efficiency correction

2. Drillrod, sampler and borehole corrections

3. Correction due to overburden pressure

1. Hammer Efficiency Correction, E_h

Different types of hammers are in use for driving the drill rods. Two types are normally used in USA. They are (Bowles, 1996)

1. Donut with two turns of manila rope on the cathead with a hammer efficiency $E_h = 0.45$.

2. Safety with two turns of manila rope on the cathead with a hammer efficiency as follows:

    Rope-pulley or cathead = 0.7 to 0.8;

    Trip or automatic hammer = 0.8 to 1.0.

2. Drill Rod, Sampler and Borehole Corrections

Correction factors are used for correcting the effects of length of drill rods, use of split spoon sampler with or without liner, and size of bore holes. The various correction factors are (Bowles, 1996).

(a) Drill rod length correction factor C_d

Length (m)	Correction factor (C_d)
> 10 m	1.0
4–10 m	0.85–0.95
< 4.0 m	0.75

(b) Sampler correction factor, C_s

Without liner $C_s = 1.00$

With liner,

Dense sand, clay = 0.80

Loose sand = 0.90

(c) Bore hole diameter correction factor, C_b

Bore hole diameter	Correction factor, C_b
60-120 mm	1.0
150 mm	1.05
200 mm	1.15

3. Correction Factor for Overburden Pressure in Granular Soils, C_N

The C_N as per Liao and Whitman (1986) is

$$C_N = \left[\frac{65.76}{p_o'} \right]^{1/2} \tag{17.5}$$

where, p_o' = effective overburden pressure in kN/m^2

There are a number of empirical relations proposed for C_N. However, the most commonly used relationship is the one given by Eq. (17.5).

N (corrected) may be expressed as

$$N = C_N N_0 E_h C_d C_s C_b \tag{17.6}$$

where N_0 is the observed value.

N is related to the standard energy ratio used by the designer. N may be expressed as N_{70} or N_{60} according to the designer's choice.

Note: N or N_{cor} is used in this text for the corrected value.

Example 17.3

The observed standard penetration test value in a deposit of fully submerged sand was 45 at a depth of 6.5 m. The average effective unit weight of the soil is 9.69 kN/m^3. The other data given are (a) hammer efficiency = 0.8, (b) drill rod length correction factor = 0.9, and (c) borehole correction factor = 1.05. Determine the corrected SPT value for standard energy (a) R_{es} = 60 percent, and (b) R_{es} = 70 %.

Solution

Per Eq. (17.6), the equation for N_{60} may be written as

(i) $N_{60} \quad = C_N N_0 E_h C_d C_s C_b$

where $N_0 \quad$ = observed SPT value

$\quad\quad C_N \quad$ = overburden correction

Per Eq. (17.5) we have

$$C_N = \left[\frac{95.76}{p_0'}\right]^{1/2}$$

where p_0' = effective overburden pressure

$$= 6.5 \times 9.69 = 63 \, kN/m^2$$

Substituting for p_0',

$$C_N = \left[\frac{95.76}{60}\right]^{1/2} = 1.233$$

Substituting the known values, the corrected N_{60} is

$N_{60} = 1.233 \times 45 \times 0.8 \times 0.9 \times 1.05 = 42$

For 70 percent standard energy

$$N_{70} = 42 \times \frac{0.6}{0.7} = 36$$

SPT Values Related to Relative Density of Cohesionless Soils

Although the SPT is not considered as refined and completely reliable method of investigation. the N values give a useful information with regards to consistency of cohesive soils and relative density of cohesionless soils. The correlation between N values (corrected) and relative density of granular soils suggested by Terzaghi and Peck is given in Table 17.3.

Before using Table 17.3 the observed N value has to be corrected for standard energy, dilatancy and overburden pressure. The correlations given in Table 17.3 is just a guide and it may vary according to the fineness of the sand and as the sand may be fine, medium or coarse.

Meyerhof (1956) has suggested the following approximate equations for computing the angle of friction from the known value of D_r.

For granular soil with fine sand and silt more than 5 per cent,

$$\phi° = 25 + 0.15 D_r \tag{17.7}$$

For granular soils with fine sand and silt less than 5 per cent,

$$\phi° = 30 + 0.15 D_r \tag{17.8}$$

where, D_r is expressed in percent.

TABLE 17.3

N and φ Related to Relative Density

N	Compactness	Relative density, D_r	$\phi°$
0–4	Very loose	0–15	< 28
4–10	Loose	15–35	28–30
10–30	Medium	35–65	30–36
30–50	Dense	65–85	36–41
> 50	Very Dense	> 85	> 41

TABLE 17.4

Relation Between N and q_u

Consistency	N	q_u, kPa
Very soft	0–2	< 25
Soft	2–4	25–50
Medium	4–8	50–100
Stiff	8–15	100–200
Very Stiff	15–30	200–400
Hard	> 30	> 400

where q_u is the unconfined compressive strength.

SPT Values Related to Consistency of Clay Soil

Peck *et al* (1974) have given for saturated cohesive soils, correlations between N value, and consistency. This correlation is quite useful but has to be used according to the soil conditions met in the field. Table 17.4 gives the correlations.

The present practice is to relate q_u with N as follows,

$$q_u = \bar{k}N \qquad (17.9a)$$

or $\quad \bar{k} = \dfrac{q_u}{N} \qquad (17.9b)$

where, $\bar{k}$ is the proportionality factor.

The author (Murthy, 1982) investigated the relationship for the clay soil met at Farakka in West Bengal, India. The soil met was preconsolidated greyish silty clay with a natural water content close to the plastic limit. The plastic limit varied from 30 to 40 per cent and the liquid limit from 50 to 100 percent. The preconsolidation ratio was of the order of 5. Undrained triaxial and unconfined compression tests were carried out on a number of undisturbed samples. The results were plotted with q_u, as abscissa and N/q_u as ordinate. There was a considerable scatter of test results. However, the trend showed that most of the points fell between two parallel lines having values $N/q_u = 5$ and

10. An average value $N/q_u = 7.5$ was assumed for this soil. As per this investigation, we may write,

$$q_u = \frac{N}{7.5} \text{kg/cm}^2 = 13.33N \text{ kPa} \qquad (17.10)$$

The value of the proportionality factor $\overline{k}$ varies from 4 for clay soil to 7.5 for silty sandy soil (Sangler *et al*, 1972)

Example 17.4

If the corrected SPT value is 35 compute the following

    (a) The relative density and ϕ as per Table 17.3

    (b) The value of ϕ for the computed relative density as per Meyerhof.

Solution

    (i) Since the SPT value is 35, the relative density and ϕ as per Table 17.3 may be taken as equal to 70% and 37° respectively.

    (ii) Since the SPT is conducted in fine silty sand, the silt content may be assumed more than 5 per cent. As such Eq. (17.7) is applicable as per Meyerhof's proposal. Therefore, $\phi° = 25 + 0.15 \times 70 = 36°$.

Example 17.5

If the observed SPT value is 45 as per Example 17.3 in a deposit of silty clay, determine

    1. The consistency of the clay deposit and the unconfined compressive strength, q_u of clay as per Table 17.4.

    2. The value of q_u as per Eq. (17.10).

Solution

    1. The consistency of clay is very hard as per Table 17.4 and the value of q_u is above 400 kN/m².

    2. The value of q_u as per Eq. (17.10): $q_u = 13.33 N = 13.33 \times 45 = 600$ kN/m².

17.6 STATIC CONE PENETRATION TEST (CPT)

Introduction

The static cone penetration test normally called as the Dutch cone penetration test (CPT) is the most popular of the penetration tests. It has gained acceptance rapidly in many countries. The method was introduced nearly 50 years ago, but its popularity remained for a long time to a small number of countries such as Holland and Belgium. One of the greatest value of the CPT consists still of its function as a small scale model pile test. Empirical correlations established over many years permit the calculations of pile bearing capacity directly from the CPT results without the use of conventional soil parameters.

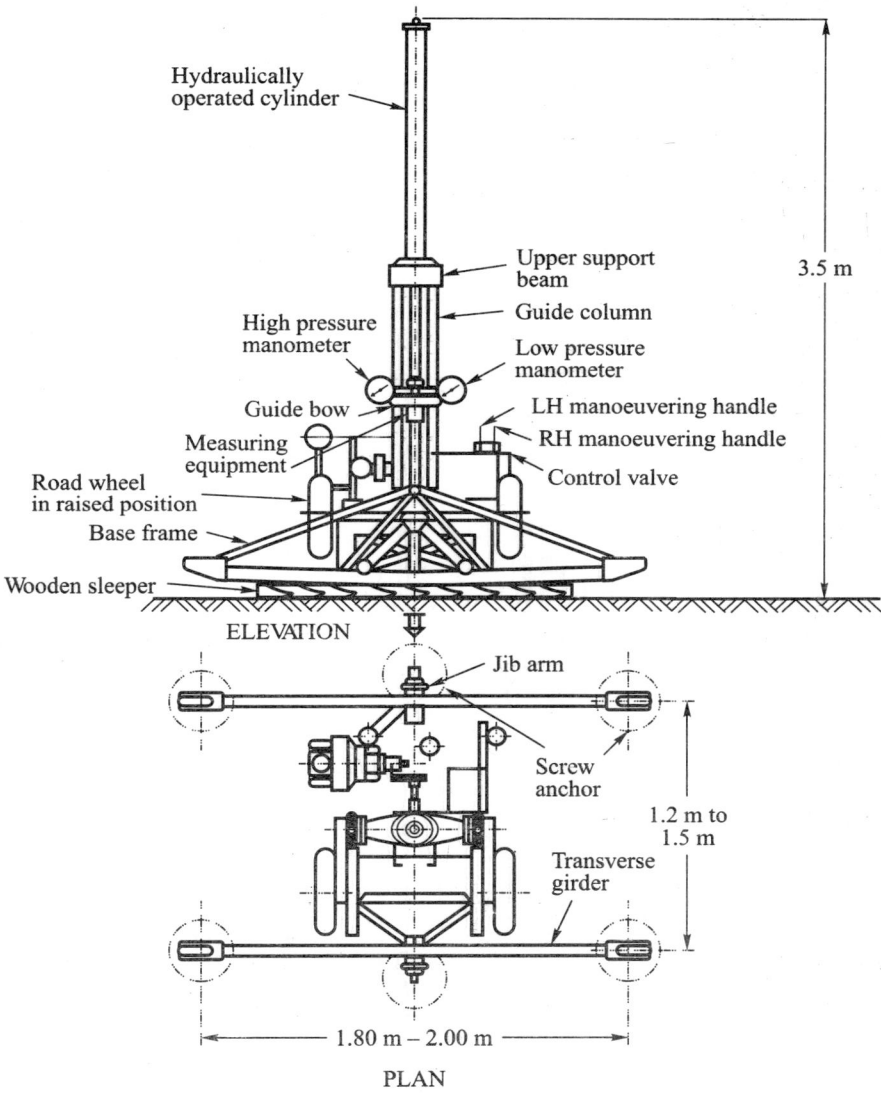

Hydraulically
operated cylinder

Upper support
beam

3.5 m

High pressure
manometer

Guide column

Low pressure
manometer

Guide bow

LH manoeuvering handle

Measuring
equipment

RH manoeuvering handle

Road wheel
in raised position

Control valve

Base frame

Wooden sleeper

ELEVATION

Jib arm

Screw
anchor

1.2 m to
1.5 m

Transverse
girder

1.80 m – 2.00 m

PLAN

Fig. 17.11(a)   Static cone penetration testing equipment

The CPT has proved valuable for soil profiling, as the soil type can be identified from the combined measurement of end resistance of cone and side friction on a jacket. The test lends itself further more for the derivation of normal soil properties such as density, friction angle and cohesion. Various theories have been developed for foundation design.

The popularity of the CPT can be attributed to the following three important factors:

1.   General introduction of the electric penetrometer providing more precise measurements. and improvements in the equipment allowing deeper penetrations.

2.   The need for penetrometer testing in-situ technique in offshore foundation investigations in view of the difficulties in achieving adequate sample quality in marine environment.

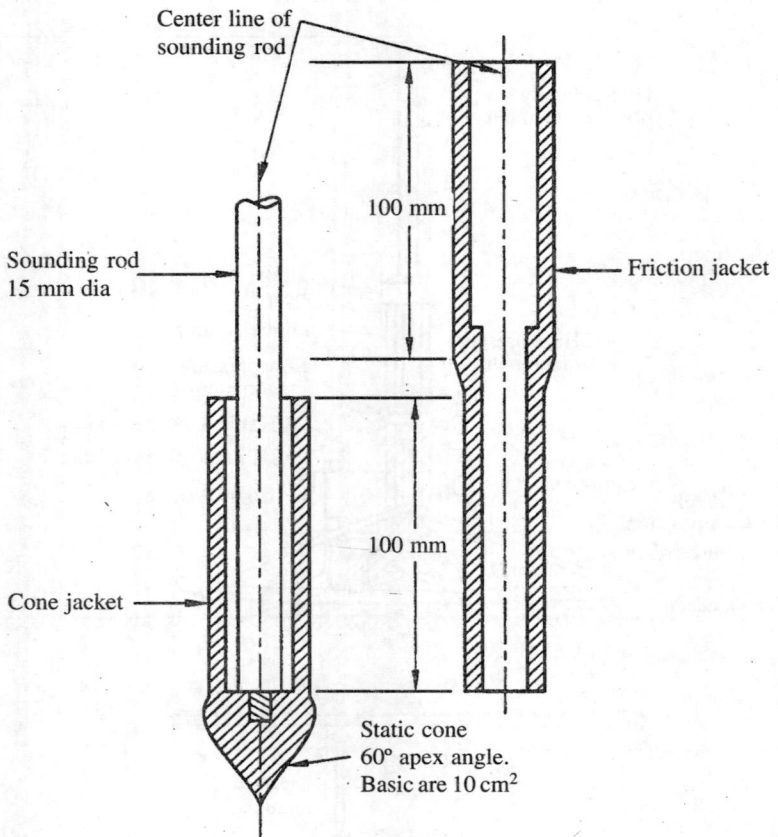

Center line of sounding rod

100 mm

Sounding rod 15 mm dia

Friction jacket

Cone jacket

100 mm

Static cone 60° apex angle. Basic are 10 cm²

Fig. 17.11(b) Static cone assembly

3. The addition of other simultaneous measurements to the standard friction penetrometer such as porewater pressure and soil temperature.

The Penetrometer

There are a variety of shapes and sizes of penetrometers being used. However, the one that is standard in most of the coues is the cone with an apex angle of 60° and a base area of 10 cm². The sleeve (jacket) has become a standard item on the penetrometer for most applications. On the 10 cm² cone penetrometer the friction sleeve should have an area of 150 cm² as per the standard practice. The ratio of side friction and bearing resistance called as *friction ratio* enables identification of the soil type (Schmertmann 1975) and provide useful information in particular when no bore hole data are available. Even when borings are made, the friction ratio supplies a check on the accuracy of the boring logs.

Static Cone Penetrometer

The Begemann Friction Cone Mechanical type penetrometer consists of a 60° cone with a base diameter of 35.6 mm (Sectional area 10 cm²). A sounding rod is screwed to the base. Additional rods of one meter length each is used. These rods are screwed or attached together to bear against each other. The sounding rods move inside mantle tubes. The inside diameter of the mantle tube is just sufficient for the sounding rods to move freely whereas the outside diameter is equal to or less than the base diameter the cone. Fig. 17.11 shows static cone penetration equipment and cone assembly.

Jacking System

The rigs used for pushing down the penetrometer consists basically a hydraulic system. The thrust capacity for cone testing on land varies from 20 to 30 kN for hand operated rigs and 100 to 200 kN for mechanically operated ones. Bourdon gauges are provided in the driving mechanism for measuring the pressures exerted by the cone and the friction jacket either individually or collectively during the operation. The rigs may be operated by placing it either on the ground or mounted on heavy duty trucks. In either case, the rig should take the necessary upthrust. For ground based rigs screw anchors are, provided to take up the reaction thrust.

Operation of Penetrometer

The point resistance of the cone (q_c) and the frictional resistance on the friction jacket can be found out by measuring the resistance offered by pushing each element independently. The cone and the jacket are pushed at a standard rate of 20 mm per second.

Effect of Rate of Penetration

Several studies have been made to determine the effect of the rate of penetration on cone bearing and side friction. Although the values tend to decrease for slower rates, the general conclusion is that the influence is insignificant for speeds between 10 and 30 mm per second. The standard rate of penetration has been generally accepted as 20 mm per second.

Cone Resistance q_c and Local Side Friction f_c

Cone penetration resistance q_c is obtained by dividing the total force Q_c acting on the cone by the base area A_c of the cone.

$$q_c = \frac{Q_c}{A_c}$$

(17.11)

In the same way, the local side friction fc is

$$f_c = \frac{Q_f}{A_f},$$

(17.12)

where, $Q_f = Q_t - Q_c$ = force required to push the friction jacket,

Q_t = the total force required to push the cone and friction jacket together in the case of a mechanical penetrometer,

A_f = surface area of the friction jacket.

Friction Ratio, R_f

Friction ratio, R_f is expressed as

$$R_f = \frac{f_c}{q_c},$$ (17.13)

where f_c and q_c are measured at the same depth. R_f is expressed as a percentage. Friction ratio is an important parameter for classifying soil.

Relationship Between q_c, Relative Density D_r and Friction Angle ϕ for Sand

Research carried out by many indicates that a unique relationship between cone resistance, relative density and friction angle valid for all sands does not exist. Robertson and Campanella (1983a) have provided a set of curves (Fig. 17.12a) which may be used to estimate D_r based on q_c and effective overburden pressure. These curves are supposed to be applicable for normally consolidated clean sand. Fig. 17.12b gives the relationship between q_c and ϕ (Robertson and Campanella, 1983b).

Relationship Between q_c and Undrained Shear Strength, c_u of Clay

The cone penetration resistance q_c and c_u may be related as

$$q_c = N_k c_u + p_o \quad \text{or} \quad c_u = \frac{q_c - p_o}{N_k}$$ (17.14)

where, N_k = cone factor,

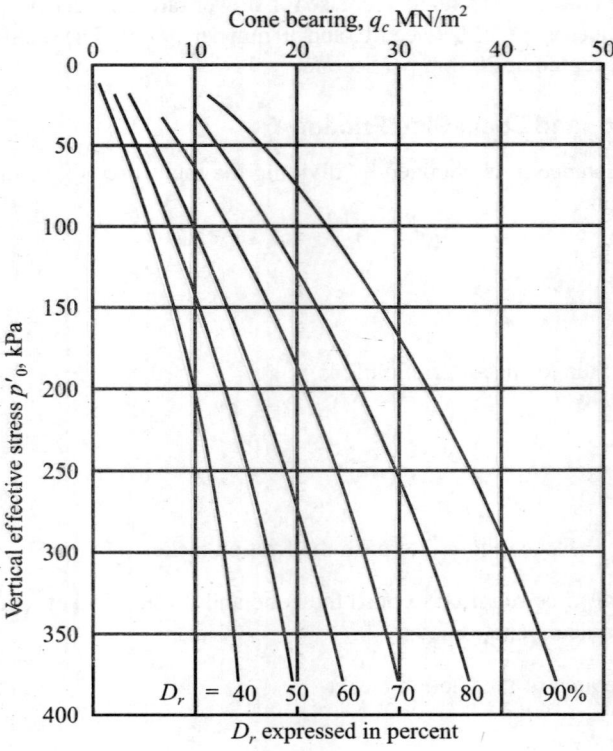

Fig. 17.12(a) Relationship between relative density D_r and penetration resistance q_c for uncemented quartz sands (Robertson and Campanella, 1983a)

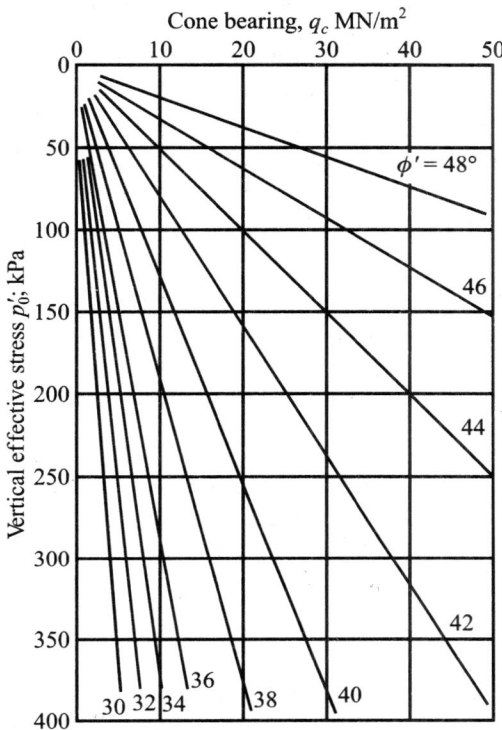

Fig. 17.12(b)   Relationship between cone point resistance q_c and the internal friction ϕ for uncemented quartz sands (Robertson and Campanella, 1983b)

$p_o = \gamma z$ = overburden pressure.

Lunne and Kelven (1981), investigated the value of the cone factor N_k for both normally consolidated and overconsolidated clays. The values of N_k as obtained are given below:

Type of clay	Cone factor
Normally consolidated	11 to 19
Overconsolidated	
At shallow depths	15 to 20
At deep depths	12 to 18

Possibly a value of 20 for N_k for both the types of clays may be alright. Sanglerat (1972) also recommends the same value for all cases where overburden correction is of negligible value. In most of the cases, the overburden pressure in Eq. (17.14) is neglected.

Correlation Between SPT and CPT

Meyerhof (1965) presented comparative data between SPT and CPT. For fine or silty medium loose to medium dense sands, he presents the correlation as

$$q_c = 400\, N \text{ kN/m}^2 = 0.4\, N \text{ MN / m}^2 \tag{17.15}$$

His findings are as given in Table 17.5

TABLE 17.5

Approximate relationship between relative density of fine sand, the SPT, the static cone resistance and the angle of internal fraction

State of sand	D_r	N	q_c MN/m^2	$\phi°$
Very loose	< 0.2	< 4	< 2.0	< 30
Loose	0.2–0.4	4–10	2–4	30–35
Medium dense	0.4–0.6	10–30	4–12	35–40
Dense	0.6–0.8	30–50	12–20	40–45
Very dense	0.8–1.0	> 50	> 20	45

TABLE 17.6

Relationship between q_c and N (After Schmertmann, 1978)

Type of soil	$q_c/N = n$
Sand and gravel mixtures	6
Sandy silts	3
Clay-silt-sand mixtures	2
Insensitive clays	1.5

The lowest values of the angle of internal friction given in Table 17.5 are conservative estimates for uniform, clean sand and they should be reduced by at least 5° for clayey sand. These values as well as the upper values of the angles of internal friction which apply to well graded sand, may be increased by 5° for gravelly sand.

Schmertmann (1978) proposed a relationship between q_c and N as a ratio $q_c/N = n$. He gave the relationship for both the Fugro electric static cone penetrometer and mechanical cone penetrometer. He found that the former gave higher values for n as compared to the latter. Table 17.6 gives values for n for mechanical penetrometer for various types of soils.

Example 17.6

If the sand deposit given in Example 17.3 is uncemented quartz sand with the correced SPT value of 35 at a depth of 6.5 m ($p'_0 = 63$ kN/m^2), determine ϕ by using Fig. 17.12b.

Solution

As per Eq. (17.15), $q_c = 0.4$ N = 14 MN/m^2. From Fig. 17.12(b) for $p'_0 = 63$ kN/m^2, $q_c = 14$ MN/m^2, the value of $\phi = 45°$.

Example 17.7

If a deposit happens to be silty clay (saturated) with a value of $q_c = 88$ kg/cm^2, determine the unconfined compressive of clay as per Eq. (17.14). Use $p_0 = 127$ kN/m^2.

Solution

As per Eq. (17.14)

$$c_u = \frac{q_c - p_o}{N_k} \text{ or } q_u = \frac{2(q_c - p_o)}{N_k}$$

Assume $N_K = 20$

Substituting for q_c, p_0 and N_K, we have,

$$q_u = \frac{2(88 \times 100 - 127)}{20} = 867 \text{ kN/m}^2$$

If we neglect the overburden pressure p_0

$$q_u = \frac{2 \times 88 \times 100}{20} = 880 \text{ kN/m}^2$$

The value of q_u is not affected much if we neglect the overburden pressure.

17.7 FIELD VANE SHEAR TEST (VST)

Vane shear test is one of the *in-situ* tests used normally for obtaining the undrained shear strengths of soft sensitive clays. It is in deep bed of such material that the vane test is most valuable for the simple reason that there is at present no other method known by which the shear strength of these clays can be measured. This test has been described in Chapter 18.

17.8 FIELD PLATE LOAD TEST (PLT)

Field plate test is the oldest of the methods for determining either the bearing capacity or settlement of footings. The details of PLT have been discussed under Shallow Foundations.

17.9 GROUND WATER CONDITIONS

Water Table Location by Hvorselev (1949) Method

As per the Hvorselev method, water table level can be located in a bore hole used for soil investigation. The bore hole should have a casing to stabilise the sides. The method normally used, is the Rising Water Level Method for determining the water table locations.

Rising Water Level Method

This method most commonly referred to as the time lag method consists of bailing the water out of the casing and observing the rate of rise of the water level in the casing at intervals of time until the rise in water level becomes negligible. The rate is observed by measuring the elapsed time and the depth of the water surface below the top of the casing. The intervals at which the readings are required will vary some what with the permeability of the soil. In no case should the total elapsed time for the readings be less than 5 minutes. In freely draining materials such as sands, gravels, etc.

the interval of time between successive readings may not exceed 1 to 2 hours, but in soils of low permeability such as fine sand silts and clays, the intervals may rise from 12 to 24 hours, and it may take a few days to determine the stabilized water table level.

Let the time be t_0, when the water table level was at depth H_0 below the normal water table level (Fig. 17.13). Let the successive rise in water levels be h_1, h_2, h_3, etc. at times t_1, t_2, t_3 respectively wherein the difference in time $(t_1 - t_0)$, $(t_2 - t_1)$, $(t_3 - t_2)$, etc. is kept constant.

Now, from Fig. 17.13

$$H_0 - H_1 = h_1, H_1 - H_2 = h_2, \ H_2 - H_3 = h_3$$

Let $(t_1 - t_0) = (t_2 - t_1) = (t_3 - t_2)$, etc. $= \Delta t$

The depths H_1, H_2, H_3 of the water level in the casing from the normal watertable level can be computed as follows

$$H_0 = \frac{h_1^2}{h_1 - h_2}, \ H_2 = \frac{h_2^2}{h_1 - h_2}, \ H_3 = \frac{h_3^2}{h_2 - h_3}$$

Let the corresponding depths of water table level below the ground surface be h_{w_1}, h_{w_2}, h_{w_3}, etc. Now, we have

$$h_{w_1} = H_w - H_0$$

$$h_{w_2} = H_w - (h_1 + h_2 + H_2)$$

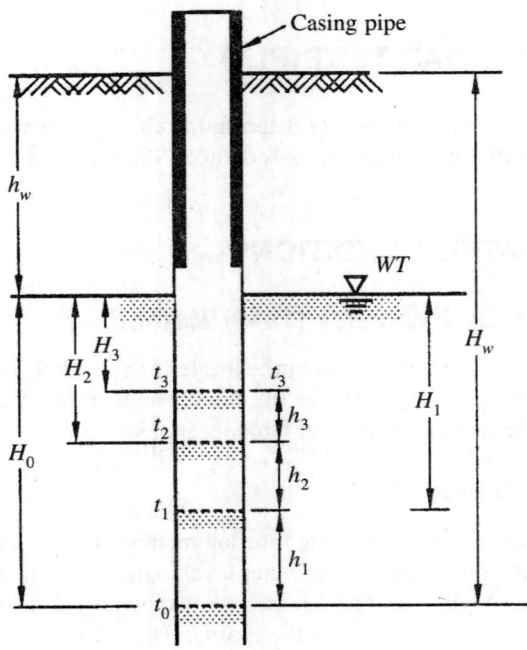

Fig. 17.13   Water table level location by rising water level method

$$h_{w_3} = H_w - \left(h_1 + h_2 + h_3 + H_3\right)$$

where, H_w, is the depth of water level in the casing from the ground surface at the start of the test. Normally $h_{w_1} = h_{w_2} = h_{w_3} = h_w$; if not an average value gives h_w.

Example *17.8*

Establish the location of ground water in a clayey strata. Water in bore hole was bailed out to a depth of 10.67 m below ground surface, and the rise of water was recorded at 24 hour intervals as follows (Fig. 17.13).

$h_1 = 64.0$ cm, $h_2 = 57.9$ cms and $h_3 = 51.8$ cm.

Solution

$$H_0 = \frac{0.64^2}{0.64 - 0.579} = 6.714 \text{ m,}$$

$$H_2 = \frac{0.579^2}{0.64 - 0.579} = 5.496 \text{ m}$$

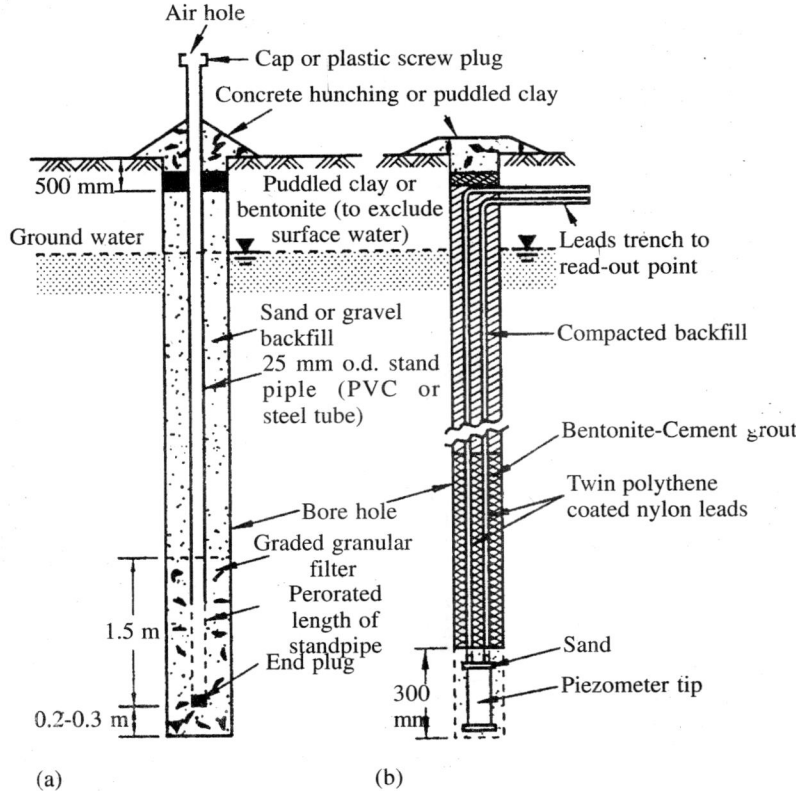

Fig. 17.14   Types of Piezometer for ground water level observations (a) Standpipe (b) Hydraulic piezometer

$$H_3 = \frac{0.518^2}{0.579 - 0.518} = 4.399 \text{ m}$$

$$h_{w_1} = 10.67 - 6.714 = 3.953 \text{ m}$$

$$h_{w_2} = 10.67 - (0.64 + 0.579 + 5.496) = 3.955$$

$$h_{w_3} = 10.67 - (0.64 + 0.579 + 0.518 + 4.399) = 4.534.$$

$$\text{Average,} = h_w = \frac{3.953 + 3.955 + 4.534}{3} = 4.147 \text{ m.}$$

Ground Water Level Observation

When measurement of water table levels has to be made over a long period of time, the best way of doing this is to install a series of stand pipes or piezometers over the area in bore holes and observe

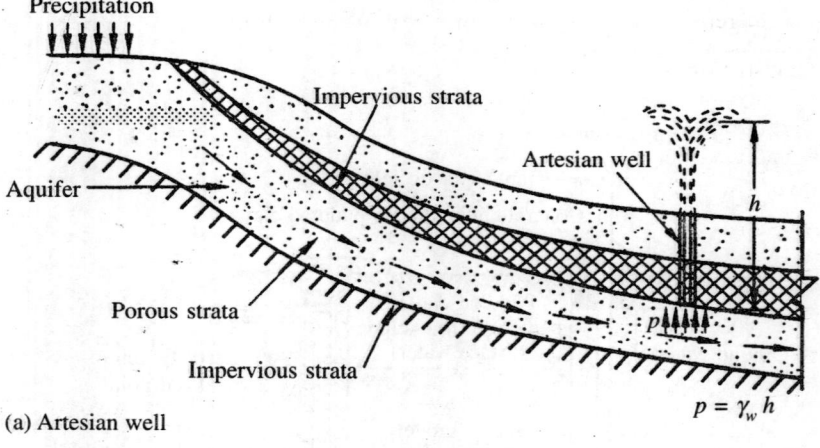

(a) Artesian well

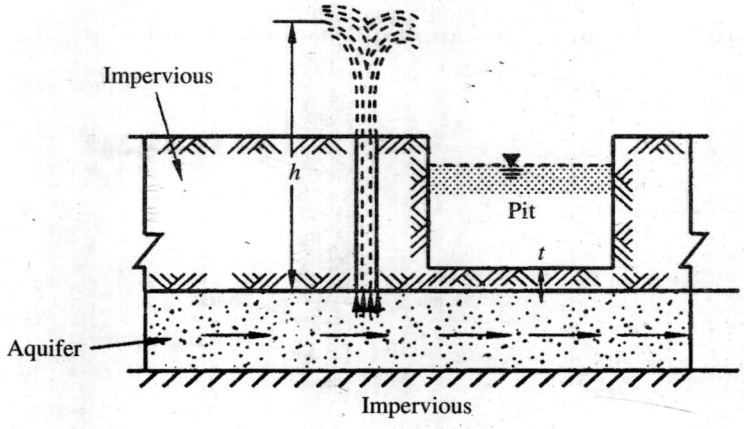

(b) Effect of artesian water on the stability of the bottom of an excavation

Fig. 17.15   Artesian ground water flow

the water levels. A simple stand pipe (Fig. 17.14a) consisting of a PVC tubing with a slotted end and surrounded by granular filter or plastic fabrics is satisfactory for granular soils or permeable rocks. In silts or clays more sensitive equipment is required. The hydraulic piezometer (Fig.17.14b) consists of a porous element connected by twin small-bore plastic tubing to a remote reading station where pressures are measured by a mercury monometer or a Bourden gauge.

Artesian Ground Water Flow

Figure 17.15a is a schematic sketch which shows the presence of artesian ground water flow and artesian pressure. If a well is made in the soil puncturing the top impermeable sratum of soil, the water in the well rises to a height h above the bottom of the well. If p is the pressure acting at the bottom of the well, the water rises to a height h, which is equal to

$$h = \frac{p}{\gamma_w}$$

Such a well is called as *artesian well*. Ground water flow under artesian pressure is quite common in valleys and in areas close to hilly tracts. In such a locality, ground water flows from higher elevations to lower elevations thereby, give rise to artesian water under high pressures. The artesian water endangers the stability of foundations if the bottom of the foundations lie close to the artesian aquifer.

Figure 17.15b indicates the possible destruction of a foundation pit excavated in a soil below which artesian condition prevails. If the impermeable soil strata below the bottom of the trench (or foundation) is of insufficient thickness, t, the *artesian water pressure* may break through the bottom of the pit inundate it, or even destroy it totally.

17.10 GEOPHYSICAL EXPLORATION

Introduction

The stratification of soils and rocks can be determined by geophysical methods of exploration which measures changes in certain physical characteristics of these materials, for example the magnetism, density, electrical resistivity, elasticity or combination of these properties. However, the utility of these methods in the field of foundation engineering is very limited since the methods do not quantify the characteristics of the various substrata. Vital information on ground water conditions is usually lacking. Geophysical methods at best provide some missing information between widely spaced bore holes but it can not replace bore holes. Two methods of exploration which are some times useful are discussed briefly in this section. They are

1. Seismic Refraction Method,

2. Electrical Resistivity Method.

Seismic Refraction Method

The method is based on the fact that seismic waves have different velocities in different types of soils (or rock) and besides the waves refract when they cross boundaries between different types of soils. If artificial impulses are produced either by detonation of explosives or mechanical blow with a heavy hammer at ground surface or at shallow depth within a hole, these shocks generate three types of waves. But, in general, only compression waves (longitudinal waves) are observed. These waves are classified as either direct, reflected or refracted. Direct waves travel in approximately straight lines from the source of impulse to the surface. Reflected or refracted waves undergo a change in direction when they encounter a boundary separating media of different seismic velocities. The seismic refraction method is more suited to shallow exploration for civil engineering purposes.

The method starts by inducing impact or shock waves into the soil at a particular location. The shock waves are picked up by geophones. In Fig. 17.16a, point A is the source of seismic impulse. The points D_1 through D_8 represent the locations of the geophones or detectors which are installed in a straight line. The spacings of the geophones is dependent on the amount of detail required and the depth of the strata being investigated. In general, the spacing must be such that the distance from D_1 to D_8 is three to four times the depth to be investigated. The geophones are connected by

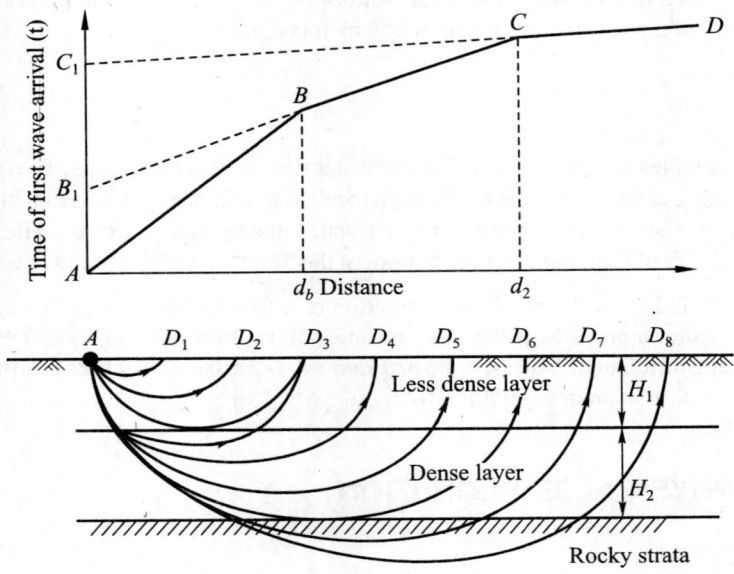

(a) Schematic representation of refraction method

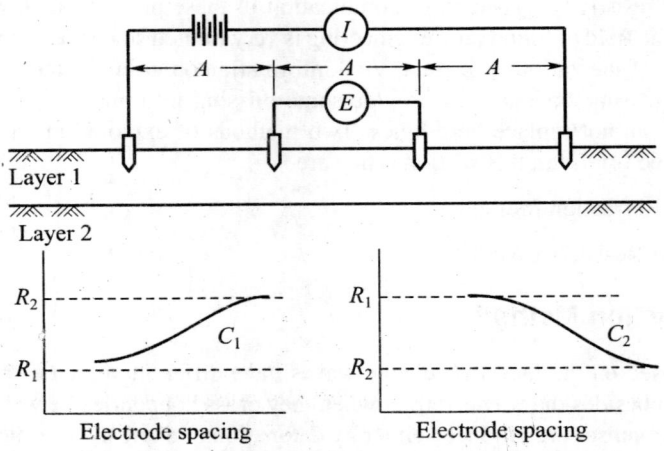

(b) Schematic representation of electrical resistivity method

Fig. 17.16 Geophysical methods of exploration

cable to a central recording device. A series of detonations or impacts are produced and the arrival time of the first wave at each geophone position is recorded in turn. When the distance between source and geophone is short, the arrival time will be that of direct wave. When the distance exceeds a certain value (depending on the thickness of the stratum), the refracted wave will be the first to be detected by the geophone. This is because, the refracted wave, although longer than that of the direct wave, passes through a stratum of higher seismic velocity.

A typical plot of test results for a three layer system is given in Fig. 17.16a with the arrival time plotted against the distance source and geophone. As in the figure, if the source-geophone spacing is more than the distance d_b, which is the distance from the source to the point B. The direct wave reaches the geophone in advance of the refracted wave and the time-distance relationship is represent by a straight line AB through the origin represented by A. If on the other hand, the source geophone distance is greater than d_b, the refracted waves arrive in advance of the direct waves and the time-distance relationship is represented by another straight line BC which will have slope different from that of AB. The slopes of the lines AB and BC are represented by $1/V_1$, and $1/V_2$ respectively, where V_1 and V_2 are the velocities of the upper and lower strata respectively. Similarly, the slope of the third line CD is represented by $1/V_3$ in the third strata.

The general types of soil or rocks can be determined from a knowledge of these velocities. The depth H_1, of the top strata (provided the thickness of the stratum is constant) can be estimated from the formula

$$H_1 = \frac{d_b}{2}\sqrt{\frac{V_2 - V_1}{V_2 + V_1}} \tag{17.16}$$

If the thickness of the strata is not constant, H_1, represents the average thickness.

Shepard and Haines (1944) have given the following equations for determining the depths H_1, and H_2 in a three layer strata;

$$H_1 = \frac{l_1 V_1}{2\cos\alpha} \tag{17.17}$$

$$H_2 = \frac{l_2 V_2}{2\cos\beta} \tag{17.18}$$

where, l_1 = AB_1, (Fig. 17.16a); The point B_1, is obtained on the vertical passing through A by extending the straight line CB,

l_2 = $(AC_1 - AB_1)$; AC_1 is the intercept on the vertical through A obtained by extending the straight line DC,

α = $\sin^{-1}(V_1/V_2)$,

β = $\sin^{-1}(V_2/V_3)$.

α and β are the angles of refraction of the first and second stratum interfaces respectively.

The formulae used to estimate the depths from seismic refraction survey data are based on the following assumptions:

1. Each strata is homogeneous and isotropic.

2. The boundaries between strata are either horizontal or inclined planes.

TABLE 17.7

Range of seismic velocities in soils near the surface or at shallow depths (Bowles 1986)

Material	Velocity m/sec
Sand	200–2000
Aluvium	500–2000
Clay	1000–2800
Loam	800–1800
Sand Stone	1400–4300
Lime Stone	1700–6400
Slate and shale	2300–4600
Granite	4000–5700
Quartzite	6100

3. Each stratum is of sufficient thickness to reflect a change in velocity on time-distance plot.

4. The velocity of wave propagation for each succeeding stratum increases with depth. Table 17.7 gives typical seismic velocities in various materials.

Electrical Resistivity Method

The method depends on differences in the electrical resistance of different soil (and rock) types. The flow of current through a soil is mainly due to electrolytic action and therefore depends on the concentration of dissolved salts in the pore water. The mineral particles of soil are poor conductors of current. The resistivity of soil, therefore, decreases as both water content and concentration of salts increase. A dense clean sand above the water table, for example, would exhibit a high resistivity due to its low degree of saturation and virtual absence of dissolved salts. A saturated clay of high void ratio, on the other hand, would exhibit a low resistivity due to the relative abundance of porewater and the free ions in that water.

There are several methods by which the field resistivity measurements are made. The most popular of the methods is the Wenner Method.

Wenner Method

The Wenner arrangement consists of four equally spaced electrodes driven approximately 20 cms in to the ground as shown in Fig. 17.16b. In this method a *dc* or very low frequency *dc* current of known magnitude is passed between, the two outer (current) electrodes, thereby producing within the soil an electric field, whose pattern is determined by the resistivities of the soils present within the field and the boundary conditions. The potential drop E, for the surface current flow lines is measured by means of the inner electrodes. The apparent resistivity, R, is given by the equation

$$R = \frac{2\pi A E}{I} \tag{17.19}$$

It is customary to express A in centimetre, E in volts, I in amperes, and R ohm-cm. The apparent resistivity represents a weighted average of true resistivity to a depth A in a large volume of soil, the ɔil close to the surface being more heavily weighted than the soil at greater depths. The presence of stratum of low resistivity, forces the current to flow closer to the surface resulting in a higher ɔltage drop and hence a higher value of apparent resistivity. The opposite is true if a stratum of low ɛsistivity lies below a stratum of high resistivity.

The method known as *sounding* is used when the variation of resistivity with depth is required. his enables rough estimates to be made of the types and depths of strata. A series of readings are ɪken, the (equal) spacing of the electrodes being increased for each successive reading. However, ɪe centre of the four electrodes remains at a fixed point. As the spacing is increased, the apparent ɛsistivity is influenced by a greater depth of soil. If the resistivity increases with the increasing ɪectrode spacings, it can be concluded that an underlym stratum of higher resistivity is beginning to ɪfluence the readings. If increased separation produces decreasing resistivity, on the other hand, a ɔwer, resistivity is beginning to influence te readings.

Apparent resistivity is plotted against spacing, perferably, on a log paper. Characteristic curves ɔr a two layer structure are shown in Fig. 17.16b. For curve C_1 the resistivity of layer 1 is lower than ɪat of 2; for curve C_2, layer 1 has a higher resistivity than that of layer. 2. The curves become ɪsymptotic to lines representing the true resistance R_1, and R_2 of the respective layers. Approximate ɪayer thickness can be obtained by comparing the observed curves of resistivity versus electrode ɔpacing with a set of standard curves.

The procedure known as profiling is used in the investigation of lateral variation of soil types. A series of readings is taken, the four electrodes being moved laterally as a unit for each successive ɛeading; the electrode spacing remains constant for each reading of the series. Apparent resistivity is ɔlotted against the centre position of the four electrodes, to natural scale; such plot can be used to ɔocate the position of soil of high or low resistivity. Contours of resistivity can be plotted over a ƺiven area.

The electrical method of exploration has been found to be not as reliable as the seismic method as the apparent resistivity of a particular method as the apparent resistivity of a particular soil or rock can vary over a wide range of values.

Representative values of resistivity are given in Table 17.8.

TABLE 17.8

Representative values of resistivity. The values are expressed in units of 10^3 ohm-cm (After Peek *et al* 1974)

Material	Resistivity Ohm-cm
Clay and saturated silt	0–10
Sandy clay and wet silty sand	10–25
Clayey sand and saturated sand	25–50
Sand	50–150
Gravel	150–500
Weathered rock	100–200
Sound rock	150–4,000

17.11 PLANNING OF SOIL EXPLORATION

Introduction

The planner has to consider the following points before making a programme:

1. Type, size and importance of the project.

2. Whether the site investigation is preliminary or detailed.

In the case of big projects, a preliminary investigation is normally required for the purpose c

1. selecting a site and making a feasibility study of the project,

2. making tentative designs and estimate of the cost of the project.

Preliminary site investigation needs only a few bore holes distributed suitably over the area f
taking samples. The data obtained from the field and laboratory tests must be adequate to provide
fairly good idea of the strength characteristics of the subsoil for making preliminary drawings an
design. In case a particular site is found unsuitable on the basis of the study, an alternate site ma
have to be chosen.

When once a site is chosen, a detailed soil investigation is undertaken. The planning of a so
investigation include the following steps:

1. A detailed study of the geographical condition of the area which include:

    (a) Collection of all the available information about the site, including the collection o
         existing topographical and geological maps,

    (b) General topographical features of the site,

    (c) Collection of the available hydraulic conditions, such as water table fluctuations
         flooding of the site etc,

    (d) Accessibility to the site.

2. Preparation of a layout plan of the project.

3. Preparation of a borehole layout plan which includes the depths and the number of bore
    holes suitably distributed over the area.

4. Marking on the layout plan any additional types of soil investigation.

5. Preparation of specifications and guidelines for the field execution of the various elements
    of soil investigation.

6. Preparation of specifications and guide lines for testing of the samples collected in the
    laboratory, presentation of field and laboratory test results, writing of report, etc.

The planner can make an intelligent, practical and pragmatic planning if he is conversant with
the various elements of soil investigation.

Depths of Bore Holes

The depth up to which bore holes should be sunk is governed by the depth of soil affected by the
foundation bearing pressures. The standard practice is to take the borings up to a depth (called as
significant depth) at which the excess vertical stress caused by a fully loaded foundation is of the
order of 20 percent or less of the net imposed vertical stress at the foundation base level. The depth
the borehole as per this practice works out to about 1.5 times the least width of the foundation from

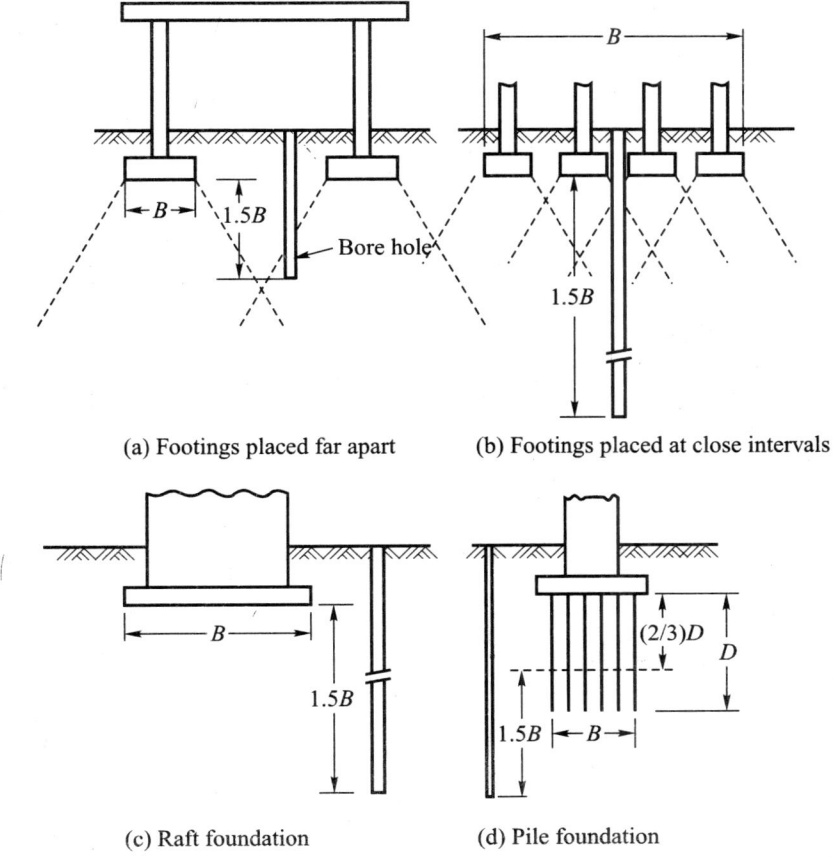

(a) Footings placed far apart

(b) Footings placed at close intervals

(c) Raft foundation

(d) Pile foundation

Fig. 17.17 Depth of Bore holes

the base level of the foundation as shown in Fig. 17.17a. Where strip or pad footings are closely spaced which results in the overlapping of the stressed zones, the whole loaded area becomes in effect a raft foundation with correspondingly deep borings as shown in Fig. 17.17b and c. In case of field or pier foundations the subsoil should be explored up to the depths required to cover the soil lying even below the tips of piles (or pile and piers which are affected by the loads transmitted to the deeper layers, Fig. 17.17b In case rock is met with at shallow depths, foundations may have to rest on rocky strata. The boring should also explore the strength characteristics of rocky strata in such cases.

Number of Bore Holes

Adequate number of bore holes is needed to

1. provide a reasonably accurate determination of the contours of the proposed bearing stratum,

2. locate any soft pockets in the supporting soil.which would adversely affect the safety and performance of the proposed design.

The number of bore holes which need to be sunk on any particular site is a difficult proble which is closely linked up with the relative cost of the investigation and the project for which it undertaken. When the soil is homogeneous over the whole area, the numbsr of bore holes could limited, but if the soil condition is erratic limiting the number would be counter productive.

17.12 EXECUTION OF SOIL EXPLORATION PROGRAMME

Introduction

The three limbs of a Soil Exploration are

1. Planning, 2. Execution, 3. Report writing.

All the three limbs are equally important for a satisfactory solution of the problem. Howeve the execution of the soil exploration programme acts as a bridge between planning and report writing and as such occupies an important place. No amount of planning would help report writing, if th field and laboratory works are not executed with Diligence and Care. It is therefore, essential tha the execution part should always be entrusted to well qualified, reliable and resourceful geotechnica consultants who will also be responsible for report writing.

Bore Logs

A detailed record of boring operations and other tests carried out in the field is an essential part o the field work. The bore hole log is made during the boring operation. The soil is classified based o the visual examination of the disturbed samples collected. A typical example of a bore hole log i given in Fig. 17.18. The log should include the difficulties faced during boring operations including the occurrence of sand boils, and the presence of artesian water conditions if any, etc.

Report

A report is the final document of the whole exercise of soil exploration. A report should be comprehensive, clear and to the point. Many can write reports, but only a very few can produce a good report. A report writer should be knowledgable, practical and pragmatic. No theory, books or codes of practice provide all the materials required to produce a good report. It is only the experience of a number of years of dedicated service in the field helps a geotechnical consultant to make the report writing an art. A good report should normally comprise of the following.

1. A general description of the nature of the project and its importance.

2. A general description of the topographical features and hydraulic conditions of the site.

3. A brief description of the various field and laboratory tests carried out.

4. Analysis and discussions of the test results.

5. Recommendations.

6. Calculations for determining safe bearing pressures, pile loads, etc.

7. Tables containing borelogs, and other field and Laboratory test results.

8. Drawings which include an index plan, a site-plan, test results plotred in the form of charts and graphs, soil profiles, etc.

Bore-hole log

Job No.	Date: 6.04.1984
Project: Farakka STPP	BH No.: 1
	GL: 64.3 m
Location: WB	WTL: 63.0 m
Boring Method: Shell and Auger	Supervisor: X
Dia of BH 15 cm	

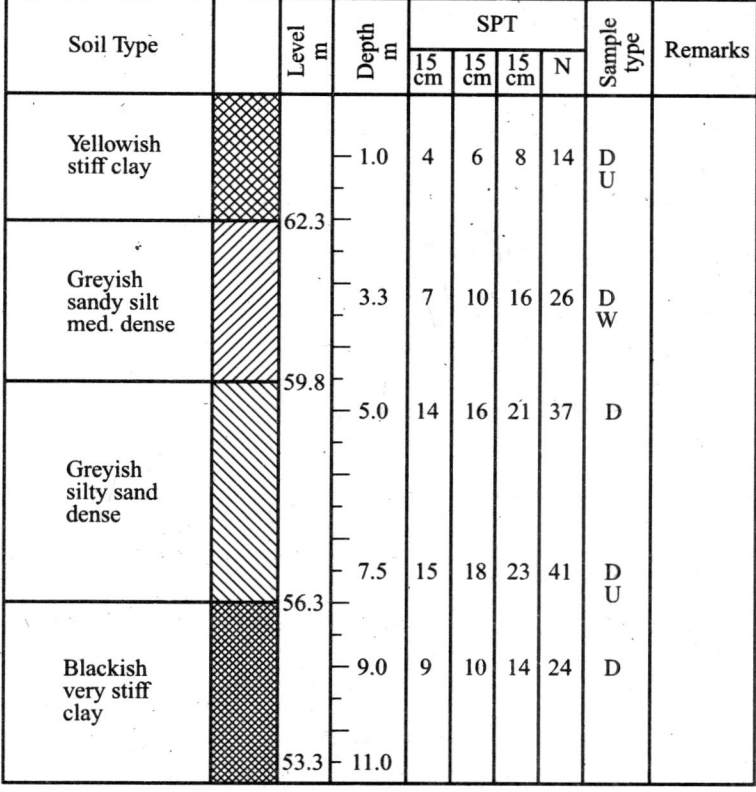

Soil Type		Level m	Depth m	SPT				Sample type	Remarks
				15 cm	15 cm	15 cm	N		
Yellowish stiff clay			1.0	4	6	8	14	D U	
		62.3							
Greyish sandy silt med. dense			3.3	7	10	16	26	D W	
		59.8							
			5.0	14	16	21	37	D	
Greyish silty sand dense									
			7.5	15	18	23	41	D U	
		56.3							
Blackish very stiff clay			9.0	9	10	14	24	D	
		53.3	11.0						

D = disturbed sample; U = undisturbed sample;
W = water sample; N = SPT value

Fig. 17.18 A typical bore-hole log

17.13 QUESTIONS AND PROBLEMS

17.1   Compute the area ratio of a sampling tube given the outside diameter = 100 mm and inside diameter = 94 mm. In what types of soil can this tube be used for sampling?

17.2   A standard penetration test was carried out at a site. The soil profile is given in Fig. Prob. 17.2 with the penetration values. The average soil data are given for each layer. Compute the corrected values of N and plot showing the

(a) variation of observed values with depth

(b) variation of corrected values with depth for standard energy 60%

Assume: $E_h = 0.7$, $C_d = 0.9$, $C_s = 0.85$ and $C_b = 1.05$

17.3   For the soil profile given in Fig. Prob 17.2, compute the corrected values of N for standard energy 70%.

17.4   For the soil profile given in Fig. Prob. 17.2, estimate the average angle of friction for the sand layers based on the following:

(a) Table 17.3

(b) Equation (17.8) by assuming the profile contains less than 5% fines (D_r may be taken from Table 17.3)

Estimate the values of ϕ and D_r for 60 percent standard energy.

Assume: $N_{cor} = N_{60}$.

17.5   For the corrected values of N_{60} given in Prob 17.2, determine the unconfined compressive strengths of clay at points C and D in Fig Prob 17.2 by making use of Table 17.4 and Eq. (17.9a). What is the consistency of the clay?

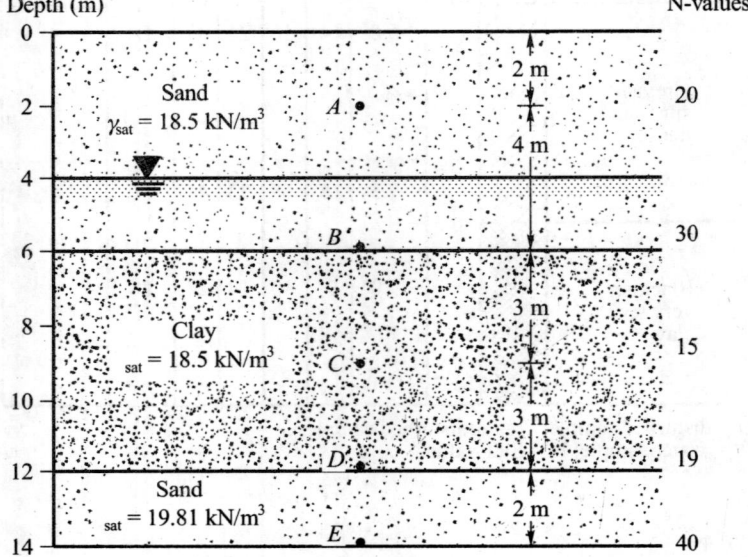

Fig. Prob. 17.2

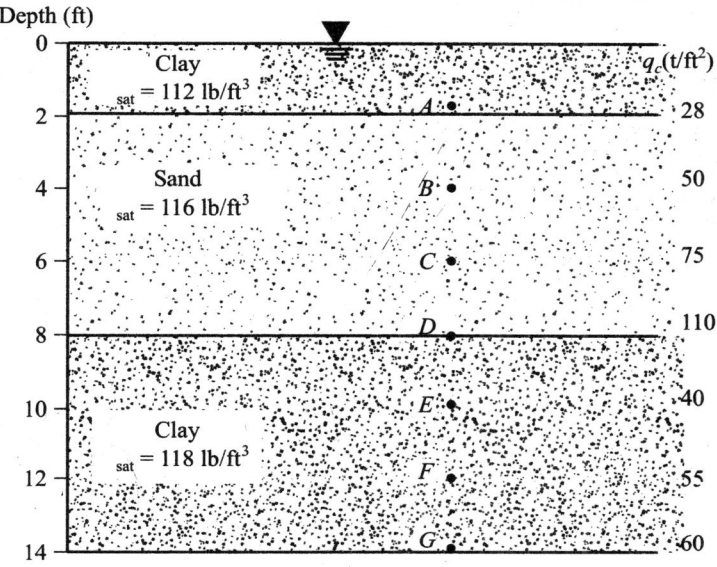

Fig. Prob. 17.6

17.6   A static cone penetration test was carried out at a site using an electric-friction cone penetrometer. Fig. Prob 17.6 gives the soil profile and values of q_c obtained at various depths.

(a)   Plot the variation of q_c with depth.

(b)   Determine the relative density of the sand at the points marked in the figure by using Fig. 17.12a.

(c)   Determine the angle of internal friction of the sand at the points marked by using Fig. 17.12b.

17.7   For the soil profile given in Fig. Prob 17.6, determine the unconfined compressive strength of the clay at the points marked in the figure using Eq. (17.14).

17.8   Determine the relative density and the friction angle if the corrected SPT value N_{60} at a site is 30 from Eq. (17.15) and Table 17.5. What are the values of D_r and ϕ for N_{70} ?

CHAPTER 18

SHALLOW FOUNDATION

18.1 INTRODUCTION

It is the customary practice to regard a foundation as shallow if the depth of the foundation is less than or equal to the width of the foundation. The different types of footings that we normally come across are given in Fig. 18.1.

A foundation is an integral part of a structure. The stability of a structure depends upon the stability of the supporting soil. Two important factors that are to be considered in this chapter are

1. The foundation must be stable against shear failure of the supporting soil.

2. The foundation must not settle beyond a tolerable limit to avoid damage to the structure.

The other factors that require consideration are the location and depth of the foundation. In deciding the location and depth, one has to consider the erosions due to flowing water, underground defects such as root holes, cavities, unconsolidated fills, ground water level, presence of expansive soils, etc.

In selecting a type of foundation, one has to consider the functions of the structure and the load it has to carry, the subsurface condition of the soil, and the cost of the superstructure.

The design loads also play an important part in the selection of the type of foundation. The various loads that are likely to be considered are (i) dead loads, (ii) live loads, (iii) wind and earthquake forces, (iv) lateral pressures exerted by the foundation earth on the embedded structural elements, and (v) the effects of dynamic loads.

In addition to the above loads, the loads that are due to the subsoil conditions are also required to be considered. They are (i) lateral or uplift forces on the foundation elements due to high water table, (ii) swelling pressures on the foundations in expansive soils, (iii) heave pressures on foundations in areas subjected to frost heave and (iv) negative frictional drag on piles where pile foundations are used in highly compressible soils.

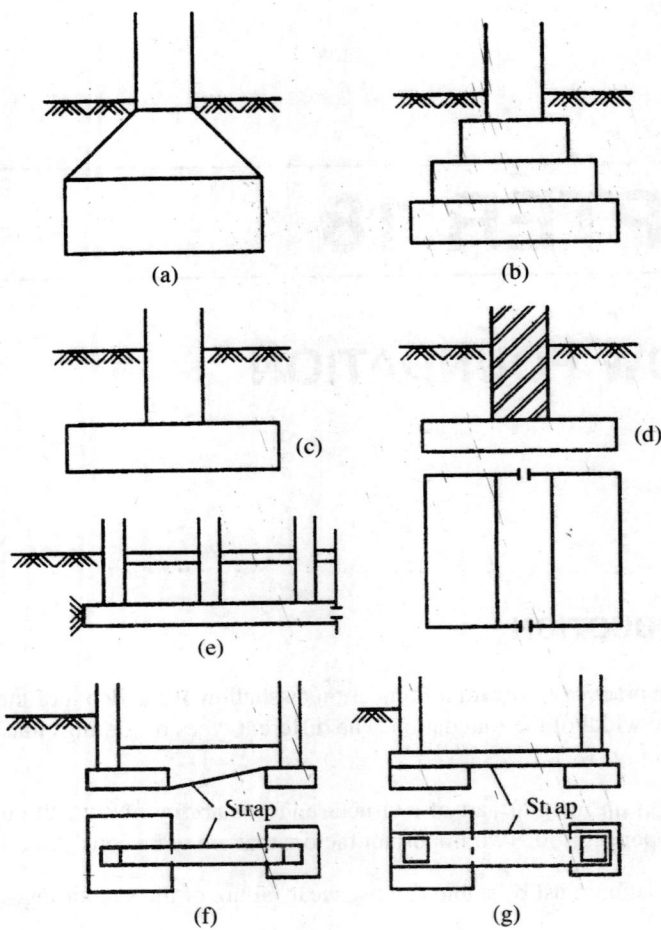

Fig. 18.1   Types of shallow foundations; (a) R.C. Foundation, (b) Stepped R.C. Foundation, (c) Plain reinforced concrete, (d) Plain concrete wall foundation, (e) Mat or raft foundation, and (f, g) Strap foundations

Steps for the Selection of the Type of Foundation

In choosing the type of foundation, the design engineer must perform five successive steps.

1. Obtain the required information concerning the nature of the superstructure and the loads to be transmitted to the foundation.

2. Obtain the subsurface soil conditions.

3. Explore the possibility of constructing any one of the types of foundation under the existing conditions by taking into account (i) the bearing capacity of the soil to carry the required load, and (ii) the adverse effects on the structure due to differential settlements. Eliminate in this way, the unsuitable types.

4.   Once one or two types of foundation are selected on the basis of preliminary studies, make more detailed studies. These studies may require more accurate determination of loads, subsurface conditions and footing sizes. It may also be necessary to make more refined estimates of settlement in order to predict the behavior of the structure.

5.   Estimate the cost of each of the promising types of foundation, and choose the type that represents the most acceptable compromise between performance and cost.

The subject matter of shallow foundations has been considered under three parts as follows.

Part A: Bearing capacity of foundations.

Part B: Safe bearing pressure and settlement of foundations

Part C: Design considerations of strap and combined footings

PART A: BEARING CAPACITY OF FOUNDATIONS

18.2   ULTIMATE BEARING CAPACITY OF SOIL

Consider the simplest case of a shallow foundation subjected to a central vertical load. The footing is founded at a depth D_f below the ground surface Fig. 18.2(a). If the settlement, S, of the footing is recorded against the applied load, Q, load-settlement curves, similar in shape to a stress-strain curve, may be obtained as shown in Fig. 18.2(b).

The shape of the curve depends generally on the size and shape of the footing, the composition of the supporting soil, and the character, rate, and frequency of loading. Normally a curve will indicate the ultimate load Q_d that the foundation can support. If the foundation soil is a dense sand or a very

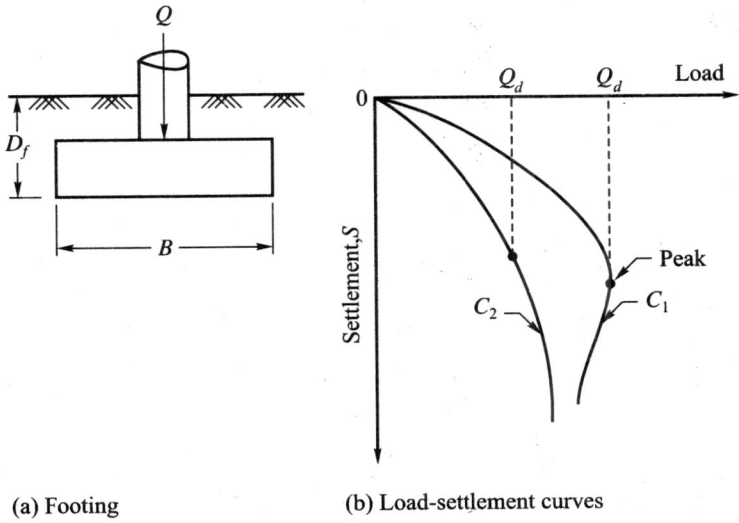

(a) Footing   (b) Load-settlement curves

Fig. 18.2   Typical load settlement curves

stiff clay, the curve passes fairly abruptly to a peak value and then drops down as shown by curve C_1 in Fig. 18.2(b). The peak load Q_d is quite pronounced in this case. On the other hand, if the soil is loose sand or soft clay, the settlement curve continues to descend on a slope as shown by curve C_2 which shows that the compression of soil is continuously taking place without giving a definite value for Q_d. On such a curve, Q_d may be taken at a point beyond which there is a constant rate of penetration.

18.3   SOME OF THE TERMS DEFINED

It will be useful to define, at this stage, some of the terms relating to bearing capacity of foundations.

(a) Total Overburden Pressure q

q_0 is the intensity of total overburden pressure due to the weight of both soil and water at the base level of the foundation.

$$q = \gamma D_{w1} + \gamma_{sat} \overline{D}_w \qquad (18.1)$$

(b) Effective Overburden Pressure q_0

q_0 is the effective overburden pressure at the base level of the foundation.

$$q_0 = \gamma D_{w1} + \gamma_b \overline{D}_w \qquad (18.2)$$

when $\overline{D}_w = 0, q_0 = \gamma D_{w1} = \gamma D_f$.

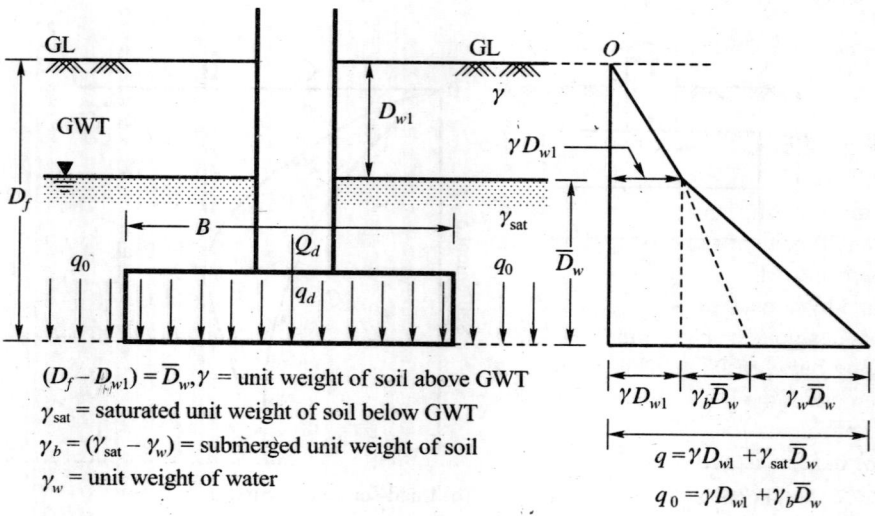

$(D_f - D_{w1}) = \overline{D}_w, \gamma = $ unit weight of soil above GWT
$\gamma_{sat} = $ saturated unit weight of soil below GWT
$\gamma_b = (\gamma_{sat} - \gamma_w) = $ submerged unit weight of soil
$\gamma_w = $ unit weight of water

$q = \gamma D_{w1} + \gamma_{sat} \overline{D}_w$
$q_0 = \gamma D_{w1} + \gamma_b \overline{D}_w$

Fig. 18.3   Total and effective over burden pressures

(c) The Ultimate Bearing Capacity of Soil, q_d

q_d is the maximum bearing capacity of soil at which the soil fails by shear.

(d) The Net Ultimate Bearing Capacity, q_{nd}

q_{nd} is the bearing capacity in excess of the effective overburden pressure q_0, expressed as

$$q_{nd} = q_d - q_0 \tag{18.3}$$

(e) Gross Allowable Bearing Pressure, q_a

q_a is expressed as

$$q_a = \frac{q_d}{F_s} \tag{18.4}$$

where F_s = factor of safety.

(f) Net Allowable Bearing Pressure, q_{na}

q_{na} is expressed as

$$q_{na} = \frac{q_d - \gamma D_f}{F_s} = \frac{q_{na}}{F} \tag{18.5}$$

(g) Safe Bearing Pressure, q_s

q_s is defined as the net safe bearing pressure which produces a settlement of the foundation which does not exceed a permissible limit.

Note: In the design of foundations, one has to use the least of the two values of q_{na} and q_s.

18.4 TYPES OF FAILURE IN SOIL

Experimental investigations have indicated that foundations on dense sand with relative density greater than 70 percent fail suddenly with pronounced peak resistance when the settlement reaches about 7 percent of the foundation width. The failure is accompanied by the appearance of failure surfaces and by considerable bulging of a sheared mass of sand as shown in Fig. 18.4(a). This type of failure is designated as general shear failure by Terzaghi (1943). Foundations on sand of relative density lying between 35 and 70 percent do not show a sudden failure. As the settlement exceeds about 8 percent of the foundation width, bulging of sand starts at the surface. At settlements of about 15 percent of foundation width, a visible boundary of sheared zones at the surface appears. However, the peak of base resistance may never be reached. This type of failure is termed local shear failure, Fig. 18.4(b), by Terzaghi (1943).

Foundations on relatively loose sand with relative density less than 35 percent penetrate into the soil without any bulging of the sand surface. The base resistance gradually increases as settlement progresses. The rate of settlement, however, increases and reaches a maximum at a settlement of about 15 to 20 percent of the foundation width. Sudden jerks or shears can be observed as soon as the settlement reaches about 6 to 8 percent of the foundation width. The failure surface, which is vertical or slightly inclined and follows the perimeter of the base, never reaches the sand surface. This type of failure is designated as punching shear failure by Vesic (1963) as shown in Fig. 18.4(c).

The three types of failure described above were observed by Vesic (1963) during tests on model footings. It may be noted here that as the relative depth/width ratio increases, the limiting relative densities at which failure types change increase. The approximate limits of types of failure to be affected as relative depth D_f/B, and relative density of sand, D_r, vary are shown in Fig. 18.5 (Vesic, 1963). The same figure shows that there is a critical relative depth below which only punching shear failure occurs. For circular foundations, this critical relative depth, D_f/B, is around 4 and for long rectangular foundations around 8.

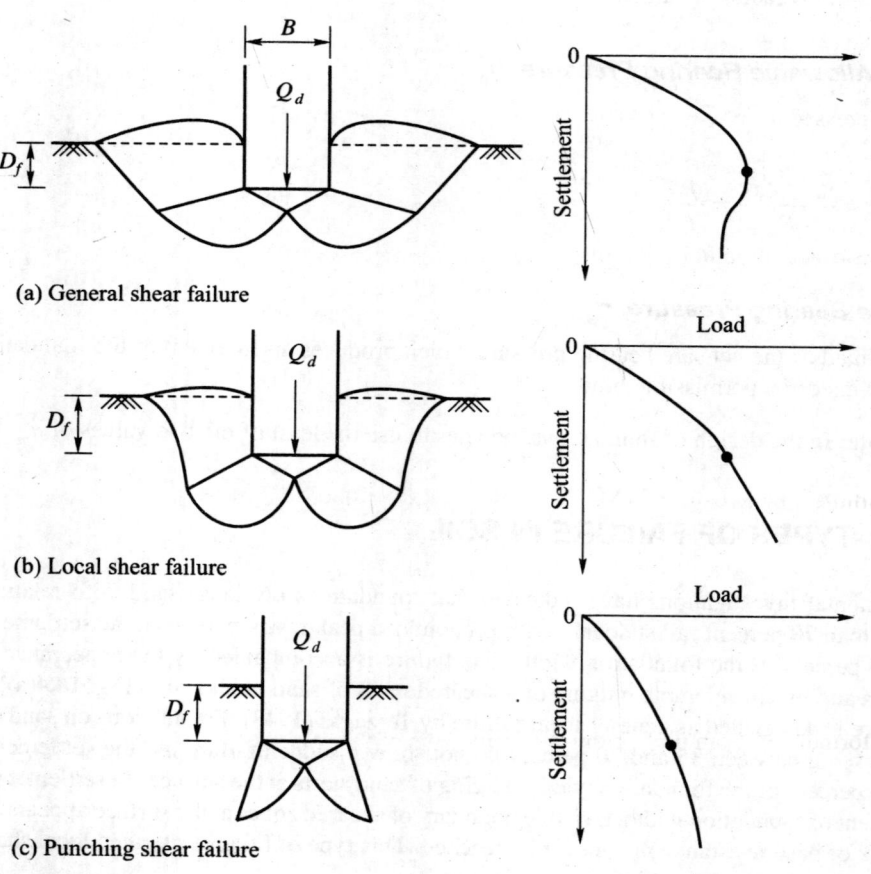

(a) General shear failure

(b) Local shear failure

(c) Punching shear failure

Fig. 18.4   Modes of bearing capacity failure (Vesic, 1963)

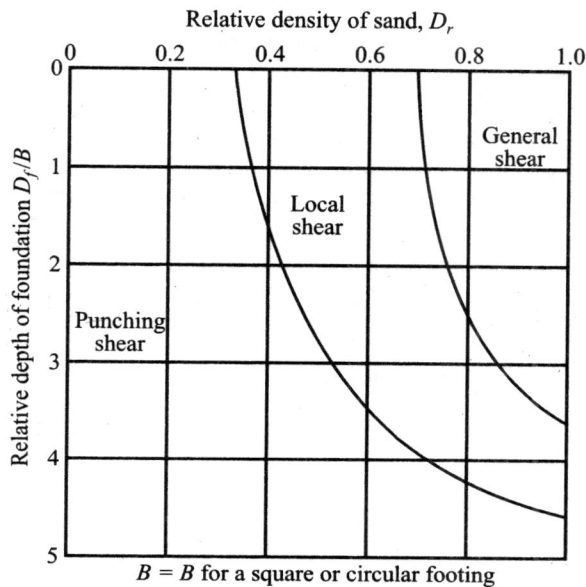

Fig. 18.5 Modes of failure of model footings in sand (After Vesic, 1963)

The surfaces of failures as observed by Vesic are for concentric vertical loads. Any small amount of eccentricity in the load application changes the modes of failure and the foundation tilts in the direction of eccentricity. Tilting nearly always occurs in cases of foundation failures because of the inevitable variation in the shear strength and compressibility of the soil from one point to another and causes greater yielding on one side or another of the foundation. This throws the center of gravity of the load towards the side where yielding has occurred, thus increasing the intensity of pressure on this side followed by further tilting.

Footings founded on precompressed clays or saturated normally consolidated clays will fail in general shear if it is loaded so that no volume change can take place and fails by punching shear if the footing is founded on soft clays.

Methods of Determining Ultimate Bearing Capacity of Soil

The ultimate bearing capacity of soil can be calculated either from some of the bearing capacity theories formulated from time to time or from some of the in-situ tests.

The methods that are considered in this chapter are:

1. General shear failure theory of Terzaghi

2. Theoritical solutions presented by Meyerhof, Brinc Hansen and Vesic

3. Solutions based on in-situ tests soils as PLT, SPT and CPT.

18.5   TERZAGHI'S METHOD OF DETERMINING BEARING CAPACITY OF SOIL

Ultimate Bearing Capacity, q_d, for a Strip Footing under General Shear Failure Criteria

Introduction

A theoritical form of equation for determining the bearing capacity of soil was first established by Prandtl (1920). He developed his theory based on the penetration of long hard metal puncher into softer material which he assumed as homogeneous, isotrophic and weightless, possessing only cohesion and friction. The form of equation he propounded did not take into account (a) the weight of the soil, and (b) the effect of soil above the base of the footing on the bearing capacity. He considered the base of the footing as smooth.

Terzaghi (1943) used the same form of equation as proposed by Prandtl (1921) and extended his theory to take into account the weight of soil and the effect of soil above the base of the foundation on the bearing capacity of soil. Terzaghi made the following assumptions for developing an equation for determining q_d, for a c-ϕ soil.

(1) The soil is semi-infinite, homogeneous and isotropic, (2) the problem is two-dimensional, (3) the base of the footing is rough, (4) the failure is by general shear, (5) the load is vertical and symmetrical, (6) the ground surface is horizontal, (7) the overburden pressure at foundation level is equivalent to a surcharge load $q_0 = \gamma D_f$, where γ is the effective unit weight of soil, and D_f, the depth of foundation less than the width B of the foundation, (8) the principle of superposition is valid, and (9) Coulomb's law is strictly valid, that is, $\sigma = c + \sigma \tan \phi$.

Mechanism of Failure

The shapes of the failure surfaces under ultimate loading conditions are given in Fig. 18.6. The zones of plastic equilibrium represented in this figure by the area $gedcf$ may be subdivided into

1.   Zone I of elastic equilibrium
2.   Zones II of radial shear state
3.   Zones III of Rankine passive state

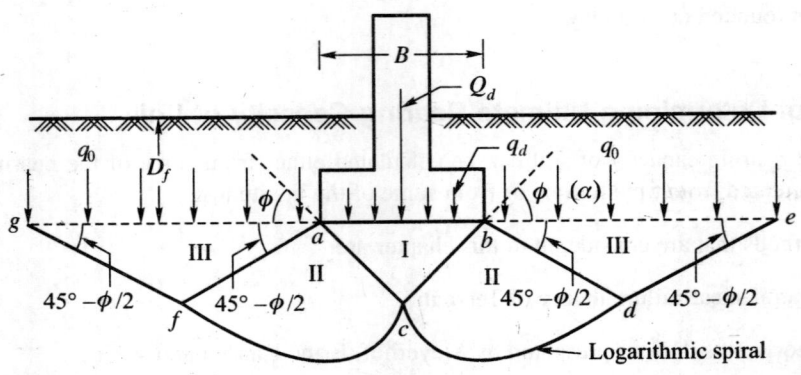

Fig. 18.6   General shear failure surface as assumed by Terzaghi for a strip footing

When load q_d per unit area acting on the base of the footing of width B with a rough base is transmitted into the soil, the tendency of the soil located within zone I is to spread but this is counteracted by friction and adhesion between the soil and the base of the footing. Due to the existence of this resistance against lateral spreading, the soil located immediately beneath the base remains permanently in a state of elastic equilibrium, and the soil located within this central Zone I behaves as if it were a part of the footing and sinks with the footing under the superimposed load. The depth of this wedge shaped body of soil abc remains practically unchanged, yet the footing sinks. This process is only conceivable if the soil located just below point c moves vertically downwards. This type of movement requires that the surface of sliding cd (Fig. 18.6) through point c should start from a vertical tangent. The boundary bc of the zone of radial shear bcd (Zone II) is also the surface of sliding. As per the theory of plasticity, the potential surfaces of sliding in an ideal plastic material intersect each other in every point of the zone of plastic equilibrium at an angle $(90° - \phi)$. Therefore the boundary bc must rise at an angle ϕ to the horizontal provided the friction and adhesion between the soil and the base of the footing suffice to prevent a sliding motion at the base.

The sinking of Zone I creates two zones of plastic equilibrium, II and III, on either side of the footing. Zone II is the radial shear zone whose remote boundaries bd and af meet the horizontal surface at angles $(45° - \phi/2)$, whereas Zone III is a passive Rankine zone. The boundaries de and fg of these zones are straight lines and they meet the surface at angles of $(45° - \phi/2)$. The curved parts cd and cf in Zone II are parts of logarithmic spirals whose centers are located at b and a respectively.

Ultimate Bearing Capacity of Soil

Strip Footings

Terzaghi developed his bearing capacity equation for strip footings by analyzing the forces acting on the wedge abc in Fig. 18.6. The theoritical analysis of the forces acting on the wedge is given in the Appendix. The equation for the ultimate bearing capacity q_d is

$$q_d = \frac{Q_d}{B} = cN_c + \gamma D_f N_q + \frac{1}{2}\gamma B N_\gamma \qquad (18.6)$$

where Q_d = ultimate load per unit length of footing, c = unit cohesion, γ the effective unit weight of soil, B = width of footing, D_f = depth of foundation, N_c, N_q and N_γ are the bearing capacity factors. They are functions of the angle of friction, ϕ.

The bearing capacity factors are expressed by the following equations

$$\left.\begin{array}{l} N_c = \left(N_q - 1\right)\cot\phi \\[2mm] N_q = \dfrac{a_\theta^2}{2\cos^2\left(45° + \phi/2\right)} \\[2mm] \text{where } a_\theta = \varepsilon^{\eta\tan\phi}, \eta = \left(0.75\pi - \phi/2\right) \\[2mm] N_\gamma = \dfrac{1}{2}\tan\phi\left(\dfrac{K_{p\gamma}}{\cos^2\phi} - 1\right) \end{array}\right\} \qquad (18.7)$$

where $K_{p\gamma}$ = passive earth pressure coefficient.

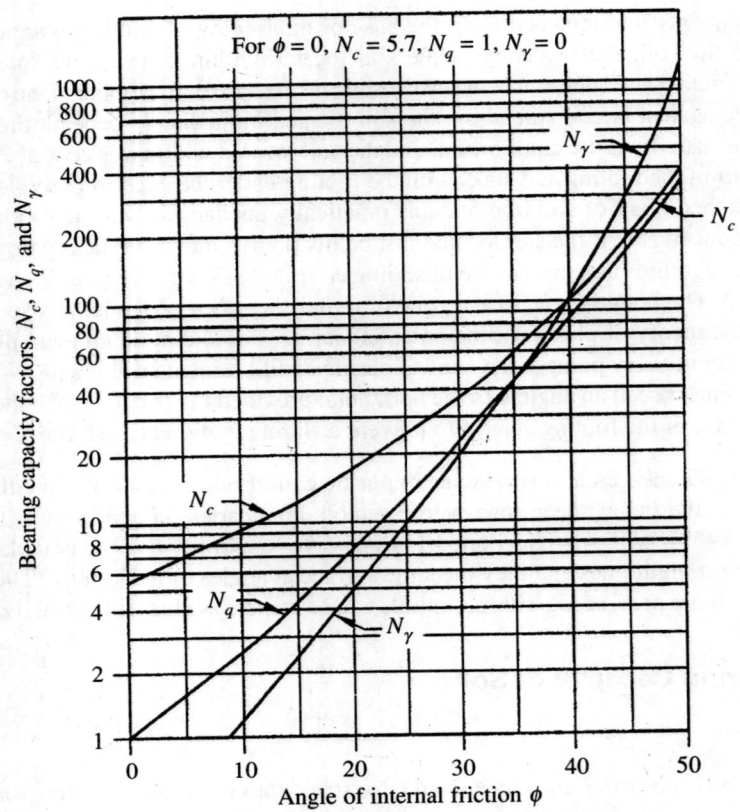

Fig. 18.7 Terzaghi's bearing capacity factors for general shear failure

Table 18.1 gives the values of N_c, N_q and N_γ for various values of ϕ and Fig. 18.7 gives the same in a graphical form.

TABLE 18.1
Bearing capacity factors of Terzaghi

ϕ	N_c	N_q	N_γ
0	5.7	1.0	0.0
5	7.3	1.6	0.5
10	9.6	2.7	1.2
15	12.9	4.4	2.5
20	17.7	7.4	5.0
25	25.1	12.7	9.7
30	37.2	22.5	19.7
35	57.8	41.4	42.4
40	95.7	81.3	100.4
45	172.3	173.3	297.5
50	347.5	415.1	1153.0

Equations for Square, Circular, and Rectangular Foundations.

Terzaghi's bearing capacity Eq. (18.6) has been modified for other types of foundations by introducing the shape factors. The equations are

Square Foundations

$$q_d = 1.3cN_c + \gamma D_f N_q + 0.4\gamma BN_\gamma \qquad (18.8)$$

Circular Foundations

$$q_d = 1.3cN_c + \gamma D_f N_q + 0.3\gamma BN_\gamma \qquad (18.9)$$

Rectangular Foundations

$$q_d = cN_c\left(1+0.3\times\frac{B}{L}\right) + \gamma D_f N_q + \frac{1}{2}\gamma BN_\gamma\left(1-0.2\times\frac{B}{L}\right) \qquad (18.10)$$

where B = width or diameter, L = length of footing.

Ultimate Bearing Capacity for Local Shear Failure

The reasons as to why a soil fails under local shear have been explained under Section 18.4. When a soil fails by local shear, the actual shear parameters c and ϕ are to be reduced as per Terzaghi (1943). The lower limiting values of c and ϕ are

$$\bar{c} = 0.67c$$

and $\tan\bar{\phi} = 0.67\tan\phi$ or $\bar{\phi} = \tan^{-1}(0.67\tan\phi)$

The equations for the lower bound values for the various types of footings are as given below.

Strip Foundation

$$q_d = 0.67c\bar{N}_c + \gamma D_f \bar{N}_q + \frac{1}{2}\gamma B\bar{N}_\gamma \qquad (18.11)$$

Square Foundation

$$q_d = 0.867c\bar{N}_c + \gamma D_f \bar{N}_q + 0.4\gamma B\bar{N}_\gamma \qquad (18.12)$$

Circular Foundation

$$q_d = 0.867c\bar{N}_c + \gamma D_f \bar{N}_q + 0.3\gamma B\bar{N}_\gamma \qquad (18.13)$$

Rectangular Foundation

$$q_d = 0.67c\left(1+0.3\times\frac{B}{L}\right)\overline{N}_c + \gamma D_f\overline{N}_q + \frac{1}{2}\gamma B\overline{N}_\gamma\left(1-0.2\times\frac{B}{L}\right) \qquad (18.14)$$

where $\overline{N}_c$, $\overline{N}_q$ and $\overline{N}_\gamma$ are the reduced bearing capacity factors for local shear failure. These factors may be obtained either from Table 18.1 or Fig. 18.7 by making use of the friction angle $\overline{\phi}$.

Ultimate Bearing Capacity q_d in Purely Cohesionless and Cohesive Soils Under General Shear Failure Criteria

Equations for the various types of footings for $(c - \phi)$ soil under general shear failure have been given earlier. The same equations can be modified to give equations for cohesionless soil (for $c = 0$) and cohesive soils (for $\phi = 0$) as follows.

It may be noted here that for $c = 0$, the value of $N_c = 0$, and for $\phi = 0$, the value of $N_c = 5.7$ for a strip footing and $N_q \doteq 1$.

(a) Strip Footing

$$\text{For } c = 0, \quad q_d = \gamma D_f N_q + \frac{1}{2}\gamma B N_\gamma$$

$$\text{For } \phi = 0, \quad q_d = 5.7c + \gamma D_f \qquad (18.15)$$

(b) Square Footing

$$\text{For } c = 0, \quad q_d = \gamma D_f N_q + 0.4\gamma B N_\gamma$$

$$\text{For } \phi = 0, \quad q_d = 7.4c + \gamma D_f \qquad (18.16)$$

(c) Circular Footing

$$\text{For } c = 0, \quad q_d = \gamma D_f N_q + 0.3\gamma B N_\gamma$$

$$\text{For } \phi = 0, \quad q_d = 7.4c + \gamma D_f \qquad (18.17)$$

(d) Rectangular Footing

$$\text{For } c = 0, \quad q_d = \gamma D_f N_q + \frac{1}{2}\gamma B N_\gamma\left(1-0.2\times\frac{B}{L}\right)$$

$$\text{For } \phi = 0, \quad q_d = 5.7c\left(1+0.3\times\frac{B}{L}\right) + \gamma D_f \qquad (18.18)$$

Similar types of equations as presented for general shear failure can be developed for local shear failure also.

Transition from Local to General Shear Failure in Sand

As already explained, local shear failure normally occurs in loose and general shear failure occurs in dense sand. There is a transition from local to general shear failure as the state of sand changes from

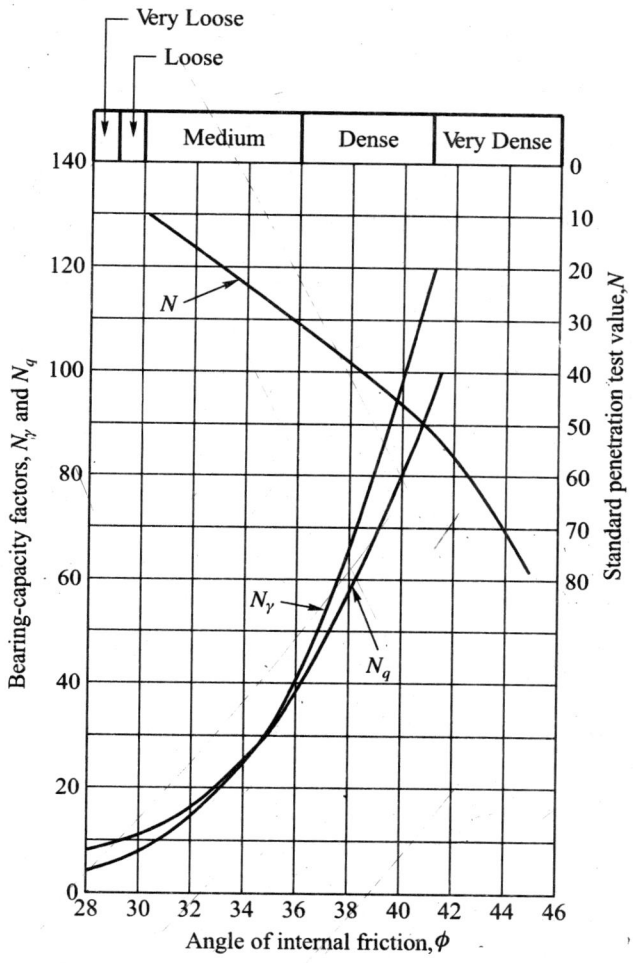

Fig. 18.8 Terzaghi's bearing capacity factors which take care of mixed state of local and general shear failures in sand

loose to dense condition. There is no bearing capacity equation to account for this transition from loose to dense state. Peck *et al.*, (1974) have given curves for N_γ and N_q which automatically incorporate allowance for the mixed state of local and general shear failures as shown in Fig. 18.8.

The curves for N_q and N_γ are developed on the following assumptions.

1. Purely local shear failure occurs when $\phi \leq 28°$.

2. Purely general shear failure occurs when $\phi \geq 38°$.

3. Smooth transition curves for values of ϕ between $28°$ and $38°$ represent the mixed state of local and general shear failures.

N_q and N_γ for values of $\phi \geq 38°$ are as given in Table 18.1. Values of $\bar{N}_q$ and $\bar{N}_\gamma$ for $\phi \leq 28°$ may be obtained from Table 18.1 by making use of the relationship $\bar{\phi} = \tan^{-1} (2/3) \tan \phi$.

In the case of purely cohesive soil local shear failure may be assumed to occur in soft to medium stiff clay with an unconfined compressive strength $q_u \leq 100$ kPa.

Figure 18.8 also gives the relationship between SPT value N and the angle of internal friction ϕ by means of a curve. This curve is useful to obtain the value of ϕ when the SPT value is known.

Net Ultimate Bearing Capacity and Safety Factor

The net ultimate bearing capacity q_{nd} is defined as the pressure at the base level of the foundation in excess of the effective overburden pressure $q_0 = \gamma D_f$ as defined in Eq. (18.3). The net q_{nd} for a strip footing is

$$q_{nd} = (q_d - \gamma D_f) = cN_c + \gamma D_f (N_q - 1) + \frac{1}{2}\gamma B N_\gamma \qquad (18.19)$$

Similar expressions can be written for square, circular, and rectangular foundations and also for local shear failure conditions.

Allowable Bearing Pressure

As per Eq. (18.4), the gross allowable bearing pressure is

$$q_a = \frac{q_a}{F_s} \qquad (18.20a)$$

In the same way the net allowable bearing pressure q_{na} is

$$q_{na} = \frac{q_d - \gamma D_f}{F_s} = \frac{q_{nd}}{F_s} \qquad (18.20b)$$

where F_s = factor of safety which is normally assumed as equal to 3.

18.6   SKEMPTON'S BEARING CAPACITY FACTOR N_C

For saturated clay soils, Skempton (1951) proposed the following equation for a strip foundation

$$q_d = cN_c + \gamma D_f \qquad (18.21a)$$

or $\quad q_{nd} = q_d - \gamma D_f = cN_c = \frac{q_u}{2} N_c \qquad (18.21b)$

where q_u = unconfined compressive strength of clay.

If we assume, $F_s = 3, \bar{N_c} = 6$ for all practical purposes, q_{na} for strip footing may be written as

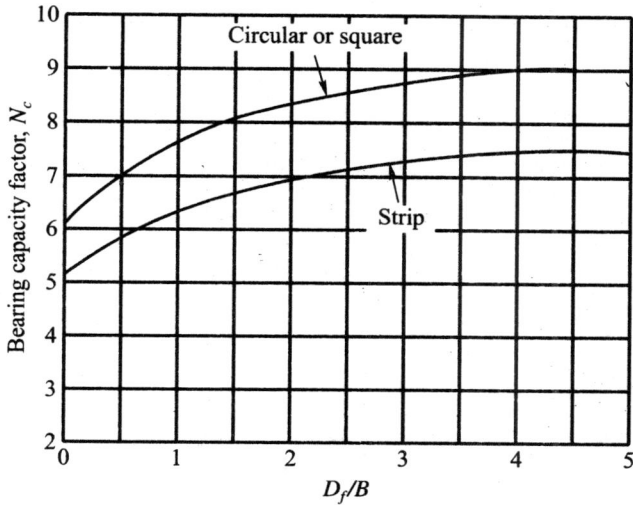

Fig. 18.9 Skempton's bearing capacity factor N_c for clay soils

$$q_{na} = \frac{q_{nd}}{F_s} = \frac{q_u \times 6}{2 \times 3} = q_u \qquad (18.21c)$$

The N_c values for strip and square (or circular) foundations as a function of the D_f/B ratio are given in Fig. 18.9. The equation for rectangular foundation may be written as follows

$$(N_c)_R = \left(0.84 + 0.16 \times \frac{B}{L}\right)(N_c)_S \qquad (18.21d)$$

where $(N_c)_R = N_c$ for rectangular foundation, $(N_c)_S = N_c$ for square foundation.

The lower and upper limiting values of N_c for strip and square foundations may be written as follows:

Type of foundation	Ratio D_f/B	Value of N_c
Strip	0	5.14
	≥4	7.5
Square	0	6.2
	≥4	9.0

18.7 EFFECT OF WATER TABLE ON BEARING CAPACITY

The theoretical equations developed for computing the ultimate bearing capacity q_d of soil are based on the assumption that the water table lies at a depth below the base of the foundation equal to or greater than the width B of the foundation or otherwise the depth of the water table from ground surface is equal to or greater than $(D_f + B)$. In case the water table lies at any intermediate depth less than the depth $(D_f + B)$, the bearing capacity equations are affected due to the presence of the water table. (Refer Fig. 18.10)

Two cases may be considered here.

Case 1. When the water table lies above the base of the foundation.

Case 2. When the water table lies within depth B below the base of the foundation.

We will consider the two methods for determining the effect of the water table on bearing capacity as given below.

Method 1

For any position of the water table within the depth $(D_f + B)$, we may write Eq. (18.6) as

$$q_d = cN_c + \gamma D_f N_q R_{w1} + \frac{1}{2} \gamma B N_\gamma R_{w2} \tag{18.22}$$

where R_{w1} = reduction factor for water table above the base level of the foundation,

R_{w2} = reduction factor for water table below the base level of the foundation,

$\gamma = \gamma_{sat}$ for all practical purposes in both the second and third terms of Eq. (18.22).

Case 1: When the water table lies above the base level of the foundation or when $D_{w1}/D_f \leq 1$

The equation for R_{w1} may be written as

$$R_{w1} = \frac{1}{2}\left(1 + \frac{D_{wl}}{D_f}\right) \tag{18.23a}$$

For $D_{w1}/D_f = 0$, we have $R_{w1} = 0.5$, and for $D_{w1}/D_f = 1.0$, we have $R_{w1} = 1.0$.

Case 2: When the water table lies below the base level or when $D_{w2}/B \leq 1$

The equation for R_{w2} is

$$R_{w2} = \frac{1}{2}\left(1 + \frac{D_{w2}}{B}\right) \tag{18.23b}$$

For $D_{w2}/B = 0$, we have $R_{w2} = 0.5$, and for $D_{w2}/B = 1.0$, we have $R_{w2} = 1.0$.

Figure 18.10 shows in a graphical form the relations D_{w1}/D_f vs R_{w1} and D_{w2}/B vs R_{w2}.

Equations (18.23a) and (18.23b) are based on the assumption that the submerged unit weight of soil is equal to half of the saturated unit weight and the soil above the water table remains saturated.

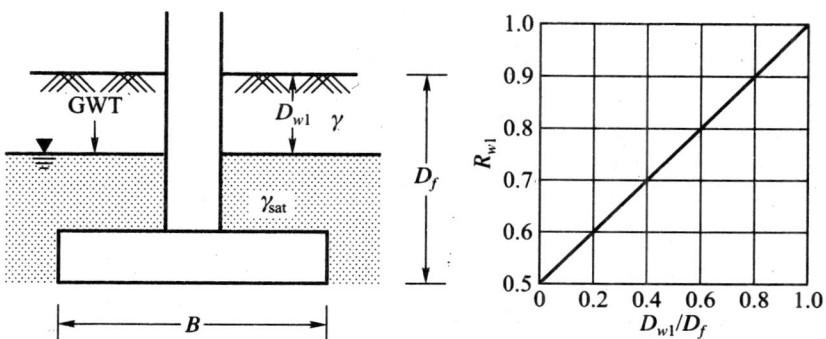

(a) Water table above base level of foundation

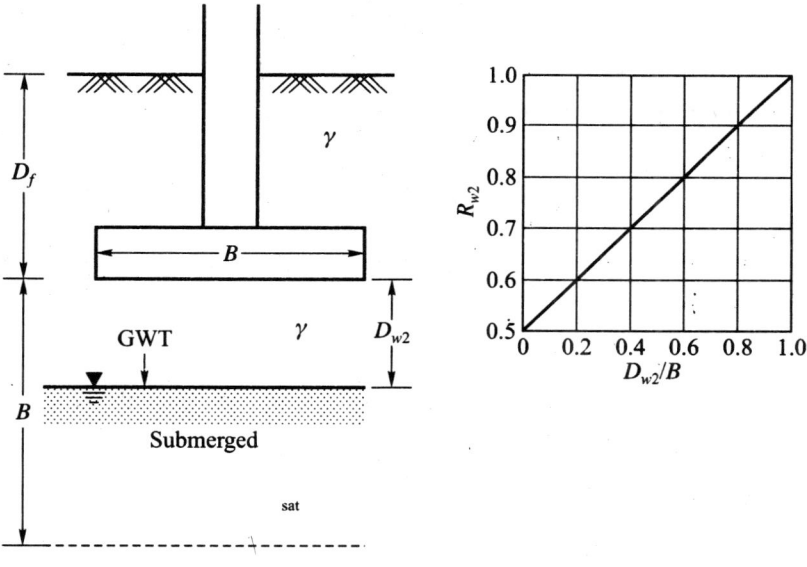

(b) Water table below base level of foundation

Fig. 18.10 Effect of WT on bearing capacity

Method 2: Equivalent effective unit weight method

Equation (18.6) for the strip footing may be expressed as

$$q_d = cN_c + \gamma_{e1}D_f N_q + \frac{1}{2}\gamma_{e2}BN_\gamma \tag{18.24}$$

where γ_{e1} = weighted effective unit weight of soil lying above the base level of the foundation

γ_{e2} = weighted effective unit weight of soil lying within the depth B below the base level of the foundation

γ_m = moist or saturated unit weight of soil lying above WT (case 1 or case 2)

γ_{sat} = saturated unit weight of soil below the WT (case 1 or case 2)

γ_b = submerged unit weight of soil = $\gamma_{sat} - \gamma_w$

Case 1

An equation for γ_{e1} may be written as

$$\gamma_{e1} = \gamma_b + \frac{D_{w1}}{D_f}(\gamma_m - \gamma_b)$$ (18.25a)

$$\gamma_{e2} = \gamma_b$$

Case 2

$$\gamma_{e1} = \gamma_m$$

$$\gamma_{e2} = \gamma_b + \frac{D_{w2}}{B}(\gamma_m - \gamma_b)$$ (18.25b)

Example 18.1

A strip footing of width 3 m is founded at a depth of 2 m below the ground surface in a $(c - \phi)$ soil having a cohesion c = 30 kN/m² and angle of shearing resistance ϕ = 35°. The water table is at a depth of 5 m below ground level. The moist weight of soil above the water table is 17.25 kN/m³. Determine (a) the ultimate bearing capacity of the soil, (b) the net bearing capacity, and (c) the net allowable bearing pressure and the load/m for a factor of safety of 3. Use the general shear failure theory of Terzaghi.

Solution

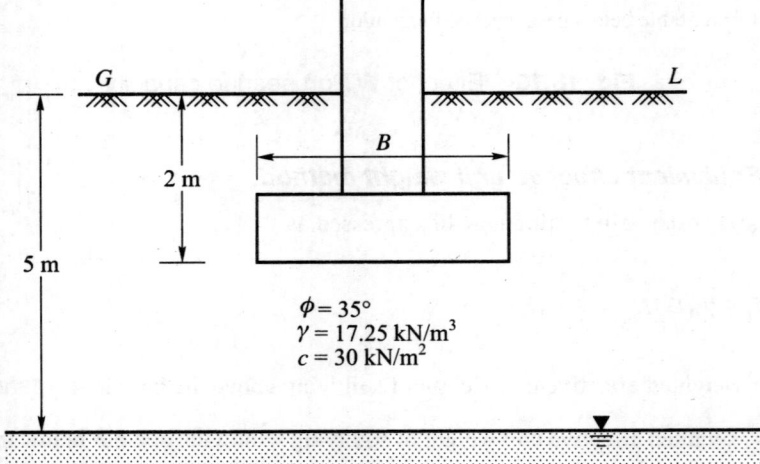

Fig. Ex. 18.1

For $\phi = 35°$, $N_c = 57.8$, $N_q = 41.4$, and $N_\gamma = 42.4$

From Eq. (18.6)

$$q_d = cN_c + \gamma D_f N_q + \frac{1}{2}\gamma B N_\gamma$$

$$= 30 \times 57.8 + 17.25 \times 2 \times 41.4 + \frac{1}{2} \times 17.25 \times 3 \times 42.4$$

$$= 4259 \text{ kN/m}^2$$

$$q_{nd} = q_d - \gamma D_f = 4259 - 17.25 \times 2 \approx 4225 \text{ kN/m}^2$$

$$q_{na} = \frac{q_{nd}}{F_s} = \frac{4225}{3} \approx 1408 \text{ kN/m}^2$$

$$Q_a = q_{na} B = 1408 \times 3 = 4224 \text{ kN/m}$$

Example 18.2

If the soil in Ex. 18.1 fails by local shear failure, determine the net safe bearing pressure. All the other data given in Ex. 18.1 remain the same.

Solution

For local shear failure:

$$\bar{\phi} = \tan^{-1} 0.67 \tan 35° = 25°,$$

$$\bar{c} = 0.67c = 0.67 \times 30 = 20 \text{ kN/m}^2$$

From Table 18.1, for $\bar{\phi} = 25°$, $\bar{N}_c = 25.1$, $\bar{N}_q = 12.7$, $\bar{N}_\gamma = 9.7$

Now from Eq. (18.11)

$$q_d = 20 \times 25.1 + 17.25 \times 2 \times 12.7 + \frac{1}{2} \times 17.25 \times 3 \times 9.7 = 1191 \text{ kN/m}^2$$

$$q_{nd} = 1191 - 17.25 \times 2 = 1156.5 \text{ kN/m}^2$$

$$q_{na} = \frac{1156.50}{3} = 385.5 \text{ kN/m}^2$$

$$Q_a = 385.5 \times 3 = 1156.5 \text{ kN/m}$$

Example 18.3

If the water table in Ex. 18.1 rises to the ground level, determine the net safe bearing pressure of the footing. All the other data given in Ex. 18.1 remain the same. Assume the saturated unit weight of the soil $\gamma_{sat} = 18.5$ kN/m³.

Solution

When the WT is at ground level we have to use the submerged unit weight of the soil.

Therefore, $\gamma_b = \gamma_{sat} - \gamma_w = 18.5 - 9.81 = 8.69$ kN/m³

$$q_{nd} = 30 \times 57.8 + 8.69 \times 2(41.4 - 1) + \frac{1}{2} \times 48.69 \times 3 \times 42.4 \approx 2992 \text{ kN/m}^2$$

$$q_{na} = \frac{2992}{3} = 997.33 \text{ kN/m}^2$$

$$Q_a = 997.33 \times 3 = 2992 \text{ kN/m}$$

Example 18.4

If the water table in Ex. 18.1 occupies any of the positions (a) 1.25 m below ground level or (b) 1.25 m below the base level of the foundation, what will be the net safe bearing pressure?

Assume $\gamma_{sat} = 18.5$ kN/m³, γ(above WT) = 17.5 kN/m³. All the other data remain the same as given in Ex. 18.1.

Solution

Method 1: By making use of reduction factors R_{w1} and R_{w2} and using Eqs. (18.19) and (12.22), we may write

$$q_{nd} = cN_c + \gamma D_f (N_q - 1) R_{w1} + \frac{1}{2} \gamma B N_\gamma R_{w2}$$

Case 1: When the WT is 1.25 m below the GL

From Eq. (18.23), we get $R_{w1} = 0.818$ for $D_{w1}/D_f = 0.625$, $R_{w2} = 0$ for $D_{w2}/B = 0$.

By substituting the known values in the equation for q_{nd}, we have

$$q_{nd} = 30 \times 57.8 + 18.5 \times 2 \times 40.4 \times 0.818 + \frac{1}{2} \times 18.5 \times 3 \times 42.5 \times 0.5 = 3547 \text{ kN/m}^2$$

$$q_{na} = \frac{3547}{3} = 1182 \text{ kN/m}^2$$

Case 2: When the WT is 1.25 m below the base of the foundation

$R_{w1} = 1.0$ for $D_{w1} / D_f = 1, R_{w2} = 0.71$ for $D_{w2} / B = 0.42$.

Now

$$q_{nd} = 30 \times 57.8 + 18.5 \times 2 \times 40.4 \times 1 \times \frac{1}{2} \times 18.5 \times 3 \times 42.5 \times 0.71 = 4029 \text{ kN/m}^2$$

$$q_{na} = \frac{4029}{3} = 1343 \text{ kN/m}^2$$

Method 2: By equivalent effective unit weight, submerged unit weight $\gamma_b = 18.5 - 9.81 = 8.69 \text{ kN/m}^3$.

As per Eq. (18.24)

$$q_{nd} = cN_c + \gamma_{e1}D_f\left(N_q - 1\right) + \frac{1}{2}\gamma_{e2}BN_\gamma$$

Case 1: When D_{w1} = 1.25 m (Fig. Ex. 18.4)

From Eq. (18.25a)

$$\gamma_{e1} = \gamma_b + \frac{D_{w1}}{D_f}\left(\gamma - \gamma_b\right)$$

where $\gamma = \gamma_{sat} = 18.5 \text{ kN/m}^3$

$$\gamma_{e1} = 8.69 + \frac{1.25}{2}\left(18.5 - 8.69\right) = 14.82 \text{ kN/m}^3$$

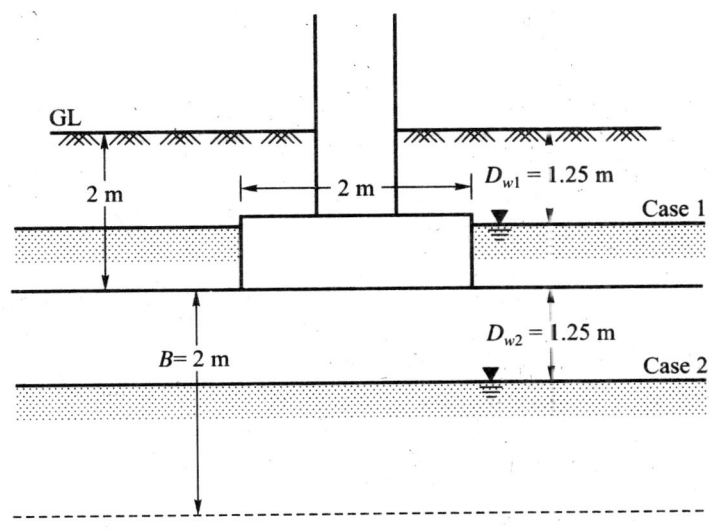

Fig. Ex. 18.4 Effect of WT on bearing capacity

$$\gamma_{e1} = \gamma_b = 8.69 \text{ kN/m}^3$$

$$q_{nd} = 30 \times 57.8 + 14.82 \times 2 \times 40.4 + \frac{1}{2} \times 8.69 \times 3 \times 42.5 = 3480 \text{ kN/m}^2$$

$$q_{na} = \frac{3480}{3} = 1160 \text{ kN/m}^2$$

Case 2: When $D_{w2} = 1.25$ m (Fig. Ex. 18.4)

From Eq. (18.25b)

$$\gamma_{e1} = \gamma = 18.5 \text{ kN/m}^3$$

$$\gamma_{e2} = 8.69 + \frac{1.25}{3}(18.5 - 8.69) = 12.78 \text{ kN/m}^3$$

$$q_{nd} = 1734 + 18.5 \times 2 \times 40.4 + \frac{1}{2} \times 12.78 \times 3 \times 42.5 = 4044 \text{ kN/m}^2$$

$$q_{na} = \frac{4044}{3} = 1348 \text{ kN/m}^2$$

Example 18.5

A square footing fails by general shear in a cohesionless soil under an ultimate load of $Q_d = 7500$ kN. The footing is placed at a depth of 2 m below ground level. Given $\phi = 35°$, and $\gamma = 17.25$ kN/m³, determine the size of the footing if the water table is at a great depth.

Solution

For a square footing Eq. (18.16) for $c = 0$, we have

$$q_d = \gamma D_f N_q + 0.4 \gamma B N_\gamma$$

For $\phi = 35°$, $N_q = 41.4$, and $N_\gamma = 42.4$

$$q_d = \frac{Q_d}{B^2} = \frac{7500}{B^2}$$

By substituting known values, we have

$$\frac{7500}{B^2} = 17.25 \times 2 \times 41.4 + 0.4 \times 17.25 \times 42.4 \, B$$

$$= 1428.3 + 292.56B$$

Simplifying and transposing, we have

$$B^3 + 4.882B^2 - 25.63 = 0$$

Solving this equation by trial and error method, we get $B = 1.95$ m.

Example 18.6

A rectangular footing of size 3×6 m is founded at a depth of 2 m below the ground surface in a homogeneous cohesionless soil having an angle of shearing resistance $\phi = 35°$. The water table is at a great depth. The unit weight of soil $\gamma = 18$ kN/m³. Determine: (1) the net ultimate bearing capacity, (2) the net allowable bearing pressure for $F_s = 3$, and (3) the safe load Q_s the footing can carry. Use Terzaghi's theory.

Solution

Using Eq. (18.18) for $c = 0$, we have

$$q_{nd} = \gamma D_f (N_q - 1) + \frac{1}{2} \gamma B N_\gamma \left(1 - 0.2 \frac{B}{L}\right)$$

By substituting the known values,

$$q_{nd} = 18 \times 2(41.4 - 1) + \frac{1}{2} \times 18 \times 3 \times 42.4 \left(1 - 0.2 \times \frac{3}{6}\right) = 2485 \text{ kN/m}^2$$

$$q_{na} = \frac{2485}{3} = 828 \text{ kN/m}^2$$

$$Q_a = (B \times L)q_{na} = 3 \times 6 \times 828 = 14904 \text{ kN}$$

Example 18.7

A rectangular footing of size 3×6 m is founded at a depth of 2 m below the ground level in a cohesive soil ($\phi = 0$) which fails by general shear. Given: $\gamma_{sat} = 18$ kN/m³, $c = 45$ kN/m². The water table is close to the ground surface. Determine q_d, q_{nd} and q_{na} by (a) Terzaghi's method, and (b) Skempton's method. Use $F_s = 3$.

Solution

(a) Terzaghi's method

Use Eq. (18.18)

For $\phi = 0°$, $N_c = 5.7$, $N_q = 1$

$$q_d = cN_c \left(1 + 0.3 \times \frac{B}{L}\right) + \gamma D_f$$

Substituting the known values,

$$q_d = 45 \times 5.7 \left(1 + 0.3 \times \frac{3}{6}\right) + 8 \times 2 = 331 \, \text{kN/m}^2$$

$$q_{nd} = \left(q_d - \gamma D_f\right) = 331 - 36 = 295 \, \text{kN/m}^2$$

$$q_{na} = \frac{q_{nd}}{F_s} = \frac{295}{3} = 98.33 \, \text{kN/m}^2$$

(b) Skempton's method

From Eqs (18.21a) and (18.21d) we may write

$$q_d = cN_{cr} + \gamma D_f$$

where N_{cr} = bearing capacity factor for rectangular foundation.

$$N_{cr} = \left(0.84 + 0.16 \times \frac{B}{L}\right) \times N_{cs}$$

where N_{cs} = bearing capacity factor for a square foundation.

From Fig.18.9, $N_{cs} = 7.4$ for $D_f / B = 0.67$.

Therefore $N_{cr} = \left(0.84 + 0.16 \times \frac{3}{6}\right) \times 7.4 = 6.8$

Now $q_d = 45 \times 6.8 + 18 \times 2 = 342 \, \text{kN/m}^2$

$$q_{nd} = \left(q_d - \gamma D_f\right) = 342 - 36 = 306 \, \text{kN/m}^2$$

$$q_{na} = \frac{q_{nd}}{F_s} = \frac{306}{3} = 102 \, \text{kN/m}^2$$

Note: Terzaghi's and Skempton's values are in close agreement for cohesive soils.

Example 18.8

If the soil in Ex. 18.6 is cohesionless ($c = 0$), and fails in local shear, determine (i) the ultimate bearing capacity, (ii) the net bearing capacity, and (iii) the net allowable bearing pressure. All the other data remain the same.

Solution

From Eq. (18.14), we have for local shear failure for $c = 0$

$$q_{nd} = \gamma D_f\left(\overline{N}_q - 1\right) + \frac{1}{2}\gamma B \overline{N}_\gamma\left(1 - 0.2 \times \frac{B}{L}\right)$$

where, $\overline{\phi} = \tan^{-1} 0.67 \tan 35° \approx 25°$, $N_q = 1.7$, and $N_\gamma = 9.7$ for $\overline{\phi} = 25°$ from Table 18.1.

By substituting known values, we have

$$q_{nd} = 18 \times 2(12.7 - 1) + \frac{1}{2} \times 18 \times 3 \times 9.7\left(1 - 0.2 \times \frac{3}{6}\right) = 657 \text{ kN/m}^2$$

$$q_{na} = \frac{657}{3} = 219 \text{ kN/m}^2$$

Example 18.9

What will be the gross and new allowable bearing pressure of sand having $\phi = 35°$ and effective unit weight of soil 18 kN/m³ under the following cases: (a) size of footing 1×1 m square, (b) circular footing of 1 m dia., and (c) 1 m wide strip footing.

The footing is placed at a depth of 1 m below ground surface and the water table is at great depth. Use $F_s = 3$. Compute by Terzaghi's general shear failure theory.

Solution

For $\phi = 35°$, $N_\gamma = 42.4$, $N_q = 41.4$

(a) From Eq. (18.16),

$$q_d(\text{sq}) = \gamma D_f N_q + 0.4 \gamma B N_\gamma$$

$$= 18 \times 1 \times 41.4 + 0.4 \times 18 \times 1 \times 42.4 = 1050.5 \text{ kN/m}^2$$

$$q_{nd} = q_d - \gamma D_f = 1050.5 - 18 \times 1 = 1032.5 \text{ kN/m}^2$$

$$q_{na} = \frac{1032.5}{3} = 344.17 \text{ kN/m}^2$$

(b) From Eq. (18.17),

$$q_d(\text{cir}) = \gamma D_f N_q + 0.3 \gamma B N_\gamma$$

$$= 18 \times 1 \times 41.4 + 0.3 \times 18 \times 1 \times 42.4 = 974.16 \text{ kN/m}^2$$

$$q_{nd} = 974.16 - 18 \times 1 = 956.16 \text{ kN/m}^2$$

$$q_{na} = \frac{956.16}{3} = 318.72 \text{ kN/m}^2$$

(c) From Eq. (18.15),

$$q_d (\text{strip}) = \gamma D_f N_q + \frac{1}{2} \gamma B N_\gamma$$

$$= 18 \times 1 \times 41.4 + \frac{1}{2} \times 18 \times 1 \times 42.4 = 1126.8 \text{ kN/m}^2$$

$$q_{nd} = 1126.8 - 18 \times 1 = 1108.81 \text{ kN/m}^2$$

$$q_{na} = \frac{1108.81}{3} = 369.6 \text{ kN/m}^2$$

Example 18.10

A strip foundation at a depth of 1.5 m below ground surface. WT is close to the ground level and the soil is cohesionless. The footing is supposed to carry a net safe load of 400 kN/m² with $F_s = 3$. Given $\gamma_{sat} = 20.85 \text{ kN/m}^3$ and $\phi = 35°$, find the width of the footing, under general failure criteria of Terzaghi.

Solution

For $\phi = 35°$, $N_\gamma = 42.4$, $N_q = 41.4$

Equation for strip footing is

$$q_{nd} = \gamma D_f (N_q - 1) + \frac{1}{2} \gamma B N_\gamma$$

Since the WT is at ground level, γ = submerged unit, γ_b, in both the terms.

$$\gamma_b = \gamma_{sat} - \gamma_w = 20.85 - 9.81 = 11.04 \text{ kN/m}^3$$

For $q_{na} = 400 \text{ kN/m}^2$, $q_{nd} = 400 \times 3 = 1200 \text{ kN/m}^2$.

we have, $1200 = 11.04 \times 1.5 (41.4 - 1) + \frac{1}{2} \times 11.04 \times B \times 42.4$

$$= 669.024 + 234.048B$$

or $B = \dfrac{530.976}{234.048} = 2.27\text{m}$

Example 18.11

At what depth should a footing of size 2 × 3 m be founded to provide a factor of safety of 3, if the soil is stiff clay having an unconfined compressive strength of 120 kN/m². The unit weight of soil is 18 kN/m². The ultimate bearing capacity of the footing is 425 kN/m². Use Terzaghi's theory. The WT is close to the ground surface.

Solution

Use Eq. (18.18) for $\phi = 0$

$$q_d = 5.7c\left(1-0.3\times\frac{B}{L}\right)+\gamma D_f$$

Since WT is close to GL, $\gamma = \gamma_b$ in the above equation.

$$\gamma_b = \gamma_{sat} - \gamma_w = 18.0 - 9.81 = 8.19 \text{ kN/m}^3$$

Substituting the known values, we have

$$q_d = 5.7\times\frac{120}{2}\left(1-0.3\times\frac{2}{3}\right)+8.19D_f = 410.4+8.19D_f$$

Since q_d (given) = 425 kN/m², we may write

$$8.19D_f + 410.4 = 425$$

$$\text{or } D_f = \frac{14.6}{8.19} = 1.78 \text{ m}$$

Example 18.12

A rectangular footing is founded at a depth of 2 m below ground level in a (c – ϕ) soil having the following properties: porosity $n = 40\%$, $G = 2.67$, $c = 15$ kN/m², and $\phi = 30°$.

The water table is close to the ground surface. If the width of the footing is 3 m, what is the length required to carry a gross allowable bearing pressure $q_a = 455$ kN/m² with a factor of safety, $F_s = 3$. Use Terzaghi's theory of general shear failure.

Solution

Use Eq. (18.10) for q_d

$$q_d = cN_c\left(1+0.3\times\frac{B}{L}\right)+\gamma D_f N_q +\frac{1}{2}\gamma BN_\gamma\left(1-0.2\times\frac{B}{L}\right)$$

Since the WT is close to the ground surface,

$$\gamma = \gamma_b = \frac{\gamma_w(G-1)}{1+e}$$

Since $n = 40\% = \dfrac{e}{1+e}$, we have $e = 0.67$. Now

$$\gamma_b = \frac{9.81(2.67-1)}{1.67} = 9.81\,\text{kN/m}^3.$$

For $\phi = 30°$, $N_c = 37.2$, $N_q = 22.5$ and $N_\gamma = 19.7$ from Table 18.1. Now substituting the known values, we have

$$q_d = 5.7\times37.2\left(1+0.3\times\frac{3}{L}\right)+9.81\times2\times22.5+\frac{1}{2}\times9.81\times3\times19.7\times\left(1-0.3\times\frac{3}{L}\right)$$

$$= 1289+\frac{328}{L}$$

Given $q_a = 455\,\text{kN/m}^2$ and $F_s = 3$, we have

$$q_d = 455\times3 = 1365\,\text{kN/m}^2$$

Now equating, we have

$$1365 = q_d = 1289+\frac{328}{L}$$

or $L = \dfrac{328}{76} = 4.32\,\text{m}$

Example 18.13

A square footing located at a depth of 1.5 m below the ground surface in cohesionless soil carries a column load of 1280 kN. The soil is submerged having an effective unit weight of 11.5 kN/m³ and an angle of shearing resistance of 30°. Find the size of the following for $F_s = 3$ by Terzaghi's theory of general shear failure.

Solution

Use Eq. (18.16) for $c = 0$

$$q_d = \gamma D_f N_q + 0.4\gamma B N_\gamma$$

Since WT is close to GL

$$\gamma = \gamma_b = 11.5\,\text{kN/m}^3$$

For $\phi = 30°$, $N_q = 22.5$ and $N_\gamma = 19.7$. Substituting the known values, we have

$q_d = 11.5 \times 1.5 \times 22.57 + 0.4 \times 11.5 \times 19.7B$

$= 388.13 + 90.62B$ (a)

Column load, $Q_a = 1280$ kN, or $q_a = \dfrac{1280}{B \times B}$ kN/m^2

Since $F_s = 3$, $q_d = \dfrac{1280 \times 3}{B^2} = \dfrac{3840}{B^2}$ (b)

Now equating Eq. (a) with (b)

$\dfrac{3840}{B^2} = 388.13 + 90.62B$

or $B^3 + 4.283B^2 - 42.37 = 0$

Solving for B, we have $B = 2.5$ m

Size of square footing $= 2.5 \times 2.5$ m.

Example 18.14

A footing of 1.5 m diameter carries a safe load (including itself weight) of 800 kN in cohesionless soil. The soil has an angle of shearing resistance $\phi = 36°$ and an effective unit weight of 12 kN/m^3. Determine the depth of foundation for $F_s = 2.5$ by Terzaghi's general shear failure criteria.

Solution

Use Eq. (18.17) for $c = 0$

$q_d = \gamma D_f N_q + 0.3\gamma BN_\gamma$

for $\phi = 36°$, $N_q = 49.38$, $N_\gamma = 54$ from Table 18.1. Substituting the known values

$q_d = 12 \times 49.38 D_f + 0.3 \times 12 \times 1.5 \times 54$

$= 592.56 D_f + 291.6$ (a)

Given $Q_a = 800$ kN, with $F_s = 2.5$, $Q_d = 2000$ kN

Therefore, $q_d = \dfrac{4Q_d}{\pi B^2} = \dfrac{4 \times 2000}{3.14 \times 1.5^2} = 1132.34$ kN/m^2 (b)

Equating Eq. (a) with (b), we have

$$592.56D_f + 291.6 = 1132.34$$

$$\text{or } D_f = \frac{840.74}{592.56} = 1.42 \text{ m}$$

Example 18.15

If the ultimate bearing capacity of a 1 m wide strip footing resting on the surface of sand is 250 kN/m², what will the net allowable pressure that a 3 × 3 m square footing resting on the surface can carry with $F_s = 3$. Assume that the soil is cohesionless. Use Terzaghi's theory.

Solution

We may write the following equations:

$$q_d(\text{square}) = 0.4\gamma B_1 N_\gamma$$

$$q_d(\text{strip}) = 0.5\gamma B_2 N_\gamma$$

where $D_f = 0$

$$\text{or } \frac{q_d(\text{square})}{q_d(\text{strip})} = \frac{0.4\gamma B_1 N_\gamma}{0.5\gamma B_2 N_\gamma} = 0.8 \times \frac{B_1}{B_2} = \frac{0.8 \times 3}{1} = 2.4$$

$$q_d(\text{square}) = 2.4 q_d(\text{strip}) = 2.4 \times 250 = 600 \text{ kN/m}^2$$

$$q_{na} = \frac{q_d(\text{square})}{3} = \frac{q_{nd}(\text{square})}{3} = \frac{600}{3} = 200 \text{ kN/m}^2$$

since $D_f = 0$.

Example 18.16

A circular plate of diameter 1.05 m was placed on a sand surface of unit weight 16.5 kN/m³ and loaded to failure. The failure load was found to give a pressure of 1500 kN/m². Determine the value of the bearing capacity factor N_γ. The angle of shearing resistance of the sand measured in a triaxial test was found to be 39°. Compare this value with the theoritical value of corresponding to N_γ. Use Terzaghi's theory.

Solution

Since the plate is placed on the surface $D_f = 0$, the equation for q_d (circular) is

$$q_d = 0.3\gamma B N_\gamma = 0.3 \times 16.5 \times 1.05 N_\gamma = 5.1975 N_\gamma$$

since $q_d = 1500$ kN/m², we have

$$N_\gamma = \frac{1500}{5.1975} = 289$$

From Table 18.1 for $\phi = 39°$, $N_\gamma = 88.8$, which is very much less than that obtained from the plate load test. This is partly due to the scale effect and partly due to sensitiveness of N_γ at higher values of ϕ.

Example 18.17

Find the net allowable bearing load per meter length of a long wall footing of 2 m wide founded on a stiff saturated clay at a depth of 1 m. The unit weight of the clay is 17 kN/m³, and the shear strength is 120 kN/m². Assume the load to be applied rapidly such that undrained conditions ($\phi = 0$) prevail. Use $F_s = 3$ and Skempton's method.

Solution

From Eq. (18.21a) we have

$$q_d = cN_c + \gamma D_f \text{ or } q_{nd} = cN_c$$

From Fig. 18.9, $N_c = 6$ for $D_f/B = 0.5$ for a stripe footing.

Therefore $q_{nd} = 120 \times 6 = 720 \text{ kN/m}^2$

$$q_{na} = \frac{720}{3} = 240 \text{ kN/m}^2$$

18.8 THE GENERAL BEARING CAPACITY EQUATION

The bearing capacity Eq. (18.6) developed by Terzaghi is for a strip footing under general shear failure. Eq. (18.6) has been modified for other types of foundations such as square, circular and rectangular by introducing shape factors. Meyerhof (1963) presented a general bearing capacity equation which takes into account the shape and the inclination of load. The general form of equation suggested by Meyerhof for bearing capacity is

$$q_d = cN_c s_c d_c i_c + q'_o N_q s_q d_q i_q + \frac{1}{2}\gamma B N_\gamma s_\gamma d_\gamma i_\gamma \tag{18.26}$$

where c = unit cohesion

q'_o = effective overburden pressure at the base level of the foundation = $\overline{\gamma} D_f$

$\overline{\gamma}$ = effective unit weight above the base level of foundation

γ = effective unit weight of soil below the foundation base

D_f = depth of foundation

$$s_c, s_q, s_\gamma = \text{shape factors}$$

$$d_c, d_q, d_\gamma = \text{depth factors}$$

$$i_c, i_q, i_\gamma = \text{load inclination factors}$$

$$B = \text{width of foundation}$$

$$N_c, N_q, N_\gamma = \text{bearing capacity factors}$$

Hansen (1970) extended the work of Meyerhof by including in Eq. (18.26) two additional factors to take care of base tilt and foundations on slopes. Vesic (1973, 1974) used the same form of equation suggested by Hansen. All three investigators use the equations proposed by Prandtl (1921) for computing the values of N_c and N_q wherein the foundation base is assumed as smooth with the angle $\alpha = 45° + \phi/2$ (Fig. 18.6). However, the equations used by them for computing the values of N_γ are different. The equations for N_c, N_q and N_γ are

$$N_q = e^{\pi\tan\phi} N_\phi \tag{18.27}$$

$$N_c = (N_q - 1)\cot\phi \tag{18.28}$$

TABLE 18.2

The values of N_c, N_q and Meyerhof (M), Hansen (H) and Vesic (V) N_γ Factors

ϕ	N_c	N_q	N_γ(H)	N_γ(M)	N_γ(V)
0	5.14	1.0	0.0	0.0	0.0
5	6.49	1.6	0.1	0.1	0.4
10	8.34	2.5	0.4	0.4	1.2
15	10.97	3.9	1.2	1.1	2.6
20	14.83	6.4	2.9	2.9	5.4
25	20.71	10.7	6.8	6.8	10.9
26	22.25	11.8	7.9	8.0	12.5
28	25.79	14.7	10.9	11.2	16.7
30	30.13	18.4	15.1	15.7	22.4
32	35.47	23.2	20.8	22.0	30.2
34	42.14	29.4	28.7	31.1	41.0
36	50.55	37.7	40.0	44.4	56.2
38	61.31	48.9	56.1	64.0	77.9
40	72.25	64.1	79.4	93.6	109.4
45	133.73	134.7	200.5	262.3	271.3
50	266.50	318.50	567.4	871.7	762.84

Note: N_c and N_q are the same for all the three methods. Subscripts identify the author for N_γ.

TABLE 18.3

Shape, depth and load inclination factors of Meyerhof, Hansen and Vesic

Factors	Meyerhof	Hansen	Vesic
s_c	$1+0.2N_\phi\left[\dfrac{B}{L}\right]$	$1+\dfrac{N_\phi}{N_c}\left[\dfrac{B}{L}\right]$	
s_q	$1+0.1N_\phi\left[\dfrac{B}{L}\right]$ for $\phi>10°$	$1+\dfrac{B}{L}\tan\phi$	
s_γ	$s_\gamma = s_q$ for $\phi>10°$ $s_\gamma = s_q =1$ for $\phi=0$	$1-0.4\dfrac{B}{L}$	The shape and depth factors of Vesic are the same as those of Hansen.
d_c	$1+0.2\sqrt{N_\phi}\left[\dfrac{D_f}{B}\right]$	$1+0.4\left[\dfrac{D_f}{B}\right]$	
d_q	$1+0.2\sqrt{N_\phi}\left[\dfrac{D_f}{B}\right]$ for $\phi>10°$	$1+2\tan\phi(1-\sin\phi)^2\left[\dfrac{D_f}{B}\right]$	
d_γ	$d_\gamma = d_q$ for $\phi>10°$ $d_\gamma = d_q =1$ for $\phi=0$	1 for all ϕ Note: Vesic's s and d factors = Hansen's s and d factors	
i_c	$\left[1-\dfrac{\alpha°}{90}\right]^2$ for any ϕ	$i_q - \dfrac{1-i_q}{N_q-1}$ for $\phi>0$	Same as Hansen for $\phi>0$
		$0.5\left[1-\dfrac{Q_h}{A_f c_a}\right]^{\frac{1}{2}}$ for $\phi=0$	$1-\dfrac{mQ_h}{A_f c_a N_c}$
i_q	$i_q = i_c$ for any ϕ	$\left[1-\dfrac{0.5Q_h}{Q_d + A_f c_a \cot\phi}\right]^5$	$\left[1-\dfrac{Q_h}{Q_d + A_f c_a \cot\phi}\right]^m$
i_γ	$\left[1-\dfrac{\alpha°}{\phi°}\right]^2$ for $\phi>0$ $i_\gamma = 0$ for $\phi=0$	$\left[1-\dfrac{0.7Q_h}{Q_d + A_f c_a \cot\phi}\right]^5$	$\left[1-\dfrac{Q_h}{Q_d + A_f c_a \cot\phi}\right]^{m+1}$

$$\left.\begin{array}{ll} N_\gamma = (N_\gamma -1)\tan(1.4\phi) & \text{(Meyerhof)} \\ N_\gamma = 1.5(N_q -1)\tan\phi & \text{(Hansen)} \\ N_\gamma = 2(N_q -1)\tan\phi & \text{(Vesic)} \end{array}\right] \qquad (18.29)$$

Table18.2 gives the values of the bearing capacity factors. Equations for shape, depth and inclination factors are given in Table 18.3. The tilt of the base and the foundations on slopes are not considered here.

In Table 18.3 The following terms are defined with regard to the inclination factors

Q_h = horizontal component of the inclined load

Q_u = vertical component of the inclined load

c_a = unit adhesion on the base of the footing

A_f = effective contact area of the footing

$$m = m_B = \frac{2 + B/L}{1 + B/L} \text{ with } Q_h \text{ parallel to } B$$

$$m = m_L = \frac{2 + B/L}{1 + B/L} \text{ with } Q_h \text{ parallel to } L$$

The general bearing capacity Eq. (18.26) has not taken into account the effect of the water table position on the bearing capacity. Hence, Eq. (18.26) has to be modified according to the position of water level in the same way as explained in Section 18.7.

Validity of the Bearing Capacity Equations

There is currently no method of obtaining the ultimate bearing capacity of a foundation other than as an estimate (Bowles, 1996). There has been little experimental verification of any of the methods except by using model footings. Up to a depth of $D_f \approx B$ the Meyerhof q_d is not greatly different from the Terzaghi value (Bowles, 1996). The Terzaghi equations, being the first proposed, have been quite popular with designers. Both the Meyerhof and Hansen methods are widely used. The Vesic method has not been much used. It is a good practice to use at least two methods and compare the computed values of q_d. If the two values do not compare well, use a third method.

18.9 BEARING CAPACITY OF FOUNDATIONS SUBJECTED TO ECCENTRIC LOADS

Foundations Subjected to Eccentric Vertical Loads

If a foundation is subjected to lateral loads and moments in addition to vertical loads, eccentricity in loading results. The point of application of the resultant of all the loads would lie outside the geometric center of the foundation, resulting in eccentricity in loading. The eccentricity e is measured from the center of the foundation to the point of application normal to the axis of the foundation. The maximum eccentricity normally allowed is $B/6$ where B is the width of the foundation. The basic problem is to determine the effect of the eccentricity on the ultimate bearing capacity of the foundation. When a foundation is subjected to an eccentric vertical load, as shown in Fig. 18.11(a), it tilts towards the side of the eccentricity and the contact pressure increases on the side of tilt and decreases on the opposite side. When the vertical load Q_d reaches the ultimate load, there will be a failure of the supporting soil on the side of eccentricity. As a consequence, settlement of the footing will be associated with tilting of the base towards the side of eccentricity. If the eccentricity is very small, the load required to produce this type of failure is almost equal to the load required for producing a symmetrical general shear failure. Failure occurs due to intense radial shear on one side of the plane of symmetry, while the deformations in the zone of radial shear on the other side are still insignificant. For this reason the failure is always associated with a heave on that side towards which the footing tilts.

Research and observations of Meyerhof (1953, 1963) indicate that effective footing dimensions obtained (Fig. 18.11) as

$$L' = L - 2e_y, \quad B' = B - 2e_x \tag{18.30}$$

should be used in bearing capacity analysis to obtain an effective footing area defined as

$$A' = B'L' \tag{18.31}$$

where e_x and e_y are the eccentricity in the direction of the axes.

The ultimate load bearing capacity of a footing subjected to eccentric loads may be expressed as

$$Q_d = q_e A' \tag{18.32}$$

where q_d = ultimate bearing capacity of the footing with the load acting at the center of the footing.

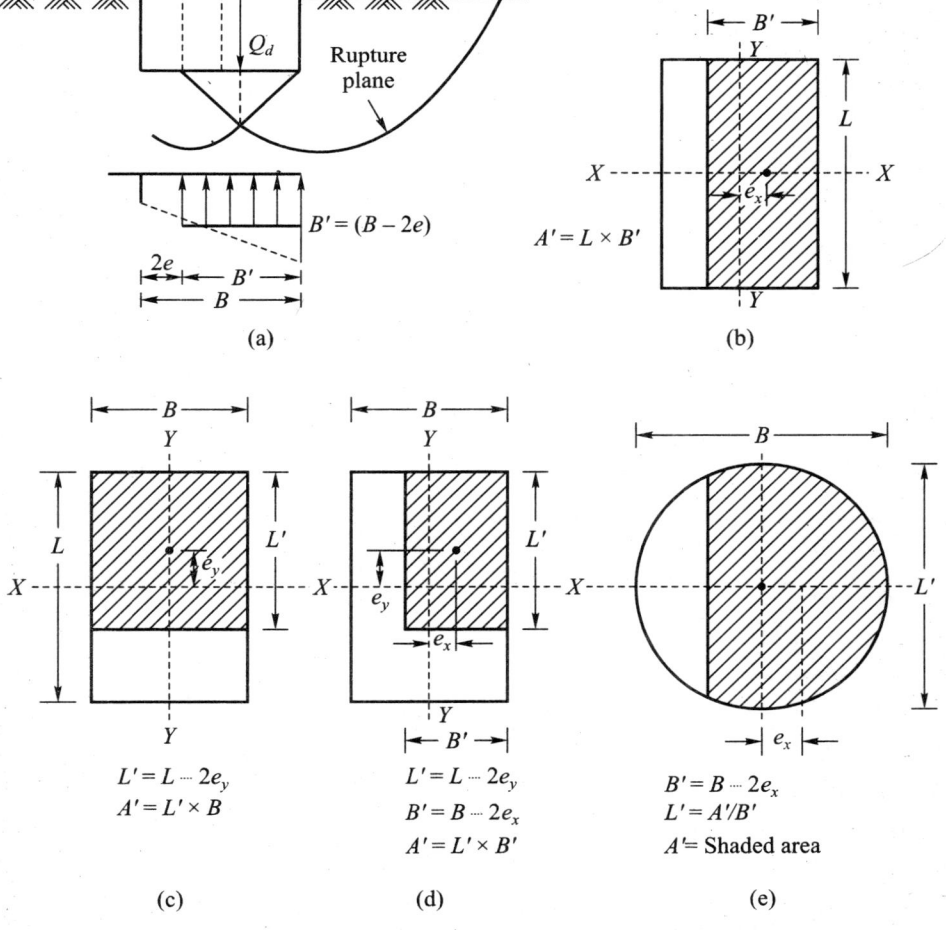

(a)

(b)

(c) $L' = L - 2e_y$
$A' = L' \times B$

(d) $L' = L - 2e_y$
$B' = B - 2e_x$
$A' = L' \times B'$

(e) $B' = B - 2e_x$
$L' = A'/B'$
$A' =$ Shaded area

Fig. 18.11 Eccentrically loaded footing (Meyerhof, 1953)

Determination of Maximum and Minimum Base Pressures Under Eccentric Loadings

The methods of determining the effective area of a footing subjected to eccentric loadings have been discussed earlier. It is now necessary to know the maximum and minimum base pressures under the same loadings. Consider the plan of a rectangular footing given in Fig. 18.12 subjected to eccentric loadings.

Let the coordinate axes XX and YY pass through the center O of the footing. If a vertical load passes through O, the footing is symmetrically loaded. If the vertical load passes through O_x on the X-axis, the footing is eccentrically loaded with one way eccentricity. The distance of O_x from O, designated as e_x, is called the eccentricity in the X-direction. If the load passes through O_y on the Y-axis, the eccentricity is e_y in the Y-direction. If on the other hand the load passes through O_{xy}, the eccentricity is called *two-way eccentricity* or *double eccentricity*.

When a footing is eccentrically loaded, the soil experiences a maximum or a minimum pressure at one of the corners or edges of the footing. For the load passing through O_{xy} (Fig. 18.12), the points C and D at the corners of the footing experience the maximum and minimum pressures respectively.

The general equation for pressure may be written as

$$q = \frac{Q}{A} \pm \frac{Qe_x}{I_y}x \pm \frac{Qe_y}{I_x}y \tag{18.33a}$$

$$\text{or } q = \frac{Q}{A} \pm \frac{M_x}{I_y}x \pm \frac{M_y}{I_x}y \tag{18.33b}$$

where q = contact pressure at a given point (x, y)

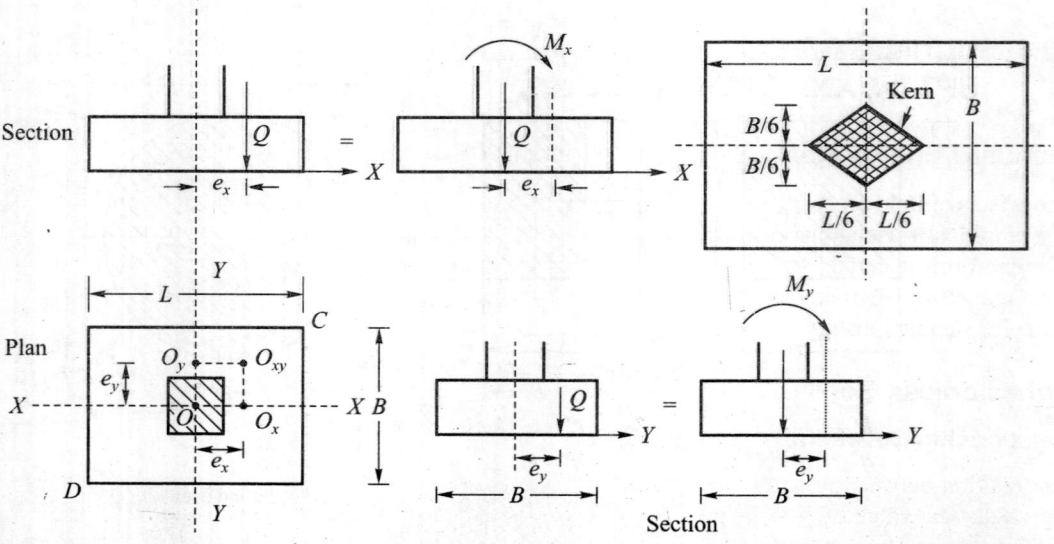

Fig. 18.12 Footing subjected to eccentric loadings

Q = total vertical load

A = area of footing

Q_{ex} = M_x = moment about axis YY

Q_{ey} = M_y = moment about axis XX

I_x, I_y = moment of inertia of the footing about XX and YY axes respectively

q_{max} and q_{min} at points C and D respectively may be obtained by substituting in Eq. (18.33) for

$$I_x = \frac{LB^3}{12}, \quad I_y = \frac{LB^3}{12}, \quad x = \frac{L}{2}, \quad y = \frac{B}{2}$$

we have

$$q_{max} = \frac{Q}{A}\left[1 + 6\frac{e_x}{L} + 6\frac{e_y}{B}\right] \tag{18.34a}$$

$$q_{min} = \frac{Q}{A}\left[1 - 6\frac{e_x}{L} - 6\frac{e_y}{B}\right] \tag{18.34b}$$

Equation (18.34) may also be used for one way eccentricity by putting either $e_x = 0$, or $e_y = 0$.

When e_x or e_y exceed a certain limit, Eq. (18.34) gives a negative value of q which indicates tension between the soil and the bottom of the footing. Eqs (18.34) are applicable only when the load is applied within a limited area which is known as the *Kern* as is shown shaded in Fig 18.12 so that the load may fall within the shaded area to avoid tension. The procedure for the determination of soil pressure when the load is applied outside the kern is laborious and as such not dealt with here. However, charts are available for ready calculations in references such as Teng (1969) and Highter and Anders (1985).

18.10 ULTIMATE BEARING CAPACITY OF FOOTINGS BASED ON THE SPT (*N*) AND CPT VALUES (*q_c*)

Standard Energy Ratio R_{es} Applicable to *N* Value

The effects of field procedures and equipment on the field values of *N* were discussed in Chapter 17. The empirical correlations established in the USA between *N* and soil properties indicate the value of *N* conforms to certain standard energy ratios. Some suggest 70% (Bowles, 1996) and others 60% (Terzaghi *et al.*, 1996). In order to avoid this confusion, the author uses *N* in this book as the corrected value for standard energy.

Cohesionless Soils

Relationship Between *N* and φ

The relation between *N* and φ established by Peck *et al.*, (1974) is given in a graphical form in Fig. 18.8. The value of *N* to be used for getting φ is the corrected value for standard energy. The angle φ obtained by this method can be used for obtaining the bearing capacity factors, and hence the ultimate bearing capacity of soil.

Cohesive Soils

Relationship Between N and q_u (Unconfined Compressive Strength)

Relationships have been developed between N and q_u (the undrained compressive strength) for the ϕ = 0 condition. This relationship gives the value of c_u for any known value of N. The relationship may be expressed as [Eq. (17.9a)]

$$q_u = 2c_u = \bar{k}N \; kPa \qquad (18.35)$$

where the value of the coefficient $\bar{k}$ may vary from a minimum of 12 to a maximum of 25. A low value of 13 yields q_u given in Table 17.4.

Once q_u is determined, the net ultimate bearing capacity and the net allowable bearing pressure can be found following Skempton's approach.

THE CPT METHOD

Cohesionless Soils

Relationship Between q_c, D_r and ϕ

Relationships between the static cone penetration resistance q_c and ϕ have been developed by Robertson and Campanella (1983b), Fig. 17.13. The value of f can therefore be determined with the known value of q_c. With the known value of ϕ, bearing capacity factors can be determined and hence the ultimate bearing capacity. Experience indicates that the use of q_c for obtaining ϕ is more reliable than the use of N.

Bearing Capacity of Soil

As per Schmertmann (1978), the bearing capacity factors N_q and N_γ for use in the Terzaghi bearing capacity equation can be determined by the use of the equation

$$N_q = N_\gamma = 1.25q_c \qquad (18.36)$$

where q_c = cone point resistance in kg/cm^2 (or tsf) averaged over a depth equal to the width below the foundation.

Undrained Shear Strength

The undrained shear strength c_u under $\phi = 0$ condition may be related to the static cone point resistance q_c as [Eq. (17.11)]

$$q_c = N_k c_u + p_o$$

$$\text{or } c_u = \frac{q_c - p_o}{N_k} = \frac{\bar{q}_c}{N_k} \qquad (18.37)$$

where N_k = cone factor, may be taken as equal to 20 (Sanglerat, 1972) both for normally consolidated and preconsolidated clays.

p_o = total overburden pressure

When once c_u is known, the values of q_{nu} and q_{na} can be evaluated as per the methods explained in earlier sections.

Example 18.18

A water tank foundation has a footing of size 6 × 6 m founded at a depth of 3 m below ground level in a medium dense sand stratum of great depth. The corrected average SPT value obtained from the site investigation is 20. The foundation is subjected to a vertical load at an eccentricity of $B/10$ along one of the axes. Figure Ex. 18.18 gives the soil profile with the remaining data. Estimate the ultimate load, Q_d, by Meyerhof's method.

Solution

From Fig. 18.8 $\phi = 33°$ for $N = 20$.

$$B' = B - 2e = 6 - 2(0.6) = 4.8 \text{ m}$$

$$L' = L = B = 6 \text{ m}$$

For $c = 0$ and $i = 1$, Eq. (18.26) reduces to

$$q_d = \gamma D_f N_q s_q d_q + \frac{1}{2} \gamma B' N_\gamma s_\gamma d_\gamma$$

From Table 18.2 for $\phi = 33°$ we have

$N_q = 26.3$, $N_\gamma = 26.55$ (Meyerhof)

From Table 18.3 (Meyerhof)

$$s_q = 1 + 0.1 N_\phi \left[\frac{B}{L} \right] = 1 + 0.1 \tan^2 \left(45^2 + \frac{33}{2} \right)(1) = 1.34$$

$$s_\gamma = s_q = 1.34 \quad \text{for } \phi > 10°$$

$$d_q = 1 + 0.1 \sqrt{N_\phi} \left[\frac{D_f}{B'} \right] = 1 + 0.1 \times 1.84 \left[\frac{3}{4.8} \right] = 1.115$$

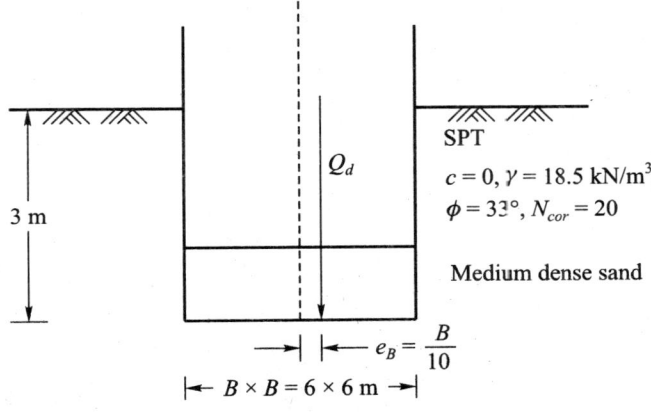

SPT

$c = 0$, $\gamma = 18.5 \text{ kN/m}^3$,
$\phi = 33°$, $N_{cor} = 20$

Medium dense sand

3 m

Q_d

$e_B = \dfrac{B}{10}$

$\longleftarrow B \times B = 6 \times 6 \text{ m} \longrightarrow$

Fig. Ex. 18.18

Substituting $d_\gamma = d_q = 1.115$ for $\phi > 10°$

$$q_d = 18.5 \times 3 \times 26.3 \times 1.34 \times 1.115 + \frac{1}{2} \times 18.5 \times 4.8 \times 26.55 \times 1.34 \times 1.115$$

$$= 2{,}181 + 1{,}761 = 3{,}942 \text{ kN/m}^2$$

$$Q_d = B \times B' \times q_d = 6 \times 4.8 \times 3942 = 113{,}530 \text{ kN} \approx 114 \text{ MN}$$

Example 18.19

Figure Ex. 18.19 gives the plan of a footing subjected to eccentric load with two way eccentricity. The footing is founded at a depth 3 m below the ground surface. Given $e_x = 0.60$ m and $e_y = 0.75$ m, determine Q_d. The soil properties are: $c = 0$, $N = 20$, $\gamma = 18.5 \text{ kN/m}^3$. The soil is medium dense sand. Use N_γ (Meyerhof) from Table 18.2 and Hansen's shape and depth factors from Table 18.3.

Solution

Figure Ex. 18.19 shows the two-way eccentricity. The effective lengths and breadths of the foundation from Eq. (18.30) is

$$B' = B - 2e_y = 6 - 2 \times 0.75 = 4.5 \text{ m}.$$

$$L' = L - 2e_x = 6 - 2 \times 0.6 = 4.8 \text{ m}.$$

Effective area, $A' = L' \times B' = 4.5 \times 4.8 = 21.6 \text{ m}^2$

As in Example 18.18

$$q_d = \gamma D_f N_q s_q d_q + \frac{1}{2} \gamma B' N_\gamma s_\gamma d_\gamma$$

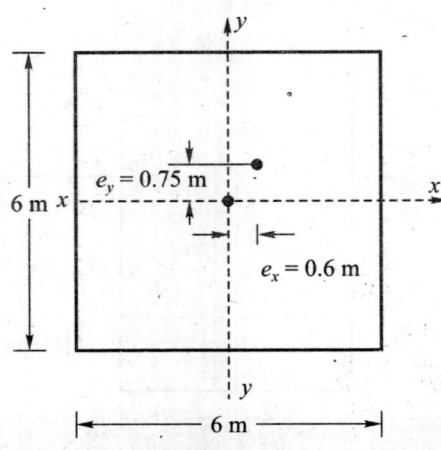

$e_y = 0.75$ m

6 m x

$e_x = 0.6$ m

y

6 m

Fig. Ex. 18.19

For $\phi = 33°$, $N_q = 26.3$ and $N_\gamma = 26.55$ (Meyerhof)

From Table 18.3 (Hansen)

$$s_q = 1 + \frac{B'}{L'}\tan 33° = 1 + \frac{4.5}{4.8} \times 0.65 = 1.61$$

$$s_\gamma = 1 + 0.4\frac{B'}{L'} = 1 - 0.4 \times \frac{4.5}{4.8} = 0.63$$

$$d_q = 1 + 2\tan 33° \left(1 - \sin 33°\right)^2 \times \frac{3}{4.5}$$

$$= 1 + 1.3 \times 0.21 \times 0.67 = 1.183$$

$$d_\gamma = 1$$

Substituting

$$q_d = 18.5 \times 3 \times 26.3 \times 1.61 \times 1.183 + \frac{1}{2} \times 18.5 \times 4.5 \times 26.55 \times 0.63 \times (1)$$

$$= 2,780 + 696 = 3,476 \text{ kN/m}^2$$

$$Q_d = A'q_d = 21.6 \times 3,476 = 75,082 \text{ kN}$$

18.11 IS CODE OF PRACTICE FOR COMPUTING BEARING CAPACITY

The IS code IS: 6403-1981 has suggested a bearing capacity similar to the one proposed by Brinch Hansen. The bearing capacity factors N_c and N_q are the same as that given in Table 18.2. Vesic's bearing capacity factor N_γ has been choosen. The values of N_γ for various values of ϕ are also given in the same table. The other details given in the code are more or less the same as that given in Table 18.3. Students may refer to the code for more information on this subject.

PART B: SAFE BEARING PRESSURE AND SETTLEMENT OF FOUNDATIONS

18.12 INTRODUCTION

Allowable and Safe Bearing Pressures

When we design a foundation, we must see that the structure is safe on two counts. They are,

1. The supporting soil should be safe from shear failure due to the loads imposed on it by the superstructure,

2. The settlement of the foundation should be within permissible limits.

Hence, we have to deal with two types of bearing pressures. They are:

1. A pressure that is safe from shear failure criteria,

2. A pressure that is safe from settlement criteria.

For the sake of convenience, let us call the first the *allowable bearing pressure* and the seco
the *safe bearing pressure*.

In all our design, we use only the net bearing pressure and as such we call q_{na} the *net allowab*
bearing pressure and q_s the net safe bearing pressure. In designing a foundation, we use the least
the two bearing pressures.

Effect of Settlement on the Structure

If the structure as a whole settles uniformly into the ground there will not be any detrimental effe
on the structure as such. The only effect it can have is on the service lines, such as water and sanita
pipe connections, telephone and electric cables etc. which can break if the settlement is considerabl
Such uniform settlement is possible only if the subsoil is homogeneous and the load distribution
uniform. Buildings in Mexico City have undergone settlements as large as 2 m. However, th
differential settlement should not exceed the permissible limits.

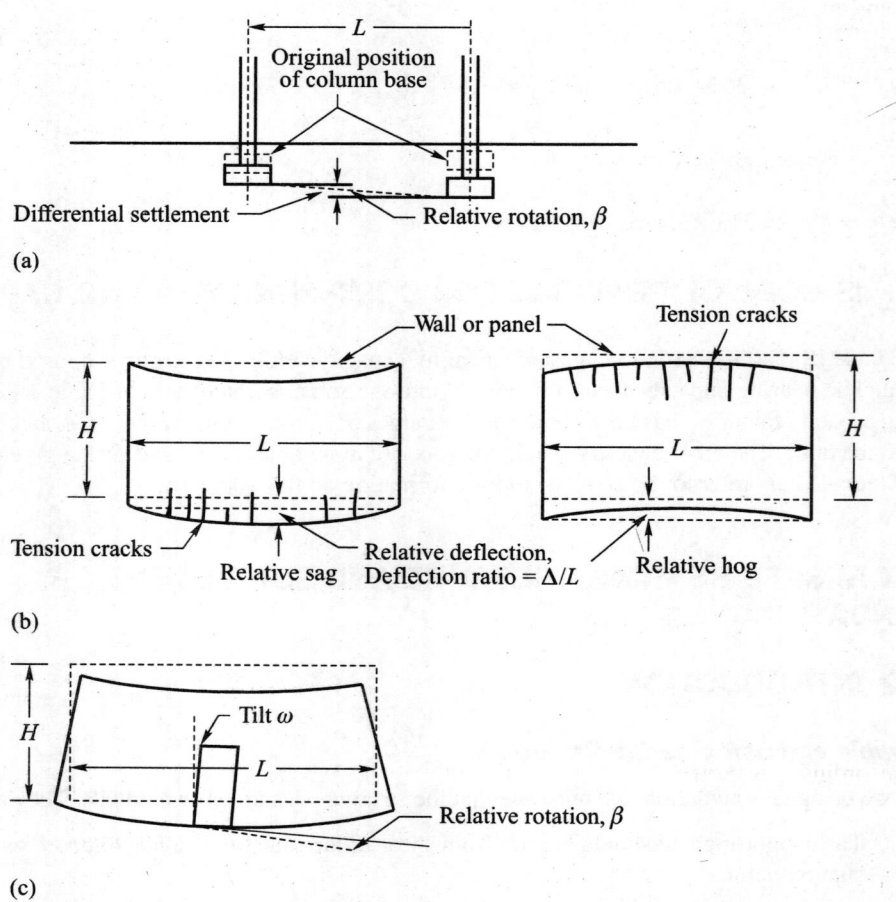

Fig. 18.13 Definition of differential settlement for framed and load-bearing wall structures (after Burland and Worth, 1974)

ABLE 18.4

Maximum settlements and differential settlements of buildings in cm. (After McDonald and Skempton, 1955)

Sl. no.	Criterion	Isolated foundations	Raft
1.	Angular distortion	1/300	1/300
2.	Greatest differential settlements		
	Clays	4.5	4.5
	Sands	3.25	3.25
3.	Maximum settlements		
	Clays	7.5	10.0
	Sands	5.0	6.25

TABLE 18.5

Permissible settlements (1955, U.S.S.R. Building Code)

Sl.no.	Type of building	Average settlement (cm)
1.	Building with plain brickwalls on continuous and separate foundations with wall length L to wall height H	
	$L/H \geq 2.5$	7.5
	$L/H \leq 1.5$	10.0
2.	Framed building	10.0
3.	Solid reinforced concrete foundation of blast furnaces, water towers etc.	30

According to experience, the differential settlement between parts of a structure may not exceed 75 percent of the normal absolute settlement. The various ways by which differential settlements may occur in a structure are shown in Fig. 18.13. Tables 18.4, 18.5, and 18.6 give the absolute and permissible differential settlements for various types of structures.

Foundation settlements must be estimated with great care for buildings, bridges, towers, power plants and similar high cost structures. The settlements for structures such as fills, earthdams, levees, etc. can be estimated with a greater margin of error.

TABLE 18.6

Permissible differential settlement (U.S.S.R. Building Code, 1955)

Sl.no.	Type of structure	Type of soil	
		Sand and hard clay	Plastic clay
1.	Steel and reinforced concrete structures	0.002L	0.002L
2.	Plain brick walls in multistory buildings		
	For $L/H \leq 3$	0.0003L	0.0004L
	$L/H \geq 5$	0.0005L	0.0007L
3.	Water towers, silos etc.	0.004L	0.004L
4.	Slope of crane way as well as track		
	for bridge crane track	0.003L	0.003L

where, L = distance between two columns or parts of structure that settle different amounts, H = height of wall.

Approaches for Determining the Net Safe Bearing Pressure

Four approaches may be considered for determining the net safe bearing pressure of soil. They are

1. from field plate load tests,

2. from charts,

3. from empirical equations,

4. from settlement consideration.

18.13 FIELD PLATE LOAD TESTS

The plate load test is a semi-direct method to estimate the allowable bearing pressure of soil to induce a given amount of settlement.

Plates, round or square, varying in size, from 30 to 60 cm and thickness of about 2.5 cm are employed for the test.

The load on the plate is applied by making use of a hydraulic jack. The reaction of the jack load is taken by a cross beam or a steel truss anchored suitably at both the ends. The settlement of the plate is measured by a set of three dial gauges of sensitivity 0.02 mm placed 120° apart. The dial gauges are fixed to independent supports which remain undisturbed during the test.

Figure 18.14 shows the arrangement for a plate load test. The method of performing the test is essentially as follows:

1. Excavate a pit of size not less than 4 to 5 times the size of the plate. The bottom of the pit should coincide with the level of the foundation.

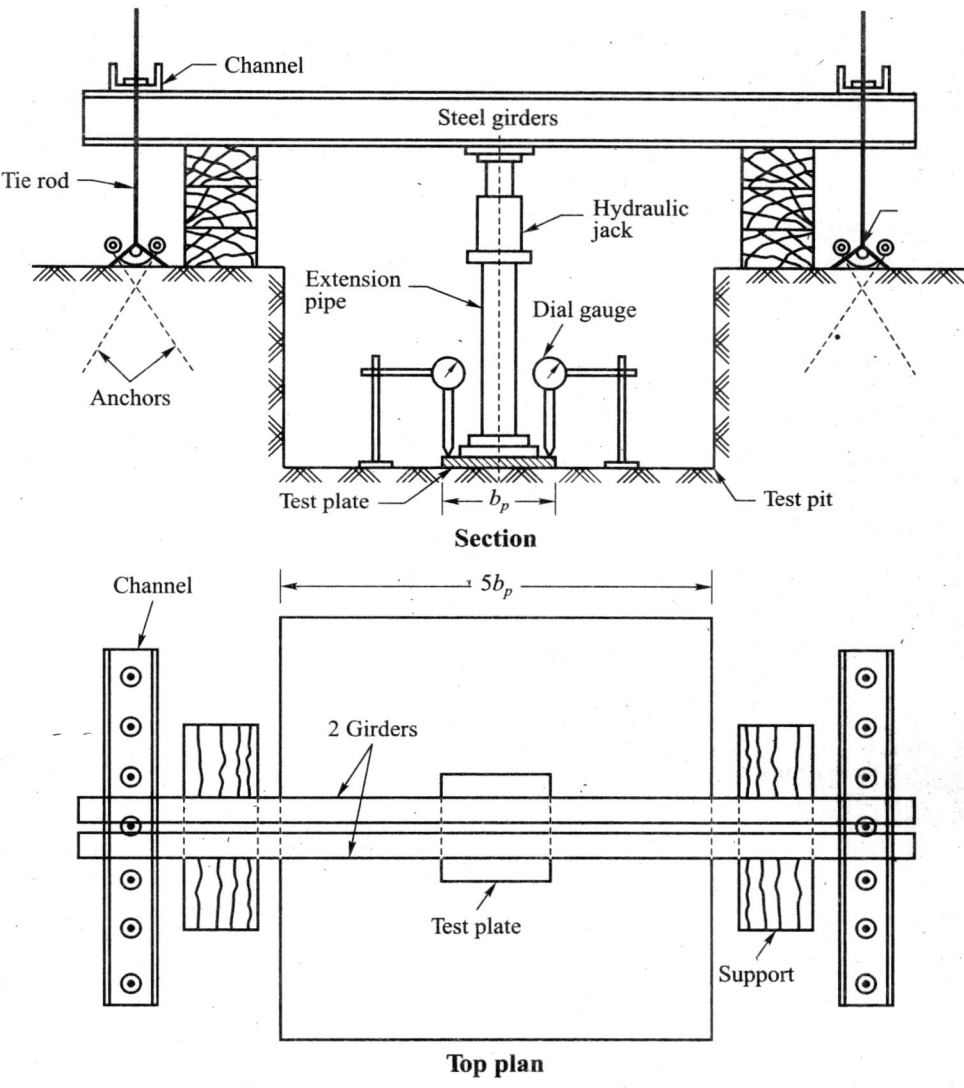

Fig. 18.14 Plate load test arrangement

2. If the water table is above the level of the foundation, pump out the water carefully and keep it at the level of the foundation.

3. A suitable size of plate is selected for the test. Normally a plate of size 30 cm is used in sandy soils and a larger size in clay soils. The ground should be levelled and the plate should be seated over the ground.

4. A seating load of about 70 gm/cm² is first applied and released after some time. A higher load is next placed on the plate and settlements are recorded by means of the dial gauges. Observations on every load increment shall be taken until the rate of settlement is less than

0.25 mm per hour. Load increments shall be approximately one-fifth of the estimated safe bearing capacity of the soil. The average of the settlements recorded by 2 or 3 dial gauges shall be taken as the settlement of the plate for each of the load increments.

5.   The test should continue until a total settlement of 2.5 cm or the settlement at which the soil fails, whichever is earlier, is obtained. After the load is released, the elastic rebound of the soil should be recorded.

From the test results, a load-settlement curve should be plotted as shown in Fig. 18.15. The allowable pressure on a prototype foundation for an assumed settlement may be found by making use of the following equations suggested by Terzaghi and Peck (1948) for square footings in granular soils.

$$S_f = S_p \left(\frac{B(b_p + 0.3)}{b_p(B + 0.3)} \right)^2 \tag{18.38}$$

for clay soils,

$$S_f = S_p \times \frac{B}{b_p} \tag{18.39}$$

where S_f = permissible settlement of foundation in mm, S_p = settlement of plate in mm, B = size of foundation in meters, b_p = size of plate in meters.

The permissible settlement S_f for a prototype foundation should be known. Normally a settlement of 2.5 cm is recommended. In Eq. 18.38 the values of S_f and b_p are therefore known. The unknowns are S_p and B. The value of S_p for any assumed size B may be found from the equation. Using the plate load settlement curve Fig. 18.15 the value of the bearing pressure corresponding to the computed

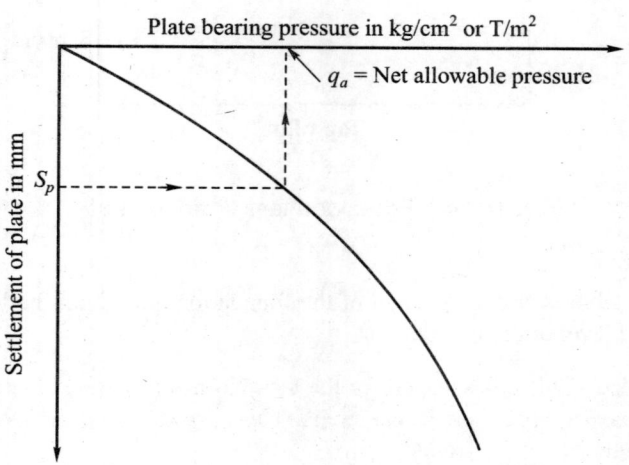

Fig. 18.15   Load settlement curve of a plate load test

value of S_p is found. This bearing pressure is the safe bearing pressure for a given permissible settlement S_f.

Since a load test is of short duration, consolidation settlements cannot be predicted. The test gives the value of immediate settlement only. If the underlying soil is sandy in nature immediate settlement may be taken as the total settlement. If the soil is a clayey type, the immediate settlement is only a fraction of the total settlement. Load tests, therefore, do not have much significance in clayey soils to determine allowable pressure on the basis of a settlement criterion.

Plate load tests should be used with caution and the present practice is not to rely too much on this test. If the soil is not homogeneous to a great depth, plate load tests give very misleading results.

Assume, as shown in Fig. 18.16, two layers of soil. The top layer is stiff clay whereas the bottom layer is soft clay. The load test conducted near the surface of the ground measures the characteristics of the stiff clay but does not indicate the nature of the soft clay soil which is below. The actual foundation of a building however has a bulb of pressure which extends to a great depth into the poor soil which is highly compressible. Here the soil tested by the plate load test gives results which are highly on the unsafe side.

A plate load test is not recommended in soils which are not homogeneous at least to a depth equal to 1½ to 2 times the width of the prototype foundation.

Plate load tests should not be relied on to determine the ultimate bearing capacity of sandy soils as the scale effect gives very misleading results. However, when the tests are carried on clay soils, the ultimate bearing capacity as determined by the test may be taken as equal to that of the foundation since the bearing capacity of clay is essentially independent of the footing size.

The plate load test is possibly the only way of determining the allowable pressures in gravelly deposits. For tests on such deposits the size of the plate should be bigger to eliminate the effect of grain size.

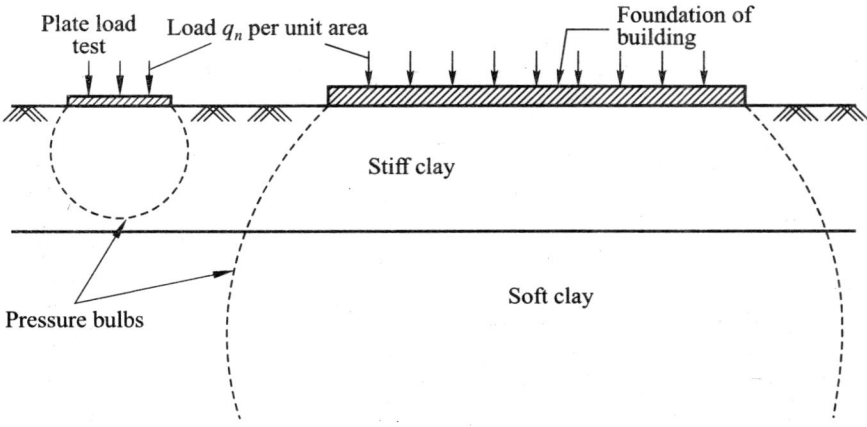

Fig. 18.16 Plate Load Test in non-homogeneous soil

Effect of Water Table on the Allowable Beating Pressure

If the plate load test is carried out above the water table, the allowable pressures as determined from the load settlement curve will be affected if the foundation gets submerged at a later date. The effect of submergence is to reduce the allowable bearing pressure by about 50 percent in granular soils.

Housel's (1929) Method of Determining Safe Bearing Pressure from Settlement Consideration

The method suggested by Housel for determining the safe bearing pressure on settlement consideration is based on the following formula

$$Q = A_p m + P_p n \tag{18.40}$$

where Q = load applied on a given plate, A_p = contact area of plate, P_p = perimeter of plate, m = a constant corresponding to the bearing pressure, n = another constant corresponding to perimeter shear.

Objective

To determine the load Q_f and the size of a foundation for a permissible settlement S_f.

Housel suggests two plate load tests with plates of different sizes, say $B_1 \times B_1$ and $B_2 \times B_2$ for this purpose.

Procedure

1. Two plate load tests are to be conducted at the foundation level of the prototype as per the procedure explained earlier.

2. Draw the load-settlement curves for each of the plate load tests.

3. Select the permissible settlement S_f for the foundation.

4. Determine the loads Q_1 and Q_2 from each of the curves for the given permissible settlement S_f.

Now we may write the following equations

$$Q_1 = mA_{p1} + nP_{p1} \tag{a}$$

for plate load test 1.

$$Q_2 = mA_{p2} + nP_{p2} \tag{b}$$

for plate load test 2.

The unknown values of m and n can be found by solving the above Eqs. (a) and (b). The equation for a prototype foundation may be written as

$$Q_f = mA_f + nP_f \tag{18.41}$$

where A_f = area of the foundation, P_f = perimeter of the foundation.

When A_f and P_f are known, the size of the foundation can be determined.

18.14 SAFE BEARING PRESSURE FROM SPT VALUES

Charts developed by Peck et al. for Cohesionless Soils

Peck at al. (1974) developed a set of charts for determining safe bearing pressures of footings on sand based on N values. These charts are applicable for a settlement of 25 mm and they are the modified version of the charts developed by Terzaghi and Peck (1948). These charts are given in Fig. 18.17. The charts are developed for depth/width ratios of $D_f/B = 0.25, 0.5$ and 1.0. Each line in the chart corresponds to a particular N value and indicates the soil pressure corresponding to a settlement of 1 inch (25 mm). The chart has been developed based on the results of tests on footings of different sizes in sand.

The curves given in Fig. 18.17 are for the condition that the water table is at a depth greater than the width of the foundation. For other positions of the water table, the necessary water table corrections have to be applied as explained earlier. But if the water table is close to the base level of the foundation, the safe bearing pressures from Fig. 18.17 should be reduced by 50 percent.

The chart applies to shallow foundations ($D_f \leq B$) resting on uniform sand for which $\gamma = 100$ lb/ft^3. The N value to be used in the chart must be the one *corrected for dilatancy and overburden pressure* and which must represent the average condition of the soil below the footing up to a depth equal to the width of the footing.

Empirical Equations from SPT Values for Footings on Cohesionless Soils

Footings on granular soils are sometimes proportioned using empirical relationships. Teng (1969) proposed an equation for a settlement of 25 mm based on the curves developed by Terzaghi and Peck (1948). The modified form of the equation is,

$$q_s = 35(N-3)\left(\frac{B+0.3}{2B}\right)^2 R_{w2}\, F_d \ \text{kN/m}^2 \tag{18.42a}$$

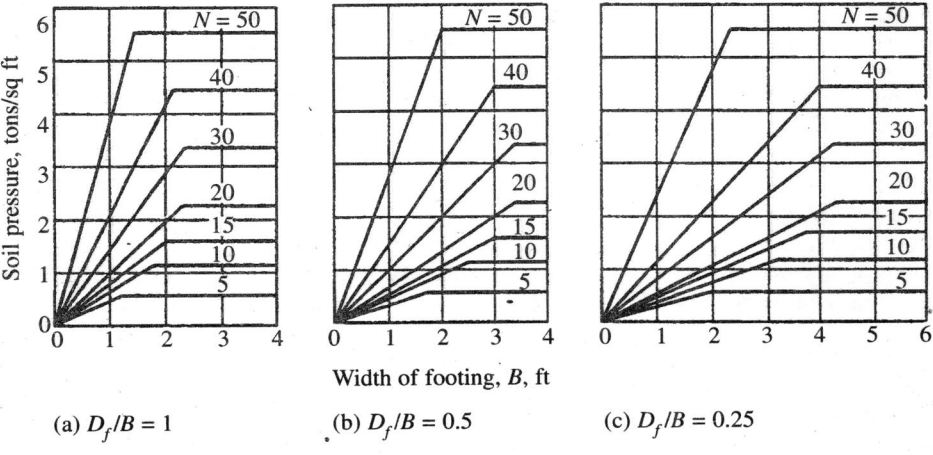

Width of footing, B, ft

   (a) $D_f/B = 1$ (b) $D_f/B = 0.5$ (c) $D_f/B = 0.25$

Fig. 18.17 Design chart for proportioning shallow footings on sand (Peck et al. 1974)

where q_s = net allowable bearing pressure for a settlement of 25 mm in kN/m^2,

$\qquad$ N = corrected standard penetration value,

$$R_{w2} = \text{water table correction factor} = \frac{1}{2}\left(1+\frac{D_{w2}}{B}\right)$$

$$F_d = \text{depth factor} = \left(1+\frac{D_f}{B}\right) \le 2.0$$

$\qquad$ B = width of footing in meters,

$\qquad$ D_f = depth of foundation in meters.

$\qquad$ D_{w2} = depth of WT below the base of foundation.

Meyerhof (1956) proposed the following equations which are slightly different from that of Teng

$$q_s = 12NR_{w2}F_d \text{ kN/m}^2 \text{ for } B \le 1.2 \text{ m} \tag{18.42b}$$

$$q_s = 8N\left(\frac{B+0.3}{2B}\right)^2 R_{w2}F_d \text{ kN/m}^2 \text{ for } B > 1.2 \text{ m} \tag{18.43}$$

where, $F_d = (1+0.33D_f / B) \le 1.33$.

Experimental results indicate that the equations presented by Teng and Meyerhof are too conservative. The values of q_s obtained by the Eqs. 18.42 and 18.43 could be increased as much as 2 to 3 times without excessive settlements occurring. However, Bowles (1996) proposes an approximate increase of 50 percent over that of Meyerhof which can also be applied to Teng's equations. Modified equations of Teng and Meyerhof are,

Teng's equation (modified),

$$q_s = 53(N-3)\left(\frac{B+0.3}{2B}\right)^2 R_{w2}F_d \text{ kN/m}^2 \tag{18.44}$$

Meyerhof's equation (modified) as per Bowles (1988),

$$q_s = 20NR_{w2}F_d \text{ kN/m}^2 \text{ for } B \le 1.2 \text{ m} \tag{18.45}$$

$$q_s = 12.5N\left(\frac{B+0.3}{B}\right)^2 R_{w2}F_d \text{ kN/m}^2 \text{ for } B > 1.2 \text{ m} \tag{18.46}$$

If the tolerable settlement is greater than 25 mm, the safe bearing pressure computed by the above equations can be increased linearly as,

$$q'_s = \frac{S'}{25}q_s \tag{18.47}$$

where q'_s = net safe bearing pressure for a settlement S' mm, q_s = net safe bearing pressure for a settlement of 25 mm.

Safe Bearing Pressures for Raft Foundation on Sand and Clay

Rafts on Sand

Because the differential settlements of a raft foundation are less than those of a footing foundation designed for the same soil pressure, it is reasonable to permit larger safe soil pressures on raft foundation. Experience has shown that a pressure approximately twice as great as that allowed for individual footings may be used because it does not lead to detrimental differential settlements. For a soil pressure that produces a differential settlement of 19 mm (3/4 in), however the maximum settlement of a raft may be about 50 mm (2 in) instead of 25 mm as for a footing foundation.

Peck *et al.* (1974) recommend the following equation for computing net safe pressure,

$$q_s \text{ (tonne/m}^2) = 2.1N, \quad \text{or} \quad q_s = 21N \text{ kN/m}^2 \tag{18.48}$$

for $5 < N < 50$

where, N is the SPT value corrected for dilatancy and overburden pressure.

Equation (18.38) gives q_s values above water table. Necessary correction factor should be used for the presence of water table.

Peck *et al.* (1974) also recommend that the q_s values as given by Eq. (18.48) may be increased some what if bedrock is encountered at a depth less than about one half the width of the raft.

The value of N to be considered is the average of the values obtained upto a depth equal to the least width of the raft. If the average value of N after correction for the influence of overburden pressure and dilatancy is less than about 5, Peck *et al.* say that the sand is generally considered to be too loose for the successful use of the raft foundation. Either the sand should be compacted or else the foundation should be established on piles or piers.

The minimum depth of foundation recommended for a raft is about 2.5 m below the surrounding ground surface. Experience has shown that if the surcharge is less than this amount, the edges of the raft settle appreciably more than the interior because of lack of confinement of the sand.

Safe Bearing Pressures for Rafts on Clay

The quantity in Eq. (18.21b) is the pressure at the elevation of the base of the raft in excess of that exened by the surrounding surcharge. Likewise, in Eq. (18.21c), q_{na} is a net soil pressure. By increasing the depth of excavation, the pressure that can safely be exerted by the building is correspondingly increased.

As for footings on clay, the factor of safety against failure of the soil beneath a raft on clay should not be less than 3 under normal loads, or less than 2 under the most extreme loads.

The settlements of the raft under the given loading condition should be calculated as per the procedures explained in this Chapter. The net safe pressure should be decided on the basis of the permissible settlement.

18.15 SAFE BEARING PRESSURE FROM CPT VALUES FOR FOOTINGS ON COHESIONLESS SOIL

The static cone penetration test in which a standard cone of 10 cm^2 sectional area is pushed into the soil without the necessity of boring provides a much more accurate and detailed variation in the soil as discussed in Chapter 17. Meyerhof (1956) suggested a set of empirical equations based on the Terzaghi and Peck curves (1948). As these equations were also found to be conservative, modified forms with an increase of 50 percent over the original values are given below.

$$q_s = 3.6q_c R_{w2} \text{kN/m}^2 \quad \text{for } B \leq 1.2 \text{ m} \tag{18.49a}$$

$$q_s = 2.1q_c\left(1+\frac{1}{B}\right)^2 R_{w2}\text{kN/m}^2 \quad \text{for } B \leq 1.2 \text{ m} \tag{18.49b}$$

An approximate formula for all widths

$$q_s = 2.7q_c R_{w2}\text{kN/m}^2 \tag{18.49c}$$

where q_c is the cone point resistance in kg/cm^2 and q_s in kN/m^2.

The above equations have been developed for a settlement of 25 mm.

Meyerhof (1956) developed his equations based on the relationship $q_c = 4N$ kg/cm^2 for penetration resistance in sand where N is the SPT value. However, this relationship varies according to the investigators.

Example 18.20

A plate load test using a plate of size 30 × 30 cm was carried out at the level of a prototype foundation. The soil at the site was cohesionless with the water table at great depth. The plate settled by 10 mm at a load intensity of 160 kN/m^2. Determine the settlement of a square footing of size 2 × 2 m under the same load intensity.

Solution

The settlement of the foundation S_f may be determined from Eq. (18.38).

$$S_f = S_p\left(\frac{B(b_p+0.3)}{b_p(B+0.3)}\right)^2 = 10\left(\frac{2(0.3+0.3)}{0.3(2+0.3)}\right)^2 = 30.24 \text{ mm}$$

Example 18.21

For Ex. 18.20 estimate the load intensity if the permissible settlement of the prototype foundation is limited to 40 mm.

Solution

In Ex. 18.20, a load intensity of 160 kN/m^2 induces a settlement of 30.24 mm. If we assume that the load-settlement is linear within a small range, we may write

$$\frac{q_2}{q_1} = \frac{S_{f2}}{S_{f1}} \text{ or } q_2 = q_1 \times \frac{S_{f2}}{S_{f1}}$$

where, $q_1 = 160$ kN/m^2, $S_{f1} = 30.24$ mm, $S_{f2} = 40$ mm. Substituting the known values

$$q_2 = 160 \times \frac{40}{30.24} = 211.64 \text{ kN/m}^2$$

Example 18.22

Two plate load tests were conducted at the level of a prototype foundation in cohesionless soil close to each other. The following data are given:

Size of plate	Load applied	Settlement recorded
0.3 × 0.3 m	30 kN	25 mm
0.6 × 0.6 m	90 kN	25 mm

If a footing is to carry a load of 100 kN, determine the required size of the footing for the same settlement of 25 mm.

Solution

Use Eq. (18.40). For the two plate load tests we may write:

$$Q_1 = A_{p1}m + P_{p1}n$$

$$Q_2 = A_{p2}m + P_{p2}n$$

PLT 1: $A_{p1} = 0.3 \times 0.3 = 0.09$ m^2; $P_{p1} = 0.3 \times 4 = 1.2$ m; $Q_1 = 30$ kN

PLT 2: $A_{p2} = 0.6 \times 0.6 = 0.36$ m^2; $P_{p2} = 0.6 \times 4 = 2.4$ m; $Q_2 = 90$ kN

Now we have

30 = 0.09 m + 1.2 n

90 = 0.36 m + 2.4 n

On solving the equations we have

$m = 166.67$, and $n = 12.5$

For prototype foundation, we may write

$$Q_f = 166.67 A_f + 12.5 P_f$$

Assume the size of the footing as $B \times B$, we have

$A_f = B^2$, $P_f = 4B$, and $Q_f = 1000$ kN

Substituting we have

$$1000 = 166.67 B^2 + 50B$$

or $B^2 + 0.3B - 6 = 0$

The solution gives $B = 2.3$ m.

The size of the footing $= 2.3 \times 2.3$ m.

Example 18.23

A square footing of size 4×4 m is founded at a depth of 2 m below the ground surface in loose to medium dense sand. The corrected standard penetration test value $N = 11$. Compute the net safe bearing pressure for a settlement of 40 mm by the use of modified equations of (a) Teng, and (b) Meyerhof.

Solution

$$q_s = 53(N-3)\left(\frac{B+0.3}{2B}\right)^2 R_{w2} F_d \ \text{kN/m}^2$$

where $R_{w2} = \dfrac{1}{2}\left(1 + \dfrac{D_{w2}}{B}\right) = 0.5$ since $D_{w2} = 0$

$$F_d = \left(1 + \frac{D_f}{B}\right) = \left(1 + \frac{2}{4}\right) = 1.5$$

$N = 11$, $B = 4$ m.

By substituting

$$q_s = 53(11-3)\left(\frac{4.3}{8}\right)^2 \times 0.5 \times 1.5 = 92 \ \text{kN/m}^2$$

(b) Meyerhof's equation (modified), Eq. (18.46)

$$q_s = 12.5N\left(\frac{B+0.3}{B}\right)^2 R_{w2} F_d$$

where $R_{w2} = 0.5$, $F_d = \left(1 + 0.33 \times \dfrac{D_f}{B}\right) = \left(1 + 0.33 \times \dfrac{2}{4}\right) = 1.165$

By substituting

$$q_s = 12.5 \times 11\left(\frac{4.3}{4}\right)^2 \times 0.5 \times 1.165 = 93 \ \text{kN/m}^2$$

Note: Both the methods give the same result.

Example 18.24

For the Ex. 18.23, compute q_s by making use of the chart of Peck $et\ al.$

Solution

Refer chart Fig. 18.17(b) for $D_f/B = 2/4 = 0.5$.

For $N = 11$, the safe pressure $q_s = 1.15$ tonne/ft$^2 = 1.15 \times 95.765 = 110$ kN/m^2

This value is quite comparable with the values obtained in Ex. 18.23.

Example 18.25

A footing of size 3×3 m is to be constructed at a site at a depth of 1.5 m below the ground surface. The water table is at the base of the foundation. The average static cone penetration resistance obtained at the site is 20 kg/cm^2. The soil is cohesive. Determine the safe bearing pressure for a settlement of 40 mm.

Solution

Use Eq. (18.49b)

$$q_s = 2.1q_c\left(1+\frac{1}{3}\right)^2 R_{w2}\text{kN/m}^2$$

where $q_c = 20$ kg/cm^2, $B = 3$ m, $R_{w2} = 0.5$.

This equation is for 25 mm settlement. By substituting, we have

$$q_s = 2.1\times 20\left(1+\frac{1}{3}\right)^2 \times 0.5 = 37.3 \text{ kN/m}^2$$

For 40 mm settlement, the value of q_s is

$$q_s = 37.3\left(\frac{40}{25}\right) \approx 60 \text{ kN/m}^2$$

18.16 FOUNDATION SETTLEMENT

Components of Total Settlement

The total settlement of a foundation comprises three parts as follows

$$S = S_e + S_c + S_s \tag{18.50}$$

where S = total settlement

S_e = elastic or immediate settlement

S_c = consolidation settlement

S_s = secondary settlement

Immediate settlement, S_e, is that part of the total settlement, S, which is supposed to take place during the application of loading. The consolidation settlement is that part which is due to the expulsion of pore water from the voids and is time-dependent settlement. Secondary settlement normally starts with the completion of the consolidation. It means, during the stage of this settlement, the pore water pressure is zero and the settlement is only due to the distortion of the soil skeleton.

Footings founded in cohesionless soils reach almost the final settlement, S, during the construction stage itself due to the high permeability of soil. The water in the voids is expelled simultaneously with the application of load and as such the immediate and consolidation settlements in such soils are rolled into one.

In cohesive soils under saturated conditions, there is no change in the water content during the stage of immediate settlement. The soil mass is deformed without any change in volume soon after the application of the load. This is due to the low permeability of the soil. With the advancement of time there will be gradual expulsion of water under the imposed excess load. The time required for the complete expulsion of water and to reach zero water pressure may be several years depending upon the permeability of the soil. Consolidation settlement may take many years to reach its final stage. Secondary settlement is supposed to take place after the completion of the consolidation settlement, though in some of the organic soils there will be overlapping of the two settlements to a certain extent.

Immediate settlements of cohesive soils and the total settlement of cohesionless soils may be estimated from elastic theory. The stresses and displacements depend on the stress-strain characteristics of the underlying soil. A realistic analysis is difficult because these characteristics are nonlinear. Results from the theory of elasticity are generally used in practice, it being assumed that the soil is homogeneous and isotropic and there is a linear relationship between stress and strain. A linear stress-strain relationship is approximately true when the stress levels are low relative to the failure values. The use of elastic theory clearly involves considerable simplification of the real soil.

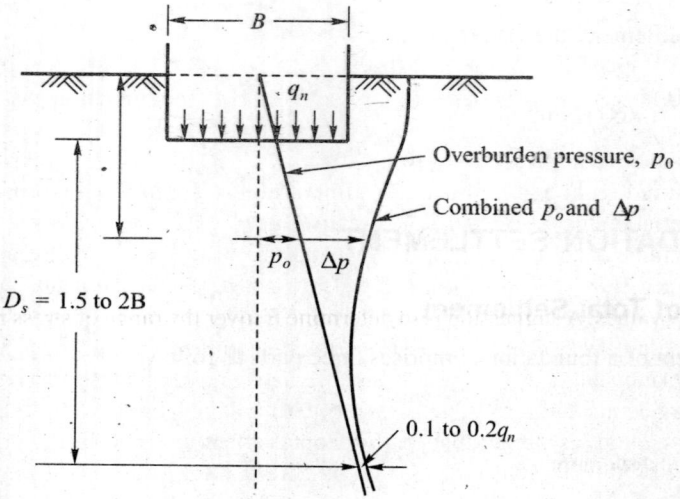

Fig. 18.18   Overburden pressure and vertical stress distribution

Some of the results from elastic theory require knowledge of Young's modulus (E_s), here called e compression or deformation modulus, E_d, and Poisson's ratio, μ, for the soil.

eat of Settlement

ootings founded at a depth D_f below the surface settle under the imposed loads due to the ompressibility characteristics of the subsoil. The depth through which the soil is compressed depends oon the distribution of effective vertical pressure p'_0 of the overburden and the vertical induced ress Δp resulting from the net foundation pressure q_n as shown in Fig. 18.18.

In the case of deep compressible soils, the lowest level considered in the settlement analysis is e point where the vertical induced stress Δp is of the order of 0.1 to $0.2q_n$, where q_n is the net ressure on the foundation from the superstructure. This depth works out to about 1.5 to 2 times the idth of the footing. The soil lying within this depth gets compressed due to the imposed foundation ressure and causes more than 80 percent of the settlement of the structure. This depth D_s is called as e *zone of significant stress*. If the thickness of this zone is more than 3 m, the steps to be followed the settlement analysis are:

1. Divide the zone of significant stress into layers of thickness not exceeding 3 m,

2. Determine the effective overburden pressure p'_0 at the center of each layer,

3. Determine the increase in vertical stress Δp due to foundation pressure q_n at the center of each layer along the center line of the footing by the theory of elasticity,

4. Determine the average modulus of elasticity and other soil parameters for each of the layers.

8.17 EVALUATION OF MODULUS OF ELASTICITY

he most difficult part of a settlement analysis is the evaluation of the modulus of elasticity E_s, that vould conform to the soil condition in the field. There are two methods by which E_s can be evaluated. They are:

1. Laboratory method,

2. Field method.

Laboratory Method

For settlement analysis, the values of E_s at different depths below the foundation base are required. One way of determining E_s is to conduct triaxial tests on representative undisturbed samples extracted from the depths required. For cohesive soils, undrained triaxial tests and for cohesionless soils drained triaxial tests are required. Since it is practically impossible to obtain undisturbed sample of cohesionless soils, the laboratory method of obtaining E_s can be ruled out. Even with regards to cohesive soils, there will be disturbance to the sample at different stages of handling it, and as such the values of E_s obtained from undrained triaxial tests do not represent the actual conditions and normally give very low values. A suggestion is to determine E_s over the range of stress relevant to the particular problem. Poulos *et al.*, (1980) suggest that the undisturbed triaxial specimen be given a preliminary preconsolidation under K_0 conditions with an axial stress equal to the effective overburden pressure at the sampling depth. This procedure attempts to return the specimen to its original state of effective stress in the ground, assuming that the horizontal effective stress in the ground was the same as that produced by the laboratory K_0 condition. Simons and Som (1970) have shown that triaxial tests on London clay in which specimens were brought back to their original *in situ* stresses gave elastic moduli which were much higher than those obtained from conventional undrained triaxial tests. This has been confirmed by Marsland (1971) who carried out 865 mm diameter plate loading

tests in 900 mm diameter bored holes in London clay. Marsland found that the average mod
determined from the loading tests were between 1.8 to 4.8 times those obtained from undrai
triaxial tests. A suggestion to obtain the more realistic value for E_s is,

1. Undisturbed samples obtained from the field must be reconsolidated under a stress syst
    equal to that in the field (K_0-condition),

2. Samples must be reconsolidated isotropically to a stress equal to 1/2 to 2/3 of the in-s
    vertical stress.

It may be noted here that reconsolidation of disturbed sensitive clays would lead to signific
change in the water content and hence a stiffer structure which would lead to a very high E_s.

Because of the many difficulties faced in selecting a modulus value from the results of laborato
tests, it has been suggested that a correlation between the modulus of elasticity of soil and the undrain
shear strength may provide a basis for settlement calculation. The modulus E_s may be expressed

$$E_s = Ac_u \tag{18.5}$$

where the value of A for inorganic stiff clay varies from about 500 to 1500 (Bjerrum, 1972) and c_u
the undrained cohesion. It may generally be assumed that highly plastic clays give lower values f
A, and low plasticity give higher values for A. For organic or soft clays the value of A may vary fro
100 to 500. The undrained cohesion c_u can be obtained from any one of the field tests mentione
below and also discussed in Chapter 17.

Field methods

Field methods are increasingly used to determine the soil strength parameters. They have been foun
to be more reliable than the ones obtained from laboratory tests. The field tests that are normall
used for this purpose are:

1. Plate load tests (PLT)

2. Standard penetration test (SPT)

3. Static cone penetration test (CPT)

4. Pressuremeter test (PMT)

5. Flat dilatometer test (DMT)

TABLE 18.7

Equations for computing E_s by making use of SPT and CPT values (in kPa)

Soil	SPT	CPT
Sand (normally consolidated)	$500 (N_{cor} + 15)$ $(35000 \text{ to } 50000) \log N_{cor}$ (U.S.S.R. Practice)	$2 \text{ to } 4 \, q_c$ $(1 + D_r^2) \, q_c$
Sand (saturated)	$250 (N_{cor} + 15)$	
Sand (overconsolidated)	–	$6 \text{ to } 30 \, q_c$
Gravelly sand and gravel	$1200 (N_{cor} + 6)$	
Clayey sand	$320 (N_{cor} + 15)$	$3 \text{ to } 6 \, q_c$
Silty sand	$300 (N_{cor} + 6)$	$1 \text{ to } 2 \, q_c$
Soft clay	–	$3 \text{ to } 8 \, q_c$

Plate load tests, if conducted at levels at which E_s is required, give quite reliable values as
npared to laboratory tests. Since these tests are too expensive to carry out, they are rarely used
:ept in major projects.

Many investigators have obtained correlations between E_s and field tests such as SPT, CPT and
IT. The correlations between E_s and SPT or CPT are applicable mostly to cohesionless soils and
some cases cohesive soils under undrained conditions. PMT can be used for cohesive soils to
:ermine both the immediate and consolidation settlements together.

Some of the correlations of E_s with N and q_c are given in Table 18.7. These correlations have
en collected from various sources.

3.18 METHODS OF COMPUTING SETTLEMENTS

any methods are available for computing elastic (immediate) and consolidation settlements. Only
ose methods that are of practical interest are discussed here. The various methods discussed in this
apter are the following:

omputation of Elastic Settlements

1. Elastic settlement based on the theory of elasticity

2. Janbu et al., (1956) method of determining settlement under an undrained condition.

3. Schmertmann's method of calculating settlement in granular soils by using CPT values.

omputation of Consolidation Settlement

1. *e*-log *p* method by making use of oedometer test data.

2. Skempton-Bjerrum method.

8.19 ELASTIC SETTLEMENT BENEATH THE CORNER OF A UNIFORMLY LOADED FLEXIBLE AREA BASED ON THE THEORY OF ELASTICITY

he net elastic settlement equation for a flexible surface footing may be written as,

$$S_e = q_n B \frac{(1-\mu^2)}{E_s} I_f \tag{18.52a}$$

vhere S_e = elastic settlement

B = width of foundation,

E_s = modulus of elasticity of soil,

μ = Poisson's ratio,

q_n = net foundation pressure,

I_f = influence factor.

In Eq. (18.52a), for saturated clays, $\mu = 0.5$, and E_s is to be obtained under undrained conditions
as discussed earlier. For soils other than clays, the value of μ has to be chosen suitably and the

corresponding value of E_s has to be determined. Table 18.8 gives typical values for μ as sugges by Bowles (1996).

I_f is a function of the L/B ratio of the foundation, and the thickness H of the compressible lay Terzaghi has a given a method of calculating I_f from curves derived by Steinbrenner (1934),

for Poisson's ratio of 0.5, $I_f = F_1$,

for Poisson's ratio of zero, $I_f = F_1 + F_2$.

where F_1 and F_2 are factors which depend upon the ratios of H/B and L/B.

For intermediate values of μ, the value of I_f can be computed by means of interpolation or the equation

$$I_f = \left[F_1 + \frac{\left(1-\mu-2\mu^2\right)F_2}{1-\mu^2} \right]$$

(18.52

The values of F_1 and F_2 are given in Fig. 18.19a. The elastic settlement at any point (Fig. 18.19b) is given by

$$S_e \text{ at point } N = \frac{q_n\left(1-\mu^2\right)}{E_s}\left[I_{f1}B_1 + I_{f2}B_2 + I_{f3}B_3 + I_{f4}B_4\right]$$

(18.52

To obtain the settlement at the center of the loaded area, the principle of superposition is followe In such a case N in Fig. 18.19b will be at the center of the area when $B_1 = B_4 = L_2 = B_3$ and $B_2 = L$ Then the settlement at the centre is equal to four times the settlement at any one corner. The curv in Fig. 18.19a are based on the assumption that the modulus of deformation is constant with dept

In the case of a rigid foundation, the immediate settlement at the center is approximately 0 times that obtained for a flexible foundation at the center. A correction factor is applied to the immedia settlement to allow for the depth of foundation by means of the depth factor d_f. Fig. 18.20

gives Fox's (1948) correction curve for depth factor. The final elastic settlement is

$$S_{ef} = C_r d_f S_e$$

(18.53

where, S_{ef} = final elastic settlement

TABLE 18.8

Typical range of values for Poisson's ratio (Bowles, 1996)

Type of soil	μ
Clay, saturated	0.4–0.5
Clay, unsaturated	0.1–0.3
Sandy clay	0.2–0.3
Silt	0.3–0.35
Sand (dense)	0.2–0.4
Coarse (void ratio = 0.4 to 0.7)	0.15
Fine grained (void ratio = 0.4 to 0.7)	0.25
Rock	0.1–0.4

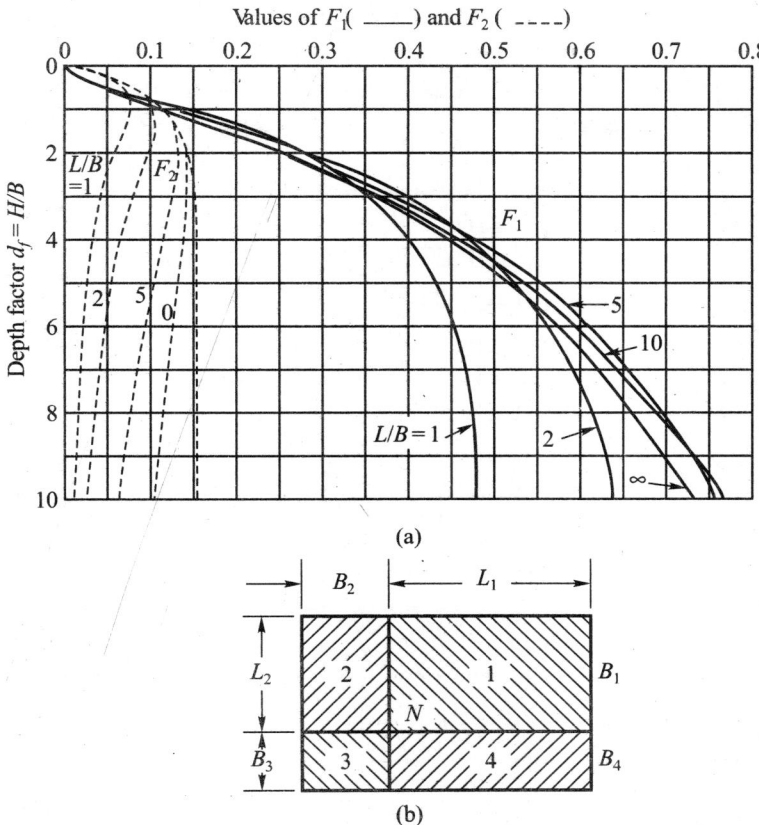

Fig. 18.19 Settlement due to load on surface of elastic layer (a) F_1 and F_2 versus H/B, (b) Method of estimating settlement (After Steinbrenner, 1934)

TABLE 18.9

Influence factor I_f (Bowles, 1988)

Shape	I_f (average values)	
	Flexible footing	Rigid footing
Circle	0.85	0.88
Square	0.95	0.82
Rectangle	1.20	1.06
L/B = 1.5	1.20	1.06
2.0	1.31	1.20
5.0	1.83	1.70
10.0	2.25	2.10
100.0	2.96	3.40

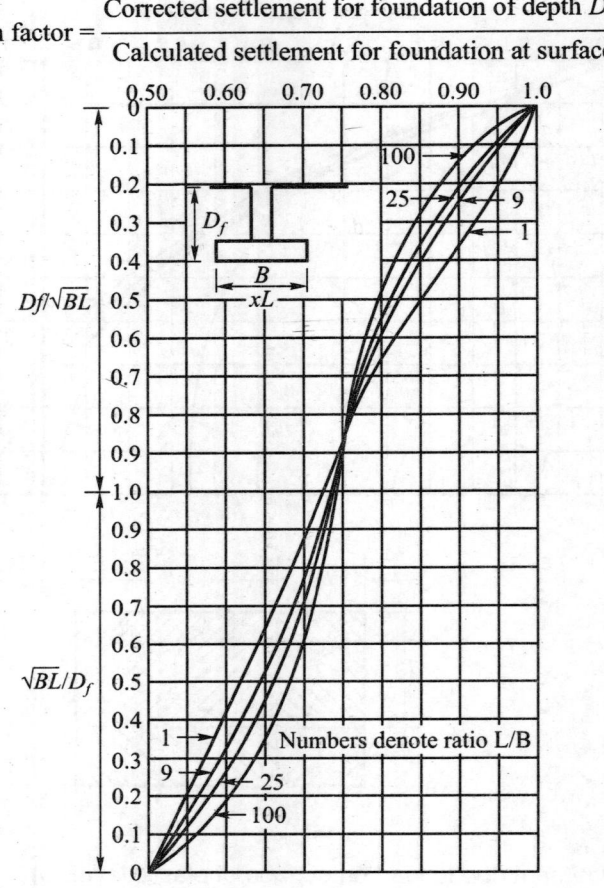

Fig. 18.20 Correction curves for elastic settlement of flexible rectangular foundations at depth (Fox, 1948)

C_r = rigidity factor taken as equal to 0.8 for a highly rigid foundation

d_f = depth factor from Fig. 18.20

S_e = settlement for a surface flexible footing

Bowles (1996) has given the influence factor for various shapes of rigid and flexible footings as shown in Table 18.9.

18.20 JANBU, BJERRUM AND KJAERNSLI'S METHOD OF DETERMINING ELASTIC SETTLEMENT UNDER UNDRAINED CONDITIONS

Probably the most useful chart is that given by Janbu et al., (1956) as modified by Christian and Carrier (1978) for the case of a constant E_s with respect to depth. The chart (Fig. 18.21) provides estimates of the average immediate settlement of uniformly loaded, flexible strip, rectangular, square or circular footings on homogeneous isotropic saturated clay. The equation for computing the settlement may be expressed as

$$S_e = \frac{\mu_0 \mu_1 q_n B}{E_s}$$

(18.54)

In Eq.(18.52), Poisson's ratio is assumed equal to 0.5. The factors μ_0 and μ_1 are related to the D_f/B and H/B ratios of the foundation as shown in Fig. 18.21. Values of μ_1 are given for various L/B ratios. Rigidity and depth factors are required to be applied to Eq. (18.54) as per Eq. (18.53). In Fig. 18.21 the thickness of compressible strata is taken as equal to H below the base of the foundation where a hard stratum is met with.

Generally, real soil profiles which are deposited naturally consist of layers of soils of different properties underlain ultimately by a hard stratum. Within these layers, strength and moduli generally increase with depth. The chart given in Fig. 18.21 may be used for the case of E_s increasing with depth by replacing the multilayered system with one hypothetical layer on a rigid base. The depth of this hypothetical layer is successively extended to incorporate each real layer, the corresponding values of E_s being ascribed in each case and settlements calculated. By subtracting the effect of the hypothetical layer above each real layer the separate compression of each layer may be found and summed to give the overall total settlement.

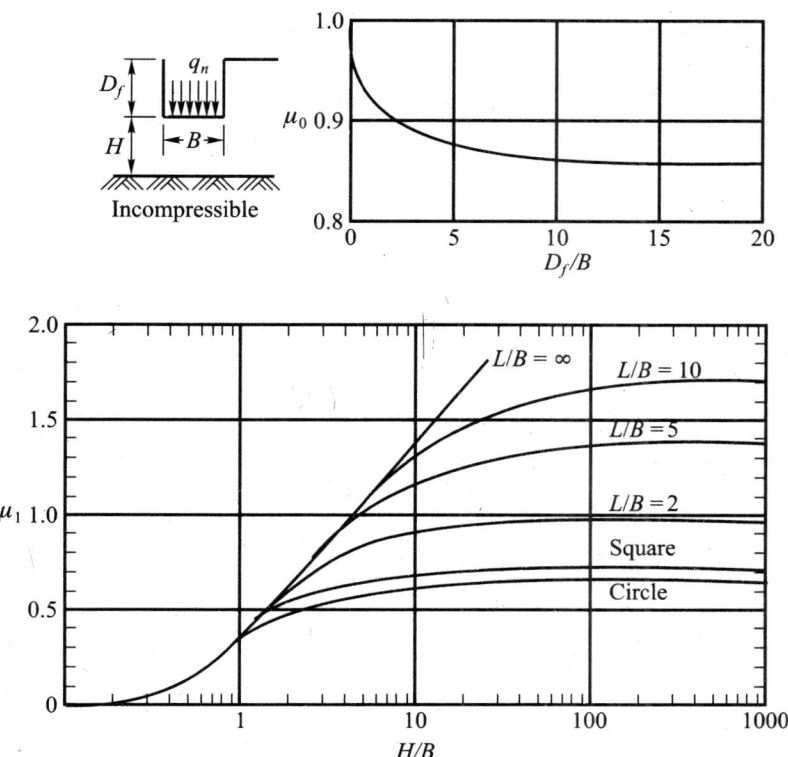

Fig. 18.21 Factors for calculating the average immediate settlement of a loaded area (after Christian and Carrier, 1978)

18.21 SCHMERTMANN'S METHOD OF CALCULATING SETTLEMENT IN GRANULAR SOILS BY USING CPT VALUES

It is normally taken for granted that the distribution of vertical strain under the center of a footing over uniform sand is qualitatively similar to the distribution of the increase in vertical stress. If true, the greatest strain would occur immediately under the footing, which is the position of the greatest stress increase. The detailed investigations of Schmertmann (1970), Eggestad, (1963) and others, indicate that the greatest strain would occur at a depth equal to half the width for a square or circular footing. The strain is assumed to increase from a minimum at the base to a maximum at $B/2$, then decrease and reaches zero at a depth equal to $2B$. For strip footings of $L/B > 10$, the maximum strain is found to occur at a depth equal to the width and reaches zero at a depth equal to $4B$. The modified triangular vertical strain influence factor distribution diagram as proposed by Schmertmann (1978) is shown in Fig. 18.22. The area of this diagram is related to the settlement. The equation (for square as well as circular footings) is

$$S = C_1 C_2 q_n \sum_0^{2B} \frac{I_z}{E_s} \Delta z \qquad (18.55)$$

where, S = total settlement,

q_n = net foundation base pressure = $(q - q'_0)$,

q = total foundation pressure,

q'_0 = effective overburden pressure at foundation level,

Δz = thickness of elemental layer,

I_z = vertical strain influence factor,

C_1 = depth correction factor,

C_2 = creep factor.

The equations for C_1 and C_2 are

$$C_1 = 1 - 0.5 \left(\frac{q'_0}{q_n} \right) \qquad (18.56)$$

$$C_2 = 1 + 0.2 \log_{10} \left(\frac{t}{0.1} \right) \qquad (18.57)$$

where t is time in years for which period settlement is required.

Equation (18.57) is also applicable for $L/B \geq 10$ except that the summation is from 0 to $4B$.

The modulus of elasticity to be used in Eq. (18.55) depends upon the type of foundation as follows:

For a square footing,

$$E_s = 2.5 q_c \qquad (18.58)$$

For a strip footing, $L/B \geq 10$,

$$E_s = 3.5 q_c \qquad (18.59)$$

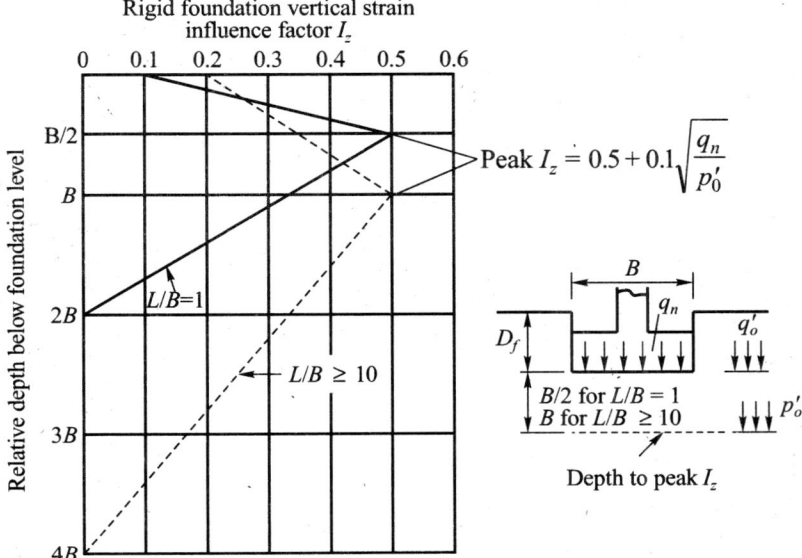

Fig. 18.22 Vertical strain influence factor diagrams (after Schmertmann et al., 1978)

Figure 18.22 gives the vertical strain influence factor I_z distribution for both square and strip foundations if the ratio $L/B \geq 10$. Values for rectangular foundations for $L/B < 10$ can be obtained by interpolation. The depths at which the maximum I_z occurs may be calculated as follows (Fig 18.22),

$$I_z = 0.5 + 0.1 \sqrt{\frac{q_n}{p_0'}} \qquad (18.60)$$

where p_0' = effective overburden pressure at depths $B/2$ and B for square and strip foundations respectively.

Further, I_z is equal to 0.1 at the base and zero at depth $2B$ below the base for square footing; whereas for a strip foundation it is 0.2 at the base and zero at depth $4B$.

Values of E_s given in Eqs (18.58) and (18.59) are suggested by Schmertmann (1978). Lunne and Christoffersen (1985) proposed the use of the tangent modulus on the basis of a comprehensive review of field and laboratory tests as follows:

For normally consolidated sands,

$$E_s = 4q_c \text{ for } q_c < 10 \qquad (18.61)$$

$$E_s = (2q_c + 20) \text{ for } 10 < q_c < 50 \qquad (18.62)$$

$$E_s = 120 \text{ for } q_c > 50 \qquad (18.63)$$

For overconsolidated sands with an overconsolidation ratio greater than 2,

$$E_s = 5q_c \text{ for } q_c < 50 \qquad (18.64a)$$

$$E_s = 250 \text{ for } q_c > 50 \qquad (18.64b)$$

where E_s and q_c are expressed in MPa.

The cone resistance diagram is divided into layers of approximately constant values of q_c and the strain influence factor diagram is placed alongside this diagram beneath the foundation which is drawn to the same scale. The settlements of each layer resulting from the net contact pressure q_n are then calculated using the values of E_s and I_z appropriate to each layer. The sum of the settlements in each layer is then corrected for the depth and creep factors using Eqs (18.56) and (18.57) respectively.

Example 18.26

Estimate the immediate settlement of a concrete footing 1.5 × 1.5 m in size founded at a depth of 1 m in silty soil whose modulus of elasticity is 90 kg/cm². The footing is expected to transmit a unit pressure of 200 kN/m².

Solution

Use Eq. (18.52a)

Immediate settlement,

$$S_e = qB\frac{(1-\mu^2)}{E_s}I_f$$

Assume $\mu = 0.35$, $I_f = 0.82$ for a rigid footing.

Given: $q = 200$ kN/m², $B = 1.5$ m, $E_s = 90$ kg/cm² ≈ 9000 kN/m².

By substituting the known values, we have

$$S_e = 200 \times 1.5 \times \frac{1-0.35^2}{9000} \times 0.82 = 0.024 \text{ m} = 2.4 \text{ cm}$$

Example 18.27

A square footing of size 8 × 8 m is founded at a depth of 2 m below the ground surface in loose to medium dense sand with $q_n = 120$ kN/m². Standard penetration tests conducted at the site gave the following corrected N_{60} values.

Depth below G.L. (m)	N_{cor}	Depth below GL (m)	N_{cor}
2	8	12	16
4	8	14	18
6	12	16	17
8	12	18	20
10	11		

The water table is at the base of the foundation. Above the water table $\gamma = 16.5$ kN/m³, and submerged $\gamma_b = 8.5$ kN/m³.

Compute the elastic settlement by Eq. (18.52a). Use the equation $E_s = 250 (N_{cor} + 15)$ for computing the modulus of elasticity of the sand. Assume $\mu = 0.3$ and the depth of the compressible layer = $2B = 16$ m (= H).

Solution

For computing the elastic settlement, it is essential to determine the weighted average value of N_{cor}. The depth of the compressible layer below the base of the foundation is taken as equal to 16 m (= H). This depth may be divided into three layers in such a way that N_{cor} is approximately constant in each layer as given below.

Layer No.	Depth (m)		Thickness	N_{cor}
	From	To	(m)	
1	2	5	3	9
2	5	11	6	12
3	11	18	7	17

The weighted average

$$N_{cor}(av) = \frac{9 \times 3 + 12 \times 6 + 17 \times 7}{16} = 13.6 \text{ or say } 14$$

From equation $E_s = 250(N_{cor} + 15)$ we have

$$E_s = 250(14 + 15) = 7250 \text{ kN/m}^2$$

The total settlement of the center of the footing of size 8 × 8 m is equal to four times the settlement of a corner of a footing of size 4 × 4 m.

In the Eq. (18.52a), $B = 4$ m, $q_n = 120$ kN/m², $\mu = 0.3$.

Now from Fig. 18.19, for $H/B = 16/4 = 4$, $L/B = 1$

$F_1 = I_f = 0.4$ for $\mu = 0.5$

$F_2 = 0.03$ for $\mu = 0.5$

Now from Eq. (18.52 b) I_f for $\mu = 0.3$ is

$$I_f = F_1 + \frac{(1 - \mu - 2\mu^2)F_2}{1 - \mu^2} = 0.40 + \frac{(1 - 0.3 - 2 \times 0.3^2)}{1 - 0.3^2} \times 0.03 = 0.42$$

From Eq. (18.52a) we have settlement of a corner of a footing of size 4 × 4 m as

$$S_e = q_n B \frac{(1 - \mu^2)}{E_s} I_f = \frac{120 \times 4(1 - 0.3^2)}{7250} \times 0.42 \times 100 = 2.53 \text{ cm}$$

With the correction factor, the final elastic settlement from Eq. (18.53) is

$$S_{ef} = C_r d_f S_e$$

where C_r = rigidity factor = 1 for flexible footing d_f = depth factor

From Fig. 18.20 for

$$\frac{D_f}{\sqrt{BL}} = \frac{2}{\sqrt{4 \times 4}} = 0.5, \quad \frac{L}{B} = \frac{4}{4} = 1 \text{ we have } d_f = 0.85$$

Now $S_{ef} = 1 \times 0.85 \times 2.53 = 2.15 \text{ cm}$

The total elastic settlement of the center of the footing is

$$S_e = 4 \times 2.15 = 8.6 \text{ cm} = 86 \text{ mm}$$

Per Table 18.4, the maximum permissible settlement for a raft foundation in sand is 62.5 mm. Since the calculated value is higher, the contact pressure q_n has to be reduced.

Example 18.28

It is proposed to construct an overhead tank at a site on a raft foundation of size 8 × 12 m with the footing at a depth of 2 m below ground level. The soil investigation conducted at the site indicates that the soil to a depth of 20 m is normally consolidated insensitive inorganic clay with the water table 2 m below ground level. Static cone penetration tests were conducted at the site using a mechanical cone penetrometer. The average value of cone penetration resistance q_c was found to be 1540 kN/m² and the average saturated unit weight of the soil = 18 kN/m³. Determine the immediate settlement of the foundation using Eq. (18.54). The contact pressure $q_n = 100$ kN/m² (= 0.1 MPa). Assume that the stratum below 20 m is incompressible.

Solution

Computation of the modulus of elasticity

Use Eq. (18.51) with $A = 500$

$$E_s = 500\, c_u$$

where c_u = the undrained shear strength of the soil

From Eq. (17.11)

$$c_u = \frac{\overline{q}_c - p_a}{N_k}$$

where　$\overline{q}_c$　=　average static cone penetration resistance = 1540 kN/m²

　　　　p_o　=　average total overburden pressure = $10 \times 18 = 180$ kN/m²

　　　　N_k　=　20 (assumed)

Therefore　$c_u = \dfrac{1540 - 180}{20} = 68\, \text{kN/m}^2$

$E_s = 500 \times 68 = 34{,}000\, \text{kN/m}^2 = 34\, \text{MPa}$

Equation (18.54) for S_e is

$$S_e = \frac{\mu_0 \mu_1 q_n B}{E_s}$$

From Fig. 18.21 for $D_f/B = 2/8 = 0.25$, $\mu_0 = 0.95$, for $H/B = 16/8 = 2$ and $L/B = 12/8 = 1.5$, $\mu_1 = 0.6$. Substituting

$$S_e(\text{average}) = \frac{0.95 \times 0.6 \times 0.1 \times 8}{34} = 0.0134\, \text{m} = 13.4\, \text{mm}$$

From Fig. 18.20 for $D_f / \sqrt{BL} = 2/\sqrt{8 \times 12} = 0.2$, $L/B = 1.5$ the depth factor $d_f = 0.94$

The corrected settlement S_{ef} is

$$S_{ef} = 0.94 \times 13.4 = 12.6\, \text{mm}$$

Example 18.29

Refer to Example 18.27. Estimate the elastic settlement by Schmertmann's method by making use of the relationship $q_c = 4\, N_{cor}$ kg/cm² where q_c = static cone penetration value in kg/cm². Assume settlement is required at the end of a period of 3 years.

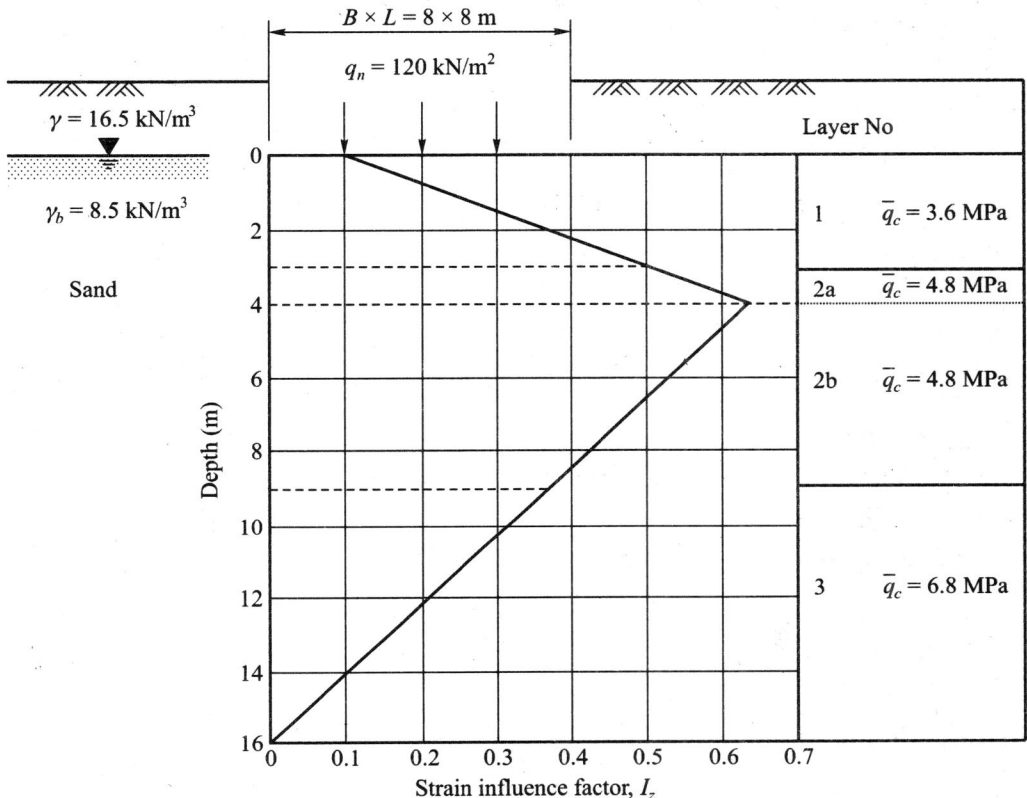

Fig. Ex. 18.29

Solution

The average value of for N_{cor} each layer given in Ex. 18.21 is given below

Layer No	Average	Average q_c	
	N_{cor}	kg/cm²	MPa
1	9	36	3.6
2	12	48	4.8
3	17	68	6.8

The vertical strain influence factor I_z with respect to depth is calculated by making use of Fig. 18.22.

At the base of the foundation $I_z = 0.1$

At depth $B/2$, $I_z = 0.5 + 0.1 \sqrt{\dfrac{q_n}{p_o'}}$

where $q_n = 120$ kPa

p_o' = effective average overburden pressure at depth = $(2 + B/2) = 6$ m below ground level.

$= 2 \times 16.5 + 4 \times 8.5 = 67$ kN/m².

$$I_z(\max) = 0.5 + 0.1\sqrt{\frac{120}{67}} = 0.63$$

$I_z = 0$ at $z = H = 16$ m below base level of the foundation. The distribution of I_z is given in Fig. Ex. 18.29. The equation for settlement is

$$S = C_1 C_2 q_n \sum_0^{2B} \frac{I_z}{E_s} \Delta z$$

where $C_1 = 1 - 0.5\left[\frac{q_0'}{q_n}\right] = 1 - 0.5\left[\frac{2\times16.5}{120}\right] = 0.86$

$$C_2 = 1 + 0.2\log\left[\frac{t}{0.1}\right] = 1 + 0.2\log\left[\frac{3}{0.1}\right] = 1.3$$

where $t = 3$ years.

The elastic modulus E_s for normally consolidated sands may be calculated by Eq. (18.61).

$$E_s = 4\overline{q}_c \text{ for } q_c < 10\text{MPa}$$

where $\overline{q}_c$ is the average for each layer.

Layer 2 is divided into sublayers 2a and 2b for computing I_z. The average of the influence factors for each of the layers given in Fig. Ex. 18.29 are tabulated along with the other calculations

Layer No.	Δz (cm)	$\overline{q}_c$ (MPa)	E_s (MPa)	I_z (av)	$\dfrac{I_z \Delta z}{E_s}$
1	300	3.6	14.4	0.3	6.25
2a	100	4.8	19.2	0.56	2.92
2b	500	4.8	19.2	0.50	13.02
3	700	6.8	27.2	0.18	4.63
				Total	26.82

Substituting in the equation for settlement S, we have

$$S = 0.86\times1.3\times0.12\times26.82 = 3.6\,\text{cm} = 36\,\text{mm}$$

18.22 ESTIMATION OF CONSOLIDATION SETTLEMENT BY USING OEDOMETER TEST DATA

Equations for Computing Settlement

Settlement calculation from e-log p curves

A general equation for computing oedometer consolidation settlement may be written as follows.

Normally consolidated clays

$$S_c = H\frac{C_c}{1+e_0}\log\frac{p_0+\Delta p}{p_0} \tag{18.65}$$

Overconsolidated clays

for $p_0 + \Delta p < p_c$

$$S_c = H \frac{C_s}{1 + e_0} \log \frac{p_0 + \Delta p}{p_0} \tag{18.66}$$

for $p_0 < p_c < p_0 + \Delta p$

$$S_c = \frac{H}{1 + e_0} \left[C_s \log \frac{p_c}{p_0} + C_c \log \frac{p_0 + \Delta p}{p_c} \right] \tag{18.67}$$

where C_s = swell index, and C_c = compression index

If the thickness of the clay stratum is more than 3 m the stratum has to be divided into layers of thickness less than 3 m. Further, e_0 is the initial void ratio and p_0, the effective overburden pressure corresponding to the particular layer; Δp is the increase in the effective stress at the middle of the layer due to foundation loading which is calculated by elastic theory. The compression index, and the swell index may be the same for the entire depth or may vary from layer to layer.

Settlement calculation from e-p curve

Equation (18.67) can be expressed in a different form as follows:

$$S_c = \Sigma H m_v \Delta p \tag{18.68}$$

where m_v = coefficient of volume compressibility

18.23 SKEMPTON-BJERRUM METHOD OF CALCULATING CONSOLIDATION SETTLEMENT (1957)

Calculation of consolidation settlement is based on one dimensional test results obtained from oedometer tests on representative samples of clay. These tests do not allow any lateral yield during the test and as such the ratio of the minor to major principal stresses, K_0, remains constant. In practice, the condition of zero lateral strain is satisfied only in cases where the thickness of the clay layer is small in comparison with the loaded area. In many practical solutions, however, significant lateral strain will occur and the initial pore water pressure will depend on the *in situ* stress condition and the value of the pore pressure coefficient A, which will not be equal to unity as in the case of a one-dimensional consolidation test. In view of the lateral yield, the ratios of the minor and major principal stresses due to a given loading condition at a given point in a clay layer do not maintain a constant K_0.

The initial excess pore water pressure at a point P (Fig. 18.23) in the clay layer is given by the expression

$$\Delta u = \Delta\sigma_3 + A(\Delta\sigma_1 - \Delta\sigma_3)$$

$$= \Delta\sigma_1 \left[A + \frac{\Delta\sigma_3}{\Delta\sigma_1}(1 - A) \right] \tag{18.69}$$

where $\Delta\sigma_1$ and $\Delta\sigma_3$ are the total principal stress increments due to surface loading. It can be seen from Eq. (18.69)

$$\Delta u > \Delta\sigma_3 \text{ if } A \text{ is positive}$$

and $\Delta v = \Delta\sigma_1$ if $A = 1$

The value of A depends on the type of clay, the stress levels and the stress system.

Figure 18.23a presents the loading condition at a point in a clay layer below the central line of circular footing. Figures 18.23 (b), (c) and (d) show the condition before loading, immediately after loading and after consolidation respectively.

By the one-dimensional method, consolidation settlement S_{oc} is expressed as

$$S_{oc} = \int_0^H m_v \Delta\sigma_1 dz \qquad (18.70)$$

By the Skempton-Bejerrum method, consolidation settlement is expressed as

$$S_{oc} = \int_0^H m_v \Delta u dz \qquad (18.71)$$

or $\quad S_{oc} = \int_0^H m_v \Delta\sigma_1 \left[A + \dfrac{\Delta\sigma_3}{\Delta\sigma_1}(1-A) \right] \qquad (18.72)$

A settlement coefficient β is used, such that $S_c = \beta S_{oc}$

The expression for β is

$$\beta = \frac{S_c}{S_{oc}} = \frac{\int_0^H m_v \Delta\sigma_1 \left[A + \dfrac{\Delta\sigma_3}{\Delta\sigma_1}(1-A) \right] dz}{\int_0^H m_v \Delta\sigma_1 dz}$$

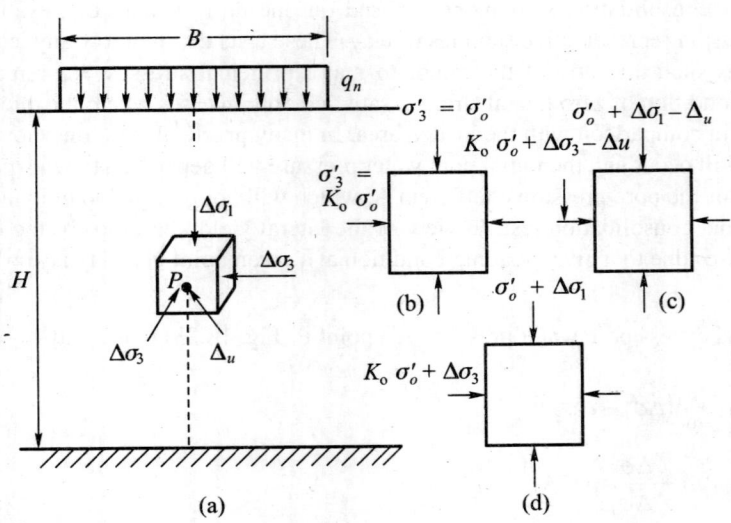

(a) Physical plane (b) Initial conditions
(c) Immediately after loading (d) After consolidation

Fig. 18.23   In situ effective stresses

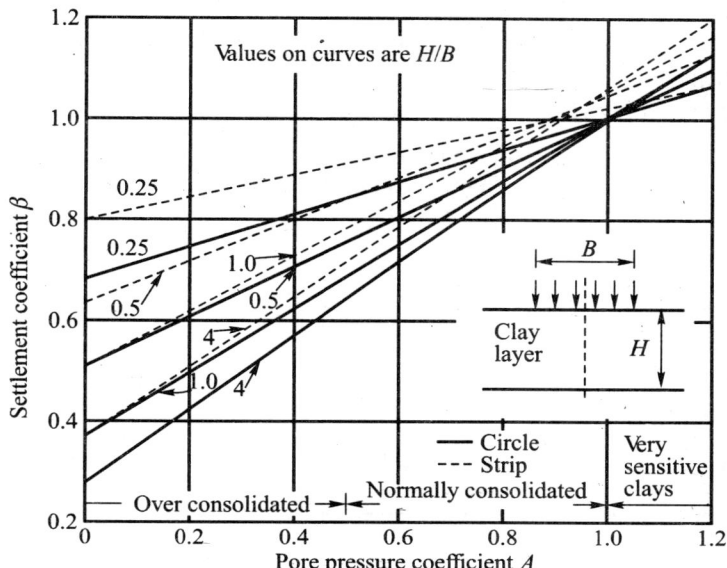

Fig. 18.24 Settlement coefficient versus pore-pressure coefficient for circular and strip footings (After Skempton and Bjerrum, 1957)

TABLE 18.10
Values of settlement coefficient β

Type of clay	β
Very sensitive clays (soft alluvial and marine clays)	1.0 to 1.2
Normally consolidated clays	0.7 to 1.0
Overconsolidated clays	0.5 to 0.7
Heavily overconsolidated clays	0.2 to 0.5

or $S_c = \beta S_{oc}$ (18.73)

where β is called the *settlement coefficient*.

If it can be assumed that m_v and A are constant with depth (sub-layers can be used in the analysis), then β can be expressed as

$$\beta = A + (1 - A)\alpha$$ (18.74)

where $\alpha = \dfrac{\int_0^H \Delta\sigma_3 dz}{\int_0^H \Delta\sigma_1 dz}$ (18.75)

Taking Poisson's ratio μ as 0.5 for a saturated clay during loading under undrained conditions, the value of β depends only on the shape of the loaded area and the thickness of the clay layer in relation to the dimensions of the loaded area and thus β can be estimated from elastic theory.

The value of initial excess pore water pressure (Δu) should, in general, correspond to the *in-situ* stress conditions. The use of a value of pore pressure coefficient A obtained from the results of a triaxial test on a cylindrical clay specimen is strictly applicable only for the condition of axial symmetry, i.e. for the case of settlement under the center of a circular footing. However, the value of A so obtained will serve as a good approximation for the case of settlement under the center of a square footing (using the circular footing of the same area).

Under a strip footing plane strain conditions prevail. Scott (1963) has shown that the value of Δu appropriate in the case of a strip footing can be obtained by using a pore pressure coefficient A_s as

$$A_s = 0.866A + 0.211 \tag{18.76}$$

The coefficient A_s replaces A (the coefficient for the condition of axial symmetry) in Eq. (18.74) for the case of a strip footing, the expression for α being unchanged.

Values of the settlement coefficient β for circular and strip footings, in terms of A and ratios H/B, are given in Fig 18.24.

Typical values of β are given in Table 18.10 for various types of clay soils.

Example 18.30

For the problem given in Ex. 18.28 compute the consolidation settlement by the Skempton-Bjerrum method. The compressible layer of depth 16 m below the base of the foundation is divided into four layers and the soil properties of each layer are given in Fig. Ex. 18.30. The net contact pressure $q_n = 100$ kN/m^2.

Solution

From Eq. (18.65), the oedometer settlement for the entire clay layer system may be expressed as

$$S_{oc} = \Sigma H_i \frac{C_c}{1+e_o} \log \frac{p_o + \Delta p}{p_o}$$

From Eq. (18.73), the consolidation settlement as per Skempton-Bjerrum may be expressed as

$$S_c = \beta S_{oc}$$

where β = settlement coefficient which can be obtained from Fig. 18.30 for various values of A and H/B.

p_o = effective overburden pressure at the middle of each layer (Fig. Ex. 13.30)

C_c = compression index of each layer

H_i = thickness of i th layer

e_o = initial void ratio of each layer

Δp = the excess pressure at the middle of each layer obtained from elastic theory (Chapter 10)

The average pore pressure coefficient is

$$A = \frac{0.9 + 0.75 + 0.70 + 0.45}{4} = 0.7$$

The details of the calculations are tabulated below.

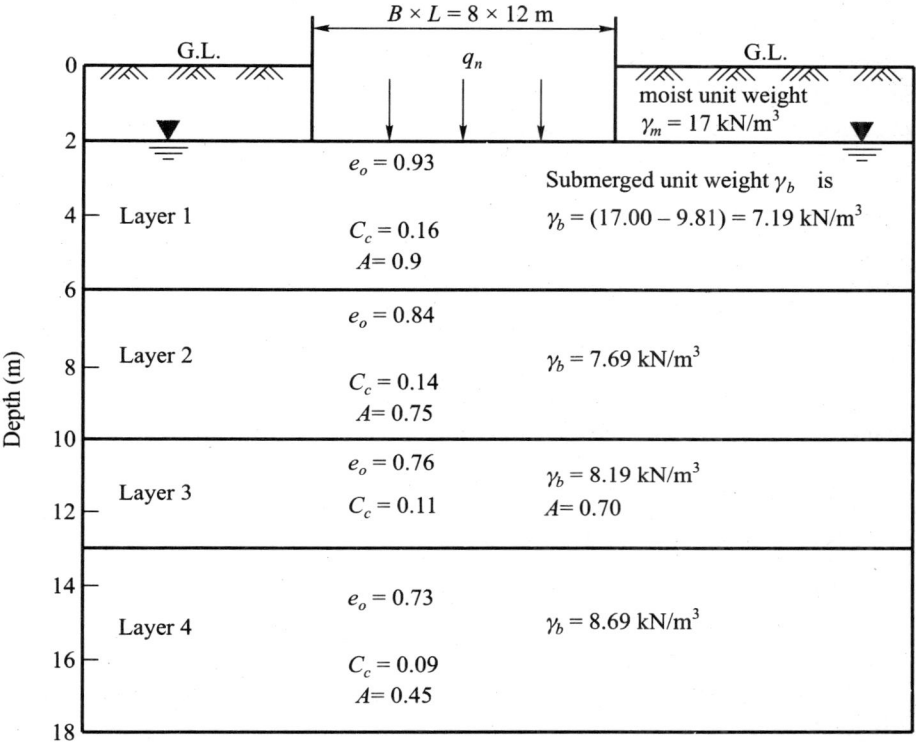

Fig. Ex. 18.30

Layer No.	H_i (cm)	p_o (kN/m²)	Δp (kN/m²)	C_c	e_o	$\log \dfrac{p_0 + \Delta p}{p_0}$	S_{oed} (cm)
1	400	48.4	75	0.16	0.93	0.407	13.50
2	400	78.1	43	0.14	0.84	0.191	5.81
3	300	105.8	22	0.11	0.76	0.082	1.54
4	500	139.8	14	0.09	0.73	0.041	1.07
						Total	21.92

For $H/B = 16/8 = 2$, $A = 0.7$, from Fig. 18.24 we have $\beta = 0.8$.

The consolidation settlement S_c is

$$S_c = 0.8 \times 21.92 = 17.536 \text{ cm} = 175.36 \text{ mm}$$

PART C: COMBINED FOOTINGS AND MAT FOUNDATIONS

18.24 INTRODUCTION

The common methods of transmitting loads to subsoil through spread footings carrying single column load have been discussed earlier. The Part C considers the following types of foundations:

1. Cantilever footings

2. Combined footings

3. Mat foundations

When a column is near or right next to a property limit, a square or rectangular footing concentrically loaded under the column would extend into the adjoining property. If the adjoining property is a public side walk or alley, local building codes may permit such footings to project into public property. But when the adjoining property is privately owned, the footings must be constructed within the property. In such cases, there are three alternatives which are illustrated in Fig. 18.25(a). These are:

1. *Cantilever footing:* A cantilever or strap footing normally comprises two footings connected by a beam called a strap. A strap footing is a special case of a combined footing.

2. *Combined footing:* A combined footing is a long footing supporting two or more columns in one row.

3. *Mat or raft foundations:* A mat or raft foundation is a large footing, usually supporting several columns in two or more rows.

The choice between these types depends primarily upon the relative cost. In the majority of cases, mat foundations are normally used where the soil has low bearing capacity and where the total area occupied by an individual footing is not less than 50 percent of the loaded area of the building.

When the distances between the columns and the loads carried by each column are not equal, there will be eccentric loading. The effect of eccentricity is to increase the base pressure on the side of eccentricity and decrease it on the opposite side. The effect of eccentricity on the base pressure of rigid footings is also considered here.

Mat Foundation in Sand

A foundation is generally termed as a mat if the least width is more than 6 meters. Experience indicates that the ultimate bearing capacity of a mat foundation on cohesionless soil is much higher than that of individual footings of lesser width. With the increasing width of the mat, or increasing relative density of the sand, the ultimate bearing capacity increases rapidly. Hence, the danger that a large mat may break into a sand foundation is too remote to require consideration. On account of the large size of mats the stresses in the underlying soil are likely to be relatively high to a considerable depth. Therefore, the influence of local loose pockets distributed at random throughout the sand is likely to be about the same beneath all parts of the mat and differential settlements are likely to be smaller than those of a spread foundation designed for the same soil pressure.

Mat Foundation in Clay

The net ultimate bearing capacity that can be sustained by the soil at the base of a mat on a deep deposit of clay or plastic silt may be obtained in the same manner as for footings on clay discussed earlier. However, by using the principle of *flotation*, the pressure on the base of the mat that induces

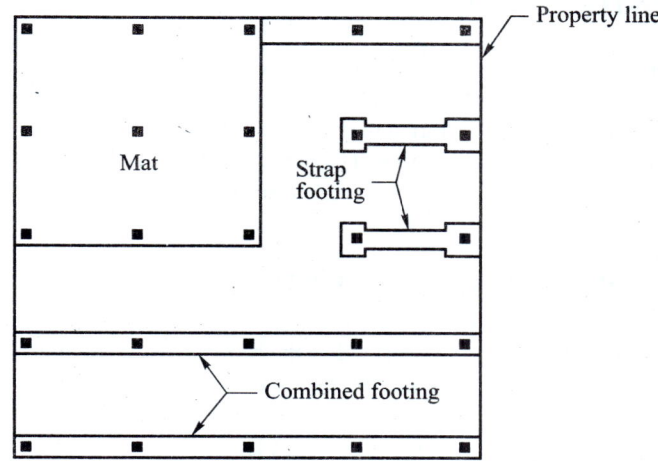

(a) Schematic plan showing mat,
strap and combined footings

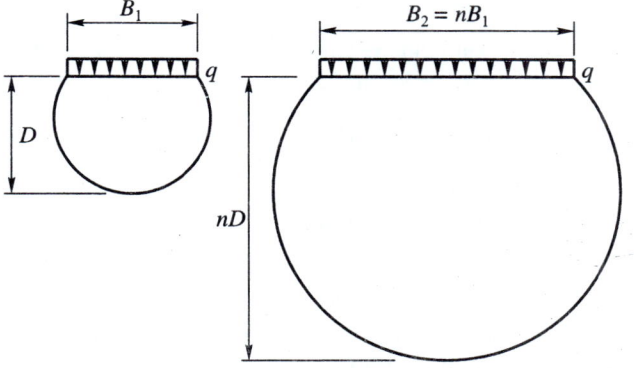

(b) Bulb of pressure for vertical stress for different beams

Fig. 18.25 (a) Types of footings, (b) beams on compressible subgrade

settlement can be reduced by increasing the depth of the foundation. A brief discussion on the principle of flotation is dealt with in this chapter.

Rigid and Elastic Foundation

The conventional method of design of combined footings and mat foundations is to assume the foundation as infinitely rigid and the contact pressure is assumed to have a planar distribution. In the case of an elastic foundation, the soil is assumed to be a truly elastic solid obeying Hooke's law in all directions. The design of an elastic foundation requires a knowledge of the subgrade reaction which is briefly discussed here. However, the elastic method does not readily lend itself to engineering applications because it is extremely difficult and solutions are available for only a few extremely simple cases.

18.25 ECCENTRIC LOADING

When the resultant of loads on a footing does not pass through the center of the footing, the footing is subjected to what is called *eccentric loading*. The loads on the footing may be vertical or inclined. If the loads are inclined it may be assumed that the horizontal component is resisted by the frictional resistance offered by the base of the footing. The vertical component in such a case is the only factor for the design of the footing. The effects of eccentricity on bearing pressure of the footings have been discussed earlier.

18.26 THE COEFFICIENT OF SUBGRADE REACTION

The coefficient of subgrade reaction is defined as the ratio between the pressure against the footing or mat and the settlement at a given point expressed as

$$k_s = \frac{q}{S},$$
(18.77)

where k_s = coefficient of subgrade reaction expressed as force/length^3 (FL^{-3}),

q = pressure on the footing or mat at a given point expressed as force/length^2 (FL^{-2}),

S = settlement of the same point of the footing or mat in the corresponding unit of length.

In other words the coefficient of subgrade reaction is the unit pressure required to produce a unit settlement. In clayey soils, settlement under the load takes place over a long period of time and the coefficient should be determined on the basis of the final settlement. On purely granular soils, settlement takes place shortly after load application. Eq. (18.77) is based on two simplifying assumptions:

1.   The value of k_s is independent of the magnitude of pressure.
2.   The value of k_s has the same value for every point on the surface of the footing.

Both the assumptions are strictly not accurate. The value of k_s decreases with the increase of the magnitude of the pressure and it is not the same for every point of the surface of the footing as the settlement of a flexible footing varies from point to point. However the method is supposed to give realistic values for contact pressures and is suitable for beam or mat design when only a low order of settlement is required.

Factors Affecting the Value of k_s

Terzaghi (1955) discussed the various factors that affect the value of k_s. A brief description of his arguments is given below.

Consider two foundation beams of widths B_1 and B_2 such that $B_2 = nB_1$ resting on a compressible subgrade and each loaded so that the pressure against the footing is uniform and equal to q for both the beams (Fig. 18.25b). Consider the same points on each beam and, let

y_1 = settlement of beam of width B_1

y_2 = settlement of beam of width B_2

Hence $k_{s1} = \dfrac{q}{y_1}$ and $k_{s2} = \dfrac{q}{y_2}$

If the beams are resting on a subgrade whose deformation properties are more or less independent of depth (such as a stiff clay) then it can be assumed that the settlement increases in simple proportion o the depth of the pressure bulb.

Then
$$y_2 = ny_1 \qquad (18.78)$$

and
$$k_{s2} = \frac{q}{ny_1} = \frac{q}{y_1}\frac{B_1}{B_2} = k_{s1}\frac{B_1}{B_2}. \qquad (18.79)$$

A general expression for k_s can now be obtained if we consider B_1 as being of unit width (Terzaghi used a unit width of one foot which converted to metric units may be taken as equal to 0.30 m). Hence by putting $B_1 = 0.30$ m, $k_s = k_{s2}$, $B = B_2$, we obtain

$$k_s = 0.3\frac{k_{s1}}{B} \qquad (18.80)$$

where k_s is the coefficient of subgrade reaction of a long footing of width B meters and resting on stiff clay; k_{s1} is the coefficient of subgrade reaction of a long footing of width 0.30 m (approximately), resting on the same clay. It is to be noted here that the value of k_{s1} is derived from ultimate settlement values, that is, after consolidation settlement is completed.

If the beams are resting on clean sand, the final settlement values are obtained almost instantaneously. Since the modulus of elasticity of sand increases with depth, the deformation characteristics of the sand change and become less compressible with depth. Because of this characteristic of sand, the lower portion of the bulb of pressure for beam B_2 is less compressible than that of the sand enclosed in the bulb of pressure of beam B_1.

The settlement value y_2 lies somewhere between y_1 and ny_1. It has been shown experimentally (Terzaghi and Peck, 1948) that the settlement, y, of a beam of width B resting on sand is given by the expression

$$y = y_1\left[\frac{2B}{B+0.3}\right]^2 \qquad (18.81)$$

where y_1 = settlement of a beam of width 0.30 m and subjected to the same reactive pressure as the beam of width B meters.

Hence, the coefficient of subgrade reaction k_s of a beam of width B meters can be obtained from the following equation

$$k_s = \frac{q}{y} = \frac{q}{y_1}\left[\frac{B+0.30}{2B}\right]^2 = k_{s1}\left[\frac{B+0.30}{2B}\right]^2 \qquad (18.82)$$

where k_{s1} = coefficient of subgrade reaction of a beam of width 0.30 m resting on the same sand.

Measurement of k_{s1}

A value for k_{s1} for a particular subgrade can be obtained by carrying out plate load tests. The standard size of plate used for this purpose is 0.30×0.30 m size. Let k_1 be the subgrade reaction for a plate of size 0.30×0.30 m size.

From experiments it has been found that $k_{s1} \approx k_1$ for sand subgrades, but for clays k_{s1} varies with the length of the beam. Terzaghi (1955) gives the following formula for clays

$$k_{s1} = k_1\left[\frac{L+0.152}{1.5L}\right] \qquad (18.83a)$$

where L = length of the beam in meters and the width of the beam = 0.30 m. For a very long beam on clay subgrade we may write

$$k_{s1} = \frac{k_1}{1.5} \qquad\qquad (18.83b)$$

Procedure to Find k_s

For sand

1. Determine k_1 from plate load test or from estimation.
2. Since $ks_{s1} \approx k_1$, use Eq. (18.82) to determine k_s for sand for any given width B meter.

For clay

1. Determine k_1 from plate load test or from estimation.
2. Determine k_{s1} from Eq. (18.83a) as the length of beam is known.
3. Determine k_s from Eq. (18.80) for the given width B meters.

When plate load tests are used, k_1 may be found by one of the two ways,

1. A bearing pressure equal to not more than the ultimate pressure and the corresponding settlement is used for computing k_1
2. Consider the bearing pressure corresponding to a settlement of 1.3 mm for computing k_1.

Estimation of k_1 Values

Plate load tests are both costly and time consuming. Generally a designer requires only the values of the bending moments and shear forces within the foundation. With even a relatively large error in the estimation of k_1, moments and shear forces can be calculated with little error (Terzaghi, 1955); an error of 100 percent in the estimation of k_s may change the structural behavior of the foundation by up to 15 percent only.

In the absence of plate load tests, estimated values of k_1 and hence k_s are used. The values suggested by Terzaghi for k_1 (converted into S.I. units) are given in Table 18.11.

TABLE 18.11a

k_1 values for foundations on sand (MN/m³)

| Relative density | Loose | Medium | Dense |
SPT Values (Uncorrected)	<10	10–30	>30
Soil, dry or moist	15	45	175
Soil submerged	10	30	100

TABLE 18.11b

k_1 values for foundation on clay (MN/m³)

Consistency	Stiff	Very stiff	Hard
c_u (kN/m²)	50–100	100–200	>200
k_1 (MN/m³)	25	50	100

Source: Terzaghi (1955)

18.27 PROPORTIONING OF CANTILEVER FOOTINGS

Strap or cantilever footings are designed on the basis of the following assumptions:

1. The strap is infinitely stiff. It serves to transfer the column loads to the soil with equal and uniform soil pressure under both the footings.

2. The strap is a pure flexural member and does not take soil reaction. To avoid bearing on the bottom of the strap a few centimeters of the underlying soil may be loosened prior to the placement of concrete.

A strap footing is used to connect an eccentrically loaded column footing close to the property line to an interior column as shown in Fig. 18.26.

With the above assumptions, the design of a strap footing is a simple procedure. It starts with a trial value of e, Fig. 18.26. Then the reactions R_1 and R_2 are computed by the principle of statics. The tentative footing areas are equal to the reactions R_1 and R_2 divided by the safe bearing pressure q_s. With tentative footing sizes, the value of e is computed. These steps are repeated until the trial value of e is identical with the final one. The shears and moments in the strap are determined, and the straps designed to withstand the shear and moments. The footings are assumed to be subjected to uniform soil pressure and designed as simple spread footings. Under the assumptions given above the resultants of the column loads Q_1 and Q_2 would coincide with the center of gravity of the two footing areas. Theoretically, the bearing pressure would be uniform under both the footings. However, it is possible that sometimes the full design live load acts upon one of the columns while the other may be subjected to little live load. In such a case, the full reduction of column load from Q_2 to R_2 may not be realized. It seems justified then that in designing the footing under column Q_2, only the dead load or dead load plus reduced live load should be used on column Q_1.

The equations for determining the position of the reactions (Fig. 18.26) are

$$R_1 = Q_1\left(1 + \frac{e}{L_R}\right), \quad R_2 = Q_2 - \frac{Q_1 e}{L_R} \tag{18.84}$$

where R_1 and R_2 = reactions for the column loads Q_1 and Q_2 respectively, e = distance of R_1 from Q_1, L_R = distance between R_1 and R_2.

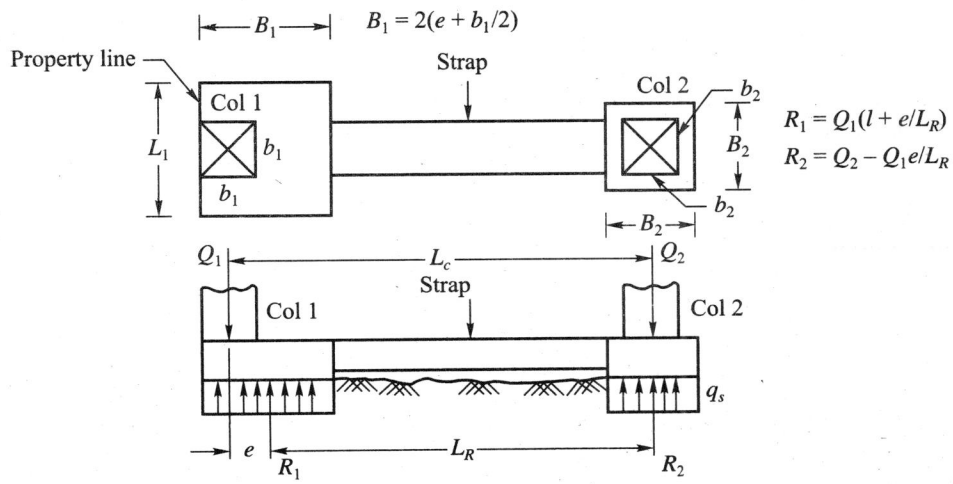

$B_1 = 2(e + b_1/2)$

$R_1 = Q_1(1 + e/L_R)$

$R_2 = Q_2 - Q_1 e/L_R$

Fig. 18.26 Principles of cantilever or strap footing design

18.28 DESIGN OF COMBINED FOOTINGS BY RIGID METHOD (CONVENTIONAL METHOD)

The rigid method of design of combined footings assumes that

1. The footing or mat is infinitely rigid, and therefore, the deflection of the footing or mat does not influence the pressure distribution,

2. The soil pressure is distributed in a straight line or a plane surface such that the centroid of the soil pressure coincides with the line of action of the resultant force of all the loads acting on the foundation.

Design of Combined Footings

Two or more columns in a row joined together by a stiff continuous footing form a combined footing as shown in Fig. 18.27a. The procedure of design for a combined footing is as follows:

1. Determine the total column loads $\Sigma Q = Q_1 + Q_2 + Q_3 + \dots$ and location of the line of action of the resultant ΣQ. If any column is subjected to bending moment, the effect of the moment should be taken into account.

2. Determine the pressure distribution q per lineal length of footing.

3. Determine the width, B, of the footing.

4. Draw the shear diagram along the length of the footing. By definition, the shear at any section along the beam is equal to the summation of all vertical forces to the left or right of the section. For example, the shear at a section immediately to the left of Q_1 is equal to the area $abcd$, and immediately to the right of Q_1 is equal to $(abcd - Q_1)$ as shown in Fig. 18.27a.

5. Draw the moment diagram along the length of the footing. By definition the bending moment at any section is equal to the summation of moment due to all the forces and reaction to the left (or right) of the section. It is also equal to the area under the shear diagram to the left (or right) of the section.

6. Design the footing as a continuous beam to resist the shear and moment.

7. Design the footing for transverse bending in the same manner as for spread footings.

It should be noted here that the end column along the property line may be connected to the interior column by a rectangular or trapezoidal footing. In such a case no strap is required and both the columns together will be a combined footing as shown in Fig. 18.27b. It is necessary that the center of area of the footing must coincide with the center of loading for the pressure to remain uniform.

18.29 DESIGN OF MAT FOUNDATION BY RIGID METHOD

In the conventional rigid method the mat is assumed to be infinitely rigid and the bearing pressure against the bottom of the mat follows a planar distribution where the centroid of the bearing pressure coincides with the line of action of the resultant force of all loads acting on the mat. The procedure of design is as follows:

1. The column loads of all the columns coming from the superstructure are calculated as per standard practice. The loads include live and dead loads.

2. Determine the line of action of the resultant of all the loads. However, the weight of the mat

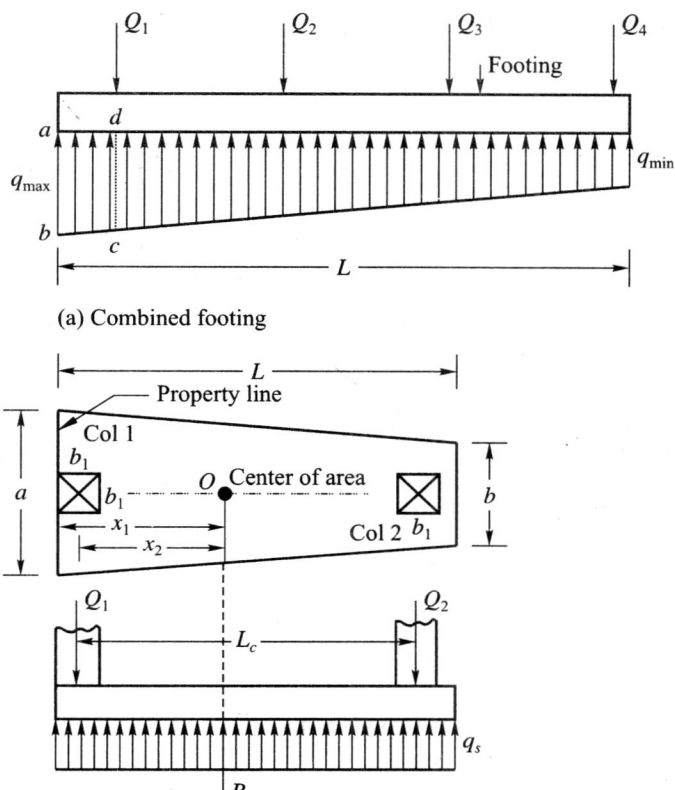

(a) Combined footing

(b) Trapezoidal combined footing

Fig. 18.27 Combined or trapezoidal footing design

is not included in the structural design of the mat because every point of the mat is supported by the soil under it, causing no flexural stresses.

3. Calculate the soil pressure at desired locations by the use of Eq. (18.33a)

$$q = \frac{Q_t}{A} \pm \frac{Q_t e_x}{I_y} x \pm \frac{Q_t e_y}{I_x} y \qquad (18.85)$$

where $Q_t = \Sigma Q$ = total load on the mat,

A = total area of the mat,

x, y = coordinates of any given point on the mat with respect to the x and y axes passing through the centroid of the area of the mat,

e_x, e_y = eccentricities of the resultant force,

I_x, I_y = moments of inertia of the mat with respect to the x and y axes respectively.

4. The mat is treated as a whole in each of two perpendicular directions. Thus the total shear force acting on any section cutting across the entire mat is equal to the arithmetic sum of all forces and reactions (bearing pressure) to the left (or right) of the section. The total bending moment acting on such a section is equal to the sum of all the moments to the left (or right) of the section.

18.30 DESIGN OF COMBINED FOOTINGS BY ELASTIC LINE METHOD

The relationship between deflection, y, at any point on an elastic beam and the corresponding bending moment M may be expressed by the equation

$$EI\frac{d^2y}{dx^2} = M \tag{18.86}$$

The equations for shear V and reaction q at the same point may be expressed as

$$EI\frac{d^3y}{dx^3} = V \tag{18.87}$$

$$EI\frac{d^4y}{dx^4} = q \tag{18.88}$$

where x is the coordinate along the length of the beam.

From the basic assumption of an elastic foundation

$$q = -yBk_s$$

where, B = width of footing, k_s = coefficient of subgrade reaction.

Substituting for q, Eq. (18.88) may be written as

$$EI\frac{d^4y}{dx^4} = -yBk_s \tag{18.89}$$

The classical solutions of Eq. (18.89) being of closed form, are not general in their application. Hetenyi (1946) developed equations for a load at any point along a beam. The development of solutions is based on the concept that the beam lies on a bed of elastic springs which is based on Winkler's hypothesis. As per this hypothesis, the reaction at any point on the beam depends only on the deflection at that point.

Methods are also available for solving the beam-problem on an elastic foundation by the method of finite differences (Malter, 1958). The finite element method has been found to be the most efficient of the methods for solving beam-elastic foundation problem. Computer programs are available for solving the problem.

Since all the methods mentioned above are quite involved, they are not dealt with here. Interested readers may refer to Bowles (1996).

18.31 DESIGN OF MAT FOUNDATIONS BY ELASTIC PLATE METHOD

Many methods are available for the design of mat-foundations. The one that is very much in use is the finite difference method. This method is based on the assumption that the subgrade can be substituted by a bed of uniformly distributed coil springs with a spring constant k_s which is called the coefficient of subgrade reaction. The finite difference method uses the fourth order differential equation

$$\nabla^4 w = \frac{q - k_s w}{D} \tag{18.90}$$

where $\nabla^4 w = \dfrac{\partial^4 w}{\partial x^4} + 2\dfrac{\partial^4 w}{\partial x^2 \partial y^2} + \dfrac{\partial^4 w}{\partial y^4}$

q = subgrade reaction per unit area,

k_s = coefficient of subgrade reaction,

w = deflection,

D = rigidity of the mat = $\dfrac{Et^3}{12(1-\mu^2)}$

E = modulus of elasticity of the material of the footing,

t = thickness of mat,

μ = Poisson's ratio.

Equation (18.90) may be solved by dividing the mat into suitable square grid elements, and writing difference equations for each of the grid points. By solving the simultaneous equations so obtained the deflections at all the grid points are obtained. The equations can be solved rapidly with an electronic computer. After the deflections are known, the bending moments are calculated using the relevant difference equations.

Interested readers may refer to Teng (1969) or Bowles (1996) for a detailed discussion of the method.

18.32 FLOATING FOUNDATION

General Consideration

A *floating foundation* for a building is defined as a foundation in which the weight of the building is approximately equal to the full weight of soil and water removed from the site of the building. This principle of flotation may be explained with reference to Fig. 18.28. Figure 18.28(a) shows a horizontal ground surface with a horizontal water table at a depth d_w below the ground surface. Figure 18.28(b) shows an excavation made in the ground to a depth D where $D > d_w$, and Fig. 18.28(c) shows a structure built in the excavation and completely filling it.

If the weight of the building is equal to the weight of the soil and water removed from the excavation, then it is evident that the total vertical pressure in the soil below depth D in Fig. 18.28(c) is the same as in Fig. 18.28(a) before excavation.

Since the water level has not changed, the neutral pressure and the effective pressure are therefore unchanged. Since settlements are caused by an increase in effective vertical pressure, if we could move from Fig. 18.28(a) to Fig. 18.28(c) without the intermediate case of 18.28(b), the building in Fig. 18.28(c) would not settle at all.

This is the principle of a floating foundation, an exact balance of weight removed against weight imposed. The result is zero settlement of the building.

However, it may be noted, that we cannot jump from the stage shown in Fig. 18.28(a) to the stage in Fig. 18.28(c) without passing through stage 18.28(b). The excavation stage of the building is the critical stage.

Cases may arise where we cannot have a fully floating foundation. The foundations of this type

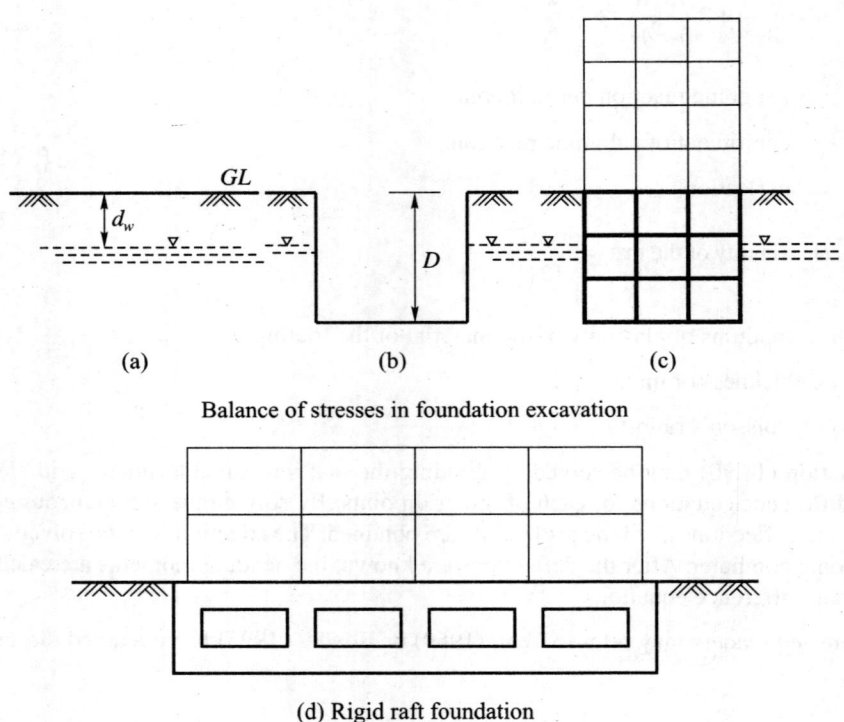

(a)   (b)   (c)

Balance of stresses in foundation excavation

(d) Rigid raft foundation

Fig. 18.28   Principles of floating foundation; and a typical rigid raft foundation

are sometimes called *partly compensated foundations* (as against *fully compensated* or *fully floating* foundations).

While dealing with floating foundations, we have to consider the following two types of soils. They are:

Type 1: The foundation soils are of such a strength that shear failure of soil will not occur under the building load but the settlements and particularly differential settlements, will be too large and will constitute *failure* of the structure. A floating foundation is used to reduce settlements to an acceptable value.

Type 2: The shear strength of the foundation soil is so low that rupture of the soil would occur if the building were to be founded at ground level. In the absence of a strong layer at a reasonable depth, the building can only be built on a floating foundation which reduces the shear stresses to an acceptable value. Solving this problem solves the settlement problem.

In both the cases, a rigid raft or box type of foundation is required for the floating foundation [Fig. 18.28(d)]

Problems to be Considered in the Design of a Floating Foundation

The following problems are to be considered during the design and construction stage of a floating foundation.

1. Excavation

The excavation for the foundation has to be done with care. The sides of the excavation should suitably be supported by sheet piling, soldier piles and timber or some other standard method.

2. Dewatering

Dewatering will be necessary when excavation has to be taken below the water table level. Care has to be taken to see that the adjoining structures are not affected due to the lowering of the water table.

3. Critical depth

In Type 2 foundations the shear strength of the soil is low and there is a theoretical limit to the depth to which an excavation can be made. Terzaghi (1943) has proposed the following equation for computing the critical depth D_c,

$$D_c = \frac{5.7s}{\gamma - (s/B)\sqrt{2}} \tag{18.91}$$

for an excavation which is long compared to its width

where γ = unit weight of soil,

   s = shear strength of soil = $q_u/2$,

   B = width of foundation,

   L = length of foundation.

   Skempton (1951) proposes the following equation for D_c, which is based on actual failures in excavations

$$D_c = N_c \frac{s}{\gamma} \tag{18.92}$$

or the factor of safety F_s against bottom failure for an excavation of depth D is

$$F_s = N_c \frac{s}{\gamma D + p}$$

where N_c is the bearing capacity factor as given by Skempton, and p is the surcharge load. The values of N_c may be obtained from Fig 18.9. The above equations may be used to determine the maximum depth of excavation.

4. Bottom heave

Excavation for foundations reduces the pressure in the soil below the founding depth which results in the heaving of the bottom of the excavation. Any heave which occurs will be reversed and appear as settlement during the construction of the foundation and the building. Though heaving of the bottom of the excavation cannot be avoided it can be minimized to a certain extent. There are three possible causes of heave:

1.   Elastic movement of the soil as the existing overburden pressure is removed.

2.   A gradual swelling of soil due to the intake of water if there is some delay for placing the foundation on the excavated bottom of the foundation.

3.   Plastic inward movement of the surrounding soil.

The last movement of the soil can be avoided by providing proper lateral support to the excavated sides of the trench.

Heaving can be minimized by phasing out excavation in narrow trenches and placing the foundation soon after excavation. It can be minimized by lowering the water table during the excavation process. Friction piles can also be used to minimize the heave. The piles are driven either before excavation commences or when the excavation is at half depth and the pile tops are pushed down to below foundation level. As excavation proceeds, the soil starts to expand but this movement is resisted by the upper part of the piles which go into tension. This heave is prevented or very much reduced.

It is only a *practical and pragmatic approach* that would lead to a safe and sound settlement free floating (or partly floating) foundation.

Example 18.31

A beam of length 4 m and width 0.75 m rests on stiff clay. A plate load test carried out at the site with the use of a square plate of size 0.30 m gives a coefficient of subgrade reaction k_1 equal to 25 MN/m^3. Determine the coefficient of subgrade reaction k_s for the beam.

Solution

First determine k_{s1} from Eq. (18.83a) for a beam of 0.30 m. wide and length 4 m. Next determine k_s from Eq. (18.80) for the same beam of width 0.75 m.

$$k_{s1} = k_1 \left[\frac{L+0.152}{1.5L} \right] = 25 \left[\frac{4+0.152}{1.5 \times 4} \right] = 17.3 \text{ MN/m}^3$$

$$k_s = 0.3 \frac{k_{s1}}{B} = \frac{0.3 \times 17.3}{0.75} \approx 7 \text{MN/m}^3$$

Example 18.32

A beam of length 4 m and width 0.75 m rests in dry medium dense sand. A plate load test carried out at the same site and at the same level gave a coefficient of subgrade reaction k_1 equal to 47 MN/m^3. Determine the coefficient of subgrade reaction for the beam.

Solution

For sand the coefficient of subgrade reaction, k_{s1}, for a long beam of width 0.3 m is the same as that for a square plate of size 0.3 × 0.3 m that is $k_{s1} = k_s$. k_s now can be found from Eq. (18.82) as

$$k_s = k_1 \left[\frac{B+0.3}{2B} \right]^2 = 47 \left[\frac{0.75 \times 0.30}{1.5} \right]^2 = 23 \text{ MN/m}^3$$

Example 18.33

The following information is given for proportioning a cantilever footing with reference to Fig. 18.26.

Column Loads: $Q_1 = 1455$ kN, $Q_2 = 1500$ kN.

Size of column: 0.5×0.5 m.

$L_c = 6.2$ m, $q_s = 384$ kN/m^2

It is required to determine the size of the footings for columns 1 and 2.

Solution

Assume the width of the footing for column $1 = B_1 = 2$ m.

First trial

Try $e = 0.5$ m. Now, $L_R = 6.2 - 0.5 = 5.7$ m.

Reactions

$$R_1 = Q_1 \left[1 + \frac{e}{L_R} \right] = 1455 \left[1 + \frac{0.5}{5.7} \right] = 1583 \text{ kN}$$

$$R_2 = \left[Q_2 - \frac{Q_1 e}{L_R} \right] = \left[1500 - \frac{1455 \times 0.5}{5.7} \right] = 1372 \text{ kN}$$

Size of footings—First trial

Col. 1. Area of footing $A_1 = \dfrac{1583}{384} = 4.122$ sq.m

Col. 2. Area of footing $A_2 = \dfrac{1372}{384} = 3.57$ sq.m

Try 1.9×1.9 m

Second trial

New value of $e = \dfrac{B_1}{2} - \dfrac{b_1}{2} = \dfrac{2}{2} - \dfrac{0.5}{2} = 0.75$ m

New $L_R = 6.20 - 0.75 = 5.45$ m

$$R_1 = 1455 \left[1 + \frac{0.75}{5.45} \right] = 1655 \text{ kN}$$

$$R_2 = \left[1500 - \frac{1455 \times 0.75}{5.45} \right] = 1300 \text{ kN}$$

$A_1 = \dfrac{1655}{384} = 4.31$ sq.m or 2.08×2.08 m

$A_2 = \dfrac{1300}{384} = 3.38$ sq.m or 1.84×1.84 m

Check $e = \dfrac{B_1}{2} - \dfrac{b_1}{2} = 1.04 - 0.25 = 0.79 \approx 0.75$ m

Use 2.08 × 2.08 m for Col. 1 and 1.90 × 1.90 m for Col. 2.

Note: Rectangular footings may be used for both the columns.

Example 18.34

Figure Ex. 18.34 gives a foundation beam with the vertical loads and moment acting thereon. The width of the beam is 0.70 m and depth 0.50 m. A uniform load of 16 kN/m (including the weight of the beam) is imposed on the beam. Draw (a) the base pressure distribution, (b) the shear force diagram, and (c) the bending moment diagram. The length of the beam is 8 m.

Solution

The steps to be followed are:

1. Determine the resultant vertical force R of the applied loadings and its eccentricity with respect to the centers of the beam.

2. Determine the maximum and minimum base pressures.

3. Draw the shear and bending moment diagrams.

$R = 320 + 400 + 16 \times 8 = 848$ kN.

Taking the moment about the right hand edge of the beam, we have,

$$Rx = 848x = 320 \times 7 + 400 \times 1 + 16 \times \frac{8^2}{2} - 160 = 2992$$

or $x = \dfrac{2992}{848} = 3.528$ m

$e = 4.0 - 3.528 = 0.472$ m to the right of center of the beam. Now from Eqs 18.34(a) and (b) using $e_y = 0$,

$$q_{min}^{max} = \frac{\Sigma Q}{A}\left[1 \pm 6\frac{e_x}{L}\right] = \frac{848}{8 \times 0.7}\left[1 \pm \frac{6 \times 0.472}{8}\right] = 205.02 \text{ or } 97.83 \text{ kN/m}^2$$

Convert the base pressures per unit area to load per unit length of beam.

The maximum vertical load $= 0.7 \times 205.02 = 143.52$ kN/m.

The minimum vertical load $= 0.7 \times 97.83 = 68.48$ kN/m.

The reactive loading distribution is given in Fig. Ex. 18.34(b).

Shear force diagram

Calculation of shear for a typical point such as the reaction point R_1 (Fig. Ex. 18.34(a)) is explained below.

Consider forces to the left of R_1 (without 320 kN).

Shear force $V =$ upward shear force equal to the area *abcd* – downward force due to distributed load on beam *ab*

$$= \frac{68.48 + 77.9}{2} - 16 \times 1 = 57.2 \text{ kN}$$

Consider to the right of reaction point R_1 (with 320 kN).

$V = -320 + 57.2 = -262.8$ kN.

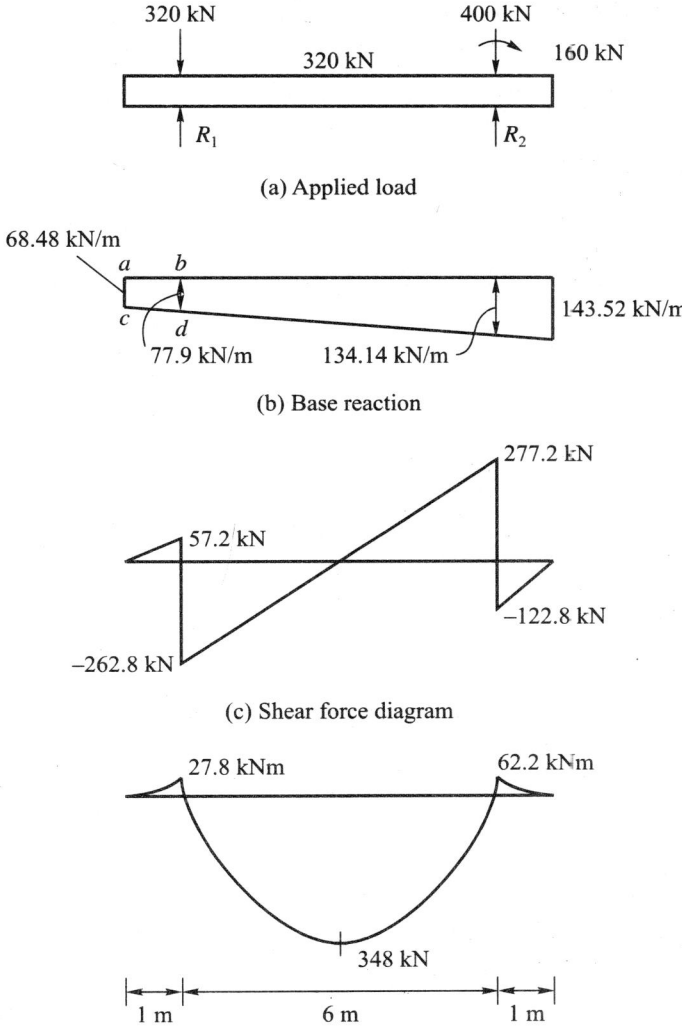

(a) Applied load

(b) Base reaction

(c) Shear force diagram

(d) Bending moment diagram

Fig. Ex. 18.34

In the same away the shear at other points can be calculated. Fig. Ex. 18.34(c) gives the complete near force diagram.

Bending moment diagram

Bending moment at the reaction point R_1 = moment due to force equal to the area *abcd* + moment due to distributed load on beam *ab*

$$= 68.48 \times \frac{1}{2} + \frac{9.42}{2} \times \frac{1}{3} - 16 \times \frac{1}{2}$$

$$= 27.8 \text{ kN - m}$$

The moments at other points can be calculated in the same way. The complete moment diagram s given in Fig. Ex. 18.34(d)

Example 18.35

The end column along a property line is connected to an interior column by a trapezoidal footin. The following data are given with reference to Fig. 18.27(b):

Column Loads: $Q_1 = 2016$ kN, $Q_2 = 1560$ kN.

Size of columns: 0.46×0.46 m.

$L_c = 5.48$ m.

Determine the dimensions a and b of the trapezoidal footing. The net allowable bearing pressur $q_{na} = 190$ kPa.

Solution

Determine the center of bearing pressure x_2 from the center of Column 1. Taking moments of all th loads about the center of Column 1, we have

$$(2016+1560)x_2 = 1560 \times 5.48$$

$$x_2 = \frac{1560 \times 5.48}{3576} = 2.39 \text{ m}$$

Now $x_1 = 2.39 + \frac{0.46}{2} = 2.62 \text{ m}$

Point O in Fig. 18.27(b) is the center of the area coinciding with the center of pressure. From the allowable pressure $q_a = 190$ kPa, the area of the combined footing required is

$$A = \frac{3576}{190} = 18.82 \text{ sq.m}$$

From geometry, the area of the trapezoidal footing (Fig. 18.27(b)) is

$$A = \frac{(a+b)L}{2} = \frac{(a+b)}{2}(5.94) = 18.82$$

or $(a+b) = 6.34 \text{ m}$

where, $L = L_c + b_1 = 5.48 + 0.46 = 5.94$ m

From the geometry of the Fig. (18.27b), the distance of the center of area x_1 can be written in terms of a, b and L as

$$x_1 = \frac{L}{3}\frac{2a+b}{a+b}$$

or $\dfrac{2a+b}{a+b} = \dfrac{3x_1}{L} = \dfrac{3 \times 2.62}{5.94} = 1.32 \text{ m}$

but $a + b = 6.32$ m or $b = 6.32 - a$. Now substituting for b we have,

$$\frac{2a+6.34-a}{6.34} = 1.32$$

and solving, $a = 2.03$ m, from which, $b = 6.34 - 2.03 = 4.31$ m.

18.33 CASE HISTORY OF FAILURE OF THE TRANSCONA GRAIN ELEVATOR

One of the best known foundation failures occurred in October 1913 at North Transcona, Manitoba, Canada. It was ascertained later on that the failure occurred when the foundation pressure at the base was about equal to the calculated ultimate bearing capacity of an underlaying layer of plastic clay (Peck and Byrant, 1953), and was essentially a shearing failure.

Fig. 18.29   The titled Transcona grain elevator (Courtesy: UMA Engineering Ltd., Manitoba, Canada)

Fig. 18.30   The straightened Transcona grain elevator (Courtesy: UMA Engineering Ltd., Manitoba, Canada)

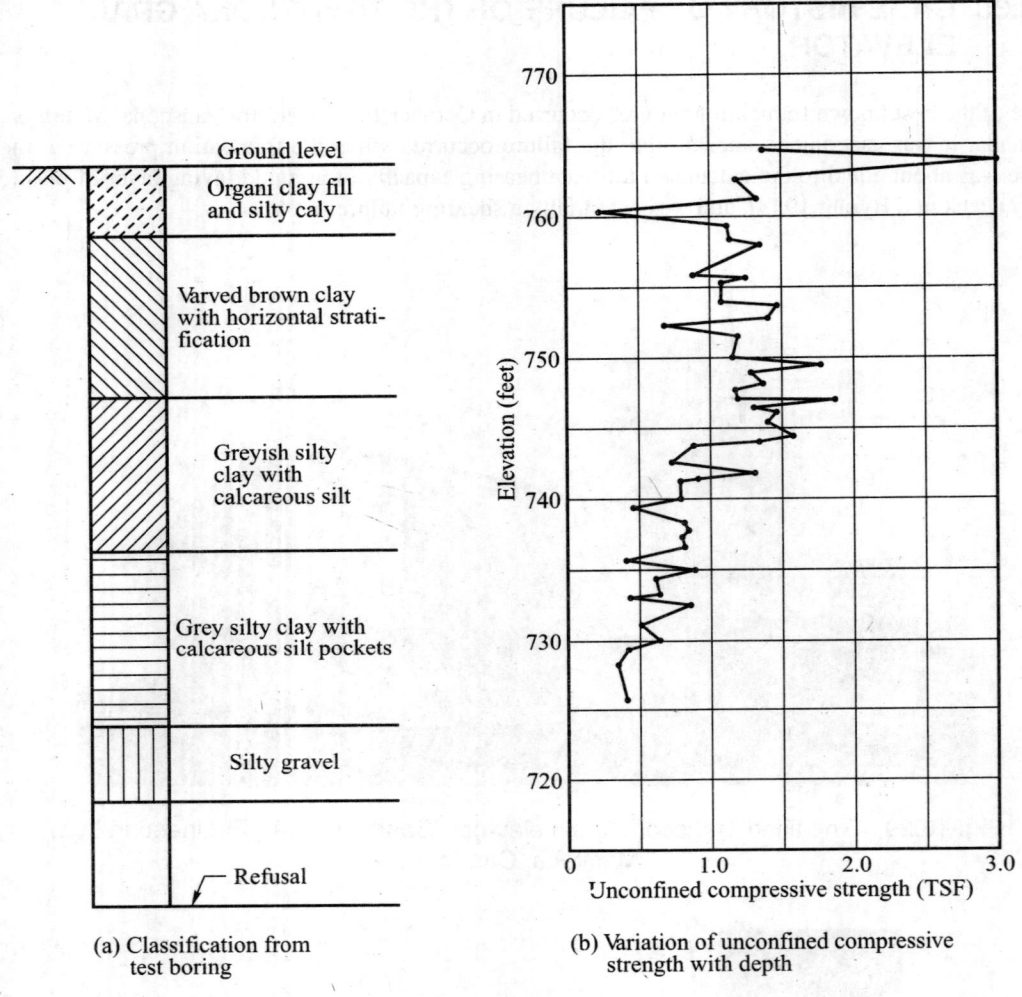

(a) Classification from
test boring

(b) Variation of unconfined compressive
strength with depth

Fig. 18.31 Results of test boring at site of Transcona grain elevator (Peck and Byrant, 1953)

The construction of the silo started in 1911 and was completed in the autumn of 1913. The silo is 77 ft by 195 ft in plan and has a capacity of 1,000,000 bushels. It comprises 65 circular bins and 48 inter-bins. The foundation was a reinforced concrete raft 2 ft thick and founded at a depth of 12 ft below the ground surface. The weight of the silo was 20,000 tons, which was 42.5 percent of the total weight, when it was filled. Filling the silo with grain started in September 1913, and in October when the silo contained 875,000 bushels, and the pressure on the ground was 94 percent of the design pressure, a vertical settlement of 1 ft was noticed. The structure began to tilt to the west and within twenty four hours was at an angle of 26.9° from the vertical, the west side being 24 ft below and the east side 5 ft above the original level (Szechy, 1961). The structure tilted as a monolith and there was no damage to the structure except for a few superficial cracks. Figure 18.19 shows a view of the tilted structure. The excellent quality of the reinforced concrete structure is shown by the fact that later it was underpinned and jacked up on new piers founded on rock. The level of the new foundation is 34 ft below the ground surface. Figure 18.20 shows the view of the silo after it was straightened in 1916.

During the period when the silo was designed and constructed, soil mechanics as a science had hardly begun. The behavior of the foundation under imposed loads was not clearly understood. It was only during the year 1952 that soil investigation was carried out close to the silo and the soil properties were analyzed (Peck and Byrant, 1953). Figure 18.21 gives the soil classification and unconfined compressive strength of the soil with respect to depth. From the examination of undisturbed samples of the clay, it was determined that the average water content of successive layers of varved clay increased with their depth from 40 percent to about 60 percent. The average unconfined compressive strength of the upper stratum beneath the foundation was 1.13 tsf, that of the lower stratum was 0.65 tsf, and the weighted average was 0.93 tsf. The average liquid limit was found to be 105 percent; therefore the plasticity index was 70 percent, which indicates that the clay was highly colloidal and plastic. The average unit weight of the soil was 120 lb/ft³.

The contact pressure due to the load from the silo at the time of failure was estimated as equal to 3.06 tsf. The theoretical values of the ultimate bearing capacity by various methods are as follows.

Methods	q_u tsf
Terzaghi [Eq. (12.19)]	3.68
Meyerhof [Eq. (12.27)]	3.30
Skempton [Eq. (12.22)]	3.32

The above values compare reasonably well with the actual failure load 3.06 tsf. Perloff and Baron (1976) give details of failure of the Transcona grain elevator.

18.34 QUESTIONS AND PROBLEMS

18.1   The total column load of a footing near ground level is 5000 kN. The subsoil is cohesionless soil with $\phi = 38°$ and $\gamma = 19.5$ kN/m³. The footing is to be located at a depth of 1.50 m below ground level. For a footing of size 3×3 m, determine the factor of safety by Terzaghi's general shear failure theory if the water table is at a depth of 0.5 m below the base level of the foundation.

18.2   What will be the factors of safety if the water table is met (i) at the base level of the foundation, and (ii) at the ground level in the case of the footing in Prob. 18.1, keeping all the other conditions the same? Assume that the saturated unit weight of soil $\gamma_{sat} = 19.5$ kN/m³, and the soil above the base of the foundation remains saturated even under (i) above.

18.3   If the factors of safety in Prob. 18.2 are other than 3, what should be the size of the footing for a minimum factor of safety of 3 under the worst condition?

18.4   If the soil in Frob. 18.1 is medium dense sand with $\phi = 34°$, and $\gamma_{sat} = 19.5$ kN/m³, determine (i) the factor of safety for the water table at ground level, and (ii) the size of footing for a minimum $F_s = 3.0$ with WT at GL. All the other conditions remain the same as in Frob. 18.1.

18.5   If the average SPT value (N) is 38 for $\phi = 38°$ in Frob. 18.1, compute the net ultimate bearing capacity at the base of the footing given in the problem by Teng (1969) method. Compare this value with the net ultimate bearing capacity by Terzaghi method. Assume water table is at ground level, and $\gamma_{sat} = 19.5$ kN/m³.

18.6   A footing of size 3×3 m is founded at a depth of 1.5 m in medium stiff clay soil having an unconfined compressive strength of 100 kPa. Determine the net safe bearing capacity of footing with the water table at ground level by Skempton's method. Assume $F_s = 3$.

18.7   If the average static cone penetration resistance, q_c in Prob. 18.6 is 10 kg/cm², determine q_{na} as per Skempton's method. The other conditions remain the same as in Prob. 18.6. Ignore the effect of overburden pressure.

18.8   For the footing given in Prob. 18.5, determine the allowable bearing pressure by Teng's method (modified).

18.9   For the footing given in Prob. 18.5, determine the allowable bearing pressure by Peck *et al.* (1974) design chart.

18.10   A raft foundation of size 12 × 8 m is founded at a depth of 2 m below ground level in cohesionless soil with water table at ground level. The average SPT (N) value beow the footing is 16. Determine the net safe bearing pressure under settlement consideration by Peck *et al.* (1974) method.

18.11   Compute the safe bearing pressure q_s of footing given in Prob. 18.1 by using average static cone penetration value q_c as 150 kg/cm², Assume that the compressible strata below the footing is equal to twice its width. The water table is at ground level.

18.12   What will be the net ultimate bearing capacity of a strip footing of width 1.5 m founded at a depth of 1.0 m in sand. The water table is at ground level. A sample of this soil under triaxial compression test failed with a lateral pressure of 50 kN/m² and a deviator stress of 150 kN/m². The saturated unit weight of the soil is 17.0 kN/m³. Use Terzaghi's theory of general shear failure.

18.13   Find the allowable net bearing load per meter length of a long wall footing 2 m wide on a stiff saturated clay. The depth of the foundation is 0.5 m. The unit weight of the clay is 17kN/m³ and its shear strength c is 120 kN/m². Assume the load to be applied rapiy such that undrained ($\phi = 0$) conditions prevail. Use a factor of safety of 3.

18.14   Determine the allowable total load for a rectangular footing 2 × 3 m for the same soil condition as in Prob. 18.13.

18.15   Determine the allowable total load for the rectangular footing in Prob. 18.14 if the eccentricity is 1/6 m in each direction.

18.16   Determine the net allowable pressure for a square footing 3 × 3 m founded at a depth of 1 m in a medium dense sand. The sand has a friction angle of 36° and has moist unit weight of 16 kN/m³, Use $F_s = 2.5$. Given $N_y = 41$, $N_q = 38$.

18.17   A footing 2.25 m square is located at a depth of 1.5 m in a sand deposit having $\phi = 38°$. Determine the net ultimate bearing capacity (a) if the water table is well below the foundation level, (b) if the water table is at the surface. The unit weight of the sand above the water table is 18 kN/m³.

18.18   A strip footing carries a load of 380 kN/m and the footing is founded at a depth of 1 m in a firm soil of unit weight 18 kN/m³. The shear strength parameters of the soil are $c_u = 15$ kN/m² and $\phi_u = 30°$. Determine the width of the footing if $F_s = 3$.

18.19   A footing 2 m square is located at a depth of 1.0 m in a stiff clay of saturated unit weight of 21 kN/m³. The undrained strength of the clay at a depth of 1 m is $c_u = 120$ kN/m² for $\phi_u = 0$. For $F_s = 3$, determine the net load the footing is allowed to carry.

18.20   A net load of 425 kN per metre length is carried on a strip footing 2 m wide at a depth of 1 m in a stiff clay of saturated unit weight 21 kN/m³, the water table being at a depth of 3 m from ground level. Determine the factor of safety with respect to general shear failure (a) when c_u = 105 kN/m² and ϕ_u= 0, (b) when c' = 10 kN/m² and ϕ' =25°.

18.21   A strip footing of 1.75 m wide is located at a depth of 1.0 m in a sand deposit of unit weight 18.5 kN/m³, the water table being at a depth of 5 m below the foundation level. The angle of internal friction of sand is 35°. The footing carries a load of 575 kN/m. Determine the factor of safety with respect to general shear failure.

18.22   A footing of size 4.0 × 2.0 m is founded at a depth of 2.0 m in a stiff clay. The saturated unit weight of the clay is 19.5 kN/m³, and c_u = 145 kN/m².

Determine the net allowable load for F_s = 3 with respect to shear failure.

18.23   The contact pressure of a footing of 2.5 m square founded at a depth of 1.25 m is 450 kN/m². The saturated unit weight of the sand is 20.5 kN/m³ and the unit weight above the water table is 17.5 kN/m³. Assuming ϕ' = 37°.5, determine the factor of safety F_s with respect to the shear failure for the following cases (a) the water table at ground level, (b) the water table at 2.5 m below ground level, and (c) the water table at 4.5 m below ground level.

18.24   A continuoutwall footing will rest at a depth of 1 m on a saturated clay that possesses an unconfined compressive strength of 110 kN/m². At a load of 145 kN/m of wall a factor of safety of 3 is required, in addition the factor of safety should not be less than 2 if the footing is subjected to a load of 195 kN/m. Determine the width of the footing.

18.25   Proportion a square footing to carry a column load 1600 kN at a factor of safety of 2.5. The base of the footing will be 1.25 below the level of the surrounding ground. The clay beneath the footing has an unconfined compressive strength of 140 kN/m².

18.26   A footing 3 m square rests at a depth of 1 m on clay that has an unconfined compressive strength 120 kN/m². If the factor of safety is not to be less than 2.5, what is the maximum column load that can be supported by the footing.

18.27   A building is to be supported on a reinforced concrete raft of size 14 × 21 m. The subsoil is clay with an unconfined compressive strength of 80 kN/m². The pressure on the soil, due to the weight of the building and the loads it will carry will be 140 kN/m² at the base of raft. If the unit weight of the excavated soil is 19.2 kN/m³, at what depth should the bottom of the raft be placed to provide a factor of safety of 3.

18.28   The foundation for a building consists of a RC raft of size 18 × 21.5 m.1t is founded at a depth of 3 m below the ground surface on a deposit of clay of unit weight 19.2 kN/m³. The cohesive strength of clay under undrained condition is 40 kN/m². What total weight of building including foundation can safely be supported by the raft with a factor of safety of 3 against bearing capacity failure?

18.29   A raft of size 15 × 20 m is founded at a depth of 2.5 m on a deposit of medium dense sand. The standard penetration tests carried out in bore holes over the area give an average value of 12 corrected for over burden pressure, and dilatancy. Compute the net pressure that the raft may carry under settlement consideration, (a) if the water table is at a depth greater than the width of the foundation, and (b) if the water table is at a depth of 5 m below the ground surface. Use Peck et al. (1974) chart.

18.30   A footing 3 × 3 m is founded in a deposit of medium dense sand at a depth of 1.5 m below ground surface. The water table is at a depth 0.5 m below the ground surface. The soil investigation at the site indicate that an average SPT value of 14 may be taken which is corrected for overburden pressure, and dilatancy. Compute the net allowable bearing pressure by Teng's modified formula for settlement consideration

18.31   Compute the safe pressure for the footing in Prob. 18.30 by the use of Peck et al. Chart, (1974).

18.32   Compute the safe pressure for the footing in Prob. 18.30 if the average static cone point penetration resistance is 56 kg/cm^2.

18.33   A footing of width 3 m and length 6 m is founded on medium dense sand at a depth 2 m below ground level. The footing is considered as rigid. The average net contact pressure at the base of the footing due to superstructure loading is 175 N/m^2. Given Poisson ratio $\mu = 0.4$, deformation modulus $E_d = 5000$ kN/m^2, and the influence factor $I_f = 1.06$ for rigid footing. Calculate the immediate settlement. The water table is assumed to be close to the base of the footing. Neglect the depth factor.

18.34   A footing of size 5 × 5 m is founded at a depth of 1.5 m on a loose to mdlium dense sand. The imposed contact pressure at the base of the footing is 80 kN/m^2. The thickness of the sandy strata below the base of the footing is 4 m. Below the sand strata lies a normally consolidated clay strata of 3 m thick overlying a sandy strata of great depth. The water table lies at the base of the footing. Sandy layer above water table is assumed as saturated. The unit weights of saturated sand and clay strata are 18.75 and 20.5 kN/m^3 respectively. The average void ratio of clay soil is taken as equal to 0.74, and its average liquid limit, w_l is 49%. Estimate the consolidation settlement of the footing. Use 2 : 1 method to determine the increase in pressure at the centre of clay layer.

18.35   For the problem given in 18.34 if the clay strata has a coefficient of permeability $k = 5 \times 10^{-8}$ cm/see and a coefficient of volume compressibility $m_v = 125 \times 10^{-6}$ cm^2/kN, determine the time required for 30% and 70% consolidation and the corresponding settlements. Assume all the other data as given in Prob. 18.34.

18.36   A 2 cm thick sample of clay was extracted from an in situ strata for laboratory compression testing. The sample was tested in a consolidometer allowing drainage both sides. As per the test 50% of the total settlement took place in 4 minutes. Find the time required in days for 50% of the total settlement of a building constructed on this clay strata of 5 m thick having drainage both sides.

18.37   A recently filled up soil was fully saturated and had an initial water content of 47% and th.specific gravity of 2.67. The fill was loaded on the surface by constructing an embankment covering a large area of the fill. Some months after the embankment was constructed, measurements of the fill indicated an average water content of 26%. Estimate the compression of the fill by assuming the initial thickness of the fill was 5 m.

18.38   It is proposed to construct an verhead tank at a site on a circular rigid raft foundation of 10 m diameter. The raft is founded at a depth of 2 m below ground level on loose sandy strata extending to a great depth. The average corrected SPT value is 10. The contact pressure at the base of the footing due to the superstructure weight is estimated to be 150 kN/m^2. Compute the factor of safety due to shear failure criteria by assuming (a) the water table is at great depth and (b) at the ground level.

18.39   For the footing in Prob. 18.38, estimate the immediate settlement with respect to sandy strata. Assume Poisson's ratio $\mu = 0.25$ and Young's modulus $E_s = 10000$ kN/m^2, and influence factor $I_f = 0.8$.

18.35 ADDITIONAL PROBLEMS

18.1   A plate load test was conducted in a medium dense sand at a depth of 5 ft below ground
        level in a test pit. The size of the plate used was 12 × 12 in. The data obtained from the test
        are plotted in Fig. Prob. 18.1 as a load-settlement curve. Determine from the curve the net
        safe bearing pressure for footings of size (a) 10 × 10 ft, and (b) 15 × 15 ft. Assume the
        permissible settlement for the foundation is 25 mm.

18.2   Refer to Prob. 18.1. Determine the settlements of the footings given in Prob 18.1. Assume
        the settlement of the plate as equal to 0.5 in. What is the net bearing pressure from Fig.
        Prob. 18.1 for the computed settlements of the foundations?

18.3   For Problem 18.2, determine the safe bearing pressure of the footings if the settlement is
        limited to 2 in.

18.4   Refer to Prob. 18.1. If the curve given in Fig. Prob. 18.1 applies to a plate test of 12 × 12 in.
        conducted in a clay stratum, determine the safe bearing pressures of the footings for a
        settlement of 2 in.

18.5   Two plate load tests were conducted in a c-ϕ soil as given below.

Size of plates (m)	Load kN	Settlement (mm)
0.3 × 0.3	40	30
0.6 × 0.6	100	30

        Determine the required size of a footing to carry a load of 1250 kN for the same settlement
        of 30 mm.

18.6   A rectangular footing of size 4 × 8 m is founded at a depth of 2 m below the ground surface
        in dense sand and the water table is at the base of the foundation. N_{cor} = 30 (Fig. Prob.
        18.6). Compute the safe bearing pressure q_s using the chart given in Fig. 18.17.

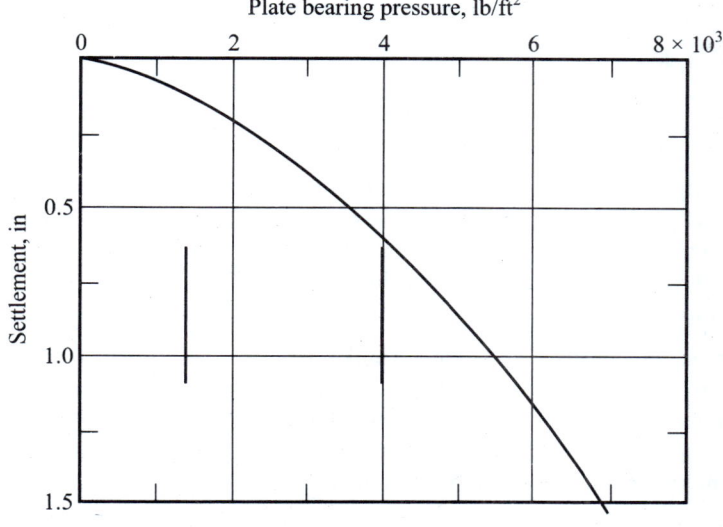

Fig. Prob. 18.1

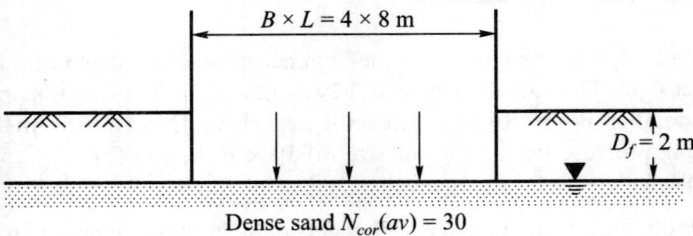

Fig. Prob. 18.6

18.7   Refer to Prob. 18.6. Compute q_s by using modified (a) Teng's formula, and (b) Meyerhof's formula.

18.8   Refer to Prob. 18.6. Determine the safe bearing pressure based on the static cone penetration test value based on the relationship given in Eq. (18.49) for $q_c = 120$ kN/m².

18.9   Refer to Prob. 18.6. Estimate the immediate settlement of the footing by using Eq. (18.52a). The additional data available are:

$\mu = 0.30$, $I_f = 0.82$ for rigid footing and $E_s = 11,000$ kN/m². Assume $q_n = q_s$ as obtained from Prob. 18.6.

18.10   Refer to Prob 18.6. Compute the immediate settlement for a flexible footing, given $\mu = 0.30$ and $E_s = 11,000$ kN/m². Assume $q_n = q_s$

18.11   If the footing given in Prob. 18.6 rests on normally consolidated saturated clay, compute the immediate settlement using Eq. (18.54). Use the following relationships.

$$\bar{q}_c = 120 \, \text{kN/m}^2$$

$$c_u = \frac{\bar{q}_c - p_0}{N_k} \text{ and } N_k = 20$$

$$E_s = 600 \, c_u \, \text{kN/m}^2$$

Given: $\gamma_{sat} = 18.5$ kN/m³, $q_n = 150$ kN/m². Assume that the incompressible stratum lies at at depth of 10 m below the base of the foundation.

18.12   A footing of size 6×6 m rests in medium dense sand at a depth of 1.5 below ground level. The contact pressure $q_n = 175$ kN/m². The compressible stratum below the foundation base is divided into three layers. The corrected N_{cor} values for each layer is given in Fig. Prob. 18.12 with other data . Compute the immediate settlement using Eq. (18.55). Use the relationship $q_c = 400 \, N_{cor}$ kN/m².

18.13   It is proposed to construct an overhead tank on a raft foundation of size 8×16 m with the foundation at a depth of 2 m below ground level. The subsoil at the site is a stiff homogeneous clay with the water table at the base of the foundation. The subsoil is divided into 3 layers and the properties of each layer are given in Fig. Prob. 18.13. Estimate the consolidation settlement by the Skempton-Bjerrum Method.

18.14   A footing of size 10×10 m is founded at a depth of 2.5 m below ground level on a sand deposit. The water table is at the base of the foundation. The saturated unit weight of soil

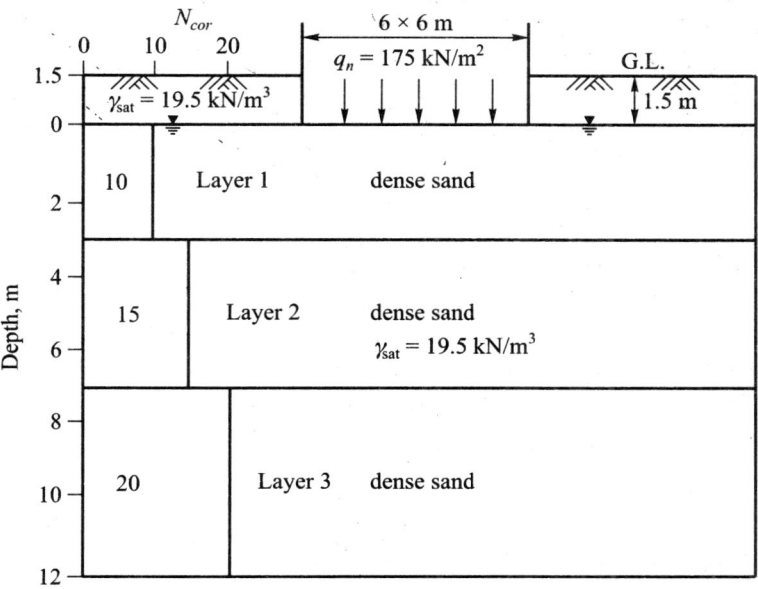

Fig. Prob. 18.12

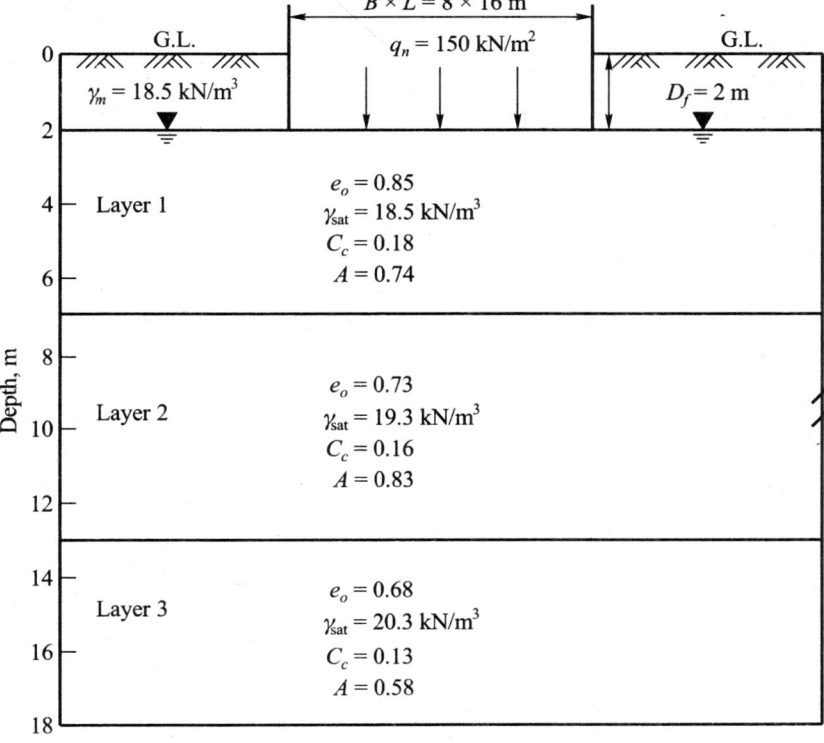

Fig. Prob. 18.13

from ground level to a depth of 22.5 m is 20 kN/m³. The compressible stratum of 20 m below the foundation base is divided into three layers with corrected SPT values (N) and CPT values (q_c) constant in each layer as given below.

Layer No	Depth from (m) foundation level		N_{cor} (av)	$q_c(av)$ MPa
	From	To		
1	0	5	20	8.0
2	5	11.0	25	10.0
3	11.0	20.0	30	12.0

Compute the settlements by Schmertmann's method.

Assume the net contact pressure at the base of the foundation is equal to 70 kPa, and $t = 10$ years.

18.15   A square rigid footing of size 10 × 10 m is founded at a depth of 2.0 m below ground level. The type of strata met at the site is

Depth below G. L. (m)	Type of soil
0 to 5	Sand
5 to 7m	Clay
Below 7m	Sand

The water table is at the base level of the foundation. The saturated unit weight of soil above the foundation base is 20 kN/m³. The coefficient of volume compressibility of clay, m_v, is 0.0001 m²/kN, and the coefficient of consolidation c_v, is 1 m²/year. The total contact pressure q_n = 100 kN/m². Water table is at the base level of foundation.

Compute primary consolidation settlement.

18.16   A circular tank of diameter 3 m is founded at a depth of 1 m below ground surface on a 6 m thick normally consolidated clay. The water table is at the base of the foundation. The saturated unit weight of soil is 19.5 kN/m³, and the *in-situ* void ratio e_0 is 1.08. Laboratory tests on representative undisturbed samples of the clay gave a value of 0.6 for the pore pressure coefficient A and a value of 0.2 for the compression index C_c. Compute the consolidation settlement of the foundation for a total contact pressure of 95 kPa. Use 2:1 method for computing Δp.

18.17   A raft foundation of size 10 × 40 m is founded at a depth of 3 m below ground surface and is uniformly loaded with a net pressure of 50 kN/m². The subsoil is normally consolidated saturated clay to a depth of 20 m below the base of the foundation with variable elastic moduli with respect to depth. For the purpose of analysis, the stratum is divided into three layers with constant modulus as given below:

Layer No	Depth from ground (m)		Elastic modulus
	From	To	E_s (MPA)
1	3	8	20
2	8	18	25
3	18	23	30

Compute the immediate settlements by using Eq. (18.52a). Assume the footing is flexible.

18.18   A beam of length 6 m and width 0.80 m is founded on dense sand under submerged conditions. A plate load test with a plate of 0.30×0.30 m conducted at the site gave a value for the coefficient of subgrade reaction for the plate equal to 95 MN/m^3. Determine the coefficient of subgrade reaction for the beam.

18.19   If the beam in Prob. 18.18 is founded in very stiff clay with the value for k_1 equal to 45 MN/m^3, what is the coefficient of subgrade reaction for the beam?

18.20   Proportion a strap footing given the following data with reference to Fig. 18.26:

$Q_1 = 580$ kN, $Q_2 = 900$ kN

$L_c = 6.2$ m, $b_1 = 0.40$ m, $q_s = 120$ kPa.

18.21   Proportion a rectangular combined footing given the following data with reference to Fig. 18.27 (the footing is rectangular instead of trapezoidal):

$Q_1 = 535$ kN, $Q_2 = 900$ kN, $b_1 = 0.40$ m,

$L_c = 4.75$ m, $q_s = 100$ kPa.

CHAPTER 19

PILE FOUNDATION

19.1 INTRODUCTION

There are three types of deep foundations. They are:

1. Pile foundations,

2. Wells or Caisson foundations,

3. Drilled pier foundations.

Piles are the long slender members either *driven* or *cast-in-situ*. They may be subjected to vertical or lateral loads or a combination of vertical and lateral loads. Caissons or Well foundations are heavier in section and they are sunk to the required depth. They are normally used to carry very heavy loads such as the loads from bridge piers or multistoried buildings.

Some times the term *drilled pier* is used for piles of bigger diameter which are normally greater then 1 m and may go upto 5 m or more. They are nothing but large diameter bored pile which may be belled at the bottom to give greater bearing area. The design principles of drilled piers are the same as that of piles of smaller diameter.

Piles are also used to take up uplift loads. Piles may be used as single piles or in groups with vertical and/or batter piles. The problem of pile foundation has been considered under the following headings in this book,

Part A Types of piles and installation,

Part B Vertical load bearing capacity of single vertical pile,

Part C Pile groups subjected to vertical loads,

Part D Behaviour of laterally loaded vertical and batter piles.

PART A: TYPES OF PILES AND INSTALLATION

19.2   CLASSIFICATION OF PILES

Piles may be classified as long or short in accordance with the L/B ratio of the pile (where L = length, B = diameter of pile). A short pile behaves as a rigid body and rotates as a unit under lateral loads. The load transferred to the tip of the pile bears a significant proportion of the total vertical load on the top. In the case of a long pile, the length beyond a particular depth loses its significance under lateral loads, but when subjected to vertical load, the frictional load on the sides of the pile bears a significant part to the total load.

Piles may further be classified as vertical piles or inclined piles. Vertical piles are normally used to carry mainly vertical loads and very little lateral load. When piles are inclined at an angle to the vertical, they are called *batter piles* or *raker piles*. Batter piles are quite effective for taking lateral loads, but when used in groups, they also can take vertical loads.

Types of Piles According to their Composition

Piles may be classified according to their composition as

1.   Timber Piles,

2.   Concrete Piles,

3.   Steel Piles.

Timber Piles: Timber piles are made of tree trunks with the branches trimmed off. Such piles shall be of sound quality and free of defects. The length of the pile may be 15 m or more. If greater lengths are required, they may be spliced. The diameter of the piles at the butt end may vary from 30 to 40 cm. The diameter at the tip end should not be less than 15 cm.

The usual maximum design load per pile does not exceed 250 kN. Timber piles work out cheaper at places where timber is available in plenty.

Concrete Piles: Concrete piles are either precast or *cast-in-situ* piles. Precast concrete piles are cast and cured in a casting yard and then transported to the site of work for driving. If the work is of a very big nature, they may be cast at the site also.

Precast piles may be made of uniform sections with pointed tips. Tapered piles may be manufactured when greater bearing resistance is required. Normally piles of square or octagonal sections are manufactured since these shapes are easy to cast in horizontal position. Necessary reinforcement is provided to take care of handling stresses. Piles may also be prestressed. Fig. 19.1 indicates maximum moments in pile, depending on the location of the pick-up points. The pick-up points on the pile should clearly be marked to avoid overstressing the pile by inadvertence. Minimum reinforcement in a pile should be at least 1 percent.

Maximum load on a prestressed concrete pile is approximately 2000 kN and on precast piles 1000 kN. The optimum load range is 400 to 600 kN.

Steel Piles: Steel piles are usually rolled H shapes or pipe piles. H-piles are proportioned to withstand large impact stresses during hard driving. Pipe piles are either welded or seamless steel pipes which may be driven either open-end or closed-end. Pipe piles are often filled with concrete after driving, although in some cases this is not necessary. The optimum load range on steel piles is 400 to 1200 kN.

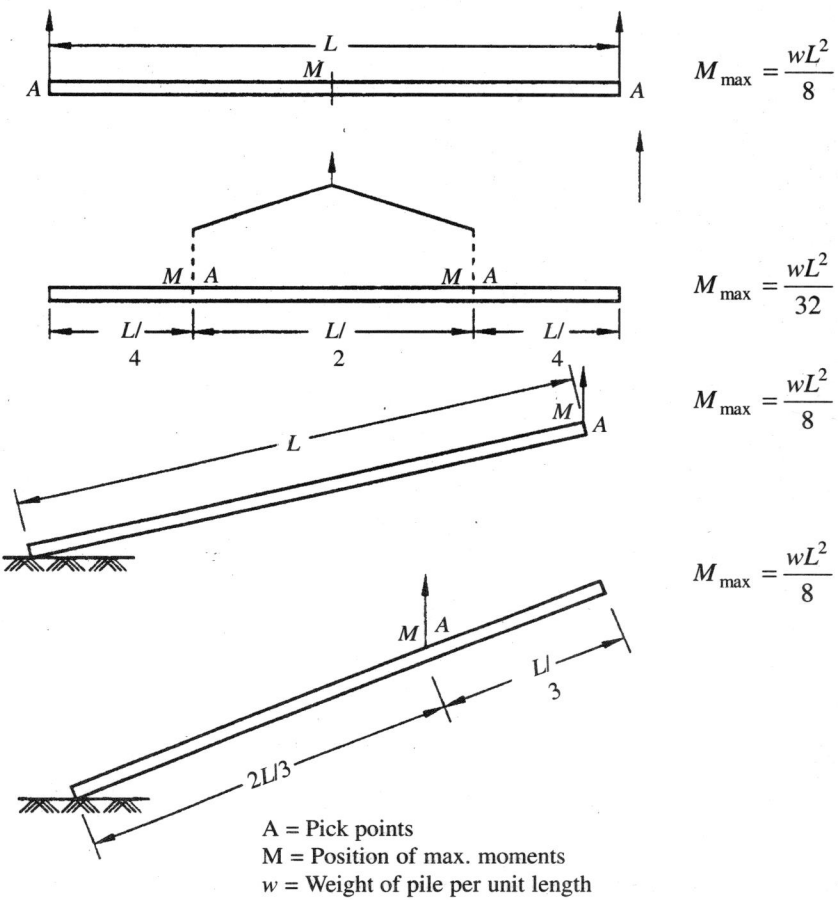

A = Pick points
M = Position of max. moments
w = Weight of pile per unit length

Fig. 19.1   Maximum moments in piles at pick-up points

Types of Piles According to the Method of Installation

According to the method of construction, there are three types of piles. They are:

1.   Driven piles,

2.   *Cast-in-situ* piles and

3.   Driven and *cast-in-situ* piles.

Driven Piles

Piles may be of timber, steel or concrete. When the piles are of concrete, they are to be precast. They may be driven either vertically or at an angle to the vertical. Piles are driven using a pile hammer. When a pile is driven into granular soil, the soil so displaced, equal to the volume of the driven pile, compacts the soil around the sides since the displaced soil particles enter the soil spaces of the adjacent mass which leads to densification of the mass. The pile that compacts the soil adjacent to it is sometimes called a *compaction pile*. The compaction of the soil mass around a pile increases its bearing capacity.

Cast-in-situ Piles

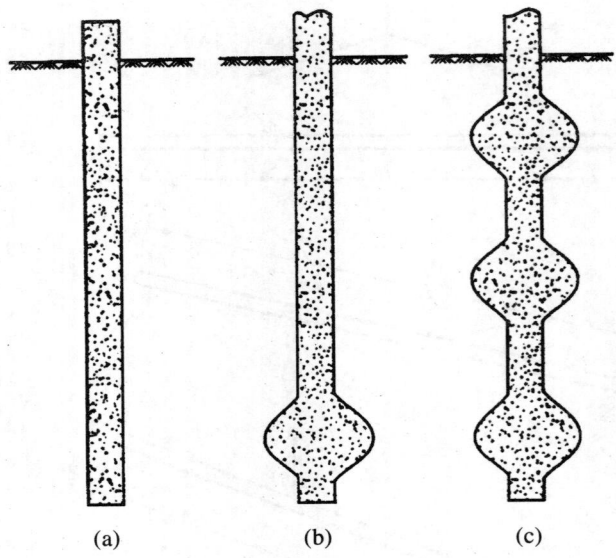

A. *Cast-in-situ* (a) straight bored, (b) single bulb, and (c) multi bulb pile

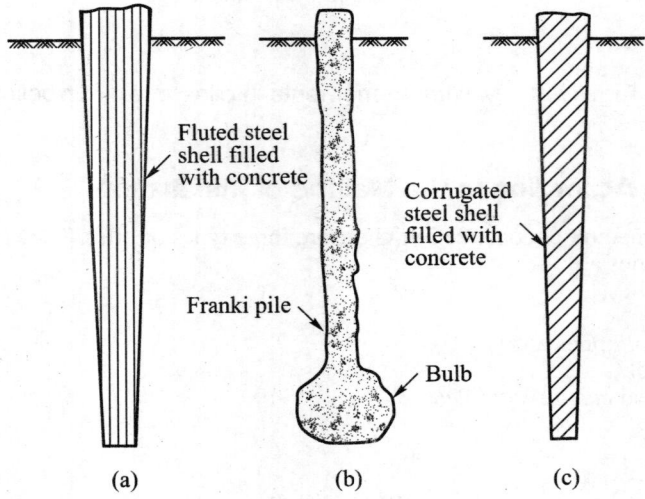

B. Driven and cast-in-situ piles

Fig. 19.2 Types of *cast-in-situ* and driven *cast-in-situ* concrete piles

Cast-in-situ piles are concrete piles. These piles are distinguished from drilled piers as small diameter piles. They are constructed by making holes in the ground to the required depth and then filling the hole with concrete. Straight bored piles or piles with one or more bulbs at intervals may be cast at the site. The latter type are called *under-reamed piles*. Reinforcement may be used as per the requirements. *Cast-in-situ* piles have advantages as well as disadvantages. The straight bored and under reamed piles are shown in Fig. 19.2(A).

Driven and Cast-in-situ Piles

This type has the advantages and disadvantages of both the driven and the *cast-in-situ* piles. The procedure of installing a *driven* and *cast-in-situ* pile is as follows:

A steel shell is driven into the ground with the aid of a mandrel inserted into the shell. The mandrel is withdrawn and concrete is placed in the shell. The shell is made of corrugated and reinforced thin sheet steel (mono-tube piles) or pipes (Armco welded pipes or common seamless pipes). The piles of this type are called a shell type. The shell-less type is formed by withdrawing the shell while the concrete is being placed. In both the types of piles the bottom of the shell is closed with a conical tip which can be separated from the shell. By driving the concrete out of the shell an enlarged bulb may be formed in both the types of piles. Franki piles are of this type. The common types of driven and cast-in-situ piles are given in Fig. 19.2(B). In some cases the shell will be left in place and the tube is concreted. This type of pile is very much used in piling over water.

Advantages and Disadvantages of the Various Types of Soils

Driven piles can be precast and installed according to specifications. Th pile work can be programmed in advance. But they cause driving during vibration which may affect the neighbouring structures. They are not suitable in soils of poor drainage quantities and this would lead to development of pore pressures or lifting to adjacent piles.

Cast-in-situ piles are suitable where the length can be adjusted according to pile conditions. They are ideal in places where vibrations are required to be avoided or where the soil has poor drainage qualities. But installation of these piles require great care and are not suitable in underground flowing water. The piles of cast-in-situ take very much less load as compared to driven piles. The advantages and disadvantages of driven and cast-in-situ piles are midway between the two piles mentioned above.

19.3 USES OF PILES

The major uses of piles are:

1. To carry vertical compression load.

2. To resist uplift load.

3. To resist horizontal or inclined loads.

Normally vertical piles are used to carry vertical compression loads coming from superstructures such as buildings, bridges etc. The piles are used in groups joined together by pile caps. The loads carried by the piles are transferred to the adjacent soil. If all the loads coming on the tops of piles are transferred to the tips, such piles are called *end-bearing* or *point-bearing piles*. However, if all the load is transferred to the soil along the length of the pile such piles are called *friction piles*. If, in the course of driving a pile into granular soils, the soil around the pile gets compacted, such piles are

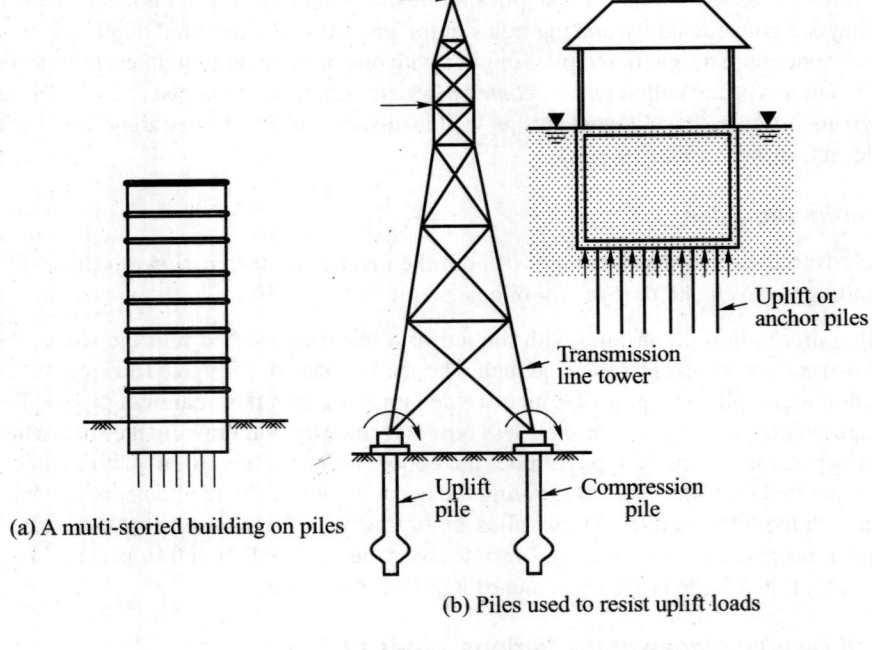

(a) A multi-storied building on piles

(b) Piles used to resist uplift loads

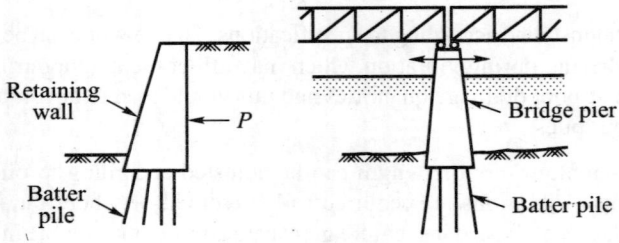

(c) Piles used to resist lateral loads

Fig. 19.3   Uses of piles

called *compaction piles*. Figure 19.3(a) shows piles used for the foundation of a multistoried building to carry loads from the superstructure.

Piles are also used to resist uplift loads. Piles used for this purpose are called *tension piles* or *uplift piles* or *anchor piles*. Uplift loads are developed due to hydrostatic pressure or overturning movement as shown in Fig. 19.3(b).

Piles are also used to resist horizontal or inclined forces. Batter piles are normally used to resist large horizontal loads. Fig. 19.3(c) shows the use of piles to resist lateral loads.

19.4   SELECTION OF PILE

The selection of the type, length and capacity is usually made from estimation based on the soil conditions and the magnitude of the load. In large cities, where the soil conditions are well known and where a large number of pile foundations have been constructed, the experience gained in the past is extremely useful. Generally the foundation design is made on the preliminary estimated values. Before the actual construction begins, pile load tests must be conducted to verify the design values. The foundation design must be revised according to the test results. The factors that govern the selection of piles are:

1. Length of pile in relation to the load and type of soil

2. Character of structure

3. Availability of materials

4. Type of loading

5. Factors causing deterioration

6. Ease of maintenance

7. Estimated costs of types of piles, taking into account the initial cost, life expectancy and cost of maintenance

8. Availability of funds

All the above factors have to be largely analyzed before deciding up on a particular type.

19.5   INSTALLATION OF PILES

The method of installing a pile at a site depends upon the type of pile. The equipment required for this purpose varies. The following types of piles are normally considered for the purpose of installation.

1. Driven piles

The piles that come under this category are:

a. Timber piles,

b. Steel piles, *H*-section and pipe piles,

c. Precast concrete or prestressed concrete piles, either solid or hollow sections.

2. Driven cast-in-situ piles

This involves driving of a steel tube to the required depth with the end closed by a detachable conical tip. The tube is next concreted and the shell is simultaneously withdrawn. In some cases the shell will not be withdrawn.

3. Bored cast-in-situ piles

Boring is done either by auguring or by percussion drilling. The bottom of the bore may be under-reamed according to requirements. After boring is completed, the bore is concreted with or without reinforcement.

A pile is driven by a pile driving rig (Fig. 19.4) which consists of a frame, a winch and impact hammers. The hammers may be the drop type, single acting or double-acting steam hammers or the vibrating type [Fig. 19.4(b)]. The weights of hammers depend upon the type of length of piles. For a single-acting hammer, weights vary from 1500 to 10,000 kg and for double acting from 350 to 2500 kg.

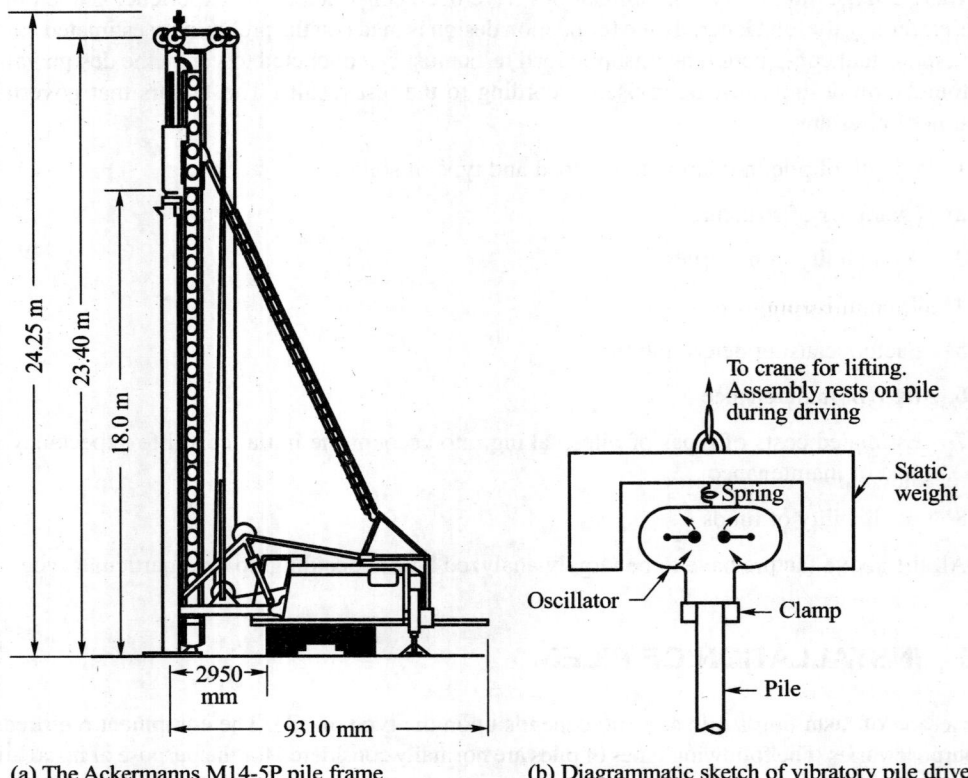

(a) The Ackermanns M14-5P pile frame   (b) Diagrammatic sketch of vibratory pile driver

Fig. 19.4   Pile driving equipment and vibratory pile driver

Sometimes water jetting is used for driving piles in firm to stiff clays or soils mixed with cobbles or boulders. Percussion drilling is some time times adopted in soils mixed with gravels and boulders.

PART B: VERTICAL LOAD BEARING CAPACITY OF SINGLE VERTICAL PILE

19.6   INTRODUCTION

The bearing capacity of groups of piles subjected to vertical or vertical and lateral loads depends upon the behaviour of a single pile. The bearing capacity of a single pile depends upon

1.   Type, size and length of pile,

2.   Type of soil,

3.   The method of installation.

The bearing capacity depends primarily on the method of installation and the type of soil met with.

In order to be able to design a safe and economical pile foundation, we have to analyze the interactions between the pile and the soil, establish the modes of failure and estimate the settlements from soil deformation under dead load, service load etc. The design should comply with the following requirements.

1. It should ensure adequate safety against failure; the factor of safety used depends on the importance of the structure and on the reliability of the soil parameters and the loading systems used in the design.

2. The settlements should be compatible with adequate behaviour of the superstructure to avoid impairing its efficiency.

Statement of the Problem

Figure 19.5(a) gives a single pile of uniform diameter and length L driven into a homogeneous mass of soil of known physical properties. It is required to determine the ultimate bearing capacity Q_u of the pile.

When the ultimate load applied on the top of the pile is Q_u, a part of the load is transmitted to the soil along the length of the pile and the balance is transmitted to the pile base. The probable load distribution along the pipe with the increase in the top load is given in Fig. 19.5(b). If the settlement of the top of the pile is measured at every stage of loading after equilibrium condition is reached a load-settlement curve of the type shown in Fig. 19.5(c) can be obtained. When the load on the pile is very much lower than the ultimate load Q_u, the load is entirely taken by friction. For example, a load Q_1 is applied on the pile Fig. 19.5(b), this load is transferred to soil within the length L_1. Only at high loads, such as loads greater than Q_2, a part of these loads is transferred to the base of the pile.

Figure 19.5(d) gives the results of an actual load test in the field. It gives an idea of the distribution of load between the shaft and the base of a pile.

When the pile reaches the ultimate load Q_u, the load transmitted to the soil along the length of the pile is called as the ultimate skin load Q_f and that transmitted to the base is called as the base or point load Q_b. The total ultimate load Q_u is a sum of these two expressed as

$$Q_u = Q_b + Q_f = q_b A_b + f_s A_s \qquad (19.1)$$

where q_b = ultimate unit bearing capacity of the pile at the base

A_b = bearing area of the base of the pile

A_s = total surface area of pile embedded below ground surface

f_s = ultimate unit skin friction

The relative proportions of the loads carried by skin friction and base resistance depend on the shear strength and elasticity of the soil. Generally the vertical movement of the pile which is required to mobilize full end resistance is much greater than that required to mobilize full skin friction. Experience indicates that in bored *cast-in-situ* piles full frictional load is normally mobilized at a settlement equal to 0.5 to 1 percent of pile diameter and the full base load Q_b at 10 to 20 percent of the diameter. But, if this ultimate load criterion is applied to piles of large diameter in clay, the settlement at the working load (with a factor of safety of 2 on the ultimate load) may be excessive.

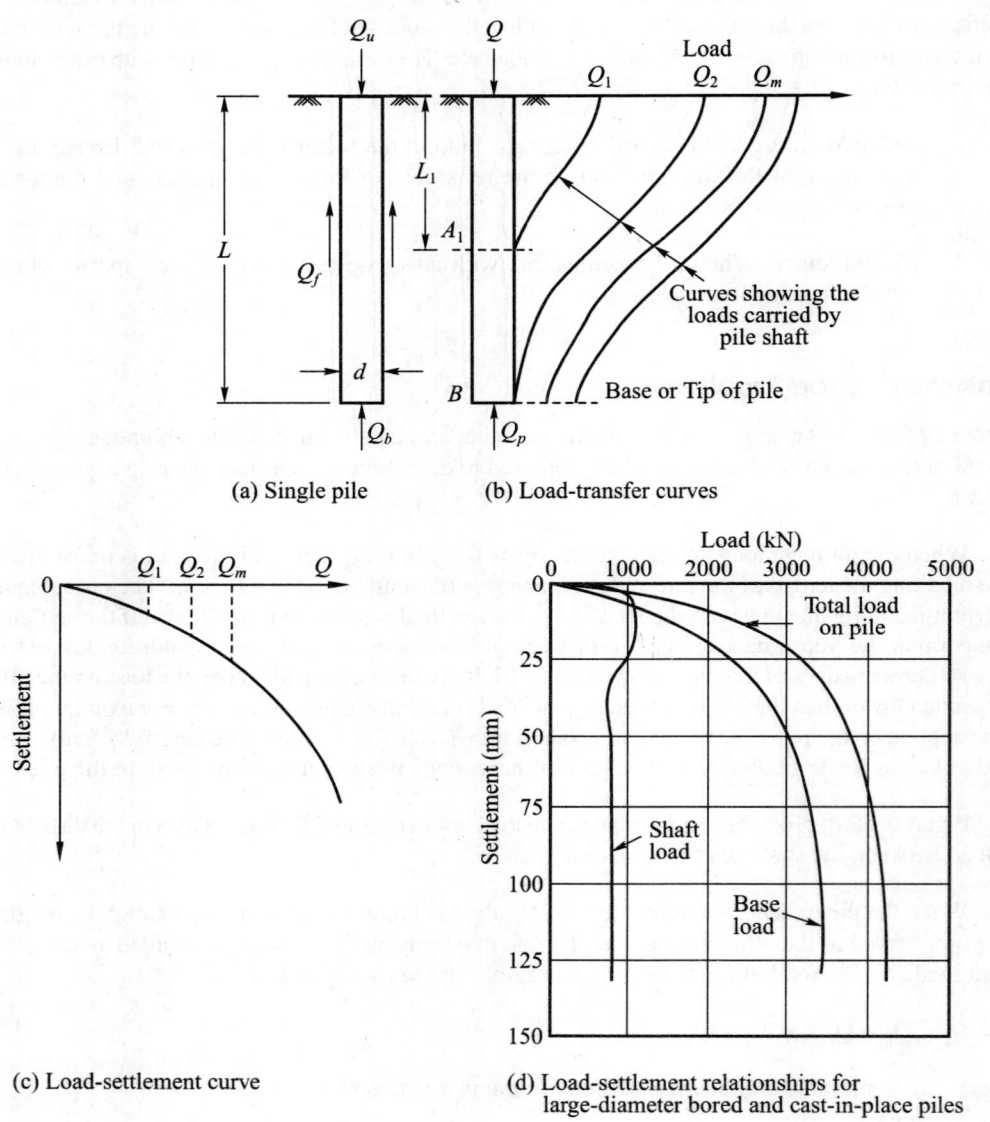

(a) Single pile

(b) Load-transfer curves

(c) Load-settlement curve

(d) Load-settlement relationships for
large-diameter bored and cast-in-place piles

Fig. 19.5   Load transfer mechanism

Definition of Failure Load

The methods of determining failure loads based on load-settlement curves are described in subsequent sections. However, in the absence of a load settlement curve, a failure load may be defined as that which causes a settlement equal to 10 percent of the pile diameter or width (as per the suggestion of Terzaghi) which is widely accepted by engineers. However, if this criterion is applied to piles of large diameter in clay and a nominal factor of safety of 2 is used to obtain the working load, then the settlement at the working load may be excessive.

Factor of Safety

In almost all cases where piles are acting as structural foundations, the allowable load is governed solely from considerations of tolerable settlement at the working load.

The working load for all pile types in all types of soil may be taken as equal to the sum of the base resistance and shaft friction divided by a suitable factor of safety. A safety factor of 2.5 is normally used. Therefore we may write

$$Q_a = \frac{Q_b + Q_f}{2.5} \qquad (19.2)$$

In case where the values of Q_b and Q_f can be obtained independently, the allowable load can be written as

$$Q_a = \frac{Q_b}{3} + \frac{Q_f}{1.5} \qquad (19.3)$$

It is permissible to take a safety factor equal to 1.5 for the skin friction because the peak value of skin friction on a pile occurs at a settlement of only 3–8 mm (relatively independent of shaft diameter and embedded length but may depend on soil parameters) whereas the base resistance requires a greater settlement for full mobilization.

The least of the allowable loads given by Eqs. (19.2) and (19.3) is taken as the design working load.

Methods of Determining Ultimate Load Bearing Capacity of a Single Vertical Pile

The ultimate bearing capacity, Q_u, of a single vertical pile may be determined by any of the following methods.

1. By the use of static bearing capacity equations.

2. By the use of SPT and CPT values.

3. By field load tests.

4. By dynamic method.

The determination of the ultimate point bearing capacity, q_b, of a deep foundation on the basis of theory is a very complex one since there are many factors which cannot be accounted for in the theory. The theory assumes that the soil is homogeneous and isotropic which is normally not the case. All the theoretical equations are obtained based on plane strain conditions. Only shape factors are applied to take care of the three-dimensional nature of the problem. Compressibility characteristics of the soil complicate the problem further. Experience and judgment are therefore very essential in applying any theory to a specific problem. The skin load Q_f depends on the nature of the surface of the pile, the method of installation of the pile and the type of soil. An exact evaluation of Q_f is a difficult job even if the soil is homogeneous over the whole length of the pile. The problem becomes all the more complicated if the pile passes through soils of variable characteristics.

19.7   GENERAL THEORY FOR ULTIMATE BEARING CAPACITY

According to Vesic (1967), only punching shear failure occurs in deep foundations irrespective of the density of the soil so long as the depth-width ratio L/D is greater than 4. The types of failure surfaces assumed by different investigators are shown in Fig. 19.6 for the general shear failure condition. The detailed experimental study of Vesic indicates that the failure surfaces do not revert back to the shaft as shown in Fig. 19.6(c).

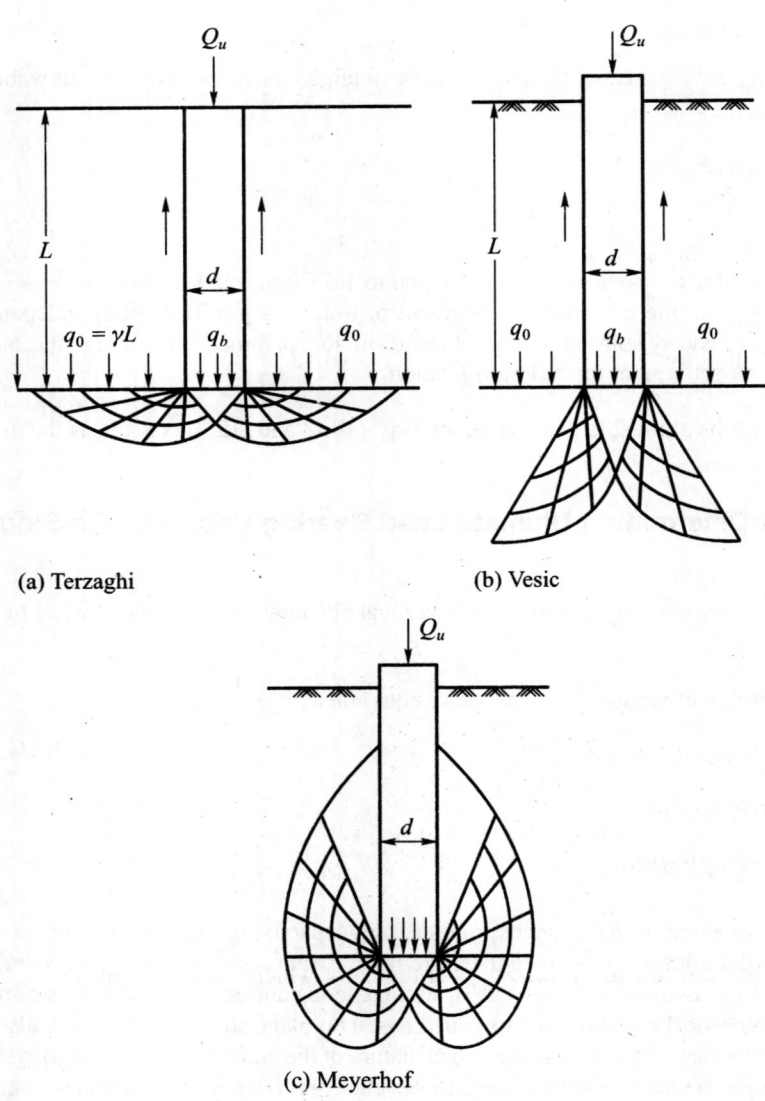

(a) Terzaghi   (b) Vesic

(c) Meyerhof

Fig. 19.6   The shapes of failure surfaces at the tips of piles as assumed by different investigators

The total failure load $\overline{Q}_u$ may be written as follows

$$\overline{Q}_u = Q_u + W_p = Q_b + Q_f + W_p$$

The net ultimate load in excess of the overburden pressure $q_0 A_b$ is

$$\overline{Q}_u - q_0 A_b = Q_u + W_p - q_0 A_b = Q_b + Q_f + W_p - q_0 A_b$$

For all practical purposes $W_p \approx q_0 A_b$. As such the expression for net ultimate load Q_u may be written as

$$Q_u = Q_b + Q_f \tag{19.4}$$

where, Q_b = the total base load of pile, Q_f = the total skin load, q_0 = the effective overburden pressure at the base level of pile, W_p = weight of the pile, and A_b = the sectional area of pile at the base level.

The general equation for the base resistance Q_b for a c-ϕ soil may be written as

$$Q_b = \left(cN_c + q_0 N_q + \frac{1}{2} \gamma d N_\gamma \right) A_b \tag{19.5a}$$

where, c = effective unit cohesion, γ = effective unit weight of soil, q_0 = effective overburden pressure at the base level of pile = γL, N_c, N_q and N_γ = the bearing capacity factors which take into account the necessary shape and depth factors, and L = depth of embedment of pile.

The term $(\frac{1}{2}) \gamma d N_\gamma$ becomes insignificant in comparison with the term $q_0 N_q$ for deep foundations and as such is normally neglected. The final equation for base resistance for a $(c - \phi)$ soil is

$$Q_b = \left(cN_c + q_0 N_q \right) A_b \tag{19.5b}$$

For a purely cohesionless soil, $c = 0$, then

$$Q_b = q_0 N_q A_b \tag{19.5c}$$

For a purely cohesive soil, $\phi = 0$, then

$$Q_b = c_b N_c A_b \tag{19.5d}$$

where, q = undrained shear strength of clay at base level.

The general equation for shaft resistance, Q_f may be written as

$$Q_f = f_s A_s \tag{19.6a}$$

where, f_s = unit shaft resistance and A_s = shaft area for which f_s is applicable.

For a pile in cohesionless soil ($c = 0$), Q_f may be written as

$$Q_f = A_s f_s = A_s \overline{q}_0 K_s \tan\delta \tag{19.6b}$$

For a pile in cohesive soil ($\phi = 0$),

$$Q_f = A_s f_s = A_s \alpha \overline{c}_u \tag{19.6c}$$

where, $\overline{q}_0$ = the average effective overburden pressure over the embedded depth of pile for which $\overline{K}_s \tan \delta$ is applicable.

$\overline{K}_s$ = lateral earth pressure coefficient,

δ = angle of wall friction,

$\overline{c}_u$ = average undrained shear strength of clay along the shaft, and

α = adhesion factor.

If f_s varies along the shaft, we may write for cohesionless soils

$$Q_f = P\sum_{0}^{L} \overline{q}_o \overline{K}_s \tan\delta\Delta L \qquad (19.6d)$$

and for cohesive soils,

$$Q_f = P\sum_{0}^{L} \alpha\overline{c}_u\Delta L \qquad (19.6e)$$

where, P = perimeter of pile, Δz = thickness of each layer for which $\overline{q}_0$, $\overline{K}_s$, δ, α, and c_u are applicable.

19.8   THE ULTIMATE LOAD BEARING CAPACITY OF A DRIVEN PILE INTO COHESIONLESS SOIL

The sum of Eqs (19.5c) and (19.6b) gives the equation for calculating the ultimate load bearing capacity Q_u for a pole driven into cohesionless soil expressed as

$$Q_u = Q_b A_b + f_s A_s = q_0 N_q A_b + \overline{q}_0 \overline{K}_s \tan\delta\, A_s \qquad (19.7a)$$

For a layered system of cohesionless soils, Q_u may be expressed as

$$Q_u = q_0 N_q A_b + P\sum_{0}^{L} \overline{q}_o K_s \tan\delta\Delta L \qquad (19.7b)$$

where P = perimeter of pile, ΔL = thickness soil layer.

Base resistance, $Q_b = q_0 N_q A_b = q_b A_b$

For a straight edged driven pile into cohesion less soil, the value of q_b is proportional to q_0 and N_q. If the soil is homogeneous, q_0 is expressed as

$$q_0 = \gamma L \qquad (19.7c)$$

where, γ = the average effective unit weight of soil at the base level of pile,

L = Length of the embedded portion of soil

Equation (19.7a) indicates that the value of q_0 increases linearly from zero at ground level to a maximum at the base level of pile. However, extensive research carried out by Vesic (1967) has indicated that q_0 increases linearly upto a limited depth only beyond which, it remains constant, irrespective of the depth of embedment of pile. This depth is called as the *critical depth*. Tomlinson

(1986) is of the opinion, that the critical depth concept is not tenable as it is not possible to duplicate the field conditions in laboratory tests. He recommends the conventional method of calculating the base resistance by limiting the value of q_b based on field experience.

The other factor that affects the base resistance q_b is the bearing capacity factor N_q which is a function of ϕ. Though pile driving into cohesionless soil increases the density of soil around and bottom of the pile, and hence the value of ϕ, Tomlinson (1986) suggests the use of the value of ϕ that exists prior to pile driving at the site.

There is great deal of variation in the values of N_q obtained by different investigators as shown in Fig. 19.7. However, Tomlinson (1986) recommends the curves developed by Berezantsev et al. (1961) as these curves conform most nearly practical criteria of pile failure. These curves are given in Fig. 19.8. It may be seen that there is a rapid increase in the values of N_q at higher values of ϕ, giving there by higher values for q_b. Tomlinson suggests to limit the value of q_b to 11000 kN/m^2 whatever may be the penetration depth of pile.

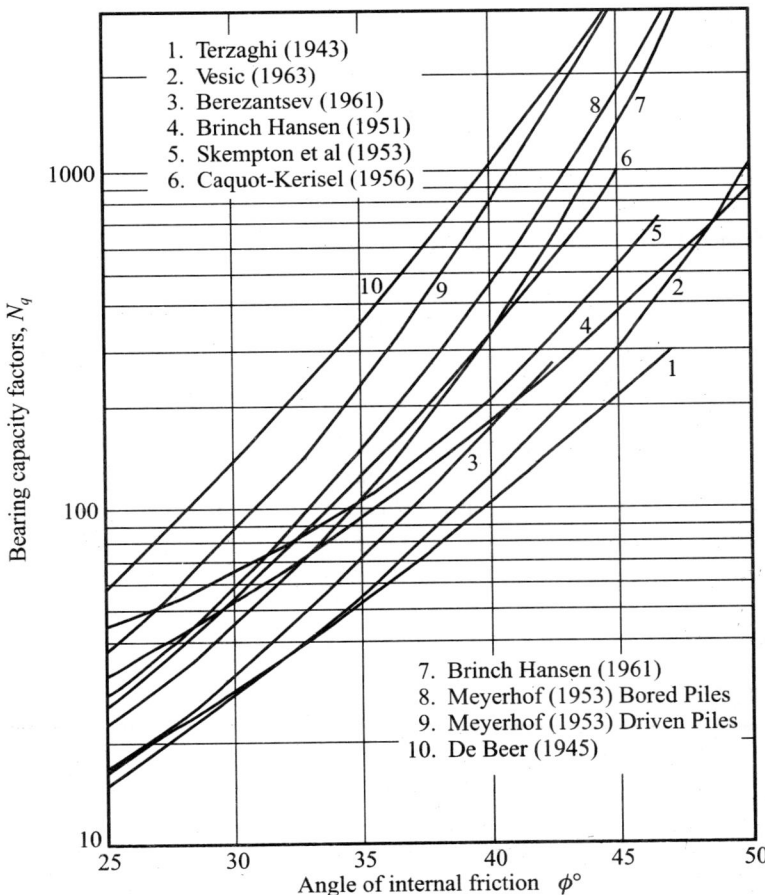

Fig.19.7 Bearing capacity factors for circular deep foundations by various investigators

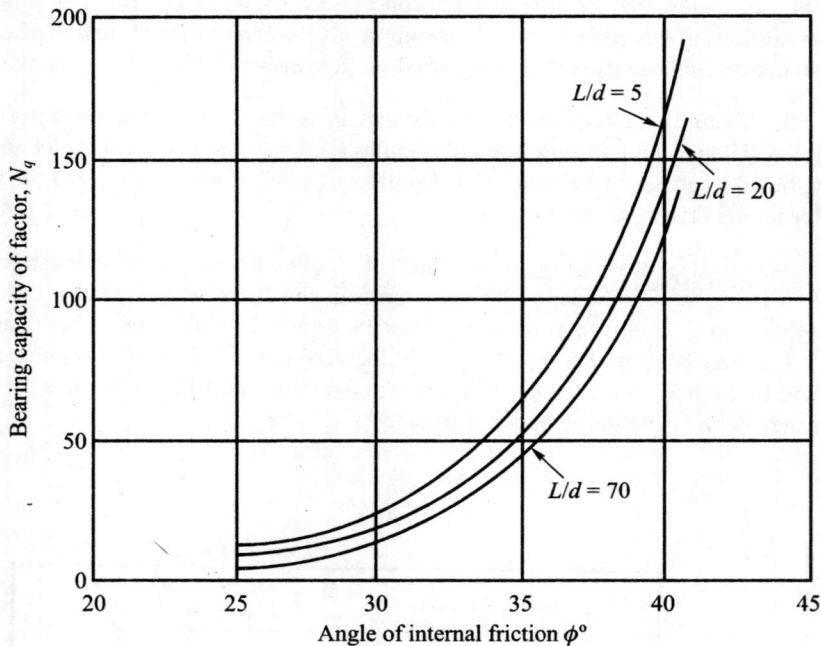

Fig. 19.8   Berezantsev's bearing capacity factor, N_q

Frictional load, $Q_f = q_0\ \bar{K}_s \tan \delta\ A_s$

The values of $\bar{K}_s$ and δ vary with the typ of material surface and density index of sand. Broms (1996) has related the values of $\bar{K}_s$ and δ to the effective angle of friction ϕ of cohesionless soil for various pile materials and density index as shown in the Table 19.1. These values are applicable to driven piles only.

Some investigators apply the same concept of *critical depth* for limiting the value of f_s. According to them the values of f_s increases linearly with depth (in a homogeneous soil) up to a certain depth and then remains constant. This depth, as per their research, may be any where between 10 and 20 pile diameter. Tomlinson argues that this depth is not a constant in all conditions of soil in the field. To overcome this problem, he recommends a maximum of 110 kN/m² for f_s.

TABLE 19.1

Values of $\bar{K}_s$ and δ

Pile material	δ	Values of $\bar{K}_s$	
		Low D_r	High D_r
Steel	20°	0.5	1.0
Concrete	3/4 ϕ	1.0	2.0
Wood	2/3 ϕ	1.5	4.0

Example 19.1

A concrete pile of 45 cm diameter was driven into sand of loose to medium density to a depth of 15 m. The following properties are known:

(a) Average unit weight of soil along the length of the pile, $\overline{\gamma} = 17.5$ kN/m³ and average value of $\phi = 30°$, (b) average $\overline{K}_s = 10$. and $\delta = 0.75\phi$.

Calculate (a) the ultimate bearing capacity of the pile, and (b) the allowable load with $F_s = 2.5$. Assume the water table is at great depth. Use Berezantsev's method.

Solution

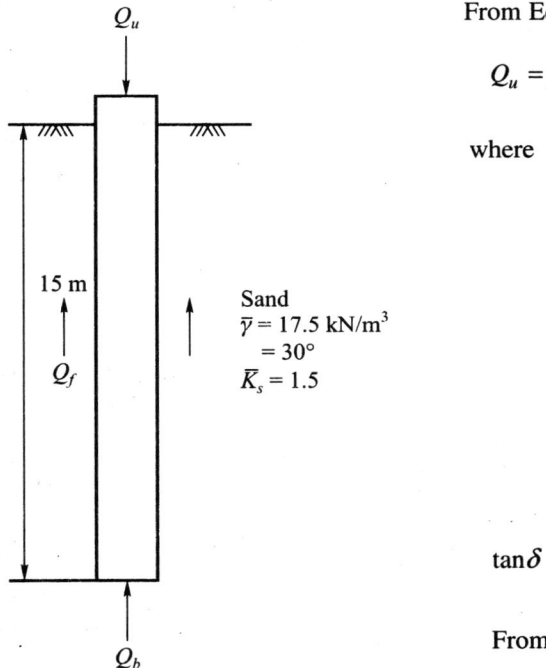

Q_u

15 m

Q_f

Q_b

Sand
$\overline{\gamma} = 17.5$ kN/m³
$= 30°$
$\overline{K}_s = 1.5$

Fig. Ex. 19.1

From Eq. (19.7a)

$$Q_u = Q_b + Q_f = q_0 A_b N_q + \overline{q} A_s \overline{K}_s \tan\delta$$

where $q_0 = \overline{\gamma}L = 17.5 \times 15 = 262.5$ kN/m²

$$\overline{q} = \frac{1}{2}\overline{\gamma}L = \frac{262.5}{2} = 131.75 \text{ kN/m}^2$$

$$A_b = \frac{3.14}{4} \times 0.45^2 = 0.159 \text{ m}^2$$

$$A_s = 3.14 \times 0.45 \times 15 = 21.195 \text{ m}^2$$

$$\delta = 0.75\phi = 0.75 \times 30 = 22°.5$$

$$\tan\delta = 0.4142$$

From Fig. 19.8, N_q for $\dfrac{L}{d} = \dfrac{15}{0.45} = 33.3$ and

$\phi = 30°$ is equal to 16.5.

Substituting the known values, we have

$Q_u = 262.5 \times 0.159 \times 16.5 + 131.75 \times 21.195 \times 1.5 \times 0.4142$

$= 688.67 + 1734.94 \approx 2424$ kN

$$Q_a = \frac{2424}{2.5} = 969.6 \text{ kN}$$

Example 19.2

Assume in Ex. 19.1 that the water table is at the ground surface and $\gamma_{sat} = 18.5$ kN/m³. All the other data remain the same. Calculate Q_u and Q_a.

Solution

Water table at the ground surface $\gamma_{sat} = 18.5$ kN/m³

$$\gamma_b = \gamma_{sat} - \gamma_w = 18.5 - 9.81 = 8.69 \text{ kN/m}^3$$

$$q_0 = 8.69 \times 15 = 130.35 \text{ kN/m}^2$$

$$\bar{q}_0 = \frac{1}{2} \times 130.35 = 65.18 \text{ kN/m}^2$$

Substituting the known values

$$Q_u = 130.35 \times 0.159 \times 16.5 + 651.8 \times 21.195 \times 1.5 \times 0.4542$$

$$= 342 + 858 = 1200 \text{ kN}$$

$$Q_a = \frac{1200}{2.5} = 480 \text{ kN}$$

Note: It may be noted here that the presence of water table up to the ground level in cohesionless soil reduces the ultimate load capacity of pile by about 50 percent.

Example 19.3

A concrete pile of 45 cm diameter is driven to a depth of 16 m through a layered system of sandy soil ($c = 0$). The following data are available.

Top layer 1: Thickness = 8 m, $\gamma_d = 16.5$ kN/m³, $c = 0.60$ and $\phi = 30°$.

Layer 2: Thickness = 6 m, $\gamma_d = 15.5$ kN/m³, $c = 0.65$ and $\phi = 35°$.

Layer 3: Extends to a great depth, $\gamma_d = 16.00$ kN/m³, $c = 0.65$ and $\phi = 38°$.

Assume that the value of δ in all the layers of sand is equal to 0.75 ϕ. The value of $\bar{K}_s$ for each layer as equal to half of the passive earth pressure coefficient. The water table is at ground level.

Calculate the values of Q_u and Q_a with $F_s = 2.5$

Solution

The soil is submerged up to GL. The specific gravity G is required for calculating γ_{sat}.

(a) Using equation, $\gamma_d = \dfrac{\gamma_w G}{1+e}$, calculate G for each layer since γ_d, γ_w and e are known.

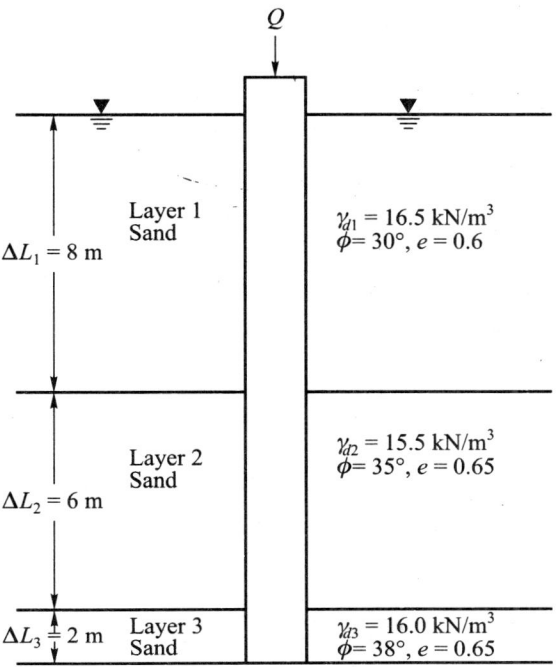

Fig. Ex. 19.3

(b) Using equation, $\gamma_{sat} = \dfrac{\gamma_w(G+e)}{1+e}$, calculate γ_{sat} for each layer and then $\gamma_b = \gamma_{sat} - \gamma_w$ for

each layer.

(c) For a layered system of soil, the ultimate load can be determined by making use of Eq. (19.7b). Now

$$Q_u = Q_b + Q_f = q_0 N_q A_b + P\sum_0^L \bar{q}_o \bar{K}_s \tan\delta\Delta L$$

(d) q_0 at the tip of the pile is

$$q_0 = \gamma_{b1}\Delta L_1 + \gamma_{b2}\Delta L_2 + \gamma_{b3}\Delta L_3$$

(e) q_0 at the middle of each layer is

$$\bar{q}_{01} = \frac{1}{2}\Delta L_1 \gamma_{b1}$$

$$\bar{q}_{02} = \Delta L_1 \gamma_{b1} + \frac{1}{2}\Delta L_2 \gamma_{b2}$$

$$\bar{q}_{03} = \Delta L_1 \gamma_{b1} + \Delta L_2 \gamma_{b2} + \frac{1}{2}\Delta L_3 \gamma_{b3}$$

(f) $N_q = 95$ for $\phi = 38°$ and $\dfrac{L}{d} = \dfrac{15}{0.45} = 33.33$ from Fig. 19.8.

(g) $A_b = 0.159\,\text{m}^2$, $P = 1.4287\,\text{m}$.

(h) $\overline{K}_s = \dfrac{1}{2}\tan^2\left(45° + \dfrac{\phi}{2}\right) = \dfrac{1}{2}K_p$. for each layer can be calculated.

(i) $\delta = 0.75\,\phi$. The values of $\tan\delta$ can be calculated for each layer.

The computed values for all the layers are given below in Table 19.2.

TABLE 19.2

Computed values for all layers

Layer no.	G	γ_b kN/m^3	$\overline{q}_0$ kN/m^2	$\overline{K}_s$	$\tan\delta$	ΔL m
1.	2.69	10.36	41.44	1.5	0.414	8
2.	2.61	9.57	111.59	1.845	0.493	6
3.	2.68	10.05	150.35	2.10	0.543	2
		Middle to tip of pile =	10.05			
		$q_0 = 160.40$				

$Q_u = 160.4 \times 95 \times 0.159 + 1.429(41.44 \times 1.5 \times 0.414$

$\qquad + 111.59 \times 1.845 \times 0.493 + 150.35 \times 2.10 \times 0.543)$

$\qquad = 2423 + 426.80 \approx 2850\,\text{kN}$

$Q_a = \dfrac{Q_u}{2.5} = \dfrac{2850}{2.5} = 1140\,\text{kN}$

19.9 CALCULATING THE ULTIMATE LOADS ON BORED AND *CAST-IN-SITU* PILES IN COHESIONLESS SOILS

Bored piles are formed in cohesionless soils by drilling with rigs. The sides of the holes might be supported by the use of casing pipes. When casing is used, the concrete is placed in the drilled hole and the casing is gradually withdrawn. In all the cases the sides and bottom if the hole will be loosened as a result of the boring operations, even though it may be initially be in a dense or medium dense state. Tomlinson suggests that the values of the parameters in Eq. (19.7) must be calculated by assuming that the ö value will represent the loose condition.

However, when piles are installed by rotary drilling under a bentonite slurry for stabilizing the sides, it may be assumed that the ϕ value used to calculate both the skin friction and base resistance will correspond to the undisturbed soil condition (Tomlinson, 1986).

The assumption of loose conditions for calculating skin friction and base resistance means that the ultimate carrying capacity of a bored pile in a cohesionless soil will be considerably lower than that of a pile driven in the same soil type. As per De Beer (1965), the base resistance q_b of a bored and *cast-in-situ* pile is about one third of that of a driven pile.

We may write,

$$q_b(\text{bored pile}) = \frac{1}{3}q_b(\text{driven pile}) \tag{19.8}$$

So far as friction load is concerned, the frictional parameter may be calculated by assuming a value of ϕ equal to 28° which represents the loose condition of the soil.

The same Eq. (19.7) may be used to compute Q_u based on the modifications explained above.

Example 19.4

If the pile in Ex. 19.2 is a bored and *cast-in-situ*, compute Q_u and Q_a. All the other data remain the same. Water table is close to the ground surface.

Solution

Per Tomlinson (1986), the ultimate bearing capacity of a bored and cast-in-situ-pile in cohesionless soil is reduced considerably due to disturbance of the soil. Per Section 19.9, calculate the base resistance for a driven pile and take one-third of this as the ultimate base resistance for a bored and *cast-in-situ* pile.

For computing δ, take $\phi = 28°$ and $\overline{K}_s = 1.0$ from Table 19.2 for a concrete pile.

Base resistance for driven pile

For $\phi = 30°$, $N_q = 16.5$ from Fig. 19.8.

$A_b = 0.159$ m²

$q_0 = 130.35$ kN/m² (From Ex. 19.2)

$Q_b = 130.35 \times 0.159 \times 16.5 = 342$ kN

For bored pile

$$Q_b = \frac{1}{3} \times 342 = 114 \text{kN}$$

Skin load

$$Q_f = A_s \overline{q} K_s \tan\delta$$

For $\phi = 28°$, $\delta = 0.75 \times 28 = 21°$, $\tan\delta = 0.384$

$A_s = 21.95$ m² (from Ex. 19.2)

$$\bar{q}_0 = 65.18\,\text{kN/m}^2\,(\text{Ex.19.2})$$

Substituting the known values,

$$Q_f = 21.95 \times 65.18 \times 1.0 \times 0.384 = 530\,\text{kN}$$

Therefore, $Q_u = 114 + 530 = 644\,\text{kN}$

$$Q_a = \frac{644}{2.5} \approx 258\,\text{kN}$$

19.10 ULTIMATE LOADS ON PILES DRIVEN INTO COHESIVE SOILS

The load bearing capacity of piles driven into cohesive soils is a sum of the base resistance load and the skin load as expressed by Eqs. (19.5d) and (19.6c) respectively. The combined equation for Q_u is

$$Q_u = q_b A_b + f_s A_s = c_b N_c A_b + \alpha \bar{c}_u A_s \tag{19.9a}$$

For a layered system of cohesive soils, Q_u may be expressed as

$$Q_u = c_b N_c A_b + P \sum_0^L \alpha \bar{c}_u \Delta L \tag{19.9b}$$

where, c_b = undrained shear strength of clay at base level of pile, $\bar{c}_u$ = average shear strength of clay under undrained condition along the length of pile considered, α = average adhesion factor considered, P = perimeter of pile, ΔL = thickness of layer.

The value of N_c that is generally accepted is 9. The adhesion factor vary with the type of material surface and the consistency of the soil. Table 19.3 gives the values of α.

TABLE 19.3

Adhesion factor, α

Material of pile	Consistency	Cohesive strength, c kN/m^2	Adhesion factor, α
Timber and concrete	Soft	0 –37.5	1–0.90
	Medium	37.5 –75.0	0.90–0.60
	Stiff	75.0 –150.0	0.60–0.45
Steel	Soft	0 –37.5	1.00–0.80
	Medium	37.5 –75.0	1.00–0.5
	Stiff		< 0.50

Example 19.5

A concrete pile 45 cm in diameter and 15 m long is driven into a homogeneous mass of clay soil of medium consistency. The water table is at the ground surface. The unit cohesion of the soil under undrained condition is 50 kN/m^2 and the adhesion factor $\alpha = 0.75$. Compute Q_u and Q_a with $F_s = 2.5$.

Solution

Given: $L = 15$ m, $D = 0.45$ m, $c_u = 50$ kN/m^2, $\alpha = 0.75$.

From Eq. (19.9a), we have

$$Q_u = Q_b + Q_f = c_b N_c A_b + A_s \alpha \bar{c}_u$$

where, $c_b = \bar{c}_u = 50 \text{ kN/m}^2$; $N_c = 9$; $A_b = 0.159 \text{ m}^2$; $A_s = 21.195 \text{ m}^2$

Substituting the known values, we have

$$Q_u = 50 \times 9 \times 0.159 + 21.195 \times 0.75 \times 50$$

$$= 71.55 + 794.81 = 866.36 \text{ kN}$$

$$Q_a = \frac{866.6}{2.5} = 346.55 \text{ kN}$$

Example 19.6

A concrete pile of 45 cm diameter is driven through a system of layered cohesive soils. The length of the pile is 16 m. The following data are available. The water table is close to the ground surface.

Top layer 1: Soft clay, thickness = 8 m, unit cohesion $c = 30$ kN/m^2 and adhesion factor $\alpha = 0.90$.

Layer 2: Medium stiff, thickness = 6 m, unit cohesion $\bar{c}_u = 50$ kN/m^2 and $\alpha = 0.75$.

Layer 3: Stiff stratum extends to a great depth, unit cohesion $\bar{c}_u = 105$ kN/m^2 and $\alpha = 0.50$.

Compute Q_u and Q_a with $F_s = 2.5$.

Solution

Here, the pile is driven through clay soils of different consistencies.

The equations for Q_u expressed as (Eq. 19.9b)

$$Q_u = 9c_b A_b + P \sum_0^L \alpha \bar{c}_u \Delta L$$

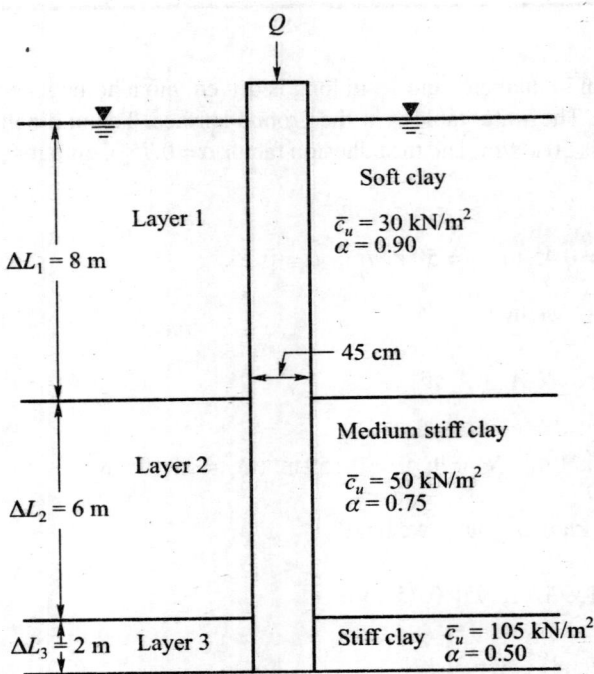

Fig. Ex. 19.6

Here, $c_b = \bar{c}_u$ of layer 3, $P = 1.429$ m, $A_b = 0.159$ m^2

Substituting the known values, we have

$$Q_u = 9 \times 105 \times 0.159 + 1.429(0.90 \times 30 \times 8$$

$$+ 0.75 \times 500 \times 6 + 0.50 \times 105 \times 2)$$

$$= 150.25 + 780.23 = 930.48 \text{ kN}$$

$$Q_a = \frac{930.48}{2.5} \approx 372 \text{ kN}$$

Example 19.7

A precast concrete pile of size 45 × 45 cm is driven into stiff clay. The unconfined compressive strength of the clay is 200 kN/m^2. Determine the length of pile required to carry a safe working load of 400 kN with $F_s = 2.5$.

Solution

The equation for Q_u is

$$Q_u = N_c c_u A_b + \alpha \bar{c}_u A_s$$

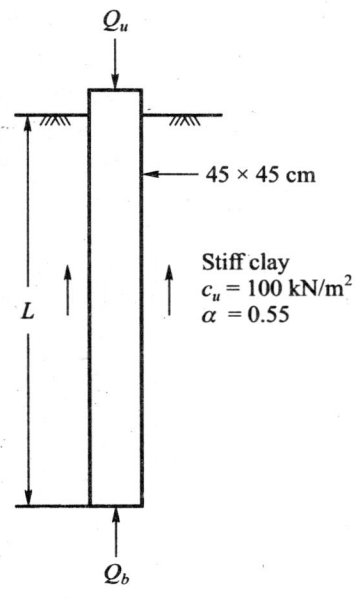

Fig. Ex. 19.7

we have

$$Q_u = 2.5 \times 400 = 1000 \text{ kN},$$

$$N_c = 9, \ c_u = 100 \text{ kN/m}^2$$

$$\alpha = 0.55 \text{ from Table 19.3},$$

$$\bar{c}_u = c_u = 100 \text{ kN/m}^2, \ A_b = 0.2025 \text{ m}^2$$

Assume the length of pile $= L$ m

Now, $A_s = 4 \times 0.45L = 1.80L$

Substituting the known values, we have

$$1000 = 9 \times 100 \times 0.2025 + 0.55 \times 100 \times 1.80L$$

or $1000 = 182 - 25 + 99L$

Simplifying, we have

$$L = \frac{1000 - 182.25}{99} = 8.26 \text{ m}$$

19.11 BEARING CAPACITY OF PILES IN GRANULAR SOILS BASED ON SPT VALUE

Meyerhof (1956) suggests the following equations for single piles in granular soils based on SPT values.

For displacement piles

$$Q_u = 400NA_b + 2\,\overline{N}A_s \tag{19.10a}$$

for H-piles

$$Q_u = 133NA_b + \overline{N}A_s \tag{19.10b}$$

for bored piles

$$Q_u = 133NA_b + 0.67\overline{N}A_s \tag{19.10c}$$

where, Q_u = ultimate total load in kN,

$\quad N$ = average corrected SPT value below pile tip

$\quad \overline{N}$ = corrected average SPT value along the pile shaft

$\quad A_b$ = base area of pile in m^2,

$\quad A_s$ = shaft surface area in m^2,

A minimum factor of safety of 4 is recommended for driven piles and 2.5 for bored piles.

Example 19.8

A reinforced concrete pile of size 30 × 30 cm and 10 m long is driven into coarse sand extending to a great depth. The average total unit weight of the soil is 18 kN/m³ and the average N-value is 15. Determine the allowable load on the pile by the static formula. Use $F_s = 2.5$. The water table is close to the ground surface.

Solution

In this example only the N-value is given. The corresponding ϕ value can be found from Fig. 18.8 which is equal to 32°.

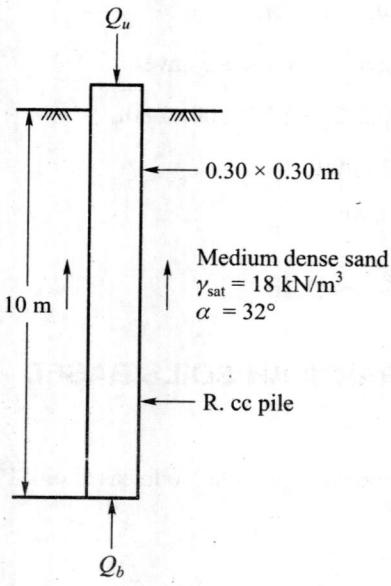

0.30 × 0.30 m

Medium dense sand
$\gamma_{sat} = 18$ kN/m³
$\alpha = 32°$

10 m

R. cc pile

Q_b

Fig. Ex. 19.8

Now from Fig. 19.8, for $\phi = 32°$, and

$$\frac{L}{d} = \frac{10}{0.3} = 33.33, \text{ the value of } N_q = 25.$$

$$A_b = 0.3 \times 0.3 = 0.09 \text{ m}^2,$$

$$A_s = 10 \times 4 \times 0.3 = 12 \text{ m}^2$$

$$\delta = 0.75 \times 32 = 24° \tan\delta = 0.445$$

The density index is low side of medium dense. From Table 19.2, we may take

$$\overline{K}_s = 1 + \frac{1}{3}(2-1) = 1.33$$

Now,

$$Q_u = q_0 N_q A_b + \overline{q}_0 \overline{K}_s \tan\delta A_s$$

$$\gamma_b = \gamma_{sat} - \gamma_w = 18.0 - 9.81 = 8.19 \text{ kN/m}^3$$

$$q_0 = \gamma_b L = 8.19 \times 10 = 81.9 \text{ kN/m}^{2}$$

$$\overline{q}_0 = \frac{q_0}{2} = \frac{81.9}{2} = 40.95 \text{ kN/m}^2$$

Substituting the known values, we have

$$Q_u = 81.9 \times 25 \times 0.09 + 40.95 \times 1033 \times 0.445 \times 12$$

$$= 184 + 291 = 475 \text{ kN}$$

$$Q_a = \frac{475}{2.5} = 190 \text{ kN}$$

Example 19.9

Determine the allowable load on the pile given in Ex. 19.8 by making use of the N-value.

Solution

As Per Ex. 19.8, $N = 15$

The expression for Q_u is

$$Q_u = 400NA_b + 2\overline{N}A_s \, (kN)$$

Here, we have to assume $N = \overline{N} = 15$

$$A_b = 0.3 \times 0.3 = 0.09 \, m^2, \, A_s = 4 \times 0.3 \times 10 = 12 \, m^2$$

Substituting, we have

$$Q_u = 400 \times 15 \times 0.09 + 2 \times 16 \times 12$$

$$= 540 + 360 = 900 \, kN$$

$$Q_a = \frac{900}{4} = 225 \, kN$$

Example 19.10

Precast concrete piles 40 cm in diameter are required to be driven for a building foundation. The design load on a single pile is 450 kN. Determine the length of the pile if the soil is loose to medium dense sand with an average N value of 15 along the pile and 21 at the tip of the pile. The water table may be taken at the ground level. The average saturated unit weight of soil is equal to 18.5 kN/m³. Use the static formula and $F_s = 2.5$.

Solution

Required to find the length of a pile to take an ultimate load of $Q_u = 2.5 \times 100 = 1125 \, kN$.

The equation for Q_u is

$$Q_u = q_0 N_q A_b + \overline{q}_0 \overline{K}_s \tan \delta A_s$$

The average value of ϕ along the pile and the value at the tip may be determined from Fig. 18.8.

For $N = 15$, $\phi = 32°$; for $N = 21$, $\phi = 34°$.

Since the soil is submerged

$$\gamma_b' = 18.5 - 9.81 = 8.69 \, kN/m^2$$

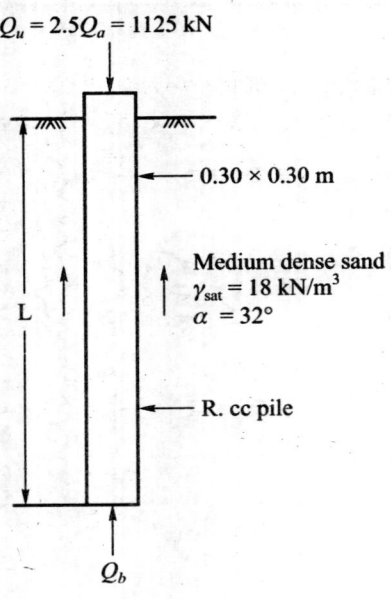

$Q_u = 2.5Q_a = 1125$ kN

0.30 × 0.30 m

Medium dense sand
$\gamma_{sat} = 18$ kN/m^3
$\alpha = 32°$

L

R. cc pile

Q_b

Fig. Ex. 19.10

Now,

$$q_0 = \gamma_b L = 8.69 L \text{ kN/m}^2$$

$$\bar{q}_0 = \frac{8.69L}{2} = 4.345 L \text{ kN/m}^2$$

$$A_b = \frac{3.14}{4} \times 0.4^2 = 0.1256 \text{ m}^2$$

$$A_s = 3.14 \times 0.4 \times L = 1.256 L \text{ m}^2$$

From Fig. 19.8, N_q = 40 for L/D = 20 (assumed)

From Table 19.2, K_s = 1.33 for the lower side of medium dense soil

$$\delta = \frac{3}{4} \times 34 = 25.5°, \tan\delta = 0.477$$

Now by substituting the known values, we have

$$1125 = (8.69L) \times 40 \times 0.1256 + (4.345L) \times 1.33 \times 0.477 \times (1.256L)$$

$$= 43.66L + 3.4622L^2$$

or $L^2 + 12.61L - 325 = 0$

The above equation when solved gives a value L = 12.80 m or we may take L = 13 m for all practical purposes.

Example 19.11

Determine the allowable load on the pile given in Ex. 19.10 by directly making use of the N-value. All the other conditions remain the same.

Solution

From Ex. 19.10, we have found L = 13 m

From Eq. (19.10a), we have

$$Q_u = 400NA_b + 2\bar{N}A_s \text{ (kN)}$$

where, $N = 21$, $\bar{N} = 15$

$A_b = 0.1256 \text{ m}^2$, $A_s = 1.256 \times 13 = 16.328 \text{ m}^2$

Substituting,

$$Q_u = 400 \times 21 \times 0.1256 + 2 \times 15 \times 16.328$$

$$= 1055 + 490 = 1545 \text{ kN}$$

Taking $F_s = 4$

$$Q_a = \frac{1545}{4} = 386 \text{ kN}$$

19.12 BEARING CAPACITY OF PILES BASED ON STATIC CONE PENETRATION TESTS (CPT)

Methods of Determining Pile Capacity

The cone penetration test may be considered as a small scale pile load test. As such the results of this test yield the necessary parameters for the design of piles subjected to vertical load.

Vander Veen's Method for Piles in Cohesionless Soils

In the Vander Veen *et al.,* (1951) method, the ultimate end-bearing resistance of a pile is taken, equal to the point resistance of the cone. To allow for the variation of cone resistance which normally occurs, the method considers average cone resistance over a depth equal to three times the diameter of the pile above the pile point level and one pile diameter below point level as shown in Fig. 19.9. Experience has shown that if a safety factor of 2.5 is applied to the ultimate end resistance as determined from cone resistance, the pile is unlikely to settle more than 15 mm under the working load (Tomlinson, 1986). The equations for ultimate bearing capacity and allowable load may be written as,

Pile base resistance, $q_b = q_p(\text{cone})$ (19.11)

Ultimate base capacity, $Q_b = A_b q_p$ (19.12)

Allowable base load, $Q_a = \dfrac{A_b q_p}{F_s}$ (19.13)

where, q_p = average cone resistance over a depth $4d$ as shown in Fig. 19.9. F_s = factor of safety.

The skin friction on the pile shaft in cohesionless soils is obtained from the relationships established by Meyerhof (1956) as follows.

For displacement piles, the ultimate skin friction, f_s, is given by

$$f_s = \frac{\bar{q}_c}{2} \text{ kN/m}^2 \tag{19.14a}$$

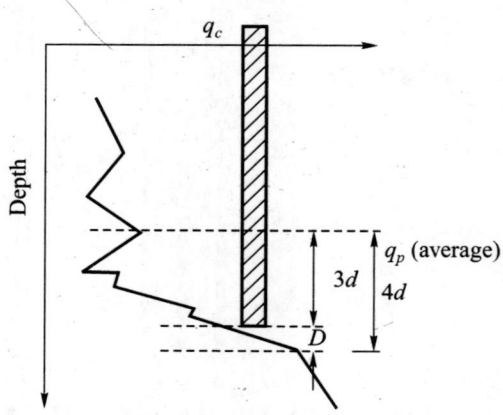

Fig. 19.9   Pile capacity by the use of CPT values by Vander Veen's method

and for *H*-section piles, the ultimate limiting skin friction is given by

$$f_s = \frac{\bar{q}_c}{4} \text{ kN/m}^2 \tag{19.14b}$$

where $\bar{q}_c$ = average cone resistance in kg/cm² over the length of the pile shaft under consideration.

Meyerhof states that for straight sided displacement piles, the ultimate unit skin friction, f_s, has a maximum value of 107 kN/m² and for *H*-sections, a maximum of 54 kN/m² (calculated on all faces of flanges and web).

The ultimate skin load is

$$Q_f = A_s f_s \tag{19.15a}$$

The ultimate load capacity of a pile is

$$Q_u = Q_b + Q_f \tag{19.15b}$$

The allowable load is

$$Q_a = \frac{Q_b + Q_f}{2.5} \tag{19.15c}$$

If the working load, Q_a, obtained for a particular position of pile in Fig. 19.9, is less than that required for the structural designer's loading conditions, then the pile must be taken to a greater depth to increase the skin friction f_s or the base resistance q_b.

Example 19.12

A concrete pile of 40 cm diameter is required to be driven into a homogeneous mass of cohesionless soil. The pile carries a safe load of 650 kN. A static cone penetration test conducted at the site indicates an average value of q_c = 40 kg/cm² along the pile and 120 kg/cm² below the pile tip.

Compute the length of the pile with $F_s = 2.5$.

Solution

$Q_u = 2.5 \times 650 = 1625$ kN

From Eq. (19.11)

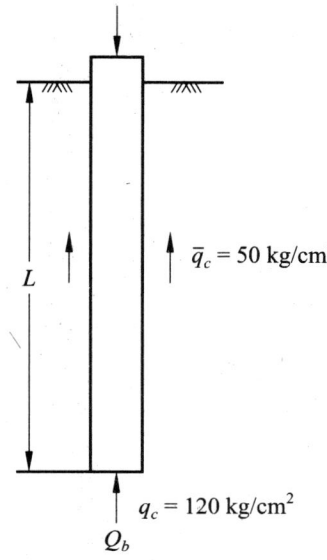

Fig. Ex. 19.12

$$q_b(\text{pile}) = q_p(\text{cone})$$

Given, $q_p = 120$ kN/m^2, therefore,

$$q_b = 120 \, \text{kg/cm}^2 = 120 \times 100 = 12000 \, \text{kN/m}^2$$

As per Section 19.8, q_b is restricted to 11,000 kN/m^2.

Therefore,

$$Q_b = A_b q_b = \frac{3.14}{4} \times 0.4^2 \times 11000 = 1382 \, \text{kN}$$

Assume the length of pile $= L$ m

The average, $\bar{q}_c = 40$ kg/cm^2

As per Eq. (19.14a),

$$f_s = \frac{\bar{q}_c}{2} \, \text{kN/m}^2 = \frac{40}{2} = 20 \, \text{kN/m}^2$$

Now, $Q_f = f_s A_s = 20 \times 3.14 \times 0.4 \times L = 25.12L$ kN

Given $Q_a = 650$ kN. with $F_s = 2.5$, $Q_u = 650 \times 2.5 = 1625$ kN.

Now, $1625 = Q_b + Q_f = (1382 + 25.12L)$ kN

or $L = \dfrac{1625 - 1382}{25.12} = 9.67$ m or say 10 m

The pile has to be driven to a depth of 10 m to carry a safe load of 650 kN with $F_s = 2.5$.

19.13 BEARING CAPACITY OF A SINGLE PILE BY LOAD TEST

Introduction

Pile load test is the most acceptable method to determine the load carrying capacity of a pile. The load test may be carried out either on a driven pile or a *cast-in-situ* pile. Load tests may be made either on a single pile or a group of piles. Load tests on a pile group are very costly and may be undertaken only in very important projects.

Pile load tests on a single pile or a group of piles are conducted for the determination of

1.   Vertical load bearing capacity,

2.   Uplift load capacity,

3.   Lateral load capacity.

Generally load tests are made to determine the bearing capacity and to establish the load settlement relationship under a compressive load. The other two types of tests may be carried out only when piles are required to resist large uplift or lateral forces.

Usually pile foundations are designed with an estimated capacity which is determined from a thorough study of the site conditions. At the beginning of construction, load tests are made for the purpose of verifying the adequacy of the design capacity. If the test results show an inadequate factor of safety or excessive settlement, the design must be revised before construction is under way.

Load tests may be carried out either on

1.   A working pile or

2.   A test pile.

A *working pile* is a pile driven or *cast-in-situ* along with the other piles to carry the loads from the superstructure. The maximum test load on such piles should not exceed one and a half times the design load.

A *test pile* is a pile which does not carry the loads coming from the structure. The maximum load that can be put on such piles may be about 2½ times the design load or the load imposed must be such as to give a total settlement not less than one-tenth the pile diameter.

Method of Carrying Out Vertical Pile Load Test

A vertical pile load test assembly is shown in Fig. 19.10. It consists of

1.   An arrangement to take the reaction of the load applied on the pile head,

2.   A hydraulic jack of sufficient capacity to apply load on the pile head, and

3.   A set of three dial gauges to measure settlement of the pile head.

Load Application

Load test may be of two types:

1.   Continuous load test.

2.   Cyclic load test.

In the case of a continuous load test, continuous increments of load are applied to the pile head. Settlement of the pile head is recorded at each load level.

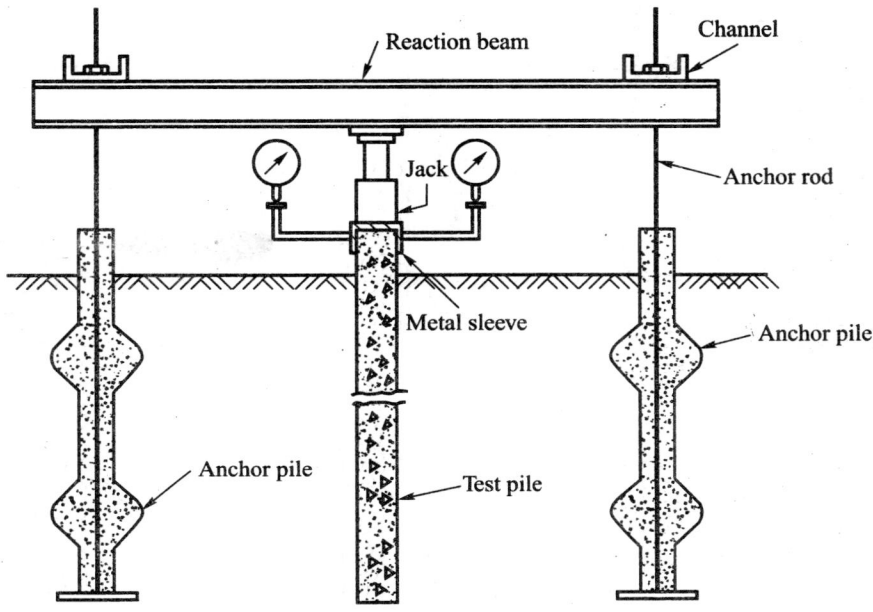

Fig. 19.10 A vertical pile load test assembly

In the case of the cyclic load test, the load is raised to a particular level, then reduced to zero, again raised to a higher level and reduced to zero. Settlements are recorded at each increment or decrement of load. Cyclic load tests help to separate frictional load from point load.

In the continuous load test, the loads are normally put in equal increments of about one-fifth of the design load. Each increment of load is kept on the pile for a sufficiently long time till the settlement rate is less than 0.25 mm per hour. The settlements are recorded with the help of dial gauges. A minimum of three dial gauges are used for recording settlements. The average of the gauge readings gives the settlement of the pile head. A load settlement curve may be plotted from the test data as shown in Figs 19.11 *a* and *b*.

Allowable Load from Single Pile Load Test Data

There are many methods by which allowable loads on a single pile may be determined by making use of load test data. If the ultimate load can be determined from load-settlement curves, allowable loads are found by dividing the ultimate load by a suitable factor of safety which varies from 2 to 3. A factor of safety of 2.5 is normally recommended. A few of the methods that are useful for the determination of ultimate or allowable loads on a single pile are given below:

1. The ultimate load, Q_u, can be determined as the abscissa of the point where the curved part of the load-settlement curve changes to a falling straight line, Fig. 19.11(a).

2. Q_u is the abscissa of the point of intersection of the initial and final tangents of the load-settlement curve, Fig. 19.11(b).

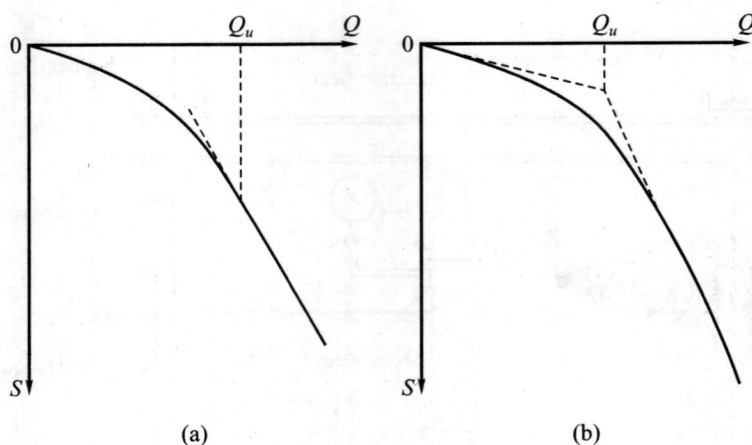

Fig. 19.11   Determination of ultimate load from load-settlement curves

3.   The allowable load Q_a is 50 percent of the ultimate load at which the total settlement amounts to one-tenth of the diameter of the pile [Indian Standard Code of Practice IS: 2911 (Part 1) - 1964], for uniform diameter piles.

4.   The allowable load Q_a is sometimes taken as equal to two-thirds of the load which causes a total settlement of 12 mm (Indian Standard Code of Practice).

5.   The allowable load Q_a is sometimes taken as equal to two-thirds of the load which causes a net (plastic) settlement of 6 mm (Indian Standard Code of Practice).

If pile groups are loaded to failure, the ultimate load of the group, Q_{gu}, may be found by any one of the first two methods mentioned above for single piles. However, if the groups are subjected to only one and a half-times the design load of the group, the allowable load on the group cannot be found on the basis of 12 or 6 mm settlement criteria applicable to single piles. In the case of a group with piles spaced at less than 6 to 8 times the pile diameter, the stress interaction of the adjacent piles affects the settlement considerably. The settlement criteria applicable to pile groups should be the same as that applicable to shallow foundations at design loads.

19.14 PILE BEARING CAPACITY FROM DYNAMIC PILE DRIVING FORMULAE

The resistance offered by a soil to penetration of a pile during driving gives an indication of its bearing capacity. Qualitatively speaking, a pile which meets greater resistance during driving is capable of carrying a greater load. A number of dynamic formulae have been developed which equate pile capacity in terms of driving energy.

The basis of all these formulae is the simple energy relationship which may be stated by the following equation

$$Wh = Q_u s \text{ or } Q_u = \frac{Wh}{s}$$

where, W = weight of the driving hammer

h = height of fall of hammer

Wh = energy of hammer blow

Q_u = ultimate resistance to penetration

s = pile penetration under one hammer blow

$Q_u s$ = resisting energy of the pile

Hiley Formula

The Hiley formula is expressed as

$$Q_u = \frac{\eta_h Wh}{s+c} \times \frac{1+C_r^2 R}{1+R}$$ (19.16)

where, $R = \dfrac{W_p}{W}$, W_p = weight of pile

C = half of the total elastic compression = $\frac{1}{2}(C_1 + C_2 + C_3)$

C_1 = elastic compression of pile cap

C_2 = elastic compression of pile

C_3 = elastic compression of soil

C_r = coefficient of restitution

η_h = efficiency of hammer

The allowable load Q_a may be obtained by dividing Q_u by a suitable factor of safety.

If the pile tip rests on rock or relatively impenetrable material, Eq. (19.16) is not valid. Chellis suggests for this condition that the use of $W_p/2$ instead of W_p may be more correct. The various coefficients used in the Eq. (19.16) are as given below:

1. *Elastic compression C_1 of cap and pile head*

Pile Material	Range of Driving Stress kg/cm^2	Range of C_1
Precast concrete pile with packing inside cap	30-150	0.12-0-50
Timber pile without cap	30-150	0.05-0.20
Steel *H*-pile	30-150	0.04-0.16

2. *Elastic compression C_2 of pile.*

This may be computed using the equation

$$C_2 = \frac{Q_u L}{AE}$$

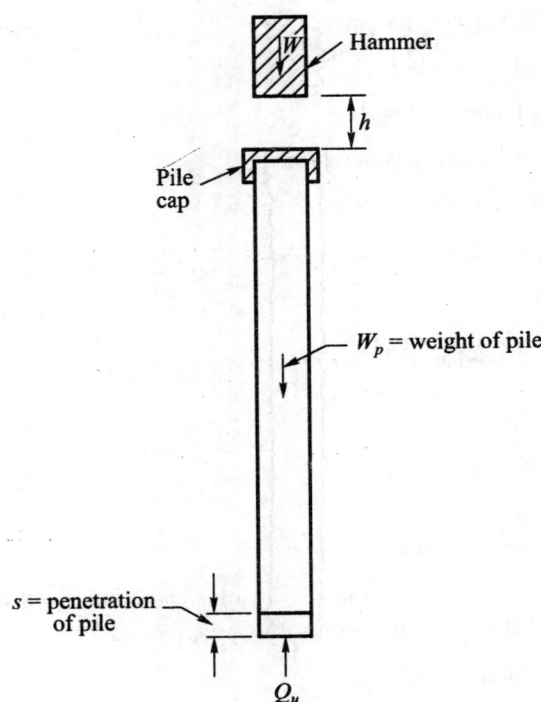

Fig. 19.12   Basic energy relationship

where, L = embedded length of the pile,

A = average cross-sectional area of the pile,

E = Young's modulus.

3. Elastic compression C_3 of soil.

The average value of C_3 may be taken as 0.1 (the value ranges from 0.0 for hard soil to 0.2 for resilient soils).

4. Pile-hammer efficiency

Hammer Type	η_h
Drop	1.00
Single acting	0.75–0.85
Double acting	0.85
Diesel	1.00

5. Coefficient of restitution C_r

Material	C_r
Wood pile	0.25
Compact wood cushion on steel pile	0.32

Cast iron hammer on concrete pile without cap   0.40

Cast iron hammer on steel pipe without cushion 0.55

Engineering News Record (ENR) Formula

The general form of the Engineering News Record Formula for the allowable load Q_a may be obtained from Eq. (19.16) by putting

$\eta_h = 1$ and $C_r = 1$ and a factor of safety equal to 6. The formula proposed by A.M. Wellington, editor of the Engineering News, in 1886, is

$$Q_a = \frac{Wh}{6(s+C)} \tag{19.17}$$

where Q_a = allowable load in kg or kN,

W = weight of hammer in kg or kN,

h = height of fall of hammer in cm,

s = final penetration in cm per blow (which is termed as set). The set is taken as the average penetration per blow for the last 5 blows of a drop hammer or 20 blows of a steam hammer,

C = empirical constant,

= 2.5 cm for a drop hammer,

= 0.25 cm for single and double acting hammers.

The equations for the various types of hammers may be written as:

1. Drop hammer

$$Q_a = \frac{Wh}{6(s+2.5)} \tag{19.18a}$$

2. Single-acting hammer

$$Q_a = \frac{Wh}{6(s+0.25)} \tag{19.18b}$$

3. Double-acting hammer

$$Q_u = \frac{W+ap}{6(s+0.25)} \tag{19.18c}$$

a = effective area of the piston in sq. cm,

p = mean effective steam pressure in kg/cm^2.

Comments on the Use of Dynamic Formulae

1.   Detailed investigations carried out by Vesic (1967) on deep foundations in granular soils indicate that the Engineering News Record Formula applicable to drop hammers,

Eq. (19.18a), gives pile loads as low as 44 % of the actual loads. In order to obtain better agreement between the one computed and observed loads, Vesic suggests the following values for the coefficient C in Eq. (19.17).

For steel pipe piles, $C = 1$ cm.

For precast concrete piles $C = 1.5$ cm.

2.   The tests carried out by Vesic in granular soils indicate that Hiley's formula does not give consistent results. The values computed from Eq. (19.16) are sometimes higher and sometimes lower than the observed values.

3.   Dynamic formulae in general have limited value in pile foundation work mainly because the dynamic resistance of soil does not represent the static resistance, and because often the results obtained from the use of dynamic equations are of questionable dependability. However, engineers prefer to use the Engineering News Record Formula because of its simplicity.

4.   Dynamic formulae could be used with more confidence in freely draining materials such as coarse sand. If the pile is driven to saturated loose fine sand and silt, there is every possibility of development of liquefaction which reduces the bearing capacity of the pile.

5.   Dynamic formulae are not recommended for computing allowable loads of piles driven into cohesive soils. In cohesive soils, the resistance to driving increases through the sudden increase in stress in pore water and decreases because of the decreased value of the internal friction between soil and pile because of pore water. These two oppositely directed forces do not lend themselves to analytical treatment and as such the dynamic penetration resistance to pile driving has no relationship to static bearing capacity.

There is another effect of pile driving in cohesive soils. During driving the soil becomes remoulded and the shear strength of the soil is reduced considerably. Though there will be a regaining of shear strength after a lapse of some days after the driving operation, this will not be reflected in the resistance value obtained from the dynamic formulae.

Example 19.13

A 40 × 40 cm reinforced concrete pile 20 m long is driven through loose sand and then into dense gravel to a final set of 3 mm/blow, using a 30 kN single-acting hammer with a stroke of 1.5 m. Determine the ultimate driving resistance of the pile if it is fitted with a helmet, plastic dolly and 50 mm packing on the top of the pile. The weight of the helmet and dolly is 4 kN. The other details are:

Weight of pile = 74 kN; weight of hammer = 30 kN; pile hammer efficiency $\eta_h = 0.80$ and coefficient of restitution $C_r = 0.40$.

Use (a) Hiley's formula and (b) Engineering News Formula. The sum of the elastic compression $\overline{C}$ of pile cap (C_1) pile material (C_2) and of soil (C_3), is $\overline{C} = C_1 + C_2 + C_3 = 19.6$ mm, $C = \overline{C}/2$.

Solution

(a) Hiley Formula

Use Eq. (19.16)

$$Q_u = \frac{\eta_h W h}{s+C} \times \frac{1+C_r^2 R}{1+R}$$

where $\eta_h = 0.80$, $W = 30$kN, $h = 1.5$m, $R = \dfrac{W_p}{W} = \dfrac{(74+4)}{30} = 2.6$, $C_r = 0.40$, $s = 0.30$cm.

Substituting we have,

$$Q_u = \frac{0.8 \times 30 \times 150}{0.3 + 1.96/2} \times \frac{1+0.4^2 \times 2.6}{1+2.6}$$

$$= 2813 \times 0.393 = 1105 \text{ kN}$$

(b) Engineering News Formula

Use Eq. (19.18b),

$$Q_u = 6Q_a = \frac{Wh}{s+0.25} = \frac{30 \times 100 \times 150 \times 10^{-3}}{0.3 + 0.25} = 818 \text{ kN}$$

19.15 UNDER-REAMED PILES

Piles under compression

The point bearing capacity of a pile can be increased by the use of an enlarged base. The enlarged base is sometimes called as a bulb or an under-ream. There could be one or more under-reams in a pile as shown in Fig. 19.13. The piles can accordingly be named as single or multi-under-reamed piles. The under-reamed piles are *cast-in-situ* concrete piles. The under-reams are made by a special equipment called as under-reaming equipment. The capacity of a pile increases as a result of the increased base area. The bearing capacity of a pile may be increased by providing more than one under-reams along the pile. The ratio of the bulb diameter to the shaft diameter may vary from 2 to 3 but normally 2.5 is provided.

Field load tests have indicated that under-reamed piles work out cheaper as compared to straight bored piles for the same design load.

Under-reamed piles may be installed both in sandy or clayey soils. The sides may be stabilised if required by the use of bentonite slurry which is also sometimes called as *drilling mud*. However, one has to be cautious while constructing under-reamed piles in granular soils as it will be very difficult to stabilise the bulbs.

The minimum distance between the centres of bulbs in a multi under-ream pile is $1.5\, d_u$, where, d_u is the diameter of the bulb. The load carrying capacity of a single under-reamed pile may be obtained from the equation.

$$Q_u = Q_b + Q_f = A_b q_b + A_s f_s \tag{19.19}$$

where, A_b = sectional area of bulb equal to $\pi d_u^2/4$,

$\quad q_b$ = base resistance per unit area of the bulb which may be calculated by anyone of the methods explained earlier,

$\quad A_s$ = surface area of the embedded shaft of the pile,

$\quad f_s$ = unit skin resistance.

The load carrying capacity of multi-under-reamed piles increases with the increase in the number of bulbs. If the distance between the centres of bulbs is not more than $1.5D_u$, the ultimate load carrying capacity of the multi-under reamed piles may be determined by making use of the equation (Fig. 19.13),

$$Q_u = A_b q_b + \overline{A}_s \overline{f}_s + A_s f_s \tag{19.20}$$

wherein, A_b = sectional area of the lowest bulb,

$\quad \overline{A}_s$ = surface area of cylinder of the diameter D_u and height equal to the distance between the centres of the extreme bulbs,

$\quad f_s$ = unit frictional resistance between pile shaft and soil,

$\quad \overline{f}_s$ = unit frictional resistance between soil to soil,

$\quad A_s$ = surface area of the embedded portion of pile shaft above the top of bulb.

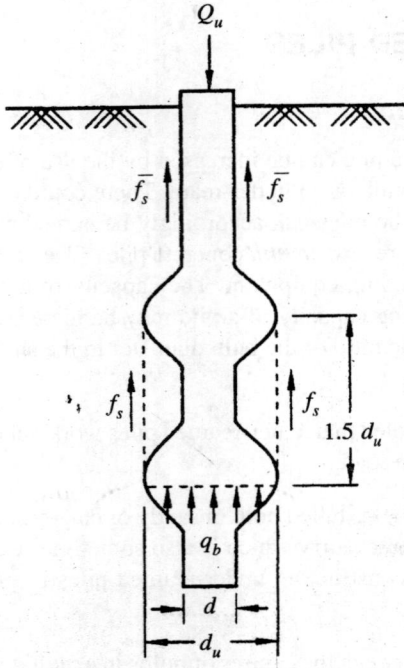

Fig. 19.13   Multi-under-reamed pile in non-expansive soil

Figure 19.13 assumes that the soil might get sheared off along the periphery of the bulbs and the soil between the bulbs might move together with the bulbs under ultimate load condition. Research work in this direction is not sufficient to substantiate this statement. The type of failure depends upon many factors such as relative density of sandy soil or consistency of clayey soil, the distance between the bulbs, the maximum settlement that can be tolerated and possibly many other factors.

Piles in Expansive soils

Under-reamed piles with a single under-ream are very much used in expansive soils to take up the uplift loads caused due to swelling of soils. The bulbs are placed in the stable zone where there will be no change in moisture content. The depth of the pile in the stable zone must be sufficient to take care of the uplift pressure exerted on the shaft in the unstable zone.

Figure 19.14 is a typical example of an under-reamed pile with a single bulb used in expansive soil for resisting uplift. Let L_1 = length of shaft in the unstable zone, L_2 = length of shaft in the stable zone, L_3 = depth of footing (bulb) of width d_u, and Q = load from the superstructure. When the soil in the unstable zone takes water during the wet season, the soil tries to expand which is partially or wholly prevented by the rough surface of the shaft of pile of length L_1. As a result of which there will be an upward force developed on the surface of shaft which tries to pull the pile out of its position. This upward liftout can be resisted in the following ways.

1. The downward dead load Q acting on the pile top.

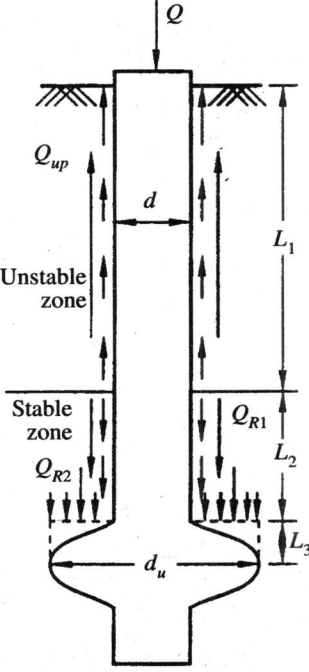

Fig. 19.14 Under-reamed pile in expansive soil

2. The resisting force provided by. the shaft of length and the bulb of diameter D_u, within the stable zone.

Two cases may have to be considered.

1. The stability of the pile when no downward dead load Q is acting on the pile top. For this condition a factor of safety of 1.2 is normally found sufficient.

2. The stability of the pile when Q is acting on the top. For this a values of $F_s = 2.0$ is used.

The equation for uplift force Q_{up} may be written as

$$Q_{up} = \pi d L_1 c_a \qquad (19.21)$$

where, c_a = adhesion between soil and shaft. The value of c_a may be taken as equal to the shear strength of clay if the wall of the shaft is rough.

The resisting forces Q_R is a sum of the resisting force Q_{R1} due to the skin load on the shaft of length L_2 and Q_{R2} due to the reaction provided by the soil above the bulb within the annular surface of diameter $(d_u - d)$. We may write

$$Q_{R1} = \pi d L_2 c_a \qquad (19.22a)$$

$$Q_{R2} = q_d \frac{\pi}{4} \left(d_u^2 - d^2 \right)$$

where, $q_d = c_b N_c = 6c_b$ for all practical purposes. Here, c_b = cohesive strength of soil at the bulb level, and $N_c = 6$. The two cases of stability are

1. Without taking the top pile load Q and using $F_s = 1.2$

$$Q_{up} = \frac{1}{1.2} \left(Q_{R1} + Q_{R2} \right) \qquad (19.22b)$$

2. By taking into account the dead load Q and using $F_s = 2.0$

$$Q_{up} - Q = \frac{1}{2.0} \left(Q_{R1} + Q_{R2} \right) \qquad (19.22c)$$

For a given shaft diameter D and bulb diameter D_u, the above equations help to determine L_2. The one that gives the maximum value for has to be used.

Plies under hydrostatic uplift

Under-reamed piles are also used as anchor piles to take up hydrostatic uplift pressures exerted on submerged structures and as tension piles to take up uplift loads under tall towers subjected to moments. The equation for uplift load for a double under-reamed pile is

$$Q_{up} = \overline{A}_s \overline{f}_s + A_s f_s + W_p \qquad (19.23)$$

where, W_p is the weight of the pile and the other terms as in Eq. (19.20). In Eq. (19.23), f_s is taken as equal to 50 percent of f_s for compression piles.

PART C: PILE GROUPS SUBJECTED TO VERTICAL LOADS

19.16 PILE SPACING AND GROUP EFFICIENCY

Spacing of Piles

The spacing of piles in a group depends on many factors such as, overlapping of adjacent piles (Fig. 19.15), cost of foundation and efficiency of pile group. When the piles are spaced closely, the soil is highly stressed in the zones of overlapping of pressures. Large spacings lead to higher costs of pipe caps etc. The spacing of pile also depends upon the method of installation. Generally, the spacing

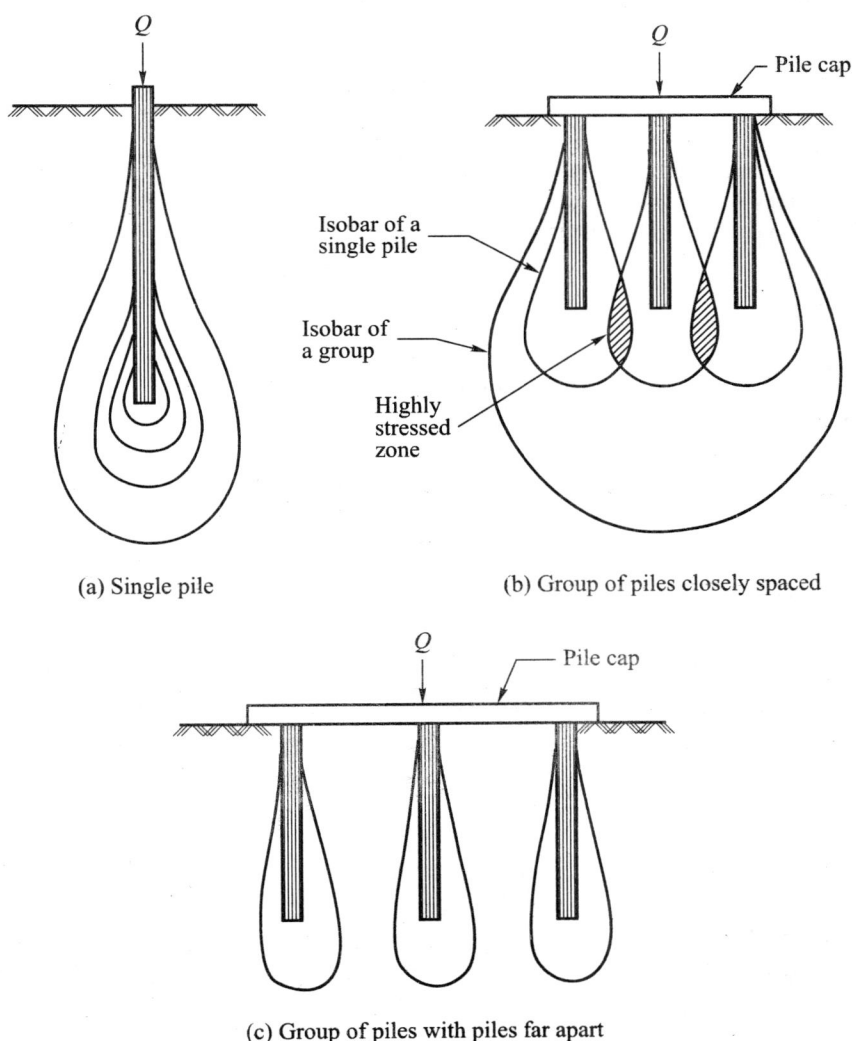

(a) Single pile   (b) Group of piles closely spaced

(c) Group of piles with piles far apart

Fig. 19.15 Pressure isobars

for point bearing piles can be much less than that required for friction piles. The spacing for straight uniform diameter piles may vary from 2 to 6 times the diameter of the shaft. For friction piles, the minimum spacing recommended is $3d$ where d is the diameter of the pile. For end bearing piles passing through relatively compressible strata, the spacing of piles shall not be less than $2.5d$. For compaction piles, the spacing may be $2d$. Typical arrangements of piles in groups are shown in Fig. 19.16.

Pile Group Efficiency

The pile group efficiency E_g is expressed as

$$E_g = \frac{Q_{gu}}{\Sigma Q_u} \tag{19.24}$$

where, Q_{gu} = ultimate load bearing capacity of a pile group;

Q_u = ultimate load bearing capacity of a single pile.

Vesic (1967) carried out tests on 4 and 9 pile groups driven into sand under controlled conditions. Piles with spacings 2, 3, 4, and 6 times pile diameter. His tests were conducted in homogeneous, and medium dense sand. His findings are given in Fig. 19.17. As per his findings the point efficiency of all

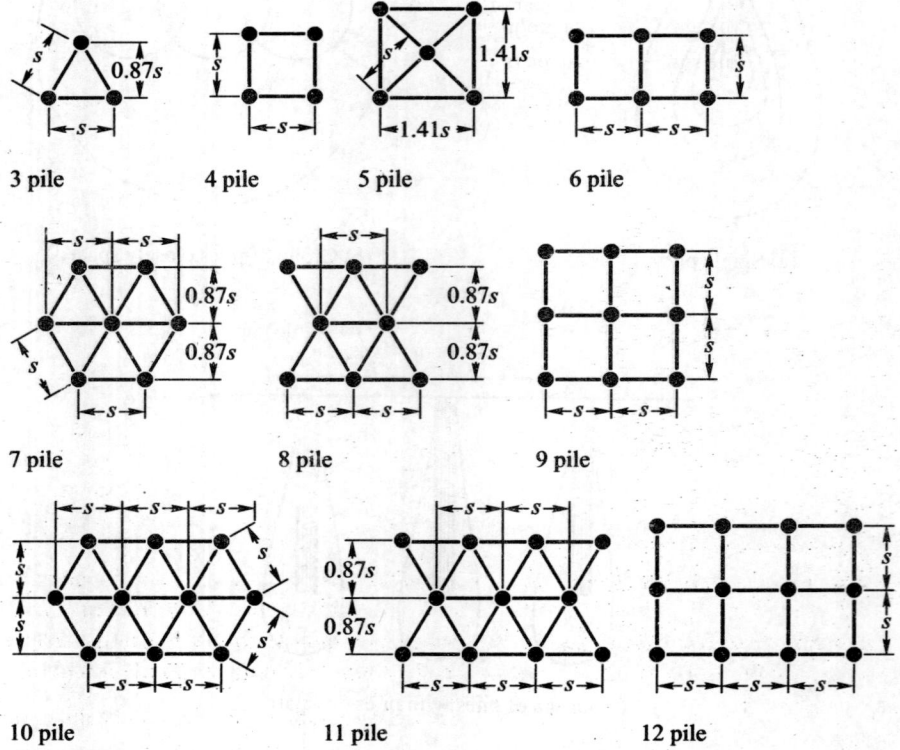

Fig. 19.16   Typical arrangements of piles in groups

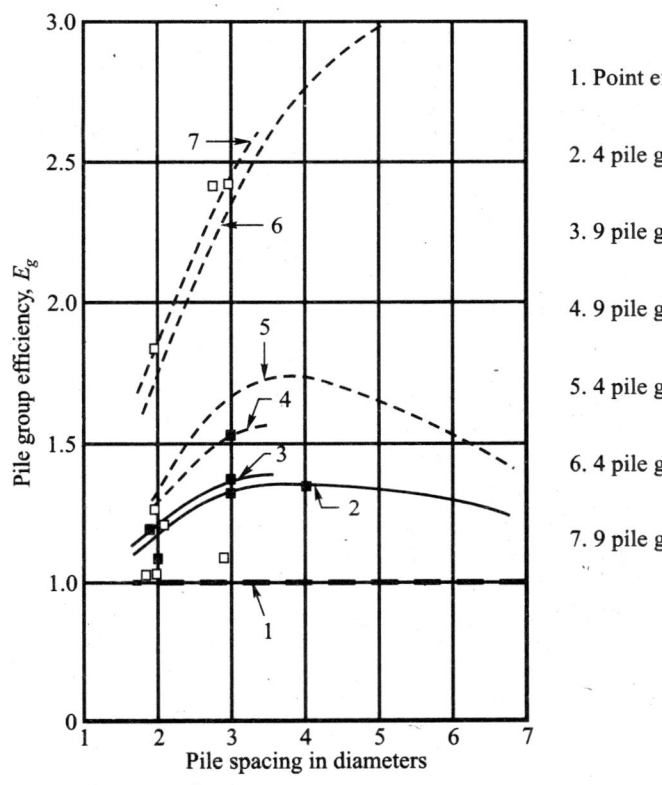

1. Point efficiency—average of all tests

2. 4 pile group—total efficiency

3. 9 pile group—total efficiency

4. 9 pile group—total efficiency with cap

5. 4 pile group—total efficiency with cap

6. 4 pile group—skin efficiency

7. 9 pile group—skin efficiency

Fig. 19.17 Efficiency of pile groups in sand Vesic, 1967)

the tests is unity. The skin efficiency increases with the increase in the pile spacing. The efficiency increases with the cap resting on the soil.

The Converse-Labarre Formula is one of the most widely used group-efficiency equations which is expressed as

$$E_g = 1 - \frac{\theta(n-1)m + (m-1)n}{90mn}$$ (19.25)

where m = number of columns of piles in a group,

n = number of rows,

$\theta = \tan^{-1} d/s$ in degrees,

d = diameter of pile,

s = spacing of piles.

There is not sufficient experimental evidence to determine efficiency of piles in clay soils.

19.17 VERTICAL BEARING CAPACITY AND SETTLEMENT OF PILE GROUPS

Driven Piles into Sand and Gravel

If piles are driven into loose to medium dense sands, the soil around the piles up to about $3d$ is supposed to get compacted. In such cases the efficiency of the pile group will be greater than unity. However, it is a conventional practice to assume

$$Q_{gn} = nQ_u \qquad (19.26)$$

where, Q_{gn} = ultimate load capacity of a pile group with n piles, and

   Q_u = the ultimate load capacity of a single pile.

Bored Piles Groups in Sand and Gravel

In the case of bored piles, the soil get loosened during boring and as such the efficiency of the pile group decreases and the efficiency factor will never be greater than unity. The load bearing capacity of a single *bored-and cast-in-situ* pile will be very much less than that of an equivalent driven for all practical purposes, the Eq. (19.26) also applies for bored piles.

Vertical Bearing Capacity of Pile Groups in Cohesive Soils

There will be considerable remoulding of soil if piles are driven into cohesive soils. This is particularly so when the soil is soft and there will be heaving of the soil if piles are driven at close intervals because of the poor drainage quality of the soil. There is every possibility of lifting of the pile also during this process of heaving of the soil. Bored piles are preferred to driven piles in cohesive soils. In case driven piles are used, the spacing of piles must be greater and the rate of driving must be such as not to allow the development of porewater pressure. A good practice is to drive piles from the centre of the group towards the edges.

Two methods are normally used for computing the ultimate bearing capacity of a pile group. They are:

1.   Assuming that the group may fail as a block called as *block failure*.

2.   Assuming that individual piles in the group will fail.

When piles are spaced at closer intervals, the soil contained between the piles move downward with the piles and at failure, piles and soil move together to give the typical *'block failure'*. Normally this type of failure occurs when piles are placed within 2 to 3 pile diameters. But for wider spacings, the piles fail individually. The efficiency ratio is less than unity at closer spacings and may reach unity at a spacing of about 8 diameters.

The equation for block failure may be written as (Fig. 19.18)

$$Q_{gu} = cN_c A_g + P_g L\bar{c} \qquad (19.27)$$

where, c = cohesive strength of clay beneath the pile group,

   $\bar{c}$ = average cohesive strength of clay around the group,

   L = length of pile,

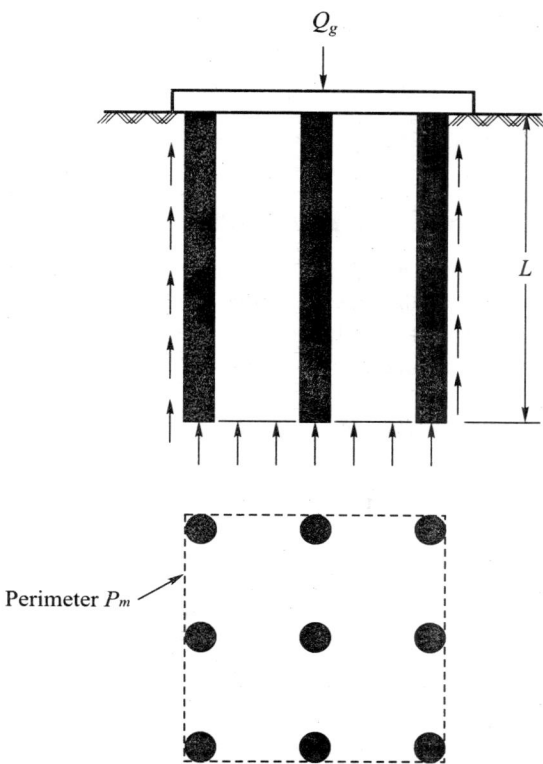

Fig. 19.18   Block failure of a pile group in clay soil

P_g = perimeter of pile group,

A_g = sectional area of group,

N_c = bearing capacity factor which may be assumed as 9 for deep foundations.

Bearing capacity of a pile group on the basis of individual pile failure may be written as

$$Q_{gu} = nQ_u \qquad\qquad\qquad (19.28)$$

where, n = number of piles in the group,

Q_u = bearing capacity of an individual pile.

Terzaghi and Peck recommend that bearing capacity of a pile group is normally taken as the smaller of the two given by Eqs. (19.27) and (19.28).

Settlement of Piles and Groups in Sands and Gravels

The present knowledge is not sufficient to evaluate the settlements of piles and pile groups. For most engineering structures, the loads to be applied to a pile group will be governed by consideration of consolidation settlement rather than by bearing capacity of the group divided by an arbitrary factor of safety of 2 or 3. It has been found from field observation that the settlement of a pile group is many times the settlement of a single pile at the corresponding working load. The settlement of a group is affected by the shape and size of the group, length of piles, method of installation of piles and possibly many other factors.

There are no equations that would satisfactorily predict the settlements of piles in sand. It is better to rely on load tests for piles in sand.

Settlement of Pile Groups in Cohesive Soils

The total settlements of pile groups may be calculated by making use of consolidation settlement equations. The problem involves evaluating the increase in stress Δp beneath a pile group when the group is subjected to a vertical load Q_g. The computation of stresses depends on the type of soil through which the pile passes. The methods of computing the stresses are explained below:

1.   The soil in the first group given in (a) of the Fig. 19.19 is homogeneous clay. The load Q_g is assumed to act on a fictitious footing at a depth $2/3L$ from the surface and distributed over the sectional area of the group. The load on the pile group acting at this level is assumed to spread out at a $2 : 1$ slope. The stress Δp at any depth z below the fictitious footing may be found as explained in Chapter 10.

2.   In the second group given in (b) of the figure, the pile passes through a very weak layer of depth L_1 and the lower portion of length L_2 is embedded in a strong layer. In this case, the load Q_g is assumed to act at a depth equal to $2/3\,L_2$ below the surface of the strong layer and spreads at a $2 : 1$ slope as before.

3.   In the third case shown in (c) of the figure, the piles are point bearing piles. The load in this case is assumed to act at the level of the firm stratum and spreads out at a $2 : 1$ slope.

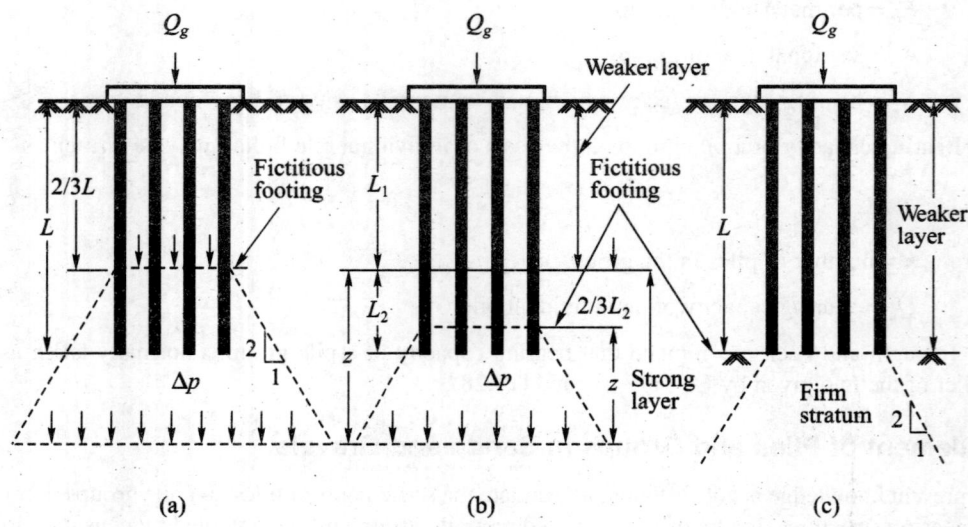

Fig. 19.19   Settlement of pile groups in clay soils

Allowable Loads on Groups of Piles

The basic criterion governing the design of a pile foundation should be the same as that of a shallow foundation, that is, the settlement of the foundation must not exceed some permissible value. The permissible values of settlements assumed for shallow foundations in Chapter 18 are also applicable to pile foundations. The allowable load on a group of piles should be the least of the values computed on the basis of the following two criteria.

1. Shear failure,

2. Settlement.

Procedures have been given in earlier chapters as to how to compute the allowable loads on the basis of a shear failure criterion. The settlement of pile groups should not exceed the permissible limits under these loads.

Example 19.14

A group of 9 piles with 3 piles in a row was driven into a soft clay extending from ground level to a great depth. The diameter and the length of the piles were 30 cm and 10 m respectively. The unconfined compressive strength of the clay is 70 kPa. If the piles were placed 90 cm centre to centre, compute the allowable load on the pile group on the basis of a shear failure criterion for a factor of safety of 2.5.

Solution

The allowable load on the group is to be calculated for two conditions: (a) block failure and (b) individual pile failure. The least of the two gives the allowable load on the group.

(a) Block failure (Fig. 19.18)

$$Q_{gu} = cN_c A_g + P_g L\bar{c} \text{ where } N_c = 9, c = \bar{c} = 70/2 = 35 \text{ kN/m}^2$$

$$A_g = 2.1 \times 2.1 = 4.4 \text{ m}^2, \quad P_g = 4 \times 2.1 = 8.4 \text{ m}, \quad L = 10 \text{ m}$$

$$Q_{gu} = 35 \times 9 \times 4.4 + 8.4 \times 10 \times 35 = 4340 \text{ kN}, \quad Q_a = \frac{4340}{2.5} = 1740 \text{ kN}$$

(b) Individual pile failure

$$Q_u = Q_b + Q_f = q_c A_b + \alpha \bar{c} A_s. \text{ Assume } \alpha = 1.$$

Now, $q_b = cN_c = 35 \times 9 = 315 \text{ kN/m}^2, \quad A_b = 0.07 \text{ m}^2,$

$$A_s = 3.14 \times 0.3 \times 10 = 9.42 \text{ m}^2$$

Substituting, $Q_u = 315 \times 0.07 + 1 \times 35 \times 0.47 = 352 \text{ kN}$

$$Q_{gu} = nQ_u = 9 \times 352 = 3168 \text{ kN}, Q_a = \frac{3168}{2.5} = 1267 \text{ kN}$$

The allowable load is 1267 kN.

Example 19.15

It is required to construct a pile foundation comprised of 20 piles arranged in 5 columns at distances of 90 cm centre to centre. The diameter and lengths of the piles are 30 cm and 9 m respectively. The bottom of the pile cap is located at a depth of 2.0 m from the ground surface. The details of the soil properties etc. are as given below with reference to ground level as the datum. The water table was found at a depth of 4 m from ground level.

Depth (m)		Soil properties
From	To	
0	2	Silt, saturated, $\gamma = 16$ kN/m³
2	4	Clay, saturated, $\gamma = 19.2$ kN/m³
4	12	Clay, saturated, $\gamma = 19.2$ kN/m³, $q_u = 120$ kN/m², $e_0 = 0.80$, $C_c = 0.23$
12	14	Clay, $\gamma = 18.24$ kN/m³, $q_u = 90$ kN/m², $e_0 = 1.08$, $C_c = 0.34$.
14	17	Clay, $\gamma = 20$ kN/m³, $q_u = 180$ kN/m², $e_0 = 0.70$, $C_c = 0.2$
17	–	Rocky stratum

Compute the consolidation settlement of the pile foundation if the total load imposed on the foundation is 2500 kN.

Solution

Assume that the total load 2500 kN acts at a depth $(2/3)L = (2/3) \times 9 = 6$ m from the bottom of the pile cap on a fictitious footing as shown in Fig. 19.19(a). This fictitious footing is now at a depth of 8 m below ground level. The size of the footing is 3.9×3.0 m. Now three layers are assumed to contribute to the settlement of the foundation. They are: *Layer 1*—from 8 m to 12 m (= 4 m thick) below ground level; *Layer 2*—from 12 m to 14 m = 2 m thick; *Layer 3*—from 14 m to 17 m = 3 m thick. The increase in pressure due to the load on the fictitious footing at the centres of each layer is computed on the assumption that the load is spread at an angle of 2 vertical to 1 horizontal [Fig. 19.19(a)] starting from the edges of the fictitious footing. The settlement is computed by making use of the equation.

$$S_t = \Sigma H_i \frac{C_c}{1+e_o} \log_{10} \frac{p_o + \Delta p}{p_o}$$

where p_o = the effective overburden pressure at the middle of each layer,

Δp = the increase in pressure at the middle of each layer

Computation of p_o

For Layer 1, $p_o = 2 \times 16 + 2 \times 19.2 + (10-4)(19.2-9.81) = 126.74$ kN/m²

For Layer 2, $p_o = 126.74 + 2(19.2 - 9.81) + 1 \times (18.24 - 9.81) = 153.95 \, \text{kN/m}^2$

For Layer 3, $p_o = 153.95 + 1(18.24 - 9.81) + 1.5 \times (20.0 - 9.81) = 177.67 \, \text{kN/m}^2$

Computation of Δp

For Layer 1

Area at 2 m depth below fictitious footing = $(3.9 + 2) \times (3 + 2) = 29.5 \, \text{m}^2$

$$\Delta p = \frac{2500}{29.5} = 84.75 \, \text{kN/m}^2$$

For Layer 2

Area at 5 m depth below fictitious footing = $(3.9 + 5) \times (3 + 5) = 71.2 \, \text{m}^2$

$$\Delta p = \frac{2500}{71.2} = 35.1 \, \text{kN/m}^2$$

For Layer 3

Area at 7.5 m depth below fictitious footing = $(3.9 + 7.5) \times (3 + 7.5) = 119.7 \, \text{m}^2$

$$\Delta p = \frac{2500}{119.7} = 20.9 \, \text{kN/m}^2$$

Settlement computation

$$\text{Layer 1} \quad S_1 = \frac{4 \times 0.23}{1 + 0.80} \log_{10} \frac{126.74 + 84.75}{126.74} = 0.114 \, \text{m}$$

$$\text{Layer 2} \quad S_2 = \frac{2 \times 0.34}{1 + 0.80} \log_{10} \frac{153.95 + 35.1}{153.95} = 0.034 \, \text{m}$$

$$\text{Layer 3} \quad S_3 = \frac{3 \times 0.2}{1 + 0.7} \log_{10} \frac{177.67 + 20.9}{177.67} = \frac{0.017 \, \text{m}}{0.165 \, \text{m}}$$

$$= 16.5 \, \text{cm}$$

Total settlement $S = 16.5$ cm.

19.18 NEGATIVE FRICTION

Figure 19.20(a) shows a single pile and (b) a group of piles passing through a recently constructed cohesive soil fill. The soil below the fill had completely consolidated under its overburden pressure.

When the fill starts consolidating under its own overburden pressure, it develops a drag on the surface of the pile. This drag on the surface of the pile is called 'negative friction'. Negative friction may develop if the fill material is loose cohesionless soil. Negative friction can also occur when fill is placed over peat or a soft clay stratum as shown in Fig. 19.20(c). The superimposed loading on such compressible stratum causes heavy settlement of the fill with consequent drag on piles.

Negative friction may develop by lowering the ground water which increases the effective stress causing consolidation of the soil with resultant settlement and friction forces being developed on the pile.

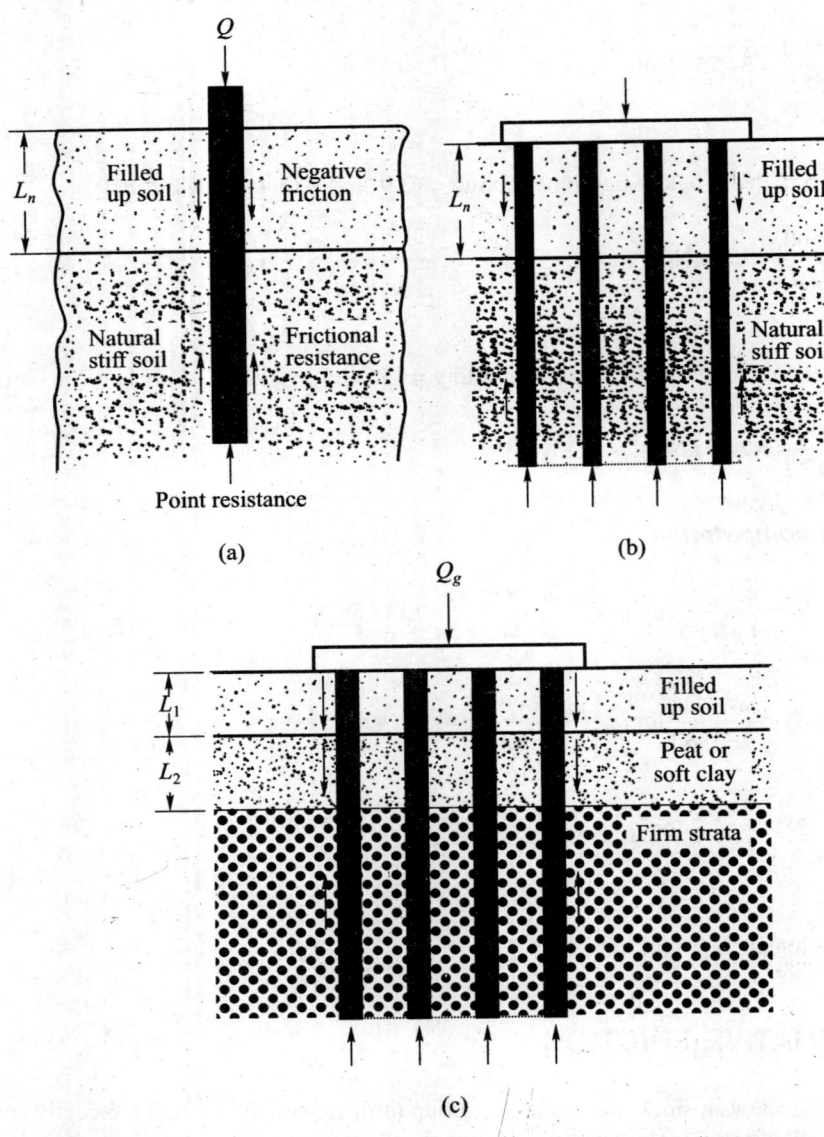

Fig. 19.20   Negative friction on piles

Negative friction must be allowed when considering the factor of safety on the ultimate carrying capacity of a pile. The factor of safety, F_s, where negative friction is likely to occur may be written as

$$F_s = \frac{\text{Ultimate carrying capacity of a single or group of piles}}{\text{Working load} + \text{Negative skin friction load}}$$

Computation of Negative Friction on a Single Pile

The magnitude of negative friction F_n for a single pile in filled up soils may be taken as [Fig. 19.20(a)].

(a) For cohesive soils

$$F_n = PL_n S \tag{19.29}$$

(b) For cohesionless soils

$$F_n = \frac{1}{2} PL_n^2 \gamma K \tan\delta \tag{19.30}$$

where, L_n = length of piles in the compressible material,

s = shear strength of cohesive soils in the fill,

P = perimeter of pile,

K = earth pressure coefficient normally lies between the active and the passive earth pressure coefficients,

δ = angle of wall friction which may vary from $\phi/2$ to ϕ.

Negative Friction on Pile Groups

When a group of piles passes through a compressible fill, the negative friction, F_{ng}, on the group may be found by any of the following methods [Fig. 19.20(b)].

(a) $F_{ng} = nF_n$ (19.31)

(b) $F_{ng} = sL_n P_g + \gamma L_n A_g$ (19.32)

where, n = number of piles in the group,

γ = unit weight of soil within the pile group to a depth L_n,

P_g = perimeter of pile group,

A_g = sectional area of pile group within the perimeter P_g,

s = shear strength of soil along the perimeter of the group.

Equation (19.31) gives the negative friction forces of the group as equal to the sum of the friction forces of all the single piles. Eq. (19.32) assumes the possibility of block shear failure along the perimeter of the group which includes the volume of the soil $\gamma L_n A_g$ enclosed in the group. The maximum value obtained from Eqs (19.31) or (19.32) should be used in the design.

When the fill is underlain by a compressible stratum as shown in Fig. 19.20(c), the total negative friction may be found as follows:

$$F_{ng} = n(F_{n1} + F_{n2}) \tag{19.33}$$

$$F_{ng} = s_1 L_1 P_g + s_2 L_2 P_g + \gamma_1 L_1 A_g + \gamma_2 L_2 A_g$$

$$= P_g(s_1 L_1 + s_2 L_2) + A_g(\gamma_1 L_1 + \gamma_2 L_2) \tag{19.34}$$

wherein, L_1 = depth of fill,

$\qquad L_2$ = depth of compressible natural soil,

$\qquad s_1, s_2$ = shear strengths of the fill and compressible soils respectively,

$\qquad \gamma_1, \gamma_2$ = unit weights of fill and compressible soils respectively,

$\qquad F_{n1}$ = negative friction of a single pile in the fill,

$\qquad F_{n2}$ = negative friction of a single pile in the compressible soil.

The maximum value of the negative friction obtained from Eqs. (19.33) or (19.34) should be used for the design of pile groups.

Example 19.16

A square pile group similar to the one shown in Fig. 19.18 passes through a recently constructed fill. The depth of fill L_n = 3 m. The diameter of the pile is 30 cm and the piles are spaced 90 cm centre to centre. If the soil is cohesive with $q_u = 60$ kN/m², and $\gamma = 15$ kN/m³, compute the negative frictional load on the pile group.

Solution

The negative frictional load on the group is the maximum of

$\qquad$ (b) $F_{ng} = nF_n$, and $\quad$ (b) $F_{ng} = sL_n P_g + \gamma L_n A_g$,

where, $P_g = 4 \times 3 = 12$ m, $A_g = 3 \times 3 = 9$ m², $c_u = 60/2 = 30$ kN/m²

$\qquad$ (a) $F_{ng} = 9 \times 3.14 \times 0.3 \times 3 \times 30 = 763$ kN

$\qquad$ (b) $F_{ng} = 30 \times 3 \times 12 + 15 \times 3 \times 9 = 1485$ kN

$\qquad$ The negative frictional load on the group = 1485 kN.

19.19 QUESTIONS AND PROBLEM: PARTS A, B AND C

19.1 $\quad$ A concrete pile of 40 cm diameter is driven into a layered system of soil as shown in Fig. Prob 19.1. The properties and other particulars are given in the figure. Estimate the ultimate load bearing capacity and the allowable load with $F_s = 3.0$ by static formula

19.2 $\quad$ Solve the Prob. 19.1 by using the standard penetration test values.

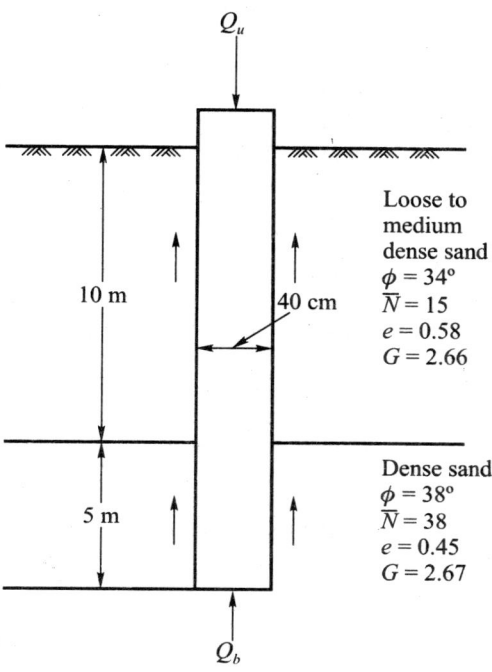

Q_u

10 m 40 cm

Loose to
medium
dense sand
$\phi = 34°$
$\overline{N} = 15$
$e = 0.58$
$G = 2.66$

5 m

Dense sand
$\phi = 38°$
$\overline{N} = 38$
$e = 0.45$
$G = 2.67$

Q_b

Fig. Prob 19.1

19.3 If the pile in Frob. 19.1 is bored and cast-in-situ, calculate Q_u and Q_a with $F_s = 3.0$ by (a) static formula and (b) by the use of N value.

19.4 A 45 cm diameter pipe pile with closed end is driven into a cohesionless soil having $\phi = 34°$. The void ratio of the soil is 0.48 and the specific gravity of the solid particles is equal to 2.64. The water table is just close to the ground surface. Estimate the uitiate pile capacity if the total length of the pile is 12 m.

19.5 If the water table in Prob. 19.4 is situated very much below the pile bottom, what is the ultimate toad capacity of the pile. Assume the soil remains dry above the water table.

19.6 A single acting steam hammer was used to drive pile with the following data: (a) weight of hammer with pile = 20 kN (b) efficiency of hammer = 75%, (c) the value of the constants (cm) $C_1 = 0.05$, $C_2 = 0.20$, $C_3 = 0.10$ (d) final set, $s = 2.5$ cm, (d) $C_r = 0.35$ and $h = 1$ m. Determine Q_u by (i) Hiley's formula, and (ii) Engineering News Record formula.

19.7 Under reamed piles with single bulbs are required for the foundation of a building an expensive soils. The cohesive strength of the soil is 50 kN/m². The shaft and bulb diameters are 45 and 115 cm respectively. If the centre of bulb is located at a depth of 4 m below the ground surface, determine the allowable load on the pile for a factor of safety of 2.5. Neglect the shaft friction.

19.8 A RCC pile of size 40 × 40 cm is driven up to a depth of 15 m into a homogenous sandy strata having a total unit weight of 19.5 kN/m³. The average corrected SPT(N) value is 16.

Water table is at the ground level. The value of $\phi = 32°$. Given: $N_q = 15$ and $\delta = 24°$. Determine the allowable pile load with $F_s = 2.5$. Assume a suitable value for coefficient of earth pressure.

19.9   A 40 cm diameter and 18 m long bored concrete pile is installed through a strata consisting of 5 m of made up cohesive soil underlain 7 m thickness of medium stiff clay having $c_u = 50$ kN/m^2. This strata is again under stiff clay having $c_u = 200$ kN/m^2. What should be the value of factor of safety if the allowable load on the pile is 400 kN. Assume that the made up soil has a cohesive strength of 30 kN/m^2 and the water table is at ground level.

19.10   For a pile designed for an allowable load of 400 kN driven by a steam hammer (single acting) with a rated energy of 2070 kN-cm, what is the approximate–terminal set of the pile using Engineering News Record formula.

19.11   A group of nine friction piles arranged in a square pattern is to be proportioned in a deposit of medium stiff clay. Assuming that the size of piles are 30×30 cm and 10 m long, find the optimum spacing for piles. Assume $a = 0.8$ and $c_u = 50$ kN/m^2.

19.12   A group of 9 piles with 3 piles in a row were driven into sand at a site. The diameter and length of the piles are 30 cm and 12 m respectively. The properties of the soil are:

$\phi = 30°$, $e = 0.7$, and $G = 2.64$.

If the spacing of the piles is 90 cm, compute the allowable load on the pile group on the basis of shear failure criteria for $F_s = 2.0$ with respect to skin resistance and $F_s = 2.5$ with respect to base resistance. For $\phi = 30°$, assume $N_q = 22.5$ and $N_\gamma = 19.7$. The water table is at ground level.

19.13   A soil profile consists of a clay layer of 8 m thick overlying dense sand extending to great depth. The value of q_u of clay strata varies from 100 (at the top) to 140 (at the bottom) kN/m^2. Estimate the load carrying capacity of a reinforced concrete pile of 40 cm diameter driven to a depth of 4 m into the sandy strata. The following properties are given.

$\alpha = 0.50$, $\phi = 36°$, and $\gamma = 18$ kN/m^3 for sandy layer

19.14   A test pile of diameter 30 cm penetrates 10 m into a stratum of clay and rests on dense sand strata. The average q_u of clay is 40 kN/m^2. Estimate how much of the ultimate load of the pile is carried by friction in the clay deposit. The following properties of the soil are:

For sand: $\phi = 38°$, $\gamma = 19.5$ kN/m^3
For clay: $\alpha = 0.75$, $\gamma_{sat} = 18$ kN/m^3
The water table is at ground level.

19.15   A bored and cast in situ concrete pile was constructed with a shaft diameter of 90 cm and a base diameter of 1.86 m. The length of shaft in clay was 14.5 m. The average unconfirmed compressive strength of a clay along the shaft was 256 kN/m^2 and at the base level 300 kN/m^2. Determine the safe working load with $F_s = 2.5$. Assume $\alpha = 0.45$. Water table is at ground level.

19.16   A ten-storied building is required to be constructed at a site where the water table is close to the surface. The foundation of the building will be supported on 30 cm diameter pipe piles. The bottom of the pile cap will be at a depth of 1.0 m below ground level. The soil investigation at the site and laboratory tests have provided. the saturated unit weights γ, the shear strength values c under undrained conditions (average), the corrected SPT values N, and the soil profile of the soil up to depth of about 40 m. The soil profile and the other details are given below.

Depth (m)		Soil	γ	N	ϕ	c (average)
From	To		kN/m³			kN/m²
0	6	Sand	19	18	33°	–
6.0	22	Med. stiff clay	18	–	–	60
22	30	sand	19.6	25	35°	–
30	38	stiff clay	18.5	–		75

Determine the ultimate bearing capacity of single pile for lengths of (a) 15 m and (b) 25 m below the bottom of the cap.

Adhesion factor: Use $\alpha = 0.50$ and $\overline{K}_s = 1.2$. Assume that the water table is at ground level.

19.17   An RCC pile of length 14.5 m and diameter 0.94 m was constructed in a deep layer of stiff clay. The bottom of the pile was belled to a diameter of 1.86 m. The average undrained shear strength, c_u, along the shaft was 128 kN/m² and that at the base was 150 kN/m². With factor of safety $F_s = 1.5$ for friction load and $F_s = 3$ the tip load, determine the safe working load. Assume the coefficient of adhesion $\alpha = 0.45$.

19.18   A bored pile with an enlarged base is to be installed in a stiff clay having undrained shear strength 220 kN/m². The saturated unit weight of the clay is 21 kN/m³. The diameters of the pile shaft and base are 1.05 m ad 3.00 m respectively. The length of the pile is 18 m. The straight portion of the pile is 16 m. The coefficient of adhesion, a for this clay may be taken as equal to 0.45. Determine the allowable load on the pile. Assume $F_s = 1.5$ for friction load and 3 for base load.

19.19   A 45 cm diameter pipe pile is to be driven through two layers of clay soil with water table close to the ground surface. The soil profile is as given below.

Type	Depth (m)		Unit cohesion c_u
	From	To	kN/m²
Soft clay	0	6	36
Medium stiff	6	38	57

Estimate the length of the piles required to carry an axial load 668 kN with $F_s = 2.5$. Use adhesion factor $\alpha = 0.8$ for soft clay and 0.6 for medium stiff clay.

19.20   A single pile of 30 cm diameter is driven to a depth of 10m into a deep layer of loose to med dense sand deposit. The soil investigation carried out at the site gives an average SPT value of 12 along the shaft and 15 at the base level. Compute the ultimate load capacity of the pile by Meyerhof's (1956) formula and the allowable load for $F_s = 4$.

19.21   If the pile in Prob. 19.20 is a bored pile, what is the ultimate load and the allowable load for
$F_s = 2.5$.

19.22   A double under-reamed pile is constructed in medium stiffclay soil with the water table
close to the ground surface. The average unconfined compressive strength of the clay along
the shaft is 60 kPa and at the bottom 90 kPa. The length of the pile is 6 m and its shaft dia
30 em. The diameter of the bulb is 75 cm. Assume any other data if required. Compute the
allowable load that the pile can carry with $F_s = 2.5$ for a bulb spacing of 112.5 cm. Assume
$\alpha = 0.6$.

19.23   Nine RCC piles of diameter 30 cm each are driven in a square pattern at 90 cm centre to
centre for a depth of 12 m into a strata of loose to medium dense sand. The average value
of ϕ may be taken as equal to 33°, The bottom of the pile cap embedding all the pile rests at
a depth of 1.5 m below ground level. The water table is close to the ground surface. At a
depth of 15 m below the ground surface there lies a clay strata of thickness 3 m and below
which lies sandy strata. The liquid limit of the clay soil has been found out to be 45%. The
saturated unit weights of sand and clay stratas are 18.5 kN/m³ and 19.5 kN/m³ respectively.
The initial void ratio of clay is 0.65. Determine (a) the ultimate bearing capacity of the pile
group and the allowable load by assuming $F_s = 2.5$, (b) Calculate the consolidation settlement
of the pile group under the allowable load.

Discuss whether the pile group is safe as per settlement consideration. Assume $\delta = 2/3\ \phi$,
and $\overline{K}_s = 2.5$. Neglect the load carried by the soil below the pile cap.

PART D: BEHAVIOUR OF LATERALLY LOADED VERTICAL AND BATTER PILES

19.20 INTRODUCTION

When a soil of low bearing capacity extends to a considerable depth, piles are generally used to
transmit vertical and lateral loads to the surrounding soil media. Piles that are used under tall chimneys,
television towers, high rise buildings, high retaining walls, offshore structures, etc. are normally
subjected to high lateral loads. These piles or pile groups should resist not only vertical movements
but also lateral movements. The requirements for a satisfactory foundation are,

1.   The vertical settlement or the horizontal movement should not exceed an acceptable
maximum value.

2.   There must not be failure by yield of the surrounding soil or the pile material.

Vertical piles are used in foundations to take normally vertical loads and small lateral loads.
When the horizontal load per pile exceeds the value suitable for vertical piles, *batter piles* are used
in combination with vertical piles. Batter piles are also called *inclined piles* or *raker piles*. The
degree of batter, is the angle made by the pile with the vertical, may up to 30°. If the lateral load acts
on the pile in the direction of batter, it is called an *in-batter* or *negative batter* pile. If the lateral load
acts in the direction opposite to that of the batter, it is called an *out-batter* or *positive batter* pile. Fig.
19.21a shows the two types of batter piles.

Extensive theoretical and experimental investigation has been conducted on single vertical piles
subjected to lateral loads by many investigators. Generalized solutions for laterally loaded vertical
piles are given by Matlock and Reese (1960). The effect of vertical loads in addition to lateral loads

has been evaluated by Davisson (1960) in terms of non-dimensional parameters. Broms (1964a, 1964b) and Poulos and Davis (1980) have given different approaches for solving laterally loaded pile problems. Brom's method is ingenious and is based primarily on the use of limiting values of soil resistance. The method of Poulos and Davis is based on the theory of elasticity.

The finite difference method of solving the differential equation for a laterally loaded pile is very much in use where computer facilities are available. Reese *et al.*, (1974) and Matlock (1970) have developed the concept of (*p-y*) curves for solving laterally loaded pile problems. This method is quite popular in the USA and in some other countries.

However, the work on batter piles is limited as compared to vertical piles. Three series of tests on single 'in' and 'out' batter piles subjected to lateral loads have been reported by Matsuo (1939). They were run at three scales. The small and medium scale tests were conducted using timber piles embedded in sand in the laboratory under controlled density conditions. Loos and Breth (1949) reported a few model tests in dry sand on vertical and batter piles. Model tests to determine the effect of batter on pile load capacity have been reported by Tschebotarioff (1953), Yoshimi (1964), and Awad and Petrasovits (1968). The effect of batter on deflections has been investigated by Kubo (1965) and Awad and Petrasovits (1968) for model piles in sand.

Full-scale field tests on single vertical and batter piles, and also groups of piles, have been made from time to time by many investigators in the past. The field test values have been used mostly to check the theories formulated for the behaviour of vertical piles only. Murthy and Subba Rao (1995) made use of field and laboratory data and developed a new approach for solving the laterally loaded pile problem.

Reliable experimental data on batter piles are rather scarce compared to that of vertical piles. Though Kubo (1965) used instrumented model piles to study the deflection behaviour of batter piles, his investigation in this field was quite limited. The work of Awad and Petrasovits (1968) was based on non-instrumented piles and as such does not throw much light on the behaviour of batter piles.

The author (Murthy, 1965) conducted a comprehensive series of model tests on instrumented piles embedded in dry sand. The batter used by the author varied from –45° to +45°. A part of the author's study on the behaviour of batter piles, based on his own research work, has been included in this chapter.

19.21 WINKLER'S HYPOTHESIS

Most of the theoretical solutions for laterally loaded piles involve the concept of *modulus of subgrade reaction* or otherwise termed as *soil modulus* which is based on Winkler's assumption that a soil medium may be approximated by a series of closely spaced independent elastic springs. Fig. 19.21(b) shows a loaded beam resting on a elastic foundation. The reaction at any point on the base of the beam is actually a function of every point along the beam since soil material exhibits varying degrees of continuity. The beam can be replaced by a beam in Fig. 19.21(c). In this figure the beam rests on a bed of elastic springs wherein each spring is independent of the other. According to Winkler's hypothesis, the reaction at any point on the base of the beam in Fig. 19.21(c) depends only on the deflection at that point. Vesic (1961) has shown that the error inherent in Winkler's hypothesis is not significant.

The problem of a laterally loaded pile embedded in soil is closely related to the beam on an elastic foundation. A beam can be loaded at one or more points along its length, whereas in the case of piles the external loads and moments are applied at or above the ground surface only.

The nature of a laterally loaded pile-soil system is illustrated in Fig. 19.21(d) for a vertical pile. The same principle applies to batter piles. A series of nonlinear springs represents the force-deformation characteristics of the soil. The springs attached to the blocks of different sizes indicate reaction increasing with deflection and then reaching a yield point, or a limiting value that depends on depth; the taper on the springs indicates a nonlinear variation of load with deflection. The gap between the pile and the springs indicates the moulding away of the soil by repeated loadings and

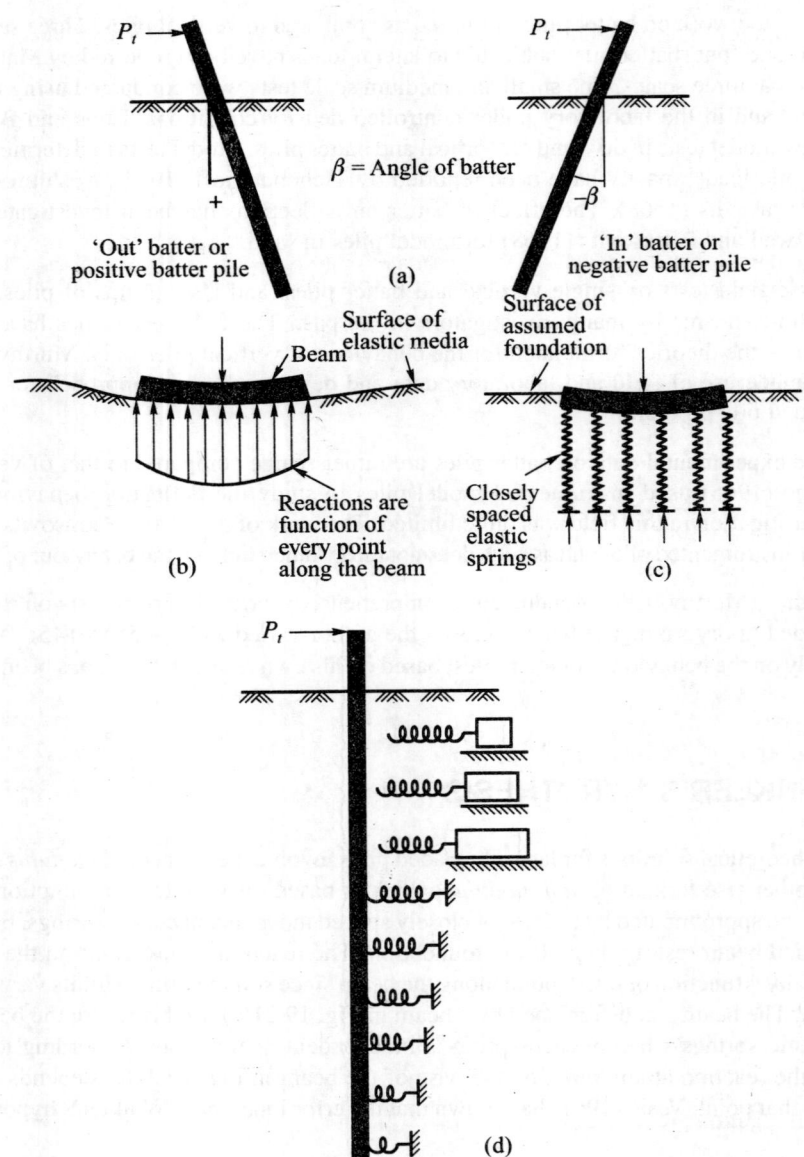

Fig. 19.21 (a) Batter piles, (b, c) Winkler's hypothesis and (d) the concept of laterally loaded pile-soil system

the increasing stiffness of the soil is shown by shortening of the springs as the depth below the surface increases.

19.22 THE DIFFERENTIAL EQUATION

Compatibility

As stated earlier, the problem of the laterally loaded pile is similar to the beam-on-elastic foundation problem. The interaction between the soil and the pile or the beam must be treated quantitatively in the problem solution. The two conditions that must be satisfied for a rational analysis of the problem are:

1. Each element of the structure must be in equilibrium.

2. Compatibility must be maintained between the superstructure, the foundation and the supporting soil.

If the assumption is made that the structure can be maintained by selecting appropriate boundary conditions at the top of the pile, the remaining problem is to obtain a solution that insures equilibrium and compatibility of each element of the pile, taking into account the soil response along the pile. Such a solution can be made by solving the differential equation that describes the pile behaviour.

The Differential Equation of the Elastic Curve

The standard differential equations for slope, moment, shear and soil reaction for a beam on an elastic foundation are equally applicable to laterally loaded piles.

The deflection of a point on the elastic curve of a pile is given by y. The x-axis is along the pile axis and deflection is measured normal to the pile-axis.

The relationships between y, slope, moment, shear and soil reaction at any point on the deflected pile may be written as follows.

deflection of the pile $= y$

slope of the deflected pile $\quad S = \dfrac{dy}{dx}$ $\hspace{5cm}$ (19.35)

moment of pile $\hspace{2cm} M = EI\dfrac{d^2y}{dx^2}$ $\hspace{4cm}$ (19.36)

shear $\hspace{3cm} V = EI\dfrac{d^3y}{dx^3}$ $\hspace{4cm}$ (19.37)

soil reaction, $\hspace{2.5cm} p = EI\dfrac{d^4y}{dx^4}$ $\hspace{4cm}$ (19.38)

where EI is the flexural rigidity of the pile material.

The soil reaction p at any point at a distance x along the axis of the pile may be expressed as

$p = -E_s y$ $\hspace{7cm}$ (19.39)

where y is the deflection at point x, and E_s is the soil *modulus*. Eqs (19.38) and (19.39) when combined gives

$$EI \frac{d^4 y}{dx^4} + E_s y = 0 \tag{19.40}$$

which is called the *differential equation* for the *elastic curve* with zero axial load.

The key to the solution of laterally loaded pile problems lies in the determination of the value of the modulus of subgrade reaction (soil modulus) with respect to depth along the pile. Fig. 19.22(a) shows a vertical pile subjected to a lateral load at ground level. The deflected position of the pile and the corresponding soil reaction curve are also shown in the same figure. The soil modulus E_s at any point x below the surface along the pile as per Eq. (19.39) is

$$E_s = -\frac{p}{y} \tag{19.41}$$

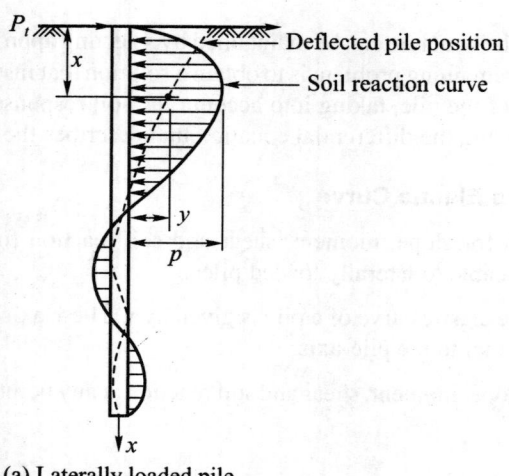

(a) Laterally loaded pile

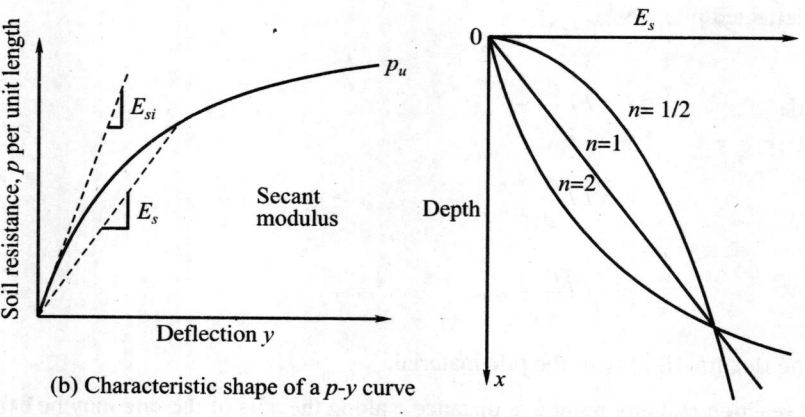

(b) Characteristic shape of a p-y curve

(c) Form of variation of E_s with depth

Fig. 19.22   The concept of (p-y) curves: (a) a laterally loaded pile, (b) characteristic shape of a p–y curve, and (c) the form of variation of E_s with depth

TABLE 19.4

Typical values of n_h for cohesive soils (Taken from Poulos and Davis, 1980)

Soil type	n_h lb/in^3	Reference
Soft NC clay	0.6 to 12.7	Reese and Matlock, 1956
	1.0 to 2.0	Davisson and Prakash, 1963
NC organic clay	0.4 to 1.0	Peck and Davisson, 1962
	0.4 to 3.0	Davisson, 1970
Peat	0.2	Davisson, 1970
	0.1 to 0.4	Wilson and Hills, 1967
Loess	29 to 40	Bowles, 1968

Table 19.4 gives some typical values for cohesive soils for n_h and Fig. 19.23 gives the relationship between n_h and the relative density of sand (Reese, 1975).

As the load P_t at the top of the pile increases the deflection y and the corresponding soil reaction p increase. A relationship between p and y at any depth x may be established as shown in Fig. 19.22(b). It can be seen that the curve is strongly non-linear, changing from an initial tangent modulus E_{si} to an ultimate resistance p_u. E_s is not a constant and changes with deflection.

There are many factors that influence the value of E_s such as the pile width d, the flexural stiffness EI, the magnitude of loading P_t and the soil properties.

The variation of E_s with depth for any particular load level may be expressed as

$$Es = n_h x^n \tag{19.42a}$$

in which n_h is termed the *coefficient of soil modulus variation*. The value of the power n depends upon the type of soil and the batter of the pile. Typical curves for the form of variation of E_s with depth for values of n equal to 1/2, 1, and 2 are given, Fig. 19.22(c). The most useful form of variation of E_s is the linear relationship expressed as

$$E_s = n_h x \tag{19.42b}$$

which is normally used by investigators for vertical piles.

19.23 NON-DIMENSIONAL SOLUTIONS FOR VERTICAL PILES SUBJECTED TO LATERAL LOADS

Matlock and Reese (1960) have given equations for the determination of y, S, M, V, and p at any point x along the pile based on dimensional analysis. The equations are

deflection,
$$y = \left[\frac{P_t T^3}{EI}\right]A_y + \left[\frac{M_t T^2}{EI}\right]B_y \tag{19.43}$$

slope,
$$S = \left[\frac{P_t T^2}{EI}\right]A_s + \left[\frac{M_t T}{EI}\right]B_s \tag{19.44}$$

moment,
$$M = [P_t T]A_m + [M_t]B_m \tag{19.45}$$

shear,
$$V = [P_t]A_v + \left[\frac{M_t}{T}\right]B_v \qquad (19.46)$$

soil reaction,
$$p = \left[\frac{P_t}{T}\right]A_p + \left[\frac{M_t}{T^2}\right]B_p \qquad (19.47)$$

where T is the relative stiffness factor expressed as

$$T = \left[\frac{EI}{n_h}\right]^{\frac{1}{5}} \qquad (19.48a)$$

for
$$E_s = n_h x$$

For a general case
$$T = \left[\frac{EI}{n_h}\right]^{\frac{1}{n+4}} \qquad (19.48b)$$

In Eqs (19.43) through (19.47), A and B are the sets of non-dimensional coefficients whose values are given in Table 19.5. The principle of superposition for the deflection of a laterally loaded

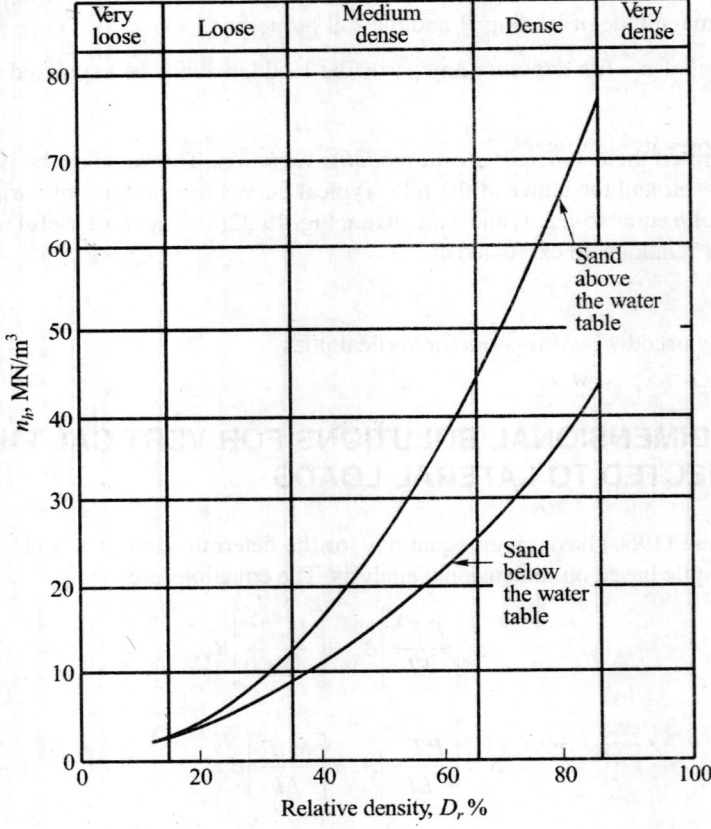

Fig. 19.23   Variation of n_h with relative density (Reese, 1975)

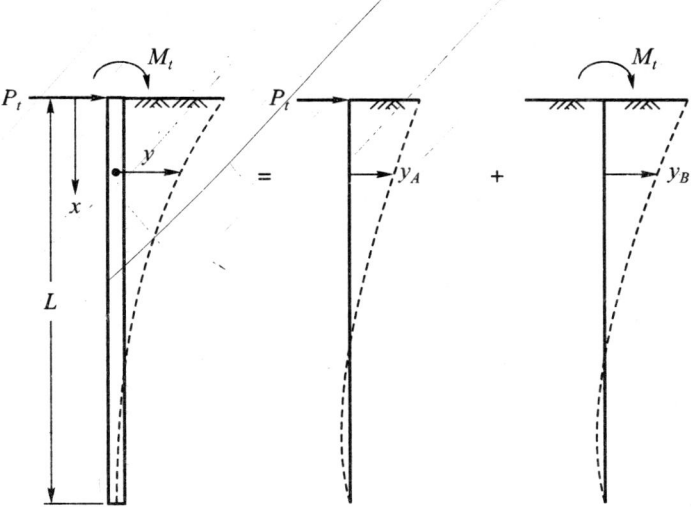

Fig. 19.24 Principle of superposition for the deflection of laterally loaded piles

pile is shown in Fig. 19.24. The A and B coefficients are given as a function of the depth coefficient, Z, expressed as

$$Z = \frac{x}{T} \tag{19.49}$$

The A and B coefficients tend to zero when the depth coefficient Z is equal to or greater than 5 or otherwise the length of the pile is more than $5T$. Such piles are called long or flexible piles. The length of a pile loses its significance beyond $5T$.

Normally we need deflection and slope at ground level. The corresponding equations for these may be expressed as

$$y_g = 2.43\frac{P_t T^3}{EI} + 1.62\frac{M_t T^2}{EI} \tag{19.50a}$$

$$S_g = 1.62\frac{P_t T^2}{EI} + 1.75\frac{M_t T}{EI} \tag{19.50b}$$

y_g for fixed head is

$$y_g = 0.93\frac{P_t T^3}{EI} \tag{19.51a}$$

Moment at ground level for fixed head is

$$M_t = -0.93[P_t T] \tag{19.51b}$$

19.24 *p-y* CURVES FOR THE SOLUTION OF LATERALLY LOADED PILES

Section 19.23 explains the methods of computing deflection, slope, moment, shear and soil reaction by making use of equations developed by non-dimensional methods. The prediction of the various

TABLE 19.5

The A and B coefficients as obtained by Reese and Matlock (1956) for long vertical piles on the assumption $E_s = n_h x$

Z	A_y	A_s	A_m	A_v	A_p
0.0	2.435	−1.623	0.000	1.000	0.000
0.1	2.273	−1.618	0.100	0.989	−0.227
0.2	2.112	−1.603	0.198	0.966	−0.422
0.3	1.952	−1.578	0.291	0.906	−0.586
0.4	1.796	−1.545	0.379	0.840	−0.718
0.5	1.644	−1.503	0.459	0.764	−0.822
0.6	1.496	−1.454	0.532	0.677	−0.897
0.7	1.353	−1.397	0.595	0.585	−0.947
0.8	1.216	−1.335	0.649	0.489	−0.973
0.9	1.086	−1.268	0.693	0.392	−0.977
1.0	0.962	−1.197	0.727	0.295	−0.962
1.2	0.738	−1.047	0.767	0.109	−0.885
1.4	0.544	−0.893	0.772	−0.056	−0.761
1.6	0.381	−0.741	0.746	−0.193	−0.609
1.8	0.247	−0.596	0.696	−0.298	−0.445
2.0	0.142	−0.464	0.628	−0.371	−0.283
3.0	−0.075	−0.040	0.225	−0.349	0.226
4.0	−0.050	0.052	0.000	−0.016	0.201
5.0	−0.009	0.025	−0.033	0.013	0.046

Z	B_y	B_s	B_m	B_v	B_p
0.0	1.623	−1.750	1.000	0.000	0.000
0.1	1.453	−1.650	1.000	−0.007	−0.145
0.2	1.293	−1.550	0.999	−0.028	−0.259
0.3	1.143	−1.450	0.994	−0.058	−0.343
0.4	1.003	−1.351	0.987	−0.095	−0.401
0.5	0.873	−1.253	0.976	−0.137	−0.436
0.6	0.752	−1.156	0.960	−0.181	−0.451
0.7	0.642	−1.061	0.939	−0.226	−0.449
0.8	0.540	−0.968	0.914	−0.270	−0.432
0.9	0.448	−0.878	0.885	−0.312	−0.403
1.0	0.364	−0.792	0.852	−0.350	−0.364
1.2	0.223	−0.629	0.775	−0.414	−0.268
1.4	0.112	−0.482	0.668	−0.456	−0.157
1.6	0.029	−0.354	0.594	−0.477	−0.047
1.8	−0.030	−0.245	0.498	−0.476	0.054
2.0	−0.070	−0.155	0.404	−0.456	0.140
3.0	−0.089	0.057	0.059	−0.0213	0.268
4.0	−0.028	0.049	0.042	0.017	0.112
5.0	0.000	0.011	0.026	0.029	−0.002

curves depends primarily on the single parameter n_h. If it is possible to obtain the value of n_h independently for each stage of loading P_t, the p-y curves at different depths along the pile can be constructed as follows:

1. Determine the value of n_h for a particular stage of loading P_t.

2. Compute T from Eq. (19.48a) for the linear variation of E_s with depth.

3. Compute y at specific depths $x = x_1$, $x = x_2$, etc. along the pile by making use of Eq. (19.43), where A and B parameters can be obtained from Table 19.5 for various depth coefficients Z.

4. Compute p by making use of Eq. (19.47), since T is known, for each of the depths $x = x_1$, $x = x_2$, etc.

5. Since the values of p and y are known at each of the depths x_1, x_2 etc., one point on the p-y curve at each of these depths is also known.

6. Repeat steps 1 through 5 for different stages of loading and obtain the values of p and y for each stage of loading and plot to determine p-y curves at each depth.

The individual p-y curves obtained by the above procedure at depths x_1, x_2, etc. can be plotted on a common pair of axes to give a family of curves for the selected depths below the surface. The p-y curve shown in Fig. 19.22b is strongly non-linear and this curve can be predicted only if the values of n_h are known for each stage of loading. Further, the curve can be extended until the soil reaction, p, reaches an ultimate value, p_u, at any specific depth x below the ground surface.

If n_h values are not known to start with at different stages of loading, the above method cannot be followed. Supposing p-y curves can be constructed by some other independent method, then p-y curves are the starting points to obtain the curves of deflection, slope, moment and shear. This means we are proceeding in the reverse direction in the above method. The methods of constructing p-y curves and predicting the non-linear behaviour of laterally loaded piles are beyond the scope of this book. This method has been dealt with in detail by Reese (1985).

Example 19.17

A steel pipe pile of 61 cm outside diameter with a wall thickness of 2.5 cm is driven into loose sand ($D_r = 30\%$) under submerged conditions to a depth of 20 m. The submerged unit weight of the soil is 8.75 kN/m³ and the angle of internal friction is 33°. The EI value of the pile is 4.35×1011 kg-cm² (4.35×10^2 MN-m²). Compute the ground line deflection of the pile under a lateral load of 268 kN at ground level under a free head condition using the non-dimensional parameters of Matlock and Reese. The n_h value from Fig. 19.23 for $D_r = 30\%$ is 6 MN/m³ for a submerged condition.

Solution

From Eq. (19.50a)

$$y_g = 2.43 \frac{P_t T^3}{EI} \text{ for } M_t = 0$$

From Eq. (19.48a),

$$T = \left[\frac{EI}{n_h} \right]^{\frac{1}{5}}$$

where, $P_t = 0.268$ MN

$EI = 4.35 \times 10^2$ MN-m^2

$n_h = 6$ MN/m^3

$$T = \left[\frac{4.35 \times 10^2}{6}\right]^{\frac{1}{5}} = 2.35 \text{ m}$$

Now $y_g = \dfrac{2.43 \times 0.268 \times (2.35)^3}{4.35 \times 10^2} = 0.0194 \text{ m} = 1.94 \text{ cm}$

Example 19.18

If the pile in Ex. 19.17 is subjected to a lateral load at a height 2 m above ground level, what will be the ground line deflection?

Solution

From Eq. (19.50a)

$$y_g = 2.43 \frac{P_t T^3}{EI} + 1.62 \frac{M_t T^2}{EI}$$

As in Ex. 19.17 $T = 2.35$ m, $M_t = 0.268 \times 2 = 0.536$ MN-m

Substituting, $y_g = \dfrac{2.43 \times 0.268 \times (2.35)^3}{4.35 \times 10^2} + \dfrac{1.62 \times 0.536 \times (2.35)^3}{4.35 \times 10^2}$

$= 0.0194 + 0.0110 = 0.0304 \text{ m} = 3.04 \text{ cm}.$

Example 19.19

If the pile in Ex. 19.17 is fixed against rotation, calculate the deflection at the ground line.

Solution

Use Eq. (19.51a)

$$y_g = \frac{0.93 P_t T^3}{EI}$$

The values of P_t, T and EI are as given in Ex. 19.17. Substituting these values

$$y_g = \frac{0.93 \times 0.268 \times (2.35)^3}{4.35 \times 10^2} = 0.0075 \text{ m} = 0.75 \text{ cm}$$

19.25 BROMS' SOLUTIONS FOR LATERALLY LOADED PILES

Broms' (1964a, 1964b) solutions for laterally loaded piles deal with the following:

1.   Lateral deflections of piles at ground level at working loads
2.   Ultimate lateral resistance of piles under lateral loads

Broms' provided solutions for both short and long piles installed in cohesive and cohesionless soils respectively. He considered piles fixed or free to rotate at the head. Lateral deflections at working loads have been calculated using the concept of subgrade reaction. It is assumed that the deflection increases linearly with the applied loads when the loads applied are less than one-half to one-third of the ultimate lateral resistance of the pile.

Lateral Deflections at Working Loads

Lateral deflections at working loads can be obtained from Fig. 19.25 for cohesive soil and Fig. 19.26 for cohesionless soils respectively. For piles in saturated cohesive soils, the plot in Fig. 19.25 gives the relationships between the dimensionless quantity βL and $(y_o kdL)/P_t$ for free-head and restrained piles, where

$$\beta = \left(\frac{kd}{4EI}\right)^{1/4}.$$

(19.52)

EI = stiffness of pile section
k = coefficient of horizontal subgrade reaction
d = width or diameter of pile
L = length of pile

A pile is considered long or short on the following conditions

Free-head pile

Long pile when $\beta L > 2.50$
Short pile when $\beta L < 2.50$

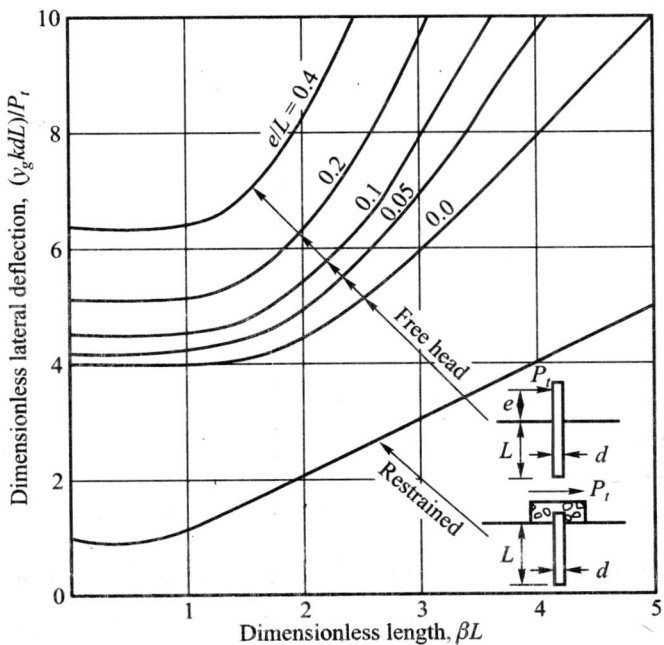

Fig. 19.25 Charts for calculating lateral deflection at the ground surface of horizontally loaded pile in cohesive soil (after Broms 1964a)

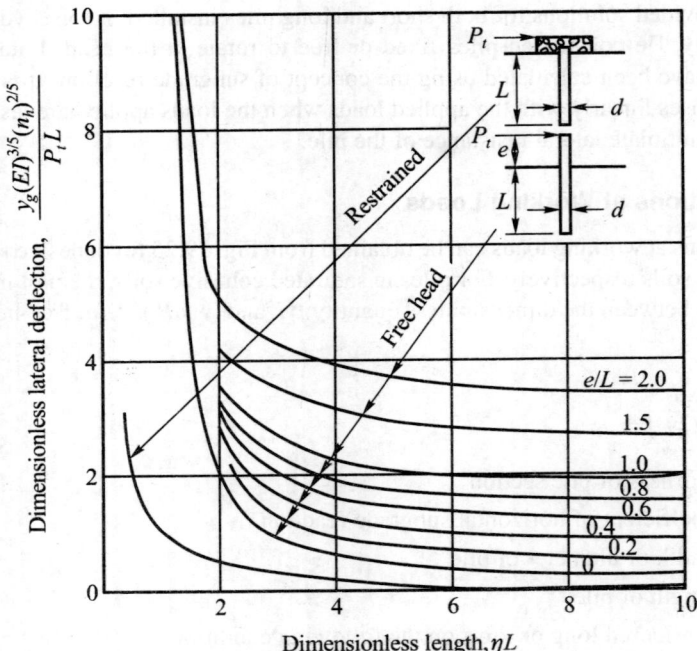

Fig. 19.26 Charts for calculating lateral deflection at the ground surface of horizontally loaded piles in cohesionless soil (after Broms 1964b)

Fixed-head Pile

Long pile when $\beta L > 1.5$

Short pile when $\beta L < 1.5$

Lateral deflections at working loads of piles embedded in cohesionless soils may be obtained from Fig. 19.26 Non-dimensionless factor $[y_g(EI)^{3/5} (n_h)^{2/5}]/P_tL$ is plotted as a function of ηL for various values of e/L

where y_g = deflection at ground level

$$\eta = \left(\frac{n_h}{EI}\right)^{1/5} \tag{19.53}$$

n_h = coefficient of soil modulus variation

P_t = lateral load applied at or above ground level

L = length of pile

e = eccentricity of load.

Ultimate Lateral Resistance of Piles in Saturated Cohesive Soils

The ultimate soil resistance of piles in cohesive soils increases with depth from $2c_u$ (c_u = undrained shear strength) to 8 to 12 c_u at a depth of three pile diameters (3d) below the surface. Broms (1964a) suggests a constant value of $9c_u$ below a depth of 1.5d as the ultimate soil resistance. Figure 19.27 gives solutions for short piles and Fig. 19.28 for long piles. The solution for long piles

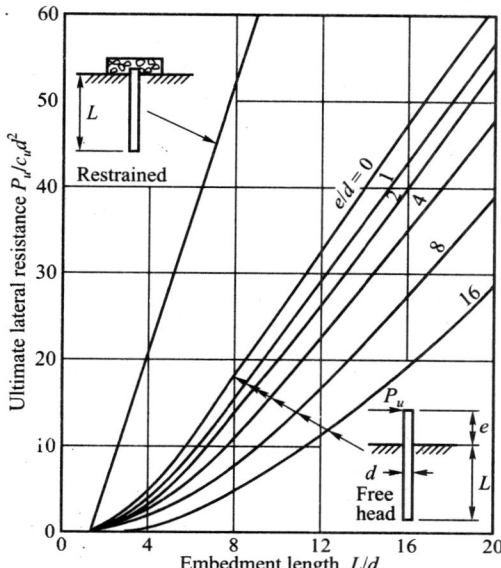

Fig. 19.27   Ultimate lateral resistance of a short pile in cohesive soil related to embedded length (after Broms (1964a))

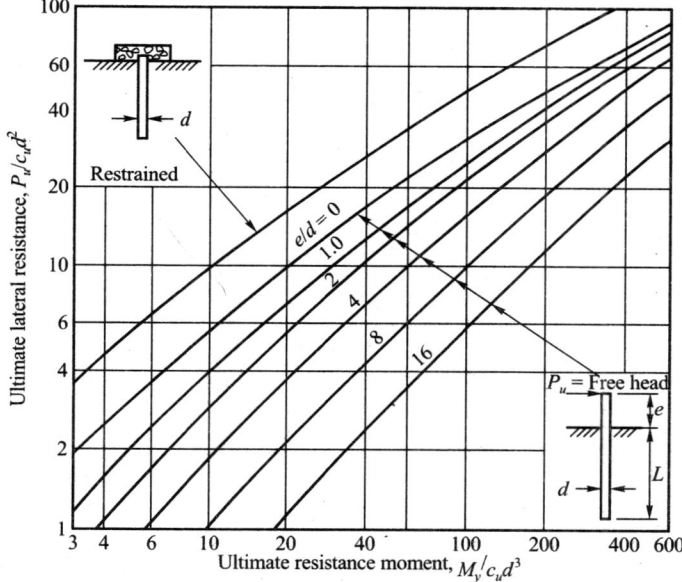

Fig. 19.28   Ultimate lateral resistance of a long pile in cohesive soil related to embedded length (after Broms (1964a))

involves the yield moment M_y for the pile section. The equations suggested by Broms for computing M_y are as follows:

For a cylindrical steel pipe section

$$M_y = 1.3 \, f_y \, Z \qquad\qquad (19.54a)$$

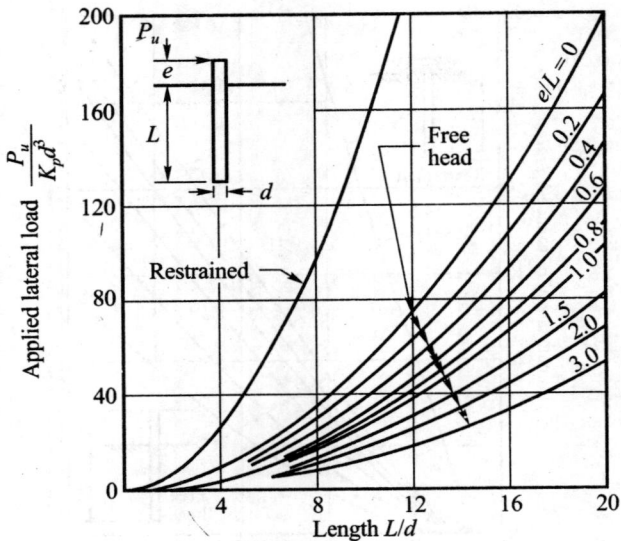

Fig. 19.29 Ultimate lateral resistance of a short pile in cohesionless soil related to embedded length (after Broms (1964b))

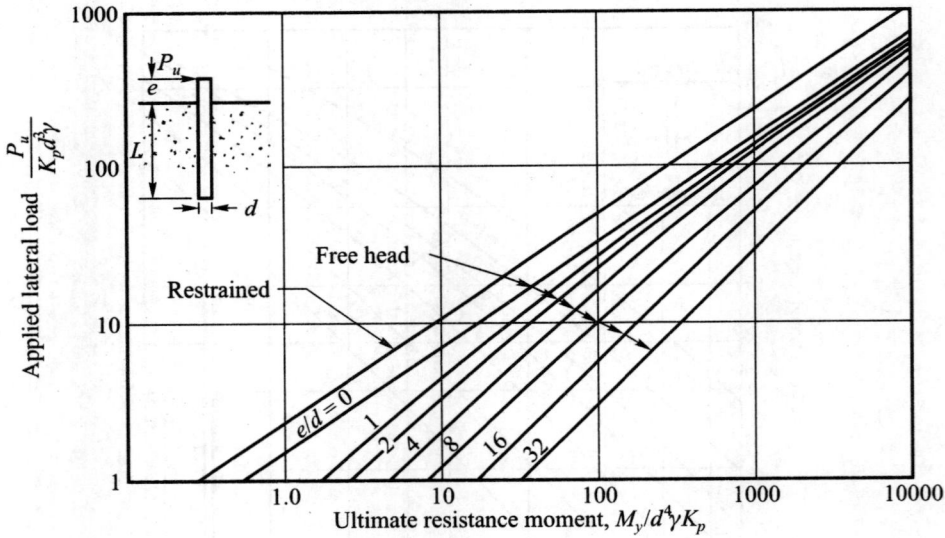

Fig. 19.30 Ultimate lateral resistance of a long pile in cohesionless soil related to embedded length (after Broms (1964b))

For an H-section

$$M_y = 1.1 f_y Z_{max} \qquad\qquad (19.54b)$$

where f_y = yield strength of the pile material

Z = section modulus of the pile section

The ultimate strength of a reinforced concrete pile section can be calculated in a similar manner.

Ultimate Lateral Resistance of Piles in Cohesionless Soils

The ultimate lateral resistance of a short piles embedded in cohesionless soil can be estimated making use of Fig. 19.29 and that of long piles from Fig. 19.30. In Fig. 19.29 the dimensionless quantity $P_u/\gamma d^3 K_P$ is plotted against the L/d ratio for short piles and in Fig. 19.30 $P_u/\gamma d^3 K_P$ is plotted against $M_y/\gamma d^4 K_P$. In both cases the terms used are

γ = effective unit weight of soil

K_P = Rankine's passive earth pressure coefficient = $\tan^2(45° + \phi/2)$

Example 19.20

A steel pipe pile of 61 cm outside diameter with 2.5 cm wall thickness is driven into saturated cohesive soil to a depth of 20 m. The undrained cohesive strength of the soil is 85 kPa. Calculate the ultimate lateral resistance of the pile by Broms' method with the load applied at ground level.

Solution

The pile is considered as a long pile. Use Fig. 19.28 to obtain the ultimate lateral resistance P_u of the pile.

The non-dimensional yield moment $= \dfrac{M_y}{c_u d^3}$

where M_y = yield resistance of the pile section

= $1.3 f_y Z$

f_y =   yield strength of the pile material

=   2800 kg/cm^2 (assumed)

Z =   section modulus $= \dfrac{\pi}{64R}[d_0^4 - d_i^4]$

I =   moment of inertia,

d_o =   outside diameter = 61 cm,

d_i =   inside diameter = 56 cm,

R =   outside radius = 30.5 cm

$$Z = \frac{3.14}{64 \times 30.5}\left[61^4 - 56^4\right] = 6{,}452.6 \text{ cm}$$

$$M_y = 1.3 \times 2{,}800 \times 6{,}452.6 = 23.487 \times 10^6 \text{ kg - cm.}$$

$$\frac{M_y}{c_u d_0^3} = \frac{23.487 \times 10^6}{0.85 \times 61^3} = 122$$

From Fig. 19.28 for $e/d = 0$, $\dfrac{M_y}{c_u d^3} = 122$, $\dfrac{P_u}{c_u d_0^2} \approx 35$

$$P_u = 35\ c_u d_0^2 = 35 \times 85 \times 0.61^2 = 1{,}107\ \text{kN}$$

Example 19.21

If the pile given in Ex. 19.20 is restrained against rotation, calculate the ultimate lateral resistance P_u.

Solution

Per Ex. 19.20 $\quad \dfrac{M_y}{c_u d_0^3} = 122$

From Fig. 19.28, for $\dfrac{M_y}{c_u d_0^3} = 122$, for restrained pile $\dfrac{P_u}{c_u d_0^2} \approx 50$

Therefore $P_u = \dfrac{50}{35} \times 1{,}107 = 1{,}581\ \text{kN}$

Example 19.22

A steel pipe pile of outside diameter 61 cm and inside diameter 56 cm is driven into a medium dense sand under submerged conditions. The sand has a relative density of 60% and an angle of internal friction of 38°. Compute the ultimate lateral resistance of the pile by Brom's method. Assume that the yield resistance of the pile section is the same as that given in Ex 19.20. The submerged unit weight of the soil $\gamma_b = 8.75\ \text{kN/m}^3$.

Solution

From Fig. 19.30

$$\text{Non - dimensional yeild moment} = \frac{M_y}{\gamma d^4 K_P}$$

where, $\quad K_p \quad = \quad \tan^2(45 + \phi/2) = \tan^2 64 = 4.20$,

$\qquad\quad M_y \quad = \quad 23.487 \times 10^6$ kg-cm,

$\qquad\quad \gamma \quad = \quad 8.75\ \text{kN/m}^3 \approx 8.75 \times 10^{-4}\ \text{kg/cm}^3$,

$\qquad\quad d \quad = \quad 61$ cm.

Substituting,

$$\frac{M_y}{\gamma d^4 K_P} = \frac{23.487 \times 10^6 \times 10^4}{8.75 \times 61^4 \times 4.2} = 462$$

From Fig. 19.30, for $\dfrac{M_y}{\gamma d^4 K_P} = 462$, for $e/d = 0$ we have $\dfrac{P_u}{\gamma d^3 K_P} \approx 80$

Therefore $P_u = 80\ \gamma d^3 K_P = 80 \times 8.75 \times 0.61^3 \times 4.2 = 667\ \text{kN}$

Example 19.23

If the pile in Ex. 19.22 is restrained, what is the ultimate lateral resistance of the pile?

Solution

From Fig.19.30, for $\dfrac{M_y}{\gamma d^4 K_P} = 462$, the value $\dfrac{P_u}{\gamma d^3 K_P} = 135$

$P_u = 135\, \gamma d^3 K_P = 135 \times 8.75 \times 0.61^3 \times 4.2 = 1{,}126\,\text{kN}.$

Example 19.24

Compute the deflection at ground level by Broms' method for the pile given in Ex. 19.17.

Solution

From Eq. (19.53)

$$\eta = \left(\frac{n_h}{EI}\right)^{1/5} = \left(\frac{6}{4.35 \times 10^2}\right)^{1/5} = 0.424$$

$\eta L = 0.424 \times 20 = 8.5$

From Fig. 19.26, for $\eta L = 8.5$, $e/L = 0$, we have

$$\frac{y_g (EI)^{3/5} (n_h)^{2/5}}{P_t L} = 0.2$$

$$y_g = \frac{0.2 P_t L}{(EI)^{3/5} (n_h)^{2/5}} = \frac{0.2 \times 0.268 \times 20}{(4.35 \times 10^2)^{3/5} (6)^{2/5}} = 0.014\,\text{m} = 1.4\,\text{cm}$$

Example 19.25

If the pile given in Ex. 19.17 is only 4 m long, compute the ultimate lateral resistance of the pile by Broms' method.

Solution

From Eq. (19.53)

$$\eta = \left(\frac{n_h}{EI}\right)^{1/5} = \left(\frac{6}{4.35 \times 10^2}\right)^{1/5} = 0.424$$

$\eta L = 0.424 \times 4 = 1.696.$

The pile behaves as an infinitely stiff member since $\eta L < 2.0$, $L/d = 4/0.61 = 6.6$.

From Fig. 19.29, for $L/d = 6.6$, $e/L = 0$, we have

$P_u / \gamma d^3 K_p = 25.$

$\phi = 33°$, $\gamma = 8.75\,\text{kN/m}^3$, $d = 61\,\text{cm}$, $K_p = \tan^2 (45° + \phi/2) = 3.4.$

Now $P_u = 25 \, \gamma d^3 \, K_p = 25 \times 8.75 \times (0.61)^3 \times 3.4 = 169$ kN

If the sand is medium dense, as given in Ex. 19.22, then $K_p = 4.20$, and the ultimate lateral resistance P_u is

$$P_u = \frac{4.2}{3.4} \times 169 = 209 \text{ kN}$$

As per Ex. 19.22, P_u for a long pile = 667 kN, which indicates that the ultimate lateral resistance increases with the length of the pile and remains constant for a long pile.

19.26 A SIMPLIFIED METHOD FOR SOLVING THE NON-LINEAR BEHAVIOUR OF LATERALLY LOADED FLEXIBLE PILE PROBLEMS

Key to the Solution

The key to the solution of a laterally loaded vertical pile problem is the development of an equation for n_h. The present state of the art does not indicate any definite relationship between n_h, the properties of the soil, the pile material, and the lateral loads. However it has been recognized that n_h depends on the relative density of soil for piles in sand and undrained shear strength c for piles in clay. It is well known that the value of n_h decreases with an increase in the deflection of the pile. It was Palmer et al (1948) who first showed that a change of width d of a pile will have an effect on deflection, moment and soil reaction even while EI is kept constant for all the widths. The selection of an initial value for n_h for a particular problem is still difficult and many times quite arbitrary. The available recommendations in this regard (Terzaghi 1955, and Reese 1975) are widely varying.

The author has been working on this problem since a long time (Murthy, 1965). An explicit relationship between n_h and the other variable soil and pile properties has been developed on the principles of dimensional analysis (Murthy and Subba Rao, 1995).

Expressions for n_h

The term n_h may be expressed as a function of the following parameters for piles in sand and clay.

For piles in sand, $\qquad n_h = \dfrac{150 C_\phi \gamma^{1.5} \sqrt{EId}}{P_e}$ $\qquad\qquad$ (19.55)

For piles in clay, $\qquad n_h = \dfrac{125 c^{1.5} \sqrt{EI\gamma d} \big/ (1 + e/d)^{1.5}}{P_e^{1.5}}$ $\qquad\qquad$ (19.56)

where

$P_e = P_t \left(1 + 0.67 \dfrac{e}{T} \right)$ an equivalent laterial load acting at ground level

P_t = Lateral load acting a height of e above ground level.

C_ϕ = correction factor for the angle of friction $\phi = 3 \times 10^{-5} \, (1.316)^{\phi°}$

Now the equation for computing ground line deflection y_g is

$$y_g = \frac{2.43 P_e T^3}{EI} \qquad\qquad (19.57)$$

It can be seen in the above equations that the numerators in both cases are constants for any given set of pile and soil properties.

The above two equations can be used to predict the non-linear behaviour of piles subjected to lateral loads very accurately.

Example 19.26

Solve the problem in Example 19.17 by the simplified method. The soil is loose sand in a submerged condition.

Given; EI $= 4.35 \times 1011$ kg- cm$^2 = 4.35 \times 10^5$ kN-m^2

   d $=$ 61 cm, $L = 20$ m, $\gamma_b = 8.75$ kN/m^3

   ϕ $=$ $33°, P_t = 268$ kN (since e = 0)

Required y_g at ground level

Solution

For a pile in sand for the case of $e = 0$, use Eq. (19.55)

$$n_h = \frac{150 C_\phi \gamma^{1.5} \sqrt{EId}}{P_e}$$

For $\phi = 33°$, $C_\phi = 3 \times 10^{-5}$ $(1.316)^{33} = 0.26$

$$n_h = \frac{150 \times 0.26 \times (8.75)^{1.5} \sqrt{4.65 \times 10^5 \times 0.61}}{P_e} = \frac{54 \times 10^4}{P_e} = \frac{54 \times 10^4}{268} = 2{,}015 \text{ kN/m}^3$$

$$T = \left[\frac{EI}{n_h}\right]^{1/5} = \left[\frac{43.5 \times 10^4}{2015}\right]^{1/5} = 2.93 \text{ m}$$

Now using Eq. (19.57)

$$y_g = \frac{2.43 \times 268 \times (2.93)^3}{4.35 \times 10^5} = 0.0377 \text{ m} = 3.77 \text{ cm}$$

It may be noted that the direct method gives a greater ground line deflection (= 3.77 cm) as compared to the 1.96 cm in Ex. 19.17.

Example 19.27

Solve the problem in Example 19.18 by the simplified method. In this case P_t is applied at a height 2 m above ground level All the other data remain the same.

Solution

From Example 19.26

$$n_h = \frac{54 \times 10^4}{P_e}$$

For $P_e = 382$ kN, $n_h = 2{,}015$ kN/m^3, $T = 2.93$ m

$$P_e = P_t\left(1+0.67\frac{E}{T}\right) = 268\left(1+0.67\times\frac{2}{2.93}\right) = 391\,\text{kN}$$

For $P_e = 391\,\text{kN}$, $n_h = \dfrac{54\times10^4}{391} = 1{,}381\,\text{kN/m}^3$

Now $T = \left[\dfrac{43.5\times10^4}{1{,}381}\right]^{1/5} = 3.16\,\text{m}$

As before $P_e = 268\left(1+0.67\times\dfrac{2}{3.16}\right) = 382\,\text{kN}$

For $P_e = 382$ kN, $n_h = 1{,}414$ kN/m^3, $T = 3.14$ m

Convergence will be reached after a few trials. The final values are

$P_e = 387$ kN, $n_h = 1718$ kN/m^3, $T = 3.025$ m

Now from Eq. (19.57)

$$y_g = \frac{2.43 P_e T^3}{EI} = \frac{2.43\times382\times(3.14)^3}{4.35\times10^5} = 0.066\,\text{m} = 6.6\,\text{cm}$$

The n_h value from the simplified method is 1,414 kN/m^3 whereas from Fig. 19.23 it is 6,000 kN/m^3. The n_h from Fig. 19.23 gives y_g which is 50 percent of the probable value and is on the unsafe side.

Example 19.28

Compute the ultimate lateral resistance for the pile given in Example 19.20 by the simplified method. All the other data given in the example remain the same.

Given: EI = 4.35×10^5 kN-m^2, $d = 61$ cm, $L = 20$ m

c_u = 85 kN/m^3, $\gamma_b = 10$ kN/m^3 (assumed for clay)

M_y = 2,349 kN-m; $e = 0$

Required: The ultimate lateral resistance P_u.

Solution

Use Eqs (19.56) and (19.48a)

$$n_h = \frac{125c^{1.5}\sqrt{EI\gamma d}}{P_t^{1.5}} \quad \text{for } e = 0$$

$$T = \left[\frac{EI}{n_h}\right]^{0.2} \tag{a}$$

Substituting the known values and simplifying

$$n_h = \frac{1{,}600\times10^5}{P_t^{1.5}} \tag{b}$$

Step 1

Let $P_t = 1,000$ kN, $n_h = \dfrac{1,600 \times 10^5}{(1000)^{1.5}} = 5,060$ kN/m^3

$$T = \left[\frac{4.35 \times 10^5}{5060} \right]^{0.2} = 2.437 \text{ m}$$

For $e = 0$, from Table 19.5 and Eq. (19.45) we may write

$M_{max} = 0.77[P_t T]$

where $A_m = 0.77$ (max) correct to two decimal places.

For $P_t = 1000$ kN, and $T = 2.437$ m

$M_{max} = 0.77 \times 1000 \times 2.437 = 1876$ kN-m $< M_y$.

Step 2

Let P_t	=	1500 kN
n_h	=	2754 kN/m^3 from Eq. (b)
and T	=	2.75 m from Eq. (a)
Now M_{max}	=	$0.77 \times 1500 \times 2.75 = 3179$ kN-m $> M_y$.

$$P_u = 1,000 + (1,500 - 1,000) \times \frac{(2,349 - 1,876)}{(3,179 - 1,876)} = 1,182 \text{ kN}$$

$P_u = 1,100$ kN by Brom's method which agrees with the simplified method.

19.27 CASE STUDIES FOR LATERALLY LOADED VERTICAL PILES IN SAND

Case 1: Mustang Island Pile LoadTest (Reese et al., 1974)

Data

Pile diameter, d	=	24 in, steel pipe (driven pile)
EI	=	4.854×10^{10} lb-in^2
L	=	69 ft
e	=	12 in.
ϕ	=	39°
γ	=	66 lb/ft^3 (= 0.0382 lb/in^3)
M_y	=	7×10^6 in-lbs

The soil was fine silty sand with WT at ground level

Required

(a) Load-deflection curve (P_t vs. y_g) and n_h vs. y_g curve
(b) Load-max moment curve (P_t Vs M_{max})
(c) Ultimate load P_u

Solutions

For pile in sand, $n_h = \dfrac{150 C_\phi \gamma^{1.5}\sqrt{EId}}{P_e}$

For $\phi = 39°$. $C_\phi = 3 \times 10^{-5}(1.316)^{39°} = 1.34$

After substitution and simplifying

$$n_h = \frac{1631 \times 10^3}{P_e} \tag{a}$$

We have $P_e = P_t\left(1 + 0.67\dfrac{e}{T}\right)$ (b)

$$T = \left[\frac{EI}{n_h}\right]^{\frac{1}{5}} \tag{c}$$

(a)   Calculation of Groundline Deflection, y_g

Step 1

Since T is not known to start with, assume $e = 0$, and $P_e = P_t = 10,000$ lbs

Now, from Eq. (a), $n_h = \dfrac{1631 \times 10^3}{10 \times 10^3} = 163 \ \text{1b/in}^3$

from Eq. (c), $T = \left[\dfrac{4.854 \times 10^{10}}{163}\right]^{\frac{1}{5}} = 49.5 \ \text{in}$

from Eq. (b), $P_e = 10 \times 10^3\left(1 + 0.67 \times \dfrac{12}{49.5}\right) = 1.624 \times 10^3 \ \text{1bs}$

Step 2

For $P_e = 11.62 \times 10^3 \ \text{1b}, \quad n_h = \dfrac{1631 \times 10^3}{11.624 \times 10^3} = 140 \ \text{1b/in}^3$

As in Step 1 $T = 51 \ \text{ins}, P_e = 12.32 \times 10^3 \ \text{lbs}$

Step 3

Continue Step 1 and Step 2 until convergence is reached in the values of T and P_e. The final values obtained for $P_t = 10 \times 10^3$ lb are $T = 51.6$ in, and $P_e = 12.32 \times 10^3$ lbs

Step 4

The ground line deflection may be obtained from

$$y_g = \frac{2.43 P_e T^3}{EI} = \frac{2.43 \times 12.32 \times 10^3 \times (51.6)^3}{4.854 \times 10^{10}} = 0.0845 \ \text{in}$$

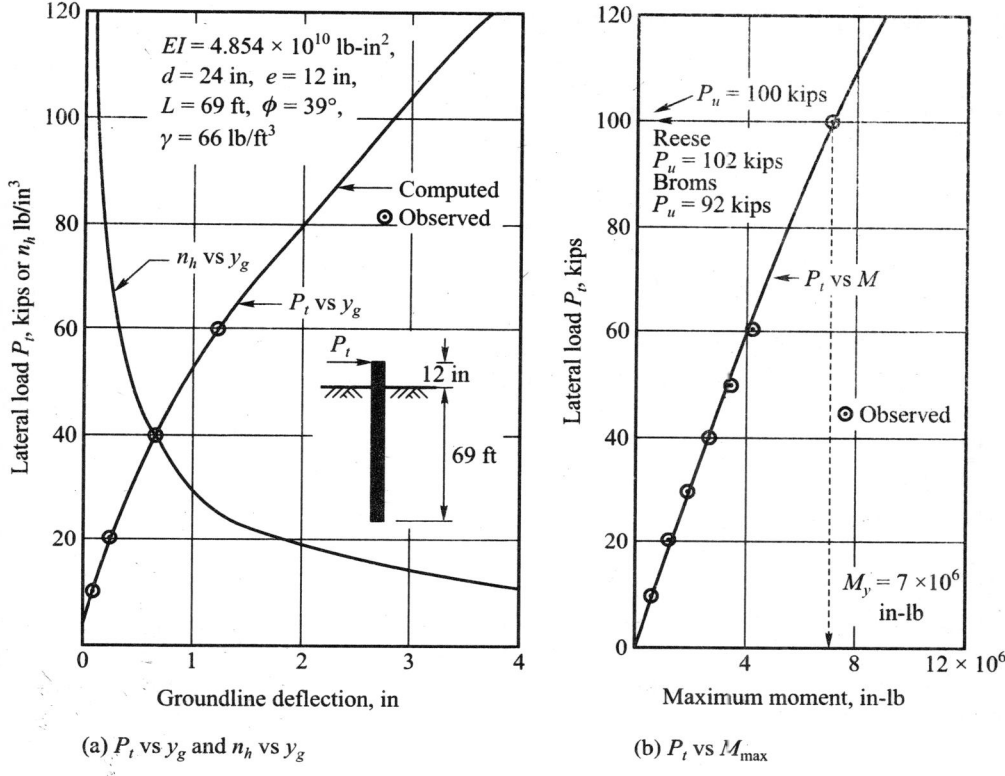

$EI = 4.854 \times 10^{10}$ lb-in^2,
$d = 24$ in, $e = 12$ in,
$L = 69$ ft, $\phi = 39°$,
$\gamma = 66$ lb/ft^3

Computed
⊙ Observed

n_h vs y_g

P_t vs y_g

P_t

12 in

69 ft

(a) P_t vs y_g and n_h vs y_g

$P_u = 100$ kips

Reese
$P_u = 102$ kips
Broms
$P_u = 92$ kips

P_t vs M

⊙ Observed

$M_y = 7 \times 10^6$
in-lb

(b) P_t vs M_{max}

Fig. 19.31 Mustang Island lateral load test

This deflection is for $P_t = 10 \times 10^3$ lbs. In the same way the values of y_g can be obtained for different stages of loadings. Fig. 19.31(a) gives a plot P_t vs. y_g. Since n_h is known at each stage of loading, a curve of n_h vs. y_g can be plotted as shown in the same figure.

b) Maximum Moment

The calculations under (a) above give the values of T for various loads P_t. By making use of Eq. (19.45) and Table 19.5, moment distribution along the pile for various loads P_t can be calculated. From these curves the maximum moments may be obtained and a curve of P_t vs. M_{max} may be plotted as shown in Fig. 19.31b.

c) Ultimate Load P_u

Figure 19.31(b) is a plot of M_{max} vs. P_t. From this figure, the value of P_u is equal to 100 kips for the ultimate pile moment resistance of 7×10^6 in-lb. The value obtained by Broms' method and by computer (Reese, 1986) are 92 and 102 kips respectively

Comments

Figure 19.31a gives the computed P_t vs. y curve by the direct approach method and the observed values. There is an excellent agreement between the two. In the same way the observed and the calculated moments and ultimate loads agree well.

Case 2: Florida Pile Load Test (Davis, 1977)

Data

Pile diameter, d = 56 in steel tube filled with concrete

EI = 132.5×10^{10} lb-in²

L = 26 ft

e = 51 ft

ϕ = 38°,

γ = 60 lb/ft³

M_y = 4630 ft-kips.

The soil at the site was medium dense and with water table close to the ground surface.

Required

(a) P_t vs. y_g curve and n_h vs. y_g curve

(b) Ultimate lateral load P_u

Solution

The same procedure as given for the Mustang Island load test has been followed for calculating the P_t vs. y_g and n_h vs. y_g curves. For getting the ultimate load P_u the P_t vs. M_{max} curve is obtained. The value of P_u obtained is equal to 84 kips which is the same as the ones obtained by Broms (1964) and Reese (1985) methods. There is a very close agreement between the computed and the observed test results as shown in Fig. 19.32.

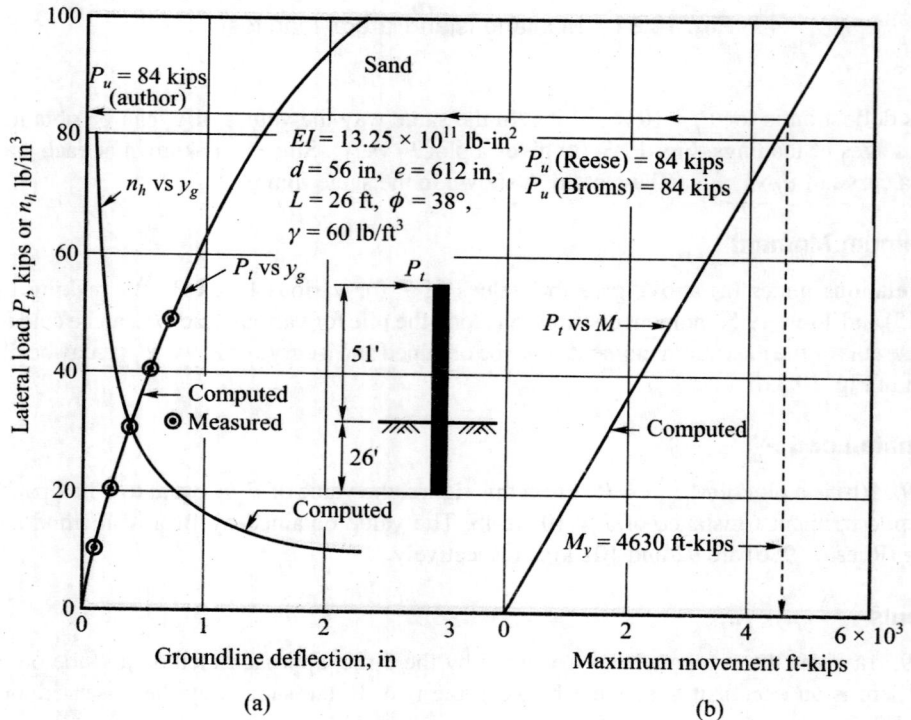

Fig. 19.32 Florida pile test (Davis, 1977)

ase 3: Model Pile Tests in Sand (Murthy, 1965)

ata

odel pile tests were carried out to determine the behaviour of vertical piles subjected to lateral ads. Aluminum alloy tubings, 0.75 in diameter and 0.035 in wall thickness, were used for the test. the test piles were instrumented. Dry clean sand was used for the test at a relative density of 67%. the other details are given in Fig. 19.33.

olution

igure 19.33 gives the predicted and observed

(a)   load-ground line deflection curve

(b)   deflection distribution curves along the pile

(c)   moment and soil reaction curves along the pile

There is an excellent agreement between the predicted and the observed values. The simplified method has been used.

9.28 CASE STUDIES FOR LATERALLY LOADED VERTICAL PILES IN CLAY

Case 1: Pile load test at St. Gabriel (Capazzoli, 1968)

Data

$$
\begin{aligned}
\text{Pile diameter, } d &= \text{10 in, steel pipe filled with concrete} \\
EI &= 38 \times 10^8 \text{ lb-in}^2 \\
L &= 115 \text{ ft} \\
e &= 12 \text{ in.} \\
c &= 600 \text{ lb/ft}^2 \\
\gamma &= 110 \text{ lb/ft}^3 \\
M_y &= 116 \text{ ft-kips}
\end{aligned}
$$

Water table was close to the ground surface.

Required

(a) P_t vs. y_g curve

(b) the ultimate lateral load, P_u

Solution

We have,

(a) $n_h = \dfrac{125c^{1.5}\sqrt{EI\gamma d}}{(1+e/d)^{1.5}}$ (b) $P_e = P_t\left(1+0.67\dfrac{e}{T}\right)$ (c) $T = \left[\dfrac{EI}{n_h}\right]^{\frac{1}{5}}$

After substituting the known values in Eq. (a) and simplifying. we have

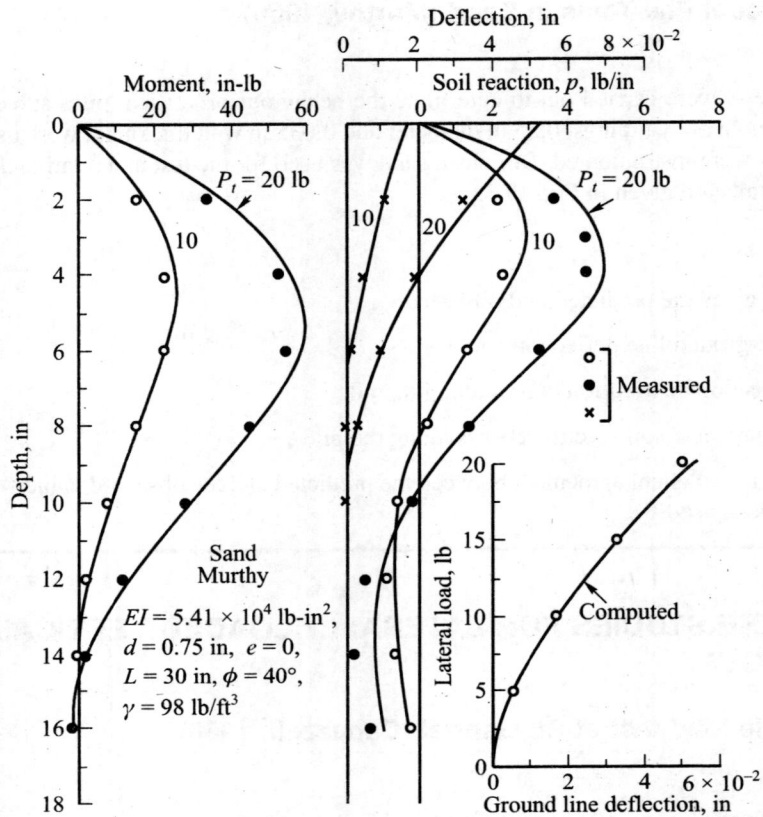

Fig. 19.33   Curves of bending moment, deflection and soil reaction for a model pile in sand (Murthy, 1965)

$$n_h = \frac{16045 \times 10^3}{P_e^{1.5}}$$

(a) Calculation of groundline deflection

1. Let $P_e = P_t = 500$ lbs

    From Eqs (a) and (c), $n_h = 45$ lb/in^3, $T = 38.51$ in

    From Eq. (d), $P_e = 6044$ lb.

2. For $P_e = 6044$ lb, $n_h = 34$ lb/in^3 and $T = 41$ in

3. For $T = 41$ in, $P_e = 5980$ lb, and $n_h = 35$ lb/in^3

4. For $n_h = 35$ lb/in^3, $T = 40.5$ in, $P_e = 5988$ lb

5. $y_g = \dfrac{2.43 P_e T^3}{EI} = \dfrac{2.43 \times 5988 \times (40.5)^3}{38 \times 10^8} = 0.25$ in

6. Continue steps 1 through 5 for computing y_g for different loads P_t. Figure 19.34 gives a plot of P_t vs. y_g which agrees very well with the measured values.

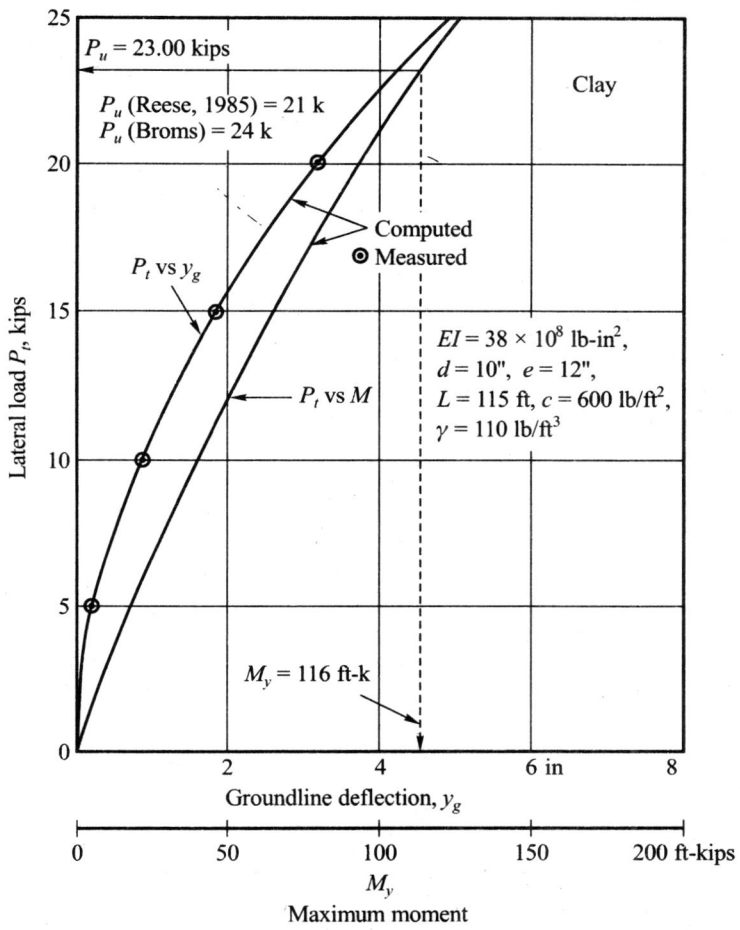

Fig. 19.34 St. Gabriel pile load test in clay

(b) Ultimate load P_u

A curve of M_{max} vs. P_t is given in Fig. 19.34 following the procedure given for the Mustang Island Test. From this curve $P_u = 23$ k for $M_y = 116$ ft kips. This agrees well with the values obtained by the methods of Reese (1985) and Broms (1964a).

Case 2: Pile Load Test at Ontario (Ismael and Klym, 1977)

Data

$$
\begin{aligned}
\text{Pile diameter, } d &= 60 \text{ in, concrete pile (Test pile 38)} \\
EI &= 93 \times 10^{10} \text{ lb-in}^2 \\
L &= 38 \text{ ft} \\
e &= 12 \text{ in.} \\
c &= 2000 \text{ lb/ft}^2 \\
\gamma &= 60 \text{ lb/ft}^3
\end{aligned}
$$

The soil at the site was heavily overconsolidated

Required

(a) P_t vs. y_g curve

(b) n_h vs. y_g curve

Solution

By substituting the known quantities in Eq. (19.56) and simplifying,

$$n_h = \frac{68495 \times 10^5}{P_e^{1.5}}, \quad T = \left(\frac{EI}{n_h}\right)^{\frac{1}{5}}, \quad \text{and} \quad P_e = P_t\left(1 + 0.67\frac{e}{T}\right)$$

Follow the same procedure as given for Case 1 to obtain values of y_g for the various loads P_t. The load deflection curve can be obtained from the calculated values as shown in Fig. 19.35. The measured values are also plotted. It is clear from the curve that there is a very close agreement between the two. The figure also gives the relationship between n_h and y_g.

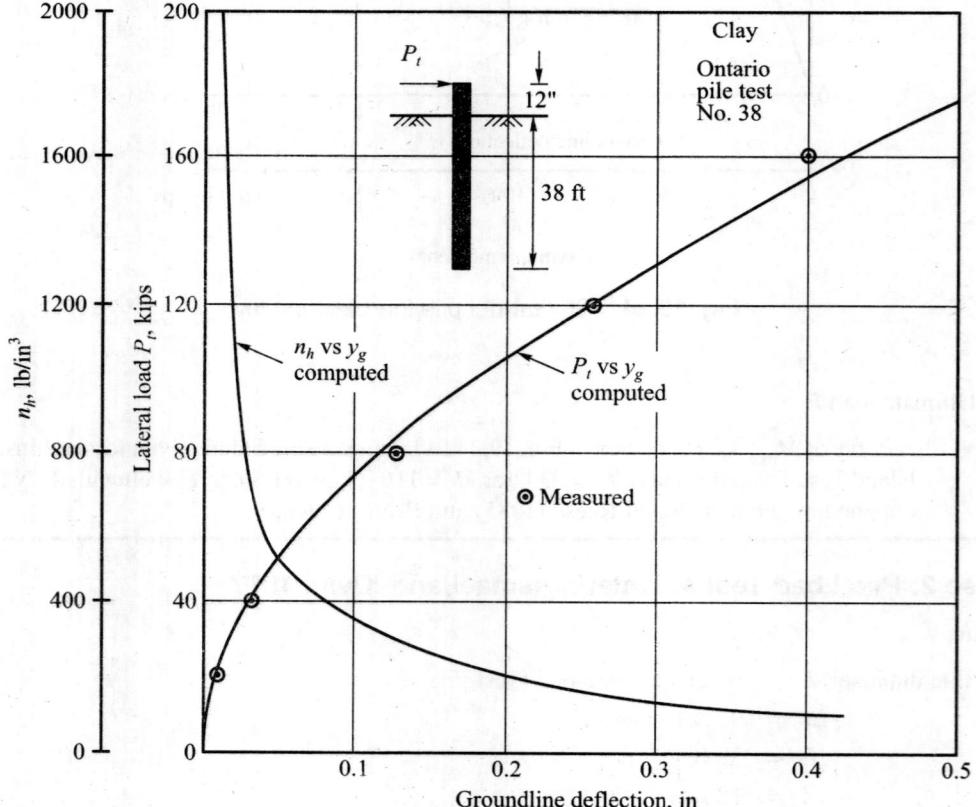

Fig. 19.35   Ontario pile load test (38)

Case 3: Restrained Pile at the Head for Offshore Structure (Matlock and Reese, 1961)

Data

The data for the problem are taken from Matlock and Reese (1961). The pile is restrained at the head by the structure on the top of the pile. The pile considered is below the sea bed. The undrained shear strength c and submerged unit weights are obtained by working back from the known values of n_h and T. The other details are

$$
\begin{aligned}
\text{Pile diameter, } d &= 33 \text{ in, pipe pile} \\
EI &= 42.35 \times 10^{10} \text{ lb-in}^2 \\
c &= 500 \text{ lb/ft}^2 \\
\gamma &= 40 \text{ lb/ft}^3 \\
P_t &= 150,000 \text{ lbs}
\end{aligned}
$$

$$
(a) \; \frac{M_t}{P_t T} = \frac{-T}{12.25 + 1.078T}, \quad (b) \; T = \left[\frac{EI}{n_h} \right]^{\frac{1}{5}}, \quad (c) \; P_e = P_t \left(1 - 0.67 \frac{e}{T} \right)
$$

Required

  (a) deflection at the pile head

  (b) moment distribution diagram

Solution

Substituting the known values in Eq (19.56) and simplifying,

$$
(d) \quad n_h = \frac{458 \times 10^6}{P_e^{1.5} (1 + e/d)^{1.5}}
$$

Calculations

1. Assume $e = 0$, $P_e = P_t = 150,000$ lb

   From Eqs (d) and (b) $n_b = 7.9$ lb / in^2, $T = 140$ in

   From Eq. (a) $\dfrac{M_y}{P_t T} = \dfrac{-140}{12.25 + 1.078 \times 140} = -0.858$

   or $M_t = -0.858 \; P_t T = P_t e$

   Therefore $e = 0.858 \times 140 = 120$ in

$$
P_e = P_t \left(1 - 0.67 \frac{e}{T} \right) = 1.5 \times 10^5 \left(1 - 0.67 \times \frac{120}{140} \right) = 63,857 \text{ lb}
$$

$$
\left(1 + \frac{e}{d} \right)^{1.5} = \left(1 + \frac{120}{33} \right)^{1.5} = 10
$$

   Now from Eq. (d), $n_h = 2.84$ lb/in^3, from Eq. (b) $T = 171.64$ in

   After substitution in Eq. (a)

$$\frac{M_t}{P_t T} = -0.875 \text{, and } e = 0.875 \times 171.64 = 150.2 \text{ in}$$

$$P_e = \left(1 - 0.67 \times \frac{150.2}{171.64}\right) \times 1.5 \times 10^5 = 62,205 \text{ lbs}$$

3.   Continuing this procress for a few more steps there will be convergence of values of n_h, T and P_e. The final values obtained are

$n_h = 2.1 \text{ lb / in}^3$, $T = 182.4 \text{ in}$, and $P_e = 62,246 \text{ lb}$

$M_t = -P_e e = -150,000 \times 150.2 = -22.53 \times 10^6 \text{ lb - in}^2$

$$y_g = \frac{2.43 P_e T^3}{EI} = \frac{2.43 \times 62,246 \times (182.4)^3}{42.35 \times 10^{10}} = 2.17 \text{ in}$$

Moment distribution along the pile may now be calculated by making use of Eq. (19.45) and Table 19.5. Please note that M_t has a negative sign. The moment distribution curve is given in Fig. 19.36. There is a very close agreement between the computed values by direct method and the Reese and Matlock method. The deflection and the negative bending moment as obtained by Reese and Matlock are

$y_m = 2.307 \text{ in and } M_t = -24.75 \times 10^6 \text{ lb-in}^2$

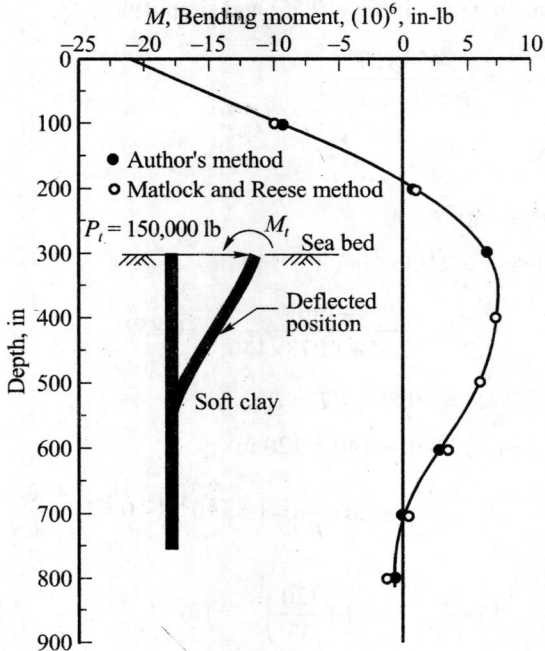

Fig. 19.36   Bending moment distribution for an offshore pile supported structure
(Matlock and Reese, 1961)

19.29 BEHAVIOUR OF LATERALLY LOADED BATTER PILES IN SAND

General Considerations

The earlier sections dealt with the behaviour of long vertical piles. The author has so far not come across any rational approach for predicting the behaviour of batter piles subjected to lateral loads. He has been working on this problem for a long time (Murthy, 1965). Based on the work done by the author and others, a method for predicting the behaviour of long batter piles subjected to lateral load has now been developed.

Model Tests on Piles in Sand (Murthy, 1965)

A series of seven instrumented model piles were tested in sand with batters varying from 0 to ±45°. Aluminum alloy tubings of 0.75 in outside diameter and 30 in long were used for the tests. Electrical resistance gauges were used to measure the flexural strains at intervals along the piles at different load levels. The maximum load applied was 20 lbs. The pile had a flexural rigidity $EI = 5.14 \times 10^4$ lb-in^2. The tests were conducted in dry sand, having a unit weight of 98 lb/ft^3 and angle of friction ϕ equal to 40°. Two series of tests were conducted-one series with loads horizontal and the other with loads normal to the axis of the pile. The batters used were 0°, ± 15°, ±30° and ±45°. Pile movements at ground level were measured with sensitive dial gauges. Flexural strains were converted to moments. Successive integration gave slopes and deflections and successive differentiations gave shears and soil reactions respectively. A very high degree of accuracy was maintained throughout the tests. Based on the test results a relationship was established between the n_h^b values of batter piles and n_h^o

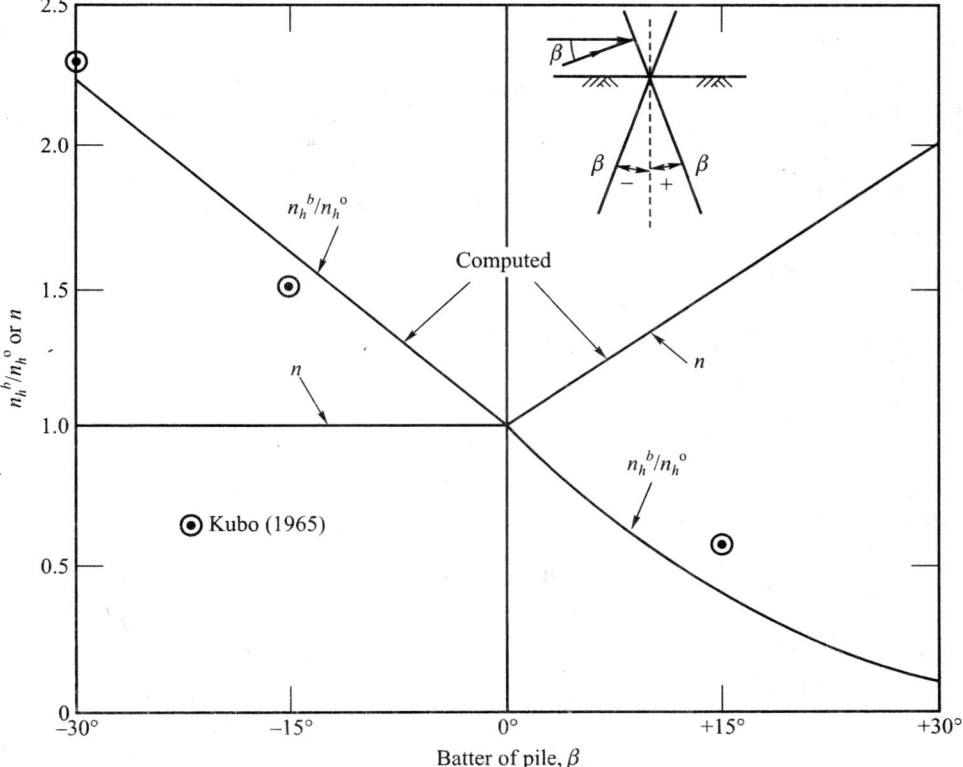

Fig. 19.37 Effect of batter on n_h^b / n_h^o and n (after Murthy, 1965)

values of vertical piles. Fig. 19.37 gives this relationship between n_h^b / n_h^o and the angle of batter β. It is clear from this figure that the ratio increases from a minimum of 0.1 for a positive 30° batter pile to a maximum of 2.2 for a negative 30° batter pile. The values obtained by Kubo (1965) are also shown in this figure. There is close agreement between the two.

The other important factor in the prediction is the value of n in Eq. (19.42a). The values obtained from the experimental test results are also given in Fig. 19.37. The values of n are equal to unity for vertical and negative batter piles and increase linearly for positive batter piles up to a maximum of 2.0 at + 30° batter.

In the case of batter piles the loads and deflections are considered normal to the pile axis for the purpose of analysis. The corresponding loads and deflections in the horizontal direction may be written as

$$\overline{P}_t(\text{Hor}) = \frac{P_t(Nor)}{\cos\beta} \tag{19.58a}$$

$$\overline{y}_g(\text{Hor}) = \frac{y_g(Nor)}{\cos\beta} \tag{19.58b}$$

where P_t and y_g are normal to the pile axis; $\overline{P}_t(\text{Hor})$ and $\overline{y}_g(\text{Hor})$ are the corresponding horizontal components.

Application of the Use of n_h^b / n_h^o and n

It is possible now to predict the non-linear behaviour of laterally loaded batter piles in the same way as for vertical piles by making use of the ratio n_h^b / n_h^o and the value of n. The validity of this method is explained by considering a few case studies.

CASE STUDIES

Case 1: Model Pile Test (Murthy, 1965).

Piles of +15° and +30° batters have been used here to predict the P_t vs. y_g and P_t vs. M_{max} relationships. The properties of the pile and soil are given below.

$EI = 5.14 \times 10^4$ lb in^2, $d = 0.75$ in, $L = 30$ in; $e = 0$

$\gamma = 98$ lb/ft^3 and $\phi = 40°$

For $\phi = 40°$, $C_\phi = 1.767$ [= 3×10–5 $(1.316)^\phi$]

From Eq. (19.55), $n_h^o = \dfrac{150C_\phi \gamma^{1.5} \sqrt{EId}}{P_t}$

After substituting the known values and simplifying we have

$$n_h^o = \frac{700}{P_t}$$

Solution: +15° batter pile

From Fig. 19.37 $n_h^b / n_h^o = 0.4$, $n = 1.5$

From Eq. (19.48b), $T_b = \left[\dfrac{EI}{n_h^b}\right]^{\frac{1}{1.5+4}} = 5.33\left[\dfrac{5.14}{n_h^b}\right]^{0.1818}$

Calculations of Deflection y_g

For $P_t = 5$ lbs, $n_h^b = 141$ lbs/in^3, $n_h^o = 141 \times 0.4 = 56$ lb/in^3 and $T_b = 3.5$ in

$$y_g = \frac{2.43 P_t^3 T_b^3}{5.14 \times 10^4} = 0.97 \times 10^{-2} \text{ in}$$

Similarly, y_g can be calculated for $P_t = 10$, 15 and 20 lbs.

The results are plotted in Fig. 19.38 along with the measured values of y_g. There is a close agreement between the two.

Calculation of Maximum Moment, M_{max}

For $P_t = 5$ lb, $T_b = 3.5$ in, The equation for M is

$$M = A_m P_t T_b = 0.77 P_t T_b$$

where $A_m = 0.77$ (max) from Table 19.5

By substituting and calculating, we have

$$M_{(max)} = 13.5 \text{ in-lb}$$

Similarly $M_{(max)}$ can be calculated for other loads. The results are plotted in Fig. 19.38 along with the measured values of $M_{(max)}$. There is very close agreement between the two.

+30° Batter Pile

From Fig 19.27, $n_h^b / n_h^o = 0.1$, and $n = 2$; $T_b = \left[\dfrac{EI}{n_h^b} \right]^{\frac{1}{2+4}} = 4.64 \left[\dfrac{5.14}{n_h^b} \right]^{0.1667}$, $n_h^o = \dfrac{700}{P_t}$

For $P_t = 5$ lbs, $n_h^o = 141$ lbs/in^3, $n_h^b = 0.1 \times 141 = 14.1$ lb/in^3, $T_b = 3.93$ in.

For $P_t = 5$ lbs, $T_b = 3.93$ in, we have, $y_g = 1.43 \times 10^{-2}$ in

As before, $M_{(max)} = 0.77 \times 5 \times 3.93 = 15$ in-lb.

The values of y_g and $M_{(max)}$ for other loads can be calculated in the same way. Fig. 19.38 gives P_t vs. y_g and P_t vs. $M_{(max)}$ along with measured values. There is close agreement up to about

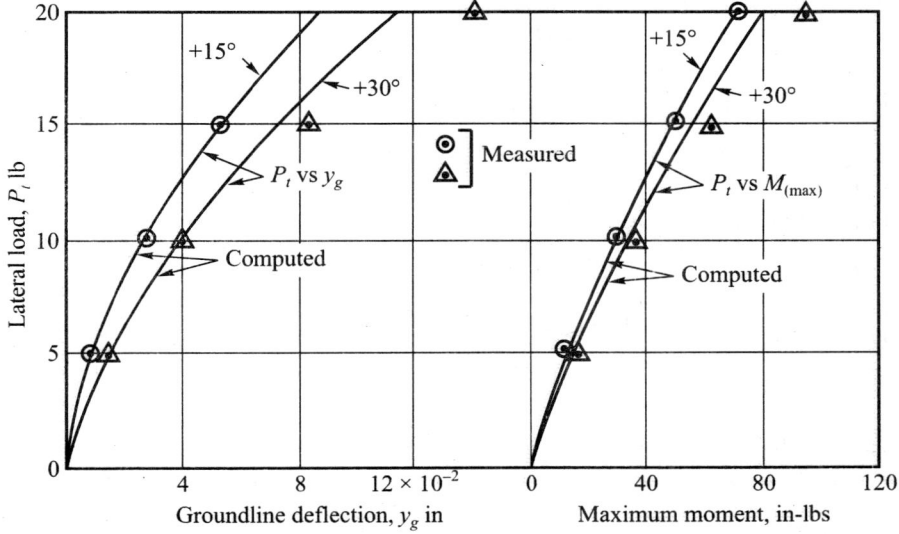

Fig. 19.38 Model piles of batter +15° and +30° (Murthy, 1965)

$P_t = 10$ lb, and beyond this load, the measured values are greater than the predicted by about 25 percent which is expected since the soil yields at a load higher than 10 lb at this batter and there is a plastic flow beyond this load.

Case 2: Arkansas River Project (Pile 12) (Alizadeh and Davisson, 1970)

Given

$EI = 278.5 \times 10^8$ lb –in^2, $d = 14$ in, $e = 0$.

$\phi = 41°$, $\gamma = 63$ lb/ft^3, $\beta = 18.4°$ (–ve)

$C_\phi = 2.33$, from Fig. 19.37 $n_h^b / n_h^o = 1.7$, $n = 1.0$

From Eq. (19.55), after substituting the known values and simplifying, we have,

(a) $n_h^o = \dfrac{1528 \times 10^3}{P_t}$, and (b) $T_b = 39.8 \left[\dfrac{278.5}{n_h^b} \right]^{0.2}$

Calculation for $P_t = 12.6^k$

From Eq. (a), $n_h^o = 121$ lb/in^3; now $n_h^b = 1.7 \times 121 = 206$ lb/in^3

From Eq. (b), $T_b = 42.27$ in

$$y_g = \frac{2.43 \times 12,600(42.27)^3}{278.5 \times 10^8} = 0.083 \text{ in}$$

$M_{(max)} = 0.77\ P_t T = 0.77 \times 12.6 \times 3.52 = 34$ ft-kips.

The values of y_g and $M_{(max)}$ for $P_t = 24.1k$, $35.5k$, $42.0k$, $53.5k$, $60k$ can be calculated in the same way the results are plotted the Fig. 19.39 along with the measured values. There is a very close agreement between the computed and measured values of y_g but the computed values of M_{max} are

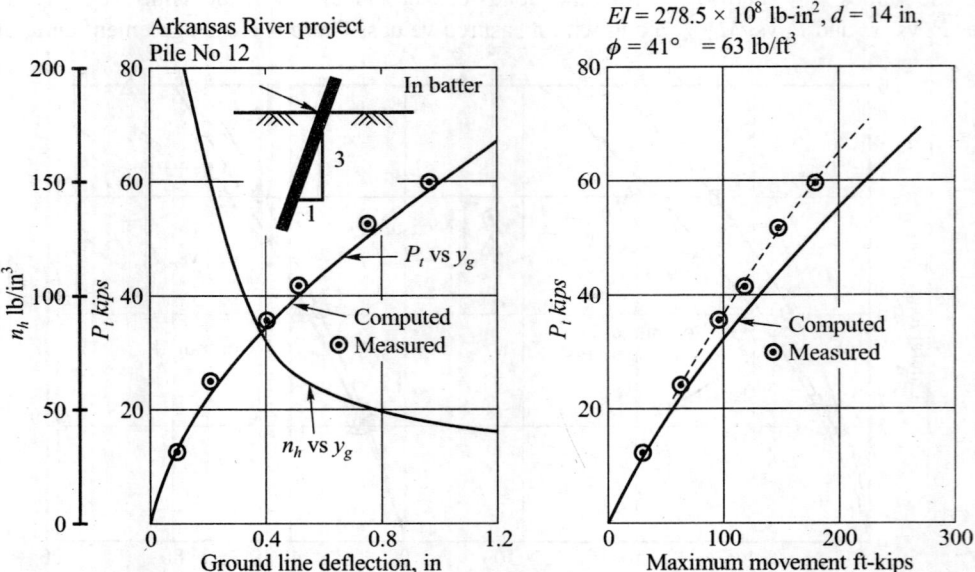

Fig. 19.39 Lateral load test-batter pile 12-Arkansas River Project (Alizadeh and Davisson, 1970)

higher than the measured values at higher loads. At a load of 60 kips, $M_{(max)}$ is higher than the measured by about 23% which is quite reasonable.

Case 3: Arkansas River Project (Pile 13) (Alizadeh and Davisson, 1970).

Given

$EI = 288 \times 108$ lb-ins, $d = 14"$, $e = 6$ in.

$\gamma = 63$ lbs/ft3, $\phi = 41°$ $(C_\phi = 2.33)$

$\beta = 18.4°$ (+ve), $n = 1.6$, $n_h^b / n_h^o = 0.3$

$$T_b = \left[\frac{EI}{n_h^b}\right]^{\frac{1}{1.6+4}} = 27\left[\frac{288}{n_h^b}\right]^{0.1786}$$

After substituting the known values in the equation for n_h^o [Eq. (19.55)] and simplifying, we have

$$n_h^o = \frac{1597 \times 10^3}{P_e}$$

Calculations for y_g for $P_t = 141.4k$

1. From Eq (b), $n_h^o = 39$ lb/in³, hence $n_h^b = 0.3 \times 39 = 11.7$ lb/in³
    From Eq. (a), $T_b \approx 48$ in.

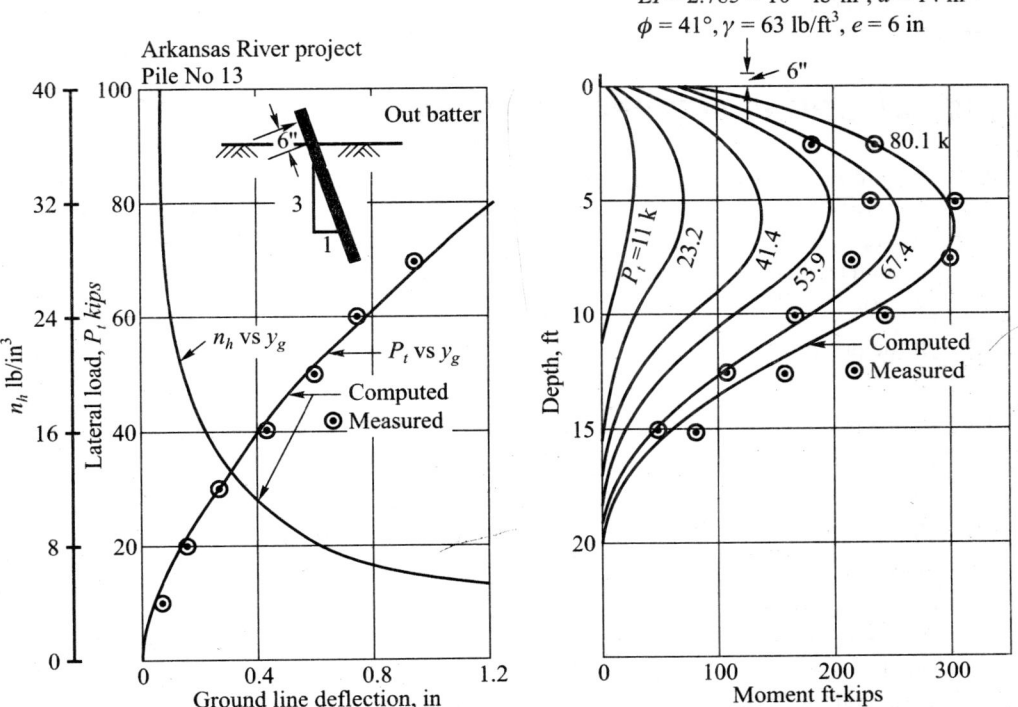

$EI = 2.785 \times 10^{10}$ lb-in², $d = 14$ in
$\phi = 41°, \gamma = 63$ lb/ft³, $e = 6$ in

Fig. 19.40 Lateral load test-batter pile 13-Arkansas River Project (Alizadeh and Davisson, 1970)

2. $P_e = P_t\left(1 + 0.67\dfrac{e}{T}\right) = 41.4\left(1 + 0.67 \times \dfrac{6}{48}\right) = 44.86 \text{ kips}$

3. For $P_e = 44.86$ kips, $n_h^o = 36$ lb / in^3, and $n_h^b = 11$ lb/in^3, $T_b = 48$ in

4. Final values: $P_e = 44.86$ kips, $n_h^b = 11$ lb / in^3, and $T_b = 48$ in

5. $y_g = \dfrac{2.43 P_e T_b^3}{EI} = \dfrac{2.43 \times 44.860 \times (48)^3}{288 \times 10^8} = 0.42 \text{ in}$

6. Follow Steps 1 through 5 for other loads. Computed and measured values of y are plotted in Fig. 19.40 and there is a very close agreement between the two. The n_h values against y_g are also plotted in the same figure.

Calculation of Moment Distribution

The moment at any distance x along the pile may be calculated by the equation

$$M = [P_t T]A_m + [M_t]B_m$$

As per the calculations shown above, the value of T will be known for any lateral load level P_t. This means $[P_t T]$ will be known. The values of A_m and B_m are functions of the depth coefficient Z which can be taken from Table 19.5 for the distance $x(Z = x/T)$. The moment at distance x will be known from the above equation. In the same way moments may be calculated for other distances. The same procedure is followed for other load levels. Fig. 19.40 gives the computed moment distribution along the pile axis. The measured values of M are shown for two load levels $P_t = 67.4$ and 80.1 kips. The agreement between the measured and the computed values is very good.

19.30 QUESTIONS AND PROBLEMS: PART D

19.1   A reinforced concrete pile 50 cm square in section is driven into a medium dense sand to a depth of 20 m . The sand is in a submerged state. A lateral load of 50 kN is applied on the pile at a height of 5 m above the ground level. Compute the lateral deflection of the pile at ground level. Given: $n_h = 15$ MN/m^3, $EI = 115 \times 10^9$ kg-cm^2. The submerged unit weight of the soil is 8.75 kN/m^3.

19.2   If the pile given in Prob. 19.1 is fully restrained at the top, what is the deflection at ground level?

19.3   If the pile given in Prob. 19.1 is 3 m long, what will be the deflection at ground level (a) when the top of the pile is free, and (b) when the top of the pile is restrained? Use Broms' method.

19.4   Refer to Prob. 19.1. Determine the ultimate lateral resistance of the pile by Broms' method. Use $\phi = 38°$. Assume $M_y = 250$ kN-m.

19.5   If the pile given in Prob. 19.1 is driven into saturated normally consolidated clay having an unconfined compressive strength of 70 kPa, what would be the ultimate lateral resistance of the soil under (a) a free-head condition, and (b) a fixed-condition? Make necessary assumptions for the yield strength of the material.

19.6   Refer to Prob. 19.1. Determine the lateral deflection of the pile at ground level by the simplified method. Assume $M_y = 250$ kN-m and $e = 0$.

19.7   Refer to Prob. 19.1. Determine the ultimate lateral resistance of the pile by the simplified method.

19.8   A precast reinforced concrete pile of 30 cm diameter is driven to a depth of 10 m in a vertical direction into a medium dense sand which is in a semi-dry state. The value of the coefficient of soil modulus variation (n_h) may be assumed as equal to 0.8 kg/cm^3. A lateral load of 40 kN is applied at a height of 3 m above ground level. Compute (a) the deflection at ground level, and (b) the maximum bending moment on the pile (assume $E = 2.1 \times 10^5$ kg/cm^2).

19.9   Refer to Prob. 19.8. Solve the problem by the direct method. All the other data remain the same. Assume $\phi = 38°$ and $\gamma = 16.5$ kN/m^3.

19.10   If the pile in Prob. 19.9 is driven at a batter of 22.5° to the vertical, and lateral load is applied at ground level, compute the normal deflection at ground level, for the cases of the load acting in the direction of batter and against the batter.

CHAPTER 20

CAISSON FOUNDATIONS

20.1 INTRODUCTION

Wells, which are also known as caissons, have been in use for foundations of bridges and other important structures since the Roman and Moghul periods in India. Moghuls in particular used wells for the foundations of their monuments, including Taj Mahal, which is a standing testimony to the skills of mankind in the earlier days. In modern times, however, one of the earliest use in India is that for an aqueduct for the upper Ganges Canal constructed in the earlier part of the 19th century. With the advent of pneumatic sinking in 1850 A.D., and discovery of better materials like reinforced concrete and steel, use of wells as foundations of bridges gained popularity.

Caisson foundations have been used for most of the major bridges in India. Materials commonly used for construction are reinforced concrete, brick or stone masonry. Use of well or caisson foundations is equally popular in the United States of America and other western countries. The size of caisson used for the San Francisco Oakland Bridge is 29.6 × 60.1 m in section and 74 m depth.

20.2 TYPES OF CAISSONS

There are three types of caissons. They are:

1. Open caissons.

2. Pneumatic caissons.

3. Box caissons.

Open Caissons

The top and bottom of the caisson (Fig. 20.1) is open during construction. They may have any shape in plan as round, oblong, oval, rectangular etc. They are of cellular construction and the provision of cells reduces the cost of construction. The open-end caisson usually has a cutting edge. The cutting edge is first fabricated at rue site and the first segment of the shaft is built on it. The soil inside the

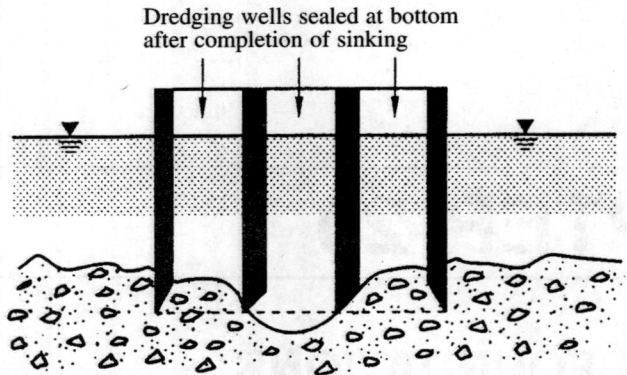

Fig. 20.1   Open caisson

shaft is removed by grab buckets and the segment is sunk vertically. Another segment is added to the top and the process of sinking is continued by excavating the soil inside. After the required depth is reached, concrete is placed under water on the open bottom as a seal to a depth that will contain the hydrostatic uplift pressure so as to avoid blowing in of the bottom when the water inside the caisson is pumped out.

When the concrete seal is completely cured, the water in the caisson can be pumped out.

Advantages of open caissons are:

1.   The caisson can be constructed to great depths.

2.   The construction cost is relatively low.

Disadvantages are:

1.   The clearing and inspection of bottom of the caisson cannot be done.

2.   Concrete seal placed in water will not be satisfactory.

3.   The rate of progress will be slowed down if boulders are met during construction.

Pneumatic Caissons

In the case of pneumatic caissons (Fig. 20.2), the working chamber at the bottom of the caisson is kept dry by forcing out water under air pressure. Air locks are provided at the top. The caisson is sunk as the excavation proceeds. Upon reaching its final depth, the working chamber is filled with concrete.

Advantages of pneumatic caisson are:

1.   Control over the work and preparation of foundation for the sinking of caisson are better since the work is done in the dry.

2.   The caisson can be sunk vertically as careful supervision is possible.

3.   The bottom of the chamber can be sealed effectively with concrete as it can be placed dry.

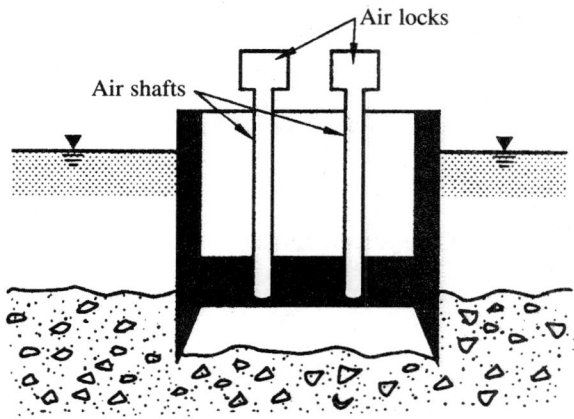

Fig. 20.2 Pneumatic caisson

4. Obstruction to sinking, such as boulders, etc. can be removed easily.

Disadvantages are:

1. Construction cost is quite high.

2. The depth of penetration below water is limited to about 35 m (3.5 kg/cm²). Higher pressures are beyond the endurance of the human body,

Box Caissons

In the case of box caissons (Fig. 20.3) the bottom is closed. This type of caisson is first cast on land and then towed to the site and then sunk on to a previously levelled foundation base, It is sunk by filling inside with sand, gravel, concrete or water. The box type of caisson is also called as floating caisson.

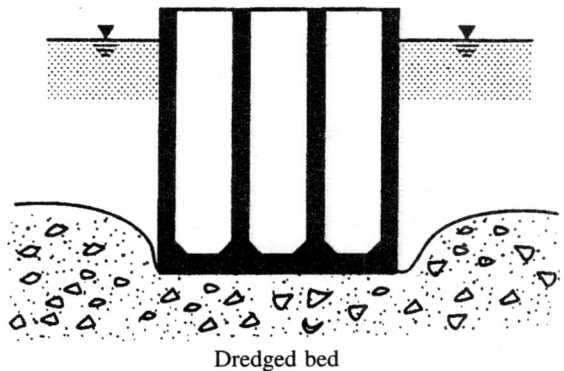

Dredged bed

Fig. 20.3 Box caisson

Advantages of box caissons are:

1. The cost of construction is relatively low.

2. It can be used where the construction of other types of caissons are not posible.

Disadvantages are:

1. The foundation base shall be prepared in advance of sinking.

2. Deep excavations for seating the caissons at the required depth is very difficult below
    water level.

3. Due care has to be taken to protect the foundation from scour.

4. The bearing capacity of the base should be assessed in advance.

20.3 STABILITY ANALYSIS OF CAISSON FOUNDATIONS

Introduction

Two types of soils are considered in the stability analysis of caisson (well) foundations. They are
cohesionless and cohesive soils. This chapter deals with the lateral stability of well foundation only.
The vertical bearing capacity of deep foundations dealt with in Chapter 19 is also applicable to well
foundations, and as such this aspect of the problem is not considered here.

Statement of the Problem

A caisson foundation used for a bridge pier shall carry both vertical and lateral loads. Vertical loads
comprise of dead and live loads. The dead loads include the weights of superstructure and substructure.
The vertical line loads are brought on to the structure due to the passing of vehicles over the bridge.
The lateral loads are caused due to braking or traction of vehicles, water current, wind, earth quakes
etc. The lateral forces might act at different points on a pier, but their effect can be simulated by
considering an equivalent force acting at bearing level.

Figure 20.4 shows a typical rectangular well foundation with all the external loads and the
resisting forces acting on the well in cohesionless soil. The external loads are,

$W_T =$ the vertical load at the bearing level of pier which includes loads of superstructure
(excluding the pier) and the live loads acting on it,

$W_s =$ weight of pier and well (considering relief due to buoyancy),

$P_u =$ equivalent lateral load acting at the bearing level at height H above the maximum scour
level under ultimate lateral load.

$[P_u = F_l P_l$ where, $F_l =$ load factor, $P_l =$ design lateral load].

The external forces are resisted by the soil surrounding the well. Since the well is a massive one
with depth/width ratio (D/B) normally not exceeding a value of 3, it is assumed to rotate as a rigid
body about a point O lying on the base of the well on the axis passing through well. When the well
rotates as a unit passive pressure develops in the front and active pressure at the back, and the active
pressure is normally neglected in the analysis since it is quite small compared to the magnitude

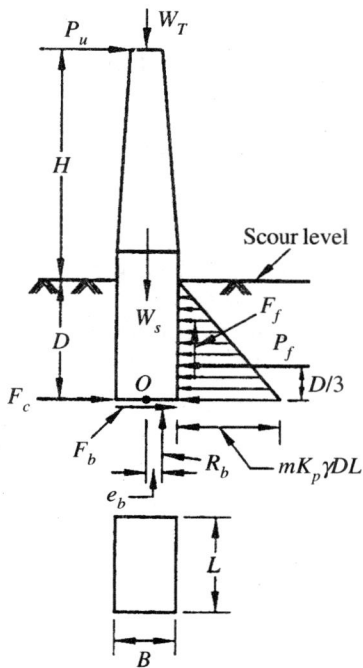

Fig. 20.4   Stability analysis of well foundation in cohesionless soils

of the passive pressure. The high lateral pressure that develops at the bottom of the back of the wall is assumed to be resisted by a line load F_c at the bottom of the well.

Two cases are considered:

1.   Stability analysis of wells in cohesionless soils.

2.   Stability analysis of wells in cohesive soils.

The principle of analysis in both the cases are based on the principles enunciated by Broms (1964) for short piles.

In both the cases it is required to determine the depth of embedment of the well (grip length) under ultimate lateral load conditions. With a suitable factor of safety, the movement or rotation of the well at the bearing level should be within the permissible limits.

The analysis is based on limit equilibrium conditions. Some experimental work has been done by Kapur on model wells (1971) in cohesionless soil on the stability analysis of wells. But there is no way of checking the validity of these methods since field tests are not available to verify his findings. However, some of his findings are briefly mentioned in this chapter. He has also developed equations based on dimensional analysis to compute slopes of wells at working loads in cohesionless soils.

Limit Equilibrium method of Determining the Grip Length of Wells in Cohesionless Soils

Experimental works of Kapur (1971) on instrumented model blocks indicate that the passive pressure distribution on the front face of the blocks is parabolic, however, under limit equilibrium condition, it is reasonable to assume that the passive pressure increases linearly with depth. It has also been found out that the passive earth pressure coefficients for blocks are greater than those applicable for plane strain conditions.

In order to simplify the procedure, Broms (1964) recommends that the maximum earth pressure which develops at failure may be taken as equal to three times the passive earth pressure calculated by Rankine's earth pressure theory. He has used this approach for the computation of ultimate lateral resistance of piles with D/B ratio greater than 3 (D = depth, B = diameter of piles). For rigid foundations with D/B ratio less than 3, Broms (1987), recommends as follows.

Let, $\overline{K}_p$ = three dimensional passive earth pressure coefficient,

K_p = Rankines passive earth pressure coefficient.

We may write

$$\overline{K}_p = mK_p \tag{20.1}$$

The value of m depends according to Broms (1987) on the D/B ratio, and his recommendations are:

D/B ratio	Value of m
≤ 1	1.0
$1< D/B <3$	D/B
≥ 3	3

The author feels that a value of 1 for m for heavy wells and a maximum of 1.5 for smaller diameter wells might be reasonable.

Figure 20.4 shows a typical rectangular caisson foundation with all the external loads and resisting forces acting on it.

The resisting forces are

P_f = passive earth pressure on the front face of the well,

F_f = vertical frictional force on the front face = $P_f \tan\delta$

δ = angle of wall friction,

R_b = base reaction acting at an eccentricity at e_b

F_b = base frictional resistance,

F_c = concentrated lateral force acting at the back of the well at base level.

The maximum passive earth pressure, P_D, at the base of the well per unit depth may be expressed as

$$P_D = mK_p \gamma DL \qquad (20.2)$$

The total passive pressure P_f is

$$P_f = \frac{1}{2} \gamma L m K_p D^2 \qquad (20.3)$$

The total frictional force F_f on the front face is

$$F_f = \frac{1}{2} \gamma L m \tan \delta D^2 \qquad (20.4)$$

where, γ = effective unit weight of soil.

Conditions for Statical Equilibrium

The magnitudes of all the resisting forces and their point of application are as given in Fig. 20.4. At limiting state, all the forces acting on the well should satisfy the following conditions of statical equilibrium. They are:

1. The sum of all the vertical forces must be equal to zero, i.e. $\Sigma V_f = 0$.

2. Sum of all the horizontal forces must be equal to zero, $\Sigma H_f = 0$.

3. Sum of he moments of all the forces about point 0 on the base (of any other point) must be equal to zero, i.e. $\Sigma M_f = 0$.

We may now write

1. $\Sigma V_f = W - R_b - P_f \tan \delta = 0 \qquad (20.5)$

2. $\Sigma H_f = P_u - P_f + F_b + F_c = 0 \qquad (20.6)$

3. $\Sigma M_f = M_{gu} + P_u D + \frac{1}{3} P_f D - R_b e_b = 0 \qquad (20.7)$

From Eq. (20.5), we have

$$R_b = W - P_f \tan \delta \qquad (20.8)$$

where, $W = W_T + W_s$.

The grip length can be found out from Eg. (20.7). Substituting in Eq. (20.7) for P_f and R_b and simplifying, we have an equation for D as

$$D^2 = \frac{12 \left(M_{gu} + P_u D - W e_b \right)}{L m K_p \gamma \left[2D + 3 \tan \delta (B - 2e_b) \right]} \qquad (20.9a)$$

Equation (20.9a) may be expressed as

$$D^2 = \frac{WD(\alpha_1\alpha_2 - \alpha_3)}{\gamma m K_p A(\alpha_4 + \alpha_5)} \tag{20.9b}$$

where $\alpha_1 = \dfrac{P_u}{W}$, $\alpha_2 = \dfrac{H}{D}+1$, $\alpha_3 = \dfrac{e_b}{D}$, $\alpha_4 = \dfrac{1}{6}\dfrac{D}{B}$, $\alpha_5 = \dfrac{1}{4}\tan\delta(B-2e_b)$

$A = L \times B$, $m =$ as per Eq.(20.1), $\gamma =$ effective unit weight of soil,

$\delta =$ angle of wall friction, $M_{gu} = P_u H$

The depth D from Eq. (20.9) can be obtained either by trial and error method or graphically.

Eccentricity of Base Reaction

Experimental investigations carried out by Kapur (1971) indicate that the eccentricity of base reaction e_b, expressed as a ratio e_b/B increases linearly with the lateral load P. The maximum value of e_b has been found out to be $B/3$ even at limit state of equilibrium. At an eccentricity of $B/3$, the maximum toe pressure q, has been found out to be about 3 times the average base pressure q, Figs. 20.5(a) and (b) give the findings of Kapur on eccentricity. However, an eccentricity of $B/6$ is suggested for computing the grip length.

Load Factor

When the load factor is 2, that is when $P_u/P_t = 2$ (where P_u the ultimate lateral load to be considered for computing the grip length D is equal to twice the maximum lateral load P_t that the well is likely to experience), the eccentricity of the reaction at the base as per Fig. 20.5(a) is $B/6$. The maximum toe pressure at this eccentricity is about $1.5q_a$ and there is no tension at the heel. If a factor of safety of 3 is used to determine the allowable vertical pressure q_a, the factor of safety of the toe pressure at

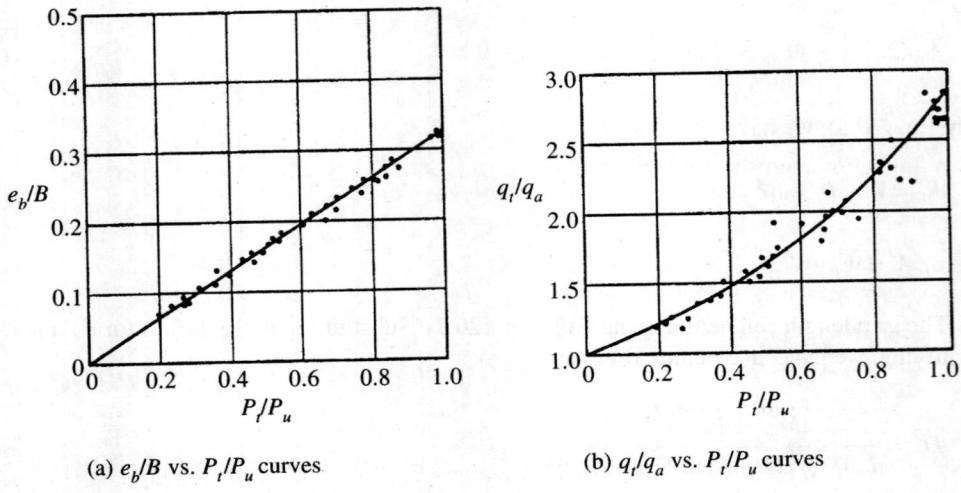

(a) e_b/B vs. P_t/P_u curves   (b) q_t/q_a vs. P_t/P_u curves

Fig. 20.5   The effect of eccentricity on toe pressure (Kapur, 1971)

a load factor of 2 is $3/1.5 = 2$ which is greater than 1.5 that is normally allowed under the maximum lateral load P_u. Load factor is not considered here for the vertical dead and live loads as these loads can be assessed fairly accurately. However, the total vertical load W may be multiplied by a suitable lo factor if required.

Shape Factor

The investigation of Kapur (1971) indicates that at equal displacements a square foundation is about 20 percent more resistant than a circular well of the same cross sectional kea. We may therefore write

$$P_{us} = 1.2 P_{uc} \tag{20.10}$$

where, P_{us} = ultimate lateral resistance of a square foundation, and P_{uc} = ultimate lateral resistance of a circular well of the same cross-sectional area.

Grip Length of Caisson in Cohesive Soils

Figure 20.6 shows the external loads and the resisting forces acting on a caisson embedded in cohesive soil. Broms (1964) assumes a maximum lateral soil reaction of $9c_u$ per unit area where, c_u is undrained shear strength, which remains constant with depth. He has assumed the soil reaction as zero upto a depth equal to 1.5 times the width of pile which is applicable only to small diameter piles. In the case of caisson foundations for bridges. the grip length is considered below the maximum scour level and as such there is no necessity to consider zero soil reaction below this depth. The forces that act on a caisson in cohesive soils are shown in Fig. 20.6.

We may write the following equations

$$P_f = 9c_u LD \tag{20.11}$$

$$F_f = \alpha c_u DL \tag{20.12}$$

$$R_b = W - F_f \tag{20.13}$$

where, c_u = undrained shear strength of soil, α = adhesions factor.

Taking moments about O we have

$$P_u D + M_{gu} = \frac{1}{2} \times 9c_u LD^2 + \frac{1}{2} \alpha c_u DLB + (W - \alpha c_u DL)e_b \tag{20.14}$$

Simplifying Eq. (20.14), we have

$$D = \frac{2(P_u D + M_{gu} - We_b)}{c_u L[9D + \alpha(B - e_b)]} \tag{20.15}$$

where, $M_{gu} = P_u H$, P_u = the ultimate lateral load.

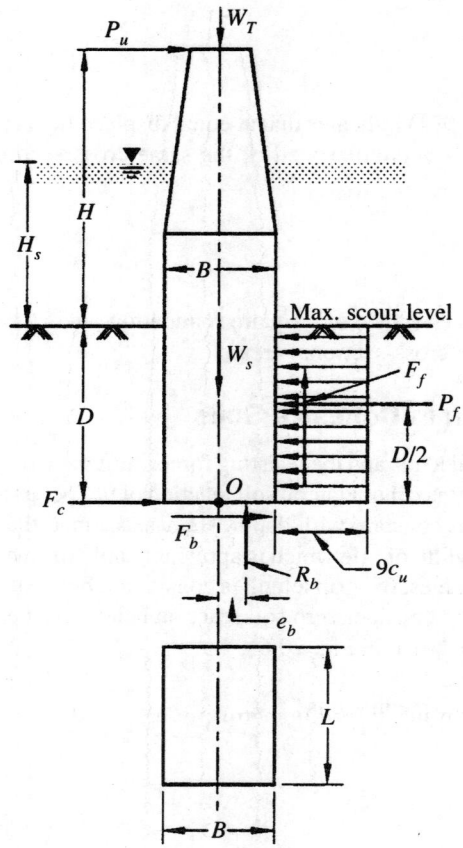

Fig. 20.6   Stability analysis of a well foundation embedded in cohesive soil

e_b may be taken as equal to $B/6$ under ultimate lateral load condition. The adhesion factor to may be selected according to consistency of the soil as explained in Chapter 19.

Equation (20.15) can be solved by choosing such value of D which balances both sides of the equation.

20.4 DETERMINATION OF SCOUR DEPTH IN COHESIONLESS SOILS

It has been stated earlier that the grip length D for the well foundation of a bridge pier is the depth below the scour level. Methods or procedures for predicting scour depth vary and one requires sound engineering judgement based on hydraulic and hydrological information, engineering geology and records of performance of adjacent structures for this purpose.

Scour that may occur at a bridge site can be categorised as follows:

1. General scour that would occur in the stream without the bridge.

2. The scour that would occur at the bridge site because of the constriction in waterway caused by the bridge and the approach embankment, and

3. The local scour that occurs because of distortion of the flow pattern in the immediate vicinity of the bridge piers and abutments.

General Scour

The depth of scour is normally measured from the high flood level (*HFL*). No reliable method is available for estimating the depth of scour. The formula that is commonly used for estimating the depth of general scour (which includes the effect of the constriction of waterway also) is the Lacey's formula which is expressed as

$$D_s = 0.473 \left(\frac{Q}{s^2 f} \right)^{1/3} \tag{20.16}$$

where, D_s = Depth of scour measured from *HFL*,

Q = maximum design discharge in cubic metre per sec,

s = $4.8\sqrt{Q/B_w}$ a factor which is limited to unity,

B_w = actual width of water spread in metre measured at *HFL*,

f = Silt factor which depends on grain size.

Local Scour

The local scour occurs at the pier points below the depth of general scour. The formula that is sometimes used for estimating the depth of local scour D_{ls} is

$$D_{ls} = 1.4 C_s B_a, \tag{20.17}$$

where, B_a = Average width of pier below the *HFL* and above the general scour level,

C_s = Coefficient which depends on the pier shape.

A value of 1.0 for cylindrical piers and 1.4 for rectangular pier is normally assumed.

The depth of local scour D_{ls} is sometimes increased by 20 percent if the discharge in the river exceeds 3000 cubic metre per second.

The total depth of scour D_{ts} be expressed as follows:

1. For discharges upto 3000 cubic metre per second

$$D_{ts} = D_s + D_{ls} \tag{20.18}$$

2. For discharges greater than 30 cubic metre per second

$$D_{ts} = D_s + 1.2D_{ls} \tag{20.19}$$

Silt Factor f

According to Lacey, the silt factor f depends on the average grain size and density of boundary materials in the river. Assuming an average specific gravity of 2.65 for the river bed material which is normally sand, Lacey gives the following formula for deteining the silt factor

$$f = 1.76\sqrt{d_m} , \tag{20.20}$$

where, d_m = average particle size in mm.

Method of Determining Average particle Size d_m

The average grain size, d_m, of any given sample of soil may be found out by carrying out sieve analysis. A typical example of determining d_m is given in Tables 20.1 and 20.2. Table 20.1 gives the percent by weight of the soil particles retained on- sieve sizes and Table 20.2 shows the method of determining d_m which is self-explanatory.

Representative samples of the bed materials at different depths upto the normal depth of scour have to be tested and the mean diameter of particle of each sample to be found out. The average of the mean diameters of all the samples tested gives the particle size which has to be used in Eq. (20.20) for computing the silt factor.

TABLE 20.1

Result of sieve analysis

Sieve designation	Sieve opening in mm	Weight of soil retained gm	Percent retained
5.60 mm	5.60	0.0	0.00
4.00 mm	4.00	0.0	0.00
2.80 mm	2.80	16.2	4.05
1.00 mm	1.00	76.5	18.30
425 micron	0.425	79.2	18.85
180 micron	0.180	150.4	35.85
75 micron	0.75	41.0	9.80
Pan	–	55.4	13.20

TABLE 20.2

Computation of average diameter of particle

Sieve no.	Average sieve size mm	Percentage weight retained	Product of columns 2 and 3
1	2	3	4
4.00 mm to 2.80	3.40	4.05	13.750
2.80 mm to 1.00 mm	1.90	18.30	34.700
1.00 mm to 425 μ	0.712	18.85	13 .400
425 μ to 180 μ	0.302	35.85	10.800
180 μ to 75 μ	0.127	9.80	1.250
75 μ and below	0.0375	13.20	0.495
	Total	100.05	74.395

$$\text{Average diameter} = \frac{74.395}{100} = 0.74395 \text{ mm.}$$

20.5 THICKNESS OF STEINING OF WELLS

The walls of wells are called as steining. The steining may be constructed of cement concrete or brick masonry. The cement concrete steining is normally reinforced to take care of stresses developed during sinking or due to changes in temperature condition. Similarly the brick masonry steining are also reinforced.

The thickness of steining is fixed by taking the following into consideration:

1. It should be possible to sink the well without excessive kentledge.

2. The wells should not get damaged during sinking.

3. If the well develops tilts and shifts during sinking it should be possible to rectify the tilts and shifts without damaging the well.

4. The well should be able to resist safely the earth pressure developed during a sand blow or other conditions like sudden drop that may be experienced during sinking.

5. Stresses at various levels of the steining should be within permissible limits under all load conditions that may be transferred to the well either during sinking or during service.

There is no recognized procedure for computing the thickness of steining. Some of the Codes of Practices on bridges propose empirical methods for computing the thickness. The method that are suggested by the Indian Roads Congress is given below.

Plain Cement Concrete Steining

The thickness, T, of the steining should not be less than 45 cm nor less than that given by the following equations:

1. For circular or dumbbell shaped wells

(i) In sandy and silty strata

$$T_s = 1.0 \left(\frac{D_f}{100} + \frac{d_0}{10} \right)$$ (20.21)

(ii) In soft clay strata

$$T_s = 1.1 \left(\frac{D_f}{100} + \frac{d_0}{10} \right)$$ (20.22)

(iii) In hard clay strata

$$T_s = 1.25 \left(\frac{D_f}{100} + \frac{d_0}{10} \right)$$ (20.23)

(iv) In soils where boulders, kankars, shale or laterite or such hard materials are met with

$$T_s = 1.25 \left(\frac{D_f}{100} + \frac{d_0}{8} \right)$$ (20.24)

2. For rectangular or double D shaped wells

(i) In sandy strata

$$T_s = 1.0 \left(\frac{D_f}{100} + \frac{L}{10} \right)$$ (20.25)

(ii) In soft clay strata

$$T_s = 1.1 \left(\frac{D_f}{100} + \frac{L}{10} \right)$$ (20.26)

(iii) In hard clay strata

$$T_s = 1.15 \left(\frac{D_f}{100} + \frac{L}{10} \right)$$ (20.27)

(iv) In soils where boulders, kankars, shale or laterite or such hard materials are met with

$$T_s = 1.20 \left(\frac{D_f}{100} + \frac{L}{10} \right)$$ (20.28)

where, D_f = fully designed depth of the well below the bed level of river existing at the time of sinking in cm

d_0 = outside diameter of the well in cm

L = length of rectangular well in cm.

The length L depends upon the number of dredging wells provided. For a single dredging well, L is the longest side. If there are more than one dredging wells, the longest side of a dredging well is to be considered.

Brick Masonry Steining

The minimum thickness, T_s of the steining of circular brick masonry wells should not be less than 45 cm and also should not be less than the values given by the following equations.

   (i) In sandy strata

$$T_s = \left(\frac{D_f}{40} + \frac{d_0}{8} \right)$$ (20.29)

   (ii) In soft clay strata

$$T_s = 1.1\left(\frac{D_f}{40} + \frac{d_0}{8} \right)$$ (20.30)

   (iii) In hard clay strata

$$T_s = 1.25\left(\frac{D_f}{40} + \frac{d_0}{8} \right)$$ (20.31)

Distance between wells

When groups of wells are near each other, special care is needed to ensure that they do not fail in the course of sinking and also do not cause disturbance to wells already sunk. The minimum clearance between the centre to centre of wells should be 1½ times the external diameter.

20.6   EXAMPLES

Example 20.1

A caisson 6.4 m external diameter has been sunk for the existing Yamuna Bridge at Agra., India. The caisson is founded in sandy soil. The following information is available. (Kapur, 1971)

Dimensions of Caisson

External diameter of caisson = 6.4 m

Cross sectional area = 32.2 sq. m

Grip length provided = 11.3 m

Height of bearing level above maximum scour level = 26.6 m

Design Loads

Type of Load caisson condition	Symbol	Seismic condition	Non-seismic
Total vertical load including weight of peir and caisson (considering buoyancy effect)	W	1930 tonnes	1930 tonnes
Total lateral load at scour level	P_t	85 tonnes	37 tonnes
Total moment at scour level	M_g	2240 tonne-metre	1050 tonne-metre

Soil properties

Relative density = 60%

Submerged unit weight of soil = 0.962 T/m^3

Angle of internal friction = 33°42′

Angle of wall friction = 27°42′

Compute the grip length as per the limit state equilibrium method and check up the length provided is adequate or not.

Solution

Compute D for Seismic condition.

Use Eq. (20:9b)

$$D^2 = \frac{WD(\alpha_1\alpha_2 - \alpha_3)}{\gamma m K_p A(\alpha_4 + \alpha_5)}$$

The caisson is a circular one. Computation is carried out on a square caisson of equivalent sectional area. The equivalent width, B, is

$$B = \sqrt{A} = \sqrt{32.2} = 5.67 \text{ m}$$

Computation may be carried out with an assumed eccentricity e_b. This can be taken normally as equal to $B/6$. However, the effect of $e_b = B/3$ on the grip length is also worth studying. The various factors are

W = 1930 tonnes = 19300 kN

With a load factor $F_l = 2$, the ultimate lateral load to be used for determining grip length D for square caisson is (as per Eq. 20.10)

$$P_u = 1.2 \times 2 \times 85 = 204 \text{ tonnes} = 2040 \text{ kN}$$

$$\alpha_1 = \frac{P_u}{W} = \frac{204}{1930} = 0.106$$

$$\alpha_2 = \frac{H}{D} + 1 = \frac{20.6}{D} + 1$$

$$\alpha_3 = \frac{e_b}{D},$$

for $e_b = \dfrac{B}{6}$, $\alpha_3 = \dfrac{5.67}{6} \times \dfrac{1}{D} = \dfrac{0.945}{D}$

for $e_b = \dfrac{B}{3}$, $\alpha_3 = \dfrac{5.67}{3} \times \dfrac{1}{D} = \dfrac{1.89}{D}$

$$\alpha_4 = \frac{1}{6}\frac{D}{B} = \frac{D}{6 \times 5.67} = \frac{D}{34}$$

$$\alpha_5 = \frac{1}{4}\tan\delta(B - 2e_b)$$

for $e_b = \dfrac{B}{6}$, $\alpha_5 = \dfrac{1}{4}\tan 27.7° \left(5.67 - 2 \times \dfrac{5.67}{6}\right) = 0.44$

for $e_b = \dfrac{B}{3}$, $\alpha_5 = 0.248$

The value of m is assumed as equal to 1 this happens to be a heavy well. Now substituting all the known factors, we have

$$K_p = \tan^2(45° + \phi/2) = \tan^2 61.85° = 3.49$$

$$D^2 = \frac{1930D\left[0.106\left(\dfrac{26.6}{D} + 1\right) - C_1\right]}{0.962 \times 1 \times 3.49 \times 32\left(\dfrac{D}{34} + C_2\right)}$$

where, $C_1 = \dfrac{0.945}{D}$, $C_2 = 0.44$ for $e_b = \dfrac{B}{6}$

$$C_1 = \frac{1.89}{D}, \quad C_2 = 0.248 \quad \text{for} \quad e_b = \frac{B}{3}$$

The value of D can be found out by trial and error method for both the cases of e_b. The value ϵ D as obtained are

for $e_b = B/6$, $D = 8.5$ m

for $e_b = B/3$, $D = 8.0$ m

The increase of eccentricity by 100 percent, decreases the depth D only by 6 percent. Therefore we can adopt $e_b = B/6$ or $B/3$ without causing significant error in the grip length.

It may also be noted here that an increase in the value of m from 1 to 1.5, decreases the value $\circ$ D from 8.0 m (for $e_b = B/3$) to 6.5, i.e. a decrease of about 20 percent. If m is increased to 2, the valu ϵ of D reduces to about 6 m which is a decrease of about 25 percent. This means an increase in the value of m by 100 percent (1 to 2) decreases the value of D by about 25 percent. This helps the designer to choose a proper value for m.

The grip length calculated by assuming $e_b = B/6$ is about 23 percent less than that provided.

Example 20.2

If the caisson foundation given in Ex. 20.1 is founded in cohesive soil with an undrained cohesive strength $c_u = 50$ kN/m², compute the grip length with all the other details as given in Ex. 20.1 for seismic condition.

Solution

Use Eq. (20.15), assume $e_b = B/6$, $\alpha = 0.5$.

$$D = \frac{2(P_u D + P_u H - We_b)}{c_u B[9D + \alpha(B - e_b)]}$$

we have, $P_u = 2040$ kN, $W = 19300$ kN.

Substituting all the values known, we have

$$D = \frac{\left(204D + 2040 \times 26.6 - 19300 \times \dfrac{5.67}{6}\right)}{5 \times 5.67\left[9D + 0.6\left(5.67 - \dfrac{5.67}{6}\right)\right]}$$

$$= \frac{2(204D + 3602)}{28.35(9D + 2.835)}$$

The solution of this equation gives $D = 6$ m. If the eccentricity e_b is taken as equal to $B/3$, the grip length $D = 4.5$ m. Possibly it is safer to use $e_b = B/6$.

CHAPTER 21

DRILLED PIER FOUNDATIONS

21.1 INTRODUCTION

Chapter 19 dealt with piles subjected to vertical loads. Drilled pier foundations, the subject matter of this chapter, belong to the same category as pile foundations. Because piers and piles serve the same purpose, no sharp deviations can be made between the two. The distinctions are based on the method of installation. A pile is installed by driving, a pier by excavating. Thus, a foundation unit installed in a drill-hole may also be called a bored cast-in-situ concrete pile. Here, distinction is made between a small diameter pile and a large diameter pile. A pile, cast-in-situ, with a diameter less than 0.75 m (or 2.5 ft) is sometimes called a small diameter pile. A pile greater than this size is called a large diameter bored-cast-in-situ pile. The latter definition is used in most non-American countries whereas in the USA, such large diameter bored piles are called drilled piers, drilled shafts, and sometimes drilled *caissons*. Chapter 19 deals with small diameter bored-cast-*in-situ* piles in addition to driven piles.

21.2 TYPES OF DRILLED PIERS

Drilled piers may be described under four types. All four types are similar in construction technique, but differ in their design assumptions and in the mechanism of load transfer to the surrounding earth mass. These types are illustrated in Fig. 21.1.

Straight-shaft end-bearing piers develop their support from end-bearing on strong soil, "hardpan" or rock. The overlying soil is assumed to contribute nothing to the support of the load imposed on the pier (Fig. 21.1(a)).

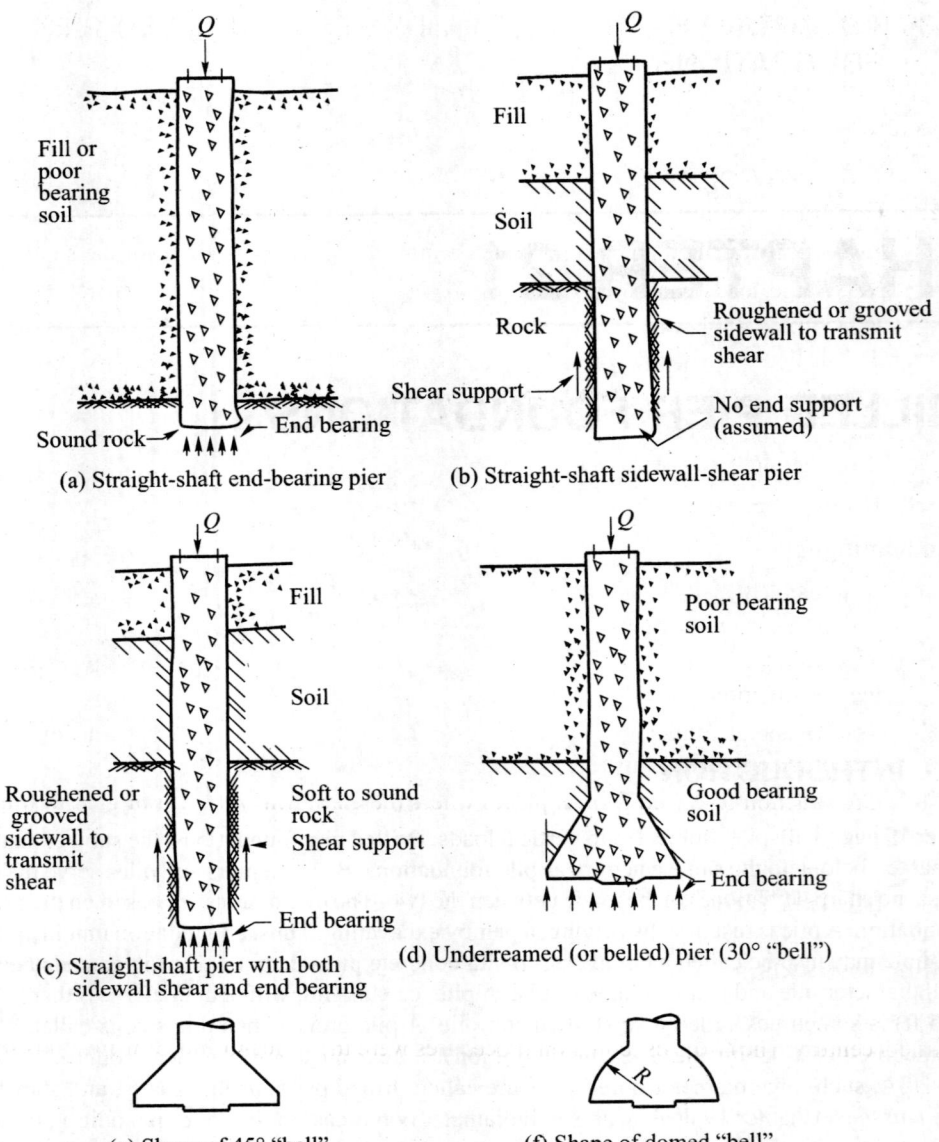

Fig. 21.1 Types of drilled piers and underream shapes (Woodward et al., 1972)

Straight-shaft side wall friction piers pass through overburden soils that are assumed to carry none of the load, and penetrate far enough into an assigned bearing stratum to develop design load capacity by side wall friction between the pier and bearing stratum (Fig. 21.1(b)).

Combination of straight shaft side wall friction and end bearing piers are of the same construction as the two mentioned above, but with both side wall friction and end bearing assigned a role in carrying the design load. When carried into rock, this pier may be referred to as a socketed pier or a "drilled pier with rock socket" (Fig. 21.1(c)).

Belled or underreamed piers are piers with a bottom bell or underream (Fig. 21.1(d)). A greater percentage of the imposed load on the pier top is assumed to be carried by the base.

21.3 ADVANTAGES AND DISADVANTAGES OF DRILLED PIER FOUNDATIONS

Advantages

1. Pier of any length and size can be constructed at the site
2. Construction equipment is normally mobile and construction can proceed rapidly
3. Inspection of drilled holes is possible because of the larger diameter of the shafts
4. Very large loads can be carried by a single drilled pier foundation thus eliminating the necessity of a pile cap
5. The drilled pier is applicable to a wide variety of soil conditions
6. Changes can be made in the design criteria during the progress of a job
7. Ground vibration that is normally associated with driven piles is absent in drilled pier construction
8. Bearing capacity can be increased by underreaming the bottom (in non-caving materials)

Disadvantages

1. Installation of drilled piers needs a careful supervision and quality control of all the materials used in the construction
2. The method is cumbersome. It needs sufficient storage space for all the materials used in the construction
3. The advantage of increased bearing capacity due to compaction in granular soil that could be obtained in driven piles is not there in drilled pier construction
4. Construction of drilled piers at places where there is a heavy current of ground water flow due to artesian pressure is very difficult

21.4 METHODS OF CONSTRUCTION

Earlier Methods

The use of drilled piers for foundations started in the United States during the early part of the twentieth century. The two most common procedures were the Chicago and Gow methods shown in Fig. 21.2. In the Chicago method a circular pit was excavated to a convenient depth and a cylindrical shell of vertical boards or staves was placed by making use of an inside compression ring. Excavation then continued to the next board length and a second tier of staves was set and the procedure continued. The tiers could be set at a constant diameter or stepped in about 50 mm. The Gow method, which used a series of telescopic metal shells, is about the same as the current method of using casing except for the telescoping sections reducing the diameter on successive tiers.

Modern Methods of Construction

Equipment

There has been a phenomenal growth in the manufacture and use of heavy duty drilling equipment in the United States since the end of World War II. The greatest impetus to this development occurred in two states, Texas and California (Woodward et al., 1972). Improvements in the machines were made responding to the needs of contractors. Commercially produced drilling rigs of sufficient size and capacity to drill pier holes come in a wide variety of mountings and driving arrangements. Mountings are usually truck crane, tractor or skid. Fig. 21.3 shows a tractor mounted rig. Drilling

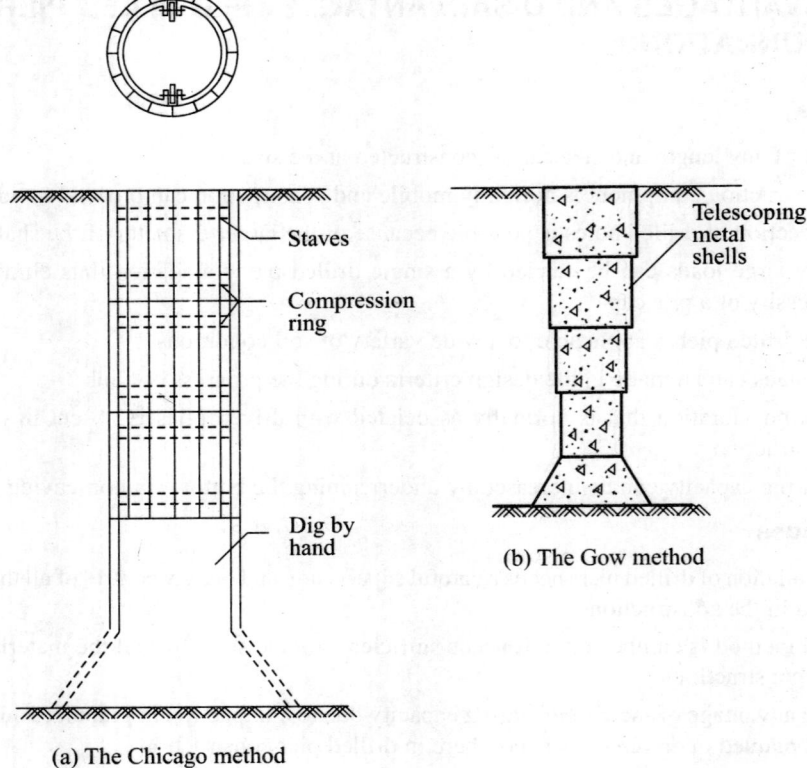

Fig. 21.2   Early methods of caisson construction

machine ratings as presented in manufacturer's catalogs and technical data sheets are usually expressed as maximum hole diameter, maximum depth, and maximum torque at some particular rpm.

Many drilled pier shafts through soil or soft rock are drilled with the open-helix auger. The tool may be equipped with a knife blade cutting edge for use in most homogeneous soil or with hard-surfaced teeth for cutting stiff or hard soils, stony soils, or soft to moderately hard rock. These augers are available in diameters up to 3 m or more. Figure 21.4 shows commercially available models.

Underreaming tools (or buckets) are available in a variety of designs. Figure 21.5 shows a typical 30° underreamer with blade cutter for soils that can be cut readily. Most such underreaming tools are limited in size to a diameter three times the diameter of the shaft.

When rock becomes too hard to be removed with auger-type tools, it is often necessary to resort to the use of a core barrel. This tool is a simple cylindrical barrel, set with tungsten carbide teeth at the bottom edge. For hard rock which cannot be cut readily with the core barrel set with hard metal teeth, a calyx or shot barrel can be used to cut a core of rock.

General Construction Methods of Drilled Pier Foundations

The rotary drilling method is the most common method of pier construction in the United States. The methods of drilled pier construction can be classified in three categories as

1.   The dry method
2.   The casing method
3.   The slurry method

Fig. 21.3   Tractor mounted hydraulic drilling rig (Courtesy: Kelly Tractor Co, USA)

Dry Method of Construction

The dry method is applicable to soil and rock that are above the water table and that will not cave or slump when the hole is drilled to its full depth. The soil that meets this requirement is a homogeneous, stiff clay. The first step in making the hole is to position the equipment at the desired location and to select the appropriate drilling tools. Fig. 17.6(a) gives the initial location. The drilling is next carried out to its fill depth with the spoil from the hole removed simultaneously.

After drilling is complete, the bottom of the hole is underreamed if required. Figure 21.6(b) and (c) show the next steps of concreting and placing the rebar cage. Figure 21.6(d) shows the hole completely filled with concrete.

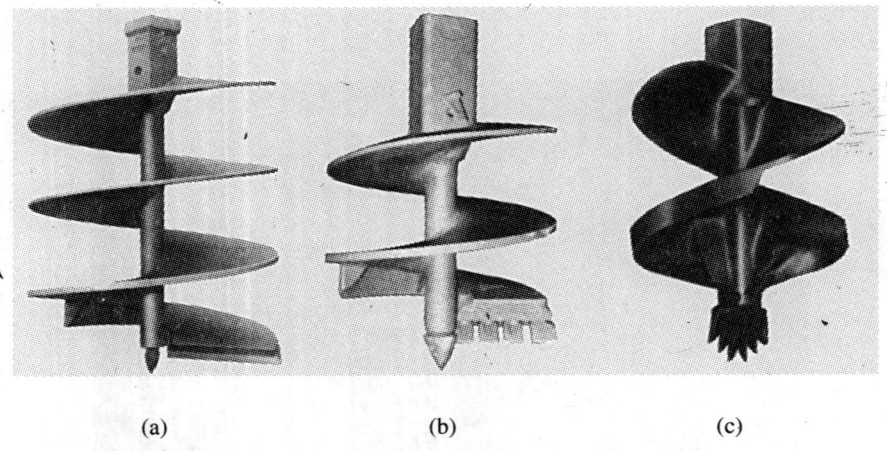

(a)   (b)   (c)

Fig. 21.4   (a) Single-flight auger bit with cutting blade for soils, (b) single-flight auger bit with hard-metal cutting teeth for hard soils, hardpan, and rock, and (c) cast steel heavy-duty auger bit for hardpan and rock (*Source:* Woodward *et al.,* 1972)

Fig. 21.5   A 30º underreamer with blade cutters for soils that can be cut readily (Source: Woodward et al., 1972)

Casing Method of Construction

The casing method is applicable to sites where the soil conditions are such that caving or excessive soil or rock deformation can occur when a hole is drilled. This can happen when the boring is made in dry soils or rocks which are stable when they are cut but will slough soon afterwards. In such a

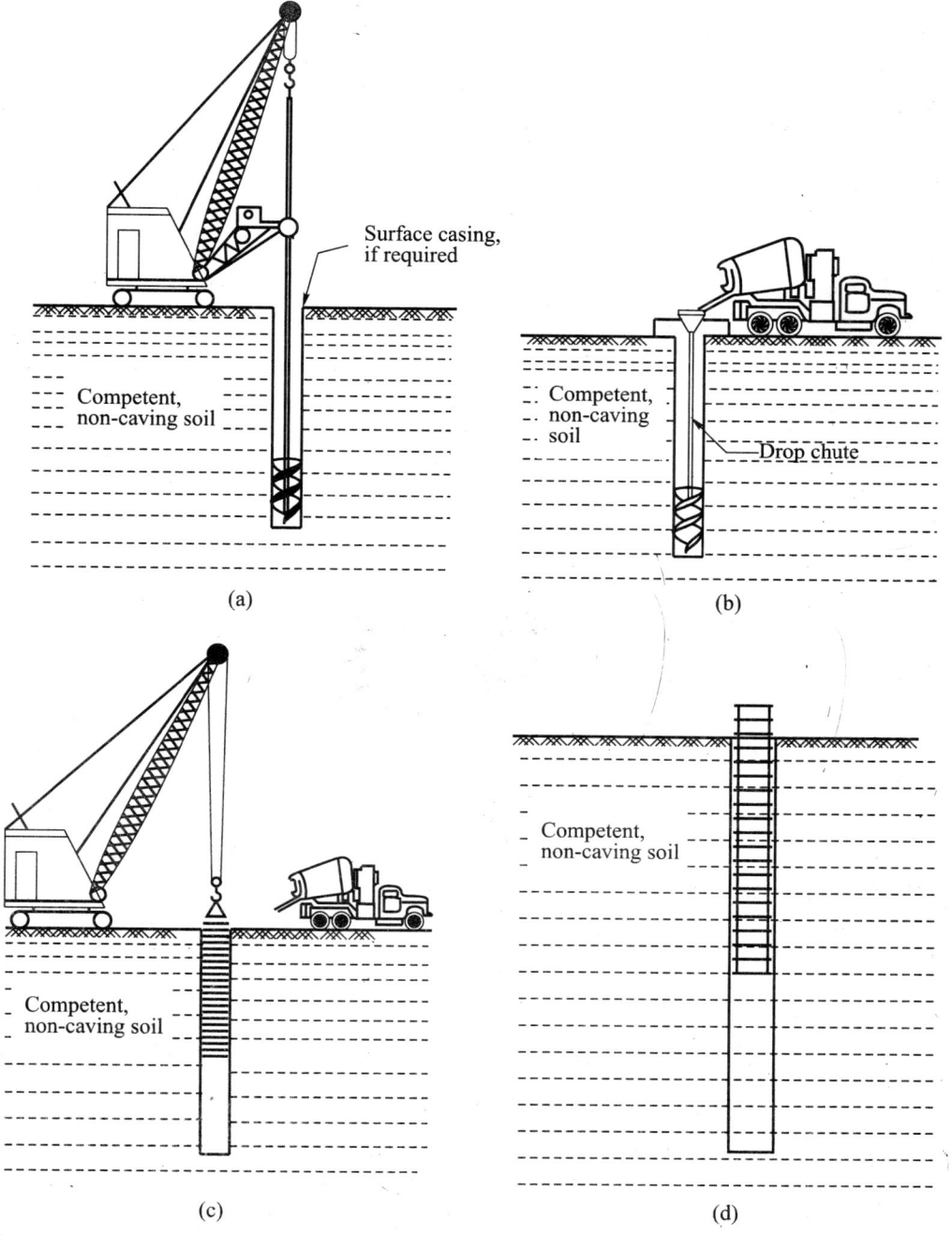

Fig. 21.6   Dry method of construction: (a) initiating drilling, (b) starting concrete pour, (c) placing rebar cage, and (d) completed shaft (O' Neill and Reese, 1999)

case, the bore hole is drilled, and a steel pipe casing is quickly set to prevent sloughing. Casing is also required if drilling is required in clean sand below the water table underlain by a layer of impermeable stones into which the drilled shaft will penetrate. The casing is removed soon after the concrete is deposited. In some cases, the casing may have to be left in place permanently. It may be noted here that until the casing is inserted, a slurry is used to maintain the stability of the hole. After the casing is seated, the slurry is bailed out and the shaft extended to the required depth. Figures 21.7(a) to (h) give the sequence of operations. Withdrawl of the casing, if not done carefully, may lead to voids or soil inclusions in the concrete, as illustrated in Fig. 21.8.

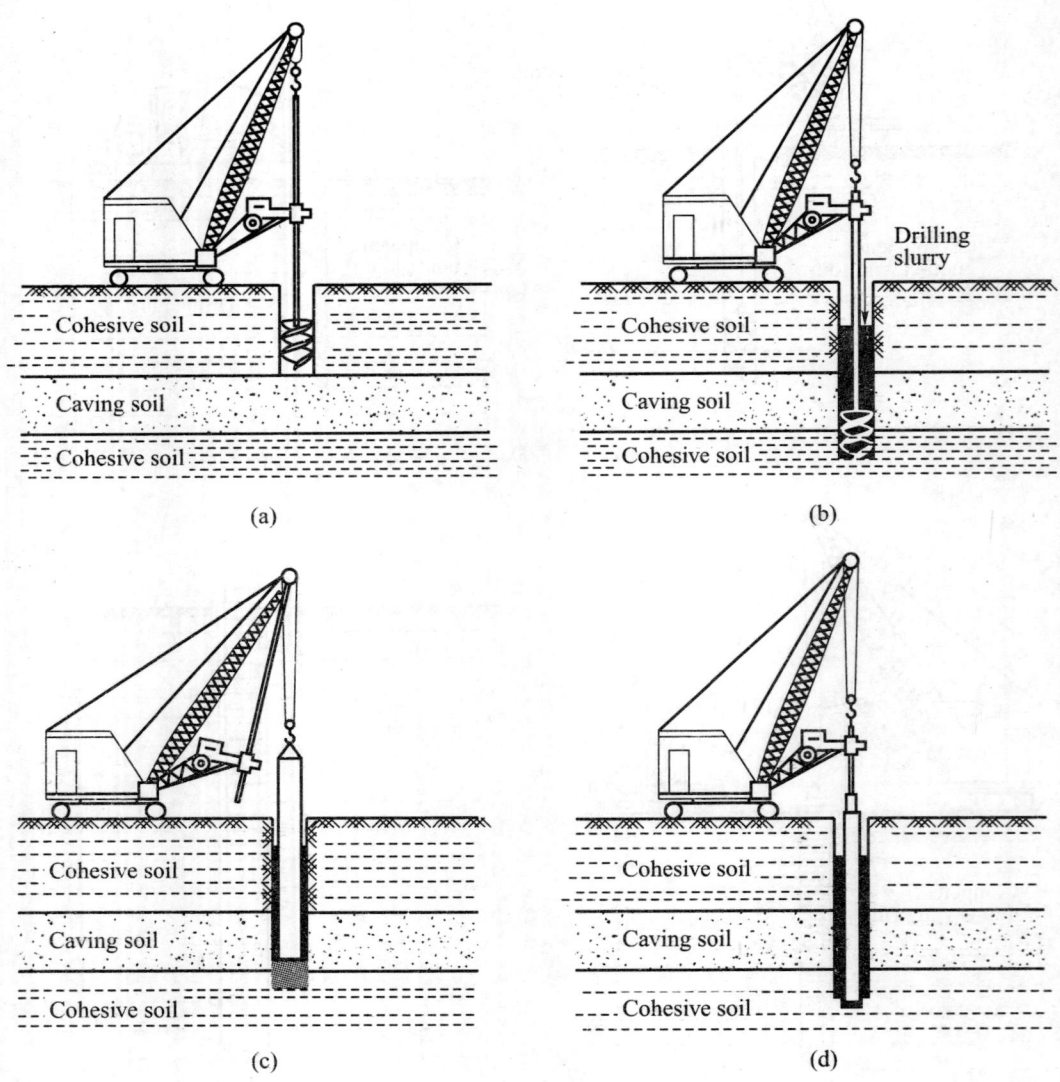

Fig. 21.7   Casing method of construction: (a) initiating drilling, (b) drilling with slurry; (c) introducing casing, (d) casing is sealed and slurry is being removed from interior of casing (continued)

Slurry Method of Construction

The slurry method of construction involves the use of a prepared slurry to keep the bore hole stable for the entire depth of excavation. The soil conditions for which the slurry displacement method is applicable could be any of the conditions described for the casing method. The slurry method is a

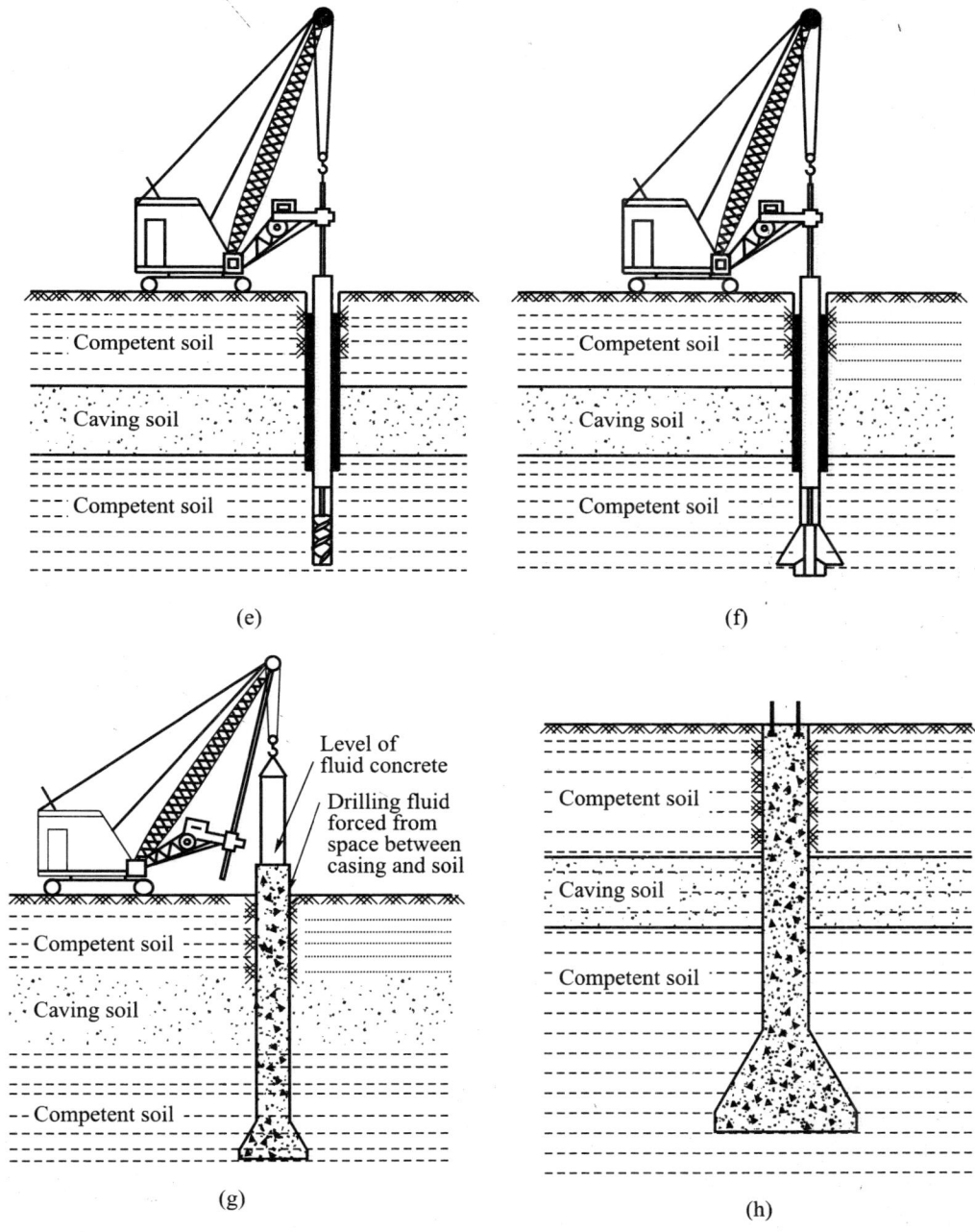

Fig. 21.7 (continued) casing method of construction: (e) drilling below casing, (f) underreaming, (g) removing casing, and (h) completed shaft (O'Neill and Reese, 1999)

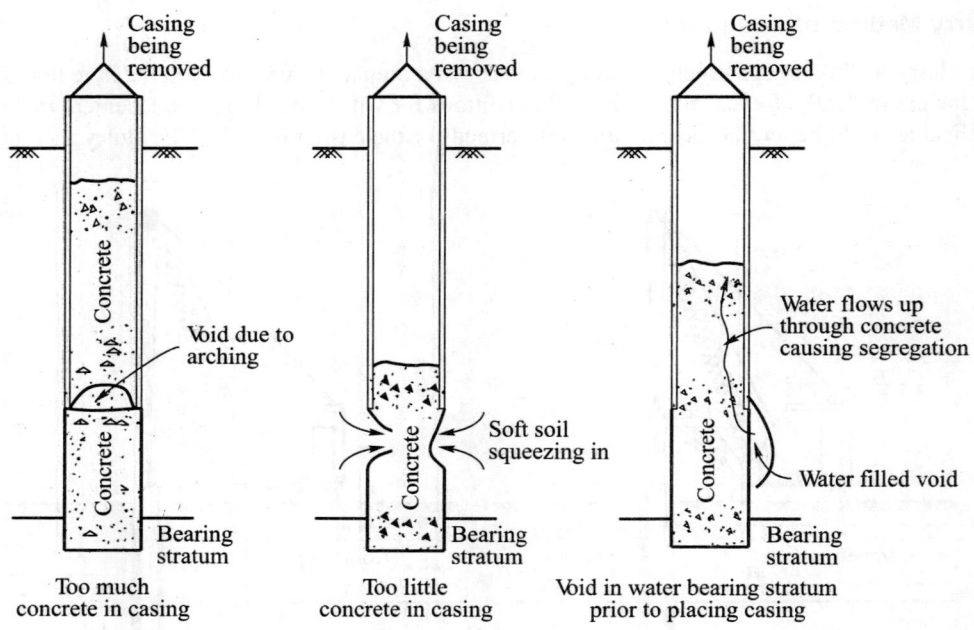

Fig. 21.8   Potential problems leading to inadequate shaft concrete due to removal of temporary casing without care (D'Appolonia, et al., 1975)

viable option at any site where there is a caving soil, and it could be the only feasible option in a permeable, water bearing soil if it is impossible to set a casing into a stratum of soil or rock with low permeability. The various steps in the construction process are shown in Fig. 21.9. It is essential in this method that a sufficient slurry head be available so that the inside pressure is greater than that from the GWT or from the tendency of the soil to cave.

Bentonite is most commonly used with water to produce the slurry. Polymer slurry is also employed. Some experimentation may be required to obtain an optimum percentage for a site, but amounts in the range of 4 to 6 percent by weight of admixture are usually adequate.

The bentonite should be well mixed with water so that the mixture is not lumpy. The slurry should be capable of forming a filter cake on the side of the bore hole. The bore hole is generally not underreamed for a bell since this procedure leaves unconsolidated cuttings on the base and creates a possibility of trapping slurry between the concrete base and the bell roof.

If reinforcing steel is to be used, the rebar cage is placed in the slurry as shown in Fig 21.9(b). After the rebar cage has been placed, concrete is placed with a tremie either by gravity feed or by pumping. If a gravity feed is used, the bottom end of the tremie pipe should be closed with a closure plate until the base of the tremie reaches the bottom of the bore hole, in order to prevent contamination of the concrete by the slurry. Filling of the tremie with concrete, followed by subsequent slight lifting of the tremie, will then open the plate, and concreting proceeds. Care must be taken that the bottom of the tremie is buried in concrete at least for a depth of 1.5 m (5 ft). The sequence of operations is shown in Fig 21.9(a) to (d).

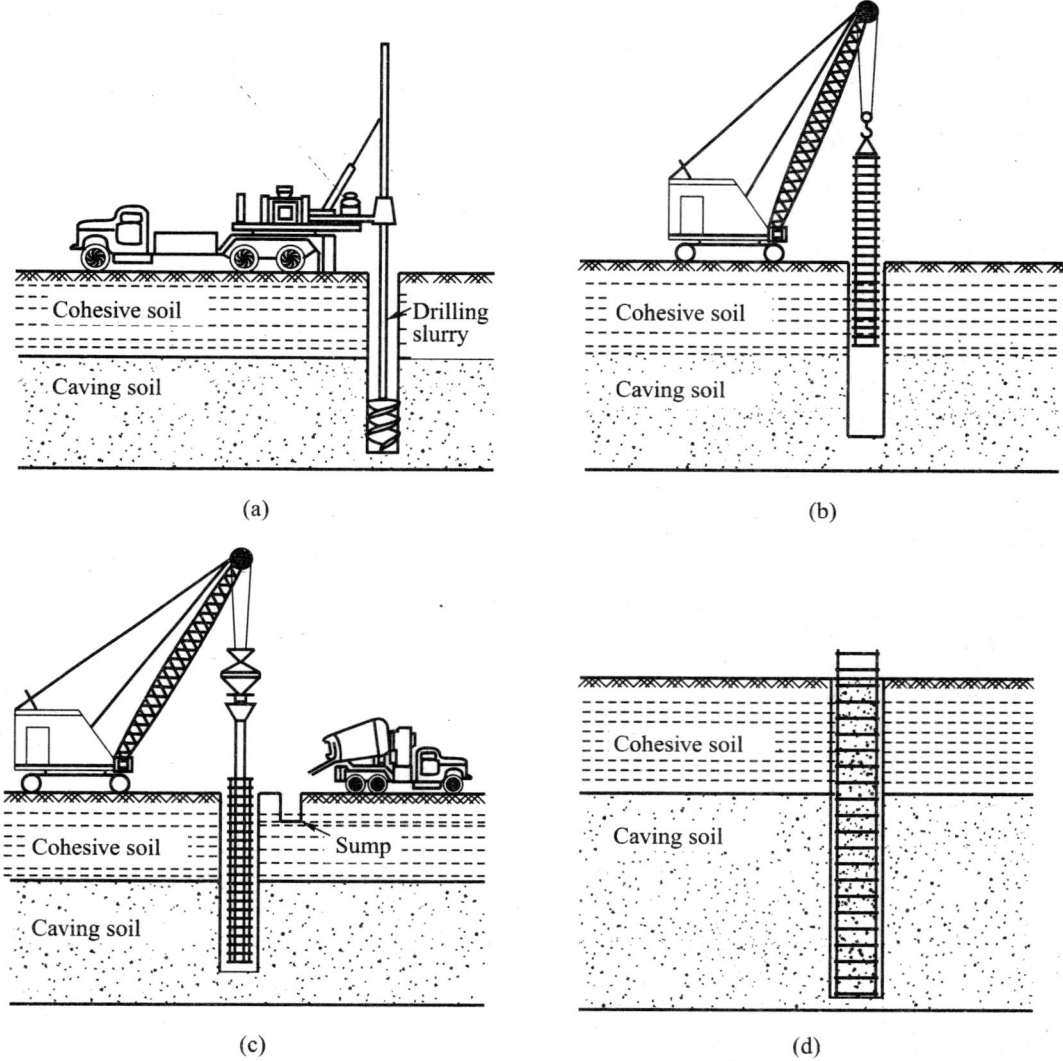

Fig. 21.9 Slurry method of construction (a) drilling to full depth with slurry, (b) placing rebar cage, (c) placing concrete, (d) completed shaft (O'Neill and Reese, 1999)

21.5 DESIGN CONSIDERATIONS

The precess of the design of a drilled pier generally involves the following:

1. The objectives of selecting drilled pier foundations for the project.
2. Analysis of loads coming on each pier foundation element.
3. A detailed soil investigation and determining the soil parameters for the design.
4. Preparation of plans and specifications which include the methods of design, tolerable settlement, methods of construction of piers, etc.
5. The method of execution of the project.

In general the design of a drilled pier may be studied under the following headings.

1. Allowable loads on the piers based on ultimate bearing capacity theories.
2. Allowable loads based on vertical movement of the piers.
3. Allowable loads based on lateral bearing capacity of the piers.

In addition to the above, the uplift capacity of piers with or without underreams has to be evaluated.

The following types of strata are considered.

1. Piers embedded in homogeneous soils, sand or clay.
2. Piers in a layered system of soil.
3. Piers socketed in rocks.

It is better that the designer select shaft diameters that are multiples of 150 mm (6 in) since these are the commonly available drilling tool diameters.

21.6   LOAD TRANSFER MECHANISM

Figure 21.10(a) shows a single drilled pier of diameter d, and length L constructed in a homogeneous mass of soil of known physical properties. If this pier is loaded to failure under an ultimate load Q_u, a part of this load is transmitted to the soil along the length of the pier and the balance is transmitted to the pier base. The load transmitted to the soil along the pier is called the ultimate *friction load or skin load*, Q_f and that transmitted to the base is the ultimate base or point load Q_b. The total ultimate load, Q_u, is expressed as (neglecting the weight of the pier)

$$Q_u = Q_b + Q_f = q_b A_b + \sum_{i=1}^{N} f_{si} P_i \Delta z_i \qquad (21.1)$$

where q_b =   net ultimate bearing pressure
         A_b =   base area
         f_{si} =   unit skin resistance (ultimate) of layer i
         P_i =   perimeter of pier in layer i
         Δz_i =   thickness of layer i
         N =   number of layers

If the pier is instrumented, the load distribution along the pier can be determined at different stages of loading. Typical load distribution curves plotted along a pier are shown in Fig 21.10(b) (O'Neill and Reese, 1999). These load distribution curves are similar to the one shown in Fig. 19.5(b). Since the load transfer mechanism for a pier is the same as that for a pile, no further discussion on this is necessary here. However, it is necessary to study in this context the effect of settlement on the mobilization of side shear and base resistance of a pier. As may be seen from Fig. 21.11, the maximum values of base and side resistance are not mobilized at the same value of displacement. In some soils, and especially in some brittle rocks, the side shear may develop fully at a small value of displacement and then decrease with further displacement while the base resistance is still being mobilized (O'Neill and Reese, 1999). If the value of the side resistance at point A is added to the value of the base resistance at point B, the total resistance shown at level D is overpredicted. On the other hand, if the designer wants to take advantage primarily of the base resistance, the side resistance at point C should be added to the base resistance at point B to evaluate Q_u. Otherwise, the designer may wish to design for the side resistance at point A and disregard the base resistance entirely.

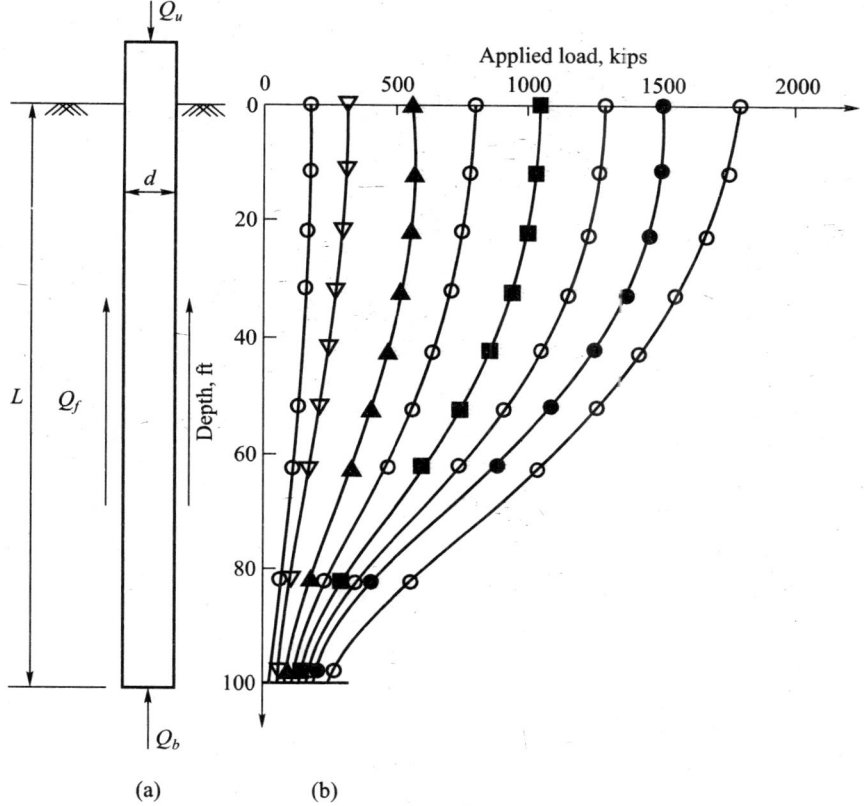

Fig. 21.10 Typical set of load distribution curves (O'Neill and Reese, 1999)

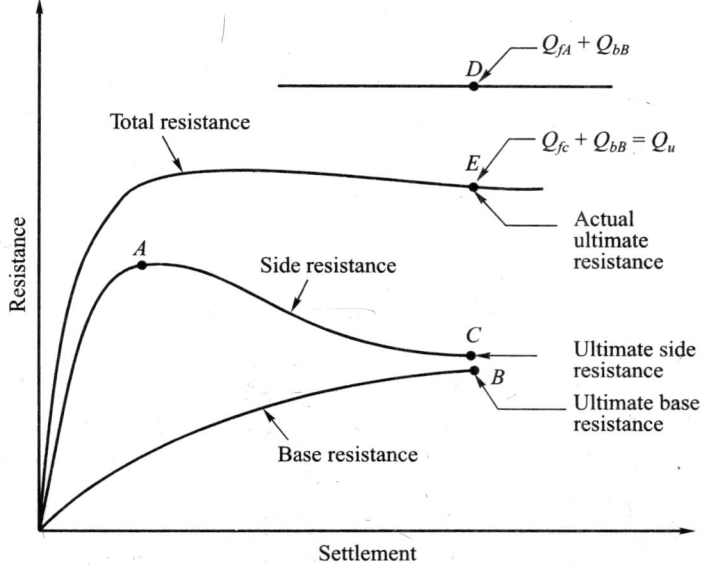

Fig. 21.11 Condition in which $(Q_b + Q_f)$ is not equal to actual ultimate resistance

21.7 VERTICAL BEARING CAPACITY OF DRILLED PIERS

For the purpose of estimating the ultimate bearing capacity, the subsoil is divided into layers (Fig. 21.12) based on judgment and experience (O'Neill and Reese, 1999). Each layer is assigned one of four classifications.

1. Cohesive soil [clays and plastic silts with undrained shear strength $c_u \leq 250$ kN/m² (2.5 t/ft²)] .

2. Granular soil [cohesionless geomaterial, such as sand, gravel or nonplastic silt with uncorrected SPT(N) values of 50 blows per 0.3/m or less].

3. Intermediate geometerial [cohesive geometerial with undrained shear strength c_u between 250 and 2500 kN/m² (2.5 and 25 tsf), or cohesionless geomaterial with SPT(N) values > 50 blows per 0.3 m].

4. Rock [highly cemented geomaterial with unconfined compressive strength greater than 5000 kN/m² (50 tsf)].

The unit side resistance f_s ($=f_{max}$) is computed in each layer through which the drilled shaft passes, and the unit base resistance q_b ($=q_{max}$) is computed for the layer on or in which the base of the drilled shaft is founded.

The soil along the whole length of the shaft is divided into four layers as shown in Fig. 21.12.

Effective Length for Computing Side Resistance in Cohesive Soil

O'Neill and Reese (1999) suggest that the following effective length of pier is to be considered for computing side resistance in cohesive soil.

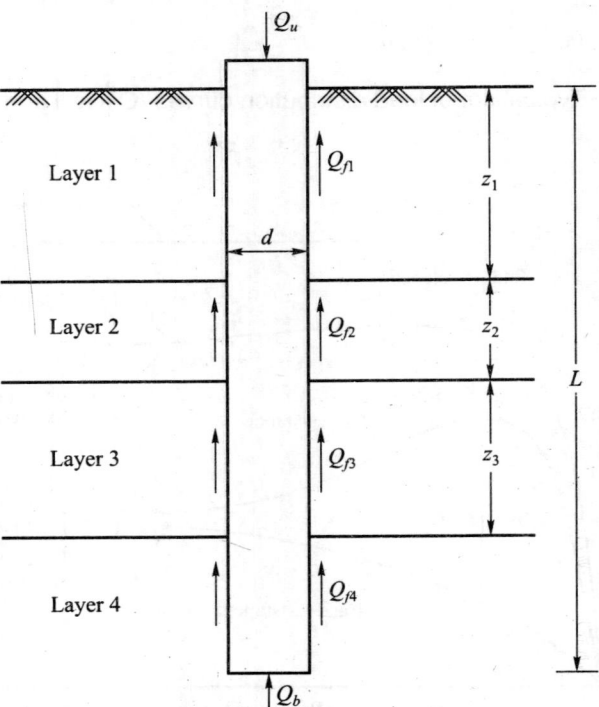

Fig. 21.12 Idealised geomaterial layering for computation of compression load and resistance (O'Neill and Reese, 1999)

Straight shaft: One diameter from the bottom and 1.5 m (5 feet) from the top are to be excluded from the embedded length of pile for computing side resistance as shown in Fig. 21.13(a) .

Belled shaft: The height of the bell (h) plus the diameter of the shaft (d) from the bottom and 1.5 m (5 ft) from the top are to be excluded as shown in Fig 21.13(b).

21.8 THE GENERAL BEARING CAPACITY EQUATION FOR THE BASE RESISTANCE q_b (= q_{max})

The equation for the ultimate base resistance may be expressed as

$$q_b = s_c d_c N_c c + s_q d_q \left(N_q - 1\right) q'_o + \frac{1}{2} \gamma d s_\gamma d_\gamma N_\gamma \qquad (21.2)$$

where N_c, N_q and N_γ = bearing capacity of factors for long footings

s_c, s_q and s_γ = shape factors

d_c, d_q and d_γ = depth factors

q'_o = effective vertical pressure at the base level of the drilled pier

γ = effective unit weight of the soil below the bottom of the drilled shaft to a depth = 1.5 d where d = width or diameter of pier at base level

c = average cohesive strength of soil just below the base.

For deep foundations the last term in Eq. (21.2) becomes insignificant and may be ignored. Now Eq. (21.2) may be written as

$$q_b = s_c d_c N_c c + s_q d_q \left(N_q - 1\right) q'_o \qquad (22.3)$$

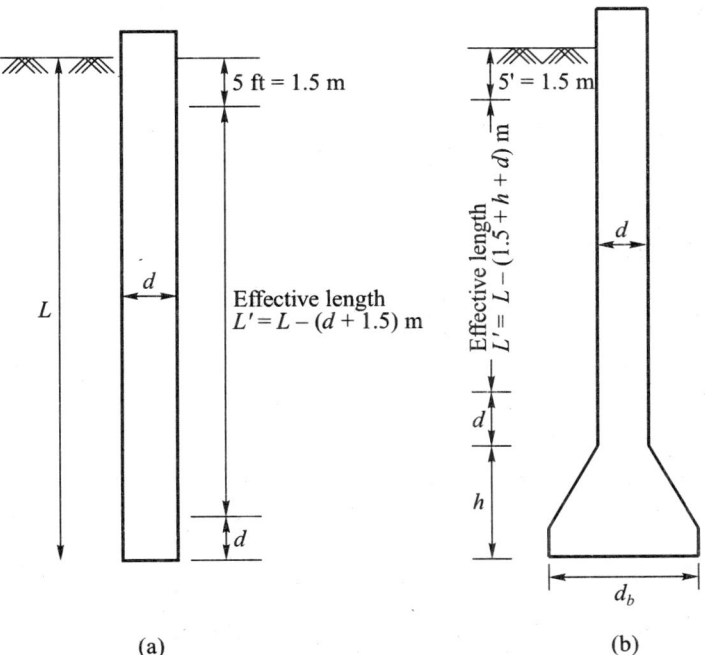

(a)   (b)

Fig. 21.13 Exclusion zones for estimating side resistance for drilled shafts in cohesive soils

21.9 BEARING CAPACITY EQUATIONS FOR THE BASE IN COHESIVE SOIL

When the Undrained Shear Strength, $c_u \leq 250$ kN/m² (2.5 t/ft²)

For $\phi = 0$, $N_q = 1$ and $(N_q - 1) = 0$, here Eq. (21.3) can be written as (Vesic, 1972)

$$q_b = N_c^* c_u \qquad (21.4)$$

in which

$$N_c^* = \frac{4}{3}(\ln I_r + 1) \qquad (21.5)$$

I_r = rigidity index of the soil

Equation (21.4) is applicable for $c_u \leq 96$ kPa and $L \geq 3d$ (where d = base width)

For $\phi = 0$, I_r may be expressed as (O'Neill and Reese, 1999)

$$I_r = \frac{E_s}{3c_u} \qquad (21.6)$$

where E_s = Young's modulus of the soil in undrained loading. Refer to Section 18.6 for the methods of evaluating the value of E_s.

Table 21.1a gives the values of I_r and N_c^* as a function of c_u.

If the depth of base $(L) < 3d$

$$q_b (= q_{max}) = \frac{2}{3}\left[1 + \frac{L}{6d}\right] N_c^* c_u \qquad (21.7)$$

When $c_u \geq 96$ kPa (2000 lb/ft²), the equation for q_b may be written as

$$q_b = 9 c_u \qquad (21.8)$$

for depth of base = $L \geq 3d$ (base width).

Note: In this text, sometimes E_c is used in place of E_s.

21.10 EQUATION BEARING CAPACITY EQUATION FOR THE BASE IN GRANULAR SOIL

Values N_c and N_q in Eq. (21.3) are for strip footings on the surface of rigid soils and are plotted as a function of ϕ in Fig. 21.14. Vesic (1977) explained that during bearing failure, a plastic failure zone develops beneath a circular loaded area that is accompanied by elastic deformation in the surrounding elastic soil mass. The confinement of the elastic soil surrounding the plastic soil has an effect on q_b $(= q_{max})$. The values of N_c and N_q are therefore dependent not only on ϕ, but also on I_r. They must be corrected for soil rigidity as given below.

TABLE 21.1a

Values of $I_r = E_s/3c_u$ and N_c^*

c_u	I_r	N_c^*
24 kPa (500 lb/ft²)	50	6.5
48 kPa (1000 lb/ft²)	150	8.0
$\geq$ 96 kPa (2000 lb/ft²)	250–300	9.0

$$N_c(\text{corrected}) = N_c C_c \left.\vphantom{\begin{matrix}a\\b\end{matrix}}\right\}$$
$$N_q(\text{corrected}) = N_q C_q \quad\quad\quad\quad\quad\quad\quad\quad (21.9)$$

where C_c and C_q are the correction factors. As per Chen and Kulhawy (1994)

Equation (21.3) may now be expressed as

$$q_b = cN_c s_c d_c C_c + (N_q - 1)q_o' s_q d_q C_q \quad\quad\quad\quad (21.10)$$

$$C_c = C_q - \frac{1-C_q}{N_c \tan\phi} \quad\quad\quad\quad\quad\quad\quad\quad (21.11a)$$

$$C_q = \exp\{[-3.8\tan\phi] + [(3.07\sin\phi)\log_{10} 2I_{rr}]/(1+\sin\phi)\} \quad\quad (21.11b)$$

where ϕ is an effective angle of internal friction. I_{rr} is the reduced rigidity index expressed as

$$I_{rr} = \frac{I_r}{1+\Delta I_r} \quad\quad\quad\quad\quad\quad\quad\quad (21.12)$$

and $I_r = \dfrac{E_d}{2(1+\mu_d)q_o'\tan\phi} \quad\quad\quad\quad\quad\quad (21.13)$

by ignoring cohesion, where,

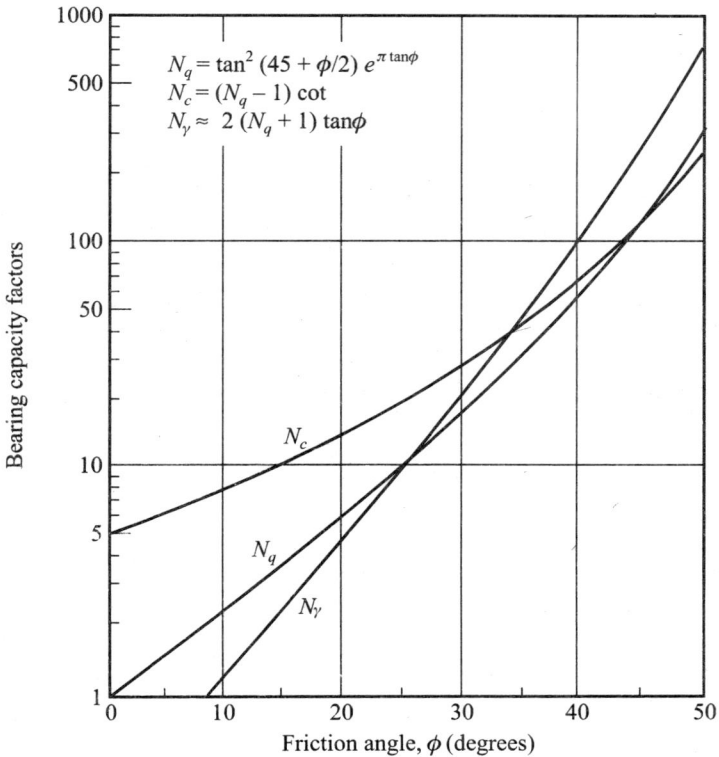

Fig. 21.14   Bearing capacity factors (Chen and Kulhawy), 1994

E_d = drained Young's modulus of the soil

μ_d = drained Poisson's ratio

Δ = volumetric strain within the plastic zone during the loading process

The expressions for μ_d and Δ may be written as (Chen and Kulhawy, 1994)

$$\mu_d = 0.1 + 0.3\phi_{rel} \tag{21.14}$$

$$\Delta = \frac{0.005(1-\phi_{rel})q_o'}{p_a} \tag{21.15}$$

where $\phi_{rel} = \dfrac{\phi° - 25°}{45° - 25°}$ for $25° \leq \phi° \leq 45°$ $\tag{21.16}$

= relative friction angle factor, p_a = atmospheric pressure = 101 kPa.

Chen and Kulhawy (1994) suggest that, for granular soils, the following values may be considered.

loose soil, E_d = 100 to 200 p_a $\tag{21.17}$

medium dense soil, E_d = 200 to 500 p_a

dense soil, E_d = 500 to 1000 p_a

The correction factors C_c and C_q indicated in Eq. (21.9) need be applied only if I_{rr} is less than the critical rigidity index $(I_r)_{crit}$ expressed as follows

$$(I_r)_{cr} = \frac{1}{2}\exp\left[2.85\cot\left(45° - \frac{\phi°}{2}\right)\right] \tag{21.18}$$

The values of critical rigidity index may be obtained from Table 21.1b for piers circular or square in section.

If $I_{rr} > (I_r)_{crit}$, the factors C_c and C_q may be taken as equal to unity.

The shape and depth factors in Eq. (21.3) can be evaluated by making use of the relationships given in Table 21.2.

TABLE 21.1b
Values of Critical Rigidity Index

Angle of shearing resistance ϕ	Critical Rigidity Index for	
	Strip foundation B/L = 0	Square foundation B/L = 1
0	13	8
5	18	11
10	25	15
15	37	20
20	55	30
25	89	44
30	152	70
35	283	120
40	592	225
45	1442	486
50	4330	1258

TABLE 21.2

Shape and depth factors (Eq. 21.3) (Chen and Kulhawy, 1994)

Factors	Value
s_c	$1 + \dfrac{N_q}{N_c}$
s_d	$dq - \dfrac{1 - d_q}{N_c \tan \phi}$
s_q	$1 + \tan \phi$
d_q	$1 + 2 \tan \phi (1 - \sin \phi)^2 \left[\dfrac{\pi}{180} \tan^{-1} \dfrac{L}{d} \right]$

Base in Cohesionless Soil

The theoretical approach as outlined above is quite complicated and difficult to apply in practice for drilled piers in granular soils. Direct and simple empirical correlations have been suggested by O'Neill and Reese (1999) between SPT N value and the base bearing capacity as given below for cohesionless soils.

$$q_b(= q_{max}) = 57.5N \text{ kPa} \leq 2900 \text{ kN/m}^2 \tag{21.19a}$$

$$q_b(= q_{max}) = 0.60 \, N \text{ tsf} \leq 30 \text{ tsf} \tag{21.19b}$$

where N = SPT value ≤ 50 blows / 0.3 m.

Base in Cohesionless IGM

Cohesionless IGM's are characterized by SPT blow counts if more than 50 per 0.3 m. In such cases, the expression for q_b is

$$q_b(= q_{max}) = 0.60 \left[N_{60} \frac{p_a}{q_o'} \right]^{0.8} q_o' \tag{21.20}$$

where
N_{60} = average SPT corrected for 60 percent standard energy within a depth of 2d (base) below the base. The value of N_{60} is limited to 100. No correction for overburden pressure

P_a = atmospheric pressure in the units used for q_o' (= 101 kPa in the SI System)

q_o' = vertical effective stress at the elevation of the base of the drilled shaft.

21.11 BEARING CAPACITY EQUATIONS FOR THE BASE IN COHESIVE IGM OR ROCK (O'NEILL AND REESE, 1999)

Massive rock and cohesive intermediate materials possess common properties. They possess low drainage qualities under normal loadings but drain more rapidly under large loads than cohesive soils. It is for these reasons undrained shear strengths are used for rocks and IGMs.

If the base of the pier lies in cohesive IGM or rock ($RQD = 100$ percent) and the depth of socket, D_s, in the IGM or rock is equal to or greater than 1.5d, the bearing capacity may be expressed as

$$q_b (= q_{max}) = 2.5 q_u \qquad (21.21)$$

where q_u = unconfined compressive strength of IGM or rock below the base

For RQD between 70 and 100 percent,

$$q_b (= q_{max}) = 4.83 (q_u)^{0.51} \text{MPa} \qquad (21.22)$$

For jointed rock or cohesive IGM

$$q_b (= q_{max}) = [s^{0.5} + (ms^{0.5} + s)^{0.5}] q_u \qquad (21.23)$$

where q_u is measured on intact cores from within $2d$ (base) below the base of the drilled pier. In all the above cases q_b and q_u are expressed in the same units and s and m indicate the properties of the rock or IGM mass that can be estimated from Tables 21.3 and 21.4.

TABLE 21.3
Descriptions of rock types

Rock type	Description
A	Carbonate rocks with well-developed crystal cleavage (eg., dolostone, limestone, marble)
B	Lithified argillaeous rocks (mudstone, siltstone, shale, slate)
C	Arenaceous rocks (sandstone, quartzite)
D	Fine-grained igneous rocks (andesite, dolerite, diabase, rhyolite)
E	Coarse-grained igneous and metamorphic rocks (amphibole, gabbro, gneiss, granite, norite, quartz-diorite)

TABLE 21.4
Values of *s* and *m* (dimensionless) based on rock classification (Carter and Kulhawy, 1988)

Quality of rock mass	Joint description and spacing	s	Value of m as function of rock type (A–E) from				
			A	B	C	D	E
Excellent	Intact (closed); spacing > 3 m (10 ft)	1	7	10	15	17	25
Very good	Interlocking; spacing of 1 to 3 m (3 to 10 ft)	0.1	3.5	5	7.5	8.5	12.5
Good	Slightly weathered; spacing of 1 to 3 m (3 to 10 ft)	4×10^{-2}	0.7	1	1.5	1.7	2.5
Fair	Moderately weathered; spacing of 0.3 to 1 m (1 to 3 ft)	10^{-4}	0.14	0.2	0.3	0.34	0.5
Poor	Weathered with gouge (soft material) spacing of 30 to 300 mm (1 in. to 1 ft)	10^{-5}	0.04	0.05	0.08	0.09	0.13
Very poor	Heavily weathered; spacing of less than 50 mm (2 in.)	0	0.007	0.01	0.015	0.017	0.025

The allowable bearing pressure q_a is also estimated based on RQD. The values of q_a for various values of RQD are given in Table 21.2.

21.12 THE ULTIMATE SKIN RESISTANCE OF COHESIVE AND INTERMEDIATE MATERIALS

Cohesive Soil

The process of drilling a borehole for a pier in cohesive soil disturbs the natural condition of the soil all along the side to a certain extent. There is a reduction in the soil strength not only due to boring but also due to stress relief and the time spent between boring and concreting. It is very difficult to quantify the extent of the reduction in strength analytically. In order to take care of the disturbance, the unit frictional resistance on the surface of the pier may be expressed as

$$f_s = \alpha c_u \qquad\qquad (21.24)$$

where α = adhesion factor

   c_u = undrained shear strength

Relationships have been developed between c_u and α by many investigators based on field load tests. Fig 21.15 gives one such relationship in the form of a curve developed by Chen and Kulhawy (1994). The curve has been developed on the following assumptions (Fig. 21.15).

$f_s = 0$ up to 1.5 m (= 5 ft) from the ground level.

$f_s = 0$ up to a height equal to $(h + d)$ as per Fig. 21.13.

O'Neill and Reese (1999) recommend the chart's trend line given in Fig. 21.15 for designing drilled piers. The suggested relationships are:

$$\alpha = 0.55 \quad \text{for } c_u / p_a \le 1.5 \qquad\qquad (21.25a)$$

$$\text{and } \alpha = 0.55 - 1\left[\frac{c_u}{p_a} - 1.5\right] \quad \text{for } 1.5 \le c_u / p_a \le 2.5 \qquad\qquad (21.25b)$$

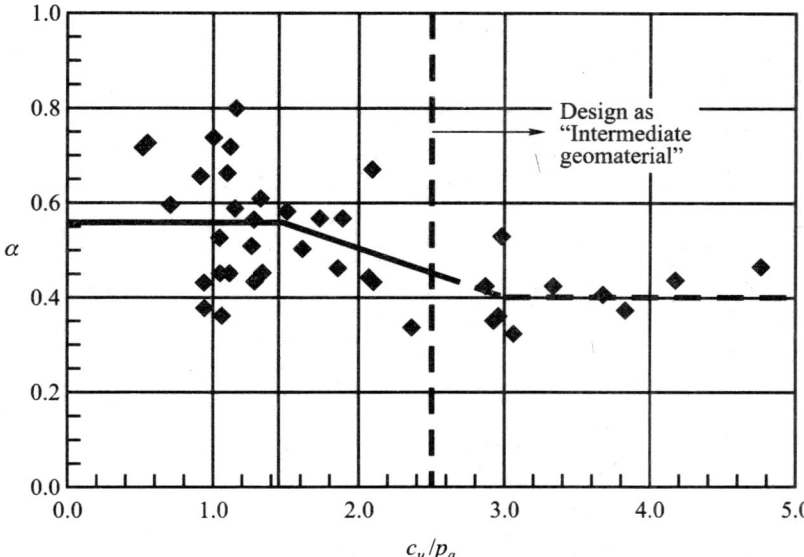

Fig. 21.15   Correlation between α and c_u/p_a

Cohesive Intermediate Geomaterials

Cohesive IGM's are very hard clay-like materials which can also be considered as very soft rock (O'Neill and Reese, 1999). IGM's are ductile and failure may be sudden at peak load. The value of f_a (please note that the term f_a is used instead of f_s for ultimate unit resistance at infinite displacement depends upon the side condition of the bore hole, that is, whether it is rough or smooth. For design purposes the side is assumed as smooth. The expression for f_a may be written as

$$f_a = \alpha q_u \tag{21.26}$$

where, q_u = unconfined compressive strength

   f_a = the value of ultimate unit side resistance which occurs at infinite displacement.
   Figure 21.16 gives a chart for evaluating α. The chart is prepared for an effective angle of friction between the concrete and the IGM (assuming that the intersurface is drained) and S_t denotes the settlement of piers at the top of the socket. Further, the chart involves the use of σ_n / p_a where σ_n is the normal effective pressure against the side of the borehole by the drilled pier and p_a is the atmospheric pressure (101 kPa).

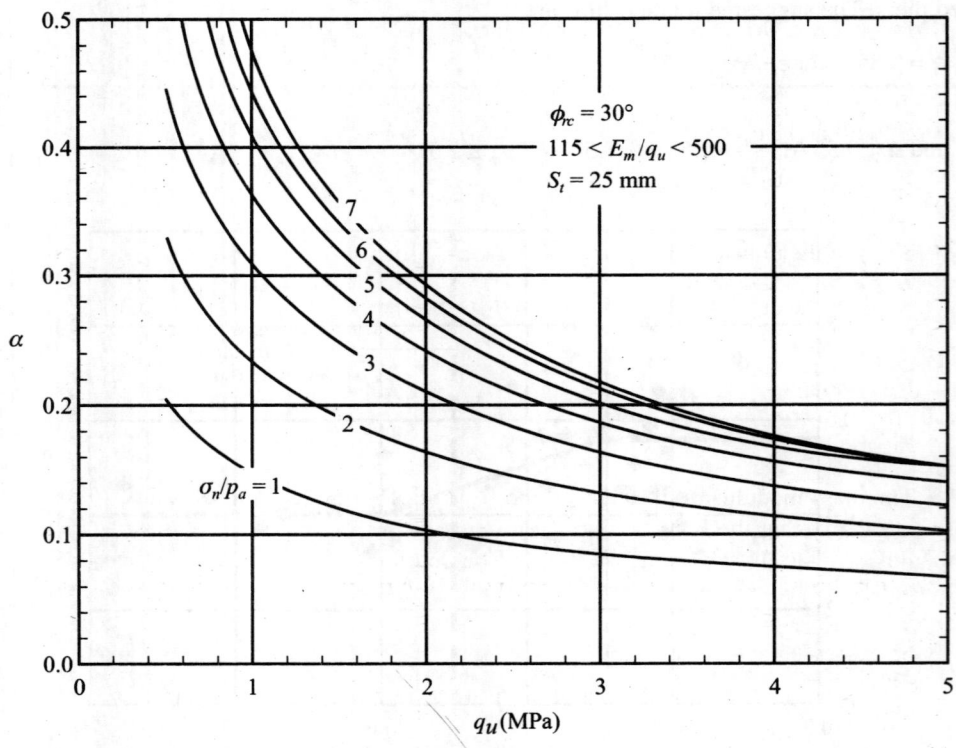

Fig. 21.16   Factor α for cohesive IGM's (O'Neill and Reese, 1999)

TABLE 21.5

Estimation of E_m/E_i based on RQD (Modified after Carter and Kulhawy, 1988)

RQD (percent)	E_m/E_i	
	Closed joints	Open joints
100	1.00	0.60
70	0.70	0.10
50	0.15	0.10
20	0.05	0.05

Note: Values intermediate between tabulated values may be obtained by linear interpolation.

TABLE 21.6

f_{aa}/f_a based on E_m/E_i (O'Neill et al., 1996)

E_m/E_i	f_{aa}/f_a
1.0	1.0
0.5	0.8
0.3	0.7
0.1	0.55
0.05	0.45

O'Neill and Reese (1999) give the following equation for computing σ_n

$$\sigma_n = M\gamma_c z_c \tag{21.27}$$

where γ_c = the unit weight of the fluid concrete used for the construction

z_c = the depth of the point at which σ_n is required

M = an empirical factor which depends on the fluidity of the concrete as indexed by the concrete slump

Figure 21.17 gives the values of M for various slumps.

The mass modulus of elasticity of the IGM (E_m) should be determined before proceeding, in order to verify that the IGM is within the limits of Fig 21.16. This requires the average Young's modulus of intact IGM core (E_i) which can be determined in the laboratory. Table 21.5 gives the ratios of E_m/E_i for various values of RQD. Values of E_m/E_i less than unity indicate that soft seams and/or joints exist in the IGM. These discontinuities reduce the value of f_a. The reduced value of f_a may be expressed as

$$f_{aa} = f_a R_a \tag{21.28}$$

where the ratio $R_a = f_{aa}/f_a$ can be determined from Table 21.6.

If the socket is classified as smooth, it is sufficiently accurate to set $f_s = f_{max} = f_{aa}$

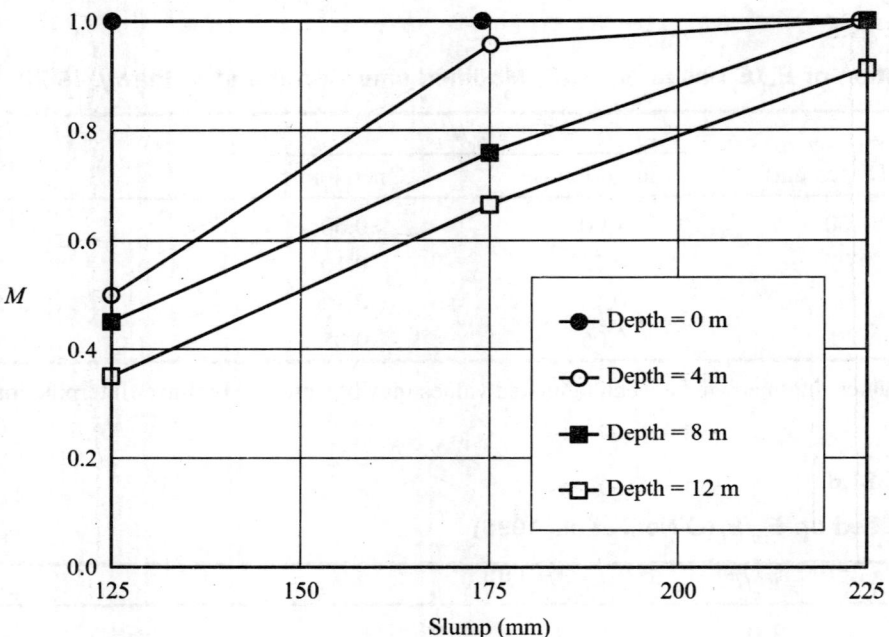

Fig. 21.17   Factor M versus concrete slump (O'Neill at al., 1996)

21.13 ULTIMATE SKIN RESISTANCE IN COHESIONLESS SOIL AND GRAVELLY SANDS (O'NEILL AND REESE, 1999)

In Sands

A general expression for total skin resistance in cohesionless soil may be written as [Eq. (21.1)]

$$Q_{fi} = \sum_{i=1}^{N} P_i f_{si} \Delta z_i = \sum_{i=1}^{N} P_i q'_{oi} K_{si} \tan\delta_i \Delta z_i \tag{21.29}$$

$$\text{or} \quad Q_{fi} = \sum_{i=1}^{N} P_i \beta_i q'_{oi} \Delta z_i \tag{21.30}$$

where $f_{si} = \beta_i\, q'_{oi}$ (21.31)

$\beta_i = K_{si} \tan\delta_i$

δ_i = angle of skin friction of the i th layer

The following equations are provided by O'Neill and Reese (1999) for computing β_i.

For SPT N_{60} (uncorrected) ≥ 15 blows / 0.3 m

$$\beta_i = 1.5 - 0.245[z_i]^{0.5} \tag{21.32}$$

For SPT N_{60} (uncorrected) < 15 blows / 0.3 m

$$\beta_i = \frac{N_{60}}{15}\left[1.5 - 0.245(z_i)^{0.5}\right] \tag{21.33}$$

In Gravelly Sands or Gravels

For SPT $N_{60} \geq 15$ blows / 0.3 m

$$\beta_i = 2.0 - 0.15[z_i]^{0.75} \tag{21.34}$$

In gravelly sands or gravels, use the method for sands if $N_{60} < 15$ blows / 0.3 m.

The definitions of various symbols used above are

β_i = dimensionless correlation factor applicable to layer i. Limited to 1.2 in sands and 1.8 in gravelly sands and gravel. Minimum value is 0.25 in both types of soil; f_{si} is limited 200 kN/m^2 (2.1 tsf)

q'_{oi} = vertical effective stress at the middle of each layer

N_{60} = design value for SPT blow count, uncorrected for depth, saturation or fines corresponding to layer i

z_i = vertical distance from the ground surface, in meters, to the middle of layer i. The layer thickness Δz_i is limited to 9 m.

21.14 ULTIMATE SIDE AND TOTAL RESISTANCE IN ROCK (O'NEILL AND REESE, 1999)

Ultimate Skin Resistance (for Smooth Socket)

Rock is defined as a cohesive geomaterial with $q_u > 5$ MPa (725 psi). The following equations may be used for computing f_s ($= f_{max}$) when the pier is socketed in rock. Two methods are proposed.

Method 1

$$f_s(= f_{max}) = 0.65 p_a \left[\frac{q_u}{p_a}\right]^{0.5} \leq 0.65 p_a \left[\frac{f_c}{p_a}\right]^{0.5} \tag{21.35}$$

where, p_a = atmospheric pressure (= 101 kPa)

q_u = unconfined compressive strength of rock mass

f_c = 28 day compressive cylinder strength of concrete used in the drilled pier

Method 2

$$f_s(= f_{max}) = 1.42 p_a \left[\frac{q_u}{p_a}\right]^{0.5} \tag{21.36}$$

Carter and Kulhawy (1988) suggested equation (21.36) based on the analysis of 25 drilled shaft socket tests in a very wide variety of soft rock formations, including sandstone, limestone, mudstone, shale and chalk.

Ultimate Total Resistance Q_u

If the base of the drilled pier rests on sound rock, the side resistance can be ignored. In cases where significant penetration of the socket can be made, it is a matter of engineering judgment to decide whether Q_f should be added directly to Q_b to obtain the ultimate value Q_u. When the rock is brittle in shear, much side resistance will be lost as the settlement increases to the value required to develop the full value of q_b ($= q_{max}$). If the rock is ductile in shear, there is no question that the two values can be added directly (O'Neill and Reese, 1999).

21.15 ESTIMATION OF SETTLEMENTS OF DRILLED PIERS AT WORKING LOADS

O'Neill and Reese (1999) suggest the following methods for computing axial settlements for isolated drilled piers:

1. Simple formulas
2. Normalized load-transfer methods

The total settlement S_t at the pier head at working loads may be expressed as (Vesic, 1977)

$$S_t = S_e + S_{bb} + S_{bs} \qquad\qquad (21.37)$$

where, S_e = elastic compression

S_{bb} = settlement of the base due to the load transferred to the base

S_{bs} = settlement of the base due to the load transferred into the soil along the sides.

The equations for the settlements are

$$S_e = \frac{L(Q_a - 0.5Q_{fm})}{A_b E} \qquad\qquad (21.38a)$$

$$S_{bb} = C_p \left(\frac{Q_{mb}}{dq_b} \right) \qquad\qquad (21.38)$$

$$S_{bs} = \left[0.93 + 0.16\sqrt{\frac{L}{d}} \right] C_p \left(\frac{Q_{fm}}{Lq_b} \right) \qquad\qquad (21.39)$$

where L = length of the drilled pier

A_b = base cross-sectional area

E = Young's modulus of the drilled pier

Q_a = load applied to the head

Q_{fm} = mobilized side resistance when Q_a is applied

Q_{bm} = mobilized base resistance

d = pier width or diameter

C_p = soil factor obtained from Table 21.7

TABLE 21.7
Values of C_p for various soils (Vesic, 1977)

Soil	C_p
Sand (dense)	0.09
Sand (loose)	0.18
Clay (stiff)	0.03
Clay (soft)	0.06
Silt (dense)	0.09
Silt (loose)	0.12

Normalized Load-Transfer Methods

Reese and O'Neill (1988) analyzed a series of compression loading test data obtained from fullsized drilled piers in soil. They developed normalized relations for piers in cohesive and cohesionless soils. Figures 21.18 and 21.19 can be used to predict settlements of piers in cohesive soils and Figs 21.20 and 21.21 in cohesionless soils including soil mixed with gravel.

The boundary limits indicated for gravel in Fig. 21.20 have been found to be approximately appropriate for cemented fine-grained desert IGM's (Walsh *et al.*, 1995). The range of validity of the normalized curves are as follows:

Figures 21.18 and 21.19

Normalizing factor	=	shaft diameter d
Range of d	=	0.46 m to 1.53 m

Figures 21.20 and 21.21

Normalizing factor	=	base diameter
Range of d	=	0.46 m to 1.53 m

The following notations are used in the figures:

S_R	=	Settlement ratio = S_a/d
S_a	=	Allowable settlement
N_{fm}	=	Normalized side load transfer ratio = Q_{fm}/Q_f
N_{bm}	=	Normalize base load transfer ratio = Q_{bm}/Q_b

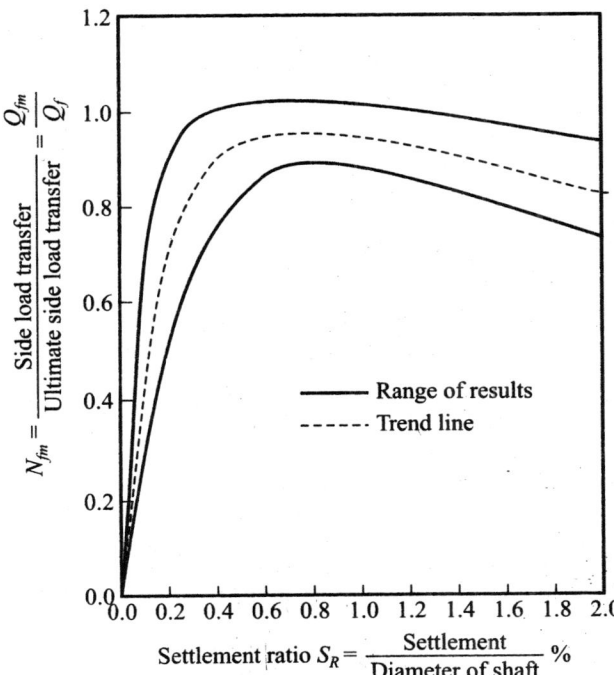

Fig. 21.18 Normalised side load transfer for drilled shaft in cohesive soil (O'Neill and Reese, 1999)

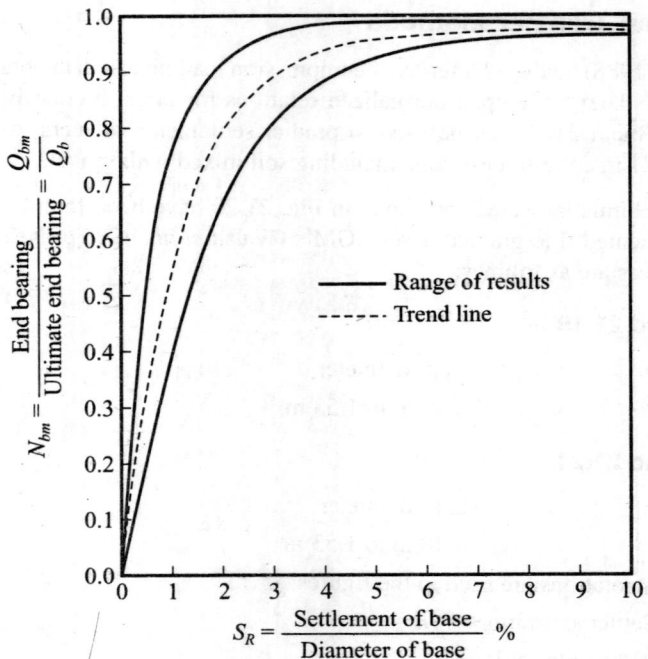

Fig. 21.19 Normalised base load transfer for drilled shaft in cohesive soil (O'Neill and Reese, 1999)

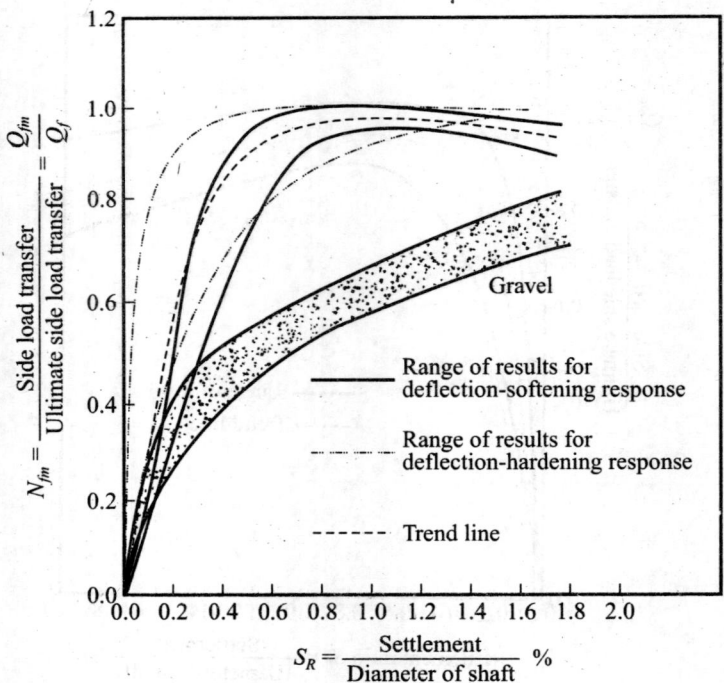

Fig. 21.20 Normalised side load transfer for drilled shaft in cohesionless soil (O'Neill and Reese, 1999)

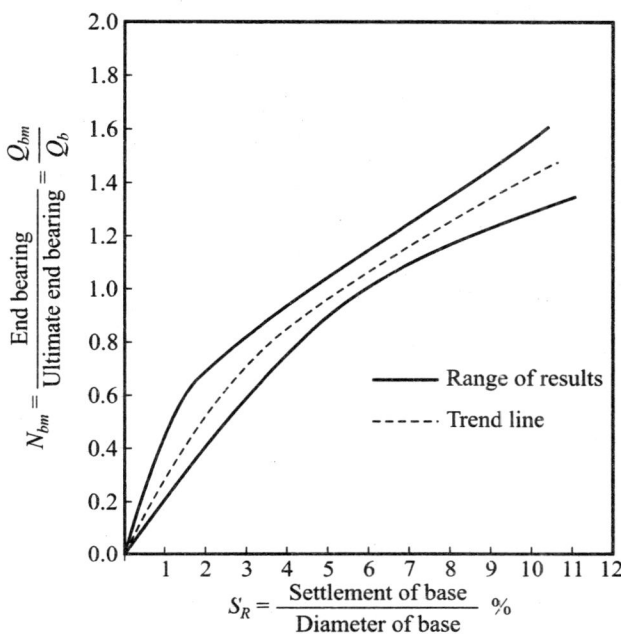

Fig. 21.21 Normalised base load transfer for drilled shaft in cohesionless soil (O'Neill and Reese, 1999)

Example 21.1

A multistory building is to be constructed in a stiff to very stiff clay. The soil is homogeneous to a great depth. The average value of undrained shear strength c_u is 150 kN/m². It is proposed to use a drilled pier of length 25 m and diameter 1.5 m. Determine (a) the ultimate load capacity of the pier, and (b) the allowable load on the pier with $F_s = 2.5$. (Fig. Ex. 21.1).

Solution

Base load

When $c_u \geq 96$ kPa (2000 lb/ft²), use Eq. (21.8) for computing q_b. In this case $c_u > 96$ kPa.

$$q_b = 9c_u = 9 \times 150 = 1350 \text{ kN/m}^2$$

$$\text{Base load } Q_b = A_b q_b = \frac{3.14 \times 1.5^2}{4} \times 1350 = 1.766 \times 1350 = 2384 \text{ kN}$$

Frictional load

The unit ultimate frictional resistance f_s is determined using Eq. (21.24)

$$f_s = \alpha c_u$$

From Fig. (21.15) $\alpha = 0.55$ for $c_u/p_a = 150/101 = 1.5$

where p_a is the atmospheric pressure = 101 kPa

Therefore $f_s = 0.55 \times 150 = 82.5$ kN/m²

The effective length of the shaft for computing the frictional load (Fig. 21.13 a) is

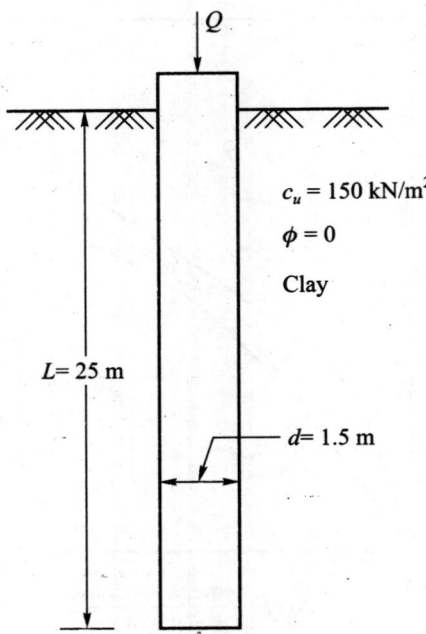

Fig. Ex. 21.1

$L' = [L - (d + 1.5)]$ m $= 25 - (1.5 + 1.5) = 22$ m

The effective surface area $A_s = \pi d L' = 3.14 \times 1.5 \times 22 = 103.62$ m^2

Therefore $\quad Q_f = f_s A_s = 82.5 \times 103.62 = 8,549$ kN

The total ultimate load is

$Q_u = Q_f + Q_b = 8,549 + 2,384 = 10,933$ kN

The allowable load may be determined by applying an overall factor of safety to Q_u. Normally $F_s = 2.5$ is sufficient.

$$Q_a = \frac{10,933}{2.5} = 4,373 \text{ kN}$$

Example 21.2

For the problem given in Ex. 21.1, determine the allowable load for a settlement of 10 mm ($= S_a$). All the other data remain the same.

Solution

Allowable skin load

Settlement ratio $S_R = \dfrac{S_a}{d} = \dfrac{10}{1.5 \times 10^3} \times 100 = 0.67\%$

From Fig. 21.18 for $S_R = 0.67\%$, $N_{fm} = \dfrac{Q_{fm}}{Q_f} = 0.95$ by using the trend line.

$Q_{fm} = 0.95 \, Q_f = 0.95 \times 8,549 = 8,122$ kN.

Allowable base load for S_a = 10 mm

From Fig. 21.19 for $S_R = 0.67\%$, $N_{bm} = \dfrac{Q_{bm}}{Q_b} = 0.4$

$Q_{bm} = 0.4\ Q_b = 0.4 \times 2{,}384 = 954$ kN.

Now the allowable load Q_{as} based on settlement consideration is

$Q_{as} = Q_{fm} + Q_{bm} = 8{,}122 + 954 = 9{,}076$ kN

Q_{as} based on settlement consideration is very much higher than Q_a (Ex. 21.1) and as such Q_a governs the criteria for design.

Example 21.3

Figure Ex. 21.3 depicts a drilled pier with a belled bottom. The details of the pile and the soil properties are given in the figure. Estimate (a) the ultimate load, and (b) the allowable load with $F_s = 2.5$.

Solution

Based load

Use Eq. (21.8) for computing q_b

$q_b = 9c_u = 9 \times 200 = 1{,}800$ kN/m^2

Base load $Q_b = \dfrac{\pi d_b^2}{4} \times q_b = \dfrac{3.14 \times 3^2}{4} \times 1{,}800 = 12{,}717$ kN

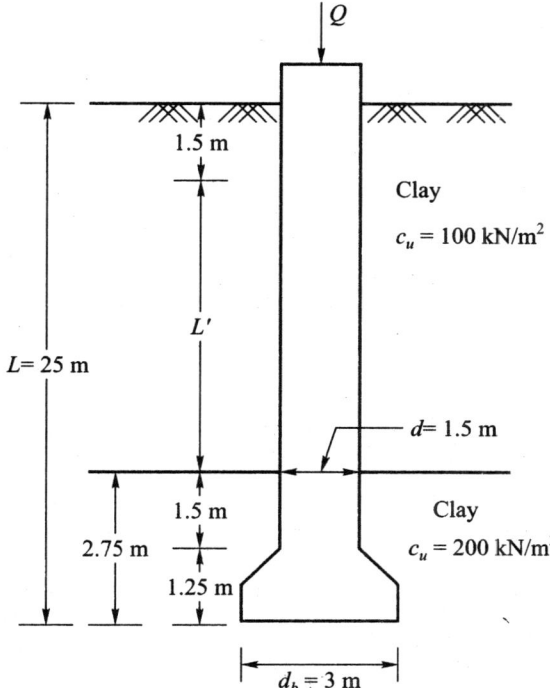

Fig. Ex. 21.3

Frictional load

The effective length of shaft $L' = 25 - (2.75 + 1.5) = 20.75$ m

From Eq. (21.24) $f_s = \alpha c_u$

For $\dfrac{c_u}{p_a} = \dfrac{100}{101} \approx 1.0$, $\alpha = 0.55$ from Fig. 21.15

Hence $f_s = 0.55 \times 100 = 55$ kN/m^2

$Q_f = PL'f_s = 3.14 \times 1.5 \times 20.75 \times 55 = 5{,}375$ kN

$Q_u = Q_b + Q_f = 12{,}717 + 5{,}375 = 18{,}092$ kN

$Q_a = \dfrac{18{,}092}{2.5} = 7{,}237$ kN

Example 21.4

For the problem given in Ex. 21.3, determine the allowable load Q_{as} for a settlement $S_a = 10$ mm.

Solution

Skin load Q_{fm} (mobilized)

Settlement ratio $S_R = \dfrac{10}{1.5 \times 10^3} \times 100 = 0.67\%$

From Fig. 21.18 for $S_R = 0.67$, $N_{fm} = 0.95$ from the trend line.

Therefore $Q_{fm} = 0.95 \times 5{,}375 = 5{,}106$ kN

Base load Q_{bm} (mobilized)

$S_R = \dfrac{10}{3 \times 10^3} \times 100 = 0.33\%$

From Fig. 21.19 for $S_R = 0.33\%$, $N_{bm} = 0.3$ from the trend line.

Hence $Q_{bm} = 0.3 \times 12{,}717 = 3815$ kN

$Q_{as} = Q_{fm} + Q_{bm} = 5{,}106 + 3{,}815 = 8{,}921$ kN

The factor of safety with respect to Q_u is (from Ex. 21.3)

$F_s = \dfrac{18{,}092}{8{,}921} = 2.03$

This is low as compared to the normally accepted value of $F_s = 2.5$. Hence Q_a rules the design.

Example 21.5

Figure Ex. 21.5 shows a straight shaft drilled pier constructed in homogeneous loose to medium dense sand. The pile and soil properties are:

$L = 25$ m, $d = 1.5$ m, $c = 0$, $\phi = 36°$ and $\gamma = 17.5$ kN/m^3

Estimate (a) the ultimate load capacity, and (b) the allowable load with $F_s = 2.5$. The average SPT value $N_{cor} = 30$ for $\phi = 36°$.

Use (i) Vesic's method, and (ii) the O'Neill and Reese method.

Solution

(i) Vesic's method

From Eq. (21.10) for $c = 0$

$$q_b = (N_q - 1)q'_o s_q d_q C_q \qquad \text{(a)}$$

$$q'_o = 25 \times 17.5 = 437.5 \text{ kN/m}^2$$

From Eq. (21.16) $\phi_{rel} = \dfrac{\phi^\circ - 25^\circ}{45^\circ - 25^\circ} = \dfrac{36 - 25}{20} = 0.55$

From Eq. (21.15) $\Delta = \dfrac{0.005(1 - 0.55) \times 437.5}{101} \approx 0.01$

From Eq. (21.14) $\mu_d = 0.1 + 0.3\phi_{rel} = 0.1 + 0.3 \times 0.55 = 0.265$

From Eq. (21.17) $E_d = 200\, p_a = 200 \times 101 = 20,200 \text{ kN/m}^2$

From Eq. (21.13) $I_r = \dfrac{E_d}{2(1 + \mu_d)q'_o \tan\phi} = \dfrac{20,200}{2(1 + 0.265) \times 437.5 \tan 36^\circ} = 25$

From Eq. (21.12) $I_{rr} = \dfrac{I_r}{1 + \Delta I_r} = \dfrac{25}{1 + 0.01 \times 25} = 20$

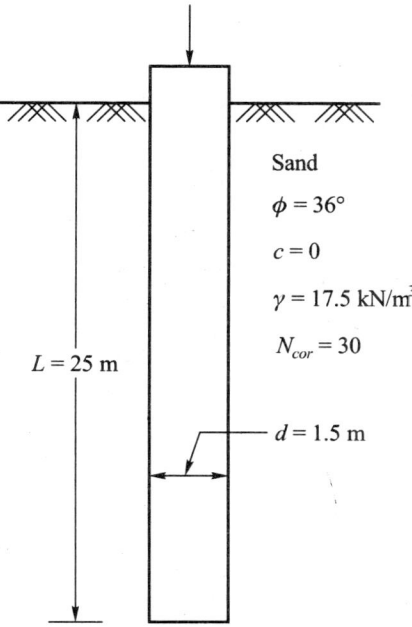

Sand

$\phi = 36^\circ$

$c = 0$

$\gamma = 17.5 \text{ kN/m}^3$

$N_{cor} = 30$

$L = 25 \text{ m}$

$d = 1.5 \text{ m}$

Fig. Ex. 21.5

From Eq. (21.11b)

$$C_q = \exp\left\{[-3.8\tan36°] + \left[\frac{3.07\sin36°\log_{10}2\times20}{1+\sin36°}\right]\right\} = \exp-(0.9399) = 0.391$$

From Fig. 21.14, $N_q = 30$ for $\phi = 36°$

From Table (21.2) $s_q = 1 + \tan36° = 1.73$

$$d_q = 1 + 2\tan36°(1-\sin36°)^2\left[\frac{3.14}{180}\times\tan^{-1}\frac{25}{1.5}\right] = 1.373$$

Substituting in Eq. (a)

$q_b = (30-1)\times437.5\times1.73\times1,373\times0.391 = 11,783\,\text{kN/m}^2 > 11,000\,\text{kN/m}^2$

As per Tomlinson (1986) the computed q_b should be less than 11,000 kN/m².

Hence $Q_b = \dfrac{3.14}{4}\times(1.5)^2\times11,000 = 19,429\,\text{kN}$

Skin load Q_f

From Eqs (21.31) and (21.32)

$$f_s = \beta q_o', \quad \beta = 1.5 - 0.24z^{0.5}, \text{ where } z = \frac{L}{2} = \frac{25}{2} = 12.5\,\text{m}$$

Substituting

$$\beta = 1.5 - 0.245\times(12.5)^{0.5} = 0.63$$

Hence $f_s = 0.63\times437.5 = 275.62\,\text{kN/m}^2$

Per Tomlinson (1986) f_s should be limited to 110 kN/m². Hence $f_s = 110\,\text{kN/m}^2$

Therefore $Q_f = \pi dL f_s = 3.14\times1.5\times25\times110 = 12,953\,\text{kN}$

Ultimate load kN $Q_u = 19,429 + 12,953 = 32,382\,\text{kN}$

$$Q_a = \frac{32,382}{2.5} = 12,953\,\text{kN}$$

O'Neill and Reese method

This method relates q_b to the SPT N value as per Eq. (21.19a)

$$q_b = 57.5N\ \text{kN/m}^2 = 57.5\times30 = 1,725\,\text{kN/m}^2$$

$Q_b = A_b q_b = 1.766\times1,725 = 3,046\,\text{kN}$

The method for computing Q_f remains the same as above.

Now $Q_u = 3,406 + 12,953 = 15,999$

$$Q_a = \frac{15,999}{2.5} = 6,400\,\text{kN}$$

Example 21.6

Compute Q_u and Q_a for the pier given in Ex. 21.5 by the following methods.

1. Use the SPT value [Eq. (19.10c)] for bored piles

2. Use the Tomlinson method of estimating Q_b and Table 19.2 for estimating Q_f. Compare the results of the various methods.

Solution

Use of the SPT value [Meyerhof Eq. (19.10c)]

$$q_b = 133N = 133 \times 30 = 3,990 \text{ kN/m}^2$$

$$Q_b = \frac{3.14 \times (1.5)^2}{4} \times 3,990 = 7,047 \text{ kN}$$

$$f_s = 0.67N = 0.67 \times 30 = 20 \text{ kN/m}^2$$

$$Q_f = 3.14 \times 15 \times 25 \times 20 = 2,355 \text{ kN}$$

$$Q_u = 7,047 + 2,355 = 9,402 \text{ kN}$$

$$Q_a = \frac{9,402}{2.5} = 3,760 \text{ kN}$$

Tomlinson Method for Q_b

For a driven pile

From Fig. 19.8 $N_q = 65$ for $\phi = 36°$ and $\dfrac{L}{d} = \dfrac{25}{1.5} \approx 17$

Hence $q_b = q'_o N_q = 437.5 \times 65 = 28,438$ kN/m²

For bored pile

$$q_b = \frac{1}{3} q_b \text{(driven pile)} = \frac{1}{3} \times 28,438 = 9,479 \text{ kN/m}^2$$

$$Q_b = A_b q_b = 1.766 \times 9,479 = 16,740 \text{ kN}$$

Q_f from Table 19.2

For $\phi = 36°$, $\delta = 0.75 \times 36 = 27$. and $\bar{K}_s = 1.5$ (for medium dense sand).

$$f_s = \bar{q}'_o \bar{K}_s \tan \delta = \frac{437.5}{2} \times 1.5 \tan 27° = 167 \text{ kN/m}^2$$

As per Tomlinson (1986) f_s is limited to 110 kN/m². Use $f_s = 110$ kN/m².

Therefore, $Q_f = 3.14 \times 1.5 \times 25 \times 110 = 12,953$ kN

$Q_u = Q_b + Q_f = 16,740 + 12,953 = 29,693$ kN

$$Q_a = \frac{29,693}{2.5} = 11,877 \text{ kN}$$

Comparison of estimated results ($F_s = 2.5$)

Example No	Name of method	Q_b (kN)	Q_f (kN)	Q_u (kN)	Q_a (kN)
21.5	Vesic	19,429	12,953	32,382	12,953
21.5	O'Neill and Reese, for Q_b and Vesic for Q_f	3,046	12,953	15,999	6,400
21.6	Meyerhof	7,047	2,355	9,402	3,760
21.6	Tomlinson for Q_b	16,740	12,953	29,693	11,877

Which method to use

The variation in the values of Q_b and Q_f are very large between the methods. Since the soils encountered in the field are generally heterogeneous in character no theory holds well for all the soil conditions. Designers have to be practical and pragmatic in the selection of any one or combination of the theoretical approaches discussed earlier.

Example 21.7

For the problem given in Example 21.5 determine the allowable load for a settlement of 10 mm. All the other data remain the same. Use (a) the values of Q_f and Q_b obtained by Vesic's method, and (b) Q_b from the O'Neill and Reese method.

Solution

(a) Vesic's values Q_f and Q_b

Settlement ratio for $S_a = 10$ mm is

$$S_R = \frac{S_a}{d} = \frac{10 \times 10^2}{1.5 \times 10^3} = 0.67\%$$

From Fig. 21.20 for $S_R = 0.67\%$ $N_{fm} = 0.96$ (approx.) using the trend line.

$$Q_{fm} = 0.96 \times Q_f = 0.96 \times 12,953 = 12,435 \text{ kN}$$

From Fig. 21.21 for $S_R = 0.67\%$

$$N_{bm} = 0.20, \text{ or } Q_{bm} = 0.20 \times 19,429 = 3,886 \text{ kN}$$

$$Q_{as} = 12,435 + 3,886 = 16,321 \text{ kN}$$

Shear failure theory give $Q_a = 12,953$ kN which is much lower than Q_{as}. As such Q_a determines the criteria for design.

(b) O'Neill and Reese $Q_b = 3,046$ kN

As above, $Q_{bm} = 0.20 \times 3,046 = 609$ kN

Using Q_{fm} in (a) above,

$$Q_{as} = 609 + 12,435 = 13,044 \text{ kN}$$

The value of Q_{as} is closer to Q_a (Vesic) but much higher than Q_a calculated by all the other methods.

Example 21.8

Figure Ex. 21.8 shows a drilled pier penetrating an IGM: clay-shale to a depth of 8 m. Joints exists within the IGM stratum. The following data are available: $L_s = 8$ m $(= z_c)$, $d = 1.5$ m, q_u (rock) $= 3 \times 10^3$ kN/m², E_i (rock) $= 600 \times 10^3$ kN/m², concrete slump $= 175$ mm, unit weight of concrete $\gamma_c = 24$ kN/m³, E_c (concrete) $= 435 \times 10^6$ kN/m², and RQD $= 70$ percent, q_u (concrete) $= 435 \times 106$ kN/m². Determine the ultimate frictional load Q_f (max).

Solution

(a) Determine α in Eq. (21.26)

$f_a = \alpha q_u$ where $q_u = 3$ MPa for rock

For the depth of socket $L_s = 8$ m, and slump $= 175$ mm

$M = 0.76$ from Fig. 21.17.

From Eq. (21.27)

$\sigma_n = M\phi_c z_c = 0.76 \times 24 \times 8 = 146$ kN/m²

$p_a = 101$ kN/m², $\sigma_n / p_a = 146/101 = 1.45$

From Fig. 21.16 for $q_u = 3$ MPa and $\sigma_n / p_a = 1.45$, we have $\alpha = 0.11$

(b) Determination of f_a

$f_a = 0.11 \times 3 = 0.33$ MPa

(c) Determination f_{aa} in Eq. (21.28)

For $RQD = 70\%$, $E_m/E_i = 0.1$ from Table 21.5 for open joints, and $f_{aa}/f_a (= R_a) = 0.55$ from Table 21.6

$f_{max} = f_{aa} = 0.55 \times 0.33 = 0.182$ MPa $= 182$ kN/m²

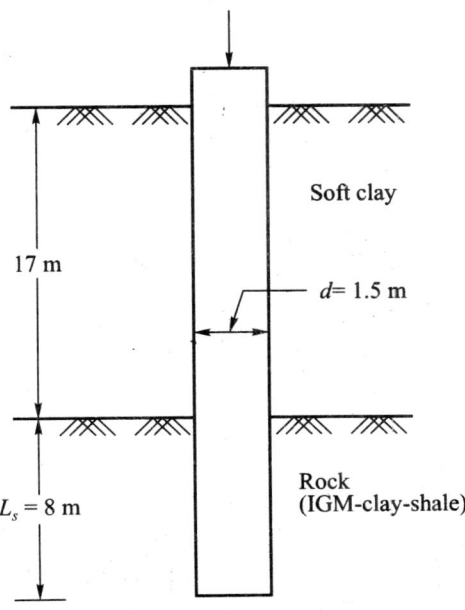

Fig. Ex. 21.8

(d) Ultimate friction load Q_f

$$Q_f = PLf_{aa} = 3.14 \times 1.5 \times 8 \times 182 = 6{,}858 \text{ kN}$$

Example 21.9

For the pier given in Ex. 21.8, determine the ultimate bearing capacity of the base. Neglect the frictional resistance. All the other data remain the same.

Solution

For RQD between 70 and 100 percent

from Eq. (21.22)

$$q_b (= q_{max}) = 4.83(q_u)^{0.5} \text{ MPa} = 4.83 \times (3)^{0.5} = 8.37 \text{ MPa}$$

$$Q_b (\text{max}) = \frac{3.14}{4} \times 1.5^2 \times 8.37 = 14.78 \text{ MN} = 14{,}780 \text{ kN}$$

21.16 UPLIFT CAPACITY OF DRILLED PIERS

Structures subjected to large overturning moments can produce uplift loads on drilled piers if they are used for the foundation. The design equation for uplift is similar to that of compression. Figure 21.22 shows the forces acting on the pier under uplift-load Q_{ul}. The equation for Q_{ul} may be expressed as

$$Q_{ul} = Q_{fr} + W_p = A_s f_r + W_p \tag{21.40}$$

where, Q_{fr} = total side resistance for uplift

W_p = effective weight of the drilled pier

A_s = surface area of the pier

f_r = frictional resistance to uplift

Uplift Capacity of Single Pier (straight edge)

For a drilled pier in cohesive soil, the frictional resistance may expressed as (Chen and Kulhawy, 1994).

$$f_r = \alpha c_u \tag{21.41a}$$

$$\alpha = 0.31 + 0.17 \frac{c_u}{p_a} \tag{21.41b}$$

where, α = adhesion factor

c_u = undrained shear strength of cohesive soil

p_a = atmospheric pressure (101 kPa)

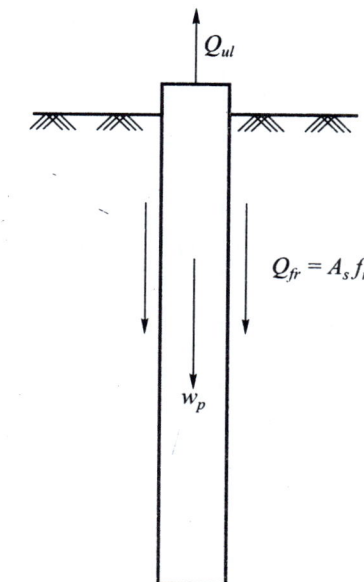

Fig. 21.22 Uplift forces for a straight edged pier

Uplift Resistance of Piers in Sand

There are no confirmatory methods available for evaluating uplift capacity of piers embedded in cohesionless soils. Poulos and Davis, (1980) suggest that the skin frictional resistance for pull out may be taken as equal to two-thirds of the shaft resistance for downward loading.

Uplift Resistance of Piers in Rock

According to Carter and Kulhawy (1988), the frictional resistance offered by the surface of the pier under uplift loading is almost equal to that for downward loading if the drilled pier is rigid relative to the rock. The effective rigidity is defined as $(E_c/E_m)(d/D_s)^2$, in which E_c and E_m are the Young's modulus of the drilled pier and rock mass respectively, d is the socket diameter and D_s is the depth of the socket. A socket is rigid when $(E_c/E_m)(d/D_s)^2 \geq 4$. When the effective rigidity is less than 4, the frictional resistance f_r for upward loading may be taken as equal to 0.7 times the value for downward loading.

Example 21.10

Determine the uplift capacity of the drilled pier given in Fig. Ex. 21.10. Neglect the weight of the pier.

Solution

From Eq. (21.40)

$$Q_{ul} = A_s f_r$$

From Eq. (21.41a) $f_r = \alpha c_u$

From Eq.(21.41b) $\alpha = 0.31 + 0.17 \dfrac{c_u}{p_a}$

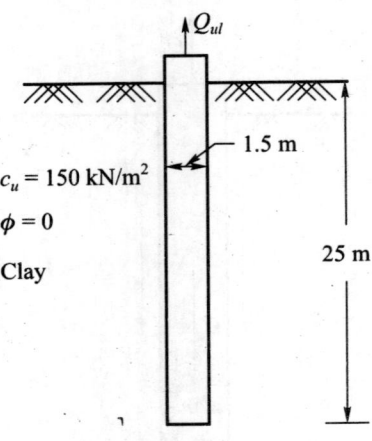

Fig. Ex. 21.10

Given: $L = 25$ m, $d = 1.5$ m, $c_u = 150$ kN/m^2

Hence $\alpha = 0.31 + 0.17 \times \dfrac{150}{101} = 0.56$

$f_r = 0.56 \times 150 = 84 \text{ kN/m}^2$

$Q_{ul} = 3.14 \times 1.50 \times 25 \times 84$

$\quad = 9{,}891$ kN

It may be noted here that $f_s = 82.5$ kN/m^2 for downward loading and $f_r = 84$ kN/m^2 for uplift. The two values are very close to each other.

21.17 QUESTIONS AND PROBLEMS

21.1   Figure Prob. 21.1 shows a drilled pier of diameter 1.25 m constructed for the foundation of a bridge. The soil investigation at the site revealed soft to medium stiff clay extending to a great depth. The other details of the pier and the soil are given in the figure. Determine (a) the ultimate load capacity, and (b) the allowable load for $F_s = 2.5$. Use Vesic's method for base load and α-method for the skin load.

21.2   Refer to Prob. 21.1. Given $d = 3$ ft, $L = 30$ ft, and $c_u = 1050$ lb/ft^2. Determine the ultimate (a) base load capacity by Vesic's method, and (b) the frictional load capacity by the α method.

21.3   Figure Prob. 21.3 shows a drilled pier with a belled bottom constructed for the foundation of a multistory building. The pier passes through two layers of soil. The details of the pier and the properties of the soil are given in the figure. Determine the allowable load Q_a for $F_s = 2.5$. Use (a) Vesic's method for the base load, and (b) the O'Neill and Reese method for skin load.

21.4   For the drilled pier given in Fig. Prob. 21.1, determine the working load for a settlement of 10 mm. All the other data remain the same. Compare the working load with the allowable load Q_a.

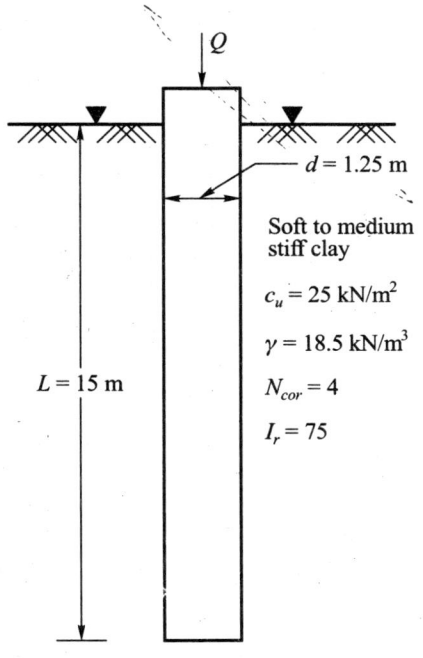

Fig. Prob. 21.1

Fig. Prob. 21.3

21.5   For the drilled pier given in Prob 21.2, compute the working load for a settlement of 0.5 in. and compare this with the allowable load Q_a.

21.6   If the drilled pier given in Fig. Prob. 21.6 is to carry a safe load of 2500 kN, determine the length of the pier for $F_s = 2.5$. All the other data are given in the figure.

21.7   Determine the settlement of the pier given in Prob. 21.6 by the O'Neill and Reese method. All the other data remain the same.

21.8   Fig. Prob. 21.8 depicts a drilled pier with a belled bottom constructed in homogeneous clay extending to a great depth. Determine the length of the pier to carry an allowable load of 3000 kN with a $F_s = 2.5$. The other details are given in the figure.

21.9   Determine the settlement of the pier in Prob 21.8 for a working load of 3000 kN. All the other data remain the same. Use the length L computed.

21.10   Figure Prob 21.10 shows a drilled pier. The pier is constructed in homogeneous loose to medium dense sand. The pier details and the properties of the soil are given in the figure. Estimate by Vesic's method the ultimate load bearing capacity of the pier.

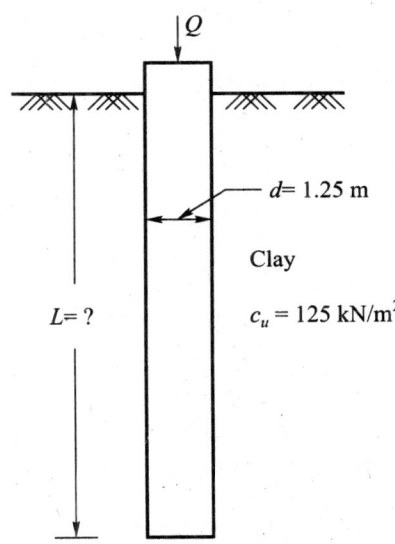

Fig. Prob. 21.6

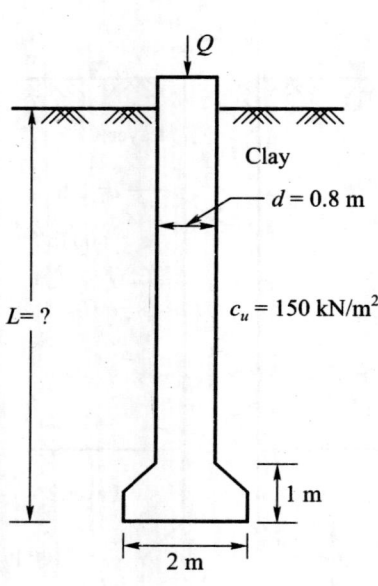

Fig. Prob. 21.8   **Fig. Prob. 21.10**

21.11   For Problem 21.10 determine the ultimate base capacity by the O'Neill and Reese method. Compare this value with the one computed in Prob. 21.10.

21.12   Compute the allowable load for the drilled pier given in Fig. 21.10 based on the SPT value. Use Meyerhof's method.

21.13   Compute the ultimate base load of the pier in Fig. Prob. 21.10 by Tomlinson's method.

21.14   A pier is installed in a rocky stratum. Fig. Prob. 21.14 gives the details of the pier and the properties of the rock materials. Determine the ultimate frictional load Q_f (max).

21.15   Determine the ultimate base resistance of the drilled pier in Prob. 21.14. All the other data remain the same. What is the allowable load with $F_s = 4$ by taking into account the frictional load Q_f computed in Prob. 21.14?

21.16   Determine the ultimate point bearing capacity of the pier given in Prob. 21.14 if the base rests on sound rock with $RQD = 100\%$.

21.17   Determine the uplift capacity of the drilled pier given in Prob. 21.1. Given: $L = 15$ m, $d = 1.25$ m, and $c_u = 25$ kN/m². Neglect the weight of the pile.

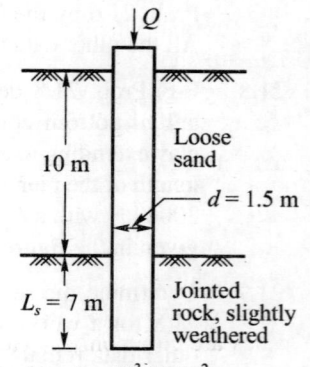

q_u (rock) = 2 × 10³ kN/m²
E_i (rock) = 40 × 10⁴ kN/m²
E_c (concrete) = 435 × 10⁶ kN/m²
RQD = 50%
q_u (concrete) = 40 × 10⁴ kN/m²
slump = 175 mm, γ_c = 23.5 kN/m³

Fig. Prob. 21.14

CHAPTER 22

BRACED CUTS AND DRAINAGE

PART A: BRACED CUTS

22.1 GENERAL CONSIDERATIONS

Shallow excavations can be made without supporting the surrounding material if there is adequate space to establish slopes at which the material can stand. The steepest slopes that can be used in a given locality are best determined by experience. Many building sites extend to the edges of the property lines. Under these circumstances, the sides of the excavation have to be made vertical and must usually be supported by bracings.

Common methods of bracing the sides when the depth of excavation does not exceed about 3 m are shown in Figs 22.1(a) and (b). The practice is to drive vertical timber planks known as sheeting along the sides of the excavation. The sheeting is held in place by means of horizontal beams called *wales* that in turn are commonly supported by horizontal *struts* extending from side to side of the excavation. The struts are usually of timber for widths not exceeding about 2 m. For greater widths metal pipes called *trench braces* are commonly used.

When the excavation depth exceeds about 5 to 6 m, the use of vertical timber sheeting will become uneconomical. According to one procedure, *steel sheet piles* are used around the boundary of the excavation. As the soil is removed from the enclosure, wales and struts are inserted. The wales are commonly of steel and the struts may be of steel or wood. The process continues until the excavation is complete. In most types of soil, it may be possible to eliminate sheet piles and to replace them with a series of *H* piles spaced 1.5 to 2.5 m apart. The *H* piles, known as *soldier piles* or *soldier beams*, are driven with their flanges parallel to the sides of the excavation as shown in Fig. 21.1(b). As the soil next to the piles is removed horizontal boards known as *lagging* are introduced as shown in the figure and are wedged against the soil outside the cut. As the general depth of excavation advances from one level to another, wales and struts are inserted in the same manner as for steel sheeting.

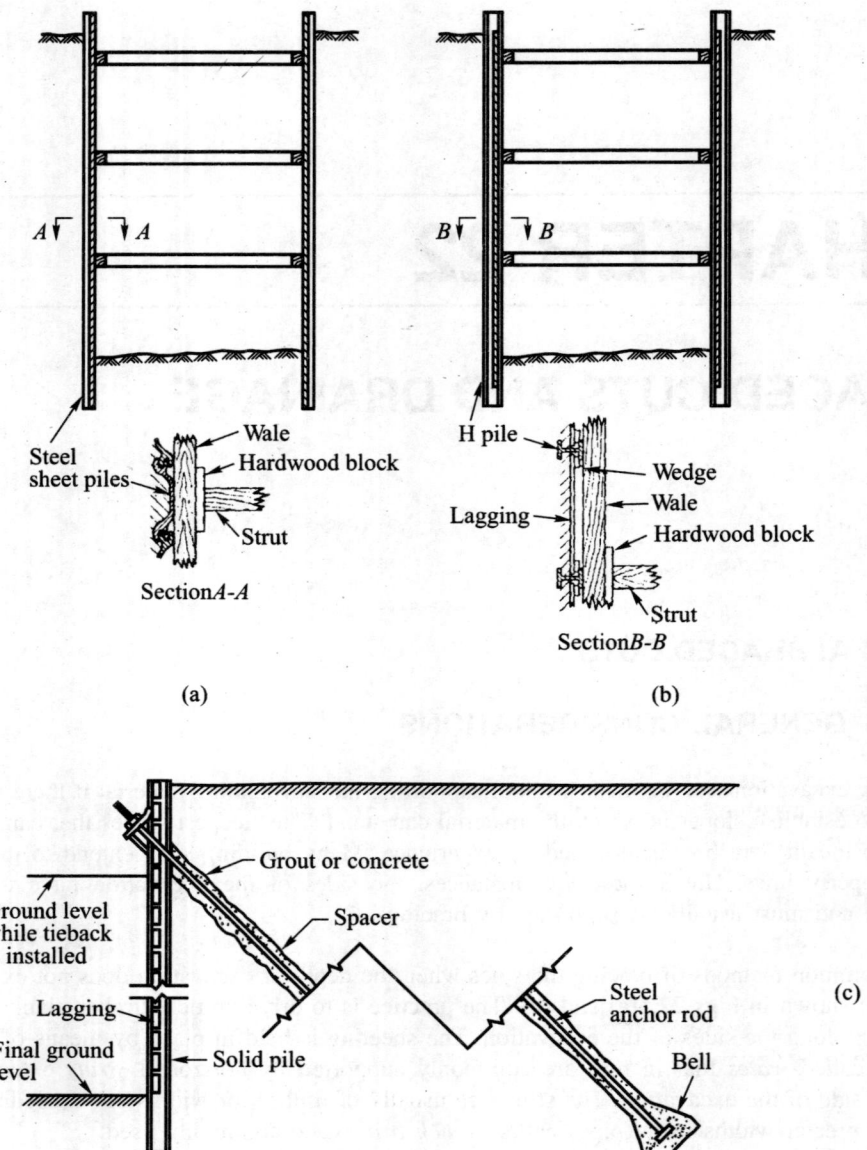

Fig. 22.1 Cross sections, through typical bracing in deep excavation. (a) sides retained by steel sheet piles, (b) sides retained by *H* piles and lagging, (c) one of several tieback systems for supporting vertical sides of open cut. several sets of anchors may be used, at different elevations (Peck, 1969)

If the width of a deep excavation is too great to permit economical use of struts across the entire excavation, *tiebacks* are often used as an alternative to cross-bracings as shown in Fig. 22.1(c). Inclined holes are drilled into the soil outside the sheeting or *H* piles. Tensile reinforcement is then inserted and concreted into the hole. Each tieback is usually prestressed before the depth of excavation is increased.

22.2 LATERAL EARTH PRESSURE DISTRIBUTION ON BRACED-CUTS

Since most open cuts are excavated in stages within the boundaries of sheet pile walls or walls consisting of soldier piles and lagging, and since struts are inserted progressively as the excavation precedes, the walls are likely to deform as shown in Fig. 22.2. Little inward movement can occur at the top of the cut after the first strut is inserted. The pattern of deformation differs so greatly from that required for Rankine's state that the distribution of earth pressure associated with retaining walls is not a satisfactory basis for design (Peck *et al*, 1974). The pressures against the upper portion of the walls are substantially greater than those indicated by the equation.

$$p_a = \frac{1 - \sin\phi}{1 + \sin\phi} p_v \tag{22.1}$$

for Rankine's condition

where

p_v = vertical pressure,

ϕ = friction angle

Apparent Pressure Diagrams

Peck (1969) presented pressure distribution diagrams on braced cuts. These diagrams are based on a wealth of information collected by actual measurements in the field. Peck called these pressure diagrams *apparent pressure envelopes* which represent fictitious pressure distributions for estimating strut loads in a system of loading. Figure 22.3 gives the apparent pressure distribution diagrams as proposed by Peck.

Deep Cuts in Sand

The apparent pressure diagram for sand given in Fig. 22.3 was developed by Peck (1969) after a great deal of study of actual pressure measurements on braced cuts used for subways.

The pressure diagram given in Fig. 22.3(b) is applicable to both loose and dense sands. The struts are to be designed based on this apparent pressure distribution. The most probable value of any individual strut load is about 25 percent lower than the maximum (Peck, 1969). It may be noted

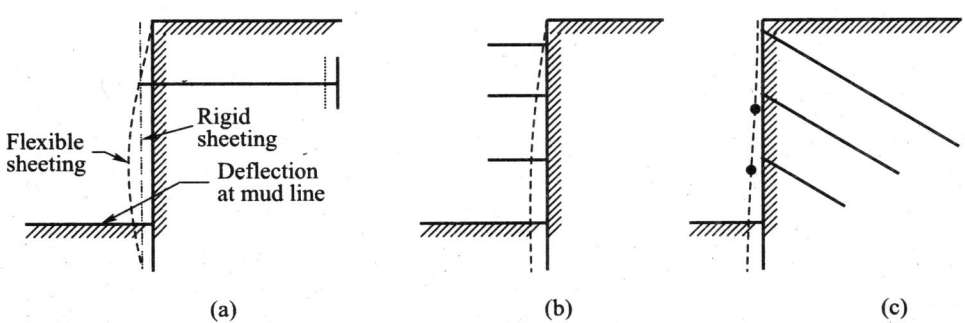

(a)   (b)   (c)

Fig. 22.2 Typical pattern of deformation of vertical walls (a) anchored bulkhead, (b) braced cut, and (c) tieback cut (Peck et al., 1974)

here that this apparent pressure distribution diagram is based on the assumption that the water table is below the bottom of the cut.

The pressure p_a is uniform with respect to depth. The expression for p_a is

$$p_a = 0.65 \, \gamma H \, K_A \tag{22.2}$$

where,

$K_A = \tan^2 (45° - \phi/2)$

γ = unit weight of sand

Cuts in Saturated Clay

Peck (1969) developed two apparent pressure diagrams, one for soft to medium clay and the other for stiff fissured clay. He classified these clays on the basis of non-dimensional factors (stability number N_s) as follows.

Stiff Fissured clay

$$N_s = \frac{\gamma H}{c} \leq 4 \tag{22.3}$$

Soft to Medium clay

$$N_s = \frac{\gamma H}{c} > 4 \tag{22.4}$$

where γ = unit weight of clay, c = undrained cohesion ($\phi = 0$)

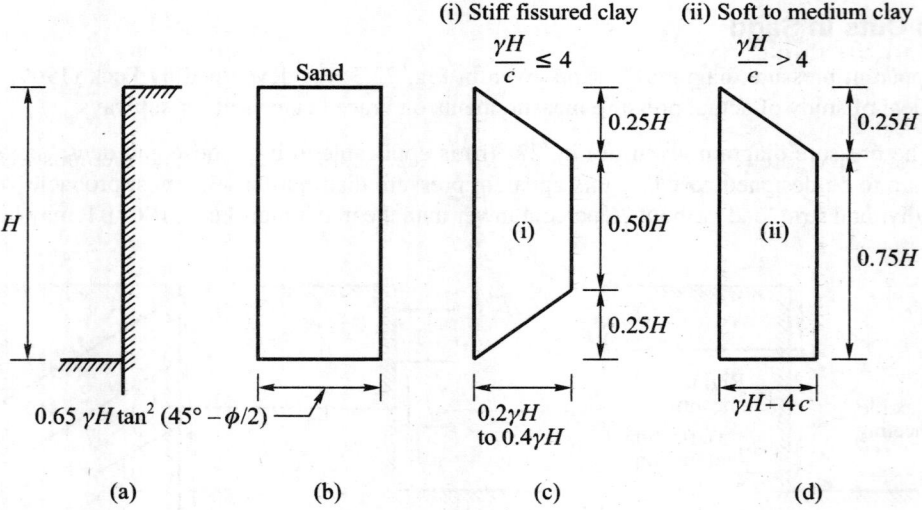

Fig. 22.3 Apparent pressure diagram for calculating loads in struts of braced cuts: (a) sketch of wall of cut, (b) diagram for cuts in dry or moist sand, (c) diagram for clays if $\gamma H/c$ is less than 4 (d) diagram for clays if $\gamma H/c$ is greater than 4 where c is the average undrained shearing strength of the soil (Peck, 1969)

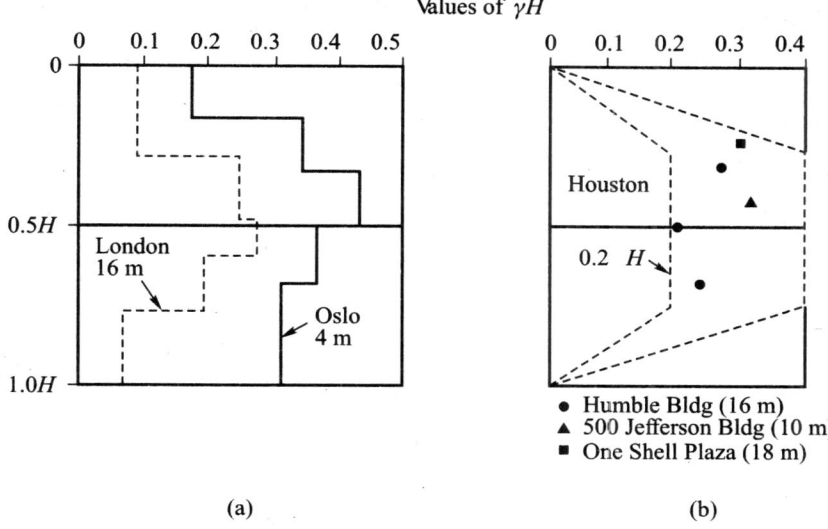

(a)   (b)

Fig. 22.4 Maximum apparent pressures for cuts in stiff clays: (a) fissured clays in London and Oslo, (b) stiff slickensided clays in Houston (Peck, 1969)

The pressure diagrams for these two types of clays are given in Fig. 22.3(c) and (d) respectively. The apparent pressure diagram for soft to medium clay (Fig. 22.3(d)) has been found to be conservative for estimating loads for design supports. Figure 22.3(c) shows the apparent pressure diagram for stiff-fissured clays. Most stiff clays are weak and contain fissures. Lower pressures should be used only when the results of observations on similar cuts in the vicinity so indicate. Otherwise a lower limit for $p_a = 0.3\ \gamma H$ should be taken. Figure 22.4 gives a comparison of measured and computed pressures distribution for cuts in London, Oslo and Houston clays.

Cuts in Stratified Soils

It is very rare to find uniform deposits of sand or clay to a great depth. Many times layers of sand and clays overlying one on another are found in nature. Even the simplest of these conditions does not lend itself to vigorous calculations of lateral earth pressures by any of the methods available.

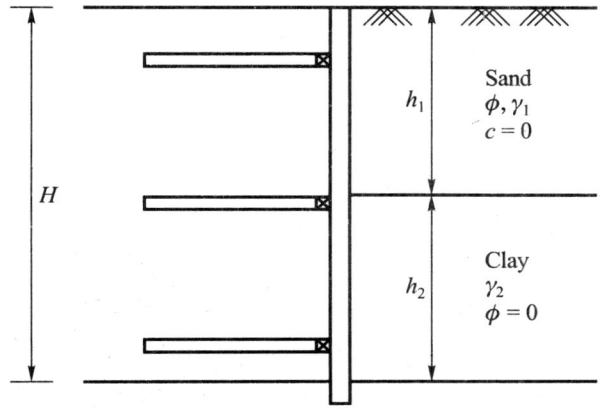

Fig. 22.5   Cuts in stratified soils

Based on field experience, empirical or semi-empirical procedures for estimating apparent pressure diagrams may be justified. Peck (1969) proposed the following unit pressure for excavations in layered soils (sand and clay) with sand overlying as shown in Fig. 22.5.

When layers of sand and soft clay are encountered, the pressure distribution shown in Fig. 22.3(d) may be used if the unconfined compressive strength q_u is substituted by the average $\overline{q_u}$ and the unit weight of soil γ by the average value $\overline{\gamma}$ (Peck, 1969). The expressions for $\overline{q_u}$ and $\overline{\gamma}$ are

$$\overline{q_u} = \frac{1}{H}[\gamma_1 K_s h_1^2 \tan\phi + h_2 n q_u] \tag{22.5}$$

$$\overline{\gamma} = \frac{1}{H}[\gamma_1 h_1 + \gamma_2 h_2] \tag{22.6}$$

where

$\quad H =$ total depth of excavation

$\quad \gamma_1, \gamma_2 =$ unit weights of sand and clay respectively

$\quad h_1, h_2 =$ thickness of sand and clay layers respectively

$\quad K_s =$ hydrostatic pressure ratio for the sand layer, may be taken as equal to 1.0 for design purposes

$\quad \phi =$ angle of friction of sand

$\quad n =$ coefficient of progressive failure varies from 0.5 to 1.0 which depends upon the creep characteristics of clay. For Chicago clay n varies from 0.75 to 1.0.

$\quad q_u =$ unconfined compression strength of clay

22.3 STABILITY OF BRACED CUTS IN SATURATED CLAY

A braced-cut may fail as a unit due to unbalanced external forces or heaving of the bottom of the excavation. If the external forces acting on opposite sides of the braced cut are unequal, the stability of the entire system has to be analyzed. If soil on one side of a braced cut is removed due to some unnatural forces the stability of the system will be impaired. However, we are concerned here about the stability of the bottom of the cut. Two cases may arise. They are

1. Heaving in clay soil
2. Heaving in cohesionless soil

Heaving in Clay Soil

The danger of heaving is greater if the bottom of the cut is soft clay. Even in a soft clay bottom, two types of failure are possible. They are

Case 1: When the clay below the cut is homogeneous at least up to a depth equal $0.7 B$ where B is the width of the cut.

Case 2: When a hard stratum is met within a depth equal to 0.7 B.

In the first case a full plastic failure zone will be formed and in the second case this is restricted as shown in Fig. 22.6. A factor of safety of 1.5 is recommended for determining the resistance here. Sheet piling is to be driven deeper to increase the factor of safety. The stability analysis of the bottom of the cut as developed by Terzaghi (1943) is as follows.

Case 1: Formation of Full Plastic Failure Zone Below the Bottom of Cut

Figure 22.6(a) is a vertical section through a long cut of width B and depth H in saturated cohesive soil ($\phi = 0$). The soil below the bottom of the cut is uniform up to a considerable depth for the formation of a full plastic failure zone. The undrained cohesive strength of soil is c. The weight of the blocks of clay on either side of the cut tends to displace the underlying clay toward the excavation. If the underlying clay experiences a bearing capacity failure, the bottom of the excavation heaves and the earth pressure against the bracing increases considerably.

The anchorage load block of soil $a\,b\,c\,d$ in Fig. 22.6(a) of width $\bar{B}$ (assumed) at the level of the bottom of the cut per unit length may be expressed as

$$Q = \gamma H\bar{B} - cH = \bar{B}H\left(\gamma - \frac{c}{B}\right) \tag{22.7}$$

The vertical pressure q per unit length of a horizontal, ba, is

$$q = \frac{Q}{B} = H\left(\gamma - \frac{c}{\bar{B}}\right) \tag{22.8}$$

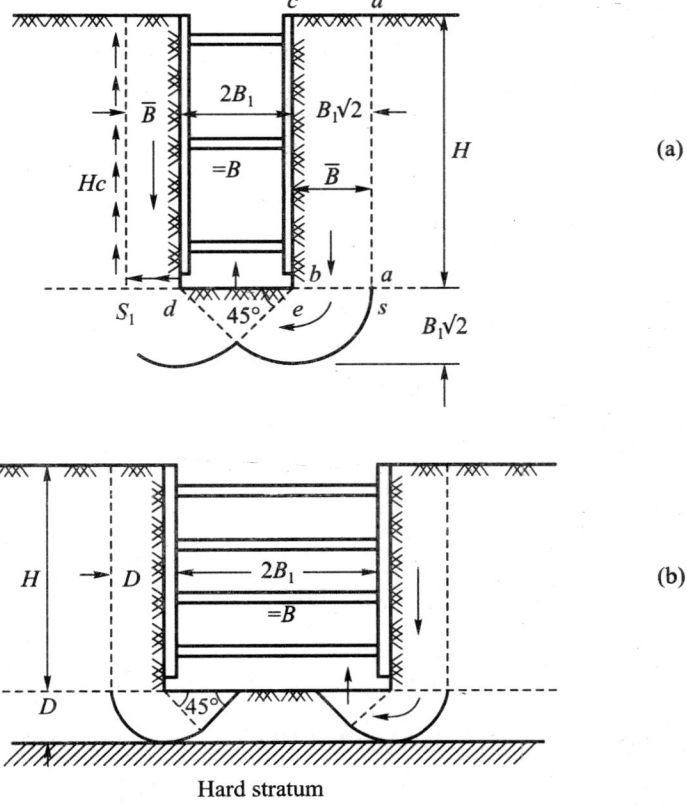

Fig. 22.6 Stability of braced cut: (a) heave of bottom of timbered cut in soft clay if no hard stratum interferes with flow of clay, (b) as before, if clay rests at shallow depth below bottom of cut on hard stratum (after Terzaghi, 1943)

The bearing capacity q_u per unit area at level ab is

$$q_u = N_c c = 5.7c \tag{22.9}$$

where $N_c = 5.7$

The factor of safety against heaving is

$$F_s = \frac{q_u}{q} = \frac{5.7c}{H\left(\gamma - \dfrac{c}{\overline{B}}\right)} \tag{22.10}$$

Because of the geometrical condition, it has been found that the width $\overline{B}$ cannot exceed $0.7\,B$. Substituting this value for $\overline{B}$,

$$F_s = \frac{5.7c}{H\left(\gamma - \dfrac{c}{0.7B}\right)} \tag{22.11}$$

This indicates that the width of the failure slip is equal to $\overline{B}\sqrt{2} = 0.7B$.

Case 2: When the Formation of Full Plastic Zone is Restricted by the Presence of a Hard Layer

If a hard layer is located at a depth D below the bottom of the cut (which is less than $0.7B$), the failure of the bottom occurs as shown in Fig. 22.6(b). The width of the strip which can sink is also equal to D.

Replacing $0.7B$ by D in Eq. (22.11), the factor of safety is represented by

$$F_s = \frac{5.7c_u}{H\left(\gamma - \dfrac{c}{D}\right)} \tag{22.12}$$

For a cut in soft clay with a constant value of c_u below the bottom of the cut, D in Eq. (22.12) becomes large, and F_s approaches the value

$$F_s = \frac{5.7c_u}{\gamma H} = \frac{5.7}{N_s} \tag{22.13}$$

where

$$N_s = \frac{\gamma H}{c_u} \tag{22.14}$$

is termed the *stability number*. The stability number is a useful indicator of potential soil movements. The soil movement is smaller for smaller values of N_s.

The analysis discussed so far is for long cuts. For short cuts, square, circular or rectangular, the factor of safety against heave can be found in the same way as for footings.

22.4   BJERRUM AND EIDE (1956) METHOD OF ANALYSIS

The method of analysis discussed earlier gives reliable results provided the width of the braced cut is larger than the depth of the excavation and that the braced cut is very long. In the cases where the braced cuts are rectangular, square or circular in plan or the depth of excavation exceeds the width of the cut, the following analysis should be used.

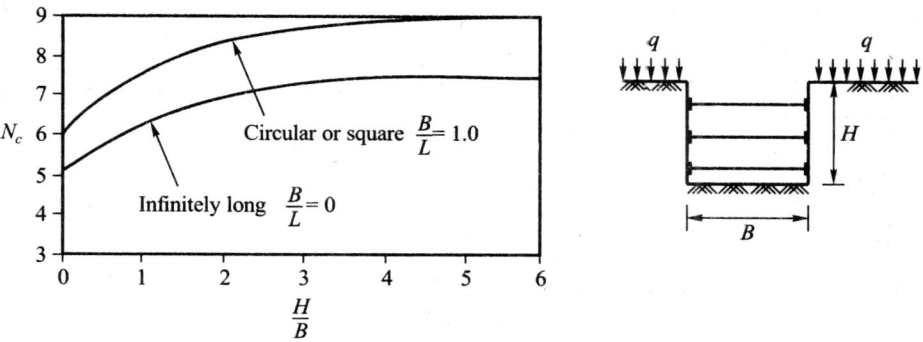

Fig. 22.7 Stability of bottom excavation (after Bjerrum and Eide, 1956)

In this analysis the braced cut is visualized as a deep footing whose depth and horizontal dimensions are identical to those at the bottom of the braced cut. This deep footing would fail in an identical manner to the bottom braced cut failed by heave. The theory of Skempton for computing N_c (bearing capacity factor) for different shapes of footing is made use of. Figure 22.7 gives values of N_c as a function of H/B for long, circular or square footings. For rectangular footings, the value of N_c may be computed by the expression

$$N_c \text{ (rect)} = (0.84 + 0.16 \ B/L) \ N_c \text{ (sq)} \tag{22.15}$$

where

L = length of excavation

B = width of excavation

The factor of safety for bottom heave may be expressed as

$$F_s = \frac{cN_c}{\gamma H + q} \geq 1.5$$

where

γ = effective unit weight of the soil above the bottom of the excavation

q = uniform surcharge load (Fig. 22.7)

Example 22.1

A long trench is excavated in medium dense sand for the foundation of a multistorey building. The sides of the trench are supported with sheet pile walls fixed in place by struts and wales as shown in Fig. Ex. 22.1. The soil properties are:

$\gamma = 18.5$ kN/m³, $c = 0$ and $\phi = 38°$

Determine: (a) The pressure distribution on the walls with respect to depth.

(b) Strut loads. The struts are placed horizontally at distances $L = 4$ m center to center.

(c) The maximum bending moment for determining the pile wall section.

(d) The maximum bending moments for determining the section of the wales.

Solution

(a) For a braced cut in sand use the apparent pressure envelope given in Fig. 22.3(b). The equation for p_a is

$$p_a = 0.65 \ \gamma H \ KA = 0.65 \times 18.5 \times 8 \ \tan^2 (45 - 38/2) = 23 \text{ kN/m}^2$$

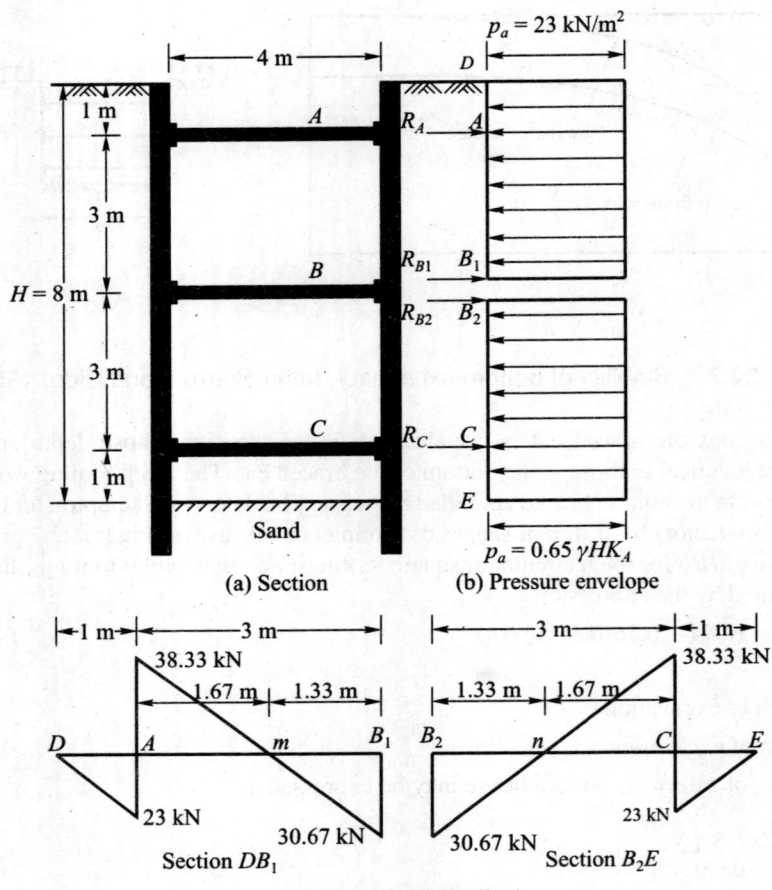

Fig. Ex. 22.1(b) shows the pressure envelope.

(b) Strut loads

The reactions at the ends of struts A, B and C are represented by R_A, R_B and R_c respectively
For reaction R_A, take moments about B

$$R_A \times 3 = 4 \times 23 \times \frac{4}{2} \quad \text{or} \quad R_A = \frac{184}{3} = 61.33 \text{ kN}$$

$R_{B1} = 23 \times 4 - 61.33 = 30.67$ kN

Due to the symmetry of the load distribution,

$R_{B1} = R_{B2} = 30.67$ kN, and $R_A = R_C = 61.33$ kN.

Now the strut loads are (for $L = 4$ m)

Strut A, $P_A = 61.33 \times 4 \approx 245$ kN

Strut B, $P_B = (R_{B1} + R_{B2}) \times 4 = 61.34 \times 4 \approx 245$ kN

Strut C, $P_C = 245$ kN

(c) Moment of the pile wall section

To determine moments at different points it is necessary to draw a diagram showing the shear force distribution.

Consider sections DB_1 and B_2E of the wall in Fig. Ex. 22.1(b). The distribution of the shear forces are shown in Fig. Ex. 22.1(c) along with the points of zero shear.

The moments at different points may be determined as follows

$$M_A = \frac{1}{2} \times 1 \times 23 = 11.5 \text{ kN - m}$$

$$M_C = \frac{1}{2} \times 1 \times 23 = 11.5 \text{ kN - m}$$

$$M_m = \frac{1}{2} \times 1.33 \times 30.67 = 20.4 \text{ kN - m}$$

$$M_n = \frac{1}{2} \times 1.33 \times 30.67 = 20.4 \text{ kN - m}$$

The maximum moment M_{max} = 20.4 kN-m. A suitable section of sheet pile can be determined as per standard practice.

(d) Maximum moment for wales

The bending moment equation for wales is

$$M_{max} = \frac{RL^2}{8}$$

where R = maximum strut load = 245 kN

L = spacing of struts = 4 m

$$M_{max} = \frac{245 \times 4^2}{8} = 490 \text{ kN - m}$$

A suitable section for the wales can be determined as per standard practice.

Example 22.2

Figure Ex. 22.2a gives the section of a long braced cut. The sides are supported by steel sheet pile walls with struts and wales. The soil excavated at the site is stiff clay with the following properties

$$c = 800 \text{ lb/ft}^2, \phi = 0, \gamma = 115 \text{ lb/ft}^3$$

Determine: (a) The earth pressure distribution envelope.

(b) Strut loads.

(c) The maximum moment of the sheet pile section.

The struts are placed 12 ft apart center to center horizontally.

Solution

(a) The stability number N_s from Eq. (22.3) is

$$N_s = \frac{\gamma H}{c} = \frac{115 \times 25}{800} = 3.6 < 4$$

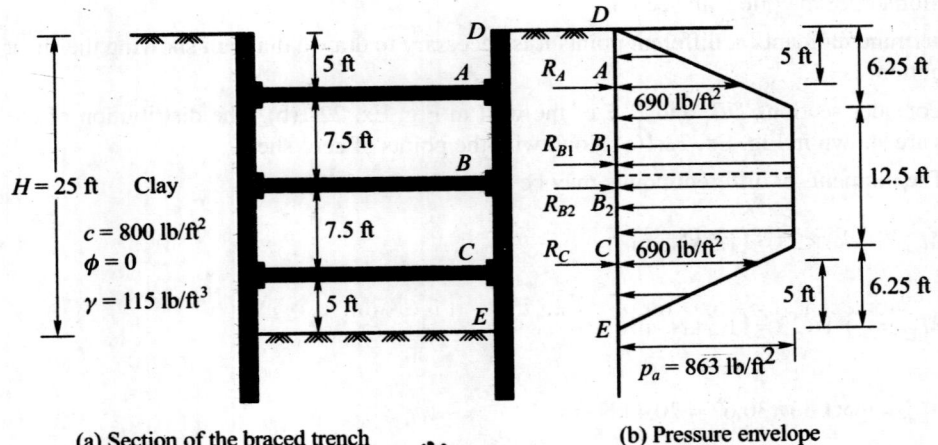

(a) Section of the braced trench   (b) Pressure envelope

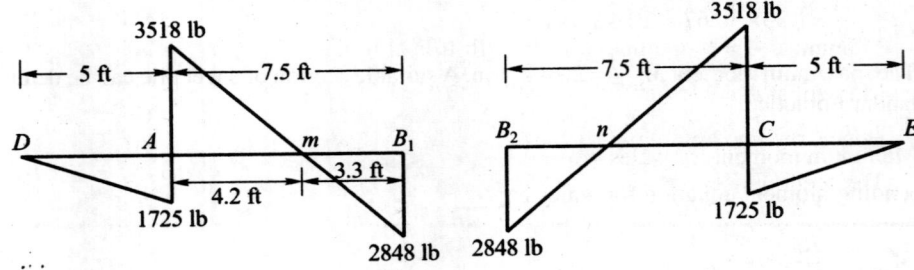

(c) Shear force diagram

Fig. Ex. 22.2

The soil is stiff fissured clay. As such the pressure envelope shown in Fig. 22.3(c) is applicable.
Assume $p_a = 0.3\,\gamma H$

$p_a = 0.3 \times 115 \times 25 = 863$ lb/ft²

The pressure envelope is drawn as shown in Fig. Ex. 22.2(b).

(b)   Strut loads

Taking moments about the strut head B_1 (B)

$$R_A \times 7.5 = \frac{1}{2}863 \times 6.25\left(\frac{6.25}{3}+6.25\right)+863 \times \frac{(6.25)^2}{2}$$

$$= 22.47 \times 10^3 + 16.85 \times 10^3 = 39.32 \times 10^3$$

$R_A = 5243$ lb/ft

$$R_{B1} = \frac{1}{2} \times 863 \times 6.25 + 8.63 \times 6.25 - 5243 = 2848 \text{ lb/ft}$$

Due to symmetry

$R_A = R_C = 5243$ lb/ft

$R_{B2} = R_{B1} = 2848$ lb/ft

Strut loads are:

$P_A = 5243 \times 12 = 62,916$ lb $= 62.92$ kips

$P_B = 2 \times 2848 \times 12 = 68,352$ lb $= 68.35$ kips

$P_C = 62.92$ kips

(c) Moments

The shear force diagram is shown in Fig. Ex. 22.2c for sections DB_1 and B_2 E

$$\text{Moment at } A = \frac{1}{2} \times 5 \times 690 \times \frac{5}{3} = 2,875 \text{ Ib - ft/ft of wall}$$

$$\text{Moment at } m = 2848 \times 3.3 - 863 \times 3.3 \times \frac{3.3}{2} = 4699 \text{ Ib - ft/ft}$$

Because of symmetrical loading

Moment at A = Moment at C = 2875 lb-ft/ft of wall

Moment at m = Moment at n = 4699 lb-ft/ft of wall

Hence, the maximum moment = 4699 lb-ft/ft of wall.

The section modulus and the required sheet pile section can be determined in the usual way.

22.5 PIPING FAILURES IN SAND CUTS

Sheet piling is used for cuts in sand and the excavation must be dewatered by pumping from the bottom of the excavation. Sufficient penetration below the bottom of the cut must be provided to reduce the amount of seepage and to avoid the danger of piping.

Piping is a phenomenon of water rushing up through pipe-shaped channels due to large upward seepage pressure. When piping takes place, the weight of the soil is counteracted by the upward hydraulic pressure and as such there is no contact pressure between the grains at the bottom of the excavation. Therefore, it offers no lateral support to the sheet piling and as a result the sheet piling may collapse. Further the soil will become very loose and may not have any bearing power. It is therefore, essential to avoid piping. For further discussions on piping, see Chapter 4 on Soil Permeability and Seepage. Piping can be reduced by increasing the depth of penetration of sheet piles below the bottom of the cut.

22.6 QUESTIONS AND PROBLEMS (PART A)

22.1 Figure Prob. 22.1 shows a braced cut in medium dense sand. Given $\gamma = 18.5$ kN/m^3, $c = 0$ and $\phi = 38°$.

(a) Draw the pressure envelope, (b) determine the strut loads, and (c) determine the maximum moment of the sheet pile section.

The struts are placed laterally at 4 m center to center.

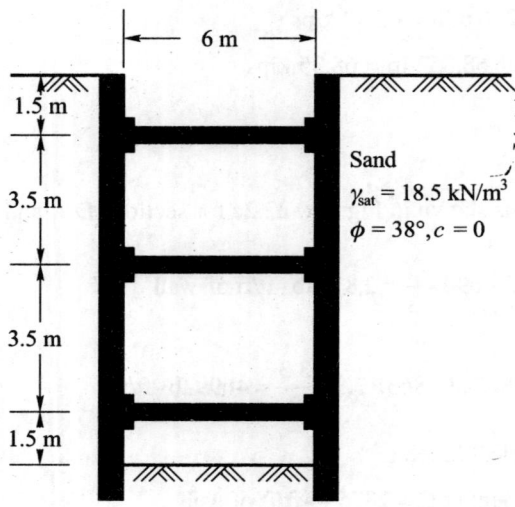

Fig. Prob. 22.1

22.2   Figure Prob. 22.2 shows the section of a braced cut in clay. Given $c = 650$ lb/ft^2, $\gamma = 115$ lb/ft^3.

(a) Draw the earth pressure envelope, (b) determine the strut loads, and (c) determine the maximum moment of the sheet pile section.

Assume that the struts are placed laterally at 12 ft center to center.

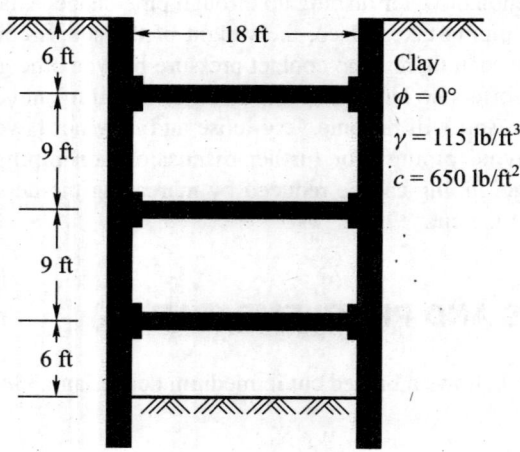

Fig. Prob. 22.2

PART B: DRAINAGE

22.7   INTRODUCTION

When the depth of excavation is greater than the distance to the free water surface in a pervious soil having a coefficient of permeability greater than about 10^{-3} cm/sec, the soil must be drained to permit construction of foundations in the dry. If the coefficient of permeability of the soil is within the range 10^{-3} to 10^{-5} cm/sec, the quantity of water that seeps into the excavation may be inconsequential but drainage may still be required to maintain the stability of the sides and bottom of the excavation. If the coefficient of permeability is smaller than about 10^{-7} cm/sec, the soil is likely to possess sufficient cohesion to overcome the influence of the seepage forces and drainage may not be required even if the excavation extends for a considerable depth below the water table.

After completion of structures with basements, it is often necessary to maintain the water level in a lowered position. This requires the installation of permanent drains.

Prior to the design and construction of a given foundation, the ground water level at the site must be reliably determined. If the ground water is high, some of the following problems are encountered:

Dewatering the site during construction

Foundation drainage

For each job it is important to determine before hand the method of dewatering and the type of foundation drains. Sometimes the cost of dewatering the site is excessive and consequently, the total construction cost may be high. In such cases, a cost comparison should be made for all feasible types of foundations or schemes of substructures from which the most economical design can be selected. Quite often the amount of free water that will flow into the excavation is difficult to predict due to the erratic pattern of waterbearing layers and pockets. Under such circumstances, the design must be made flexible enough so that it can be adjusted to suit the actual conditions as the excavation proceeds or when the excavation is completed.

22.8   METHODS OF DEWATERING

When construction is made below the ground water level, the site must be dewatered feir the following purposes:

1. To provide a suitable working surface at the bottom of the excavation.

2. To stabilize the banks of the excavation thus avoiding the hazards of slides and sloughing.

3. To prevent disturbance of the soil at the bottom of excavation caused by boils or piping. Such disturbance may reduce the bearing power of the soil.

The amount of water to be removed from the site varies from a trivial quantity to large volumes, depending upon the height of water head, the permeability of the soil below the water level, and the size of the area to be dewatered. Extensive dewatering is necessary for deep excavation in permeable soils (sand, gravel, or soils containing such seams) whereas little dewatering is required for shallow excavations or excavations in impervious soils (clays).

A successful dewatering job depends upon the proper selection of the method and the constant vigil on the operation. The surface water should be diverted away from the excavation. Furthermore, the possibility of piping or boil should be analyzed. The dewatering may be done in one or a

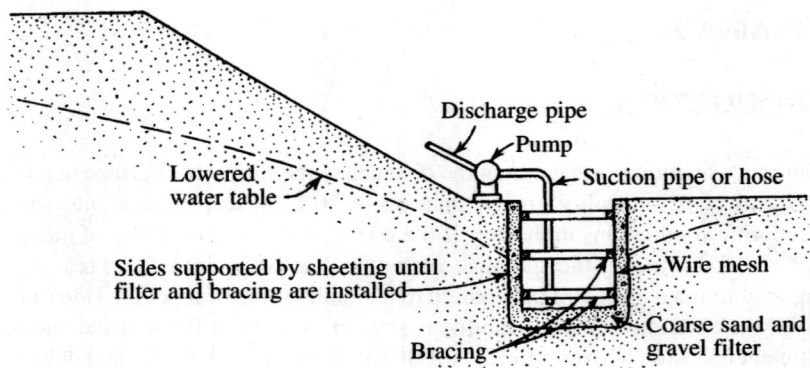

Fig. 22.8 Pump sump

combination of the following methods:

A. Sumps: A sump is merely a hole in the ground from which water is being pumped for the purpose of removing water from the adjoining area. This method is most commonly used for removal of surface water but is also useful where the amount of water to be removed is small.

One sump may be sufficient for a small area, whereas several sumps with ditches leading to them are necessary for dewatering a large area. If the soil is predominantly sand or gravel and if the excavation exceeds several feet below the ground water level, the pump sump method may become inadequate, and another method should be used.

In any dewatering operation it is important to guard against the danger of carrying away the fine particles from granular soils. As fine particles are carried away by the flowing water, the bearing capacity of the soil may be impaired. If existing foundations are in the vicinity, pumping may cause settlement of these foundations. To avoid such difficulties, the sump should be lined with a filter material which has grain-size gradations determined. Generally, the filter material is installed in the following manner,

1. Drive sheeting around the sump for the full depth of the sump.

2. Install a cage inside the sump. The cage may be made of wire mesh with internal strutting or a perforated pipe.

3. Fill the filter material in the space outside the cage and at the bottom of the cage.

4. Withdraw the sheeting.

The relative amount of soil particles carried away by pumping can be determined by visual examination of the water discharged from the end of the hose. For a prolonged pumping, a bucket of discharged water should be collected periodically, and the water allowed to set for several hours. The amount of soil particles settled at the bottom of the bucket can be observed visually.

B. Well points: A well point is a two to three inch diameter pipe two to four feet long which is perforated and covered with a screen. The lower end of the pipe has a driving head with water holes for jetting.

Well points are connected to two to three inch diameter pipes known as riser pipes and are inserted into the ground by driving or jetting. The upper ends of riser pipes lead to a header pipe

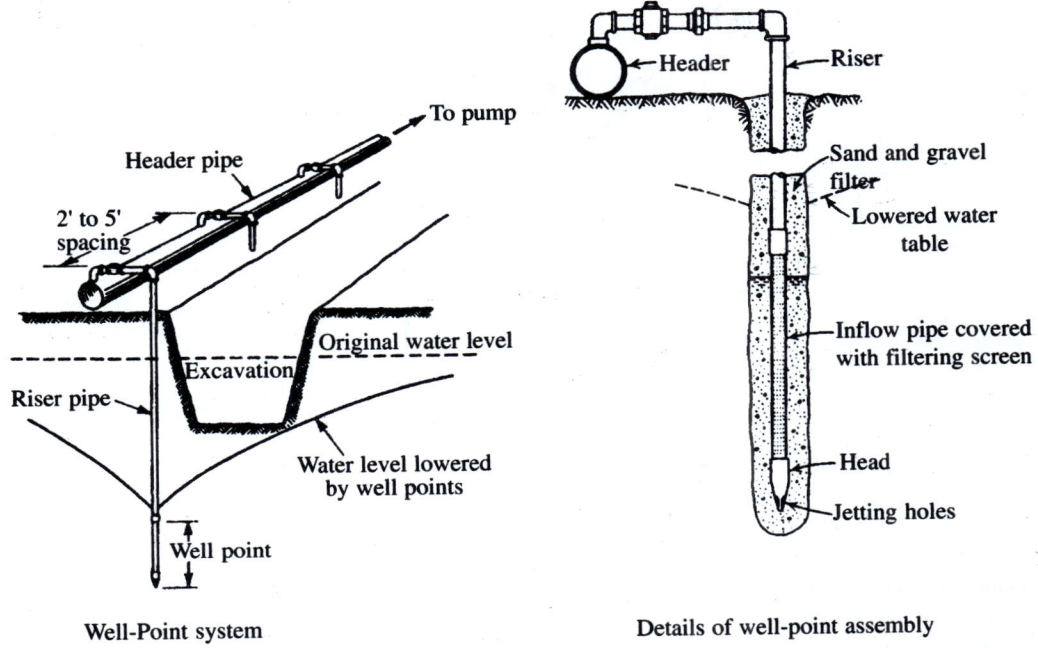

Well-Point system

Details of well-point assembly

Fig. 22.9   Well-point

which, in turn, is connected to a pump. The ground water is drawn by the pump into the well points through the header pipe and discharged. The well points are installed with two to five foot spacing. Figure 22.9 shows the details of a well-point assembly and a well-point system. This type of dewatering system is effective in soils constituted primarily of sand fraction or other soils containing seams of such materials. In highly pervious soils such as coarse gravels, the spacing required to handle the water may be so small that well points become impracticable. They are not useful to draw water out of clays because of the slow process of water seepage. In silt strata, well points may be used if the upper two to three feet of the riser pipes are encased in a tamped clay seal and if pumping is maintained for a period of several weeks. By so doing, a vacuum pressure is created in the silt.

In stratified soils, the screened portion of the well point does not draw water from all the strata above it. In order to facilitate the dewatering in all strata and thus cut down the cost, vertical sand drains may be provided within the influence area of the well points. These sand drains are usually 12 to 24 inches diameter at 15 ft spacings.

The well points can lower a water level to a maximum of 18 ft below the center-line of the header. Under ideal conditions and using special high vacuum equipment, the depth of lowering has been increased to as much as 25 ft. For lowering water level to a greater depth, the multiple stage system of well points must be used which employs two or more tiers of well points. Under average conditions, any number of stages can be used, each stage lowering the water level about 15 ft. A typical setup for a multi-stage system is shown in Fig. 22.10. However, multiple stage system requires additional footage of header pipes and additional pumps. It also increases the width of excavation due to the berms required for headers. Therefore, for dewatering a large head of water, other methods should be considered. The selection of dewatering method should be made on the basis of total cost including initial cost and the cost of operation.

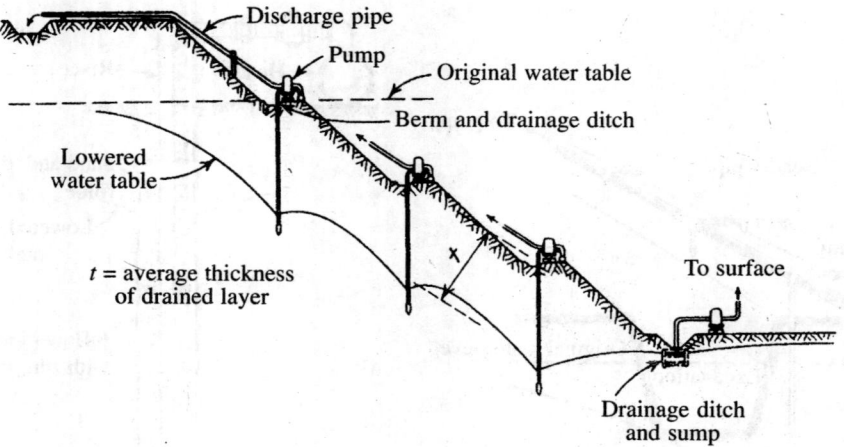

Fig. 22.10 Multiple-stage well-point setup

C. Deep well pumps: To install a deep well pump, a well of 6 in. to 2 ft or larger diameter, is bored to the desired depth; then a deep well turbine, a submersible pump or a water ejector is lowered to the bottom of each well. Such wells are capable of lowering a large head of water and are spaced at 25 ft to more than 120 ft apart depending upon the depth of water to be lowered and other conditions. Filter material should be provided in the well to prevent loss of fine particles in the adjacent ground and clogging of the system. The filter material should be selected according to standard rules.

Although the cost of installation and operation of deep wells is high, this system can be less expensive than the multiple-stage well point system. Furthermore, the wells can be located at some considerable distance from the edge of excavation thus causing very little interference with other con-struction activities. In large jobs where a number of construction equipment are in use, this factor may be of decisive advantage.

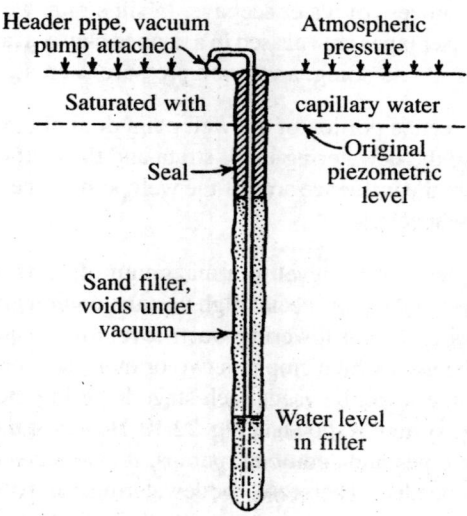

Fig. 22.11 Vaccum well-point installation

D. Vaccum Well-Point: If the permeability is less than about 10^{-4} cm/sec, drainage cannot be accomplished simply by pumping from well points because the capillary forces prevent the flow of water from the pores of the soil. However, drainage can be accomplished by consolidation. This may be done by means of the vacuum method of well-point operation (Fig. 22.11). In this method, the well points are set in holes about 8 in. in diameter made by either augering or jetting. A filter of medium to coarse sand is then placed around the point and pipe to within about 2 or 3 ft of the surface. Above the filter an impervious material such as clay is tamped to form a seal. Special techniques may be required in holes that fail to stand open.

The pumps for such an installation must be capable of maintaining a vacuum in the well points and the surrounding filter. The pressure around the well points is thereby reduced to a small fraction of atmospheric pressure whereas the surface of the ground is acted upon by the weight of the atmosphere. Thus, the soil becomes consolidated under a pressure of about 1 ton/sq ft.

The vacuum process is highly effective in silts and organic silts, but the time required to achieve consolidation and stability is likely to be several weeks.

CHAPTER 23

CELLULAR COFFERDAMS

23.1 INTRODUCTION

A cofferdam is a temporary structure. It is normally constructed to divert water in a river or to keep away water from an enclosed area in order to construct a permanent structure. A bridge pier in a standing water may be constructed by first constructing cofferdam around the site and then pumping out water from the site to enable the construction work to go on in a dry condition. During the entire period of construc-tion a certain amount of pumping is constantly needed to keep away the water that leaks through the dam and foundation. Cofferdams are of many types (Refer Chapter 15). The relative merits of the various types are:

1. Cantilever sheet pile cofferdams [Fig. 15.4(a)]. They are suitable in cases where the height of dam is very small. Such dams are normally subjected to large leakage and flood damage.

2. Braced cofferdams [Fig. 15.4(c)]. They are economical for small to moderate heights. They are also susceptible to floods damage.

3. Earth embankments. There is no limitation to height but the construction occupies more time and space.

4. Double-wall cofferdams [(Fig. 15.4(g)]. They are suitable for moderate heights.

5. Cellular cofferdams [(Fig. 15.4 (e)]. These dams are suitable from moderate to large hclghts. They can constantly be used in excavation in water. When the excavation is in large area overlain by water, as in a river or lake botton, cellular cofferdams are generally used to provide a water barrier. This type of structure is widely used to provide a dry work area where dams are constructed in rivers, and for waterfront construction.

The cellular structure shown in Fig. 23.1 is economical since stability is achieved by using a soil cell fill for mass, which is relatively cheap. The internal bracings which would obstruct the work area are avoided. The sheet piling may be pulled out and reused. It is not usually necessary to drive

the piling to great depths in the natural soil, thus avoiding damage to the piles during driving. T
achieve a cell which is stable against bursting it is necessary that the sheet piling be driven so th:
continuity of interlocks is maintained. The cellular type is more water tight than the braced cofferdan
The construction problem is relatively simple.

The design procedures of cantilever and braced piles are already discussed. The design of eart
embankment will be discussed in the next chapter. The design procedures of double-wall cofferdam
and cellular cofferdams are fundamentally the same. The basic principles of design of these dams ar
discussed in this chapter.

23.2 CELLULAR COFFERDAMS

Cellular cofferdams are basically of two types as shown in Fig. 23.1. They are :

1. Circular type.

2. Diaphragm type.

Circular Type

Circular type consists of individual large diameter circles connected together by arcs of smalle
diameter. These arcs usually intersect the circles at a point at 30° or 45° with the longitudinal axis o
the cofferdam. They are often perpendicular to the circle, but ocassionally different angles may be
used.

N.B.: Dimensions used in the design equations

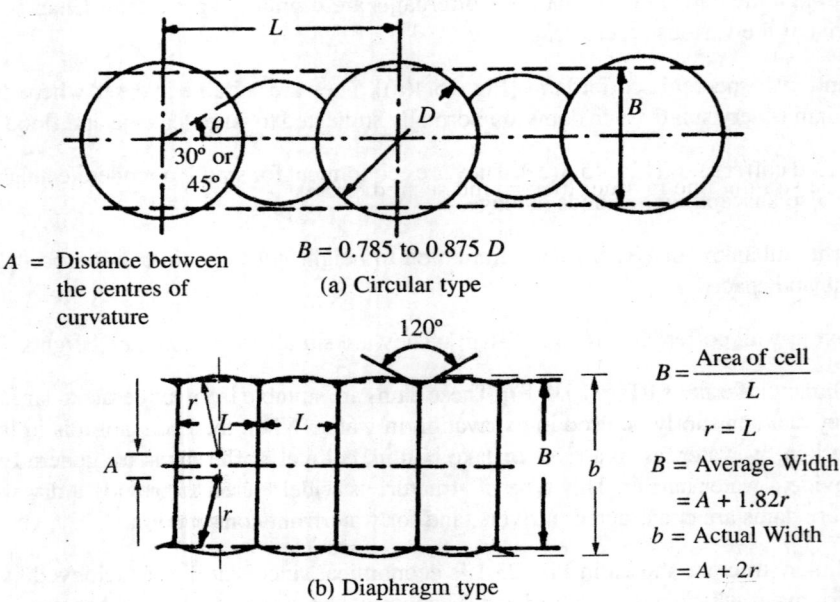

A = Distance between
the centres of
curvature

$B = 0.785$ to 0.875 D

(a) Circular type

$B = \dfrac{\text{Area of cell}}{L}$

$r = L$

B = Average Width

$= A + 1.82r$

b = Actual Width

$= A + 2r$

(b) Diaphragm type

Fig. 23.1 Cellular cofferdams

Advantages of the Curcular Type

1. The circular type can be used singularly in a group or at end.

2. It will not collapse in the event of failure of adjoining cells (due to interlock damage, sudden floods, etc.)

3. Each cell can be filled independent of the other without hampering the progress of work.

4. It requires less number of piles per lineal metre of cofferdam as compared with the diaphragm type of an equal design.

Diaphragam Type

It consists of two series of arcs connected together by diaphragms perpendicular to the axis of the cofferdam. Generally the radii of these arcs are made equal to the distances between the diaphragms. At the intersection point the two arcs and the diaphragm make an angle of 120° between each other.

Advantages of the Diaphragm Type

1. It has uniform interlock stress throughout the section at any given level. The stress is smaller than at the point of circular cell of comparable design.

2. It can be widened readily by increasing the length of the diaphragm if it is required by stability. This will not increase the interlock stress which is a function of the radius of the arc.

23.3 COMPONENTS OF CELLULAR COFFERDAMS

A cellular cofferdam is made of:

1. Steel cells

2. Cell fill

Steel Cells

Steel cells are fabricated out of steel sheet piles. In large diameter cells any two adjacent piles are almost on a straight line. In smaller cells, however, each sheet pile must deflect at a relatively large angle from the straight line in order to form the desired circle.

Cell Fills

The material used for the cell fill should have the following properties.

1. The fill should be free-draining granular soil with little fine particles.

2. It should possess high angle of friction.

3. The fill should be as dense as possible.

4. It should possess large resistance to scour and leakage. Well graded soils are most suitable.

Normally, natural deposits of mixed sand and gravel possess all of these desirable properties and as such are the best materials for the cell fill.

23.4 DIMENSIONS OF COFFERDAM

The height of a cofferdam is fixed on the basis of the maximum flood level in the river where it is to

be constructed. Its diameter is based on the safety requirements. The desigh of the cofferdam begi
with a tentative proportion which is subsequently analysed for stability and other safety requiremen
The design is usually made on the basis of a section one metre long with a uniform average width
which is used in the design calculations. For the design perposes, a simple procedure may be us
whereby the average width is determined such that the area in the rectangular section and that in ti
actual cofferdam are equal.

Let, B = Average width

L = distance centre to centre of cell

Then, $B = \dfrac{\text{Area of main cell} + \text{Area of connecting cell}}{L}$

The average width B of cells ranges from $0.785D$ to $0.875D$ where D is the diameter of cell.

23.5 STABILITY OF CELLULAR COFFERDAMS

There are a few methods of design of cellular cofferdams available in technical literature. The metho
that is described in this book is the one that is largely used by TVA (1957) engineers. The tentativ
dimensions that are initially assumed are to be checked for their safety requirements. The preliminar
dimensions assumed for cofferdams should satisfy the following stability requirements.

Cofferdams on Rock

1. Resistance to sliding

The lateral and vertical forces that act on a cellular cofferdam are shown in Fig. 23.2(a). Th
forces are

P = The lateral pressure due to water and the submerged weight of soil below the river bed
This pressure tends to push the cell away from its position.

W = Effective weight of the fill material which is a sum of the submerged weight below th
saturation line (assumed at 2 : 1 slope) and the total weight above the saturation line.

P_p = Passive pressure. If there is no berm on the wall side, the passive pressure which resist:
the movement of the cell is due to the soil below the river bed level. If a berm is provided
to add to stability, the passive pressure due to the berm should also be taken into account

F_f = The frictional resistance that is developed at the base of the cell. This is equal to $W \tan \phi$
for soil to soil sliding at the base.

A cofferdam should provide adequate resistance to sliding on the base caused by the lateral
pressure P (taken per unit length of cell) acting on the cell face on the river side. The factor of safety
F_s, against sliding may be written as

$$F_s = \frac{P_p + F_f}{P} \tag{23.1}$$

The factor of safety should be greater than 1. A maximum value of 1.25 may be used in the
analysis.

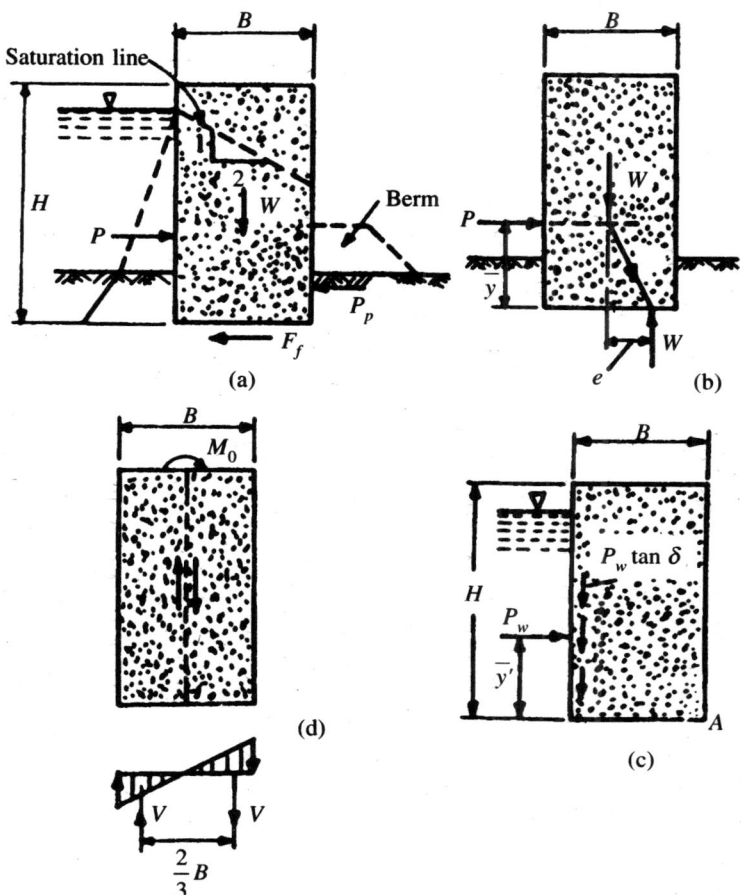

(a) Resistance to sliding (b) Resistance to overturning (c) Resistance for friction of the steel piling on the cell fill material (d) Resistance for shear along the centre line of the cell.

Fig. 23.2   Stability of Cofferdams

2.   Resistance to overturning

The cofferdam should be stable against overturning. The stability to overturning may be checked by two methods.

(a) *Check for the resultant weight to fall within the middle-third of the base of cell [Fig. 23.2(b)].*

The overturning moment M_0 due to the lateral pressure P is

$$M_0 = P\bar{y}$$ (23.2)

Where $\bar{y}$ is the point of application of P above the base.

The resisting Moment M_r due to reaction at the base is

$$M_r = We = \gamma HBe$$ (23.3)

where

γ = effective unit weight of the fill material

H = Height of fiil

B = Average width of Cell

e = Eccentricity

When $M_0 = M_r$, we have

$$e = \frac{P\bar{y}}{\gamma HB} \tag{23.4}$$

Since soil cannot take tension, the safety requirement is

$$e = \frac{P\bar{y}}{\gamma HB} \le \frac{B}{6} \tag{23.5}$$

Equation (23.5) indicates that the width of the cell should be increased as the height of the cell increases so that the reaction may fall within the middle third.

(b) *Check for frictional resistance of the steel piling on the cell fill material.*

The resistance to overturning may be analysed in a different way as shown in Fig. 23.2(c). One may argue that as the cell tends to tip over, the soil will pour out at the heel. For this to occur the frictional resistance of the steel piling on the cell fill is to be fully developed. On this side of the cell the water pressure P_w is pushing the piling against the fill so that the friction force per unit length of cell is $P_w \tan \delta$ where δ is the angle of wall friction between soil and the pile. The resisting shear force acts downward on the wall of the pile as shown in the Fig. 23.2(c). The overturning moment about the toe of the cell (point A in the figure) is

$$M_0 = P_w \bar{y}' \tag{23.6}$$

The resisting moment is

$$M_r = P_w B \tan\delta \tag{23.7}$$

If the Cofferdam is to be stable, the resisting moment should be more than the overturning moment or factor of safety F_s may be written as

$$F_s = \frac{P_w B \tan\delta}{P_w \bar{y}'} = \frac{B \tan\delta}{\bar{y}'} \tag{23.8}$$

A factor of safety of 1.1 to 1.25 is generally suggested.

3. Cell shear

Overturning moments on a cell develops shear stress on a vertical plane through the centre line of the cell as shown in Fig. 23.2(d). For stability the shearing resistance along this plane which is thf sum of soil shear resistance and resistance in the interlocks, must be equal to or greater than the shear due to overturning effects. Referring to Fig. 23.2(d), and assuming a linear pressure distribution across the base of the cell, the overturning moment due to the overturning shear force developed at the base is

$$M_0 = \frac{2}{3} BV \qquad (23.9)$$

Where $2/3\, B$ = lever arm of the shearing force V acting as a couple.

Solving for the overturning shear on the plane through the centre line, we have

$$V = 1.5 \frac{M_0}{B} \qquad (23.10)$$

The total resisting shear, F_{cr} = resisting shear S_r in the soil

$$+ \text{ Resisting shear in the locks } F_c$$

The resisting shear in the soil S_r, is given by the expression

$$S_r = \frac{1}{2} \gamma H^2 \overline{K}_A \tan\phi \qquad (23.11)$$

here $\overline{K}_A$ = Coefficient of earth pressure which is found to be greater than the active earth pressure coefficient K_A. The expression for determining $\overline{K}_A$ as proposed by TVA (1957) is

$$\overline{K}_A = \frac{\cos^2 \phi}{2 - \cos^2 \phi} \qquad (23.12)$$

where ϕ = angle of the internal friction of the fill material.

The frictional force in the interlock is computed as follows :

Hoop tension per unit depth of wall in the interlock at any depth from the surface *is* $T = p_a r$

where,

p_a = lateral pressure on the wall at any depth z from the surface of fill.

r = radius of the cell.

Since p_a increases with the depth, the total hoop-tension force for cell of total depth H is

$$T = \frac{1}{2} \gamma H^2 K_A r \qquad (23.13)$$

where,

γ = effective unit weight of fill material

K_A = Coefficient of earth pressure due to Rankine or Coulomb.

If a sheet pile rests on a rock bed or is embedded in soil, a restraining force preventing the lateral movement of the cell is developed at the base. This effect causes the maximum pressure to be developed at a depth $H' \approx 0.75\, H_c$ = height of the sheet piling above the point of fixty, or embedment. The depth qf the point of fixity from the top is taken. for all practical purposes as the average of the total height of the cell and the depth of water above 0.75 H_c, the river bed.

The triangular distribution of earth pressure p_a for the entire height of sheet piling is shown in Fig. 23.3(a). When fixity condition prevails, the soil pressure distribution will be as shown in Fig. 23.3(b). The area of this triangle gives the total lateral earth pressure acting on the cell. If we assume the pressure distribution on the entire height H of piling with the fixity condition prevailing at depth, the total lateral earth pressure is

$$P_a = \frac{1}{2}\gamma H(0.75H_c)K_A$$

(23.1·)

Therefore, the adjusted hoop-tension force is

$$T = P_a r = \frac{1}{2}\gamma H(0.75H_c)K_a r$$

$$= \frac{3}{8}\gamma H H_c K_A r$$

(23.1!)

The friction force in the interlock is therefore

$$\overline{F_i} = Tf = \frac{3}{8}\gamma H H_c K_A rf$$

(23.16)

where f = Coefficient of interlock friction which may be taken as 0.3.

Since the analysis is usually performed on a unit length of cell (say 1 metre length), the shea
resistance per unit length of cell is

$$F_i = \frac{\overline{F_i}}{L} = \frac{3}{8}\gamma H H_c K_A f \frac{r}{L}$$

(23.17

where L = distance between cross walls for diaphragm type cellular dams, and $L = r$, the
radius, for circular cells.

The hoop tension per unit length of circular cell is therefore

$$F_i = \frac{3}{8}\gamma H H_c K_A f$$

(23.18

The total cell shear resistance, F_{cr} is, therefore, for a circular cell type cofferdam

$$F_{cr} = \frac{1}{2}\gamma H^2 \overline{K}_A \tan\phi + \frac{3}{8}\gamma H H_c K_A f$$

(23.19)

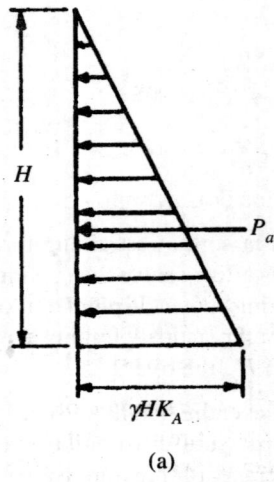

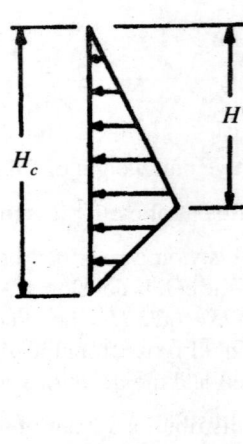

(a)   (b)

Fig. 23.3   Earth pressure distribution on the sheet-piling wall

The factor of safety, F_s for the overturning shear force may be written as

$$F_s = \frac{F_{cr}}{V}$$

(23.20a)

or $F_s = F_{cr} \dfrac{2B}{3M_0}$

(23.20b)

The factor of safety recommended varies from 1.1 to 1.25.

4. Resistance to bursting

The cell should be stable against bursting pressure. The critical locations are in the interlock points and in the T's and Y's used for the connecting area. The bursting pressure is the maximum hoop-tension developed in the interlock at a depth H' from the surface of fill. The maximum pressure p_a at depth $H' \approx 0.75H_c$ as shown in Fig. 23.3(b) is equal to

$$p_a = \gamma H' K_A$$

The bursting pressure T at depth H', is therefore

$$T = p_a r + \gamma_w H'', \text{ per unit depth of wall}$$

(23.21)

where

γ_w = Unit weight of water

H'' = Depth of water above the point of maximum pressure p_a

The allowable value for the interlock tension T, depends on the size and shape of the rolled steel pile sections. The computed maximum interlock tension, should not exceed the maximum stress specified by the manufacturers of steel sheet piles.

According to the Tennesse Valley Authority (TVA), the stress T' in a 90°-Tee, used for the connecting arc, can be obtained as an approximation from

$$T' = \frac{T}{\cos \theta}$$

(23.22)

where T is obtained from Eq. (23.21), and θ is the angle of intersection of the connecting arcs as shown in Fig. 23.1(a).

Cofferdams on Deep Layers of Sand or Clay

The principle of analysis cofferdams founded on deep layers of sand or clay is the same as that applied to cofferfams on rock. In addition, the following requirements must be satisfied.

1. The sheet piling must be driven to such a depth at which level the bearing capacity of the soil should be at least 1.5 times the maximum vertical pressure transmitted to the soil by the cellfill.

The maximum pressure transmitted to the soil by the cellfill when the sheet pile is subjected to lateral force may be written as

$$F_v = \frac{1}{2} \gamma H^2 K_A \tan \delta$$

(23.23)

where

F_v = Vertical pressure per unit length of sheet of piling,

γ = Effective unt weight of cell fill

H = Height of cell

K_A = Coefficient of active earth pressure.

δ = Angle of friction between cell fill and piling.

2. Cellular cofferdams to be founded on sand bed should be designed to prevent boiling at the toe due to seepage of water.

Figure 23.4 gives the section of a cellular cofferdam founded on sand with the flow nets drawn. Due to high permeability of the soil, water retained behind the dam percolates below its base at a relatively large velocity and rise up in front of the toe. If the seepage pressure in front of the toe is more than the buoyant weight of the soil, boiling of sand or quick.sand condition develops. The danger of boiling can (readily be eliminated by the use of loaded filter as explained in Sec. 7.11. As an alternative, the sheet piling may be driven to a great depth in order to eliminate the boiling condition. The depth of sheet piling below the bed level required for this purpose is approximately equal to 0.67 H where H is the height sheet pile above bed.

3. Cellular Cofferdams founded on clay should be investigated for the bearing capacity of the clay at the toe end of the dam. The maximum height of the Cofferdam above the bed level is a function of tlte undrained shear strength of clay. The equation for the critical height $\overline{H}_c$ may be written as

$$\overline{H}_c = 5.7 \frac{c_u}{\gamma} \tag{23.24}$$

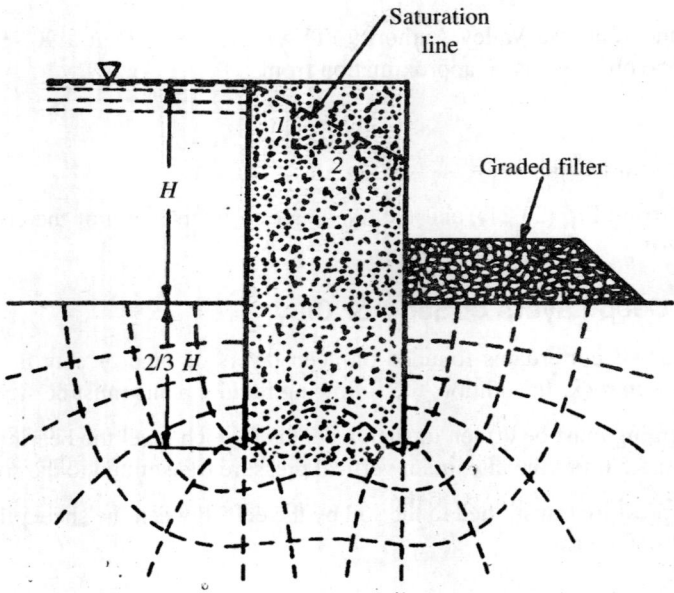

Fig. 23.4 Seepage in cellular cofferdam in sand

where

c_u = undrained shear strength of clay

γ = effective unit weight of cell fill.

If a minimum factor of safety of 1.5 is used, the allowable height of cofferdam above the clay bed is

$$H = 3.8 \frac{c_u}{\gamma} \qquad\qquad\qquad (23.25)$$

23.6 EXAMPLES

Example 23.1 Design a cellular cofferdam. diaphragm type, with the dimensions given in Fig. Ex. 23.1. The other available data are

(i) Unit weight of fill above saturation line, $\gamma = 1.7$ t/m³, (ii) Submerged unit weight of fill and soil outside the cell, $\gamma = 1.1$ t/m³, (iii) Angle of internal friction of fill and soil, $\phi = 30°$, (iv) Frictional coefficient of fill on rock, $f = 0.57$, (v) Interlock friction, $f = 0.3$, (vi) Interlock tension allowed, $T = 1450$ kg/cm, (vii) Frictional coefficient of steel on fill, $f = 0.4$, (viii) Allowable steel tensile stress, $f_t = 1500$ kg/cm².

Solution

Let b = actual width of cell, B = average width of cell = $A + 1.82r$, A = distance between the centres of curvature, r = radius of the curved portion of cell.

1. Sliding stability

The average width b is used in the stability analysis. Weight of fill above rock level per umt length of cell, $W = W_1 + W_2$

where,

W_1 = weight of fill above saturation line,

W_2 = Weight of fill below saturation line,

$$W_1 = \frac{1}{2} \times 1.7(1+1+B/2)B \text{ tonnes}$$

$$W_2 = \frac{1}{2} \times 1.1(15+15+B/2)B \text{ tonnes}$$

Therefore $W = 18.2\,B - 0.025\,B^2$

The frictional resistance due to the cell weight at the rock level (neglecting the weight of the steel pile wall) is

$$F_f = FW = 0.57(18.2B - 0.025B^2)$$

Passive pressure

$$P_p = \frac{1}{2}\gamma_b h^2 K_p = \frac{1}{2} \times 1.1 \times 6^2 \times \tan^2(45° + 30/2)$$

$$= 59 \text{ tonnes}$$

The driving force P_d is

$$P_d = P_w + P_a$$

where, P_w = Total pressure due to water

P_a = Active pressure due to soil

$$P_w = \frac{1}{2} \times 1 \times 15^2 = 112.5 \text{ tonnes}$$

$$P_a = \frac{1}{2} \times 1.1 \times 6^2 \times 0.33 = 6.5 \text{ tonnes}$$

$$P_d = 112.5 + 6.5 = 119.0 \text{ tonnes}$$

Let the factor of safety $F_s = 1.25$. Therefore

$$1.25 = \frac{F_f + P_p}{P_d} = \frac{0.57(18.2B - 0.025B^2) + 59}{119}$$

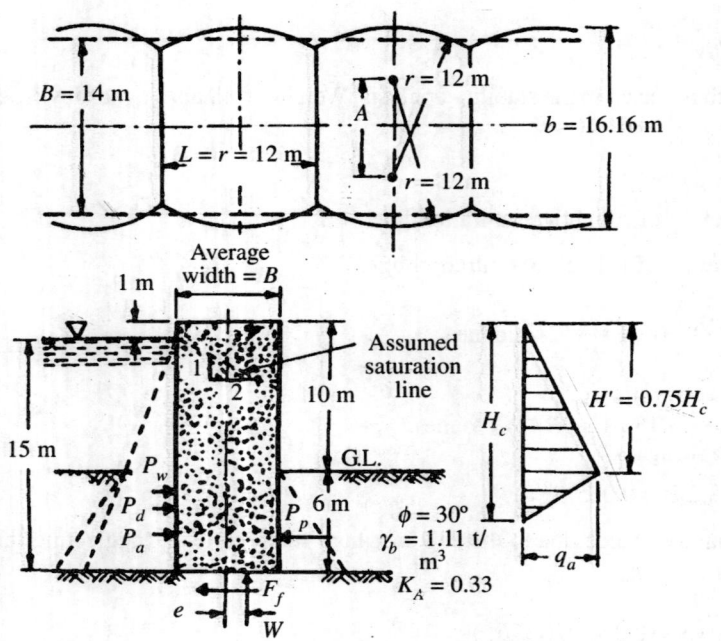

Fig. Ex. 23.1

The quadratic equation in B may be written as

$$0.014B^2 - 10.4B + 90 = 0$$

Solving for B, we have

$$B = 8.8 \text{ m}$$

. *Width to satisfy overturning with a factor of safety* 1.25

Overturning moment M_0,

$$M_0 = P_w \times \frac{15}{3} + P_a \times \frac{6}{3} - P_p \times \frac{6}{3}$$

$$= 112.5 \times 5 + 6.5 \times 2 - 59 \times 2$$

$$= 457.5 \text{ m - tonnes}$$

The maximum allowable eccentricity is

$$e = \frac{B}{6}$$

For stability, we may write the equation

$$We = M_0 F_s$$

Therefore,

$$(18.2B - 0.025B^2)\frac{B}{6} = 457.5 \times 1.25 = 572$$

The value of $B = 14$ m by trial and error method.

Next, check overturning from shear of piling on cell fill.

Summing moments about the toe, we obtain

$$fBP_w = M_0 F_s$$

or $B = \dfrac{M_0 F_s}{f P_w} = \dfrac{457.5 \times 1.25}{0.4 \times 112.5} = 12.7 \text{ m}$

The controlling width, $B = 14$ m.

3. *Check shear along centre line of cell and interlock friction. Assume the radius of Cell $r = L$, the distance between diaphragms.*

    The total weight of soil in the cell per metre length

    $$= 18.2 \times 14 - 0.025 \times 14^2 = 250 \text{ tonnes}$$

    The average unit weight of the fill in the cell is

    $$\gamma_a = \frac{250}{16 \times 14} = 1.15 \text{ tonnes/m}^3$$

The lateral pressure coefficient for $\phi = 30°$ is

$$\overline{K}_A = \frac{\cos^2\phi}{2-\cos^2\phi} = \frac{0.75}{2-0.75} = 0.6$$

where, $\tan\phi = 0.577$

The soil shear resistance S_r along the centre line of the cell is (assuming average unit weight γ

$$S_r = \frac{1}{2}\gamma_a H^2 \overline{K}_A \tan\phi = \frac{1}{2}\times 1.15 \times 16^2 \times 0.6 \times 0.577 = 51 \text{ tonnes}$$

For computing interlock shear F_c, assume $H_c = \frac{16+9}{2} = 12.5$ m

$$F_c = \frac{3}{8}\gamma_a H H_c K_A f = \frac{3}{8}\times 1.15 \times 16 \times 12.5 \times 0.33 \times 0.3 = 8.6 \text{ tonnes.}$$

The shear on the centre line of the cell due to overturning is

$$V = \frac{1.5M_0}{B} = \frac{1.5\times 457.5}{14} = 49 \text{ tonnes}$$

The safety factor is $F_s = \dfrac{S_r + F_c}{V} = \dfrac{51+8.6}{49} = 1.22 \approx 1.25$

4.   *Check for interlock tension*

The depth at which maximum pressure occurs on the cell wall is

$$H' = \frac{3}{4}H_c = \frac{3}{4}\times 12.5 = 9.37 \text{ m};$$

$$H'_w = 9.0 - (10-9.37) = 8.37 \text{ m}$$

$$q_a = \gamma_a H'K_A + \gamma_w H'_w = 1.15\times 9.37 \times 0.33 + 1\times 8.37 = 3.56 + 8.37 = 11.93 \text{ tonnes/m}^2$$

Interlock tension, $T = \dfrac{q_a r}{100}$ tonnes/cm,

$$= \frac{11.93\,r}{100} = 0.1193\,r \text{ tonnes/cm}$$

where r is in metres

For T less than or equal to 1.45 tonnes/cm

$$r = \frac{1.45}{0.1193} = 12.1 \text{ m} \approx 12 \text{ m}$$

The interlock tension $T = 0.0093 \times 12 = 1.43\, T/cm$

Assuming thickness of web = 1.25 cm, the web stress

$$= \frac{1.43}{1.25} = 1.14 \text{ tonnes/sq.cm} < 1.5 \text{ tonnes/cm}^2$$

The final cell dimensions are (Fig. Ex. 23.1)

The distance between diaphragm $L = r = 12$ m

Actual width of cell,

$$B = 14 = A + 2r = A + 24$$

Average width of cell,

$$b = 14 = A + 1.82r$$

$$= A + 21.84$$

Solving for B we have $B = 16.16$ m

With $B = 16.16$ m, the distance $A = -7.84$ m

Example 23.2. Design a circular cofferdam by making use of the data given in Ex.23.1.

Solution

All computations remain the same as in Ex. 23.1. The diameter D of the cell is obtained from

$$D = \frac{b}{0.875} = \frac{14}{0.875} = 16 \text{ m}$$

Hoope tension $T = \dfrac{q_a r}{100} = \dfrac{11.93 \times 16}{100 \times 2}$

$$= 0.95 \text{ tonnes/cm} < 1.45 \text{ T/cm}$$

The stress in the Tee is

$$T' = \frac{0.95}{\cos 45°} = 1.39 \text{ tonnes/cm}$$

The final dimensions for a circular cell are

$$b = 14 \text{ m}, \qquad D = 16 \text{ m}$$

23.7   QUESTIONS AND PROBLEMS

23.1   What are the different types of cellular cofferdams? Discuss their advantages and disadvantages.

23.2   What are the components of a cellular cofferdam? What are the desirable properties of fill in a cell?

23.3   Find the depth of embedment of a diaphragm type cofferdam of total width $B = 7$ m and height above ground level = 7 m. Provide a factor of safety of 1.2. Use the other data given in Ex. 23.1. The dam retains water on one side up to the top.

FOUNDATIONS ON COLLAPSIBLE AND EXPANSIVE SOILS

24.1 GENERAL CONSIDERATIONS

The structure of soils that experience large loss of strength or great increase in compressibility with comparatively small changes in stress or deformations is said to be *metastable* (Peck *et al.*, 1974). Metastable soils include (Peck *et al.*, 1974):

1. Extra-sensitive clays such as quick clays,

2. Loose saturated sands susceptible to liquefaction,

3. Unsaturated primarily granular soils in which a loose state is maintained by apparent cohesion, cohesion due to clays at the intergranular contacts or cohesion associated with the accumulation of soluble salts as a binder, and

4. Some saprolites either above or below the water table in which a high void ratio has been developed as a result of leaching that has left a network of resistant minerals capable of transmitting stresses around zones in which weaker minerals or voids exist.

Footings on quick clays can be designed by the procedures applicable for clays as explained in Chapter 12. Very loose sands should not be used for support of footings. This chapter deals only with soils under categories 3 and 4 listed above.

There are two types of soils that exhibit volume changes under constant loads with changes in water content. The possibilities are indicated in Fig. 24.1 which represent the result of a pair of tests in a consolidation apparatus on identical undisturbed samples. Curve *a* represents the *e*-log *p* curve for a test started at the natural moisture content and to which no water is permitted access. Curves *b* and *c*, on the other hand, correspond to tests on samples to which water is allowed access under all loads until equilibrium is reached. If the resulting *e*-log *p* curve, such as curve *b*, lies entirely below curve *a*, the soil is said to have *collapsed*. Under field conditions, at present overburden pressure p_1

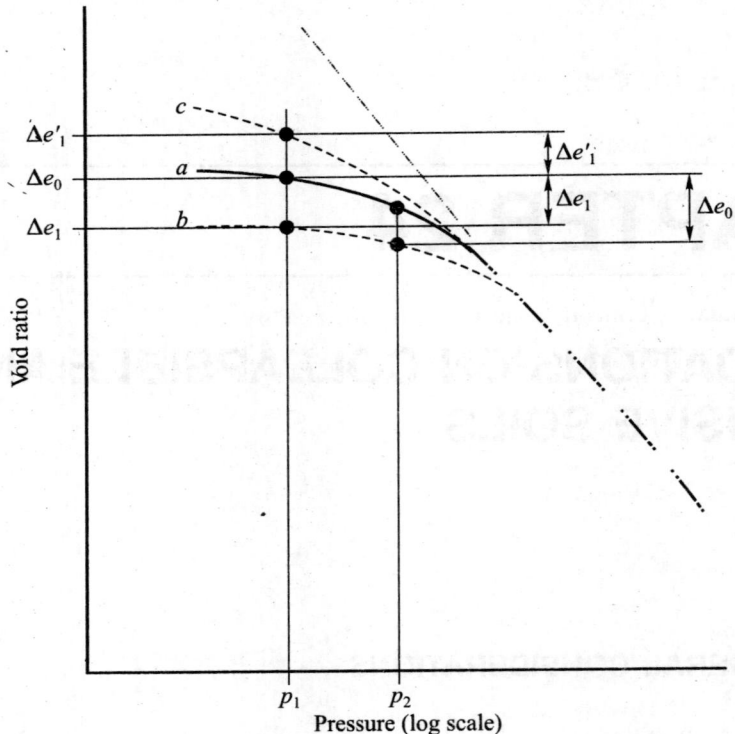

Fig. 24.1 Behaviour of soil in double oedometer or paired confined compression te
(a) relation between void ratio and total pressure for sample to which no water is adde
(b) relation for identical sample to which water is allowed access and which experienc
collapse, (c) same as (b) for sample that exhibits swelling (after Peck *et al.*, 1974)

and void ratio e_0, the addition of water at the commencement of the tests to sample 1, causes the
ratio to decrease to e_1. The collapsible settlement S_c may be expressed as

$$S_c = \frac{H\Delta e_1}{1+e_0}$$

(24

where H = the thickness of the stratum in the field.

Soils exhibiting this behaviour include true loess, clayey loose sands in which the clay se
merely as a binder, loose sands cemented by soluble salts, and certain residual soils such as th
derived from granites under conditions of tropical weathering.

On the other hand, if the addition of water to the second sample leads to curve c, loc
entirely above a, the soil is said to have *swelled*. At a given applied pressure p_1, the void r
increases to e'_1, and the corresponding rise of the ground is expressed as

$$S_c = \frac{H\Delta e'_1}{1+e_0}$$

(24.

Soils exhibiting this behaviour to a marked degree are usually montmorillonitic clays with h
plasticity indices.

T A: COLLAPSIBLE SOILS

GENERAL OBSERVATIONS

rding to Dudley (1970), and Barden *et al.,* (1973), four factors are needed to produce collapse oil structure:

. An open, partially unstable, unsaturated fabric

. A high enough net total stress that will cause the structure to be metastable

. A bonding or cementing agent that stabilizes the soil in the unsaturated condition

. The addition of water to the soil which causes the bonding or cementing agent to be reduced, and the interaggregate or intergranular contacts to fail in shear, resulting in a reduction in total volume of the soil mass.

Collapsible behaviour of compacted and cohesive soils depends on the percentage of fines, the l water content, the initial dry density and the energy and the process used in compaction.

Current practice in geotechnical engineering recognizes an unsaturated soil as a four phase rial composed of air, water, soil skeleton, and contractile skin. Under the idealization, two es can flow, that is air and water, and two phases come to equilibrium under imposed loads, that e soil skeleton and contractile skin. Currently, regarding the behaviour of compacted collapsing , geotechnical engineering recognized that:

1. Any type of soil compacted at *dry of optimum* conditions and at a low dry density may develop a collapsible fabric or metastable structure (Barden *et al.,* 1973).

2. A compacted and metastable soil structure is supported by microforces of shear strength, that is bonds, that are highly dependent upon capillary action. The bonds start losing strength with the increase of the water content and at a critical degree of saturation, the soil structure collapses (Jennings and Knight 1957; Barden *et al.,* 1973).

3. The soil collapse progresses as the degree of saturation increases. There is, however, a critical degree of saturation for a given soil above which negligible collapse will occur regardless of the magnitude of the prewetting overburden pressure (Jennings and Burland, 1962; Houston *et al.,* 1989).

4. The collapse of a soil is associated with localized shear failures rather than an overall shear failure of the soil mass.

5. During wetting induced collapse, under a constant vertical load and under K_o-oedometer conditions, a soil specimen undergoes an increase in horizontal stresses.

6. Under a triaxial stress state, the magnitude of volumetric strain resulting from a change in stress state or from wetting, depends on the mean normal total stress and is independent of the principal stress ratio.

The geotechnical engineer needs to be able to identify readily the soils that are likely to collapse to determine the amount of collapse that may occur. Soils that are likely to collapse are loose ., altered windblown sands, hillwash of loose consistency, and decomposed granites and acid eous rocks.

Some soils at their natural water content will support a heavy load but when water is provided y undergo a considerable reduction in volume. The amount of collapse is a function of the relative portions of each component including degree of saturation, initial void ratio, stress history of the

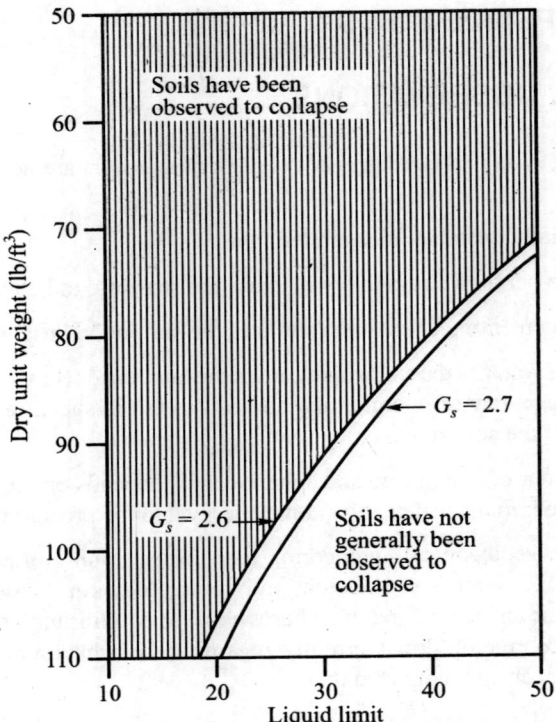

Fig. 24.2   Collapsible and noncollapsible loess (after Holtz and Hilf, 1961)

materials, thickness of the collapsible strata and the amount of added load.

Collapsing soils of the loessial type are found in many parts of the world. Loess is foun many parts of the United States, Central Europe, China, Africa, Russia, India, Argentina and elsewh

Holtz and Hilf (1961) proposed the use of the natural dry density and liquid limit as criteri predicting collapse. Figure 24.2 shows a plot giving the relationship between liquid limit and unit weight of soil, such that soils that plot above the line shown in the figure are susceptibl collapse upon wetting.

24.3 COLLAPSE POTENTIAL AND SETTLEMENT

Collapse Potential

A procedure for determining the collapse potential of a soil was suggested by Jennings and Kni (1975). The procedure is as follows:

A sample of an undisturbed soil is cut and fit into a consolidometer ring and loads are appl progressively until about 200 kPa (4 kip/ft^2) is reached. At this pressure the specimen is flooded v water for saturation and left for 24 hours. The consolidation test is carried on to its maximum loadi The resulting e-log p curve plotted from the data obtained is shown in Fig. 24.3.

The collapse potential C_p is then expressed as

$$C_p = \frac{\Delta e_c}{1 + e_0}$$

(24.

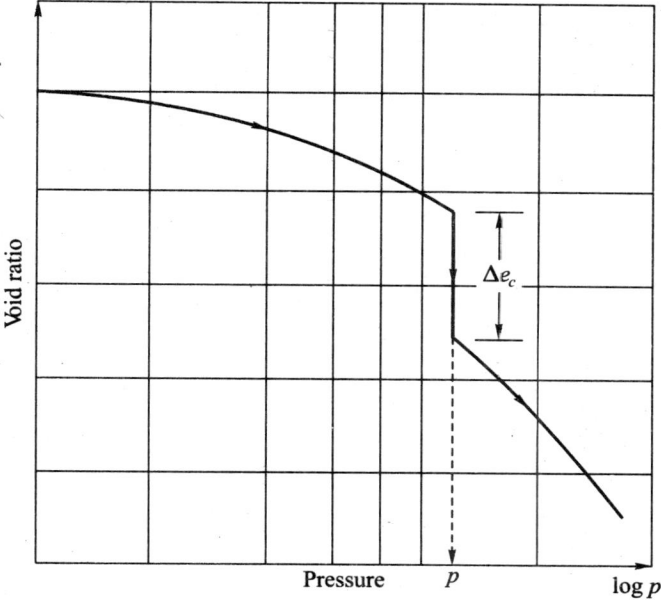

Fig. 24.3 Typical collapse potential test result

TABLE 24.1

Collapse potential values

$C_p\%$	Severity of problem
0–1	No problem
1–5	Moderate trouble
3–10	Trouble
10–20	Severe trouble
> 20	Very severe trouble

in which Δe_c = change in void ratio upon wetting, e_o = natural void ratio.

The collapse potential is also defined as

$$C_p = \frac{\Delta H_o}{H_c}$$ (24.2b)

where, ΔH_c = change in the height upon wetting, H_c = initial height.

Jennings and Knight have suggested some values for collapse potential as shown in Table 24.1. These values are only qualitative to indicate the severity of the problem.

24.4 COMPUTATION OF COLLAPSE SETTLEMENT

The double oedometer method was suggested by Jennings and Knight (1975) for determining a quantitative measure of collapse settlement. The method consists of conducting two consolidation

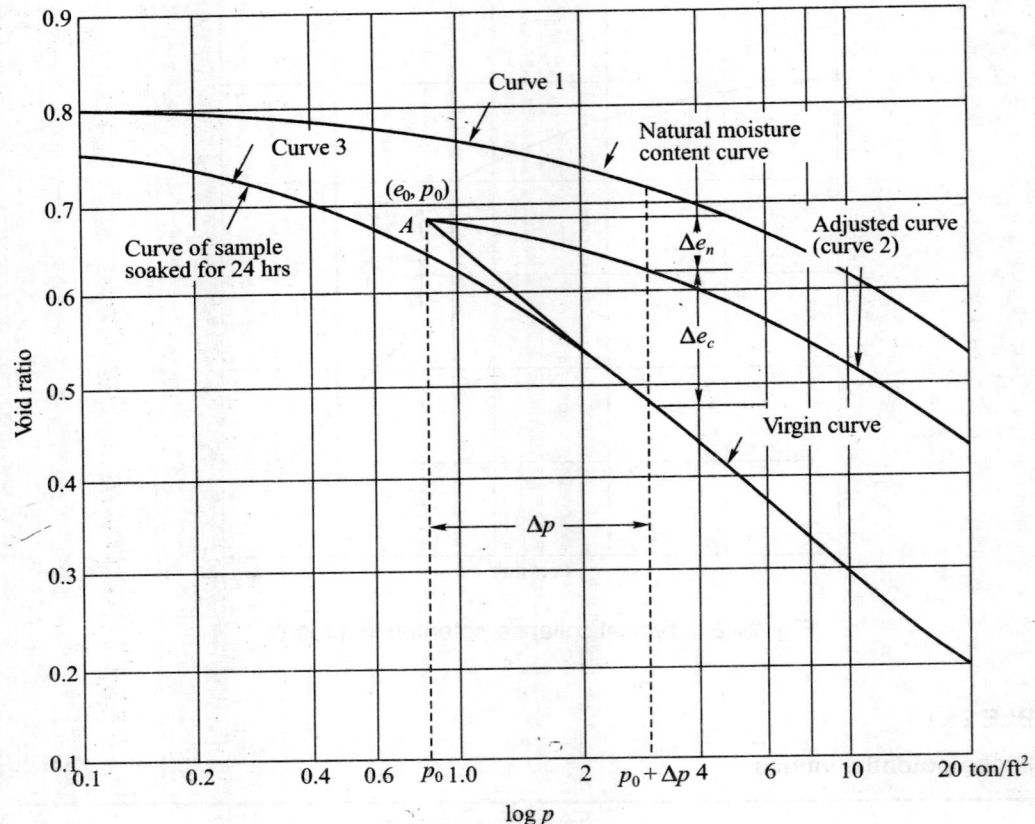

Fig. 24.4 Double consolidation test and adjustments for normally consolidated soil
(Clemence and Finbarr, 1981)

tests. Two identical undisturbed soil samples are used in the tests. The procedure is as follows:

1. Insert two identical undisturbed samples into the rings of two oedometers.

2. Keep both the specimens under a pressure of 1 kN/m^2 (= 0.15 lb/in^2) for a period of 24 hours.

3. After 24 hours, saturate one specimen by flooding and keep the other at its natural moisture content.

4. After the completion of 24 hour flooding, continue the consolidation tests for both the samples by doubling the loads. Follow the standard procedure for the consolidation test.

5. Obtain the necessary data from the two tests, and plot e-log p curves for both the samples as shown in Fig. 24.4 for normally consolidated soil.

6. Follow the same procedure for overconsolidated soil and plot the e-log p curves as shown in Fig. 24.5.

From e-log p plots, obtain the initial void ratios of the two samples after the first 24 hour of loading. It is a fact that the two curves do not have the same initial void ratio. The total overburden pressure p_0 at the depth of the sample is obtained and plotted on the e-log p curves in Figs 24.4 and 24.5. The preconsolidation pressures p_c are found from the soaked curves of Figs 24.4 and 24.5 and plotted.

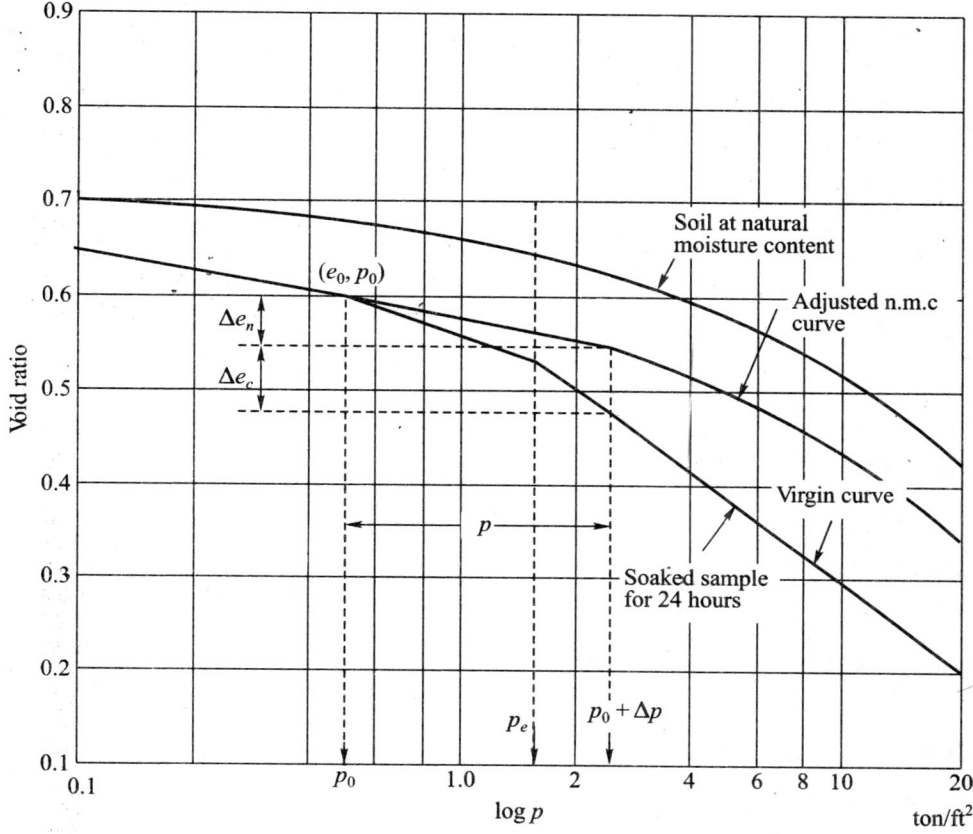

Fig. 24.5   Double consolidation test and adjustments for overconsolidated soil
(Clemence and Finbarr, 1981)

Normally Consolidated Case

For the case in which p_c/p_0 is about unity, the soil is considered normally consolidated. In such a case, compression takes place along the virgin curve. The straight line which is tangential to the soaked e-log p curve passes through the point (e_0, p_0) as shown in Fig. 24.4. Through the point (e_0, p_0) a curve is drawn parallel to the e-log p curve obtained from the sample tested at natural moisture content. The settlement for any increment in pressure Δp due to the foundation load may be expressed in two parts as

$$S_1 = \frac{\Delta e_n H_c}{1 + e_0} \tag{24.3a}$$

$$S_2 = \frac{\Delta e_c H_c}{1 + e_0} \tag{24.3b}$$

where Δe_n = change in void ratio due to load Δp as per the e-log p curve without change in moisture content

   Δe_c = change in void ratio at the same load Δp with the increase in moisture content (settlement caused due to collapse of the soil structure)

H_c = thickness of soil stratum susceptible to collapse.

From Eqs (24.3a) and (24.3b), the total settlement due to the collapse of the soil structure is

$$S_c = S_1 + S_2 = \frac{H_c}{1+e_0}(\Delta e_n + \Delta e_c) \tag{24.4}$$

Overconsolidated Case

In the case of an overconsolidated soil the ratio p_c/p_0 is greater than unity. Draw a curve from the point (e_0, p_0) on the soaked soil curve parallel to the curve which represents no change in moisture content during the consolidation stage. For any load $(p_0 + \Delta p) > p_c$, the settlement of the foundation may be determined by making use of the same Eq. (24.4). The changes in void ratios Δe_n and Δe_c are defined in Fig. 24.5.

Example 24.1

A footing of size 10×10 ft is founded at a depth of 5 ft below ground level in collapsible soil of the loessial type. The thickness of the stratum susceptible to collapse is 30 ft. The soil at the site is normally consolidated. In order to determine the collapse settlement, double oedometer tests were conducted on two undisturbed soil samples as per the procedure explained in Section 24.3. The $e \log p$ curves of the two samples are given in Fig. 24.4. The average unit weight of soil $\gamma = 106.6$ lb/ft^3 and the induced stress Δp, at the middle of the stratum due to the foundation pressure, is 4,400 lb/ft^2 (= 2.20 t/ft^2). Estimate the collapse settlement of the footing under a soaked condition.

Solution

Double consolidation test results of the soil samples are given in Fig. 24.4. Curve 1 was obtained with natural moisture content. Curve 3 was obtained from the soaked sample after 24 hours. The virgin curve is drawn in the same way as for a normally loaded clay soil (Fig. 7.9a).

The effective overburden pressure p_0 at the middle of the collapsible layer is

$p_0 = 15 \times 106.6 = 1,599$ lb/ft^2 or 0.8 ton/ft^2

A vertical line is drawn in Fig. 24.4 at $p_0 = 0.8$ ton/ft^2. Point A is the intersection of the vertical line and the virgin curve giving the value of $e_0 = 0.68$. $p_0 + \Delta p = 0.8 + 2.2 = 3.0$ t/ft^2. At $(p_0 + \Delta p) = 3$ ton/ft^2, we have (from Fig. 24.4)

$\Delta e_n = 0.68 - 0.62 = 0.06$

$\Delta e_c = 0.62 - 0.48 = 0.14$

From Eq. (24.3)

$$S_1 = \frac{\Delta e_n H_c}{1+e_0} = \frac{0.06 \times 30 \times 12}{1+0.68} = 12.86 \text{ in.}$$

$$S_2 = \frac{\Delta e_c H_c}{1+e_0} = \frac{0.14 \times 30 \times 12}{1+0.68} = 30.00 \text{ in.}$$

Total settlement $S_c = 42.86$ in.

The total settlement would be reduced if the thickness of the collapsible layer is less or the foundation pressure is less.

Example 24.2

Refer to Example 24.1. Determine the expected collapse settlement under wetted conditions if the soil stratum below the footing is overconsolidated. Double oedometer test results are given in Fig. 24.5. In this case $p_0 = 0.5$ ton/ft2, $\Delta p = 2$ ton/ft^2, and $p_c = 1.5$ ton/ft^2.

Solution

The virgin curve for the soaked sample can be determined in the same way as for an overconsolidated clay. Double oedometer test results are given in Fig. 24.5. From this figure:

$$e_0 = 0.6, \Delta e_n = 0.6 - 0.55 = 0.05, \Delta e_c = 0.55 - 0.48 = 0.07$$

As in Ex. 24.1

$$S_1 = \frac{\Delta e_n}{1+e_0} H_c = \frac{0.05 \times 30 \times 12}{1+0.6} = 11.25 \text{ in.}$$

$$S_2 = \frac{\Delta e_c}{1+e_0} H_c = \frac{0.7 \times 30 \times 12}{1+0.6} = 15.75 \text{ in.}$$

Total $S_c = 27.00$ in.

24.5 FOUNDATION DESIGN

Foundation design in collapsible soil is a very difficult task. The results from laboratory or field tests can be used to predict the likely settlement that may occur under severe conditions. In many cases, deep foundations, such as piles, piers etc, may be used to transmit foundation loads to deeper bearing strata below the collapsible soil deposit. In cases where it is feasible to support the structure on shallow foundations in or above the collapsing soils, the use of continuous strip footings may provide a more economical and safer foundation than isolated footings (Clemence and Finbarr, 1981). Differential settlements between columns can be minimized, and a more equitable distribution of stresses may be achieved with the use of strip footing design as shown in Fig. 24.6 (Clemence and Finbarr, 1981).

Load-bearing beams

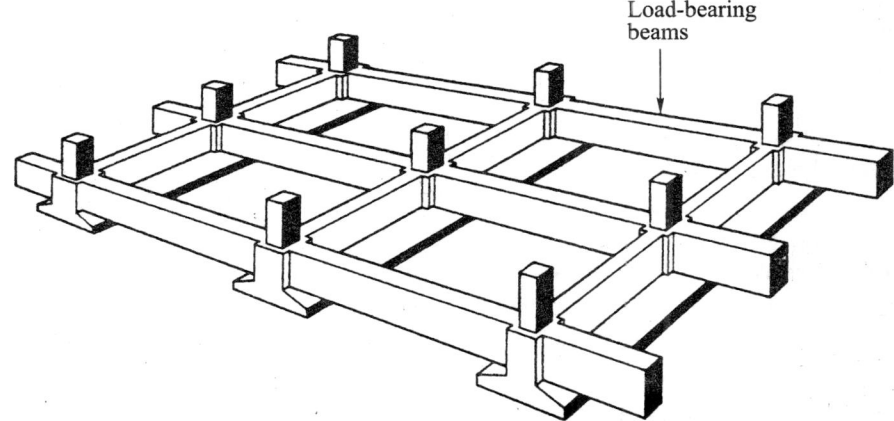

Fig. 24.6 Continuous footing design with load-bearing beams for collapsible soil (after Clemence and Finbarr, 1981)

24.6   TREATMENT METHODS FOR COLLAPSIBLE SOILS

On some sites, it may be feasible to apply a pretreatment technique either to stabilize the soil or cause collapse of the soil deposit prior to construction of a specific structure. A great variety of treatment methods have been used in the past. Moistening and compaction techniques, with either conventional impact, or vibratory rollers may be used for shallow depths up to about 1.5 m. For deeper depths, vibroflotation, stone columns, and displacement piles may be tried. Heat treatment to solidify the soil in place has also been used in some countries such as Russia. Chemical stabilization with the use of sodium silicate and injection of carbon dioxide have been suggested (Semkin et al., 1986).

.   . Field tests conducted by Rollins *et al.,* (1990) indicate that dynamic compaction treatment provides the most effective means of reducing the settlement of collapsible soils to tolerable limits. Prewetting, in combination with dynamic compaction, offers the potential for increasing compaction efficiency and uniformity, while increasing vibration attenuation. Prewetting with a 2 percent solution of sodium silicate provides cementation that reduces the potential for settlement. Prewetting with water was found to be the easiest and least costly treatment, but it proved to be completely ineffective in reducing collapse potential for shallow foundations. Prewetting must be accompanied by preloading, surcharging or excavation in order to be effective.

PART B: EXPANSIVE SOILS

24.7   DISTRIBUTION OF EXPANSIVE SOILS

The problem of expansive soils is widespread throughout world. The countries that are facing problems with expansive soils are Australia, the United States, Canada, China, Israel, India, and Egypt. The clay mineral that is mostly responsible for expansiveness belongs to the montmorillonite group. The major concern with expansive soils exists generally in the western part of the United States. In the northern and central United States, the expansive soil problems are primarily related to highly overconsolidated shales. This includes the Dakotas, Montana, Wyoming and Colorado (Chen, 1988). In Minneapolis, the expansive soil problem exists in the Cretaceous deposits along the Mississippi River and a shrinkage/swelling problem exists in the lacustrine deposits in the Great Lakes Area. In general, expansive soils are not encountered regularly in the eastern parts of the central United States.

In eastern Oklahoma and Texas, the problems encompass both shrinking and swelling. In the Los Angeles area, the problem is primarily one of desiccated alluvial and colluvial soils. The weathered volcanic material in the Denver formation commonly swells when wetted and is a cause of major engineering problems in the Denver area.   :

The six major natural hazards are earthquakes, landslides, expansive soils, hurricane, tornado and flood. A study points out that expansive soils tie with hurricane wind/storm surge for second place among America's most destructive natural hazards in terms of dollar losses to buildings. According to the study, it was projected that by the year 2000, losses due to expansive soil would exceed 4.5 billion dollars annually (Chen, 1988).

24.8   GENERAL CHARACTERISTICS OF SWELLING SOILS

Swelling soils, which are clayey soils, are also called expansive soils. When these soils are partially saturated, they increase in volume with the addition of water. They shrink greatly on drying and develop cracks on the surface. These soils possess a high plasticity index. Black cotton soils found in many parts of India belong to this category. Their colour varies from dark grey to black. It is easy to recognize these soils in the field during either dry or wet seasons. Shrinkage cracks are visible on the ground surface during dry seasons. The maximum width of these cracks may be up to 20 mm or more and they travel deep into the ground. A lump of dry black cotton soil requires a hammer to break. During rainy seasons, these soils become very sticky and very difficult to traverse.

Expansive soils are residual soils which are the result of weathering of the parent rock. The depths of these soils in some regions may be up to 6 m or more. Normally the water table is met at great depths in these regions. As such the soils become wet only during rainy seasons and are dry or partially saturated during the dry seasons. In regions which have well-defined, alternately wet and dry seasons, these soils swell and shrink in regular cycles. Since moisture change in the soils bring about severe movements of the mass, any structure built on such soils experiences recurring cracking and progressive damage. If one measures the water content of the expansive soils with respect to depth during dry and wet seasons, the variation is similar to the one shown in Fig. 24.7.

During dry seasons, the natural water content is practically zero on the surface and the volume of the soil reaches the shrinkage limit. The water content increases with depth and reaches a value w_n

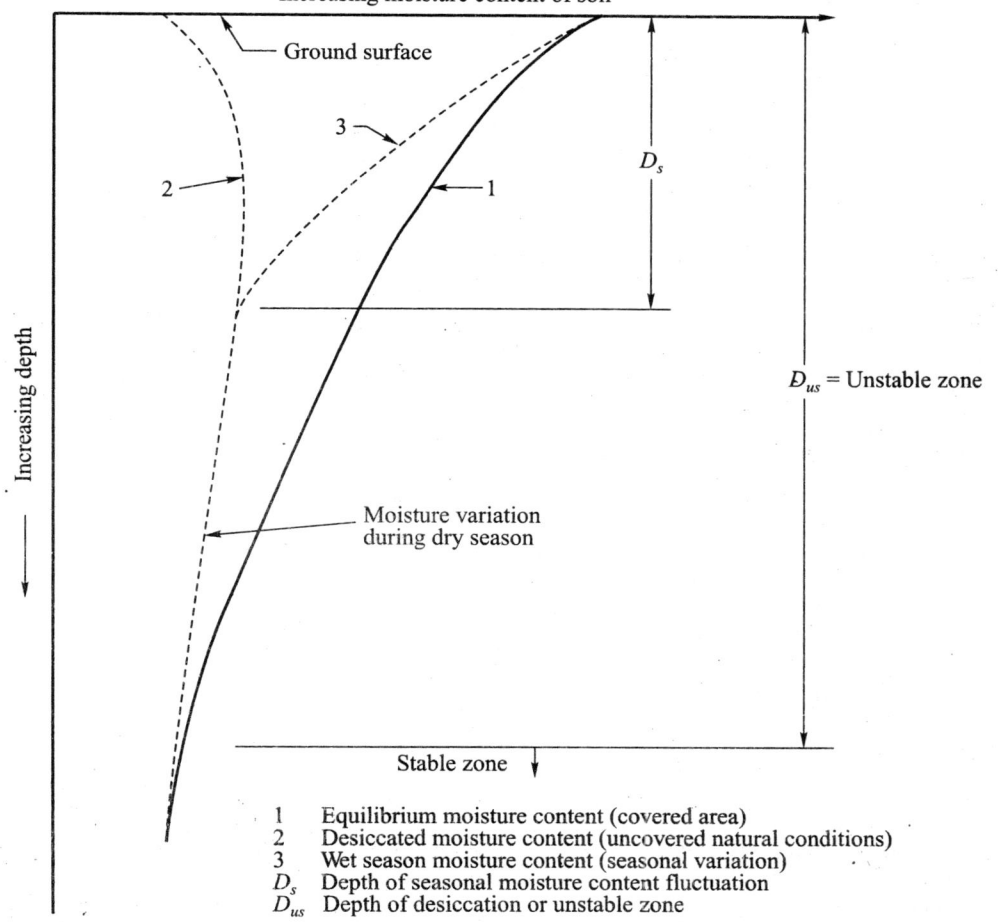

Fig. 24.7   Moisture content variation with depth below ground surface (Chen, 1988)

at a depth D_{us}, beyond which it remains almost constant. During the wet season the water content increases and reaches a maximum at the surface. The water content decreases with depth from a maximum of w_n at the surface to a constant value of w_n at almost the same depth D_{us}. This indicates that the intake of water by the expansive soil into its lattice structure is a maximum at the surface and nil at depth D_{us}. This means that the soil lying within this depth D_{us} is subjected to drying and wetting and hence cause considerable movements in the soil. The movements are considerable close to the ground surface and decrease with depth. The cracks that are developed in the dry seasons, close due to lateral movements during the wet seasons.

The zone which lies within the depth D_{us} may be called the *unstable zone* (or active zone) and the one below this the *stable zone*. Structures built within this unstable *zone* are likely to move up and down according to seasons and hence suffer damage if differential movements are considerable.

If a structure is built during the dry season with the foundation lying within the unstable zone, the base of the foundation experiences a *swelling pressure* as the partially saturated soil starts taking in water during the wet season. This swelling pressure is due to constraints offered by the foundation for free swelling. The maximum swelling pressure may be as high as 2 MPa (20 tsf). If the imposed bearing pressure on the foundation by the structure is less than the swelling pressure, the structure is likely to be lifted up at least locally which would lead to cracks in the structure. If the imposed bearing pressure is greater than the swelling pressure, there will not be any problem for the structure. If on the other hand, the structure is built during the wet season, it will definitely experience settlement as the dry season approaches, whether the imposed bearing pressure is high or low. However, the imposed bearing pressure during the wet season should be within the allowable bearing pressure of the soil. The *better practice* is to *construct a structure during the dry season and complete it before the wet season.*

In covered areas below a building there will be very little change in the moisture content except due to lateral migration of water from uncovered areas. The moisture profile is depicted by curve 1 in Fig. 24.7.

24.9 CLAY MINERALOGY AND MECHANISM OF SWELLING

Clays can be divided into three general groups on the basis of their crystalline arrangement. They are:

1. Kaolinite group

2. Montmorillonite group (also called the smectite group)

3. Illite group.

The kaolinite group of minerals are the most stable of the groups of minerals. The kaolinite mineral is formed by the stacking of the crystalline layers of about 7 Å thick one above the other with the base of the silica sheet bonding to hydroxyls of the gibbsite sheet by hydrogen bonds. Since hydrogen bonds are comparatively strong, the kaolinite crystals consists of many sheet stackings that are difficult to dislodge. The mineral is, therefore, stable and water cannot enter between the sheets to expand the unit cells.

The structural arrangement of the montmorillonite mineral is composed of units made of two silica tetrahedral sheets with a central alumina-octahedral sheet. The silica and gibbsite sheets are combined in such a way that the tips of the tetrahedrons of each silica sheet and one of the hydroxyl layers of the octahedral sheet form a common layer. The atoms common to both the silica and gibbsite layers are oxygen instead of hydroxyls. The thickness of the silica-gibbsite-silica unit is about 10 Å. In stacking of these combined units one above the other, oxygen layers of each unit are adjacent to oxygen of the neighboring units, with a consequence that there is a weak bond and excellent cleavage between them. Water can enter between the sheets causing them to expand significantly and thus the structure can break into 10 Å thick structural units. The soils containing a considerable amount of montmorillonite minerals will exhibit high swelling and shrinkage characteristics. The illite group of minerals has the same structural arrangement as the montmorillonite group. The presence of potassium as the bonding materials between units makes the illite minerals swell less.

24.10 DEFINITION OF SOME PARAMETERS

Expansive soils can be classified on the basis of certain inherent characteristics of the soil. It is first necessary to understand certain basic parameters used in the classification.

Swelling Potential

Swelling potential is defined as the percentage of swell of a laterally confined sample in an oedometer test which is soaked under a surcharge load of 7 kPa (1 lb/in^2) after being compacted to maximum dry density at optimum moisture content according to the AASHTO compaction test.

Swelling Pressure

The swelling pressure p_s, is defined as the pressure required for preventing volume expansion in soil in contact with water. It should be noted here that the swelling pressure measured in a laboratory oedometer is different from that in the field. The actual field swelling pressure is always less than the one measured in the laboratory.

Free Swell

Free swell S_f is defined as

$$S_f = \frac{V_f - V_i}{V_i} \times 100 \qquad (24.5)$$

where V_i = initial dry volume of poured soil

V_f = final volume of poured soil

According to Holtz and Gibbs (1956), 10 cm^3 (V_i) of dry soil passing thorough a No. 40 sieve is poured into a 100 cm^3 graduated cylinder filled with water. The volume of settled soil is measured after 24 hours which gives the value of V_f. Bentonite-clay is supposed to have a free swell value ranging from 1200 to 2000 percent. The free swell value increases with plasticity index. Holtz and Gibbs suggested that soils having a free-swell value as low as 100 percent can cause considerable damage to lightly loaded structures and soils heaving a free swell value below 50 percent seldom exhibit appreciable volume change even under light loadings.

24.11 EVALUATION OF THE SWELLING POTENTIAL OF EXPANSIVE SOILS BY SINGLE INDEX METHOD (CHEN, 1988)

Simple soil property tests can be used for the evaluation of the swelling potential of expansive soils (Chen, 1988). Such tests are easy to perform and should be used as routine tests in the investigation of building sites in those areas having expansive soil. These tests are:

1. Atterberg limits tests
2. Linear shrinkage tests
3. Free swell tests
4. Colloid content tests

Atterberg Limits

Holtz and Gibbs (1956) demonstrated that the plasticity index, I_p, and the liquid limit, w_l, are useful indices for determining the swelling characteristics of most clays. Since the liquid limit and the swelling of clays both depend on the amount of water a clay tries to absorb, it is natural that they are related. The relation between the swelling potential of clays and the plasticity index has been established as given in Table 24.2.

Linear Shrinkage

The swell potential is presumed to be related to the opposite property of linear shrinkage measured in a very simple test. Altmeyer (1955) suggested the values given in Table 24.3 as a guide to the determination of potential expansiveness based on shrinkage limits and linear shrinkage.

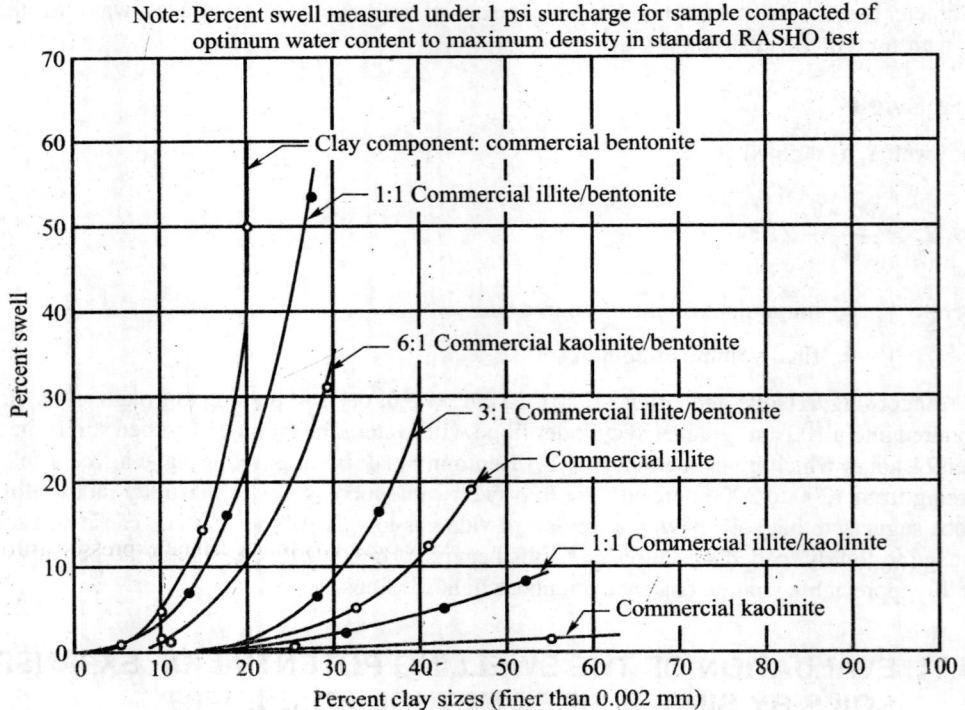

Note: Percent swell measured under 1 psi surcharge for sample compacted of optimum water content to maximum density in standard RASHO test

Fig. 24.8   Relationship between percentage of swell and percentage of clay sizes for experimental soils (after Seed *et al.*, 1962)

TABLE 24.2

Relation between swelling potential and plasticity index, I_p

Plasticity index I_p	(%) Swelling potential
0 –15	Low
10 –35	Medium
20 –55	High
35 and above	Very high

TABLE 24.3

Relation between swelling potential, shrinkage limits, and linear shrinkage

Shrinkage limit %	Linear shrinkage %	Degree of expansion
< 10	> 8	Critical
10–12	5–8	Marginal
> 12	0–5	Non-critical

Colloid Content

There is a direct relationship between colloid content and swelling potential as shown in Fig. 24.8 (Chen, 1988). For a given clay type, the amount of swell will increase with the amount of clay present in the soil.

24.12 CLASSIFICATION OF SWELLING SOILS BY INDIRECT MEASUREMENT

By utilizing the various parameters as explained in Section 24.11, the swelling potential can be evaluated without resorting to direct measurement (Chen, 1988).

USBR Method

Holtz and Gibbs (1956) developed this method which is based on the simultaneous consideration of several soil properties. The typical relationships of these properties with swelling potential are shown in Fig. 24.9. Table 24.4 has been prepared based on the curves presented in Fig. 24.9 by Holtz and Gibbs (1956).

The relationship between the swell potential and the plasticity index can be expressed as follows (Chen, 1988)

$$S_p = Be^{A(I_p)} \tag{24.6}$$

where, $A = 0.0838$

$B = 0.2558$

$I_p =$ plasticity index.

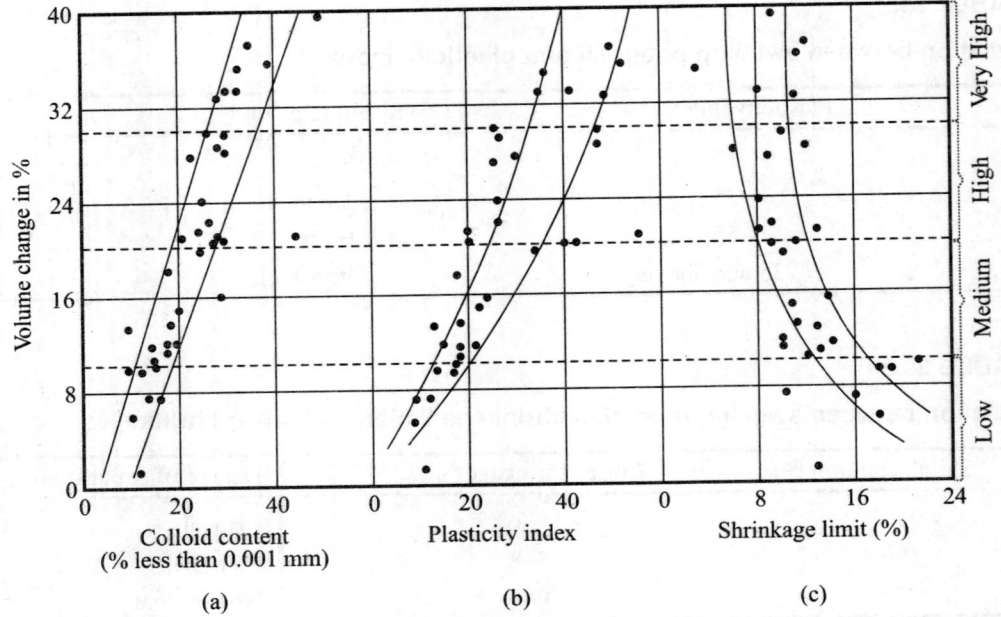

Fig. 24.9 Relation of volume change to (a) colloid content, (b) plasticity index, and (c) shrinkage limit (air-dry to saturated condition under a load of 1 lb per sq in) (Holtz and Gibbs, 1956)

TABLE 24.4

Data for making estimates of probable volume changes for expansive soils (Source: Chen, 1988)

Data from index tests*			Probable expansion,	
Colloid content, per-cent minus 0.001 mm	Plasticity index	Shrinkage limit	percent total vol. change	Degree of expansion
> 28	> 35	< 11	> 30	Very high
20–13	25–41	7–12	20–30	High
13–23	15–28	10–16	10–30	Medium
< 15	< 18	> 15	<10	Low

*Based on vertical loading of 1.0 psi. (after Holtz and Gibbs, 1956)

Figure 24.10 shows that with an increase in plasticity index, the increase of swelling potential is much less than predicted by Holtz and Gibbs. The curves given by Chen (1988) are based on thousands of tests performed over a period of 30 years and as such are more realistic.

Activity Method

Skempton (1953) defined activity by the following expression

$$A = \frac{I_p}{C} \tag{24.7}$$

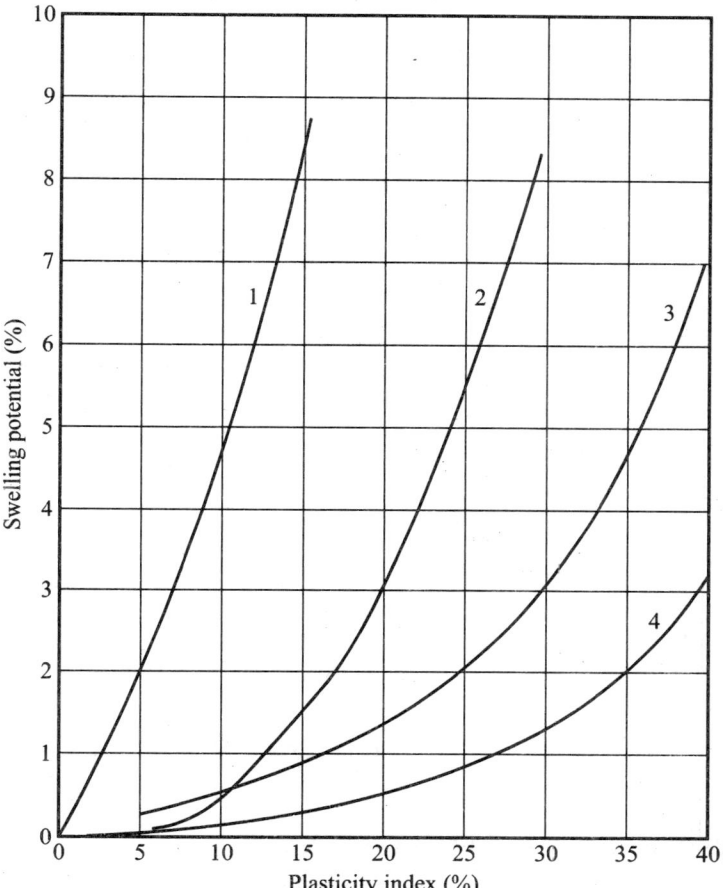

1. Holtz and Gibbs (Surcharge pressure 1 psi)
2. Seed, Woodward and Lundgren (Surcharge pressure 1 psi)
3. Chen (Surcharge pressure 1 psi)
4. Chen (Surcharge pressure 6.94 psi)

Fig. 24.10 Relationships of volume change to plasticity index (*Source:* Chen, 1988)

where I_p = plasticity index

C = percentage of clay size finer than 0.002 mm by weight.

The activity method as proposed by Seed, Woodward, and Lundgren, (1962) was based on remoulded, artificially prepared soils comprising of mixtures of bentonite, illite, kaolinite and fine sand in different proportions. The activity for the artificially prepared sample was defined as

$$\text{activity } A = \frac{I_p}{C-n} \tag{24.8}$$

where $n = 5$ for natural soils and, $n = 10$ for artificial mixtures.

The proposed classification chart is shown in Fig. 24.11. This method appears to be an improvement over the USBR method.

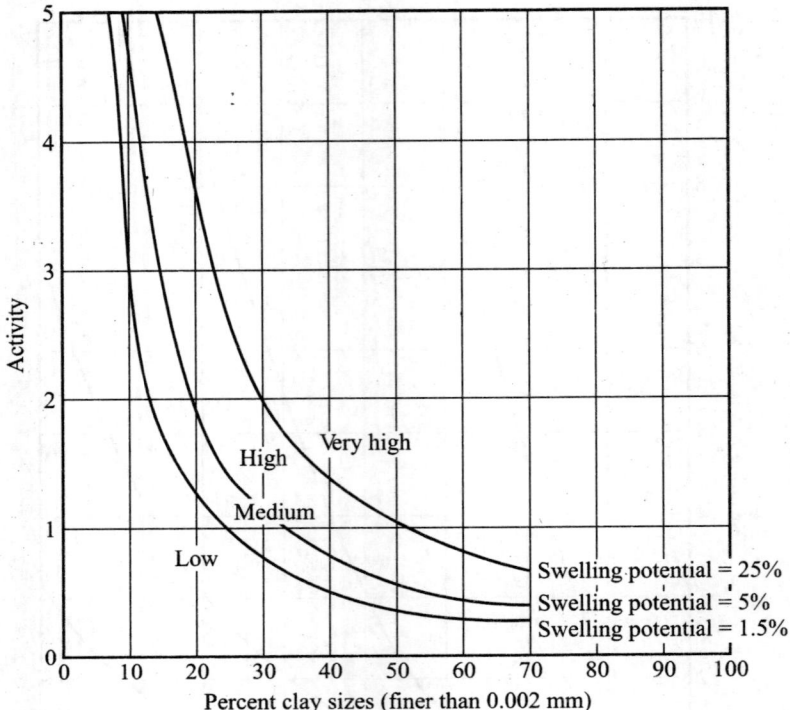

Fig. 24.11   Classification chart for swelling potential (after Seed, Woodward, and Lundgren, 1962)

The Potential Volume Change Method (PVC)

A determination of soil volume change was developed by Lambe under the auspices of the Federal Housing Administration (*Source:* Chen, 1988). Remoulded samples were specified. The procedure is as given below.

The sample is first compacted in a fixed ring consolidometer with a compaction effect of 55,000 ft-lb per cu ft. Then an initial pressure of 200 psi is applied, and water added to the sample which is partially restrained by vertical expansion by a proving ring. The proving ring reading is taken at the end of 2 hours. The reading is converted to pressure and is designated as the *swell index*. From Fig. 24.12, the swell index can be converted to potential volume change. Lambe established the categories of PVC rating as shown in Table 24.5.

The PVC method has been widely used by the Federal Housing Administration as well as the Colorado State Highway Department (Chen, 1988).

TABLE 24.5
Potential volume change rating (PVC)

PVC rating	Category
Less than 2	Non-critical
2–4	Marginal
4–6	Critical
> 6	Very critical

(*Source:* Chen, 1988)

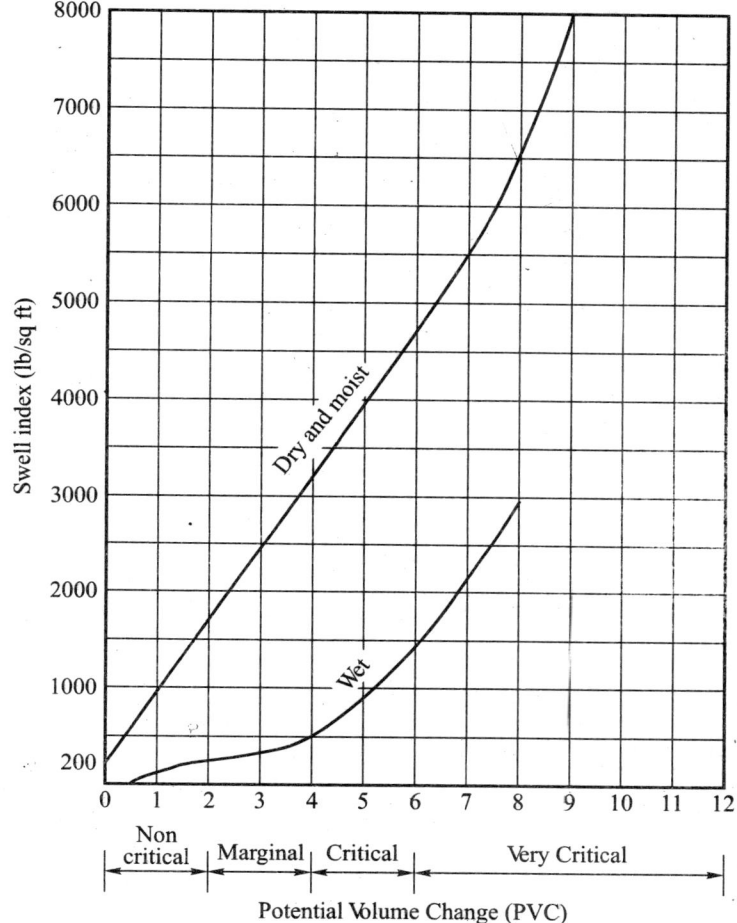

Fig. 24.12   Swell index versus potential volume change (from 'FHA soil PVC meter publication,' Federal Housing Administration Publication no. 701) (*Source:* Chen, 1988)

Figure 24.13(a) shows a soil volume change meter (ELE International Inc). This meter measures both shrinkage and swelling of soils, ideal for measuring swelling of clay soils, and fast and easy to operate.

Expansion Index (EI)—Chen (1988)

The ASTM Committee on Soil and Rock suggested the use of an *Expansion Index* (EI) as a unified method to measure the characteristics of swelling soils. It is claimed that the EI is a basic index property of soil such as the liquid limit, the plastic limit and the plasticity index of the soil.

The sample is sieved through a No 4 sieve. Water is added so that the degree of saturation is between 49 and 51 percent. The sample is then compacted into a 4 inch diameter mould in two layers to give a total compacted depth of approximates 2 inches. Each layer is compacted by 15 blows of 5.5 lb hammer dropping from a height of 12 inches. The prepared specimen is allowed to consolidate under 1 lb/in^2 pressure for a period of 10 minutes, then inundated with water until the rate of expansion ceases.

TABLE 24.6

Classification of potentially expansive soil

Expansion Index,	EI Expansion potential
0–20	Very low
21–50	Low
51–90	Medium
91–130	High
> 130	Very high

(Source: Chen, 1988)

The expansion index is expressed as

$$EI = \frac{\Delta h}{h_i} \times 1000 \qquad (24.9)$$

where Δh = change in thickness of sample, in.

h_i = initial thickness of sample, in.

The classification of a potentially expansive soil is based on Table 24.6.

This method offers a simple testing procedure for comparing expansive soil characteristics.

Figure 24.13(b) shows an ASTMD-829 expansion index test apparatus (ELE International Inc). This is a completely self-contained apparatus designed for use in determining the expansion index of soils.

Swell Index

Vijayvergiya and Gazzaly (1973) suggested a simple way of identifying the swell potential of clays, based on the concept of the swell index. They defined the swell index, I_s, as follows

$$I_s = \frac{w_n}{w_l} \qquad (24.10)$$

where w_n = natural moisture content in percent

w_l = liquid limit in percent

The relationship between I_s and swell potential for a wide range of liquid limit is shown in Fig. 24.14. Swell index is widely used for the design of post-tensioned slabs on expansive soils.

Prediction of Swelling Potential

Plasticity index and shrinkage limit can be used to indicate the swelling characteristics of expansive soils. According to Seed at al., (1962), the swelling potential is given as a function of the plasticity index by the formula

$$S_p = 60kI_p^{2.44} \qquad (24.11)$$

(a)   (b)

Fig. 24.13 (a) Soil volume change meter, and (b) Expansion index test apparatus
(Courtesy: Soiltest)

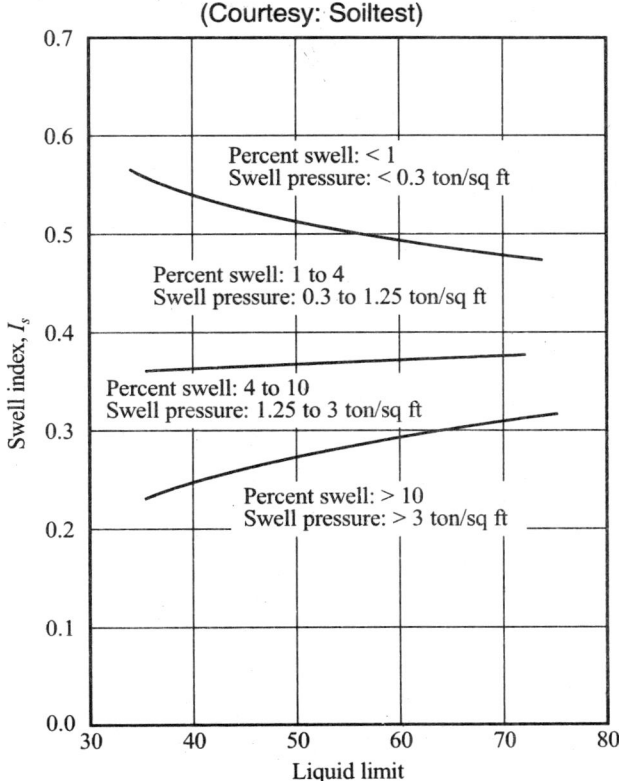

Fig. 24.14 Relationship between swell index and liquid limit for expansive clays
(*Source:* Chen, 1988)

where S_p = swelling potential in percent

   I_p = plasticity index in percent

   k = 3.6×10^{-5}, a factor for clay content between 8 and 65 percent.

24.13 SWELLING PRESSURE BY DIRECT MEASUREMENT

ASTM defines swelling pressure which prevents the specimen from swelling or that pressure which is required to return the specimen to its original state (void ratio, height) after swelling. Essentially, the methods of measuring swelling pressure can be either stress controlled or strain controlled (Chen, 1988).

   In the stress controlled method, the conventional oedometer is used. The samples are placed in the consolidation ring trimmed to a height of 0.75 to 1 inch. The samples are subjected to a vertical pressure ranging from 500 psf to 2000 psf depending upon the expected field conditions. On the completion of consolidation, water is added to the sample. When the swelling of the sample has ceased the vertical stress is increased in increments until it has been compressed to its original height. The stress required to compress the sample to its original height is commonly termed the *zero volume change swelling pressure*. A typical consolidation curve is shown in Fig. 24.15.

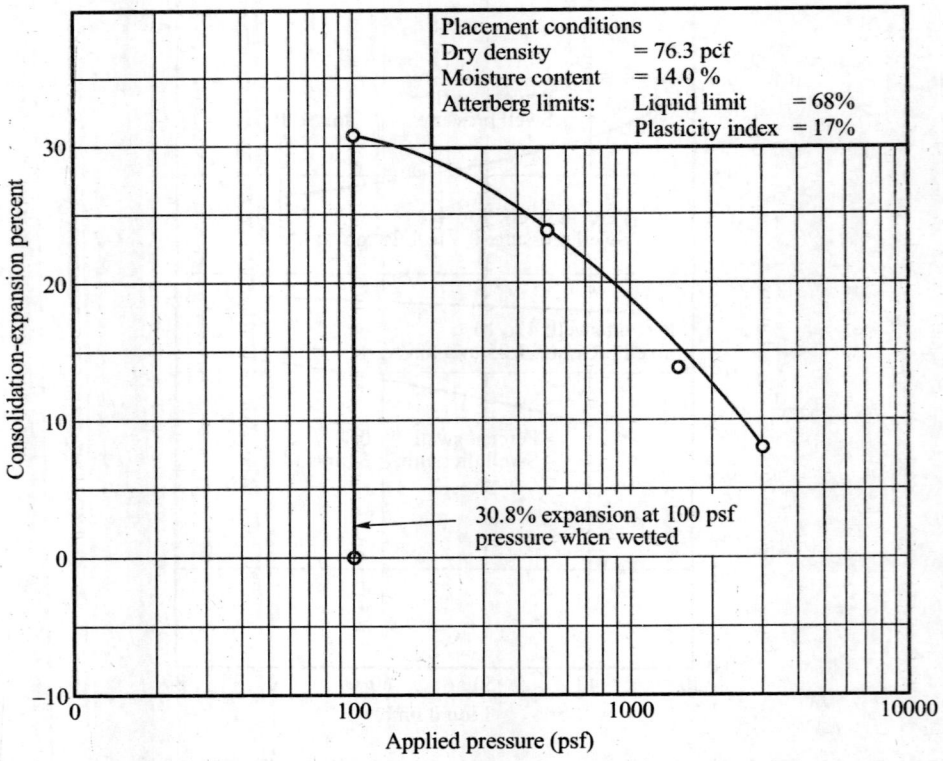

Fig. 24.15   Typical stress controlled swell-consolidation curve

Prediction of Swelling Pressure

Komornik et al., (1969) have given an equation for predicting swelling pressure as

$$\log p_s = 2.132 + 0.0208 w_l + 0.00065 \gamma_d - 0.0269 w_n \qquad (24.12)$$

where p_s = swelling pressure in kg/cm^2

w_l = liquid limit (%)

w_n = natural moisture content (%)

γ_d = dry density of soil in kg/cm^3

24.14 EFFECT OF INITIAL MOISTURE CONTENT AND INITIAL DRY DENSITY ON SWELLING PRESSURE

The capability of swelling decreases with an increase of the initial water content of a given soil because its capacity to absorb water decreases with the increase of its degree of saturation. It was found from swelling tests on black cotton soil samples, that the initial water content has a small effect on swelling pressure until it reaches the shrinkage limit, then its effect increases (Abouleid, 1982). This is depicted in Fig. 24.16(a).

The effect of initial dry density on the swelling percent and the swelling pressure increases with an increase of the dry density because the dense soil contains more clay particles in a unit volume and consequently greater movement will occur in a dense soil than in a loose soil upon wetting (Abouleid, 1982). The effect of initial dry density on swelling pressure is shown in Fig. 24.16(b).

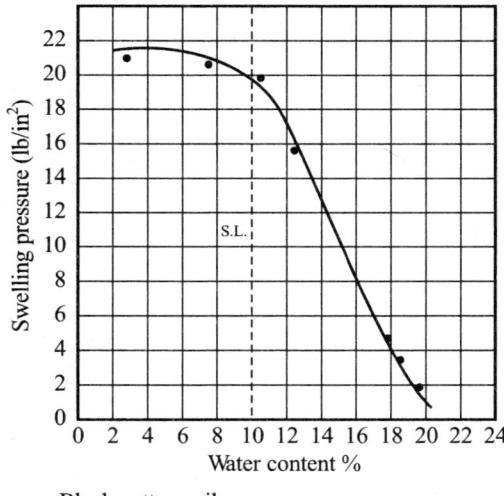

Black cotton soil
$w_l = 90$, $w_p = 30$, $w_s = 10$

Mineralogy: Largely sodium, Montmorillonite

(a)

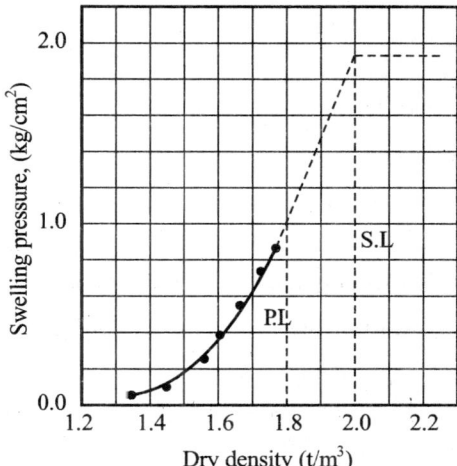

(b)

Fig. 24.16 (a) effect of initial water content on swelling pressure of black cotton soil, and (b) effect of initial dry density on swelling pressure of black cotton soil
(*Source:* Abouleid, 1982)

24.15 ESTIMATING THE MAGNITUDE OF SWELLING

When footings are built in expansive soil, they experience lifting due to the swelling or heaving the soil. The amount of total heave and the rate of heave of the expansive soil on which a structure founded are very complex. The heave estimate depends on many factors which cannot be readi determined. Some of the major factors that contribute to heaving are:

1. Climatic conditions involving precipitation, evaporation, and transpiration affect the moistu in the soil. The depth and degree of desiccation affect the amount of swell in a given so horizon.

2. The thickness of the expansive soil stratum is another factor. The thickness of the stratum controlled by the depth to the water table.

3. The depth to the water table is responsible for the change in moisture of the expansive so lying above the water table. No swelling of soil takes place when it lies below the wate table.

4. The predicted amount of heave depends on the nature and degree of desiccation of the so immediately after construction of a foundation.

5. The single most important element controlling the swelling pressure as well as the swe potential is the in-situ density of the soil. On the completion of excavation, the stress conditio in the soil mass undergoes changes, such as the release of stresses due to elastic rebound o the soil. If construction proceeds without delay, the structural load compensates for th stress release.

6. The permeability of the soil determines the rate of ingress of water into the soil either b gravitational flow or diffusion, and this in turn determines the rate of heave.

Various methods have been proposed to predict the amount of total heave under a given structura load. The following methods, however, are described here.

1. The Department of Navy method (1982)

2. The South African method [also known as the Van Der Merwe method (1964)].

The Department of Navy Method

Procedure for Estimating Total Swell under Structural Load

1. Obtain representative undisturbed samples of soil below the foundation level at intervals of depth. The samples are to be obtained during the dry season when the moisture contents are at their lowest.

2. Load specimens (at natural moisture content) in a consolidometer under a pressure equal to the ultimate value of the overburden plus the weight of the structure. Add water to saturate the specimen. Measure the swell.

3. Compute the final swell in terms of percent of original sample height.

4. Plot swell versus depth.

5. Compute the total swell which is equal to the area under the percent swell versus depth curve.

Procedure for Estimating Undercut

The procedure for estimating undercut to reduce swell to an allowable value is as follows:

1. From the percent swell versus depth curve, plot the relationship of total swell versus depth at that height. Total swell at any depth equals area under the curve integrated upward from the depth of zero swell.

2. For a given allowable value of swell, read the amount of undercut necessary from the total swell versus depth curve.

Van Der Merwe Method (1964)

Probably the nearest practical approach to the problem of estimating swell is that of Van Der Merwe. This method starts by classifying the swell potential of soil into very high to low categories as shown in Fig. 24.17. Then assign potential expansion (PE) expressed in in./ft of thickness based on Table 24.7.

TABLE 24.7

Potential expansion

Swell potential	Potential expansion (PE) in./ft
Very high	1
High	1/2
Medium	1/4
Low	0

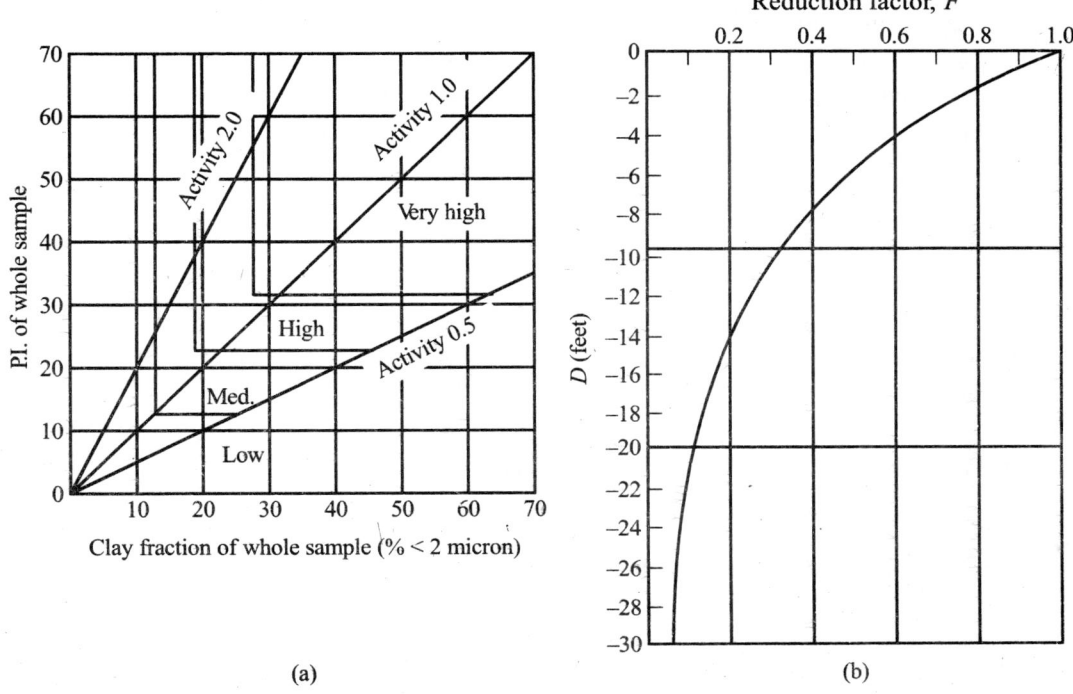

(a)

(b)

Fig. 24.17 Relationships for using Van Der Merwe's prediction method: (a) potential expansiveness, and (b) reduction factor (Van der Merwe, 1964)

Procedure for Estimating Swell

1. Assume the thickness of an expansive soil layer or the lowest level of ground water.

2. Divide this thickness (z) into several soil layers with variable swell potential.

3. The total expansion is expressed as

$$\Delta H_e = \sum_{i=1}^{i=n} \Delta_i \tag{24.13}$$

where ΔH_e = total expansion (in.)

$$\Delta_i = (PE)_i (\Delta D)_i (F)_i \tag{24.14}$$

$$(F)_i = \log^{-1}\left(-\frac{D_i}{20}\right) = \text{reduction factor for layer } i.$$

z = total thickness of expansive soil layer (ft)

D_i = depth to midpoint of i th layer (ft)

$(\Delta D)_i$ = thickness of i th layer (ft)

Figure 24.17(b) gives the reduction factor plotted against depth.

24.16 DESIGN OF FOUNDATIONS IN SWELLING SOILS

It is necessary to note that all parts of a building will not equally be affected by the swelling potential of the soil. Beneath the centre of a building where the soil is protected from sun and rain the moisture changes are small and the soil movements the least. Beneath outside walls, the movement are greater. Damage to buildings is greatest on the outside walls due to soil movements.

Three general types of foundations can be considered in expansive soils. They are

1. Structures that can be kept isolated from the swelling effects of the soils

2. Designing of foundations that will remain undamaged in spite of swelling

3. Elimination of swelling potential of soil.

All three methods are in use either singly or in combination, but the first is by far the most widespread. Figure 24.18 show a typical type of foundation under an outside wall. The granular fill provided around the shallow foundation mitigates the effects of expansion of the soils.

24.17 DRILLED PIER FOUNDATIONS

Drilled piers are commonly used to resist uplift forces caused by the swelling of soils. Drilled piers, when made with an enlarged base, are called, *belled piers* and when made without an enlarged base are referred to as *straight-shaft piers*.

Woodward, *et al.*, (1972) commented on the empirical design of piers: "Many piers, particularly where rock bearing is used, have been designed using strictly empirical considerations which are derived from regional experience". They further stated that "when surface conditions are well established and are relatively uniform, and the performance of past constructions well documented, the design by experience approach is usually found to be satisfactory."

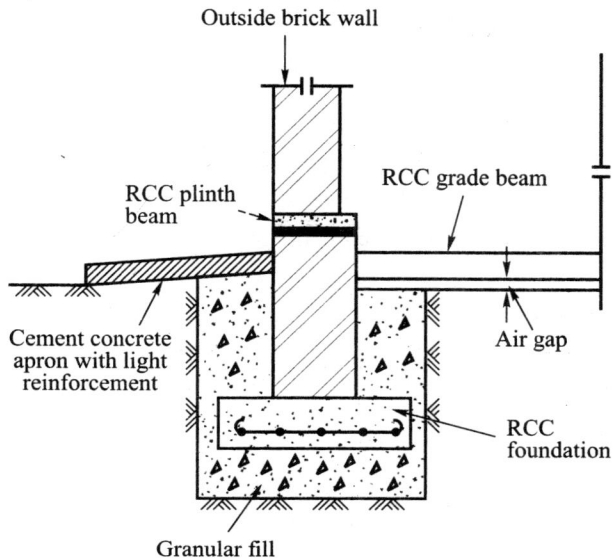

Fig. 24.18 Foundation in expansive soil

The principle of drilled piers is to provide a relatively inexpensive way of transferring the structural loads down to stable material or to a stable zone where moisture changes are improbable. There should be no direct contact between the soil and the structure with the exception of the soils supporting the piers.

Straight-shaft Piers in Expansive Soils

Figure 24.19(a) shows a straight-shaft drilled pier embedded in expansive soil. The following notations are used.

L_1 = length of shaft in the unstable zone (active zone) affected by wetting.

L_2 = length of shaft in the stable zone unaffected by wetting

d = diameter of shaft

Q = structural dead load = qA_b

q = unit dead load pressure and

A_b = base area of pier

When the soil in the unstable zone takes water during the wet season, the soil tries to expand which is partially or wholly prevented by the rough surface of the pile shaft of length L_1. As a result there will be an upward force developed on the surface of the shaft which tries to pull the pile out of its position. The upward force can be resisted in the following ways.

1. The downward dead load Q acting on the pier top

2. The resisting force provided by the shaft length L_2 embedded in the stable zone.

Two approaches for solving this problem may be considered. They are:

1.   The method suggested by Chen (1988)

2.   The O'Neill (1988) method with belled pier.

Two cases may be considered. They are:

1.   The stability of the pier when no downward load Q is acting on the top. For this condition a factor of safety of 1.2 is normally found sufficient.

2.   The stability of the pier when Q is acting on the top. For this a value of $F_s = 2.0$ is used.

Equations for Uplift Force Q_{up}

Chen (1988) suggested the following equation for estimating the uplift force Q_{up}

$$Q_{up} = \pi d \alpha_u p_s L_1 \tag{24.15}$$

where d = diameter of pier shaft

α_u = coefficient of uplift between concrete and soil = 0.15

p_s = swelling pressure

= 10,000 psf (480 kN/m^2) for soil with high degree of expansion

= 5,000 (240 kN/m^2) for soil with medium degree of expansion

The depth (L_1) of the unstable zone (wetting zone) varies with the environmental conditions. According to Chen (1988) the wetting zone is limited only to the upper 5 feet of the pier. It is possible for the wetting zone to extend beyond 10–15 feet in some countries and limiting the depth of unstable zone to a such a low value of 5 ft may lead to unsafe conditions for the stability of structures. However, it is for designers to decide this depth L_1 according to local conditions. With regards to swelling pressure p_s, it is unrealistic to fix any definite value of 10,000 or 5,000 psf for all types of expansive soils under all conditions of wetting. It is also not definitely known if the results obtained from laboratory tests truly represent the in situ swelling pressure. Possibly one way of overcoming this complex problem is to relate the uplift resistance to undrained cohesive strength of soil just as in the case of friction piers under compressive loading. Equation (24.15) may be written as

$$Q_{up} = \pi d \alpha_s c_u L_1 \tag{24.16}$$

where α_s = adhesion factor between concrete and soil under a swelling condition

c_u = unit cohesion under undrained conditions

It is possible that the value of α_s may be equal to 1.0 or more according to the swelling type and environmental conditions of the soil. Local experience will help to determine the value of α_s. This approach is simple and pragmatic.

Resisting Force

The length of pier embedded in the stable zone should be sufficient to keep the pier being pulled out of the ground with a suitable factor of safety. If L_2 is the length of the pier in the stable zone, the resisting force Q_R is the frictional resistance offered by the surface of the pier within the stable zone. We may write

$$Q_R = \pi d L_2 \alpha c_u \tag{24.17}$$

here $\quad \alpha =$ adhesion factor under compression loading

$\quad\quad c_u =$ undrained unit cohesion of soil

The value of α may be obtained from Fig. 21.15.

Two cases of stability may be considered:

. Without taking into account the dead load Q acting on the pier top, and using $F_s = 1.2$

$$Q_{up} = \frac{Q_R}{1.2} \qquad\qquad\qquad (24.18a)$$

2. By taking into account the dead load Q and using $F_s = 2.0$

$$(Q_{up} - Q) = \frac{Q_R}{2.0} \qquad\qquad\qquad (24.18b)$$

For a given shaft diameter d equations (24.18a) and (24.18b) help to determine the length L_2 of the pier in the stable zone. The one that gives the maximum length L_2 should be used.

Belled Piers

Piers with a belled bottom are normally used when large uplift forces have to be resisted. Figure 24.19(b) shows a belled pier with all the forces acting.

The uplift force for a belled pier is the same as that applicable for a straight shaft. The resisting force equation for the pier in the stable zone may be written as (O'Neill, 1988)

$$Q_{R1} = \pi d L_2 \alpha c_u \qquad\qquad\qquad (24.19a)$$

$$Q_{R2} = \frac{\pi}{4}\left[d_b^2 - d^2\right]\left[cN_c + \gamma L_2\right] \qquad\qquad\qquad (24.19b)$$

where

$\quad d_b \;=$ diameter of the underream

$\quad N_c \;=$ bearing capacity factor

$\quad c \;\;\;=$ unit cohesion under undrained condition

$\quad \gamma \;\;\;=$ unit weight of soil

The values of N_c are given in Table 24.8 (O'Neill, 1988)

Two cases of stability may be written as before.

1. Without taking the dead load Q and using $F_s = 1.2$

$$Q_{up} = \frac{1}{2}\left[Q_{R1} + Q_{R2}\right]$$

2. By taking into account the dead load and $F_s = 2.0$

$$(Q_{up} - Q) = \frac{1}{2.0}\left[Q_{R1} + Q_{R2}\right]$$

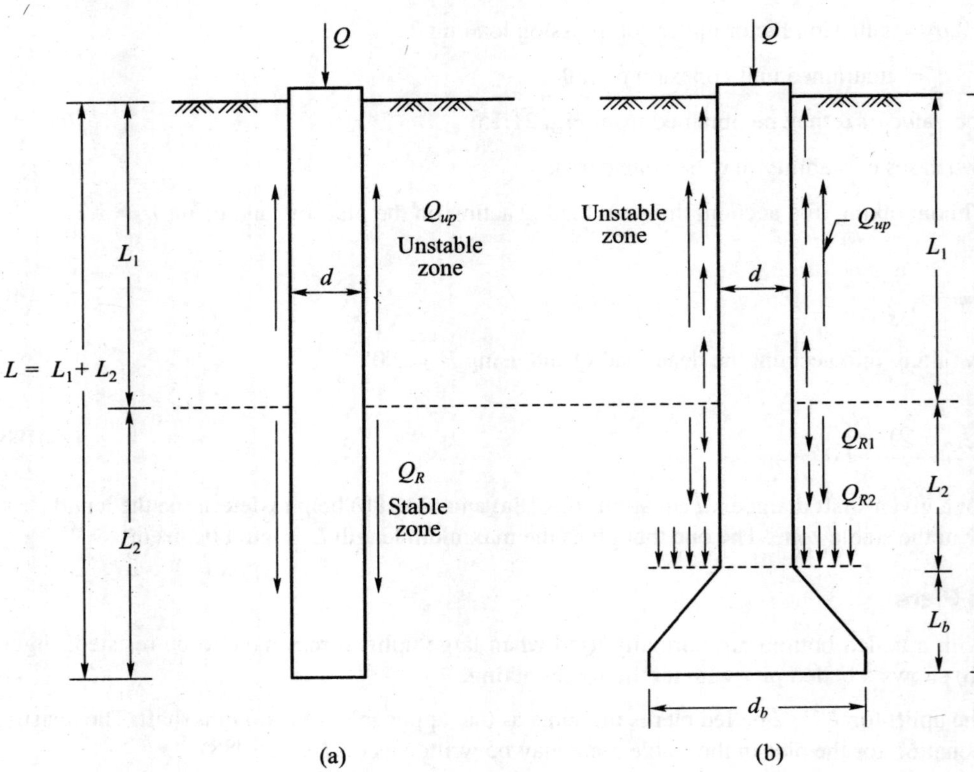

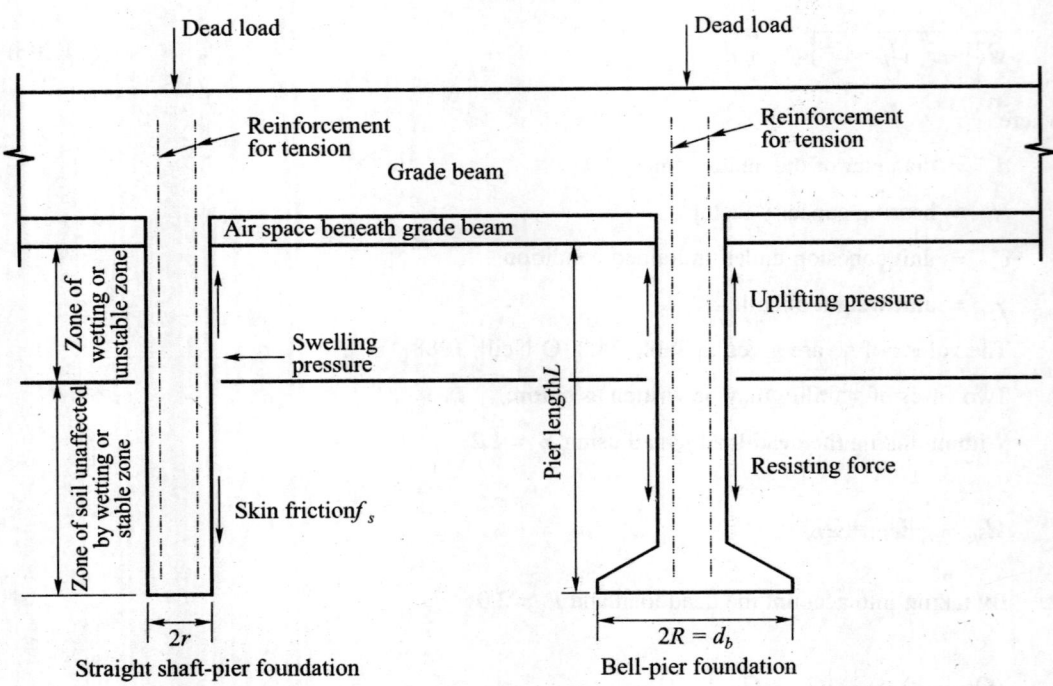

Fig. 24.19 Drilled pier in expansive soil

Fig. 24.20 Grade beam and pier system (Chen, 1988)

TABLE 24.8
Values of N_c

L_2/d_b	N_c
1.7	4
2.5	6
≥ 5.0	9

For a given shaft diameter d and base diameter d_b, the above equations help to determine the value of L_2. The one that gives the maximum value for L_2 has to be used in the design.

Figure 24.20 gives a typical foundation design with grade beams and drilled piers (Chen, 1988). The piers should be taken sufficiently below the unstable zone of wetting in order to resist the uplift forces.

Example 24.3

A footing founded at a depth of 1 ft below ground level in expansive soil was subjected to loads from the superstructure. Site investigation revealed that the expansive soil extended to a depth of 8 ft below the base of the foundation, and the moisture contents in the soil during the construction period were at their lowest. In order to determine the percent swell, three undisturbed samples at depths of 2, 4 and 6 ft were collected and swell tests were conducted per the procedure described in Section 24.16. Figure Ex. 24.3a shows the results of the swell tests plotted against depth. A line passing through the points is drawn. The line indicates that the swell is zero at 8 ft depth and maximum at a base level equal to 3%. Determine (a) the total swell, and (b) the depth of undercut necessary for an allowable swell of 0.03 ft.

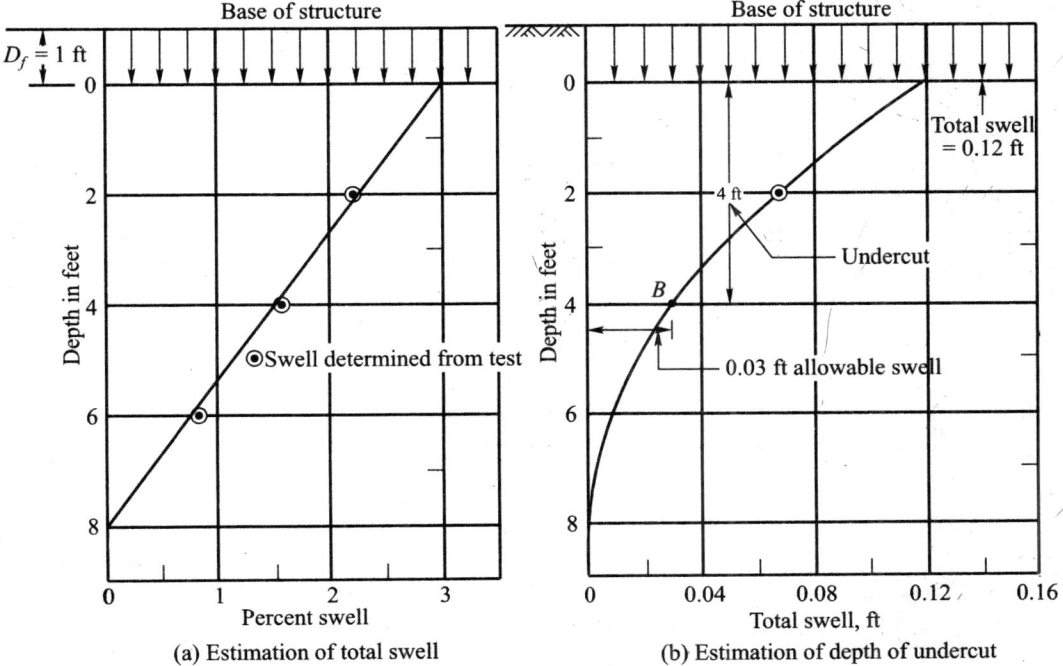

(a) Estimation of total swell   (b) Estimation of depth of undercut

Fig. Ex. 24.3

Solution

(a)   The total swell is equal to the area under the percent swell versus depth curve in Fig. Ex. 24.3a.

$$\text{Total swell} = \frac{1}{2} \times 8 \times 3 \times 1/100 = 0.12 \, \text{ft}$$

(b)   Depth of undercut

From the percent swell versus depth relationship given by the curve in Fig. 24.3a, total swell at different depths are calculated and plotted against depth in Fig. 24.3b. For example the total swell at depth 2 ft below the foundation base is

Total swell = 1/2 (8 – 2) × 2.25 × 1/100 = 0.067 ft plotted against depth 2 ft. Similarly total swell at other depths can be calculated and plotted. Point B on curve in Fig. 24.3b gives the allowable swell of 0.03 ft at a depth of 4 ft below foundation base. That is, the undercut necessary in clay is 4 ft which may be replaced by an equivalent thickness of nonswelling compacted fill.

Example 24.4

Figure Ex. 24.4 shows that the soil to a depth of 20 ft is an expansive type with different degrees of swelling potential. The soil mass to a 20 ft depth is divided into four layers based on the swell potential rating given in Table 24.7. Calculate the total swell per the Van Der Merwe method.

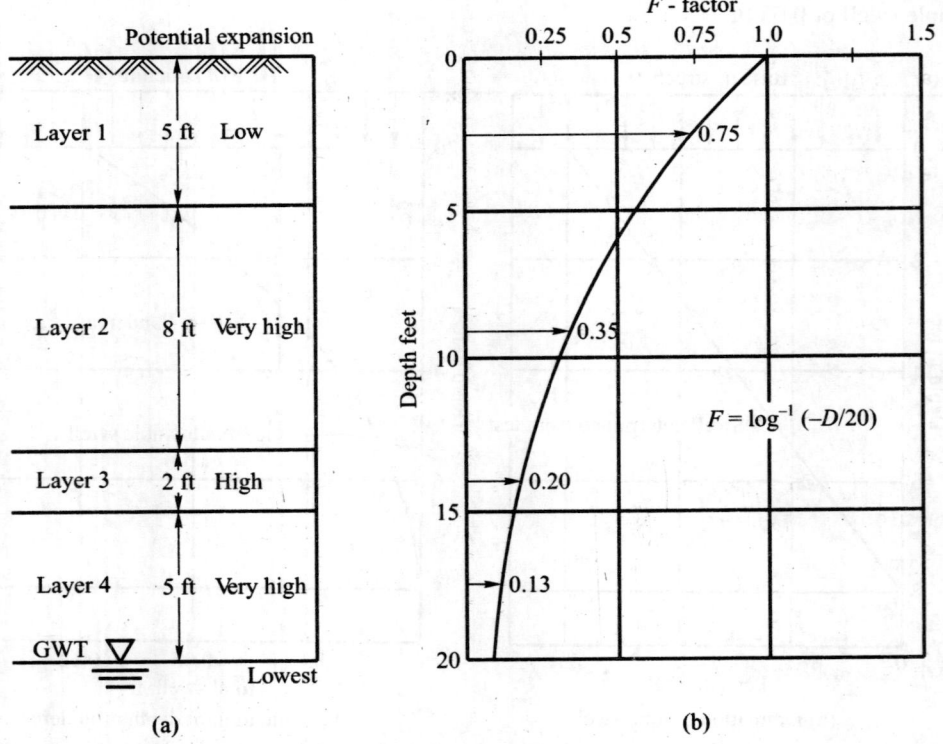

Fig. Ex. 24.4

Solution

The procedure for calculating the total swell is explained in Section 24.15. The details of the calculated results are tabulated below.

The details of calculated results

Layer No.	Thickness ΔD(ft) ft	PE	D	F	ΔH_s (in.)
1	5	0	2.5	0.75	0
2	8	1.0	9.0	0.35	2.80
3	2	0.5	14.0	0.20	0.20
4	5	1.0	17.5	0.13	0.65
				Total	3.65

In the table above D = depth from ground level to the mid-depth of the layer considered. F = reduction factor.

Example 24.5

A drilled pier [refer to Fig. 24.19(a)] was constructed in expansive soil. The water table was not encountered. The details of the pier and soil are:

$L = 20$ ft, $d = 12$ in., $L_1 = 5$ ft, $L_2 = 15$ ft, $p_s = 10,000$ lb/ft², $c_u = 2089$ lb/ft², SPT(N) = 25 blows per foot,

Required:

(a) total uplift capacity Q_{up}

(b) total resisting force due to surface friction

(c) factor of safety without taking into account the dead load Q acting on the top of the pier

(d) factor of safety with the dead load acting on the top of the pier

Assume $Q = 10$ kips. Calculate Q_{up} by Chen's method (Eq. 24.15).

Solution

(a) Uplift force Q_{up} from Eq. (24.15)

$$Q_{up} = \pi d \alpha_u p_s L_1 = \frac{3.14 \times (1) \times 0.15 \times 10,000 \times 5}{1000} = 23.55 \text{ kips}$$

(b) Resisting force Q_R

From Eq. (24.17)

$$Q_R = \pi d (L - L_1) \alpha c_u$$

where $c_u = 2089$ lb/ft² ≈ 100 kN/m²

$$\frac{c_u}{p_a} = \frac{100}{101} \approx 1.0 \text{ where } p_a = \text{atmospheric pressure} = 101 \text{ kPa}$$

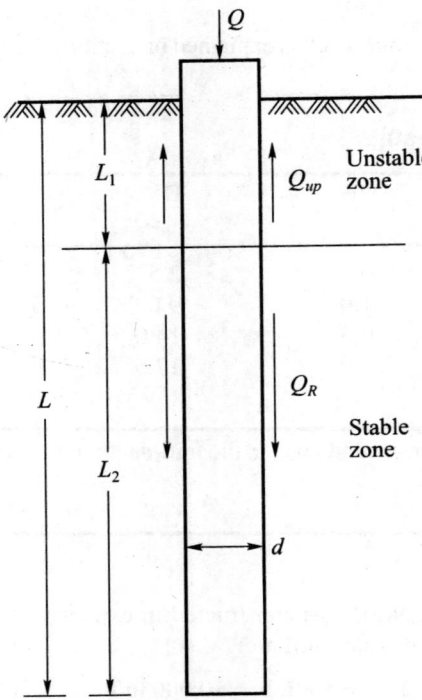

Fig. Ex. 24.5

From Fig. 21.15, $\alpha = 0.55$ for $c_u/p_a \approx 1.0$

Now substituting the known values

$Q_R = 3.14 \times 1 \times (20 - 5) \times 0.55 \times 2000 = 51{,}810$ lb ≈ 52 kips

(c)   Factor of safety with $Q = 0$

From Eq. (24.18a)

$$F_s = \frac{Q_R}{Q_{up}} = \frac{52}{23.5} = 2.2 > 1.2 \text{ required - -OK}$$

(d)   Factor of safety with $Q = 10$ kips

From Eq. (24.18b)

$$F_s = \frac{Q_R}{(Q_{up} - Q)} = \frac{52}{(23.5 - 10)} = 3.9 > 2.0 \text{ required - -OK.}$$

Example 24.6

Solve Example 24.5 with $L_1 = 10$ ft. All the other data remain the same.

Solution

(a)   Uplift force Q_{up}

$Q_{up} = 23.5 \times (10/5) = 47.0$ kips

where $Q_{up} = 23.5$ kips for $L_1 = 5$ ft

(b)   Resisting force Q_R

$Q_R = 52 \times (10/15) \approx 34.7$ kips

where $Q_R = 52$ kips for $L_2 = 15$ ft

(c)   Factor of safety for $Q = 0$

$$F_s = \frac{34.7}{47.0} = 0.74 < 1.2 \text{ as required - -not OK}$$

(d)   Factor of safety for $Q = 10$ kips

$$F_s = \frac{34.7}{(47-10)} = \frac{34.7}{37} = 0.94 < 2.0 \text{ as required - -not OK}$$

The above calculations indicate that if the wetting zone (unstable zone) is 10 ft thick the structure will not be stable for $L = 20$ ft.

Example 24.7

Determine the length of pier required in the stable zone for $F_s = 1.2$ where $Q = 0$ and $F_s = 2.0$ when $Q = 10$ kips. All the other data given in Example 24.6 remain the same.

Solution

(a)   Uplifting force Q_{up} for L_1 (10 ft) = 47 kips

(b)   Resisting force for length L_2 in the stable zone.

$Q_R = \pi_d \alpha c_u L_2 = 3 \ 14 \times 1 \times 0.55 \times 2000 L_2 = 3454 L_2$ Ib/ft^2

(c) $Q = 0$, minimum $F_s = 1.2$

or $1.2 = \dfrac{Q_R}{Q_{up}} = \dfrac{3,454 L_2}{47,000}$

solving we have $L_2 = 16.3$ ft.

(d) $Q = 10$ kips. Minimum $F_s = 2.0$

$$F_s = 2.0 = \frac{Q_R}{(Q_{up} - Q)} = \frac{3,454 L_2}{(47,000 - 10,000)} = \frac{3,454 L_2}{37,000}$$

Solving we have $L_2 = 21.4$ ft.

The above calculations indicate that the minimum $L_2 = 21.4$ ft or say 22 ft is required fo
the structure to be stable with $L_1 = 10$ ft. The total length $L = 10 + 22 = 32$ ft.

Example 24.8

Figure Ex. 24.8 shows a drilled pier with a belled bottom constructed in expansive soil. The wate
table is not encountered. The details of the pier and soil are given below:

$$L_1 = 10 \text{ ft}, L_2 = 10 \text{ ft}, L_b = 2.5 \text{ ft}, d = 12 \text{ in.}, d_b = 3 \text{ ft}, c_u = 2000 \text{ lb/ft}^2, p_s = 10,000 \text{ lb/ft}^2, \gamma = 110 \text{ lb/ft}^3$$

Required

(a) Uplift force Q_{up}

(b) Resisting force Q_R

(c) Factor of safety for $Q = 0$ at the top of the pier

(d) Factor of safety for $Q = 20$ kips at the top of the pier

Solution

(a) Uplift force Q_{up}

As in Ex. 24.6 $Q_{up} = 47$ kips

(b) Resisting force Q_R

$$Q_{R1} = \pi d L_2 \alpha c_u$$

$a = 0.55$ as in Ex. 24.5

Substituting known values

$$Q_{R1} = 3.14 \times 1 \times 10 \times 0.55 \times 2000 = 34540 \text{ lbs} = 34.54 \text{ kips}$$

$$Q_{R2} = \frac{\pi}{4}\left[d_b^2 - d^2\right]\left[cN_c + \gamma L_2\right]$$

where $d_b = 3$ ft, $c = 2000$ lb/ft^2, $N_c = 7.0$ from Table 24.8 for $L_2/d_b = 10/3 = 3.33$

Substituting known values

$$Q_{R2} = \frac{3.14}{4}\left[3^2 - (1)^2\right]\left[2000\times7.0+110\times10\right]$$

$$= 6.28 [15,100] = 94,828 \text{ lbs} = 94.8 \text{ kips}$$

$$Q_R = Q_{R1} + Q_{R2} = 34.54 + 94.8 = 129.3 \text{ kips}$$

(c) Factor of safety for $Q = 0$

$$F_s = \frac{Q_R}{Q_{up}} = \frac{129.3}{47.0} = 2.75 > 1.2 \text{ --OK}$$

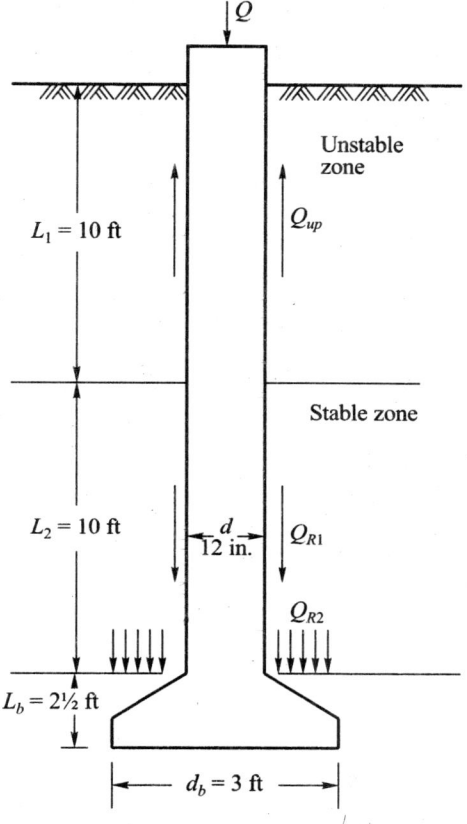

Fig. Ex. 24.8

(d) Factor of safety for $Q = 20$ kips

$$F_s = \frac{Q_R}{(Q_{up} - Q)} = \frac{129.3}{(47 - 20)} = 4.79 > 2.0 \text{ -- as required OK.}$$

The above calculations indicate that the design is over conservative. The length L_2 can be reduced to provide an acceptable factor of safety.

24.18 ELIMINATION OF SWELLING

The elimination of foundation swelling can be achieved in two ways. They are:

1.   Providing a granular bed and cover below and around the foundation (Fig. 24.18)

2.   Chemical stabilization of swelling soils

Figure 24.18 gives a typical example of the first type. In this case, the excavation is carried out up to a depth greater than the width of the foundation by about 20 to 30 cms. Freely draining soil, such as a mixture of sand and gravel, is placed and compacted up to the base level of the foundation. A reinforced concrete footing is constructed at this level. A mixture of sand and gravel is filled up loosely over the fill. A reinforced concrete apron about 2 m wide is provided around the building to prevent moisture directly entering the foundation. A cushion of granular soils below the foundation absorbs the effect of swelling, and thereby its effect on the foundation will considerably be reduced.

A foundation of this type should be constructed only during the dry season when the soil has shrunk to its lowest level. Arrangements should be made to drain away the water from the granular base during the rainy seasons.

Chemical stabilization of swelling soils by the addition of lime may be remarkably effective if the lime can be mixed thoroughly with the soil and compacted at about the optimum moisture content. The appropriate percentage usually ranges from about 3 to 8 percent. The lime content is estimated on the basis of pH tests and checked by compacting, curing and testing samples in the laboratory. The lime has the effect of reducing the plasticity of the soil, and hence its swelling potential.

24.19 QUESTIONS AND PROBLEMS

24.1 A building was constructed in a loessial type normally consolidated collapsible soil with the foundation at a depth of 1 m below ground level. The soil to a depth of 6 m below the foundation was found to be collapsible on flooding. The average overburden pressure was 56 kN/m^2. Double consolidometer tests were conducted on two undisturbed samples taken at a depth of 4 m below ground level, one with its natural moisture content and the other under soaked conditions per the procedure explained in Section 24.4. The following data were available.

Applied pressure, kN/m^2	10	20	40	100	200	400	800
Void ratios at natural moisture content	0.80	0.79	0.78	0.75	0.725	0.68	0.61
Void ratios in the soaked condition	0.75	0.71	0.66	0.58	0.51	0.43	0.32

Plot the e-log p curves and determine the collapsible settlement for an increase in pressure $\Delta p = 34$ kN/m^2 at the middle of the collapsible stratum.

24.2 Soil investigation at a site indicated overconsolidated collapsible loessial soil extending to a great depth. It is required to construct a footing at the site founded at a depth of 1.0 m below ground level. The site is subject to flooding. The average unit weight of the soil is 19.5 kN/m^3. Two oedometer tests were conducted on two undisturbed samples taken at a depth of 5 m from ground level. One test was conducted at its natural moisture content and the other on a soaked condition per the procedure explained in Section 24.4.

The following test results are available.

Applied pressure, kN/m^2	10	20	40	100	200	400	800	2000
Void ratio under natural moisture condition	0.795	0.79	0.787	0.78	0.77	0.74	0.71	0.64
Void ratio under soaked condition	0.775	0.77	0.757	0.730	0.68	0.63	0.54	0.37

The swell index determined from the rebound curve of the soaked sample is equal to 0.08. Required:

(a) Plots of e-log p curves for both tests.

(b) Determination of the average overburden pressure at the middle of the soil stratum.

(c) Determination of the preconsolidation pressure based on the curve obtained from the soaked sample.

(d)   Total collapse settlement for an increase in pressure $\Delta p = 710$ kN/m²?

24.3   A footing for a building is founded 0.5 m below ground level in an expansive clay stratum which extends to a great depth. Swell tests were conducted on three undisturbed samples taken at different depths and the details of the tests are given below.

Depth (m) below GL	Swell %
1	2.9
2	1.75
3	0.63

Required:

(a)   The total swell under structural loadings

(b)   Depth of undercut for an allowable swell of 1 cm

24.4   Figure Prob. 24.4 gives the profile of an expansive soil with varying degrees of swelling potential. Calculate the total swell per the Van Der Merwe method.

24.5   Figure Prob. 24.5 depicts a drilled pier embedded in expansive soil. The details of the pier and soil properties are given in the figure.

Determine:

(a)   The total uplift capacity.

(b)   Total resisting force.

(c)   Factor of safety with no load acting on the top of the pier.

(d)   Factor of safety with a dead load of 100 kN on the top of the pier.

Calculate Q_{up} by Chen's method.

24.6   Solve problem 24.5 using Eq. (24.16)

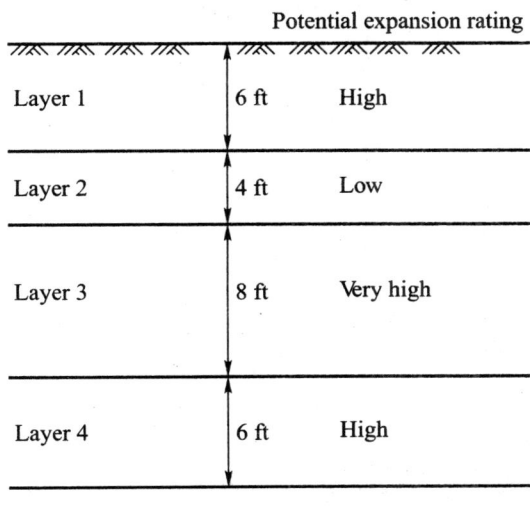

Fig. Prob. 24.4

24.7 Figure Prob. 24.7 shows a drilled pier with a belled bottom. All the particulars of the pier and soil are given in the figure.

Required:

(a) The total uplift force.

(b) The total resisting force

(c) Factor of safety for $Q = 0$

(d) Factor of safety for Q = 200 kN.

Use Chen's method for computing Q_{up}.

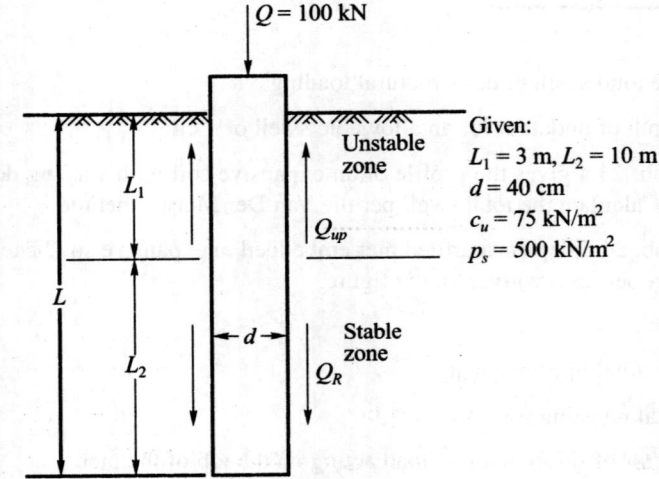

Given:
$L_1 = 3$ m, $L_2 = 10$ m
$d = 40$ cm
$c_u = 75$ kN/m^2
$p_s = 500$ kN/m^2

Fig. Prob. 24.5

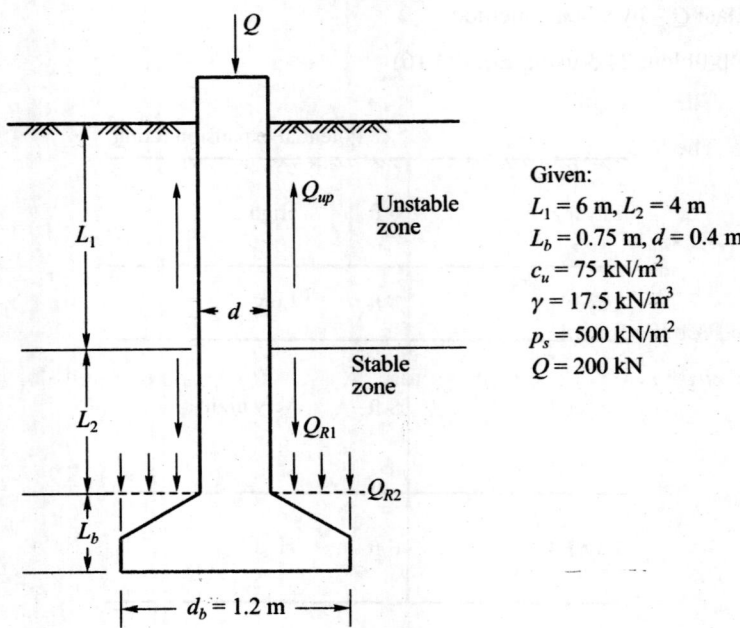

Given:
$L_1 = 6$ m, $L_2 = 4$ m
$L_b = 0.75$ m, $d = 0.4$ m
$c_u = 75$ kN/m^2
$\gamma = 17.5$ kN/m^3
$p_s = 500$ kN/m^2
$Q = 200$ kN

$d_b = 1.2$ m

Fig. Prob. 24.7

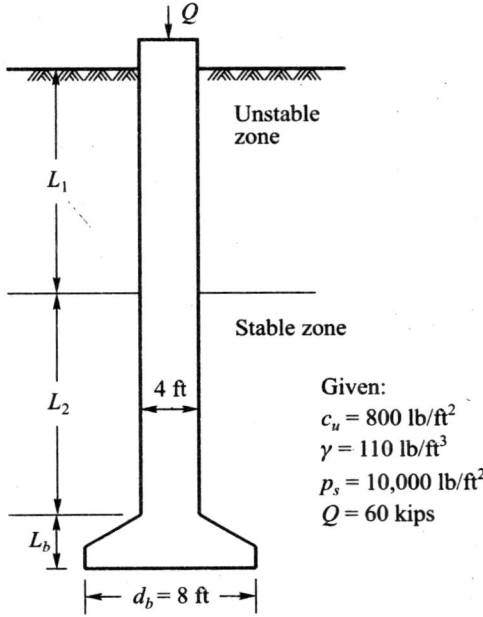

Fig. Prob. 24.9

Given:
$c_u = 800 \text{ lb/ft}^2$
$\gamma = 110 \text{ lb/ft}^3$
$p_s = 10,000 \text{ lb/ft}^2$
$Q = 60 \text{ kips}$

24.8   Solve Prob. 24.7 by making use of Eq. (24.16) for computing Q_{up}.

24.9   Refer to Fig. Prob. 24.9. The following data are available:

$L_1 = 15$ ft, $L_2 = 13$ ft, $d = 4$ ft, $d_b = 8$ ft and $L_b = 6$ ft.

All the other data are given in the figure.

Required:

(a)   The total uplift force

(b)   The total resisting force

(c)   Factor of safety for $Q = 0$

(d)   Factor of safety for $Q = 60$ kips.

Use Chen's method for computing Q_{up}.

24.10   Solve Prob. 24.9 using Eq. (24.16) for computing Q_{up}.

24.11   If the length L_2 is not sufficient in Prob. 24.10, determine the required length to get $F_s = 3.0$.

Fig. Prob. 24.5

CHAPTER 25

CONCRETE RETAINING WALLS

25.1 INTRODUCTION

The common types of concrete retaining walls and their uses were discussed in Chapter 14. The lateral pressure theories and the methods of calculating the lateral earth pressures were described in detail in the same chapter. The two classical earth pressure theories that have been considered are those of Rankine and Coulomb. In this chapter we are interested in the following:

1. Conditions under which the theories of Rankine and Coulomb are applicable to cantilever and gravity retaining walls under the active state.

2. The common minimum dimensions used for the two types of retaining walls mentioned above.

3. Use of charts for the computation of active earth pressure.

4. Stability of retaining walls.

5. Drainage provisions for retaining walls.

25.2 CONDITIONS UNDER WHICH RANKINE AND COULOMB FORMULAS ARE APPLICABLE TO RETAINING WALLS UNDER ACTIVE STATE

Conjugate Failure Planes Under Active State

When a backfill of cohesionless soil is under an active state of plastic equilibrium due to the stretching of the soil mass at every point in the mass, two failure planes called *conjugate rupture planes* are formed. These are further designated as the *inner failure plane* and the *outer failure plane* as shown in Fig. 25.1. These failure planes make angles of α_i and α_0 with the vertical. The equations for these angles may be written as (for a sloping backfill)

$$\alpha_i = \frac{90-\phi}{2} + \frac{\varepsilon-\beta}{2} \tag{25.1a}$$

$$\alpha_0 = \frac{90-\phi}{2} - \frac{\varepsilon-\beta}{2} \tag{25.1b}$$

where $\quad \sin\varepsilon = \dfrac{\sin\beta}{\sin\phi}$ \hfill (25.2)

when $\beta = 0$, $\quad \alpha_i = \dfrac{90-\phi}{2} = 45° - \dfrac{\phi}{2}, \quad \alpha_0 = \dfrac{90-\phi}{2} = 45° - \dfrac{\phi}{2}$

The angle between the two failure planes $= 90 - \phi$.

Conditions for the Use of Rankine's Formula

1.   Wall should be vertical with a smooth pressure face.

2.   When walls are inclined, it should not come in the way of the formation of the outer failure plane. Figure 25.1 shows the formation of failure planes. Since the sloping face AB' of the retaining wall makes an angle a_w greater than a_o, the wall does not interfere with the formation of the outer failure plane. The plastic state exists within wedge ACC'.

The method of calculating the lateral pressure on AB' is as follows.

1.   Apply Rankine's formula for the vertical section AB.

2.   Combine P_a with W_s, the weight of soil within the wedge ABB', to give the resultant P_R.

Let the resultant P_R in this case make an angle δ_r with the normal to the face of the wall. Let the maximum angle of wall friction be δ_m. If $\delta_r > \delta_m$, the soil slides along the face AB' of the wall.

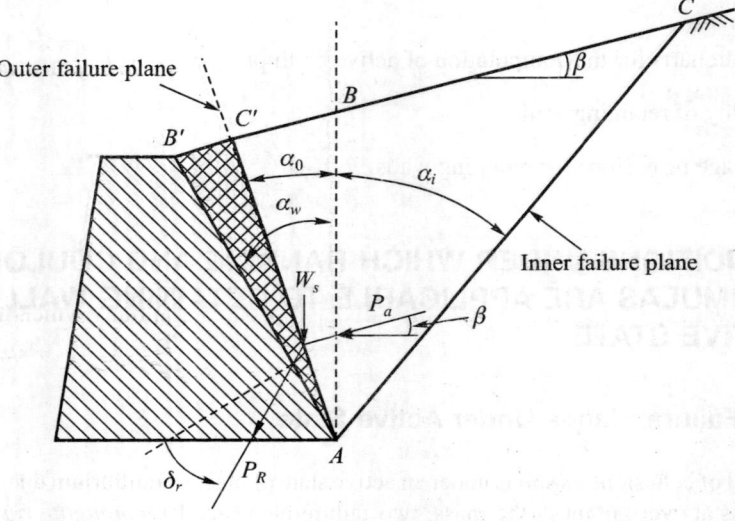

Fig. 25.1   Application of Rankine's active condition to gravity walls

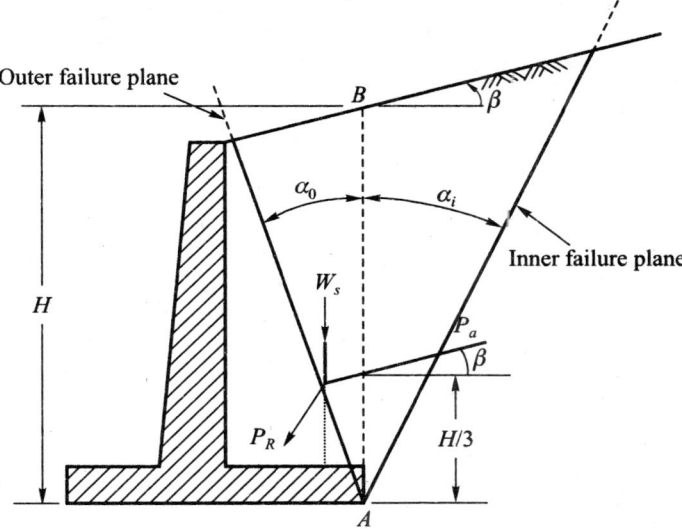

Fig. 25.2 Lateral earth pressure on cantilever walls under active condition

In such an eventuality, the Rankine formula is not recommended but the Coulomb formula may be used.

Conditions for the Use of Coulomb's Formula

1. The back of the wall must be plane or nearly plane.
2. Coulomb's formula may be applied under all other conditions where the surface of the wall is not smooth and where the soil slides along the surface.

In general the following recommendations may be made for the application of the Rankine or Coulomb formula without the introduction of significant errors:

1. Use the Rankine formula for cantilever and counterfort walls.
2. Use the Coulomb formula for solid and semisolid gravity walls.

In the case of cantilever walls (Fig. 25.2), P_a is the active pressure acting on the vertical section AB passing through the heel of the wall. The pressure is parallel to the backfill surface and acts at a height $H/3$ from the base of the wall where H is the height of the section AB. The resultant pressure P_R is obtained by combining the lateral pressure P_a with the weight of the soil W_s between the section AB and the wall.

25.3 PROPORTIONING OF RETAINING WALLS

Based on practical experience, retaining walls can be proportioned initially which may be checked for stability subsequently. The common dimensions used for the various types of retaining walls are given below.

Gravity Walls

A gravity walls may be proportioned in terms of its height given in Fig. 25.3(a). The minimum top width suggested is 0.30 m. The tentative dimensions for a cantilever wall are given in Fig. 25.3(b) and those for a counterfort wall are given in Fig. 25.3(c).

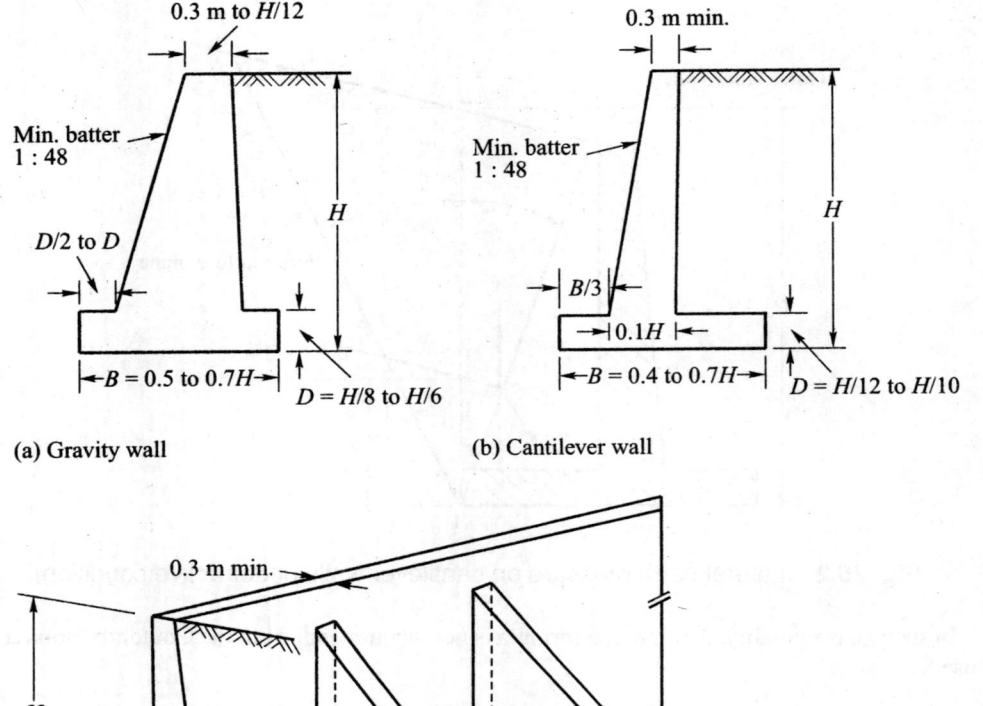

Fig. 25.3   Tentative dimensions for retaining walls

25.4   EARTH PRESSURE CHARTS FOR RETAINING WALLS

Charts have been developed for estimating lateral earth pressures on retaining walls based on certain assumed soil properties of the backfill materials. These semi empirical methods represent a body of valuable experience and summarize much useful information. The charts given in Fig. 25.4 are meant to produce a design of retaining walls of heights not greater than 6 m. The charts have been developed for five types of backfill materials given in Table 25.1. The charts are applicable to the following categories of backfill surfaces. They are

1.   The surface of the backfill is plane and carries no surcharge
2.   The surface of the backfill rises on a slope from the crest of the wall to a level at some elevation above the crest.

The chart is drawn to represent a concrete wall but it may also be used for a reinforced soil wall. All the dimensions of the retaining walls are given in Fig. 25.4. The total horizontal and vertical pressures on the vertical section of $A B$ of height H are expressed as

$$P_h = \frac{1}{2} K_h H^2 \qquad (25.3)$$

TABLE 25.1

Types of backfill for retaining walls

Type	Backfill material
1	Coarse-grained soil without admixture of fine soil particles, very permeable (clean sand or gravel)
2	Coarse-grained soil of low permeability due to admixture of particles of silt size
3	Residual soil with stones, fine silty sand, and granular materials with conspicuous clay content
4	Very soft or soft clay, organic silts, or silty clays
5	Medium or stiff clay

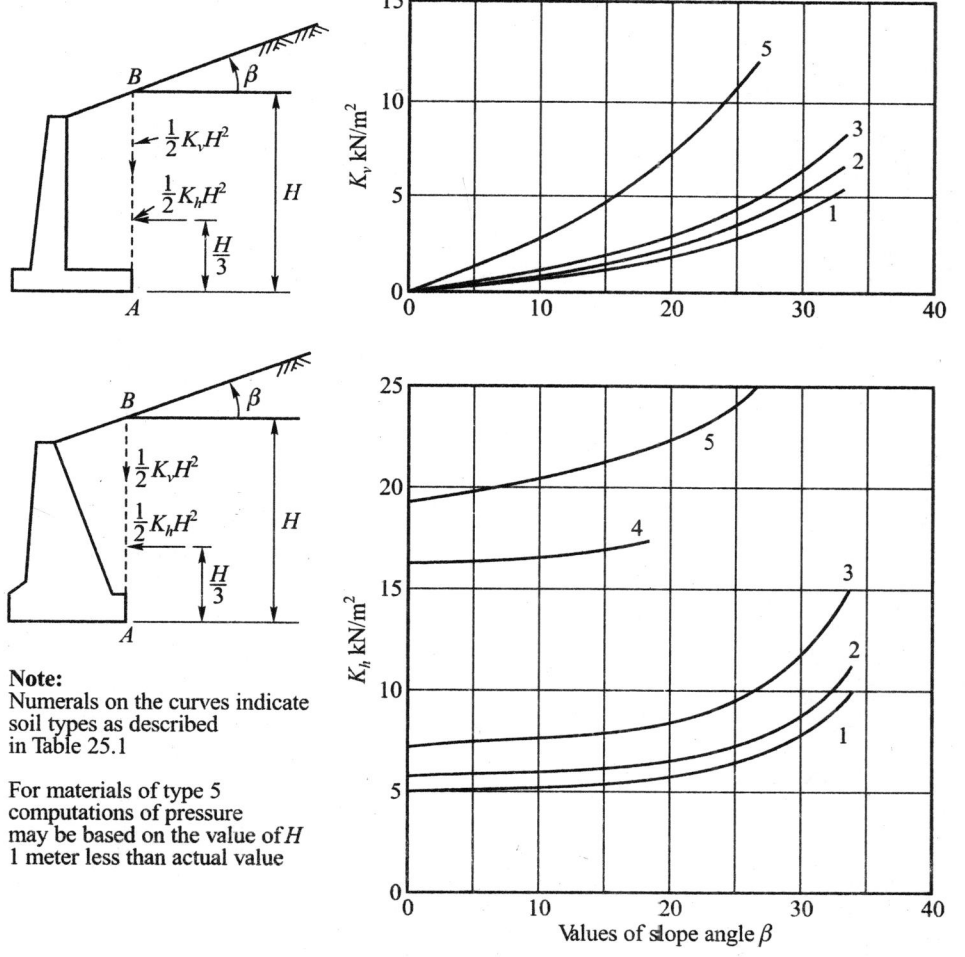

Note:
Numerals on the curves indicate
soil types as described
in Table 25.1

For materials of type 5
computations of pressure
may be based on the value of H
1 meter less than actual value

Fig. 25.4   Chart for estimating pressure of backfill against retaining walls supporting backfills with a plane surface. (Terzaghi, Peck, and Mesri, 1996)

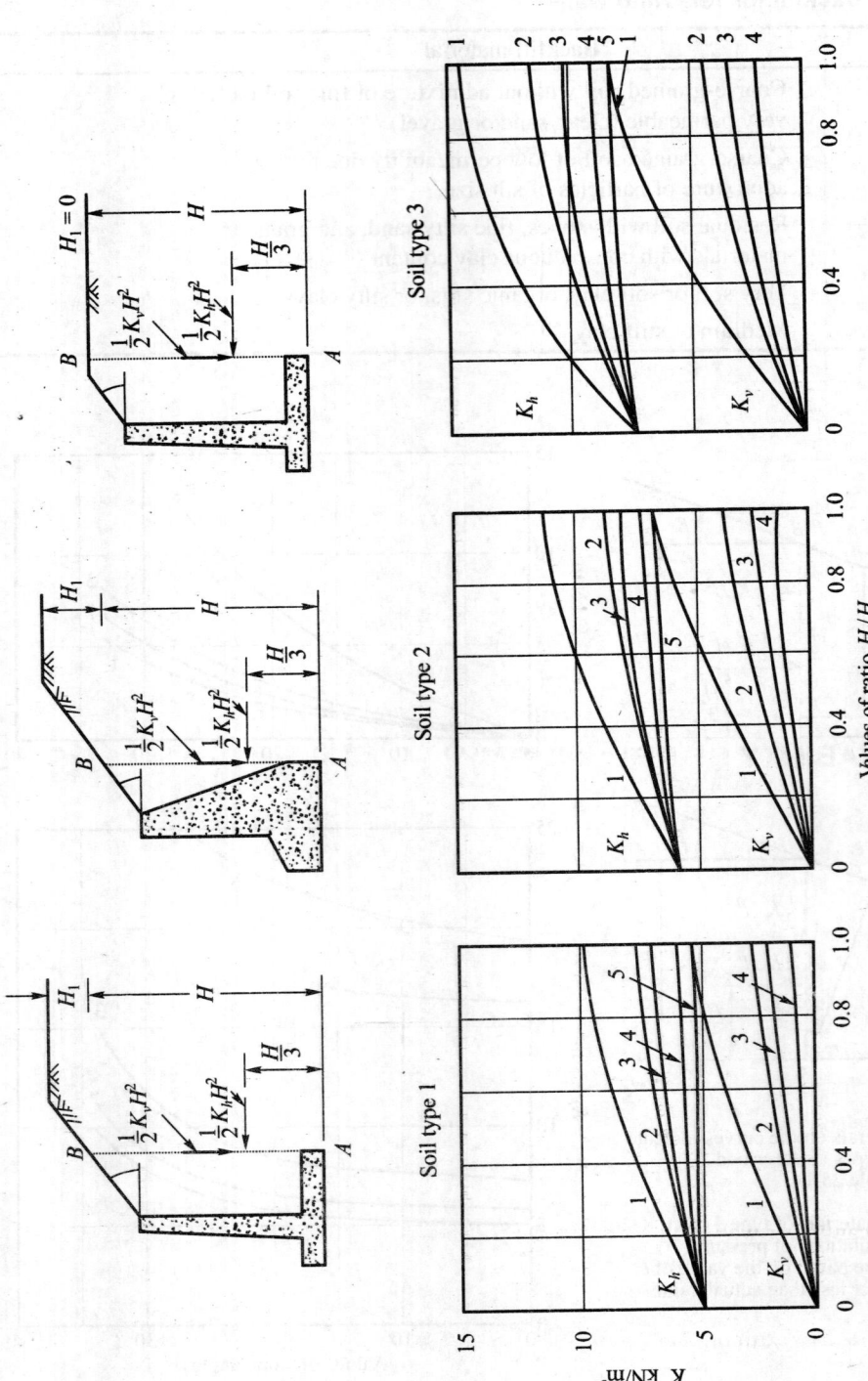

Fig. 25.5 Chart for estimating pressure of backfill against retaining walls supporting backfills with a surface that slopes upward from the crest of the wall for limited distance and then becomes horizontal. (Terzaghi et al., 1996)

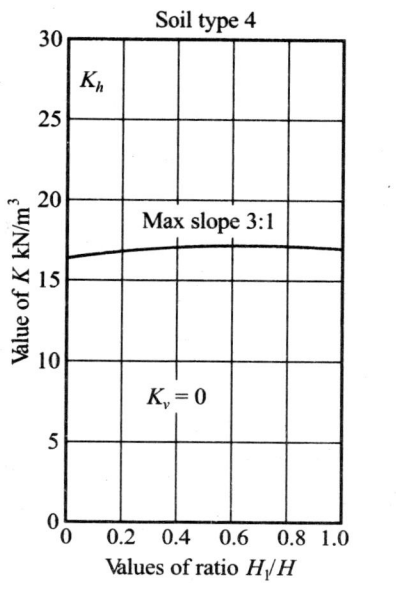
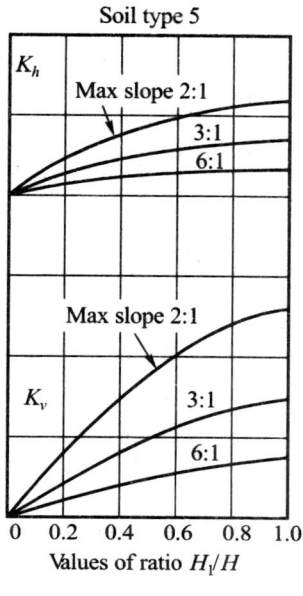

Fig. 25.5 (Continued)

$$P_v = \frac{1}{2} K_v H^2 \qquad (25.4)$$

Values of K_h and K_v are plotted against slope angle β in Fig. 25.4 and the ratio H_1/H in Fig. 25.5.

25.5 STABILITY OF RETAINING WALLS

The stability of retaining walls should be checked for the following conditions:

1. Check for sliding
2. Check for overturning
3. Check for bearing capacity failure
4. Check for base shear failure

The minimum factors of safety for the stability of the wall are:

1. Factor of safety against sliding = 1.5
2. Factor of safety against overturning = 2.0
3. Factor of safety against bearing capacity failure = 3.0

Stability Analysis

Consider a cantilever wall with a sloping backfill for the purpose of analysis. The same principle holds for the other types of walls.

Figure 25.6 gives a cantilever wall with all the forces acting on the wall and the base, where

P_a = active earth pressure acting at a height $H/3$ over the base on section AB

P_h = $P_a \cos \beta$

P_v = $P_a \sin \beta$

β = slope angle of the backfill

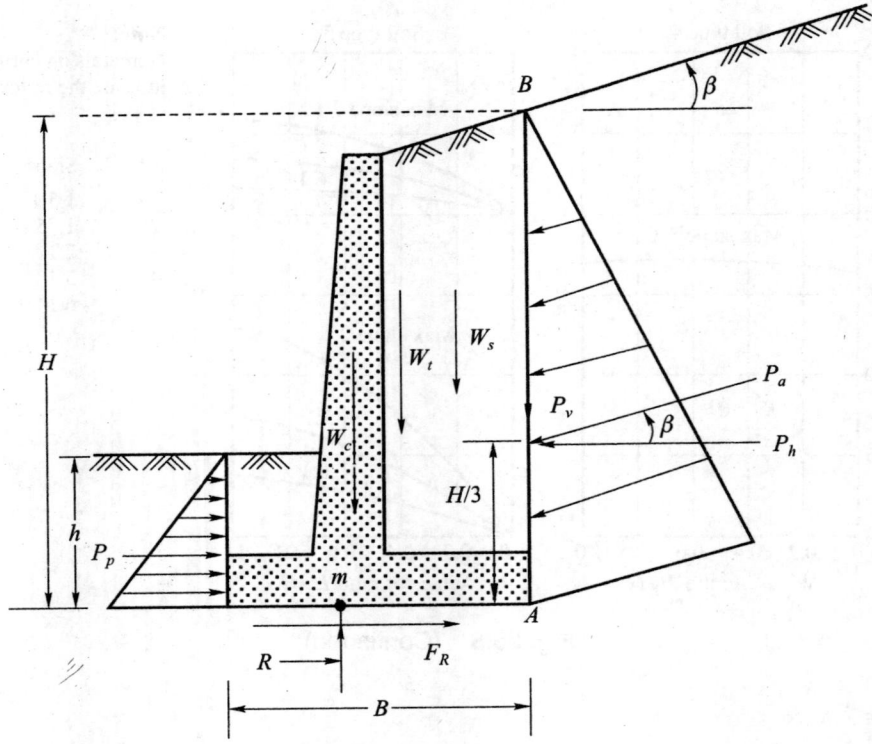

(a) Forces acting on the wall

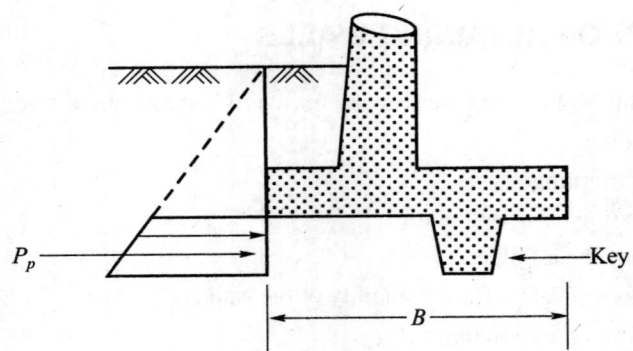

(b) Provision of key to increase sliding resistance

Fig. 25.6 Check for sliding

W_s = weight of soil
W_c = weight of wall including base
W_t = the resultant of W_s and W_c
P_P = passive earth pressure at the toe side of the wall.
F_R = base sliding resistance

Check for Sliding (Fig. 25.6)

The force that moves the wall = horizontal force P_h

The force that resists the movement is

$$F_R = c_a B + R \tan \delta + P_p \tag{25.5}$$

R = total vertical force = $W_s + W_c + P_v$,

δ = angle of wall friction

c_a = unit adhesion

If the bottom of the base slab is rough, as in the case of concrete poured directly on soil, the coefficient of friction is equal to $\tan \phi$, ϕ being the angle of internal friction of the soil.

The factor of safety against sliding is

$$F_s = \frac{F_R}{P_h} \geq 1.5 \tag{25.6}$$

In case $F_s < 1.5$, additional factor of safety can be provided by constructing one or two keys at the base level shown in Fig. 25.6b. The passive pressure P_p (Fig. 25.6a) in front of the wall should not be relied upon unless it is certain that the soil will always remain firm and undisturbed.

Check for Overturning

The forces acting on the wall are shown in Fig. 25.7. The overturning and stabilizing moments may be calculated by taking moments about point O. The factor of safety against overturning is therefore

$$F_o = \frac{\text{Sum of moment that resist overturning}}{\text{Sum of overturning moments}} = \frac{M_R}{M_o} \tag{25.7a}$$

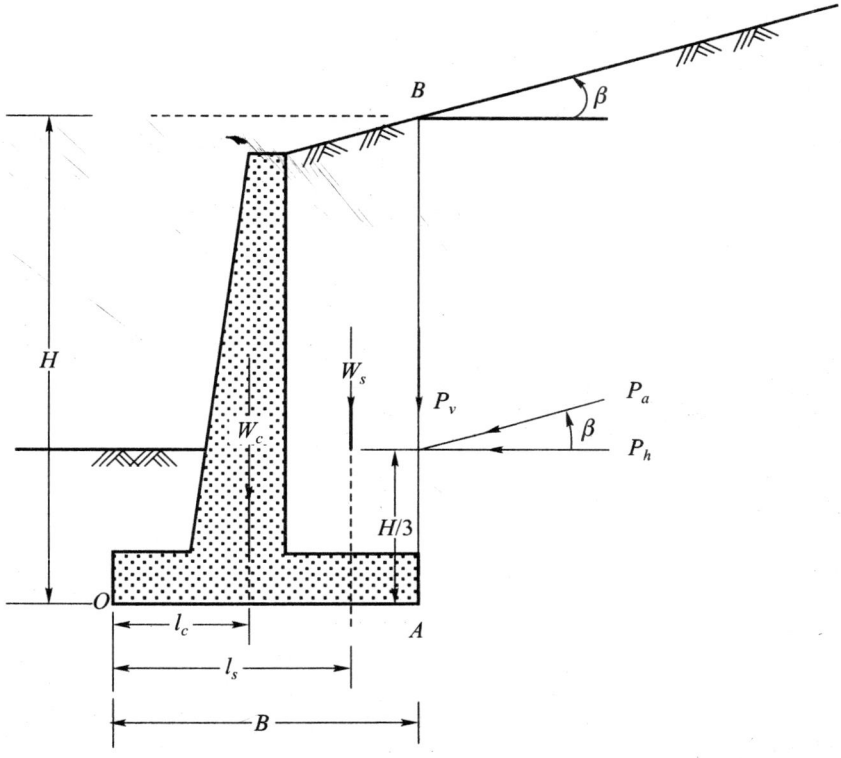

Fig. 25.7 Check for overturning

we may write (Fig. 25.7)

$$F_o = \frac{W_c l_c + W_s l_s + P_v B}{P_h (H/3)} \tag{25.7b}$$

where F_o should not be less than 2.0.

Check for Bearing Capacity Failure (Fig. 25.8)

In Fig. 25.8, W_t is the resultant of W_s and W_c. P_R is the resultant of P_a and W_t and P_R meets the base at m. R is the resultant of all the vertical forces acting at m with an eccentricity e. Fig. 25.8 shows the pressure distribution at the base with a maximum q_t at the toe and a minimum q_h at the heel.

An expression for e may be written as

$$e = \frac{B}{2} - \frac{(M_R - M_o)}{\Sigma V} \tag{25.8a}$$

where $R = \Sigma V =$ sum of all vertical forces

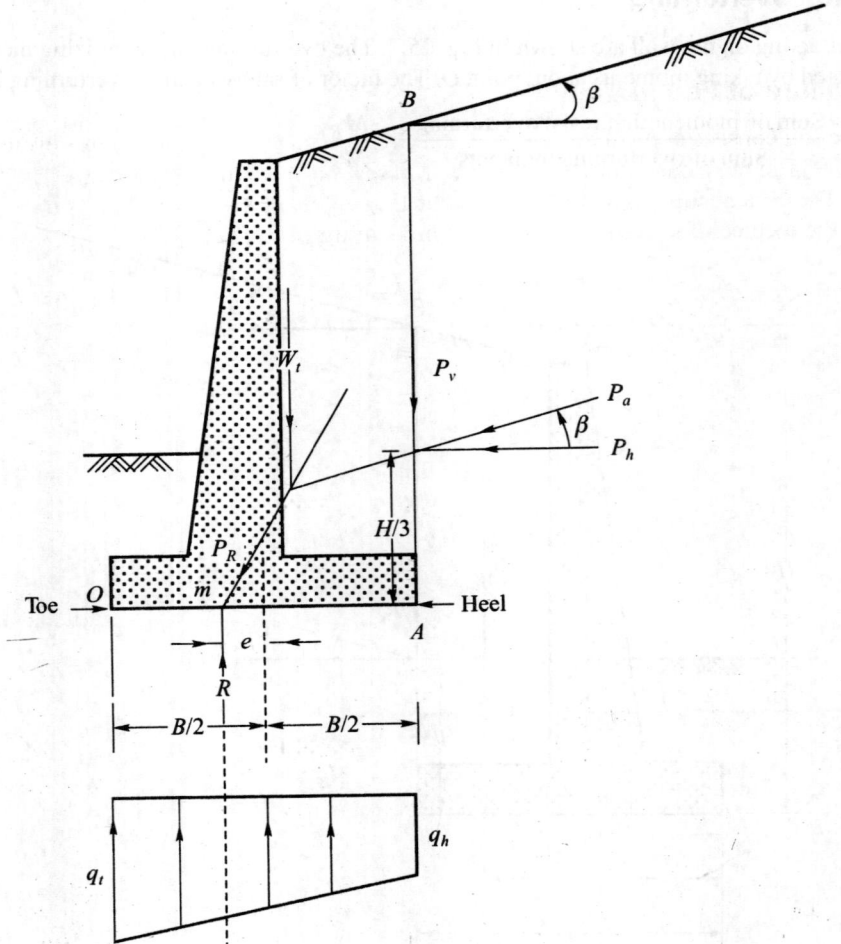

Fig. 25.8 Stability against bearing capacity failure

The values of q_t and q_h may be calculated by making use of the equations

$$q_t = \frac{R}{B}\left[1+\frac{6e}{B}\right] = q_a\left[1+\frac{6e}{B}\right]$$

(25.8b)

$$q_t = q_a\left[1+\frac{6e}{B}\right]$$

(25.8c)

where, $q_a = R/B$ = allowable bearing pressure.

Equation (25.8) is valid for $e \leq B/6$. When $e = B/6$, $q_t = 2q_a$ and $q_h = 0$. The base width B should be adjusted to satisfy Eq. (25.8). When the subsoil below the base is of a low bearing capacity, the possible alternative is to use a pile foundation.

The ultimate bearing capacity q_d may be determined as in Chapter 18 by taking into account the eccentricity. It must be ensured that

$$q_t \leq \frac{q_d}{F_s} = \frac{q_d}{3}$$

Base Failure of Foundation (Fig. 25.9)

If the base soil consists of medium to soft clay, a circular slip surface failure may develop as shown in Fig. 25.9. The most dangerous slip circle is actually the one that penetrates deepest into the soft material. The critical slip surface must be located by trial. Such stability problems may be analyzed either by the method of slices or any other method discussed in Chapter 16.

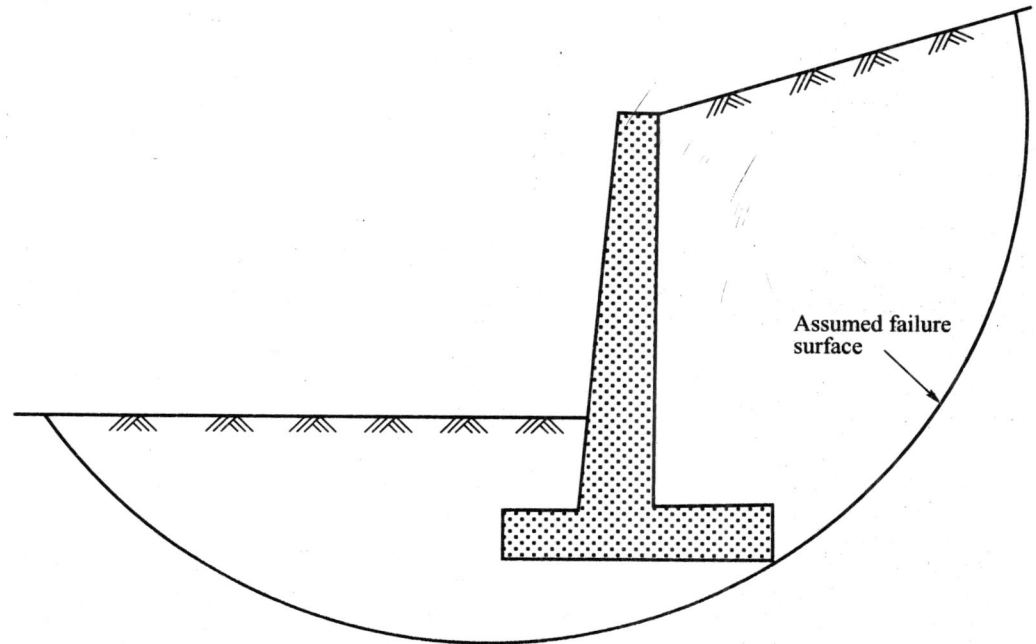

Assumed failure surface

Fig. 25.9   Stability against base slip surface shear failure

Drainage Provision for Retaining Walls (Fig. 25.10)

The saturation of the backfill of a retaining wall is always accompanied by a substantial hydrostatic pressure on the back of the wall. Saturation of the soil increases the earth pressure by increasing the unit weight. It is therefore essential to eliminate or reduce pore pressure by providing suitable drainage. Four types of drainage are given in Fig. 25.10. The drains collect the water that enters the backfill and this may be disposed of through outlets in the wall called weep holes. The graded filter material should be properly designed to prevent clogging by fine materials. The present practice is to use geotextiles or geogrids.

The weep holes are usually made by embedding 100 mm (4 in.) diameter pipes in the wall as shown in Fig. 25.10. The vertical spacing between horizontal rows of weep holes should not exceed 1.5 m. The horizontal spacing in a given row depends upon the provisions made to direct the seepage water towards the weep holes.

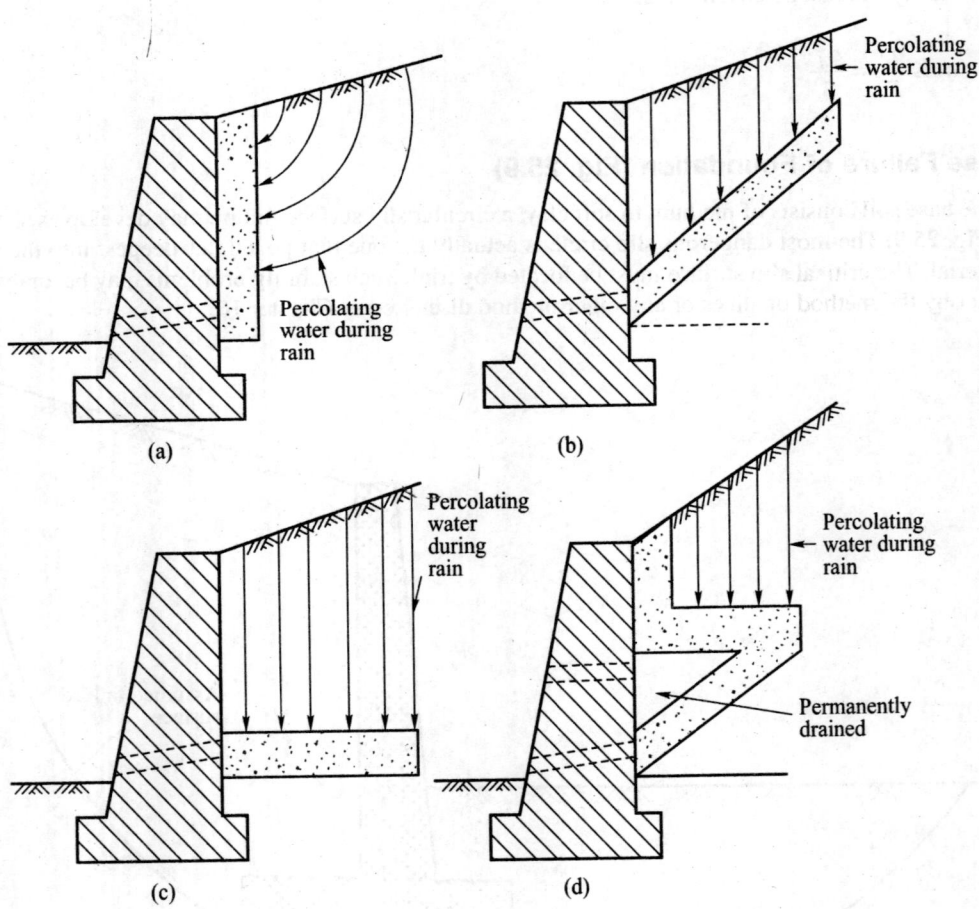

Fig. 25.10   Diagram showing provisions for drainage of backfill behind retaining walls: (a) vertical drainage layer, (b) inclined drainage layer for cohesionless backfill, (c) bottom drain to accelerate consolidation of cohesive back fill, (d) horizontal drain and seal combined with inclined drainage layer for cohesive backfill (Terzaghi et al., 1996)

Example 25.1

Figure Ex. 25.1(a) shows a section of a cantilever wall with dimensions and forces acting thereon. Check the stability of the wall with respect to (a) overturning, (b) sliding, and (c) bearing capacity.

Solution

Check for Rankine's condition

From Eq. (25.1b)

$$\alpha_0 = \frac{90-\phi}{2} - \frac{\varepsilon-\beta}{2}$$

where $\sin\varepsilon = \dfrac{\sin\beta}{\sin\phi} = \dfrac{\sin15°}{\sin30°} = 0.5176$

or $\varepsilon \approx 31°$

$$\alpha_0 = \frac{90-30}{2} - \frac{31-15}{2} = 22°$$

The outer failure line AC is drawn making an angle $22°$ with the vertical AB. Since this line does not cut the wall Rankine's condition is valid in this case.

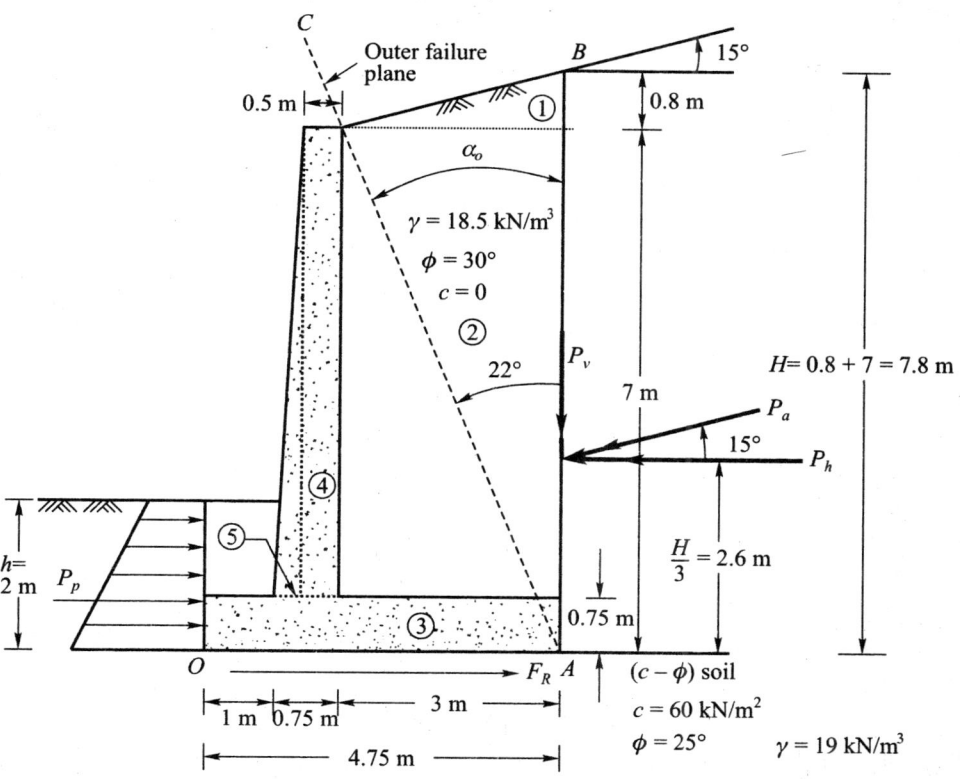

Fig. Ex. 25.1(a)

Rankine active pressure

Height of wall $= AB = H = 7.8$ m (Fig. Ex. 25.1(a))

$$P_a = \frac{1}{2}\gamma H^2 K_A$$

where $K_A = \tan^2(45° - \phi/2) = \frac{1}{3}$

substituting

$$P_a = \frac{1}{2} \times 18.5 \times (7.8)^2 \times \frac{1}{3} = 187.6 \text{ kN/m of wall}$$

$$P_h = P_a\cos\beta = 187.6\cos 15° = 181.2 \text{ kN/m}$$

$$P_v = P_a\sin\beta = 187.6\sin 15° = 48.6 \text{ kN/m}$$

Check for overturning

The forces acting on the wall in Fig. Ex. 25.1(a) are shown. The overturning and stabilizing moments may be calculated by taking moments about point O.

The whole section is divided into 5 parts as shown in the figure. Let these forces be represented by $w_1, w_2, \ldots w_5$ and the corresponding lever arms as $l_1, l_2, \ldots l_5$. Assume the weight of concrete $\gamma_c = 24$ kN/m³. The equation for the resisting moment is

$$M_R = w_1l_1 + w_2l_2 + \ldots w_5l_5$$

The overturning moment is

$$M_o = P_h\frac{H}{3}$$

The details of calculations are tabulated below.

Section No.	Area (m²)	Unit weight kN/m³	Weight kN/m	Lever arm(m)	Moment kN-m
1	1.20	18.5	22.2	3.75	83.25
2	18.75	18.5	346.9	3.25	1127.40
3	3.56	24.0	85.4	2.38	203.25
4	3.13	24.0	75.1	1.50	112.65
5	0.78	24.0	18.7	1.17	21.88
			$P_v = 48.6$	4.75	230.85
			$\Sigma_v = 596.9$		$\Sigma_M = 1{,}779.3 = M_R$

$$M_O = 181.2 \times 2.6 = 471.12 \text{ kN–m}$$

$$F_s = \frac{M_R}{M_o} = \frac{1{,}779.3}{471.12} = 3.78 > 2.0 \text{ --- OK}$$

Check for sliding (Fig. 25.1a)

The force that resists the movement as per Eq. (25.5) is

$$F_R = c_aB + R\tan\delta + P_p$$

where B = width = 4.75 m

$c_a = \alpha c_u$, α = adhesion factor = 0.55 from Fig. 21.15

R = total vertical force Σ_V = 596.9 kN

For the foundation soil:

δ = angle of wall friction $\approx \phi$ = 25°

From Eq. (14.45)

$$P_p = \frac{1}{2}\gamma h^2 K_p + 2ch\sqrt{K_p}$$

where h = 2 m, γ = 19 kN/m³, c = 60 kN/m²

$K_p = \tan^2 (45° + \phi/2) = \tan^2 (45° + 25/2) = 2.46$

substituting

$$P_p = \frac{1}{2}\times 19 \times 2^2 \times 2.46 + 2\times 60 \times 2 \times \sqrt{2.46} = 470 \text{ kN/m}$$

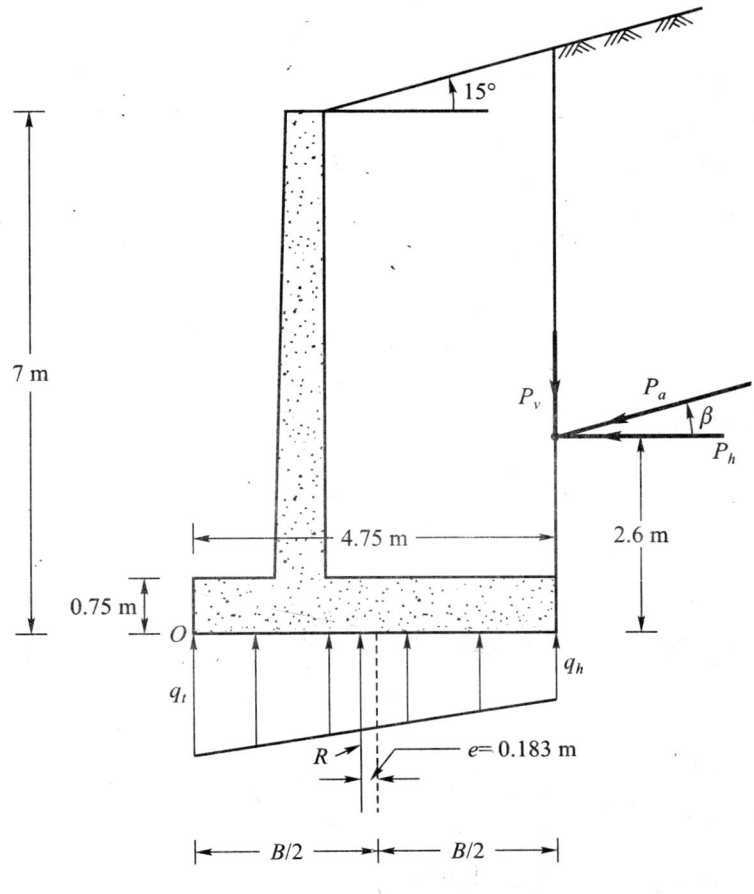

Fig. Ex. 25.1(b)

$$F_R = 60 \times 4.75 + 596.9 \tan 25° + 470 = 285 + 278 + 470 = 1033 \text{ kN/m}$$

$$P_h = 181.2 \text{ kN/m}$$

$$F_s = \frac{F_R}{P_h} = \frac{1033}{181.2} = 5.7 > 1.5 \text{---OK}$$

Normally the passive earth pressure P_p is not considered in the analysis. By neglecting P_p, the factor of safety is

$$F_s = \frac{1033 - 470}{181.2} = \frac{563}{181.2} = 3.1 > 1.5 \text{---OK}$$

Check for bearing capacity failure (Fig. Ex. 25.1b)

From Eq. (25.8b and c), the pressures at the toe and heel of the retaining wall may be written as

$$q_t = \frac{R}{B}\left[1 + \frac{6e}{B}\right]$$

$$q_h = \frac{R}{B}\left[1 - \frac{6e}{B}\right]$$

where e = eccentricity of the total load $R\ (= \Sigma V)$ acting on the base. From Eq. (25.8a), the eccentricity e may be calculated.

$$e = \frac{B}{2} - \frac{(M_R - M_o)}{R} = \frac{4.75}{2} - \frac{(1,779.3 - 471.12)}{596.9} = 0.183 \text{ m}$$

$$\text{Now } q_t = \frac{596.9}{4.75}\left[1 + \frac{6 \times 0.183}{4.75}\right] = 154.7 \text{ kN/m}^2$$

$$q_h = \frac{596.9}{4.75}\left[1 - \frac{6 \times 0.183}{4.75}\right] = 96.6 \text{ kN/m}^2$$

The ultimate bearing capacity q_d may be determined by the procedure explained in Chapter 18. It has to be ensured that

$$q_t \le \frac{q_d}{F_s}$$

where $F_s = 3$

25.6   QUESTIONS AND PROBLEMS

25.1   Figure Prob. 25.1 gives a section of a cantilever wall. Check the stability of the wall with respect to (a) overturning, (b) sliding, and (c) bearing capacity.

25.2   Check the stability of the wall given in Prob. 25.1 for the condition that the slope is horizontal and the foundation soil is cohesionless with $\phi = 30°$. All the other data remain the same.

25.3   Check the stability of the cantilever wall given in Fig. Prob. 25.3 for (a) overturning, (b) sliding, and (c) bearing capacity failure.

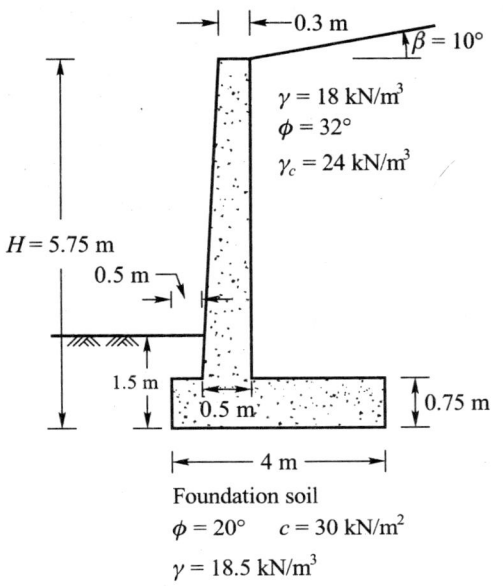

$H = 5.75$ m

Foundation soil
$\phi = 20°$ $c = 30$ kN/m^2
$\gamma = 18.5$ kN/m^3

Fig. Prob. 25.1

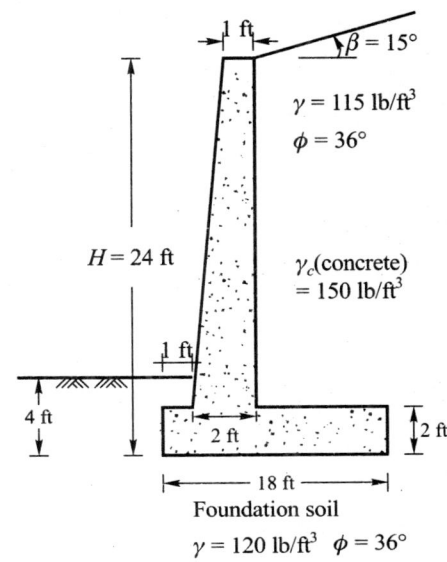

$H = 24$ ft

Foundation soil
$\gamma = 120$ lb/ft^3 $\phi = 36°$

Fig. Prob. 25.3

25.4   Check the stability of the wall in Prob. 25.3 assuming (a) $\beta = 0$, and (b) the foundation soil
has $c = 300$ lb/ft^2, $\gamma = 115$ lb/ft^3, and $\phi = 26°$.

25.5   Figure Prob. 25.5 depicts a gravity retaining wall. Check the stability of the wall for sliding,
and overturning.

25.6   Check the stability of the wall given in Fig. Prob. 25.6. All the data are given on the figure.

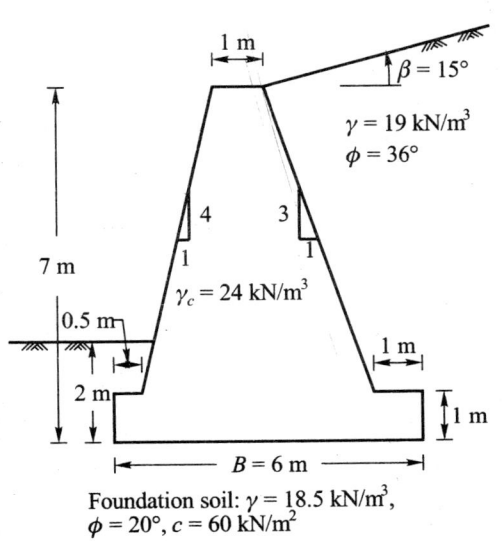

Foundation soil: $\gamma = 18.5$ kN/m^3,
$\phi = 20°$, $c = 60$ kN/m^2

Fig. Prob. 25.5

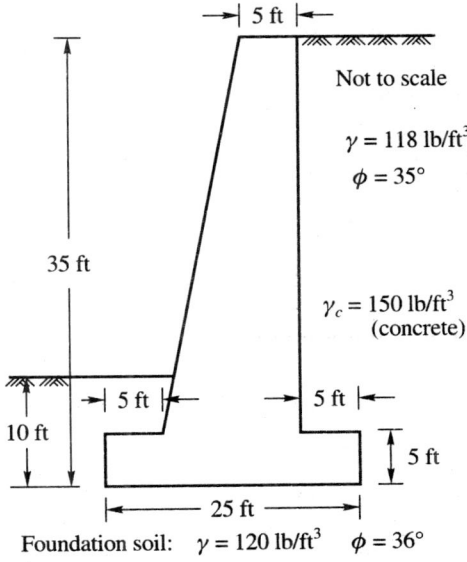

Foundation soil: $\gamma = 120$ lb/ft^3 $\phi = 36°$

Fig. Prob. 25.6

25.7   Check the stability of the gravity wall given in Prob. 25.6 with the foundation soil having properties $\phi = 30°$, $\gamma = 110$ lb/ft^3 and $c = 500$ lb/ft^2. All the other data remain the same.

25.8   Check the stability of the gravity retaining wall given in Fig. Prob. 25.8.

25.9   Check the stability of the gravity wall given in Prob. 25.8 for Coulomb's condition. Assume $\delta = 2/3\phi$.

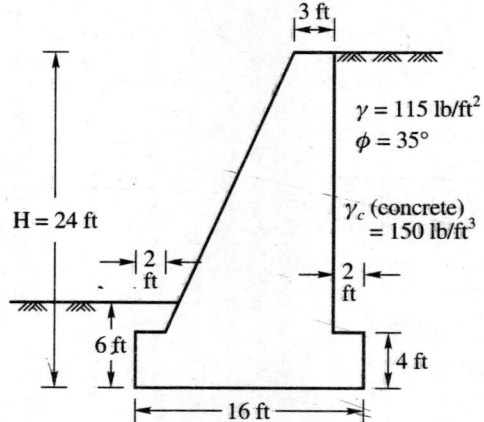

Foundation soil: $\gamma = 120$ lb/ft^3, $\phi = 36°$

Fig. Prob. 25.8

CHAPTER 26

MECHANICALLY STABILISED EARTH RETAINING WALLS

26.1 GENERAL CONSIDERATIONS

Reinforced earth is a construction material composed of soil fill strengthened by the inclusion of rods, bars, fibers or nets which interact with the soil by means of frictional resistance. The concept of strengthening soil with rods or fibers is not new. Throughout the ages attempts have been made to improve the quality of adobe brick by adding straw. The present practice is to use thin metal strips, geotextiles, and geogrids as reinforcing materials for the construction of reinforced earth retaining walls. A new era of retaining walls with reinforced earth was introduced by Vidal (1969). Metal strips were used as reinforcing material as shown in Fig. 26.1(a). Here the metal strips extend from the panel back into the soil to serve the dual role of anchoring the facing units and being restrained through the frictional stresses mobilized between the strips and the backfill soil. The backfill soil creates the lateral pressure and interacts with the strips to resist it. The walls are relatively flexible compared to massive gravity structures. These flexible walls offer many advantages including significant lower cost per square meter of exposed surface. The variations in the types of facing units, subsequent to Vidal's introduction of the reinforced earth walls, are many. A few of the types that are currently in use are (Koerner, 1999)

1. Facing panels with metal strip reinforcement

2. Facing panels with wire mesh reinforcement

3. Solid panels with tie back anchors

4. Anchored gabion walls

5. Anchored crib walls

6. Geotextile reinforced walls

7. Geogrid reinforced walls

975

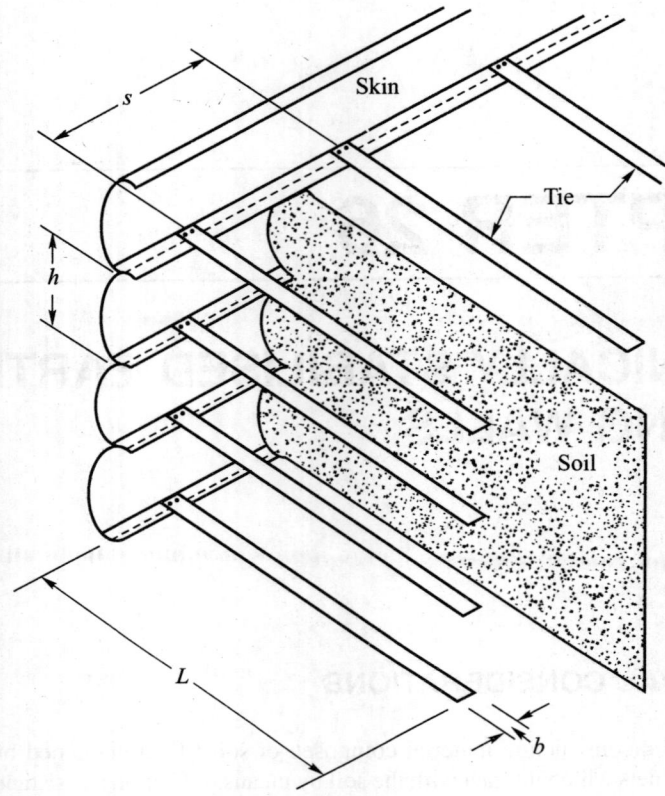

Fig. 26.1(a)   Component parts and key dimensions of reinforced earth wall (Vidal, 1969)

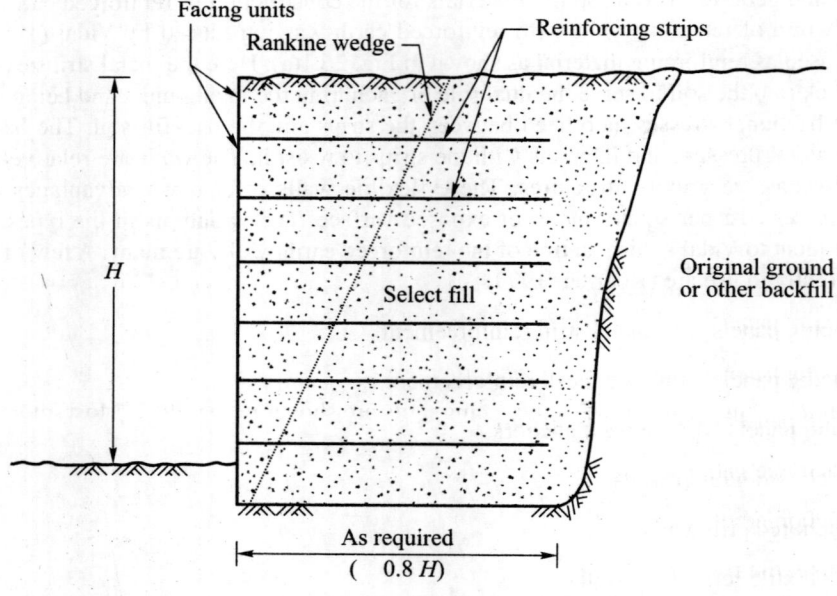

(b) Line details of a reinforced earth wall in place

Fig. 26.1(b)   Reinforced earth walls (Bowles, 1996)

(c) Front face of a reinforced earth wall under construction for a brdge approach fill using patented precast concrete wall face units

Fig. 26.1(c) Reinforced earth walls (Bowles, 1996)

In all cases, the soil behind the wall facing is said to be *mechanically stabilized earth* (MSE) and the wall system is generally called an MSE wall.

The three components of a MSE wall are the facing unit, the backfill and the reinforcing material. Figure 26.1(b) shows a side view of a wall with metal strip reinforcement and Fig. 26.1(c) the front face of a wall under construction (Bowles, 1996).

Modular concrete blocks, currently called segmental retaining walls (SRWS, Fig. 26.2(a)) are most common as facing units. Some of the facing units are shown in Fig. 26.2. Most interesting in regard to SRWS are the emerging block systems with openings, pouches, or planting areas within them. These openings are soil-filled and planted with vegetation that is indigenous to the area (Fig. 26.2(b)). Further possibilities in the area of reinforced wall systems could be in the use of polymer rope, straps, or anchor ties to the facing in units or to geosynthetic layers, and extending them into the retained earth zone as shown in Fig. 26.2(c).

A recent study (Koerner 2000) has indicated that geosynthetic reinforced walls are the least expensive of any wall type and for all wall height categories (Fig. 26.3).

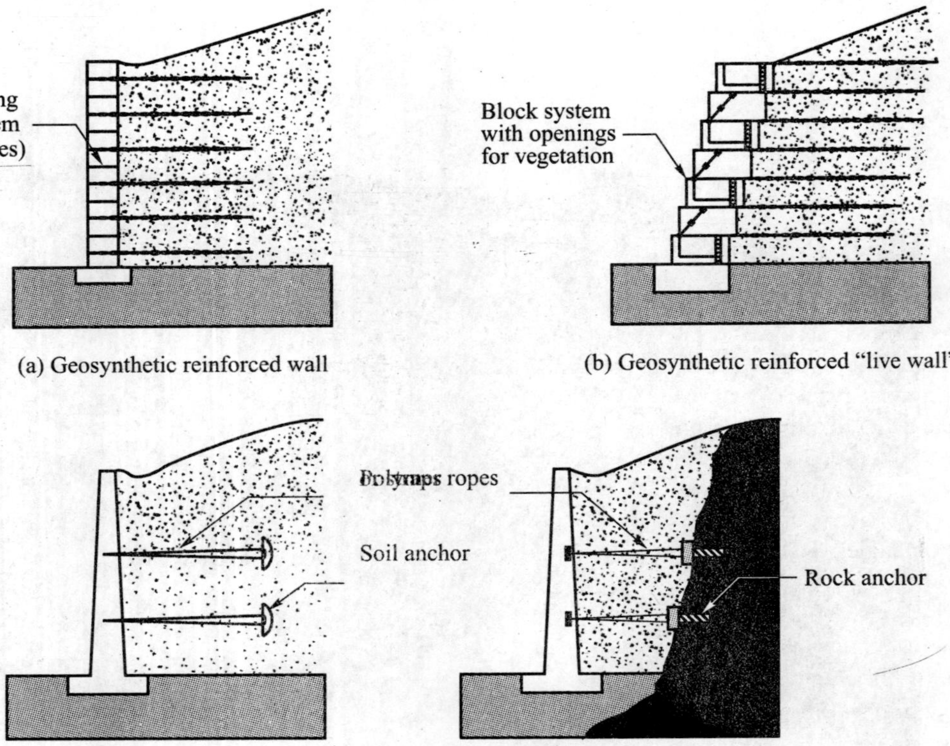

Facing
system
(varies)

(a) Geosynthetic reinforced wall

Block system
with openings
for vegetation

(b) Geosynthetic reinforced "live wall"

Polymps ropes

Soil anchor

Rock anchor

(c) Future types of geosynthetic anchorage

Fig. 26.2 Geosynthetic use for reinforced walls and bulkheads (Koerner, 2000)

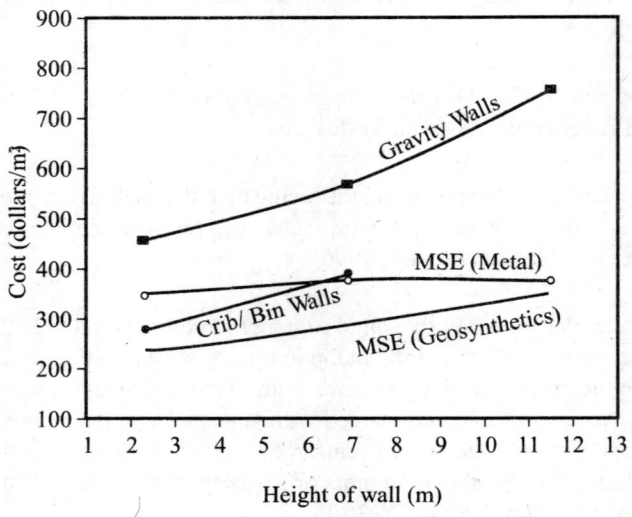

Fig. 26.3 Mean values of various categories of retaining wall costs
(Koerner, 2000)

26.2 BACKFILL AND REINFORCING MATERIALS

Backfill

The backfill, is limited to cohesionless, free draining material (such as sand), and thus the key properties are the density and the angle of internal friction.

Reinforcing material

The reinforcements may be strips or rods of metal or sheets of geotextile, wire grids or geogrids (grids made from plastic).

Geotextile is a permeable geosynthetic comprised solely of textiles. Geotextiles are used with foundation soil, rock, earth or any other geotechnical engineering-related material as an integral part of a human made project, structure, or system (Koerner, 1999). AASHTO (M288-96) provides (Table 26.1) geotextile strength requirements (Koerner, 1999). The tensile strength of geotextile varies with the geotextile designation as per the design requirements. For example, a woven slitfilm polypropylene (weighing 240 g/m^2) has a range of 30 to 50 kN/m. The friction angle between soil and geotextiles varies with the type of geotextile and the soil. Table 26.2 gives values of geotextile friction angles (Koerner, 1999).

The test properties represent an idealized condition and therefore result in the maximum possible numerical values when used directly in design. Most laboratory test values cannot generally be used directly and must be suitably modified for in-situ conditions. For problems dealing with geotextiles the ultimate strength T_u should be reduced by applying certain reduction factors to obtain the allowable strength T_a as follows (Koerner, 1999).

$$T_a = T_u \left(\frac{1}{RF_{1D} \times RF_{CR} \times RF_{CD} \times RF_{BD}} \right) \tag{26.1}$$

where

$\quad T_a \quad = \quad$ allowable tensile strength
$\quad T_u \quad = \quad$ ultimate tensile strength
$\quad RF_{ID} \quad = \quad$ reduction factor for installation damage
$\quad RF_{CR} \quad = \quad$ reduction factor for creep
$\quad RF_{BD} \quad = \quad$ reduction factor for biological degradation and
$\quad RF_{CD} \quad = \quad$ reduction factor for chemical degradation

Typical values for reduction factors are given in Table 26.3.

TABLE 26.2
Peak soil-to-geotextile friction angles and efficiencies in selected cohesionless soils*

Geotextile type	Concrete sand ($\phi = 30°$)	Rounded sand ($\phi = 28°$)	Silty sand ($\phi = 26°$)
Woven, monofilament	26° (84 %)	–	–
Woven, slit-film	24° (77%)	24° (84 %)	23° (87 %)
Nonwoven, heat-bonded	26° (84 %)	–	–
Nonwoven, needle-punched	30° (100 %)	26° (92 %)	25° (96 %)

* Numbers in parentheses are the efficiencies. Values such as these should not be used in final design. Site specific geotextiles and soils must be individually tested and evaluated in accordance with the particular project conditions: saturation, type of liquid, normal stress, consolidation time, shear rate, displacement amount, and so on (Koerner, 1999).

TABLE 26.1

AASHTO M288-96 geotextile strength property requirements

Test methods	Units	Geotextile Classification* † ‡					
		Case 1		Case 2		Case 3	
		Elongation <50 %	Elongation ≥50 %	Elongation <50 %	Elongation ≥50 %	Elongation <50 %	Elongation ≥50 %
Grab strength	ASTM D4632	N	1400	900	1100	700	800	500
Sewn seam strength ‡	ASTM D4632	N	1200	810	990	630	720	450
Tear strength	ASTM D4533	N	500	350	400	250	300	180
Puncture strength	ASTM D4833	N	500	350	400	2505	300	180
Burst strength	ASTM D3786	kPa	3500	1700	2700	1300	2100	950

† As measured in accordance with ASTM D4632. Woven geotextiles fail at elongations (strains)< 50%, while nonwovens fail at elongation (strains) > 50%.

‡ When sewnseams are required. Overlap seam requirements are application specific.

§ The required MARV tear strength for woven monofilament geotextiles is 250 N.

TABLE 26.3

Recommended reduction factor values for use in [Eq. (26.1)]

Application Area	Range of Reduction Factors			
	Installation Damage	Creep*	Chemical Degradation	Biological Degradation
Separation	1.1 to 2.5	1.5 to 2.5	1.0 to 1.5	1.0 to 1.2
Cushioning	1.1 to 2.0	1.2 to 1.5	1.0 to 2.0	1.0 to 1.2
Unpaved roads	1.1 to 2.0	1.5 to 2.5	1.0 to 1.5	1.0 to 1.2
Walls	1.1 to 2.0	2.0 to 4.0	1.0 to 1.5	1.0 to 1.3
Embankments	1.1 to 2.0	2.0 to 3.5	1.0 to 1.5	1.0 to 1.3
Bearing capacity	1.1 to 2.0	2.0 to 4.0	1.0 to 1.5	1.0 to 1.3
Slope stabilization	1.1 to 1.5	2.0 to 3.0	1.0 to 1.5	1.0 to 1.3
Pavement overlays	1.1 to 1.5	1.0 to 2.0	1.0 to 1.5	1.0 to 1.1
Railroads (filter/sep.)	1.5 to 3.0	1.0 to 1.5	1.5 to 2.0	1.0 to 1.2
Flexible forms	1.1 to 1.5	1.5 to 3.0	1.0 to 1.5	1.0 to 1.1
Silt fences	1.1 to 1.5	1.5 to 2.5	1.0 to 1.5	1.0 to 1.1

* The low end of the range refers to applications which have relatively short service lifetimes and / or situations where creep deformations are not critical to the overall system performance. (Koerner, 1999)

TABLE 26.4

Recommended reduction factor values for use in Eq. (26.2) for determining allowable tensile strength of geogrids

Application Area	RF_{ID}	RF_{CR}	RF_{CD}	RF_{BD}
Unpaved roads	1.1 to 1.6	1.5 to 2.5	1.0 to 1.5	1.0 to 1.1
Paved roads	1.2 to 1.5	1.5 to 2.5	1.1 to 1.6	1.0 to 1.1
Embankments	1.1 to 1.4	2.0 to 3.0	1.1 to 1.4	1.0 to 1.2
Slopes	1.1 to 1.4	2.0 to 3.0	1.1 to 1.4	1.0 to 1.2
Walls	1.1 to 1.4	2.0 to 3.0	1.1 to 1.4	1.0 to 1.2
Bearing capacity	1.2 to 1.5	2.0 to 3.0	1.1 to 1.6	1.0 to 1.2

Geogrid

A geogrid is defined as a geosynthetic material consisting of connected parallel sets of tensile ribs with apertures of sufficient size to allow strike-through of surrounding soil, stone, or other geotechnical material (Koerner, 1999).

Geogrids are matrix like materials with large open spaces called apertures, which are typically 10 to 100 mm between the ribs, called *longitudinal* and *transverse* respectively. The primary function of geogrids is clearly reinforcement. The mass of geogrids ranges from 200 to 1000 g/m^2 and the open area varies from 40 to 95 %. It is not practicable to give specific values for the tensile strength of geogrids because of its wide variation in density. In such cases one has to consult manufacturer's literature for the strength characteristics of their products. The allowable tensile strength, T_a, may be determined by applying certain reduction factors to the ultimate strength T_u as in the case of geotextiles.

The equation is

$$T_a = T_u \left(\frac{1}{RF_{1D} \times RF_{CR} \times RF_{BD} \times RF_{CD}} \right) \qquad (26.2)$$

The definition of the various terms in Eq. (26.2) is the same as in Eq. (26.1). However, the reduction factors are different. These values are given in Table 26.4 (Koerner, 1999).

Metal Strips

Metal reinforcement strips are available in widths ranging from 75 to 100 mm and thickness on the order of 3 to 5 mm, with 1 mm on each face excluded for corrosion (Bowles, 1996). The yield strength of steel may be taken as equal to about 35000 lb/in2 (240 MPa) or as per any code of practice.

26.3   CONSTRUCTION DETAILS

The method of construction of MSE walls depends upon the type of facing unit and reinforcing material used in the system. The facing unit which is also called the *skin* can be either flexible or stiff, but must be strong enough to retain the backfill and allow fastenings for the reinforcement to be attached. The facing units require only a small foundation from which they can be built, generally consisting of a trench filled with mass concrete giving a footing similar to those used in domestic housing. The segmental retaining wall sections of dry-laid masonry blocks, are shown in Fig. 26.2(a). The block system with openings for vegetation is shown in Fig. 26.2(b).

The construction procedure with the use of geotextiles is explained in Fig. 26.3(a). Here, the geotextile serve both as a reinforcement and also as a facing unit. The procedure is described below (Koerner, 1985) with reference to Fig. 26.3(a).

1.   Start with an adequate working surface and staging area (Fig. 26.3a).

2.   Lay a geotextile sheet of proper width on the ground surface with 4 to 7 ft at the wall face draped over a temporary wooden form (b).

3.   Backfill over this sheet with soil. Granular soils or soils containing a maximum 30 percent silt and /or 5 percent clay are customary (c).

4.   Construction equipment must work from the soil backfill and be kept off the unprotected geotextile. The spreading equipment should be a wide-tracked bulldozer that exerts little pressure against the ground on which it rests. Rolling equipment likewise should be of relatively light weight.

5.   When the first layer has been folded over, the process should be repeated for the second layer with the temporary facing form being extended from the original ground surface or the wall being stepped back about 6 inches so that the form can be supported from the first layer. In the latter case, the support stakes must penetrate the fabric.

6.   This process is continued until the wall reaches its intended height.

7.   For protection against ultraviolet light and safety against vandalism the faces of such walls must be protected. Both shotcrete and gunite have been used for this purpose.

Figure 26.3(b) shows complete geotextile walls (Koerner, 1999).

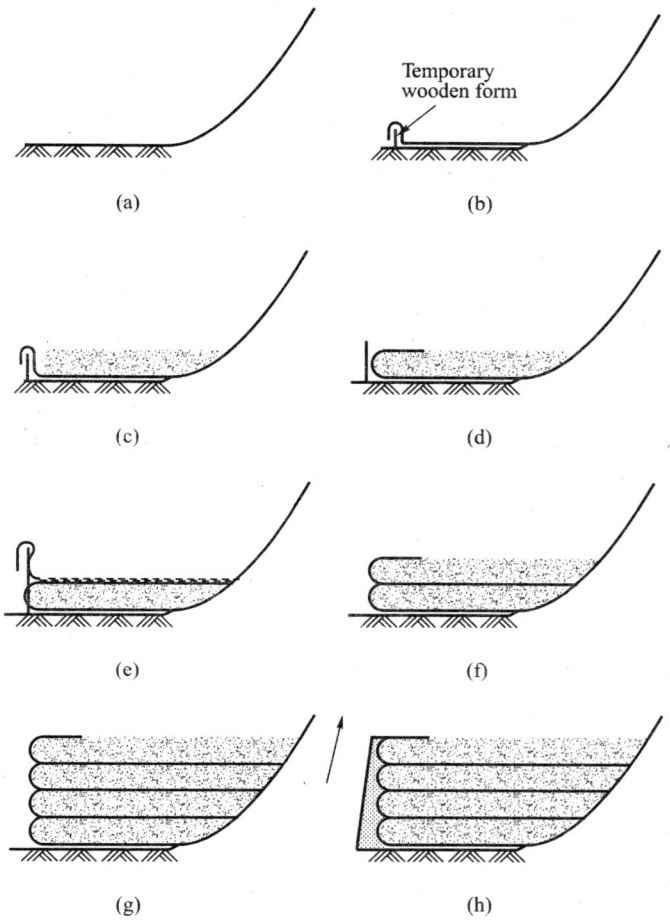

Fig. 26.3(a)   General construction procedures for using geotextiles in fabric wall construction (Koerner, 1985)

Fig. 26.3(b)   Geotextile walls (Koerner, 1999)

26.4   DESIGN CONSIDERATIONS FOR A MECHANICALLY STABILIZED EARTH WALL

The design of a MSE wall involves the following steps:

1.   Check for internal stability, addressing reinforcement spacing and length.

2.   Check for external stability of the wall against overturning, sliding, and foundation failure.

The general considerations for the design are:

1.   Selection of backfill material: granular, freely draining material is normally specified. However, with the advent of geogrids, the use of cohesive soil is gaining ground.

2.   Backfill should be compacted with care in order to avoid damage to the reinforcing material.

3.   Rankine's theory for the active state is assumed to be valid.

4.   The wall should be sufficiently flexible for the development of active conditions.

5.   Tension stresses are considered for the reinforcement outside the assumed failure zone.

6.   Wall failure will occur in one of three ways

   (a)   tension in reinforcements

   (b)   bearing capacity failure

   (c)   sliding of the whole wall soil system.

7.   Surcharges are allowed on the backfill. The surcharges may be permanent (such as a roadway) or temporary.

   (a)   Temporary surcharges within the reinforcement zone will increase the lateral pressure on the facing unit which in turn increases the tension in the reinforcements, but does not contribute to reinforcement stability.

   (b)   Permanent surcharges within the reinforcement zone will increase the lateral pressure and tension in the reinforcement and will contribute additional vertical pressure for the reinforcement friction.

   (c)   Temporary or permanent surcharges outside the reinforcement zone contribute lateral pressure which tends to overturn the wall.

8.   The total length L of the reinforcement goes beyond the failure plane AC by a length L_e. Only length L_e (effective length) is considered for computing frictional resistance. The length L_R lying within the failure zone will not contribute for frictional resistance (Fig. 26.4a).

9.   For the propose of design the total length L remains the same for the entire height of wall H. Designers, however, may use their discretion to curtail the length at lower levels. Typical ranges in reinforcement spacing are given in Fig. 26.5.

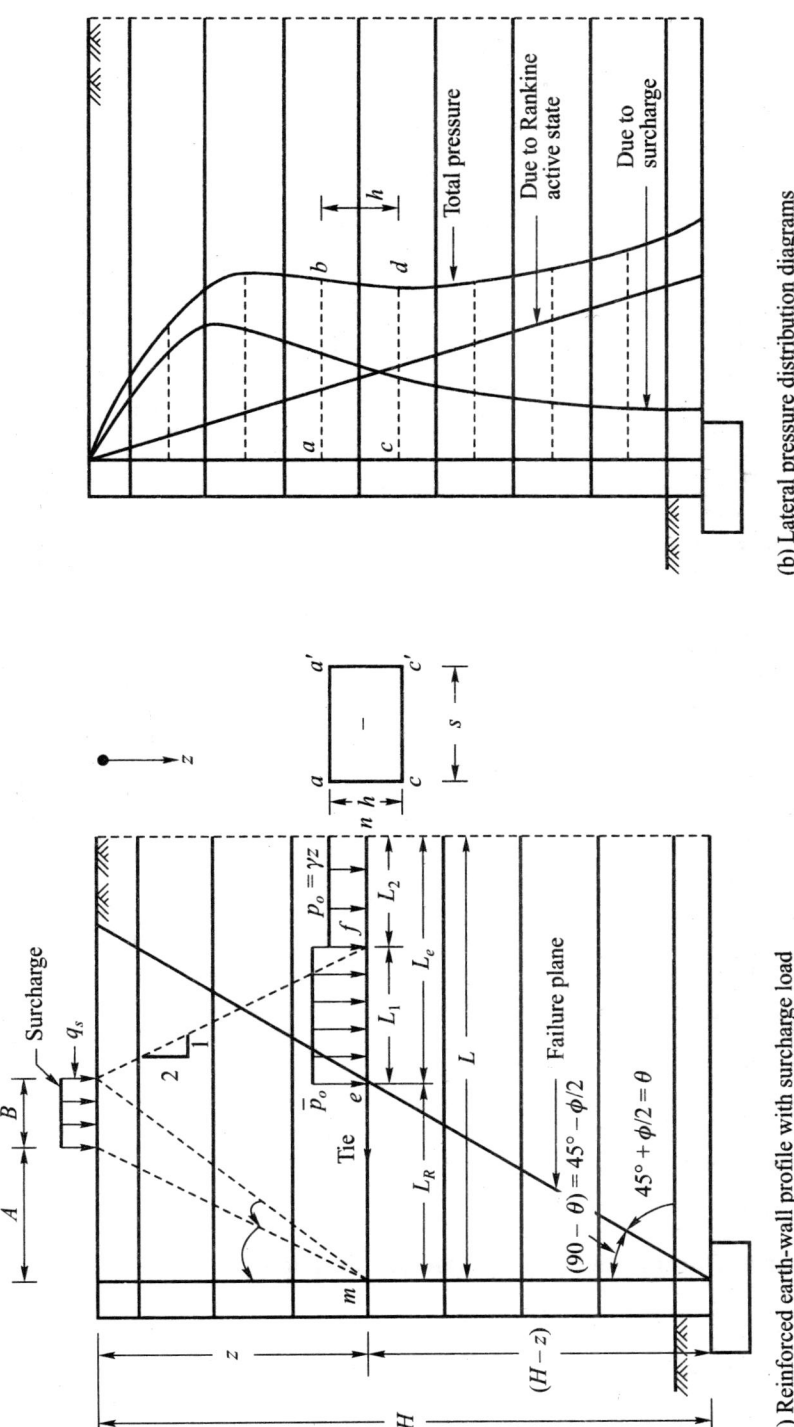

(b) Lateral pressure distribution diagrams

(a) Reinforced earth-wall profile with surcharge load

Fig. 26.4 Principles of MSE wall design

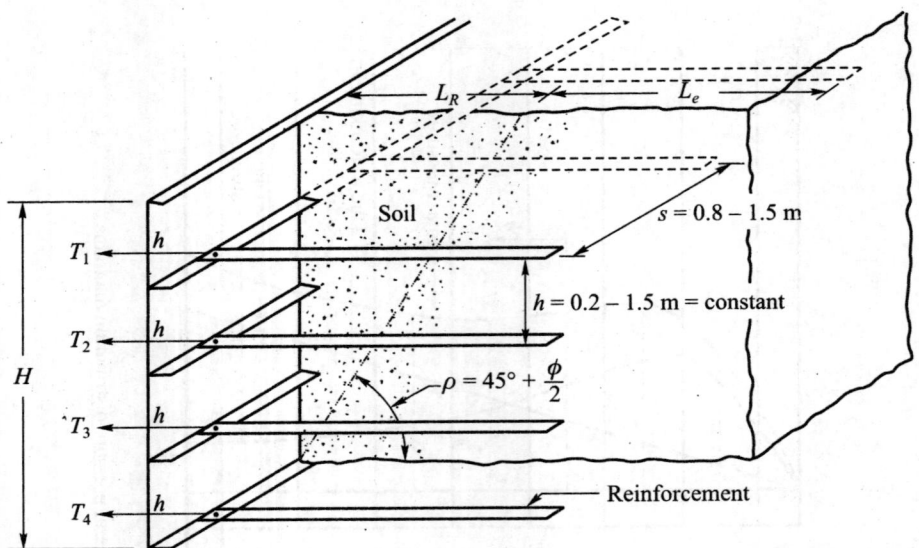

Fig. 26.5 Typical range in strip reinforcement spacing for reinforced earth
walls (Bowles, 1996)

26.5 DESIGN METHOD

The following forces are considered:

1. Lateral pressure on the wall due to backfill
2. Lateral pressure due to surcharge if present on the backfill surface.
3. The vertical pressure at any depth z on the strip due to
    (a) overburden pressure p_o only
    (b) overburden pressure p_o' and pressure due to surcharge.

Lateral Pressure

Pressure due to Overburden

Lateral earth pressure due to overburden

$$\text{At depth } z \quad p_a = p_{oz}K_A = \gamma z K_A \tag{26.3}$$

$$\text{At depth } H \quad p_a = p_{oH}K_A = \gamma H K_A \tag{26.4}$$

Total active earth pressure

$$P_a = \frac{1}{2}\gamma H^2 K_A \tag{26.5}$$

Pressure Due to Surcharge (a) of Limited Width, and (b) Uniformly Distributed

(a) From Eq. (14.77)

$$q_h = \frac{2q_s}{\pi}(\beta - \sin\beta\cos 2\alpha) \tag{26.6}$$

(b) $q_h = q_s K_A$ \hfill (26.7)

Total lateral pressure due to overburden and surcharge at any depth z

$$p_h = p_a + q_h = (\gamma z K_A + q_h) \tag{26.8}$$

Vertical pressure

Vertical pressure at any depth z due to overburden only

$$p_o = \gamma z \qquad (26.9)$$

due to surcharge (limited width)

$$\Delta q = \frac{q_s B}{B + z} \qquad (26.10)$$

where the 2:1 (2 vertical : 1 horizontal) method is used for determining Δq at any depth z.

Total vertical pressure due to overburden and surcharge at any depth z.

$$\bar{p}_o = p_o + \Delta q \qquad (26.11)$$

Reinforcement and Distribution

Three types of reinforcements are normally used. They are

1. Metal strips
2. Geotextiles
3. Geogrids.

Galvanized steel strips of widths varying from 5 to 100 mm and thickness from 3 to 5 mm are generally used. Allowance for corrosion is normally made while deciding the thickness at the rate of 0.001 in. per year and the life span is taken as equal to 50 years. The vertical spacing may range from 20 to 150 cm (8 to 60 in.) and can vary with depth. The horizontal lateral spacing may be on the order of 80 to 150 cm (30 to 60 in.). The ultimate tensile strength may be taken as equal to 240 MPa (35,000 lb/in^2). A factor of safety in the range of 1.5 to 1.67 is normally used to determine the allowable steel strength f_a.

Figure 26.5 depicts a typical arrangement of metal reinforcement. The properties of geotextiles and geogrids have been discussed in Section. However, with regards to spacing, only the vertical spacing is to be considered. Manufacturers provide geotextiles (or geogrids) in rolls of various lengths and widths. The tensile force per unit width must be determined.

Length of Reinforcement

From Fig. 26.4(a)

$$L = L_R + L_e = L_R + L_1 + L_2 \qquad (26.12)$$

where L_R = $(H - z) \tan (45° - \phi/2)$

L_e = effective length of reinforcement outside the failure zone

L_1 = length subjected to pressure $(p_o + \Delta q) = \bar{p}_o$

L_2 = length subjected to p_o only.

Strip Tensile Force at any Depth z

The equation for computing T is

$$T = p_h \times h \times s / \text{strip} = (\gamma z K_A + q_h) h \times s \qquad (26.13)$$

The maximum tie force will be

$$T(\max) = (\gamma z K_A + q_{hH})h \times s \qquad (26.14)$$

where p_h $= \gamma z K_A + q_h$

q_h = lateral pressure at depth z due to surcharge

q_{hH} = q_h at depth H

h = vertical spacing

s = horizontal spacing

$$T = p_a + p_q \qquad (26.15)$$

where P_a $= 1/2 \gamma H^2 K_A$—Rankine's lateral force

P_q = lateral force due to surcharge

Frictional Resistance

In the case of strips of width b both sides offer frictional resistance. The frictional resistance F_R offered by a strip at any depth z must be greater than the pullout force T by a suitable factor of safety. We may write

$$F_R = 2b[(p_o + \Delta q)L_1 + p_o L_2]\tan\delta \le TF_s \qquad (26.16)$$

or $$F_R = 2b[\bar{p}_o L_1 + p_o L_2]\tan\delta \le TF_s \qquad (26.17)$$

where F_s may be taken as equal to 1.5.

The friction angle δ between the strip and the soil may be taken as equal to ϕ for a rough strip surface and for a smooth surface δ may lie between 10 to 25°.

Sectional Area of Metal Strips

Normally the width b of the strip is assumed in the design. The thickness t has to be determined based on T (max) and the allowable stress f_a in the steel. If f_y is the yield stress of steel, then

$$f_a = \frac{f_y}{F_s(\text{steel})} \qquad (26.18)$$

Normally F_s (steel) ranges from 1.5 to 1.67. The thickness t may be obtained from

$$t = \frac{T(\max)}{bf_a} \qquad (26.19)$$

The thickness of t is to be increased to take care of the corrosion effect. The rate of corrosion is normally taken as equal to 0.001 in/yr for a life span of 50 years.

Spacing of Geotextile Layers

The tensile force T per unit width of geotextile layer at any depth z may be obtained from

$$T = p_h h = (\gamma z K_A + q_h)h \qquad (26.20)$$

where q_h = lateral pressure either due to a stripload or due to uniformly distributed surcharge.

The maximum value of the computed T should be limited to the allowable value T_a as per Eq. (26.1). As such we may write Eq. (26.20) as

$$T_a = TF_s = (\gamma z K_A + q_h)hF_s \tag{26.21}$$

or $\quad h = \dfrac{T_a}{(\gamma z K_A + q_h)F_s} = \dfrac{T_a}{p_h F_s} \tag{26.22}$

where F_s = factor of safety (1.3 to 1.5) when using T_a.

Equation (26.22) is used for determining the vertical spacing of geotextile layers.

Frictional Resistance

The frictional resistance offered by a geotextile layer for the pullout force T_a may be expressed as

$$F_R = 2[(\gamma z + \Delta q)L_1 + \gamma z L_2]\tan\delta \ge T_a F_s \tag{26.23}$$

Equation (26.23) expresses frictional resistance per unit width and both sides of the sheets are considered.

Design with Geogrid Layers

A tremendous number of geogrid reinforced walls have been constructed in the past 10 years (Koerner, 1999). The types of permanent geogrid reinforced wall facings are as follows (Koerner, 1999):

1. *Articulated precast panels* are discrete precast concrete panels with inserts for attaching the geogrid.

2. *Full height precast panels* are concrete panels temporarily supported until backfill is complete.

3. *Cast-in-place concrete panels* are often wrap-around walls that are allowed to settle and, after 1/2 to 2 years, are covered with a cast-in-place facing panel.

4. *Masonry block facing walls* are an exploding segment of the industry with many different types currently available, all of which have the geogrid embedded between the blocks and held by pins, nubs, and/or friction.

5. *Gabion facings* are polymer or steel-wire baskets filled with stone, having a geogrid held between the baskets and fixed with rings and/or friction.

The frictional resistance offered by a geogrid against pullout may be expressed as (Koerner, 1999)

$$F_R = 2C_i C_r L_e p_o \tan\phi \ge TF_s \tag{26.24}$$

where $\quad C_i$ = interaction coefficient = 0.75 (may vary)

$\qquad C_r$ = coverage ratio = 0.8 (may vary)

All the other notations are already defined. The spacing of geogrid layers may be obtained from

$$h = \frac{T_a C_r}{p_h} \tag{26.25}$$

where p_h = lateral pressure per unit length of wall.

26.6   EXTERNAL STABILITY

The MSE wall system consists of three zones. They are:

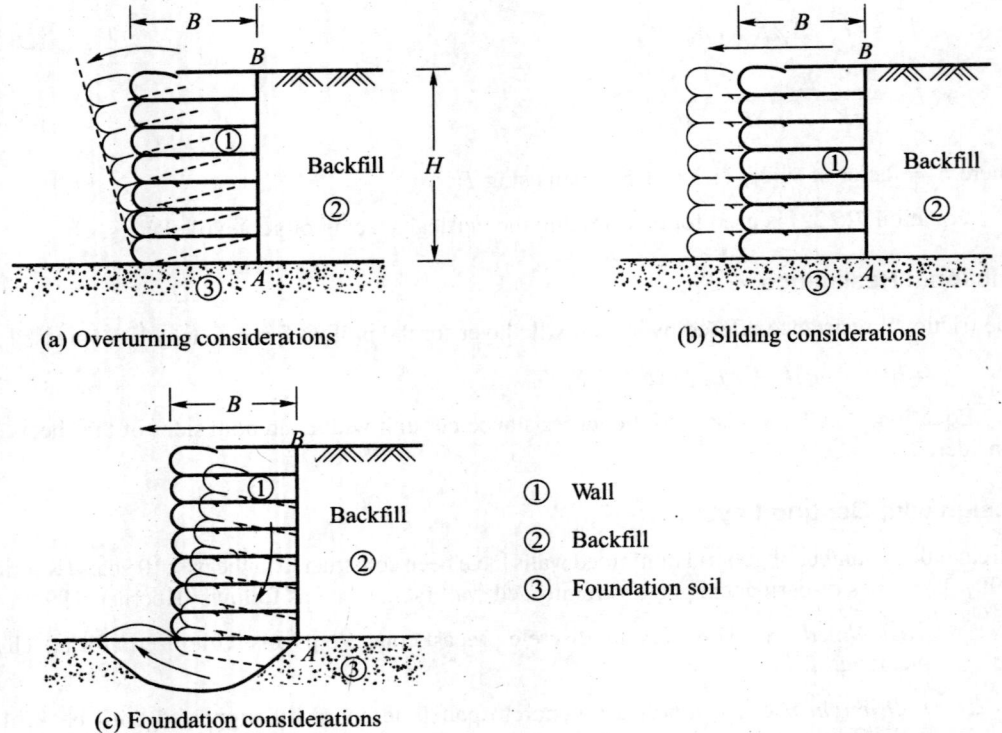

(a) Overturning considerations

(b) Sliding considerations

(c) Foundation considerations

① Wall

② Backfill

③ Foundation soil

Fig. 26.6 External stability considerations for reinforced earth walls

1. The reinforced earth zone.
2. The backfill zone.
3. The foundation soil zone.

The reinforced earth zone is considered as the wall for checking the internal stability whereas all three zones are considered for checking the external stability. The soils of the first two zones are placed in layers and compacted whereas the foundation soil is a normal one. The properties of the soil in each of the zones may be the same or different. However, the soil in the first two zones is normally a free draining material such as sand.

It is necessary to check the reinforced earth wall (width = B) for external stability which includes overturning, sliding and bearing capacity failure. These are illustrated in Fig. 26.6. Active earth pressure of the backfill acting on the internal face AB of the wall is taken in the stability analysis. The resultant earth thrust P_a is assumed to act horizontally at a height $H/3$ above the base of the wall. The methods of analysis are the same as for concrete retaining walls.

Example 26.1

A typical section of a retaining wall with the backfill reinforced with metal strips is shown in Fig. Ex. 26.1. The following data are available:

Height $H = 9$ m; $b = 100$ mm; $t = 5$ mm; $f_y = 240$ MPa; F_s for steel = 1.67; F_s on soil friction = 1.5; $\phi = 36°$; $\gamma = 17.5$ kN/m³; $\phi = 25°$; $h \times s = 1 \times 1$ m.

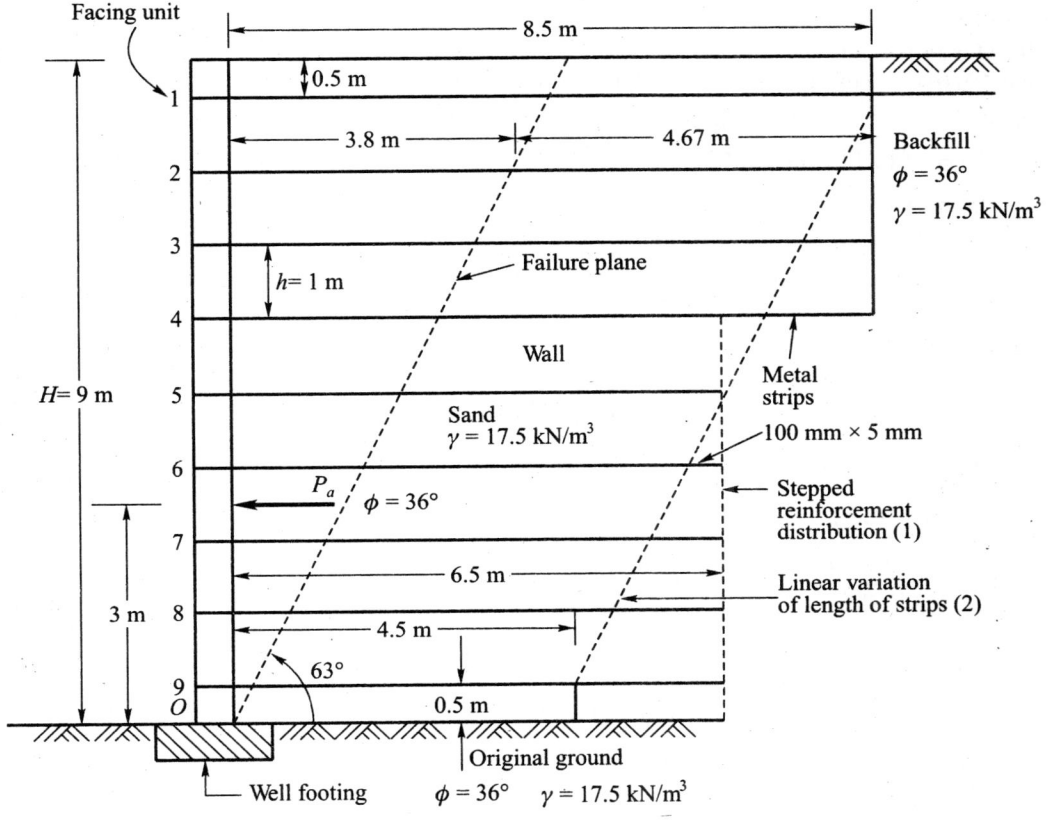

Fig. Ex. 26.1

Required:

(a)   Lengths L and L_e at varying depths.

(b)   The largest tension T in the strip.

(c)   The allowable tension in the strip.

(d)   Check for external stability.

Solution

From Eq. (26.13), the tension in a strip at depth z is

$$T = \gamma z K_A sh \text{ for } q_h = 0$$

where $\gamma = 17.5$ kN/m^3, $K_A = \tan^2(45° - 36/2) = 0.26$, $s = 1$m; $h = 1$ m.

Substituting

$T = 17.5 \times 0.26 \ (1) \ [1] \ z = 4.55z$ kN/strip.

$$L_e = \frac{F_s T}{2\gamma z b \tan\delta} = \frac{1.5 \times 4.55z}{2 \times 17.5 \times 0.1 \times 0.47 \times z} = 4.14 \text{ m}$$

This shows that the length $L_e = 4.14$ m is a constant with depth. Fig. Ex. 26.1 shows the positions of L_e for strip numbers 1, 2 ... 9. The first strip is located 0.5 m below the backfill surface and the 9th at 8.5 m below with spacings at 1 m apart. Tension in each of the strips may be obtained by using the equation $T = 4.55$ z. The total tension ΣT as computed is

$\Sigma T = 184.29$ kN/m since $s = 1$ m.

As a check the total active earth pressure is

$$P_a = \frac{1}{2}\gamma H^2 K_A = \frac{1}{2}\times 17.5 \times 9^2 \times 0.26 = 184.28 \text{ kN/m} = \Sigma T$$

The maximum tension is in the 9th strip, that is, at a depth of 8.5 m below the backfill surface. Hence

$$T = \gamma z K_A sh = 17.5 \times 8.5 \times 0.26 \times 1 \times 1 = 38.68 \text{ kN/strip}$$

The allowable tension is

$$T_a = f_a tb$$

where $f_a = \dfrac{240 \times 10^3}{1.67} = 143.7 \times 10^3 \text{ kN/m}^2$

Substituting $T_a = 143.7 \times 10^3 \times 0.005 \times 0.1 = 72 \text{ kN} > T - \text{OK}$.

The total length of strip L at any depth z is

$$L = L_R + L_e = (H - z)\tan(45 - \phi/2) + 4.14 = 0.51(9 - z) + 4.14 \text{ m}$$

where $H = 9$ m.

The lengths as calculated have been shown in Fig. Ex. 26.1. It is sometimes convenient to use the same length L with depth or stepped in two or more blocks or use a linear variation as shown in the figure.

Check for External Stability

Check of bearing capacity

It is necessary to check the base of the wall with the backfill for the bearing capacity per unit length of the wall. The width of the wall may be taken as equal to 4.5 m (Fig. Ex. 26.1). The procedure as explained in Chapter 18 may be followed. For all practical purposes, the shape, depth, and inclination factors may be taken as equal to 1.

Check for sliding resistance

$$F_s = \frac{\text{Sliding resistance } F_R}{\text{Driving force } P_a}$$

where $F_R = W \tan\delta = \dfrac{4.5 + 8.5}{2} \times 17.5 \times 9 \tan 36°$

$= 1024 \times 0.73 = 744 \text{ kN}$

where $\delta = \phi = 36°$ for the foundation soil, and W = weight of the reinforced wall

$P_a = 184.28 \text{ kN}$

$$F_s = \frac{744}{184.28} = 4 > 1.5 - \text{-OK}$$

(a) Check for overturning

$$F_s = \frac{M_R}{M_o}$$

From Fig. Ex. 26.1 taking moments of all forces about O, we have

$$M_R = 4.5 \times 9 \times 17.5 \times \frac{4.5}{2} + \frac{1}{2} \times 9 \times (8.5 - 4.5)(4.5 + \frac{4}{3}) \times 17.5$$

$$= 1595 + 1837 = 3432 \text{ kN - m}$$

$$M_o = P_a \times \frac{H}{3} = 184.28 \times \frac{9}{3} = 553 \text{ kN - m}$$

$$F_s = \frac{3432}{553} = 6.2 > 2 - \text{OK}.$$

Example 26.2

A section of a retaining wall with a reinforced backfill is shown in Fig. Ex. 26.2. The backfill surface is subjected to a surcharge of 30 kN/m². Required:

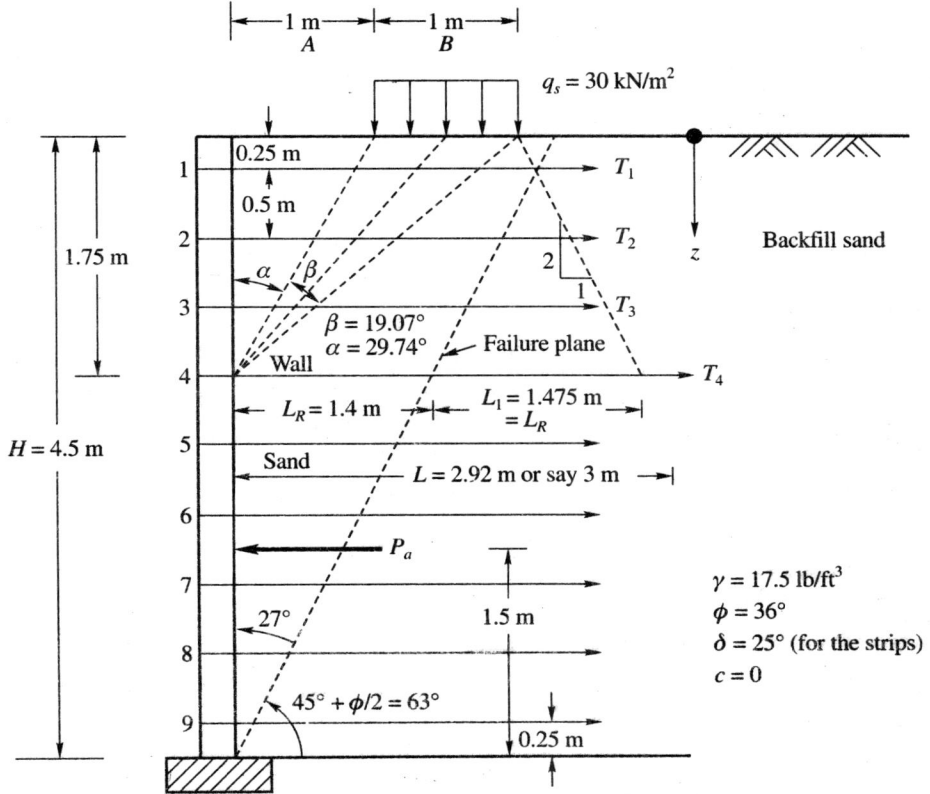

Fig. Ex. 26.2

(a)   The reinforcement distribution.

(b)   The maximum tension in the strip.

(c)   Check for external stability.

Given: $b = 100$ mm, $t = 5$ mm, $f_a = 143.7$ MPa, $c = 0$, $\phi = 36°$, $\delta = 25°$, $\gamma = 17.5$ kN/m^3, $s = 0.5$ m, and $h = 0.5$ m.

Solution

From Eq. (26.13)

$$T = (\gamma z K_A + q_h)h \times s = (p_o + q_h)A_c$$

where $\gamma = 17.5$ kN/m^3, $K_A = 0.26$, $A_c = h \times s = (0.5 \times 0.5)$ m^2

From Eq. (26.6)

$$q_h = \frac{2q_s}{\pi}[\beta - \sin\beta \cos2\alpha]$$

Refer to Fig. Ex. 26.2 for the definition of α and β.

$q_s = 30$ kN/m^2

The procedure for calculating length L of the strip for one depth $z = 1.75$ m (strip number 4) is explained below. The same method is valid for the other strips.

Strip No. 4. Depth $z = 1.75$ m

$$p_a = \gamma z\, K_A = 17.5 \times 1.75 \times 0.26 = 7.96 \text{ kN/m}^2$$

From Fig. Ex. 26.2, $\beta = 19.07° = 0.3327$ radians

$$\alpha = 29.74°$$

$$q_s = 30 \text{ kN/m}^2$$

$$q_h = \frac{2q_s}{\pi}[\beta - \sin\beta \cos2\alpha] = \frac{2 \times 30}{3.14}[0.3327 - \sin 19.07° \cos 59.5°] = 3.19 \text{ kN/m}^2$$

Figure Ex. 26.2 shows the surcharge distribution at a 2 (vertical) to 1 (horizontal) slope. Per the figure at depth $z = 1.75$ m, $L_1 = 1.475$ m from the failure line and $L_R = (H - z) \tan(45° - \phi/2) = 2.75 \tan(45° - 36°/2) = 1.4$m from the wall to the failure line. It is now necessary to determine L_2 (Refer to Fig. 26.4a).

Now $T = (7.96 + 3.19) \times 0.5 \times 0.5 = 2.79$ kN/strip.

The equation for the frictional resistance per strip is

$$F_R = 2b(\gamma z + \Delta q)L_1 \tan\delta + (\gamma z L_2 \tan\delta)2b$$

From the 2:1 distribution Δq at $z = 1.75$ m is

$$\Delta q = \frac{Q}{B+z} = \frac{30 \times 1}{1+1.75} = 10.9 \text{ kN/m}^2$$

$$p_o = 17.5 \times 1.75 = 30.63 \text{ kN/m}^2$$

Hence $\bar{p}_o = 10.9 + 30.63 = 41.53 \text{ kN/m}^2$

Now equating frictional resistance F_R to tension in the strip with $F_s = 1.5$, we have

$F_R = 1.5\ T$. Given $b = 100$ mm; from Eq. (26.17)

$F_R = 2b\tan\delta(\bar{p}_o L_1 + p_o L_2) = 1.5T$

Substituting and taking $\delta = 25°$, we have

$2 \times 0.1 \times 0.47[41.53 \times 1.475 + 30.63L_2] = 1.5 \times 2.79$

Simplifying

$\qquad L_2 = -0.546 \text{ m} \approx 0$

Hence $\qquad L_e = L_1 + 0 = 1.475 \text{ m}$

$\qquad L = L_R + L_e = 1.4 + 1.475 = 2.875 \text{ m}$

L can be calculated in the same way at other depths.

Maximum tension T

The maximum tension is in strip number 9 at depth $z = 4.25$m

Allowable $T_a = f_a bt = 143.7 \times 10^3 \times 0.1 \times 0.005 = 71.85$ kN

$T = (\gamma z K_A + q_h)sh$

where $\gamma z K_A = 17.50 \times 4.25 \times 0.26 = 19.34 \text{ kN/m}^2$

$\qquad q_h = 0.89 \text{ kN/m}^2$ from equation for q_h at depth $z = 4.25$m.

Hence $T = (19.34 + 0.89) \times 1/2 \times 1/2 = 5.05$ kN/strip < 71.85 kN – OK

Example 26.3 (Koerner, 1999)

Figure Ex. 26.3 shows a section of a retaining wall with geotextile reinforcement. The wall is backfilled with a granular soil having $\gamma = 18 \text{ kN/m}^3$ and $\phi = 34°$.

A woven slit-film geotextile with warp (machine) direction ultimate wide-width strength of 50 kN/m and having $\delta = 24°$ (Table 26.2) is intended to be used in its construction.

The orientation of the geotextile is perpendicular to the wall face and the edges are to be overlapped to handle the weft direction. A factor of safety of 1.4 is to be used along with sitespecific reduction factors (Table 26.3).

Required:

(a)   Spacing of the individual layers of geotextile.

(b)   Determination of the length of the fabric layers.

(c)   Check the overlap.

(d)   Check for external stability.

The backfill surface carries a uniform surcharge dead load of 10 kN/m².

Solution

(a) The lateral pressure p_h at any depth z is expressed as

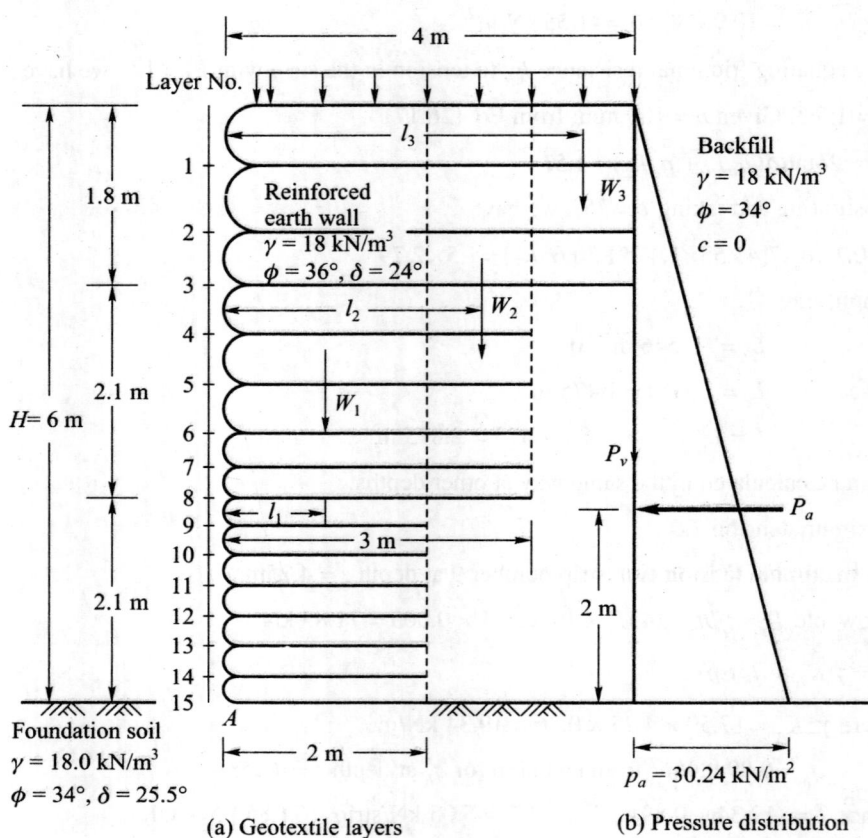

Layer No.

Fig. Ex. 26.3

(a) Geotextile layers (b) Pressure distribution

$p_h = p_a + q_h$

where $p_a = \gamma z K_A$, $q_h = q K_A$, $K_A = \tan^2(45° - 36/2) = 0.26$

Substituting

$p_h = 18 \times 0.26\ z + 0.26 \times 10 = 4.68\ z + 2.60$

From Eq. (26.1), the allowable geotextile strength is

$$T_a = T_u \left[\frac{1}{RF_{ID} \times RF_{CR} \times RF_{CD} \times RF_{BD}} \right]$$

$$= 50 \left[\frac{1}{1.2 \times 2.5 \times 1.15 \times 1.1} \right] = 13.2\ \text{kN/m}$$

From Eq. (26.13), the expression for allowable stress in the geotextile at any depth z may be expressed as

$T = T_a = p_h\ h\ F_s$

$$h = \frac{T_a}{p_h F_s}$$

where h = vertical spacing (lift thickness)

T_a = allowable stress in the geotextile

p_h = lateral earth pressure at depth z

F_s = factor of safety = 1.4

Now substituting

$$h = \frac{13.2}{[4.68(z)+2.60]1.4} = \frac{13.2}{6.55(z)+3.64}$$

At z = 6 m, $h = \dfrac{13.2}{6.55 \times 6 + 3.64} = 0.307$ m or say 0.30 m

At z = 3.3 m, $h = \dfrac{13.2}{6.55 \times 3.3 + 3.64} = 0.52$ m or say 0.50 m

At z = 1.3 m, $h = \dfrac{13.2}{6.55 \times 1.3 + 3.64} = 1.08$ m, but use 0.65 m for a suitable distribution.

The depth 3.3 m or 1.3 m are used just as a trial and error process to determine suitable spacings. Figure Ex. 26.3 shows the calculated spacings of the geotextiles.

(b) Length of the fabric layers

From Eq. (26.23) we may write

$$L_e = \frac{TF_s}{2\gamma z \tan\delta} = \frac{p_h h F_s}{2\gamma z \tan\delta} = \frac{h(4.68z + 2.06)1.4}{2 \times 18 z \tan 24°} = L_e = \frac{h(6.55z + 3.64)}{16z}$$

From Fig. (26.12) the expression for L_R is

$$L_R = (H - z)\tan(45° - \phi/2) = (H - z)\tan(45° - 36/2) = (6.0 - z)(0.509)$$

The total length L is

$$L = L_R + L_e$$

The computed L and suggested L are given in a tabular form below.

Layer No	Depth z (m)	Spacing h (m)	L_e (m)	L_e (min) (m)	L_R (m)	L (cal) (m)	L (suggested) (m)
1	0.65	0.65	0.49	1.0	2.72	3.72	4.0
2	1.30	0.65	0.38	1.0	2.39	3.39	–
3	1.80	0.50	0.27	1.0	2.14	3.14	–
4	2.30	0.50	0.26	1.0	1.88	2.88	3.0
5	2.80	0.50	0.25	1.0	1.63	2.63	–
6	3.30	0.50	0.24	1.0	1.37	2.37	–
7	3.60	0.30	0.14	1.0	1.22	2.22	–
8	3.90	0.30	0.14	1.0	1.07	2.07	–
9	4.20	0.30	0.14	1.0	0.92	1.92	2.0
10	4.50	0.30	0.14	1.0	0.76	1.76	–
11	4.80	0.30	0.14	1.0	0.61	1.61	–
12	5.10	0.30	0.14	1.0	0.46	1.46	–
13	5.40	0.30	0.14	1.0	0.31	1.31	–
14	5.70	0.30	0.14	1.0	0.15	1.15	–
15	6.00	0.30	0.13	1.0	0.00	1.00	–

It may be noted here that the calculated values of L_e are very small and a minimum value of 1.0 m should be used.

(c) Check for the overlap

When the fabric layers are laid perpendicular to the wall, the adjacent fabric should overlap a length L_o. The minimum value of L_o is 1.0m. The equation for L_o may be expressed as

$$L_o = \frac{hp_hF_s}{2\times 2\gamma z\tan\delta} = \frac{h[4.68(z)+2.60]1.4}{4\times 18(z)\tan 24°}$$

The maximum value of L_o is at the upper layer at $z = 0.65$. Substituting for z we have

$$L_o = \frac{0.65[4.68(0.65)+2.60]1.4}{4\times 18(0.65)\tan 24°} = 0.25 \text{ m}$$

Since this value of L_o calculated is quite low, use $L_o = 1.0$m for all the layers.

(d) Check for external stability

The total active earth pressure P_a is

$$P_a = \frac{1}{2}\gamma H°K_A = \frac{1}{2}\times 18\times 6^2\times 0.28 = 90.7 \text{ kN/m}$$

$$F_s = \frac{\text{Resisting moment } M_R}{\text{Driving moment } M_o} = \frac{W_1l_1 + W_2l_2 + W_3l_3 + P_vl_4}{P_a(H/3)}$$

where $W_1 = 6\times 2\times 18 = 216$ kN and $l_1 = 2/2 = 1$m

$W_2 = (6-2.1)\times(3-2)(18) = 70.2$ kN, and $l_2 = 2.5$m

$W_3 = (6-4.2)(4-3)(18) = 32.4$ kN and $l_3 = 3.5$m

$$F_s = \frac{213\times(1) + 70.2\times(2.5) + 32.4\times(3.5)}{90.7\times(2)} = 2.78 > 2 \text{ - OK}$$

Check for sliding

$$F_s = \frac{\text{Total resisting force } F_R}{\text{Total driving force } F_d}$$

$$F_R = (W_1 + W_2 + W_3)\tan\delta$$

$$= (216 + 70.2 + 32.4)\tan 25.5°$$

$$= 318.6\times 0.477 = 152 \text{ kN}$$

$$F_d = P_a = 90.7 \text{ kN}$$

Hence $F_s = \dfrac{152}{90.7} = 1.68 > 1.5 \text{ - OK}$

Check for a foundation failure

Consider the wall as a surface foundation with $D_f = 0$. Since the foundation soil is cohesionless, we may write

$$q_u = \frac{1}{2}\gamma B N_\gamma$$

Use Terzaghi's theory. For $\phi = 34°$, $N_\gamma = 38$, and $B = 2m$

$$q_u = \frac{1}{2}\times 18\times 2\times 38 = 684 \text{ kN/m}^2$$

The actual load intensity on the base of the backfill
q(actual) $= 18 \times 6 + 10 = 118 \text{ kN/m}^2$

$$F_s = \frac{684}{118} = 5.8 > 3 \text{ which is acceptable}$$

Example 26.4 (Koerner, 1999)

Design a 7m high geogrid-reinforced wall when the reinforcement vertical maximum spacing must be 1.0 m. The coverage ratio is 0.80 (Refer to Fig. Ex. 26.4). Given: $T_u = 156 \text{ kN/m}$, $C_r = 0.80$, $C_i = 0.75$. The other details are given in the figure.

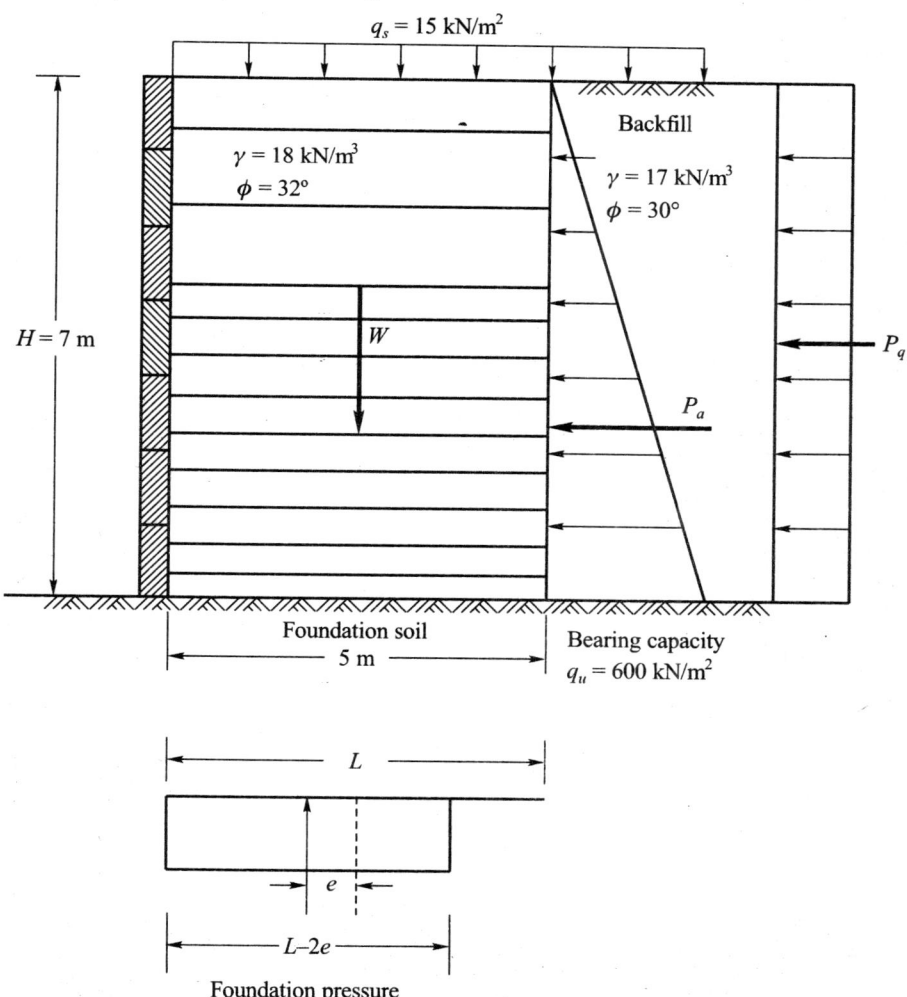

Fig. Ex. 26.4

Solution

Internal Stability

From Eq. (26.8)

$$p_h = (\gamma z K_A + q_h) = \gamma z K_A + q_s K_A$$

$$K_A = \tan^2(45° - \phi/2) = \tan^2(45° - 32/2) = 0.31$$

$$p_h = (18 \times z \times 0.31) + (15 \times 0.31) = 5.58z + 4.65$$

1. For geogrid vertical spacing.

Given $T_u = 156$ kN/m

From Eq. (26.2) and Table 26.4, we have

$$T_a = T_u \left[\frac{1}{RF_{ID} \times RF_{CR} \times RF_{BD} \times RF_{CD}} \right]$$

$$T_a = 156 \left[\frac{1}{1.2 \times 2.5 \times 1.3 \times 1.0} \right] = 40 \text{ kN/m}$$

But use $T_{design} = 28.6$ kN/m with $F_s = 1.4$ on T_a

From Eq. (26.25)

$$T_{design} = \frac{h p_h}{C_r}$$

$$28.6 = h \frac{5.58z + 4.65}{0.8}$$

or $h = \dfrac{22.9}{5.58z + 4.65}$

Maximum depth for $h = 1$m is

$$1.0 = \frac{22.9}{5.58z + 4.65} \text{ or } z = 3.27\text{m}$$

Maximum depth for $h = 0.5$m

$$0.5 = \frac{22.9}{5.58z + 4.65} \text{ or } z = 7.37\text{m}$$

The distribution of geogrid layers is shown in Fig. Ex. 26.4.

2. Embedment length of geogrid layers.

From Eqs (26.24) and (26.21)

$$2C_1 C_r L_e p_o \tan\phi = T_H F_s = p_h h F_s$$

Substituting known values

$$2 \times 0.75 \times 0.8 \times (L_e) \times 18 \times (z) \tan 32° = h(5.58z + 4.65) 1.5$$

Simplifying $L_e = \dfrac{(0.62z + 0.516)h}{z}$

The equation for L_R is

$$L_R = (H - z)\tan(45° - \phi/2) = (7 - z)\tan(45° - 32/2)$$

$$= 3.88 - 0.554(z)$$

From the above relationships the spacing of geogrid layers and their lengths are given below.

Layer No	Depth (m)	Spacing (m)	L_e (m)	L_e (min) (m)	L_R (m)	L (cal) (m)	L (required) (m)
1	0.75	0.75	0.98	1.0	3.46	4.46	5.0
2	1.75	1.00	0.92	1.0	2.91	3.91	5.0
3	2.75	1.00	0.81	1.0	2.36	3.36	5.0
4	3.25	0.50	0.39	1.0	2.08	3.08	5.0
5	3.75	0.50	0.38	1.0	1.80	2.80	5.0
6	4.25	0.50	0.37	1.0	1.52	2.52	5.0
7	4.75	0.50	0.36	1.0	1.25	2.25	5.0
8	5.25	0.50	0.36	1.0	0.97	1.97	5.0
9	5.75	0.50	0.36	1.0	0.69	1.69	5.0
10	6.25	0.50	0.35	1.0	0.42	1.42	5.0
11	6.75	0.50	0.35	1.0	0.14	1.14	5.0

External Stability

(a) Pressure distribution

$$P_a = \frac{1}{2}\gamma H^2 K_A = \frac{1}{2}\times 17 \times 7^2 \tan(45° - 30/2) = 138.8 \text{ kN/m}$$

$$P_q = q_s K_A H = 15 \times 0.33 \times 7 = 34.7 \text{ kN/m}$$

Total ≈ 173.5 kN/m

1. Check for sliding (neglecting effect of surcharge)

$$F_R = W\tan\delta = \gamma \times H \times L\tan 25° = 18 \times 7 \times 5.0 \times 0.47 = 293.8 \text{ kN/m}$$

$$P = P_a + P_q = 173.5 \text{ kN/m}$$

$$F_s = \frac{293.8}{173.5} = 1.69 > 1.5 \quad \text{OK}$$

2. Check for overturning

Resisting moment $\quad M_R = W \times \dfrac{L}{2} = 18 \times 7 \times 5 \times \dfrac{5}{2} = 1575 \text{ kN-m}$

Overturning moment $\quad M_O = P_a \times \dfrac{H}{3} + P_q \times \dfrac{H}{2}$

$$\text{or} \quad M_O = 138.8 \times \frac{7}{3} + 34.7 \times \frac{7}{2} = 445.3 \text{ kN-m}$$

$$F_s = \frac{1575}{445.3} = 3.54 > 2.0 \text{ OK}$$

3. Check for bearing capacity

$$\text{Eccentricity } e = \frac{M_O}{W + q_s L} = \frac{445.3}{18 \times 7 \times 5 + 15 \times 5} = 0.63$$

$$e = 0.63 < \frac{L}{6} = \frac{5}{6} = 0.83 \text{ OK}$$

Effective length $= L - 2e = 5 - 2 \times 0.63 = 3.74\text{m}$

$$\text{Bearing pressure } = [18 \times 74 + 15]\left(\frac{5}{3.74}\right) = 189 \text{ kN/m}^2$$

$$F_s = \frac{600}{189} = 3.17 > 3.0 \text{ OK}$$

26.7 EXAMPLES OF MEASURED LATERAL EARTH PRESSURES

Backfill Reinforced with Metal Strips

Laboratory tests were conducted on retaining walls with backfills reinforced with metal strips (Lee *et al.*, 1973). The walls were built within a 30 in. × 48 in. × 2 in. wooden box. Skin elements were made from 0.012 in aluminum sheet. The strips (ties) used for the tests were 0.155 in wide and 0.0005 in thick aluminum foil. The backfill consisted of dry Ottawa No. 90 sand. The small walls built of these materials in the laboratory were constructed in much the same way as the larger walls in the field. Two

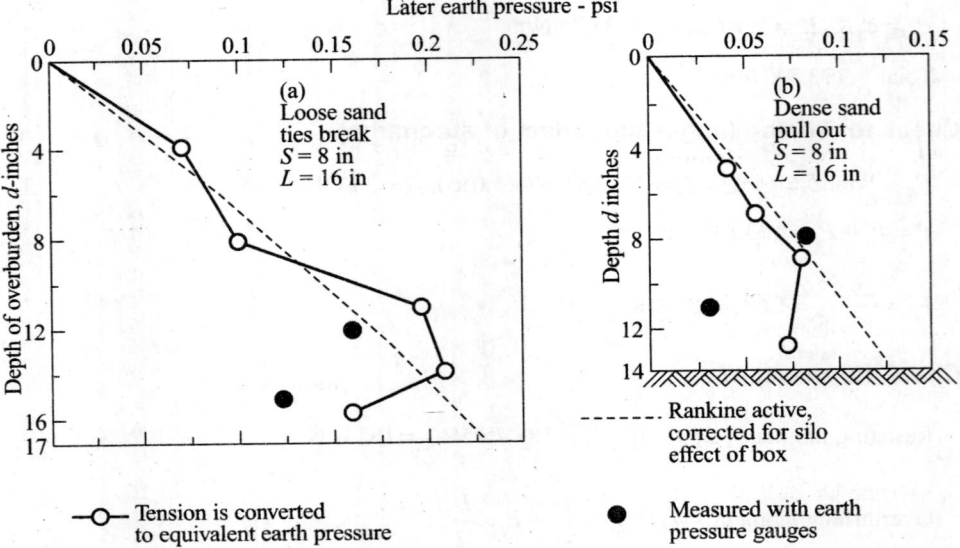

Fig. 26.7 Typical examples of measured lateral earth pressures just prior to wall failure
(1 in. = 25.4 mm; 1 psi = 6.89 kN/m²) (Lee et al., 1973)

different sand densities were used: loose, corresponding to a relative density, $D_r = 20\%$, and medium dense, corresponding to $D_r = 63\%$, and the corresponding angles of internal friction were 31° and 44° respectively. SR-4 strain gages were used on the ties to determine tensile stresses in the ties during the tests.

Examples of the type of earth pressure data obtained from two typical tests are shown in Fig. 26.7. Data in Fig. 26.7(a) refer to a typical test in loose sand whereas data in 26.7(b) refer to test in dense sand. The ties lengths were different for the two tests. For comparison, Rankine lateral earth pressure variation with depth is also shown. It may be seen from the curve that the measured values of the earth pressures follow closely the theoretical earth pressure variation up to two thirds of the wall height but fall off comparatively to lower values in the lower portion.

Field Study of Retaining Walls with Geogrid Reinforcement

Field studies of the behaviour of geotextile or geogrid reinforced permanent wall studies are fewer in number. Berg et al., (1986) reported the field behaviour of two walls with geogrid reinforcement. One wall in Tucson, Arizona, 4.6 m high, used a cumulative reduction factor of 2.6 on ultimate strength for allowable strength T_a and a value of 1.5 as a global factor of safety. The second wall was in Lithonia, Georgia, and was 6 m high. It used the same factors and design method. Fig. 26.8 presents the results for both the walls shortly after construction was complete. It may be noted that the horizontal pressures at various wall heights are overpredicted for each wall, that is, the wall designs that were used appear to be quite conservative.

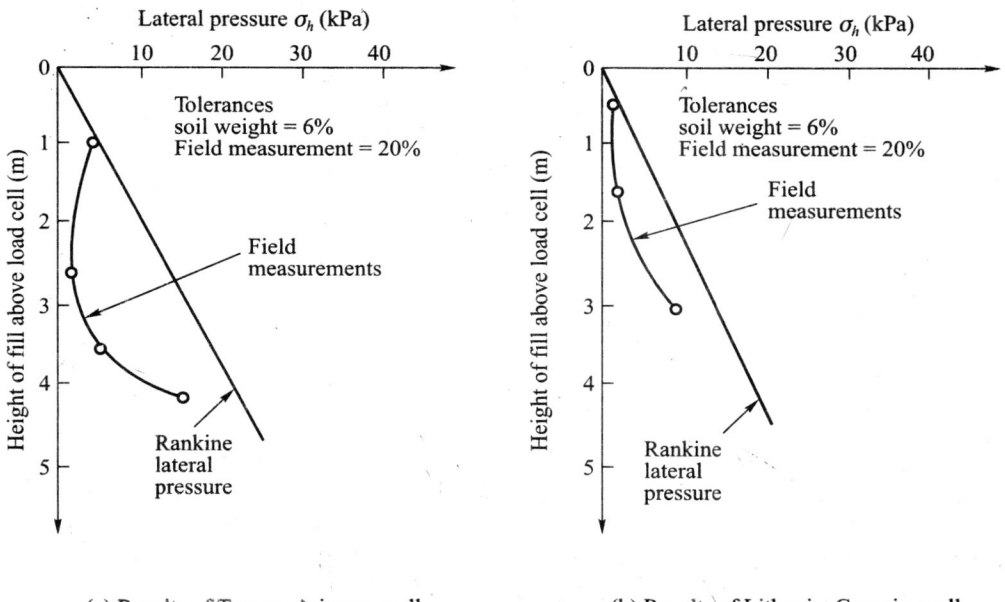

(a) Results of Tucson, Arizona, wall   (b) Results of Lithonia, Georgia, wall

Fig. 26.8   Comparison of measured stresses to design stresses for two geogrid reinforced walls (Berg *et al.*, 1986)

26.8 QUESTIONS AND PROBLEMS

26.1 A typical section of a wall with granular backfill reinforced with metal strips is given in Fig. Prob. 26.1. The following data are available.

$H = 6$ m, $b = 75$ mm, $t = 5$ mm, $f_y = 240$ MPa, F_s for steel $= 1.75$, F_s on soil friction $= 1.5$.

The other data are given in the figure. Spacing: $h = 0.6$ m, and $s = 1$ m.

Required

(a) Lengths of tie at varying depths

(b) Check for external stability

26.2 Solve the Prob. 26.1 with a uniform surcharge acting on the backfill surface. The intensity of surcharge is 20 kN/m^2.

26.3 Figure Prob. 26.2 shows a section of a MSE wall with geotextile reinforcement.

Required:

(a) Spacing of the individual layers of geotextile

(b) Length of geotextile in each layer

(c) Check for external stability

26.4 Design a 6 m high geogrid-reinforced wall (Fig. Prob. 26.4), where the reinforcement maximum spacing must be at 1.0 m. The coverage ratio $C_r = 0.8$ and the interaction coefficient $C_i = 0.75$, and $T_a = 26$ kN/m. (T_{design})

Given : Reinforced soil properties : $\gamma = 18$ kN/m^3 $\phi = 32°$

Foundation soil : $\gamma = 17.5$ kN/m^3 $\phi = 34°$

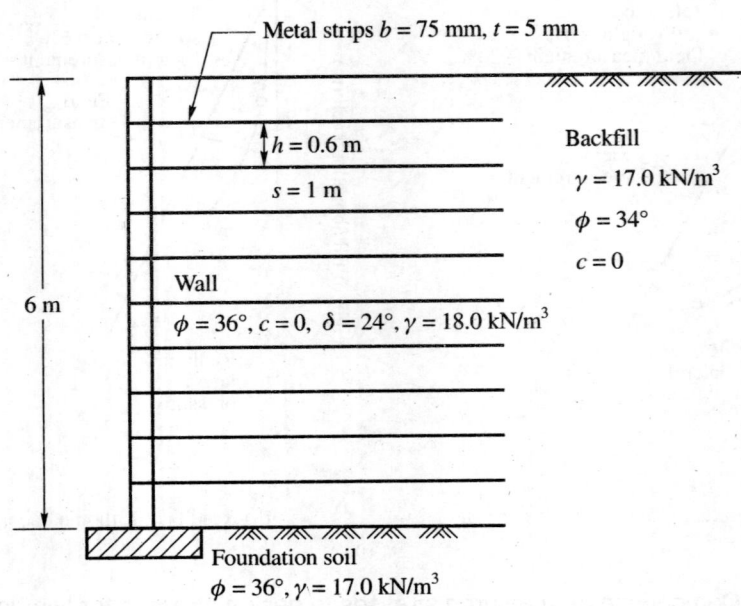

Metal strips $b = 75$ mm, $t = 5$ mm

$h = 0.6$ m

$s = 1$ m

Backfill
$\gamma = 17.0$ kN/m^3
$\phi = 34°$
$c = 0$

Wall
$\phi = 36°$, $c = 0$, $\delta = 24°$, $\gamma = 18.0$ kN/m^3

6 m

Foundation soil
$\phi = 36°$, $\gamma = 17.0$ kN/m^3

Fig. Prob. 26.1

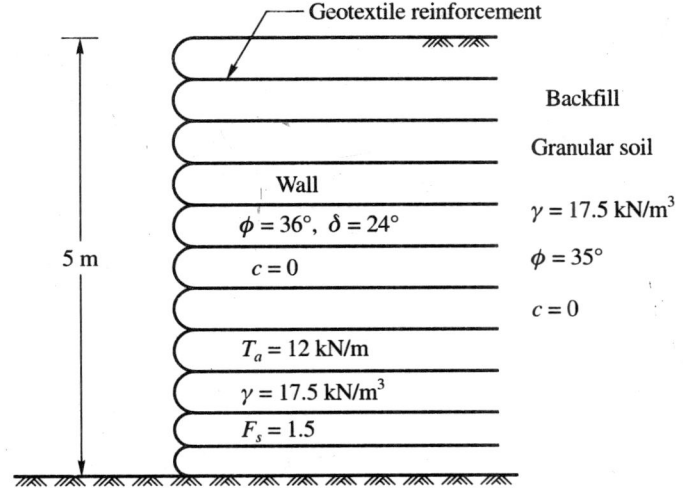

Foundation soil: $\gamma = 18.5$ kN/m^3, $\phi = 36°$.

Fig. Prob. 26.2

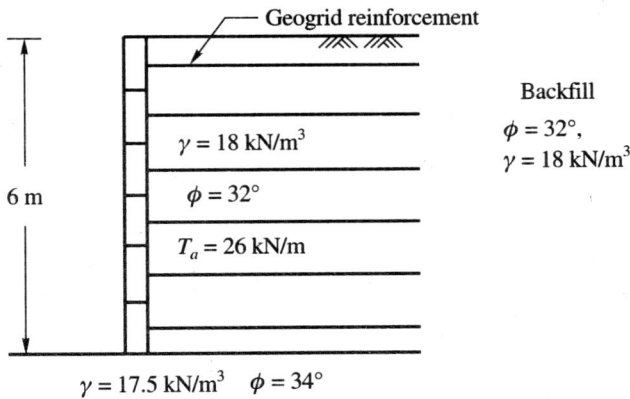

Fig. Prob. 26.4

APPENDIX A

SI UNITS IN GEOTECHNICAL ENGINEERING

Introduction

There has always been some confusion with regards to the system of units to be used in engineering practices and other commercial transactions. FPS (Foot-pound-second) and MKS (Metre-Kilogram-second) systems are still in use in many parts of the world. Sometimes a mixture of two or more systems are in vogue making the confusion all the greater. Though the SI (Le System International d'Unites or the International System of Units) units was first conceived and adopted in the year 1960 at the Eleventh General Conference of Weights and Measures held in Paris, the adoption of this coherent and systematically constituted system is still slow because of the past association with the FPS system. The conditions are now gradually changing and possibly in the near future the SI system will be the only system of use in all academic institutions in the world over. It is therefore essential to understand the basic philosophy of the SI units.

The Basics of the SI System

The SI system is a fully coherent and rationalized system. It consists of six basic units and two supplementary units, and several derived units. (Table A.1)

TABLE A.1

Basic units of interest in geotechnical engineering

	Quantity	Unit	SI symbol
1.	Length	Metre	m
2.	Mass	Kilogram	kg
3.	Time	Second	S
4.	Electric current	Ampere	A
5.	Thermodynamic temperature	Kelvin	K

TABLE A.2

Derived units

Quantity	Unit	SI symbol	Formula
acceleration	metre per second squared	–	m/sec^2
area	square metre	–	m^2
density	kilogram per cubic metre	–	kg/m^3
force	newton	N	$kg\text{-}m/s^2$
pressure	pascal	Pa	N/m^2
stress	pascal	Pa	N/m^2
moment or torque	newton-metre	N-m	$kg\text{-}m^2/s^2$
unit weight	newton per cubic metre	N/m^3	kg/s^2m^2
frequency	hertz	Hz	cycle/sec
volume	cubic metre	m^3	–
volume	litre	L	$10^{-3}m^3$
work (energy)	joule	J	N-m

Supplementary Units

The supplementary units include the *radian* and *steradian*, the units of plane and solid angles, respectively.

Derived Units

The derived units used by geotechnical engineers are tabulated in Table A.2.

Prefixes are used to indicate *multiples* and *submultiples* of the basic and derived units as given below.

Factor	Prefix	Symbol
10^6	mega	M
10^3	kilo	k
10^{-3}	milli	m
10^{-6}	micro	μ

Mass

Mass is a measure of the amount of matter an object contains. The mass remains the same even if the object's temperature and its location change. Kilogram, kg, is the unit used to measure the quantity of mass contained in an object. Sometimes *Mg (megagram)* and gram (*g*) are also used as a measure of mass in an object.

Time

Although the second (s) is the basic SI time unit, minutes (min), hours (h), days (d) etc. may be used as and where required.

Force

As per Newton's second law of motion, force, F, is expressed as $F = Ma$, where, $M =$ mass expressed in kg, and a is acceleration in units of m/sec^2. If the acceleration is g, the standard value of which is $9.80665 \ m/sec^2 \approx 9.81 \ m/s^2$, the force F will be replaced by W, the weight of the body. Now the above equation may be written as $W = Mg$.

The correct unit to express the weight W, of an object is the *newton* since the weight is the gravitational force that causes a downward acceleration of the object.

Newton, N, is defined as the force that causes a 1 kg mass to accelerate 1 m/s²

or $1N = 1\dfrac{kg \text{-} m}{s^2}$

Since, a *newton*, is too small a unit for engineering usage, multiples of newtons expressed as *kilonewton, kN,* and *meganewton, MN,* are used. Some of the useful relationships are

1 kilonewton, kN = 10^3 newton = 1000 N

meganewton, MN = 10^6 newton = 10^3 kN = 1000 kN

Stress and Pressure

The unit of *stress* and *pressure* in SI units is the *pascal (Pa)* which is equal to 1 newton per square metre (N/m^2). Since a *pascal* is too small a unit, multiples of pascals are used as *prefixes* to express the unit of stress and pressure. In engineering practice kilopascals or megapascals are normally used. For example,

1 kilopascal = 1 kPa = 1 kN/m² = 1000 N/m²

1 megapascal = 1 MPa = 1 MN/m² = 1000 kN/m²

Density

Density is defined as mass per unit volume. In the SI system of units, mass is expressed in kg/m^3. In many cases, it may be more convenient to express density in megagrams per cubic metre or in gm per cubic centimetre. The relationships may be expressed as

1 g/cm³ = 1000 kg/m³ = 10^6 g/m³ = 1 Mg/m³

It may be noted here that the density of water, ρ_w, is exactly 1.00 g/cm³ at 4 °C, and the variation is relatively small over the range of temperatures in ordinary engineering practice. It is sufficiently accurate to write

$\rho_w = 1.00$ g/cm³ $= 10^3$ kg/m³ $= 1$ Mg/m³

Unit weight

Unit weight is still the common measurement in geotechnical engineering practice. The relationship between unit weight, γ, and density ρ, may be expressed as $\gamma = \rho g$.

For example, if the density of water, $\rho_w = 1000$ kg/m³, then

$$\gamma_w = \rho_w g = 1000\frac{kg}{m^3} \times 9.81\frac{m}{s^2} = 9810\frac{kg}{m^3} \cdot \frac{m}{s^2}$$

Since $1N = 1\dfrac{Kg\,m}{s^2}$, $\gamma_w = 9810\dfrac{N}{m^3} = 9.81\,kN/m^3$

TABLE A.3

Conversion factors

To convert	SI to FPS			FPS to SI		
	From	To	Multiply by	From	To	Multiply by
Length	m	ft	3.281	ft	m	0.3048
	m	in	39.37	in	m	0.0254
	cm	in	0.3937	in	cm	2.54
	mm	in	0.03937	in	mm	25.4
Area	m^2	ft^2	10.764	ft^2	m^2	929.03×10^{-4}
	m^2	in^2	1550	in^2	m^2	6.452×10^{-4}
	cm^2	in^2	0.155	in^2	cm^2	6.452
	mm^2	in^2	0.155×10^{-2}	in^2	mm^2	645.16
Volume	m^3	ft^3	35.32	ft^3	m^3	28.317×10^{-3}
	m^3	in^3	61,023.4	in^3	m^3	16.387×10^{-6}
	cm^3	in^3	0.06102	in^3	cm^3	16.387
Force	N	lb	0.2248	lb	N	4.448
	kN	lb	224.8	lb	kN	4.448×10^{-3}
	kN	kip	0.2248	kip	kN	4.448
	kN	US ton	0.1124	US ton	kN	8.896
Stress	N/m^2	lb/ft^2	20.885×10^{-3}	lb/ft^2	N/m^2	47.88
	kN/m^2	lb/ft^2	20.885	lb/ft^2	kN/m^2	0.04788
	kN/m^2	$US ton/ft^2$	0.01044	$US ton/ft^2$	kN/m^2	95.76
	kN/m^2	kip/ft^2	20.885×10^{-3}	kip/ft^2	kN/m^2	47.88
	kN/m^2	lb/in^2	0.145	lb/in^2	kN/m^2	6.895
Unit weight	kN/m^3	lb/ft^3	6.361	lb/ft^3	kN/m^3	0.1572
	kN/m^3	lb/in^3	0.003682	lb/in^3	kN/m^3	271.43
Moment	N-m	lb-ft	0.7375	lb-ft	N-m	1.3558
	N-m	lb-in	8.851	lb-in	N-m	0.11298
Moment in inertia	mm^4	in^4	2.402×10^{-6}	in^4	mm^4	0.4162×10^6
	m^4	in^4	2.402×10^6	in^4	m^4	0.4162×10^{-6}
Section modulus	mm^3	in^3	6.102×10^{-5}	in^3	mm^3	0.16387×10^5
	m^3	in^3	6.102×10^4	in^3	m^3	0.16387×10^{-4}
Hydraulic conductivity	m/min	ft/min	3.281	ft/min	m/min	0.3048
	cm/min	ft/min	0.03281	ft/min	cm/min	30.48
	m/sec	ft/sec	3.281	ft/sec	m/sec	0.3048
	cm/sec	in/sec	0.3937	in/sec	cm/sec	2.54
Coefficient of consolidation	cm^2/sec	in^2/sec	0.155	in^2/sec	cm^2/sec	6.452
	$m^2/year$	in^2/sec	4.915×10^{-5}	in^2/sec	$m^2/year$	20.346×10^3
	cm^2/sec	ft^2/sec	1.0764×10^{-3}	ft^2/sec	cm^2/sec	929.03

TABLE A.4

Conversion factors—general

To convert from	To	Multiply by
Angstrom	inches	$3.9370079 \ 10^{-9}$
	feet	3.28084×10^{-10}
	units millimetres	1×10^{-7}
	centimetres	1×10^{-8}
	metres	1×10^{-10}
Microns	inches	3.9370079×10^{-5}
US gallon (gal)	cm^3	3785
	m^3	3.785×10^{-3}
	ft^3	0.133680
	litres	3.785
Pounds	dynes	4.44822×10^5
	grams	453.59243
	kilograms	0.45359243
Tons (short or US tons)	kilograms	907.1874
	pounds	2000
	kips	2
Tons (metric)	grams	1×10^6
	kilograms	1000
	pounds	2204.6223
	kips	2.2046223
	tons (short or US tons)	1.1023112
kips/ft²	lbs/in²	6.94445
	lbs/ft²	1000
	US tons/ft²	0.5000
	kg/cm²	0.488244
	metric ton/ft²	4.88244
Pounds/in³	gms/cm³	27.6799
	kg/m³	27679.905
	lbs/ft³	1728
Poise	kN-sec/m²	10^{-4}
	poise	10^{-3}
millipoise	kN-sec/m²	10^{-7}
	gm-sec/cm²	10^{-6}
ft/min	ft/day	1440
	ft/year	5256×10^2
ft/year	ft/min	1.9025×10^{-6}
cm/sec	m/min	0.600
	ft/min	1.9685
	ft/year	1034643.6

APPENDIX B

SLOPE STABILITY CHARTS AND TABLES

As per Eq.(16.43), the factor of safety F_s is defined as

$$F_s = m - nr_u$$

where, m, n = stability coefficients, and r_u = pore pressure ratio. The values of m and n may be obtained from Figs. B.1 to B.14

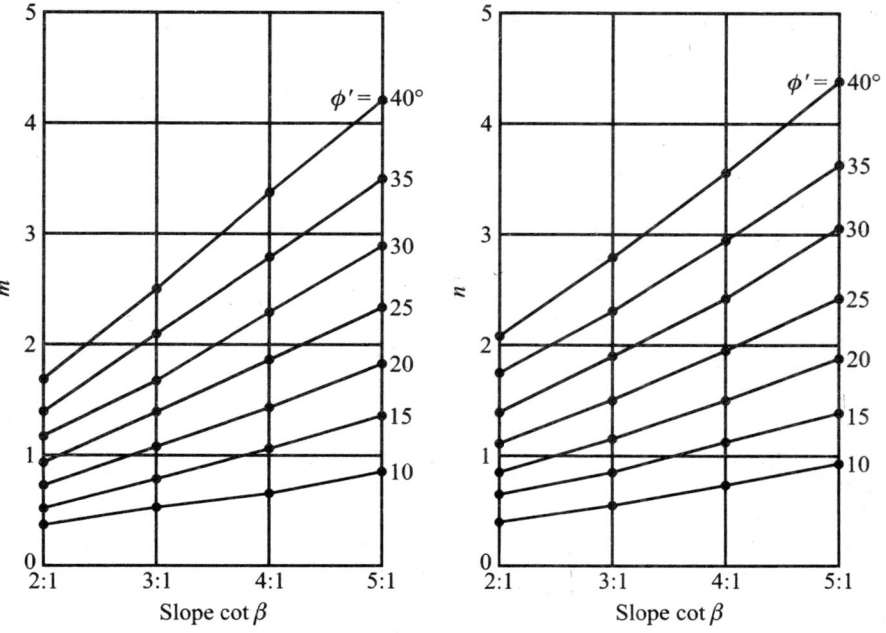

Fig. B.1   Stability coefficients m and n for $c'/\gamma H = 0$ (Bishop and Morgenstern, 1960)

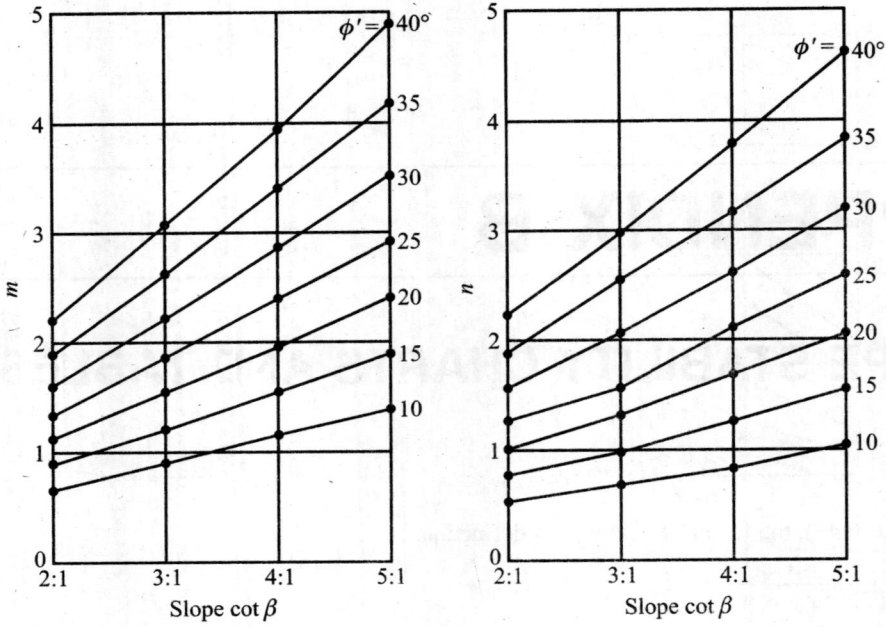

Fig. B.2   Stability coefficients for $c'/\gamma H = 0.025$ and $n_d = 1.00$
(Bishop and Morgenstern, 1960)

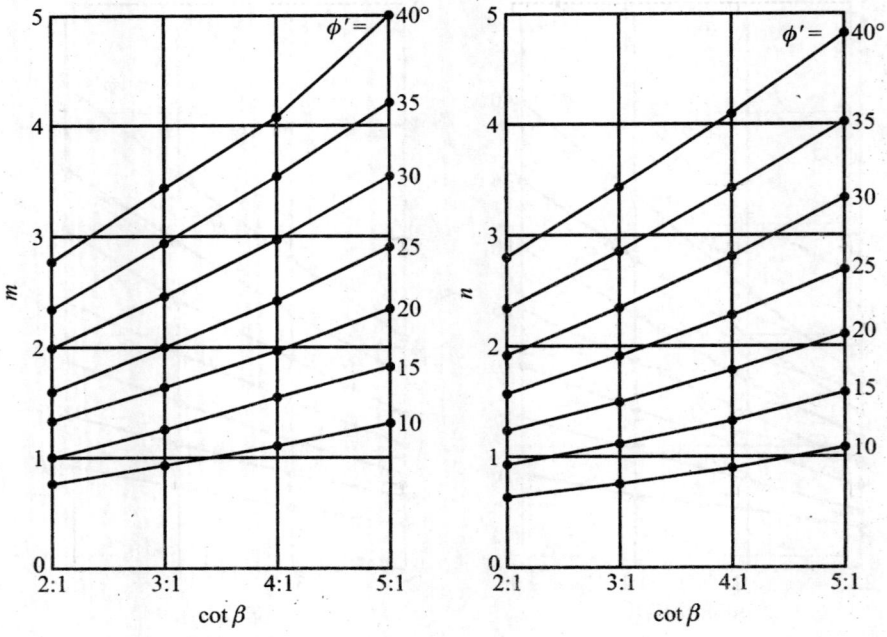

Fig. B.3   Stability coefficients m and n for $c'/\gamma H = 0.025$ and $n_d = 1.25$
(Bishop and Morgenstern, 1960)

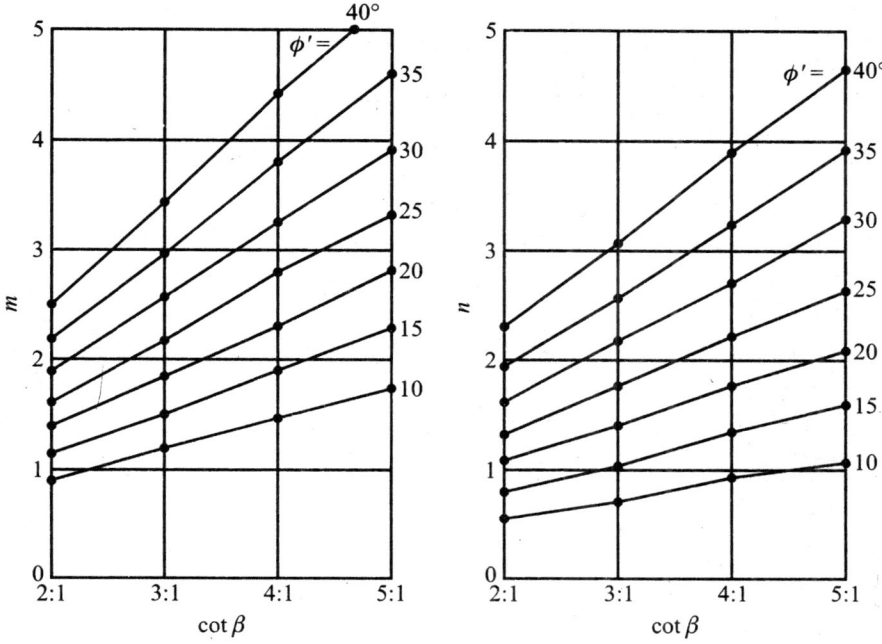

Fig. B.4   Stability coefficients m and n for $c'/\gamma H = 0.05$ and $n_d = 1.00$
(Bishop and Morgenstern, 1960)

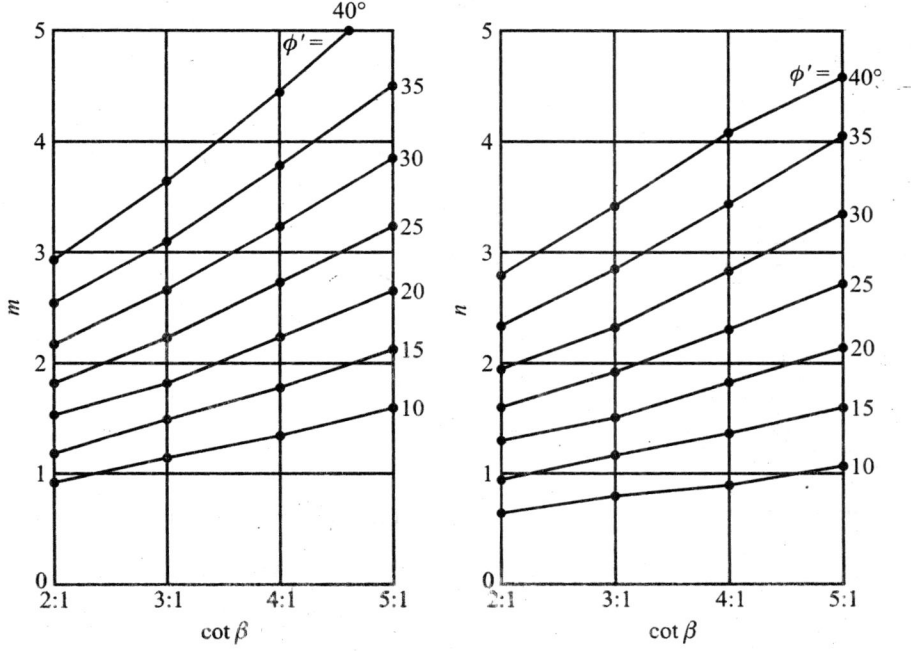

Fig. B.5   Stability coefficients m and n for $c'/\gamma H = 0.05$ and $n_d = 1.25$
(Bishop and Morgenstern, 1960)

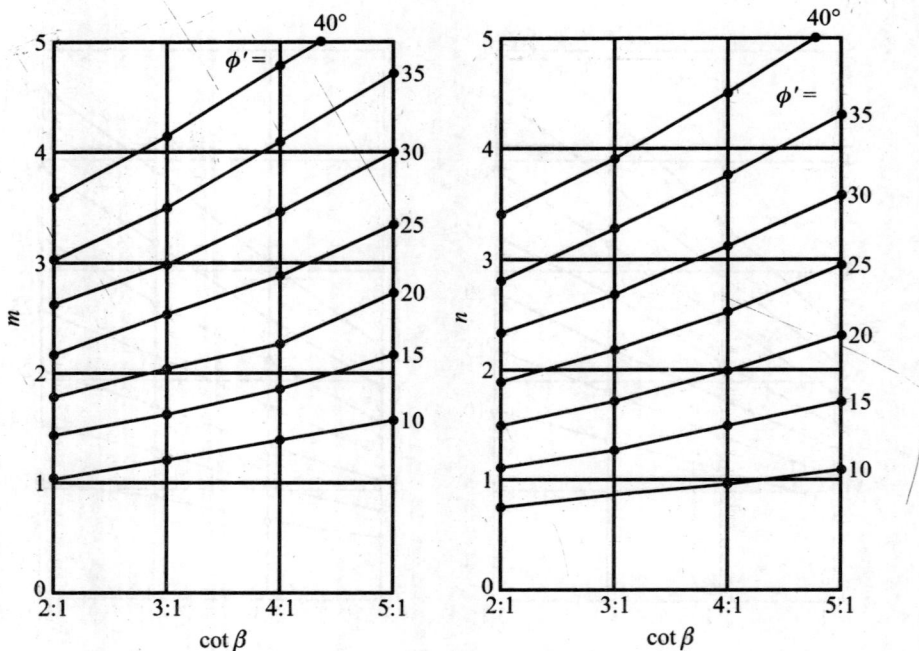

Fig. B.6   Stability coefficients m and n for $c'/\gamma H = 0.05$ and $n_d = 1.50$
(Bishop and Morgenstern, 1960)

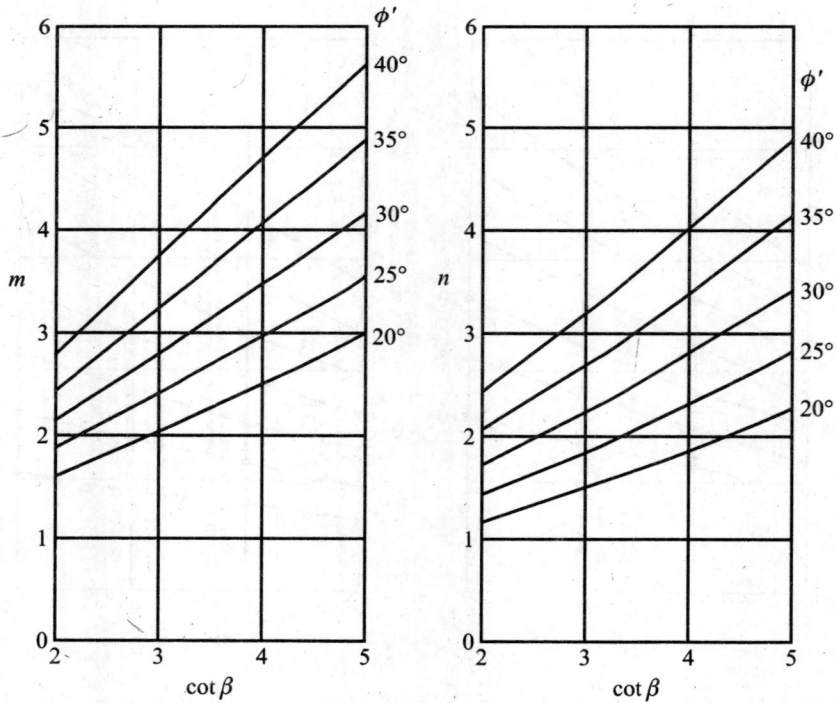

Fig. B.7   Stability coefficients m and n for $c'/\gamma H = 0.075$ toe circles
(O'Connor and Mitchell, 1977)

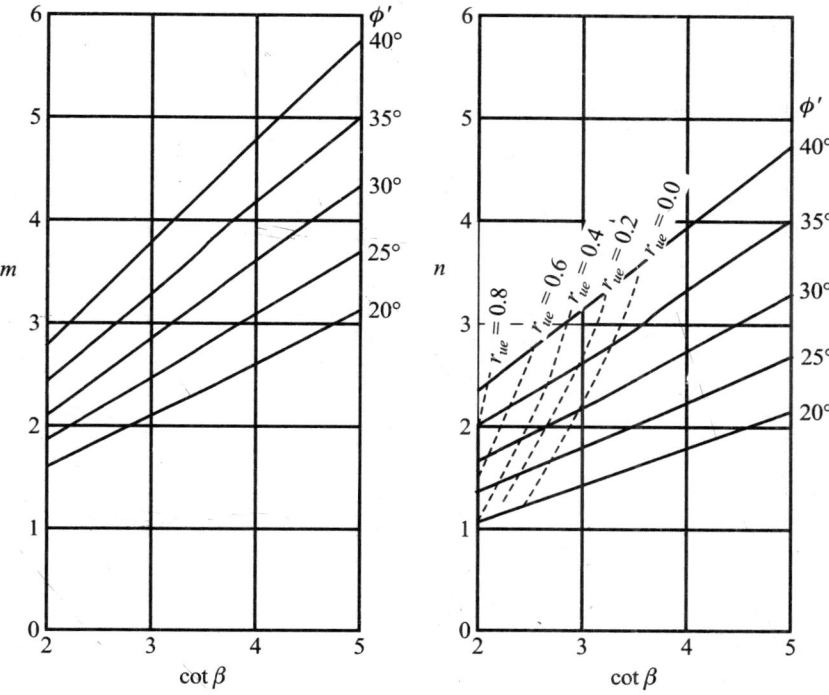

Fig. B.8 Stability coefficients m and n for $c'/\gamma H = 0.075$ and $n_d = 1.00$
(O'Connor et al., 1977)

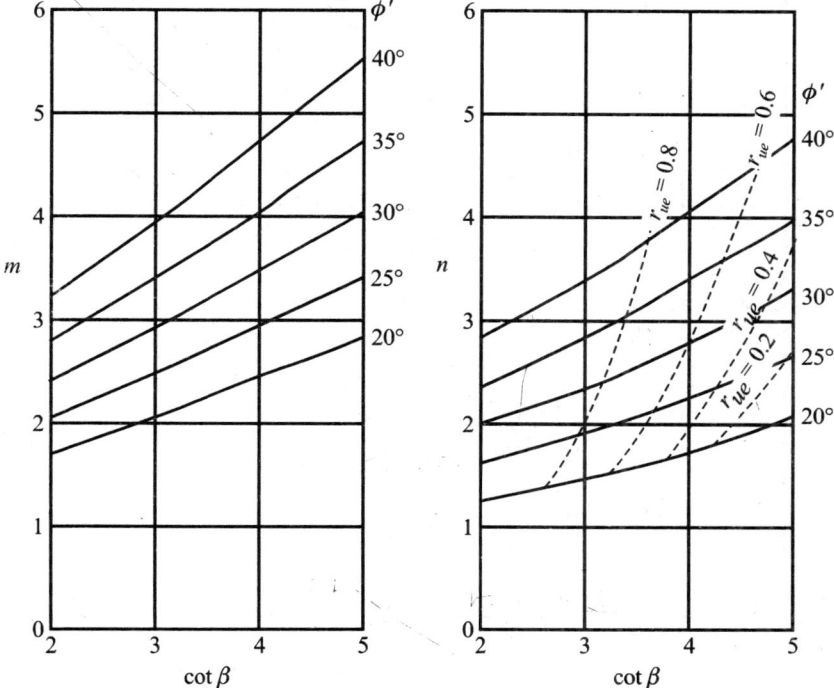

Fig. B.9 Stability coefficients m and n for $c'/\gamma H = 0.075$ and $n_d = 1.25$
(O'Connor and Mitchell, 1977)

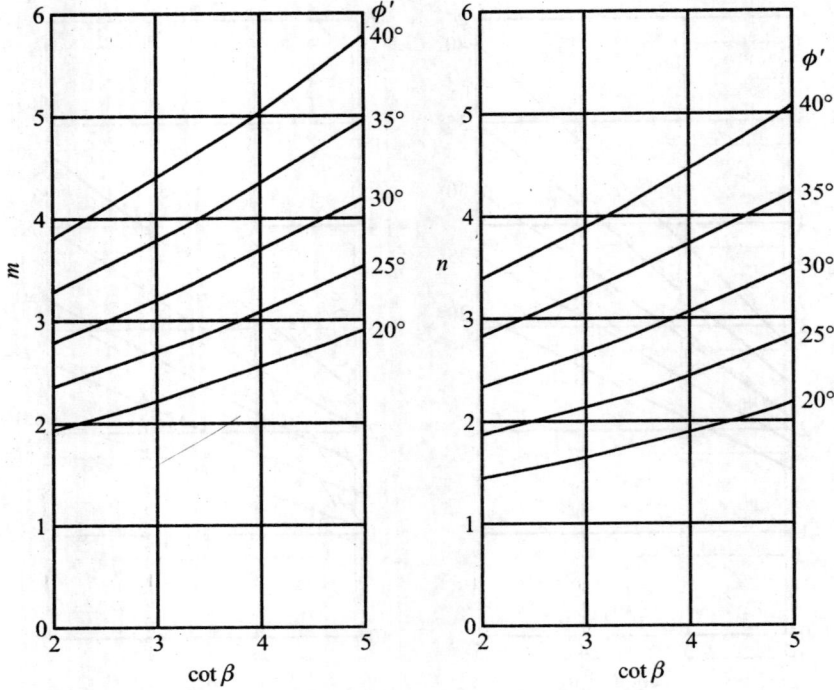

Fig. B.10   Stability coefficients m and n for $c'/\gamma H = 0.075$ and $n_d = 1.50$
(O'Connor and Mitchell, 1977)

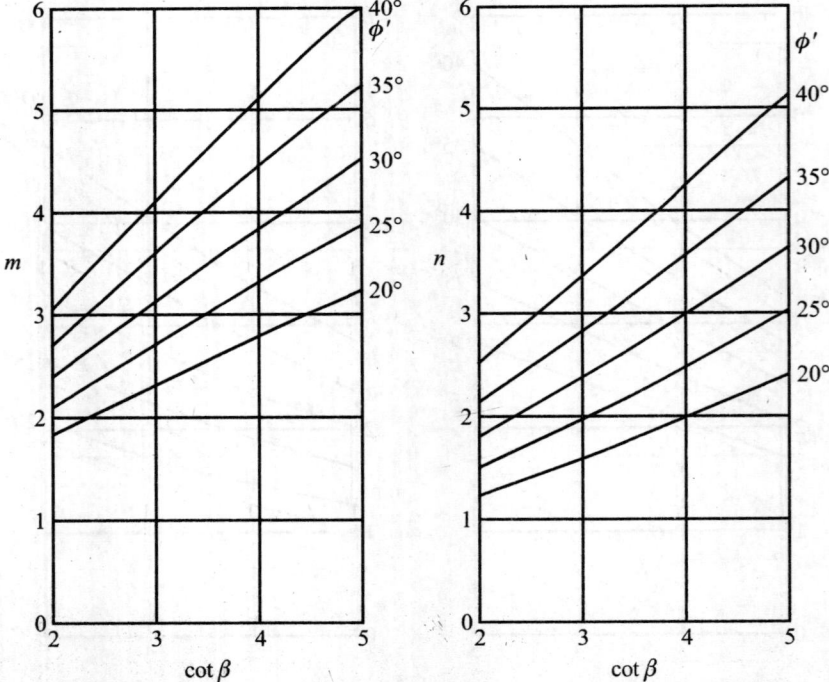

Fig. B.11   Stability coefficients m and n for $c'/\gamma H = 0.100$ toe circles
(O'Connor and Mitchell, 1977)

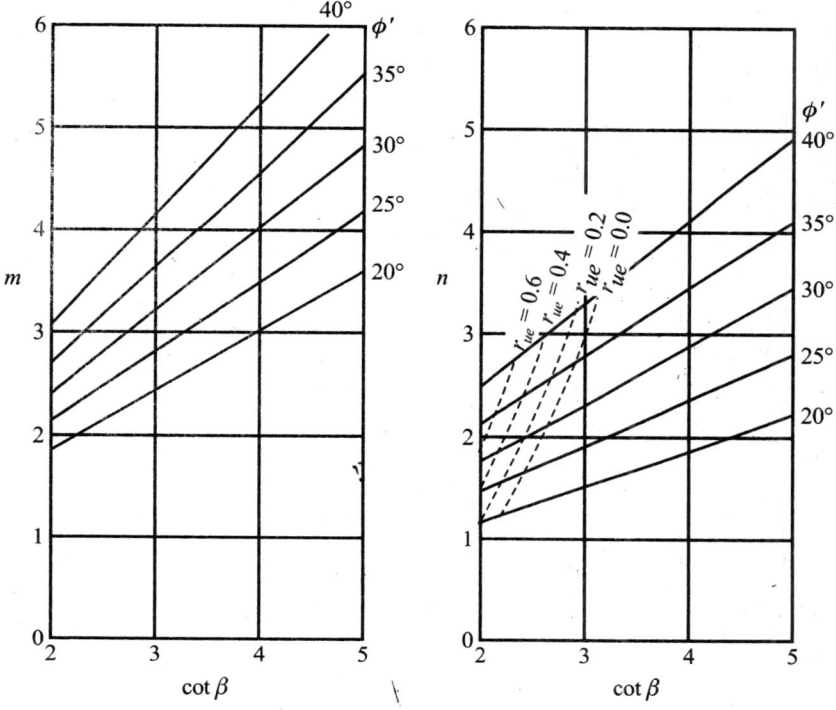

Fig. B.12 Stability coefficients m and n for $c'/\gamma H = 0.100$ and $n_d = 1.00$
(O'Connor and Mitchell, 1977)

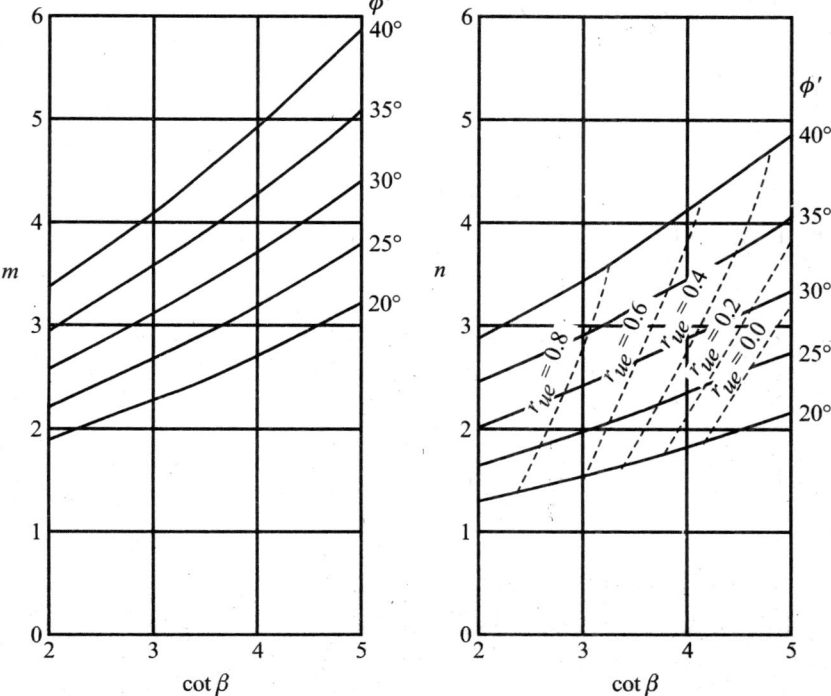

Fig. B.13 Stability coefficients m and n for $c'/\gamma H = 0.100$ and $n_d = 1.25$
(O'Connor and Mitchell, 1977)

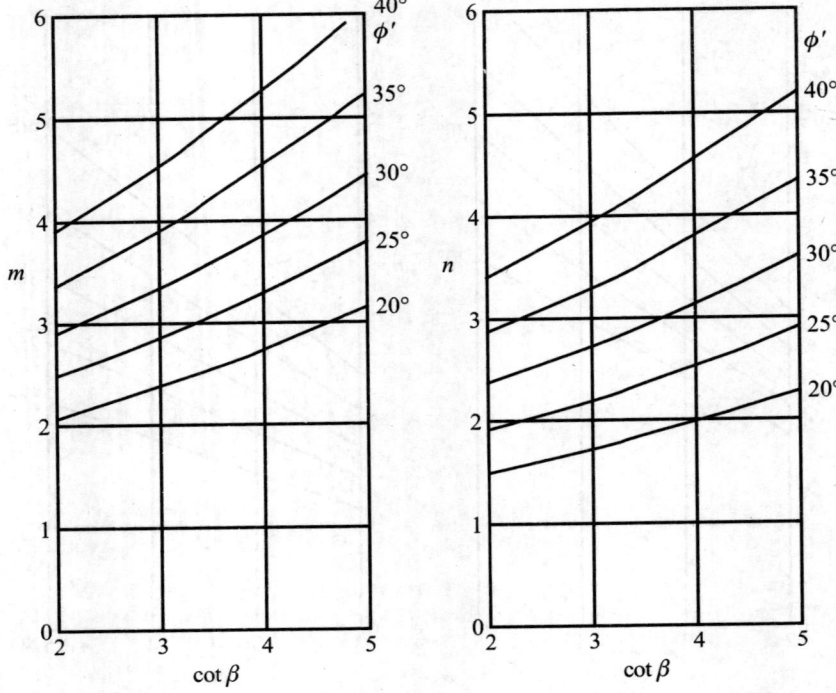

Fig. B.14   Stability coefficients m and n for $c'/\gamma H = 0.100$ and $n_d = 1.50$
(O'Connor and Mitchell, 1977)

Bishop and Morgenstern (1960) Stability Coefficients are Presented in Tabular Form

$$F_s = m - n.r_u$$

TABLE B.1

Stability coefficients m and n for $\dfrac{c'}{\gamma H} = 0$

	Stability coefficients for earth slopes							
ϕ'	Slope 2:1		Slope 3:1		Slope 4:1		Slope 5:1	
	m	m	m	m	n	n	n	n
10.0	0.353	0.441	0.529	0.588	0.705	0.749	0.882	0.917
12.5	0.443	0.554	0.665	0.739	0.887	0.943	1.109	1.153
15.0	0.536	0.670	0.804	0.893	1.072	1.139	1.340	1.393
17.5	0.631	0.789	0.946	1.051	1.261	1.340	1.577	1.639
20.0	0.728	0.910	1.092	1.213	1.456	1.547	1.820	1.892
22.5	0.828	1.035	1.243	1.381	1.657	1.761	2.071	2.153
25.0	0.933	1.166	1.399	1.554	1.865	1.982	2.332	2.424
27.5	1.041	1.031	1.562	1.736	2.082	2.213	2.603	2.706
30.0	1.155	1.444	1.732	1.924	2.309	2.454	2.887	3.001
32.5	1.274	1.593	1.911	2.123	2.548	2.708	3.185	3.311
35.0	1.400	1.750	2.101	2.334	2.801	2.977	3.501	3.639
37.5	1.535	1.919	2.302	2.588	3.069	3.261	3.837	3.989
40.0	1.678	2.098	2.517	2.797	3.356	3.566	4.196	4.362

TABLE B.2

Stability coefficients m and n for $\dfrac{c'}{\gamma H} = 0.025$ and $n_d = 1.00$

ϕ'	Slope 2:1		Slope 3:1		Slope 4:1		Slope 5:1	
	m	m	m	m	n	n	n	n
10.0	0.678	0.534	0.906	0.683	1.130	0.846	1.365	1.031
12.5	0.790	0.655	1.066	0.849	1.337	1.061	1.620	1.282
15.0	0.901	0.776	1.224	1.014	1.544	1.273	1.868	1.534
17.5	1.012	0.898	1.380	1.179	1.751	1.485	5.121	1.789
20.0	1.124	1.022	1.542	1.347	1.962	1.698	2.380	2.050
22.5	1.239	1.150	1.705	1.518	2.177	4.916	2.646	2.317
25.0	1.356	1.282	1.875	1.696	2.400	2.141	2.921	2.596
27.5	1.478	1.421	2.050	1.882	2.631	2.375	3.207	2.880
30.0	1.606	1.567	2.235	2.078	2.873	2.622	3.508	3.191
32.5	1.739	1.721	2.431	2.285	3.127	2.883	3.823	3.511
35.0	1.880	1.885	2.635	2.505	3.396	3.160	4.156	3.849
37.5	2.030	2.060	2.855	2.741	3.681	3.458	4.510	4.209
4.0	2.190	2.247	3.090	2.933	3.984	3.778	4.885	4.592

TABLE B.3

Stability coefficients m and n for $\dfrac{c'}{\gamma H} = 0.025$ and $n_d = 1.25$

ϕ'	Slope 2:1		Slope 3:1		Slope 4:1		Slope 5:1	
	m	m	m	m	n	n	n	n
10.0	0.737	0.614	0.901	0.726	1.085	0.867	1.285	1.014
12.5	0.878	0.759	1.076	0.908	1.299	1.089	1.543	1.278
15.0	1.019	0.907	1.253	1.093	1.515	1.312	1.803	1.545
17.5	1.162	1.059	1.433	1.282	1.736	1.541	2.065	1.814
20.0	1.309	1.216	1.618	1.478	1.961	1.775	2.344	2.090
22.5	1.461	1.379	1.808	1.680	2.194	2.017	2.610	2.373
25.0	1.619	1.547	2.007	1.891	2.437	2.269	2.897	2.669
27.5	1.783	1.728	2.213	2.111	2.689	2.531	3.196	2.976
30.0	1.956	1.915	2.431	2.342	2.943	2.806	3.511	2.299
32.5	2.139	2.112	2.659	2.588	3.231	3.095	3.841	2.638
35.0	2.331	2.321	2.901	2.841	3.524	3.404	4.191	3.998
37.5	2.536	2.541	3.158	3.112	3.835	3.723	4.563	4.379
40.0	2.753	2.775	3.431	3.399	4.164	4.064	4.988	4.784

TABLE B.4

Stability coefficients m and n for $\dfrac{c'}{\gamma H} = 0.05$ and $n_d = 1.00$

ϕ'	Slope 2:1		Slope 3:1		Slope 4:1		Slope 5:1	
	m	m	m	m	n	n	n	n
10.0	0.913	0.563	1.181	0.717	1.469	0.910	1.733	1.069
12.5	1.030	0.690	1.343	0.878	1.688	1.136	1.995	1.316
15.0	1.145	0.816	1.506	1.043	1.904	1.353	2.256	1.576
17.5	1.262	0.942	1.671	1.212	2.117	1.565	2.517	1.825
20.0	1.380	1.071	1.840	1.387	2.333	1.776	2.783	2.091
22.5	1.500	1.202	2.014	1.568	2.551	1.989	3.055	2.365
25.0	1.624	1.338	2.193	1.757	2.778	2.211	3.336	2.651
27.5	1.753	1.480	2.380	1.952	3.013	2.444	3.628	2.948
30.0	1.888	1.630	2.574	2.157	3.261	2.693	3.934	3.259
32.5	2.029	1.789	2.777	2.370	3.523	2.961	4.256	3.585
35.0	2.178	1.958	2.990	2.592	3.803	3.253	4.597	3.927
37.5	2.336	2.138	3.215	2.826	4.103	3.574	4.959	4.288
40.0	2.505	2.332	3.451	3.071	4.425	3.926	5.344	4.668

TABLE B.5

Stability coefficients m and n for $\dfrac{c'}{\gamma H} = 0.05$ and $n_d = 1.25$

ϕ'	Slope 2:1		Slope 3:1		Slope 4:1		Slope 5:1	
	m	m	m	m	n	n	n	n
10.0	0.919	0.633	1.119	0.766	1.344	0.886	1.594	1.042
12.5	1.065	0.792	1.294	0.941	1.563	1.112	1.850	1.300
15.0	1.211	0.950	1.471	1.119	1.782	1.338	2.109	1.562
17.5	1.359	1.108	1.650	1.303	2.004	1.567	2.373	1.831
20.0	1.509	1.266	1.834	1.493	2.230	1.799	2.643	2.107
22.5	1.663	1.428	2.024	1.690	2.463	2.038	2.921	2.392
25.0	1.822	1.595	2.222	1.897	2.705	2.287	3.211	2.690
27.5	1.988	1.769	2.428	2.113	2.957	2.546	3.513	2.999
30.0	2.161	1.950	2.645	2.342	3.221	2.819	3.829	3.324
32.5	2.343	2.141	2.873	2.583	3.500	3.107	4.161	3.665
35.0	2.535	2.344	3.114	2.839	3.795	3.413	4.511	4.025
37.5	2.738	2.560	3.370	3.111	4.109	3.740	4.881	4.405
40.0	2.953	2.791	3.642	3.400	4.442	4.090	5.273	4.806

TABLE B.6

Stability coefficients m and n for $\dfrac{c'}{\gamma H} = 0.05$ and $n_d = 1.50$

ϕ'	Slope 2:1		Slope 3:1		Slope 4:1		Slope 5:1	
	m	m	m	m	n	n	n	n
10.0	1.022	0.751	1.170	0.828	1.343	0.974	1.547	1.108
12.5	1.202	0.936	1.376	1.043	1.589	1.227	1.829	1.399
15.0	1.383	1.122	1.583	1.260	1.835	1.480	2.112	1.690
17.5	1.565	1.309	1.795	1.480	2.084	1.734	2.398	1.983
20.0	1.752	1.501	2.011	1.705	2.337	1.993	2.690	2.280
22.5	1.943	1.698	2.234	1.937	2.597	2.258	2.990	2.585
25.0	2.143	1.903	2.467	2.179	2.867	2.534	3.302	2.902
27.5	2.350	2.117	2.709	2.431	3.148	2.820	3.626	3.231
30.0	2.568	2.342	2.964	2.696	3.443	3.120	3.967	3.577
32.5	2.798	2.580	3.232	2.975	3.753	3.436	4.236	3.940
35.0	3.041	2.838	3.515	3.269	4.082	3.771	4.707	4.325
37.5	3.299	3.102	3.817	3.583	4.431	4.128	5.112	4.735
40.0	3.574	3.389	4.136	3.915	4.803	4.507	5.543	5.171

TABLE B.7

Stability coefficients m and n for $\dfrac{c'}{\gamma H} = 0.075$ and toe circle

ϕ'	Slope 2:1		Slope 3:1		Slope 4:1		Slope 5:1	
	m	m	m	m	n	n	n	n
20	1.593	1.158	2.055	1.516	2.498	1.903	2.934	2.301
25	1.853	1.430	2.426	1.888	2.980	2.361	3.520	2.861
30	2.133	1.730	2.826	2.288	3.496	2.888	4.150	3.461
35	2.433	2.058	3.253	2.730	4.055	3.445	4.846	4.159
40	2.773	2.430	3.737	3.231	4.680	4.061	5.609	4.918

TABLE B.8

Stability coefficients m and n for $\dfrac{c'}{\gamma H} = 0.075$ and $n_d = 1.00$

ϕ'	Slope 2:1		Slope 3:1		Slope 4:1		Slope 5:1	
	m	m	m	m	n	n	n	n
20	1.610	1.100	2.141	1.443	2.664	1.801	3.173	2.130
25	1.872	1.386	2.502	1.815	3.126	2.259	3.742	2.715
30	2.142	1.686	2.884	2.201	3.623	2.758	4.357	3.331
35	2.443	2.030	3.306	2.659	4.177	3.331	5.024	4.001
40	2.772	2.386	3.775	3.145	4.785	3.945	5.776	4.759

TABLE B.9

Stability coefficients m and n for $\dfrac{c'}{\gamma H} = 0.075$ and $n_d = 1.25$

ϕ'	Slope 2:1		Slope 3:1		Slope 4:1		Slope 5:1	
	m	m	m	m	n	n	n	n
20	1.688	1.285	2.071	1.543	2.492	1.815	2.954	2.173
25	2.004	1.641	2.469	1.975	2.792	2.315	3.523	2.730
30	2.352	2.015	2.888	2.385	3.499	2.857	4.149	3.357
35	2.782	2.385	3.357	2.870	4.079	3.457	4.831	4.043
40	3.154	2.841	3.889	3.428	4.729	4.128	5.063	4.830

TABLE B.10

Stability coefficients m and n for $\dfrac{c'}{\gamma H} = 0.075$ and $n_d = 1.50$

ϕ'	Slope 2:1		Slope 3:1		Slope 4:1		Slope 5:1	
	m	m	m	m	n	n	n	n
20	1.918	1.514	2.199	1.728	2.548	1.985	2.931	2.272
25	2.308	1.914	2.660	2.200	3.083	2.530	2.552	2.915
30	2.735	2.355	3.158	2.714	3.659	3.128	4.218	3.585
35	3.211	2.854	3.708	3.285	4.302	3.786	4.961	4.343
40	3.742	3.397	4.332	3.926	5.026	4.527	5.788	5.185

TABLE B.11

Stability coefficients m and n for $\dfrac{c'}{\gamma H} = 0.100$ and toe circle

ϕ'	Slope 2:1		Slope 3:1		Slope 4:1		Slope 5:1	
	m	m	m	m	n	n	n	n
20	1.804	1.201	2.286	1.588	2.748	1.974	3.190	2.361
25	2.076	1.488	2.665	1.945	3.246	2.459	3.796	2.959
30	2.362	1.786	3.076	2.359	3.770	2.961	4.442	3.576
35	2.673	2.130	3.518	2.803	4.339	3.518	5.146	4.249
40	3.012	2.486	4.008	3.303	4.984	4.173	5.923	5.019

TABLE B.12

Stability coefficients m and n for $\dfrac{c'}{\gamma H} = 0.100$ and $n_d = 1.00$

ϕ'	Slope 2:1		Slope 3:1		Slope 4:1		Slope 5:1	
	m	m	m	m	n	n	n	n
20	1.841	1.143	2.421	1.472	20982	1.815	3.549	2.157
25	2.102	1.430	2.785	1.845	3.458	2.303	4.131	2.743
30	2.378	1.714	3.183	2.258	3.973	2.830	4.751	3.372
35	2.692	2.086	3.612	2.715	4.516	3.359	5.426	4.059
40	3.025	2.445	4.103	3.230	5.144	4.001	6.187	4.831

TABLE B.13

Stability coefficients m and n for $\dfrac{c'}{\gamma H} = 0.100$ and $n_d = 1.25$

ϕ'	Slope 2:1		Slope 3:1		Slope 4:1		Slope 5:1	
	m	m	m	m	n	n	n	n
20	1.874	1.301	2.283	1.558	2.751	1.843	3.253	2.158
25	2.197	1.642	2.681	1.972	3.233	2.330	3.833	2.758
30	2.540	2.000	3.112	2.415	3.753	2.858	4.451	3.372
35	2.922	2.415	3.588	2.914	4.333	3.458	5.141	4.072
40	3.345	2.855	4.199	3.457	4.987	4.142	5.921	4.872

TABLE B.14

Stability coefficients m and n for $\dfrac{c'}{\gamma H} = 0.100$ and $n_d = 1.50$

ϕ'	Slope 2:1		Slope 3:1		Slope 4:1		Slope 5:1	
	m	m	m	m	n	n	n	n
20	2.079	1.528	2.387	1.742	2.768	2.014	3.158	2.285
25	2.477	1.942	2.852	2.215	3.297	2.542	3.796	2.927
30	2.908	2.385	3.349	2.728	3.881	3.143	4.468	3.614
35	3.385	2.884	3.900	3.300	4.520	3.800	5.211	4.372
40	3.924	3.441	4.524	3.941	5.247	4.542	6.040	5.200

REFERENCES

Abouleid, A.F. (1982). "Measurement of Swelling and Collapsible Soil Properties," *Foundation Engineering,* Edited by George Pilot. Presses de le cole nationale des Ponts et chaussees, Paris, France.

Alpan, I. *"The Empirical Evaluation of the coefficient K_o and K_{or} Soil* And Foundation" (*Jap. Soc. Soil Mech. Found. Eng*), Jan 1967.

Altmeyer, W.T. (1955). "Discussion of Engineering properties of expansive clays," *Proc. ASCE,* Vol. 81, No. SM 2.

American Society for Testing and Materials (1994). *Annual Book of ASTM Standards,* Vol. 04.08, Philadelphia, Pa.

Amos Komornik, and David David, *"Prediction of Swelling Pressures of Clays",* Pro. ASCE, *SM & Found Div. SMI,* 1969.

Anderson, P., *Substructure Analysis and Design,* Second Ed., The Rotand Press Co., New York, 1956.

Aterberg, A., *Uber die Physikalische Bodenuntersuchung Und Uber die Plastizitat der Tone* Int. Mitt. Bodenkunde, Vol. 1, Berlin, 1911.

Azzouz, A.S., Kriezek, R.J., and Corotis, R.B., *Regression Analysis of soil compressibility,* Soils Found., Tokyo, Vol. 16, 1976. 4

Awad, A. and G. Petrasovits, *Consideration on The Bearing Capacity of Vertical and Batter Piles Subjected to Forces Acting in Different Directions,* Proc. 3rd Budapest Conf. SM and FE., Budapest, 1968.

Baver, L.D., *Soil Physics* Third Ed., John Wiley and Sons, 1906.

Baguelin, F., J.F. Jezequel and D.H. Shields, *The Pressure Meter and Foundation Engineering,* Trans Tech. Publications, Clausthal, Germany, 1978.

Barden, L., McGown, A., and Collins, K. (1973). "The Collapse Mechanism in Partly Saturated Soil," *Engineering Geology,* Vol. 7.

Bazara, A.R., *Use of Standard Penetration Test for Estimating Settlements of Shallow Foundations on Sand,* PhD Thesis, Univ of Illinois, U.S.A., 1967.

Bell, A.L., *"The Lateral Pressure and Resistance of Clay and Supporting Power of Clay Foundations",* Century of Soil Mechanics, ICF, London, 1915.

Berg, R.R., Bonaparte, R., Anderson, R.P., and Chouery, V.E. (1986). "Design Construction and Performance of Two Tensar Geogrid Reinforced Walls" *Proc. 3rd Intl. Conf. Geotextiles,* Vienna: Austrian Society of Engineers.

Berezantsev, V.G., V. Khristoforov, and V. Golubkov, *Load Bearing Capacity and Deformation of Piled Foundations,* Proc. 5th Int. Conf. SM & FE, Vol. 2, 1961.

Boit, M.A, and Clingam, F.M, *General Theory of Three-Dimensional Consolidation,* J. Applied Physics, Vol. 12, 1941.

Bishop, A.W, *The use of Pore Pressure coefficients in Practice,* Geotechnique, Vol. 4, No. 4, 1954, London.

Bishop, A.W. (1955). "The Use of Slip Circle in the Stability Analysis of Earth Slopes," *Geotechnique,* Vol. 5, No. I.

Bishop, A.W, *"The Measurement of Pore Pressure in The Triaxial Tests",* Conf. Pore Pressure and Suction in Soils, Butterworths, London, 1960-61.

Bishop, A.W., and Henkel, D.J. (1962). *The Measurement of Soil Properties in the Triaxial Test,* Second Ed., Edward Arnold, London.

Bishop, A.W, and Bjerrum, L, *The Relevance of the Triaxial Test to the Solution of Stability Problems,* Proc. Res. Conf. Shear Strength of Soils, ASCF, 1960.

Bishop, A.W, and Henkel, D.J, *The Measurement of Soil Properties in the Triaxial Test,* Second Ed., Edward Arnold, London, 1969.

Bishop, R.F., R Hill and N.F. Mott, *The Theory of Indentation and Hardness Tests,* Proc of' The Physical Sociey, Vol. 57, Part 3, 1945.

Bishop, A.W, and Morgenstern, N, *Stability Coefficients for Earth Slopes* Geotechnique, Vol. 10, No.4, 1960, London.

Bjerrum, L, et al., *From Theory to Practice in Soil Mechanics,* John Wiley and Sons, 1960.

Bjerrum, L., and N.E. Simons, *Comparison of Shear Strength Characteristics of Normally Consolidated Clay* Proc. Res. COTA. Shear Strength Cohesive Soils, ASCE, 1960.

Bjerrum, L. (1973). "Problems of soil mechanics and construction on soft days," *Proc. 8th Int. Conf. on Soil Mech. and Found. Eng.,* Moscow, 3.

Bjerrum, L., and Eide, O. (1956). "Stability of Strutted Excavations in Clay," *Geotechnique,* Vol. 6, No. 1.

Bjerrum, L., and Simons, N.E. (1960). "Comparison of Shear Strength Characteristics of Normally Consolidated Clay," *Proc. Res. Conf. Shear Strength Cohesive Soils,* ASCE.

Bowles, J.E, *Foundation Analysis and Design,* McGraw-Hill, 1977, New York.

Bowles, J.E. (1996). *Foundation Analysis and Design,* McGraw-Hill, New York.

Borrowicka, H, *Influence of Rigidity of a Circular Foundation Slab on the Distribution of Pressures over the Contact Surface,* Proc. First Int. Conf. Soil Mech. Found. Eng., Vol. 2, 1936.

Borrowicka, H, *The Distribution of Pressure under 2 Uniformly Loaded Elastic Strip Resting on Elastic-Isotropic Ground,* Second Cont. Int. Assoc. Bridge Struct, Eng., Final Report, Vol 8, 1938, Berlin.

Boussinesq, J, *Application des Potentials a L' Erade de L'Equilibre Elastiques,* Gauthier-Villars, 1883, Paris.

Briad, J.L., T.D. Smith, and L.M. Tacker, *A Pressure meter Method for Laterally Loaded Piles,* Inst. Conf. of SM and F.Engg., San Francisco, 1985.

Broms, B.B., *Lateral Resistance of Piles in Cohesionless Soils,* JSMFD, ASCE, Vol. 90, SM 2, 1964 a.

Broms, B.B., *Lateral Resistance of Piles in Cohesionless Soils,* JSMFD, ASCE, Vol. 90, SM 3, 1964 b.

Brooker, E.W., and Ireland, H.O., *Earth Pressures at Rest Related to Stress History,* Canadian Geotech. Jou, Vol 2, No.1, 1965.

Brown, R.E., *Drill Rod Influence on Standard Penetration Test,* JGED, ASCE, Vol. 103, SM 3, 1977.

Brown, R.E. (1977). "Vibroflotation Compaction of Cohesioinless Soils," *J. of Geotech. Engg. Div.,* ASCE, Vol. 103, No. GT12.

Burmister, D.M, *The Theory of Stresses and Displacements in Layer Systems and Application to Design of Airport Runways,* Proc. Highway Res. Board, Vol 23, 1943.

Burland, J.B. and M.C. Burbridge, *Settlement of Foundations on Sand and Gravel,* Proc. ICE, Vol. 78, 1985.

Burland, J.B., *Shaft Friction of Piles in Clay-A Simple Fundamental Approach,* Ground Eng, Vol. 6   1973.

Bustamante, X., and L. Gianeselli, *Portance Rielle et Portance Calculee des Pieux Isoles, Sollicites Verticalment,* Revue Francaise de Geotechnique, No. 16, 1981.

Butler, F.G, *Heavily, over-Consolidated Clays-a Review,* Proc. of the Cont. on Settlement of Structures, Pentech press, Cambridge, 1974.

Button, S.J., *The Bearing Capacity of Footings on a two-Layer Cohesive Sub-soil,* 3rd ICEMFE, Vol.1, 1953.

Caron, P.C., and Thomas, F.B. (1975). "Injection," *Foundation Engineering Hand book,* Edited by Winterkorn and Fang, Van Nastrand Reinhold Company, New York.

Carter, J.P., and Kulhawy, F.H. (1988). "Analysis and Design of Drilled Shaft Foundations Socketed into Rock," Final Report Project 1493–4, EPRI EL–5918, Geotechnical Group, Cornell University, Ithaca.

Casagrande, A, *The Hydrometer Method for Mechanical Analysis of Soils and other Granular Materials,* Cambridge, Massachusetts, 1931.

Casagrande, A, *Classification and Identification of Soils,* Proc. ASCE, Vol 73, No.6, Part I, June 1967, New York.

Casagrande, A, *Contribution to Soil Mechanics,* 1925 to *1940* Boston Soc. of Civil Engineers.

Casagrande, A, *The determination of the Preconsolidation Load and its Practical Significance.* Proc. First Int. Cont. Soil. Mech. Found Eng, 1936.

Casagrande, A, *Review of Past and Current Works on Electro-osmotic Stabilisation of Soils,* Harvard Soil Mech. Series No. 45, Harvard "Univ., Cambridge, Mass-1958 and 1959.

Casagrande, L, *Naeherungsmethoden zur Bestimmurg von Art'und Menge der sickerung durch geschuettete Daemme,* Thesis, Techniche Hochschule, Vienna, 1932.

Caquot, A., and K'erisel, J, *Tables for the Calculation of Passive Pressure, Active Pressure, and Bearing Capacity of Foundations* (Translated by M.A. Bec, London), Gauthier-Villars, Paris.

Carrillo, N, *Simple Two and Three-Dimensional cases in the Theory of consolidation of Soils,* J. Math. Phys., Vol 21, 1942.

Chen, F.H. (1988). *Foundations on Expansive Soils.* Elsevier Science Publishing Company Inc., New York.

Chen, Y.J., and Kulhawy, F.H. (1994). "Case History Evaluation of the Behavior of Drilled Shaft Under Axial and Lateral Loading," Final Report, Project 1493-04, EPRI TR-104601, Geotechnical Group, Cornell University, Ithacka.

Christian, J.T., and W.D. Carrier, *Janbu, Bjerrum and Kjaernsli's Chart Reinterpreted,* Canadian Geotechnical Journal, Vol. 15, 1978.

Clemence, S.P., and Finbarr, A.O. (1981). "Design Considerations for Collapsible Soils," *J. of the Geotech. Eng. Div.,* ASCE, Vol. 107, No. GT3.

Coulomb, C.A, *Essai Sur Une Application des regles des Maximis et Minimis a Queiques Problems des Statique Relatifs a LArchitecture,* Mem. Acad Roy. Pres. Divers Savants, Paris, Vol 7, 1776.

Coyle, H.M., and L.C. Reese, *Load Transfer for Axially Loaded Piles in Clay,* JSMFD, ASCE, Vol. 92, SM 2, 1966.

Coyle, H.M., and I.H. Sulaiman, *Skin Friction for Steel Piles in Sand,* JSMFD, ASCE, Vol. 93, SM 6, 1967.

Craig, RF, *Soil Mechanics* Van Nostrand Reinhold Co., 1980, London.

Culmann, C., *Graphische Statik, Zurich,* 1875.

Cummings, A.E., *Foundation Stresses in an Elastic Solid with a Rigid underlying Boundary, Civil* Eng., Vol 11, 1941.

D'Appolonia, E., Ellison, R.D., and D'Appolonia, D.J. (1975). "Drilled Piers," *Foundation Engineering Handbook,* Edited: H.F. Winterkorn and H.Y. Fang, Van Nostrand Reinhold Company, NewYork.

Darcy, H., *Les Fontaines Publiques de la Ville de Dijon,* Dalmot, 1986, Paris.

Das, B.M., *Advanced Soil Mechanics,* McGraw-Hill Book Co., 1983.

Davisson, M.T., *Behaviour of Flexible Vertical Piles Subjected to Moments, Shear and Axial Load,* PhD Thesis, Univ. of Illinois, Urbana, U.S.A., 1960.

De Ruiter, J., and F.L. Beringen, *Pile Foundations for Large North Sea Structures.* Marine Geotechnology, Vol. 3, No.3, 1979.

Desai, C.S., *Numerical Design Analysis for Piles in Sand.* JQED, ASCE, Vol. 100, No. 676, 1974.

Douglas, B.J., *The Electric Cone Penetrometer Test: A User's Guide to Contracting for Services. Quality Assurance, Data Analysis,* The Earth Technology Corporation, Long Beach, California, U.S.A., 1984.

Douglas, B.J., and R.S. Olsen, *Soil Classification Using the Electric Cone Penetrometer,* Cone Penetration Testing and Experience. ASCE Fall Convention, 1981.

Dunn, I.S., Anderson, I.R. and Kiefer, F.W., *Fundamentals of Geotechnical Analysis,* John Wiley and Sons, 1980.

Dupuit, J., *Etudes Theoriques et practiques sur Ie Mouvement des eaux das les Canaux Decouverts et a travers les Terrains Permeables,* Dunod, 1863, Paris.

Fadum, R.E., *Influence Values for Estimating Stresses in Elastic Foundations,* Proc. Second. Int. Cool. Soil Mech. Found Eng., Vol 3, 1948.

Fancher, G.H., Lewis, J.A., and Barnes, K.B., *Mineral Industries Experiment Station, Bulletin* 12, Pennsylvania State Univ., University Part, Pa., 1933.

Fellenius, W., *Erdstatische Berechnungen mit Reibung and Kohasion and unter Annahme keissy Undrischer Gleitflachen,* Wilhelm Ernt and Sohn, II Ed., 1947, Berlin.

Forchheimer, P., *General Hydraulics,* Third Ed.. Leipzig, 1930.

Fox, E.N., *The mean Elastic Settlements of Uniformly Loaded Area at Depth Below the Ground Surface,* 2nd ICSMEF, Vol. 1, 1948.

Frohllck, O.K., *General Theory of Stability of Slopes.* Geotechnique, Vol 5, 1955.

Gibson, R.E., Lo, K.Y., *A Theory of Consolidation for Soils Exhibiting Secondary Compression,* Norwegian Geotechnical Institute, Publication No.41, 1961.

Gibson, R.E., *Experimental Determination of the True Cohesion and True angle of Internal Friction in Clays,* Proc. Third Int. Conf. Soil Mech. Found Eng. (Zurich), Vol 1, 1953.

Gibbs, H.J., and W.G. Holtz, *Research on Determining The Density of Sands by Spoon Penetration Testing,* 4th ICSMEF, Vol. 1, 1957.

Gilboy, G, *Mechanics of Hydraulic Fill Dams,* Journal Boston Society of Civil Engineers, Boston, 1934.

Glesser, S.M., *Lateral Load Tests, on Vertical Fixed-Head and Free-Head Pile,* ASTM, STP 154, 1953.

Gray, D.H, and Mitchell, J.K, *Fundamental Aspects of Electro-osmosis in Soils* J. Soil Mech. Found. Div., ASCE, Vol 93, 1967.

Grim, R.E., *Applied Clay Mineralogy,* McGraw Hill, 1962.

Hanna, T.H., *Foundations in Tension - Ground Anchors,* Trans Tech. Publications and McGraw-Hill   Company, 1982.

Hansen, J.B., *A Revised and Extended Formula for Bearing Capacity,* Danish Geotechnical Institute Bul. No. 28, Copenhagen, 1970.

Harr, M.E, *Ground Water and Seepage,* McGraw Hill, 1962, New York.

Harr, M.E., *Foundations of Theoretical Soil Mechanics.* McGraw Hill, 1966, New York.

Hausmann, M.R. (1990). *Engineering Principles of Ground Modification,* McGraw Hill, New York.

Hazen, A, *Discussion of Dams and Sand Foundations* by A.C. Koenig, Trans ASCE, Vol 73, 1911.

Hazen, A, *Hydraulics* Third Ed., Leipzig, 1930.

Hazen, A, *Some Physical Properties of sands and gravels with special reference to their use in filtration* 24th Annual Report, Massachusetts State Board of Health, 1892.

Henkel, D.J., *The Shear Strength of Saturated Remoulded Clays.* Proc. Re. Conf. Shear Strength Cohesive Soils. ASCE, 1960.

Heteyni, M., *Beams on Elastic Foundations,* The Univ. of Michigan Press, Ann Arbor, MI. 1946.

Holl, D.L., *Stress Transmission on Earths,* Proc, Highway Res. Board, Vol. 20, 1940.

Holl, D.L., *Plane-Strain Distribution of Stress in Elastic Media,* Iowa Eng. Expts Station, Iowa, 1941.

Holtz, R.D., and Kovacs, W.D. (1981). *An Introduction to Geotechnical Engineering,* Prentice Hall Inc., New Jersey.

Holtz, W.G. and Gibbs, H.J. (1956). "Engineering Properties of Expansive Clays." Trans ASCE. 121.

Holtz, W. G. and Hilf, J. W. (1961). "Settlement of Soil Foundations Due to Saturation," *Proc. 5th International SMFE,* Paris, Vol 1.

Hough, B.K., *Basic Soil Engineering,* Ronald Press, 1957, New York.

Huntington, W.C., *Earth Pressures and Retaining Walls,* John Wiley and Sons, New York.

Hvorslev, M.J., *Physical Components of The Shear Strength of Saturated Clays,* Proc. Res. Conf. on Shear Strength of Cohesive Soils, ASCE 1960, Boulder, Colorado.

Hvorslev, M.J., *Surface Exploration and Sampling of Soils for Civil Engineering Purposes,* Water ways Experimental Station, Engineering Foundations, New York, 1949.

Hrennikoff, A., *Analysis of Pile Foundations with Batter Piles,* Proc. ASCE, Vol. 75, 1949.

I.S: 2720 (Part V), *Determining Liquid Limit by the use of Static Cone Penetrometer* -1965.

Jaky, J., *The Coefficient of Earth Pressure at Rest,* J. Soc. Hungarian Arch. Eng., 1944

Janbu, N., L. Bjerrum, and B. Kjaernsli, *Veiledning Veldlqsning afundamenteringsoppgarer.* Pub. No.6, Norwegian Geotechnical Institute, 1956.

Jennings, J.E., and Knight, K. (1957). "An Additional Settlement of Foundation Due to a Collapse Structure of Sandy Subsoils on Wetting," *Proc. 4th Int. Conf on S.M. and F. Eng.,* Paris, Vol. 1.

Jennings, J.E., and Knight, K. (1975). "A Guide to Construction on or with Materials Exhibiting Additional Settlements Due to 'Collapse' of Grain Structure," *Proc. Sixth Regional Conference for Africa on S.M and F. Engineering,* Johannesburg.

Jumikis, A.R., *Soil Mechanics,* D. Van Nostrand Co. inc., 1962, New York.

Kaldjian, M.J., *Discussion of Design Procedures for Dynamically Loaded Foundations by ,* R. V. Whitman and F. E. Richart, Jr., JSMFD, Proc. ASCE, Vol. 95, SM 1, 1969.

Kapur, R., *Lateral stability Analysis of Well Foundations* Ph.D. thesis, Univ. Roorkee, 1971.

Karlsson, R., Viberg, L. (1967). "Ratio c_u/p in relation to liquid limit and plasticity index with special reference to Swedish clays," *Proc. Geotech. Conf,* Olso, Norway, Vol. 1.

Kishida, H., *Ultimate Bearing Capacity of Piles driven into Loose Sand, Soil and Fndns,* Vol. 7, 1967.

Koerner, R.M., *Construction and Geotechnical Methods in Foundation Engineering,* McGraw-Hill Book Company, New York, 1985.

Koerner, R.M. (1999). *Design with Geosynthetics.* Fourth Edition, Prentice Hall, New Jersey.

Koerner, R.M. (2000). "Emerging and Future Developments of Selected Geosynthetic Applications" Thirty-Second Terzaghi Lecture, Drexel University, Philadelphia.

Koglar, F., and Scheidig, A., *Soil Mechanics* by A.R. Jumikis, D. Van Nostrand, 1962.

Kolbuszewski, J.J., *An Experimental Study of the Maximum Porosities of Sands,* Proc. Second Int. Conf. Soil Mech. Found Eng., Vol 1, 1948.

Kovacs, W.D., and Salomone, *SPT Hammer Energy Measurement,* JGED, ASCE, GT4, 1982.

Kubo, J., *Experimental Study of the Behaviour of Laterally Loaded Piles,* Proc. 6th Int. Conf. SM & FE., vol. 2, 1965.

Ladd, C.C., and Foott, R., *New Design Procedure for Stability of Soft Clays,* J. Geotech Eng. Div.., ASCE, Vol. 100, 1974.

Lambe, T.W., *The Structure of Compacted Clay and The Engineering Behaviour of Compacted Clay,* J. Soil Mech, Found, Div, ASCE, Vol. 84, 1958.

Lambe, T.W., and Whitman, R.V., *Soil Mechanics,* John Wiley and Sons, Inc, 1969.

Lee, K.L, and Singh, A., *Relative Density and Relative Compaction,* J. Soil Mech., Found. Div, ASCE, Vol. 97, SM 7, 1971.

Lee, K.L., Adams, B.D., and Vagneron, J.J. (1973). "Reinforced Earth Retaining Walls," *Jou. of SMFD.,* ASCE, Vol. 99, No. SM 10.

Leonards, G.A., *Foundation Engineering,* McGraw Hill, 1962.

Liao, S.S., and R.V. Whitman, *Overburden Correction factors for sand,* JGED, Vol. 112, No.3, 1986.

Littlejohn, G.S., and D.A. Bruce, *Rock Anchors: State of the Art, Part I - Design,* Ground Engineering, Vol. 8, No.3, 1975.

Loos, W., and H. Breth, *Modellversuche uber Biege Beanspruch - ungen Von Pfahlen and Spunwenden,* Der Bauingenieur, Vol. 28, 1949.

Lowe, J., *New Concepts in Consolidation and settlement Analysis.* J. Geotech. Eng. Div., ASCE Vol. 100, 1974.

Lunne, T., and H.P. Christoffersen, *Interpretation of Cone Penetrometer data for Offshore Sands,* Norwegian Geotechnical Institute, Pub. No. 156, 1985.

Matlock, H., and LC. Reese, *Generalized Solutions for Laterally Loaded Piles,* JSMFD, ASCE, Vol. 89, 1960.

Matlock, H., *Correlations for Design of Laterally Loaded Piles in soft Clay,* Proc. 2nd Offshore Tech. Conf. Houston, Vol. 1, 1970.

Matlock, H., and L.C. Reese, *Foundation Analysis of offshore Supported Structures,* Proc. 5th Tnt. Conf. SM & FD, Vol. 2, 1961.

Matso, H., *Tests on the Lateral Resistance of Piles (in Japanese).* Research Inst. of Civ. Bugg., Min. of Home Affairs, Report No. 46, 1939.

Means, RE, and Parcher, J.V., *Physical Properties of Soils.* Prentice-Hall ofIndia, 1965. New Delhi.

Menard, L., *An Apparatus for Measuring the Strength of Soils in Place,* MSc Thesis, Univ. Illinois, Urbana, U.S.A., 1957.

Menard, L., *Interpretation and Application of pressure meter Test Results to Foundation Design,* Soils - Soils, No. 26, 1976.

Mesri, G., *Coefficient of Secondary Compression,* J. Soil Mech, Found. Div., ASCE, Vol 99, SM 1, 1973.

Meyerhof, G.G., *Shallow Foundations* JSMFD, ASCE, Vol. 901, SM 2, 1965.

Meyerhof, G.G., *The Ultimate Bearing Capacity of Foundations, Geotechnique,* Vol. 2, 1951.

Meyerhof, G.G., *Some Recent Research an Bearing Capacity of Foundations,* CGL Ottawa, Vol. 1, 1963.

Meyerhof, G.G., *Discussions on Sand Density by Spoon Penetration.* 4th ICSMFE, Vol. 3, 1957.

Meyerhof, G.G., *The ultimate bearing capacity of foundations of slopes,* 4th ICSMFE, 1957, London.

Meyerhof, G.G., *The Bearing Capacity of Foundations under Eccentric and Inclined Loads,* 3rd ICSMFE, Vol. 1, 1953.

Meyerhof, G.G., *Bearing Capacity and Settlement of Pile Foundations,* JGED, ASCE, Vol. 102, GT 3, 1976.

Meyerhof, G.G., *Penetration Tests and Bearing Capacity of Cohesion less Soils.* JSMFD, ASCE, Vol. 82, SM 1, 1956.

Meyerhof, G.G., *Penetration Testing Outside Europe, General Report,* Proc. of the European Symp. on Penetration Testing, Vol. 2, Stockholm, 1975.

Mindlin, R.D., *Stress Distribution around a Tunnel,* Trans ASCE. Am. Soc. Civil Engrs. Vol. 104, 1939.

Mindlin, R.D, *Forces at a point in the Interior of a Semi-Infinite Solid,* Physics, Vol. 7, 1936.

Mitchell, J.K., and Freitag, D.R. (1959). "A Review and Evaluation of Soil-Cement Pavements," *J. of SM and FD,* ASCE, Vol. 85, No. SM6.

Mohr, 0., *Die Elastizitatsgrenze Und Bruch eines Materials* (The elastic limit and the failure of a Material), Zeitschrift Veneins Deuesche Ingenieure, Vol 4, 1900.

Morgenstern, N.R. (1963). "Stability Charts for Earth slopes During Rapid Drawdown," *Geotechnique,* Vol. 13, No. 2.

Murphy, V.A. *The effect of ground water characteristics on the seismic design of structures* Proc. 2nd World conf on Earthquake Eng., Tokyo, Japan, 1960 and *Foundation Engineering Hand Book* Edited by Winterkorn, H.F. *et al,* Van Nostrand company.

Murthy, V.N.S., *Behaviour of Batter Piles Subjected to Lateral Loads,* PhD Thesis, Indian Institute of Technology, Kharagpur, India, 1965.

Murthy, V.N.S., *Report on Soil Investigation for the construction of cooling water system, Part II of   Farakha Superthermal Power Project,* National Thermal Power Corporation, New Delhi, 1982.

Murthy, V.N.S., and Subba Rao, K.S. (1995) *"Prediction of Nonlinear Behaviour of Laterally Loaded Long Piles",* Foundation Engineer, Vol. 1, No. 2, New Delhi.

Muskat, M. *The Flow of homogeneous fluids through porous media* McGraw Hill Co., 1946.

Nagaraj, T.S., and Srinivasa Murthy, *Prediction of the Pre-consolidation Pressure and Recompression Index of Soils,* Geotech, Testing Journal, ASTM Vol. 8, 1985.

Newmark, N.M., *Influence Charts for Computation of Stresses in* Elastic Soils, Univ. of Illinois Expt. Stn., Bulletin No.338, 1942.

Nordlund, R.L., *Bearing Capacity of Piles in Cohesionless Soils,* JSMFD, ASCE, Vol. 89, SM 3, 1963.

Norris, G.M., and R.D. Holtz, *Cone penetration Testing and Experience,* Proc. ASCE National Convention, St. Louis, Missouri 1981 - Published by ASCE, NY.

O' Connor, M.J., and Mitchell, R.J. (1977). "An Extension of the Bishop and Morgenstern Slope Stability Charts," *Canadian Geotech. J.* Vol. 14.

O' Neill, M.W., and Reese, L.C. (1999). "Drilled Shafts: Construction Procedures and Design Methods" Report No. FHWA-IF-99-025, Federal Highway Administration Office of Infrastructure/Office of Bridge Technology, HIBT, Washington D.C.

O'Neill, M.W., Townsend, F. C., Hassan, K. H., Buttler A., and Chan, P. S. (1996). "Load Transfer for Intermediate Geomaterials," Publication No. FHWA-RD-95-171, Federal Highway Administration, Office of Enggniering and Highway Operations R & D, McLean, V. A.

O'Neill, M.W. (1988). "Special Topics in Foundations," *Proc. Geotech. Eng. Div.,* ASCE National Convention, Nashville.

Osterberg, J.O., *Influence Values for Vertical Stresses in Semi-Infinite Mass due to Embankment Loading,* Proc. Fourth Int. Conf. Soil Mech. Found. Eng., Vol. 1, 1957.

Palmer, L.A. and J.B. Thompson, *The Earth Pressure and Deflection along Embedded Lengths of Piles Subjected to Lateral Thrusts,* Proc. 2nd Int. Conf. SM and FE, Rotterdam, Vol. 5, 1948.

Peck, R.B. (1969). "Deep excavations and Tunneling in Soft Ground," *Proc., 7th Int. Conf. on SMFE,* Mexico city.

Peck, R.B., Hanson, W.E., and Thornburn, T.H., *Foundation Engineering* John Wiley and Sons, Inc., New York, 1974.

Peck, R.B, and Terzaghi, K., *Soil Mechanics* in *Engineering Practice,* John Wiley and Sons.

Petterson, K.E., *The Early history of circular sliding surfaces,* Geotechnique, The Institution of Engineers, London, Vol. 5, 1955.

Poulos, H.G., *Elastic Solutions for Soil and Rock Mechanics,* Wiley and Sons, 1974, New York.

Poulos, H.G., and E.H. Davis, *Pile Foundation Analysis and Design,* John Wiley & Sons, New York, 1980.

Rankine, W.J.M., *On the Stability of Loose Earth Dams,* Phil. Trans. Royal Soc., Vol. 147, 1857, London.

Reese, L.C., and H. Matlock, *Non-dimensional solutions for Laterally Loaded Piles with Soil Modulus assumed Proportional to depth,* Proc. 8th Texas Conf. S.M. and EE., Spec. Pub 29, Bureau of Eng. Res., Univ. of Texas, Austin, 1956.

Reese, L.C. and R.C. Welch, *Lateral Loadings of Deep Foundations,* in *Stiff Clay,* JGED, ASCE, Vol. 101, GT 7, 1975.

Reese, L., W.R. Cox, and F.D. Koop, *Analysis of Laterally Loaded Piles* in *Sand,* Proc. 6th Offshore Tech. Conf. Houston, 1974.

Reese, L.C., and O'Neill, M.W. (1988). "Field Load Tests of Drilled Shafts," *Proc., International Seminar on Deep Foundations and Auger Piles,* Van Impe (ed.), Balkema, Rotterdam.

Reltov, B.E., *Foundation Engineering* Edited by G. Leonards, McGraw Hill, 1962.

Rendulic, L., *Der Hydrodynamische spannungsaugleich in zentral entwasserten Tonzylindern,* Wasserwirtsch. u. Technik, Vol. 2, 1935.

Reynolds, O., *An Experimental Investigation of the circumstances which determine whether the motion of water shall be Direct or Sinuous and the Law of Resistance* in *parallel Channel,* Trans. Royal Soc. Vol. 174, 1883, London.

Robertson, P.K., and R.G. Campanella, *Interpretation of Cone Penetration Tests, Part 1 - Sand,* CGJ, Ottawa, Vol. 20, 1983.

Robertson, P.K. and R.G. Campanella, *SPT-CPT Correlations,* JGED, ASCE, Vol. 109, 1983 a and b.

Sanglerat, G., *The Penetrometer and Soil Exploration,* Elsevier Publishing Co., Amsterdam, 1972.

Schmertmann, J.H., *Guidelines for Cone Penetration Test: Performance and Design,* U.S. Deptt. of Transportation, 1978.

Schmertmann, J.H., *Static Cone to Compute Static Settlement over sand,* JSMFD, ASCE, Vol. 96, SM 3, 1970.

Scott, R.F., *Principles of Soil Mechanics,* Addison - Wesley Publishing Co., Inc., London, 1963.

Seed, H.B., Woodward, R.J., and Lundgren, R. (1962). "Prediction of Swelling Potential for Compacted Clays," *Journal ASCE,* SMFD, Vol. 88, No. SM 3.

Seed H.B., and Whitmann, R.V., *Design of Earth Retaining Structures for dynamic loads, ASCE, spec. conf Lateral stresses* in *the ground and design of earth retaining structures,* 1970.

Simons, N.E., *Comprehensive Investigation of the Shear Strength of an Undisturbed Drammen Clay,* Proc. Res. Conf. Shear Strength Cohesive Soils, ASCE, 1960.

Simons, N.E. and N.N. Som, *Settlement of Structures on Clay with Particular emphasis on London Clay,* Industry Research Institute, Assoc. Report 22, 1970.

Skempton, A.W., *The Bearing Capacity of Clays,* Proc. Building Research Congress, Vol. 1, 1951.

Skempton, A.W., *Cast-in-situ* Bored Piles in London Clay, Geot., Vol. 9, 1959.

Skempton, A.W., *The Pore Pressure Coefficients A and B, Geotechnique,* Vol. 4, 1954.

Skempton, A.W., and Northey, R.D., *Sensitivity of Clays,* Geotechnique, Vol. 3, 1952-53, London.

Skempton, A.W., and L. Bjerrum, *A Contribution to the Settlement Analysis, of Foundations on Clay,* Geotechnique, 7, 1957.

Smit, R.E., and Wahls, H.E., *Consolidation under Constant Rate of Strain* J. Soil Mech. Found. Div, ASCE, Vol. 95, SM 2, 1969.

Spencer, E. (1967). "A Method of Analysis of the Stability of Embankments, Assuming Parallel Inter-slice Forces," *Geotechnique,* Vol. 17, No. 1.

Steinbrenner, W., *Taflen zur Setzungsberechnung,* Die Strasse, Vol. 1, 1935.

Taylor, D.W. (1937). "Stability of Earth Slopes," *J. of Boston Soc. of Civil Engineers,* Vol. 24.

Taylor, D.W., *Fundamentals of Soil Mechanics,* John Wiley and Sons, 1948, New York.

Terzaghi, K. (1943). *Theoretical Soil Mechanics,* Wiley & Sons, New York.

Terzaghi, K., *Evaluation of Co-efficient of Sub-grade Reaction,* Geotechnique, London, Vol. 5, 1955.

Terzaghi, K., and R.B. Peck, *Soil Mechanics in Engineering Practice,* John Wiley & Sons, N.Y., 1967.

Terzaghi, K., Peck, R.B., and Mesri, G. (1996). *Soil Mechanics in Engineering Practice,* John Wiley & Sons, Inc., Third Edition, New York.

Timoshenko, S.P., and Goodier, J.N. *Theory of Elasticity,* Third Ed., McGraw Hill, 1970, New York.

Tschebotarioff, G.P., *The Resistance of Lateral Loading of Single Piles and of Pile Groups,* ASTM Spl. Pub. No. 154, 1953.

Tschebotarioff, G.P., *Soil Mechanics, Foundation and Earth Structures,* McGraw Hill, 1958, New York.

Tsytovich, N., *Soil Mechanics,* Mir Publishers, Moscow, 1986.

U.S. Army Engineer Water Ways Experiments Station, *The Unified soil classification system* Technical memo, No. 3-357, Corps Engineers, Vicksburg, U.S.A., 1957.

Van Der Merwe, D.H. (1964). "The Prediction of Heave from the Plasticity Index and Percentage clay Fraction of Soils," *Civil Engineer in South Africa,* Vol. 6, No. 6.

Van Der Veen, C., *The Bearing Capacity of a Pile Predetermined by a Cone Penetration Test,* Proc. 4th ICSMF, London, 1957.

Vesic, A.S., *Contribution a L'Extude des Foundations Sur Pieux Verticaux et Inclines,* Annales des Travaux Publics de Belgique, No.6, 1956.

Vesic, A.S., *A Study of Bearing Capacity of Deep Foundations,* Final Report, School of Civil Eng., Georgia Inst. Tech., Atlanta, U.S.A., 1967.

Vesic, A.S. (1972). "Expansion of cavities in Infinite Soil Mass," *J.S.M.F.D.,* ASCE, Vol. 98.

Vesic, A.S., *Analysis of Ultimate Loads of Shallow Foundations,* JSMFD, ASCE, Vol. 99, SM 1, 1973.

Vesic, A.S., *Bearing Capacity of Shallow Foundations,* Foundation Engineering Hand Book, Van Nostrand Reinhold Book Co., N.Y., 1974.

Vesic, A.S., *Tests on Instrumented Piles, Ogeechee River Site,* JSMFD, ASCE, Vol. 96, SM 2, 1970.

Vesic, A.S. (1977). *Design of Pile Foundations,* Synthesis of Highway Practice 42, Res. Bd., Washington D.C.

Vidal, H. (1969). "The Principle of Reinforced Earth," HRR No. 282.

Vijayvergiya, V.N., and Focht, J.A. Jr., (1972). *A New way to Predict the Capacity of Piles in Clay,* 4th Annual Offshore Tech Conf., Houston, Vol. 2.

Vijayvergiya, V.N., and J.A. Focht, Jr., *A New Way to Predict The Capacity, of Piles in Clay,* 4th Annual Offshore Tech Conf., Houston, Vol. 2, 1972.

Westergaard, H.M., *Stresses in Concrete Pavements Computed by Theoretical Analysis Public Roads,* Vol. 7, No.2, 1926, Washington.

Westergaard, H.M., *The Resistance of a group of Piles,* J. Western Society of Engineers, Vol. 22, 1917.

Whitman R.V., and F.E. Richart, Jr., *Design Procedures for Dynamically Loaded Foundations,* JSMFD, Proc. ASCE, Vol. 93, SM 6, 1967.

Winterkom, H.F., & Hsai Yang Faud, *Foundation Engineering hand book* Van Nostrand Reinhold company, New York, 1975.

Woodward, Jr., R.J., Gardner, W.S., and Greer, D.M. (1972). *Drilled Pier Foundations,* McGraw Hill, New York.

INDEX